T0314155

UNIVERSITY OF WASHINGTON PUBLICATIONS IN BIOLOGY, VOLUME 17

VASCULAR PLANTS

OF THE PACIFIC NORTHWEST

By

C. Leo Hitchcock

Arthur Cronquist

Marion Ownbey

J. W. Thompson

PART 3: SAXIFRAGACEAE TO ERICACEAE

By C. Leo Hitchcock and Arthur Cronquist

Illustrated by Jeanne R. Janish

University of Washington Press

SEATTLE AND LONDON

Copyright © 1961 by the University of Washington Press
Second printing, 1971
Third printing, 1977
Fourth printing, 1984
Fifth printing, 1990
Sixth printing, 1994
Seventh printing, 2001
Printed in the United States of America

Library of Congress Catalog Card Number 56-62679
ISBN 0-295-73985-1

The paper used in this publication meets the minimum require-
ments of American National Standard for Information Sciences—
Permanence of Paper for Printed Library Materials, ANSI Z39.48-
1984. ♾

Vascular Plants of the Pacific Northwest

PART 3: SAXIFRAGACEAE TO ERICACEAE

Introduction*

Part III of the Vascular Plants of the Pacific Northwest has been written in similar fashion to Parts IV and V, in that the manuscript prepared by one person has been sent to the other authors of the Flora for review. Hitchcock and Cronquist are primarily responsible for the preparation of this volume, the latter submitting the treatment of the genus *Rosa* and the family Umbelliferae. The horticultural notes to be found throughout the book were written in collaboration with B. O. Mulligan and Carl S. English, Jr. Mr. R. C. Barneby was freely consulted concerning the genera *Astragalus* and *Oxytropis;* Lyman Benson reviewed the Cactaceae; George Eiten advised concerning the Oxalidaceae. The assistance of these and other workers is acknowledged in the text. It is a pleasure to mention the improvement of the manuscript as the result of correction and additions from Dr. Ownbey. Mrs. Janish has been responsible for the elimination of many inconsistencies through her insistence that material sent her for illustration harmonize completely with the keys and descriptions of the text. Mr. Thompson has done the lettering for the plates and aided in other ways in bringing this part to publication.

As one of the two persons primarily responsible for the preparation of the Flora and the writer of the greater portion of this particular volume, it would be at least immodest of me not to mention that Dr. Cronquist has contributed much more than the proportionate authorship would indicate. Although each of us has, from the start of the work on the Flora, reviewed the manuscript of the other, testing the adaptability of the keys, descriptions, and other data to the material in the herbaria of our respective institutions, for various reasons Dr. Cronquist has contributed much more to the completion of that portion under my authorship than I to his. For his extensive assistance, always freely and cheerfully given, I wish to express my appreciation. C. L. H.

SAXIFRAGACEAE Saxifrage Family

Flowers usually several to many but sometimes few or only 1, borne in simple to compound, open to congested, basically cymose but often apparently racemose inflorescences, mostly perfect, regular or occasionally slightly irregular, from very slightly perigynous to epigynous; calyx well developed, greenish or white to strongly cyanic and sometimes petaloid, usually at least partially adnate to the pistil at the base, frequently with a free, connate, and flared to tubular or campanulate portion (hypanthium), the lobes usually 5 (4-6), sometimes slightly unequal or the calyx otherwise somewhat irregular; petals mostly equal to and alternate with the calyx lobes, rarely fewer or even lacking, usually borne (with or slightly above the stamens) toward the top of the hypanthium (if any) or at the edge of the ovary where it is freed from the calyx, often at the edge of a disclike nectary, frequently smaller and less showy than the sepals; stamens usually 5 (either alternate with or opposite the petals, and in Parnassia alternate with simple to branched or fimbriate staminodia), 10, or occasionally only 3 *(Tolmiea),* the filaments nearly always distinct, often flattened, the anthers 2-celled, longitudinally dehiscent; carpels usually 2 or 3 (varying sometimes to 4, 5, or 6), from completely fused except for the separate stigmas to almost or quite distinct and the bases then sometimes more or less joined only by the adnate calyx, the ovary from (occasionally) superior to (less commonly) completely inferior, when compound either 1-celled and with usually 2-3 parietal placentae or 2- to 3 (rarely 4 or 5)-celled and with axile placentation, often tapered above into beaklike, usually sterile portions below the solid styles (if any); stigmas equal to the carpels, mostly capitate or discoid but sometimes more elongate, in *Parnassia* sessile or subsessile; fruit capsular or (if the carpels distinct) more truly follicular; seeds (1) few-many per carpel, with fleshy endosperm, smooth to muricate, the testa tight to loose-fitting, occasionally winged; annual or more commonly perennial herbs, only slightly if at all woody, with alternate (ours) to very rarely opposite, mostly exstipulate but sometimes conspicuously stipulate, sessile to long-petiolate, simple to compound leaves.

About 50 genera and nearly 800 species, mostly in the temperate and colder areas of N. Am. (especially) and Eurasia, but also in various parts of the Southern Hemisphere.

The family, as here considered, includes herbaceous 4-merous members often referred to the Par-

*This investigation was supported in part by funds provided for biological and medical research by the State of Washington Initiative Measure No. 171 and in part (the illustrations) by a grant from the National Science Foundation.

nassiaceae, but excludes certain genera often placed here, including some opposite-leaved shrubs with mostly numerous stamens (referred to the Hydrangeaceae) and baccate fruited, inferior ovaried, alternate-leaved shrubs here placed in the Grossulariaceae.

Many of the genera are of economic importance, chiefly because of their horticultural value, including *Saxifraga, Heuchera, Boykinia, Parnassia, Tiarella,* and *Tolmiea* of our local flora and such extraterritorial but well-known genera as *Astilbe* and *Rodgersia.*

1 Pistil 4-carpellary, with 4 parietal placentae; stigmas almost or quite sessile; stamens 5,
    alternate with simple to fimbriate staminodia   PARNASSIA
1 Pistil usually 2 or 3 (4 or 5)-carpellary and mostly with 2-3 axile or parietal placentae;
    stigmas usually borne on well-developed styles; stamens (3) 5-10; staminodia lacking
    2 Petals lacking; sepals 4; stamens 4-8; flowers greenish and inconspicuous   CHRYSOPLENIUM
    2 Petals present, or the sepals 5 and stamens (3) 5 or 10; flowers often showy
        3 Leaf blades nearly orbicular, peltate, mostly at least 1 dm. broad; petioles 2-15
            dm. long, developing later than the long naked flowering scapes   PELTIPHYLLUM
        3 Leaf blades not orbicular and peltate, mostly less than 1 dm. broad, usually
            developing with or before the flowers
            4 Ovary 1-celled and with usually 2 (less commonly 3) parietal to semibasal placentae
                5 Fruit consisting of 2 very unequal dehiscent parts (valves); stamens 10; petals
                    entire, linear   TIARELLA
                5 Fruit consisting of 2 (or 3) essentially equal parts; stamens 5 or 10; petals entire
                  to laciniate
                    6 Styles or style branches 3; petals white or pinkish, laciniate; plants usually
                      bulblet-bearing in the leaf axils or on the underground parts   LITHOPHRAGMA
                  6 Styles or style branches normally 2; petals mostly either greenish or not lacini-
                  ate; plants usually not bulblet-bearing
                    7 Petals entire
                      8 Flowers paniculate or thyrsoid, at least in the lower part of the often al-
                        most spicate inflorescence; seeds finely echinulate in longitudinal rows;
                        stamens usually 5; calyx never narrowly turbinate   HEUCHERA
                    8 Flowers racemose; seeds rarely echinulate but if so then the calyx
                      narrowly turbinate; stamens sometimes 3 or 10
                      9 Stamens regularly 3; calyx greenish-purple, nearly cylindric, 7-10
                        mm. long, oblique at base and irregular above, split much more
                        deeply on one side than between the rest of the lobes   TOLMIEA
                    9 Stamens 5 or 10; calyx mostly less than 7 mm. long, neither oblique
                      nor irregular, usually saucer-shaped to campanulate-turbinate
                      10 Seeds finely echinulate; racemes loosely 5- to 12-flowered, not
                        secund; leaves reniform, 1-4 cm. broad, usually ciliate   CONIMITELLA
                      10 Seeds not echinulate; racemes closely 10- to 45-flowered, strongly
                        secund; leaves usually cordate, 2-8 cm. broad, commonly hirsute
                        on one or both surfaces   MITELLA
                7 Petals trifid to pectinately lobed
                  11 Calyx (including that portion adnate to the ovary) usually only 2-4
                    mm. long, in 1 species up to 4-6 mm. long but then the petals tri-
                    lobed, the styles less than 1 mm. long, and the flowering stems
                    leafless   MITELLA
                11 Calyx (including the adnate portion) (6) 7-10 mm. long; petals usually
                  5- to many-lobed; styles over 1 mm. long; flowering stems more or
                    less leafy
                    12 Stamens 10   TELLIMA
                  12 Stamens 5   ELMERA
        4 Ovary 2 (rarely 3-5)-celled, the placentation axile
            13 Stamens 10
                14 Carpels distinct almost to the base, adnate to the calyx for less than 1/5
                  their length; leaves leathery, broadly ovate-elliptic to elliptic-obovate,
                  mostly 3-6 cm. long, crenate but not lobed; plants rhizomatous   LEPTARRHENA
                14 Carpels usually fused or adnate to the calyx for at least 1/5 their length,
                  if not then the plants otherwise not as in *Leptarrhena*

15 Styles partially connate; petals pink to deep red; calyx campanulate, usually reddish,
   (5) 6-10 mm. long; leaves alternate, petiolate   TELESONIX
15 Styles free above the ovuliferous portion of the ovary; petals usually white, but if (as
   rarely) pink or red (to purple) then either the calyx not at once campanulate, red,
   and so much as 6 mm. long, or the leaves sessile and opposite   SAXIFRAGA
13 Stamens 5
   16 Calyx free of the pistil, campanulate, (10) 12-16 mm. long, the sepals lanceolate-acu-
      minate; cauline leaves with prominent, toothed stipules   BOLANDRA
   16 Calyx usually adnate to at least the base of the pistil, various as to size and shape;
      cauline leaves, if any, often without toothed stipules
      17 Plants bulbiferous at the crown, neither stoloniferous nor conspicuously rhizomatous,
         the rootstocks, if any, filiform; flowering stems rarelv over 2 dm. tall; upper
         cauline leaves conspicuously stipulate; petals white or violet   SUKSDORFIA
      17 Plants not bulbiferous at the crown, usually either stoloniferous or with evident
         rhizomes; flowering stems either well over 3 dm. tall or else the upper cauline leaves
         without conspicuous stipules; petals white
         18 Petals 1.5-2.5 mm. long, withering persistent; calyx mostly 2.5-3.5 mm. long;
            stems rarely as much as 2.5 dm. tall; plant stoloniferous   SULLIVANTIA
         18 Petals mostly 4-7 mm. long, deciduous; calyx rarely less than 4 mm. long;
            stems (1.5) 2-8 (10) dm. tall; plant rhizomatous but never stoloniferous   BOYKINIA

Bolandra Gray

Flowers rather few in open, conspicuously bracteate, terminal panicles, complete; calyx tubular-campanulate, with 5 slenderly lanceolate, spreading lobes, greenish but usually purplish-tinged at least on the lobes; petals and stamens 5, borne at the summit of the calyx tube, the petals linear, nearly erect or only slightly spreading, reddish-purple, considerably exceeding (and alternate with) the stamens which are opposite the sepals; pistil 1, the 2 carpels fused only 1/4 to 1/5 their length, entirely free of the calyx, tapered rather uniformly upward and the nonovuliferous or stylar portion hollow almost to the nearly globose stigmas; ovary 2-celled basally, the placentation axile; fruit a loc-ulicidal capsule; perennial herbs with very short, bulbiferous rootstocks, the lower leaves with more or less reniform, palmately veined blades and long petioles and small stipules, the upper cauline leaves reduced but with the stipules becoming large and similar to the bladeless bracts of the inflorescence.

One Californian species besides the following. (Named for Dr. Henry N. Bolander, 1831-1897, connected with the State Geological Survey of California.)

Bolandra oregana Wats. Proc. Am. Acad. 14:292. 1879. (Howell, "On wet rocky banks of the Willamette River, near Oregon City, Oregon," June, 1877)
   B. imnahaensis Peck, Rhodora 36:266. 1934. (Peck 17495, small canyon along the Imnaha R., 3 mi. above Imnaha, Wallowa Co., Oreg., July 4, 1933)
Weakly glandular-pubescent, herbaceous perennial with numerous bulblets along the very short, horizontal rootstocks, the stems mostly single, (1.5) 2-4 (6) dm. tall; basal and lower cauline leaves with slender petioles up to 15 cm. long, the blades reniform, (2) 3-7 cm. broad, shallowly lobed and with 9-13 acutely dentate or usually somewhat serrate-dentate segments; petioles much shortened on the upper leaves and the stipules much more conspicuous and leaflike; bracts of the inflorescence somewhat clasping, 1-3 cm. long, deeply crenate-dentate; panicle branches (1) 2-7, remote, spreading, 1- to 7-flowered; calyx accrescent and eventually 14-18 mm. long, the linear-lanceolate, usually purplish lobes equaling or slightly exceeding the campanulate-tubular portion; petals purplish, linear, about equal to the calyx lobes, the stamens about 1/3 as long, the filaments reddish-purple; capsule about 1 cm. long, the carpels fused only 1/5-1/4 their length. N = 7.

Moist, mossy rocks, usually near waterfalls, on both sides of the lower Columbia R. Gorge, and along the Snake R. and its tributaries in s.e. Wash., n.e. Oreg., and adj. Ida. May-early June.

Boykinia Nutt. Nom. Conserv.

Flowers nearly or quite regular, complete, borne in terminal and subterminal cymose panicles; calyx turbinate to campanulate, adnate to the ovary for 1/3 to 1/2 the length and prolonged above as a short tubular hypanthium, the 5 lobes equal, lanceolate; petals white, spatulate or obovate to oblong-ovate, exceeding the sepals, short-clawed, deciduous; stamens 5, inserted with the petals at the top of the hypanthium, opposite to (and usually shorter than) the calyx segments; pistil 2-carpellary, the ovary 2-celled, with axile placentation, at least half inferior, tapered into the distinct, somewhat beaklike styles; seeds many, minutely tuberculate; glandular-pubescent and often brownish-pilose herbaceous perennials from rootstocks, with alternate, cordate to reniform, several times shallowly cleft and toothed leaves and with stipules from well developed and sometimes even leafletlike, to much reduced and represented only by long brownish bristles.

Seven or 8 species of N. Am., mostly in w. U.S. (In honor of Dr. Samuel Boykin, 1786-1848, a naturalist of Georgia.)

The plants are easily introduced and grown in the wild garden, preferring a moist or shady location where they will blossom throughout the summer.

1 Stipules, even of the upper leaves, reduced to no more than a slight membranous expansion on the petiole and several slender, usually brownish bristles   B. ELATA
1 Stipules (at least the upper ones) conspicuous, more or less leafletlike, often clasping or connate   B. MAJOR

Boykinia elata (Nutt.) Greene, Fl. Fran. 190. 1891.

  Saxifraga elata Nutt. in T. & G. Fl. N. Am. 1:575. 1840. *B. occidentalis* var. *elata* Gray, Proc. Am. Acad. 8:383. 1872. *Therofon elatum* Greene, Man. S. F. Bay Reg. 121. 1894. *B. nuttallii* Macoun, Can. Rec. Sci. 6:408. 1896. *(Nuttall,* "near Chenook Point at the estuary of the Oregon")

  B. occidentalis T. & G. Fl. N. Am. 1:577. 1840. *Therofon occidentalis* Kuntze, Rev. Gen. 1:227. 1891. *Therophon occidentale* Rydb. N. Am. Fl. 22²:124. 1905. *(Douglas,* Oregon)

  Therophon cincinnatum Rosend. & Rydb. N. Am. Fl. 22²:124. 1905, at least as to our plants. *B. occidentalis* var. *cincinnata* Rosend. Engl. Bot. Jahrb. 37, Beibl. 83:61. 1905. *B. circinnata* Henry, Fl. So. B. C. 164. 1915. *(Pringle,* Santa Cruz, Calif., in 1882)

  Therophon vancouverense Rydb. N. Am. Fl. 22²:125. 1905. *B. vancouverensis* Fedde, Just Bot. Jahresb. 33¹: 607. 1906. *(Harry Edwards,* Vancouver I., B. C., Aug. 1, 1874)

  Stems rather slender, 1.5-6 dm. tall, pilose throughout with brownish to reddish, usually glandular hairs, or glandular-pubescent only above; leaves long-petioled below, becoming nearly sessile above, the stipules reduced to a short dilation and several long brownish bristles; blades cordate to somewhat reniform, the larger ones 2-6 (8) cm. broad, 5- to 7-cleft nearly half their length, and 2-3 times dentate into ultimately slightly bristle-tipped teeth, gradually reduced to the nearly sessile main bracts; inflorescence an open, loose panicle of somewhat secund cymes, densely glandular and usually reddish or purplish; flowers slightly irregular; calyx turbinate-campanulate, 4-5 mm. long, somewhat accrescent, lobed slightly over 1/3 the length into narrowly lanceolate glabrous segments, the adnate portion glandular; petals spatulate-oblanceolate, 5-6 mm. long, short-clawed, deciduous; stamens much shorter than the sepals, the filaments about equaling the anthers; seeds ellipsoid, purplish-black, 0.5-0.6 mm. long, finely tuberculate. N=7.

Moist woods and along streams, from the coast to the lower w. slopes of the Cascades, B. C. southward, through Wash. and Oreg., and along the coast and in the Sierra Nevada of Calif. June-Aug.

The species is extremely variable, especially in the type, color, and amount of pubescence, and not separable into natural infraspecific taxa in our area.

Boykinia major Gray, Bot. Calif. 1:196. 1876.

  Therofon major Kuntze, Rev. Gen. 1:227. 1891. *Therofon majus* Wheelock, Bull. Torrey Club 23:70. 1896. (Four collections from Calif. and Oreg. were cited)

  B. major var. *intermedia* Piper, Erythea 7:172. 1899. *Therofon intermedium* Heller, Muhl. 1:53. 1904. *Therofon majus* ssp. *intermedium* Piper, Contr. U. S. Nat. Herb. 11:311. 1906. *B. intermedia* G. N. Jones, U. Wash. Pub. Biol. 5:168. 1936. *(F. H. Lamb 1267,* New London, Chehalis Co., Wash., June 10, 1897)

  Stems stout, from brownish-pilose and more or less glandular to glandular-pubescent and often also finely puberulent (at least in the inflorescence), (2) 3-8 (10) dm. tall; lower leaves long-petioled (up to 2 dm.) and with brownish, membranous stipules, upwards the petioles becoming greatly shortened (lacking) and the stipules usually more pronounced, leafletlike, and often clasping or connate; blades reniform, the basal ones up to 20 cm. broad, unequally 3- to 7-lobed usually at least halfway to the cordate base and again unequally 2-3 times toothed or lobed; inflorescence a many-flowered, rather compact, fairly flat- or round-topped, cymose panicle; calyx 4-6 mm. long at anthesis, campanulate, slightly accrescent, the lobes lanceolate and usually somewhat attenuate, about equal to the hemispheric based, flaring, tubular portion; petals 4-6 (7) mm. long, rather firm, oblong-ovate to obovate-oblanceolate, gradually or abruptly narrowed to a short, broad claw; stamens not quite equaling the sepals; free hypanthium considerably flared, yellowish-glandular lined; seeds irregularly ellipsoid, 0.6-0.8 mm. long, finely tuberculate, black.

In meadows and along streams; w. Wash. to Calif., and n.e. through Oreg. to Ida. and Mont. June-Sept.

This attractive and easily cultivated perennial is represented by the following fairly well-marked and apparently disjunct varieties, the var. *intermedia* being somewhat intermediate to *B. elata:*
1 Petals usually undulate-margined, narrowed abruptly to very short claws, the blades mostly
   ovate to nearly orbicular; inflorescence closely flowered, the branches crowded, the
   whole rounded to flattened across the top; w. Mont., across Ida. and n.e. Oreg. to
   w. Oreg., s. to Calif.   var. major
1 Petals more nearly obovate to spatulate, plane-margined, attenuate to the claws; inflorescence
   more open, the branches few-flowered, the whole more pyramidal than flat-topped; Olympic
   Peninsula, Wash., s. to Tillamook Co., Oreg.   var. intermedia Piper

Chrysosplenium L. Golden Carpet

Flowers greenish and inconspicuous, solitary or in few-flowered terminal or apparently axillary cymes, regular, perfect, perigynous; calyx adnate to the lower half of the ovary, broadly campanulate, the lobes 4, spreading, rounded; petals lacking; stamens 4 or 8 (ours), alternating with small fleshy glands that form a minute disclike margin surrounding the ovary; filaments very short, usually no longer than the anthers; carpels 2, the ovary 1-celled with 2 parietal placentae, more than half inferior, the 2 short styles protruding through the disc, in development the capsule elongating and ultimately about half inferior, 2-lobed and dehiscent throughout the superior portion; seeds several to numerous, smooth, plump, the raphe forming an angled ridge; perennial, glabrous (ours) herbs with succulent, more or less stolonous stems and small, opposite or alternate, exstipulate, petiolate, crenate leaves.

Forty or more species, mostly in moist areas of temperate to arctic N.Am., S.Am., and Eurasia. (From the Greek *chrysos,* gold, and *splen,* spleen, the name popularly believed to have been coined because of some fancied medicinal value of the plants.)
1 Leaves alternate, with 3-7 broad crenations; stamens 4   C. TETRANDUM
1 Leaves (except the uppermost) opposite, with 15-20 crenate-serrations; stamens 8
                                                                      C. GLECHOMAEFOLIUM

Chrysosplenium glechomaefolium Nutt. in T. & G. Fl. N.Am. 1:589. 1840.
   C. oppositifolium var. *scouleri* Hook. Fl. Bor. Am. 1:242. 1833. *C. scouleri* Rose, Bot. Gaz. 23:
   277. 1897. *(Dr. Scouler,* "Columbia River on the North-West coast")
   Glabrous perennial, the stems usually dichotomously branched, prostrate and freely rooting to
ascending at the tips, up to 2 dm. long, forming mats; leaves opposite (except for the uppermost),
with petioles up to 1 cm. long, the blades from oval or broadly ovate and with truncate to cuneate base,
to reniform, (3) 5-15 mm. broad, up to 2 cm. long, shallowly crenate-serrate with about 15-20 teeth,
usually minutely glandular in the sinuses; flowers solitary in the upper axils, short-pedicellate, about
3 mm. broad; sepals 4, rounded; stamens 8; capsules ultimately about 3-4 mm. long, adnate about
half their length; seeds about 0.5 mm. long, ovoid-ellipsoid, dark purplish-black, smooth.

Moist, usually springy or swampy places; B.C. southward, on the w. side of the Cascades, and mostly along the coast, to n.w. Calif. Late Mar.-May.

Chrysosplenium tetrandum (Lund) Fries, Bot. Notiser 1858:193. 1858.
   C. alternifolium of R. Br. Chlor. Melv. 17. 1823 but not of L. in 1753. *C. alternifolium* var. *tetran-*
   dum Lund, Bot. Notiser 1846:39, 78. 1846, nom. nud., the first valid publication not ascertained by
   us. *C. alternifolium* ssp. *tetrandum* Hultén, Fl. Aleut. Is. 217. 1937. (Typification doubtful)
   Slenderly stoloniferous, the stems as much as 1.5 dm. long, with erect tips up to 10 cm. tall, simple below but usually 2-3 times branched near the ends to form few-flowered, leafy cymes; leaves alternate, chiefly basal or clustered near the stem ends, with slender petioles up to 2 cm. long, the
blades oval to reniform, mostly 5-10 (12) mm. broad, with 3-7 broad, fairly deep crenations; flowers
about 3 mm. broad; calyx lobes oval; stamens 4; free portion of the capsule shorter than the adnate

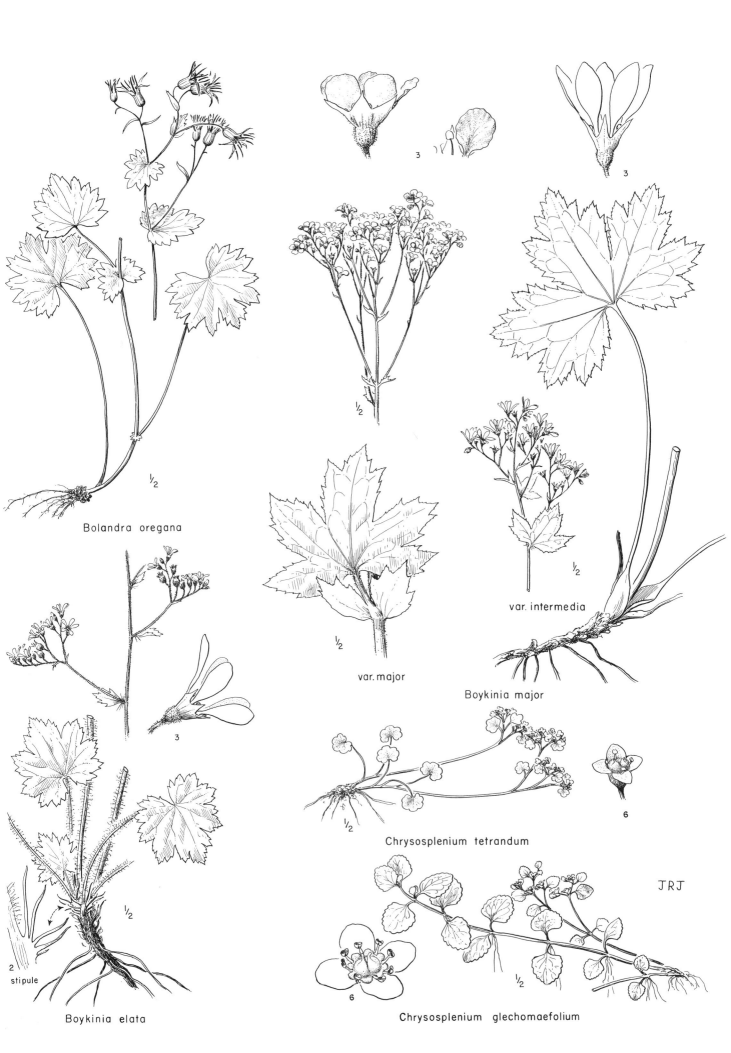

Bolandra oregana

3

3

var. intermedia

var. major

Boykinia major

3

2
stipule

Boykinia elata

Chrysosplenium tetrandum

6

6

Chrysosplenium glechomaefolium

JRJ

portion, the whole about 4 mm. long, after dehiscence the remnants 4-lobed; seeds about 0.5 mm. long, lenticular, dark brown, smooth. N=12.

More or less circumpolar, southward in N. Am., in rock crevices and on wet banks, etc., from Alas. to B. C. and Okanogan Co., Wash., and in the Rocky Mts. to Colo. June.

Conimitella Rydb.

Flowers in a terminal, inconspicuously bracteate raceme, regular, complete; calyx turbinate-ob-conic, adnate to the ovary for about 1/2-1/3 its length, the free, tubular portion (hypanthium) about equaling the 5 erect to slightly spreading lobes; petals 5, white, with an entire blade and slender claw, inserted with the 5 stamens near the base of the hypanthium; stamens opposite the calyx lobes, the short filaments about equaling the anthers; pistil 2 (3-5?)-carpellary, 1-celled with 2 parietal placentae, almost completely inferior, the apices short and hollow and with almost no true style; stigmas nearly sessile, capitate; seeds numerous; perennial scapose herbs with short rhizomes and rather leathery reniform leaves.

Only the one species. (Latin *conus*, a cone, and *Mitella*, to which the plant is closely related, because of the elongate, cone-shaped hypanthium.)

Conimitella williamsii (D. C. Eat.) Rydb. N. Am. Fl. 22[2]:97. 1905.
  Heuchera williamsii D. C. Eat. Bot. Gaz. 15:62. 1890. *Lithophragma williamsii* Greene, Erythea 3: 102. 1895. *(R. S. Williams,* Belt Mts., Mont., July, 1882)
  Tellima nudicaulis Greene, Pitt. 2:162. 1891. *(F. D. Kelsey,* near Deer Lodge, Mont., May 30, 1889)
Perennial from short ascending rootstocks; flowering stems leafless, 2-4 (6) dm. tall, finely glandular-puberulent; leaves with slender petioles (1) 1.5-6 (10) cm. long, the blades reniform, rather leathery, usually purplish on the lower surface, 1-4 cm. broad, shallowly and broadly bicrenate, stiffly ciliate, otherwise usually glabrous; flowers 5-10 (12) in a loose, upwardly blossoming, finely glandular-puberulent raceme, the bracts more or less minutely laciniate; pedicels 1-5 mm. long; calyx narrowly turbinate, 4-6 mm. long (accrescent and up to 9 mm. long in fruit), the lobes ovate-oblong, about 1 mm. long; petals 4-5 mm. long, the rhombic to oblanceolate, entire blades subequal to the slender claws; stamens usually about equaling the hypanthium; capsule dehiscent along the inner (ventral) side of the free, stylelike portion; seeds numerous, black, about 1 mm. long, shallowly muricate in longitudinal lines.

Rock crevices, moist cliffs, and open montane slopes, often on limestone; Rocky Mts. and adjacent ranges of e. Mont., southward to n. w. Wyo. and e. Ida. June-July.

Elmera Rydb.

Flowers in minutely bracteate, simple, terminal racemes on leafy (occasionally leafless) flowering stems; calyx cup-shaped, adnate to the ovary only at the base, the free hypanthium not at all flared, lined with a thin glandular disc, regular to very slightly lopsided, about as long as the calyx lobes; petals white, with a short claw about equal to the 3- to 7-cleft (entire) blade; stamens 5, opposite to, and shorter than, the calyx lobes, inserted slightly lower than the petals on the hypanthium; pistil 2-carpellary; ovary 1-celled, with 2 parietal placentae, adnate for scarcely 1/4 its length, tapered above into erect, conical, hollow, ventrally dehiscent beaks ending in 2 short, thick, solid styles; stigmas capitate; fruit an ovoid, many-seeded capsule dehiscent by inner (ventral) sutures on the beaklike upper portion; seeds ovoid-oblong, with a conspicuous raphe, more or less evenly verrucose in longitudinal rows; herbaceous perennials with slender rootstocks and long-petioled, reniform, palmately lobed and once or twice crenate leaves, strongly pubescent, glandular above.

Only the one species. (Named for A. D. E. Elmer, 1870-1942, an American botanist who collected in the Northwest from 1896 to 1900.)

Elmera is closely related to *Heuchera* and *Tellima,* its one species having been referred to each. From *Heuchera* it differs the more strongly because of the leafy flowering stems, racemose inflorescence, less inferior ovary, and cleft petals. There is probably as much reason for combining *Elmera* and *Tellima* as for maintaining them separately since they differ mainly in stamen number, in the degree of dissection of the petals, in habit, and in other minor ways more usually regarded as specific, rather than generic, distinctions.

Elmera racemosa (Wats.) Rydb. N.Am. Fl. 22^2:97. 1905.

> *Heuchera racemosa* Wats. Proc. Am. Acad. 20:365. 1885. *Tellima racemosa* Greene, Erythea 3:55.
> 1895. *(Suksdorf,* Mt. Adams, Wash., July, 1883)
> ELMERA RACEMOSA var. PUBERULENTA C. L. Hitchc. hoc loc. A var. *racemosa* differt par-
> tibus inferioribus caulorum glandulo-puberulentis, non hirsutis; petiolis glandulo-puberulentis vel
> puberulentis et sparse hirsutis. (Type: *J. W. Thompson 8739,* ledges of cliffs at head of Beverly
> Creek, Wenatchee Mts., Kittitas Co., Wash., in the U. of Wash. Herb.) Other collections from
> the same general region include: *Thompson 5793, 6841, 9510, 10764,* and *12631, Morrill 372,* and
> *English & English 2347.*

Rootstocks slender, horizontal, the flowering stems erect, 1-2.5 dm. tall; basal leaves with the
blades reniform, (2) 3-5 cm. broad, considerably broader than long, the stipules large, membranous,
brownish; cauline leaves usually (0-1) 2-3; raceme loosely 10- to 35-flowered, the bracts small, fimbriate, brownish; calyx greenish-yellow, the hypanthium distinctly tubular, 3-4 mm. long, about
equaling the triangular-oblong, erect to (allegedly) spreading lobes; capsule 4-5 mm. long, ovoid, the
beaks scarcely 0.5 mm. long; seeds dark brown, about 0.6 mm. long.

In rock crevices and on rocky ledges and talus slopes, montane to subalpine; Wash., from Hart's
Pass, Okanogan Co., southward to the s. side of Mt. Adams, also in the Olympic and the Wenatchee
mts. June-Aug.

The species is represented by these two clearly marked varieties:

1 Petioles and lower stems from only strongly hirsute-glandular to both hirsute and puberulent;
   inflorescence glandular-puberulent and less conspicuously hirsute; Cascade Mts., from the
   vicinity of Mt. Rainier s. to Mt. Adams, and in the Olympic Mts.   var. racemosa
1 Petioles and lower stems glandular-puberulent, the stems without any, and the petioles with
   comparatively few (if any), longer stiff hairs; inflorescence sometimes with longer hairs
   as well as with glandular puberulence but the pubescence always shorter than in var. *race-*
   mosa; Cascade Mts., Snohomish Co. (Columbia Peak), and the Mt. Stuart area of the
   Wenatchee Mts., in n. Kittitas Co., n. to Hart's Pass, Wash.   var. puberulenta C. L. Hitchc.

Heuchera L. Alumroot

Flowers regular to irregular, usually complete, borne in an open to congested and more or less
spikelike, membranous-bracteate thyrse, the inflorescence sometimes racemose above; calyx greenish, yellowish, or reddish-tinged, from shallowly saucer-shaped to conic or tubular-campanulate,
basally adnate almost to the top of the ovary but with a short to well-developed, free hypanthium, the
lobes generally 5; petals usually 5 but sometimes fewer (lacking), white to greenish-yellow, mostly
distinctly clawed and with an ovate to spatulate or linear, entire blade; stamens normally 5, one or
more often rudimentary, opposite the calyx lobes, borne with the petals at or near the top of the hypanthium; ovary from about half to nearly completely inferior, 2-carpellary, 1-celled with 2 parietal
placentae, projecting above into hollow, conical, stylelike, ventrally dehiscent beaks, true styles
from well developed to almost or entirely lacking; stigmas from capitate to slightly lunate, but not
lobed; capsule many-seeded, dehiscent along the beaks; seeds spinulose in longitudinal rows (all ours)
to warty or nearly smooth; herbaceous perennials with branched crowns and usually with thick scaly
rootstocks and erect, naked to bracteate flowering stems, more or less glandular; leaves basal, long-
petioled, palmately lobed and usually deeply once or twice crenate-dentate, the stipules membranous
and fused with the petioles.

Probably about 35 species of N.Am., ranging from s. Mex. to the Arctic, preponderantly in w. U.S.
(Named for Johann Heinrich von Heucher, 1677-1747, Professor of Medicine at Wittenberg.)

One species, *H. sanguinea* (coral-bells), of Ariz. and Mex., is a general favorite in cultivation,
especially in rockeries; several of our native species, notably *H. glabra* and *H. micrantha,* are well
worth a place in the native garden, preferably in a moist spot. *H. cylindrica* has considerable merit
as a rock garden or dry-wall plant, suitable for either side (but more easily grown to the e.) of the
Cascade Mts.

The genus is a difficult one, taxonomically, because of the variability of the species and the amount
of hybridization between taxa, the two conditions being closely related. In the treatment presented
here, whereby closely similar populations (which are evidently interfertile) are treated as infraspecific rather than specific taxa, there is a considerable reduction in the number of species recognized.

References:

Calder, J. A. and D. B. O. Savile. Studies in Saxifragaceae-I. The Heuchera cylindrica complex in and adjacent to British Columbia. Brittonia 11:49-67. 1959.

Rosendahl, Carl Otto, Frederick K. Butters, and Olga Lakela. A monograph on the genus Heuchera. Minn. Stud. Pl. Sci. 2:1-180. 1936.

1 Stamens strongly exserted; styles slender, 1.5-3 mm. long; flowers in open to diffuse panicles
 2 Calyx (as measured from the tip of the pedicel) 1.5-3 (3.5) mm. long and nearly as broad, the free hypanthium rarely much over 0.5 mm. long, considerably shorter than the adnate portion; panicle diffuse
  3 Seeds slenderly ellipsoid, slightly curved, 3-4 times as long as broad, medium brown, prominently echinulate; basal portion of the flowering stem and the petioles glabrous or occasionally glandular-puberulent, the stipules ciliate; leaf blades (as measured from the basal sinus) nearly always broader than long, acutely lobed 1/3-1/2 their length, the sinuses narrow **H. GLABRA**
  3 Seeds broadly ellipsoid-oblong, not curved, less than twice as long as broad, nearly black, finely echinulate to tessellate; basal portion of the flowering stem and the petioles usually conspicuously villous with white or brownish hairs, rarely glandular-puberulent to glabrous but then the stipules long-villous margined; leaf blades often longer than broad and with rounded, shallow lobes **H. MICRANTHA**
 2 Calyx (measured from the tip of the pedicel) (3) 3.5-10 mm. long, usually not nearly so broad, the free hypanthium well over 0.5 mm. long, often equaling or longer than the adnate portion; panicle open to contracted but not diffuse
  4 Calyx (3) 3.5-5 mm. long, pinkish; hypanthium 0.5-1.5 mm. long; extreme s.e. Oreg. to s. Calif., e. to s.w. Ida., Utah, Colo., and Ariz.; not known from our region *H. rubescens* **Torr.**
  4 Calyx 5-10 mm. long, greenish; hypanthium 1-5 mm. long; Alta. to Colo., mostly e. of the Rockies, but entering our range in c. Mont. **H. RICHARDSONII**
1 Stamens usually shorter than the calyx lobes; styles short and thick, rarely as much as 1 mm. long; flowers usually in congested panicles
 5 Calyx at anthesis mostly 2-3 mm. long, regular, turbinate at the adnate base, the hypanthium flared and somewhat saucer-shaped, nearly as long as the spreading lobes, 1-1.3 mm. long, lined with a thin glandular disc that more or less covers the almost completely inferior ovary; petals broadly elliptic to ovate **H. PARVIFOLIA**
 5 Calyx at anthesis usually well over 3 mm. long, often oblique, generally campanulate at base and the hypanthium cup-shaped, the lobes usually erect, mostly 1.5 mm. long; hypanthium not gland-lined or if so the disc not covering the top of the ovary, the ovary usually no more than half inferior; petals often lanceolate to oblanceolate or wanting
  6 Petioles and the lower portions of the stems densely villous with brownish (dried), mostly eglandular hairs 2-5 mm. long; filaments often over twice as long as the dehisced anthers; Cascade Mts. and w. from s. B.C. to s. Oreg., w. of the crest in Wash. except up the Columbia R. Gorge, sometimes eastward in Oreg. **H. CHLORANTHA**
  6 Petioles and the lower portion of the stems glabrous to glandular-pubescent or somewhat glandular-villous, the hairs rarely as much as 2 mm. long and then mostly whitish when dried and usually glanduliferous; filaments mostly less than twice as long as the dehisced anthers; nearly entirely e. of the Cascade crest
   7 Calyx 4-6 (6.5) mm. long at anthesis; petals 5 and present on all (or nearly all) flowers, from 1/2 as long as the calyx lobes to nearly twice as long, the blade oblanceolate to spatulate; stamens not over 1.5 mm. long; leaves usually broader than long **H. GROSSULARIIFOLIA**
   7 Calyx (4.5) 6-8 mm. long at anthesis; petals usually not over half as long as the calyx lobes, mostly linear and fewer than 5, or even lacking; stamens commonly over 1.5 mm. long; leaves generally longer than broad **H. CYLINDRICA**

Heuchera chlorantha Piper, Contr. U.S. Nat. Herb. 16:206. 1913. *(Suksdorf 1739,* Falcon Valley, Klickitat Co., Wash., June 28, 1892)
 Strong perennial with a branched crown and short thick rootstocks; flowering stems 4-10 dm. tall, the lower portion conspicuously villous-hirsute with brownish (usually whitish before drying), spreading to retrorse hairs 2-5 mm. long, gradually becoming glandular-hirsute to glandular-pubescent

with mostly whitish hairs above, bractless or with 1-2 small brownish-membranous bracts below tne panicle; leaf blades cordate-ovate to nearly reniform, (3) 4-8 cm. broad and somewhat shorter, with 5-9 shallow lobes that are broadly rounded, more or less doubly crenate, and slightly dentate-mucronulate; inflorescence contracted and dense, up to 15 cm. long; calyx turbinate-campanulate, greenish to cream, 7-9 mm. long at anthesis, up to 12 mm. long in fruit, the hypanthium slightly lopsided, about as long as the adnate portion of the ovary, lined with a thick glandular disc, the lobes somewhat unequal, oblong, obtuse to rounded, equaling or slightly longer than the rest of the calyx; petals various, often in part or entirely lacking, not over half so long as the sepals, linear to broadly spatulate; anthers about 1. 5 mm. long before dehiscence, the filaments then not so long but elongating rapidly after anthesis and becoming 2-3 times as long as the slightly shrunken (0. 8-1 mm. long) anthers; ovary about 4/5 inferior at anthesis, projecting into conical, hollow, stylelike beaks as long as the fertile portion, the closed styles 0. 2-0. 5 mm. long; capsule 9-11 mm. long; seeds dark brown, ellipsoid-ovoid, 0. 7-0. 8 mm. long, echinulate with longitudinal rows of slender, slightly tapered spines.

On gravelly prairies to wooded hillsides; w. of the Cascade summit from Queen Charlotte Is. , B. C., to Douglas Co. , Oreg. , but up the Columbia R. Gorge to the e. base of the Cascades, and s. along the Cascade Mts. , in Oreg. , to Klamath Co. May-Aug.

Heuchera cylindrica Dougl. ex Hook. Fl. Bor. Am. 1:236. 1833.
  Yamala cylindrica Raf. Fl. Tellur. 2:75. 1837. *(Douglas,* "On the declivities of low hills and on the steep banks of streams on the west side of the Rocky Mountains")
  H. ovalifolia Nutt. in T. & G. Fl. N.Am. 1:581. 1840. *H. cylindrica* var. *ovalifolia* Wheelock, Bull. Torrey Club 17:203. 1890. *(Nuttall,* Blue Mountains, Oreg.) = var. *alpina.*
  H. glabella T. & G. Fl. N.Am. 1:581. 1840. *H. cylindrica* var. *glabella* Wheelock, Bull. Torrey Club 17:203. 1890. *(Nuttall,* "Rocky Mountains towards Oregon")
  H. cylindrica var. *alpina* Wats. Bot. King Exp. 5:96. 1871. *H. alpina* Blank. Mont. Agr. Stud. Bot. 1:62. 1905. *H. ovalifolia* var. *alpina* Rosend. Engl. Bot. Jahrb. 37, Beibl. 83:81. 1905. *(Watson,* Clover Mts. , Nev. , Sept. , 1868)
  H. saxicola E. Nels. Bot. Gaz. 30:118. 1900. *(A. & E. Nelson 5687,* Undine Falls, Yellowstone Park, July 6, 1899) = var. *cylindrica.* This taxon, treated at the specific level by Calder and Savile, is usually maintained on the basis of three alleged peculiarities: hirsute pubescence, bracteate stems, and unusually short filaments. The first of these distinctions applies equally well to var. *cylindrica,* especially in the n. Cascades; the third seems not to exist. If recognized at the varietal level, the name to be used is "suksdorfii."
  H. columbiana Rydb. N.Am. Fl. 22²:116. 1905. *H. glabella* ssp. *columbiana* Piper, Contr. U.S. Nat. Herb. 11:321. 1906. *(J. B. Winston,* near Swan Lake [error for Loon Lake, Stevens Co., acc. Piper], Wash. , in 1897) = var. *cylindrica.*
  H. suksdorfii Rydb. N.Am. Fl. 22²:116. 1905. *H. glabella* var. *suksdorfii* Rosend. Engl. Bot. Jahrb. 37, Beibl. 83:81. 1905. *(Suksdorf,* Falcon Valley, Klickitat Co. , Wash. , June 28, 1882) = those plants, otherwise referable to var. *cylindrica, glabella,* or *alpina,* that have unusually large bracts on the inflorescence. Such specimens are largely confined to the Wenatchee Mts. , but are not maintainable as a taxonomic unit without obscuring whatever naturalness there is to the other three varieties.
  H. cylindrica var. *glabella* f. *valida* R. B. & L. Minn. Stud. Pl. Sci. 2:138. 1936. *(Leiberg 1541,* N. Fork Coeur d'Alene R. , Ida.) = var. *glabella.*
  H. cylindrica var. *septentrionalis* R. B. & L. Minn. Stud. Pl. Sci. 2:138. 1936. *(Butters & Holway 404,* Avalanche Crest, Glacier, B.C.) = transitional between var. *glabella* and var. *cylindrica* although closer to the latter.
  H. ovalifolia var. *orbicularis* R. B. & L. Minn. Stud. Pl. Sci. 2:143. 1936. *H. cylindrica* var. *orbicularis* Calder & Savile, Brittonia 11:58. 1959. *(A. Isabel Mulford,* Silver City, Owyhee Co., Ida. , July 4, 1892) = intergradient between var. *cylindrica* and var. *glabella,* with the base of the leaves more or less cordate.
  H. ovalifolia var. *thompsonii* R. B. & L. Minn. Stud. Pl. Sci. 2:144. 1936. *(J. W. Thompson 4833,* cliff south of Elgin, Union Co. , Oreg.) = a slightly aberrant form of var. *cylindrica,* approaching var. *alpina* but with the leaves more hirsute.
  H. suksdorfii f. *ribesoides* R. B. & L. Minn. Stud. Pl. Sci. 2:147. 1936. *(Elmer 1197,* Mt. Stuart, Kittitas Co. , Wash.) = var. *alpina,* but intermediate to var. *glabella;* perhaps unique, but if so because of the leaves rather than because of the bracts.
  H. suksdorfii f. *sandbergii* R. B. & L. Minn. Stud. Pl. Sci. 2:148. 1936. *(Sandberg & Leiberg,*

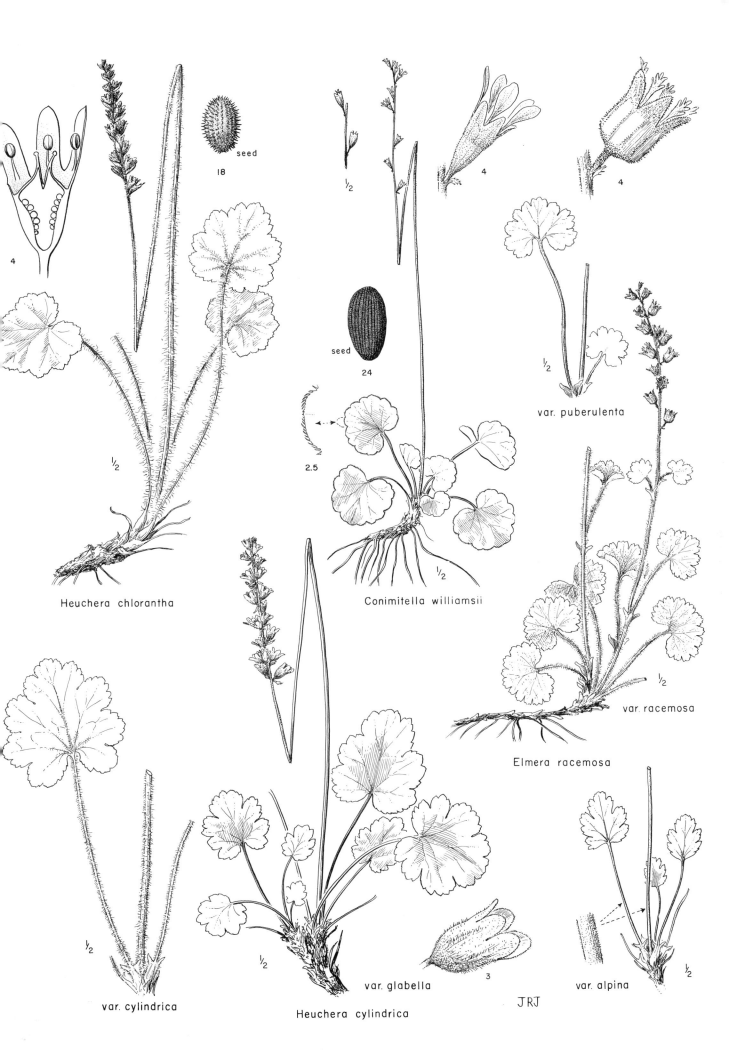

seed 18

4

½

seed 24

2.5

4

4

½

var. puberulenta

Heuchera chlorantha

Conimitella williamsii

var. racemosa

Elmera racemosa

var. cylindrica

½

½

var. glabella

3

Heuchera cylindrica

var. alpina

½

JRJ

Nason City, Wash., Aug., 1893) = var. *glabella,* to which category other collections of a similar nature from the Wenatchee Mts. were referred by Lakela.

Strong perennial with a branching crown and short thick rhizomes; flowering stems (1) 1.5-9 dm. tall, glabrous to glandular-puberulent or hirsute below, always rather thickly glandular-puberulent to glandular-hirsute above, naked or with 1-3 brownish (and membranous) to greenish, simple or 3-lobed bracts; leaf blades ovate or ovate-oblong to cordate-ovate or nearly reniform, mostly 1-6 (up to 14) cm. broad, and from somewhat shorter to considerably longer than broad, from glabrous to copiously glandular-puberulent or hirsute; panicle ultimately open, mostly 3-12 (20) cm. long, the peduncles and pedicels of the lateral cymes together usually shorter than the flowers; calyx (5) 6-8 mm. long at anthesis, up to 10 mm. long shortly afterward, cream-colored to greenish-yellow, turbinate-campanulate to tubular-campanulate, the hypanthium equaling or somewhat longer than the adnate portion of the ovary, slightly to considerably longer on one side than the other and thus irregular, lined with a thin and delicate to thick, more or less glandular disc that partially covers the top of the ovary, the 5 sepals erect, oblong-lanceolate to obovate and overlapping, considerably longer than the hypanthium; petals linear and much shorter than the calyx lobes, sometimes 5, but usually fewer, often lacking; stamens shorter than the calyx lobes, the anthers mostly 1-1.3 mm. long just before anthesis and equaling or exceeding the broad-based filaments, but following anthesis the filaments usually elongating to equal or considerably exceed the anthers; ovule-bearing portion of the ovary 2/3-4/5 inferior, the ovary tapered above to conical, hollow, empty stylar beaks and very short true styles 0.1-0.5 mm. long; stigmas capitate to somewhat crescent-shaped but not lobed; capsule 6-10 mm. long; seeds dark brown, 0.6-0.9 mm. long, oblong-ellipsoid, longitudinally echinulate with slender, straight to curved, conical spines. N=7.

Rocky soil, cliffs, and talus slopes; B.C. southward in and on the e. side of the Cascades, to c. Oreg. and n.e. Calif., e. to Alta., Mont., n.w. Wyo., and n. Nev. Apr.-Aug.

There has been little agreement among various workers concerning the taxa included here. There are several features of the plant that show great variation, and to which much taxonomic significance has been attributed: 1) the leaves (especially the petioles) and the lower portion of the stems range from glabrous to glandular-hirsute; 2) the leaf blades vary in lobation and particularly in the shape of the base, from rounded to deeply cordate; 3) the flowers differ in size, but the difference is greatly accentuated by the fact that there is a rapid growth of the calyx during and after anthesis; 4) this sudden growth change is even more marked in the stamens where just before anthesis the anthers are at their greatest size, always equaling to greatly exceeding the filaments; during and following anthesis the filaments lengthen and eventually usually equal or exceed the anthers; 5) the relative degree of development of the bracts of the flowering stems, in some cases these are large and greenish, more frequently they are much reduced (even absent) and membranous; 6) the degree of development of the disc which lines the hypanthium, in some cases this is very thin and almost indiscernible, in others it may be much thickened; 7) the relative length and degree of divergence of the beaklike styles of the fruit, dependent largely upon the age of the flowers. There is very little correlation among these several variable characters and it is possible to recognize a great many taxa if all the above tendencies are considered important. Regardless of which variations are selected for emphasis, there is essentially complete gradation between the largely sympatric taxa thus delimited.

The most workable treatment of the species was that proposed by Piper (Contr. U.S. Nat. Herb. 16: 206. 1913) where the more conspicuous variants were considered as intergradient and largely artificial infraspecific taxa.

Very recently, Calder & Savile have recognized 6 taxa in the complex, namely *H. saxicola* and *H. cylindrica,* with 5 vars. under the latter. Their interpretation of the group is based on the premise that several populations now recognizable as geographic races became established in refugia separated largely by north-south valley glaciers during Pleistocene time. In our area, at least, three of these taxa are not sufficiently clearly defined to be recognizable.

1 Leaves and lower portion of the stems finely glandular-pubescent (but varying to hirsute), the blades thick, mostly 1-2.5 cm. broad and slightly longer, usually rounded (or somewhat acute) to moderately cordate, varying (especially in the n. part of the range) to cordate and then merging with var. *cylindrica* and with var. *glabella;* Wenatchee Mts., s. Chelan Co., to Grant Co., Wash., s. to Jackson Co., Oreg., and n.e. Calif., and e. through the Steens, Blue, and Wallowa mts., Oreg., to s.w. Ida. and n. Nev.   var. alpina Wats.

1 Leaves and lower part of the stems variously pubescent, the blades frequently over 2.5 cm. broad, usually cordate and with a definite sinus

  2 Lower part of the stems and the leaves, at least the petioles, glabrous or sparsely glandu-

lar-pubescent; with the var. *cylindrica* in much of its range and completely transitional to
it, but less common, or perhaps lacking, in most of B.C. and in Alta., the most common
phase in most of Mont. and much of Ida. and Wash., infrequent in Oreg., transitional to
var. *alpina* in c. Wash. and n.e. Oreg.   var. glabella (T. & G.) Wheelock
2 Lower part of the stems and the petioles glandular-pubescent to hirsute; bracts of the stems
   rather conspicuous in many cases, especially in the n. Cascades (and constituting the basis on which *H. suksdorfii* is sometimes maintained) and occasionally in the Rocky Mts.,
   where considered especially significant if *H. saxicola* is recognized; the common phase of
   the species in the n. Cascades and in B.C. to Mont. and s. to n.w. Wyo.; occurring also in Ida. and s. in Wash. to Oreg. where usually more or less intermediate to var. *glabella* and var. *alpina*   var. cylindrica

Heuchera glabra Willd. ex R. & S. Syst. Veg. 6:216. 1820. ("In Americae borealis plagis occidenta-
   libus")
   Tiarella colorans Grah. Edinb. New Phil. Journ. 7:349. 1829. (Plants cultivated at Edinburgh from
     seeds collected by Drummond in the Rocky Mts.)
   H. divaricata Fisch. ex Ser. in DC. Prodr. 4:51. 1830. *(Langsdorff,* "in Asiae borealis insulis Ka-
     diak et Sitka")
Rootstocks well developed, horizontal to ascending; flowering stems 1-several, (1.5) 2.5-6 dm.
tall, glabrous to sparsely glandular-puberulent below but becoming rather strongly glandular-puberulent to -pubescent in the inflorescence; leaves mainly basal, the blades cordate-ovate, mostly 3-9 cm.
broad and not quite so long, sparsely glandular-puberulent on the lower surface but otherwise usually
glabrous, 5 (7)-lobed 1/3-1/2 way to the base, the lobes again shallowly lobed and coarsely crenate-
dentate, ultimately with acute teeth; cauline leaves 1 or 2, greatly reduced and the uppermost usually
more or less bractlike; inflorescence a large, open, linear-bracteate, many-flowered thyrse, the
branches and pedicels filiform; calyx obconic-cyathiform, (1.5) 2-3 (3.5) mm. long, glandular-pubescent, the 5 lobes ovate-oblong, about 2/3 as long as the almost completely adnate lower portion, the
free hypanthium less than 0.5 mm. long; petals white, 2-4 times as long as the sepals, with slender
claws equal to, or somewhat longer than, the narrowly rhombic to spatulate, entire blades; stamens
5, strongly exserted, about equaling the petals; ovary almost completely inferior at anthesis, the free
portion shortly conical below the slender styles; stigmas slightly enlarged, capitate; capsule at length
exserted from the calyx, 5-6 mm. long, ovoid, slenderly beaked by the persistent styles; seeds medium brown, 0.7-0.8 mm. long, narrowly ellipsoid, 3-4 times as long as broad, somewhat lunate,
prominently echinulate with longitudinal rows of about 20 spines that are approximately 1/4-1/6 as
long as the seed is thick.
   Stream banks and crevices of moist rocks, from the coast to above timber line, Alas. southward,
in both the Cascade and Olympic mts. in Wash., to Mt. Hood, Oreg., e. in Wash. to the Wenatchee
Mts. and the Entiat Valley, and in B.C. to the Selkirk Mts. and possibly to the Rocky Mts. Early
June-late Aug.

Heuchera grossulariifolia Rydb. Mem. N.Y. Bot. Gard. 1:196. 1900.
   H. hallii var. *grossulariifolia* Rosend. Engl. Bot. Jahrb. 37, Beibl. 83:81. 1905. *(Rydberg & Bes-
     sey 4288,* Pony, Mont., July 6, 1897, is the first collection cited)
   HEUCHERA GROSSULARIIFOLIA var. TENUIFOLIA (Wheelock) C. L. Hitchc. hoc loc. *H. cylindrica*
     var. *tenuifolia* Wheelock, Bull. Torrey Club 17:204. 1890. *H. tenuifolia* Rydb. N.Am. Fl. 22[2]:116.
     1905. *(Thos. Howell,* near the Dalles, Oreg., is the first of 2 collections cited)
   H. gracilis Rydb. N.Am. Fl. 22[2]:114. 1905. *(Henderson 3159,* Mt. Idaho, Camas Prairie, Ida.) =
     var. *grossulariifolia.*
   H. cusickii R. B. & L. Minn. Stud. Pl. Sci. 2:157. 1936. *(Cusick 2539,* Cliffs of N. Pine, Oreg.)=
     var. *grossulariifolia.*
Caudex thick, branched, the flowering stems leafless, 1.5-6.5 (8) dm. tall, glandular-puberulent
throughout to glandular-hirsute above, but sometimes eglandular or even glabrous near the base;
leaves glabrous to glandular-puberulent, the blades cordate-orbicular to cordate-reniform, 1-7 cm.
broad, always slightly to considerably shorter as measured from the open to rather narrow sinus,
(3) 5- to 7-lobed for 1/5-1/3 of the length, the lobes sometimes overlapping, coarsely 2-3 times
crenate-dentate, the margins usually coarsely ciliate-cuspidate; inflorescence 1-6 cm. long and generally tightly congested at anthesis, in fruit usually open and 5-12 (20) cm. long, the branches ascending, up to 1.5 (8) cm. long, 2- to 6-flowered; calyx 4-6 (6.5) mm. long at anthesis, oblique,

one side considerably longer than the other, tubular-campanulate, the basal adnate portion about equal to the shorter side of the scarcely flared hypanthium, lobes oblong to oval, erect, equaling or somewhat longer than the hypanthium; petals white, from considerably shorter than the sepals to half again as long, narrowly clawed and with an oblanceolate to spatulate blade; stamens shorter than the sepals, incurved, the stout filaments from only half as long to nearly as long as the cordate, oval anthers; ovary about 3/4 inferior at anthesis, tapered above to stout hollow beaks and very short thick styles not over 1 mm. long; stigmas enlarged, capitate but with a slight lobing; capsules 4.5-7 mm. long; seeds 0.6-0.8 mm. long, dark brown, ovoid-ellipsoid, closely covered with short conical spines.

Grassy hillsides and rocky canyon walls to alpine scree and talus slopes; s. w. Mont. and c. Ida. to the Wallowa Mts. of n.e. Oreg., and in the Columbia R. Gorge of Oreg. and Wash. May-early Aug.

This is a rather variable taxon that has previously been treated as three separate species, the many intermediates being accounted for as postulated hybrids; actually, with the range of the species more completely understood, it appears to be a natural entity consisting of 2 well-marked but completely intergradient varieties.

1 Leaves mostly 1-2.5 (3.5) cm. broad; petals usually as long to half again as long as the sepals; calyx 4-5 (6) mm. long; flowering stems rarely over 4.5 dm. tall; in Ravalli, Beaverhead, Deerlodge, and Madison cos., Mont., in Lemhi, Custer, Blaine, Elmore, Boise, and Valley cos., and in the Snake R. Canyon from Payette to Idaho Co., Ida., and in Wallowa and Baker cos., Oreg.   var. grossulariifolia
1 Leaves mostly (2.5) 3-7 cm. broad; petals usually shorter than the sepals; calyx 5-6.5 mm. long; flowering stems up to 8 dm. tall, the plants tending to be larger in nearly all respects (except the petals) than the more eastern variety; Payette and Salmon R. drainages in n. Elmore and Boise cos., and in Valley and Idaho cos., Ida., and in the Columbia R. Gorge of Wasco and Hood R. cos., Oreg., and adj. Wash.   var. tenuifolia (Wheelock) C. L. Hitchc.

Heuchera micrantha Dougl. ex. Lindl. Bot. Reg. 15:pl. 1302. 1830. (*Douglas,* "near the Grand Rapids of the Columbia")

    H. longipetala Moc. ex Ser. in DC. Prodr. 4:52. 1830. (Moc. pl. notk. ined. icon. "in Americae bor. plagis occid.," described from a plate of Mocino's of a plant from Nootka, Vancouver I.)

    H. barbarossa Presl, Rel. Haenk. 2:56. 1835. (*Haenke,* Nootka Sound) = var. *diversifolia.*

    H. diversifolia Rydb. N. Am. Fl. 22²:102. 1905. *H. micrantha* var. *diversifolia* R. B. & L. Minn. Stud. Pl. Sci. 2:42. 1936. (*F. H. Lamb 1305,* Baldy Peak, Wash.)

    H. glaberrima Rydb. N. Am. Fl. 22²:103. 1905. *H. micrantha* f. *glaberrima* R. B. & L. Minn. Stud. Pl. Sci. 2:42. 1936. *H. micrantha* var. *glaberrima* W. H. Baker, Am. Midl. Nat. 46:154. 1951. (*Howell,* Oregon) = var. *micrantha.*

    H. nuttallii Rydb. N. Am. Fl. 22²:103. 1905. *H. micrantha* var. *nuttallii* Rosend. Eng. Bot. Jahrb. 37, Beibl. 83:77. 1905. (*Nuttall,* "Among rocks on the Oregon, near the mouth of the Willamette") = var. *micrantha.*

    H. lloydii Rydb. N. Am. Fl. 22²:113. 1905. (*F. E. Lloyd,* on the 45th parallel in the Cascade Mts., in 1895) = var. *micrantha.*

    H. micrantha var. *diversifolia* f. *acuta* R. B. & L. Minn. Stud. Pl. Sci. 2:43. 1936. (*Rosendahl 1812,* Cowichan Lake, Vancouver I., B.C., June 15, 1907) = var. *diversifolia.*

Very similar to *H. glabra* but differing in that the petioles and lower portions of the stems are usually long-villous with whitish hairs that turn brownish when dried, much less commonly the villosity scant but the stipules with pectinate-villous marginal hairs 1-3 mm. long; leaf blades reniform to cordate-ovate or cordate-oblong, (2) 3-8 cm. broad, mostly either shallowly lobed much less than 1/3 their length with 5-7 rounded lobes and distinctly broader than long, or more acutely lobed up to 1/3 their length and then usually longer than broad, commonly more or less strigose on the lower (and sometimes on the upper) surface; calyx generally villous, more campanulate than cyathiform, 1.5-3 mm. long, the hypanthium about 0.5 mm. long; seeds deep brownish-purple, ovoid-oblong, 0.6-0.7 mm. long, at least 1/2 as thick, finely echinulate with longitudinal rows of 25-30 spines that are about 1/10 as long as the thickness of the seed. N=7.

Gravelly stream banks and rock crevices from near sea level to subalpine, where often on talus slopes; B.C. southward, in the Cascades and w., to Monterey Co. and the Sierra Nevada of Calif., e. in Oreg. to the Blue and Wallowa mts., and into adj. Ida. May-Aug.

Although *H. micrantha* and *H. glabra* are usually recognizable with no difficulty, there is sufficient intermediacy between them to indicate that they interbreed to some extent. There are several geographic races of *H. micrantha,* two of which, perhaps the least clearly marked, occur in our area.

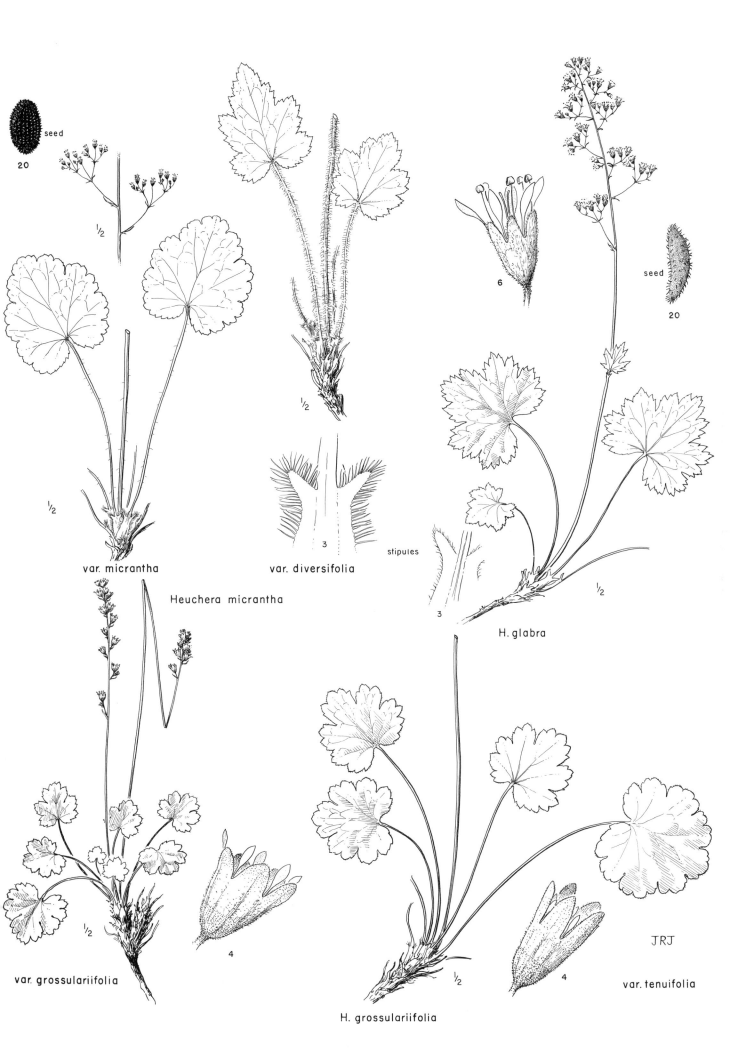

seed

20

½

½

var. micrantha

½

var. diversifolia

stipules

3

Heuchera micrantha

6

seed

20

3

½

H. glabra

½

var. grossulariifolia

4

H. grossulariifolia

½

4

JRJ

var. tenuifolia

1 Leaf blades at least as broad as long, shallowly lobed much less than 1/3 their length, the
      lobes rounded; petioles and lower stems villous to puberulent or subglabrous; occasional
      in Wash. , but becoming the common phase from the Columbia R. Gorge s. to Marion Co. ,
      Oreg. , e. in Oreg. to Ida.   var. micrantha
1 Leaf blades often longer than broad, more deeply and acutely lobed (up to 1/3 their length);
      petioles and lower stems usually strongly villous; the common phase in Wash. and B. C.
                                                                var. diversifolia (Rydb.) R. B. & L.

Heuchera parvifolia Nutt. ex T. & G. Fl. N.Am. 1:581. 1840. *(Dr. James,* Rocky Mts. , is the first
   of 2 collections cited, the second, *Nuttall,* "Blue Mountains of Oregon," was probably collected in
   Wyo.)
   H. utahensis Rydb. N.Am. Fl. 22^2:114. 1905. *H. parvifolia* var. *utahensis* Garrett, Spr. Fl. Wa-
      satch Reg. 39. 1911. *(M. E. Jones 9458,* City Creek Canyon, Utah, in 1880)
   H. parvifolia var. *dissecta* M. E. Jones, Bull. U. Mont. Biol. no. 15:32. 1910. *(M. E. Jones,* "An-
      aconda and Durant," according to R. B. & L. the type was collected in Deer Lodge Valley, Mont.,
      July 19, 1905)
   H. flabellifolia Rydb. N. Am. Fl. 22^2:115. 1905. *(Rydberg & Bessey 4292,* Bridger Mts. , Gallatin
      Co. , Mont. , in 1897) = var. *dissecta.*
   H. flabellifolia var. *subsecta* R. B. & L. Minn. Stud. Pl. Sci. 2:173. 1936. *(Heller 11107,* Ruby
      Mts. , near Blaine Post Office, Nev.) = var. *utahensis,* at least as to the plants from Ida.
      Perennial from a large branched crown, more or less glandular-puberulent throughout or some-
   times slightly hirsute, occasionally nearly glabrous below; leaves all basal, the blade cordate-orbicular to reniform, (1) 2-6 cm. broad and considerably shorter, shallowly to deeply 5- to 7-lobed, the
   segments often again less deeply lobed and strongly and broadly crenulate; flowering stems 1. 5-6 dm.
   tall; inflorescence at first tightly congested, at length open and up to 2. 5 dm. long, the branches 1-3
   (6) cm. long; calyx greenish, mostly 2-3 mm. long, turbinate at base where adnate to the ovary, the
   free hypanthium flared considerably, and nearly as long as the triangular to ovate, spreading lobes,
   the hypanthium lined with a conspicuous glandular disc which more or less completely covers the top
   of the ovary, the whole appearance of the calyx at anthesis very quickly changed by the accrescence of
   the lower adnate portion; petals white, about 1. 5 times as long as the calyx lobes, the blade elliptic or
   rhombic to broadly ovate, slightly longer than the narrow claw; stamens shorter than the sepals, incurved, the filaments about equaling the broadly cordate anthers; ovary at anthesis nearly completely
   inferior, often almost flat-topped, with only the carpel tips and the short (0. 2-0. 4 mm. long) styles
   projecting through, or the disc less complete and the upper conical portions of the ovary projecting
   through, this condition largely dependent upon the stage of development of the flower, the ovary growing rapidly and the adnate portion of the mature calyx becoming more campanulate and 2. 5-3. 5 mm.
   long; capsule 4-7 mm. long; seeds ovoid-ellipsoid, dark brown, 0. 8-1 mm. long, finely echinulate
   with longitudinal rows of short conical spines.
      Montane, on granitic and limestone cliffs and gravelly slopes and talus; Rocky Mts. from Alta. to N.
   M. , w. to c. Ida. , Nev. , and Ariz. June-Aug.
      The taxon is represented by several geographic races, two of which occur in our range. These have
   previously been maintained as distinct species largely on the basis of the degree of development of the
   perigynous glandular disc. Flowers of the two taxa, at comparable stages of development, appear not
   to differ in the way alleged, and since the two intergrade freely, they are here treated as varieties.
   Other varieties, including var. *parvifolia,* occur to the south of our range.
1 Flowering stems mostly at least 2 dm. tall; leaves usually 3-6 cm. broad, deeply cordate with
      a narrow sinus, lobed 1/5-1/3 of the length; Wyo. to Colo. and Utah, w. to c. Ida. and Nev.
                                                                var. utahensis (Rydb.) Garrett
1 Flowering stems mostly less than 2 dm. tall; leaves (1) 2-3 (4) cm. broad, from deeply cor-
      date and with a narrow sinus to shallowly cordate, mostly lobed 1/3-1/2 of the length; the
      common phase of the plant in Mont. and Alta. , also in c. Ida.; freely intergradient with
      var. *utahensis*   var. dissecta M. E. Jones

Heuchera richardsonii R. Br. in Frankl. Journ. 766. 1823.
   H. hispida richardsonii Rydb. in Britt. Man. 483. 1901. *(John Richardson,* British N.Am.)
   H. ciliata Rydb. Mem. N.Y. Bot. Gard. 1:196. 1900. *(Tweedy 259,* Mill Creek, Mont. , in 1887)
      Perennial from a (usually) branched caudex, only very slightly or not at all rhizomatous; leaves all
   basal, the flowering stems leafless, 4-7 dm. tall, moderately hispid with whitish hairs below, becom-

ing glandular-puberulent above; leaf blades (2) 3-6 cm. broad, cordate-reniform, shallowly (5) 7- to 9-lobed and prominently crenate-dentate, usually glabrous on the upper surface and somewhat whitish-hirsute on the lower; flowers in a rather narrow thyrse; calyx greenish, strongly glandular-puberulent, (5) 6-9 mm. long, campanulate-cylindric, the base obconic, adnate to the ovary, the hypanthium lopsided, 2-4 mm. long on one side but scarcely half as long on the other, the lobes oblong, erect, about as long as the hypanthium; petals clawed, spatulate, equaling or slightly exceeding the sepals; stamens more or less exserted; ovary more than half inferior; styles stout, 1.5-2.5 mm. long; stigmas small; seeds dark brown, about 0.7 mm. long, ovoid, finely echinulate.

Rock crevices and stream banks to moist sandy prairie, from Mackenzie, and probably n. B. C., southward, almost entirely on the e. side of the Rockies, to Colo., e. to Sask. and Man., Minn., Wisc., and Ind. June-July.

This species, which barely enters our area in the vicinity of Bozeman, Mont., is composed of several geographic races, our material, as described above, being referable to the var. *richardsonii*.

Leptarrhena R. Br.

Flowers numerous in a rather tight, several-branched thyrse; calyx deeply saucer shaped, the 5 lobes erect, longer than the adnate lower portion, a free hypanthium lacking; petals white, small, persistent; stamens 10, inserted with the petals at the edge of the ovary; carpels 2, fused only at the base of the very slightly (not over 1/6) inferior ovary, tapered above to hollow, ventrally dehiscent, beaklike apices, true styles almost lacking, the stigmas no broader than the style, capitate; fruit 2 ventrally dehiscent, lance-ovoid, membranous follicles; seeds numerous, erect, much elongate and more or less lunate, the testa loose, plainly cellular, attenuate to a tubular tail at each end; strongly rhizomatous perennial herbs with leathery, persistent, short-petiolate, toothed leaves and sparsely leafy, simple flowering stems that are more or less glandular-pubescent.

Only the one species. (Greek *leptos*, fine or slender, and *arrhen*, male, in reference to the slender filaments.)

Leptarrhena pyrolifolia (D. Don) R. Br. ex Ser. in DC. Prodr. 4:48. 1830.
   Saxifraga pyrolifolia D. Don, Trans. Linn. Soc. 13:389. 1822. (*Merk*, Kamtschatka, is the first of 2 collections cited)
   Saxifraga amplexifolia Sternb. Rev. Saxifr. Suppl. 1:2. 1822. *L. amplexifolia* R. Br. ex Ser. in DC. Prodr. 4:48. 1830. (*Fischer, Langsdorf*, and *Chamisso*, separate collections all from Unalaska)
   L. amplexifolia var. *laxiflora* Ser. in DC. Prodr. 4:48. 1830. (No specimens cited)
   L. amplexifolia var. *glomerata* Ser. loc. cit. (No specimens cited)
Rootstocks widespreading, horizontal; leaves glabrous, mostly basal, narrowly obovate to elliptic or ovate-oblong, 3-15 cm. long, narrowed to a short broad petiole less than 1/2 as long as the blade, prominently once (and often much less strongly secondarily) crenate-serrate, bright green on the upper surface, pale green beneath; stipules membranous, fused with the petioles; cauline leaves (except the crowded basal ones) 1-3, remote, sessile and amplexicaul, mostly oblong-oblanceolate and much smaller than the basal blades; flowering stems simple, (1) 2-4 (5) dm. tall; inflorescence strongly congested at early anthesis but becoming much more open, the bracts minute; calyx 2-3 mm. broad, the sepals about 1 mm. long; petals spatulate to oblanceolate, usually unequal, up to twice as long as the sepals; stamens equaling or half again as long as the petals; follicles 5-6 mm. long; seeds light brown, 2-3.5 mm. long, the empty tail-like ends of the testa 2-3 times as long as the rest of the seed.

Stream banks and wet meadows to moist subalpine slopes; Alas. southward, along the coast and in the Olympic and Cascade mts., to Mt. Adams, Wash., and in the high Cascades of Oreg. to the Sisters, e. in B. C. and n. Wash. to n. Ida., w. Mont., and s.w. Alta. June-Aug.

The plant takes well to cultivation and, because of its deep green leathery leaves and usually reddish follicles, is well worth a place in the native garden or in a moist spot in the rockery.

Lithophragma Nutt. Prairie Star; Woodland Star

Flowers rather showy, often partially or wholly replaced by small purplish bulblets, borne in terminal, simple or (especially when represented in part by bulblets) somewhat compound racemes; calyx narrowly cyathiform-obconic to campanulate or cup-shaped, shallowly 5-lobed, partially adnate to the ovary but with a well-developed hypanthium; petals 5, white to pink or purplish-tinged, often somewhat unequal, narrowly clawed and with a large, expanded, usually digitately (pinnately) cleft or divided (ours) to shallowly lobed or sometimes entire blade; stamens 10, included in the hypanthium or partially exserted, the filaments about equaling or shorter than the oblong-cordate anthers, inserted with the petals somewhat below the top of the hypanthium; pistil 3-carpellary; ovary 1-celled with 3 parietal placentae, 1/5-5/6 inferior; styles 3, forming short beaks to the 3-valved capsule; stigmas rounded; seeds numerous, from slightly wrinkled but otherwise smooth to irregularly reticulate, verrucose, or muricate; perennial, usually glandular-pubescent herbs with simple, sparingly leafy flowering stems and slender rootstocks bearing numerous rice-grain bulblets and often with bulblets in the axils of the cauline leaves and the bracts of the inflorescence; leaves mostly basal, orbicular-reniform to reniform, palmately parted or cleft and less deeply laciniately lobed to shallowly lobed or deeply crenate, the petioles slender, with dilated, membranous stipular bases, the cauline leaves 1-several, reduced, sometimes sessile.

Eight or 9 species of temperate w. N. Am. (Greek *lithos*, stone, and *phragma*, wall, supposedly in reference to the habitat of the plant.)

The plants are well worth a place in the woodland or rock garden, but of brief seasonal duration, dying to the ground by midsummer. The bulbs must be kept as dry as possible during the dormant period for best results.

1 Seeds muricate; basal leaves glabrous or very sparsely pubescent; flowers mostly 2-5 (7),
    more or less corymbose at anthesis, later becoming "racemose," but the lower pedicels
    usually elongate and mostly 1.5-3 times as long as the calyx; cauline leaves often bulbiferous; petals usually 5-lobed
    2 Cauline leaves bulbiferous
    2 Cauline leaves not bulbiferous   L. BULBIFERA
                                           L. GLABRA
1 Seeds from nearly smooth (slightly wrinkled) to verrucose or reticulate but never muricate;
    basal leaves moderately to copiously pubescent at least on the lower surface; flowers (sometimes as few as 4 or 5) usually 6-11, racemose, the lower pedicels mostly subequal to (to
    half again as long as) the calyx; cauline leaves not bulbiferous; petals often only 3-lobed
    3 Calyx obconic-cyathiform, (3) 4-6 mm. long at anthesis, mostly 6-10 mm. long in fruit;
       ovary at least 2/3 inferior; petals commonly 3-lobed   L. PARVIFLORA
    3 Calyx more nearly campanulate, 2-3 (3.5) mm. long at anthesis and 3.5-5 mm. long in
       fruit; ovary about 1/2 inferior; petals mostly 5 (7)-lobed   L. TENELLA

Lithophragma bulbifera Rydb. N. Am. Fl. 22²:86. 1905.
    Tellima bulbifera Fedde, Just Bot. Jahresb. 33¹:614. 1906. *L. glabra* var. *bulbifera* Jeps. Fl. Calif.
      2:130. 1936. (*F. Tweedy 4411*, Battle, Carbon Co., Wyo.)
    L. tenella var. *ramulosa* Suksd. West Am. Sci. 15:61. 1906. (*Suksdorf 4013*, near Bingen, Wash.,
      Apr., 1898-1904)

Very similar to *L. glabra*, but the cauline leaves more numerous (up to 5), bearing in their axils 1-several small reddish-purple bulblets; inflorescence often more elongate and not infrequently compound, with most of the flowers of the lateral branches replaced by bulblets; flowers perhaps slightly smaller, the calyx 2.5-3.5 mm. long at anthesis; entire plant, but especially the upper stems and the inflorescence, often more or less reddish-purple.

Grassy hillsides and sagebrush desert to ponderosa pine and Douglas fir forest; e. side of the Cascade Mts. from s. B.C. to Calif., e. to Alta., the Dakotas, and Colo. Mar.-early July.

Because of the transition between *L. bulbifera* and *L. glabra* there is perhaps good basis for their treatment as varieties of the latter species, a procedure followed by Jepson. The earliest varietal name available for this taxon is "ramulosa" rather than "bulbifera." No name change is intended here.

Lithophragma glabra Nutt. in T. & G. Fl. N. Am. 1:584. 1840.
    Tellima glabra Steud. Nom. 2nd ed. 2:665. 1841. (*Nuttall*, "Blue Mountains of the Oregon")
    L. tenella var. *florida* Suksd. West Am. Sci. 15:61. 1906. (*Suksaorf 4011*, Bingen, Wash., Apr. 15,
      1904)

Flowering stems 5-25 (30) cm. tall, sparsely to thickly glandular-pubescent throughout, the hairs usually purple-tipped and the plants usually somewhat purplish-tinged, especially above; basal leaves with petioles 1-2.5 (7.5) cm. long, blades (0.5) 1-2 (40) cm. broad, usually cleft to the base into (mostly) 5 cuneate segments that are more or less tricrenately lobed shallowly to deeply, glabrous or only very sparingly pubescent, often bulblet bearing in their axils; cauline leaves usually 2 (1-3), similar to the basal or more narrowly segmented, becoming subsessile; inflorescence 2- to 5 (rarely as many as 7)-flowered, at anthesis contracted, flat-topped, and corymbose, later becoming more open

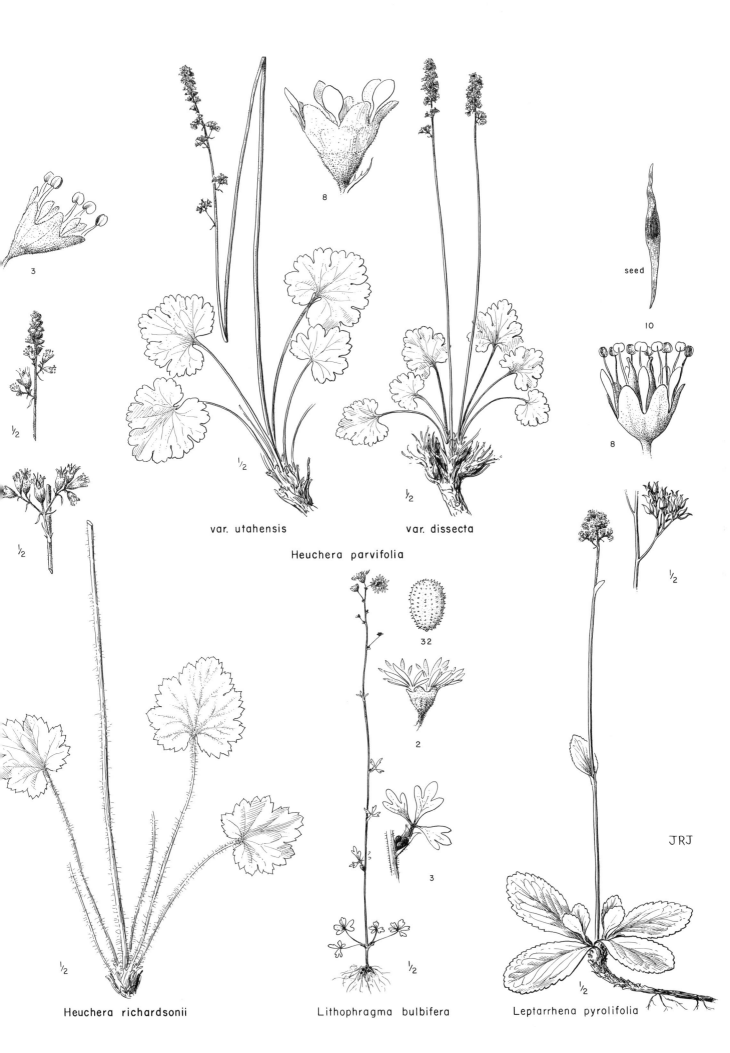

3

½

½

8

var. utahensis   var. dissecta

Heuchera parvifolia

seed

10

8

½

32

2

3

½

½

JRJ

Heuchera richardsonii   Lithophragma bulbifera   Leptarrhena pyrolifolia and racemose, the lower pedicels usually elongating to 1. 5-3 times the length of the calyx (up to 15 mm. long); calyx campanulate to almost cup-shaped, 3-4 mm. long at anthesis, up to 4-6 mm. long in fruit, about as wide at the top (in pressed specimens), adnate to the ovary for not more than 1 mm., the free hypanthium flared only slightly, the lobes broadly triangular, obtuse to rounded, 0. 5-1 mm. long; petals (white to) pinkish- or purplish-tinged, the narrow claw about equaling the calyx lobes, the blade cuneate-obovate, up to 8 mm. long, from merely 3-cleft to more usually 5-cleft and the segments again more or less deeply divided and often with smaller basal teeth; ovary much less than half inferior; seeds brownish, about 0. 5 mm. long, minutely muricate with sharp-pointed spines.

Grasslands and sagebrush plains to ponderosa pine or oak woodland; e. base of the Cascades from Chelan and Kittitas cos., Wash., to the Columbia R. Gorge, e. to n.e. Oreg., s.e. Wash., and across n. Ida. to n.w. Mont. Mar.-May.

Lithophragma parviflora (Hook.) Nutt. ex T. & G. Fl. N.Am. 1:584. 1840.
  Tellima parviflora Hook. Fl. Bor. Am. 1:239, pl. 78A. 1833. *Pleurendotria parviflora* Raf. Fl.
    Tellur. 2:73. 1837. *(Menzies,* "North California," is the first of 3 specimens cited)
  Pleurendotria reniformis Raf. Fl. Tellur. 2:74. 1837. *(Nuttall,* Oregon Mts.)
  L. parviflora var. *micrantha* T. & G. Fl. N.Am. 1:584. 1840. *(Nuttall,* "Dry hills on the Flat-head
    River near the Rocky Mountains")
  Flowering stems (5) 10-30 (50) cm. tall, rather densely glandular-pubescent throughout and commonly distinctly purplish above, often canescent; petioles of the basal leaves (1) 2-6 (8) cm. long, the blades 1-3 cm. broad, divided nearly or quite to the base into (3) 5 main divisions that are more or less biternately to ternately cleft and lobed; cauline leaves usually 2 (1-3), often cleft into narrower segments, the upper ones nearly sessile; racemes at first congested, then becoming open and up to as much as 15 cm. long, (2) 5- to 11-flowered; pedicels ascending to erect, from shorter to slightly longer than the fruiting calyx; calyx at anthesis distinctly obconic-cyathiform, acute at base and attenuate gradually to the pedicel, (3) 4-6 mm. long, in fruit becoming clavate-obconic and up to 10 mm. long, the 5 lobes triangular-ovate, 1-2 mm. long, and about as broad, slightly flared; petals white to pinkish or somewhat purplish, usually slightly unequal, 5-10 mm. long, the blade cuneate-obovate, usually digitately 3 (to 5)-cleft, narrowed abruptly to a slender claw about as long as the calyx lobes; ovary nearly completely inferior; styles about 1 mm. long; seeds ellipsoid, brown, about 0.5 mm. long, irregularly reticulate and longitudinally ridged.

Prairies and grassland to sagebrush desert and lower montane forest; B. C. southward, on both sides of the Cascades, to n. Calif., e. to Alta. and S. D. and s. to Colo. Mar.-June.

The species is variable in nearly all respects, but especially in stature, leaf size, and pubescence. Although of wide distribution it is apparently not differentiated into geographic races, even though the plants along the coast of Wash. and Oreg. appear to have distinctly pinkish- to purplish-tinged stems and flowers.

Lithophragma tenella Nutt. in T. & G. Fl. N.Am. 1:584. 1840.
  Tellima tenella Walpers Rep. 2:371. 1843. *(Nuttall,* "In the central range of the Rocky Mountains, on
    the banks of the Big Sandy and Siskadee Rivers of the Colorado of the West")
  L. australis Rydb. N.Am. Fl. 22²:86. 1905. *(J. S. Newberry,* Cedar Creek, Camp 79 of the Ives'
    Expedition, probably in Ariz., Apr. 24, 1858)
  LITHOPHRAGMA TENELLA var. THOMPSONII (Hoover) C. L. Hitchc. *L. thompsonii* Hoover,
    Leafl. West. Bot. 4:38. 1944. *(Hoover 5745,* South Fork of Wide Hollow Creek, Yakima Co., Wash.,
    May 2, 1942)
  Flowering stems 10-25 (30) cm. tall, rather copiously glandular-pubescent with yellowish-tipped hairs, the plants seldom at all purplish; leaves not bulbiferous, the basal with petioles mostly 1-2 times as long as the blades, the blades more or less reniform to suborbicular, 5-15 mm. wide, usually hirsute on both surfaces, from ternately divided (with ternately deep-lobed segments) to 3 (5)-lobed half (or less) of their length and with shallowly toothed lobes, or merely with 3-7 coarse crenate-dentations; cauline leaves usually 2-3, becoming sessile above, more finely dissected than the basal, usually bi- to triternately divided and lacerate into linear segments or teeth; flowers 5-10, the inflorescence at first congested but very soon elongate and plainly racemose, the pedicels slender to stout and usually equaling or shorter than, but sometimes half again as long as, the fruiting calyx; calyx 2-3 (3.5) mm. long at anthesis, up to 5 mm. long in fruit, rather narrowly companulate, the lobes broadly triangular, scarcely 1 mm. long; petals white or somewhat (but not at all deeply) pinkish, the claw equaling the calyx lobes, the blade 3-5 mm. long, mostly with 3 large upper lobes and 2 (4) much smaller basal

ones; stamens inserted slightly above midlength and included in the free hypanthium; ovary 1/2-3/5 inferior at anthesis, less than half inferior in fruit; styles about 0.5 mm. long; capsule about equaling the calyx at maturity; seeds brown, ellipsoid, about 0.5 mm. long, from irregularly reticulate or slightly verrucose to nearly smooth except for some wrinkling.

Sagebrush desert to pine forest, along the e. side of the Cascades in Wash. and in s. Oreg., eastward, but only occasional, through s. Ida. and Mont. to Wyo., and in slightly modified form s. to Nev., Ariz., and Colo. May-June in our area.

To the south of our area occurs the very slightly larger flowered, somewhat coarsely hairy phase of this species which has been called *L. australis;* whether or not this is a distinctive geographic race is not certain, although it would appear that a bulbiferous phase ranging from c. Nev. to extreme s. Ida. *is* significant. The species is seldom collected and hence not very well understood, at least in our area, where there appear to be two recognizable varieties:

1 Pedicels rarely over 1 1/4 times as long as the fruiting calyx; basal leaves usually divided or
  lobed nearly or quite to the petiole; Klamath Co., Oreg., probably into n. Calif. and Nev.,
  s. to s. Mont. and Wyo.   var. tenella
1 Pedicels more slender, the lower ones often 1 1/2 times as long as the fruiting calyx; basal leaves (in part at least) lobed no more than half their length; e. side of the Cascades in
  Wash., known from Okanogan and Grant to Yakima cos.; a very similar plant occurs in
  Lemhi Co., Ida.   var. thompsonii (Hoover) C. L. Hitchc.

Mitella L. Mitrewort

Flowers complete, regular, borne in elongate, membranous-bracteate, simple racemes; calyx saucer-shaped to turbinate-campanulate, adnate for some length to the ovary, the free portion deeply 5-lobed but usually with a connate, often thinly gland-lined and usually more or less flared hypanthium; petals 5, alternate with the sepals and borne with the stamens at, or near the top of, the hypanthium, greenish, white, or pinkish- to purplish-tinged, slenderly clawed and with a usually filiformly dissected to trilobed (rarely entire) blade; stamens 10, or 5 and then either opposite or alternate with the calyx lobes, the filaments from much shorter to slightly longer than the cordate or subreniform to oblong-ovate anthers; pistil 2-carpellary, ovary 1-celled with 2 parietal placentae, from less than half to nearly completely inferior, bilobed and projecting into short, hollow or solid, terminal, stylelike or stylar portions; stigmas flattened, rounded to strongly bilobed; fruit a capsule, dehiscent by adaxial (ventral) sutures on the free, lobed portion, the dehiscent fruit appearing almost circumscissile; seeds numerous, generally almost or quite black and shining, obscurely reticulate-pitted; perennial, rhizomatous (sometimes stoloniferous in late summer), glandular-puberulent and often somewhat hirsute herbs with leafless or 1- to 3-foliate flowering stems.

Twelve species of (mostly w.) N. Am. (n. of Mex.), Japan, and n.e. Asia. (A diminutive form of the Latin *mitra*, cap, because of the shape of the fruit.)

The plants are easily grown, either from seeds or transplants and are an attractive element in the native garden.

1 Stamens 10; ovary less than 1/2 inferior   M. NUDA
1 Stamens 5; ovary at least 1/2 inferior
  2 Stamens alternate with the calyx lobes (opposite the petals); anthers cordate-reniform,
    much broader than long; petals greenish   M. PENTANDRA
  2 Stamens opposite the calyx lobes (alternate with the petals); anthers usually at least as
    long as broad; petals often white
    3 Flowering stems with 1-3 leaves; flowers blossoming from the top downward (plant
      unique in this respect); calyx 5-6 mm. broad; styles about 1 mm. long   M. CAULESCENS
    3 Flowering stems usually leafless or if leaf-bearing then the calyx less than 5 mm.
      broad and the styles much less than 1 mm. long; racemes blossoming from the bot-
      tom upward
      4 Calyx shallowly saucer-shaped, considerably broader than long, the lobes triangular,
        spreading-recurved; anthers cordate; stigmas bilobed; petals usually greenish-yel-
        low
        5 Leaf blades mostly 4-8 cm. broad, cordate to reniform, always shorter (as meas-
          ured from the basal sinus) than broad, slightly if at all white-hirsute; pedicels 1-
          2 (in fruit up to 5) mm. long   M. BREWERI
        5 Leaf blades mostly 1.5-3.5 cm. broad, cordate-ovate to cordate-oblong, always

longer (as measured from the basal sinus) than broad, the upper surface usually copiously hirsute with coarse hairs; pedicels mostly 0.5-1.5 mm. long   M. OVALIS
4 Calyx cup-shaped to campanulate, usually considerably longer than broad, the lobes ovate to
   oblong, often erect or with only the tips spreading; anthers ovate to oblong; stigmas not bilobed; petals usually whitish or tinged with pink to purple
      6 Leaf blades cordate-ovate to cordate-triangular, about as long (measured from the basal
         sinus) as broad, with 5 (7) distinctly angular lobes, the terminal lobe often acute; petals
         apically 3- or 5-lobed; flowering stems often with 1 or 2 leaves; Cascade Mts. from s.
         Wash. southward
                                                                                        M. DIVERSIFOLIA
      6 Leaf blades reniform to cordate-ovate, always shorter (as measured from the basal sinus)
         than broad, shallowly rounded-lobed, the terminal segment not acute; petals never more
         than 3-lobed; flowering stems rarely with any true leaves; general over our area
            7 Racemes strongly secund, 10- to 45-flowered; calyx (3) 4-6 mm. long, the lobes with a
               simple central vein and branched lateral veins; petals mostly (1.5) 2-4 mm. long, the
               blade trilobed into divaricately spreading to ascending, mostly filiform segments; ex-
               treme e. Wash. and Oreg. to Mont. and southward, common in Ida.   M. STAUROPETALA
            7 Racemes weakly or not at all secund, (4) 10- to 20-flowered; calyx 1.5-3.5 mm. long,
               the lobes with a branched central vein and usually simple lateral veins; petals (1) 1.5-
               2.5 (3.5) mm. long, the blades digitately cleft into 3 ascending to erect, narrow (but
               not filiform) segments; B.C. to Calif., e. to Mont., rare or lacking in Ida.   M. TRIFIDA

Mitella breweri Gray, Proc. Am. Acad. 6:533. 1865.
   Pectiantia breweri Rydb. N.Am. Fl. 22^2:93. 1905. *(Brewer,* Mt. Hoffman, Calif.)
   M. breweri f. *denticulata* Rosend. Engl. Bot. Jahrb. 50, Suppl.:385. 1914. *(Butters & Holway 142,*
   "Prospector's Valley, in the Canadian Rockies and Silkirk Mountain, B.C.")
   Perennial with leafless, or occasionally membranous-bracteate flowering stems (1) 1.5-3 (4) dm.
tall, glabrous or sparsely brownish arachnoid-pilose below, becoming copiously glandular-puberulent
in the inflorescence; leaf blades cordate to reniform, (2) 4-8 (11) cm. broad, glabrous to sparsely
whitish- or brownish-hirsute, very shallowly and indistinctly 7- to 11-lobed, and lightly once or twice
crenate-dentate; racemes 20- to 60-flowered; flowers greenish-yellow, blossoming upward; pedicels
stout, 1-2 (or in fruit rarely up to 5) mm. long; calyx broadly saucer-shaped, becoming more cup-
shaped in fruit, 3-3.5 mm. broad, the hypanthium about equal to the 5 spreading to recurved, triangular lobes; petals 1.5-2 mm. long, pinnatisect into 5-9 filiform, often paired segments; stamens 5,
opposite to, and nearly as long as, the calyx segments; disc thin, covering the ovary; ovary nearly
completely inferior; styles about 0.2 mm. long, the two branches stigmatic at the tip (stigmas commonly described as bilobed). N=7.
   Moist mt. valleys and open to wooded slopes, ranging upward to timber line; B.C. southward, in the
Cascade and Olympic mts., to the Sierra Nevada of c. Calif., e. to Alta., n.w. Mont., and n. Ida.
May-Aug.

Mitella caulescens Nutt. in T. & G. Fl. N.Am. 1:586. 1840.
   Mitellastra caulescens Howell, Fl. N.W.Am. 201. 1898. *(Nuttall,* "shady woods of the Oregon, near
   the mouth of the Wahlamet")
   Perennial, often stoloniferous; flowering stems (1.5) 2-4 (4.5) dm. tall, glandular-puberulent,
sparsely leafy; leaf blades cordate, (3) 5 (7)-lobed 1/4-1/3 of their length and crenate-dentate, rather
sparsely hirsute with fairly coarse hairs, at least on the upper surface, (2) 3-7 cm. broad, the cauline
leaves 1-3 in number and considerably reduced; racemes elongate, loose, up to 25-flowered and as
much as 20 cm. long, blossoming from the top downward; pedicels 2-8 mm. long; calyx broadly campanulate, greenish, 5-6 mm. broad, adnate to the ovary for less than 1/3 its length, the free hypanthium about equal to the 5 deltoid-ovate, spreading lobes; petals 3-4 mm. long, greenish but often pur-
plish-based, pectinately dissected into 8 (4-6) filiform and usually minutely glandular lateral segments;
stamens 5, opposite (and shorter than) the calyx segments; disc lacking; ovary about 1/2 inferior at
anthesis; styles about 1 mm. long, the stigmas very small, capitate.
   Deep woods to meadowland and swampy ground, from the seacoast to middle elevations in the mts.; B.C.
southward to n.w. Calif., e. to n. and w.c. Ida. and n.w. Mont. Apr.-June.

Mitella diversifolia Greene, Pitt. 1:32. 1887.
   Ozomelis diversifolia Rydb. N.Am. Fl. 22^2:94. 1905. *(C. C. Marshall,* Trinity Mts., Calif., July, 1886)

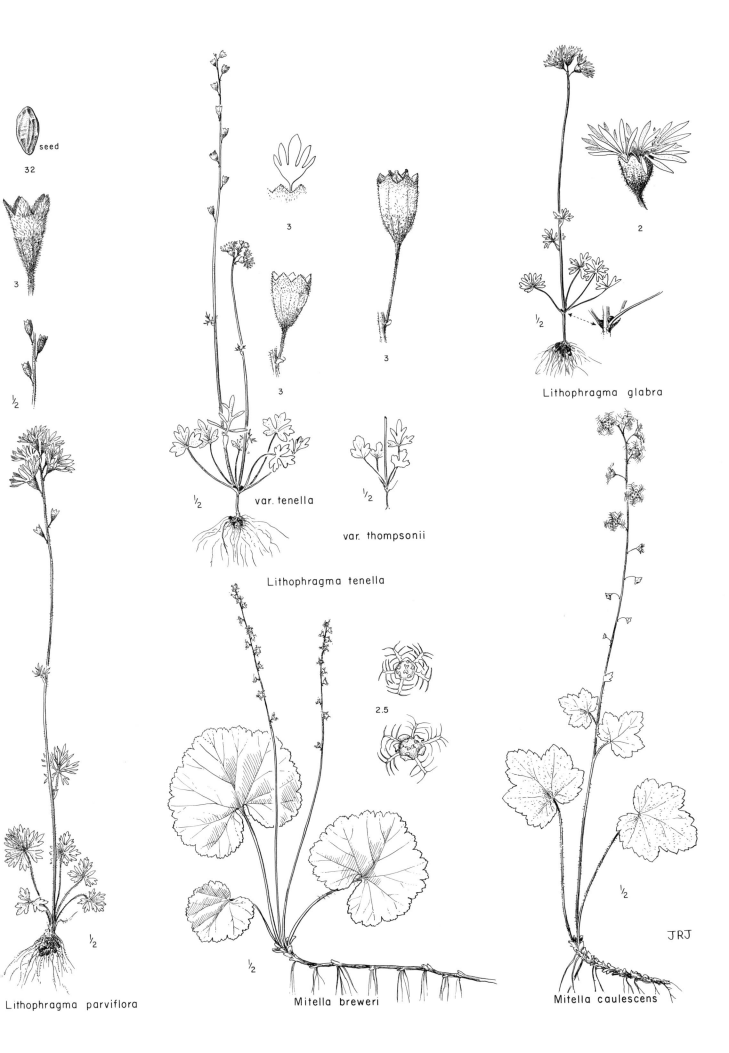

seed

32

½

Lithophragma parviflora

3

3

var. tenella

½   var. thompsonii

Lithophragma tenella

3

3

Lithophragma glabra

½

2.5

Mitella breweri

Mitella caulescens

½

JRJ

Plants more or less densely glandular-puberulent throughout, and pubescent with long, coarse, spreading to retrorse hairs; flowering stems 1.5-5 dm. tall, leafless or with 1 (very rarely 2) reduced leaf; leaf blades cordate-ovate, mostly 3-6 cm. broad and about as long, 5 (7)-lobed and otherwise entire or undulate to shallowly crenate; racemes 8- to 35-flowered, blossoming upward; pedicels stout, 0.5-1.5 mm. long; calyx obconic-campanulate, 2.5-3.5 mm. long, the lobes about 1 mm. long, whitish, oblong-ovate, the tubular portion adnate to the ovary for slightly more than half its length, the free hypanthium slightly or not at all flared, gland-lined; petals about twice as long as the calyx lobes, white (pinkish- or purplish-tinged) with a long slender claw and an oval, apically 3 (5)-lobed blade; stamens 5, opposite the sepals; ovary at least 3/4 inferior; styles (other than the empty terminal portions of the ovary) nearly lacking, the thick, somewhat discoid or very indistinctly lobed stigmas subsessile.

Usually in moist woods or on stream banks; Cascade Mts., from Mt. Adams southward to the Trinity Mts. of n.w. Calif. May-June.

Mitella nuda L. Sp. Pl. 406. 1753. (N. Asia)

Rhizomatous and usually stoloniferous, the erect scapes 3-20 cm. tall, leafless or with a reduced, sessile leaf near the base, finely glandular-puberulent; leaf blades cordate to reniform, 1-3 cm. long, mostly doubly crenate, sparsely hirsute at least on the upper surface; racemes 3- to 12-flowered, blossoming upward, very rarely with 2- to 3-flowered lateral branches; pedicels 2-6 mm. long; calyx broadly campanulate-saucer-shaped, lobed at least 1/2 the length, the segments ovate, obtuse to acute, spreading, about 1.5 mm. long, the tubular portion adnate to the ovary, a free hypanthium almost lacking; petals 5, greenish-yellow, about 4 mm. long, filiformly dissected into usually 8 lateral divisions; stamens 10; ovary 1/3-1/2 inferior; styles somewhat divergent, about 0.5 mm. long.

In damp woods and along stream banks and in bogs; Alas. southward to the Cascades of n.w. Wash., e. to Lab. and Newf., and s. to n.c. Mont., Minn., and Pa.; e. Asia. June-Aug.

Mitella ovalis Greene, Pitt. 1:32. 1887.

Pectiantia ovalis Rydb. N.Am. Fl. 22²:94. 1905. *(Bolander,* Mendocino Co., Calif.)
M. hallii Howell, Erythea 3:33. 1895. *(Elihu Hall 164,* Cascade Mts. of Oreg. and Wash., in 1871)

Coarsely brownish- to whitish-hirsute on the foliage and lower part of the stems, but becoming glandular and more puberulent above; scapes leafless, but sometimes membranous-bracteate below, 1.5-3 (3.5) dm. tall; leaf blades cordate-ovate to cordate-oblong, 1.5-3.5 (4.5) cm. broad and up to 1/2 again as long, shallowly and indistinctly 5- to 9-lobed and lightly bicrenate-serrate or -dentate; racemes rather closely 20- to 60-flowered, eventually open, blossoming upward; flowers greenish-yellow, with short stout pedicels rarely as much as 2 mm. long; calyx shallowly saucer-shaped, becoming slightly more cup shaped in fruit, 2.5-3.5 mm. broad, adnate to the ovary for about 1/3 the length, the lobes shallow, triangular, spreading-recurved, about equaling the free hypanthium; petals about 1.5 mm. long, pinnatifid into 4-7 linear, unpaired segments; stamens 5, opposite the calyx lobes; ovary nearly completely inferior; styles about 0.3 mm. long, the stigmas bilobed.

In deep moist woods, creek bottoms, and wet banks, from the w. side of the Cascades to the coast, B.C. southward to Marin Co., Calif. Late Mar.-May.

Mitella pentandra Hook. in Curtis' Bot. Mag. 56:pl. 2933. 1829.

Drummondia mitelloides DC. Prodr. 4:50. 1830. *Pectiantia mitelloides* Raf. Fl. Tellur. 2:72. 1837.
Mitellopsis drummondia Meisn. Pl. Vasc. Gen. Comm. 100. 1838. *Mitellopsis pentandra* Walpers Rep. 2:370. 1843. *Pectiantia pentandra* Rydb. N.Am. Fl. 22²:93. 1905. (Cultivated plant, raised from seed collected by Drummond in the Rocky Mts.)
Pectiantia latiflora Rydb. N.Am. Fl. 22²:93. 1905. *(Mrs. Bailey Willis,* Palace Camp, Wash.)
Merely a large-flowered, more pilose phase of the species, especially common in the Washington Cascades and in Alas. and Ida.
Mitella pentandra f. *stolonifera* Rosend. Engl. Bot. Jahrb. 50, Suppl.:386. 1914. *(C. Dallen 5,* upper valley of the Nisqually, Mt. Rainier, Wash.)
Mitella pentandra f. *maxima* Rosend. Engl. Bot. Jahrb. 50, Suppl.:387. 1914. *(E. L. Greene,* Selkirk Mts., B.C., in 1890)

Occasionally stoloniferous, the scapes (1) 2-3 (4) dm. tall, naked or with 1 or 2 membranous bracts or even with 1 (2) reduced, petiolate or sessile leaves, from glandular-puberulent or glandular-pubescent to nearly glabrous; leaf blades ovate-cordate, (1.5) 2-5 (8) cm. broad and usually somewhat longer, subglabrous to coarsely hirsute on both surfaces, shallowly and indistinctly 5- to 9-lobed and

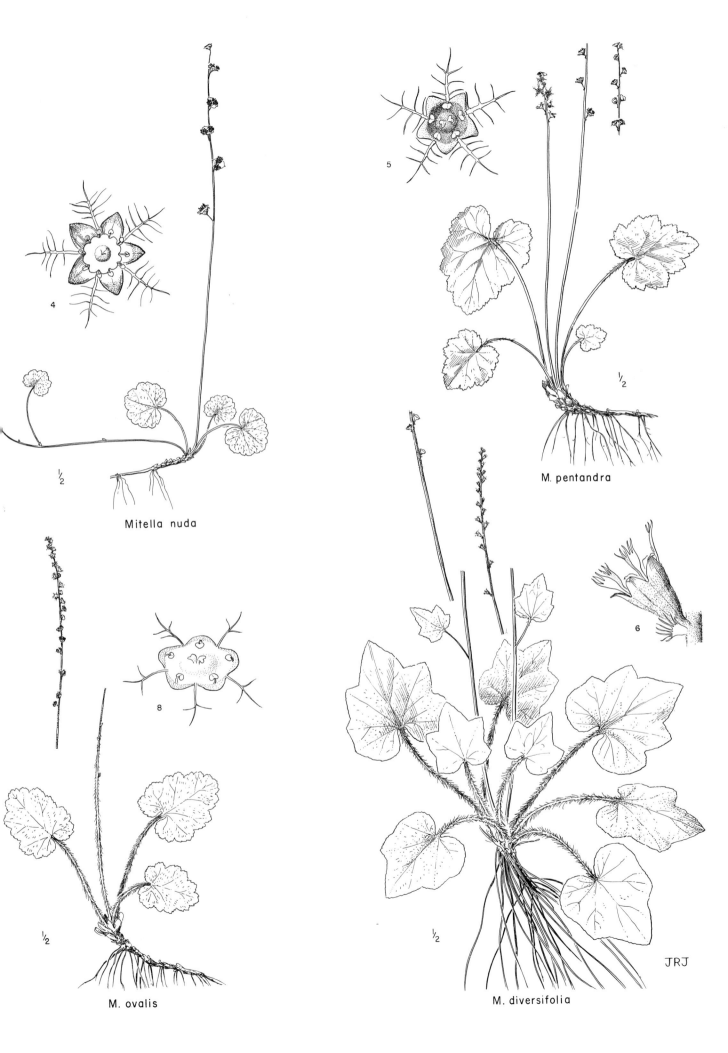

4

5

Mitella nuda

½

M. pentandra

½

8

6

½

M. ovalis

½

M. diversifolia

JRJ

doubly crenate-dentate; flowers 6-25, racemose to somewhat paniculate, blossoming upward; pedicels 2-7 mm. long; calyx broadly saucer-shaped, becoming more cup-shaped in fruit, 3-4 (5) mm. broad, the lobes triangular, spreading to recurved; petals greenish, 2-3 mm. long, pectinately dissected into 8 (4-10) filiform lateral segments; stamens 5, opposite the petals; ovary almost completely inferior; styles almost lacking, the 2 stigmas nearly sessile, bilobed and cordate in outline as viewed from above.

Moist woods, especially along streams, to wet mt. meadows; Alas. southward, along the coast, to n.e. Calif. and into the Sierras, e. to Alta. and Colo. June-Aug.

Mitella stauropetala Piper, Erythea 7:161. 1899.
    Ozomelis stauropetala Rydb. N.Am. Fl. 22²:95. 1905. *(Sandberg, McDougal, & Heller 232,* Craig Mts., Nez Perce Co., Ida.)
    M. stenopetala Piper, Erythea 7:161. 1899. *Ozomelis stenopetala* Rydb. N.Am. Fl. 22²:96. 1905. *M. stauropetala* var. *stenopetala* Rosend. Engl. Bot. Jahrb. 50, Suppl.:380. 1914. *(Watson 365,* Wahsatch Mts., Utah, May, 1869)
    M. stenopetala var. *parryi* Piper, Erythea 7:162. 1899. *Ozomelis parryi* Rydb. N.Am. Fl. 22²:96. 1905. *(Parry 102,* Striking [error for Stinking] Water, Wyo., in 1873) = var. *stenopetala.*

Scapes usually several, up to 5 dm. tall, leafless, but mostly with 1-2 tiny, membranous bracts, glandular-puberulent above and often more or less coarsely hirsute; leaf blades cordate to reniform, often purplish-tinged, 2-8 cm. broad and somewhat shorter, inconspicuously to plainly 5- to 7-lobed and broadly once or twice crenate, sparsely hirsute at least on the upper surface; racemes closely 10- to 35 (45)-flowered, strongly secund, blossoming upward; pedicels 0.5-1.5 (in fruit up to 3) mm. long; calyx turbinate-campanulate, (3) 4-6 mm. long, the lobes oblong to oblong-obovate, erect but with spreading tips at least as long as the connate portion, greenish-white to purplish-tinged, each with usually 2 branched lateral veins and a simple median nerve; petals white or purplish, (1.5) 2-4 mm. long, spreading, usually trilobed into divaricately spreading to ascending, filiform to very narrowly linear segments about as long and as broad as the claw, occasionally merely trilobed or sometimes even entire; stamens 5, opposite the sepals; ovary about 2/3 (3/4) inferior; stigmas broad and flattened, subsessile.

Open to dense woods, where more or less moist, from extreme e. Wash. (Spokane to Asotin cos.) and the Blue, Wallowa, and Ochoco mts., Oreg., eastward to the Rockies in Mont., and s. to Colo. and Utah. May-June.

Most if not all plants of our area are referable to var. *stauropetala,* characterized by a calyx often over 3 mm. long, and petals that are filiformly dissected. Southward, in s.e. Ida., Wyo., e. Utah, and n. Colo., var. *stauropetala* is replaced by var. *stenopetala* (Piper) Rosend., which has the calyx rarely over 3 mm. long, and the petals less deeply dissected (to entire) and with much broader segments. In many respects this phase of the species tends to resemble *M. trifida.*

Mitella trifida Grah. Edinb. New Phil. Journ. 1829:185. 1829.
    Ozomelis varians Raf. Fl. Tellur 2:73. 1837. *Mitellopsis hookeri* Meisn. Pl. Vas. Gen. Comm. 2: 100. 1838. *Mitellopsis trifida* Walpers Rep. 2:370. 1843. *Ozomelis trifida* Rydb. N.Am. Fl. 22²: 95. 1905. (Specimens cultivated at Edinburgh from seeds collected by Drummond in w. N.Am.)
    Mitella violacea Rydb. Bull. Torrey Club 24: 248. 1897. *Ozomelis violacea* Rydb. N.Am. Fl. 22²: 95. 1905. *Mitella trifida* var. *violacea* Rosend. Engl. Bot. Jahrb. 50, Suppl.:381. 1914. *(Flodman 527,* Spanish Basin, Madison Range, Mont., July 11, 1896)
    Mitella micrantha Piper, Erythea 7:162. 1899. *Ozomelis micrantha* Rydb. N.Am. Fl. 22²:96. 1905. *Mitella trifida* f. *micrantha* Rosend. Engl. Bot. Jahrb. 50, Suppl.:382. 1914. *(Watson 135,* Fort Colville, Wash., Sept. 29, 1880)
    Ozomelis pacifica Rydb. N.Am. Fl. 22²:95. 1905. *(Elmer 2778,* Olympic Mts., Clallam Co., Wash., in 1900)

Scapes usually clustered, leafless or with a much reduced leaf near the base, often with 1 or 2 membranous bracts, 1.5-3.5 (4.5) dm. tall, from glandular-puberulent (especially above) to glandular-pilose or somewhat hirsute below; leaf blades cordate to cordate-ovate, 2-6 (8) cm. broad, usually slightly shorter, rather indistinctly 5- to 7-lobed and very lightly to strongly crenate; racemes rather closely (4) 10- to 20-flowered but eventually elongate and open, slightly if at all secund, blossoming upward; pedicels thick, 0.5-2 mm. long; calyx 1.5-3.5 mm. long, broadly campanulate, the lobes oblong to ovate, mostly erect to spreading, thin, whitish or purplish-tinged, with (1) 3 main nerves, the middle nerve branched, the lateral nerves usually unbranched; petals white to strongly purplish-tinged, (1) 1.5-2.5 (3.5) mm. long, the blade narrowly to broadly rhombic or oblanceolate, entire to (more commonly) trifid, cuneate-based and narrowed to a long claw about as wide as the segments of the blade; stamens 5, opposite (and much shorter than) the calyx lobes; ovary about half inferior; stigmas subsessile, broadly flattened.

Deep forest to moist montane slopes; B.C. southward, in the Cascade and Olympic mts., to n. Calif., e. in Wash. and in Can. to Alta., and s. in the Rockies to s. Mont. and in Oreg. to Grant Co. and (probably) the Blue Mts., not presently known from Ida. although probably to be found in the n. part of the state. May-July.

The species is slightly variable in flower size, pubescence, and degree of lobing of the petals. Small-flowered plants from Mont., with entire to shallowly lobed and often purplish petals, have been called var. *violacea* (Rydb.) Rosend. but probably do not merit such distinction, as they occur sporadically elsewhere, especially in B.C. and in the Cascades of Wash.

Parnassia L. Grass-of-Parnassus

Flowers erect, single and terminal on leafless or 1-bracteate peduncles, regular, often rather showy; calyx with a short, broadly obconic, basal connate portion adnate (in ours) to the ovary, or sometimes partially free, deeply 5-lobed, the segments acute to rounded, 3- to many-nerved; petals 5, white (ours) or pale yellow, spreading, (1) 3- to many-nerved, rounded to clawed at the base, entire to pectinate-fimbriate on the basal half, deciduous or withering persistent; fertile stamens 5, inserted on the calyx alternate with the petals, the filaments usually tapered to the rather large anthers; staminodia always present, opposite the petals, in our species from (very rarely) entire and filament-like to broadly scalelike and usually oblong but flaring above and tipped with (3) 5-numerous fingerlike to filamentlike, mostly linear segments usually ending in capitate, glandular knobs; pistil 4-carpellary, the ovary from nearly entirely superior to about 1/4 inferior in our species, 1-celled, with 4 parietal placentae; styles nearly or quite lacking, the 4 stigmas sessile or subsessile; fruit an apically dehiscent, loculicidal, usually ovoid, membranous capsule; seeds numerous, irregularly angled, the testa loose from the rest of the seed and more or less inflated, strongly cellular-reticulate; endosperm lacking; glabrous, scapose, perennial herbs from short, usually erect and simple rootstocks, the leaves (except for the sessile bract of the peduncle) basal, petiolate, and entire.

About 15 species of temperate and arctic N.Am. and Eurasia, often montane. (Named after Mt.Parnassus in Greece.)

The genus is sometimes treated as a separate family, Parnassiaceae, but the morphological features of *Parnassia,* namely, the glabrous entire basal leaves, pedunculate solitary flowers, the staminodia, and the 4-merous pistil with parietal placentation and sessile stigmas, are all to be found in other members of the Saxifragaceae, although not in the same combination. The staminodia are commonly supposed to represent the fused vestiges of several stamens, but there seems to be as good reason to believe that they are derived from single stamens, and that *Parnassia* has a basic number of 10, rather than numerous, stamens.

The species usually grow in boggy soil and several of them do well in cultivation if given sufficient water, especially *P. fimbriata,* the most attractive of the several native species. They are interesting and attractive wild-garden subjects.

1 Petals fimbriate-pectinate on the lower half   P. FIMBRIATA
1 Petals not fimbriate-pectinate on the lower half
  2 Flowering stems usually bractless or with a near-basal bract; petals 1- to 3-nerved, from slightly shorter to slightly longer than the calyx lobes; staminodia with (1) 3-5 segments, usually only about half as long as the filaments; anthers less than 1 mm. long   P. KOTZEBUEI
  2 Flowering stems usually with a bract above the level of the basal leaves; petals 5- to 13- veined, generally considerably longer than the calyx lobes; staminodia with 5-numerous segments, usually well over half as long as the filaments; anthers at least 1 mm. long
    3 Leaf blades elliptic to elliptic-ovate, neither truncate nor cordate at the base; bract never at all clasping; petals usually 5-veined, mostly 4-7 mm. long, rarely over 1.5 times as long as the calyx lobes   P. PARVIFLORA
    3 Leaf blades from lanceolate to broadly ovate, often cordate; bract often clasping; petals 7- to 13-veined, mostly (6) 7-12 mm. long, usually over 1.5 times as long as the calyx lobes   P. PALUSTRIS

Parnassia fimbriata Konig. Ann. Bot. 1:391. 1805. *(Menzies,* "on the coast of Northwest America")
PARNASSIA FIMBRIATA var. INTERMEDIA (Rydb.) C. L. Hitchc. hoc loc. *P. intermedia* Rydb.
N. Am. Fl. 22[1]: 78. 1905. *(Watson 370,* in part, E. Humboldt Mts., Nev.)
PARNASSIA FIMBRIATA var. HOODIANA C. L. Hitchc. hoc loc. A var. *intermedia* differt corpore
staminodiorum angustiore, segmentis longioribus. (Type: *J. W. Thompson 11195,* moist meadow
by Lost Lake, Mt. Hood, Hood River Co., Oreg., July 31, 1934, in U. of Wash. Herb.)
Rootstock short, rather stout, from slightly ascending to nearly erect; flowering stems 1-several,
mostly 1.5-3 (5) dm. tall, the bract cordate and more or less clasping, mostly 5-15 (20) mm. long,
borne from slightly below to considerably above midlength of the scape; petioles (1) 3-10 (15) cm.
long; leaf blades (1.5) 2-4 (5) cm. broad, mostly reniform or somewhat reniform-auriculate and
broader than long, but not uncommonly more nearly cordate or truncate at base, and sometimes
slightly cuneate and somewhat longer than broad; calyx fused with the ovary for only about 1 mm., the
segments oblong-ovate to elliptic-oval, 4-7 mm. long, usually 5 (7)-veined, entire or more commonly crenulate-fimbriate, at least toward the rounded tip; petals white, 5- to 7-veined, 8-12 mm. long
(about twice as long as the calyx lobes), more or less cuneate-obovate in general appearance but clawlike at the base and with numerous long filiform-linear, plainly cellular-verrucose fimbriae, becoming more or less erose to entire on the upper half; staminodia thickened and scalelike, flared above
the middle and usually with a central, subterminal, larger lobe and 7-9 marginal, short, thick, rounded lobes, but sometimes with 5-many elongate, slender, capitate-tipped segments; filaments stout,
about equaling the calyx segments, anthers 2-2.5 mm. long; capsule ovoid, about 1 cm. long.
    Bogs, wet meadows, and stream banks, lower montane to arctic-alpine; Alas. southward to Calif.,
e. to w. Alta., Mont., Wyo., Colo., and N. M. July-Sept.
    On the whole *P. fimbriata* is rather stable. Although perhaps it undergoes less variation in the lob-
ing of the staminodia than do other species, variants differing almost exclusively in staminodial structure have become established in at least two areas. These have more commonly been treated collectively as a separate species, but they would appear to be described more accurately as two well-
marked, rather local varieties, differing from the main race of the species as follows:
1 Staminodia short and thick, the marginal segments short and rounded, not at all filamentlike;
    range of the species as a whole but possibly not sympatric with the other vars.   var. fimbriata
1 Staminodia ending in longer, more slender, filamentlike, usually capitate segments
    2 Segments of the staminodia mostly less than 10, slender, strongly capitate, all margin-
        al, equaling (or longer than) the rather narrow basal scale; Cascade Mts. of n. Oreg.,
        mostly collected on Mt. Hood, but known from Washington, Clackamas, and Hood River
        cos.   var. hoodiana C. L. Hitchc.
    2 Segments of the staminodia mostly 12 or more, 1 or 2 median and subterminal, all most-
        ly shorter than the basal scale which is considerably flared toward the top; E. Humboldt
        Mts., Nev., n. to the mts. of Ida. in Owyhee and Bannock cos., also collected in Sho-
        shone Co., the range expected to prove more continuous than presently known in Ida.
                                                                var. intermedia (Rydb.) C. L. Hitchc.

Parnassia kotzebuei Cham. in Spreng. Syst. 1:951. 1825. ("Ad sinum Escholzii sub circulo arctico
Amer. bor. occid.")
PARNASSIA KOTZEBUEI var. PUMILA Hitchc. & Ownbey hoc loc. A var. *kotzebuei* differt petalis
uninervis, corpore staminodiorum angustiore, integro vel 1- aut 2-lobato. (Type: *Hitchcock &
Muhlick 21605,* moist, rocky slope near entrance to Crescent Mine, 1 mi. s. of Gilbert, R. 18 E.,
T. 34 N., Okanogan Co., Wash., in U. of Wash. Herb.)
Rootstocks very short, erect; flowering stems mostly single (2-3 or rarely 6 or 7), usually bractless but sometimes with a near-basal, ovate to lanceolate, nonclasping bract up to 15 mm. long; petioles equaling, to considerably longer than the blades; leaf blades ovate or deltoid-ovate to nearly elliptic, 5-15 (20) mm. long, from nearly truncate to considerably tapered at the base; calyx adnate to
the ovary for 1-4 mm. (proportionately more than in our other species), the segments narrowly oblong-lanceolate, up to 7 mm. long, usually 3-nerved; petals more or less elliptic-lanceolate or elliptic-ovate, mostly about equaling the calyx lobes, 1- or 3-veined, or in particularly vigorous plants
sometimes with 2 additional weak marginal nerves; staminodia with a thin, oblong scale ending in
mostly 5 (4 or 6) short, inconspicuously capitate-tipped segments, or reduced to a single, linear, entire to bilobed scale; filaments rather slender, mostly about equaling the sepals and considerably longer than the staminodia; anthers 0.4-1.0 mm. long; capsule up to 1 cm. long. N=9.
    On the arctic tundra from Alas. to Great Slave Lake, southward in the Rocky Mts., to Alta. and

B. C. , and from a few localities in Mont. and Wyo. , the Ruby Mts. of Nev. , and the Cascades of Okanogan Co. , Wash. , also in n. e. N. Am. , Lab. , and the Gaspe Peninsula to Greenl. ; Asia. Probably more common in our area than collections would indicate. July-Sept.

Although plants of Mont. and Nev. appear to be much more dwarfed than those from farther north, they are not otherwise dissimilar, whereas a more distinctive population is known from Okanogan Co. , Wash.

1 Staminodia consisting of an oblong scale, much broader than the fertile filaments, with 4-6
    terminal filamentlike segments ending in capitate knobs; petals 3 (5)-nerved, mostly 3.5-
    7 mm. long; range of the species except not in Wash.   var. kotzebuei
1 Staminodia consisting of a short, linear scale often no broader than the fertile filaments,
    the tip entire or irregularly lacerate into 1-3 teeth; petals 1 (3) -nerved; plants consider-
    ably dwarfed, the sepals and petals 3-4 mm. long; known only from near Gilbert, Okanogan
    Co. , Wash.   var. pumila Hitchc. & Ownbey

Parnassia palustris L. Sp. Pl. 273. 1753. ("Habitat in Europae uliginosis")
    P. multiseta sensu Fern. Rhodora 28:211. 1926, not *P. palustris* β *multiseta* Ledeb. in 1842. *P.
    palustris* var. *neogaea* Fern. Rhodora 39:311. 1939. *P. palustris* ssp. *neogaea* Hultén, Fl. Alas. 5:
    956. 1945. (*Fernald & Gilbert 28481*, Little Quirpon, Newf. , Aug. 8, 1925)
    PARNASSIA PALUSTRIS var. MONTANENSIS (Fern. & Rydb.) C. L. Hitchc. hoc loc. *P. montanen-
    sis* Fern. & Rydb. N. Am. Fl. 22[1]:79. 1905. (*F. L. Scribner 54*, Sixteen-Mile Creek, Mont. , July
    12, 1883)

Rootstock usually very short and nearly erect, but occasionally longer and ascending; flowering stems (7) 10-25 (30) cm. tall, the bract lanceolate to broadly ovate or even cordate and clasping, up to 3 cm. long and 2. 5 cm. broad, usually borne below (but occasionally at or above) midlength of the scape; petioles mostly 1. 5-4 times as long as the blades; leaf blades deltoid-ovate, ovate, or elliptic-ovate, (5) 10-20 (30) mm. long, abruptly tapered, truncate, or somewhat cordate or reniform at base; calyx adnate to the ovary for (1. 5) 2-3 mm. , the lobes lanceolate to narrowly oblong-lanceolate, entire, (4) 5-8 (9) mm. long, 5- to 9-veined; petals narrowly to broadly ovate to elliptic-obovate, (6) 7-12 mm. long, 7- to 13-veined, sessile to very slightly clawed; staminodia with a thickened, scale-like, cuneate or oblong base flared into a broadened, generally obovate upper portion that is divided into (5) 7-11 (many) slender filamentlike segments ending in more or less capitate knobs; filaments broad, from shorter to considerably longer than the (1. 5) 2-3 (3. 5) mm. anthers; capsule ovoid, (8) 10-12 mm. long. 2N=16, 18, 27, 36, 54.

Arctic tundra to moist, often shaded areas in the mountains where usually along streams or around springs; arctic Am. southward to Que. , Minn. , B. C. , and in the Rocky Mts. to Colo. and Utah and w. to the Charleston Mts. of Nev. and the s. Sierra Nevada (Mono and Inyo cos.), Calif. ; Eurasia. July-Aug.

Parnassia palustris is more variable than our other species in general size and in the structure of the leaves and staminodia. Our material is in general referable to the var. *montanensis*. Other clearly marked geographic races do not exist in w. U. S. , except perhaps for the plant of the c. Sierra Nevada, called *P. palustris* var. *californica* by Gray (Bot. Calif. 1:202. 1876), which is characterized by bracts usually well above midlength of the flowering stems and staminodia with numerous (15-27) filiform segments. Variety *montanensis* differs from the other N. Am. races as follows:

1 Staminodia mostly with (7) 9-27 segments; petals usually over 1. 5 times as long as the calyx
    lobes; bracts often clasping
    2 Bract usually well above midlength of the scape, very small, not at all clasping; stamino-
        dia with mostly 17-27 very slender segments; c. Sierra Nevada, Calif. , s. as far as
        Mono Co.   var. californica Gray
    2 Bract usually at or below the middle of the scape, often as large as the basal leaves, fre-
        quently clasping; staminodia various
        3 Petals mostly with (11) 13 veins, rather quickly deciduous; staminodia (and petals) with
            short broad claws; Eurasian, also in arctic Am. , where less common than the next
                                                                                      var. palustris
        3 Petals mostly with 7-11 veins, usually withering persistent; staminodia with slender
            clawlike bases; Alas. to e. N. Am. , s. to B. C. , probably entirely n. or e. of our
            area, but transitional to var. *montanensis*   var. neogaea Fern.
1 Staminodia mostly with 5-7 (9) segments; petals usually not over 1. 5 times as long as the
    calyx lobes; bracts seldom at all clasping; variable and almost exactly intermediate be-

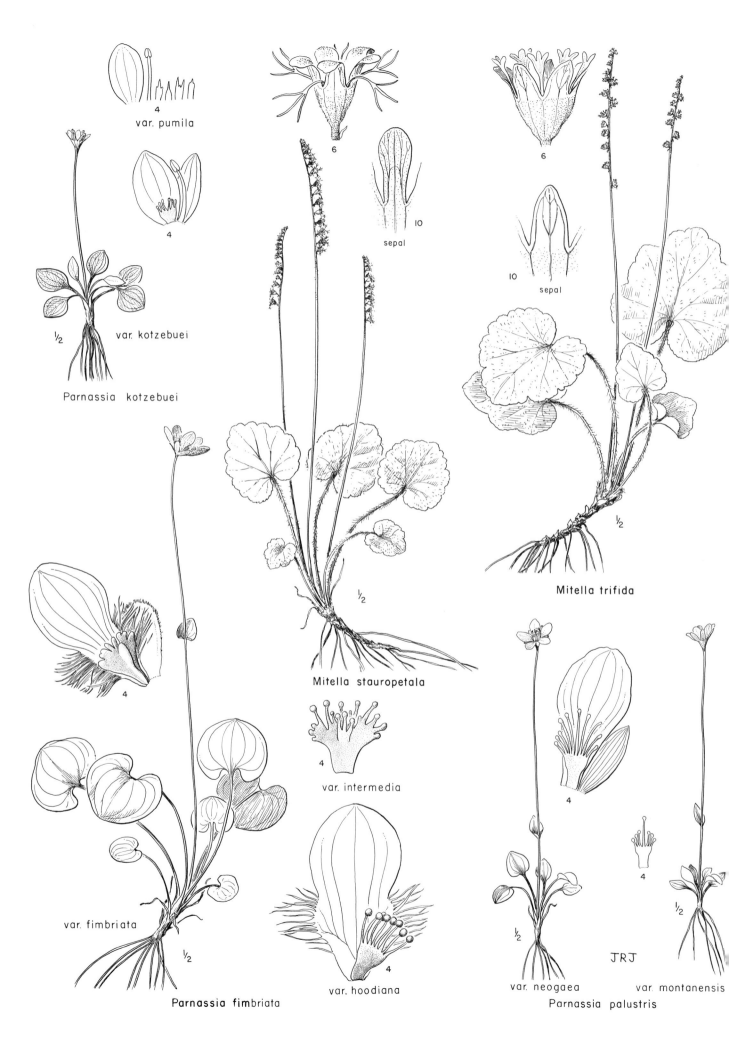

4
var. pumila

4

var. kotzebuei

½

Parnassia kotzebuei

6

10
sepal

6

10
sepal

Mitella trifida

Mitella stauropetala

4
var. intermedia

4

var. fimbriata

½

4
var. hoodiana

Parnassia fimbriata

4

½

4

½
var. neogaea

JRJ

var. montanensis

Parnassia palustris

tween *P. palustris* var. *palustris* and *P. parviflora;* Rocky Mts., s. Alta. to Colo. and
Utah, w. to s. Nev. and the s. Sierra Nevada in Mono and Inyo cos., Calif., where transitional to var. *californica* in the structure of the staminodia, but with bracts usually below midlength of the scape   var. montanensis (Fern. & Rydb.) C. L. Hitchc.

Parnassia parviflora DC. Prodr. 1:320. 1824. ("Amer. bor.")
   Rootstock short, erect or ascending; flowering stems 10-25 (30) cm. tall, the bract narrowly ovate
to lanceolate, not clasping, 7-16 mm. long, usually borne slightly to considerably below midstem; petioles from shorter to much longer than the blades; leaf blades ovate to elliptic, basally tapered, 1-2
(2.5) cm. long; calyx adnate to the ovary for about 1 mm., the lobes linear-lanceolate to very narrowly oblong, (3) 4-6 (7) mm. long, usually 5-nerved, the outer nerves less prominent than the other 3;
petals mostly 4-7 (10) mm. long, rarely as much as twice as long as the calyx lobes, cuneate-obovate
to elliptic-oblong, not pectinate, usually 5- to 7-nerved; scale of the staminodia with an oblong clawlike base, flared above only slightly, usually somewhat longer than the 5-7 (9) marginal, capitate-tipped
segments; filaments from shorter to somewhat longer than the staminodia; anthers 1-1.5 mm. long;
capsule ovoid, 7-11 mm. long.
   Bogs, wet meadows, and stream banks; B.C. to Que., s. to n. Ida., c. Mont., S.D., and Minn.
July-Sept.
   In many respects this appears to be only a very small-flowered phase of *P. palustris* (as was suggested by DeCandolle at the time of original publication). *P. parviflora* is replaced to a large extent
(if not entirely), to the s. of Ida. and Mont., by *P. palustris,* with which it tends to merge, through
the var. *montanensis.*

Peltiphyllum Engl.

   Flowers in large, bractless, paniculate-corymbose cymes, rather showy, regular; calyx adnate only to the base of the ovary, deeply 5-lobed, a free hypanthium lacking; petals white to bright pink, entire; stamens 10; carpels usually 2, free above the point of adnation with the calyx, tapered to the dis-
coid-capitate stigma; fruit follicular, dehiscent the full length; seeds cellular-rugulose; perennial
herbs from thick rhizomes with large peltate leaves that develop somewhat later than the tall, leafless,
flowering stems.
   Only the one species. (Greek *pelte,* shield, and *phyllon,* leaf, because of the peltate leaves.)

Peltiphyllum peltatum (Torr.) Engl. in E. & P. Nat. Pflanzenf. 3^{2a}:61. 1890.
   Saxifraga peltata Torr. ex Benth. Pl. Hartw. 311. 1849. *(Hartweg,* "in montibus Sacramento," acc.
   Jepson the type is *Hartweg 246,* from Pine Creek, Sierra Nevada foothills, Butte Co., Calif.)
   Rhizomes fleshy but tough, up to 5 cm. thick; leaves all basal, with membranous-stipulate, rough-
hirsute petioles up to 1 (1.5) m. long, the blades nearly orbicular, peltate, 5-40 cm. broad, 10- to
15-lobed and again more shallowly lobed and serrate-dentate, cupped in the center; scapes stout, up
to 1.5 (2) m. tall, strongly hirsute-glandular, naked or with a small bract; calyx lobes oblong-ovate,
2.5-3.5 mm. long, reflexed; petals oblong-elliptic to oblong-obovate, 4.5-7 mm. long, spreading,
deciduous; filaments flattened and broad at the base, tapered to the nearly oval anthers, usually erect;
follicles purplish, 6-10 mm. long, fused with the ovary for 1-2 cm.; seeds about 1 mm. long. N=17.
Umbrella plant.
   In, and at the margins of, cold mountain streams, usually very firmly anchored among rocks; c.
Sierra Nevada to Siskiyou and Humboldt cos., Calif., into s.w. Oreg. and extending n. as far as Benton Co. Apr.-June.
   This is indeed an unusual plant because of its size and habitat, taking readily to cultivation and well
meriting a place in the moist native garden. Except for its larger size it is somewhat suggestive of the
more commonly grown Bergenias, and in floral character actually not materially different from many
of the species of *Saxifraga,* with which genus *Bergenia* is often combined.

Saxifraga L. Saxifrage

   Flowers regular to irregular, perfect, usually complete (apetalous), perigynous to more or less
completely epigynous, generally cymose (solitary) but often pseudoracemose or in open to diffuse cymose panicles; calyx saucer-shaped to conic or campanulate, often with a free connate portion (hypanthium) between the adnate base and the (usually) 5 erect to reflexed, persistent lobes; petals common-

ly 5, white (sometimes spotted or flecked with yellow or reddish-purple) to greenish or violet-purple (1 species), usually alike but not rarely unlike in size or shape, clawless or with a distinct and often slender claw, deciduous or persistent; stamens 10, inserted with the petals on or at the top of the hypanthium (if any) or on the calyx around the ovary, the filaments from slender and more or less subulate to distinctly clavate and somewhat petaloid, deciduous or persistent, the anthers oval; carpels usually 2 (3, 4, or even 5) from distinct and joined at base only by the adnate calyx, to connate well above the ovuliferous portion, generally tapered above into slender, often recurving and beaklike portions that vary from completely dehiscent to at least partially indehiscent, commonly becoming solid and truly stylar above; stigmas usually capitate and enlarged, but sometimes smaller in diameter than the style, and not infrequently decurrent on the ventral (inner) surface of the style; fruit from plainly capsular and dehiscent across the top by the ventral sutures of the stylar beaks (the placentation axile), to follicular and dehiscent the full length; seeds numerous, from smooth to variously wrinkled, winged, crested, muricate, or tuberculate, the testa sometimes loose and inflated; occasionally annual but usually perennial (ours) herbs with leafless or leafy flowering stems and entire to toothed, lobed, or pinnatifid, simple, alternate (rarely opposite), petiolate to sessile leaves, glabrous to (usually) glandular-hairy, often with bulbils in the leaf axils or in the inflorescence.

About 300 species, widely distributed but primarily of the temperate or arctic regions of the Northern Hemisphere, many of the species circumboreal. (Latin *saxum*, rock, and *frangere*, to break.) It is sometimes assumed that the plant was so named because of the habitat of many of the species, but it is more commonly believed that the name was coined because of its usage by the herbalists in the treatment of "stones" of the urinary tract.

Many of the species of *Saxifraga* are choice ornamentals and among our species *S. oppositifolia* and *S. bronchialis* are considered to be nice rock garden subjects, although our native plants, especially of the former species, respond very sulkily, if at all, even to pampering care. The plant in successful cultivation is of European origin.

Although opinions differ as to the generic bounds for *Saxifraga,* some authors distributing our species among several segregate genera, many of our taxa that have been accorded generic rank are members of circumboreal species-complexes and do not stand out so sharply when considered in the over-all picture of the genus.

References:

Calder, J. A. and D. B. O. Savile. Studies in Saxifragaceae - II. Saxifraga Sect. Trachyphyllum in North America. Brittonia 11:228-249. 1959.

Engler, A. and E. Irmscher. Saxifragaceae—Saxifraga I. Das Pflanzenreich IV, 117, 1:1-709. 1916.

Johnson, A. M. A revision of the North American species of the section Bora   ..gler of the genus Saxifraga. Minn. Stud. Pl. Sci. 4:1-109. 1923.

————. Studies in Saxifraga. Am. Journ. Bot. 14:38-43, 323-326. 19. , 18:797-802. 1931; 21:109-112. 1934.

In the following keys and in the description of species, the length of the flowering stems as given includes the inflorescence, and the length of the leaf blade is measured from the point of attachment of the petiole even when the leaves are cordate or reniform.

1 Flowers purple (very rarely albino), borne singly on short leafy stems; leaves opposite, decussate, entire, imbricate, strongly marcescent   S. OPPOSITIFOLIA
1 Flowers white or yellow, rarely pinkish- or purplish-tinged, usually more than 1 per stem;
  cauline leaves (if any) alternate and often toothed or lobed
    2 Plant a delicate rhizomatous perennial (often apparently annual) with slender, succulent,
      trailing to erect, glabrous to sparsely glandular-puberulent stems; leaves mainly cau-
      line, entire to 3 (5)-toothed, strongly petiolate; calyx slightly keeled below the sinuses
      of the acute lobes; seeds pectinate-spinulose in longitudinal rows   S. NUTTALLII
    2 Plant of varied habit, but usually with the leaves either all entire or all basal; calyx not
      keeled below the usually rounded lobes; seeds not spinulose-pectinate in rows
        3 Leaves entire, mostly linear to lanceolate, usually 1 or more of them borne on the
          flowering stems below the inflorescence
            4 Filaments clavate; leaves fleshy, only 1-3 (0) on the flowering stems, rounded at the
              tip, glabrous or with only a few basal cilia; petals white; testa of the seeds loose
              and inflated   S. TOLMIEI
            4 Filaments not clavate; leaves often more than 3 on the flowering stems, sometimes

ciliate full length, mostly acute; petals often yellow; testa of the seeds not loose and inflated

  5 Petals yellow; leaves usually not rigidly pungent

    6 Plants with numerous leafless, filiform, axillary stolons; ovary half inferior at an-
thesis; inflorescence pubescent with purplish, gland-tipped hairs   S. FLAGELLARIS

    6 Plants not stoloniferous; ovary usually much less than half inferior at anthesis; in-
florescence often not purplish-hairy

      7 Petals ovate to obovate, abruptly narrowed to a short but distinct claw; leaves ob-
tuse to rounded, glabrous; calyx lobes reflexed   S. CHRYSANTHA

      7 Petals either not abruptly narrowed to a claw or the leaves acute or strongly cil-
iate; calyx lobes slightly spreading to erect

        8 Basal leaves mostly 1.5-5 cm. long, slender-petiolate; flowers single; calyx
and peduncle rusty-pilose, eglandular; circumboreal, southward from Alas.
to n. B.C., also in the Rocky Mts. of Colo. and to be expected in Mont. and
Ida., but not presently known from our area   *S. hirculus* L.

        8 Basal leaves mostly less than 1.5 cm. long, not slender-petiolate; flowers of-
ten more than 1 per stem; calyx and peduncle pubescent or puberulent and usually glandular, never brownish-pilose; Canadian Rockies, extending southward to s. B.C. and Alta.   S. AIZOIDES

  5 Petals white, usually purplish-spotted; leaves mostly rigidly pungent   S. BRONCHIALIS

3 Leaves toothed to lobed, or if entire then all basal

  9 Leaves (or some of them) with 3 apical cuspidate teeth, otherwise entire; petals white,
usually reddish- to purplish-spotted; ovary not more than 1/5 inferior

    10 Most of the leaves 3-toothed and over 10 mm. long; circumboreal, Alas. to Lab., s.
to B.C., reported (but not seen) from our area   *S. tricuspidata* Rottb.

    10 Most of the leaves entire, rarely as much as 10 mm. long; s.w. Wash. and n.w.
Oreg.   S. BRONCHIALIS

  9 Leaves various, but not with 3 apical cuspidate teeth; petals various; ovary often more
than 1/5 inferior

    11 Flowers rarely more than 10; ovary usually at least 1/2 inferior at anthesis, if less
than half inferior then most of the flowers replaced by bulbils; leaf blade generally
not more than 2 cm. in width or length; at least 2 or 3 of the leaves usually cauline

      12 Leaves mostly narrowly oblong-cuneate to oblanceolate-obovate, apically 3 (5-7)
-toothed or -lobed into sometimes linear segments; some of the leaves (especially the basal) usually entire, the petiole (if any) generally broad, no longer than
(and not clearly distinguishable from) the blade; bulbils lacking

        13 Plants strongly caespitose; leaves usually lobed (rather than toothed), those of
the simple flowering stems generally less deeply lobed than the basal (sometimes entire); petals gradually narrowed to a broad base and only slightly or
not at all clawed; filaments longer than the sepals   S. CAESPITOSA

        13 Plants not strongly caespitose, usually single-stemmed from a basal rosette of
more nearly toothed than lobed leaves, those of the (often basally branched)
flowering stems commonly more prominently toothed than the basal ones; petals abruptly narrowed to a short but distinct claw; filaments shorter than the
sepals   S. ADSCENDENS

      12 Leaves (at least in part) reniform and rather evenly lobed, the basal ones with a
slender petiole at least as long as the blade; bulbils usually present in the inflorescence or in the axils of the basal leaves

        14 Bulbils borne in the axils of the upper cauline leaves and in place of at least
the lower flowers; flowering stems commonly at least 10 cm. tall; ovary
scarcely 1/4 inferior   S. CERNUA

        14 Bulbils lacking in the axils of the cauline leaves and in the inflorescence; flow-
ering stems mostly less than 10 cm. tall; ovary about 1/2 inferior   S. DEBILIS

    11 Flowers usually more than 10; ovary often less than 1/2 inferior at anthesis; flower-
ing stems (except in *S. mertensiana*) usually leafless although sometimes bracteate;
leaf blades usually well over 2 cm. in length or width

      15 Leaves more or less orbicular, or reniform or cordate at base, usually broader
than long; filaments clavate

16 Leaves with thin, membranous, mostly sheathing stipules, the petioles and the flowering
   stem usually pilose; leaf blades commonly shallowly lobed, the primary segments 3-
   toothed; flowering stems generally with 1 or more leaves or scales below the inflorescence; flowers often partially replaced by bulbils, and bulblets generally present
   among the basal leaves; capsules reflexed   S. MERTENSIANA
16 Leaves with narrow, nonsheathing stipular margins, the petioles and stems glabrous to
   pubescent but rarely pilose; leaf blades mostly simply toothed; flowering stems usually
   leafless below the nonbulbiferous inflorescence; capsules erect
   17 Inflorescence glandular-pubescent to glandular-puberulent, the hairs mostly 1- to
      3-celled, the gland often reddish-purple; petals dissimilar, the larger ones with
      an oblong to oval blade not over 1.5 times as long as wide, often truncate to slight-
      ly cordate at base and narrowed abruptly to a slender claw   S. ARGUTA
   17 Inflorescence usually somewhat pilose with often wavy, several-celled hairs, not
      conspicuously glandular; petals mostly alike, the blades more or less oblong-el-
      liptic, usually well over 1.5 times as long as broad, rounded to cuneate at base
      and narrowed gradually to a rather wide claw   S. PUNCTATA
15 Leaves never reniform, mostly flabellate to lanceolate or obovate, usually much longer
   than broad; filaments often not clavate
   18 Seeds with several longitudinal rows of closely set, flattened, ribbonlike papillae; pet-
      als white, abruptly narrowed to a short, slender claw; leaves flabellate to cuneate-
      oblanceolate, with 3-11 (17) prominent teeth on the apical portion and tapered rather
      uniformly to a slender to broadly winged petiole
      19 Filaments clavate; plants very sparsely pubescent with uniseriate hairs; leaves fla-
         bellate, narrowed to eventually slender petioles; flowers all normal   S. LYALLII
      19 Filaments slender, not clavate; plant usually copiously pubescent, at least the
         larger hairs multiseriate (composed of more than 1 row or series of cells); leaves
         more cuneate-oblanceolate, narrowed very gradually to broadly winged petioles;
         flowers usually partially replaced by bulbils   S. FERRUGINEA
   18 Seeds variously wrinkled or reticulate but not with ribbonlike ridges of papillae; petals
      white to greenish (lacking), usually rounded to a broad, clawed or clawless base;
      leaves ovate to elliptic, entire to shallowly toothed most of the length of the blade; pet-
      ioles usually not broadly winged
      20 Ovary less than half inferior at anthesis, the stamens inserted at the edge of a nar-
         row bandlike gland or nectary surrounding but not covering the top of the ovary;
         plants mostly 1-2.5 (rarely to 3) dm. tall; leaves regularly crenate-serrate; fil-
         aments often clavate
         21 Filaments always strongly clavate and petaloid; leaves abruptly rounded to sub-
            cordate at base; sepals strongly reflexed; ovuliferous portion of the ovary near-
            ly entirely superior; inflorescence open and diffuse; petals yellow-spotted; gland
            or nectary visible on the maturing or ripe fruit; n. Calif. and coastal Oreg. to
            near the Columbia R., where transitional to the next   S. MARSHALLII
         21 Filaments from strongly clavate to subulate; leaves mostly tapered at the base;
            sepals spreading to reflexed; ovuliferous portion of the ovary mostly 1/5-1/3
            inferior; inflorescence diffuse to narrow and congested; petals seldom spotted;
            gland or nectary not visible on the maturing fruit; range mostly n. and e. of
            S. marshallii   S. OCCIDENTALIS
      20 Ovary either over half inferior at anthesis or plants averaging over 3 dm. tall; sta-
         mens inserted at the edge of a flattened, lobed disc that more or less covers the
         greater part of the ovuliferous portion of the pistil; leaves usually entire to sinuate
         or remotely denticulate, rather than regularly crenate-serrate; filaments not cla-
         vate
         22 Flowering stems mostly 3-12 dm. tall; inflorescence elongate and narrow to open-
            ly corymbose, rarely less than 1 dm. long in late anthesis; leaves usually well
            over 5 cm. long, slenderly ovate to oblanceolate, narrowed gradually to a ses-
            sile base or to a broadly winged petiole, neither the leaves nor the bracts ever
            reddish pilose-lanate   S. OREGANA
         22 Flowering stems mostly not over 3 dm. tall; inflorescence often congested and

less than 1 dm. long even in fruit; leaves various, frequently with narrow petioles, they
 and the bracts sometimes reddish-pilose
23 Leaves with deltoid to rhombic-deltoid, plainly crenate-serrate to crenate-dentate blades
 and short, broadly winged petioles, glabrous on the upper surface but usually rusty-
 arachnoid-pilose on the lower surface; inflorescence closely congested and glomerate
 or interrupted-glomerate; flowering stems mostly 1-2 (0.5-3) dm. tall, copiously
 glandular-pubescent above; Rocky Mts., w. only to c. Ida.   S. RHOMBOIDEA
23 Leaves usually entire or at most denticulate, pubescent to glabrous; inflorescence usually more open, sometimes diffusely cymose-paniculate; flowering stems 1-3 (4) dm.
 tall; c. Ida. (and possibly w. Mont.) w. to coastal B.C., Wash., and Oreg.
                                                                              S. INTEGRIFOLIA

Saxifraga adscendens L. Sp. Pl. 405. 1753.
 Muscaria adscendens Small, N.Am. Fl. 22²:129. 1905. ("Habitat in Pyrenaeis, Baldo, Tauro Ras-
  tadiensi")
 S. petraea sensu Hook. Fl. Bor. Am. 1:245. 1833, but not of L. in 1753. *Ponista oregonensis* Raf.
  Fl. Tellur. 2:66. 1837. *S. oregonensis* A. Nels. Bot. Gaz. 42:52. 1906. *S. tridactylites* ssp.
  adscendens var. *normalis* f. *americana* Engl. & Irmsch. Pflanzenr. IV, 117, 1:221. 1916. *S. ad-
  scendens* ssp. *oregonensis* Bacigalupi in Abrams, Ill. Fl. Pac. St. 2:359. 1944. *S. adscendens*
  var. *oregonensis* Breit. Can. Field-Nat. 71:56. 1957. *(Drummond,* "Alpine Rivulets upon the
  Rocky Mountains")
 S. incompta Peck, Rhodora 36: 267. 1934. *(Peck 18034,* N. slope of Peet's [often "Pete's"] Point,
  Wallowa Mts., Oreg., July 29, 1933) = var. *oregonensis.*
  Strongly glandular-pubescent, short-lived perennial, usually with a simple caudex, the flowering
stem from simple to freely branched, (3) 5-10 cm. tall; leaves 5-15 mm. long, the blades from entire to apically 3 (5)-toothed or shallowly lobed, obovate, gradually narrowed to a broad or narrow
petiolelike base, the cauline leaves 3-15; calyx usually reddish-purple, campanulate, 2.5-3.5 mm.
long at anthesis, the ovate-triangular lobes nearly erect, about equal to the connate portion which
in fruit becomes considerably accrescent, turbinate, and up to 4 mm. long; petals white, deciduous,
2-3 times as long as the calyx lobes, oblanceolate to (more frequently) obovate or even somewhat flabellate, narrowed abruptly to a claw about half as long as the calyx lobes; stamens inserted at the edge
of a narrow disc surrounding the almost completely inferior ovary, the filaments slender, somewhat
shorter than the sepals; styles slender, about 1 mm. long; stigmas capitate, slightly decurrent; capsule 3.5-5 mm. long; seeds obovoid, about 0.5 mm. long, dark brownish-black, nearly smooth, with
a prominent raphe. N=11.
  Rock crevices, glacial moraines, and alpine gravelly meadows; n. Rocky Mts., southward in B.C.
to the n. Cascades of Wash. and to Utah and Colo., w. to c. Ida. and the Wallowa Mts. of n.e. Oreg.;
Europe. July-Aug.
  Our material, as described above, is referable to the var. *oregonensis* (Raf.) Breit. It differs
from the European var. *adscendens* in its small stature, smaller flowers, and broader cauline
leaves.

Saxifraga aizoides L. Sp. Pl. 403. 1753.
 Leptasea aizoides Haw. Saxifr. Enum. 40. 1821. ("Habitat in Alpibus, Lapponicis, Styriacis, West-
  morlandicis, Baldo")
 S. van-bruntiae Small, Bull. Torrey Club 25:316. 1898. *Leptasea van-bruntiae* Small, N.Am. Fl.
  22²:153. 1905. *S. aizoides* var. *euaizoides* f. *flava* subf. *van-bruntiae* Engl. & Irmsch. Pflanzenr.
  IV, 117, 1:466. 1916. *(Mr. & Mrs. Cornelius Van Brunt,* Sulphur Mt., near Banff, B.C.)
  More or less matted perennial with numerous ascending to erect, leafy flowering stems 5-12 cm.
tall, glandular-puberulent at least above; leaves sessile, linear, those of the trailing stems up to
about 10 mm. long, about 1 mm. broad, glabrous or sparsely ciliate, those of the erect stems usually
broader (to 2 mm.), glabrous to sparsely ciliate, acute or apiculate but not rigidly pungent; flowers
1-several in racemelike cymes, the pedicels mostly several-bracteate; calyx adnate to the ovary for
about 1 mm., without a free hypanthium, the lobes triangular-ovate, 3-4 mm. long, spreading; petals
yellow, orange-dotted, elliptic-oblong, clawless, slightly longer than the sepals, usually 3-nerved,
persistent; stamens (often 11 or 12) of 2 lengths, from shorter to longer than the calyx lobes; pistil
often 3-carpellary, the carpels distinct almost to the adnate, very slightly inferior basal portion, tap-

ered above into short thick styles; stigmas very slightly decurrent; capsule about 5 mm. long; seeds cellular-rugose. N=13.

Along streams and on moist gravel banks and talus slides; Rocky Mts. of Alta. and B.C. to the Selkirk Mts., B.C. (just entering our area), e. N. Am.; Europe. June-Aug.

Although several races of the species are recognized in Europe, plants of e. N. Am. and those of the Canadian Rockies are not sufficiently dissimilar to warrant their inclusion in separate taxa.

Saxifraga arguta D. Don, Trans. Linn. Soc. 13:356. 1822.[*]
    Micranthes arguta Small, N.Am. Fl. 22²:147. 1905. *S. punctata* var. *arguta* Engl. & Irmsch. Pflanzenr. IV, 117, 1:11. 1916, in large part. *S. punctata* ssp. *arguta* Hultén, Fl. Alas. 5:932. 1945. *(Menzies,* "ad oras occidentales Americae septentrionalis")
    S. punctata var. *acutidentata.* Engl. Verh. Zool.-Bot. Ges. Wien 19:548. 1869. *(Lyall,* South Clear-Creek, Cascade Mts., Wash.)
    S. odontophylla of Piper, Contr. U.S. Nat. Herb. 11:314. 1906, but not of Wall. in 1834. *S. odontoloma* Piper, Smith. Miscell. Coll. 50:200. 1907. *Micranthes odontoloma* Heller, Muhl. 8:60. 1912. *(Sandberg & Leiberg 570,* Mt. Stuart, Wash.)
    S. punctata var. *arguta* f. *piperiana* Engl. & Irmsch. Pflanzenr. IV, 117, 1:11. 1916. *(Piper 2213,* Olympic Mts., Wash., is the first specimen cited that fits the description)
    S. punctata var. *arguta* f. *minor* Engl. & Irmsch. loc. cit. (Typification obscure)
Perennial with well-developed horizontal rootstocks and usually a single naked flowering stem (1.5) 2-6 dm. tall, mostly glabrous below except for the somewhat ciliate leaves but glandular-puberulent to glandular-pubescent above and throughout the inflorescence, the hairs short, tipped with reddish to purplish (yellowish) glands; petioles usually several times as long as the leaf blades, somewhat widened to nonsheathing stipular bases; blades reniform to (very occasionally) more nearly flabelliform, (1.5) 2-8 cm. broad, usually about 3/4 as long, coarsely crenate- or crenate-dentate, the teeth mostly 15-29, up to 7 mm. long; inflorescence an open, spreading, cymose panicle as much as 30 cm. long; calyx cleft nearly the full length, the lobes oblong-ovate to oblong-lanceolate, 1.5-2 mm. long, sparsely glandular-ciliolate or glabrous, often pinkish or purplish, a free hypanthium lacking; petals deciduous, white, 3-4 mm. long, equal or 2 or 3 somewhat broader than the others, the blade narrowed abruptly (and often truncate or somewhat sagittate) at base; stamens about equal to the petals, the filaments very strongly clavate, white, more or less petaloid, usually persistent; ovary about 9/10 superior, the 2 (3-4) carpels fused 1/3-3/5 their length, narrowed to slender styles up to 1 mm. long; stigmas capitate, slightly decurrent; capsule (5) 6-10 mm. long, usually purplish; seeds brown, nearly 1 mm. long, slenderly fusiform, longitudinally ridged with several rows of rufflelike, contiguous, cellular protuberances.

Lower montane to alpine, along streams, lakes, and wet meadows; s. Alas. and B.C. southward, through the Cascade and Olympic mts., and the Sierras, to s. Calif., e. to Alta., and in the Rockies to N.M. and Ariz., common in Ida., Mont., and n.e. Oreg. Early July-Sept.

Although sometimes considered to be a race of *S. punctata*, *S. arguta* seems amply distinct from that taxon and scarcely at all intergradient with it.

Saxifraga bronchialis L. Sp. Pl. 400. 1753. (Siberia)
    S. austromontana Wieg. Bull. Torrey Club 27:389. 1900. *Leptasea austromontana* Small, N.Am. Fl. 22²:153. 1905. *S. bronchialis* ssp. *austromontana* Piper, Contr. U.S. Nat. Herb. 11:313. 1906. *S. bronchialis* var. *austromontana* attributed to Piper by G. N. Jones, U. Wash. Pub. Biol. 5:167. 1936. *(Lyall,* Cascade Mts., is the first of 6 specimens cited)
    S. cognata E. Nels. Bot. Gaz. 30:118. 1900. *(Nelson & Nelson 5551,* Yellowstone Park, June 28, 1899) = var. *austromontana.*
    Leptasea vespertina Small, N.Am. Fl. 22²:153. 1905. *S. bronchialis* var. *vespertina* Rosend. Engl. Bot. Jahrb. 37, Beibl. 83:73. 1905. *S. bronchialis* ssp. *vespertina* Piper, Contr. U.S. Nat. Herb. 11:313. 1906. *S. vespertina* Fedde, Just Bot. Jahresb. 33¹:613. 1906. *(Frank H. Lamb 1312,* Baldy Peak, Chehalis Co. [now Grays Harbor Co. ?], Wash., July 24, 1897)

---

[*]Calder & Savile (Can. Journ. Bot. 38:409-39. [May] 1960) have adopted *S. odontoloma* Piper as the proper specific name for this taxon, relegating (apparently with good reason) the name, *S. arguta* D. Don, to synonymy under *S. punctata* L.

Parnassia parviflora

Saxifraga
adscendens

Peltiphyllum peltatum

Saxifraga arguta

capsule

var. austromontana

var. vespertina

Saxifraga bronchialis

Saxifraga aizoides

leaf

leaf

leaf

JRJ

S. tricuspidata var. *micrantha* of Engl. & Irmsch. Pflanzenr. IV, 117, 1:464. 1916, in part (the
   Douglas specimen from Mt. Hood), but not of Sternb. Suppl. II:62. 1831.

Caespitose, usually pulvinate perennial, up to 3 dm. broad, the leaves rigid, closely crowded and
imbricate, somewhat mosslike, linear-lanceolate to oblong or spatulate, marcescent for many years,
(3) 5-15 (18) mm. long, 1-2.5 (3) mm. broad, entire or very rarely with 2 tiny, subapical, lateral
lobes, coarsely and sometimes harshly ciliate, rounded to acute, but always with a small spinose tip,
otherwise glabrous; flowering stems 5-13 (16) cm. tall, with several reduced, sessile, spinose leaves,
more or less glandular-pubescent, especially above; flowers (1) 2-10 (15) in an open racemelike or
paniculate, but usually rather flat-topped, much modified cymose inflorescence; calyx saucer-shaped,
without a free hypanthium, the adnate portion 1/2-1/3 as long as the lobes, the sepals 1.5-2.5 (3) mm.
long, triangular to oval, glabrous or sparingly ciliate, spreading but not recurved; petals deciduous,
white, usually strongly purple-spotted above the middle, mostly 5-7 mm. long, oblong to oblong-oval,
narrowed slightly at the base but without a claw; filaments somewhat shorter than the petals, inserted
at the edge of the ovary, not clavate; ovary very slightly (1/5-1/10) inferior at anthesis, projecting upward into short hollow stylar beaks 1-2 mm. long, true solid styles lacking; stigmas slightly decurrent; capsule usually purplish, 4-5 mm. long exclusive of the somewhat divergent beaks; seeds dark
brown, 0.7-0.9 mm. long, oblanceolate-ellipsoid, nearly smooth, the testa tight to the rest of the
seed. 2N=48, 49, about 150.

From sea level to arctic-alpine, in rock crevices and rock slides to open slopes and alpine scree;
circumboreal, in N. Am. from Alas. to Greenl., southward to n. Oreg., in both the Cascade and
Olympic mts., and in the Rockies to N. M., common in Ida., and in the Blue and Wallowa mts. of n.e.
Oreg. June-Aug.

Plants from the cordilleran region of Can. and the U.S. differ in minor ways from the rather highly variable Siberian *S. bronchialis,* and have usually been segregated from it, either at the specific or infraspecific level, under the epithet "austromontana." Until the taxonomy can be clarified for the species as a whole, our plants may be maintained as two vars. separable as follows:

1 Leaves lanceolate to linear, always entire, acute, usually averaging over 4 times as long as
      broad, with coarse cilia usually less than 0.3 mm. long; B.C. to most of montane Wash.,
      southward, especially in the Rocky Mts., through much of Ida. and w. Mont., to n.e.
      Oreg., Wyo., Colo., and N. M.   var. austromontana (Wieg.) G. N. Jones
1 Leaves oblong to spatulate, very occasionally slightly 3-lobed near the tip, mostly not over
      4 times as long as broad, rounded to obtuse, with numerous cilia 0.3-0.5 mm. long; Co-
      lumbia R. Gorge and the adj. Cascade Mts. of Wash. and Oreg. (n. to Mt. Rainier), Sad-
      dle Mt., Clatsop Co., Oreg., and on Mt. Baldy, Olympic Mts., Grays Harbor Co., Wash.;
      perhaps with good reason maintained at the specific level by Calder and Savile.
                                                                           var. vespertina (Small) Rosend.

Saxifraga caespitosa L. Sp. Pl. 404. 1753.
   Muscaria caespitosa Haw. Saxifr. Enum. 37. 1821. *S. decipiens* var. *caespitosa* Engl. Monog. Saxi-
      fr. 188. 1872. *Dactyloides caespitosa* Nieuwl. Am. Midl. Nat. 4:91. 1915. ("Habitat in Alpibus
      Lapponicis, Helveticis, Tridentinis, Monspelii")
   S. exarata Hook. Fl. Bor. Am. 1:244. 1833, but not of Vill. in 1779. *Muscaria monticola* Small,
      N.Am. Fl. 22[2]:130. 1905. *S. monticola* Fedde, Just Bot. Jahresb. 33[1]:613. 1906. *S. caespitosa*
      ssp. *exaratoides* var. *drummondii* Engl. & Irmsch. Pflanzenr. IV, 117, 1:375. 1916. (*Drummond,*
      "Rocky Mountains between latitudes 52° and 56°") = var. *minima.*
   Muscaria delicatula Small, N.Am. Fl. 22[2]:129. 1905. *S. delicatula* Fedde, Just Bot. Jahresb. 33[1]:
      613. 1906. *S. caespitosa* ssp. *exaratoides* var. *delicatula* Engl. & Irmsch. Pflanzenr. IV, 117, 1:
      377. 1916. (*Rydberg,* Gray's Peak, Colo., Aug. 23, 1895) = var. *minima.*
   Muscaria micropetala Small, N.Am. Fl. 22[2]:129. 1905. (*Frank Tweedy 118,* Buffalo Creek, Teton
      Forest Reserve, Wyo., Aug., 1897) = var. *minima?*
   Muscaria emarginata Small, N.Am. Fl. 22[2]:130. 1905. *S. caespitosa* var. *emarginata* Rosend. Engl.
      Bot. Jahrb. 37, Beibl. 83:65. 1905. *S. emarginata* Fedde, Just Bot. Jahresb. 33[1]:613. 1906. (*El-
      mer 2649,* Olympic Mts., Wash., June, 1900)
   S. caespitosa var. *minima* Blank. Mont. Agr. Coll. Stud. Bot. 1:64. 1905. (*R. S. Williams,* Single-
      Shot Mountain, Teton Co., Wyo., July 4, 1897, is the first collection cited)
   S. caespitosa ssp. *exaratoides* var. *purpusii* Engl. & Irmsch. Pflanzenr. IV, 117, 1:377. 1916.
      (*Purpus 6642,* Mt. Tomasaki [Tomasak], Utah) = var. *minima.*
   SAXIFRAGA CAESPITOSA var. SUBGEMMIFERA (Engl. & Irmsch.) C. L. Hitchc. hoc loc. *S. caes-*

pitosa ssp. *subgemmifera* Engl. & Irmsch. Pflanzenr. IV, 117, 1:377. 1916. *(Th. Howell,* Cascades of the Columbia R., Oreg., is the first of 3 collections cited)

Loosely to densely caespitose, often mat-forming perennial with numerous short leafy branches and erect, glandular-pubescent, sparingly leafy flowering stems 3-15 cm. tall; leaves with (2) 3 (5-7) linear to lanceolate lobes, narrowed to broad petioles, more or less chartaceous, weakly glandular-ciliate to strongly pilose-arachnoid, crowded on the sterile shoots; cauline leaves (1) 2-5 below the inflorescence, often entire, the lower ones sometimes with more or less well-developed foliar buds or with the branches developing; flowers (1) 2-10 in a loose cymose (falsely racemose to paniculate) inflorescence, the pedicels up to 4-5 cm. long and with 1-3 bractlets; calyx turbinate to broadly campanulate and 2.5-5.5 mm. long at anthesis, with somewhat spreading, deltoid or ovate to oblong-lanceolate lobes from equal to, to only half as long as, the completely adnate lower portion which is considerably accrescent and up to 6 mm. long in fruit, a free hypanthium lacking; petals white, deciduous, 2-4 times as long as the calyx lobes, cuneate-oblanceolate to obovate, rounded to slightly retuse, narrowed to a broad, sometimes slightly clawed base; stamens up to twice as long as the calyx lobes, the filaments slender, not clavate; ovary covered by the disc, completely inferior at anthesis and about 4/5 inferior in fruit, narrowed to short hollow conical apices and true solid styles 1-1.5 mm. long; stigmas obovate, slightly decurrent; capsule (4) 5-7 mm. long; seeds more or less ovoid, dark brown, about 0.5 mm. long, minutely tuberculate, with a prominent raphe, the testa tight to the rest of the seed. N=40.

Cliffs, rock crevices, and rocky slopes, sea level to arctic-alpine; circumboreal and across arctic Am. to the Gaspe Peninsula, southward from Alas. to the Cascade and Olympic mts., the Columbia R. Gorge, and n.w. Oreg., and in the Rocky Mts. of Mont. to c. Ida., Utah, n.e. Nev., N.M., and Ariz. Apr.-Sept., dependent upon elevation and latitude.

The species is an extremely variable one, due, in part at least, to local ecological conditions. When the plants are growing at higher elevations in exposed places they are apt to be very compact and short-stemmed, when at lower elevations or in sheltered rock crevices, much more sprawly, longer stemmed, and more loosely branched, the lower cauline leaves often bearing rudimentary to well-developed axillary buds.

Small was of the opinion that there were 6 distinct N.Am. species in this complex, but Engler and Irmscher included Small's taxa under three elements of the 4 subspecies, 14 varieties, 4 subvarieties, 21 forms, and 7 subforms which they took to constitute the one species, *S. caespitosa,* and they proposed two additional new varieties and one subspecies for our material.

Although there seems to be good reason to treat the complex as one large polymorphic species, it is very difficult to determine what the relationship of our material is to that of arctic America and Eurasia and, therefore, what names are appropriate for the three rather clearly marked geographic races found in our area, for which the above description was designed.

1 Pubescence of the leaves and stems meager, not at all pilose; petals more nearly oblanceolate
    than obovate, rarely over 2.5 times as long as the usually triangular to ovate calyx lobes;
    Can. Rockies, s. through Mont. and c. Ida. to Colo., Utah, Nev., N.M., and Ariz.; s. of
    our range seemingly dwarfed and possibly meriting recognition as a separate variety; to the
    north apparently intergradient with the arctic race usually referred to as ssp. *sileniflora*
    (Sternb.) Hult.; in B.C. mergent with the var. *emarginata*   var. minima Blank.
1 Pubescence more abundant, that of the leaves often pilose-arachnoid; petals more obovate
    than oblanceolate, usually over 2.5 times as long as the oblong-lanceolate calyx lobes
  2 Leaves with short to long cilia, seldom pilose, closely imbricate on the usually tightly
     compacted sterile shoots; subalpine in the Olympic Mts., and from Mt. Rainier n. to
     c. B.C.   var. emarginata (Small) Rosend.
  2 Leaves long-pilose, often somewhat arachnoid, not closely imbricate; plants tending to
     be sprawly and diffusely branched, the lower cauline leaves frequently with buds in their
     axils (due probably to the lax habit of these lowland plants, and matched in this respect
     by such specimens as *Hitchcock & Muhlick 11981,* from Fergus Co., Mont.); Columbia
     R. Gorge, Saddle Mt., and coastal Oreg. s. to Lincoln Co., n. along the coast of Wash.
     to Island Co., and in the Olympic Mts., where mergent with var. *emarginata*
                var. subgemmifera (Engl. & Irmsch.) C. L. Hitchc.

Saxifraga cernua L. Sp. Pl. 403. 1753.

Lobaria cernua Haw. Saxifr. Enum. 20. 1821. ("Habitat in Alpibus Lapponicis frequens")

S. simulata Small, N.Am. Fl. 22²:128. 1905. *(Rydberg 681,* Sylvan Lake and Harney Peak, Black

Hills, S.D., July 21, 1892)

S. cernua f. *simplicissima* Ledeb. Fl. Alt. 2:122. 1830. (Typification obscure)

Perennial, single-stemmed to more usually somewhat caespitose and with 2-several simple (ours) to branched leafy flowering stems (8) 10-15 (20) cm. tall, rather thickly glandular-pubescent to grayish glandular-pilose or rusty-lanate below; basal leaves usually several, slender-petiolate, often bearing numerous ricelike bulblets in their axils, the blades mostly reniform, (5) 10-15 (20) mm. broad, with 5-7 (9) rather prominent rounded teeth or shallow lobes; cauline leaves usually several, the lower ones like the basal but with much shorter petioles, the upper ones fewer-lobed to entire; inflorescence false-ly racemose to paniculate, the lowermost 1 or 2 (sometimes all) flowers replaced by small reddish-purple bulbils; calyx turbinate to broadly campanulate, 3-4.5 mm. long at anthesis, usually purplish or purplish-mottled, the ovate to oblong-ovate, erect lobes 2.5-4 times as long as the adnate lower portion, a free hypanthium lacking; petals white, the 3 nerves often purplish near the base; up to 12 mm. long, 2-5 times as long as the calyx lobes, obovate to cuneate-obovate, retuse, not clawed, de-ciduous; stamens usually exceeding the calyx lobes, the filaments not clavate; ovary about 1/4 inferior at anthesis, less so at maturity, the calyx not accrescent; styles 1-1.5 mm. long; stigmas slightly de-current; mature capsules and seeds not seen. N=25, 30, 32, 35; 2N=36, 50, 60, 64.

Circumboreal, Alas. to Lab., southward in moist rock crevices, along stream banks, and in glacial detritis to the Cascade Mts. of n. Wash., and in the Rocky Mts. to N.M., w. to c. Ida. and Elko Co., Nev., e. to S.D. July-Aug.

The species is apparently abundant in the Arctic but only rarely collected in the U.S. and s. Can.

Saxifraga chrysantha Gray, Proc. Am. Acad. 12:83. 1877.

Leptasea chrysantha Small, N.Am. Fl. 22[2]:152. 1905. *(Parry 164,* Colorado, is the first of 3 col-lections cited)

Perennial with slender rootstocks and few to numerous tufted-leafy offsets and glabrous to sparsely glandular-pubescent, simple, leafy flowering stems 2-6 cm. tall, from basal rosettes of fleshy, gla-brous, linear-oblanceolate to somewhat spatulate, entire leaves (3) 5-10 mm. long; leaves of the flow-ering stems (1) 2-5, linear to narrowly oblong, 3-7 mm. long, glabrous or sometimes sparsely glan-dular-pubescent; flowers single, or occasionally paired; calyx with ovate to oblong-ovate, sharply re-flexed lobes 2-3 mm. long and 4-5 times as long as the saucer-shaped adnate basal portion, the free hypanthium barely 0.5 mm. long; petals deciduous, yellow, finely cross-rugulose with orange below the middle, broadly obovate to ovate, 5-7 mm. long, 5- to 9-nerved, narrowed very abruptly at base to a short but clearly defined claw; filaments slender, not at all clavate, inserted with the petals around the ovary rather than at the outer edge of the hypanthium; ovary less than 1/5 inferior, syncar-pous nearly to the sessile, broad, capitate stigmas, true styles lacking; capsule ovoid, 6-8 mm. long; seeds brown, about 1 mm. long, narrowly fusiform-oblong, lightly reticulate-wrinkled.

Open rocky slopes and moraines, often near snowbanks or seepage; Uinta Mts. of Utah and Wyo. and in the Rocky Mts. from n. N.M. to the Beartooth Mts. in n. Park Co., Wyo., and probably to be found in adjacent Mont. July-Aug.

Saxifraga debilis Engelm. ex Gray, Proc. Acad. Phila. 1863:62. 1863.

S. cernua var. *debilis* Engl. Monog. Saxifr. 107. 1872. *(Hall & Harbour 198,* "Alpine," Colo., in 1862)

S. rivularis of various authors with respect to our material.

Tufted perennial forming small patches up to 3-8 cm. broad, from glabrous throughout to more usu-ally glandular-pubescent and often pilose, at least on the bases of the lower leaves; flowering stems usually several, leafy, 1-10 cm. tall; leaves mostly basal, often bulbiferous, the petioles slender, frequently more or less brownish-pilose, margined with membranous, usually brownish, papery stip-ules, the blades generally more or less reniform, 5-15 (18) mm. broad, coarsely and shallowly 3- to 5 (7)-lobed; cauline leaves 1-3 (0), similar to the basal or entire; flowers 1-2 (3); calyx turbinate-campanulate, usually purplish, 2.5-3.5 mm. long at anthesis, the lobes erect, ovate to ovate-oblong, rounded, from as long to half again as long as the completely adnate lower portion, a free hypan-thium lacking; petals deciduous, white but more or less pinkish-veined, narrowly oblong to cuneate-oblong, 2-3 times as long as the calyx lobes, rounded at base and scarcely or not at all clawed; sta-mens equaling the calyx lobes, the filaments not clavate; ovary nearly half inferior at anthesis, slight-ly less than half inferior when ripe, the superior portion of the carpels free nearly half their length, tapered and ventrally dehiscent to the sessile, slightly decurrent stigmas, true solid styles lacking;

capsule 4-6 mm. long; seeds medium brown, about 0.5 mm. long, ellipsoid-prismatic, very minutely rugulose.

Damp cliffs, rock crevices, and talus near snowbanks; B.C. southward to the Cascades of Wash. (as far as Mt. Rainier), the Blue and Wallowa mts. of Oreg., the c. Sierra Nevada of Calif., and the San Francisco Mts., Ariz., e. in B.C. to the Rocky Mts., and southward through Mont. to e. Utah and Colo., probably also in c. Ida. July-Aug.

Plants of the Cascades and n. Rocky Mts. have often been called *S. rivularis* L. and might, with considerable reason, be regarded as a well-marked variety of that arctic species, differing in their somewhat smaller flowers and more numerous teeth on the leaves, and in the development of basal bulbils. Although they have been considered specifically distinct from plants of the s. Rocky Mts., there is no basis upon which such separation can be maintained even though material from Utah seems to be more nearly glabrous and to have flowers with narrower, proportionately longer sepals than those from our range.

Saxifraga ferruginea Grah. Edinb. New Phil. Journ. 1829:349. 1829.

Hexaphoma ferruginea Raf. Fl. Tellur. 2:66. 1837. *Spatularia ferruginea* Small, N.Am. Fl. 22^2:150. 1905. *Hydatica ferruginea* Small, N.Am. Fl. 22^6:554. 1918. (Described from plants grown at Edinburgh from seeds presented by Richardson)

Saxifraga stellaris var. *brunoniana* Bong. Mém. Acad. St. Pétersb. VI, 2:140. 1832. *Saxifraga leucanthemifolia* var. *brunoniana* Engl. Monog. Saxifr. 135. 1872. *Spatularia brunoniana* Small, N. Am. Fl. 22^2:149. 1905. *(Mertens,* Sitka Isl.) Probably = var. *ferruginea.*

Saxifraga nutkama Howell, Fl. N.W.Am..194. 1898. *Saxifraga nootkana* Moc. ex Rosend. Engl. Bot. Jahrb. 37, Beibl. 83:65. 1905. *Saxifraga ferruginea* var. *macounii* Engl. & Irmsch. Pflanzenr. IV, 117, 1:70. 1916. (Alas. to Oreg.)

Spatularia newcombei Small, Torreya 2:55. 1902. *Saxifraga newcombei* Engl. & Irmsch. Pflanzenr. IV, 117, 1:70. 1916. *Hydatica newcombei* Small, N.Am. Fl. 22^6:554. 1918. *(Dr. C. F. Newcombe,* Queen Charlotte Is., B.C., summer, 1901) probably = var. *grandiflora* Johnson.

Spatularia vreelandii Small, N.Am. Fl. 22^2:149. 1905. *Saxifraga ferruginea* var. *vreelandii* Engl. & Irmsch. Pflanzenr. IV, 117, 1:70. 1916. *Hydatica vreelandii* Small, N.Am. Fl. 22^6:554. 1918. *Saxifraga ferruginea* f. *vreelandii* St. John & Thayer, Mazama 11:77. 1929. *(F. K. Vreeland 1048,* Glacier Basin, Mont., Aug. 5, 1901) = var. *macounii.*

Saxifraga ferruginea var. *foliacea* Johnson, Minn. Stud. Pl. Sci. 4:62. 1923. *(Sandberg,* Palouse County and Lake Coeur d'Alene, Ida., Aug., 1892, is the first of 6 collections cited) = var. *macounii.*

Saxifraga ferruginea var. *grandiflora* Johnson, Minn. Stud. Pl. Sci. 4:62. 1923. ("Windham Bay, June 11, 1905, *C. F. Baker, no. 4921, J. D. Culbertson")* Probably a recognizable large-flowered race from s. Alas. to the Queen Charlotte Is.

Saxifraga ferruginea var. *cuneata* Johnson, Minn. Stud. Pl. Sci. 4:63. 1923. *(W. H. Osgood,* Moresby I., Queen Charlotte Is., July 2, 1900) = same as *S. newcombei,* probably = var. *grandiflora* Johnson.

Saxifraga ferruginea var. *diffusa* Johnson, Minn. Stud. Pl. Sci. 4:63. 1923. *(Th. Howell 1621,* near Yes Bay, Alas., Aug. 21, 1895) = var. *ferruginea.*

Saxifraga ferruginea var. *stellariformis* Johnson, Minn. Stud. Pl. Sci. 4:64. 1923. *(Th. Howell,* Mt. Hood, Oreg., Aug., 1881) = var. *macounii.*

Saxifraga ferruginea var. *nivea* Johnson, Minn. Stud. Pl. Sci. 4:64. 1923. *(Henderson 288,* Mt. Adams, Wash., Aug. 10, 1882) = var. *macounii.*

Perennial with a short erect caudex or with a thick rootstock; basal leaves 2-10 (14) cm. long, (3) 5-15 (30) mm. broad, tapered gradually to broad, winged petioles not clearly delimited from the bluntly to sharply (5) 7- to 17-toothed, oblanceolate to cuneate-obovate blades, grayish-pilose to thinly hirsute or merely sparsely ciliate, at least the longer hairs multiseriate (composed of more than 1 row of cells); flowering stems 1-several, (1) 1.5-3.5 (6) dm. tall, usually freely branched and floriferous to near the base, pilose to pubescent below, always glandular-pubescent above, leafless but the main bracts from leaflike to greatly reduced and linear; inflorescence mostly openly and diffusely paniculate, the flowers numerous and from all normal to nearly all modified into more or less well developed, leafy bulblets or propagules that often drop from the plant before the fruits mature; calyx nearly completely free of the ovary, lobed almost to the base, the segments 1.5-2.5 mm. long, oblong-ovate, sharply reflexed, usually reddish or purplish; petals white (purplish), (3) 4-6 (8) mm. long, curved to one side of the flower, with a distinct slender claw 0.5-1 mm. long, usually dimorphic, 3 of them with considerably broader and more nearly oblong-ovate, truncate- to slightly cordate-based, basally yellow-bimaculate blades, the other 2 more nearly elliptic-lanceolate to elliptic-ovate and with cuneate-based blades; stamens slightly shorter than the petals, the filaments white, not clavate; ovary almost completely superior, the ovuliferous portion entirely above the level of the calyx, the 2 carpels fused for over half their length, tapered to very short styles less than 0.5 mm. long, the stigmas very small; capsule 4-6 mm. long, the tips slightly divergent; seeds brownish, fusiform-ellipsoid, about 0.7 mm. long, with several longitudinal rows of flattened, comblike papillae.

Alaska southward in the Coastal and Cascade mts., through Wash. and Oreg. to n.w. Calif., e. in B.C. to Ida. and w. Mont. June-Aug.

Saxifraga ferruginea is closely related to the circumboreal, polymorphic *S. stellaris* L. The phase of *S. stellaris* in Alas. and n. B.C. is referred to as var. *comosa* Poir (in Lam. Encycl. 6:680. 1804), but has also been called *S. foliolosa* R. Br. (Chlor. Melv. 17. 1823). It differs from our plants in having smaller leaves (mostly less than 15 mm. long and with usually only 3-5 small, rounded teeth) and an almost completely bulbiferous inflorescence, the lateral branches being reduced to short divaricate stubs. *S. stellaris* reportedly ranges s. to the Selkirk Mts. of B.C. and an isolated station in Colo.; it is closely approached by occasional specimens of *S. ferruginea* var. *macounii* from the Olympic Mts., Wash.

There are 2 genetic races of *S. ferruginea* in our area, and there is apparently another distinctive phase of the species in the Queen Charlotte Is. and vicinity.

1 Flowers partially replaced by bulblets; with the more southern range of the species, gradually replaced northward by var. *ferruginea*   var. macounii Engl. & Irmsch.
1 Flowers all (or nearly all) normal; Alas. southward, along the coast, to Nanaimo, Vancouver
   I., very occasionally elsewhere to the south   var. ferruginea

Saxifraga flagellaris Willd. in Sternb. Rev. Saxifr. 25. 1810.

Hirculus flagellaris Haw. Saxifr. Enum. 41. 1821. *Leptasea flagellaris* Small, N. Am. Fl. 22^2:154. 1905. *(Adams,* Caucasus region)

S. setigera Pursh, Fl. Am. Sept. 312. 1814. *S. flagellaris* var. *setigera* Engl. Monog. Saxifr. 225. 1872. *(Nelson,* "On the north-west coast")

S. flagellaris ssp. *euflagellaris* var. *platysepala* f. *uniflora* Engl. & Irmsch. Pflanzenr. IV, 117, 1: 161. 1916. *(Moseley,* Jong's [Long's] Peak, Colo., is the first collection cited)

S. flagellaris ssp. *euflagellaris* var. *platysepala* f. *minor* Engl. & Irmsch. Pflanzenr. IV, 117, 1: 161. 1916. *(G. Engelmann,* Rocky Mts., Colo., is the first collection cited)

S. flagellaris ssp. *euflagellaris* var. *platysepala* f. *major* Engl. & Irmsch. Pflanzenr. IV, 117, 1: 161. 1916. *(W. W. Eggleston 6031,* Clover Mt., above Garfield, Colo., is the first collection cited)

Perennial with slender rhizomes and numerous filiform, naked stolons up to 10 cm. long, from the tips of which small buds and eventually new crowns develop, the individual crowns small, with a basal tuft of oblanceolate to linear-oblanceolate leaves 10-20 mm. long and 2-4 mm. broad, strongly glandular-ciliate and acute-acicular at the tip; flowering stems (3) 5-10 (12) cm. tall, with usually overlapping leaves that are gradually reduced upward, strongly purplish glandular-pubescent upward and throughout the inflorescence; flowers 1-3 (4); calyx turbinate-campanulate, the lobes triangular to oblong-lanceolate, 2.5-3 mm. long, ascending, as long to nearly half again as long as the adnate lower portion, a free hypanthium lacking; petals yellow, 6-9 mm. long, cuneate-obovate, strongly 5- to 7-nerved, narrowed rather gradually to a broad, somewhat clawlike base; stamens inserted at the edge of the ovary, the filaments not clavate, equaling or slightly longer than the calyx lobes; ovary at least half inferior, projecting into 2 short, hollow, stylelike beaks, the stigmas sessile, large, slightly decurrent; capsule 4-5 mm. long; seeds brownish, ovoid-lanceolate, about 0.6 mm. long, nearly smooth, the cellular testa closely investing the rest of the seed. N=16.

Circumboreal, in N. Am. from Alas. to Ellesmere I., and southward, on alpine scree and moist rocks, to B.C., and in the Rocky Mts. from s. Mont. to n.e. Utah, Colo., N.M., and Ariz.; to be expected in other parts of Mont. July-Aug.

Saxifraga integrifolia Hook. Fl. Bor. Am. 1:249, pl. 86. 1833.

Micranthes integrifolia Small, N. Am. Fl. 22^2:137. 1905. *S. cephalantha* Heller ex Engl. & Irmsch. Pflanzenr. IV, 117, 1:59. 1916, in synonymy. *(Dr. Scouler,* "Near the mouth of the Columbia, North-West coast of America")

S. fragosa Suksd. ex Small, Bull. Torrey Club 23:363. 1896. *S. integrifolia* var. *fragosa* Rosend. Engl. Bot. Jahrb. 37, Beibl. 83:68. 1905. *Micranthes fragosa* Small, N. Am. Fl. 22^2:137. 1905.

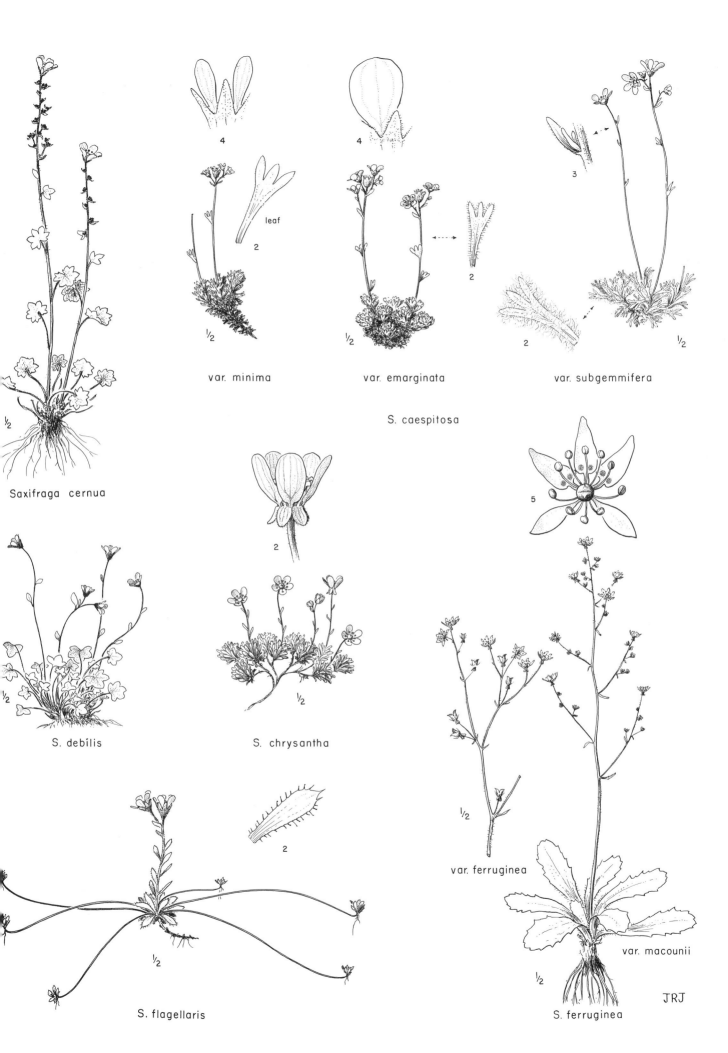

var. minima

var. emarginata

var. subgemmifera

S. caespitosa

Saxifraga cernua

S. debilis

S. chrysantha

var. ferruginea

var. macounii

S. flagellaris

S. ferruginea

JRJ

(Suksdorf 1727, near the Columbia R. , w. Klickitat Co. , Wash.) = var. *claytoniaefolia.*

S. claytoniaefolia Canby ex Small, Bull. Torrey Club 23:365. 1896. *S. integrifolia* var. *claytoniae-folia* Rosend. Engl. Bot. Jahrb. 37, Beibl. 83:67. 1905. *Micranthes claytoniaefolia* Small, N. Am. Fl. 22²:141. 1905. *S. fragosa* f. *claytoniifolia* Engl. & Irmsch. Pflanzenr. IV, 117, 1:58. 1916. *S. fragosa* ssp. *claytoniaefolia* Bacigalupi in Abrams, Ill. Fl. Pac. St. 2:363. 1944. *(Frank Tweedy,* The Dalles, Oreg. , May, 1883)

S. plantaginea Small, Bull. Torrey Club 23:366. 1896. *Micranthes plantaginea* Small, N. Am. Fl. 22²:135. 1905. *(Sandberg & Leiberg,* Spokane, Wash. , May, 1893) = between var. *columbiana* and var. *leptopetala.*

S. bracteosa Suksd. Deuts. Bot. Monats. 18:27. 1900. *(Suksdorf,* "Auf steinigen, meistens ebenen Plätzen bei Bingen, Klickitat-County," Wash. , Mar. 21, May, 1892) = var. *integrifolia.*

S. bracteosa var. *angustifolia* Suksd. Deuts. Bot. Monats. 18:28. 1900. *S. aphanostyla* Suksd. loc. cit. , in synonymy. *(Suksdorf,* "Auf Wiesenrändern im Falkenthal, Klickitat-County," Wash. , May 12, June 24, 1803) = var. *integrifolia.*

SAXIFRAGA INTEGRIFOLIA var. COLUMBIANA (Piper) C. L. Hitchc. hoc loc. *S. columbiana* Piper, Bull. Torrey Club 27:393. 1900. *Micranthes columbiana* Small, N. Am. Fl. 22²:135. 1905. *(Piper 1508,* Pullman, Wash.)

S. apetala Piper, Bull. Torrey Club 27:393. 1900. *Micranthes apetala* Small, N. Am. Fl. 22²:135. 1905. *S. integrifolia* var. *apetala* M. E. Jones, Bull. U. Mont. Biol. 15:32. 1910, as to synonymy, but not as to plants intended. *S. columbiana* var. *apetala* Engl. & Irmsch. Pflanzenr. IV, 117, 1:62. 1916. (Type cited as *Vasey 358,* Wash. , in 1889; however, *Vasey 358,* in the Wash. State University Herb. , is *Gaultheria ovatifolia,* but another sheet, which definitely fits the description of *S. apetala,* namely *Vasey 293,* is marked "Dup. type" in Piper's handwriting, the label reading "collected by G. R. Vasey, in 1889, in and near the Cascade Mountains of Kittitas, Chelan and King Counties. ")

Micranthes bidens Small, N. Am. Fl. 22²:137. 1905. *S. bidens* Fedde, Just Bot. Jahresb. 33¹:613. 1906. *(John Macoun,* Cedar Hill, Vancouver I. , May 12, 1887) = var. *integrifolia.*

S. bracteosa var. *leptopetala* Suksd. W. Am. Sci. 15:60. 1906. *S. integrifolia* var. *leptopetala* Engl. & Irmsch. Pflanzenr. IV, 117, 1:60. 1916. *(Suksdorf 4014,* Bingen, Klickitat Co. , Wash. , Apr. , 1899 and 1904)

S. bracteosa var. *micropetala* Suksd. W. Am. Sci. 15:60. 1906. *S. integrifolia* var. *micropetala* Engl. & Irmsch. Pflanzenr. IV, 117, 1:60. 1916. *(Suksdorf 4016,* Bingen, Wash. , Apr. , 1904) = var. *integrifolia* toward var. *leptopetala.*

S. fragosa var. *leucandra* Suksd. W. Am. Sci. 15:60. 1906. *S. fragosa* f. *leucandra* Engl. & Irmsch. Pflanzenr. IV, 117, 1:59. 1916. *(Suksdorf 4015,* Bingen, Wash. , Apr. , May, 1904) = between vars. *integrifolia* and *claytoniaefolia.*

S. laevicarpa Johnson, Minn. Stud. Pl. Sci. 4:46. 1923. *(Wm. Moodie,* Friday Harbor, San Juan Co. , Wash. , Apr. 3, 1910) = var. *integrifolia.*

Perennial, sometimes with an apparently simple caudex but commonly with short to rather extensive, ascending to horizontal, usually brittle rhizomes and mostly with short, lateral, cormlike propagules; flowering stems mostly single, leafless, 1-3 (4) dm. tall, from coarsely pilose to pubescent or subglabrous below, generally copiously glandular-pubescent above and in the inflorescence, with mostly reddish or purplish glands; leaves often bulbiferous in their axils, from more or less elliptic to lanceolate and subsessile or with a broadly winged but short petiole, to rhombic-ovate or ovate to deltoid and narrowed abruptly to a slender petiole, the blade mostly (1) 2-4 (but as much as 12) cm. long and (0. 5) 1-2 (but up to 6) cm. broad, usually entire or sinuate, but varying to remotely or (occasionally) regularly denticulate to dentate, sometimes glabrous, but usually sparsely to conspicuously ciliate, and often more or less rusty with slender, appressed, tangled hairs on the lower surface, rarely also coarsely hirsute; inflorescence loosely to rather closely and compactly cymose-paniculate, the lowest bract sometimes leaflike, but usually linear to linear-lanceolate, up to 2 cm. long, some or all of the bracts often rusty-floccose; calyx broadly conic and adnate to the pistil at base, without a free hypanthium, the lobes deltoid to oblong-lanceolate, 1-2 mm. long, spreading or (more commonly) reflexed; petals (rarely lacking) white to greenish-white or yellowish, sometimes purplish- or pinkish-tinged, deciduous, varying from ovate or broadly elliptic to obovate, rounded to retuse, (1. 5) 2-3 (4. 5) mm. long, and narrowed abruptly to a very slightly clawed base, to spatulate or narrowly oblanceolate, often less than 1. 5 mm. long, and narrowed gradually to a clawlike base; stamens usually shorter than the calyx lobes, the filaments broadly subulate, not at all clavate; anthers yellow or orange to greenish; carpels (often 3 or 4) nearly or quite distinct, but at anthesis largely inferior, the short, thick stylar beak almost immersed under a broad, lobed disc and only the rather large capitate stigmas visible, but in development the fruits growing above the disc and at maturity the follicles 3.5-5 mm. long, often reddish or purplish, and 1/2-2/3 superior; seeds brown, about 0.6 mm. long, lightly wrinkled lengthwise. N=about 28.

Prairies, grassy slopes, and wet banks to subalpine meadows; B.C. to Calif., e. to c. Ida. (and probably w. Mont.) and Nev. Late Mar.-July.

Saxifraga integrifolia is an extremely variable species, believed to consist of several major geographic races as well as numerous local variants that are not being accorded names in this treatment. Although the races here treated as varieties have often been given specific status, they are largely intergradient, and in many areas (but especially in the Columbia R. Gorge) populations are mostly of an intermediate nature. Not only do the varieties of *S. integrifolia* appear to be interfertile, but in various parts of their range they apparently cross with, and therefore show transition to, such species as *S. marshallii*, *S. occidentalis*, *S. rhomboidea*, and *S. oregana*.

1 Petals white, 1.5-3 (4.5) mm. long, ovate, oval, or obovate, usually at least half as broad as long
  2 Leaf blades mostly 2-5 (6) cm. long, narrowly rhombic to ovate-lanceolate or rhombic-ovate and narrowed gradually (or rather abruptly) to broad petioles rarely as long, mostly strongly ciliate-pilose and usually slightly to densely coarse-hirsute as well as rusty-arachnoid on the lower surface, the bracts also usually rusty-arachnoid; w. side of the Cascades from B.C. southward to Lincoln Co., Oreg., up the Columbia R. Gorge to Wasco Co., and completely transitional to var. *claytoniaefolia;* not uniform, plants from the "prairies" of w. Wash. often more pubescent (especially the leaves more hirsute) than those from along the coast and on the islands of the Puget Sound   var. integrifolia
  2 Leaf blades deltoid to ovate or lanceolate but narrowed abruptly to slender petioles often as long or longer, mostly glabrous except for some weak ciliation and sometimes some loose rusty floccosity on the lower surface; e. side of the Cascades from (B.C.?) n. Okanogan Co., Wash., to s. Oreg., and probably n.e. Calif., down the Columbia R. Gorge and into the Willamette Valley where intermediate to *S. marshallii;* e., in n. Oreg., to Grant Co., and to s.e. Wash. and w. Ida., where there is complete transition to var. *columbiana* and an approach to *S. occidentalis* var. *idahoensis;* south of our area this taxon is closely simulated by the plant usually called *S. nidifica* Greene, which is very doubtfully specifically distinct   var. claytoniaefolia (Canby) Rosend.
1 Petals commonly yellowish or greenish-white, often tinged with purple or pink (sometimes lacking), 1-2.5 mm. long, usually spatulate or oblanceolate, mostly (2) 2.5-3.5 times as long as broad
  3 Leaves mostly rhombic-lanceolate, narrowed gradually to broad, often conspicuously ciliate-pilose petioles; stems mostly strongly hirsute at base and copiously glandular-pubescent above; inflorescence sometimes glomerate
    4 Inflorescence densely congested, more or less capitate, very rarely with any of the lower branches evident; petals usually lacking; anthers yellow; occasional along the e. base of the Cascades from Yakima Co. to Okanogan Co., Wash., sometimes to the exclusion of other varieties, but often with var. *claytoniaefolia* which tends to blossom somewhat later   var. apetala (Piper) M. E. Jones
    4 Inflorescence open, at least the lower branches somewhat distant (1-2 cm. apart); petals nearly always present; anthers usually orange; s. B.C. southward, along the e. base of the Cascades, to n. Oreg., and down the Columbia R. Gorge, where intergradient with var. *integrifolia*, e. to w. Ida. and perhaps to w. Mont.
        var. leptopetala (Suksd.) Engl. & Irmsch.
  3 Leaves mostly rhombic-ovate, narrowed abruptly to more slender, weakly ciliate to glabrous petioles; stems sparsely pilose to nearly glabrous at base and usually only moderately glandular-pubescent above; inflorescence not glomerate; very similar to var. *claytoniaefolia* but with less pubescence and with much smaller petals; Blue and Wallowa mts., Oreg. (where often intergradient to *S. occidentalis* var. *idahoensis),* to adj. s.e. Wash. and to c. and s.w. Ida., s. to s.e. Oreg., Elko Co., Nev., and n.e. Calif.
        var. columbiana (Piper) C. L. Hitchc.

Saxifraga lyallii Engl. Verh. Zool.-Bot. Ges. Wien 19:542. 1869.
  Micranthes lyallii Small, N.Am. Fl. 22^2:143. 1905. *(Lyall,* Fort Colville, in 1861; acc. Savile & Calder [Can. Journ. Bot. 38:415. 1960] really in the Rocky Mts.)
  S. lyallii var. *laxa* Engl. Monog. Saxifr. 142. 1872. *(Lyall,* mixed with the type of *S. lyallii)*

S. lyallii var. *hultenii* Calder & Savile, Can. Journ. Bot. 38:418. 1960. *(Calder, Savile, & Taylor 23106, Moresby I., Queen Charlotte Is., B.C.)*

Perennial with well-developed rootstocks, often forming small mats; leaves all basal, the blades cuneate to flabellate or flabellate-obovate, mostly (5) 10-25 (50) mm. long and nearly as broad, prominently dentate or dentate-serrate with (3-5) 7-9 (11) teeth 2-4 mm. long, narrowed almost uniformly to rather slender petioles about as long as the blades, glabrous or sparsely brownish-pilose above, the inflorescence, calyx, and fruits often bright red; flowering stems (5) 8-25 (30) cm. tall; inflorescence rarely over 15-flowered, from simply cymose-racemose (the top flower blossoming first) to branched and less obviously cymose, the peduncles slender and often bracteate, the main bracts linear and usually entire; calyx lobed almost to the base, the segments oblong-lanceolate, (1.5) 2-3 (3.5) mm. long, sharply reflexed, a free hypanthium lacking; petals white, deciduous, often aging to pink, 2.5-4 mm. long, the blade oblong-oval, rounded to a short broad claw barely 0.5 mm. long, the tip from subacute to rounded or even slightly retuse; stamens persistent, about equaling the petals, the filaments white, distinctly clavate; carpels, especially of the oldest (top) flowers, often 3-5 (7), fused for only 0.5-1 mm., the ovuliferous portion of the ovary superior, the loculi barely extending below the point of insertion of the stamens, the upper portion of the carpels tapered to short (later divergent) styles less than 1 mm. long and the stigmas decurrent; fruit essentially follicular, 7-12 mm. long exclusive of the slender, divergent, mostly stylar beaks; seeds brownish, ellipsoid, about 0.7 mm. long, with several rows of thin, more or less scalelike, flattened papillae.

Usually along streams or ponds or in wet gravelly meadows; montane to alpine in our area; Alas. southward to the n. Cascades (Whatcom and Okanogan cos.) of Wash., e. in B.C. (and perhaps in n. Wash.) to extreme w. Alta., and s. in the Rocky Mts. to n. Ida. and w. Mont. (Beaverhead Co.). July-Aug.

Our material, as described above, is referable chiefly to the var. *lyallii,* although the recently described var. *hultenii* approaches our range on the north.

Saxifraga marshallii Greene, Pitt. 1:159. 1888.

Micranthes marshallii Small, N.Am. Fl. 22²:145. 1905. *S. marshallii* f. *crenata* Engl. & Irmsch. Pflanzenr. IV, 117, 1:36. 1916. *(C. C. Marshall,* Hoopa Valley, Humboldt Co., Calif., Apr., 1887)

S. hallii Johnson, Minn. Stud. Pl. Sci. 4:24. 1923. *(Elihu Hall 151.* Oreg., in 1871)

S. laevicaulis Johnson, Minn. Stud. Pl. Sci. 4:26. 1923. *(Piper 5061,* Mt. Grayback, Oreg., June 15, 1904)

S. petiolata Johnson, Minn. Stud. Pl. Sci. 4:29. 1923. *(Th. Howell,* Woodville, Oreg., May, 1889)

Perennial with a short, erect or ascending, mostly simple caudex and usually a single (2-3) leafless flowering stem 1-3 dm. tall, pubescent (throughout or at least above) with multicellular, reddish-purplish, glandular hairs; leaves several in a rosette, the small ones usually sparsely to copiously rusty-tomentose on the lower surface, the larger ones with the blades from ovate and with a cuneate to rounded or nearly truncate base, to elliptic-oblong, 2-5 cm. long, (0.5) 1-3 cm. broad, coarsely serrate or crenate-serrate with about 13-25 teeth, lanate to merely sparsely ciliate, narrowed abruptly or gradually to broadly winged or scarcely winged petioles from shorter than, to nearly 3 times as long as, the blades; inflorescence an open, often diffuse, cymose panicle, the branches slender, spreading, the bracts mostly small and linear, but the lowermost sometimes more leaflike and toothed; calyx adnate only to the base of the carpels, with a very short (0.1-0.3 mm. long), free hypanthium, the lobes ovate-lanceolate to oblong-lanceolate, (1) 1.5-2.5 mm. long, reflexed, glabrous or sparsely glandular-puberulent or somewhat rufous-pilose; petals deciduous, (2) 2.5-3.5 mm. long, the blade with 2 yellow spots on the lower half, abruptly rounded to a short but distinct claw; stamens persistent, inserted at the edge of the ovary and on the base of the hypanthium, the filaments white, spreading, about equaling the petals, strongly clavate and somewhat petaloid; ovary almost completely superior, at least the ovuliferous portion above the point of adnation, the 2 (3) carpels distinct to this point or below, surrounded by a ribbonlike nectary the upper margin of which is plainly visible; styles very short, the carpels dehiscent to the capitate but usually slightly decurrent stigmas; fruit more nearly follicular than capsular, much contracted at base, 3.5-5 mm. long, the stylar tips divaricate; seeds fusiform, about 0.7 mm. long, brownish, longitudinally wrinkled.

Moist or wet banks and rock crevices from n.e. Calif. northward, mostly near the coast, to Lane Co., Oreg., then passing gradually into *S. occidentalis* in the range n. to the Columbia R. Mar.-June.

It is highly questionable that this species should be maintained separately from *S. occidentalis,* since the two appear to intergrade. Their mergence would necessitate even additional nomenclatural changes, however, since the name *S. marshallii* antedates *S. occidentalis.*

Saxifraga mertensiana Bong. Mém. Acad. St. Pétersb. VI, 2:141. 1832.

Heterisia mertensiana Small, N. Am. Fl. 22²:156. 1905. *Micranthes mertensiana* Rosend. Engl.
Bot. Jahrb. 37, Beibl. 83:71. 1905. *Saxifraga mertensiana* var. *bulbillifera* Engl. & Irmsch.
Pflanzenr. IV, 117, 1:15. 1916. *(Mertens,* "Sitka Island")

Saxifraga heterantha Hook. Fl. Bor. Am. 1:252. 1833. *Steiranisia heterantha* Raf. Fl. Tellur. 2:
69. 1837. *(Douglas,* "Common on moist rocks of the River Columbia")

Heterisia eastwoodiae Small, N. Am. Fl. 22²:156. 1905. *Saxifraga mertensiana* var. *eastwoodiae*
Engl. & Irmsch. Pflanzenr. IV, 117, 1:15. 1916. *(Eastwood,* South Fork of Smith R., Del Norte
Co., Calif., May 1, 1905)

Saxifraga mertensiana var. *grandipilosa* St. John & Hardin, Mazama 11:77. 1929. *(Edith Hardin 526,*
Mt. Hermann, Whatcom Co., Wash., July 20, 1928)

Perennial with short thick rhizomes, tending to form large clumps, very succulent and often brittle,
the flowering stems (1) 1.5-4 dm. tall, usually with 1-3 leaves near their base, glabrous or more
commonly pilose and often glandular below, becoming pilose-pubescent and purplish-glandular above,
at least the longer hairs multiseriate (composed of more than 1 row of cells); leaves with petioles up
to 4 times as long as the blades and with well-developed, membranous and connate-sheathing stipules,
the basal ones usually with well-developed bulblets in their axils, the blades reniform to orbicular,
(2) 3-10 cm. broad and nearly as long, coarsely crenate-lobate and secondarily crenate-dentate, usually sparsely hirsute at least on the lower surface; inflorescence an open, cymose, many-flowered
panicle, generally with at least some of the flowers replaced by pinkish bulbils; calyx cleft nearly to
the base, the lobes reflexed, lanceolate to lance-oblong, mostly sparsely glandular-puberulent; petals
white, 3-5 mm. long, oblong-elliptic to somewhat obovate, narrowed abruptly to a clawless or short-
clawed base, withering persistent; stamens about equaling the petals, the filaments white, clavate,
somewhat petaloid, persistent, anthers pink; carpels 2-3 (4), the ovuliferous portion of the ovary
fused nearly the full length and not more than 1/10 inferior, surrounded by a bright yellow, bandlike
nectary (disc) and tapered above into conical stylelike portions and slender styles 1-2 mm. long; stigmas slightly decurrent; capsule about 5 mm. long, the dehiscent conical portion recurved and the
beaklike stylar tips divaricate; seeds light brown, slightly more than 1 mm. long, narrowly fusiform,
lightly wrinkled, the raphe winglike.

From lowland to montane, chiefly on wet banks and along gravelly streams; s. Alas. southward to
n.w. Calif. and the c. Sierra Nevada, in our area from the Cascades to the coast, e. through s. B.C.
to n.w. Mont., c. Ida., and n.e. Oreg. Apr.-early Aug.

There is much variation from plant to plant in the relative number of normal and bulbiliferous flowers. Therefore, the var. *eastwoodiae* (Small) Engl. & Irmsch., the normal-flowered phase, found occasionally throughout much (if not all) of the range, seems not particularly significant.

Saxifraga nuttallii Small, Bull. Torrey Club 23:368. 1896.

S. elegans of Nutt. in T. & G. Fl. N. Am. 1:573. 1840, but not of Sternb. in 1831. *Cascadia nuttallii*
Johnson, Am. Journ. Bot. 14:(printed corrigendum distributed with reprint). 1927. *(Nuttall,* "banks
of the Oregon below and near the Wahlamet")

S. nuttallii var. *macrophylla* Engl. & Irmsch. Pflanzenr. IV, 117, 1:231. 1916. *(Howell,* Silverton,
Oreg., is the first collection cited) Merely the more lush form of the species.

Perennial by very slender, brittle and fragile rhizomes, the stems weak, trailing to erect, freely
branching to simple, often with stolonous basal shoots, very slender, 5-40 cm. long, glabrous or
very sparsely pubescent with purplish-glandular hairs; leaves glabrous, succulent, the lower ones mostly
entire, with obovate to elliptic blades 1.5-4 mm. long and slender petioles usually as long, the numerous cauline leaves from similar to considerably larger, the blades sometimes more oblong-obovate to nearly flabellate, up to 19 mm. long and 11 mm. broad, more commonly with (2) 3 (4-5) shallow to rather deep teeth or lobes across the rounded tip and with very slender petioles; flowers (1) 2-
many in racemelike or paniclelike, loose, simple to compound, scorpioid, prominently bracteate
cymes, often the entire stem with lateral floriferous branches, the individual peduncles (pedicels) filiform, up to 3 cm. long, bractless or with 1-3 slender bracteoles; calyx usually glabrous (very
sparsely glandular-pubescent), at anthesis turbinate-campanulate, 2.5-3.5 mm. long, the lobes ascending-erect, triangular to ovate-triangular, acute, equaling to slightly longer than the connate portion which is about 2/3 adnate to the ovary and 1/3 free as a distinct hypanthium, slightly keeled
lengthwise below the sinuses of the lobes; petals white, more or less prominently pinkish-veined, oblong to oblong-oblanceolate, 2-4 times as long as the calyx lobes, deciduous; stamens equaling or exceeding the calyx lobes, inserted with the petals at or a little above midlength of the hypanthium; fila-

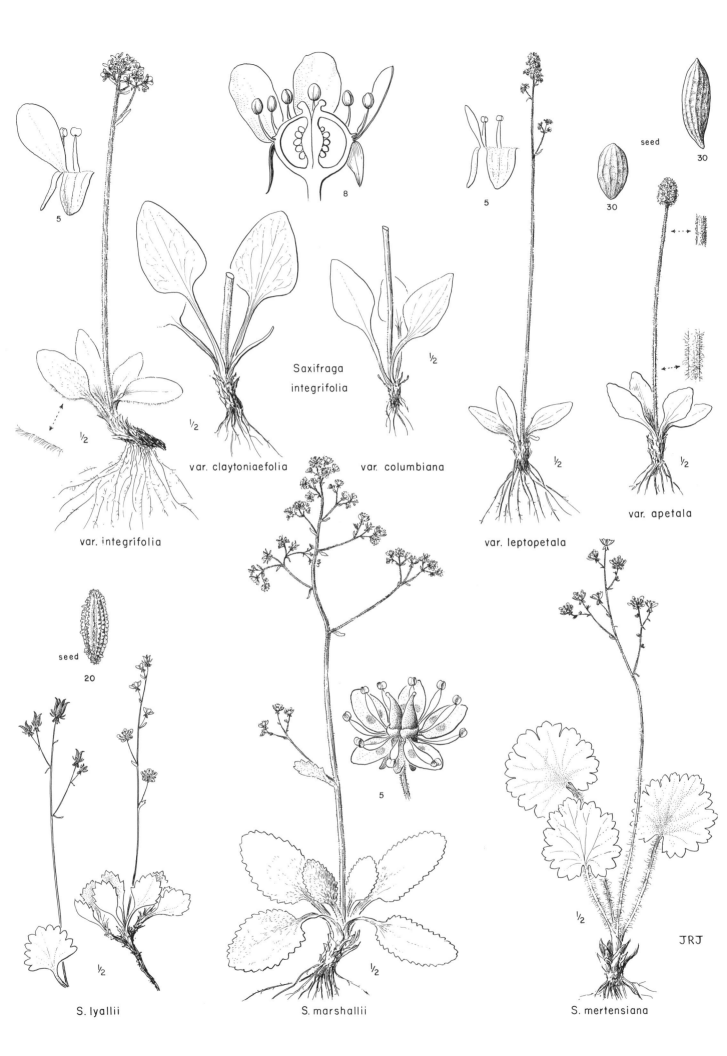

5

8

5

seed

30

30

seed

var. claytoniaefolia

var. columbiana

Saxifraga
integrifolia

½

½

½

½

var. integrifolia

var. leptopetala

½

var. apetala

½

seed

20

5

S. lyallii

S. marshallii

S. mertensiana

½

½

½

½

JRJ

ments not clavate; anthers orange; ovary at anthesis about 2/3 inferior, the 2 carpels free above the adnate portion, prolonged into slender solid styles nearly 1 mm. long, the stigmas small, capitate, the adnate portion of the ovary (chiefly) undergoing growth during the maturation of the fruit; mature capsule 3.5-5 mm. long and 2/3-3/4 inferior; seeds dark brown, about 0.6 mm. long, more or less crescent-shaped, with about 8 longitudinal rows of prominent comblike spines.

Wet banks and near waterfalls or trickling water, usually growing in moss; Columbia R. Gorge and the lower Willamette R. Valley, Saddle Mt., and southward along the coast of Oreg. to Curry and Coos cos., and from near Montesano, Grays Harbor Co., Wash. Mar.-July.

In general, this taxon is unlike our other species of *Saxifraga*, because of the spiny seeds, well-developed hypanthium, and triangular sepals. In comparison with certain of the Eurasian species it seems not so unique, however, and there is certainly as much reason for its maintenance in *Saxifraga* as for its placement in the monotypic genus *Cascadia*, especially since the latter genus was created at least partially because of the mistaken idea that the only species was an annual.

Saxifraga occidentalis Wats. Proc. Am. Acad. 23:264. 1888.
  Micranthes occidentalis Small, N.Am. Fl. 22²:144. 1905. (*Drummond*, Rocky Mountains of British Columbia, is the first collection cited)
  S. saximontana E. Nels. Erythea 7:168. 1899. *Micranthes saximontana* Small, N.Am. Fl. 22²:145. 1905. (*A. & E. Nelson 5917*, Yancey's, Yellowstone Nat. Park, July 17, 1899) = var. *occidentalis*.
  SAXIFRAGA OCCIDENTALIS var. IDAHOENSIS (Piper) C. L. Hitchc. hoc loc. *S. idahoensis* Piper, Bull. Torrey Club 27:394. 1900. *S. marshallii* var. *idahoensis* Engl. & Irmsch. Pflanzenr. IV, 117, 1:37. 1916. (*Henderson 4588*, Kendrick, Ida., Apr. 26, 1896)
  SAXIFRAGA OCCIDENTALIS var. RUFIDULA (Small) C. L. Hitchc. hoc loc. *Micranthes rufidula* Small, N.Am. Fl. 22²:140. 1905. *S. rufidula* Macoun, Ott. Nat. 20:62. 1906. *S. rufidula* f. *minor* Engl. & Irmsch. Pflanzenr. IV, 117, 1:39. 1916. *S. occidentalis* ssp. *rufidula* Bacigalupi in Abrams, Ill. Fl. Pac. St. 2:366. 1944. (*John Macoun*, Mt. Finlayson, Vancouver I., May 17, 1887)
  SAXIFRAGA OCCIDENTALIS var. ALLENII (Small) C. L. Hitchc. hoc loc. *Micranthes allenii* Small, N.Am. Fl. 22²:144. 1905. *S. allenii* Fedde, Just Bot. Jahresb. 33¹:613. 1906. (*O. D. Allen 242*, Goat Mountains, Wash., June 27, 1896)
  Micranthes lata Small, N.Am. Fl. 22²:145. 1905. *S. lata* Fedde, Just Bot. Jahresb. 33¹:613. 1906. (*John Macoun*, Lytton, B.C., Apr. 16, 1889) = var. *allenii*.
  Micranthes aequidentata Small, N.Am. Fl. 22²:145. 1905. *S. aequidentata* Rosend. Engl. Bot. Jahrb. 37, Beibl. 83:70. 1905. *S. rufidula* f. *major* Engl. & Irmsch. Pflanzenr. IV, 117, 1:39. 1916. (*Suksdorf 967*, Lower Cascades, Skamania Co., Wash., May 29, 1886) = var. *rufidula*.
  SAXIFRAGA OCCIDENTALIS var. DENTATA (Engl. & Irmsch.) C. L. Hitchc. hoc loc. *S. marshallii* f. *dentata* Engl. & Irmsch. Pflanzenr. IV, 117, 1:36. 1916. (*Heller 10059*, Elk Rock, near Oswego, Clackamas Co., Oreg., is the first collection cited)
  S. klickitatensis Johnson, Minn. Stud. Pl. Sci. 4:25. 1923. (*Suksdorf*, Klickitat Co., Wash., Apr. 9 and May, 1883) = var. *rufidula*.
  S. microcarpa Johnson, Minn. Stud. Pl. Sci. 4:28. 1923. (*M. F. Elrod 98a*, Missoula, Mont.) = var. *occidentalis*.
  S. gormani Suksd. Torreya 23:106. 1923. (*M. W. Gorman 4081*, Elk Rock, Multnomah Co., Oreg., June 2, 1917) = var. *dentata*.
  S. chelanensis Johnson, Am. Journ. Bot. 21:109. 1934. (*J. W. Thompson 5979*, 10 mi. n. of Entiat, Chelan Co., Wash., Apr. 18, 1931) = intermediate between *S. occidentalis* var. *idahoensis* and *S. integrifolia* var. *claytoniaefolia*, perhaps as close to the former, but collected just to the north of its range.
  S. marshallii var. *divaricata* Peck, Leafl. West. Bot. 5:59. 1947. (*Peck 18158*, Snake River Canyon, near mouth of Battle Creek, Wallowa Co., Oreg., Mar. 28, 1934) = var. *idahoensis*.
  S. occidentalis var. *wallowensis* Peck, Leafl. West. Bot. 5:60. 1947. (*Peck 18542*, above Ice Lake, Wallowa Mts., Wallowa Co., Oreg., July 4, 1934) = var. *occidentalis*.
  SAXIFRAGA OCCIDENTALIS var. LATIPETIOLATA C. L. Hitchc. hoc loc. Planta var. *dentatis* similis sed petiolis latis et multo brevioribus quam laminis; floris pubescentissimis; pilis undatis, flavo-glandularibus. (Type: *W. M. Gorman 3561*, Saddle Mt., Clatsop Co., Oreg., June 20, 1915, in U. of Wash. Herb.)
Perennial with short horizontal rhizomes, usually forming small clumps but often apparently with a short simple caudex and 1 or sometimes 2-3 leafless flowering stems, (0.5) 1-2.5 (3) dm. tall, vari-

ously pubescent but usually reddish-glandular; leaf blades from ovate to elliptic, up to 6 cm. long and 3 cm. broad, coarsely crenate-serrate with 15-30 teeth, the smaller ones usually with at least some reddish tomentum on the lower surface, the petioles slender to slightly winged, from shorter than the blades to 2 or 3 times their length; inflorescence cymose-paniculate, from considerably contracted and with the branches nearly erect to more open and with divaricate branches, pyramidal to rounded or flat-topped; calyx divided 1/2-3/4 of its length into ovate to oblong-lanceolate, spreading to re-flexed lobes 1-2.5 mm. long, the connate portion from about equally adnate to the ovary and free as a reflexed hypanthium 0.1-0.3 mm. long, to entirely adnate and without a free hypanthium; petals white (pinkish- or purplish-tinged), deciduous, 1.5-3.5 mm. long, the blade with 2 basal yellowish spots, or emaculate, ovate to oblong, narrowed abruptly to a short, distinct claw or nearly clawless; sta-mens persistent, about equaling the petals, inserted around the ovary, the filaments white to reddish, clavate (sometimes very strongly so) to subulate; carpels almost completely distinct, the ovary ad-nate to the calyx for 1/10-1/3 of its length, with a broad glandular band around the superior base (but the gland completely fused with, and apparently part of, the carpels), tapered to very short, ultimate-ly divergent styles; stigmas capitate and slightly decurrent; fruit more nearly follicular than capsular, contracted at base and into the styles, greenish to reddish-purple, 2.5-6 mm. long; seeds brownish, fusiform, about 0.7 mm. long, longitudinally wrinkled.

Moist banks and meadows to subalpine rocky slopes; B.C. southward, in the Cascades and Olympic Mts., to n.w. Oreg., e. to s.e. Alta., Ida., Mont., and n.w. Wyo., and as far s. as Elko Co., Nev. Apr.-Aug.

Saxifraga occidentalis is a complex, variable species consisting of numerous local races nearly all of which have been recognized at more than one taxonomic level. Several of these are more striking than others, but without exception they are interfertile, as judged by their complete intergradation. Probably because of the degree of isolation, the plants from west of the Cascades appear sharply de-limited from those of the east of this range, but through the Columbia Gorge they blend completely and even show some tendency toward intermediacy with *S. integrifolia* and *S. marshallii*. Perhaps the two most puzzling populations are those found in the foothills of the Wenatchee Mts. of s. Chelan and n. Kittitas cos., Wash., and in the Clearwater and adjacent Snake R. canyons in e. Oreg. and w. Ida. They are characterized chiefly by broad, petaloid, clavate filaments, and are not noticeably different from one another. They are freely transitional, however, to other races of *S. occidentalis* in their vicinity, and therefore must be included with that species. The taxon was described as *S. idahoensis*, but, because of its very close resemblance to *S. marshallii*, has more usually been treated as a vari-ety of the latter species, from which it is apparently well isolated geographically at the present time. Although the maintenance of *S. marshallii* as a separate species admittedly is in part for the sake of expediency, nevertheless it is felt that such treatment is about as natural as possible under the cir-cumstances. The several varieties are separable fairly satisfactorily on the following basis:

1 Filaments petaloid, very strongly clavate, 3-4 times as broad above the middle as at the base; petals sometimes with 2 near-basal yellow spots; foothills of the Wenatchee Mts., chiefly in and near Tumwater Canyon, s. Chelan and n. Kittitas cos., Wash., and in the lower Clearwater and adj. Snake R. canyons, Ida., and the lower tributaries of the Snake in n.e. Oreg., completely transitional to var. *occidentalis* in the latter locality and to var. *allenii* and possibly to var. *occidentalis* in the former   var. idahoensis (Piper) C. L. Hitchc.
1 Filaments not petaloid although sometimes clavate, commonly more nearly subulate; petals not yellow-spotted
   2 Inflorescence usually rather small and somewhat compact, rounded to pyramidal, mostly less than 5 cm. long at anthesis and less than 10 cm. long in fruit, the branches ascend-ing to erect; filaments clavate; B.C. s. in the Cascades to Whatcom and Chelan cos., Wash., e. through B.C. to s.e. Alta., and s. through Mont and Ida. to Wyo., and to n.e. Oreg. and s.e. Wash., where considerably modified by crossing with var. *idaho-ensis* with which it is partially sympatric   var. occidentalis
   2 Inflorescence open, often flat-topped, the branches erect to spreading or divaricate, often over 5 cm. long at anthesis and 10 cm. long in fruit; filaments clavate to subulate. The following 4 taxa are not very clearly marked, and in an ultraconservative treatment might all be merged into a single variety ("rufidula" the oldest available name); it is be-lieved, however, that they represent definite evolutionary tendencies even if their de-limitation cannot be made clearly and concisely.
      3 Inflorescence usually definitely flat-topped; bracts and calyces often reddish pilose-la-nate; filaments only slightly or not at all clavate; ovary mostly 1/3-1/4 inferior at an-

thesis; scapes, sepals, ovaries, filaments (and sometimes the petals) often strongly purple-
tinged; mostly w. of the Cascades and near the coast, B. C. s. to n. w. Oreg., and up the
Columbia R. Gorge to Wasco Co., Oreg.; completely transitional to var. *allenii* and to var.
dentata in the Gorge and in adj. Oreg., and not uniform even elsewhere; in particular,
plants from Mt. Olympus and vicinity, Clallam Co., Wash., are apetalous or else have
small purplish petals; and those from the e. end of the Columbia R. Gorge (*S. rufidula* var.
major of Engl. & Irmsch.) usually lack the brownish pilosity characteristic of the more
coastal material   var. rufidula (Small) C. L. Hitchc.

3 Inflorescence usually pyramidal; bracts and calyces seldom reddish-pilose; ovary mostly
  less than 1/3 inferior at anthesis; plants often without any purple tinge

  4 Filaments clavate; inflorescence not strongly diffuse, the branches mostly ascending, in
    aspect more or less intermediate between vars. *occidentalis* and *rufidula;* s. B. C. south-
    ward mostly along the e. side of the Cascades but at fairly high elev. (to 8000 ft.), to
    Mt. Rainier and the s. side of Mt. Adams, and mergent with var. *rufidula* and var. *den-
    tata* in the Columbia R. Gorge   var. allenii (Small) C. L. Hitchc.

  4 Filaments only slightly if at all clavate; inflorescence strongly diffuse, the branches often
    stiffly divaricate; plants usually of lowland valleys or foothills

    5 Leaves mostly narrowed to a distinct petiole nearly or quite as long as the blade; pubes-
      cence of the inflorescence mostly short, with purplish- or reddish-glandular tips; low-
      er Columbia R. Gorge s. to Tillamook Co., Oreg.

                                                var. dentata (Engl. & Irmsch.) C. L. Hitchc.

    5 Leaves gradually narrowed to a broad petiole much shorter than the blade; inflores-
      cence densely pubescent with rather wavy, inconspicuously yellow-glandular hairs;
      known only from Saddle Mt., Clatsop Co., Oreg.   var. latipetiolata C. L. Hitchc.

Saxifraga oppositifolia L. Sp. Pl. 402. 1753.

  Antiphylla oppositifolia Fourr. Ann. Soc. Linn. Lyon II, 16:386. 1868. ("Habitat in rupibus Alpium
    Spitzbergensium, Lapponicarum, Pyrenaicarum, Helveticarum")
  S. oppositifolia var. *albiflora* Lange, Consp. Fl. Groenl. 66. 1880. *S. oppositifolia* f. *albiflora* Fern.
    Rhodora 38:233. 1936. (*Vahl*, Frederikshaab, Greenl.) The rare albino variant.

  Caespitose, pulvinate perennial, forming thick cushionlike plants up to 2 dm. broad and 2-4 cm.
high, glabrous except for cilia on the foliage and the calyx, or slightly pilose along the stem, not glandular, often purplish-tinged throughout; leaves indefinitely marcescent, opposite and decussate, oblong
to obovate, 2.5-5 mm. long, sessile, entire but coarsely ciliate, closely crowded and strongly imbricate on the numerous sterile shoots, but fewer, often not overlapping, and occasionally alternate on
the upright, 1-4 cm. long, simple, 1-flowered stems; calyx cup-shaped, lobed for at least 3/4 the
length, the lobes oblong-oval, 2.5-3.5 mm. long, fleshy, coarsely ciliate, erect or only slightly
spreading, a free hypanthium lacking; petals purple, erect, 7-9 mm. long, spatulate-obovate to ovate-
oblong, narrowed to short and broad claws; stamens nearly twice as long as the calyx lobes, inserted
at the edge of the ovary; ovary about 1/5 inferior at anthesis, projecting into 2 tapered, hollow, sterile portions about as long as the true, solid, 1-2 mm. styles; stigmas slightly enlarged, capitate; capsule 6-8 mm. long; seeds light brown, ellipsoid, about 0.6 mm. long, lightly reticulate. N=13, 26.

  Tundra and coastal bluffs in the n., to alpine scree and rock crevices in our area; circumboreal, in
N. Am. from Alas. to Lab., Newf., and Vt., s. to the Olympic and c. Cascade mts. of Wash., the
Wallowa Mts., Oreg., and the Rocky Mts. of Mont., c. Ida., and n. Wyo. June-early Aug.

  Illustrated, out of sequence, on page 57.

Saxifraga oregana Howell, Erythea 3:34. 1895.

  Micranthes oregana Small, N. Am. Fl. 22[2]:138. 1905. ("mountain marshes of Oregon and Washing-
    ton," probably *Thos. Howell 1498*, Lake Labish, near Salem, Oreg., June 28, 1893)
  S. integrifolia var. *sierrae* Cov. Proc. Biol. Soc. Wash. 7:78. 1892. *S. sierrae* Small, Bull. Tor-
    rey Club 23: 366. 1896. *Micranthes sierrae* Heller, Muhl. 2:52. 1905. *S. oregana* var. *sierrae*
    Engl. & Irmsch. Pflanzenr. IV, 117, 1:63. 1916. (*Coville 1705*, 8 mi. n. w. of Whitney Meadows,
    Tulare Co., Calif., Aug. 25, 1891) = var. *oregana*, as reported for our area. True var. *sierrae*
    Cov., of the Sierra Nevada, Calif., has slightly smaller petals than var. *oregana*.
  SAXIFRAGA OREGANA var. MONTANENSIS (Small) C. L. Hitchc. hoc loc. *S. montanensis* Small,
    Bull. Torrey Club 23:367. 1896. *Micranthes montanensis* Small, N. Am. Fl. 22[2]:139. 1905. (*F.
    Tweedy 58*, s. w. Montana, July, 1888)

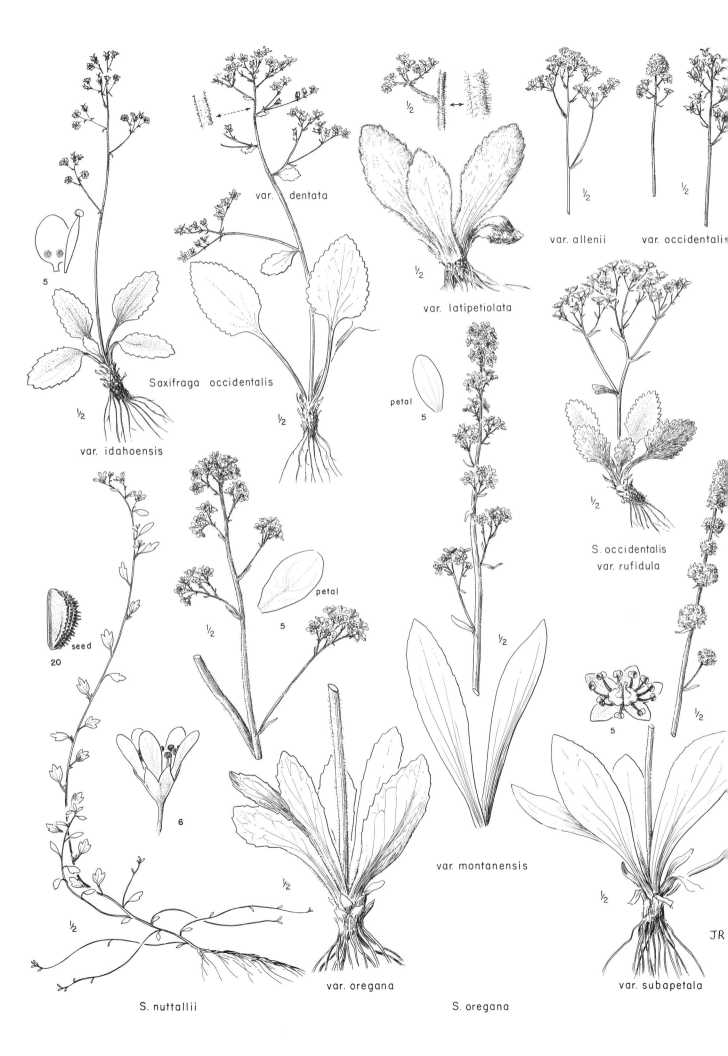

var. dentata

var. latipetiolata

var. allenii

var. occidentalis

5

Saxifraga occidentalis

var. idahoensis

petal
5

S. occidentalis
var. rufidula

seed
20

petal
5

var. montanensis

5

6

S. nuttallii

var. oregana

S. oregana

var. subapetala

JR

SAXIFRAGA OREGANA var. SUBAPETALA (E. Nels.) C. L. Hitchc. hoc loc. *S. subapetala* E. Nels.
Erythea 7:169. 1899. *Micranthes subapetala* Small, N.Am. Fl. 22[2]:139. 1905. *S. montanensis* var.
subapetala Engl. & Irmsch. Pflanzenr. IV, 117, 1:64. 1916. *(Nelson & Nelson 6089,* Obsidian
Creek, Yellowstone Nat. Park, July 24, 1899)
Micranthes arnoglossa Small, N.Am. Fl. 22[2]:138. 1905. *S. arnoglossa* Fedde, Just Bot. Jahresb.
33[1]:613. 1906. *(C. F. Baker 509,* Marshall Pass, Colo., July 19, 1901) = var. *montanensis.*
S. subapetala var. *normalis* A. Nels. Bot. Gaz. 42:53. 1906. (No collection cited) Probably = var.
montanensis.

Strong perennial from an erect, simple (or sometimes several-branched) caudex; flowering stems
leafless, 3-12 dm. tall, glabrous to pilose at base but becoming copiously glandular-pubescent to
glandular-pilose above, the glands yellow, pink, or purple; leaves (6) 10-20 (35) cm. long, rather
uniformly contracted to a broadly winged (very short to 3-6 cm.) petiole, the blades oblanceolate or
elliptic-oblanceolate to ovate-lanceolate or narrowly obovate, entire to sinuate-denticulate or occasionally prominently serrate, glabrous to sparsely hirsute and usually strongly ciliate-pilose at least
along the petiole; inflorescence from much elongate and narrowly cymose-paniculate to open and
somewhat diffuse, not uncommonly 2-3 dm. long in fruit, copiously glandular-hairy; bracts from
rather leaflike to linear; calyx conic and adnate to the ovary for 0.5-1 mm., without a free hypanthium, the lobes oblong-ovate to oblong-lanceolate, (1.5) 2-3 mm. long, usually reflexed; petals white
to greenish-white, often unequal in size and sometimes 1 or more lacking, usually narrowed to a
clawless or very slightly clawed base, narrowly obovate to narrowly or broadly oblong or ovate-oblong, mostly (1) 2-4 mm. long and 1-2.5 mm. broad, rounded to slightly acute, deciduous; filaments
broadly subulate, commonly 1-2 mm. long, greenish-white or occasionally pinkish, inserted at the
edge of a lobed, fleshy disc which at anthesis more or less covers the pistil except for the styles;
carpels not rarely 3-4, distinct to below the point of adnation with the calyx, the ovary 1/2-1/5 inferior at anthesis, in fruit often apparently (but not completely) superior, the calyx adnate to about the
base of the placentae; styles very short and thick, the carpels dehiscent almost to the discoid stigmas;
fruit follicular, green (to reddish), (3) 4-5 mm. long, the tips spreading; seeds brownish, 0.8-1.2
mm. long, more or less fusiform, rather prominently wrinkled lengthwise.

Bogs, stream banks, and wet meadowland along the w. slope of the Cascade Mts. from Snohomish
Co., Wash., southward, through the Willamette Valley and the Cascade Mts. of Oreg. to the Sierra Nevada of Calif., e. in Oreg. to much of montane Ida. and Mont., and s. to Wyo. and Colo. Apr.-July.

It has been a rather general assumption that the plants of the Rocky Mts. are specifically distinct
from those of the Cascade Mts., supposedly because of their smaller, narrower, greenish (rather
than white) petals, and more nearly inferior ovary. In general, the plants of the two areas do *tend* to
differ, but many plants from Ida. and Mont. (especially in Granite Co.) cannot be separated from the
plants of the Cascades by any such criteria, and it is more realistic to treat *S. oregana* as a wide
ranging species with 3 fairly well-marked, but completely intergradient varieties, separable as follows:
1 Petals white, usually at least 3 (to 5) mm. long and half as broad; inflorescence broadly py-
    ramidal at early anthesis, later open and with remote, long, often spreading lower branches;
    ovary usually not over 1/3 inferior at anthesis, in fruit apparently almost superior; flow-
    ering stems often conspicuously long-hirsute; range of the species in w. Wash. and Oreg.,
    and in the Cascade Mts., southward gradually transitional to the slightly smaller-petaled
    var. *sierrae* of Calif., ranging eastward to e. Oreg. (where completely transitional with
    var. *montanensis)* and occasional as far e. as w. Mont.; plants of the Cascades just to the
    south of Mt. Hood tend to have very conspicuously toothed leaves and perhaps are geneti-
    cally significant   var. oregana
1 Petals from white to greenish-white or greenish, mostly 2-3 (4) mm. long, usually less
    than half (to over half) as broad, occasionally some or all lacking; inflorescence narrow
    and elongate, the branches nearly erect; ovary usually at least 1/3 (to 1/2) inferior at an-
    thesis, in fruit generally at least 1/3 inferior; flowering stems sparsely hirsute to glan-
    dular-pubescent
    2 Filaments, fruits, and calyx lobes usually greenish; petals mostly at least 2 mm. long
        but not rarely lacking altogether; Ida. and Mont. to Colo., w. to e. Oreg.
                                                                var. montanensis (Small) C. L. Hitchc.
    2 Filaments (always), fruits (usually), and calyx lobes (often) purplish-tinged to deeply
        reddish-purple; petals either completely lacking or mere vestiges barely 1 mm. long;
        c. Mont. to Wyo., sometimes with var. *montanensis* but occurring to its exclusion in
        much of the range in s. c. Mont.   var. subapetala (E. Nels.) C. L. Hitchc.

Saxifraga punctata L. Sp. Pl. 401. 1753. (Siberia)

S. aestivalis F. & M. Ind. Sem. Hort. Petrop. 1:37. 1835. *Micranthes aestivalis* Small, N. Am. Fl. 22²:147. 1905. (Typification obscure, but acc. Hultén [Fl. Alas. 5:932. 1945], this name must be considered synonymous with *S. punctata)*

S. nelsoniana sensu Piper, Contr. U.S. Nat. Herb. 11:314. 1906, but not of D. Don, Trans. Linn. Soc. 13:355. 1822.

SAXIFRAGA PUNCTATA var. CASCADENSIS (Calder & Savile) C. L. Hitchc. hoc loc. *S. paddoensis* Suksd. W. Am. Sci. 15:59. 1906. *S. punctata* ssp. *cascadensis* Calder & Savile, Can. Journ. Bot. 38:425. 1960. *(Suksdorf 2504,* Mt. Paddo [Adams], Wash., Sept. 17, 1894)

S. punctata ssp. *pacifica* Hultén, Fl. Alas. 5:928. 1945. *(Anderson 6270,* Juneau, lectotype by Calder & Savile)

Perennial with a branched caudex and moderately long, horizontal or ascending rootstocks, mostly glabrous below or the leaves somewhat hairy, the upper portion of the flowering stems and the inflorescence always moderately to heavily pilose-arachnoid with slender, multicellular, sinuous to curled or matted, mostly eglandular hairs; flowering stems usually single, 1-3 dm. tall, simple and leafless below the inflorescence; basal leaves several, the petioles 1-5 times as long as the blades, with a slightly widened, nonsheathing, stipular-margined base, the blades glabrous or ciliate, reniform, (1.5) 2-6 (8) cm. broad, 1/2-4/5 as long, coarsely (5) 7- to 21 (25)-toothed, the teeth up to 1 cm. long, ovate and abruptly dentate to crenate-dentate; inflorescence a rather close, rounded, many-flowered, cymose panicle, becoming loose but narrow and up to 15 cm. long in fruit, the lowest branch often from a leaflike bract; calyx cleft nearly to the base, only very slightly adnate to the ovary, the lobes oblong-ovate to lance-oblong, about 1.5 mm. long, reflexed, usually reddish, ciliolate but otherwise glabrous, a free hypanthium lacking; petals white, 2.5-3.5 mm. long, narrowly to rather broadly oblong or oblong-elliptic, usually well over 1.5 times as long as broad, often rounded to retuse, narrowed abruptly and rounded or cuneate to a short broad claw, deciduous; stamens not quite so long as the petals, the filaments clavate, white, somewhat petaloid; carpels (sometimes 3 or 4) fused for less than half the length of the almost (about 9/10) superior ovary, sometimes nearly or quite distinct, the fruit from capsular to follicular, usually purplish, about 6 mm. long, the tips spreading to sharply reflexed; styles very short, thick; stigmas capitate; seeds brownish, fusiform, about 0.8 mm. long, longitudinally ridged with several rows of rufflelike, contiguous, cellular protuberances. N=24.

Montane in our area and usually on moist banks or along streams; Alas. southward, through the Olympic and Cascade mts. of Wash., to Oreg., e. in B.C. to Alta.; Eurasia. Late June-Aug.

This is a variable species consisting, in N. Am., of several geographic races. It is closely related to *S. arguta* and sympatric (but not intergradient) with it in Wash. and s. B.C. Our material, as described above is mostly referable to var. *cascadensis* (Calder & Savile) C. L. Hitchc., which occurs from s. w. B.C. to n. w. Oreg. Calder & Savile recognizes 5 other vars. in N. Am., all n. of our range.

Saxifraga rhomboidea Greene, Pitt. 3:343. 1898.

Micranthes rhomboidea Small, N. Am. Fl. 22²:136. 1905. *S. integrifolia* var. *rhomboidea* M. E. Jones, Bull. U. Mont. Biol. 15:32. 1910. (No specimens cited)

S. rydbergii Small, Mem. N. Y. Bot. Gard. 1:194. 1900. *Micranthes rydbergii* Small, N. Am. Fl. 22²:134. 1905. *S. hieracifolia* var. *rydbergii* Engl. & Irmsch. Pflanzenr. IV, 117, 1:27. 1916. *(Rydberg & Bessey 1268,* Electric Peak, Yellowstone Park, Aug. 18, 1897)

Micranthes crenatifolia Small, N. Am. Fl. 22²:134. 1905. *S. rhomboidea* var. *crenatifolia* Engl. & Irmsch. Pflanzenr. IV, 117, 1:29. 1916. *(F. W. Traphagen,* Deer Lodge, Mont., May, 1888)

S. greenei Blank. Mont. Agr. Coll. Stud. Bot. 1:65. 1905. *Micranthes greenei* Small, N. Am. Fl. 22²:137. 1905. *S. rhomboidea* var. *typica* f. *greenei* Engl. & Irmsch. Pflanzenr. IV, 117, 1:28. 1916. *(Blankinship,* Mt. Hyalite, Mont., Aug. 1, 1902)

Perennial with a usually simple, short, erect caudex; flowering stems single (very rarely 2), leafless, (0.5) 1-2 (3) dm. tall, glabrous to glandular-pubescent at the base and always rather copiously glandular-pubescent above with whitish hairs tipped with a yellowish to reddish gland; leaves mostly with thick, deltoid to rhombic-deltoid blades 1-4 (5) cm. long, usually fairly prominently crenate-dentate to crenate-serrate, ciliate, generally glabrous on the upper surface but sparsely to moderately rusty-arachnoid or -pilose (glabrous) on the lower surface, narrowed more or less abruptly to a broad, winged petiole usually shorter than the blade; inflorescence cymose-paniculate, from congested and nearly globose to more open and with the lower 1-4 branches up to 4 cm. apart but bearing secondary, tightly congested glomerules; bracts lacking or linear and not over 1 cm. long; calyx with an adnate, broadly conic base about equal to the deltoid to oblong-ovate, spreading to reflexed, 1-2 mm. lobes, a free hypanthium lacking; petals persistent, white or cream, not spotted, from about as long to nearly twice as long as the calyx lobes, spatulate to oblong or oblong-ovate and narrowed somewhat abruptly to a broad, sometimes clawlike base, rounded to slightly retuse; stamens about equaling the calyx lobes, filaments broadly subulate, inserted at the edge of a lobed gland; carpels (often 3 or even 4) distinct nearly to the base but at anthesis the ovary well over half inferior and partially covered by the gland, tapered to broad, beaklike, fully dehiscent tips; true styles lacking; stigmas enlarged and capitate; fruit more nearly follicular than capsular, the ovuliferous portion 4-6 mm. long, ending in recurved stylar beaks, usually purplish or reddish; seeds brownish, about 0.6 mm. long, lightly wrinkled longitudinally.

Moist places in sagebrush-covered slopes to subalpine meadows in the Rocky Mts., from s. B.C. and Alta. to Colo., w. to Juab Co., Utah, and c. Ida. Late May-Aug.

Saxifraga rhomboidea appears to intergrade slightly with *S. integrifolia* in the western limits of its range, but on the whole is more distinctive than most of the related taxa and recognizable because of the largely inferior ovary, conspicuous gland, subulate filaments, toothed leaves, and congested inflorescence, as well as by the usually dense, whitish, glandular-pubescence of the upper stems.

Saxifraga tolmiei T. & G. Fl. N.Am. 1:567. 1840.
 Leptasea tolmiei Small, N.Am. Fl. 22^2:155. 1905. (*Tolmie*, "North West Coast")
 S. ledifolia Greene, Pitt. 2:101. 1890. *Leptasea ledifolia* Small, N.Am. Fl. 22^2:155. 1905. *S. tolmiei* var. *ledifolia* Engl. & Irmsch. Pflanzenr. IV, 117, 1:88. 1916. (*Sonne,* near Truckee, Calif.)

Low, mat-forming perennial with numerous sterile leafy branches; leaves strongly marcescent, entire, glabrous or the bases sparsely long-ciliate, fleshy and more or less terete or slightly revolute, 3-10 (12) mm. long, pyriform to spatulate or oblanceolate and narrowed to broad, petiolar, exstipulate bases; flowering stems erect, 3-8 cm. tall, glabrous or pubescent with purplish-tipped hairs, leafless or with 1-3 somewhat reduced leaves, the flowers either single or 1-4 in a loose cymose inflorescence, the lower branches usually with 1-3 leaflike bracteoles; calyx glabrous, often purplish-tinged, saucer-shaped, the oval to oblong-ovate lobes 2-3 mm. long, spreading, 3-4 times as long as the basal adnate portion of the calyx, a free hypanthium lacking; petals white, up to twice as long as the calyx lobes, elliptic-obovate to broadly oblanceolate, gradually narrowed to the shortly clawed or clawless base; stamens inserted with the petals at the edge of the ovary, about equaling the calyx lobes, the filaments distinctly clavate and somewhat petaloid; carpels sometimes 3 or more, the fertile portion of the ovary about 1/4 inferior at anthesis, the hollow, beaklike, free, stylar portions about 1/2 as long as the ovary itself, true styles lacking; stigmas small, capitate; capsule ovoid, often purplish-mottled, 8-12 mm. long; seeds somewhat prismatic, light brown, about 0.8 mm. long, the testa plainly cellular, loose from (and much larger than) the rest of the seed.

In mountain meadows, usually near streams or in moist alpine talus, scree, or rock crevices; Alas. southward through the Cascades to the Sierra Nevada of c. Calif., w. to Vancouver I. and the Olympic Peninsula, the e. range uncertain, but known from the Bitterroot Mts. of Mont. and Ida. July-Aug.

A very attractive (but not easily grown) subject for the rock garden, with the following 2 varieties in our area:

1 Leaves with a few long cilia at base; flowering stems glandular-pubescent; range of the species in our area except for the Bitterroot Mts., rare or lacking in Calif.   var. tolmiei
1 Plant entirely glabrous; presently known in our area only from St. Mary's Peak, Bitterroot
  Mts., Ravalli Co., Mont., and Hazard Lakes, Idaho Co., Ida., the common plant in Calif.
                                                       var. ledifolia (Greene) Engl. & Irmsch.

The only 2 collections of the var. *ledifolia* seen from Mont.-Ida. include smaller leaved and more dwarfish plants than those studied from Calif. Whether or not this is a significant tendency is not known, however, and the recognition of only the one taxon seems at least expedient, the range being unusual but by no means unique.

Suksdorfia Gray Nom. Conserv.

Flowers few to many in a greatly modified (although still often somewhat flat-topped), cymose (but apparently racemose to paniculate) inflorescence, borne opposite (rather than axillary to) the bracts, if any; calyx lobes 5, erect to spreading, as long as the adnate portion and the free tubular portion (hypanthium) combined; petals 5, white or purplish-violet, erect to spreading, entire or sometimes bilobed; stamens 5, opposite the calyx lobes; pistil 2-carpellary; ovary 2-celled, with axile placentation, from slightly more than half to nearly completely inferior, prolonged into 2 hollow, tapered, nonovuliferous, stylelike beaks; stigmas capitate, nearly sessile; capsule dehiscent along the ventral sutures of the beaks; seeds many, somewhat prismatic, faintly to prominently warty; herbaceous perennials with very short, sparsely to copiously bulbiferous rootstocks, leafy flowering stems, and crenate to deeply divided, cordate to reniform basal leaves and strongly stipulate cauline leaves, glandular-pubescent at least in the inflorescence.

Only the two species. (Named for Wilhelm N. Suksdorf, 1850-1932, of Bingen, Wash., during his time one of the foremost collectors and students of the northwest flora.)

Both species are rather attractive and well worth a place in the rock garden, but neither has been introduced into the trade. They differ from one another in many significant details and have usually been treated in separate genera as *Hemieva* (or *Saxifraga) ranunculifolia* and *Suksdorfia violacea*.

1 Petals violet (rarely white), erect; calyx narrowly turbinate-campanulate; hypanthium not
　　disc-lined   S. VIOLACEA
1 Petals white, spreading; calyx shallowly and broadly campanulate; hypanthium lined with a
　　thick disc which partially covers the ovary   S. RANUNCULIFOLIA

Suksdorfia ranunculifolia (Hook.) Engl. in E. & P. Nat. Pflanzenf. 3²ᵃ:52. 1890.

Saxifraga ranunculifolia Hook. Fl. Bor. Am. 1:246. 1833. *Hemieva ranunculifolia* Raf. Fl. Tellur. 2:70. 1837. *Boykinia ranunculifolia* Gray, Am. Journ. Sci. 42:21. 1842. *(Douglas,* "high grounds around the Kettle Falls of the Columbia, and on the Rocky Mountains")

Stem usually single, simple or occasionally branching from the base, 1-3.5 dm. tall, glabrous to sparsely pubescent below, becoming strongly glandular-pubescent above; leaves light green, rather fleshy, the basal usually several, with a rather stout petiole 3-11 cm. long, the blade (1) 2-4 cm. broad, divided nearly to the petiole into 3 cuneate-obovate, entire to crenately lobed segments; stipules membranous; cauline leaves 4-9, the petiole reduced but the stipules enlarged upward; flowers numerous in a somewhat flat-topped, several-branched "panicle"; calyx shallowly and broadly campanulate, the lobes triangular-lanceolate, 1.5-2.5 mm. long, spreading, the tips slightly recurved; petals white or purplish-tinged at base, spreading, persistent, (2.5) 3-4 mm. long, obovate to oblanceolate, entire, not clawed; stamens about equaling the calyx lobes, the filaments equaling or slightly longer than the anthers; ovary about 2/3 inferior; capsule about 4 mm. long (including the 1 mm. long beaks); seeds brown, about 0.6 mm. long.

On wet mossy rocks from the foothills to subalpine slopes, often in areas that become dry by midsummer; B.C. southward, in (and mostly on the e. side of) the Cascade Mts., to n. Calif. and the Sierra Nevada, e. to Alta., n. Mont., c. Ida., and n.e. Oreg. May-Aug.

Suksdorfia violacea Gray, Proc. Am. Acad. 15:42. 1879.

Hemieva violacea Wheelock, Bull. Torrey Club 23:71. 1896. *(Suksdorf,* Columbia River, near the junction of the White Salmon River, Klickitat Co., Wash.)

Stem simple below the inflorescence, 1-2 dm. tall, from glandular-pubescent throughout to somewhat pilose and less glandular; basal leaves 1-3, usually withered by anthesis, slender-petiolate, the blade (1) 1.5-2.5 cm. broad and considerably shorter, coarsely crenate-lobate; cauline leaves 3-5, the petioles reduced, but the stipules enlarged, upward; flowers (1) 2-10 in an open, loose, simple "raceme" or a 2- to 3-branched "panicle"; calyx 4.5-6 mm. long at anthesis, narrowly turbinate-campanulate, the 5 lobes linear-lanceolate, nearly erect, the tips slightly spreading, about equal to the rest of the calyx, in fruit the calyx 6-9 mm. long and more campanulate; petals suberect, violet (occasionally almost white), 6-9 mm. long, slightly unequal, one often shallowly bilobed, spatulate to oblanceolate, the blade not sharply delimited from the short broad claw; anthers nearly sessile, 1-1.5 mm. long, the filaments scarcely 1/4 as long; ovary at anthesis nearly completely inferior, only the beaklike portions (about 1 mm. long) free; capsule 4-6 mm. long; seeds brownish, about 0.5 mm. long.

Rock crevices, mossy banks, cliffs, and sandy shaded areas, usually where wet (at least early in the season) but sometimes where completely dry by late May, widely but mostly intermittently distributed, B.C. and n.e. Wash. to n. Ida. and n.w. Mont., s. on the e. side of the Cascades in Wash. (s. Chelan Co.) to the Columbia R. Gorge. Mar. (in the Columbia R. Gorge) to June.

Sullivantia T. & G.

Flowers usually numerous in an apparent "panicle" or rarely in a "raceme" (really a much modi-

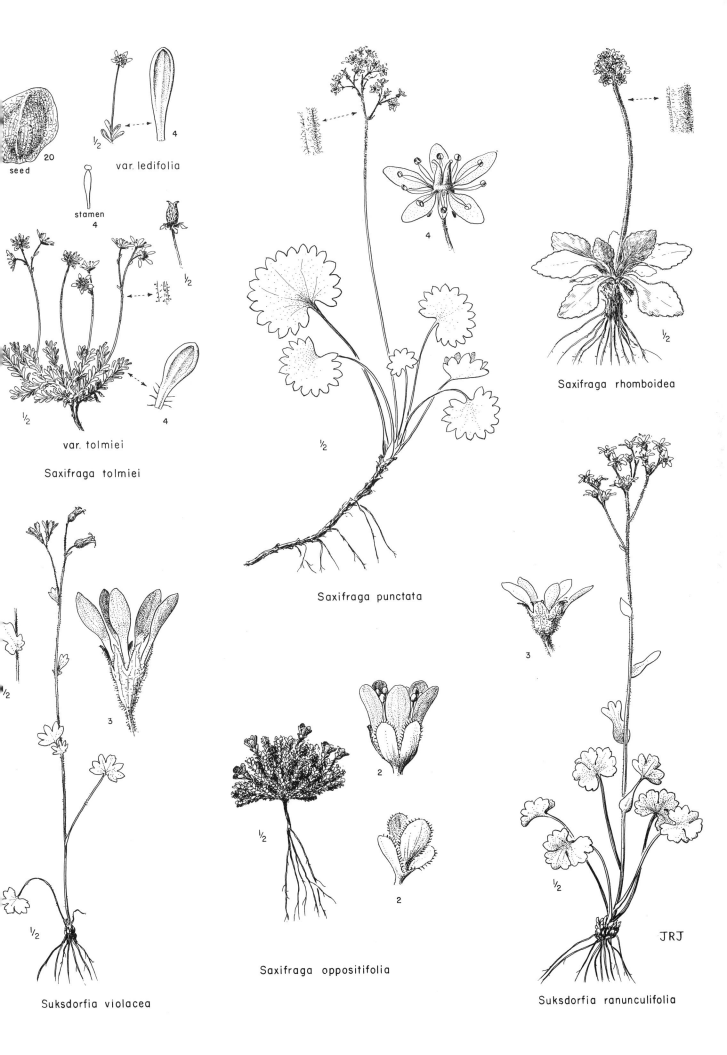

seed

20

1/2

var. ledifolia

4

stamen

4

1/2

var. tolmiei

1/2

4

Saxifraga tolmiei

4

1/2

Saxifraga punctata

1/2

Saxifraga rhomboidea

3

3

1/2

Suksdorfia violacea

2

1/2

2

Saxifraga oppositifolia

1/2

JRJ

Suksdorfia ranunculifolia

fied, compound cyme); calyx turbinate, the 5 lobes triangular, about as long (at anthesis) as the lower adnate portion and the upper free portion (hypanthium) combined, the hypanthium lined with a thin membranous disc; petals white, persistent; stamens 5, opposite the calyx lobes; pistil 2-carpellary, the ovary 2-celled, with axile placentation, the fertile portion about 3/4 inferior, the sterile upper fourth of the carpels tapered and beaklike, true styles lacking; stigmas sessile, capitate; capsule dehiscent along the ventral suture of the sterile portion of the 2 carpels; seeds several, linear-fusiform, narrowly wing-margined, lightly reticulate-pitted; perennial, moderately glandular-pubescent, stoloniferous herbs, with leafy flowering stems and cordate-reniform, incised-lobed and sharply toothed leaves.

Five or 6 species, confined to the U.S., the others from Wyo. and Colo. to Minn. and Ohio. (Named for W. S. Sullivant, 1803-1873, an American moss specialist.)

Reference:

Rosendahl, C. Otto. A revision of the genus Sullivantia. Minn. Stud. Biol. Sci. no. 6:401-427. 1927.

Sullivantia oregana Wats. Proc. Am. Acad. 14:292. 1879. *(J. Howell,* banks of the Willamette River, near Oregon City, Oreg., in 1877)

Delicate, yellowish-green perennial spreading by long slender stolons, nearly or quite glabrous except for some glandular pubescence on the upper portion of the flowering stems and on the inflorescence, the hairs mostly purplish-tipped; basal leaves long-petiolate, the blades reniform, 1-10 cm. broad, incisely lobed 1/3-1/2 their length into 7-9 cuneate segments and again once or twice sharply toothed; flowering stems 5-20 (25) cm. tall, with 1-3 leaves that are greatly reduced upward; flowers erect, but becoming sharply reflexed in fruit; calyx glabrous, pale green, 2.5-3.5 mm. long, more or less campanulate; petals slightly longer than the calyx lobes, the blade oval to obovate-oblanceolate, narrowed to a very short, broad claw; stamens shorter than the sepals, the cordate anthers about equaling the slender filaments; capsule about 4 mm. long; seeds brown, about 1.5 mm. long.

On moist cliffs, especially near waterfalls; Columbia R. Gorge and lower Willamette R., Oreg. Late May-Aug.

Telesonix Raf.

Flowers showy, reddish-purple, complete, regular, borne in compact, terminal, rather few-flowered, bracteate panicles; calyx turbinate-campanulate, adnate to the lower part of the ovary, and with a somewhat expanded, tubular, free hypanthium, the 5 lobes ovate-lanceolate; petals ovate to spatulate, inserted with the 10 stamens at the top of the hypanthium, from barely as long to nearly twice as long as the calyx lobes; filaments slender, about equaling the anthers but shorter than the petals; pistil 2-carpellary, the ovary with axile placentation, about half inferior, tapered above into the somewhat beaklike, free or partially connate styles; glandular-pubescent, perennial herbs from short, thick rootstocks, with reniform, doubly crenate leaves and slightly expanded, membranous stipules.

Only the one species. (Derivation uncertain.)

Telesonix jamesii (Torr.) Raf. Fl. Tellur. 2:69. 1837.

Saxifraga jamesii Torr. Ann. Lyc. N.Y. 2:204. 1827. *Saxifraga jamesiana* Engl. Monog. Saxifr. 109. 1872. *Boykinia jamesii* Engl. in E. & P. Nat. Pflanzenf. 3^{2a}:51. 1890. *Therofon jamesii* Wheelock, Bull. Torrey Club 23:70. 1896. *(Dr. James,* Rocky Mts.)

Therofon heucheraeforme Rydb. Bull. Torrey Club 24:247. 1897. *Telesonix heucheriformis* Rydb. N. Am. Fl. 22^2:126. 1905. *Boykinia heucheriformis* Rosend. in Engl. Bot. Jahrb. 37, Beibl. 83:64. 1905. *Saxifraga heucheriformis* M. E. Jones, Bull. U. Mont. Biol. 61:32. 1910. *Boykinia jamesii* var. *heucheriformis* Engl. in E. & P. Nat. Pflanzenf. 2nd ed. 18a:120. 1930. *Telesonix jamesii* var. *heucheriforme* Bacigalupi, Leafl. West. Bot. 5:71. 1947. *(Flodman 514,* Bridger Mts., Mont., July 28, 1896, is the first collection cited)

Stems 1-several, 5-15 (20) cm. tall, usually glandular-pubescent throughout with the hairs considerably longer above, the upper stems and the inflorescence often light to deep reddish-purple; leaves mostly basal, with slender petioles, the blades reniform, doubly crenate to shallowly lobed and doubly crenate-dentate, (1) 2-5 (6) cm. broad; cauline leaves (only 1 or 2 below the first flowers) considerably reduced; panicle leafy-bracteate, 5- to 25-flowered, often secund; calyx 9-13 mm. long, slightly accrescent, the lobes ovate-lanceolate to deltoid, about 2/5 the total length of the calyx; petals reddish-purple, up to about 3 mm. long; ovary about 1/2 inferior (in fruit somewhat less), the styles ultimately surpassing the calyx lobes; seeds brown, shining, more or less oblong, 1-2 mm. long. N=7.

Moist rock crevices and talus slopes, usually (but not always) on limestone; Alta. to S.D., southward, through the higher mts. of Mont. and Fremont Co., Ida., to Wyo., Colo., and e. Utah and s. Nev. July-Aug.

Our material, as described above, (with the range given above except for Colo.) is referable to the var. *heucheriformis* (Rydb.) Bacigalupi; the var. *jamesii* of Colorado has larger, more nearly orbicular petals (3-5.5 mm. long) and more completely connate stylar beaks on the capsule.

Although the species is grown successfully in England, it is difficult if not impossible of transplantation from the wild. It is indeed a beautiful species that has real potential for the rock garden, especially e. of the Cascades.

Tellima R. Br.

Flowers in elongate, open, minutely bracteate, terminal racemes; hypanthium well developed, campanulate-tubular, scarcely flared, regular, much longer than the adnate, somewhat turbinate calyx base and the 5 triangular-ovate, erect lobes; petals short-clawed, the blade pinnately divided, spreading; stamens 10, borne slightly below the petals near the top of the hypanthium, included, but the filaments equaling, to considerably longer than, the anthers; pistil 2-carpellary, ovary 1-celled with 2 parietal placentae, about 1/4 inferior, tapered above to 2 hollow, sterile, beaklike, ventrally dehiscent portions nearly as long as the ovuliferous, syncarpous ovary, the styles (proper) very short and thick; stigmas semicapitate; capsule dehiscent along the sutures of the beaks, many-seeded; coarsely hirsute and more or less glandular perennial with sparsely leafy flowering stems and cordate-ovate, more or less lobed leaf blades, long petioles, and rather small membranous stipules.

Only the one species. (The name an anagram of *Mitella*, under which genus the species was first published.)

Tellima grandiflora (Pursh) Dougl. in Lindl. Bot. Reg. 14:pl. 1178. 1828.
   Mitella grandiflora Pursh, Fl. Am. Sept. 314. 1814. *(Menzies,* "On the north-west coast")
   Tiarella alternifolia Fisch. ex Ser. in DC. Prodr. 4:50. 1830. *(Langsdorff,* Sitka)
   Tellima odorata Howell, Fl. N.W. Am. 199. 1898. *(Howell,* "along the Columbia river near the Cascades . . .") Well within the range of variation of the species.
   Tellima breviflora Rydb. N.Am. Fl. 22[2]:90. 1905. *(E. W. Hammond 128,* Sykes Creek, near Wisner, Oreg., in 1892)

Flowering stems up to 8 dm. tall, from a decumbent and somewhat rhizomatous base, sparingly leafy, copiously hirsute below, becoming glandular and less hirsute upward; basal leaves with very strongly hirsute petioles 5-20 cm. long, the blades cordate-triangular or cordate-ovate to more nearly reniform, 3-8 (10) cm. broad and about as long, shallowly (3) 5- to 7-lobed and irregularly once or twice crenate-dentate; cauline leaves 1-3, reduced; racemes loosely 10- to 35-flowered; pedicels much shorter than the flowers; calyx greenish, (5) 6-8 mm. long at anthesis and up to 11 mm. in fruit; petals greenish-white to deep reddish, often coloring with age; filaments 1-2.5 times as long as the anthers; capsule about equaling the calyx; seeds brown, narrowly ellipsoid-ovoid, 0.8-1 mm. long, rather prominently wrinkled-warty in longitudinal rows. N=7.

Common along streams, in woods, and (in general) on rich soil, from the seacoast to moderately high in the mountains; s. Alas. southward, along the coast, to s. of San Francisco Bay, inland in B.C. to the Selkirk Mts. and to n. Ida. and n.e. Wash., but otherwise usually w. of the Cascade crest in Wash. and Oreg., except in the Columbia R. Gorge. Apr.-July.

This plant is worthy of a place in the wild garden but apt to prove too aggressive in the average planting.

Tiarella L. False Mitrewort

Flowers in elongate leafless panicles (ours) or racemes, perfect; calyx irregular, the upper lobe usually the largest, the hypanthium 1/2-1/4 as long as the lobes, campanulate, nearly or quite free of the ovary, bearing the stamens and 5 petals at its summit; petals white, linear to subulate (ours), very similar to the filaments; stamens 10, considerably longer than the calyx, the filaments slender, the anthers oval; pistil 2-carpellary; ovary superior, 1-celled with parietal placentation but divided well over half its length into 2 erect, sterile, usually unequal, hornlike extensions ending in filiform styles (1) 1.5-2.5 mm. long (in ours); stigmas very small, capitate; fruit a few-seeded capsule, dehiscent along the unequal sterile valves above the fertile basal portion; seeds nearly black, shining and almost smooth, the raphe rather prominent; perennial, rhizomatous herbs (ours) with cordate and palmately lobed to 3-foliolate, variously toothed to cleft leaves and 2- to 3-foliate flowering stems.

About 6 species, with 1 in Asia, the others in w. and e. N.Am. (Latin, diminutive of the Greek *tiara,* an ancient Persian headdress, from the appearance of the fruit.)

Reference:

Lakela, Olga. A monograph of the genus Tiarella in North America. Am. Journ. Bot. 24:344-351. 1937.

Our 3 taxa are remarkably alike in all but the degree of leaf dissection and even though they are largely sympatric, with one exception they do not appear to intergrade. It therefore seems mandatory that all 3 be treated at the specific level.

The plants take well to cultivation but are (perhaps erroneously) not very highly regarded as garden subjects. They have rather attractive foliage, flower for much of the early summer, and are not such potential weeds as several other of our native members of the Saxifragaceae.

1 Leaves simple (very rarely some of the upper cauline leaves 3-foliolate), shallowly to rather
    deeply 3- to 5-lobed   T. UNIFOLIATA
1 Leaves 3-foliolate, the leaflets petiolulate
  2 Leaflets usually lobed no more than half their length, not laciniately cleft into ultimately
     narrow segments   T. TRIFOLIATA
  2 Leaflets cleft or divided nearly their full length and more or less laciniate into ultimately
     narrow oblong segments   T. LACINIATA

Tiarella laciniata Hook. Fl. Bor. Am. 1:239. 1833.

Petalosteira laciniata Raf. Fl. Tellur. 2:74. 1837. *T. trifoliata laciniata* Wheelock, Bull. Torrey
    Club 23:72. 1896. *(Menzies,* "North-West coast of America")

Rootstocks slender, horizontal to ascending; basal leaves more or less hirsute and usually glandular, with slender petioles mostly 2-3 times as long as the blades; leaf blades cordate, 1.5-4 cm. long, up to 6 cm. broad, trifoliolate; leaflets petiolulate, the lateral ones again nearly divided into 2 unequal segments, the upper lobe divided less deeply into 3 segments, the secondary lobes again deeply incised and finally prominently crenate-dentate and bristly-apiculate; flowering stems 2-3.5 (4) dm. tall, glandular-hirsute; panicle up to 15 cm. long, narrow; calyx finely glandular-puberulent, 1.5-2.5 mm. long, campanulate, lobed about 1/2-2/3 the length; hypanthium free of the ovary; petals linear, similar to and about equaling the filaments; filaments 1-3 times as long as the calyx lobes, those opposite the petals much shorter than the alternate 5; styles slender, 1.5-3 mm. long; valves of the capsule mostly 3-5 and 7-10 mm. long respectively; seeds about 1.5 mm. long.

In damp woods; Vancouver and adjacent islands of the Puget Sound, also in Skamania Co., Wash. May-July.

Tiarella trifoliata L. Sp. Pl. 406. 1753.

Blondia trifoliata Raf. Fl. Tellur. 2:75. 1837. *(G. Demidoff,* "Habitat in Asia boreali") Hultén (Fl.
    Alas. 5:943. 1945) states that this was probably an error and he postulates that the actual type was:
    Steller, Cape St. Elias (Kayok I.), Alas.
T. stenopetala Presl, Rel. Haenk. 2:55. 1831. *(Haenke,* Nootka-S[o]und?)

Plant very similar to *T. laciniata* in habit, habitat, and floral characters, but the leaves 1.5-7 cm. long and up to 9 cm. broad, the three petiolulate leaflets lobed no more than half their length, the lobes secondarily deeply crenate-dentate and bristly-apiculate; flowering stems (1.5) 2-5.5 (6) dm. tall.

In moist woods especially on stream banks; Aleutian Is. and s. Alas. to n. Oreg., from the coast to the w. slopes of the Cascades up to about 3500 ft. elevation, e. in c. B.C. to the Rocky Mts., s. to n. Mont. and in w. Ida. to Idaho Co., not known from e. of the Cascade Mts. in Wash. or Oreg. May-Aug.

Tiarella unifoliata Hook. Fl. Bor. Am. 1:238. 1833.

Petalosteira unifoliata Raf. Fl. Tellur. 2:74. 1837. *T. trifoliata* var. *unifoliata* Kurtz. Bot. Jahrb.
    19:378. 1894. *(Drummond,* "Height of land in the Rocky Mountains, near the source of the Columbia, and at Portage River")
T. unifoliata f. *trisecta* Lakela, Am. Journ. Bot. 24:350. 1937. *(Butters & Holway 216c,* Beaver
    Valley, Alta.)

Very similar to *T. laciniata* and *T. trifoliata* in most respects but the basal leaves up to 12 cm.

3

4

5

1/2

1/2   Telesonix jamesii

1/2   Tiarella laciniata

1/2

Sullivantia oregana

10
sepal tip

3

1/2

10

2

Tellima grandiflora

1/8

JRJ

broad and 8 cm. long, simple, broadly cordate in outline, palmately 5-lobed for 1/4-1/2 (5/6) their length, rarely divided, the lobes unequal, the lower 2 considerably smaller, all secondarily lobed less deeply and crenate-toothed to more or less doubly crenate-dentate and bristly-apiculate; cauline leaves similar or relatively more deeply 3-lobed; flowering stems (1.5) 2-4.5 dm. tall; calyx mostly 1.5-3 mm. long, the upper lobe usually 3-4 times as long as the hypanthium.

Moist woods and stream banks; s. Alas. southward to Santa Cruz Co., Calif., in both the Cascades and the Olympic Mts., mostly above 2000 ft. elevation, e. to s.w. Alta., w. Mont., n. Ida., and n.e. Oreg. June-Aug.

Although the species varies considerably in the depth of leaf lobation, plants with unusually deeply lobed (divided) leaves are at least occasional in extreme s.e. B.C. and s.w. Alta. (Selkirk Mts.) and in w. Ida., in an area where *T. unifoliata* and *T. trifoliata* occur more or less in proximity. Such plants were called *T. unifoliata* f. *trisecta* by Lakela. That they are hybrids can only be surmised. Although it would seem possible that they are sufficiently well established to be ranked as a geographic race, no change in their nomenclatural status is here proposed. They merit careful study.

Tolmiea T. & G. Nom. Conserv.

Flowers racemose, irregular, complete; calyx free of the ovary, with a tubular, oblique-based hypanthium longer than the 3 larger and 2 smaller lobes, cleft nearly to the base between the two smaller lobes; petals usually (always?) 4, linear-subulate, persistent, borne at the top of the hypanthium; stamens only 3, opposite the larger (upper) calyx lobes, borne somewhat below the top of the hypanthium; pistil 2-carpellary, the ovary 1-celled, with parietal placentation, superior, prolonged above the ovuliferous portion into 2 divergent, obconic, hollow, ventrally dehiscent beaks that taper to slender styles; fruit a membranous capsule dehiscent along the divergent beaks; seeds ovoid, finely spinulose; perennial herbs with cordate, palmately veined and shallowly lobate, long-petiolate leaves, and sparingly leafy flowering stems with terminal racemes.

Only the 1 species. (Named for Dr. William Fraser Tolmie, 1812-1886, surgeon for the Hudson's Bay Company at Ft. Vancouver.)

Tolmiea menziesii (Pursh) T. & G. Fl. N. Am. 1:582. 1840.

Tiarella menziesii Pursh, Fl. Am. Sept. 313. 1814. *Heuchera menziesii* Hook. Fl. Bor. Am. 1: 237. 1833. *Leptaxis menziesii* Raf. Fl. Tellur. 2:76. 1837. *(Menzies,* "On the north-west coast")

Hirsute throughout and somewhat glandular at least in the inflorescence, with well-developed rootstocks; flowering stems up to 8 dm. tall; leaf blades up to 10 cm. broad and nearly as long, shallowly 5- to 7-lobed and irregularly once or twice crenate-dentate, reduced upwards on the stem; stipules well developed, membranous; raceme loosely many-flowered, 1-3 dm. long; calyx greenish-purple to chocolate-colored, the hypanthium 5-9 mm. long, the lobes 3-5 mm. long; petals chocolate, up to twice as long as the calyx lobes; stamens unequal, the lower two longer than the upper one; capsule slender, 9-14 mm. long, eventually exceeding the calyx and protruding sidewise in the slit between the two lowest (smallest) lobes, the beaks equal; seeds about 0.5 mm. long. N=14. Youth-on-age, pig-a-back plant, thousand mothers.

Moist woods, especially along streams; s. Alas. southward, through B.C., Wash., and Oreg., from the lower levels of the Cascade Mts. (rarely e. of the divide) to the coast, extending to Santa Cruz Co., Calif. May-Aug.

An unusual plant, reproducing vegetatively by the development of buds at the base of the leaf blade (hence the common names), and often sold as a house plant. It takes well to the moist woodland garden where it merits a place in almost any company.

GROSSULARIACEAE Currant or Gooseberry Family

Flowers mostly 2-several (rarely 1) in bracteate (bractless) racemes, usually perfect (ours) but sometimes imperfect and rarely the plants dioecious; calyx generally at least partially adnate to the ovary and with a short to well-developed, tubular to saucer-shaped, free hypanthium, the lobes mostly 5 (4-6), valvate to imbricate, usually persistent, mostly larger and more showy than the 5 (4-6) petals; stamens 5 (4-6), alternate with the petals and with them inserted near, or at, the top of the hypanthium; pistil 2 (1-6)-carpellary, the ovary inferior (ours) to sometimes superior, 1-celled with 2 parietal placentae (ours) or 2- to 6-celled and with axile placentation, the styles from free to fully connate; fruit few- to many-seeded, baccate (ours) to capsular, the seeds with abundant endosperm; deciduous, spinose to unarmed shrubs (ours) or unarmed trees, with simple, alternate, mostly exstipulate leaves.

About a dozen genera and 150 species. Most of the genera, with the notable exception of *Ribes*, occur chiefly or entirely in the Southern Hemisphere; all are included in the Saxifragaceae by some authors, or, with the exception of *Ribes*, sometimes relegated to a third family, the Escalloniaceae.

Ribes L. Currants and Gooseberries

Flowers (1) 2 to many in mostly bracteate racemes (umbels) terminating short lateral branches, complete (ours), regular to slightly irregular, epigynous, greenish-white, white, yellow, or pink to red or purple, sometimes showy; pedicels often jointed immediately below the ovary; free portion of the calyx consisting of 5 erect or spreading to sharply reflexed lobes (sepals) and a tubular to saucer-shaped or rotate hypanthium, often lined internally with an inconspicuous to prominent, sometimes strongly lobed or, rarely, somewhat tubular disc; petals 5, always smaller than the sepals, generally erect, from oblong to cuneate, flabellate, or reniform, usually narrowed to a somewhat clawlike base, inserted at (or near) the top of the hypanthium or at the edge of the disc; stamens 5, alternate with the petals, the filaments subterete to greatly flattened, the anthers mostly oval, less commonly somewhat sagittate or cordate, rarely apiculate or tipped with a small cuplike gland; carpels 2, the ovary 1-celled with 2 parietal placentae, nearly or quite inferior at anthesis, always inferior in fruit; styles 2, from connate nearly to the stigmas to almost fully distinct; stigmas usually capitate; fruit a globose or subglobose, many-seeded, yellowish, reddish, bluish, or black berry, often glaucous, generally crowned with the persistent floral parts, glabrous to pubescent and often glandular, bristly, or spiny, usually more or less palatable; glabrous or (more commonly) pubescent and often glandular, erect to spreading or prostrate shrubs, the glands from sessile and then usually yellowish and crystalline in appearance to stalked; branches unarmed or with 1-several nodal spines and often also with few to many (sometimes very sharp) internodal bristles; leaves alternate, the blades palmately veined and mostly shallowly to deeply 3- to 5 (7)-lobed and variously toothed, usually more or less rotund but with a reniform or cordate to truncate or (occasionally) cuneate base.

Well over 100 species, mostly in the temperate and colder regions of the Northern Hemisphere, well represented in the Andes of S. Am.; especially abundant in w. N. Am. from Mex. to Alas. (From *ribas*, the Arabic name for the plant.)

There is a fairly clear line of demarcation between the armed species, most of which are commonly called gooseberries and often referred to a distinct genus *Grossularia*, and the predominantly unarmed currants *(Ribes* proper).

Several of the more showy species, including *R. aureum* and especially *R. sanguineum*, are well known and mostly easily grown ornamental shrubs. *R. lobbii* has real potentiality as an ornamental and *R. menziesii* slightly less, but neither takes well to cultivation w. of the Cascades. *R. cereum* has much to recommend it to the gardener e. of the Cascades, but is difficult in the Puget Sound area. Several species are attractive because of the berries, including *R. niveum*.

The cultivated currants are derived mostly from *R. sativum* Syme, and the gooseberries from *R. reclinatum* L., but several of our native species are utilized for the making of jam, jelly, and pie. Unfortunately many of the native and cultivated species are susceptible to the fungus which causes the white pine blister rust and serve as an avenue for the spread of the disease through the forest. As a control measure, efforts have been made to eradicate the known host species of *Ribes* in proximity to valuable stand of white pine.

Reference:

Berger, Alwin. A taxonomic review of currants and gooseberries. N. Y. State Agr. Exp. Sta. Tech. Bull. no. 109:1-118. 1924.

1 Plants with spines or prickles at the nodes and often also along the internodes and on the fruit (these sometimes very few in number and reduced in size); all species, except *R. lacustre* and *R. montigenum*, with the pedicels not jointed below the ovary, and with the flowers no more than 5 per raceme

  2 Free hypanthium shallowly cup-shaped or saucer-shaped; pedicels jointed below the ovary; racemes 3- to 15-flowered *(Lacustria* or *Grossularioides)*

    3 Leaves copiously pubescent and more or less glandular, 1-2.5 (4) cm. broad; berries reddish; pedicels stout, rarely as much as twice as long as the bracts   R. MONTIGENUM

    3 Leaves glabrous or only sparsely pubescent, never glandular, mostly 2-5 (7) cm. broad;

berries deep purple; pedicels slender, often at least twice as long as the bracts
                                                                                                    R. LACUSTRE
2 Free hypanthium tubular or campanulate, rarely cup-shaped but never saucer-shaped;
      pedicels not jointed near the ovary; racemes (1) 2- to 5-flowered *(Grossularia*—wild
      gooseberries)
   4 Styles shorter than the stamens, connate their full length; stigma enlarged, bilobed, ob-
         long; flowers less than 8 mm. long*, copiously crisp-pubescent externally; leaf blades
         3 (5)-cleft for often well over half their length, rarely as much as 15 mm. broad
                                                                                                    R. VELUTINUM
   4 Styles mostly (not always) exceeding the stamens, not completely connate but usually
         distinct for at least 1/4 of the length; stigma small, capitate-discoid; flowers general-
         ly over 8 mm. long, but if smaller then never crisp-pubescent (although rarely sparse-
         ly pilose) externally; leaf blades seldom cleft more than half their length, often much
         more than 15 mm. broad
      5 Styles hairy on the lower half; if sepals crimson then the berry neither prickly nor
            glandular
         6 Stamens at least twice as long as the petals, sometimes exceeding the extended
               sepals and conspicuously exserted
            7 Calyx lobes white or slightly greenish, rarely pinkish-tinged; sepals 5-8 mm.
                  long, usually at least twice the length of the 2-2.5 (3) mm. hypanthium; sta-
                  mens generally exceeding the extended calyx lobes, the anthers hairy; leaves
                  pubescent on the lower surface at least; styles connate much more than half
                  their length   R. NIVEUM
            7 Calyx lobes greenish to pinkish or purplish; sepals sometimes less than twice
                  as long as the hypanthium; stamens not rarely shorter than the calyx lobes,
                  the anthers glabrous; leaves sometimes glabrous except for the marginal cilia;
                  styles mostly connate only half their length
               8 Calyx lobes usually purplish or strongly purplish-tinged, 5-7 mm. long; al-
                     most entirely w. of the Cascades in our area   R. DIVARICATUM
               8 Calyx lobes usually greenish or only slightly purplish-tinged, 3-4 (5) mm.
                     long; e. of the Cascades in our area
                  9 Leaves commonly glabrous on one or both surfaces; calyx usually not hairy
                        externally; hypanthium (2) 2.5-3 (3.5) mm. long but not so broad, mostly
                        from 3/4 as long to as long as the calyx lobes; stamens usually not ex-
                        ceeding the extended sepals   R. INERME
                  9 Leaves commonly thickly pubescent on both surfaces; calyx usually hairy
                        externally; hypanthium 1.2-2 mm. long and as broad, rarely over half as
                        long as the calyx lobes; stamens usually slightly exceeding the extended
                        sepals   R. KLAMATHENSE
         6 Stamens usually about as long as the petals, rarely as much as twice their length,
               never as long as the extended sepals and not conspicuously exserted even with the
               sepals reflexed
            10 Hypanthium flared, more or less campanulate, 2-3.5 (4) mm. long, about as
                  broad (at the top) as long; flowers usually less than 10 (to 11) mm. long with
                  the sepals extended; anthers barely 1 mm. long
               11 Flowers mostly 8-11 mm. long with the sepals extended, the calyx lobes
                     mostly about half again as long as the hypanthium; styles often connate to
                     above midlength; plants usually erect, 0.5-2 m. tall, the branches rather
                     slender, rarely at all bristly when young   R. IRRIGUUM
               11 Flowers mostly not over 8 mm. long with the sepals extended, the calyx
                     lobes little if any longer than the hypanthium; styles connate only to mid-
                     length or less; plants often either low and intricately branched or with the
                     young growth bristly
                  12 Styles from shorter than the hypanthium to about equaling the petals,

---

*In fruiting specimens the calyx (especially the hypanthium) may be shrunken to not more than half
its size at anthesis. Measurements are recorded for fresh or boiled specimens.

connate for less than 1 mm.; petals 1-1. 5 mm. long, about half as long as the calyx lobes; anthers as long as the filaments; young branches sparsely or not at all bristly but the nodal spines very closely crowded, the plants gnarled and intricately branched   R. HENDERSONII

12 Styles longer than the petals, often about equaling the extended calyx lobes, connate for about half their length; petals 2-2. 5 mm. long, nearly as long as the calyx lobes; anthers only half as long as the filaments; plants usually armed with more internodal bristles than nodal spines but highly variable and often with only nodal spines   R. OXYACANTHOIDES

10 Hypanthium tubular, only slightly (or not at all) flared, (4) 5-6 mm. long, not so broad; flowers mostly well over 10 mm. long with the sepals extended; anthers usually more than 1 mm. long

13 Calyx glabrous externally; flowers 10-13 mm. long, the hypanthium 4-5 mm. long
                                                                                      R. SETOSUM

13 Calyx more or less finely pilose externally; flowers 11-16 mm. long, the hypanthium 5-6 mm. long   R. COGNATUM

5 Styles glabrous; sepals often crimson and then the berry prickly or glandular

14 Anthers lanceolate, at least 2 mm. long, broadest at the base and tapered to indehiscent mucronulate tips, smooth on the back; calyx lobes crimson

15 Plant in general nearly glabrous, the leaves never glandular; pedicels shorter than the subtending, (3) 4-5 mm. bracts; anthers reddish; branches not at all bristly, becoming reddish-brown   R. CRUENTUM

15 Plant rather heavily pubescent, especially on the glandular lower surface of the leaves and on the calyx; pedicels 2-3 times as long as the 2-3 mm. bracts; anthers white; branches very bristly, becoming grayish or straw-colored   R. MENZIESII

14 Anthers more or less oval, rarely as much as 2 mm. long, broadest near the middle, not mucronulate but dehiscent to the tip; calyx lobes mostly greenish or if crimson then the anthers papillate or warty on the back

16 Calyx greenish or only slightly reddish-tinged; stamens much shorter than the extended sepals; anthers white, smooth on the back; berry strongly bristly with spines 2-4 mm. long   R. WATSONIANUM

16 Calyx crimson at least on the inner surface of the lobes; stamens about equaling the extended sepals; anthers reddish or purple, warty or capitate-papillate on the back; berry stipitate-glandular rather than spiny   R. LOBBII

1 Plants without prickles or spines; pedicels jointed below the ovary; flowers usually more than 5 (rarely as few as 2) per raceme (wild currants)

17 Free hypanthium campanulate to cylindric, never saucer-shaped or shallowly cup-shaped, at least as long as broad; ovary never with sessile, yellow, crystalline glands; flowers sometimes bright yellow

18 Flowers bright yellow, glabrous *(Symphocalyx*–golden currants)

19 Hypanthium over 10 mm. long, about twice as long as the sepals; e. of the Rocky Mts. and probably not occurring natively in our area, although possibly to be found as an escape   *R. odoratum* Wendl. *(R. longiflorum* Nutt.)

19 Hypanthium less than 10 mm. long and usually less than twice as long as the sepals   R. AUREUM

18 Flowers other than bright yellow, usually glandular or pubescent or both

20 Plant (except for the ovary) more or less generally sprinkled with sessile, yellowish, crystalline glands; flowers greenish-white, the hypanthium usually equaling or slightly exceeding the sepals; petals white, 2/3-4/5 as long as the sepals
                                                                                      R. AMERICANUM

20 Plant eglandular or with noncrystalline glands, never bearing sessile, yellowish, crystalline glands; flowers various *(Calobotrya)*

21 Anthers not gland-tipped; flowers (at least in dried material) pale to deep rose, the hypanthium shorter than the sepals

22 Hypanthium 3-5 mm. long, tubular-campanulate; calyx lobes spreading, only slightly if at all longer than the hypanthium; from the e. slopes of the Cascade Mts. westward   R. SANGUINEUM

22 Hypanthium 1. 5-2 (2. 5) mm. long, bowl-shaped; calyx lobes usually ex-

ceeding, and sometimes nearly twice as long as, the hypanthium; s.e. Wash. (and
n.e. Oreg. ?) and w.c. Ida. in our range   R. MOGOLLONICUM
21 Anthers with a small cuplike gland at the tip (visible under 10X magnification); flowers
    white or pinkish to green, if tinged with red then the hypanthium usually at least twice
    as long as the sepals
    23 Hypanthium twice as long as the sepals; berry red   R. CEREUM
    23 Hypanthium about equaling the sepals; berry bluish to black   R. VISCOSISSIMUM
17 Free hypanthium saucer-shaped or very shallowly cup-shaped, broader than long; ovary
    sometimes with sessile, yellowish, crystalline glands; flowers never bright yellow
    24 Ovary (and usually the young herbage) sprinkled with sessile, yellow, crystalline glands
        25 Racemes drooping; hypanthium more or less cup-shaped; sepals usually purplish
            tinged; one of the domesticated black currants, seldom cultivated but occasionally
            escaping and persisting in the wild   *R. nigrum* L.
        25 Racemes erect or spreading; hypanthium saucer-shaped; sepals often whitish or
            green
            26 Hypanthium and sepals nearly pure white; racemes mostly 4-10 (17) cm. long,
                the bracts very narrowly linear-lanceolate, 1-3 mm. long, all shorter than
                the pedicels, not at all greenish; leaves primarily 3-lobed less than half their
                length, with the lower segments again much less deeply and very unequally
                lobed; entirely e. of the Cascades in our area   R. HUDSONIANUM
            26 Hypanthium and sepals greenish and usually strongly purplish-brown tinged;
                racemes mostly (10) 15-30 cm. long, the lower bracts usually more or less
                leaflike and exceeding the pedicels, the upper ones reduced but usually green-
                ish-tipped, rarely less than 4 mm. long; leaves primarily 5-lobed often for
                well over half their length, the lower segments usually again less deeply divid-
                ed into unequal to nearly equal lobes, the blades then nearly equally 7-lobed;
                almost entirely w. of the Cascades   R. BRACTEOSUM
    24 Ovary glandless or with stipitate, nonyellow, noncrystalline glands; if herbage (as rare-
      ly) with sessile crystalline glands, then the ovary always stipitate-glandular
        27 Ovary glabrous and smooth; ripe fruits red; sepals usually at least as broad as long
            28 Calyx greenish-yellow; petals yellowish-red; plants usually erect; anthers dumb-
              bell shaped, the sacs separated by almost the width of the filament   R. SATIVUM
            28 Calyx deep purplish or purplish-tinged; petals reddish-purple; plants often de-
              cumbent and rooting at the nodes; anthers broadly cordate and retuse, the sacs
             almost contiguous although slightly separated   R. TRISTE
        27 Ovary pubescent or stipitate-glandular or both; ripe fruits mostly dark bluish to
          black; sepals longer than broad
           29 Racemes drooping to pendent; filaments much broadened basally, borne on a low,
             coronalike disc projecting upward in the center of the flower; petals pink,
             spreading with the sepals; bracts 3-5 mm. long, usually equaling the pedicels
                                                    R. HOWELLII
          29 Racemes spreading to erect; filaments neither flattened nor borne on an erect,
             coronalike disc; petals and bracts various
            30 Hypanthium about 2 mm. long, nearly as long as the sepals, bowl-shaped or
              shallowly campanulate; flowers red or deep pink; s. Oreg. to Nev. and
              Calif., not believed to reach our area, although reported for Mont. by
              Rydberg   *R. nevadense* Kell.
            30 Hypanthium less than 2 mm. long and considerably shorter than the spread-
              ing sepals, shallowly saucer-shaped; flowers various
              31 Bracts of the racemes oblong to obovate, 3-4 mm. long, at least half as
                long as the pedicels
                32 Flowers greenish-white or yellowish-green; leaves usually lobed less
                    than half their length, the lobes triangular; berry black; known only
                    from the Blue Mts., Wash., and Seven Devils Mts., Ida., in our
                    area, otherwise in the Great Basin and Rocky Mts.   R. MOGOLLONICUM
                32 Flowers yellowish or pinkish; leaves usually lobed well over half their
                    length, the lobes oblong-ovate or oblong-obovate, rounded; berry red;
                    rather local in the Cascade Mts. of s. Douglas and Klamath and Jack-

son cos., Oreg., and not believed to extend as far n. as our area
R. erythrocarpum Cov. & Leib.

31 Bracts of the racemes narrowly lanceolate to ovate, mostly 1-2 mm. long, much less than half as long as the pedicels

   33 Petals with a reniform or crescent-shaped blade and a short claw, at least as broad as long, red to purplish; disc brownish to red, nearly flat; stalked glands of the ovary less than 0.5 mm. long; coastal from Alas. to n. Calif., e. in B.C. to the Rocky Mts.
R. LAXIFLORUM

   33 Petals cuneate to flabellate, longer than broad, whitish to pink; disc pale pink to greenish-white, saucer-shaped; stalked glands of the ovary up to 1.2 mm. long; e. Alas. to Lab., s. to n. B.C., Minn., Me., and in the Appalachians to N.C., the southern limits in w. N.Am. uncertain, but not believed to occur in our area   *R. glandulosum* Grauer

Ribes americanum Mill. Gard. Dict. 8th ed. Ribes no. 4. 1768. (Pennsylvania)

   R. floridum L. 'Her. Stirp. Nov. 4. 1785. *Coreosma florida* Spach, Ann. Sci. Nat. II, 4:22. 1835. *(Collinson,* Pennsylvania)

Erect to somewhat spreading, unarmed shrubs mostly about 1 m. tall, the young branches crisp-puberulent and more or less thickly dotted with sessile, yellowish, crystalline glands, aging to gray and eventually to nearly black; leaves shallowly cordate, sessile-glandular and more or less thickly coarse-pubescent at least on the lower surface, (2) 3-8 cm. broad, not quite so long, 3-lobed nearly half the length (or with 3 prominent and 2 much smaller basal lobes), the segments coarsely bicrenate-serrate; petioles equaling or shorter than the blades, with a few slender processes along the lower margins; flowers 6-15 in spreading to drooping, pubescent racemes about as long as the leaves; bracts narrowly lanceolate, up to 10 mm. long, from shorter to longer than the jointed pedicels; ovary glabrous (very sparsely hairy); calyx cream to greenish-white, usually more or less hairy, the hypanthium broadly tubular-campanulate, (3) 3.5 (4) mm. long, not quite so broad at the top, the lobes narrowly oblong-spatulate to nearly oblong, mostly not quite so long as the hypanthium, reflexed; petals whitish, oblong to oblong-obovate, mostly 2.5-3 mm. long (2/3-4/5 as long as the sepals); stamens subequal to the petals, the filaments glabrous, broad-based, abruptly narrowed just below the oval (1 mm. long) anthers; styles glabrous, about equaling the stamens, connate (and unusually thick) almost to the stigmas; berry ovoid, about 1 cm. long, smooth, not very palatable. Black currant.

Swamps, stream banks, and moist ravines and canyons, from the e. edge of the Rocky Mts., Alta. to N.M., e. to N.S. and Va., entering our area in c. Mont., extending w. almost to Missoula. May-June.

Ribes aureum Pursh, Fl. Am. Sept. 164. 1814.

   R. aureum var. *praecox* Lindl. Trans. Hort. Soc. Lond. 6:240. 1828. *(Lewis,* "On the banks of the rivers Missouri and Columbia")

   R. jasmiflorum Agardh, Svensk. Landtbr. Akad. Ann. 9:143. 1823.

   R. flavum of Colla, Mem. Accad. Torino 33:114. 1828, but not of Berlandier in 1826.

   R. tenuiflorum Lindl. Trans. Hort. Soc. Lond. 7:242. 1828 (1830?).

   R. oregoni Herincq. Hort. Fr. 1872:225, pl. 8. 1872. (Horticultural plants of doubtful origin)

Erect or rounded, unarmed shrubs 1-3 m. tall, the branches reddish and glabrous to finely puberulent when young, glabrous and dark gray with age; leaves finely pubescent to glabrous when young, but with age thick, pale green, and usually glabrous except for some ciliation, sometimes with a few nearly sessile glands, broadly deltoid-ovate to ovate, with a broadly cuneate to somewhat cordate base, 2-5 cm. broad, mostly 3-lobed less than half their length, the segments entire or with 2-5 rounded teeth but sometimes the lower ones again shallowly lobed; flowers fragrant, 5-18 in ascending to reflexed racemes equaling or longer than the leaves; pedicels up to 8 mm. long, jointed under the ovary; calyx glabrous, golden yellow, the hypanthium cylindric, (5) 6-8 (10) mm. long; calyx lobes oblong-elliptic, spreading but not usually reflexed, (4) 5-7 mm. long; ovary half as long as the hypanthium; petals yellow to orange or reddish, oblong-obovate, erect; stamens about equaling the petals, the filaments about equal to the oblong, minutely apiculate anthers; styles connate almost to the stigmas, glabrous, from slightly shorter to somewhat longer than the extended calyx lobes; berry glabrous, globose, about 7 mm. long, red to black (yellow), palatable. N=8. Golden currant.

Stream banks and flood plains in grasslands and sagebrush desert to ponderosa pine forest, from the e. slope of the Cascades, n. c. Wash. to Calif., eastward to the e. side of the Rocky Mts. from Sask. and S.D. to N.M. Apr.-May.

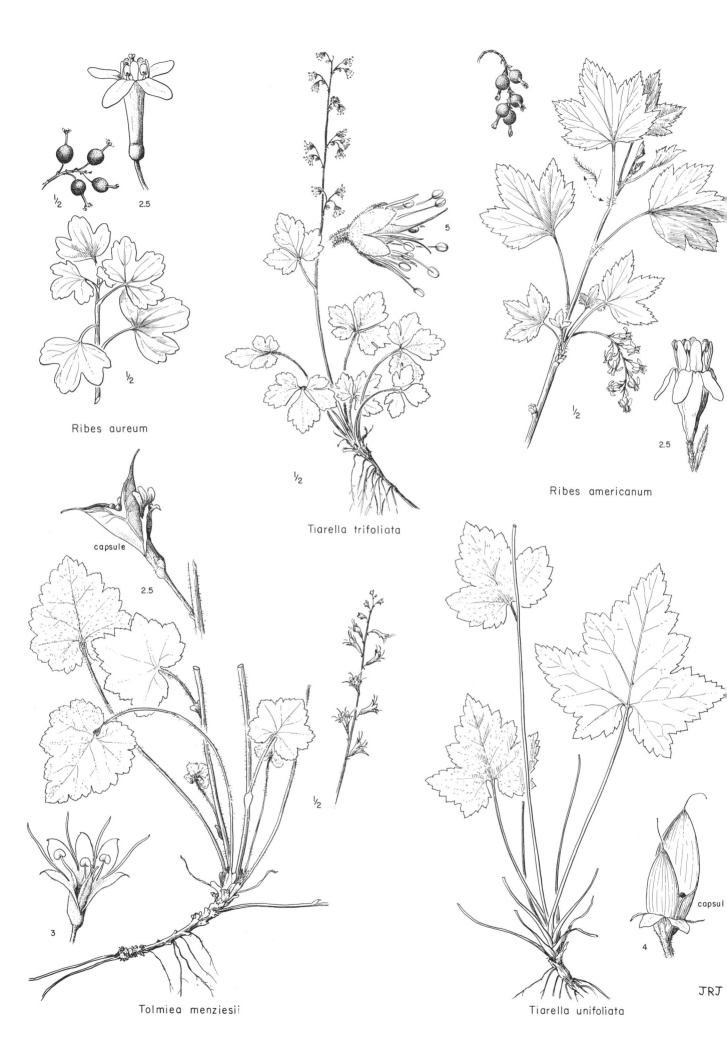

Ribes aureum

Tiarella trifoliata

Ribes americanum

capsule

Tolmiea menziesii

Tiarella unifoliata

capsul

JRJ

Variant plants, such as those with yellow fruit or with more reddish perianths or with greater puberulence, occur, but their distribution appears to be sporadic rather than geographically significant.

Ribes bracteosum Dougl. ex Hook. Fl. Bor. Am. 1:233. 1832. ("North-West coast of America, at the confluence of the Columbia with the ocean," *Dr. Scouler, Douglas*)
   R. bracteosum var. *viridiflorum* Jancz. Mém. Soc. Phys. Nat. Genèv. 35:339. 1907. (No type given)
   Erect, often straggly, unarmed shrubs, (1) 1.5-3 m. tall, more or less closely sprinkled with round, sessile, yellowish, shining, crystalline glands, and with a rather characteristic sweetish but somewhat disagreeable odor; leaf blades (3) 4-12 (24) cm. broad, usually not quite so long, cordate, from sparsely pubescent to glabrous except for the glands of the lower (paler) surface, deeply 5- to 7-lobed mostly over half the length, the main segments ovate-lanceolate, shallowly lobate and once or twice serrate; petioles from shorter to much longer than the blades, sparsely pubescent and with several acicular, mostly persistent processes near the base along the slightly winged margins; flowers numerous in ascending to erect racemes up to (10) 15-30 cm. long, the pedicels slender, 5-12 mm. long, jointed, often with 1-3 tiny bracteoles immediately under the flower; bracts conspicuous, the lower ones often leaflike, gradually reduced upward, the upper ones becoming very narrowly oblong and only (3) 4-5 mm. long, but usually with a greenish, often somewhat expanded tip, at least the lower ones usually exceeding the pedicels; ovary very thickly glandular and usually somewhat hairy, the free calyx brownish-purple to greenish or sometimes nearly white; hypanthium widely flared and deeply saucer-shaped, about 1.5 mm. long, more or less pubescent and sparsely glandular externally, lined internally with a thick, lobed disc that covers and submerges the ovary; calyx lobes ovate-lanceolate to oblong-lanceolate, 3-3.5 (4) mm. long, spreading; petals white, cuneate-flabelliform, with an oblong basal claw; stamens inserted slightly below, and about equal to, the petals; filaments glabrous; anthers about 0.5 mm. long; styles about equaling the petals, connate and tapered upward for 1/4-1/2 their length; berry subglobose, about 1 cm. long, glaucous-black, glandular and with a disagreeable taste. N=8.
   Stream banks and moist woods, usually where there is seepage; Alas. southward to n.w. Calif., common from the Cascades (nearly always w. of the crest) to the coast, occasionally e. of the Cascades, as in Okanogan Co., Wash. May-June.

Ribes cereum Dougl. Trans. Hort. Soc. Lond. 7:512. 1830. *(Douglas,* "river Columbia from the Great Falls 45° 46' 17" N. Lat. to the source of that stream in the Rocky Mts.")
   RIBES CEREUM var. INEBRIANS (Lindl.) C. L. Hitchc. hoc loc. *R. inebrians* Lindl. Bot. Reg. 17:pl. 1471. 1832. *Cerophyllum inebrians* Spach, Hist. Veg. 6:154. 1838. *R. inebrians* var. *maius* Jancz. Mém. Soc. Phys. Nat. Genèv. 35:336. 1907. (Described from cultivated plants of unstated origin)
   R. reniforme Nutt. Journ. Acad. Phila. 7:25. 1834. *(Wyeth,* "Sources of the Columbia") = var. *cereum.*
   R. cereum var. *pedicellare* Gray, Bot. Calif. 1:207. 1876. (Montana, no collections cited) = var. *cereum.*
   R. spathianum Koehne, Gartenfl. 48:338. 1899. (Garden plants, originally collected in Colorado by Purpus) = var. *cereum.*
   R. pumilum Nutt. ex Rydb. Fl. Colo. 178. 1906. (On dry hills from Mont. to N.M. and Ariz., no collectors cited) = var. *cereum.*
   R. cereum var. *farinosum* Jancz. Mém. Soc. Phys. Nat. Genèv. 35:338. 1907. *(Douglas,* mountains near the Columbia River) = var. *cereum.*
   RIBES CEREUM var. COLUBRINUM C. L. Hitchc. hoc loc. Planta var. *inebrians* similis, sed foliis, calycis, stylibusque glabris vel subglabris. (Type: *Cronquist 6181,* French Cr., 2 mi. s. of Salmon R., 20 mi. e. of Riggins, Idaho Co., Ida., Apr. 21, 1950, in U. of Wash. Herb.)
   Spreading or rounded to erect, unarmed shrub (0.2) 0.5-1.5 (2) m. tall, the new branches finely puberulent and often sparsely to copiously short-stipitate-glandular, turning grayish-brown or reddish-brown; leaves from almost reniform to broadly cuneate-flabellate, from quite glabrous or only sparsely stipitate-glandular to downy and often conspicuously stipitate-glandular on both surfaces, usually 1.5-2.5 (0.5-4) cm. broad and from rounded and almost equally coarsely crenate-dentate to (more commonly) shallowly 3- or 5-lobed much less than half the length and closely crenate-dentate; flowers 2-8, capitate-racemose at the ends of peduncles much shorter than the leaves, the entire inflorescence usually both finely pubescent and more or less sticky with short-stalked to subsessile glands; pedicels jointed under the ovary, usually shorter than the bracts; calyx greenish-white to white or faintly to strongly pinkish-tinged, from nearly glabrous to rather strongly pubescent as well as

stipitate-glandular, the hypanthium nearly cylindric, (5) 6-8 (9) mm. long, the lobes deltoid-ovate, spreading-recurved, 1.5-3 mm. long but never so much as 1/2 as long as the hypanthium; petals flabellate to spatulate-obovate, 1-2 mm. long, usually about 1/2 exserted, equaling to considerably exceeding the stamens; anthers equaling or shorter than the filaments, 0.7-1.5 mm. long, oval, tipped with a small cup-shaped gland; styles sometimes connate nearly or quite to the stigmas, glabrous or sparsely to thickly pubescent, included within, or slightly exserted from, the hypanthium; fruit ovoid, 6-8 mm. long, sparingly glandular (glabrous), dull to bright red, unpalatable. N=8. Squaw currant.

Common on the e. slope of the Cascades from B.C. through Oreg. and southward to s. Calif., e. to Mont., Neb., Colo., N.M., and Ariz., from sagebrush desert to subalpine ridges. Apr.-June.

The variation in nearly all features of the plant seems to be largely fortuitous and often due to local conditions of exposure, moisture, and soil. *R. inebrians* Lindl. and *R. reniforme* Nutt. have usually been recognized as separate species, closely related to *R. cereum*. The former is freely intergradient with *R. cereum* and would appear to be no more than a rather poorly differentiated geographic race. The latter is not recognizable, even at that level. Our three vars. are separable as follows:

1 Bracts of the inflorescence usually more or less flabellate, truncate to broadly rounded and several-lobed or very prominently toothed; leaves from glabrous to copiously pubescent and more or less glandular on both surfaces; B.C. s. to Ariz. and s. Calif., e. to c. Mont. and Ida. and w. Nev., almost entirely replaced eastward by var. *inebrians* with which it is freely intergradient   var. cereum
1 Bracts ovate to obovate, usually pointed, entire to sharply denticulate or with 2 or 3 shallow lobes
  2 Plants, especially the leaves, mostly strongly pubescent; calyx pubescent as well as glandular; leaves mostly less than 15 (20) mm. broad; c. Ida. to c. Mont. (Fergus Co.) and e. and s. (to the exclusion of var. *cereum*) to Neb., N.M., Utah, and e. Nev.
                                                                          var. inebrians (Lindl.) C. L. Hitchc.
  2 Plants mostly almost glabrous on the leaves and calyx; leaves generally 15-30 mm. broad; confined to the Snake R. Canyon and its tributaries in w.c. Ida. and adj. Wash. (Asotin Co.) and Wallowa Co., Oreg.; freely intergradient with var. *cereum* and possibly with var. *inebrians*, to which it is perhaps more closely related   var. colubrinum C. L. Hitchc.

Ribes cognatum Greene, Pitt. 3:115. 1896.
    R. palousense Elmer in Jancz. Mém. Soc. Phys. Nat. Genèv. 35:382. 1907, not validly published. *Grossularia cognata* Cov. & Britt. N.Am. Fl. 22³:222. 1908. (*Th. Howell,* Pendleton, Oreg., May 17, 1896)

An erect to clambering shrub up to 3.5 m. tall, the branches slender, arching, becoming straw-colored, from finely pubescent to semilanate and moderately to not at all bristly, but the basal sprouts yellowish and often densely bristly; nodal spines mostly 1 (2 or 3), slender, up to 1 cm. long, straight; leaf blades broadly ovate, more or less truncate to cordate at the base, 1.5-4 (7) cm. broad, sparsely to densely pubescent and often grayish on both surfaces, usually also more or less glandular-puberulent, from nearly equally 5-lobed to 3-lobed with the basal segments again shallowly cleft, the lobes oblong-rounded, with 5-9 crenate-dentations; petioles mostly shorter than the blades, densely pubescent, often glandular-puberulent and usually more or less glandular-setose; flowers (1) 2-3 (5), borne in short, spreading to drooping racemes usually shorter than the leaves; pedicels slender, not jointed, 1-2 times as long as the ovate, glandular-ciliate, 1.5-2.5 mm. bracts; flowers 11-16 mm. long, finely pilose externally except on the glabrous ovary, the hypanthium nearly cylindric, considerably constricted above the ovary, 5-6 mm. long, the lobes from as long to about 2/3 as long, oblong, obtuse to rounded, spreading to slightly reflexed, greenish-white or somewhat pinkish; petals (2.5) 3-3.5 mm. long, oblong-oblanceolate, cuneate-based, whitish; stamens equaling or slightly longer than the petals; anthers about 1.3 mm. long; filaments glabrous, 1.5-2 times as long as the anthers; styles connate and pilose to midlength or slightly above, from about equal to the stamens to almost as long as the extended calyx lobes; fruit globular, reddish but drying to bluish-black, about 1 cm. long, palatable.

Dry to moist stream banks and lower hillsides, from Umatilla Co., Oreg., and s.e. Wash., e. to n.c. Ida., possibly n. to s. B.C. Late Mar.-May.

Ribes cruentum Greene, Pitt. 4:35. 1899.
    R. amictum var. *cruentum* Jancz. Mém. Soc. Phys. Nat. Genèv. 35:366. 1907. *Grossularia cruenta* Cov. & Britt. N.Am. Fl. 22³:215. 1908. *Ribes roezlii* var. *cruentum* Rehd. in Bailey, Cycl. Am.

Hort. 2962. 1916. ("Common in the Californian Coast Range, from Sonoma Co. northward into southern Oregon," no collection cited)

Grossularia cruenta var. *oregonensis* Berger, N.Y. Agr. Exp. Sta. Tech. Bull. no. 109:80. 1924. (*E. A. Walpole 182,* Glendale and Douglas counties, Oreg., Apr. 20, 1899)

Rounded, armed, freely branched shrubs 0.5-1.5 (2) m. tall , the young stems minutely puberulent but not at all bristly, soon glabrous and reddish-brown; nodal spines mostly 3 (1), slightly curved, 5-10 (15) mm. long; leaves oblong-ovate, truncate or somewhat cuneate to slightly cordate at base, mostly 1-1.5 (2.5) cm. broad, nearly or quite glabrous except for the soft ciliation, shallowly 3 (5) -lobed much less than half their length, the lobes rounded, crenate-denticulate; flowers single or in pairs on glabrous, drooping peduncles shorter than the leaves; pedicels glabrous or sparsely setose-glandular, not jointed, at anthesis shorter than the glabrous, (3) 4-5 mm. bracts; ovary glandular-bristly; calyx glabrous to sparsely pubescent, crimson; hypanthium nearly cylindric, (4.5) 5-6 mm. long, only half as broad at the top; calyx lobes narrowly lanceolate, somewhat acute, 8-11 mm. long, reflexed; petals white to pink, 3.5-5 mm. long, cuneate-obovate, deeply erose; filaments stout, glabrous, slightly exceeding the petals; anthers reddish, glabrous, lanceolate, somewhat sagittate, 2.5-3 mm. long, with a short, indehiscent, apiculate tip; styles equaling or exceeding the extended calyx lobes, glabrous, connate to well above midlength; berry globose, about 1 cm. long, reddish to deep purple, strongly spinose and with short-stalked glands or slight pubescence, not palatable.

Lowland gravel bars, riverbanks, and canyon hillsides to subalpine ridges; s. Lane Co. to Jackson, Josephine, and Curry cos., Oreg., and s. to Napa Co., Calif. Mar.-May.

The species is so variable that it is doubtful if the somewhat more pubescent phase, "var. *oregonensis* Berger," is taxonomically significant.

Ribes divaricatum Dougl. Trans. Hort. Soc. Lond. 7:515. 1830.

R. divaricatum var. *douglasii* Jancz. Mém. Soc. Phys. Nat. Genèv. 35:391. 1907. *Grossularia divaricata* Cov. & Britt. N. Am. Fl. 22³:224. 1908. (*Douglas,* "bank of streams near Indian villages, on the North West Coast of America, from the 45° to the 52° N. Lat.")

R. villosum Nutt. in T. & G. Fl. N. Am. 1:547. 1840, not of Roxb. in 1824. *R. tomentosum* Koch, Wochenschr. Gärtn. Pflanzenk. 2:138. 1859. *R. divaricatum villosum* Zabel, Handb. Laubh. Deuts. Dendr. Ges. 137. 1903. (*Nuttall,* "St. Barbara, California, common near the village on the plain")

R. divaricatum var. *glabriflorum* Koehne, Deuts. Dendr. 200. 1893. (No type given)

R. divaricatum var. *pubiflorum* Koehne, loc. cit. (No type given)

R. suksdorfii Heller, Muhl. 3:11. 1907. (*Suksdorf,* "somewhere in Washington in 1897," probably from Klickitat Co.)

An erect to spreading shrub mostly 1.5-3 m. tall, more or less pubescent throughout, the branches tending to arch, with gray to somewhat brownish bark, generally unarmed (very occasionally bristly) except for the usual 1-3 stout, chestnut-colored, (5) 10-20 mm. spines at the nodes; leaves broadly ovate, (1) 2-6 cm. wide, rounded to truncate or slightly cordate at base, prominently crenate-serrate, usually trilobed over half their length, the sinuses open, the lower segments often again shallowly cleft into 2 unequal lobes; petioles from shorter to slightly longer than the blades; racemes slender, drooping, (1) 2- to 3 (4)-flowered, mostly shorter than the leaves; pedicels slender, (3) 5-10 mm. long, not jointed below the flowers; bracts oval, 1-1.5 (2) mm. long, glabrous or ciliate; calyx usually red or reddish-green, glabrous to copiously pubescent, the free hypanthium narrowly campanulate or conic, about 2.5 (1.5-3) mm. long; calyx lobes narrowly oblong, obtuse or rounded, (3) 5-7 mm. long and at least twice (to 3 times) as long as the hypanthium, recurved; petals cuneate-lunate to obovate, white to red, 1.5-2.5 mm. long, slightly less than half as long as the calyx lobes; stamens exceeding the extended calyx lobes by 1-2 mm., the filaments glabrous, the anthers oval, about 1 mm. long; styles about equaling the stamens, copiously pilose-villous to above midlength, connate from slightly less to slightly more than half their length; berry globose, smooth, purplish-black, about 1 cm. long, palatable. N=8.

Open woods, prairies, and moist hillsides; B.C. southward to Ventura Co., Calif., from the lower levels on the w. side of the Cascades to the coast, and up the Columbia R. Gorge to Klickitat Co., Wash. Apr.-May.

The species is extremely variable in pubescence, flower size, and leaf size and lobation, with several infraspecific variants described. In our area geographic races do not appear to be established, although plants from the islands and coast of Whatcom and Island cos., Wash., appear to have flowers that are more pubescent than average, and one very peculiar collection from San Juan I. (*Barbara*

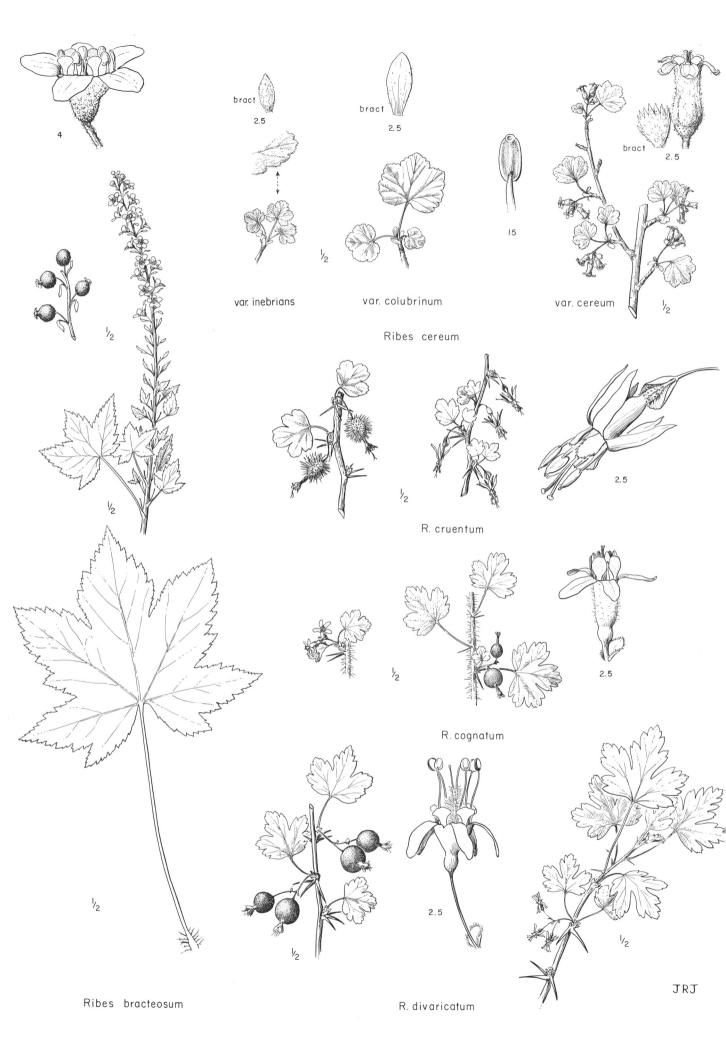

4

bract
2.5

bract
2.5

bract
2.5

15

var. inebrians

var. colubrinum

var. cereum

1/2

Ribes cereum

R. cruentum

R. cognatum

Ribes bracteosum

R. divaricatum

JRJ

Blanchard 4, at U. of Wash.) has unusually small flowers (hypanthium 1.5 mm. long, calyx lobes only 3 mm. long) and deeply lobed leaves mostly less than 1 cm. broad.

Hybrids of the species with *R. niveum (R. succirubrum* Zabel) and with *R. lobbii (R. knightii* Rehd.) have been detected (the first in cultivation, the second in the wild) and propagated for the horticultural trade.

RIBES HENDERSONII C. L. Hitchc. hoc loc.
> *Grossularia neglecta* Berger, N.Y. Agr. Exp. Sta. Tech. Bull. no. 109:106. 1924. *R. neglectum* Standl. Field Mus. Pub. Bot. 8:140. 1930, but not of Rose in 1905. *(Henderson 4048,* Lost River Mts., Ida., Aug. 14, 1895.)

A low, rounded to sprawling, intricately branched shrub 3-7 (10) dm. tall, the stems slightly puberulent and sometimes sparingly bristly, quickly glabrous and yellowish but ultimately gnarled and crooked and grayish-barked, the nodal spines 1-7, very conspicuous, yellow, the larger ones rigid, spreading, 7-10 mm. long; leaf blades 7-12 (15) mm. broad, cordate-reniform, 5 (3)-lobed about half the length or less, the lobes more or less cuneate-rounded, with 3-5 (7) unequal crenate-dentations, finely and softly pubescent and also glandular-puberulent on both surfaces but more glandular beneath; petioles mostly shorter than the blades, glandular-pubescent and usually also with a few glandular setae; racemes reduced, 1- to 2-flowered, much shorter than, and nearly hidden by, the leaves; pedicels very short, often equaled by the ovate, glandular-ciliate, 1-2 mm. bracts; flowers 6-8 mm. long, sparingly pilose externally; hypanthium broadly campanulate, about 2.5 mm. long and broad, the calyx lobes about as long, ovate-oblong, spreading but not strongly reflexed, pale greenish-white or with a slight reddish tinge; petals 1-1.5 mm. long, about half as long as the calyx lobes, white, broadly obovate-rhombic to flabelliform-reniform and cuneate-based, the upper margin broadly rounded and slightly erose; stamens barely equaling (or shorter than) the petals, the anthers oblong, nearly 1 mm. long, at least as long as the stout filaments; styles shorter than the petals, barely exceeding the hypanthium, connate for less than 1 mm., pilose to above midlength; berry glabrous, "reddish when ripe, about 10 mm. across."

On limestone cliffs and talus of the Lost River, Lemhi, and other ranges of Custer and Lemhi cos., Ida., at 6000 ft. and upward, and in the Anaconda Range, Deerlodge Co., Mont., on granite (?); probably the range more continuous. June-July.

Ribes hendersonii is a very distinctive, easily recognized, local species that is undoubtedly closely related to the more northern *R. oxyacanthoides,* from which it differs in general appearance and in its smaller petals, shorter filaments, and shorter and less connate styles.

Ribes howellii Greene, Erythea 4:57. 1896.
> *R. acerifolium* Howell, Erythea 3:34. 1895, but not of Koch in 1869. ("On Mounts Hood and Adams near the snow line; also at the mouth of the Columbia River," presumably collected by Th. Howell)

A spreading to erect shrub generally not over 1 m. tall, the whole plant more or less finely puberulent and often thickly sprinkled with small, nearly sessile to rather strongly stipitate glands, leaf blades deeply cordate, mostly 3-8 cm. broad and not quite so long, lobed about half the length into 3 main and 2 smaller, ovate-deltoid, bicrenate-serrate segments; racemes pendent, shorter than the leaves, 7- to 15-flowered; pedicels jointed under the ovary and usually minutely 1- to 3-bracteolate at the tip; ovary finely crisp-puberulent, the hypanthium also somewhat glandular, shallowly bowl-shaped, barely 1.5 mm. long, lined internally for about half the length with a thick disc from the rim of which the very slightly connate petals and stamens arise, the whole forming a short tube in the center of the flower; calyx lobes broadly oblong-ovate to more nearly deltoid-obovate, about 3 mm. long and 1/2-3/4 as broad, spreading, the tips usually recurved; petals red, 1-1.5 mm. long, obovate-cuneate, inwardly pouched and nearly keeled in the center, spreading with the sepals; filaments strongly flattened, glabrous, usually equaling the petals; styles glabrous, about equaling the stamens, connate nearly to the stigmas; berry glaucous-black, globular, up to 1 cm. long, sparsely glandular and slightly pubescent.

Montane to alpine stream banks, meadowland thickets, and open ridges and rock slides (to tree line); s. B.C. southward, in the Cascade and Olympic mts., to n. Oreg., also in the mts. just e. of Priest Lake, Boundary and Bonner cos., Ida. June-Aug.

Ribes hudsonianum Richards. in Frankl. App. Frankl. Journ. 2nd ed. 6. 1824.
> *R. petiolare* Dougl. Trans. Hort. Soc. Lond. 7:514. 1830. *R. hudsonianum* var. *petiolare* Jancz.

Mém. Soc. Phys. Nat. Genèv. 35:346. 1907. *(Douglas,* "On the western base of the Rocky Mts. from 48° to the 52° N. Lat.")

An erect, unarmed shrub mostly 0. 5-1. 5 (2) m. tall, more or less glandular all over with sessile, round, yellow, crystalline and shining glands, and with a characteristic strong, sweetish, rather unpleasant odor; leaf blades broadly cordate, (2) 3. 5-9 (12) cm. broad, not quite so long, usually glandular but otherwise glabrous to rather copiously hairy (and pale) on the lower surface and sparsely hairy on the upper, primarily 3-lobed less than half the length into broadly deltoid, and coarsely bicrenate-dentate segments, the two lower segments less deeply and very unequally lobed, very exceptionally, the blades almost equally 5-lobed; racemes many-flowered, ascending to spreading, up to 17 cm. long; pedicels slender, jointed below the flowers, 3-8 mm. long; ovary from (very rarely) glabrous to closely covered with sessile glands, or even very sparsely hairy, about 2/3 inferior at anthesis, the superior portion conical; free portion of the calyx white, more or less thickly crisp-pubescent, the hypanthium widely flared and somewhat saucer-shaped, 1-1. 5 mm. long, lined internally with a thin disc which partially covers the ovary; sepals triangular to ovate-lanceolate, 3-4 (5) mm. long, widely spreading; petals white, about 1. 5 mm. long, oblong, becoming cuneate-flabelliform upward; stamens inserted slightly below (but about equaling) the petals; filaments glabrous; anthers oval, about 0. 7 mm. long; styles about 2 mm. long, connate slightly more than half their length at anthesis, usually as long as the conical, superior portion of the ovary; berry subglobose, 7-12 mm. long, black and more or less glaucous, glabrous or more commonly sessile-glandular, bitter and not at all palatable.

Stream banks, moist woods, and thickets at the edge of mt. meadows; Alas. to Hudson's Bay, southward to n. Calif., Utah, Wyo., and Minn. May-July.

The species consists of a northern, more pubescent phase and a southern, nearly glabrous phase, both of which have usually been recognized at the specific level. The variation in pubescence appears to be of a clinal nature and the two races are completely intergradient over a wide area; they are distinguishable as follows:

1 Plant rather generally hairy, the leaf blades pubescent over the entire lower surface and usually with some pubescence on the upper surface; ovary often without glands; Alas. to Hudson's Bay, s. to s. B.C. and occasional to Okanogan Co., Wash., n. Ida. and Mont., and Minn.

var. hudsonianum

1 Plant from almost totally glabrous (except for the sessile glands) to lightly pubescent on the calyces, young stems, petioles, and at least along the veins of the lower surface of the leaves; ovary glandular, s. B.C. southward, in the Cascade Mts., to s.w. Oreg. and n. Calif., e. to Ida., Mont., Utah, and n. Wyo.   var. petiolare (Dougl.) Jancz.

Ribes inerme Rydb. Mem. N.Y. Bot. Gard. 1:202. 1900.

Grossularia inermis Cov. & Britt. N. Am. Fl. 22³:224. 1908. *(Tweedy 830,* Slough Creek, Yellowstone Park, in 1885)

R. hirtellum purpusii Koehne in Späth, Kat. 119. 1899-1900. *R. purpusii* Koehne ex Blank. Mont. Agr. Coll. Stud. Bot. 1:64. 1905. *R. oxyacanthoides* var. *purpusii* Jancz. Mém. Soc. Phys. Nat. Genèv. 35:388. 1907.

Grossularia inermis var. *pubescens* Berger, N.Y. Agr. Exp. Sta. Tech. Bull. no. 109:105. 1924. (North Colorado, no specimens cited) A hairy leaved phase.

Erect to sprawling shrubs mostly 1-2 (3) m. tall, the branches rather slender, usually glabrous, sometimes sparsely retrorse-bristly when young, but usually smooth except for the occasional single (3), straight to recurved, mostly 4-10 mm. nodal spines; bark grayish, flaky, deep red underneath; leaf blades mostly broadly ovate, rounded to (more commonly) cordate at the base, generally 2-5 (8) cm. broad, glabrous (except for the soft ciliation) or occasionally sparingly pubescent, 3 (5) -lobed not quite half the length, the segments deeply crenate-serrate, occasionally the lateral lobes again less deeply cut into larger upper and smaller lower segments; petioles from shorter to longer than the blades, mostly slightly pubescent and with a few longer, slender, stipitate-glandular hairs; racemes drooping, glabrous, shorter than the leaves, (1) 2- to 4-flowered; pedicels slender, 3-7 mm. long, not jointed; bracts 1-2 mm. long, ovate, ciliolate; hypanthium tubular-campanulate (2) 2. 5-3 (3. 5) mm. long and not so broad, glabrous or sparsely hirsute, greenish to purplish- or reddish-tinged, the lobes oblong-lanceolate to narrowly lanceolate, obtuse to rounded, spreading to curved-reflexed, equaling to half again as long as the hypanthium; petals cuneate-obovate or oblong, 1-1. 5 (2) mm. long, 1/2-3/5 as long as the calyx lobes, white or pinkish; stamens equaling, or subequal to, the extended calyx lobes; filaments glabrous; anthers ovate, about 1 mm. long, glabrous; styles connate no more than half their length, about equaling the stamens, pilose-villous to midlength or above; berry smooth, reddish-purple, 7-9 mm. long, palatable.

Stream banks and thickets at the edge of meadows, to open or wooded mt. ridges; B.C. southward, on the e. slope of the Cascade Mts., to the n. Coast Range and s. Sierra Nevada of Calif., e. to Mont., Wyo., Colo., and N.M. May-June.

Ribes inerme is most apt to be confused with *R. irriguum* or *R. niveum;* from the former it differs in having a less campanulate (more slender) hypanthium, petals less than half as long as the calyx lobes, and stamens at least twice as long as the petals; from *R. niveum* it can be told because of its smaller, glabrous anthers and smaller flowers. Although the species is normally glabrous-leaved, hairy leaved plants occur sporadically in our range but do not appear to constitute a distinctive race.

Ribes irriguum Dougl. Trans. Hort. Soc. Lond. 7:516. 1830.
> *R. divaricatum* var. *irriguum* Gray, Am. Nat. 10:273. 1876. *R. oxyacanthoides* var. *irriguum* Jancz. Mém. Soc. Phys. Nat. Genèv. 35:388. 1907. *Grossularia irrigua* Cov. & Britt. N.Am. Fl. 22^3:222. 1908. *(Douglas,* ". . . On the Blue Mountains in 46° 33' . . ., also on the hills and banks of the Spokan River")
> *R. leucoderme* Heller, Bull. Torrey Club 24:93. 1897. *R. oxyacanthoides* var. *leucoderme* Jancz. Mém. Soc. Phys. Nat. Genèv. 35:388. 1907. *(Heller & Heller 3175,* Lake Waha, Craig Mts., Nez Perce Co., Ida., June 2, 1896)

An erect to spreading shrub (0.5) 1-2.5 (3) m. tall, the branches slender, arched, finely puberulent, not bristly, grayish-brown but soon turning gray; nodal spines mostly 1 (2 or 3), straight, up to 1 cm. long; leaf blades from ovate and with a nearly truncate base to cordate or somewhat reniform, 1.5-3.5 (5 or even 7 on young shoots) cm. broad, usually 5 (3)-lobed up to half the length, the lobes rounded, irregularly once or twice shallowly incised and crenate-serrate, from (usually) finely pubescent on both surfaces to glabrous above; racemes spreading to drooping, generally shorter than the leaves, with usually paired (1-3) flowers; pedicels short, 1-3 (4) mm. long, not jointed; flowers 8-11 mm. long; calyx glabrous, the hypanthium campanulate, 2-4 mm. long and commonly as broad at the top; calyx lobes oblong, rounded to obtuse, up to 5 (7) mm. long, as much as half again as long as the hypanthium, spreading but usually not sharply reflexed, pale greenish or with some reddish or purplish tinge, at least near the base; petals obovate-cuneate, rounded and often erose, 2-3 mm. long, usually well over half as long as the extended sepals, white or slightly pinkish; stamens equaling or slightly exceeding the petals, considerably shorter than the extended sepals, the filaments stout, glabrous; anthers oval, about 1 mm. long; styles slightly exserted from the extended sepals, connate to near the tips, pilose on the lower half; berry globose, deep bluish-purple, about 1 (to 1.3) cm. long, palatable?

Moist to dry canyons and open to wooded hillsides; extreme n.e. Oreg. and adj. Wash., e. to w. Mont., n. to s.e. B.C., never e. of the continental divide. Apr.-June.

Although *R. irriguum* is really a distinctive species, it is often confused with at least 3 others, even though there is no indication that it intergrades with any of them; these are *R. inerme* (see above), *R. hendersonii* (which has much smaller flowers, glandular leaves, and more heavily armed, intricately branched, glabrous stems), and *R. oxyacanthoides* (which has smaller flowers, with a more slender, proportionately longer hypanthium, and usually glandular leaves).

Ribes klamathense (Cov.) Fedde, Just Bot. Jahresb. 36^2:519. 1910.
> *Grossularia klamathensis* Cov. N.Am. Fl. 22^3:225. 1908. *R. inerme* var. *klamathense* Jeps. Man. Fl. Pl. Calif. 472. 1925. *(Applegate 2008,* Keno, Klamath Co., Oreg., May 10, 1898)

An erect shrub as much as 2 m. tall, the branches slender, arching, glabrous, maturing to dark gray, very sparingly armed, seldom if ever at all bristly and with only single weak nodal spines rarely over 6 mm. long, or some nodes unarmed; leaf blades broadly cordate, finely pubescent on both surfaces, (1.5) 2-5 cm. broad, generally 3 (5)-lobed (mostly less than half the length) and coarsely round-toothed; petioles usually longer than the blades, pubescent only along the upper side, sparsely setose-glandular near the base; racemes glabrous, usually drooping, shorter than the leaves, (2) 4- to 5-flowered; pedicels slender, up to 10 mm. long, not jointed; bracts glabrous, 1-1.5 (2) mm. long; calyx glabrous or more commonly sparsely pilose externally, mostly 5-7 mm. long; hypanthium greenish, 1.5-2 mm. long, about as broad at the top; calyx lobes often slightly pinkish- or purplish-margined, narrowly oblong, pointed to rounded, 3-4 mm. long, reflexed; petals white, broadly cuneate-flabellate, truncate to slightly rounded across the top, about 1.5 mm. long; stamens equaling or slightly exceeding the extended calyx lobes, the filaments glabrous; anthers oval, less than 1 mm.

long; styles connate and pilose for half their length, about equaling the stamens; berry glabrous, nearly globose, glaucous-black, about 8 mm. long.

Mostly at the edge of streams and marshes, from s. Jefferson Co., Oreg., southward, on the e.
edge of the Cascade Mts., to n. w. Calif. Apr.-May.

This species is very closely related to *R. inerme* and perhaps only a variety thereof, but our material
does not include intergradients between the two.

Ribes lacustre (Pers.) Poir. in Lam. Encyc. Suppl. 2:856. 1812.
  R. oxyacanthoides var. *lacustre* Pers. Syn. Pl. 1:252. 1805. *R. grossularioides* Michx. in Steud.
    Nom. 691. 1821. *Limnobotrya lacustris* Rydb. Fl. Rocky Mts. 296, 1062. 1917. ("Hab. ad lacus
    mistassinis," Canada)
  R. echinatum Dougl. Trans. Hort. Soc. Lond. 7:517. 1830. *(Douglas,* "Grand Rapids, on the Colum-
    bia, and mountains of northern California")
  R. lacustre var. *parvulum* Gray, Bot. Calif. 1:206. 1876. *R. parvulum* Rydb. Mem. N. Y. Bot. Gard.
    1:203. 1900. *Limnobotrya parvula* Rydb. Fl. Rocky Mts. 297, 1062. 1917. (Sierra Nevada, "east
    to the Rocky Mountains and north to British Columbia," no collections cited)
An erect to spreading shrub mostly 1-1.5 (2) m. tall, the young branches finely puberulent (but not
glandular) and very thickly to remotely bristly with slender, sharp prickles and larger nodal spines
up to 12 mm. long, rarely the plant with only a few nodal spines; leaf blades cordate, (1) 2-5 (7) cm.
broad, about as long, glabrous or sparsely puberulent at least along the veins, never glandular, mostly 5-lobed half the length or more (the lowest lobes much the smallest) and again irregularly shallowly lobate and deeply once or twice crenate-dentate; racemes sometimes longer than the leaves, more
or less drooping, 7- to 15 (18)-flowered, usually (reddish or purplish) stipitate-glandular and puberulent; pedicels slender, 3-7 mm. long, jointed just below the flowers; calyx pale yellowish-green to
dull reddish-brown, the pinkish or reddish color usually deepening with age, the hypanthium scarcely
1 mm. long, shallowly saucer-shaped to crateriform, lined with a prominent pinkish disc, the petals
inserted at the edge of the disc, the stamens on the disc itself; sepals very broadly ovate-oblong, 2.5-
3 mm. long, often at least as broad; petals pinkish, cuneate-flabellate, 1/2-2/3 as long as the sepals,
narrowed to a slight claw; stamens from shorter to longer than the petals; ovary (glabrous) sparsely
to thickly stipitate-glandular with slender, usually reddish- to purplish-tipped hairs; berry 6-8 mm.
long, dark purple, usually slenderly stipitate-glandular, somewhat palatable. N=8. Swamp goosebe r-
ry, swamp black currant, prickly currant.

Moist woods and stream banks to drier forest slopes or subalpine ridges; Alas. to Newf., southward to Calif., Utah, Colo., S.D., Mich., and Pa. Apr.-July.

The leaves and the pubescence of the plant vary (apparently) fortuitously, and geographic races are
not recognizable, at least in our area, although there seems to be some indication that plants of the
Olympic Peninsula have less bristly fruits than average and that plants from n.c. Ida. may be a little
less spinose than average.

Ribes laxiflorum Pursh, Fl. Am. Sept. 731. 1814. *(Menzies,* "On the north-west coast")
  R. affine Dougl. ex Bong. Mém. Acad. St. Pétersb. VI, 2:138. 1833. *(Douglas,* Columbia River)
  R. laxiflorum var. *californicum* Jancz. Bull. Acad. Sci. Crac. 1913:728. 1913.
  R. laxiflorum var. *japonicum* Jancz. loc. cit.
Usually decumbent or spreading shrubs less than 1 m. high, but in dense woods sometimes semi-
scandent and up to 7 m. tall, the young branches finely crisp-puberulent and sparsely subsessile-
glandular, becoming deep reddish-brown; leaf blades mostly 4-8 (12) cm. broad, usually not quite so
long, glabrous on the upper surface, paler on the lower surface, more or less crisp-puberulent (at
least on the veins), and usually with a sprinkling of relatively few sessile, crystalline glands and more
numerous very short-stipitate, noncrystalline glands, deeply cordate, usually 5-lobed nearly half the
length, the segments ovate-triangular and shallowly to coarsely bicrenate-serrate; racemes (6) 8- to
18-flowered, floriferous full length, erect or ascending, usually shorter than the leaves, rather copiously crisp-puberulent and moderately (usually reddish) stipitate-glandular throughout; pedicels up
to 1 cm. long, jointed (but not bracteolate) immediately below the ovary; ovary finely puberulent and
prominently reddish stipitate-glandular, the glandular hairs up to 0.5 mm. long; free portion of the
calyx from greenish-white with only a pinkish tinge to deep red or purplish, more or less pubescent
and sparsely glandular externally; hypanthium shallowly bowl-shaped, about 1 mm. long, lined with a
prominent, thick, flat, 5-lobed, reddish to brownish disc that completely covers the top of the ovary;
calyx lobes 2.5-3 mm. long, from deltoid-ovate and as broad as long to oblong-obovate and not much

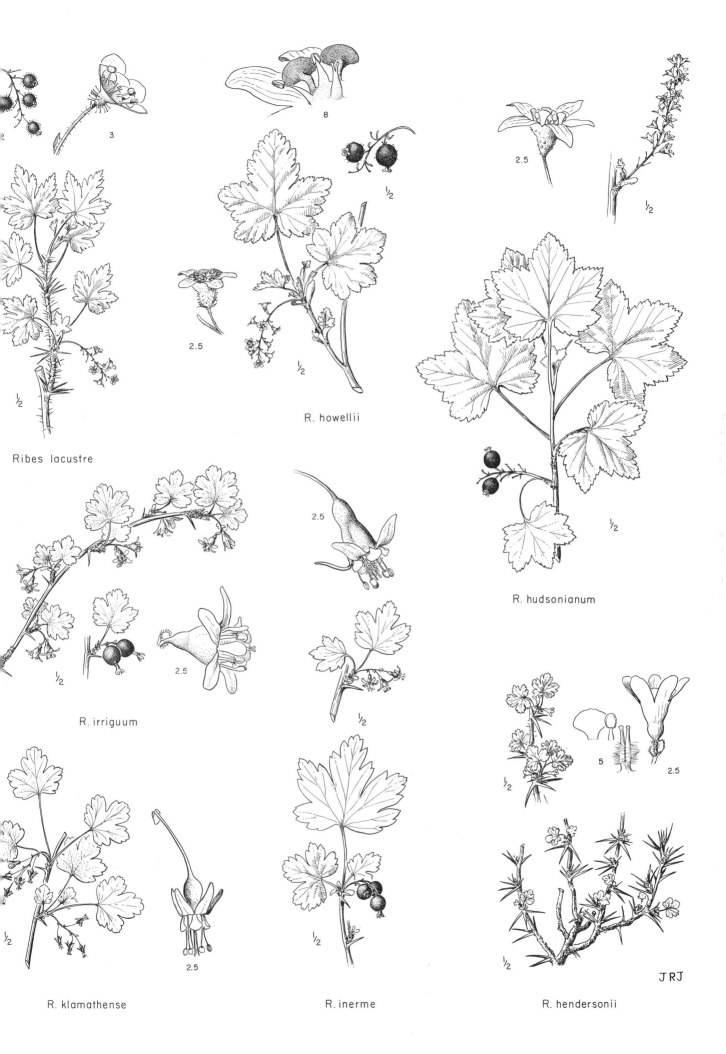

3

8

2.5

½

½

2.5

½

R. howellii

Ribes lacustre

2.5

½

½

R. hudsonianum

2.5

½

R. irriguum

½

5

2.5

½

½

2.5

½

R. klamathense

2.5

R. inerme

½

R. hendersonii

JRJ

over half as broad as long, spreading and the tips usually slightly recurved; petals red to purplish, 1-1. 5 mm. long and at least as broad, reniform or crescent-shaped with a cuneate, clawlike base, inserted at the top of the hypanthium and at the tips of the lobes of the disc; stamens inserted slightly below the petals and at the sinuses between the lobes of the disc or on the disc itself, about equaling the petals, the filaments reddish, glabrous, semiterete, the anthers oval, strongly extrorse to inverted, about 0. 7 mm. long; styles about equaling the stamens, glabrous, connate 1/3-2/3 the length; berry glaucous, purplish-black, ovoid, up to 1 cm. long, glandular-bristly.

Wet coastal woods to montane slopes; Alas. southward along the coast to Calif. (in Oreg. apparently always w. of the Cascade crest), e. through B.C. and n. Wash. to s.w. Alta. and n. Ida. Apr. -July.

The plants of the rain forests in the lowlands from the Olympic Peninsula, Wash., s. to Clatsop Co., Oreg., tend to become lax and more or less scandent, and appear to have deep reddish-purple flowers; perhaps they constitute a distinctive race.

Ribes lobbii Gray, Am. Nat. 10:274. 1876.

R. subvestitum sensu Hook. in Curtis' Bot. Mag. 82:pl. 4931. 1856, but not of H. & A. in 1838.
Grossularia lobbii Cov. & Britt. N.Am. Fl. 22³:217. 1908. (Garden plants, from specimens collected by Lobb, supposedly in California)

A freely branched, spreading shrub mostly 5-10 (15) dm. tall, the stems finely pubescent, not bristly, turning brownish and ultimately deep gray-reddish; nodal spines usually 3, slender, straight, 7-12 (15) mm. long; leaf blades ovate, usually shallowly cordate, mostly 1. 5-2. 5 cm. broad and as long or slightly longer, sparsely pubescent (glabrous) on the upper surface, paler and usually pubescent as well as glandular beneath, 3 (5)-lobed less than half the length, the lobes rounded, shallowly cleft and deeply crenate-dentate; flowers 1 or 2, pendent from stipitate-glandular peduncles shorter than the leaves, the pedicels slender, not jointed, copiously setose-glandular; ovary closely stipitate-glandular; calyx greenish with a strong reddish tinge on the outside, sparsely pilose; hypanthium narrowly campanulate, 3. 5-5. 5 mm. long, nearly as broad at the top; calyx lobes narrowly oblong, pointed, sharply·reflexed and bright red on the inner surface, 10-13 mm. long; petals white or pinkish, broadly cuneate-flabelliform but usually inrolled on the edges, 4-5 (7) mm. long; stamens considerably longer than the petals and subequal to the extended calyx lobes; filaments white; anthers purple or reddish, oval, about 1. 2 (up to 2) mm. long, rounded at the tip, warty or capitate-papillate on the back; styles about equaling the stamens, connate for 2/5-3/5 their length, glabrous; berry globose-ellipsoid, up to 15 mm. long, reddish-brown, coarsely setose-glandular but not prickly, unpalatable.

Creek banks and lowland valleys to open or forested montane slopes; B.C. s. to n.w. Calif., from the Cascade Mts. to the coast, more abundant on the e. side of the Cascades in Wash., but occasional on the Olympic Peninsula. Apr. -early June.

This is a very attractive shrub, well worth propagation as an ornamental.

Ribes menziesii Pursh, Fl. Am. Sept. 732. 1814.

R. menziesianum R. & S. Syst. Veg. 5:507. 1819. *Grossularia menziesii* Cov. & Britt. N.Am. Fl. 22³:213. 1908. *(Menzies,* "On the north-west coast, near Fort Trinidad")
R. subvestitum H. & A. Bot. Beechey Voy. 346. 1838. *(Douglas,* California)

An erect shrub 1-2 m. tall, the branches stout, slightly arched, very thickly (glandular) bristly, pubescent, becoming grayish or straw-colored; nodal spines usually 3, spreading, 1-1. 5 (2) cm. long; leaf blades broadly ovate with mostly a semitruncate (slightly cordate) base, 1. 5-4 cm. broad, 3 (5)-lobed less than half the length and with rather numerous crenate dentations, deep green, rugose, and usually glabrous on the upper surface, grayish, strongly pubescent, and glandular-setose on the lower surface; racemes short, pendent, only 1-or 2-flowered; pedicels slender, setose-glandular and grayish-pubescent, not jointed; ovary strongly (purplish) glandular-bristly and somewhat pubescent; calyx crimson, more or less hairy, the hypanthium somewhat conic, 2. 5-3. 5 mm. long, almost or quite as broad at the top, the lobes oblong-lanceolate, rounded, 2-3 times as long as the hypanthium, reflexed; petals white or pinkish to yellow, 3-4 mm. long, broadly flabellate-cuneate, often with inrolled margins, coarsely erose; filaments stout, glabrous, equaling or as much as 1 mm. longer than the petals, the anthers white, glabrous, lanceolate, semisagittate at base, mucronate and indehiscent at the tip, about 2. 5 mm. long; styles equaling or slightly exceeding the extended sepals, connate to above midlength, glabrous; berry ellipsoid-globose, 10-13 mm. long, reddish-purple, pubescent and shortly glandular-bristly, not palatable.

Coastal ranges from c. Calif. northward to Lane Co., Oreg. Late Mar. -early May.

Although the species is represented by several races in California, our material, as described above, is all referable to the var. *menziesii*.

Ribes mogollonicum Greene, Bull. Torrey Club 8:121. 1881. *(Greene,* Mogollon Mts., N.M., in 1881)
  R. wolfii Rothrock, Am. Nat. 8:358. 1874, as to specimen mentioned *(Prof. John Wolf,* Twin Lakes and Mosquito Pass, Colorado Territory), but not as to synonym cited, namely *R. sanguineum* var. *variegatum* Wats. Bot. King Exp. 100. 1871 (which was based on the specimen: *Dr. Anderson 381,* Washoe Mts., near Carson City, Nev.).

Glandular, unarmed shrub 0.5-3 m. tall, low and spreading to erect, the young branches puberulent, ultimately glabrous and white-barked, but brownish-red beneath the thin outer bark; leaves bright green and glabrous on the upper surface, slightly paler and somewhat puberulent and sessile-glandular beneath, 2.5-6 (9) cm. broad, deeply cordate, rather shallowly (less than half the length) (3) 5-lobed, the lobes nearly triangular, acute to obtuse, finely biserrate-dentate, the petioles subequal to the blade, puberulent and often sparsely glandular; racemes spreading to erect, 2-5 cm. long, usually no longer than the leaves, the peduncles, rachis, and pedicels strongly crisp-puberulent and more or less stipitate-glandular, not jointed; bracts oblong-spatulate, (3) 4-5 mm. long, equaling to twice as long as the pedicels; flowers (4) 10-20, crowded; ovary puberulent and strongly stipitate-glandular; calyx greenish-white or yellowish-green, the hypanthium flared and saucer-shaped, 1-1.5 mm. long, crisp-puberulent, the lobes spreading, oblong, 2.5-3.5 (4) mm. long, 3 (5)-veined; petals also spreading or semi-erect, whitish-green or yellowish-green, flabelliform, less than half as long as the calyx lobes; stamens about equaling the petals, barely exserted, the oval anthers barely 0.5 mm. long; style 1.5-2 mm. long, bifid less than half the length, glabrous, thickened basally; berry ovoid, black, glandular, about 10 mm. long.

Chiefly in the mountains of Utah and Colo. to N.M. and Ariz., but also in the Blue Mts. of s.e. Wash., and the Seven Devils Mts., Idaho Co., Ida. June-July.

As far as is known, the species occurs, in our area, only in the Seven Devils Mts., and at or near Wenatchee Guard Station, about 20 mi. s.w. of Asotin, Asotin Co., Wash., where it has been collected several times *(G. N. Jones 1889, L. N. Goodding 404,* and *Hitchcock & Muhlick 21817).*

The flowers have no trace of pink when fresh or when dried quickly, although they apparently sometimes turn slightly pinkish under certain drying conditions. It may have been this peculiarity that led Rothrock to believe that his Coloradan plants were the same as the pinkish-flowered currant from Nevada, earlier described by Watson as *R. sanguineum* var. *variegatum*. It is obvious Rothrock was strongly influenced by the description supplied by Watson, to the extent that the perianth of *R. wolfii* was described as red, and elsewhere in the discussion it is plain not only that Rothrock considered his species identical with Watson's variety but that his intent was to raise the latter to specific status, expanding the description. Several later workers have noted the constant differences between the Rocky Mt. and the Sierran plants that warrant their recognition as separate species, and have amended the description of *R. wolfii* better to fit the former. However, the name *R. wolfii* must be included with *R. sanguineum* var. *variegatum* (because of its typification), leaving *R. mogollonicum* as the only name available for our plant, which is identical with those of the Rocky Mts.

Ribes montigenum McClatchie, Erythea 5:38. 1897.
  R. nubigenum McClatchie, Erythea 2:80. 1894, but not of Philippi in 1857. *Limnobotrya montigena* Rydb. Fl. Rocky Mts. 397, 1062. 1917. *(Mr. & Mrs. A. J. McClatchie,* Mt. San Antonio, Calif., Aug. 16, 1893)
  R. lacustre var. *molle* Gray, Bot. Calif. 1:206. 1876. *R. molle* Howell, Fl. N. W. Am. 209. 1898, but not of Poepp. in 1858. ("In the Sierra Nevada at 6,000 to 10,000 feet, from Mariposa Co. northward," no collector mentioned)
  R. lacustre var. *lentum* M. E. Jones, Proc. Calif. Acad. Sci. II, 5:681. 1895. *R. lentum* Cov. & Rose, Proc. Biol. Soc. Wash. 15:28. 1902. *(M. E. Jones 56950,* Bromide Pass, Henry Mts., Utah, July 27, 1894, is the first of 3 collections cited)

A spreading, freely branched shrub mostly only 2-5 (7) dm. tall; herbage copiously pubescent, puberulent, and also setulose-glandular almost throughout, the branches more or less bristly and with (1) 3-5 mostly slender nodal spines 4-6 (but up to well over 10) mm. long; leaf blades usually 1-2.5 (4) cm. broad, 5-cleft about 2/3-3/4 of their length and irregularly lobed and cut into crenate-dentate teeth; racemes axillary, pendent, usually little if any longer than the leaves, mostly 4- to 7 (10)-flowered; pedicels 1-3 (5) mm. long, jointed; hypanthium very short, saucer-shaped (crateriform), less than 1 mm. long, lined with a thin yellowish-pinkish disc; sepals broadly ovate to obovate, yellowish-green to pinkish, 2.5-3 mm. long, spreading; petals scarcely half as long as the sepals, pinkish or purplish, broadly fan-shaped or cuneate-lunate, narrowed to a very short claw inserted at the edge of the disc; stamens yellow, about equaling the petals and inserted slightly below them and on the disc; ovary sparsely to thickly glandular-bristly, the glands usually purplish; berry 5-7 mm. long, obovoid-spherical, reddish, setulose-glandular, somewhat palatable. Alpine prickly currant.

Subalpine to alpine talus slopes, ridges, and rock crevices; s. B.C. southward, on the e. slope of the Cascades, to the Sierra Nevada and the mountains of s. Calif., e. to the Rocky Mts. from Mont. to N.M. Late June-Aug.

Undoubtedly very closely related to *R. lacustre* but seldom if ever crossing with it and apparently amply distinct therefrom, both morphologically and genetically.

Ribes niveum Lindl. Bot. Reg. 20:pl. 1692. 1834.
    Grossularia nivea Spach, Hist. Veg. 6:179. 1838. (Cultivated plants, originally collected by Douglas in northwest America)
    R. niveum f. *pilosum* St. John, Fl. S.E. Wash. 188. 1937. *(St. John 6794*, Lewiston Grade, n. side Snake R. Canyon, Nez Perce Co., Ida.)

Erect shrub (0.5) 1.5-3 m. tall, the long slender branches mostly grayish- or reddish-brown, not at all bristly but with mostly 1 (2 or 3) stout nodal spines 7-15 mm. long; leaf blades from broadly reniform to more usually very broadly ovate, with a truncate to rounded-cuneate base, (1) 2-4 (5) cm. broad and about as long, finely pubescent on one or both surfaces or only between the veins on the lower surface (rarely glabrous), 3- to 5-lobed about half the length or less, the lobes broad, usually more or less ternately once or twice crenate-dentate; racemes slender, drooping, glabrous, usually somewhat longer than the leaves, (2) 3- to 4-flowered; pedicels slender, (7) 10-15 mm. long, not jointed; calyx glabrous (rarely very sparsely hirsute), the free hypanthium narrowly campanulate, 2-2.5 (3) mm. long, about as broad at the top, white, pale greenish or ochroleucous; sepals narrowly oblong to narrowly oblanceolate, white or very slightly pinkish, (5) 6-7 (8) mm. long, mostly rounded (acute), sharply reflexed in anthesis; petals white, erect, oblong to cuneate-obovate, rounded, about 2 mm. long; stamens usually exceeding the extended calyx lobes by 1-3 mm., the filaments finely pilose, the anthers also pilose or arachnoid, about 1.5 mm. long (about 1 mm. long when dried), ovate-oblong, greenish; styles connate well over half their length and villous-pilose on the lower 1/2-3/4, equaling or slightly exceeding the stamens; berry globose, 8-12 mm. long, bluish-black, palatable but sour. N=8.

Mostly in thickets along intermittent or permanent streams, but sometimes on open hillsides; n.c. Oreg. (Wasco to Gilliam, Wheeler, and Grant cos.), e. to s.e. Wash. and w. Ida. (Latah to Owyhee cos.), s. to Nev. and Klamath cos., Oreg. Apr.-May.

In most of its range the species is not particularly variable, but plants from the eastern part of the range (Grant Co., Oreg., e.), and notably in w. Ida., tend to have more strongly pubescent leaves than those from farther w. and s.; they have appropriately been called f. *pilosum* St. John.

In cultivation *R. niveum* has been crossed with *R. inerme* (or *R. oxyacanthoides?*) resulting in a hybrid, *R. robustum* Jancz., that is cultivated for its fruit. *R. succirubrum* Zabel is another hybrid of *R. niveum* (with *R. divaricatum*).

Ribes oxyacanthoides L. Sp. Pl. 201. 1753.
    Grossularia oxyacanthoides Mill. Gard. Dict. 8th ed. no. 4. 1768. (Canada)
    R. saxosum sensu Rydb. Mem. N.Y. Bot. Gard. 1:202. 1900, but not of Lindl.

An ascending to sprawling shrub 6-15 dm. tall, the branches rather stout, yellowish, strongly puberulent and very thickly bristly when young, aging to gray but usually remaining bristly and with 1-3 (5) nodal spines up to 1 cm. long; leaf blades broadly ovate and with truncate to somewhat cordate base, (1) 1.5-4 cm. broad, 5 (3)-lobed about half their length, the segments cuneate-rounded and irregularly crenate-dentate into prominent, unequal teeth, more or less pubescent and usually strongly glandular-puberulent at least on the lower surface; racemes much shorter than the leaves, 1- to 2 (3)-flowered; pedicels not jointed; calyx glabrous externally, the hypanthium campanulate, 2.5-3.5 mm. long and nearly or quite as broad, the lobes broadly to narrowly oblong-ovate, rounded, about equaling the hypanthium, spreading to somewhat reflexed, greenish-yellow; petals obovate to oblong-obovate and slightly cuneate, 2-2.5 mm. long, nearly as long as the calyx lobes, rounded and erose; stamens about equaling the petals, the anthers oblong-oval, 1 (1.2) mm. long, only about half as long as the stout, glabrous filaments; styles from as long as the petals to as long as the extended calyx lobes, connate and pilose for about half their length; berry deep bluish-purple, globose, 10-12 mm. long, palatable. N=8.

Prairies and lower mountains; e. B.C. eastward to Newf., s. to N.D. and Mich., not s. of the Canadian border in our area. May-June.

Ribes sanguineum Pursh, Fl. Am. Sept. 164. 1814.

Calobotrya sanguinea Spach, Ann. Sci. Nat. II, 4:21. 1835. *Coreosma sanguinea* Spach, Hist. Veg. 6:155. 1838. *(Lewis, "On the Columbia river")*

An erect, unarmed shrub 1-3 m. tall, finely pubescent throughout on the young growth and usually somewhat stipitate-glandular at least on the petioles and ovary, the bark reddish-brown; leaf blades broadly reniform or cordate-orbicular to deltoid-ovate, the bases sometimes nearly truncate, mostly 2.5-6 cm. broad, much paler and more densely hairy on the lower than on the upper surface, from nearly equally to irregularly 5-lobed with the lobes deltoid to rounded and finely 2-3 times crenate and denticulate; racemes stiffly ascending to erect, equaling or exceeding the leaves, (5) 10- to 20 (30)-flowered, somewhat stipitate-glandular and crisp-pubescent throughout, the pedicels jointed; calyx pale to deep rose, more or less stipitate-glandular and finely pubescent, the hypanthium tubular to slightly campanulate, 3-5 mm. long and scarcely as broad (even in pressed specimens), the lobes ovate-elliptic to oblong or oblanceolate, subequal to the hypanthium, spreading; petals white to light rose, obovate-spatulate, entire (slightly erose), 2.5-3.5 mm. long; stamens about equaling the petals, the filaments pinkish, glabrous, the anthers less than 1 mm. long; styles glabrous, connate almost to the stigmas, usually not quite equaling the extended sepals; berry nearly globose, usually about 7-9 mm. long, glaucous-black, somewhat stipitate-glandular, unpalatable. N=8. Red currant, red flowering currant, blood currant.

In open to wooded, moist to rather dry valleys and lower mountains; B.C. to the Coast Range of Calif. s. of San Francisco, from the coast to the e. slope of the Cascades in Wash. and n. Oreg. but only w. of the crest of the Cascades southward. Mar.-June, dependent upon elevation.

The species is one of our most beautiful ornamental shrubs and one which varies greatly in most characters, the most noticeable being the color of the flowers (from pale pink to very deep red). Several of these color phases have been propagated and are sold commercially under descriptive horticultural names, such as *albescens, atrorubens,* and *splendens,* as well as *flore-pleno.*

Along the coast of Calif. the plants tend to have somewhat less hairy leaves (with proportionately broader and shorter terminal lobes) and drooping racemes; such specimens have been called *R. glutinosum* Benth. (Trans. Hort. Soc. Lond. II, 1:476. 1835) and *R. sanguineum* var. *glutinosum* Loud. (Arb. 988. 1844). The latter disposition more nearly represents their proper status. Less hairy phases of *R. sanguineum* in Oreg. have been referred erroneously to the var. *glutinosum.*

Ribes sativum (Reichb.) Syme, Engl. Bot. 4th ed. 42, pl. 520. 1865.

R. rubrum var. *sativum* Reichb. Fl. Germ. Excurs. 562. 1830-32. (No type given)

Erect, unarmed shrubs up to 1.5 m. tall, more or less crisp-puberulent and somewhat stipitate-glandular on the young growth, the old bark dark reddish-brown; leaf blades cordate, mostly 4-8 cm. broad, not quite so long, sparsely hairy at least along the veins of the undersurface, 3-lobed or more commonly unequally 5-lobed about half the length, the lobes once or twice crenate-serrate; racemes ascending to drooping, 8- to 20-flowered, nearly or quite glabrous, about equaling the leaves; pedicels slender, up to 6 mm. long, jointed under the glabrous and smooth ovary; calyx glabrous, ochroleucous or greenish, the free hypanthium saucer-shaped, about 1 mm. long, lined to about midlength with a prominent, 5-angled disc which completely covers the top of the ovary; calyx lobes about 2 mm. long, usually at least as wide, broadly deltoid-ovate, abruptly narrowed to a slender base, spreading and usually somewhat revolute at the tips; petals scarcely 1 mm. long, cream to pinkish, cuneate-flabellate, inserted with the stamens at the top of the hypanthium; filaments about equaling the petals; anthers broader than long, more or less dumbbell-shaped, with a connective at least as broad as the filament; styles glabrous, barely equaling the stamens, connate half their length or more; berry globose, red, up to 1 cm. long, sour but palatable. Currant.

Native of Europe, but frequently cultivated and sometimes escaped or bird-disseminated, perhaps more commonly in e. U.S., but known from supposedly wild collections in B.C., Wash., and Oreg. Apr.-May.

This species is the parent of some of the more commonly cultivated horticultural varieties of red currants.

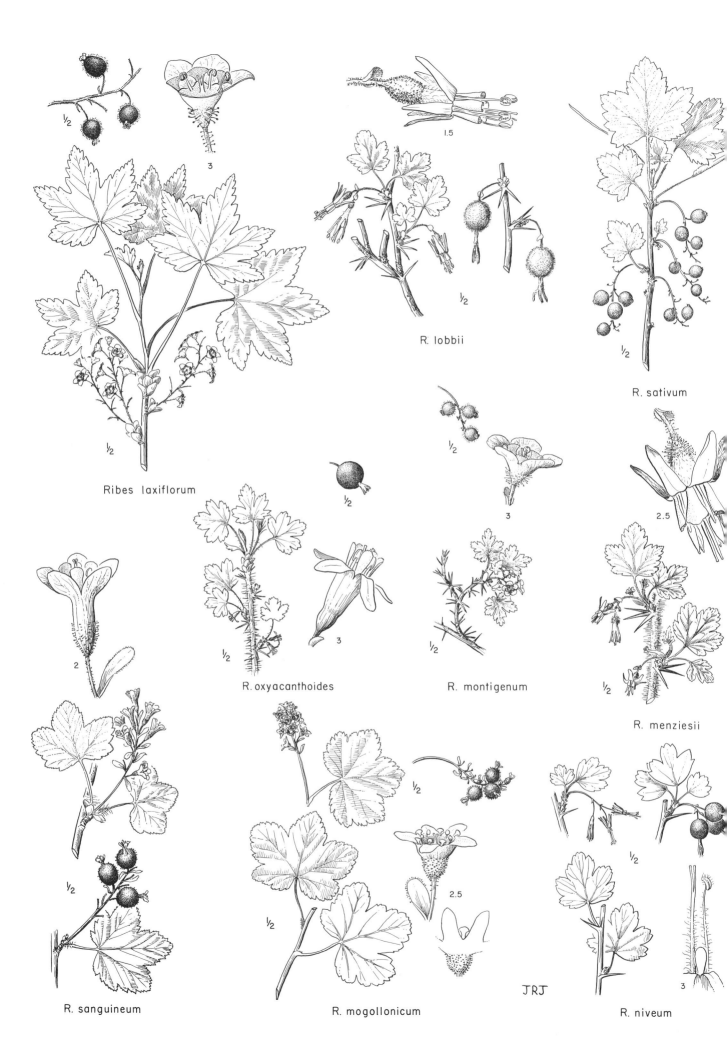

1.5

R. lobbii

R. sativum

Ribes laxiflorum

3

2.5

R. oxyacanthoides

R. montigenum

R. menziesii

2

R. sanguineum

R. mogollonicum

2.5

JRJ

R. niveum

Ribes setosum Lindl. Trans. Hort. Soc. Lond. 7:243. 1828.
  Grossularia setosa Cov. & Britt. N. Am. Fl. 223:222. 1908. (Cultivated plants of American origin)
  R. saximontanum E. Nels. Bot. Gaz. 30:119. 1900. (*E. Nelson 5542*, Golden Gate, Yellowstone
    Park, June 28, 1899)
  R. camporum Blank. Mont. Agr. Coll. Stud. Bot. 1:63. 1905. (*Blankinship 9*, Big Horn River, 7 mi.
    s. of Custer Station, May 3, 1890)
  A spreading shrub mostly 0.5-1 (3) m. tall, the branches slender, arched to recurved, finely puberulent, from not at all to densely bristly, changing to straw-colored or reddish beneath the thin grayish
bark; nodal spines 1-5, stout, straight to recurved, up to 1.5 cm. long; leaf blades mostly broadly
cordate, (1) 1.5-4 cm. wide, glabrous to finely pubescent on the upper surface, usually finely pubescent and often glandular-puberulent on the lower surface, (3) 5-lobed half their length or less, the
lobes oblong-cuneate, with (3) 5-7 (9) irregular crenate-dentations; racemes usually much shorter
than the leaves, (1) 2- to 3-flowered, the pedicels not jointed; calyx externally glabrous, the hypanthium nearly cylindric, 4-5 mm. long, the lobes spreading to somewhat recurved, oblong to slightly
obovate, obtuse to rounded, greenish-white or pinkish-tinged, subequal to the hypanthium; petals
white to pinkish, obovate-cuneate, rounded and slightly erose, 2-3 mm. long; stamens about equaling
the petals, the anthers ovate-oblong, 1-1.2 mm. long, shorter than the glabrous, often purplish or
reddish filaments; styles from shorter than the petals to slightly longer than the extended calyx lobes,
pilose and connate to midlength or slightly above; berry globose, glabrous, deep purplish-black, (8)
10-12 mm. long, of excellent flavor. N=8.
  Valleys and hillsides, most commonly along watercourses; c. and e. Ida., n. through Mont. to Assiniboia, and s. to Wyo., e. to Neb., the Dakotas, and Mich. May-July.
  In many respects *R. setosum* is similar to *R. cognatum* but it constantly differs as indicated in the
key and apparently its range is entirely to the e. of that species.

Ribes triste Pall. Nov. Act. Acad. Sci. Petrop. 10:378. 1797. (E. Siberia)
  R. albinervium Michx. Fl. Bor. Am. 1:110. 1803. ("Hab. in Canada; ad amnem Mistassin")
  R. ciliosum Howell, Fl. N.W. Am. 208. 1898. ("Mount Hood on the south side," *[Th. Howell?])*
  R. migratorium Suksd. Deuts. Bot. Monats. 18:86. 1900. (*Suksdorf*, near Bingen, Klickitat Co.,
    Wash., Mar. 21 and May, 1892)
  An unarmed, decumbent and nodally rooting to spreading shrub rarely as much as 1 m. tall, the
young twigs glabrous or sparsely crisp-puberulent and with scattered short-stipitate glands, the older
branches straw-colored, becoming purplish-brown; leaf blades broadly cordate, up to 10 cm. broad,
usually considerably shorter, glabrous above but more or less hairy beneath (at least along the veins),
usually 3 (5)-lobed less than half their length, the segments broadly triangular to ovate-triangular,
coarsely bicrenate-dentate; racemes short, drooping, glabrous to sparsely pubescent and more or less
stipitate-glandular, shorter than the leaves, 6- to 13-flowered; pedicels up to 4 mm. long, jointed under the glabrous and smooth ovary; calyx glabrous, mostly dark reddish-purple or greenish-white but
strongly purplish-maculate or -tinged; hypanthium saucer-shaped, scarcely 1 mm. long, lined to about
midlength with a prominent, reddish-purple, 5-lobed disc which covers the top of the ovary; calyx
lobes broadly cuneate-rhombic, about 2 mm. long, and as broad, spreading; petals broadly cuneate,
scarcely 1 mm. long, reddish-purple; filaments scarcely equaling the petals; anthers broader than
long, deeply cordate-based and retuse at the top, the pollen sacs slightly separated by the connective;
styles glabrous, about equaling the stamens, connate 1/3-3/4 their length; berry ovoid, less than 1 cm.
long, bright red, smooth, sour.
  Moist woods and around springs or seepage to montane rock slides; Alas. to Newf., southward to
the Cascades and Columbia R. Gorge of Wash. and n. Oreg., the Blue Mts., Wash. and Oreg., and to
S.D. and Va.; Asia. Mar. (in the Columbia R. Gorge) to June.
  Several varieties of this widespread species have been described but it is not known whether any of
the epithets proposed apply to our material, all of which seems to differ from the plants of n. B.C.
and Alas. in having more nearly 5-lobed than 3-lobed leaves, with the lobes somewhat narrower and
more deeply toothed.

Ribes velutinum Greene, Bull. Calif. Acad. Sci. 1:83. 1885.
  Grossularia velutina Cov. & Britt. N. Am. Fl. 223:220. 1908. ("Northern part of California and re-
    gions adjacent," no collector given)
  R. glanduliferum Heller, Muhl. 2:56. 1905. *Grossularia glandulifera* Berger, N.Y. Agr. Exp. Sta.
    Tech. Bull. no. 109:91. 1924. *R. velutinum* var. *glanduliferum* Jeps. Man. Fl. Pl. Calif. 472.

1925. *(Heller 8005,* Yreka, Siskiyou Co. , Calif. , June 9, 1905) = var. *velutinum.*
RIBES VELUTINUM var. GOODDINGII (Peck) C. L. Hitchc. hoc loc. *R. gooddingi* Peck, Torreya
28:54. 1928. *(William Sherwood 407,* 5 mi. w. of Imnaha, Wallowa Co. , Oreg. , June 11, 1922)
A spreading, often rounded shrub mostly 1-2 m. tall, the branches usually arching, slender, glabrous to copiously pubescent when young, never bristly, aging to dark grayish; nodal spines usually
solitary (3), straight, 1-2 cm. long; leaf blades from nearly orbicular to cordate or reniform, (2)
10-15 mm. broad, thickish, glabrous to finely pubescent and often slightly glandular-puberulent on
both surfaces, cleft often well over half their length into 3 or 5 more or less cuneate, entire to 2- to
3-toothed (rounded) segments; racemes much shorter than the leaves, 1- to 2 (3)-flowered; pedicels
not jointed, very short; ovary glabrous to crisp-puberulent and often also glandular; calyx rather copiously crisp-puberulent to more or less crisp-hirsute externally, whitish or yellowish, sometimes
tinged slightly with pink; hypanthium 1. 5-2 (2. 5) mm. long, about as broad, tubular or slightly campanulate, lined with an unusually thick glandular disc; calyx lobes oblong, 1-1. 5 times as long as the
hypanthium, spreading to nearly erect; petals white or yellowish, oblong-obovate, about 1. 5 (2) mm.
long, half as long as the calyx lobes or slightly longer; stamens about equaling the petals, the anthers
pink, about 1 mm. long, subequal to the very stout, glabrous filaments; styles connate full length,
glabrous to finely pubescent, about equaling the stamens (in dried flowers the anthers usually stuck to
the rather large, ovoid, bilobed stigma); berry globose, 6-8 mm. long, apparently reddish (but recorded as purplish), dry and not palatable.
Desert canyons and hillsides to ponderosa pine forest, from Whitman Co. , Wash. , through s. w. Ida.
and e. Oreg. to the desert of s. Calif. , e. to Utah, Nev. , and n. Ariz. Mar. -May.
The plants are extremely variable in leaf lobation and in the amount and type of pubescence on the
leaves, ovary, and calyx, and the species is probably differentiated into more than the two races found
in our area. Most of our material has deeply cleft, nearly glabrous leaves, glabrous fruits, and long
slender spines. It is not uniform in this respect, however, as plants from Whitman Co. , Wash. , have
hairy ovaries, and in general the amount of pubescence varies. It is obvious that the plants of n. e.
Oreg. and adj. Wash. and Ida. are not sharply delimited from, but rather that they gradually blend
with, those of s. c. Oreg. and southward. The following key, therefore, overemphasizes the degree of
distinction between the two races:
1 Leaves usually glabrous, mostly cleft more than half the length into 3 (5) slender, cuneate,
    entire to 1- to 3-toothed segments; ovary mostly glabrous, sometimes sparsely crisp-
    hairy, never glandular; Snake R. Canyon and its tributaries from s. Whitman Co. , Wash. ,
    up the Salmon R. to Lemhi Co. , Ida. , and through n. e. and e. Oreg. to Klamath Co. ,
    where there is a blending with var. *velutinum;* not at all uniform, even along the Snake R.
                                                                        var. gooddingii (Peck) C. L. Hitchc.
1 Leaves densely pubescent and often glandular, mostly cleft no more than half the length into
    broader, mostly 2- to 3-toothed segments; ovary usually pubescent to glandular (glabrous);
    even more variable than var. *gooddingii,* with leaves sometimes (e. g. *Cronquist 7176,* from
    Crook Co. , Oreg.) cleft nearly the full length; Crook Co. , Oreg. , s. and e. to Calif. and
    Utah
                                                                        var. velutinum

Ribes viscosissimum Pursh, Fl. Am. Sept. 163. 1814.
    Coreosma viscosissima Spach, Ann. Sci. Nat. II, 4:23. 1835. *R. viscosissimum* var. *purshii* Jancz.
    Mém. Soc. Phys. Nat. Genèv. 35:328. 1907. *(Lewis,* "On the Rocky-mountain in the interior of
    North America")
    R. hallii Jancz. Bull. Acad. Sci. Crac. 1906:9. 1906. *R. viscosissimum* var. *hallii* Jancz. Mém. Soc.
    Phys. Nat. Genèv. 35:328. 1907. *(Hall & Babcock 4533* [acc. Jepson], Lake Independence, Calif.)
An erect to spreading and usually rather straggly, unarmed shrub mostly about (1) 2 m. tall, rather
generally soft-pubescent (downy) and more or less thickly stipitate-glandular, the old branches becoming dark reddish-brown; leaf blades mostly (2) 3-6 (10) cm. broad, from sparsely stipitate-glandular on both surfaces (but otherwise nearly glabrous) to downy along the veins, copiously soft-pubescent
(as well as glandular) on both surfaces, deeply cordate, 3- or 5-lobed much less than half their length
and the lobes rounded, irregularly once or twice coarsely crenate-dentate; racemes erect to somewhat drooping, copiously pubescent and stipitate-glandular, usually shorter than the leaves, (4) 6- to
12 (17)-flowered; pedicels jointed, mostly exceeding the bracts; ovary glabrous to strongly stipitate-
glandular and usually finely pubescent; calyx greenish, greenish-yellow, or yellowish-white and sometimes strongly pinkish- or purplish-tinged; the hypanthium tubular-campanulate, (5) 6-7 (8) mm. long,
the lobes oblong, usually pointed, spreading, mostly subequal to the hypanthium; petals broadly ovate

or oval, narrowed abruptly to a short broad claw, 2.5-4 mm. long, white or cream, about half as long as the calyx lobes; stamens inserted at about the same level as the petals and equaling or slightly exceeding them; anthers oblong-oval, 1-1.5 mm. long, tipped by a small cuplike gland; styles equal or subequal to the stamens, connate almost to the stigmas, glabrous; berry ovoid, 10-12 mm. long, deep bluish-black, glabrous to stipitate-glandular and more or less hairy, rather dry and with a disagreeable taste and smell. Sticky currant.

Along creeks and on open to heavily timbered, moist to rather dry slopes, even to timber line; Cascade Mts. (mostly on the e. side) from B.C. through Wash. and Oreg. to n.w. and Sierran Calif. and n. Ariz., e. to Mont., w. Wyo., and n.w. Colo. May-June.

The species consists of the following clearly marked geographic races:

1 Ovary glabrous or only very sparsely stipitate-glandular; berry glaucous; Cascade Mts. of s. Oreg., occasional n. to Mt. Rainier; the more common phase in Calif., where many plants are transitional to the next   var. hallii Jancz.
1 Ovary rather strongly glandular and usually also soft-pubescent; berry only slightly or not at all glaucous; Cascade Mts. from B.C. s. to Mt. Rainier, Wash., e. through Wash., Ida., and n.e. Oreg., to Mont. and Colo., and in Utah, Nev., Ariz., and the southern Sierra Nevada   var. viscosissimum

The plants comprising the var. *viscosissimum* are not uniform, the leaves, especially, varying from glandular but scarcely at all downy to (more commonly) strongly downy, and the fruits from slightly or not at all pubescent but strongly glandular, to downy and glandular. The less downy plants are much more common in the Washington Cascades than elsewhere.

Ribes watsonianum Koehne, Deuts. Dendrol. 197. 1893.

R. ambiguum Wats. Proc. Am. Acad. 18:193. 1883, but not of Maxim. in 1874. *Grossularia watsoniana* Cov. & Britt. N.Am. Fl. 22³:218. 1908. (*Suksdorf*, Mt. Adams, in 1882.

An ascending to erect shrub up to 2 m. tall, the branches rather slender although not arching, copiously grayish-hairy and abundantly setiferous-glandular but very slightly if at all bristly, aging to straw-color and ultimately grayish-brown; nodal spines 1 or 3, somewhat curved, mostly 3-7 (10) mm. long; leaf blades thin, broadly ovate, with a truncate or (more often) cordate base, thickly puberulent on both surfaces or more especially beneath, often also somewhat glandular, 2-5 cm. broad, 3 (5)-lobed less than half the length, the lobes broad, rounded, again cleft into 2-5 coarsely crenate-denticulate segments; racemes 1- to 2 (3)-flowered, pubescent and glandular, the peduncles shorter than the leaves, drooping; pedicels rather stout, not jointed; ovary copiously glandular-bristly; calyx finely pilose and often sparingly glandular, greenish or with some reddish tinge, the hypanthium tubular-campanulate, 2.5-3 mm. long, about as broad at the top; calyx lobes narrowly oblong, (5.5) 6-7 (9) mm. long; petals oblong, white, 3.5-4 mm. long, rounded, entire to slightly erose; stamens about equaling the petals, the filaments 2-3.5 times as long as the oval, rounded, white, 0.8-1.2 mm. anthers; styles mostly slightly shorter than the extended calyx lobes, connate about half their length, glabrous; berry subglobose, about 1 cm. long, reddish, densely yellowish-spiny.

Canyons and ridges on both sides of the Cascade Mts. and in the Wenatchee Mts., from Chelan Co., Wash., s. to the e. side of Mt. Hood, Wasco and Hood River cos., Oreg. Late May-early July.

HYDRANGEACEAE Hydrangea Family

Flowers inconspicuous to showy, cymose in flat-topped to elongate inflorescences, or occasionally apparently racemose (solitary), perfect or sometimes the lowermost or marginal ones sterile and then often with the calyx enlarged and particularly showy; calyx more or less adnate to the ovary, usually without a free hypanthium (ours), the lobes 4-6 (up to 10); petals 4-6 (10), mostly white but sometimes colored; stamens usually numerous or at least twice as many as the petals, but as few as 5, the filaments often flattened and sometimes slightly connate; pistil (2) 3- to 5 (up to 10)-carpellary, the ovary partially to completely inferior, the styles distinct to completely connate, the stigmas varied; fruit usually a subglobose to fusiform or conic, often woody capsule; seeds from numerous to only one per locule (ours) or even solitary, with abundant endosperm; trees or shrubs (sometimes climbing) with deciduous to sometimes persistent, opposite, simple, exstipulate leaves.

About a dozen genera and perhaps 75 species of the subtropics and N. Temp. Zone, with several genera endemic to s.w. U.S.

Because of the woody habit, opposite exstipulate leaves, usually numerous stamens, and mostly 3- to 5-locular ovary, the family constitutes a distinct, natural unit, although it is sometimes treated as part of the Saxifragaceae. Such genera as *Philadelphus, Deutzia, Hydrangea,* and *Carpenteria* are considered choice ornamental shrubs.

1 Stamens 25-50; petals never less than 5 mm. long; capsule many-seeded, loculicidal

PHILADELPHUS

1 Stamens 8-12; petals less than 5 mm. long; capsule septicidal, with only 1 seed per locule

WHIPPLEA

Philadelphus L. Mock Orange; Syringa

Flowers complete, rather showy, often fragrant, borne at the ends of short lateral branches, (1) 3-11 or more in false racemes (ours) or cymes or panicles; calyx adnate to the ovary up to the level of the 4 (5) lobes, a free hypanthium lacking; petals 4 (5), white or rarely somewhat reddish or purplish at the base; stamens numerous, free (ours) or the filaments more or less connate; pistil 4 (3 or 5)-carpellary; ovary nearly completely inferior, 4-locular, with axile placentation, the ovules numerous, suspended; styles (rarely free) connate half their length or more, the 4 (5) stigmas from linear, clavate, or oar-shaped, to capitate, in ours stigmatic for considerably more of their length on the inner (adaxial) margin than on the outer edge; fruit a rather woody, loculicidal capsule; erect to spreading shrubs with opposite, deciduous (ours) to evergreen, exstipulate, petiolate, entire to toothed leaves, the axillary buds naked or protected.

About 50 species of wide geographic distribution, mostly in temp. and subtropical N. Am. and Asia. (From the Greek *philos,* love, and *delphos,* brother, said to commemorate Ptolemy Philadelphus, king of Egypt in the pre-Christian era.)

Several of the species are very highly valued as hardy ornamental shrubs because of their large, showy, and more or less fragrant blossoms. The commonly grown plant in the Pacific Northwest is the European *P. coronarius* L. Although there is no record of its having become established without human care, it can be told from our sparingly cultivated native species by the (usually) 20-25 stamens, more fragrant flowers, more numerous teeth of the leaves, and shorter appendage of the seeds, the embryo being about equal to the tail.

Reference:

Hu, Shiu-Ying. A monograph of the genus Philadelphus. Journ. Arn. Arb. 35:275-333. 1954; 36:52-109. 1955; 37:15-90. 1956.

Philadelphus lewisii Pursh, Fl. Am. Sept. 329. 1814. *(Lewis,* "on the waters of Clarck's river")

P. *gordonianus* Lindl. Bot. Reg. 24:Misc. notes 21. 1838. *P. lewisii* var. *gordonianus* Jeps. Man. Fl. Pl. Calif. 466. 1925. (Cultivated plant grown from seed collected by Douglas along the banks of the Columbia)

P. *cordatus* Petz. & Kirchn. Arb. Muscav. 203. 1864, as a synonym, only.

P. *grahami* Petz. & Kirchn. loc. cit., nom. subnud.

P. *columbianus* Koehne, Gartenfl. 45:542. 1896. *P. gordonianus* var. *columbianus* Rehd. Journ. Arn. Arb. 1:196. 1919. (Cultivated plants, thought possibly to have come from British Columbia)

P. *confusus* Piper, Bull. Torrey Club 29:225. 1902. *(O. D. Allen 221,* Tum Tum Mt., Wash., Aug. 13 and Sept. 17, 1896)

P. *angustifolius* Rydb. N.Am. Fl. 22 2:166. 1905. *P. lewisii* var. *angustifolius* Hu, Journ. Arn. Arb. 36:79. 1955. *(Mrs. Bailey Willis,* Palace Camp, Wash., in 1883)

P. *helleri* Rydb. N.Am. Fl. 22 2:166. 1905. *P. lewisii* var. *helleri* Hu, Journ. Arn. Arb. 36:79. 1955. *(Heller & Heller 3374,* Lake Waha, Nez Perce Co., Ida., in 1896)

P. *platyphyllus* Rydb. N.Am. Fl. 22 2:167. 1905. *P. lewisii* var. *platyphyllus* A. H. Moore, Rhodora 16:77. 1914. *(H. E. Brown 561,* s. side of Mt. Shasta, Calif., in 1897)

P. *intermedius* A. Nels. Bot. Gaz. 42:53. 1906. *P. lewisii* var. *intermedius* Hu, Journ. Arn. Arb. 36:81. 1955. *(Macbride & Payson 3037,* Martin, Blaine Co., Ida., lectotype by Hu) A small-leaved form.

P. *lewisii* var. *parvifolius* Hu, Journ. Arn. Arb. 36:77. 1955. *(Kirkwood 1151,* Missoula, Mont.) A small-leaved form.

P. *lewisii* var. *oblongifolius* Hu, Journ. Arn. Arb. 36:78. 1955. *(Piper 3838,* Wawawai, Wash.)

P. *lewisii* var. *ellipticus* Hu, Journ. Arn. Arb. 36:81. 1955. *(Coville & Applegate 269,* Klamath Co., Oreg.)

P. *trichothecus* Hu, Journ. Arn. Arb. 36:84. 1955. *(W. R. Carter,* Oak Bay, Victoria, B.C.) Petals and top of the ovary glabrous, the anthers hairy.

P. oreganus Nutt. ex Hu, Journ. Arn. Arb. 36:85. 1955. *(Nuttall,* Oregon woods) Petals and anthers
glabrous, the top of the ovary hairy.

P. zelleri Hu, Journ. Arn. Arb. 36:87. 1955. *(Zeller & Zeller 974,* Friday Harbor, San Juan I.,
Wash.) Top of the ovary, petals, and anthers all hairy.

Rounded to erect shrub mostly 1.5-2.5 (3) m. tall; leaf blades ovate, ovate-lanceolate, or narrowly
to broadly elliptic, entire to remotely denticulate on the older branches, often strongly but remotely
serrate-dentate on the vigorous new shoots, from rather strongly strigose or strigose-villous on both
surfaces to glabrous above and sometimes glabrous below except along the nerves or at the nerve angles, ciliate, (1) 2.5-7 (9) cm. long, (0.5) 1-4 (6) cm. broad on the floriferous branches, sometimes
slightly larger on the new shoots; flowers fragrant, 3-11 in terminal racemes on lateral branches, the
lower 2, 4, or 6 from the axils of gradually reduced leaves, the upper ones inconspicuously bracteate
or bractless; adnate portion of the calyx usually glabrous but sometimes sparsely strigose toward the
top, turbinate-campanulate, up to 6 mm. long, the lobes ovate to ovate-lanceolate, more or less acuminate, mostly 5-6 mm. long; petals white, (ovate) oblong to nearly obovate, (5) 10-20 (25) mm. long,
obtuse or rounded and often emarginate; stamens 25-40 (50), unequal, the filaments glabrous or occasionally with 1 or more hairs above, the anthers ovate-oblong, glabrous or not infrequently with from
1 to several straight to crisped hairs; styles about as long as the stamens, from nearly fully connate
to distinct half the length, glabrous, the branches often stigmatic on the inner (adaxial) surface for the
full length; ovary nearly completely inferior at anthesis, the top glabrous to hairy, at first merely
rounded but enlarging and becoming more pointed in fruit, the capsule ovoid-elliptic and pointed at the
ends, about 3/5 inferior, 6-10 mm. long; seeds about 3 mm. long, linear-fusiform, the tip finely
pointed and the funicular end with an unequally erose-coronate caruncle about twice as long as the embryo.

In gullies and watercourses and on rocky cliffs, talus slopes, and rocky hillsides of sagebrush desert and ponderosa or lodgepole pine to coastal Douglas fir and redwood forest; B.C. to n. Calif., e.
through s. B.C., Wash., and all but s.e. Oreg., to the continental divide in Mont., and from the
Snake R. northward in Ida., from sea level along the coast up to as much as 7000 ft. elevation on the
e. side of the Cascades. May-July.

The species is extremely variable in both vegetative and floral characters and appears to be particularly responsive to local ecological conditions. Although there might be some reason for the naming
of horticultural strains that can be propagated vegetatively, ecological or geographic races are not
detectable, with 2 possible exceptions. There may be statistical basis for the claim that the plants
from west of the Cascades *[P. gordonianus* Lindl., or var. *gordonianus* (Lindl.) Jeps.] have more
hairy leaves and/or more completely connate styles than those from east of the mountains. It is also
possible (but not considered probable) that the plants of the dry interior lowlands, especially those of
desert scabland and sagebrush, may constitute an ecologic race characterized chiefly by smaller,
more elliptic leaves (var. *parvifolius* Hu; *P. intermedius* A. Nels.)

Whipplea Torr.

Flowers complete, inconspicuous, borne in small, pedunculate, congested and capitate to somewhat
open cymose panicles; calyx fused half its length or less with the ovary, a free hypanthium lacking,
the lobes (4) 5 or 6; petals white, (4) 5 or 6; stamens usually equal in number to the sepals and petals
and borne opposite them, didymous, the filaments flattened, the anthers dumbbell-shaped, broader
than long; pistil (3) 4- or 5-carpellary, the styles connate only at the base or entirely free; ovary
about half inferior, (3) 4- or 5-celled, the ovules 1 per cell, suspended; fruit a subglobose, septicidal
capsule; trailing to spreading shrubs with opposite, nearly sessile, exstipulate leaves.

Only the one species. (Named for Lieut. A. W. Whipple, 1818-1863, who commanded the U.S. Railroad Expedition of 1853-1854.)

Whipplea modesta Torr. Pac. R. R. Rep. 4:90, pl. 7. 1857. *(Bigelow,* Redwood, Calif., Apr. 12,
1854)

Main stems slender, trailing and freely rooting, up to 1 m. long, with numerous short, erect, terminally floriferous shoots, rather strongly coarse-pubescent throughout; leaves ovate or ovate-elliptic, 1-2.5 cm. long, mostly 5-15 mm. broad, subsessile to short-petiolate, remotely crenate-serrulate, semideciduous but often more or less marcescent; peduncles 2-5 cm. long; flowers 5-10; calyx lobes narrowly oblong or oblong-lanceolate, about 1.5-2 mm. long, erect, eventually deciduous;
petals about twice as long as the calyx lobes, narrowly rhombic or rhombic-obovate; stamens about

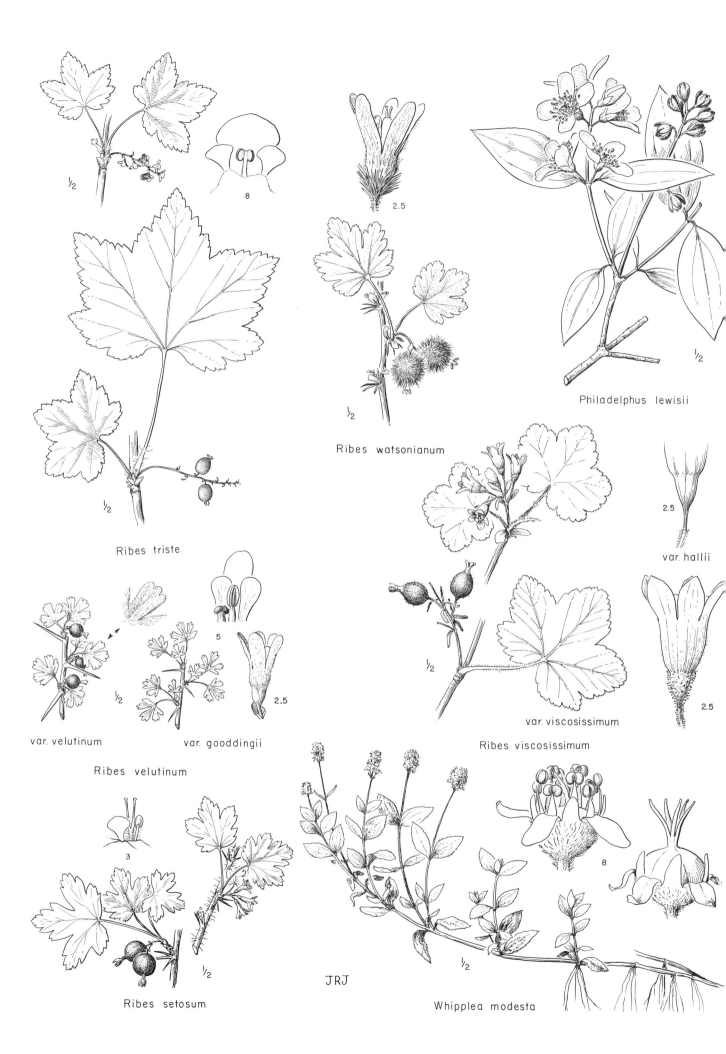

8

2.5

Philadelphus lewisii

Ribes watsonianum

Ribes triste

var. hallii

var. velutinum

5

2.5

var. viscosissimum

2.5

var. gooddingii

Ribes viscosissimum

Ribes velutinum

3

8

Ribes setosum

JRJ

Whipplea modesta

equaling the petals, the anthers about 0.6 mm. broad and half as long; capsule depressed-globose, separating into (3) 4-5 leathery, ventrally dehiscent, 1-seeded segments; seeds nearly filling the cavities, plump, reticulate-alveolate, with a small membranous caruncle.

Usually where rather dry and rocky, in open to lightly forested areas on the w. side of the Cascades, from the Olympic Peninsula, Wash., s. as far as Monterey Co., Calif. Apr.-June.

This rather attractive plant is to be considered of fair value as a ground cover, especially on drier areas.

ROSACEAE Rose Family

Flowers from single to numerous and from racemose or paniculate to corymbose, umbellate, or cymose, complete to occasionally apetalous or imperfect, mostly regular, perigynous to epigynous; calyx usually 5 (4-10)-merous, frequently bearing bracteoles alternate with the lobes, connate basally, sometimes adnate to the pistil, but usually free as a saucerlike to campanulate, turbinate, or urceolate, generally disc-lined hypanthium, persistent or the lobes (and sometimes also the hypanthium) deciduous; petals 5 (4-10), often showy, occasionally much reduced or lacking, usually deciduous; stamens very rarely only 1-4, occasionally 5 (6-8), frequently 10, but most commonly from 15 to numerous, usually in indistinct series of 5 per whorl, borne with the petals on the hypanthium generally near the outer margin at the edge of the disc, the anthers 2-celled, mostly dehiscent by slits on the margin or adaxial face (or transversely) or rarely by pores; pistils 1-several, mostly simple, free of the hypanthium, and (if more than 1) borne cyclically or spirally on an enlarged receptacle, sometimes when single composed of 2-5 carpels, the ovary partially to completely inferior, the placentation parietal, the styles free and equal to the number of carpels, or partially to completely connate but the stigmas equaling the carpels in number; fruit an achene, a follicle dehiscent adaxially or on both sutures, or fleshy and then of 1 or 5 drupes, a single pome, or in a few genera an aggregation of tiny drupelike fruits in which the receptacle may be fleshy (blackberry), dry (raspberry), or fleshy and the true fruits dry (strawberry); seeds usually exalbuminous; annual to perennial herbs, shrubs, or trees, sometimes armed, with alternate or basal (rarely opposite), simple to ternately, pinnately, or occasionally palmately compound, deciduous to evergreen, stipulate (rarely exstipulate) leaves, the stipules from free to completely adnate to the petiole, sometimes caducous.

A cosmopolitan family of perhaps 100 genera and 2000 species, most common in north temperate and boreal regions.

The group is one of great economic importance, producing most of our common fleshy fruits *(Pyrus, Prunus, Fragaria, Rubus)* and many plants of great ornamental value, including such familiar examples as *Cotoneaster, Pyracantha, Crataegus, Prunus, Spiraea, Dryas,* and *Rosa.* It is a rather heterogeneous assemblage of genera held together chiefly because of their general mutual resemblances, but the family is rather readily divisible, largely on the basis of the fruit, into several subgroups. These taxa usually are treated as tribes, although not rarely as separate families, especially the two characterized by drupaceous (Drupaceae, Amygdalaceae, or Prunaceae) or by pomaceous (Pomaceae or Malaceae) fruits.

1 Plants prickle-bearing; leaves pinnate; carpels several, enclosed within the globose to urceolate calyx (in fruit the calyx usually reddish and fleshy but the carpels themselves dry)   ROSA
1 Plants not at once prickle-bearing, pinnate-leaved, and with several carpels enclosed within a globose to urceolate and ultimately fleshy calyx
  2 Calyx adnate to the compound, 2- to 5-carpellary, more or less completely inferior ovary; ripe fruit fleshy; deciduous trees or shrubs, never dioecious
    3 Leaves pinnate, the leaflets 5-17   SORBUS
    3 Leaves simple
      4 Plants usually armed with strong sharp thorns; carpels with a hardened, shelly covering, each 1-seeded   CRATAEGUS
      4 Plants unarmed; carpels with papery to cartilaginous walls, 2-seeded
        5 Leaves oblanceolate, 1-4 cm. long, entire to finely serrulate, without well-differentiated blade and petioles although narrowed to a slender base; petals pink   PERAPHYLLUM
        5 Leaves with a well-differentiated petiole and blade, usually from ovate or oblong to obovate, serrate to strongly toothed or lobed; petals white or pink
          6 Flowers racemose; carpels falsely divided, each cavity 1-seeded; leaves never lobed   AMELANCHIER

    6 Flowers corymbose; carpels not falsely partitioned, each cavity usually 2-seeded; leaves
        sometimes lobed   PYRUS

2 Calyx sometimes enclosing (but not adnate to) the 1 to several superior, 1-carpellary ovaries;
   ripe fruit dry to fleshy; plants herbaceous, shrubby, or rarely arborescent *(Aruncus* and *Osmaronia* dioecious)

    7 Plant a shrub or small tree with large, deciduous, simple leaves and corymbose or umbellate to racemose flowers; pistils 1-5; fruit fleshy, drupaceous

        8 Flowers imperfect, pistils about 5; leaves entire   OSMARONIA

        8 Flowers perfect; pistil 1; leaves toothed   PRUNUS

    7 Plant an herb or shrub, if shrubby the leaves often persistent, or lobed to compound; pistils
        1-many, if 1 or 5 the fruit not drupaceous; flowers usually perfect

        9 Petals lacking; pistils 1-2 (3), the ovary completely enclosed in the calyx

          10 Plant a shrub; leaves entire to toothed, persistent   CERCOCARPUS

          10 Plant an herb; leaves pinnate or palmately lobed to dissected

             11 Leaves odd-pinnate; style terminal; flowers spicate   SANGUISORBA

             11 Leaves palmately lobed or dissected; style basal; flowers not spicate   ALCHEMILLA

        9 Petals present, or if (as only abnormally) lacking then the pistils more than 3, or the
        ovary not enclosed by the calyx

          12 Hypanthium crowned with a ring of hooked bristles, completely enclosing the 1-2
             achenes; pinnate-leaved herbs   AGRIMONIA

          12 Hypanthium not hooked-bristly above, often not enclosing the ovary; plants various

             13 Leaves deciduous, simple, cuneate, deeply 3-toothed at the tip, otherwise entire, revolute-margined, greenish above, tomentose beneath; pistils 1 (2); erect, rigidly branched shrubs   PURSHIA

             13 Leaves various but if cuneate and 3-toothed then the pistils more than 2, or plants herbaceous or evergreen

               14 Leaves bi- to triternately dissected into linear segments, persistent

                 15 Plant an erect herb from a taproot; flowers in open or congested cymes; fruit an achene   CHAMAERHODOS

                 15 Plant a trailing semishrub; flowers racemose; fruit a follicle   LUETKEA

             14 Leaves various, often deciduous, never bi- to triternately dissected

               16 Leaves pinnate-pinnatifidly dissected into ultimate segments 0.5-1.5 mm. long and about 0.5 mm. broad; plant a low, glandular, stellate-pubescent shrub   CHAMAEBATIARIA

               16 Leaves not pinnate-pinnatifid into tiny ultimate segments or plant not a low, glandular, stellate-pubescent shrub

                 17 Plants prostrate, matted or cushion-forming, woody shrubs with persistent, more or less marcescent, entire to merely coarsely crenate leaves; flowers single at the branch ends or in pedunculate spikelike racemes

                   18 Flowers numerous in pedunculate racemes; perianth 5-merous, small, the petals white, 1.5-2.5 mm. long; leaves entire

                                                 PETROPHYTUM

                 18 Flowers single at the branch ends or on scapes; perianth either 8- to 10-merous or the petals pink and 2-12 mm. long; leaves entire to crenate

                   19 Perianth mostly 8- to 10-merous; petals 8-12 mm. long, white or yellow; styles plumose in fruit; flowers pedunculate; leaves 10-40 mm. long   DRYAS

                   19 Perianth 5-merous; petals 2-3 mm. long, pink; styles not at all plumose; flowers subsessile; leaves 2.5-4 mm. long KELSEYA

               17 Plants often herbaceous, if shrubby then not trailing, the leaves lobed to compound, or the flowers neither single nor racemose

                 20 Plant an erect, unarmed shrub with toothed or lobed (but never compound) leaves; pistils (1) 2-7

                   21 Leaves palmately 3- to 5-lobed; calyx usually stellate; fruit a somewhat inflated, several-seeded follicle, dehiscent on both sutures   PHYSOCARPUS

21 Leaves pinnately many-lobed to toothed; calyx not stellate; fruit an achene or a follicle
    dehiscent only on the adaxial suture
    22 Fruit a 2- to several-seeded follicle; stamens 25-50; leaves usually merely toothed
        (rarely shallowly lobed); petals white, pink, red, or purplish   SPIRAEA
    22 Fruit an achene; stamens usually 20; leaves mostly shallowly lobed; petals white
                                                    HOLODISCUS
20 Plant often herbaceous but if an erect shrub then with compound leaves, or with prickles,
    or with numerous (more than 7) pistils
    23 Plants dioecious, rhizomatous herbs 1-2. 5 m. tall, with ternate- to triternate-pin-
        natisect leaves; panicles 1-5 dm. long; petals white, scarcely 1 mm. long   ARUNCUS
    23 Plants various but if herbaceous then the leaves not ternate, or the flowers perfect
        and the petals usually well over 1 mm. long
        24 Leaves pinnate or pinnately divided, the terminal segment much the largest,
            broadly ovate, palmately (3) 5- to 7-cleft, and doubly serrate, 8-20 cm. broad;
            plant a strongly rhizomatous herb 1-2 m. tall; petals white, about 6 mm. long;
            achenes strongly flattened, with a stipe 2-3 mm. long   FILIPENDULA
        24 Leaves never at once pinnate and with a large palmately-divided terminal seg-
            ment; plants rarely rhizomatous herbs over 1 m. tall
            25 Hypanthium narrowly obconic, completely enclosing the 2-6 canescent achenes;
                filaments persistent and stiffly erect in fruit; low, rhizomatous herb with
                cordate-orbicular, shallowly lobed to toothed leaves   WALDSTEINIA
            25 Hypanthium usually shallow, saucer-shaped to campanulate, rarely enclos-
                ing the usually numerous fruits; filaments not stiffly erect; plants various
                26 Stamens in ours 5; pistils usually 2-15
                    27 Leaves ternate; style laterally attached to the ovary   SIBBALDIA
                    27 Leaves pinnately dissected; style subterminal   IVESIA
                26 Stamens 10 or more; pistils usually numerous
                  28 Stamens 10; anthers dehiscing lengthwise by slits on the inner (adaxial)
                    face; perennials with pinnate leaves, the leaflets (5) 7-19, some-
                    times dissected   HORKELIA
                28 Stamens either more than 10 or, if 10, then the plants annuals or
                  biennials with nondissected pinnae
                  29 Calyx ebracteolate; plants rhizomatous, stoloniferous, or trail-
                    ing to erect, sometimes armed shrubs; mature fruit an aggrega-
                    tion of weakly coherent druplets   RUBUS
                  29 Calyx bracteolate; plant various, but mostly herbaceous, never
                    armed; mature fruit an achene, often borne partially embedded
                    in the fleshy receptacle
                    30 Receptacle enlarged and hemispheric, in fruit becoming spongy
                      or fleshy; leaves usually ternate; stems strongly stolonous,
                      freely rooting at the nodes
                    31 Petals yellow; receptacle spongy to fleshy but not juicy   DUCHESNEA
                    31 Petals white or pinkish; receptacle fleshy and juicy   FRAGARIA
                  30 Receptacle not becoming spongy or fleshy and strawberrylike;
                    leaves various but if ternate the stems rarely if ever stolonous
                    32 Style slender, apical, straight to bent or geniculate, per-
                      sistent in fruit, although (if jointed) the upper segment
                      usually deciduous; leaves usually lyrately pinnatifid   GEUM
                  32 Style various, apically to laterally or almost basally in-
                    serted, straight, of only 1 segment, jointed to the achene
                    and usually deciduous (or readily broken off); leaves var-
                    ious, but rarely if ever lyrate-pinnatifid   POTENTILLA

Agrimonia L.

Flowers complete, perigynous, rather small, borne in long, bracteate, spikelike racemes; calyx
with a conic (ours) to hemispheric, eventually 10-grooved and much indurated hypanthium often exceeding the 5 lobes, marked externally with a ring of hooked bristles, and projecting upward (inside the

petals and stamens) to form a domed covering through which the styles project; sepals spreading at anthesis, becoming incurved, persistent; petals 5, yellowish, rather small, deciduous, borne with the 5-15 stamens just within and above the sepals on the base of the constricted inward extension of the hypanthium; pistils 2 (3), enclosed by (but free of) the hypanthium; styles terminal, projecting above the hypanthium; stigmas rather large, somewhat bilobed; fruit achenial, often only 1 developing; rhizomatous perennial herbs with large, unequally odd-pinnate leaves and foliaceous stipules.

Perhaps 20 species, widely distributed, chiefly in the N. Temp. Zone, but also in S. Am. and Africa. (Supposedly a variant of *Argemone.)*

Besides the following species, *A. gryposepala* Wallr. is said to occur in e. B.C. near (if not actually within) our limits. It differs in having a more glandular rachis in the inflorescence and a less hairy hypanthium.

Agrimonia striata Michx. Fl. Bor. Am. 1:287. 1803. ("Hab. in Canada")

Fibrous-rooted, strongly rhizomatous perennial, 5-10 (15) dm. tall, papillate-hirsute below, becoming both hirsute and puberulent (and sometimes very slightly glandular) above; leaves with 5-13 very unequal main leaflets, the upper (usually largest) ones up to 6 cm. long, strongly serrate, dark green and more or less strigose on the upper surface, but much paler, pubescent and abundantly sessile-glandular beneath; stipules up to 2 cm. long; racemes 5-20 cm. long; hypanthium about 3 mm. long at anthesis, equaling the ovate-lanceolate lobes, in fruit accrescent and up to 5 mm. long, crowned with 3-4 rows of subterminal, ascending, hooked bristles nearly as long as the sepals, the sides strongly 10-furrowed and lightly strigose, especially between the ribs.

Dry to moist, often sandy to rocky soil; e. B.C. to Que. and N.S., southward, on the e. side of the Rocky Mts., to N.M., Ia., and N.Y., skirting and perhaps entering our area in s.e. B.C. and c. Mont. June-July.

Alchemilla L.

Flowers inconspicuous, greenish, apetalous, perigynous, borne in axillary clusters or in terminal, branched cymes or panicles; calyx free of the ovary, with a tubular to campanulate hypanthium, completely enclosing the achenes and in ours almost closed by a circular disc at the throat, 4 (5)-lobed and usually with as many alternating bracteoles; stamens 1-4 (5), inserted at the edge of the disc; anthers opening by a transverse slit; pistils 1 (2), 1-carpellary, the style basal; stigma capitate; ovule 1; fruit an achene; annual or perennial herbs with alternate, palmately lobed to dissected leaves and connate stipules.

About 100 species, mostly of the N. Temp. Zone of Eurasia, but well represented from S. Am. n. to Mex., and in Africa. (Said to be from *Alkemelych,* the Arabic name for the plant, or from *alchemy,* in reference to its use.)

Reference:

Walters, S. M. Alchemilla vulgaris in Britain. Watsonia 1:6-18. 1949.

1 Flowers in clusters along the stem in the axils of the connate stipules; leaf blades ternately
    dissected   A. OCCIDENTALIS
1 Flowers in terminal branching cymes or panicles; leaf blades cordate-reniform, shallowly
    round-lobed   A. VULGARIS

Alchemilla occidentalis Nutt. in T. & G. Fl. N.Am. 1:432. 1840.

Alchemilla arvensis var. *occidentalis* Piper in Piper & Beattie, Fl. Palouse Region 96. 1901. *Aphanes occidentalis* Rydb. N.Am. Fl. 22[4]:380. 1908. *(Nuttall,* "Rocky plains of the Oregon, towards the sea")

Alchemilla cuneifolia Nutt. in T. & G. Fl. N.Am. 1:432. 1840. *Aphanes cuneifolia* Rydb. N.Am. Fl. 22[4]:380. 1908. *(Nuttall,* "Dry plains, St. Barbara, California")

Alchemilla microcarpa Boiss. & Reut. Diagn. Hisp. 1:11. 1842. *Aphanes microcarpa* Rothm. in Fedde Rep. Sp. Nov. 42:172. 1937. *(Reuter,* near San Rafael in the Sierra de Guadarrama, Spain)

Aphanes macrosepala Rydb. N.Am. Fl. 22[4]:380. 1908. *(E. W. Hammond 116,* Wimer, Oreg., in 1892)

Low, spreading, more or less villous-hirsute annual, the freely branched stems mostly 5-10 (up to 20) cm. long; leaves short-petiolate, with connate-sheathing, deeply toothed stipules and cuneate-obovate or flabelliform, more or less biternately lobed blades usually 4-8 mm. long; flowers borne along most of the length of the stems, commonly 5-15 at each node in the axils of the sheathing stipules (opposite the petiole); pedicels from shorter to longer than the stipules; calyx 1-1.5 mm. long, the hypanthium narrowly ellipsoid-campanulate, hirsute-villous, 3-4 times as long as the 4 (5) triangular-ovate, erect or only slightly spreading lobes; stamens usually 1 (2), opposite a sepal, extrorse; achene glabrous, ellipsoid.

Open (often waste) fields to wooded slopes, definitely weedy; s. B.C. to Calif. and to e. Wash., e. Oreg., and possibly in Ida. and Mont., common in e. U.S.; probably native to Europe. Apr.-May.

Our plant is sometimes said to be a local native species but it seems more likely that it is an introduced weed. It is closely related to *A. arvensis* (L.) Scop., a larger plant with leaf blades mostly 7-12 mm. long, hypanthium (1.5) 2-2.5 mm. long and usually ribbed, and more hairy calyx lobes.

Alchemilla vulgaris L. Sp. Pl. 123. 1753. (Europe)

A. pratensis F. W. Schmidt, Fl. Boëm. 3:88. 1794.

Perennial with extensive thick rhizomes, rather copiously villous-hirsute throughout, 1.5-5 dm. tall; basal leaves short-petiolate, the blades cordate-reniform, mostly 3-8 (10) cm. broad, with mostly (5-7) 9 shallow, rounded, sharply serrate-dentate lobes; stipules up to 15 mm. long, toothed; flowers numerous, clustered in terminal, conspicuously bracteate, branching cymes or panicles; hypanthium 1.5-2 mm. long, conic-campanulate, sparsely pilose, the 4 lobes ovate-deltoid, nearly as long as the hypanthium; stamens 4, alternate with the calyx lobes. 2N=96, 91-191. Lady's mantle.

This Eurasian weedy species is known from Meagher Co., Mont. *(Hitchcock & Muhlick 12081)*, but it probably occurs in other parts of our area. It is common in e. U.S. May-July.

Alchemilla vulgaris is a complex of apomictic races, many of which have been named at both the varietal and specific level; if the segregates were to be recognized, our plants probably would be called *A. pratensis* F. W. Schmidt.

Amelanchier Medic. Serviceberry; Shadbush

Flowers in small terminal racemes at the end of the branches, perfect, regular, usually appearing with, or slightly before, the leaves; calyx more or less campanulate, adnate at base to the ovary but with a free, somewhat flared hypanthium lined internally with a thin to thickish glandular disc, the 5 lobes triangular to lanceolate, persistent; petals 5, white (pink), rather showy, linear to oblanceolate; stamens somewhat variable in number, usually about (10, 15 or) 20, inserted with the petals at the edge of the disc at the top of the hypanthium; filaments slender, broadened and sometimes very slightly connate at base; pistil 2- to 5-carpellary, the ovary compound, 2- to 5-locular, nearly or quite inferior, the styles 2-5, from semidistinct to connate almost up to the small capitate stigmas; fruit a dry to somewhat fleshy, reddish to purplish, baccate pome, usually with 2 exalbuminous seeds per carpel that are separated by false partitions; unarmed shrubs or small trees with alternate, deciduous, simple, serrate to subentire leaves and small, linear, quickly caducous stipules.

About 12 species, mostly of temperate N.Am., also in temperate Eurasia. (Derivation uncertain, but believed to be from the French name for a European species.)

Several of the species are considered worthful ornamentals; the fruits are edible, although seldom used at the present time. The plants are browsed by both domestic and wild animals.

Reference:

Jones, George Neville. American species of Amelanchier. Ill. Biol. Monog. 20:1-126. 1946.

The genus is well known for the degree of intergradation between taxa most of which are here treated as varieties rather than as species.

1 Leaves ultimately glabrous (or sparsely sericeous on the lower surface); petals mostly 10-20 (5-25) mm. long; fruit glabrous; styles usually 5 (4)   A. ALNIFOLIA
1 Leaves usually permanently lanate at least on the lower (and often on the upper) surface; petals 5-10 mm. long; fruit frequently pubescent; styles mostly (2-3) 4, very rarely 5
                                                                                                A. UTAHENSIS

Amelanchier alnifolia Nutt. Journ. Acad. Phila. 7:22. 1834.

Aronia alnifolia Nutt. Gen. Pl. 1:306. 1818. *Pyrus alnifolia* Lindl. Trans. Linn. Soc. 13:98. 1821. *Amelanchier canadensis* var. *alnifolia* T. & G. Fl. N.Am. 1:473. 1840. *(Nuttall,* "In ravines and on the elevated margins of small streams from Fort Mandan to the Northern Andes")

Amelanchier florida Lindl. Bot. Reg. 19:pl. 1589. 1833. *A. alnifolia* var. *florida* Schneider, Ill. Handb. Laubh. 1:739, fig. 411. 1906. (Cultivated specimens, from seed collected by Douglas in "Northwest America") = var. *semiintegrifolia*.

AMELANCHIER ALNIFOLIA var. SEMIINTEGRIFOLIA (Hook.) C. L. Hitchc. hoc loc. *A. ovalis* var. *semiintegrifolia* Hook. Fl. Bor. Am. 1:202. 1833. *A. canadensis* var. *semiintegrifolia* Farw. Rep. Mich. Acad. Sci. 17:174. 1916. *(Douglas,* "Plentiful about the Grand Rapids, and at Fort Vancouver, on the Columbia, and on the high ground of the Multnomak River")

A. parvifolia Hort. ex Loud. Arb. & Frut. Brit. 2:877. 1838. *A. florida* var. *parvifolia* Loud. loc. cit. (Cultivated specimen from the Garden of the Royal Horticultural Society) = var. *semiintegrifolia.*

A. canadensis var. *pumila* Nutt. in T. & G. Fl. N. Am. 1:474. 1840. *A. pumila* M. Roem. Syn. Mon. 3:145. 1847. *A. alnifolia* var. *pumila* Nels. in Coult. & Nels. New Man. Bot. Rocky Mts. 266. 1909. *(Nuttall,* "near the sources of the Platte in the Rocky Mountains")

A. glabra Greene, Fl. Fran. 52. 1891. *(Rev. Dr. Bonté,* Donner Lake region, Calif., June, 1888) = var. *pumila.*

AMELANCHIER ALNIFOLIA var. CUSICKII (Fern.) C. L. Hitchc. hoc loc. *A. cusickii* Fern. Erythea 7:121. 1899. *(Cusick 1858,* Union Co., Oreg., May 5 and June, 1898)

A. cuneata Piper, Bull. Torrey Club 27:392. 1900. *(Piper 2173,* Ellensburg, Wash., May 20, 1897) = var. *pumila.*

A. polycarpa Greene, Pitt. 4:127. 1900. *(C. F. Baker,* Piedra, Colo., July, 1899) = var. *pumila.*

A. gormani Greene, Pitt. 4:129. 1900. *(Gorman,* Yes Bay, Alas., June 16 and Sept. 6, 1895) = var. *semiintegrifolia.*

A. basalticola Piper in Piper & Beattie, Fl. Palouse Reg. 100. 1901. *(Prof. Byron Hunter,* Bluffs of Snake River, Whitman Co., Wash., near Lewiston, Ida.) = var. *pumila.*

A. oxyodon Koehne, Gartenfl. 51:609, fig. 126b. 1902. *(Purpus 104,* near Yale, B. C., is the first specimen cited) = var. *semiintegrifolia.*

A. ephemerotricha Suksd. Werdenda 1:20. 1927. *(Suksdorf 10026,* Falcon Valley, Klickitat Co., Wash., May 23, June 20, and Sept. 20, 1918) = var. *semiintegrifolia.*

A. ephemerotricha var. *silvicola* Suksd. Werdenda 1:20. 1927. *(Suksdorf 10494,* moist woods, near Bingen, Klickitat Co., Wash., June 19 and Aug. 18, 1920) = var. *semiintegrifolia.*

A. vestita Suksd. Werdenda 1:22. 1927. *(Suksdorf 10025,* Falcon Valley, Klickitat Co., Wash., May 23, June 20, and Sept. 20, 1918) = var. *semiintegrifolia.*

AMELANCHIER ALNIFOLIA var. HUMPTULIPENSIS (G. N. Jones) C. L. Hitchc. hoc loc. *A. florida* var. *humptulipensis* G. N. Jones, U. Wash. Pub. Biol. 5:181. 1936. *(G. N. Jones 4565* and *5819,* Humptulips Prairie, Grays Harbor Co., Wash.)

A. florida f. *tomentosa* Sealy in Curtis' Bot. Mag. 160:pl. 9496, figs. a-e. 1937. ("North-west North America, from Alaska to California," described from cultivated plants raised in Scotland, the source of the seeds unknown) = var. *semiintegrifolia.*

Low and spreading to erect shrubs or sometimes small trees, mostly 1-5 (0. 5-10) m. tall; young branches reddish-brown, glabrous or more commonly sparsely to thickly sericeous or grayish-tomentose, usually rather quickly glabrate, eventually grayish barked; leaves with slender petioles (5) 10-20 (25) mm. long, the blades thin to coriaceous, often glaucous, from oval to oblong or elliptic-oblong, (1.5) 2-4 (5) cm. long, from 1/2-5/6 as broad, cuneate to rounded or subcordate at the base, rounded to semitruncate at the tip, from (very occasionally) glabrous to sparsely or copiously sericeous or grayish-pubescent at least on the lower surface, nearly entire to (usually) sharply serrate from almost the full length to only across the tip; flowers 3-20 in short racemes, the lowest ones from the axils of scarcely reduced leaves, the upper ones from linear, sericeous or woolly (glabrous), quickly deciduous bracts; pedicels slender, ascending, mostly 5-10 (15) mm. long; calyx from glabrous within as well as outside to sparsely or densely floccose or lanate, especially on the inner surface of the lobes; hypanthium 1-2 mm. long, the lobes acute to acuminate, spreading to recurved, (1) 1.5-3.5 (5) mm. long; petals white (occasionally pinkish), linear to rather broadly linear-oblanceolate, (5) 10-20 (25) mm. long, 2-6 (8.5) mm. broad; stamens 12-15 or more commonly about 20, the filaments (1.5) 2-3 (4) mm. long, the anthers 0.5-1 (1.5) mm. long; styles usually 5, very occasionally 4, from connate nearly full length to distinct almost to the base, equaling or shorter than the stamens; ovary almost completely inferior, the rounded top (inside the hypanthium) glabrous to densely grayish-tomentose; fruit usually glabrous and more or less glaucous, nearly globose, (6) 10-14 mm. long, generally dark purplish, rather juicy and palatable.

Open woods, canyons, and hillsides, from near sea level to subalpine; s. Alas. southward to Calif., e. to Alta., the Dakotas, Neb., Colo., N.M., and Ariz. Apr.-July.

This complex has usually been separated into from 2 to several species by other workers but there is at least as much reason to treat it as a single polymorphic species, since the segregate taxa are rather completely intergradient and therefore not at all clearly delimitable.

The morphological features that show particular variation, and which have been believed to be more or less important are both floral and vegetative. Although pubescence has been considered of diagnostic significance, the quantity and, to some extent, the type vary greatly. In particular, its presence on certain parts of the flower, such as the calyx lobes, top of the ovary, etc., is usually considered to be of genetic import. The relative length of the calyx lobes, petals, styles, and anthers, the number of styles, and the shape and texture of the leaves have all been emphasized in delimiting taxa. Unfortunately, most of these features tend to vary independently rather than in combination, so that the number of taxa recognizable depends largely upon what particular characters are selected for emphasis. The following variants are the more outstanding, but their distinctness is unduly if not falsely emphasized by the mechanics of a key:

1 Top of the ovary (inside the hypanthium) glabrous or at most with only a few scattered hairs; petals rarely over 15 mm. long; leaves rather thick and somewhat coriaceous, sparsely pubescent to glabrous; s.e. Wash. to w. Mont., s. to s.e. Oreg., n.e. Calif., Utah, and Colo., freely transitional to vars. *cusickii* and *alnifolia;* closely approached by small-flowered plants of var. *cusickii* characteristically found in the foothills on the e. side of the Cascades from B.C. southward; ordinarily two taxa are recognized here, *A. pumila* (entirely glabrous) and *A. basalticola* (nearly glabrous but with slightly pubescent sepals), although the distinction is not genuine as plants cannot be separated geographically or ecologically on such a basis   var. pumila (Nutt.) A. Nels.

1 Top of the ovary usually rather strongly hairy or (if the ovary glabrous or subglabrous on top) the petals well over 15 mm. long; leaves commonly thin, often copiously hairy

  2 Petals usually less than 12 mm. long; top of the ovary mostly strongly pubescent

    3 Flowers in part generally with only 4 styles; leaves elliptic-oblong, usually 2-3 cm. long and only slightly more than half as broad, subentire or with a few tiny teeth well above midlength; always w. of the Cascades where occasional from s. B.C. to s.w. Wash.; on the whole one of the most clearly marked of the several vars.   var. humptulipensis (G. N. Jones) C. L. Hitchc.

    3 Flowers rarely with other than 5 styles; leaves various but usually strongly toothed for most of the upper half; s. Alas. and B.C. s. to s. Oreg., with few exceptions entirely e. of the Cascades, becoming more common e. to Alta., the Dakotas, Utah, Colo., and Neb.; morphologically not separable from occasional specimens from w. of the Cascades (e.g. *Piper,* May 2, 1888, Seattle, Wash.)
                                            var. alnifolia

  2 Petals usually well over 12 mm. long; top of the ovary often only weakly hairy or even glabrous

    4 Top of the ovary concealed by the copious, more or less tomentose pubescence; petals generally less than 16 mm. long and less than 4 mm. broad; calyx lobes averaging less than 3 (but up to 3.5) mm. long; sometimes a small tree with an erect trunk (unlike all the more e. varieties in this respect); the common phase in and w. of the Cascade Mts. from s. Alas. to n. Calif., but abundant e. through s. B.C. and n. Wash. to Mont., occasional in Oreg. and c. Ida., although the material from e. of the Cascades is often arbitrarily referred to other taxa
                            var. semiintegrifolia (Hook.) C. L. Hitchc.

    4 Top of the ovary from glabrous to rather copiously hairy, but usually not grayish-tomentose; petals mostly over 16 (to 25) mm. long and up to 8.5 mm. broad; calyx lobes averaging at least 3 (to 5) mm. long; the largest flowered of the varieties, rather clearly marked but completely transitional to other vars., especially to var. *alnifolia* in the s.e. Wash.-w. Ida. region, to var. *semiintegrifolia* in the Cascades, and apparently even sometimes to var. *pumila;* most common in s.e. Wash., n.e. Oreg., and adj. w. Ida., but large-flowered plants from the e. slopes of the Cascades in Wash. and Oreg., in s. B.C., and even occasionally in Mont., w. Wyo., and n. Utah, are usually referred here, even though the transition to var. *alnifolia* is complete in the Rocky Mts.   var. cusickii (Fern.) C. L. Hitchc.

Amelanchier utahensis Koehne, Wissen. Beil. Progr. Falk-Realgym. Berl. 95:25, pl. 2, fig. 20c. 1890.

  A. alnifolia var. *utahensis* M. E. Jones, Proc. Calif. Acad. Sci. 5:679. 1895. *(M. E. Jones 1716,*

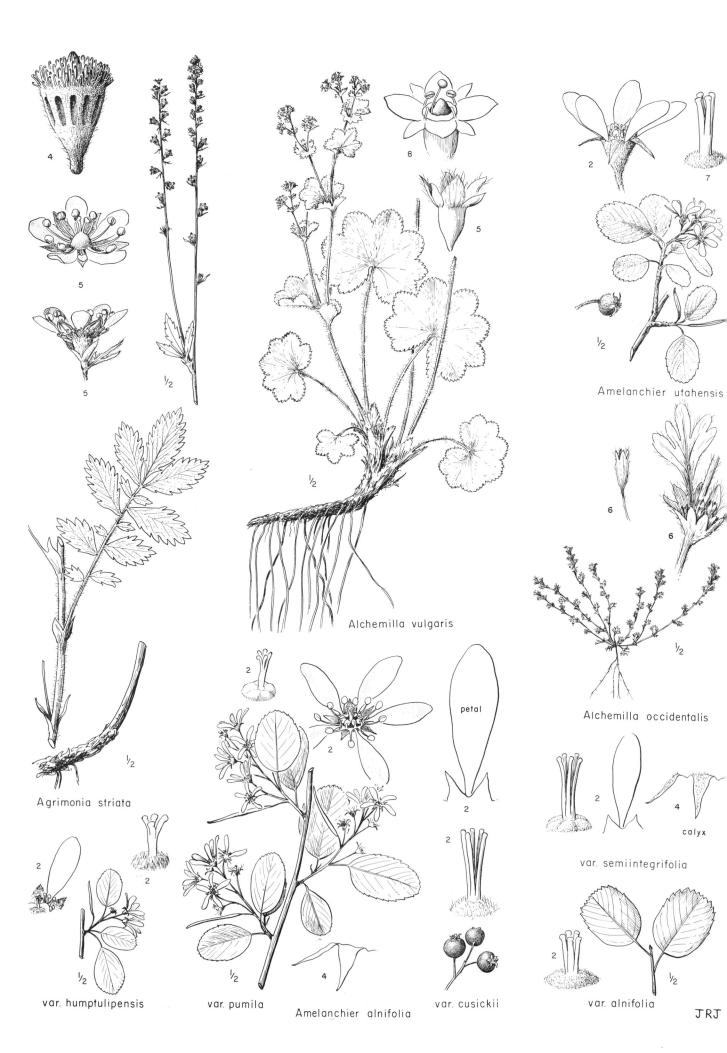

4

5

5

1/2

Agrimonia striata

2

2

1/2

var. humptulipensis

2

1/2

var. pumila

2

2

5

8

petal

2

2

4

Amelanchier alnifolia

var. cusickii

Alchemilla vulgaris

2

7

1/2

Amelanchier utahensis

6

6

1/2

Alchemilla occidentalis

2

2

4

calyx

var. semiintegrifolia

2

1/2

var. alnifolia

J R J

"at Leeds, S. Utah," in 1880, acc. Jones)

A. oreophila A. Nels. Bot. Gaz. 40:65. 1905. *A. utahensis* var. *oreophila* Clokey, Madroño 8:57.
    1945. *A. alnifolia* var. *oreophila* Davis, Madroño 11:144. 1951. *(L. N. Goodding 1456,* Camp Cr.,
    Routt Co., Colo., is the first of several collections cited)

Much like *A. alnifolia,* but usually low and irregularly branched, 0.5-3 (5) m. tall; young branches
tomentose or sericeous, glabrate after 1-2 years; leaf blades yellowish-green, rather coriaceous,
usually permanently cinereous or tomentose on both surfaces, oval or oblong-elliptic to broadly obovate, (1) 1.5-3 (3.5) cm. long, serrate or crenate-serrate (sometimes to below midlength); flowers
3-6 (10) in short, permanently sericeous or tomentose racemes; calyx tomentose to glabrous, the
lobes narrowly lanceolate, 2.5-3.5 mm. long; petals white or pinkish-blushed in the bud, cuneate-oblanceolate, (5) 6-9 (10) mm. long, up to 4 mm. broad; stamens mostly about 15 (10-17); styles 4 (2,
3, or even sometimes 5), usually connate for less than half their length; top of the ovary densely tomentose to (rarely) glabrous; fruit subglobose-pyriform, up to 10 mm. long, dark purple and fleshy
or dry and more reddish.

Rimrock, valleys, gullies, and hillsides, from sagebrush desert to middle elevations in the mountains; c. Ida. southward to Baja Calif. and Sonora, w. to s.e. Oreg., Nev., and s.e. Calif., e. to
s. Mont., Wyo., Colo., N.M., and w. Tex., occasional in c. Oreg. (Wheeler Co.) and known from
Yakima Co., Wash., on the basis of *Cotton 365* which was referred to *A. alnifolia* by Jones. May-
June.

Aruncus L. Goatsbeard

Flowers rather small, imperfect, borne in large terminal panicles of numerous spikelike, racemose branches; calyx broadly and shallowly campanulate to saucer-shaped, the hypanthium free of the
ovaries, lined internally with a prominent glandular disc, about equal to the 5 more or less triangular
lobes; petals 5, white, deciduous, borne with the rather numerous stamens (or the staminal vestiges
in the pistillate flowers) at the top of the hypanthium; pistils 3 (4-5), distinct, the ovary superior, the
style terminal but oblique and divergent, the stigma capitate; fruit a small, few-seeded follicle, dehiscent on the ventral suture; seeds (1) 2-4, erect, attenuate at each end, the seed coat loose, cel-
lular-alveolate; endosperm scanty; tall, dioecious, rhizomatous, perennial herbs with alternate, exstipulate, ternate-pinnatisect to triternate-pinnatisect leaves.

Probably only 2 or 3 species. (From the Greek *aryngos,* meaning goat's beard, in reference to the
white inflorescence.)

Reference:
Fernald, M. L. Memoranda on Aruncus. Rhodora 38:179-182, 237. 1936.

Aruncus sylvester Kostel. in Ind. Hort. Prag. 15. 1844.
    Spiraea aruncus L. Sp. Pl. 490. 1753. *Ulmaria aruncus* Hill, Hort. Kew. 214. 1768. *Astilbe aruncus*
        Trev. Bot. Zeit. 13:819. 1855. *Aruncus aruncus* Karst. Deutsch. Fl. 779. 1882. ("Habitat in Aus-
        triae, Alvorniae montanis")
    Spiraea acuminata Dougl. ex Hook. Fl. Bor. Am. 1:173. 1833, in synonymy. *Aruncus acuminatus*
        Rydb. N.Am. Fl. 22³:255. 1908. *(Drummond,* "near the source of the Columbia," is the first of 3
        collections cited)

Stems several, 1-2 (2.5) m. tall, usually glabrous; leaves numerous, mostly cauline, long-petiolate, reduced upward, from imperfectly triternate-pinnatisect below to ternate-pinnatisect above, the
leaflets mostly ovate to oblong-lanceolate and usually acuminate, as much as 15 cm. long and 8 cm.
broad, sharply twice-serrate, dark green and usually glabrous on the upper surface, hairy and paler
beneath; panicles 1-5 dm. long; calyx 1.5-2 mm. broad, the lobes spreading, about equaling the hypanthium; petals white, about 1 mm. long or slightly smaller in the pistillate flowers; staminate flowers with 15-20 stamens and tiny vestiges of pistils; pistillate flowers with vestigial stamens; follicles
about 3 mm. long, erect except for the divergent, persistent, 0.5 mm. style; seeds 2-2.5 mm. long.

Moist woods, especially along streams; Alas. southward to n.w. Calif., from the Cascades to the
coast in Wash. and much of Oreg., but e. in B.C. to the Selkirk Mts.; Eurasia, Japan, Kamchatka.
Late May-early July.

Although our plants have often been treated as a distinct species under the name *A. acuminatus*
Rydb., the basis for their separation from the Eurasian material is obscure.

The species is attractive because of the fernlike foliage and filmy sprays of flowers. It is easily established by seedage, but transplanting from the wild is to be recommended, as staminate plants have

two advantages over the pistillate; they produce larger flowers and do not become pesty from prolific seed dispersal. If pistillate plants are grown, the flower clusters should be cut before seed ripens.

Cercocarpus H. B. K. Mountain Mahogany

Flowers perfect, apetalous, sessile, 1-several in axillary clusters terminal on short lateral branches; calyx free of the ovary, the hypanthium trumpetlike, the tube persistent around (but free of) the developing ovary, the limb bearing the numerous (at least 15) stamens at several levels, deciduous as the fruits mature, the lobes 5, deltoid to lanceolate-acuminate; filaments slender; anthers oval to oblong, deeply emarginate at both tips, attached well above the base; pistil single, 1-carpellary; ovary superior; style terminal, exserted, stigmatic at the tip; fruit a terete, hardened, pilose-villous achene, usually included in the persistent hypanthium tube but the style greatly elongate and plumose; shrubs or small trees with alternate, persistent, simple, entire to toothed leaves and small stipules adnate to the petiole.

Possibly 10 species of w. U.S. and Mex. (From the Greek *kerkos*, tail, and *carpos*, fruit, because of the long, persistent styles.)

Reference:

Martin, F. L. A revision of Cercocarpus. Brittonia 7:91-111. 1950.

Both species, but especially *C. ledifolius*, make attractive shrubs or small trees, adaptable even to the Puget Sound area, in spite of their xeric native habitat. They are completely hardy, and are to be recommended for wider use e. of the Cascades.

1 Leaves entire, the margins usually strongly revolute; anthers glabrous   C. LEDIFOLIUS
1 Leaves toothed, slightly if at all revolute; anthers pubescent   C. MONTANUS

Cercocarpus ledifolius Nutt. in T. & G. Fl. N. Am. 1:427. 1840. *(Nuttall,* "Rocky Mountains, in alpine situations on the summits of the hills of Bear River of Timpanagos; near the celebrated 'Beer Springs'")

    C. ledifolius var. *intercedens* C. K. Schneider, Mitt. Deuts. Dendrol. Ges. 1905:128. 1905. *C. ledifolius* var. *intercedens* f. *subglaber* C. K. Schneider, loc. cit. *(M. E. Jones 5615,* Provo Slate Canyon, Utah, in 1894)

    C. ledifolius var. *intercedens* f. *hirsutus* C. K. Schneider, Mitt. Deuts. Dendrol. Ges. 1905:128. 1905. *(Pammel & Blackwood 3726,* Ogden, Utah, is the first of 2 specimens cited) = var. *intercedens.*

    C. hypoleucus Rydb. N. Am. Fl. 22⁵:424. 1913. *C. ledifolius* var. *hypoleucus* Peck, Man. High. Pl. Oreg. 407. 1941. *(Rydberg 2695,* Melrose, Mont., July 7, 1895) = var. *intercedens.*

Intricately branched shrubs or small trees 1-6 (10) m. tall, the younger branches pubescent and reddish but soon glabrous and eventually grayish- or somewhat reddish-barked; leaves persistent, the blades narrowly elliptic to elliptic-lanceolate, but often with the margins strongly revolute and then more nearly linear in outline, 1-3 (3.5) cm. long, up to 10 mm. broad, acute at both ends, dark green and sparsely pubescent or somewhat lanate (but often glabrate and shining) on the upper surface, paler and sparsely to densely grayish-lanate on the lower surface, more or less resinous; petioles (1) 2-5 mm. long; flowers 1-3 in the leaf axils, sessile, bracteate; calyx lanate, the tube of the hypanthium 4-7 (9) mm. long, the limb broader than long, the lobes ovate, 1.5-2 mm. long, recurved; stamens 20-30, the anthers glabrous; achene 5-7 mm. long, the stylar tail 5-8 cm. long.

Desert foothills to mountain slopes, usually in rocky soil; s.e. Wash. to the Rocky Mts. of Mont., s. to s.w. Oreg., and to n. Sierran and s. Calif., n. Ariz., and w. Colo. Apr.-June.

Our material is readily separable into 2 geographic races as follows:

1 Leaves with strongly revolute margins, linear or linear-lanceolate and mostly less than 3 mm. broad; s.e. Wash. and adj. Oreg. across montane c. Ida. to Mont. and n. Wyo., occasional s. to Colo. and Ariz.   var. intercedens C. K. Schneider
1 Leaves with only slightly revolute margins, narrowly elliptic to lanceolate-elliptic, 4-8 (10) mm. broad; with var. *intercedens* in s.e. Wash. but only occasional eastward; the common phase s. through Nev. and Calif.   var. ledifolius

Cercocarpus montanus Raf. Atl. Journ. 146. 1832. *(Dr. James,* Rocky Mountains in the summer of 1820)

    C. betuloides Nutt. in T. & G. Fl. N. Am. 1:427. (June) 1840. *C. betulaefolius* Nutt. ex Hook. Ic. Pl. pl. 322. (Oct.) 1840. *C. parvifolius* var. *glaber* Wats. Bot. Calif. 1:175. 1876. *C. parvifolius* var. *betuloides* Sarg. Silva 4:66. 1892. *C. montanus* var. *glaber* Martin, Brittonia 7:101. 1950. *(Nuttall,* Santa Barbara, Calif.)

Shrub or small tree to 7 m. tall, with smooth, grayish to brown bark; leaves short-petiolate, the blades lanceolate to deltoid-rotund, mostly 1-3 (4) cm. long and (0.5) 1-2.5 cm. broad, shallowly crenate-serrate, dark green and glabrous above, pale and often sparsely hairy beneath, the lateral veins prominent; flowers 1-3 in the axils of short spur shoots; calyx short-lanate or sericeous to sub-glabrous, the tube 3-6 mm. long at anthesis (lengthening to 8-14 mm. in fruit), the limb shallow, about 6 mm. broad, the lobes recurved; stamens 25-40, the anthers hairy; achenes up to 12 mm. long, the stylar tail mostly 3-6 cm. long. N=9.

Desert foothills and mountains to coastal chaparral and ponderosa pine forest; s.w. Oreg. to Baja Calif., and in the Rocky Mts. and Great Basin from Wyo. through Utah and Colo. to c. Mex. Feb.-June.

Our material, as described above, is representative of the var. *glaber* (Wats.) Martin, which ranges from Baja Calif., through Ariz. and much of Calif., to s.w. Oreg., and northward (much less commonly) as far as Lane and Deschutes cos.

Chamaebatiaria (Porter) Maxim. Fern Bush; Desert-sweet

Flowers numerous in somewhat leafy-bracteate panicles terminating the branches, fairly showy, complete, perigynous; calyx turbinate-campanulate, free of the pistils, the 5 lobes imbricate in bud, erect in flower, the hypanthium lined internally with a thin, glandular, lobed disc; petals 5, white, deciduous, borne with the rather numerous (about 50) stamens at the edge of the disc; carpels usually 5 (4-6), the ovaries apparently more or less connate in flower but distinct in fruit, superior; styles terminal, distinct; stigmas capitate; fruit of 5 (4-6) distinct, rather coriaceous follicles, fully dehiscent on the ventral suture and very slightly on the dorsal suture; seeds several, erect, the embryo straight; low, glandular and stellate-pubescent shrubs with alternate, deciduous, finely pinnate-pinnatifid, stipulate leaves.

Only the one species. (Resembling *Chamaebatia.*)

Chamaebatiaria millefolium (Torr.) Maxim. Acta Hort. Petrop. 6:225. 1879.

Spiraea millefolium Torr. Pac. R. R. Rep. 4:83. 1857. *Sorbaria millefolium* Focke in E. & P. Nat. Pflanzenf. 3³:16. 1888. *Basilima millefolium* Greene, Fl. Fran. 57. 1891. *(Bigelow,* "near William's mountain," Ariz., Jan. 5, 1854)

C. glutinosa Rydb. N. Am. Fl. 22³:258. 1908. *Spiraea glutinosa* Fedde, Just Bot. Jahresb. 36²:489. 1910. *(M. B. Howard,* Mammoth Range, Nye Co., Nev., in 1868)

Spreading, aromatic, glandular, stellate shrub mostly 1-2 m. tall; stipules linear-spatulate, usually persistent; leaf blades oblong-lanceolate, 2.5-4 (5) cm. long, finely divided into 16-30 narrow pinnae and secondarily pinnatifid into ultimate tiny lobes 0.5-1 (1.5) mm. long and about 0.5 mm. broad; panicles (5-10) 20 cm. long, leafy-bracteate; calyx 4-6 mm. long, the lobes about equaling the hypanthium; petals slightly longer than the sepals, broadly obovate-cuneate; stamens and styles about equaling (or shorter than) the sepals; follicles 5-6 mm. long, finely pubescent; seeds 3-7, linear, somewhat flattened at each end, terete in the embryo-bearing central portion, 3-4 mm. long.

Desert and semidesert canyons and mountain sides, especially in lava beds; s. Deschutes Co., Oreg., southward, mostly on the e. side of the Sierra Nevada, to s.e. Calif., Nev., and Ariz., e. across the Snake R. plains of Ida. to Utah and possibly to w. Wyo. June-Aug.

An aromatic shrub with beautiful foliage, and not unattractive flowers. Its horticultural potentiality for the area w. of the Cascades has not been explored fully, but it is well worth a trial, especially on well-drained, sunny banks.

Chamaerhodos Bunge

Flowers inconspicuous, complete, perigynous, borne in dichotomous, diffusely branching to rather congested, bracteate cymes; calyx free of the pistils, turbinate, the 5 lobes erect to convergent, about equaling the hypanthium; petals 5, deciduous; stamens 5, borne with (and opposite) the petals at the top of the hypanthium, the anthers opening by a continuous ringlike slit; pistils 5-10 (20), 1-carpellary, inserted on a short axis above the base of the hypanthium; styles slender, basally attached to the superior, 1-ovulate ovaries; stigmas capitate; fruit a membranous, ovoid-pyriform achene; seed ascending, the embryo straight; small, biennial or perennial, glandular-pubescent herbs with bi- to triternately divided leaves.

About five species, all except ours strictly Siberian. (Greek *chamai*, on the ground, and *rhodon*, rose, that is ground rose.)

Chamaerhodos erecta (L.) Bunge in Ledeb. Fl. Alt. 1:430. 1829.

Sibbaldia erecta L. Sp. Pl. 284. 1753. *(D. Gmelin,* "Habitat in Sibira")

CHAMAERHODOS ERECTA var. PARVIFLORA (Nutt.) C. L. Hitchc. hoc loc. *Sibbaldia erecta* var. *parviflora* Nutt. Gen. Pl. 1:207. 1818. *C. erecta* var. *nuttallii* Pickering ex T. & G. Fl. N. Am. 1:433. 1840. *C. nuttallii* Rydb. N. Am. Fl. 22⁴:377. 1908. *C. erecta* ssp. *nuttallii* Hultén, Fl. Alas. 6:1035. 1946. *(Nuttall,* "Missouri, near the Mandan villages")

Strongly glandular-pubescent and hirsute, short-lived perennial (biennial) from a strong taproot, often strongly reddish- or purplish-tinged; stem mostly one, usually freely branched, at least above, 1-2.5 (3) dm. tall; leaves numerous, the basal ones rosulate, marcescent, 1.5-3 (4) cm. long, slender-petiolate, the blades 2-3 times ternately dissected into linear segments; flowers many in a large, freely (and in part dichotomously) branched, more or less flat-topped, paniculate cyme; calyx 4-5 mm. long, the hypanthium hirsute-hispid at least at the sinuses of the lobes, slightly constricted about midlength, hirsute within; petals white (purplish-tinged), obovate-cuneate, about equaling the sepals; filaments subulate; styles about equaling the stamens; achenes 1.5 mm. long, brownish, pyriform-flattened.

Plains and arid hills; Yukon, Lake Athabasca, and s.w. Alta. to Sask. and Man., southward, almost entirely e. of the continental divide, to Colo., in our area in s.w. Mont.; Asia. Late June-July. Usually in more or less denuded areas.

The N.Am. material, as described above, is referable to the var. *parviflora* (Nutt.) C. L. Hitchc., differing from the Asiatic chiefly in having somewhat smaller flowers.

Crataegus L. Haw; Hawthorn; Thorn Apple

Flowers in axillary and terminal caducous-bracteate corymbs, rather showy, complete, epigynous; calyx with a short, disc lined, saucer shaped to campanulate, free hypanthium above the ovary, the 5 lobes persistent or deciduous; petals 5, deciduous, white (pink); stamens (5) 10-25 (or more), the filaments slender; pistil (1) 2- to 5-carpellary, the ovary partially to completely inferior; styles (1) 2-5, distinct; stigmas capitate; fruit yellow, red, purple, or black, pomaceous, with the (1) 2-5 bony, 1-seeded pyrenes or nutlets embedded in considerable flesh; shrubs or small trees, usually well armed with strong thorns (spinose branches), and with alternate, deciduous, petiolate, toothed to lobed (ours) or pinnatifid leaves and small deciduous stipules.

Perhaps 300 species of the Northern Hemisphere, most abundant in N.Am. e. of the Rocky Mts., many in Europe and adj. Asia and Africa. (From the Greek *kratos*, strength, the wood noted for its great strength.)

In c. and e. U.S. this is probably the most "difficult" genus encountered, with hundreds of taxa described but few easily recognizable, apparently largely because of the freedom with which nearly all intercross.

Our species are of rather slight horticultural value, although *C. columbiana* has very attractive fruit and might well rate a place in the larger garden. Several non-native species are of much greater merit and frequently escape from cultivation to be mistaken for native elements of the vegetation. Two of these are included in the following key:

1 Leaves deeply 3- to 7-lobed for usually over half the width; styles 1-2 (3); fruits red or yellow, usually containing only 1 or 2 "stones"; ornamental shrubs or small trees escaped from cultivation and often animal-disseminated; found chiefly w. of the Cascades
  2 Style 1; fruit with only 1 stone; leaves 3- to 7-lobed   *C. monogyna* Jacq.
  2 Styles 2 (3); fruit usually with 2 stones; leaves 3- to 5-lobed   *C. oxyacantha* L.
1 Leaves subentire to shallowly lobed (less than 1/2 the width); styles 2-5; fruit usually with more than 2 stones; native species
  3 Styles usually 5; thorns mostly 1-2 (rarely to 3) cm. long; ovary rarely at all tomentose; fruit black   C. DOUGLASII
  3 Styles usually 2-4 (5); thorns (2) 4-7 cm. long; ovary often copiously hairy to tomentose; fruit dark red   C. COLUMBIANA

Crataegus columbiana Howell, Fl. N.W. Am. 163. 1898. ("Columbia river and its tributaries east of the Cascade Mountains," no collection cited)

 C. *piperi* Britt. Torreya 1:55. 1901. *C. columbiana* var. *piperi* Eggleston, Rhodora 10:79. 1908. (*Piper 1535*, Pullman, Wash.)

 C. *williamsii* Eggleston, Bull. Torrey Club 36:641. 1909. (*R. S. Williams*, Columbia Falls, Mont., Sept. 14, 1892) = var. *columbiana*.

 Reddish-barked, straggling shrubs or small trees 1-3 (5) m. tall, armed with slender, straight to recurved thorns (2) 4-7 cm. long; leaves short-petiolate, the blades broadly ovate to oblong or obovate, 2.5-6 (7) cm. long, once or twice serrate and usually very slightly lobed, generally permanently pubescent on both surfaces; ovary and hypanthium crisp-pubescent to grayish-lanate, the calyx lobes narrowly lanceolate, 3-5 mm. long, strongly glandular-denticulate, pubescent on both surfaces; petals white, 5-8 mm. long; stamens 10; styles 2-4 (5); fruit pubescent or lanate to glabrous, dark red, ovoid, about 1 cm. long.

 Meadows and stream courses to dry hillsides; e. side of the Cascade Mts. from s. B.C. e. to Alta. and s. to n. Oreg., Ida., and Mont. Apr.-June.

 Plants of s.e Wash. and adj. Ida and n.e. Oreg. appear to be varietally distinct from the rest of the species, as follows:

1 Inflorescence hairy to tardily glabrous; calyx and ovary (glabrous) pubescent but not lanate; fruits usually glabrous; widespread but lacking or scarce within the range of var. *piperi*
                                                                                         var. columbiana
1 Inflorescence grayish-hairy; calyx and ovary lanate; fruit more or less lanate; Whitman Co., Wash., to immediately adj. Ida. and n.e. Oreg.   var. piperi (Britt.) Eggleston

Crataegus douglasii Lindl. Bot. Reg. 21:pl. 1810. 1835.

 C. *punctata* var. *brevispina* Dougl. in Hook. Fl. Bor. Am. 1:201. 1833. *C. sanguinea* var. *douglasii* T. & G. Fl. N.Am. 1:464. 1840. *Anthomeles douglasii* Roem. Syn. Rosifl. 3:140. 1847. *Mespilus douglasii* Aschers. & Graebn. Syn. Mitteleur. Fl. 6^2:24. 1906. *C. brevispina* Heller, Cat. N.Am. Pl. 2nd ed. 98. 1900. (*Douglas, Scouler*, "banks of streams in the North-West Coast of America")

 C. *rivularis* Nutt. in T. & G. Fl. N.Am. 1:464. 1840. *C. douglasii* var. *rivularis* Sarg. Gard. & For. 2:401. 1889. (*Nuttall*, "Oregon, along rivulets in the Rocky Mountains")

 C. *douglasii* var. *suksdorfii* Sarg. Bot. Gaz. 44:65. 1907. (*Suksdorf 4034*, West Klickitat Co., Wash., is the first of 5 *Suksdorf* collections cited)

 C. *douglasii* f. *badia* Sarg. Bot. Gaz. 44:65. 1907. (*Piper 2358*, Union Flat, 6 mi. s. of Pullman, Wash., is the first of 5 *Piper* collections cited)

 Large shrubs or small trees mostly 1-4 (up to 8) m. tall, armed with stout, straight or slightly curved thorns 1-2 (rarely to 3) cm. long; leaf blades (2) 3-6 (9) cm. long, from as broad to (rarely) slightly over half as broad, simply serrate or more commonly biserrate or shallowly lobed and then once or twice serrate, pubescent to glabrate on both surfaces; ovary and hypanthium glabrous to crisp-pubescent, the sepals triangular, reflexed, entire to somewhat glandular-denticulate, 1.5-2.5 mm. long, slightly villous on the upper surface or along the margins toward the tip; petals white, nearly orbicular, (4) 5-7 mm. long, somewhat crenulate; stamens 10-20; styles normally 5; fruit blackish, glabrous, about 1 cm. long. 2N=about 51.

 Coastal bluffs, valleys, meadowland thickets, and in the sagebrush and ponderosa pine areas where usually along watercourses; s. Alas. southward, on both sides of the Cascades, to Calif., e. to Alta., the Dakotas, and n. Wyo., also in n. Mich. and s.w. Ont. May-June.

 As pointed out by others, material from the two sides of the Cascades differs slightly as follows:

1 Flowers mostly with 10 stamens; ovary often slightly hairy; leaves tending to be broader above the middle than below, often weakly lobed; e. of the Cascades from B.C. to Calif., and e. to the limits of the range, occasional w. of the Cascades, at least in the Puget Trough   var. douglasii
1 Flowers mostly with 20 stamens; ovary usually glabrous; leaves more nearly elliptic or oblong than obovate, generally merely serrate or biserrate; B.C. to s. Oreg., entirely w. of the Cascades except inland somewhat in the Fraser R. Valley and Columbia R. Gorge   var. suksdorfii Sarg.

 A third race, var. *rivularis* (Nutt.) Sarg., ranges from Wyo to Colo., Utah, and e. Nev., possibly entering our range in s.e. Ida. It has somewhat longer thorns (2-3 cm. long) and narrower, longer leaf blades (1.5-2 times as long as broad).

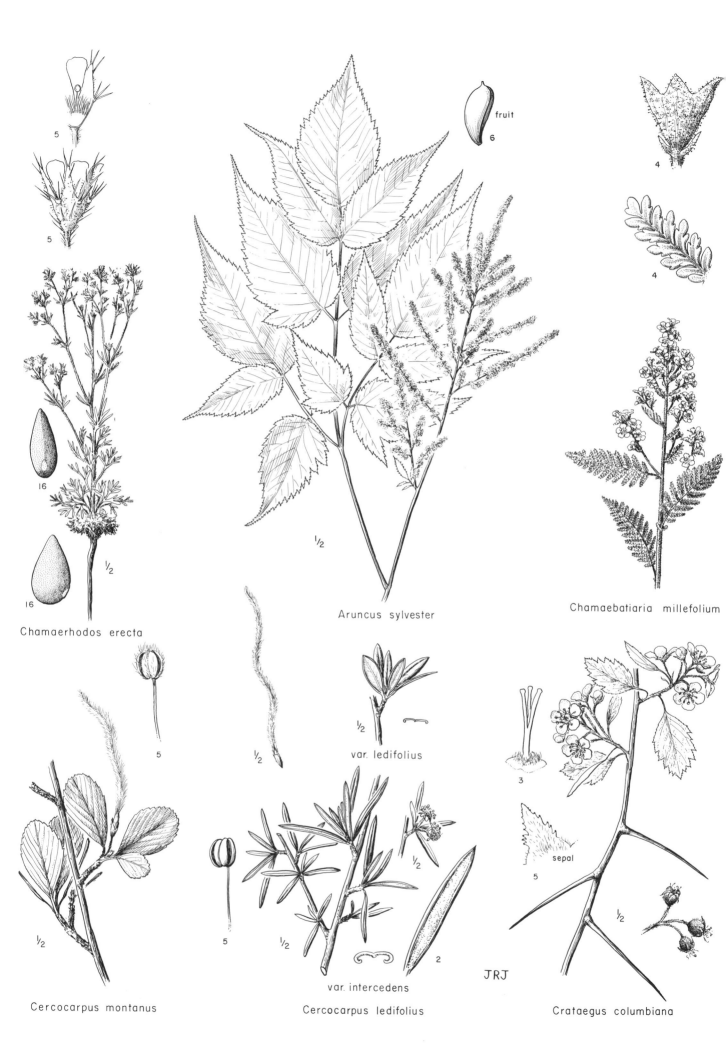

5

5

fruit

6

4

4

16

16

Chamaerhodos erecta

Aruncus sylvester

Chamaebatiaria millefolium

5

½

var. ledifolius

3

sepal

5

½

Cercocarpus montanus

5

½

2

var. intercedens

JRJ

Cercocarpus ledifolius

Crataegus columbiana

Dryas L. Mountain Avens

Flowers solitary on naked or weakly bracteate scapes, perigynous, complete (rarely imperfect), rather showy; calyx usually glandular-stipitate and tomentose, the hypanthium saucerlike, disc lined within, the lobes mostly 8-10, ovate to narrowly oblong, persistent; petals usually 8-10, white to yellow, elliptic to obovate, erect to spreading, deciduous; stamens numerous, borne with the petals at the edge of the disc at the outer edge of the hypanthium, the filaments slender, the anthers dehiscent by lateral slits; pistils numerous, the styles persistent, elongating and plumose in fruit; fruit an achene, tipped with the plumose style; prostrate shrubs with usually freely rooting woody branches, often forming large patches; leaves slender-petiolate, stipulate, evergreen and generally marcescent, the blades simple, crenate-serrate to entire, usually revolute, greenish on the upper surface and commonly white-tomentose on the lower, at least along the midrib.

Perhaps a half dozen species of Eurasia and N. Am., primarily of arctic or arctic-alpine habitats. (Latin *dryas*, a wood nymph.)

Our species are highly valued rock garden plants, propagated by layering or by cuttings, but in our area only *D. drummondii* can be brought in from the wild and cultivated successfully. It is slower growing than *D. octopetala* which was domesticated from European plants and is now very widely grown in the rock garden.

References:

Hultén, Eric. Dryas. Fl. Alas. 6:1043-1050. 1945.

Juzepczuk, S. V. Beitrag zur Systematik der Gattung Dryas L. Bull. Jard. Bot. Princip. U.R.S.S. 28:306-327. 1929.

Porsild, A. E. The genus Dryas in North America. Can. Field Nat. 61:175-192. 1947.

Porsild agreed in most respects with the conclusions of Juzepczuk, that there are 8 or 9 distinct, intersterile species of the genus in N. Am. Hultén, on the other hand, felt that several of these were interfertile races of *D. octopetala,* a belief that seems to be warranted on the basis of herbarium material and field observations.

1 Sepals ovate; petals yellow, ascending rather than spreading; filaments hairy near the base; receptacle flattened; leaf blades cuneate-based; peduncles with 1-4 tiny bracts   D. DRUMMONDII
1 Sepals narrowly oblong-lanceolate; petals white or cream, spreading; filaments glabrous; receptacle convex; leaf blades mostly cordate to rounded at the base; peduncle bractless or with only 1 bract
   2 Leaves entire or at most only crenate on the lower half of the blade, the midrib non-glandular on the lower surface   D. INTEGRIFOLIA
   2 Leaves crenate the entire length of the blade, the midrib often glandular on the lower surface   D. OCTOPETALA

Dryas drummondii Richards. ex Hook. in Curtis' Bot. Mag. 57:pl. 2972. 1830.

Dryas octopetala var. *drummondii* Wats. Bibl. Ind. 281. 1878. *Dryadaea drummondii* Kuntze, Rev. Gen. 1:215. 1891. (Said to have been described partially from cultivated plants and partially from several specimens collected by Richardson and by Drummond in the Rocky Mts. and northward)
Dryas tomentosa Farr, Ott. Nat. 20:110. 1906. *Dryas drummondii* var. *tomentosa* Williams, Ann. Mo. Bot. Gard. 23:452. 1936. *(Farr,* "near the summit of the Pass leading from Emerald Lake into the Yoho Valley, " B. C.)

Leaf blades oblong-elliptic to somewhat obovate, (1) 1.5-3 (3.5) cm. long, up to 2 cm. broad, more or less cuneate at base, coarsely once or twice crenate-serrate, dark green and glabrous to (more commonly) sparsely to moderately tomentose on the upper surface, white-tomentose beneath; scapes up to 20 (25) cm. tall, more or less tomentose, sparingly stipitate-glandular near the tip; calyx strongly stipitate-glandular, the hypanthium strongly villous within, the lobes ovate, 4-6 mm. long; petals pale to rather deep yellow, ascending (rather than spreading) in flower, elliptic to nearly obovate, 8-12 mm. long, the outer ones, and sometimes all, usually stipitate-glandular and often also somewhat tomentose in a median line; filaments hairy, at least near their base; styles often yellow-plumose.

From well up in the mountains, frequently above timber line, in cirques, on rocky ridges and talus slopes, and along the streams, to (sometimes) the foothills on gravel bars along the rivers; Alas. s. to all but s.w. B.C., and in the Rocky Mts. to Mont., in Pend Oreille Co., Wash., and Wallowa Co., Oreg., but not known from Ida., e. to Mack. and along the n. shore of L. Superior and the St. Lawrence R. May-early July.

Dryas integrifolia Vahl, Skriv. Nat. Hist. Selsk. Kjobenh. 4:171. 1798.

D. octopètala var. *integrifolia* Hook. f. Journ. Linn. Soc. 5:83. 1860. (Greenland)

D. tenella Pursh, Fl. Am. Sept. 350. 1814. *(Peck,* "on the white hills of New Hampshire")

Leaf blades 1-2 (2.5) cm. long, usually 1/3-1/5 as broad, narrowly oblong to somewhat lanceolate, generally broadest below the middle, entire or very slightly coarse-crenate on the lower half, rounded to cordate at base, the margins revolute, the upper surface dark green, shiny, and usually glabrous, the lower surface white-tomentose; scapes mostly less than 10 (15) cm. tall, tomentose and usually somewhat stipitate-glandular; calyx strongly stipitate-glandular, the hypanthium glabrous within, the lobes narrowly oblong-lanceolate or narrowly oblong, 4-6 mm. long; petals white, spreading, about 1 cm. long, elliptic; filaments glabrous. N=9.

Alpine gravel bars, talus, and arctic tundra, high montane in our area; Alas. to Greenl., s. in the Canadian Rockies to near the international border, and across much of Can. to N. H. June-July.

Although Porsild reports the species for Mont., we have seen no material from within our range.

Dryas octopetala L. Sp. Pl. 701. 1753.

Dryadaea octopetala Kuntze, Rev. Gen. 1:215. 1891. (Lapland)

Dryas hookeriana Juz. Bull. Jard. Bot. Princip. U.R.S.S. 28: 325. 1929. *Dryas octopetala* ssp. *hookeriana* Hultén, Fl. Alas. 6:1046. 1945. *Dryas octopetala* var. *hookeriana* Breit. Can. Field Nat. 71:57. 1957. *(Drummond,* Rocky Mts.)

Dryas crenulata Juz. Bull. Jard. Bot. Princip. U.R.S.S. 28:325. 1929. (E. Siberia)

DRYAS OCTOPETALA var. ANGUSTIFOLIA C. L. Hitchc. hoc loc. Foliis 1-4 mm. latis, saepe inferne eglandulosis. (Type: *Hitchcock & Muhlick 11945,* 3 mi. from mouth of Half Moon Canyon, Fergus Co., Mont., July 5, 1945, in U. of Wash. Herb.)

Leaf blades linear-lanceolate to oblong, cuneate to rounded or semicordate at base, 1-3 cm. long, (3) 5-10 (12) mm. broad, the margins coarsely crenate, often strongly revolute, the upper surface deep green, strongly rugose and minutely warty-glandular (ours), but otherwise glabrous or sparsely hairy, the lower surface tomentose or without hairs but usually with sessile to stalked glands on the midvein (rarely eglandular); scapes 3-15 (20) cm. tall, stipitate-glandular and more or less tomentose; calyx tomentose and with prominent purplish-black stipitate glands, the lobes narrowly oblong-lanceolate, (4) 5-7 mm. long, less than half as broad; petals cream to nearly pure white, oblong-obovate to elliptic, 10-12 mm. long, glabrous; filaments glabrous; achenes about 3 mm. long. N=9, 18.

From mid-elevations in the mountains (where usually on gravel bars) to above timber line, on talus, exposed ridges, and open meadows; Alas. southward to the n. Cascade Mts. of Wash., e. along the Arctic coast to Lab. and the Gaspé Peninsula, s. in the Rocky Mts. to Mont., n.e. Oreg., Ida., and Colo.; Asia. Late June-Aug.

This is a polymorphic complex with several distinct races (some of which occur to the n. of our area) that have variously been given varietal, subspecific, or specific status. It appears that there is considerable intergradation between the races, and in our area, at least, it does not seem that the taxa are delimited clearly enough to be recognizable as species. Our material consists of two phases:

1 Leaves tending to average at least 5 mm. broad, always glandular on the lower surface along the midrib and often on the lateral nerves; s. Alas. s. to the Cascades of Wash., and in the Rocky Mts. to Ida., n. e. Oreg., and Wyo.   var. hookeriana (Juz.) Breit.
1 Leaves tending to average less than 5 mm. broad, often not at all glandular on the lower surface; in our area confined to c. Ida. (Custer Co.) and to c. Mont. (Stillwater, Fergus, Beaverhead, Madison, and Carbon cos.). In general, the taxon is characterized chiefly by the narrow leaves. Plants (even from one locality) may be glandular or eglandular on the lower surfaces of the leaves. For example, *Hitchcock & Muhlick 11945,* from the Big Snowy Mts., is glandular (and referable to *D. hookeriana* Juz.), whereas *Hitchcock 16055* from the same area is eglandular and therefore referable to *D. crenulata* Juz. in Porsild's key.   var. angustifolia C. L. Hitchc.

Duchesnea Smith

Flowers single on axillary peduncles, complete, perigynous; calyx with broadly obovate, 3 (5)-lobed bracteoles alternating with the 5 sepals, the hypanthium very shallowly saucer-shaped, much shorter than the bracteoles; petals 5, deciduous, yellow; stamens 20-25, the filaments slender, inserted at the edge of the hypanthium; pistils numerous, the styles slender, attached slightly above midlength of the ovary; achenes usually reddish; receptacle becoming enlarged, hemispheric, and spongy or semifleshy (but not very palatable) in fruit; herbaceous perennials with ternately compound, stipulate leaves borne in basal rosettes and alternately along the trailing stems.

Two species of s. Asia, closely related to *Fragaria* and *Potentilla;* ours an introduced weed. (In commemoration of the French botanist, A. N. Duchesne, 1747-1827.)

Duchesnea indica (Andr.) Focke in Engl. & Prantl, Nat. Pflanzenf. 3^3:33. 1888.

Fragaria indica Andr. Bot. Rep. pl. 479. 1807. (India)

Coarsely strigose to sericeous perennial, the stems trailing and freely rooting at the nodes and producing offset plants; leaves slender-petiolate, the stipules usually toothed to lobed; leaflets 3, elliptic or ovate-elliptic, 2-4 cm. long, crenate-serrate; calyx 4-6 mm. long at anthesis, the lobes considerably shorter than the 3- to 5-lobed bracteoles but ultimately as much as 10 mm. long and in fruit sometimes equal to the bracteoles; petals 3-5 mm. long; stamens shorter than the petals; fruit red, more or less globose, up to 1 cm. broad, strawberrylike in appearance but of very poor flavor, the achenes superficial. N=42. Indian strawberry.

This species, native to India, is sometimes grown as an ornamental chiefly because of the red strawberrylike fruits which are eaten by birds; it is occasionally reported, usually as a garden escape, w. of the Cascades in B.C., Wash., and Oreg., as well as in many other parts of the U.S. where it is often much more strongly established. May-June.

Filipendula Mill.

Flowers numerous in large cymose panicles, complete to sometimes (in part) imperfect, perigynous; calyx deeply 5 (4-7)-lobed, the hypanthium nearly flat, much shorter than the reflexed sepals; petals 5 (4-7), deciduous, borne at the outer edge of the hypanthium; stamens 20-40, in 2-4 series of 10 (8-14) on small ridges opposite the petals and sepals, only slightly perigynous, the innermost series almost hypogynous; filaments slender; anthers broad, the two halves almost free except at the submedian point of attachment to the filament; pistils 5-15, distinct, erect; styles rather thick, about equal to the ovary; stigmas large, more or less lunate-reniform and deeply grooved; ovules 1-2; fruit dry, indehiscent, flattened, 1-seeded; perennial, caulescent, rhizomatous herbs with palmately lobed to pinnatisect, prominently stipulate leaves.

Perhaps 10 species of temperate Eurasia (chiefly) and N.Am., usually in moist areas. (Latin *filum,* thread, and *pendulus,* hanging, in reference to the tubers found hanging to the roots of one species.)

The plants take well to cultivation and most of them are highly regarded ornamentals, suggesting *Spiraea* and *Astilbe.*

Filipendula occidentalis (Wats.) Howell, Fl. N.W. Am. 185. 1898.

Spiraea occidentalis Wats. Proc. Am. Acad. 18:192. 1883. (*T. Howell;* and *L. F. Henderson,* Rocky banks of the Trask R., Tillamook Co., Oreg., July, 1882)

Strongly rhizomatous perennial herb with erect, mostly simple stems up to 2 m. tall, glabrous or slightly pubescent below, becoming finely pubescent above; leaves pubescent on both surfaces at least on the veins, pinnately divided, the rachis petiolelike, with 1-3 (4) pairs of remote, linear to ovate, serrate to doubly toothed leaflets 5-15 mm. long, the terminal leaflet broadly ovate, 8-20 cm. broad and nearly as long, palmately (3) 5- to 7-cleft into ovate to oblong-lanceolate, acute to acuminate, doubly serrate lobes; flowers numerous in a nearly flat-topped, open, freely branched panicle of cymes; sepals reflexed, narrowly lanceolate, (2) 3-4 mm. long; petals white, about 6 mm. long, elliptic-oblong; stamens white, about equaling the petals; fruit brownish-hairy, with a slender stipe 2-3 mm. long, the body of the achene about 4 mm. long, strongly flattened.

Northwestern Oreg., where known only from the Trask, Wilson, and Tillamook rivers in Tillamook Co.; usually growing in rock crevices just above the high water level. June-July.

This plant seems not to have been introduced into the horticultural trade, although it has excellent qualities, both floral and foliar. It is known to take well to moist areas in full sun or partial shade.

Fragaria L. Strawberry

Flowers often rather showy, usually several (1 or 2 to 25) in open and repeatedly dichotomous to reduced and often racemelike, conspicuously bracteate cymes on short to well-developed scapes; calyx with a short, spreading to saucerlike hypanthium and 5 sepaloid bracteoles alternate with the 5 ovate to lanceolate, usually ascending lobes; petals white or pinkish, deciduous; stamens mostly 20

var. suksdorfii

Dryas integrifolia

var. douglasii

Crataegus douglasii

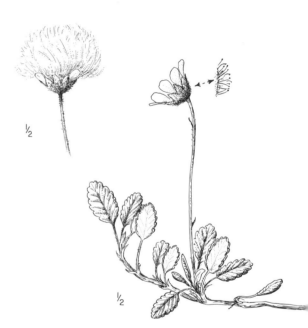

Dryas drummondii

petal

Duchesnea indica

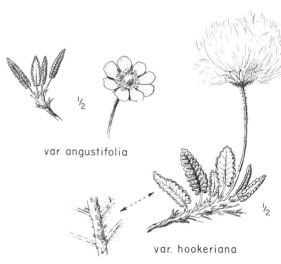

var. angustifolia

var. hookeriana

Dryas octopetala

J

but up to 25, sometimes partially or totally sterile, borne in 4 or 5 series at the outer edge of the hypanthium, the filaments rather short, broadened basally; pistils numerous, borne on an elongate, more or less hemispheric, glabrous or (more commonly) sparsely to copiously villous-lanate receptacle that gradually enlarges and becomes fleshy and usually juicy and palatable with the ripening of the fruits; ovaries from superficial and not at all imbedded to partially sunken below the surface of the receptacle in small pitlike depressions, the styles slender, 1.5-3 times as long as the ovaries on which they are laterally inserted; achenes rather small, more or less pyriform; perennial herbs with stolonous, nodally rooting stems and usually scaly rootstocks, the leaves basal, long-petiolate, stipulate, usually (ours) trifoliolate and coursely crenate-serrate.

As usually considered, perhaps 30 species of temperate Eurasia and N. Am., with a few in S. Am., although most (perhaps all) might be referred to as few as three wide-ranging specific complexes. (From *fraga,* the Latin name of the strawberry.)

Reference:

Rydberg, P. A. A monograph of the North American Potentilleae. Fragaria. Mem. Dept. Bot. Columbia Univ. 2:165-185. 1898.

Whereas Rydberg believed that our western species (of which he recognized 8) were closely related to, but distinct from, others of e. N. Am. and Europe, most workers have come to the conclusion that our taxa are merely races of two or three very widespread and polymorphic species. The species are more readily distinguishable in the field than they are after preservation because the contrasting color and texture of the leaves of living plants are largely lost in drying, whereas the differences in the teeth of the leaflets and in the relative length of the peduncles are not so fully diagnostic.

1 Plants maritime; leaves thickened, deep green, strongly reticulate-veiny beneath, rugose
    on the upper surface   F. CHILOENSIS
1 Plants not maritime; leaves either thin or not reticulate-veiny, usually bright yellow-green
    or bluish-green
    2 Leaves bright yellow-green and not glaucous on the upper surface, relatively thin and
       more or less veiny, the upper surface slightly bulged between the primary lateral
       veins and almost always somewhat pilose-silky; terminal tooth of the leaflets rel-
       atively well developed, usually projecting beyond the uppermost lateral teeth; in-
       florescence commonly equaling or surpassing the leaves   F. VESCA
    2 Leaflets glaucous and somewhat bluish-green on the upper surface, rather thick and
       not very prominently veiny, the upper surface not bulged between the veins and
       nearly always glabrous; terminal tooth of the leaflets small, usually surpassed by
       the adjacent lateral teeth; inflorescence commonly shorter than the leaves   F. VIRGINIANA

Fragaria chiloensis (L.) Duchesne, Hist. Nat. Frais. 165. 1766.
    F. vesca var. *chiloensis* L. Sp. Pl. 495. 1753. *F. chilensis* Molina, Sagg. Chile 134. 1784. *F. vir-*
    giniana chiloensis Johow. Rev. Chil. Hist. Nat. 49-50:504. 1948. (S. Am.)
    F. chilensis var. *scouleri* Wats. Bibl. Ind. 282. 1878. *F. chiloensis scouleri* Rydb. Monog. Potent.
    170. 1898. (*Scouler,* Ft. Vancouver, Wash.)
    F. cuneifolia Nutt. ex T. & G. Fl. N. Am. 1:448. 1840, in synonymy under *F. chilensis* γ; ex Howell,
    Fl. N. W. Am. 174. 1898. (*Nuttall,* "Western Coast from Puget Sound to California")

Petioles, peduncles, and the thick, usually reddish- or purplish (brownish)-tinged stolons pubescent with silky, spreading to somewhat retrorse hairs; basal leaves numerous, the petioles usually purplish-reddish (brownish-) tinged, 4-10 (20) cm. long; leaflets evidently petiolulate, the terminal one with a stalk (2) 3-10 mm. long, thick, coriaceous, obovate to cuneate-obovate, 1.5-4 cm. long, rounded to truncate, coarsely crenate-serrate mostly above midlength and sometimes only across the tip, the upper surface deep green, shining, and glabrous, the lower surface strongly reticulate and grayish-silky to somewhat tomentose; scapes 3-10 (15) cm. long, usually shorter than the leaves, even in fruit; flowers 5-15 (25) in well-developed cymes; calyx abundantly silky; bracteoles more or less elliptic, 5-8 mm. long, the sepals lanceolate, acute or acuminate, up to 10 mm. long; petals white, obovate-orbicular, (10) 12-16 mm. long; fruit usually at least 1.5 cm. broad, rather strongly pilose-lanate with hairs as long as the ovaries; achenes about 2 mm. long and about 1/3 immersed in the receptacle. N=28.

Coastal, from Alas. to Calif., also on the shores of some of the islands of Puget Sound; along most of the coast of S. Am.; Hawaii. Apr.-June.

This is one of the parents of many of the varieties of cultivated strawberries. It is often used as a ground cover for which it is ideally suited as it does not tend to spread as aggressively as our other species and is a far more attractive plant. The species is strictly maritime and reports of its natural occurrence at any distance from the coast are probably erroneous.

Fragaria vesca L. Sp. Pl. 494. 1753.
  Potentilla vesca Scop. Fl. Carn. 2nd ed. 1:363. 1772. (Europe)
  FRAGARIA VESCA var. CRINITA (Rydb.) C. L. Hitchc. hoc loc. *F. californica* Newberry, Pac. R. R. Rep. 6[2]:73. 1857, but not of Cham. & Schlecht. in 1827. *F. crinita* Rydb. Monog. Potent. 171. 1898. *F. californica* var. *crinita* Hall, U. Calif. Pub. Bot. 4:198. 1912. *(J. S. Newberry,* Williamson's Exp., Willamette Valley, Oreg.)
  F. helleri Holz. Bot. Gaz. 21:36. 1896. *(J. H. Sandberg,* Pine Creek, near Farmington Landing, Latah Co., Ida., in 1892) = var. *bracteata,* a pink-flowered phase that is sporadic in distribution.
  F. bracteata Heller, Bull. Torrey Club 25:194. 1898. *F. vesca* var. *bracteata* Davis, Madroño 11: 144. 1951. *(Heller & Heller 3615,* along Santa Fe Creek, 9 mi. e. of Santa Fe, N. M., May 29, 1897)
  F. retrorsa Greene, Ott. Nat. 18:216. 1905. *(Macoun 34336,* Chilliwack Valley, B. C.)
  Very strongly stoloniferous, the trailing stems, petioles, and peduncles greenish or very lightly tinged with reddish-purple, lightly to densely pubescent with fine, spreading to slightly ascending hairs; petioles mostly (3) 5-10 (15) cm. long; leaflets thin to rather thick, broadly elliptic to obovate-oblong, coarsely crenate-serrate most of the length, bright yellow-green and usually very sparsely hairy on the upper surface, pale green, finely pilose-silky, and usually slightly glaucous but not strongly reticulate on the lower surface, subsessile or with stalks as much as 3 mm. long; scapes ultimately often equaling or exceeding the leaves; cyme usually diffuse, 3- to 11 (15)-flowered; calyx finely silky, the bracteoles linear-elliptic, about 4 mm. long; sepals 4-5 mm. long, acuminate and sometimes very shortly setose-caudate, spreading to erect in fruit; petals white- to pinkish-tinged, 8-11 mm. long; fruit succulent and palatable, up to 1 cm. broad, the achenes about 1.3 mm. long, sunken up to 1/3-3/4 of their thickness in shallow pits in the receptacle. N=7.
  Moist woods, stream banks, and sandy meadows; Europe and N. Am. Apr. -June.
  Our material, as described above, is referable to two varieties separable from var. *vesca* as follows:
1 Achenes almost completely superficial on the receptacle; Europe, occasional as an
    escape in e. N. Am., not known from our area   var. vesca
1 Achenes partially sunken (up to 3/4 their thickness) in shallow pits in the receptacle
  2 Peduncles in fruit usually exceeding the leaves; leaflets thin, the lower surface not
      more than half-covered by the pubescence; B. C. southward, (chiefly in, and on the
      e. side of, the Cascades) to Calif., e. to Alta., Mont., Wyo., and N.M., occa-
      sional in the Puget Trough, Wash., and the Willamette Valley, Oreg., where tran-
      sitional to the next   var. bracteata (Heller) Davis
  2 Peduncles in fruit often shorter than the leaves; leaflets rather thick and often fairly
      evidently hairy, the lower surface nearly concealed by the copious silky pubescence;
      B. C. southward to Calif., chiefly in the valleys on the w. side of the Cascades, but
      also in both the Cascade and Olympic mts., up to 4000 ft. elev., occasional farther
      e.; often mistaken for *F. chiloensis,* but probably well isolated from it both ecolog-
      ically and genetically and apparently never a strand plant; freely transitional to var.
      bracteata   var. crinita (Rydb.) C. L. Hitchc.

Fragaria virginiana Duchesne, Hist. Nat. Frais. 204. 1766.
  Potentilla ovalis Lehm. Delect. Sem. Hort. Hamb. 1849:9. 1849. *F. virginiana* var. *ovalis* Davis, Madroño 11: 144. 1951. ("Habitat in terris Mexicanis")
  F. virginiana var. (?)*glauca* Wats. Bot. King Exp. 85. 1871. *F. vesca* var. *americana* of Rydb. Contr. U. S. Nat. Herb. 3:496. 1896, but not of Porter in 1890. *F. glauca* Rydb. Monog. Potent. 183. 1898. *F. ovalis glauca* Nels. in Coult. & Nels. New Man. Bot. Rocky Mts. 252. 1909. *(Watson 322,* "In the Wahsatch and Uintas")
  F. platypetala Rydb. Monog. Potent. 177. 1898. *F. virginiana* var. *platypetala* Hall, U. Calif. Pub. Bot. 4:198. 1912. *(John Macoun,* Spout, B. C., in 1890)
  F. pauciflora Rydb. Monog. Potent. 183. 1898. *(J. H. Flodman 591,* Mont., in 1896) = var. *glauca.*
  F. latiuscula Greene, Ott. Nat. 18:216. 1905. *(Macoun 34337, 34330,* and *34339,* all from Chilliwack Valley, B. C.) = var. *platypetala.*

F. suksdorfii Rydb. N.Am. Fl. 22⁴:361. 1908. *(Suksdorf 486,* Falcon Valley, Klickitat Co. , Wash. ,
   in 1883) = var. *platypetala.*

Freely stoloniferous and weakly to strongly pubescent with appressed to spreading hairs along the
greenish or slightly reddish-tinged stolons, petioles, and scapes; petioles up to 15 cm. long; leaflets
narrowly to broadly obovate or cuneate-obovate to elliptic-obovate, mostly 2-7 (10) cm. long, rather
thick, glabrous and usually glaucous and bluish-green on the upper surface, sparsely to abundantly
silky-villous on the lower surface, coarsely crenate-serrate most of the length, usually distinctly petiolulate, the stalk of the terminal leaflet 2-7 mm. long; peduncles from (usually) shorter than, to
sometimes as long as, the leaves; flowers 2-15, mostly in open (but sometimes in reduced and falsely
racemose) cymes; calyx sparsely to copiously villous-silky, the bracteoles 4-7 mm. long, narrower
and shorter than the lanceolate-elliptic, spreading to ascending, acuminate, 5-8 mm. sepals; petals
white (pinkish), 6-13 mm. long; fruit about 1 cm. broad, succulent and palatable, the achenes about
1. 5 mm. long, partially sunken (up to 3/4 their thickness) in shallow pits in the receptacle. N=28.

Open woods to sandy or gravelly meadows and stream banks in the plains and lower mountains;
Alas. to Calif. , e. to the Atlantic coast and s. to Colo. and Ga. May-Aug.

Our material, as described above, appears to be only slightly different from the var. *virginiana* of
the Mississippi Valley and eastward, but consists of two phases that are sometimes regarded as separate species.

1 Pubescence rather scanty, that of the petioles and scapes appressed; petals mostly 5-10
   mm. long, narrowly obovate; B.C. e. to Mack. and Alta. , s. to the Dakotas, Mont. ,
   e. Ida. , and N.M. , in the Rocky Mt. states perhaps more common on the e. side of
   the continental divide; apparently lacking in Wash. and n. Ida.   var. glauca Wats.
1 Pubescence usually rather abundant, that of the petioles and scapes spreading; petals
   mostly broadly obovate-orbicular, 8-12 mm. long; mostly e. of the crest of the Cascades, but occasional on the w. side, Alas. to Calif. , e. to w. Mont. , Wyo. , Colo. ,
   and Utah   var. platypetala (Rydb.) Hall

Geum L.

Flowers complete, perigynous, mostly several in bracteate, usually open (but sometimes reduced)
cymes, occasionally single; calyx with simple to sometimes cleft bracteoles alternate with the 5 erect
to reflexed lobes, the hypanthium free of the pistils, saucer-shaped to cup-shaped; petals 5, usually
yellow but sometimes strongly pinkish-tinged to red or purplish, shorter to longer than the sepals;
stamens numerous, borne in several series above (outside of) a (usually) ringlike or cuplike disc which
is fused with, or free of, the hypanthium; pistils numerous; style straight to bent or strongly geniculate and jointed, often considerably elongate in fruit and then mostly with a (sometimes tardily) deciduous terminal segment, at least the lower segment persistent on the achene; seed erect; herbaceous,
generally more or less rhizomatous herbs with usually numerous irregularly to lyrately pinnate or pinnatifid basal leaves and few, alternate or opposite, mostly 3-lobed cauline leaves.

About 50 species, mostly in the N. Temp. (to Arctic) Zone; a few in S. Am. , and one in s. Africa.
(The ancient Latin name for some one of the species.)

Although none of our species can be said to have choice ornamental qualities, *G. rivale* is known in
the trade and is easily grown in moist areas, *G. rossii* is also listed and should be rated as perhaps
best of our species, at least for e. of the Cascades, and *G. triflorum* is well worth a place in a wild
garden, especially the var. *campanulatum* for the Puget Sound area.

1 Sepals reflexed in flower; hypanthium saucer-shaped, lined with a glandular disc at least
   on the lower half; styles strongly geniculate and jointed, the persistent (longer) portion
   hooked at the tip
   2 Lower (persistent) segment of the style eglandular, glabrous or slightly hirsute near the
      base; terminal segment of the basal leaves somewhat larger than the main lateral lobes
      but similarly cuneate-based   G. ALEPPICUM
   2 Lower (persistent) segment of the style somewhat glandular-pubescent; terminal seg-
      ment of the basal leaves many times larger than the main lateral lobes and usually
      rounded to subcordate at base   G. MACROPHYLLUM
1 Sepals ascending to erect; hypanthium turbinate to bowl-shaped, the lower half usually
   not disc-lined; styles often neither geniculate nor jointed and hooked on the persistent
   portion
   3 Petals erect to convergent, the flower somewhat vase-shaped; cauline leaves 2(4),

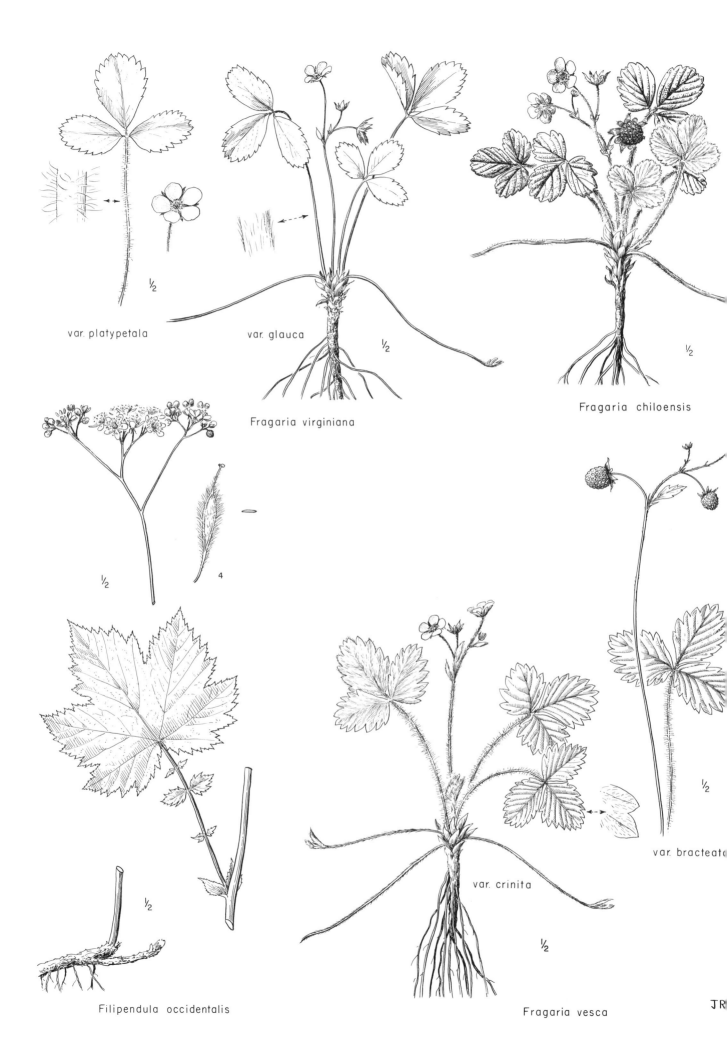

var. platypetala

var. glauca

½

½

Fragaria virginiana

Fragaria chiloensis

½

½

4

½

½

var. bracteata

Filipendula occidentalis

½

var. crinita

½

Fragaria vesca

JR

opposite, their bases more or less sheathing   G. TRIFLORUM
3 Petals spreading, or at least not erect or convergent, the flower more nearly rotate;
    cauline leaves 1 to several, alternate
    4 Style strongly geniculate and jointed, the terminal segment ultimately deciduous, the
        persistent lower segment hooked at the tip; hypanthium bowl shaped; basal leaves
        lyrate-pinnatifid, the segments 7-15; plants mostly (3) 4-6 (8) dm. tall   G. RIVALE
    4 Styles straight or only slightly bent, not jointed, persistent on the achene; hypan-
        thium shallowly funnelform; basal leaves interruptedly pinnatifid, with 9-31 seg-
        ments; plants rarely over 3 dm. tall   G. ROSSII

Geum aleppicum Jacq. Ic. Pl. Rar. 1:pl. 93. 1781-1786. (No type given)
    G. canadense of Murr. Novi Comm. Gott. 5:33. 1783, but not of Jacq. in 1773. *G. strictum* Ait.
    Hort. Kew. 2:217. 1787, nom. illegit. *G. aleppicum* var. *strictum* Fern. Rhodora 37:294. 1935.
    G. aleppicum ssp. *strictum* Clausen, Cornell Agr. Exp. Sta. Mem. 291:9. 1949. (Cultivated plants
    originating in N. Am.)
Perennial with a short rootstock and 1-several simple stems up to 1 m. tall, puberulent below and
becoming spreading-hirsute above; basal leaves several, slightly rosulate, the blades oblong-obovate
in outline, up to 15 (20) cm. long, interruptedly lyrate-pinnatifid, the main segments 5-9, cuneate-
obovate, strongly cleft and usually doubly toothed, the terminal segment lobed well over half the
length; cauline leaves several, the lower ones pinnatifid but with large leafletlike stipules, the upper
ones becoming only 3-lobed; flowers several in a leafy-bracteate, unsymmetrical cyme; hypanthium
saucer-shaped, 3-4 mm. long; sepals reflexed, 5-8 mm. long; petals yellow, spreading, equaling or
slightly exceeding the sepals; stamens 60-100; achenes flattened, elliptic in outline, 3-4 mm. long;
style strongly geniculate above midlength, of 2 distinct segments, the lower segment glabrous or
slightly hirsute (but eglandular) near the base, brownish (reddish), about 4. 5 mm. long, persistent
and hooked, the upper segment hirsute, about 1. 5 mm. long, deciduous. N=21.
    Along streams, or in marshy or damp woods; occasional e. of the Cascade Mts. from B. C. to
Calif. , e. to the Rocky Mts. from Alta. to N. M. , perhaps more common through s. Can. and n. U. S.
to Que. and Newf. ; Eurasia. June-July.
    Our plant, as described above, is said to differ from the Eurasian counterpart in having less hairy
achenes and it is often recognized at either the specific or infraspecific level (var. *strictum* Fern.).
However, the alleged difference is not readily apparent, if actual.

Geum macrophyllum Willd. Enum. Pl. Hort. Berol. 557. 1809. (Garden plants originally from Kam-
    tchatka)
    G. urbanum ssp. *oregonense* Scheutz, Nova Acta Soc. Sci. Upsal. III, 7[6]:26. 1870. *G. oregonense*
    Rydb. Bull. Torrey Club 25:56. 1898. *G. macrophyllum* var. *rydbergii* Farw. Pap. Mich. Acad.
    Sci. 23:129. 1938. *(Waerngren,* "in regione Oregonense") = var. *perincisum.*
    G. perincisum Rydb. N. Am. Fl. 22[5]:405. 1913. *G. macrophyllum* var. *perincisum* Raup, Rhodora
    33:176. 1931. *G. macrophyllum* ssp. *perincisum* Hultén, Fl. Alas. 6:1040. 1945. *(W. C. McCalla*
    2074, Cave Avenue, Banff, Alta.)
Perennial with a short rootstock; basal leaves several, rosulate, the blades interruptedly lyrate-
pinnatifid, up to 3 dm. long; leaflets 9-23, the terminal much the largest, broadly triangular-ovate to
cordate, up to 15 cm. in width; stems 1-several, mostly 3-7 (10) dm. tall, spreading-hirsute or
somewhat hispid, usually also finely pubescent and often strongly glandular at least above; cauline
leaves 2-5, with large leafletlike stipules; flowers several in a somewhat asymmetrical cyme; hypanthium saucer-shaped, mostly 3-4 mm. long; sepals reflexed, (3) 4-5 mm. long; petals yellow, 4-6 (7)
mm. long; mature achenes compressed, elliptic in outline, about 3 mm. long, the style geniculate
above midlength and in 2 distinct segments, the lower segment sparsely glandular-puberulent, about
4 mm. long, persistent, usually reddish, hooked at the tip, the upper segment 1-1. 5 mm. long, usually pubescent, yellowish, tardily deciduous. N=21.
    Common in moist woods or meadowland or along stream banks from near sea level to subalpine
areas; Alas. s. to Baja Calif. and Mex. , e. through Can. to the Great Lakes, St. Lawrence R. , and
N. S. ; Asia. Apr. -Aug.
    The species is differentiated into 2 well-marked varieties:
1 Terminal segment of the basal leaves very shallowly rounded-lobed and minutely once or
    twice serrate-dentate; leaflets of the cauline leaves usually more nearly toothed (once or
    twice) than cleft; peduncles and pedicels slightly if at all glandular; faces of the achenes

pubescent; Alas. to s. Calif., mostly w. of the Cascades in our area, less common e.
to s.e. B.C. and w. Mont.; Great Lakes to Newf., Lab., Me., and Vt.; Asia

var. macrophyllum

1 Terminal segment of the basal leaves lobed up to 1/3 or 1/2 the length and again coarsely
once or twice toothed or cleft; leaflets of the cauline leaves usually shallowly cleft to
deeply toothed; peduncles and pedicels rather strongly glandular; achenes sparsely puberulent to glabrous on the faces; Alas. s., mainly e. of the crest of the Cascades, to
Calif., and Ariz., e. to Mich., and common in the Rocky Mt. states, s. to N.M.

var. perincisum (Rydb.) Raup

Geum rivale L. Sp. Pl. 501. 1753.

Carophyllata rivalis Scop. Fl. Carn. 2nd ed. 1:365. 1772. (Europe)

G. aurantiacum Fries in Scheutz, Nova Acta Soc. Sci. Upsal. 7[6]:30. 1870. (Described from seeds
that supposedly came from the w. coast of N.Am.) The yellow-petaled phase.

G. pulchrum Fern. Rhodora 8:11. 1906. *(Williams, Collins, & Fernald,* by the St. Lawrence, Bic,
Que., July 6, 1905) The phase with yellow petals, believed by some to be a hybrid of *G. rivale* and
G. macrophyllum.

Perennial with short to rather extensive, scaly rootstocks, the basal leaves several, somewhat rosulate, up to 3 dm. long, lyrate-pinnatifid, the leaflets mostly 7-15, once or twice crenate-serrate,
1-3 of them much larger than the others, the terminal one cuneate-obovate, up to 10 cm. long; flowering stems mostly (3) 4-6 (8) dm. tall, hirsute but becoming puberulent above; cauline leaves 2-5, alternate, reduced, the stipules leafletlike, the blades pinnatifid below to deeply 3-lobed above; inflorescence open, (1) 3- to 7 (9)-flowered, cymose although alternately branched, the flowers nodding in
the bud but becoming erect; calyx reddish-purple, the lobes lanceolate, acute or acuminate, about 10
mm. long, erect, the hypanthium broadly cup-shaped, shorter than the sepals; petals yellow to pinkish,
mostly (1) 2-3 mm. shorter than the sepals; stamens 100 or more; styles strongly geniculate; mature
achenes elliptic in outline, 3-4 mm. long, strongly hirsute, the lower (persistent) joint of the style 6-
8 mm. long, hirsute below, glabrous above, hooked at the tip, the upper segment ultimately deciduous,
sparsely hirsute, 3-4 mm. long. N=21.

Stream banks, lake shores, bogs, and wet meadows; B.C. and Okanogan Co., Wash., to Alta., s.
in the Rocky Mts. to N.M., e. to Mo., Ind., N.J., Newf., and Que.; Eurasia. Late June-July.

Geum rossii (R. Br.) Ser. in DC. Prodr. 2:553. 1825.

Sieversia rossii R. Br. Chlor. Melv. 18:pl. 100. 1823. *Acomastylis rossii* Greene, Leafl. 1:174.
1906. *(Ross,* Melville I.)

GEUM ROSSII var. TURBINATUM (Rydb.) C. L. Hitchc. hoc loc. *Potentilla nivalis* Torr. Ann. Lyc.
N.Y. 1:32. 1827, but not of Lapeyr in 1782. *G. turbinatum* Rydb. Bull. Torrey Club 24:91. (Feb.
28) 1897. *Sieversia turbinata* Greene, Pitt. 4:50. 1899. *Acomastylis turbinata* Greene, Leafl. 1:
174. 1906. *(James,* James' Peak, Colo.)

G. sericeum Greene, Pitt. 3:172. (May 19) 1897. *Sieversia sericea* Greene, Pitt. 4:50. 1899. *Acomastylis sericea* Greene, Leafl. 1:174. 1906. *(Watson 320,* East Humboldt Mts., Nev.) = var.
turbinatum, merely a somewhat more hairy phase, sporadic in occurrence.

Potentilla gracilipes Piper, Bull. Torrey Club 27:392. 1900. *Sieversia gracilipes* Greene, Leafl. 1:
4. 1903. *Acomastylis gracilipes* Greene, Leafl. 1:174. 1906. *(Cusick 2246,* head of Anthony's
Creek, Blue Mts., Baker Co., Oreg., July, 1899) = var. *turbinatum*—a rather robust plant, usually 15-30 cm. tall, and perhaps more silvery than average, but apparently well matched by Rocky
Mt. material, in spite of its pronounced geographic isolation.

GEUM ROSSII var. DEPRESSUM (Greene) C. L. Hitchc. hoc loc. *Acomastylis depressa* Greene,
Leafl. 1:174. 1906. *(Elmer 1182,* Mt. Stuart, Chelan Co., Wash., Aug., 1898)

Perennial with thick scaly rootstocks, forming dense clumps up to 3 dm. broad; basal leaves numerous, strongly marcescent, the blades oblong in outline, (3) 4-10 (12) cm. long, interruptedly pinnate or pinnatifid, the main segments (9) 15-25 (31), entire (the lower ones) and linear or elliptic to
broadly cuneiform (the upper ones) and 3- to 5 (7)-toothed to -cleft, from nearly glabrous and greenish, to silvery-sericeous or -villous, sparingly glandular-pubescent; cauline leaves several, alternate, much reduced; flowering stems simple, (5) 8-20 (30) cm. tall, sparsely pubescent to villous;
flowers 1-4; calyx often purplish-tinged, from nearly glabrous to pubescent (hirsute), the hypanthium
shallowly funnelform, 3-5 mm. long, equaled or exceeded by the bracteoles, the sepals 6-10 mm.
long; petals spreading, yellow, obovate, sometimes retuse, 6-12 mm. long; stamens 50-70, inserted

just below the petals near the tip of the hypanthium; pistils few to many on a short columnar receptacle; mature achenes fusiform-lanceolate, 2.5-4 mm. long, hairy, the styles persistent, straight, glabrous, about as long as the achene. N=28.

Arctic tundra from Alas. southward, in the mountains, on gravelly meadows and scree slopes, to n.c. Wash., n.e. Oreg., Nev., Ariz., and N.M.; Asia. June-July.

In the s. and w. limits of its range in N.Am. the species is represented by numerous disjunct populations. One such, in c. Wash., differs in almost intangible ways from the Alaskan and from the Rocky Mt. plants, but is considered to be no more than varietally distinct, in spite of its geographic isolation. Our races are not nearly so distinctive as the following key might indicate:

1 Calyx usually green or only slightly purplish-tinged, the lobes mostly strongly veined; hypanthium (when pressed) generally over twice as broad as long; segments of the leaflets mostly 3-6 mm. broad, tending to be merely toothed or lobed less than half their length; flowers mostly single and terminal, less commonly 2 or 3
    2 Petals usually (8) 10-12 mm. long; leaflets greenish, subglabrous to lightly pubescent or puberulent; arctic-alpine, across n. Asia, and from Alas. and Yukon s. to Vancouver I., s.e. B.C., and to the Mission Mts., Lake Co., Mont., where intergradient with var. *turbinatum*   var. rossii
    2 Petals mostly 6-10 mm. long; leaflets glandular-pubescent and grayish, villous-silky; local in the Wenatchee Mts., Chelan Co., Wash.; most similar to var. *rossii* but completely disjunct from it   var. depressum (Greene) C. L. Hitchc.
1 Calyx usually strongly purplish-tinged, the lobes not heavily veined; hypanthium (when pressed) generally less than twice as broad as long; segments of the leaflets commonly lobed at least half their length, tending to be narrow, averaging 2-4 (5) mm. broad; flowers often several; alpine talus slides and cirques to subalpine wind-swept ridges, Rocky Mts., from c. and w. Mont. southward to N.M., disjunctly w. to the Wallowa Mts. of n.e. Oreg. and to Nev. and Ariz.; not uniform, the plants of n.e. Oreg. tending to be much taller, and perhaps more grayish-pubescent than those of the n. Rocky Mts., but no more silvery than those of the s. Rocky Mts. and Nev.
                          var. turbinatum (Rydb.) C. L. Hitchc.

Geum triflorum Pursh, Fl. Am. Sept. 736. 1814.
  Sieversia triflora R. Br. in Richards. Bot. App. Frankl. Journ. 2nd ed. 21. 1823. *Erythrocoma triflora* Greene, Leafl. 1:175. 1906. *G. ciliatum* var. *triflorum* Jeps. Fl. Calif. 2:206. 1936. (*Bradbury*, "In Upper Louisiana")
  G. ciliatum Pursh, Fl. Am. Sept. 352. 1814. *Sieversia ciliata* G. Don, Gen. Hist. Pl. 2:528. 1832. *Erythrocoma ciliata* Greene, Leafl. 1:177. 1906. *G. triflorum* var. *ciliatum* Fassett, Rhodora 30:207. 1928. (*Lewis*, "On the banks of the Kooskoosky")
  Sieversia rosea Grah. Edinb. New Phil. Journ. 11:193. 1831. (Cultivated plants from seeds coll. by Drummond in the Rocky Mts.)
  Erythrocoma affinis Greene, Leafl. 1:175. 1905. (*W. C. McCalla 2073*, near Banff, Alta.) = var. *triflorum*.
  Erythrocoma flavula Greene, Leafl. 1:177. 1906. *Sieversia flavula* Rydb. N.Am. Fl. 22⁵:410. 1913. *G. triflorum* var. *ciliatum* f. *flavulum* Fassett, Rhodora 30:207. 1928. (*Nelson 829*, Wind River Mts., Wyo.) = var. *ciliatum*, a phase thereof with yellowish calyces.
  Erythrocoma ciliata var. *ornata* Greene, Leafl. 1:178. 1906. (*Heller*, Ida., is the first of 3 collections cited) = var. *ciliatum*, the (entirely sporadic) variant with cleft calyx bracteoles.
  GEUM TRIFLORUM var. CAMPANULATUM (Greene) C. L. Hitchc. hoc loc. *Erythrocoma campanulata* Greene, Leafl. 1:178. 1906. *Sieversia campanulata* Rydb. N.Am. Fl. 22⁵:409. 1913. *G. campanulatum* G. N. Jones, U. Wash. Pub. Biol. 5:178. 1936. (*Elmer 2529*, Olympic Mts., Wash., July, 1900)
  Erythrocoma canescens Greene, Leafl. 1:178. 1906. *Sieversia canescens* Rydb. N.Am. Fl. 22⁵:409. 1913. (*Brewer*, in 1863, above Ebbett's Pass, Sierra Nevada, Calif., is the first of several collections cited) = var. *ciliatum*, a variant with short calyx bracteoles and abundant pubescence.
  Erythrocoma grisea Greene, Leafl. 1:178. 1906. *Sieversia grisea* Rydb. N.Am. Fl. 22⁵:409. 1913. *G. ciliatum* var. *griseum* Kearney & Peebles, Journ. Wash. Acad. Sci. 29:481. 1939. (*Leiberg*, San Francisco Mts., Ariz., June 25, 1901) = var. *ciliatum*, at least in the sense in which the name has been applied to our plants; merely a broader-leafleted form with linear calyx bracteoles.
Perennial with thick scaly rootstocks, forming clumps up to 3 dm. or more broad; leaves mainly

Geum aleppicum

6

G. rivale

1/2

var. depres[sum]

var. depressum

1/2

6

var. turbina[tum]

var. rossii

1/2

var. macrophyllum

var. perincisum

G. macrophyllum

G. rossii

J

basal, the blades (3) 5-15 cm. long, oblong to obovate, unequally and interruptedly pinnate to pinnatifid or lyrate above, the segments up to 30, unequal, the main ones from deeply cleft into linear or
oblong ultimate divisions to cleft much less than half their length and again 2- to 3-toothed, from puberulent to hirsute or pilose, usually somewhat grayish; flowering stems up to 3 (4) dm. tall, bearing,
about midlength, a pair of opposite, much reduced leaves with leafletlike stipules; flowers 1-9, mostly
cymose; calyx narrowly to broadly turbinate or campanulate to cup-shaped, reddish-purple to pink or
nearly yellow and only reddish-veined, the hypanthium broad, almost hemispheric, about 4-5 mm.
long, bracteoles mostly linear to narrowly elliptic, usually somewhat spreading, simple to 2- to 3-
cleft, from somewhat shorter to considerably longer than the erect to convergent, valvate, 8-12 mm.
sepals; petals valvate, erect to convergent, light yellow to strongly pinkish- or reddish-purple-tinged,
narrowly to broadly elliptic or elliptic-obovate, from 1-3 mm. longer to as much as 5 mm. shorter
than the calyx bracteoles; achenes pyriform, about 3 mm. long; lower part of the style strongly plumose, (2.5) 3-5 cm. long, straight or tortuous, purplish, the terminal glabrous segment only 3-6
mm. long (not lengthening in fruit), often slightly geniculate at the point of juncture with the lower
segment and tardily deciduous from (or persistent on) it. N=21. Purple avens, lion's-beard, old man's
whiskers, grandfather's-beard.

Moister spots in the sagebrush plains and lower foothills to subalpine ridges and talus; B. C. southward, mostly on the e. side of the Cascades, to the Sierra Nevada and e. through Can. to Newf., s.
to N.Y., Ill., Neb., and in the Rocky Mts. to N.M., Utah, and Nev. Apr.-early Aug., dependent
upon elevation.

There has been much diversity of opinion regarding both the generic status of, and the significance
of the variation in, this complex. In general the several taxa that have been recognized at the specific
level are largely sympatric and completely transitional and there seems to be no good reason to recognize more than 3 races for our area, separable as follows:

1 Terminal segment of the style usually persistent on the ripe fruit, the lower (plumose)
      segment up to 5 cm. long; larger leaflets of the basal leaves cleft or toothed less
      than half their length; the common phase from e. Alta. to Newf. and s., occasional
      but sporadic in the Rocky Mt. states on the e. side of the continental divide   var. triflorum
1 Terminal segment of the style usually ultimately deciduous from the fruit, the lower
      (plumose) segment rarely over 3.5 cm. long; larger leaflets of the basal leaves
      commonly much more deeply and repeatedly cleft or divided into ultimately linear
      or narrowly oblong segments
   2 Petals usually longer than the subequal bracteoles and sepals, otherwise similar
        to the next, from which it is not always separable as indicated; apparently a mon-
        tane ecotype of the Olympic Mts., and of Saddle Mt., Clatsop Co., Oreg., but not
        satisfactorily separable from a race on Whidbey I., Wash., which is transitional
        to var. *ciliatum*   var. campanulatum (Greene) C. L. Hitchc.
   2 Petals usually shorter or no longer than the sepals and bracteoles; general in and e.
        of the Cascades from B. C. to Calif., e. to Alta. and the Rocky Mt. states, show-
        ing great variation, and overlapping both var. *triflorum* and var. *campanulatum*
        in character   var. ciliatum (Pursh) Fassett

Holodiscus Maxim. Nom. Conserv.

Flowers numerous, perfect, perigynous, whitish (ours) to pink, borne in diffuse, pubescent panicles
(ours) or racemes on short, bracteolate pedicels; calyx deeply 5-lobed, ebracteolate, the lobes valvate in bud, ascending to erect in anthesis, persistent, the hypanthium shallowly saucer-shaped; petals
oval to elliptic, rounded to short-clawed, deciduous; stamens usually 20, inserted just above (outside)
a ringlike, entire (nonlobed) disc lining the hypanthium, the filaments slightly dilated at base; pistils
generally 5, the ovary superior, usually copiously hirsute, bearing 2 pendulous ovules; fruit a somewhat flattened, short-stipitate, usually 1-seeded achene, obliquely elliptic-ovate in outline, the inner
(adaxial) margin straight, the outer (abaxial) margin strongly convex; style persistent, terminal; small
to large shrubs (ours) or small trees with alternate, exstipulate, toothed to shallowly lobed, deciduous
leaves.

Perhaps half a dozen species of w. N. Am. and n. S. Am., although specific delimitation is more than
ordinarily controversial in the group. (From the Greek *holo*, whole, and *diskos*, disc, in reference to
the unlobed or "whole" disc lining the hypanthium.)

Reference:

Ley, Arline. A taxonomic revision of the genus Holodiscus (Rosaceae). Bull. Torrey Club 70:275-288. 1943.

1 Leaf blades usually over 3 cm. long, not decurrent along the mostly 10-15 mm. petioles, the shallow lobes or main teeth generally with several secondary teeth, the lower surface more or less tomentose or crisply pilose   H. DISCOLOR
1 Leaf blades usually only 1-2 cm. long, somewhat decurrent and forming slight wings to the mostly 2-4 mm. petioles, shallowly to coarsely toothed, sometimes secondarily mucronulate, the lower surface sessile-glandular to strongly pubescent but never at all tomentose, the hairs straight or only slightly crisped   H. DUMOSUS

Holodiscus discolor (Pursh) Maxim. Acta Hort. Petrop. 6:254. 1879.

Spiraea discolor Pursh, Fl. Am. Sept. 342. 1814. *Schizonotus discolor* Raf. New Fl. N. Am. 3:75. 1836. *Schizonotus argenteus* var. *discolor* Kuntze, Rev. Gen. 1:225. 1891. *Schizonotus discolor* var. *purshianus* Rehd. in Bailey, Cycl. Am. Hort. 4:1627. 1902. *Sericotheca discolor* Rydb. N. Am. Fl. 22³:262. 1908. *(Lewis,* "On the banks of the Kooskoosky," Ida.)

Spiraea ariaefolia Smith in Rees, Cycl. 33:no. 16. 1819. *Spiraea discolor* var. *ariaefolia* Wats. Bot. Calif. 1:170. 1876. *Schizonotus argenteus* var. *ariaefolius* Kuntze, Rev. Gen. 1:225. 1891. *Schizonotus ariaefolius* Greene, Fl. Fran. 58. 1891. *(Menzies,* "northwest coast of America")

A more or less erect shrub (0. 5) 1-3 m. tall, the branches rather slender, often arching, angled by the decurrent petioles, the bark deep grayish-red; petioles mostly 10-15 mm. long; leaf blades broadly ovate to ovate-lanceolate, rarely less than 3 (and mostly 4-7, or even 10) cm. long, greenish and more or less hirsute on the upper surface, paler and strongly pilose-lanate to lanate and usually also puberulent and sometimes sessile-glandular on the lower surface, rounded to cuneate at base, with 15-25 shallow lobes to deep teeth, the main primary segments with 3-9 secondary mucronulate serrulations; panicles diffuse, (7) 10-17 cm. long; flowers about 5 mm. broad; petals slightly exceeding the sepals; styles about 1. 5 mm. long; achenes hirsute, about 2 mm. long. Ocean spray, Indian arrow wood.

Open gravelly to rocky soil, coastal bluffs to open dry to moist woods and lower mts. ; B. C. southward, along the coast, to s. Calif., e. to w. Mont., n. Ida., and n. e. Oreg., most common in our area on the w. side of the Cascade summit. June-Aug.

Although there are other varietal races of the species south of our area, all of our material appears to be referable to the var. *discolor*. It is possible, however, that the occasional specimen from along the Columbia R. and lower Snake R. canyons, with small, hirsute leaves, may be found to be other than mere badly stunted plants.

By some this is rated as one of our finest native shrubs. However, familiarity breeds contempt, and all too often the plant is ripped out as one of the first steps in home landscaping.

Holodiscus dumosus (Hook.) Heller, Cat. N. Am. Pl. 4. 1898.

Spiraea dumosa Nutt. ex T. & G. Fl. N. Am. 1:416. 1840, as a synonym; Hook. in Lond. Journ. Bot. 6:217. 1847. *Spiraea discolor* var. *dumosa* Wats. Bot. Calif. 1:170. 1876. *Schizonotus argenteus* var. *dumosus* Kuntze, Rev. Gen. 1:226. 1891. *H. discolor* var. *dumosa* Dippel, Handb. Laubh. 3: 508. 1893. *Schizonotus dumosus* Koehne, Deuts. Dendr. 265. 1893. *Schizonotus discolor* var. *dumosus* Rehd. in Bailey, Cycl. Am. Hort. 1629. 1902. *Sericotheca dumosa* Rydb. N. Am. Fl. 22³: 263. 1908. *(Nuttall,* stony and sandy places of the Platte River, lectotype by Ley, the second of 3 collections mentioned by T. & G.)

HOLODISCUS DUMOSUS var. GLABRESCENS (Greenm.) C. L. Hitchc. hoc loc. *Spiraea discolor* var. *glabrescens* Greenm. Erythea 7:116. 1899. *H. glabrescens* Heller, Muhl. 1:40. 1904. *Sericotheca glabrescens* Rydb. N. Am. Fl. 22³:265. 1908. *H. discolor* var. *glabrescens* Jeps. Man. Fl. Pl. Calif. 479. 1925. *(Cusick 1253,* "Stein's Mountains," Oreg., June, 1885)

H. microphyllus Rydb. Bull. Torrey Club 31:559. 1904. *Sericotheca microphylla* Rydb. N. Am. Fl. 22³: 264. 1908. *H. discolor* var. *microphyllus* Jeps. Fl. Calif. 2:166. 1936. *(M. E. Jones 1142,* Alta, Wasatch Mts., Utah, in 1879) = var. *dumosus.*

Sericotheca concolor Rydb. N. Am. Fl. 22³:264. 1908. *(J. Torrey 134,* Mt. Davidson, Nev., in 1865) = var. *dumosus.*

Low, intricately branched, spreading shrub (1) 5-15 (20) dm. tall with often reddish twigs lightly ridged by the decurrent petioles, the bark of the old branches deep grayish-red; leaves in fascicles of up to 6 or 7 at the ends of spur branches and alternate on the new growth, the blades (0. 5) 1-2 cm.

long, greenish and glabrous to copiously puberulent on the upper surface, but paler, prominently veined, and villous-pubescent or puberulent and usually more or less densely glandular to glabrate or glabrous beneath (except for the spherical, sessile, yellow glands), from ovate-elliptic to obovate with a cuneate base, with 6-15 shallow to prominent teeth that are sometimes secondarily mucronulate-denticulate; inflorescence 3-10 cm. long, usually a few- to several-branched panicle but sometimes reduced to a simple raceme; flowers 4-5 mm. broad; petals elliptic-oval, about 2 mm. long; stamens about equaling the petals; ovary villous, the style 1-1.5 mm. long; achenes about 2 mm. long.

Dry, rocky desert valleys and hillsides to well up in the mountains; n.c. Oreg. southward, on the e. side of the Cascades, to n.w. Calif. and the c. Sierra Nevada, e. and s. to c. Ida. and n.w. Wyo., Colo., N.M., Utah, Ariz., and Nev. June-Aug.

Although not so floriferous as *H. discolor*, in some horticultural respects this is the nicer of the two species, especially for areas e. of the Cascades.

Our plants are referable to two taxa, here treated as members of the same complex because of their intergradiency, although they are more usually referred to separate species, *H. dumosus* and *H. microphyllus* (or *H. discolor*).

1 Leaves copiously villous-pubescent (and usually glandular) on the lower surface, more
   nearly ovate or elliptic than obovate, and often toothed to below the middle; the common
   phase from c. Ida. (Blaine, Custer, Elmore cos.) southward; transitional to other vars.
   in Wyo., and Utah, but occasional w. to Siskiyou Co., Calif.   var. dumosus
1 Leaves sparsely pubescent to glabrous but obviously glandular on the lower surface, more
   nearly obovate than elliptic or ovate and generally toothed only above the middle; c. and
   e. Oreg., from extreme s.e. Marion Co. (where possibly intergradient to *H. discolor*)
   southward to n.e. Calif. and n.w. Nev.   var. glabrescens (Greenm.) C. L. Hitchc.

Horkelia Cham. & Schlecht.

Flowers complete, perigynous, borne in usually congested, terminal, cymose clusters; calyx bracteolate, with a saucer-shaped to cup-shaped hypanthium, the 5 lobes ascending to spreading; petals white (cream colored) to strongly pinkish-tinged, deciduous; stamens 10, biseriate; filaments flattened and dilated downward (ours) to subulate, mostly shorter than the petals; anthers dehiscing lengthwise by slits on the inner (adaxial) face; pistils mostly 5-30 (up to 250), borne on a short, rounded receptacle, the style subterminal, straight or nearly so, generally enlarged and glandular below, deciduous from the ripening, eventually ovoid, nearly (or quite) smooth achenes; perennial herbs, usually strongly pubescent and often glandular, with alternate, pinnate leaves, the leaflets toothed or divided into mostly linear segments, the upper ones generally confluent, the stipules usually laciniately cleft.

Seventeen species, all from w. N.Am., mostly in California. (Named for Johann Horkel, a German physiologist, 1769-1846.)

Reference:
Keck, David D. Revision of Horkelia and Ivesia. Lloydia 1:75-111. 1938.
1 Basal leaves with filiform-lobed stipules, the segments of the blades mostly 5-11 (13),
   narrowly oblanceolate to linear-cuneate, 2- to 3- toothed at the apex; petals cream; w.
   side of the Cascades in our area   H. CONGESTA
1 Basal leaves mostly with unlobed, wholly adnate stipules, the segments of the blades
   mostly (9) 11-19, cuneate to obovate-flabellate, several-lobed or -cleft; petals white
   or pink, often pinkish-veined; e. of the Cascade summits in our area   H. FUSCA

Horkelia congesta Dougl. ex Hook. in Curtis' Bot. Mag. 56:pl. 2880. 1829.
   Sibbaldia congesta Dietr. Syn. Pl. 2:1020. 1840. *Potentilla congesta* Baill. Hist. Pl. 1:369. 1867-
   69. (Described from cultivated specimens grown by the Horticultural Society of London, from seeds
   collected by Douglas, "detected at Cape Mendocena, and on the low hills of the Umptqua River upon
   the North-west coast of America")
   H. hirsuta Lindl. Bot. Reg. 23:pl. 1997. 1837. *(Douglas,* "California")

Taprooted perennial with a simple to branched caudex, the stems usually several, 2-4 (5) dm. tall, pilose-hirsute below, becoming glandular-puberulent above; basal leaves several, pilose-hirsute to silky, the stipules divided into 2-3 filiform segments, the blades 3-8 cm. long, with (5) 7-13 mostly narrowly oblanceolate or linear-cuneate, apically 2- to 3-toothed segments up to 2 cm. long; cauline leaves several, transitional to the several-lobed bracts; hypanthium cup-shaped, pilose-hirsute and glandular-puberulent, yellowish, not quite so long as the greenish-tinged, lanceolate-acute, spreading,

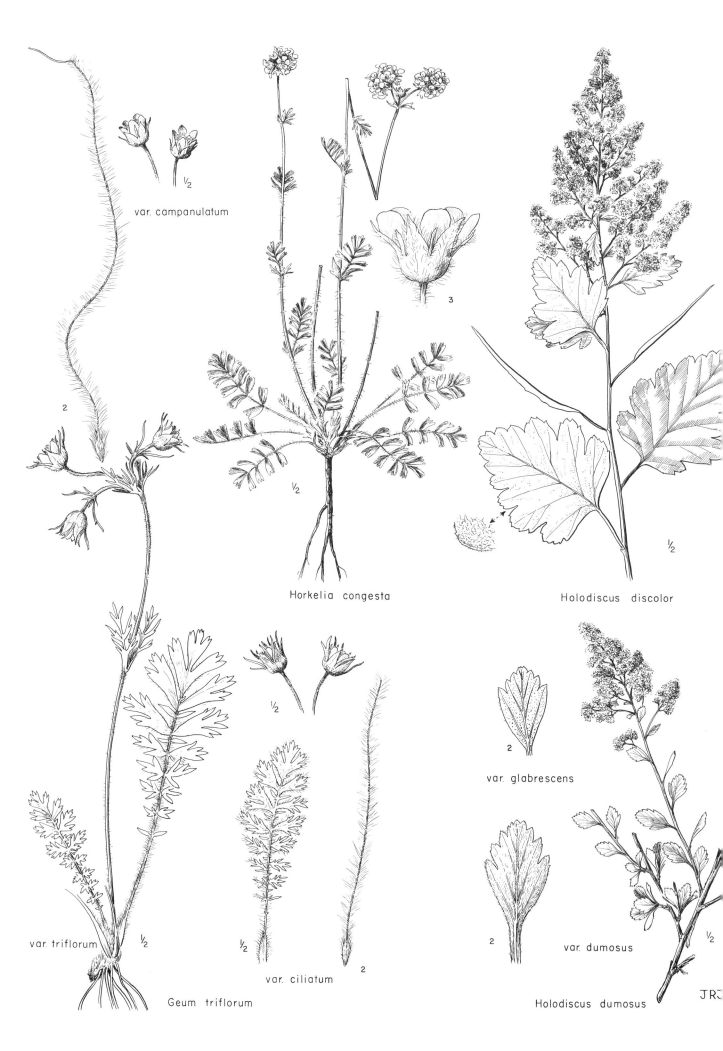

var. campanulatum

2

Horkelia congesta

½

3

Holodiscus discolor

var. triflorum   ½

Geum triflorum

½

var. ciliatum

2

2

var. glabrescens

2

var. dumosus   ½

Holodiscus dumosus

JRJ

4-5 mm. calyx lobes; petals cream, 5-6 mm. long, obovate-flabellate, slightly retuse, distinctly clawed; pistils mostly 10-20, the styles about equaling the stamens; mature achenes about 2 mm. long.

Open, sandy or rocky flats or sparsely wooded areas; Willamette Valley, Oreg., southward to the Umpqua and Rogue R. valleys, Oreg. Apr.-June.

Our material, as described above, is referable to the var. *congesta*, which ranges from Washington Co. to Douglas Co., Oreg.

Horkelia fusca Lindl. Bot. Reg. 23:pl. 1997. 1837.
  Potentilla douglasii Greene, Pitt. 1:103. 1887. *(Douglas,* "California")
  H. capitata Lindl. Bot. Reg. 23:pl. 1997. 1837. *Potentilla capitata* Greene, Pitt. 1:104. 1887. *H. fusca* ssp. *capitata* Keck, Lloydia 1:97. 1938. *H. fusca* var. *capitata* Peck, Man. High. Pl. Oreg. 398. 1941. *(Douglas,* "California")
  H. parviflora Nutt. ex H. & A. Bot. Beechey Voy. 338. 1838. *Potentilla andersonii* Greene, Pitt. 1:104. 1887. *H. fusca* ssp. *parviflora* Keck, Lloydia 1:99. 1938. *H. fusca* var. *parviflora* Peck, Man. High. Pl. Oreg. 399. 1941. *(Nuttall,* "R. Mts. U. Calif.")
  H. pseudocapitata Rydb. ex Howell, Fl. N.W. Am. 180. (Apr. 1) 1898. ("Southern base of the Blue Mountains in Oregon") = var. *capitata.*
  H. pseudocapitata Rydb. Monog. Potent. 134. (Nov. 25) 1898. *Potentilla douglasii* var. *pseudocapitata* Jeps. Fl. Calif. 2:203. 1936. *H. fusca* ssp. *pseudocapitata* Keck, Lloydia 1:99. 1938. *H. fusca* var. *pseudocapitata* Peck, Man. High. Pl. Oreg. 398. 1941. *(T. S. Brandegee,* Tanesville [=Janesville], Lassen Co., Calif., in 1892) Not the same as *H. pseudocapitata* Rydb. ex Howell.
  H. brownii Rydb. N.Am. Fl. 22[3]:276. 1908. *(H. E. Brown 530,* s. side of Mt. Shasta, Siskiyou Co., Calif., in 1897) = var. *pseudocapitata.*
  H. tenuisecta Rydb. N.Am. Fl. 22[3]:278. 1908. *Potentilla douglasii* var. *tenuisecta* Crum, Leafl. West. Bot. 1:100. 1934. *(Suksdorf 2492,* Falcon Valley, Klickitat Co., Wash., in 1882) = var. *fusca.*
  H. caeruleomontana St. John, Fl. S. E. Wash. 199. 1937. *(St. John & Smith 8329,* Wildcat Springs, Columbia Co., Wash., July 5, 1927) = var. *capitata,* the extreme phase of the variety, with large, unusually glandular calyx and long petals.

Taprooted perennial with a simple to branched caudex; stems usually several, (1) 1.5-6 (7) dm. tall, glabrous to pubescent or finely pilose below, becoming hirsute to pilose and glandular-pubescent above; basal leaves several, up to 2 dm. long, the stipules wholly adnate and forming a slight wing to the slender petioles, the blades with (9) 11-19 ovate or cuneate to obovate-flabellate, deeply crenate to linear-lobate or -cleft segments (0.5) 1-2 (2.5) cm. long; cauline leaves gradually reduced upward, with much larger, several-lobed stipules; calyx usually reddish- or purplish-tinged, the hypanthium cup-shaped, the segments triangular to lanceolate-acuminate, 2-4.5 mm. long, ascending; petals cuneate to obovate, rounded or (more commonly) retuse to somewhat obcordate, 2.5-6 mm. long, white to rather deep pink, usually pinkish- or reddish-lined; pistils up to 25, the style about twice as long as the ovary; mature achenes about 2 mm. long, nearly smooth.

Damp meadows to open forest and rocky slopes; c. Wash. to the Sierra Nevada of Calif., e. to Ida., w. Wyo., and n. Nev. June-July.

The species is notably variable, but several well-defined geographic races are readily recognizable in the more western part of the range, as follows:

1 Leaflets divided usually at least half their length into linear segments; herbage greenish, puberulent to pubescent but not grayish-hairy; n. Kittitas Co., Wash., s. along the Cascade foothills to n. Hood River and Wasco cos., Oreg.   var. fusca
1 Leaflets usually merely toothed or lobed much less than half their length, if more deeply divided then usually grayish-hairy, often silky
  2 Leaflets mostly broadly ovate to obovate; petals usually over 4 mm. long; n.w. Ida. southward, in w. Ida. and the Blue and Steens mts., Oreg., to n.w. Calif.
                                                                            var. capitata (Lindl.) Peck
  2 Leaflets usually more cuneate-oblanceolate or cuneate-obovate; petals rarely over 4 mm. long
    3 Lower surface of the leaves not at all grayish, puberulent and usually glandular; common in the Sierra Nevada, n. on the e. side of the Cascades (chiefly) to about Deschutes Co., Oreg., and very occasional farther n. to Wash.
                                                                            var. parviflora (Nutt.) Peck
    3 Lower surface of the leaves strongly pubescent and usually grayish, slightly if at all

glandular; from n. Deschutes Co. , Oreg. , southward (becoming more abundant in s.
Oreg. , and n. Calif.) to the c. Sierra Nevada   var. pseudocapitata (Rydb.) Peck

Ivesia T. & G.

Flowers perfect, perigynous, borne in crowded to somewhat diffuse cymes; calyx with a saucer-
shaped to turbinate or campanulate, prominently to obscurely disc-lined hypanthium, the lobes often
erect, persistent, alternate with, and usually longer than, the linear calyx bracteoles; petals yellow
(ours) or white or purplish, deciduous, usually more or less clawed; stamens 5 (ours) or sometimes
10, 15, or 20, the filaments commonly slender, inserted near the outer margin of the hypanthium, the
anthers laterally dehiscent; pistils 1-15 or more (mostly 2-6), borne on a low, usually hirsute receptacle, the style straight, subterminal, deciduous; fruit an achene, generally smooth; perennial, often
glandular herbs with pinnate, mostly basal leaves and usually dissected leaflets.
    Twenty-two species of w. N. Am. (Commemorating Dr. Eli Ives, 1779-1861, American physician
and botanist.)
    Reference:
    Keck, David D. Revision of Horkelia and Ivesia. Lloydia 1:111-142. 1938.
    The species have a certain attractiveness, chiefly because of the foliage, and perhaps could be
grown successfully in rockeries in drier areas. They are almost impossible, except for the specialist,
w. of the Cascades.
1 Leaflets mostly less than 20; inflorescence open, dichotomously branched, the flowers on
    filiform pedicels, recurved in fruit; Harney Co. , Oreg. (mostly in the Steens Mts.), n.c.
    Nev. , and Twin Falls Co. , Ida. , presently not known from quite within our area   *I. baileyi* Wats.
1 Leaflets mostly over 20; inflorescence congested, often subcapitate, the pedicels usually
    stout and erect
    2 Petals longer than the calyx lobes; hypanthium shallowly bowl-shaped, broader than
        deep   I. TWEEDYI
    2 Petals no longer than the sepals; hypanthium turbinate-conic, the width at the sum-
        mit less than the depth   I. GORDONII

Ivesia gordonii (Hook.) T. & G. in Newberry, Pac. R. R. Rep. 6[3]:72. 1857.
    Horkelia gordoni Hook. in Journ. Bot. & Kew Misc. 5:341, pl. 12. 1853. *(Gordon,* "Upper Platte
    River")
    I. alpicola Rydb. ex Howell, Fl. N. W. Am. 182. (Apr. 1) 1898. *Horkelia gordonii* var. *alpicola*
    Rydb. Monog. Potent. 152. (Nov. 25) 1898. ("On Mount Adams, Washington, at 5000-6000 feet
    elevation," collector not specified, but probably Th. Howell, Aug. 15, 1882)
    Strong perennial with a taproot and a simple or (more commonly) branched caudex covered with
blackish leaf bases, from very slightly to strongly glandular-puberulent and usually also finely spreading-hirsute; flowering stems 5-15 (21) cm. tall; basal leaves numerous, the blades 2-7 (10) cm. long;
leaflets usually more than 20, deeply divided into 3-5 (or more) obovate to oblong-cuneate segments
3-8 mm. long; cauline leaves usually 1, greatly reduced (sometimes lacking), the stipules usually several-lobed; calyx yellowish, the hypanthium turbinate-conic, (2. 5) 3-4 mm. long, equaled or slightly
exceeded by the deltoid-lanceolate, erect sepals; petals yellow, obovate to spatulate, from 2/3 as long
to nearly as long as the sepals; pistils mostly 2-4 (6); achenes ovoid-flattened, about 2 mm. long.
    Rocky alpine and subalpine ridges and talus slopes and along flood plains and riverbanks at lower
elevations; Table Mt. , Chelan Co. , and Goat Rocks area (Lewis-Yakima cos.) and Mt. Adams, Cascade Mts. , Wash. , from the Blue and Wallowa mts. of s. e. Wash. and n. e. Oreg. , e. to Ravalli and
Madison cos. , Mont. , c. and s. Ida. , Wyo. , Utah, and n. e. Colo. , and from s. c. Oreg. to n. w. and
Sierra Nevadan Calif. Late June-Aug.

Ivesia tweedyi Rydb. N. Am. Fl. 22[3]:288. 1908.
    Horkelia tweedyi Nels. & Macbr. Bot. Gaz. 61:31. 1916. *(F. Tweedy,* Yakima Region, Wash. , Aug. ,
    1883)
    Strong perennial with a large taproot and usually a freely branched caudex, the stems 5-20 cm. tall,
from nearly glabrous to glandular-puberulent or glandular-pubescent below, becoming more glandular
and usually somewhat pilose above, often reddish or purplish; basal leaves numerous, the blades 3-8
cm. long; leaflets 19-35, mostly less than 1 cm. long, dissected into many filiform to linear segments,
glabrous or more generally finely glandular-puberulent, sometimes sparsely pilose-hirsute; cauline

leaves commonly 1-3, greatly reduced, the stipules well developed, usually dissected; calyx yellowish-green, the hypanthium shallowly bowl-shaped, shorter than the deltoid-ovate, ascending, yellowish, 2.5-3 mm. lobes; petals yellow, from broadly elliptic to spatulate, slightly longer than the calyx lobes and the stamens; pistils (2) 4-6 (9), the styles about 3 mm. long; achenes about 2 mm. long.

Dry, open to wooded, usually rocky slopes or alpine ridges, often on serpentine; e. side of the Cascade Mts. from s. Chelan to n. Yakima Co., Wash., also from the Coeur d'Alene Mts., Ida. June-Aug.

Kelseya (Wats.) Rydb.

Flowers single and terminal, subsessile, complete, perigynous; calyx ebracteolate, 5-merous, campanulate, the lobes nearly twice as long as the hypanthium; petals 5, pink; stamens (7 to) 10, inserted with the petals at the outer margin of the disc-lined hypanthium; pistils usually 3 (4-5 reported), free of the calyx, the style slender, mostly deciduous from the mature fruit; stigma truncate, entire, not enlarged; ovules about 3 (2-5); fruit usually 1 (2-4)-seeded, leathery, folliclelike but completely dehiscent on the 2 almost equally prominent sutures; caespitose, pulvinate shrubs with alternate, entire, greatly crowded and imbricate, persistent, exstipulate leaves.
Only the one species. (Named for its discoverer, Rev. F. D. Kelsey, 1849-1905.)

Kelseya uniflora (Wats.) Rydb. Mem. N.Y. Bot. Gard. 1:207. 1900.
   Eriogynia uniflora Wats. Proc. Am. Acad. 25:130. 1890. *Luetkea uniflora* Kuntze, Rev. Gen. 1:217.
      1891. *Spiraea uniflora* Piper, Erythea 7:172. 1899. *(F. D. Kelsey, Gate of the Mountains, Mont.,*
      in 1888 and 1892)
Intricately branched, solid-cushioned shrubs up to 4-6 dm. broad but less than 1 dm. tall, the leafy branches densely crowded, mostly 1-3 (8) cm. long, covered with marcescent, often hardened and crustose leaves, the functional foliage crowded, imbricate, elliptic-oblanceolate to spatulate-obovate, 2.5-4 mm. long, finely sericeous, grayish-green; calyx campanulate in flower, more nearly globose in fruit, strongly tinged with pinkish-purple, the lobes oblong-oblanceolate, about 1.5 mm. long, the hypanthium purplish within; petals pinkish with a purple tinge, fading to brown, oblong-elliptic, 2-3 mm. long, persistent; stamens usually slightly longer than the petals, reddish-purple; styles about equaling the stamens; follicles ellipsoid, about 3 mm. long, the seeds fusiform.
In limestone rock crevices, at 6500 to 11,500 ft. elevation; known only from a few localities in Lewis & Clark and Meagher cos., Mont., Custer and Butte cos., Ida., and the Big Horn Mts., Wyo. Late May-early July.
This would undoubtedly be one of the most highly prized rock garden plants known if it could be introduced into cultivation, but unfortunately its domestication does not seem possible. Reports that the plant is white-flowered are erroneous.

Luetkea Bong.

Flowers racemose, complete, perigynous, borne in leafy-bracteate racemes; calyx ebracteolate, the hypanthium obconic, lined internally with a shallowly lobed disc, the sepals spreading to erect; petals white, tardily deciduous; stamens about 20; pistils (4) 5 (6), free of the calyx, the style rather short and slender, the stigma small, not lobed; fruit a several-seeded follicle with a stipelike base, dehiscent primarily on the very conspicuous raised ventral suture and secondarily on the indistinct dorsal suture; seeds fusiform; trailing evergreen semishrubs with numerous, biternate, exstipulate leaves.
Only the one species. (Named for Count F. P. Luetke, 1797-1882, a Russian commander and Arctic explorer whose expedition circumnavigated the earth.)

Luetkea pectinata (Pursh) Kuntze, Rev. Gen. 1:217. 1891.
   Saxifraga pectinata Pursh, Fl. Am. Sept. 312. 1814. *Eriogynia pectinata* Hook. Fl. Bor. Am. 1:255.
      1833. *Spiraea pectinata* T. & G. Fl. N. Am. 1:417. 1840. *(Menzies, "On the north-west coast")*
   L. sibbaldioides Bong. Mém. Acad. St. Pétersb. VI, 2:130. 1832. *(Mertens, Sitka Island)*
Rhizomatous and stoloniferous semishrubs forming extensive mats, the upright leafy flowering stems mostly (5) 10-15 cm. tall; leaves mostly crowded in thick basal tufts, often marcescent, the petiole narrowly wing-margined, 5-10 mm. long, about equaling the usually biternately dissected, linear-lobed blade; calyx glabrous, the lobes triangular, about 2 mm. long, nearly twice as long as

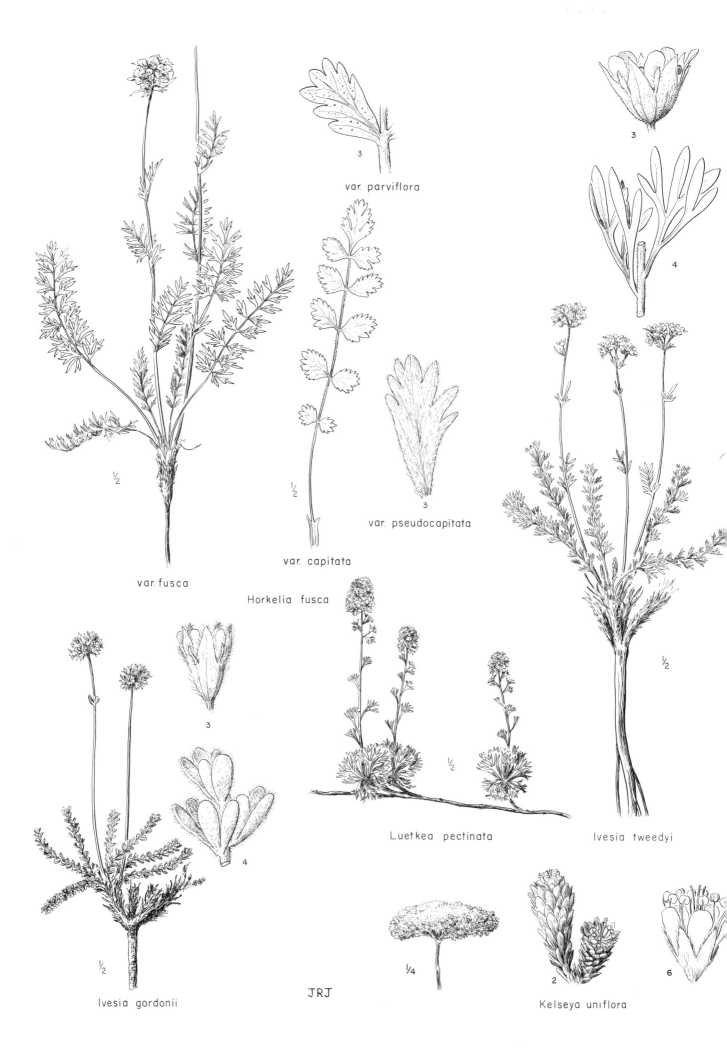

var. parviflora

3

var. pseudocapitata

3

var. capitata

var. fusca

Horkelia fusca

½

½

½

½

3

4

Ivesia gordonii

Luetkea pectinata

Ivesia tweedyi

JRJ

¼

2

6

Kelseya uniflora

the hypanthium; petals spatulate to obovate, about 3-3. 5 mm. long; stamens somewhat shorter than the corolla, the slender filaments slightly connate basally; styles shorter than the stamens; follicles about 5 mm. long, slightly sericeous on the ventral suture. Partridge foot.

Moist or shaded, usually sandy soil, often where snow persists until late in the season, subalpine to well above timber line in the mts. of w. N.Am., from Alas. and Yukon southward in the Cascades (and w.) to n. Calif., and in the Rocky Mts. to s. B.C., s.w. Alta., and the Bitterroot Mts. of w. Mont. and e. Ida. *(Hitchcock & Muhlick 15331)*. June-Aug.

This is one of our most attractive native ground covers for semishady areas, as around dwarf conifers or on exposed (if well irrigated) banks and rockeries. Although particularly appealing when in flower, it rates as one of our best foliage plants. It is easily propagated by cuttings and divisions.

Osmaronia Greene Indian Plum; Osoberry

Flowers imperfect, perigynous, borne in axillary, bracteate racemes; calyx turbinate-campanulate, ebracteolate (but the pedicels bracteolate), the hypanthium gland-lined, persistent in the staminate flowers, quickly circumscissile on the pistillate flowers near the base and leaving a collarlike remnant under the developing fruit; sepals 5; petals 5, white, deciduous; staminate flowers without vestiges of pistils, the stamens 15, borne in 3 distinct series; pistillate flowers with the normal number of stamens, but the filaments very short and the anthers small and probably never functional; pistils usually 5, distinct, the styles included in the hypanthium, deciduous; stigmas capitate; ovules 2; fruit thin-fleshed, 1-seeded drupes, 5 (or through abortion only 1-4) per flower; endosperm lacking, the cotyledons folded; deciduous, bitter-barked shrubs with alternate, entire (undulate) leaves and quickly deciduous stipules.

Only the 1 species. (Greek *osme,* smell, fragrant, plus the generic name *Aronia;* the flowers are fragrant and the crushed foliage has a fresh cucumberlike smell.)

Osmaronia cerasiformis (T. & G.) Greene, Pitt. 2:191. 1891.
   Nuttallia cerasiformis T. & G. in H. & A. Bot. Beechey Voy. 337. 1838. *(Nuttall,* "on the Columbia")
   Exochorda davidiana Baill. Adansonia 9:149. 1868-70. *Nuttallia davidiana* Baill. Adansonia 11:328. 1875. (Typification obscure)
   O. cerasiformis var. *lancifolia* Greene, Pitt. 5:309. 1905. *(J. W. Macoun 34367,* Chilliwack Valley, Can.)
   O. cerasiformis var. *nigra* Greene, Pitt. 5:309. 1905. *(O. D. Allen,* "Nesqually" Valley, Wash.)
Shrub to small tree 1.5-3 (5) m. tall, with purplish-brown bark; petioles 5-10 (15) mm. long, the leaf blades narrowly oblong-lanceolate or -elliptic to -obovate, 5-12 cm. long and up to 4 cm. broad, glabrous above, slightly paler and often pubescent (glabrate) on the lower surface; racemes axillary, usually pendent, the bracts deciduous; flowers greenish-white, blossoming as the leaves develop; calyx 6-7 mm. long, the lobes about equaling the turbinate hypanthium; petals 5-6 mm. long (smaller on the pistillate flowers), elliptic-obovate, very short-clawed, ascending; drupes 8-10 mm. long, bluish-black, bitter. N=8.

Stream banks, roadsides, and moist to fairly dry, open woods; B.C. southward, from the coast to the w. slope of the Cascades, through Wash. and Oreg. to n. Calif. and the w. side of the Sierra Nevada. Mostly Mar.-Apr.

The plant is a cheerful harbinger of spring, the foliage is a fresh, cool green, and the fruits are attractive if the birds allow them to ripen. Nevertheless, it can scarcely be considered a choice ornamental.

Peraphyllum Nutt. in T. & G.

Flowers 1-3 at the ends of short lateral branches, appearing with the fasciculate leaves, perfect, epigynous; calyx turbinate, the 5 lobes spreading to reflexed, persistent, about equal to the combined adnate portion and the disc-lined, free hypanthium; petals 5, deciduous; stamens about 15 (20), inserted at the top of the hypanthium at the edge of the disc; pistil one, 2 (3)-carpellary, the ovary inferior, completely 2 (3)-celled but falsely 4 (6)-celled by the intrusion of 2 (3) false parietal septa; styles 2 (3), free, hollow below; stigma capitate-discoid; fruit a rather cartilaginous, semiglobose, several-seeded pome; unarmed, deciduous shrubs with simple, minutely stipulate leaves.

Only the 1 species. (Greek *pera,* leather pouch, and *phyllon,* leaf.)

Peraphyllum ramosissimum Nutt. in T. & G. Fl. N. Am. 1:474. 1840. (Nuttall, "Dry hill-sides near
the Blue Mountains of the Oregon")

An intricately and rigidly branched, dark grayish-barked shrub 0.5-2 m. tall; leaves fascicled at
the ends of short spurs, narrowly oblanceolate, abruptly acute, (1) 1.5-4 cm. long, narrowed to very
short petioles, more or less appressed-puberulent, minutely serrulate (entire), the tips of the teeth
(and the tiny reddish stipules) usually quickly deciduous; calyx lobes triangular, 3.5-5 mm. long,
somewhat lanate and caducous-serrulate; petals pale pinkish to rose, spreading, obovate-oblong to
obovate, 6-7 mm. long; stamens and styles well exserted; pome yellowish to reddish, bitter, 8-10
mm. long.

Sagebrush desert to juniper or ponderosa pine woodland; n. c. Oreg., from Grant and Baker cos.
s. to n. e. Calif., e. through s. Ida. to Utah and Colo. May-June.

An attractive shrub but not believed to be adaptable to the climate w. of the Cascades.

Petrophytum (Nutt.) Rydb.

Flowers small, complete, perigynous, borne in crowded, spikelike, simple (sometimes compound),
bracteate racemes on erect, bracteate-leafy peduncles; calyx ebracteolate, free of the pistils, the
hypanthium turbinate to hemispheric, lined internally with a disc that is thickened or rimlike at the
base of the petals and stamens; sepals 5, erect to reflexed, persistent; petals 5, white, equaling to
twice as long as the sepals, mostly withering persistent; stamens 20-40, the filaments slender; pistils 5 (3-7), distinct, the ovary 2- to 4-ovulate, the style 2-4 times as long as the ovary, pilose to short-
pubescent near the base, the stigma not at all enlarged; fruit follicular, dehiscent on the ventral and
(tardily) on the dorsal sutures; seeds usually 1 or 2, linear, exalbuminous; prostrate, matted, woody
shrubs, with alternate (but closely crowded and tufted), simple, entire, more or less spatulate leaves
that are persistent and usually marcescent and cushion-forming.

Only the three species. (Greek petros, rock, and phyton, plant, the plants growing in rock crevices.)

The three species comprising this genus form a very natural group, all remarkably alike in habit
and general appearance, yet each clearly distinct from the others. They are sometimes included in
Spiraea, but differ not only in habit and leaf character from that genus, but also because of the racemose inflorescence, more prominent disc, and dorsally as well as ventrally dehiscent follicles.

Although they are not easily maintained w. of the Cascades, all are growable with at least encour-
aging success in dry, sunny areas, such as a rock wall.

1 Leaves 1-nerved on the lower surface, sericeous; styles about 3 mm. long   P. CAESPITOSUM
1 Leaves 3-nerved, glabrous to sericeous; styles 1-2 mm. long
　　2 Stamens 20-25; styles about 2 mm. long; hypanthium turbinate, grayish-sericeous,
　　　　the sepals lanceolate-triangular, erect; peduncles 5-15 cm. tall; racemes often
　　　　compound; pedicels 0.5-2 mm. long   P. CINERASCENS
　　2 Stamens 35-40; styles 1-1.5 mm. long; hypanthium hemispheric, sparsely hairy to
　　　　glabrous, the sepals oblong, reflexed; peduncles 1-5 (6) cm. tall; racemes simple;
　　　　pedicels 2-3 (5) mm. long   P. HENDERSONII

Petrophytum caespitosum (Nutt.) Rydb. Mem. N. Y. Bot. Gard. 1:206. 1900 (as Petrophyton).
Spiraea caespitosa Nutt. in T. & G. Fl. N. Am. 1:418. 1840. Eriogynia caespitosa Wats. Bot. Gaz.
15:242. 1890. Luetkea caespitosa Kuntze, Rev. Gen. 1:217. 1891. (Nuttall, "Rocky Mountains,
towards the sources of the Platte")

Strongly caespitose, forming cushions or mats up to 1 m. broad; leaves spatulate to oblanceolate,
(5) 7-12 (14) mm. long, (1) 1.5-4 mm. broad, 1-nerved, strongly sericeous and grayish-green on
both surfaces; peduncles (1) 2-8 cm. tall, with several reduced, bractlike leaves; racemes closely
crowded, spikelike (pedicels 0.5-1.5 [3] mm. long), sometimes more or less compound at the base,
(1) 2-5 cm. long; calyx sericeous to somewhat hirsutulous, the hypanthium turbinate, about 1 mm.
long, slightly exceeded by the lanceolate-triangular, erect lobes, the disc with a prominent, entire
margin projecting above the point of adnation of the petals and stamens; petals white, spatulate-oblanceolate, from as long to half again as long as the sepals; stamens 20, the filaments slender, glabrous,
about twice as long as the petals; pistils 5 (3-6), the styles slender, nearly 3 mm. long, slightly pilose; follicles glabrous to sparsely pilose, about 2 mm. long.

From the foothills to alpine summits, almost entirely on shelving rocks (granitic or limestone), rooting in the crevices; n. e. Oreg. to Calif., e. to Ida., Mont., and S.D., and s. to Ariz., N.M., and Tex. Late June-Aug.

Material from the limestone ranges of c. Ida. seems to be slightly more spreading-hairy than that from other areas, but is probably not varietally separable.

Petrophytum cinerascens (Piper) Rydb. N.Am. Fl. 22[3]:253. 1908.

Spiraea cinerascens Piper, Erythea 7:171. 1899. *Luetkea cinerascens* Heller, Muhl. 1:53. 1904. (*Elmer 853*, bluffs of the Columbia R., 12 mi. s. of Chelan, Chelan Co., Wash.)

Prostrate, matted shrublet; leaves oblanceolate-spatulate, (10) 15-25 mm. long, 2-4 (5) mm. broad, 3-nerved, sparsely appressed-pubescent; peduncles 5-15 cm. tall, with scattered bractlike leaves below, but with more numerous, linear, acute bracts bearing rudimentary buds below the simple or compound, tight, (1) 2-6 cm. racemes; pedicels 0.5-2 mm. long; calyx grayish-sericeous and crisp-puberulent, the hypanthium turbinate, scarcely 1 mm. long, the internal disc with a barely free edge; sepals lanceolate-triangular, about 1.5 mm. long, erect; petals narrowly cuneate-oblanceolate, scarcely exceeding the sepals; stamens 20-25, the filaments sparsely pilose, about twice as long as the sepals; pistils 5 (4-7), the style slender, about 2 mm. long, sparsely pubescent; follicles nearly 3 mm. long.

On basaltic cliffs along the Columbia R. in Chelan Co., Wash. June-Aug.

Petrophytum hendersonii (Canby) Rydb. N.Am. Fl. 22[3]:253. 1908.

Eriogynia hendersoni Canby, Bot. Gaz. 16:236. 1891. *Luetkea hendersonii* Greene, Pitt. 2:219. 1892. *Spiraea hendersoni* Piper, Erythea 7:172. 1899. (*Henderson*, Olympic Mts., Wash., July 15, 1890)

Caespitose, prostrate shrublet; leaves oblanceolate to spatulate-obovate, 10-20 (25) mm. long, 2-6 mm. broad, 3-nerved, sparsely to rather copiously pilose-sericeous (glabrate); peduncles 1-5 (6) cm. long, with several bractlike leaves; racemes simple, 2-4 (5) cm. long, tight, the pedicels mostly 2-3 (5) mm. long; calyx sparsely hairy, the hypanthium hemispheric, scarcely 1 mm. long, slightly more than half as long as the oblong, reflexed, ciliate lobes, the internal disc thickened at the outer edge but not projecting past the point of adnation of the petals and stamens; petals oblong to oblong-obovate, 2-2.5 mm. long; stamens 35-40, equaling or slightly exceeding the petals; pistils usually 5 (4-7), the styles stout, 1-1.5 mm. long, sparsely pilose; follicles barely 2 mm. long.

Rocky cliffs and talus slopes; Olympic Mts., Wash., at 4500-7000 ft. elevation. July-Sept.

Physocarpus Maxim. Nom. Conserv. Ninebark

Flowers rather numerous in terminal corymbs, complete, regular, perigynous; calyx ebracteolate, free of the pistils, persistent, the hypanthium hemispheric or campanulate to turbinate, disc lined, the 5 lobes spreading to reflexed; petals 5, white, spreading, deciduous; stamens 20-40, inserted with the petals at the edge of the disc; pistils 1-5, ovaries rather weakly connate, styles slender, stigmas capitate; fruit follicular, several-seeded, more or less inflated but usually also compressed, dehiscent on two sutures; seeds (1) 2-4, pyriform, hardened, shining; shrubs with exfoliating bark and alternate, deciduous, usually palmately 3- to 5-lobed leaves and deciduous stipules, generally stellate-pubescent at least on the calyx.

About a dozen species of w. N.Am. and (one) of Asia. (Greek *physa*, bellows or bladder, and *carpos*, fruit, because of the inflated follicles.)

The plants are easily cultivated and not unattractive (but hardly choice) ornamentals.

1 Pistil usually 1; leaf blades mostly less than 2 cm. long; filaments of 2 alternating and markedly unequal lengths; e. Calif., Nev., and Utah; reported from Ida., but probably s. of our range   *P. alternans* (M. E. Jones) J. T. Howell
1 Pistils 2-5; leaf blades mostly well over 2 cm. long; filaments from equal to somewhat unequal but not alternately longer and markedly shorter
  2 Pistils usually 2 (3-5), copiously stellate, connate at least half the length of the ovaries; entirely e. of the Cascades   P. MALVACEUS
  2 Pistils 3-5 (never 2), glabrous to sparsely stellate along the ventral suture, connate only at the base; mostly w. of the Cascades in our area except also in n. Ida. P. CAPITATUS

Physocarpus capitatus (Pursh) Kuntze, Rev. Gen. 1:219. 1891.

Spiraea capitata Pursh, Fl. Am. Sept. 342. 1814. *Spiraea opulifolia* var. *mollis* T. & G. Fl. N.Am.

1:414. 1840. *Neillia opulifolia* var. *mollis* Brew. & Wats. Bot. Calif. 1:171. 1876. *Neillia capitata*
Greene, Pitt. 2:28. 1889. *Opulaster capitatus* Kuntze, Rev. Gen. 2:949. 1891. *Opulaster opulifolius*
var. *capitatus* Jeps. Fl. W. Middle Calif. 276. 1901. *(Menzies, "On the north-west coast")*
Spiraea opulifolia var. *tomentella* Ser. in DC. Prodr. 2:542. 1825. *Physocarpa tomentosa* Raf. New
Fl. N. Am. 3:74. 1836. (No specimens cited)

Spreading to erect shrub 2-4 (6) m. tall, the branches glabrous to minutely stellate, angled; petioles
slender, 1-3 (4) cm. long; leaf blades ovate to cordate, 3- or 5-lobed less than half the length and
again irregularly biserrate to incised and biserrate, rounded to acute, (3) 4-8 (10) cm. long, nearly
as broad, sparsely stellate to glabrous and dark green above, paler and usually copiously stellate on
the lower surface; calyx finely stellate, the lobes ovate-lanceolate, about 3 mm. long, slightly longer
than the hemispheric hypanthium, somewhat reflexed; petals suborbicular, about 4 mm. long; stamens
about 30, equaling or exceeding the petals; pistils 3-5, the ovaries weakly connate about half their
length, glabrous or stellate above on the ventral suture; styles about equaling the stamens; mature
follicles usually glabrous, 7-11 mm. long, turgid; seeds usually 2.3-2.8 mm. long, pale, smooth,
somewhat obliquely pyriform. N=9.

Stream banks, swamps, and lake margins to moist woods in the lower mountains; s. Alas. southward, almost entirely on the w. side of the Cascade Mts., to the c. Sierras and coastal s. Calif., but
also in n. Ida. May-June.

This is one of the several species of plants occurring mainly on the lowlands w. of the Cascade Mts.
and in the moister part of n. Ida.; in this case there are no known intervening stations. It is closely
related to *P. opulifolius* (L.) Raf. of n.e. and n.c. N.Am. (sometimes cultivated in our area), which
has smaller seeds (generally less than 2 mm. long) and usually shorter sepals and proportionately
longer follicles.

Physocarpus malvaceus (Greene) Kuntze, Rev. Gen. 1:219. 1891.
Neillia malvacea Greene, Pitt. 2:30. 1889. *Opulaster malvaceus* Kuntze, Rev. Gen. 2:949. 1891.
Neillia monogyna var. *malvacea* M. E. Jones, Zoë 4:43. 1893. *(Greene, "Lake Pend d'Oreille in
northern Idaho," Aug. 9, 1889)*
Spiraea opulifolia var. *pauciflora* T. & G. Fl. N.Am. 1:414. 1840. *Spiraea pauciflora* Nutt. ex T. &
G. loc. cit., as a synonym. *Opulaster pauciflorus* Heller, Bull. Torrey Club 25:581, 626. 1898.
P. pauciflorus Piper in Piper & Beattie, Fl. Palouse Region 94. 1901. *Neillia torreyi* Hook. f. in
Curtis' Bot. Mag. 127:pl. 7758. 1901, but not *N. torreyi* of Wats. in 1876, which = *P. monogynus*.
(Dr. James, "Rocky Mountains in about lat. 40°," is the first of 3 collections cited)
Opulaster pubescens Rydb. Mem. N.Y. Bot. Gard. 1:205. 1900. *(Scribner 35, Hound Creek, Mont.,
in 1883, is the first of 3 collections cited)*
Opulaster cordatus Rydb. N.Am. Fl. 22³:242. 1908. *(Sandberg, MacDougal, & Heller 575, in part,
Farmington Landing, Coeur d'Alene, Ida., in 1892) A plant with more than the normal number of
pistils.

Very similar to *P. capitatus*, but smaller (0.5-2 m. tall) and usually growing in drier habitats, the
leaves averaging slightly smaller and the corymbs generally fewer-flowered; pistils commonly only
2 (fairly frequently 3, but very rarely 4 or 5), the ovaries densely grayish-stellate and connate at
least to midlength.

Canyon bottoms and rocky hillsides to ponderosa pine and Douglas fir forest; entirely to the e. of the
Cascades, from s.c. B.C. through c. and e. Wash. to c. and e. Oreg., e. to s.w. Alta., Mont.,
Wyo., and Utah. June-July.

Occasional sporadic plants from Wash., Ida., and Mont. have been seen to bear flowers with a variable number of pistils, usually 2 or 3, but sometimes 4 or 5. They seem not to constitute a geographic race.

Potentilla L. Cinquefoil; Five-finger

Flowers from single and terminal to numerous in bracteate, sometimes leafy, simple to branching
cymes, complete but occasionally functionally imperfect; calyx with a basal saucer-shaped to cupuliform hypanthium, usually gland lined within, the 5 lobes spreading to erect, alternating with usually
narrower and shorter, entire to rarely toothed bracteoles; petals generally yellow, sometimes white,
occasionally reddish or purple, from shorter to longer than the sepals, deciduous; stamens usually
10-30 (in ours), occasionally fewer or more numerous, inserted at the outer margin of the hypanthium,
the anthers basifixed, more or less dorsiventrally compressed, dehiscent along the outer margin; pis-

2

3

½

Osmaronia cerasiformis

½

Peraphyllum ramosissimum

2

½

Petrophytum caespitosum

6

2

6

2

½

½

Petrophytum hendersonii

6

2

3

2

2

Petrophytum cinerascens

2

½

Physocarpus capitatus

6

½

Physocarpus malvaceus

JRJ

tils numerous, the style apically to laterally or almost basally inserted, from filiform (and usually smooth at the base) to considerably thickened at (or somewhat above) the base (and then usually slightly glandular-warty or papillose below midlength), joined with the ovary and usually deciduous; achenes smooth to strongly reticulate, sometimes enclosed by the accrescent calyx; annual to perennial herbs or (1 species of ours) shrubs, often with well-developed rootstocks, and pinnately to digitately compound, toothed to dissected, alternate, stipulate leaves.

Possibly 200 species of the Northern Hemisphere, chiefly in the Temperate Zone, but n. into the Arctic; many montane. (From the Latin *potens,* powerful, in reference to supposed medicinal properties of some of the species.)

Several species have horticultural promise if not proven worth, *P. fruticosa* being considered choice. *P. villosa* is decidedly attractive and perhaps manageable, *P. gracilis* is easily grown, but not particularly attractive, and *P. flabellifolia* is well worth a trial.

The genus is a notoriously difficult one, taxonomically, chiefly because of the apomictic reproduction and free hybridization of many of the species. It is not very sharply delimited from the related *Horkelia* and *Ivesia,* and as treated here includes the segregate genera *Comarum, Drymocallis, Dasiphora,* and *Argentina.*

References:

Clausen, Jens, Keck, David D. , and William M. Hiesey. Experimental studies of the nature of species, II, Potentilla glandulosa and its allies. III, Potentilla gracilis and its allies. IV, Potentilla drummondii and Potentilla breweri. Carn. Inst. Wash. Pub. no. 520:26-195. 1940.

Rydberg, P. A. A monograph of the North American Potentilleae. Mem. Dept. Bot. Columbia Univ. 2:1-223. 1898.

1 Flowers purple to deep red   P. PALUSTRIS
1 Flowers white, cream, or yellow
  2 Plant shrubby; ovaries and achenes strongly hirsute   P. FRUTICOSA
  2 Plant herbaceous; ovaries and achenes glabrous
    3 Plants annuals or biennials, or perhaps sometimes short-lived perennials but without rootstocks, the leaves never white-tomentose on the lower surface; stamens 10-20
      4 Leaflets less than 1 cm. long; stems prostrate to ascending, rarely over 3 dm. long   P. NEWBERRYI
      4 Leaflets mostly 1-5 cm. long; stems usually erect and often over 3 dm. long
        5 Mature achenes with a thickened, wedge-shaped appendage on the inner (adaxial) margin that is nearly or quite as large as the rest of the fruit; stamens seldom fewer than 20; lower leaves mostly 5- to 9-foliolate   P. PARADOXA
        5 Mature achenes without a prominent appendage on the inner margin; stamens 10-20; lower leaves usually 3 (5)-foliolate
          6 Stems stiffly hirsute below with unicellular, more or less pustular based, spreading hairs; achenes usually strongly undulate-corrugate longitudinally; stamens (15) 20; petals mostly at least 3/4 as long as the sepals
                                                                P. NORVEGICA
          6 Stems soft-pubescent below, often with glandular or with multicellular or semitomentose pubescence; achenes smooth or very slightly striate; stamens mostly 10-15; petals usually less than 3/4 as long as the sepals
            7 Basal portion of the stems soft-pubescent, often more or less lanate, eglandular, the hairs unicellular; lower cauline leaves frequently 5-foliolate; calyx eglandular   P. RIVALIS
            7 Basal portion of the stems in part pubescent with multicellular, more or less moniliform, often glandular hairs; lower cauline leaves 3-foliolate; calyx mealy-glandular   P. BIENNIS
    3 Plants perennials, usually with well-developed rootstocks; leaves often white-tomentose on the lower surface; stamens 20-30 (40)
      8 Flowers single on naked peduncles; plants strongly stoloniferous and with multipinnate leaves
        9 Mature achenes only slightly if at all wrinkled on the back, without a pedicel-like base; stolons, petioles, and leaf rachises from sparsely appressed-hairy to glabrous; coastal   P. PACIFICA
        9 Mature achenes rather corky and with 1 to several longitudinal ridges or

wrinkles, often with a persistent basal stalk; stolons, petioles, and leaf rachises densely to sparsely silky with usually spreading hairs; mostly e. of the Cascades in our area   P. ANSERINA

8 Flowers usually several on more or less leafy flowering stems; plants not stoloniferous; leaves often ternate to palmate

  10 Style slenderly fusiform, usually roughened at or below midlength, attached somewhat below midlength of the ovary; leaves pinnate, with 5-11 leaflets; stamens about 25, sometimes up to 40 *(Drymocallis)*

    11 Calyx shallowly bowl-shaped, the sepals ascending at anthesis; petals 2-4 mm. longer than the sepals, cupped upward, never spreading   P. CAMPANULATA

    11 Calyx usually nearly rotate or if with ascending sepals then the petals either from shorter to barely longer than the sepals or spreading at anthesis

      12 Cymes narrow and strict (often much elongate), the lateral branches almost erect; sepals mostly 6-8 mm. long at anthesis; petals pale yellow ("white"), usually equaling to slightly exceeding (by 1-2 mm.) the sepals; plants mostly over 4 dm. tall   P. ARGUTA

      12 Cymes usually open to diffuse, sometimes glomerate, the lateral branches not tightly appressed; sepals often less than 6 mm. long; petals pale to deep yellow, if pale yellow ("white") then usually shorter than the sepals or the plants mostly less than 4 dm. tall or the inflorescence open   P. GLANDULOSA

  10 Style usually tapered from the base, or filiform or attached near the top of the ovary; leaves palmate or if pinnate often with more than 11 leaflets; stamens often 20

    13 Basal leaves predominantly odd-pinnate, at least some with 5 or more leaflets, few (if any) digitate although occasionally a few ternate

      14 Plant a short-lived perennial, eglandular, grayish-hirsute, low and spreading from a simple or branched crown and a taproot, but without rhizomes; petals 4-6 mm. long, nearly white, the inflorescence very leafy; style slender but thickened and more or less glandular at the base   P. NEWBERRYI

      14 Plant either a strong upright perennial, or glandular, or with petals over 6 mm. long or yellow, or with styles not thickened at the base; inflorescence usually not strongly leafy

        15 Plants finely glandular-puberulent, 5-10 (15) cm. tall; style slender, attached about midway between midlength and the tip of the ovary; leaflets (3) 5 (7), closely crowded, orbicular-flabellate, shallowly cleft and crenate, mostly less than 10 mm. long   P. BREVIFOLIA

        15 Plants with most if not all of the hairs eglandular, usually well over 10 cm. tall; style usually attached terminally; leaflets various

          16 Styles thickened and glandular-roughened or papillate at the base, tapered to the tip, shorter or only very slightly longer than the mature achenes; leaflets oblong to oblanceolate-obovate, toothed or lobed 1/2-4/5 of the way to the midvein; stipules usually deeply cleft

            17 Plants mostly 2-5 dm. tall; leaflets 5-11, more than 5 on at least some of the leaves   P. PENSYLVANICA

            17 Plants mostly less than 2 dm. tall; leaves usually at least in part ternate, the leaflets never more than 5   P. QUINQUEFOLIA

          16 Styles usually slender, only slightly tapered, not glandular-roughened at the base, considerably longer than the mature achenes; leaflets various; stipules mostly entire or only shallowly lobed or toothed

            18 Leaflets obovate-cuneate, (1) 2-5 cm. long, greenish, dissected usually about halfway to the midvein, crowded; plants mostly over 3 dm. tall; Cascades and w. from s. B.C. to n. Calif.   P. DRUMMONDII

            18 Leaflets various, but usually smaller and narrower, often dissected nearly to the midvein, frequently white-hairy; plants often much less than 3 dm. tall

              19 Leaflets usually oblong, (1) 2-5 cm. long, grayish-tomentose beneath and often grayish-strigose on the upper surface, rarely toothed over halfway to the midvein; plants often over 3 (and rarely less than 2) dm. tall

20 Anthers 0.5-0.7 mm. long; leaves usually unmistakably pinnate   P. HIPPIANA
20 Anthers over 0.8 mm. long; leaves subdigitate   P. GRACILIS
19 Leaflets usually less than 2 cm. long or otherwise not as above, often either green-
    ish or dissected much more than halfway to the midvein; if gray or tomentose then
    the plants usually less than 2 dm. tall
    21 Pinnae mostly 5-7, crowded, often in part digitate, obovate to oblong, mostly 1-3
        (5) cm. long, usually greenish on both surfaces (never tomentose); plants (1) 1.5-
        4.5 dm. tall; Rocky Mts.   P. DIVERSIFOLIA
    21 Pinnae either more numerous, or tomentose at least beneath; plants often less than
        1.5 dm. tall
        22 Leaflets usually 5-7 (9); plant low and spreading, with an upright, often simple
            crown, the stems rarely over 1.5 dm. long   P. CONCINNA
        22 Leaflets (7) 9-21; plant often erect, or with a branched crown and short root-
            stocks, the stems often over 1.5 dm. tall
            23 Ultimate segments of the leaflets mostly over 1.5 mm. broad; leaflets
                usually (7) 9-11 (13), usually sparsely to densely lanate on the lower
                surface, 1-2 (2.5) cm. long; Cascades and Sierra Nevada to s.e. Oreg.
                and c. Nev.   P. BREWERI
            23 Ultimate segments of the leaflets mostly about 1 (1.5) mm. broad; leaflets
                (7) 9-21, rarely over 1 cm. long, strigose-sericeous to grayish-sericeous
                but rarely grayish-lanate beneath; Rocky Mts.   P. OVINA
13 Basal leaves predominantly digitate, although sometimes with only 3 leaflets (ternate)
    24 Leaflets of basal leaves mostly 3, very rarely 5
        25 Leaves thin, glabrous or very sparsely pubescent, greenish on both surfaces;
            styles not warty at the base   P. FLABELLIFOLIA
        25 Leaves mostly thick and leathery, at least the lower surface grayish or whitish
            from the pubescence; styles usually warty or papillate at base
            26 Basal leaves in part 5-foliolate   P. QUINQUEFOLIA
            26 Basal leaves all 3-foliolate
               27 Calyx bracteoles broadly elliptic to oval, mostly 1.5-2 (rarely to 3) times
                as long as broad   P. VILLOSA
              27 Calyx bracteoles linear to narrowly elliptic or narrowly oblong, commonly
                at least 3 times as long as broad
                  28 Petioles and at least the lower portion of the stem without any tomentum,
                  covered with short, straight (or nearly straight) puberulence and with
                  much longer, coarse, straight, spreading to antrorse hairs P. HOOKERIANA
                28 Petioles and lower stems either lanate or pilose with slightly wavy
                  hairs, but without any fine puberulence
                    29 Petioles and the lower portion of the stems finely tomentose   P. NIVEA
                    29 Petioles and lower part of the stems pilose   P. UNIFLORA
    24 Leaflets of basal leaves mostly 5 or more
        30 Basal leaves in part usually 3-foliolate, all grayish-lanate beneath, the leaflets
            generally toothed 1/2-3/4 the way to the midvein with regular, oblong-rounded
            teeth; stems rarely over 2 dm. tall; styles warty-papillose at the base
                                                P. QUINQUEFOLIA
        30 Basal leaves rarely if ever less than 5-foliolate, often neither grayish nor tomentose
            beneath, the leaflets sometimes toothed less than halfway to the midvein; stems
            sometimes over 2 dm. tall; styles not always warty-papillose at the base
            31 Plant an erect, hirsute-hispid and finely pubescent (but not at all tomentose),
              short-lived perennial with mainly cauline leaves and leafy, flat-topped cymes;
              mature achenes prominently reticulate and with a slight dorsal keel; stamens
              usually 25 (30?), the anthers at least 1 mm. long   P. RECTA
            31 Plant not as above, often tomentose or with mainly basal leaves; mature achenes
              often smooth; stamens usually 20, the anthers often less than 1 mm. long
              32 Flowers mostly well under 1 cm. broad, borne on nearly filiform pedi-
                cels in leafy-bracteate cymes; plants with numerous, ascending, con-
                spicuously leafy stems; leaflets 0.5-2.5 cm. long, greenish above,
                whitish-lanate beneath, toothed well over halfway to the midvein; style

thickened and glandular-warty at the base   P. ARGENTEA
32 Flowers mostly over 1 cm. broad, borne in inconspicuously bracteate cymes; plants
      usually with most of the leaves basal rather than cauline; leaflets various, but never
      greenish above and lanate beneath if the flowers are less than 1 cm. broad; styles
      often filiform and not warty-glandular at the base
      33 Plant low and spreading, mostly less than 1 dm. tall but the stems to 2 dm. long;
            leaves whitish-tomentose beneath   P. CONCINNA
      33 Plant either erect or averaging well over 2 dm. tall, the leaves not white-tomen-
            tose beneath
            34 Plants alpine (or subalpine), (1) 1.5-4.5 dm. tall; leaflets mostly 1-3.5 (to 5)
                  cm. long, usually greenish or nearly equally grayish-sericeous on both sur-
                  faces; anthers mostly 0.4-0.6 mm. long   P. DIVERSIFOLIA
            34 Plants mostly from the lowlands to medium elevations in the mountains, as
                  much as 8 dm. tall; leaflets (2) 3-8 (12) cm. long, often tomentose beneath
                  or much paler on the lower than on the upper surface; anthers mainly 0.8-
                  1.3 mm. long   P. GRACILIS

Potentilla anserina L. Sp. Pl. 495. 1753.
    P. argentina Huds. Fl. Angl. 195. 1762. *Fragaria anserina* Crantz, Stirp. Austr. 2:9. 1763. *Dac-
    tylophyllum anserina* Spenn, Fl. Frib. 1084. 1829. *Argentina anserina* Rydb. Monog. Potent. 159.
    1898. (Europe)
    P. anserina var. *sericea* Hayne, Arzn. Gerv. 4:31. 1816. (Europe) Leaves white-hairy on both sides.
    P. anserina var. *concolor* Ser. in DC. Prodr. 2:582. 1825. (No collector or locality mentioned)
    Argentina argentea Rydb. Bull. Torrey Club 33:143. 1906. (Numerous specimens cited from most of
    w. N.Am.) Leaves white-hairy on both surfaces.
    Grayish, silky-tomentose, strawberrylike perennial, widely spreading by long, prostrate, freely
rooting stolons; stipules prominent, those of the stolons connate-sheathing and deeply linear-lobed;
leaf blades from whitish-silky-lanate on both surfaces to greenish above, (0.5) 1-3 dm. long, pinnate;
leaflets (7) 15-25 (29), obovate to oblong, rounded, sharply and coarsely serrate, (0.5) 1-3.5 cm.
long, interspersed with much smaller, usually entire, leaflets; flowers solitary at the nodes of the stolons on naked peduncles 3-10 (15) cm. long; calyx silky, the hypanthium shallowly bowl-shaped; sepals
ovate-triangular, (3) 4-6 mm. long, and spreading to reflexed at anthesis, but up to 12 mm. long and
erect in fruit; petals yellow, oblong-obovate to oval, (6) 8-12 mm. long, rounded; stamens 20-25; filaments flattened and broadened at base, 1-3 times as long as the anthers; pistils numerous; style slender, smooth, midlaterally attached to the ovary; mature achenes light brown, plump, obliquely ovoid,
about 2 mm. long, usually with 1 to 3 (several) longitudinal "corky" ridges or wrinkles. N=14, 21.
    Meadowland, stream banks, pond margins, and mud flats, from Alas. southward, mostly e. of the
Cascades in our area, to s. Calif., e. to the coast of the Atlantic; Eurasia. May-Aug.
    The degree of pubescence of the leaves is highly variable and not a valid basis for the delimitation of
infraspecific taxa.

Potentilla argentea L. Sp. Pl. 497. 1753.
    Fragaria argentea Crantz, Inst. 2:177. 1766. *Hypargyrium argenteum* Fourr. Ann. Soc. Linn. Lyon
    II, 16:371. 1868. ("Habitat in Europae ruderatis")
    Perennial with a heavy woody caudex, grayish (on the lower surface of the leaves and on the stems)
with a mixture of tomentum and silky to stiff hairs, the latter pubescence, only, present on the greenish upper surface of the leaves; stems numerous, (1) 1.5-3 (5) dm. tall; leaves mainly cauline (5-10
per stem); leaflets usually 5, digitate, oblanceolate, (0.5) 1-2 (2.5) cm. long, lobate-serrate 1/2-3/4
the way to the midvein into 5-9 (rarely more) lanceolate to narrowly oblong, more or less revolute-
margined teeth; stipules lanceolate, entire, 4-8 mm. long; cymes compound, many-flowered, leafy-
bracteate at the lower nodes; calyx cupuliform, sericeous, 4-6 mm. broad; sepals ovate-lanceolate,
2-3 mm. long; petals yellow, obovate-cuneate, rounded to very slightly emarginate, equaling or barely exceeding the sepals; stamens usually 20; pistils numerous, the styles thickened and glandular-
warty at base, considerably tapered upward, apical on, and little if any longer than, the lightly reticulate, 0.6-0.8 mm. achenes. 2N=14, 18, 35, 42, 56.
    This European plant is established in various parts of s. Can. and e. U.S., and is known from scat-
tered localities in our region, as from Stevens Co., Wash. (where it occurs both in ponderosa pine

forest and along the gravelly bank of the Columbia R.), and from Idaho Co. , Ida. (on sandy meadow-land). June-July.

Potentilla arguta Pursh, Fl. Am. Sept. 736. 1814.

   P. pensylvanica var. *arguta* Ser. in DC. Prodr. 2:581. 1825. *P. agrimonoides* var. *arguta* Farw. Asa Gray Bull. 3:7. 1895. *Drymocallis arguta* Rydb. Monog. Potent. 192. 1898. *(Bradbury, "Upper Louisiana")*

   Geum agrimonoides Pursh, Fl. Am. Sept. 351. 1814. *P. agrimonoides* Farw. Asa Gray Bull. 3:7. 1895, but not of Bieb. in 1808. *Drymocallis agrimonioides* Rydb. N. Am. Fl. 22⁴:368. 1908. ("On the rocky banks of the Susquehanna, Pensylvania, and on the upper parts of the Missouri") = var. *arguta*.

   P. convallaria Rydb. Bull. Torrey Club 24:249. 1897. *Drymocallis convallaria* Rydb. Monog. Potent. 193. 1898. *P. arguta* var. *convallaria* Th. Wolf, Bibl. Bot. 16, Heft 71:134. 1908. *P. arguta* ssp. *convallaria* Keck, Carn. Inst. Wash. Pub. 520:39. 1940. *(Rydberg & Flodman 604, near Bozeman, Mont., in 1896)*

   Drymocallis corymbosa Rydb. N. Am. Fl. 22⁴:369. 1908. *P. corymbosa* Fedde, Just Bot. Jahresb. 36²:494. 1910. *(Rydberg & Bessey 4348, Spanish Basin, Mont., in 1897)* = var. *convallaria*.

   Perennial from a simple or (usually) branched caudex and mostly rather short rootstocks, the stems (3) 4-8 (10) dm. tall, commonly anthocyanous, strongly pilose or villous with multicellular, moniliform (when dried), glandular (at least above), often brownish hairs; basal leaves several, weakly rosulate, pinnate, the leaflets mostly (5) 7-9 (rarely 11), ovate to obovate, oblong, or elliptic, (1) 1.5-4 (6) cm. long, deeply serrate-dentate to doubly serrate or even shallowly incised, usually copiously short-hirsute as well as glandular-puberulent but sometimes sparsely hairy or even glabrate; inflorescence cymose, narrow and tight, the lateral branches often numerous but almost strictly erect, the whole rather flat topped, sometimes the flowers semiglomerate; calyx glandular, the hypanthium saucer shaped, the lobes oblong-lanceolate, mostly (5) 6-8 mm. long and flared at anthesis, but up to 12 (15) mm. long and erect in fruit; petals spreading, "pale yellow," "lemon," "cream," "clear cream," "creamy-white," or "white," oblong-obovate to obovate, from slightly shorter to as much as 2 mm. longer than the sepals; stamens usually 25; pistils numerous, the styles slenderly fusiform, glandular-roughened near the middle, inserted below midlength of the ovary; achenes 1-1.3 mm. long, very slightly beaked. N=7.

   Alaska s. e. to Que., s. to Oreg., Ariz., N. M., the central states as far s. as Okla., and in e. U.S. to Pa. and N.J. May-July.

   Our material, as described above, is referable to the var. *convallaria* (Rydb.) Th. Wolf, which ranges from s. Alas. southward, on the e. side of the Cascades, to s. Oreg., and e. to the Rocky Mts. in B.C., Mont., and Wyo., and into n.e. Nev., Utah, and Ariz. The var. *arguta* is a larger flowered, brownish-pilose plant found almost entirely e. of the Rocky Mts.

   As a whole, *P. arguta* is fairly distinctive because of the tall habit, very narrow inflorescence, large sepals, and light-colored (but never truly white?) petals. Many specimens are more or less intermediate to *P. glandulosa* var. *intermedia,* however, and it is not possible to draw a clear line of demarcation between that taxon and *P. arguta* var. *convallaria*.

Potentilla biennis Greene, Fl. Fran. 1:65. 1891.

   Tridophyllum bienne Greene, Leafl. 1:189. 1905. ("Butte Co. to Kern and San Luis Obispo," Calif., no collections cited)

   P. lateriflora Rydb. Bull. Torrey Club 23:261. 1896. *(Henry Engelmann, Simpson's Exp., Utah, in 1859, is the first of several collections cited)*

   P. kelseyi Rydb. N. Am. Fl. 22⁴:306. 1908. *(F. D. Kelsey, near Helena, Mont., in 1891)*

   Annual or biennial with a slender taproot and usually a simple (branched) caudex, pubescent with a mixture of fine, slender, spreading to somewhat appressed or tomentose hairs and thicker, multicellular, glandular hairs; stems mostly single, ascending to erect, 1-6 dm. tall, the usually numerous branches strongly ascending, ending in leafy-bracteate, many-flowered, rather open cymes; leaves mostly cauline, reduced upward, with well-developed, oblong-lanceolate, usually entire stipules, and 3 (4 or 5), rotund-obovate to oblanceolate, coarsely crenate-serrate leaflets 1-4 (5) cm. long; calyx glandular-puberulent and often appressed-hirsute, shallowly cup shaped, 5-8 mm. broad at anthesis, considerably accrescent in fruit, the sepals erect, ovate-triangular, much longer than the hypanthium; petals yellow, cuneate-obovate, about half as long as the sepals; stamens usually either 10 or 15; pistils numerous, the style basally thickened, terminal; achenes yellow, about 0.8 mm. long, smooth.

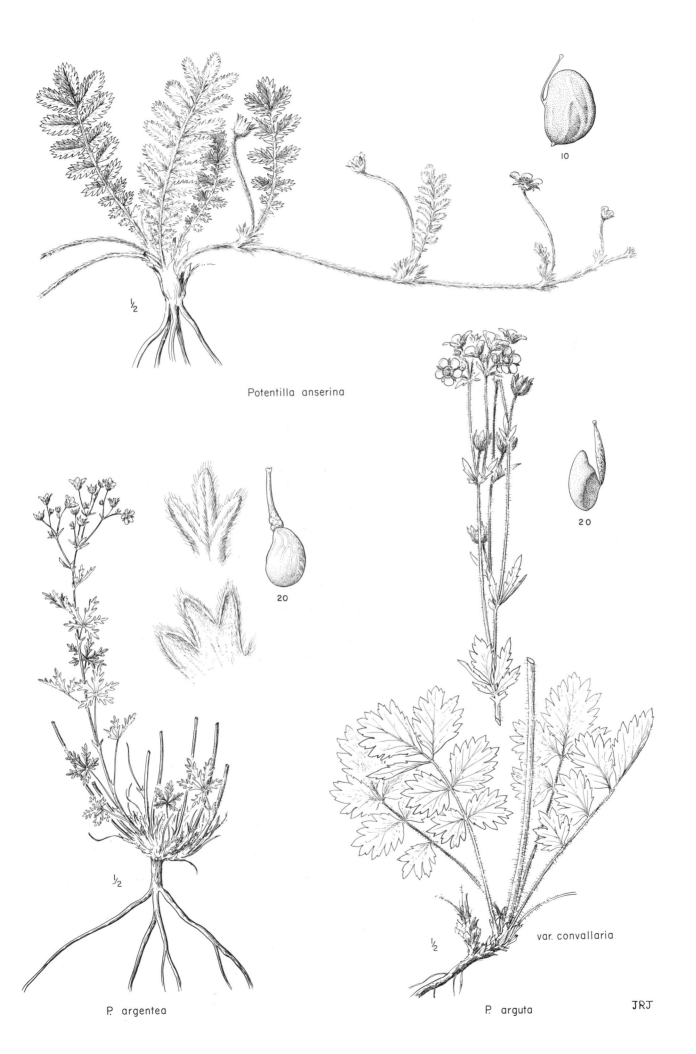

10

Potentilla anserina

20

P. argentea

20

var. convallaria

P. arguta

JRJ

Waste places, along roadsides, and (especially) in sandy soil along streams, ponds, lakes, and moist meadows, B.C. southward, mostly e. of the Cascades in Wash. and Oreg., to Ariz. and Baja Calif., e. (but less common) to Sask., the Dakotas, and Colo. May-Aug.

Potentilla brevifolia Nutt. in T. & G. Fl. N.Am. 1:442. 1840. (Nuttall, ". . . near Goodier River of the Oregon")
  P. *brevifolia* var. *perseverans* A. Nels. Am. Journ. Bot. 18:432. 1931. (F. M. Hurdle 34, Twin Peaks, Challis Nat. Forest, Ida., Aug., 1928, is the first of 2 collections cited)
  Somewhat yellowish-green perennial from a freely branched caudex with trailing rhizomelike scaly branches, forming mats up to 2 dm. in diameter, finely glandular-puberulent but otherwise usually glabrous or occasionally with sparse pilosity above; flowering stems numerous, mostly 5-10 (20) cm. tall, sparingly leafy; basal leaves several, pinnate; petioles slender, mostly considerably longer than the blade; leaflets (3) 5 (7), closely crowded, usually 6-10 mm. long, orbicular (with a broadly cuneate base) to flabellate-obovate, shallowly incised once or twice and crenate; cymes open and branching, the bracts rather inconspicuous; calyx shallowly cupped, the sepals ovate to oblong-lanceolate, spreading or slightly ascending, 3-5 mm. long, sometimes sparsely hirsute as well as glandular; petals yellow, broadly obovate, rounded to retuse, spreading, mostly 2-3 mm. longer than the sepals; stamens usually 20; pistils numerous, the style slender, inserted somewhat above midlength of the ovary; mature achenes smooth, greenish, about 1.3 mm. long.
  High mountains from n.e. Oreg. (Grant Co.) through c. Ida. to Wyo., s. to n. Nev. July-Aug.

Potentilla breweri Wats. Proc. Am. Acad. 8:555. 1873. (Brewer 1720, Mono Pass, Calif.)
  P. *breweri* var. *expansa* Wats. Bot. Calif. 1:179. 1876. (J. G. Lemmon, Sierra Co., Calif.)
  P. *plattensis* var. *leucophylla* Greene, Erythea 1:5. 1893. (C. F. Sonne, Independence Lake, Nevada Co., Calif., June 26, 1892)
  Perennial with a simple to sparingly branched crown and often with short rootstocks, rather copiously sericeous and usually also more or less lanate (at least on the lower surface of the leaves), distinctly grayish, eglandular; stems (0.5) 1-2.5 (3.5) dm. tall; basal leaves pinnate, the leaflets (7) 9-11 (13), crowded, broadly obovate-cuneate, 1-2 (2.5) cm. long, usually cleft considerably more than halfway to the midvein into linear or linear-lanceolate divisions; stipules ovate to lanceolate, 0.5-1.5 cm. long, membranous and entire below to foliaceous and often toothed above; cauline leaves 2-3, greatly reduced; cyme several-flowered, open; calyx sericeous to lanate, shallowly cupulate, as much as 1.5 cm. broad in fruit, the bracteoles linear to lanceolate, shorter than the usually lanceolate lobes; petals yellow, obovate to obcordate, exceeding the sepals; stamens usually 20; pistils numerous; style slender, subapical, exceeding the smooth, greenish, 1.3 mm. achene. 2N=about 72-73, 100, 102.
  Moist meadows and stream banks to open exposed slopes, alpine to midmontane; Sierra Nevada, Calif., n. to the Cascades and Siskiyou Mts. of n. Calif. and s. Oreg., and intermittent n. in the Cascade Mts. of Oreg. at least to n. Lane Co., also in the mountains of s.e. Oreg. and c. Nev. June-Aug.
  Specimens from Klickitat Co., Wash., which were referred to P. *drummondii* by Keck, seem to show some evidence of introgression with P. *breweri*, although the latter is not known from n. of Oreg.

Potentilla campanulata, provisional name.
  NOTE: It was my belief that D. D. Keck would have published the above epithet (which will be found in herbaria on several Cronquist collections) while it was still possible to include the appropriate citation here. Since it is expected that Dr. Keck eventually will provide valid publication for the species, its listing in this fashion was contrived so as not to substitute for that event. Had I intended publication myself, the taxon would have been treated as an additional infraspecific element of P. *glandulosa*. C. L. H. Aug. 20, 1960.
  Very similar to P. *glandulosa*, the branches of the caudex up to 3 dm. long; flowering stems 1.5-4 dm. tall, usually strongly anthocyanous, glandular-pilose throughout with moniliform (mostly), purple-tipped hairs and sometimes with finer, pointed, 1-celled, eglandular hairs that are spreading (on the stems) to more or less appressed (on the leaves); pinnae 5-9 (11), ovate-oblong to obovate, (0.5) 1-3 (or the terminal to 4) cm. long, sharply dentate to bidentate, light green; inflorescence cymose, eventually open, the branches slender, spreading, leafy-bracteate; calyx shallowly cupulate, the lobes broadly lanceolate, (5) 6-8 (in fruit to 10) mm. long, ascending in flower, erect in fruit, the bracteoles only about half as long, narrow; corolla cupulate, the petals ascending, light to "butter" yellow, 2-4 mm. longer than the sepals, oblong-obovate, rounded to slightly obtuse; stamens usually 25; pistils numerous, the style slenderly fusiform, 2-2.5 mm. long, attached below midlength of the ovary; achenes about 1mm. long, not beaked.

15

fruit

1.5

Potentilla   brevifolia

15

3

Potentilla   camparulata

P. breweri

P. biennis

JRJ

Basaltic cliffs and talus to canyon slopes and washes; John Day Valley, c. Oreg., chiefly in Grant, Wheeler, and Sherman cos. May-June.

Potentilla concinna Richards. in Frankl. 1st Journ. 739. 1823.

    P. humifusa Nutt. Gen. Pl. 1:310. 1818, but not of Willd. in 1813. *Tormentilla humifusa* G. Don, Gen. Hist. Pl. 2:562. 1832. *P. concinna* var. *humifusa* Lehm. Stirp. Pug. 9:49. 1851. *P. concinna* var. *humistrata* Rydb. Contr. U.S. Nat. Herb. 3:497. 1896. *P. concinna* var. *typica* f. *humifusa* Th. Wolf, Bibl. Bot. 16, Heft 71:250. 1908. *(Nuttall,* "near Fort Mandan, Missouri" [= N. Dak.])

    P. nivea var. *dissecta* Wats. Proc. Am. Acad. 8:559. 1873, not *P. dissecta* Pursh in 1814. *P. concinna* var. *divisa* Rydb. Bull. Torrey Club 23:431. 1896. *P. divisa* Rydb. N.Am. Fl. 22⁴:330. 1908. *(Drummond 368* and *Douglas,* "In the Rocky Mts. of British America and Montana") Possibly referable to *P. nivea.*

    P. saximontana Rydb. Bull. Torrey Club 23:399. 1896. *(John Wolf 366,* Colo., in 1873, is the first of 3 collections cited) See discussion following key to vars.

    POTENTILLA CONCINNA var. RUBRIPES (Rydb.) C. L. Hitchc. hoc loc. *P. rubricaulis* Rydb. Monog. Potent. 101. 1898, but not of Lehm. in 1830. *P. rubripes* Rydb. Bull. Torrey Club 33:143. 1906. (Several specimens are cited, from Colo., Alta., and Wyo., *F. Tweedy 209,* Colo., in 1896, being the first; however, in the N.Am. Fl. 22⁴:337. 1908, Rydberg cited the type as: *Biltmore Colorado Expedition 1425,* Pike's Peak, Colo., June 25, 1896)

    POTENTILLA CONCINNA var. MACOUNII (Rydb.) C. L. Hitchc. hoc loc. *P. macounii* Rydb. Monog. Potent. 101. 1898. *(John Macoun 16709,* Crow's nest Pass, Alta., in 1897, is the first of 2 collections cited)

    P. intermittens Rydb. N.Am. Fl. 22⁴:318. 1908. *(C. F. Baker 25,* Cameron Pass, Colo., in 1896) = var. *rubripes.*

Perennial with a stout taproot and a sparingly branched, thick caudex, rarely at all rhizomatous, the stem low and spreading to ascending, mostly 4-12 (20) cm. long, but usually not over half that tall, copiously hirsute-strigose and more or less grayish, to sparsely strigose and greenish; under surface of the leaves whitish-tomentose and usually also strongly strigose-hirsute (to sericeous), the upper surface tomentose and sericeous or hirsute to merely (as commonly) strigose or hirsute and grayish to pale green; basal leaves numerous, mostly pinnately 5- or 7 (9)-foliolate, the leaflets from closely crowded and more or less pseudodigitate, to somewhat scattered, oblong to oblanceolate or cuneate-oblanceolate, (0.5) 1-3 cm. long, from deeply cleft (nearly to the midvein) into linear segments, to shallowly few-toothed only near the tip; stipules lanceolate, 3-10 mm. long, entire to cleft, the cauline more or less connate-sheathing; stem leaves 1 or 2, not much reduced; cymes small, rather closely (2) 3- to 7-flowered, inconspicuously to (less commonly) prominently bracteate; calyx shallowly cupuliform, sericeous-villous, 8-15 mm. broad, the lobes lanceolate, 3-6 mm. long; petals yellow, obovate, shallowly emarginate, half again as long as the sepals; stamens usually 20, the anthers oval, about 0.7 mm. long; pistils numerous; style slender, subapically attached, equaling or slightly exceeding the smooth, 1.5 mm. long achene.

Sandy prairies and foothills to alpine ridges, in the Rocky Mts. from Alta. s. to Utah and N.M., also in Nev., Ariz., and Ida., e. to Sask. and the Dakotas. May-June or July, dependent upon altitude.

As here considered, *P. concinna* is a wide ranging, low growing species characterized by the tomentose leaves which vary in an almost continuous series from truly digitate to plainly pinnate, the leaflets ranging from merely shallowly toothed to deeply dissected. This pattern of leaf variation is closely paralleled in *P. diversifolia,* although on the whole there appears to be little if any intergradation between that strigose to sericeous (but never at all tomentose) species and *P. concinna.* There is a strong tendency for some of the phases included here to merge with *P. ovina,* but in general that species not only has more numerous leaflets, but also lacks tomentum.

The several freely intergradient races of the species in our area are separable on the following basis:

1 Leaves usually more nearly digitate than pinnate, the leaflets mostly only 5

  2 Leaflets generally toothed not over 1/2 the way to the midrib; Alta. to Man., s. through e. Mont. and the Dakotas to N.M. and s.e. Utah, occasionally into Ida. and Nev.
                                                                 var. concinna

  2 Leaflets dissected over halfway to the midrib; with the range of the var. *concinna* and completely mergent with it, but perhaps farther w. (into Ida. and Nev.)   var. divisa Rydb.

1 Leaves pinnate, the leaflets often 7 (9), from closely crowded and subdigitate to

plainly pinnate, the lower leaflets often scattered or reduced and somewhat distant
from the main ones

3 Leaflets usually grayish and slightly to heavily tomentose on the upper surface; Alta.
to c. Mont. , mostly on the e. slope of the Rocky Mts. ; perhaps the most clearly
marked race of the complex   var. macounii (Rydb.) C. L. Hitchc.

3 Leaflets usually greenish on the upper surface, often strongly hirsute or strigose,
but not tomentose; chiefly in Colo. but occasional n. to Alta.
                                                                var. rubripes (Rydb.) C. L. Hitchc.

It seems probable that one or more additional taxa to the south of our area, which are similar to,
and intergradient with, *P. concinna*, might reasonably be considered as geographic races of it. One
such phase consists of plants that differ from var. *rubripes* only in having somewhat broader and
shorter sepals and perhaps more loosely scattered leaflets. At various times Rydberg, who described
the taxon under the name *P. saximontana*, pointed out its great similarity both to *P. rubripes* and to
the complex including *P. pinnatisecta*, *P. ovina*, etc. *P. saximontana* is not known to enter our range,
so its reduction to varietal status is not formally proposed.

Potentilla diversifolia Lehm. Stirp. Pug. 2:9. 1830. (No specimens cited in the original description,
but in Hook. Fl. Bor. Am. 1:190. 1833 cited as: *Drummond*, "Alpine Prairies, as well as on the
higher summits of the Rocky Mountains, between lat. 52° and 56°")

P. dissecta Nutt. Journ. Acad. Phila. 7:21. 1834, but not of Pursh in 1814. *(Wyeth*, "Kamas Prairie towards the sources of the Columbia") = var. *diversifolia*.

P. glaucophylla Lehm. Delect. Sem. Hort. Hamb. 1836:7. 1836. *P. diversifolia* var. *glaucophylla*
Lehm. Stirp. Pug. 9:44. 1851. *P. dissecta* var. *glaucophylla* Wats. Proc. Am. Acad. 8:556. 1873.

P. diversifolia var. *multisecta* Wats. Bot. King Exp. 86. 1871. *P. dissecta* var. *multisecta* Wats.
Proc. Am. Acad. 8:557. 1873. *P. multisecta* Rydb. Bull. Torrey Club 23:397. 1896. *(Watson*,
East Humboldt Mts. , Nev.)

POTENTILLA DIVERSIFOLIA var. PERDISSECTA (Rydb.) C. L. Hitchc. hoc loc. *P. decurrens*
Rydb. Bull. Torrey Club 23:396. 1896. *P. perdissecta* Rydb. N.Am. Fl. 22⁴:327. 1908, not *P.
dissecta* var. *decurrens* Wats. in 1873. *(Flodman 572*, Spanish Peaks, Madison Range, Mont. , July, 1896)

P. vreelandii Rydb. N.Am. Fl. 22⁴:325. 1908. *(Vreeland 1092*, Sperry Glacier, Mont. , in 1901) =
var. *diversifolia*.

Perennial with a branching caudex and short thick rootstocks, from sparsely hirsute-strigose and
greenish, to rather grayish-sericeous, at least on the lower surface of the leaflets; stems usually
several, spreading to erect, (1) 1. 5-4. 5 dm. tall; leaves mainly basal, the blades with 5 (7) main
leaflets, mostly digitate, but not rarely semipinnate or truly pinnate, often with 1 or 2 (3-4) much reduced (sometimes entire) leaflets more or less distant from the main ones; leaflets (oblong) oblanceolate to broadly obovate, mostly 1-3 (5) cm. long, from shallowly triangular-toothed to dissected (almost to the midvein) into narrowly oblong to linear segments; cauline leaves mostly 1 or 2 below the
inflorescence; stipules ovate-lanceolate, 1-2 cm. long, usually entire; cymes open, many-flowered;
calyx saucer-shaped, villous-sericeous, up to 1. 5 cm. broad in fruit, the lobes triangular-lanceolate,
(3) 4-6 mm. long; petals yellow, obcordate, (4) 6-9 mm. long; stamens usually 20; pistils numerous;
style slender, equaling or exceeding the fruit and subapically attached to it; achenes 1. 3-1. 6 mm. long,
ultimately weakly reticulate. 2N=90-91, about 101.

From alpine, where most common in meadows and on ledges and rocky slopes, to subalpine or montane, where chiefly along stream banks; Yukon and B. C. southward, in the higher mountains of Wash.
and Oreg. , to Calif. and Nev. , e. to Sask. , and s. in the Rocky Mts. to Utah and N. M. June-Aug.

This is an extremely variable species, especially in leaf character, as the specific name would indicate. Perhaps the greater part of the material, here referred to var. *diversifolia*, has 5 or 7 digitate or subdigitate leaflets that may vary from shallowly toothed to deeply dissected and from very
sparsely to densely sericeous-strigose to sericeous (material from the Olympic Mts. , Wash. , is more
copiously hairy on both surfaces of the leaves than that from any other region). At the other extreme,
the leaves are plainly pinnate, with the 5-7 leaflets (and not rarely 1-4 much smaller leaflets scattered below) deeply dissected (nearly to the base or midvein) into linear or narrowly oblong, subglabrous
to moderately sericeous (or strigose) segments. The strigose-leaved plants of this nature are the var.
multisecta, the glabrate to sericeous plants are var. *perdissecta;* there is every degree of intermediacy between these varieties. A fourth taxon, var. *glaucophylla* Lehm. , is frequently recognized for
our area, characterized by somewhat thickish, glaucous leaflets. It may represent a montane ecotype,

but is so weakly defined that it is omitted from formal recognition in the following key:
1 Leaves primarily digitate, the leaflets usually merely deeply toothed to shallowly dissected; with the range of the species, and by far the most common of our vars. var. diversifolia
1 Leaves subpinnate to pinnate, the leaflets dissected almost to the base or to the midvein into linear or narrowly oblong segments
  2 Leaves grayish and coarsely strigose, the ultimate segments mostly linear, rarely
    over 1. 5 mm. broad; primarily of the Great Basin, from Utah and Nev. n. (but
    much less commonly) to Ida. and Mont.   var. multisecta Wats.
  2 Leaves greenish, usually slightly to moderately sericeous, often glabrate, the seg-
    ments linear to oblong, frequently up to 3-4 mm. broad; with the var. *diversifolia*
    in much of its range, but especially common in Mont. , s. e. B. C. , and s. e. Alta. ,
    elsewhere usually more intermediate to var. *diversifolia*
                                                        var. perdissecta (Rydb.) C. L. Hitchc.
     Potentilla diversifolia tends to merge with the *P. plattensis-pinnatisecta-wyomingensis* com-
plex through the vars. *multisecta* and *perdissecta,* but in general is a taller species, with fewer,
greener, and less finely divided leaflets. It also simulates the *P. concinna* complex (especially the
var. *rubripes),* but never has the undersurface of the leaves tomentose as in that species.

Potentilla drummondii Lehm. Stirp. Pug. 2:9. 1830.
    P. dissecta var. *drummondii* Kurtz, Engl. Bot. Jahrb. 19:374. 1894. *(Drummond,* "Rocky Moun-
      tains north of the Smoking River, in lat. 56°, " fide Hooker)
    P. cascadensis Rydb. Monog. Potent. 109. 1898. *P. drummondii* var. *cascadensis* Th. Wolf, Bibl.
      Bot. 16, Heft 71:492. 1908. *(Suksdorf 2165,* Chiquash Mts. , Skamania Co. , Wash. , in 1892)
    P. anomalofolia Peck, Proc. Biol. Soc. Wash. 49:110. 1936. *(Peck 16819,* near the Klamath Agency,
      Klamath Co. , Oreg. , July 10, 1933) An unusually luxuriant specimen.
    Perennial with a branching crown and short thick rootstocks, from sparingly to rather copiously
hirsute-strigose but always greenish, often glabrate, eglandular; stems (1. 5) 2. 5-4. 5 (6) dm. tall;
leaves pinnate, the leaflets mostly 5-9 (11), closely crowded, obovate-cuneate, (1) 2-5 cm. long, the
upper 3 often somewhat confluent, all cleft about halfway to the midvein into linear or lanceolate teeth;
stipules 1-2 cm. long, ovate to lanceolate, usually entire; cauline leaves (1) 2-3; cymes many-flowered, open; calyx sparingly hirsute, shallowly saucer-shaped, accrescent in fruit and up to 1. 5 cm.
broad, the bracteoles lanceolate and slightly shorter than the more ovate lobes; petals yellow, obovate to obcordate, 6-11 mm. long; stamens about 20; pistils numerous; style slender, subterminal,
considerably exceeding the smooth, 1. 5 mm. achene. 2N=various numbers between about 64 and 108.
    Alpine to subalpine wet meadows to open slopes; s. w. Alta. and s. B. C. , southward in the Cascade
Mts. of Oreg. and Wash. to n. Calif. , also in the Olympic Mts. , Wash. , and in the mountains of s. e.
Oreg. June-Aug.
    The species apparently is sympatric with, and somewhat transitional to, both *P. breweri* and *P.
ovina* in s. e. Oreg.

Potentilla flabellifolia Hook. ex T. & G. Fl. N.Am. 1:442. 1840. *(Douglas,* "Summit of Mount Rainier,
    Oregon")
    P. gelida sensu Wats. Proc. Am. Acad. 8:559. 1873, not of Meyer in 1831.
    Perennial with a branched crown and well-developed rootstocks, forming large clumps, subglabrous
to moderately puberulent or crisp-pubescent, especially above, but greenish throughout, often sessile-
glandular at least on the calyx; stems (0. 7) 1. 5-2. 5 (3) dm. tall; basal leaves numerous, long-petiolate, the blades ternate, the leaflets cuneate-obovate to flabelliform, (1) 1. 5-2. 5 (3. 5) cm. long,
deeply crenate-dentate to lacerate and often secondarily toothed; stipules membranous below, ovate-
rounded, 5-15 mm. long; cauline leaves usually only 1 or 2; cymes few-flowered, leafy-bracteate;
calyx saucer-shaped, 1-1. 5 cm. broad, the bracteoles elliptic to oval, obtuse to rounded, entire to 2-
or 3-toothed at the tip, from much shorter to only slightly shorter than the deltoid-lanceolate lobes;
petals yellow, up to 10 mm. long, obcordate; stamens usually 20, inserted just outside a deeply colored (usually blackish-purple) disc; pistils numerous; style slender, subapical on, and longer than,
the smooth, 1. 5 mm. achene.
    Wet meadows and stream banks to alpine or subalpine ridges and talus slopes; B. C. , southward in
the Cascades (and in the Olympic Mts.) to the Sierra Nevada of Calif. , e. in B. C. to the Rocky Mts.
of s. e. Alta. , and through most of montane Ida. and Mont. , also in the Blue and Wallowa mts. of n. e.
Oreg. June-Aug.

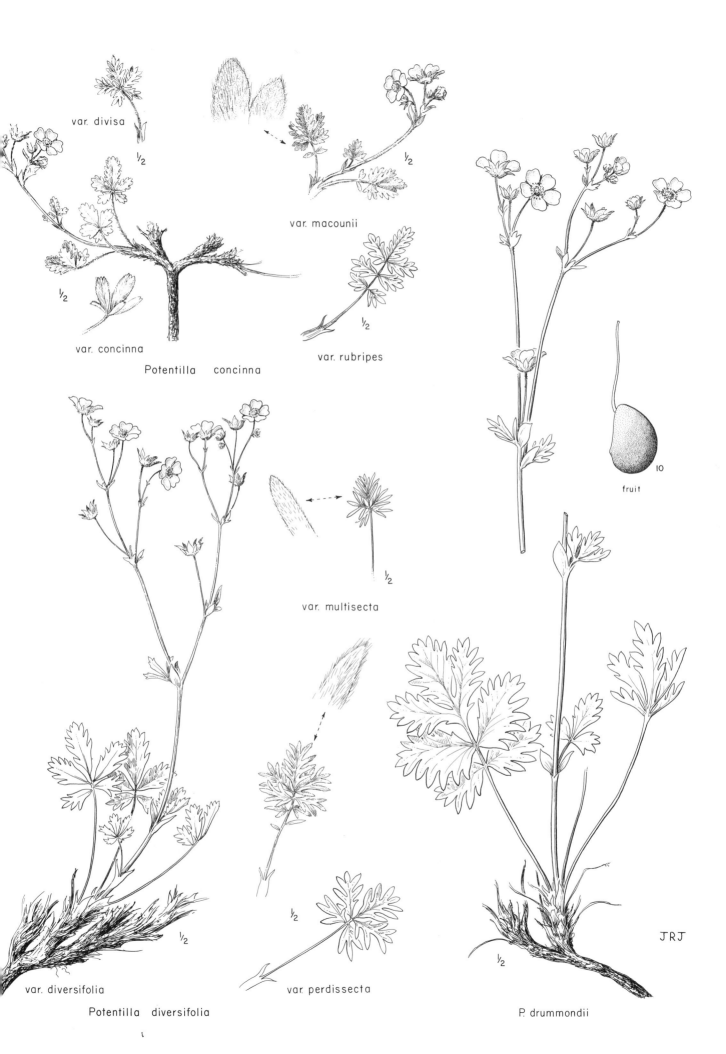

var. divisa

½

var. macounii

½

var. concinna

½

½

Potentilla   concinna

var. rubripes

½

fruit

10

var. multisecta

½

var. diversifolia

½

var. perdissecta

½

½

JRJ

Potentilla   diversifolia

P. drummondii

Potentilla fruticosa L. Sp. Pl. 495. 1753.
Fragaria fruticosa Crantz, Inst. 2:176. 1766. *Dasiphora riparia* Raf. Antik. Bot. 167. 1840. *Dasi-
phora fruticosa* Rydb. Monog. Potent. 188. 1898. ("Habitat in Eboraco, Anglia, Oelandia australi,
Sibiria")
Potentilla floribunda Pursh, Fl. Am. Sept. 355. 1814. *Pentaphylloides floribunda* A. Löve, Svensk
Bot. Tidskr. 48:224. 1954. ("Canada and on the mountains of New York and New Jersey")
Spreading to erect shrub 1-10 (16) dm. tall, the young branches silky-pilose but soon glabrate and
with reddish-brown, shredding bark; leaves numerous, the stipules brownish-scarious, sheathing,
projecting into ovate to lanceolate tips; leaf blades pinnately (3) 5 (7)-foliolate, the leaflets crowded,
(5) 10-20 mm. long, linear to narrowly elliptic-oblong, entire, often revolute, appressed-silky and
more or less grayish, especially beneath; flowers single in the leaf axils or 3-7 (9) and clustered in
small, rather open, terminal cymes; calyx eglandular, soft-hairy, the hypanthium saucer-shaped,
much shorter than the ovate-triangular and more or less acuminate, spreading, 4-6 mm. sepals; calyx bracteoles narrowly lanceolate, from slightly shorter to more usually longer than the sepals; petals yellow, spreading, oval to ovate-oblong, (6) 8-13 mm. long; stamens 25-30; pistils numerous;
style midlaterally attached, slender at base but thickened upward (clavate), up to twice as long as the
strongly hirsute ovary; mature achenes ovoid, light brown, about 1.5-1.8 mm. long, strongly whitish-
hirsute. N=7, 14. Shrubby cinquefoil.
Alaska southward through the Cascade and Olympic mts. (where mostly subalpine) to the Sierra Nevada, e. on the plains and lower foothills to subalpine slopes, to Lab. and N.S., N.J., Pa., Iowa,
Colo., and N.M.; Eurasia. Late June-Aug.
The plant takes well to cultivation and has produced numerous variants, several of which have been
selected for horticultural usage, including dwarf, slow-growing, semiprostrate, and unusually large-
flowered forms.

Potentilla glandulosa Lindl. Bot. Reg. 19:pl. 1583. 1833.
P. arguta var. *glandulosa* Cockerell, W. Am. Sci. 5:11. 1888. *Drymocallis glandulosa* Rydb. Monog.
Potent. 198. 1898. *(Douglas,* California)
P. wrangelliana Fisch. & Avé-Lall. in Ind. Sem. Hort. Petrop. 7:54. 1840. *Drymocallis wrangel-
liana* Rydb. Monog. Potent. 199. 1898. *P. glandulosa* var. *wrangelliana* Th. Wolf, Bibl. Bot. 16,
Heft 71:137. 1908. (Fort Ross, Calif.) = var. *glandulosa.*
P. glandulosa var. *nevadensis* Wats. Bot. Calif. 1:178. 1876, not *P. nevadensis* Boiss. *P. hanseni*
Greene, Pitt. 3:20. 1896. *Drymocallis glandulosa* var. *monticola* Rydb. Monog. Potent. 199. 1898.
P. glandulosa var. *genuina* f. *monticola* Th. Wolf, Bibl. Bot. 16, Heft 71:136. 1908. *Drymocallis
monticola* Rydb. N. Am. Fl. 22⁴:370. 1908. *P. glandulosa* var. *monticola* Jeps. Man. Fl. Calif.
487. 1925. *P. glandulosa* ssp. *nevadensis* Keck, Carn. Inst. Wash. Pub. 520:42. 1940. *(Rothrock,*
South Fork Kern R., Calif.)
P. glandulosa var. *reflexa* Greene, Fl. Fran. 65. 1891. *P. reflexa* Greene, Pitt. 3:19. 1896. *Dry-
mocallis reflexa* Rydb. Monog. Potent. 203. 1898. *(Greene,* near Calaveras Big Trees, Calif.,
June, 1889, acc. Keck)
P. valida Greene, Pitt. 3:20. 1896. *Drymocallis valida* Piper, Contr. U.S. Nat. Herb. 11:342. 1906.
(Greene, vicinity of Victoria, Vancouver I., in 1890) = var. *glandulosa.*
P. rhomboidea Rydb. Bull. Torrey Club 23:248. 1896. *Drymocallis rhomboidea* Rydb. Monog.
Potent. 203. 1898. *(Th. Howell,* Deer Creek Mts., Oreg., July 5, 1887) = var. *glandulosa.*
P. pseudorupestris Rydb. Bull. Torrey Club 24:250. 1897. *Drymocallis pseudorupestris* Rydb.
Monog. Potent. 194. 1898. *P. rupestris* var. *americana* Th. Wolf, Bibl. Bot. 16, Heft 71:129. 1908,
in part. *P. glandulosa* ssp. *pseudorupestris* Keck, Carn. Inst. Wash. Pub. 520:41. 1940. *P. glan-
dulosa* var. *pseudorupestris* Breit. Can. Field-Nat. 71:58. 1957. *(Rydberg & Flodman 598,* Little
Belt Mts., Mont., is the first of several collections cited)
P. ciliata Howell, Fl. N.W. Am. 175. 1898, but not of Greene in 1887. *P. ashlandica* Greene, Pitt.
3:248. 1897. *Drymocallis ashlandica* Rydb. Monog. Potent. 200. 1898. *P. glandulosa* var. *fissa* f.
ashlandica Th. Wolf, Bibl. Bot. 16, Heft 71:137. 1908. *P. glandulosa* ssp. *ashlandica* Keck, Carn.
Inst. Wash. Pub. 520:43. 1940. *(Th. Howell,* Siskiyou Mts. near Ashland Butte, July 8, 1897) =
var. *nevadensis.*
Drymocallis glabrata Rydb. Monog. Potent. 201. 1898. *P. glandulosa* var. *glutinosa* f. *glabrata* Th.
Wolf, Bibl. Bot. 16, Heft 71:137. 1908. *P. glandulosa* ssp. *glabrata* Keck, Carn. Inst. Wash. Pub.
520:36. 1940. *(Elmer 412,* Ellensburg, Wash., in 1897) = var. *intermedia.*

POTENTILLA GLANDULOSA var. INTERMEDIA (Rydb.) C. L. Hitchc. hoc loc. *Drymocallis pseudorupestris* var. *intermedia* Rydb. Mem. N.Y. Bot. Gard. 1:220. 1900. *(Flodman 597,* Spanish Basin, Mont., in 1896, is the first of 2 collections cited)

Drymocallis foliosa Rydb. N.Am. Fl. 22⁴:371. 1908. *(J. H. Flodman 596,* Bridger Mts., Mont., in 1896) = var. *intermedia.*

Drymocallis viscosa Rydb. N.Am. Fl. 22⁴:372. 1908. *P. viscosa* Fedde, Just Bot. Jahresb. 36²: 494. 1910. *(Elmer 564,* Mt. Chapaca, Okanogan Co., Wash., in 1897) = var. *pseudorupestris.*

Drymocallis pumila Rydb. N.Am. Fl. 22⁴:372. 1908, not *P. pumila* Poir. *P. glandulosa* var. *pumila* Jeps. Fl. Calif. 2:181. 1936. *(Cusick 2571,* "Stein's Mts.," Oreg., in 1901) = var. *nevadensis.*

Drymocallis amplifolia Rydb. N.Am. Fl. 22⁴:373. 1908. *P. amplifolia* Fedde, Just Bot. Jahresb. 36²:494. 1910. *(Suksdorf 1761,* near the Columbia R., western Klickitat Co., Wash., in 1894) = var. *glandulosa.*

Drymocallis oregana Rydb. N.Am. Fl. 22⁴: 374. 1908. *(Nuttall,* Columbia River) = var. *glandulosa.*

Drymocallis albida Rydb. N.Am. Fl. 22⁴:375. 1908. *P. albida* Fedde, Just Bot. Jahresb. 36²:494. 1910. *(Suksdorf 2209,* Bingen, Klickitat Co., Wash., in 1893) = var. *glandulosa.*

Glandular perennial with a branched crown and short to well-developed rootstocks, the individual stems simple below the inflorescence, (1) 1.5-4 (7) dm. tall, often anthocyanous, from glabrous below to glandular-puberulent and hirsute or densely pilose-glandular throughout, the hairs from 1-celled to moniliform; leaves pinnate; leaflets 5-9, flabellate-cuneate to rhombic, obovate, or oblong-obovate, 1-3 (5) cm. long, once or twice sharply serrate, glandular-pubescent to short-hirsute and weakly or not at all glandular, sometimes glabrate; cauline leaves few, reduced upward; inflorescence cymose, few-flowered, contracted to diffuse, the branches widely spreading to erect, the bracts from large and leafletlike to much reduced, simple, and ovate to lanceolate; calyx from rotate to shallowly bowl-shaped, the segments 4-8 (accrescent to as much as 12) mm. long, lanceolate to ovate-oblong, spreading to erect in fruit, the bracteoles from as long to (usually) shorter; petals from deep to very pale yellow (often said to be white), spreading to nearly erect, oval to broadly or narrowly obovate, from slightly shorter to somewhat longer than the sepals; stamens usually 25 (to 40); pistils numerous; style thickened above the base and tapered to each end, somewhat glandular-roughened, exceeding the ovary to which it is attached well below the middle; achenes attached somewhat laterally, erect, (0.8) 1 (1.3) mm. long. N=7.

Common from s.w. B.C. to n. Baja Calif., e. to s.w. Alta., c. Mont., Wyo., Colo., Utah, and Ariz. May-July.

Although the species undergoes great variation in practically all floral and vegetative characters, certain geographical races are detectable. These are so completely intergradient and often sporadic in occurrence that they are not always easily categorized, even with living material. The following varieties represent the races recognized by Keck, but there are several others that are slightly more nebulous and not accorded nomenclatural status:

1 Petals from shorter to very slightly (less than 0.5 mm.) longer than the sepals, often
   narrowly obovate to oblanceolate; stems and leaves usually with all hairs glandular
    2 Petals spreading to reflexed, up to 1 mm. shorter than the sepals; inflorescence usual-
      ly with reduced, nonleafy bracts; common from s. Calif. n. through the Sierras to n.
      Calif. and s. Oreg., sporadic farther northward, on the e. side of the Cascades, to
      Wasco Co., Oreg.   var. reflexa Greene
    2 Petals usually ascending to erect, about equaling the sepals; inflorescence usually
      leafy-bracteate; typically coastal, from s. B.C. to Baja Calif., but fairly common
      farther e. in the n. part of the range, and merging with other varieties in c. and e.
      Wash., w. Ida., and n.e. Oreg.   var. glandulosa
1 Petals usually 0.5-1.5 mm. longer than the sepals, oval to broadly obovate; stems and
   leaves often with many nonglandular hairs
    3 Petals lemon to butter-yellow; plants mostly 3-6 dm. tall, with few if any nonglandular
      hairs; inflorescence leafy-bracted, rather congested, the branches few, often erect;
      mostly from Mont. to Colo. and n. Utah, where transitional chiefly to var. *pseudoru-
      pestris,* but w. to the foothills of the Cascades in Wash., and common in n.e. Oreg.
      and much of Ida.   var. intermedia (Rydb.) C. L. Hitchc.
    3 Petals ochroleucous (nearly white, to cream or light canary yellow); plants often less
      than 3 dm. tall, frequently with much nonglandular pubescence; inflorescence seldom
      leafy-bracteate, the branches often more openly flowered
        4 Plant glandular almost throughout, the leaves and stems not noticeably pilose or

hirsute; with about the same range as var. *intermedia* and often transitional to it

                                                              var. pseudorupestris (Rydb.) Breit

4 Plant with many eglandular hairs on the leaves and stems, mostly more s. and w. than

     var. *pseudorupestris*   var. nevadensis Wats.

     Two other taxa, *P. arguta* and *P. campanulata,* are very closely related to *P. glandulosa* and often included in it. Although they are scarcely more distinctive than some of the varieties of *P. glandulosa,* they have apparently attained greater genetic isolation and are somewhat precariously maintainable at the specific level.

Potentilla gracilis Dougl. ex Hook. in Curtis' Bot. Mag. 57:pl. 2984. (May) 1830. (Cultivated specimen, derived from seeds collected by Douglas, "on the banks of the Columbia and the plains of the Multnomah rivers")

    P. pulcherrima Lehm. Stirp. Pug. 2:10. (Aug.) 1830. *P. pennsylvanica* var. *pulcherrima* T. & G. Fl. N.Am. 1:438. 1840. *P. hippiana* var. *pulcherrima* Wats. Proc. Am. Acad. 8:555. 1873. *P. gracilis* var. *pulcherrima* Fern. Rhodora 42:213. 1940. (No specimens cited)

    P. flabelliformis Lehm. Stirp. Pug. 2:12. (Aug.) 1830. *P. gracilis* var. *flabelliformis* Nutt. ex T. & G. Fl. N.Am. 1:440. 1840. (No specimens cited)

    P. chrysantha Lehm. in Hook. Fl. Bor. Am. 1:193. 1833, but not of Trev. in 1828. *P. gracilis* var. *chrysantha* Rydb. Fl. Nebr. 21:16. 1895. (*Drummond,* "Moist Prairies, near the Rocky Mountains") = var. *glabrata.*

    P. rigida Nutt. Journ. Acad. Phila. 7:20. 1834, but not of Wall. in 1828. *P. nuttallii* Lehm. Stirp. Pug. 9:44. 1851. *P. gracilis* var. *rigida* Wats. Proc. Am. Acad. 8:557. 1873. *P. gracilis* var. *nuttallii* Sheld. Bull. Geol. & Nat. Hist. Surv. Minn. 9:71. 1894. *P. gracilis* ssp. *nuttallii* Keck, Carn. Inst. Wash. Pub. 520:134. 1940. (*Wyeth,* "Towards the sources of the Missouri, and as far down as the old Arikaree village") = var. *glabrata.*

    P. fastigiata Nutt. in T. & G. Fl. N.Am. 1:440. 1840. *P. holopetala* var. *fastigiata* Lehm. Stirp. Pug. 9:46. 1851. *P. gracilis* var. *fastigiata* Wats. Proc. Am. Acad. 8:557. 1873. (*Nuttall,* "Plains of the Rocky Mountains") = identity uncertain, probably the same as var. *pulcherrima* but possibly the same as (and therefore representing the appropriate name for) what is here called var. *glabrata.*

    P. blaschkeana Turcz. ex Lehm. Hamb. Gart. & Blumenz. 9:506. 1853. *P. gracilis* var. *blaschkeana* Jeps. Man. Fl. Pl. Calif. 489. 1925. (*Blaschke,* "in colonis Americanis Rossicis") = var. *glabrata.*

POTENTILLA GRACILIS var. GLABRATA (Lehm.) C. L. Hitchc. hoc loc. *P. nuttallii* var. *glabrata* Lehm. Rev. Potent. 89. 1856. *P. glabrata* Rydb. N.Am. Fl. 22⁴:313. 1908. (*Burke,* Oregon)

    P. flabelliformis var. *tenuior* Lehm. Rev. Potent. 108. 1856. *P. flabelliformis* var. *typica* f. *tenuior* Th. Wolf, Bibl. Bot. 16, Heft 71:214. 1908. (No specimens cited) = var. *flabelliformis.*

    P. candida Rydb. Bull. Torrey Club 24:6. 1897. (*Watson 337,* Nevada, in 1868) = var. *elmeri.*

    P. flabelliformis var. *ctenophora* Rydb. Bull. Torrey Club 24:7. 1897. *P. ctenophora* Rydb. Monog. Potent. 75. 1898. *P. gracilis* var. *ctenophora* Boiv. Phytologia 4:90. 1952. (Wyo. and Calif. to B.C. and Sask., no collections cited) = var. *flabelliformis.*

    P. pectinisecta Rydb. Bull. Torrey Club 24:7. 1897. (Specimens from Ariz., Wyo., Mont., and Utah cited, the first being *E. Palmer 145,* Ariz., in 1877) = var. *elmeri.*

    P. longipedunculata Rydb. Monog. Potent. 39. 1898. *P. gracilis* var. *longipedunculata* Th. Wolf, Bibl. Bot. 16, Heft 71:211. 1908. (*Susie Howell,* Monmouth, Polk Co., Oreg., in 1893) = intermediate between var. *permollis* and var. *gracilis.*

    P. viridescens Rydb. Monog. Potent. 69. 1898. *P. gracilis* var. *viridescens* Th. Wolf, Bibl. Bot. 16, Heft 71:211. 1908. (*John Macoun 14447,* Manitoba, in 1896, is the first of many collections cited) = var. *glabrata.*

    P. glomerata A. Nels. Bull. Torrey Club 26:480. 1899. *P. blaschkeana* var. *glomerata* Th. Wolf, Bibl. Bot. 16, Heft 71:212. 1908. (*A. Nelson 4115,* Bear River, Evanston, Wyo., July 27, 1897) = var. *glabrata.*

    P. jucunda A. Nels. Bull. Torrey Club 27:32. 1900. *P. diversifolia* var. *jucunda* Th. Wolf, Bibl. Bot. 16, Heft 71:502. 1908. (*A. Nelson 3223a,* Green Top, Wyo., June 28, 1897) = var. *glabrata.*

POTENTILLA GRACILIS var. BRUNNESCENS (Rydb.) C. L. Hitchc. hoc loc. *P. brunnescens* Rydb. Bull. Torrey Club 28:173. 1901. (*F. Tweedy 212,* Spread Creek, Teton Forest Reserve, Wyo., in 1897)

    P. filipes Rydb. Bull. Torrey Club 28:174. 1901. *P. pulcherrima* var. *filipes* Th. Wolf, Bibl. Bot. 16, Heft 71:209. 1908. *P. gracilis* var. *filipes* Boiv. Phytologia 4:90. 1952. (*Rydberg & Vreeland*

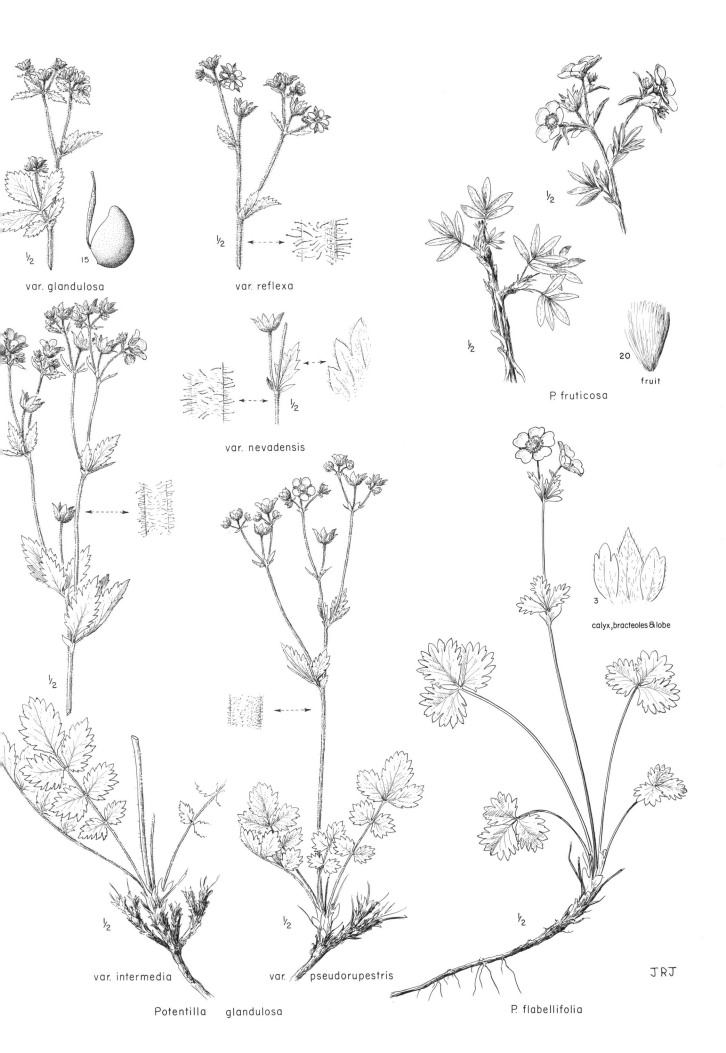

var. glandulosa

15

var. reflexa

½

P. fruticosa

20

fruit

var. nevadensis

½

3

calyx, bracteoles & lobe

½

var. intermedia

½

var. pseudorupestris

½

P. flabellifolia

Potentilla glandulosa

JRJ

6039, Wahatoya Canyon, Spanish Peaks, Colo., in 1900) = var. *pulcherrima.*

POTENTILLA GRACILIS var. PERMOLLIS (Rydb.) C. L. Hitchc. hoc loc. *P. permollis* Rydb. Bull. Torrey Club 28:175. 1901. *P. blaschkeana* var. *permollis* Th. Wolf, Bibl. Bot. 16, Heft 71:212. 1908. *(Elmer 1830,* Endicott, Whitman Co., Wash., in 1898)

P. hallii Rydb. Bull. Torrey Club 28:176. 1901. *P. gracilis* var. *hallii* Th. Wolf, Bibl. Bot. 16, Heft 71:211. 1908. *(Hall & Chandler 182,* Pine Ridge, Fresno Co., Calif., in 1900) = var. *permollis,* but intermediate to var. *glabrata.*

P. grosseserrata Rydb. N. Am. Fl. 22⁴:312. 1908. *(G. R. Vasey 322,* Washington, in 1899) = the very coarse-toothed phase, referable chiefly to var. *glabrata.*

P. rectiformis Rydb. N. Am. Fl. 22⁴:312. 1908. *(Elmer 69,* Pullman, Wash., in 1896) = var. *glabrata.*

P. dascia Rydb. N. Am. Fl. 22⁴:313. 1908. *(Harford & Dunn 1144,* The Dalles, Oreg., in 1869) = var. *permollis.*

P. macropetala Rydb. N. Am. Fl. 22⁴:313. 1908. *(F. E. Lloyd,* Tillamook, Oreg., in 1894) = var. *gracilis.* This is a large-flowered, large-leafleted phase that is not significant enough to recognize, even though it has been re-collected at Tillamook.

P. pecten Rydb. N. Am. Fl. 22⁴:315. 1908. *(Rydberg & Bessey 4377,* Bridger Mts., Mont., in 1897) = var. *elmeri.*

P. elmeri Rydb. N. Am. Fl. 22⁴:315. 1908. *P. gracilis* var. *elmeri* Jeps. Fl. Calif. 2:189. 1936. *(Elmer 4009,* Ventura Co., Calif., in 1902)

P. longiloba Rydb. N. Am. Fl. 22⁴:317. 1908. *(Elrod et al. 110,* Lo-Lo, w. Mont., in 1897) = var. *flabelliformis.*

P. dichroa Rydb. N. Am. Fl. 22⁴:319. 1908. *P. gracilis* var. *dichroa* Peck, Man. High. Pl. Oreg. 393. 1941. *(D. T. MacDougal 185,* Old Sentinel Mt., near Missoula, Mont., in 1901) = var. *glabrata.*

P. indiges Peck, Torreya 32:150. 1932. *(Peck 16034,* John Day River, 15 mi. above Dayville, Grant Co., Oreg., June 17, 1927) = var. *glabrata,* toward var. *elmeri.*

Perennial with a heavy, branched, erect or ascending caudex, exceedingly variable as to pubescence; stems usually several, ascending to more commonly erect, (3) 4-8 dm. tall, sparsely to thickly spreading-hirsute to puberulent, somewhat lanate, or perhaps most commonly chiefly strigose or strigillose; basal leaves numerous, variable, the petioles up to 3 dm. long, usually several times as long as the leaflets, the blades commonly digitate but very occasionally semipinnate; leaflets (of the basal leaves) (5) 7-9, varying from nearly glabrous (or almost equally strigose) to puberulent or hirsute, more or less glandular, and generally concolorous (on the one extreme) to much more heavily strigose or tomentose and much lighter on the lower surface, usually cuneate-oblanceolate to broadly oblanceolate or oblong-elliptic, (2) 3-8 (12) cm. long, plane to occasionally folded, from evenly crenate-dentate with 5-6 teeth per cm. to very deeply dissected almost to the midvein into segments that vary from lanceolate and as much as 1 cm. broad at base to narrowly linear and less than 2 mm. broad at base, the margins plane to slightly revolute; stipules lanceolate, up to 2.5 cm. long, entire to toothed or lacerate; cauline leaves 1-2 (3); cymes mostly large and many-flowered, open, conspicuously bracteate, usually more or less flat-topped; calyx cupuliform, 6-10 mm. broad, but accrescent and up to 12 mm. broad and nearly as high in fruit, from sparsely pubescent and glandular or more commonly eglandular, to hirsute, sericeous, or strigose, the bracteoles narrowly lanceolate, slightly to considerably shorter than the lanceolate to ovate-lanceolate and usually acuminate, 4-10 mm. lobes; petals yellow, obovate-obcordate, slightly to considerably longer than the sepals; stamens usually 20; pistils numerous; style subapically attached, slender but very slightly enlarged and somewhat glandular-verrucose near the base, 1.5-2 mm. long, usually about equaling the mature, nearly or quite smooth, usually greenish achene. 2N= 52-109. Cinquefoil.

Alaska southward, along the coast and inland, through B.C., Wash., and Oreg., to n. Calif. and the Sierra Nevada, to Baja Calif., e. to Sask., the Dakotas, Neb., N.M., and Ariz., in varied habitats, often on moderately saline soil, in grasslands, moister areas in the sagebrush desert, montane forest, and subalpine meadows. June-Aug.

Potentilla gracilis is the most variable of our species and the most difficult to treat taxonomically. In the complex there is great diversity, at least in vegetative features, and many specific or infraspecific segregates have been proposed on the basis of various peculiarities. Perhaps the most obvious differences concern the type and depth of indentation of the leaf margins. The leaflets vary from crenate-serrate or coarsely few-toothed, to the condition where they are dissected nearly to the midvein into linear, plane or slightly revolute-margined segments. The leaves also vary in vesture, apparently nearly always independently of indentation, ranging from nearly glabrous to heavily strigose or hirsute, or densely tomentose, and from equally hairy on both surfaces to more commonly with tomentum only beneath. Mixed with the other pubescence there may be few to numerous (usually short) glandular hairs, especially on the calyx. In general the hairs of the stem and petioles are appressed, but on some plants (notably those from along the base of the Cascades and in and near the Sierras), there is a marked tendency for the pubescence to be spreading. Since pubescence, leaf lobation, and stature of the plant vary almost if not entirely independently of one another, it is not surprising that numerous taxa have been described in the complex.

Aside from the fact that many plants are partially male-sterile and (as would be expected) have somewhat abnormally small stamens, there is a slight tendency for coastal and Cascadian plants to have smaller anthers (0. 7-1. 2 mm. long) as contrasted with the more eastern plants in which the anthers are rarely less than 1 mm. long.

It is not uncommon to find small, local populations of a rather uniform composition in one locality, although nearby the plants may be much more variable. Keck (Carn. Inst. Wash. Pub. 520:125-175. 1940) has pointed out that the complex is to a considerable extent apomictic. He recognized 4 species, based on leaf lobation and type of pubescence, as distinct from *P. gracilis* proper. Although these are perhaps the most conspicuous of the many variants in the complex, they are completely intergradient and largely sympatric and are here treated at the varietal level along with two other variants nearly as clearly marked.

1 Calyx sparsely hirsute but finely glandular-pubescent; leaves greenish and finely glandular-pubescent (as well as hirsute) on both surfaces, the leaflets mostly very coarsely toothed about halfway to the midvein, but varying from less than half this deeply lobed to lobed nearly to the midvein; from Stevens Co., Wash., and the Blue and Wallowa mts. of s. Wash. and n. e. Oreg., through c. and s. Ida. to e. Mont., s. to Wyo., Utah, and n. e. Nev.; intergradients to vars. *pulcherrima* and *flabelliformis* are common in the
above range   var. brunnescens (Rydb.) C. L. Hitchc.
1 Calyx usually abundantly strigose to hirsute, generally eglandular; leaves various, but seldom glandular
  2 Leaflets dissected at least 2/3 of the way to the midvein, the segments usually linear and grayish- or whitish-hairy beneath
    3 Leaflets white-tomentose beneath but usually subglabrous (to sericeous and greenish) on the upper surface; s. c. B. C., s. along the e. base of the Cascades to n. e. Calif., e. through much of Ida. and n. e. Oreg. to s. e. Alta. and e. Mont., where intergradient to var. *brunnescens (Hitchcock 15891),* and in general completely transitional to vars. *gracilis* and *elmeri;* 2N=56-65
                                        var. flabelliformis (Lehm.) Nutt. ex. T. & G.
    3 Leaflets silky (or silky and tomentose) beneath, but sericeous and grayish (although somewhat greenish) on the upper surface; n. c. Oreg. s. to e. and s. Calif., e. to Missoula and Granite cos., and s. c. Mont., w. Wyo., Colo., and N. M.; closely approximated by occasional plants from n. c. Wash. and c. B. C. that are here referred to var. *glabrata;* intergradient with var. *brunnescens* in w. Ida. *(Hitchcock & Muhlick 14023);* N=21   var. elmeri (Rydb.) Jeps.
  2 Leaflets dissected no more than 2/3 of the way to the midvein, the segments usually somewhat lanceolate and often greenish (although generally pubescent) beneath
    4 Leaflets usually grayish beneath, finely, evenly, and deeply serrate, the teeth extending 1/5 to nearly 1/2 the way to the midvein; blades often semipinnate; chiefly Rocky Mts., from B. C. and Alta. s., through Mont. and c. and e. Ida., to e. Nev., Ariz., N. M., and n. Mex., but also occasional in e. Wash. and n. e. Oreg.; freely transitional to var. *brunnescens* in c. Mont. and Wyo., where often glandular and sparsely pubescent, and to var. *glabrata;* only arbitrarily separable in many cases from var. *flabelliformis;* 2N=70, 71, 108
                                      var. pulcherrima (Lehm.) Fern.
    4 Leaflets either greenish beneath, or more coarsely and/or sharply serrate
      5 Stems and petioles hirsute, the plants usually hoary, the hairs spreading; B. C. s., along the e. base of the Cascade Mts., through c. Wash. and Ida. to e. Calif. and Nev., only occasional e. to Mont. and Colo.
                                    var. permollis (Rydb.) C. L. Hitchc.

5 Stems and petioles strigose to appressed-silky, the hairs slightly if at all spreading
  6 Lower surface of the leaves white-woolly, much lighter than the upper surface; pre-
     dominently coastal, and usually the only phase w. of the Cascades, from Alas. to
     n.w. Calif., not rarely e. of the Cascades in s. B.C. and c. Wash. and Oreg.,
     occasional throughout the range of the species and transitional to vars. *permollis* and
     glabrata (especially) but also to both var. *pulcherrima* and var. *flabelliformis*   var. gracilis
  6 Lower surface of the leaves variously pubescent but rarely woolly, usually green-
     ish, not much lighter in color than the upper surface; occasional throughout the
     range of the species, but especially common northward (where there is a marked
     tendency for the leaves to become glabrous or subglabrous), common from Alas.
     through B.C. and w. Alta. to e. Wash., Ida., and Mont., and (in a somewhat
     more heavily pubescent phase) chiefly e. of the Cascades to n. Sierran and s. Cal-
     if., e. to Neb., Utah, and N.M.   var. glabrata (Lehm.) C. L. Hitchc.

Potentilla gracilis is usually easily distinguishable from *P. diversifolia* and *P. concinna,* in
general consisting of taller plants, with larger and more constantly digitate leaflets. However, it
shows some tendency to merge with each, to the elimination of sharp lines of demarcation, so that
separation of the 3 by means of a key will not always be possible. Even the larger anthers of *P. gracilis* (0.7 to more commonly 1-1.3 mm. long, in contrast to 0.5-0.8 in the others) do not always serve
to distinguish it.

Potentilla hippiana Lehm. Stirp. Pug. 2:7. 1830.
  P. leucophylla Torr. Ann. Lyc. N.Y. 2:197. 1827, not of Pall. in 1773. *Potentilla pennsylvanica* var.
    hippiana T. & G. Fl. N.Am. 1:438. 1840. *Pentaphyllum hippianum* Lunell, Am. Midl. Nat. 4:416.
    1916. *(James 130,* sources of the Platte)
  Potentilla leneophylla Torr. & James ex Eaton, Man. Bot. 5th ed. 344. 1829. (No specimens cited)
    This, an earlier name than *Potentilla hippiana,* is being rejected on the basis that its application is am-
    biguous; some taxonomists contend the name was a misprint for *Potentilla leucophylla,* others that it
    is a valid name, in reference to the woolly leaves.
  Potentilla effusa Dougl. ex Lehm. Stirp. Pug. 2:8. 1830. *Pentaphyllum effusum* Lunell, Am. Midl.
    Nat. 4:416. 1916. ("On elevated grounds of the Assinaboyne, and the higher parts of the Red Riv-
    ers," probably collected by Douglas)
  Potentilla effusa var. *filicaulis* Nutt. in T. & G. Fl. N.Am. 1:437. 1840. *Potentilla filicaulis* Rydb.
    Bull. Torrey Club 24:2. 1897. *(Nuttall,* "Rocky Mountains towards the sources of the Platte")
  Potentilla effusa var. *gossypina* Nutt. in T. & G. Fl. N.Am. 1:437. 1840. *(Nuttall,* location not spec-
    ified)
  Potentilla diffusa Gray, Pl. Fendl. 41. 1849, not of Willd. in 1809. *Potentilla hippiana* var. *dif-
    fusa* Lehm. Del. Sem. Hort. Hamb. 1849:8. 1849. *Potentilla hippiana* var. *propinqua* Rydb. Bull.
    Torrey Club 24:3. 1897. *Potentilla propinqua* Rydb. Bull. Torrey Club 28:176. 1901. *(Fendler 198,*
    Santa Fe Creek, N.M.)
  Potentilla argyrea Rydb. N.Am. Fl. 224:341. 1908. *Potentilla hippiana* var. *argyrea* Boiv. Phytolo-
    gia 4:90. 1952. *(John Macoun 14441,* Moose Jaw, Assiniboia, Sask., in 1896)
    Perennial with a heavy branching crown, more or less grayish-hirsute and tomentose throughout
or the upper surface of the leaves green, eglandular; stems (1.5) 2-5 (in nondepauperate plants usually
at least 3) dm. tall, generally freely branched and floriferous to or below midlength; leaves pinnate;
leaflets (5) 7-11 (13), rather crowded, oblong or oblong-oblanceolate, mostly 2-5 cm. long, grayish
on both surfaces or greenish above, sharply lanceolate-toothed about halfway (or less) to the midvein;
stipules ovate to lanceolate, up to 3 cm. long, entire; cauline leaves 2-5; inflorescence strongly bracteate, ultimately cymose but freely branched, the branches strongly ascending to erect; calyx saucer-
shaped, grayish-tomentose and hirsute; petals yellow, narrowly obovate, rounded to retuse, (4) 5-7
(8) mm. long (slightly longer than the sepals); stamens usually 20; pistils numerous; style slender,
longer than (and subapically attached to) the smooth, 1.4-1.6 mm. achene.

    Open grassland and sagebrush (where often on saline soil) to juniper scabland and ponderosa pine
forest of the foothills and lower levels of the mountains; Rocky Mts., especially on the e. slope, from
Alta. to N.M. and Ariz., e. to Sask., the Dakotas, and Neb. Late June-Aug.

    Potentilla hippiana is variable in many respects, such as the height of the plant, size of flowers,
and the amount of pubescence on the upper surface of the leaves, ranging from greenish to white. The
calyx bracteoles vary from linear and scarcely half as long, to nearly as broad and as long as the sepals. *P. effusa* has usually been maintained as distinct from *P. hippiana* almost entirely on the basis

var. gracilis

var. glabrata

var. pulcherrima

calyx, bracteoles & lobe

var. flabelliformis

var. permollis

var. elmeri

var. brunnescens

Potentilla gracilis

JRJ

of its small bracteoles, even though the two taxa were known to be sympatric and completely mergent.

Potentilla hookeriana Lehm. Del. Sem. Hort. Hamb. 1849:10. 1849.
  P. nivea var. *hookeriana* Th. Wolf, Bibl. Bot. 16, Heft 71:240. 1908. ("Habitat in America septentrionali")

Perennial with a branched crown and short rootstocks, grayish-tomentose on the lower surface of the leaves but otherwise in general finely puberulent and hirsute with coarse, straight, spreading to antrorse-spreading (or the puberulence of the petioles retrorse) hairs, the puberulence often lengthening and becoming crisp upward; stems numerous, 1-2 (2.5) dm. long; leaf blades ternate; leaflets usually greenish above, white beneath, oblong-obovate, 7-20 mm. long, coarsely toothed about halfway to the midvein, the mostly 5-7 teeth triangular-oblong; cymes (3) 5- to 7-flowered, loose and open; calyx shallowly cupuliform, mostly 6-8 mm. broad, the bracteoles linear to narrowly lanceolate, at least 2.5 times as long as broad, shorter than the 3-4 mm. lobes; petals yellow, obcordate, 4-6 mm. long; stamens 20, the anthers very small, scarcely 0.4 mm. long; pistils numerous; style subapically attached, thickened at base and papillate on the lower half, shorter than the mature achene which is smooth and about 1 mm. long.

Alpine slopes and glacial moraines to montane river bars; Alas. to Sask., s. to c. B.C. and in the Rocky Mts. to s.w. Alta. and (reportedly) to n. Mont.; Kamtchatka and Siberia. June-July.

Although closely related to *P. nivea, P. hookeriana* seems readily separable from it.

Potentilla newberryi Gray, Proc. Am. Acad. 6:532. 1865.
  Ivesia gracilis Gray in Newberry, Pac. R. R. Rep. 6[3]:72. 1857, not *P. gracilis* Dougl. in 1829. (*Newberry*, Rhett Lake, Calif. or Oreg., in 1855)
  P. newberryi var. *arenicola* Rydb. Monog. Potent. 112. 1898. (*Th. Howell*, Wallula, Wash., in 1896, is the first of 2 collections cited)

Grayish-silky to -hirsute, short-lived perennial from a taproot and simple to branched (but non-rhizomatous) crown, the stems numerous, prostrate to ascending (ours) or erect, leafy throughout, 0.5-3 (5) dm. long; basal leaves few, mostly 2-4 cm. long; leaflets (5) 7-15 (21), crowded, pinnately dissected into 3-7 (9) linear or narrowly oblong to elliptic or spatulate segments 3-5 (7) mm. long; inflorescence usually freely branched and leafy bracteate, the fragrant flowers in large part hidden by the foliage; calyx shallowly bowl shaped; sepals lanceolate, spreading, (3) 4-5 (6) mm. long; petals spreading, cream or "white," obovate, rounded to slightly retuse and obcordate, (4) 5-6 mm. long; stamens 20; pistils rather numerous; style elongate, tapered from a glandular-roughened base, subapically inserted upon the ovary; mature achenes strongly reticulate, brownish, about 1.3 mm. long.

Dry lake shores and vernal pools and water holes; s.c. Wash., southward to n.e. Calif. Apr.-July.

Potentilla nivea L. Sp. Pl. 499. 1753.
  Fragaria nivea Crantz, Inst. 2:179. 1766. ("Habitat in Alpibus Lapponiae, Sibiriae")
  P. nivea var. *macrophylla* Ser. in DC. Prodr. 2:571. 1825. (No specimens cited, but American authors have considered this phase to include N.Am. plants)
  P. nipharga Rydb. N.Am. Fl. 22[4]:332. 1908. (*I. S. Onion*, Fort Good Hope, Mackenzie River)
  P. nivea ssp. *fallax* Porsild, Bull. Nat. Mus. Can. 121:226. 1951. (*A. E. Porsild 9476*, Lower Lapie crossing, Yukon)

Perennial with a branched crown and short rootstocks, more or less tomentose and usually grayish throughout except for the often greenish and strigose-hirsute upper surface of the leaves; stems 3-15 (20) cm. tall; leaves ternate; leaflets oblong-obovate to more or less oval, 5-15 (35) mm. long, pinnately (5) 7- to 11-toothed from less to slightly more than halfway to the midvein; inflorescence from much contracted and 1- to 2-flowered to open and 3- to 9-flowered, evidently bracteate; calyx silky to somewhat tomentose, saucer shaped, 8-12 mm. broad, the bracteoles linear to very narrowly lanceolate, shorter than the lanceolate, 3-4 mm. lobes; petals yellow, obcordate, 1-2 mm. longer than the sepals; stamens usually 20; pistils numerous; style subapically attached to the nearly smooth, 1-1.5 mm. achene, slightly thickened and weakly papillate near the base. 2N=56, 63, 70.

Arctic tundra and gravel bars to alpine slopes and meadows; Alas. to Que., southward to B.C., and in the Rocky Mts. to Colo. and e. Utah, also in Nev.; Greenl., Eurasia. June-Aug.

The species is variable in pubescence, flower- and leaf-size, and general habit, with many infraspecific taxa named. Only monographic work on this complex (including *P. villosa, P. hookeriana, P. quinquefolia,* and *P. uniflora*) can clarify the relationship of the American and Eurasian plants. *P. ni-*

pharga, the phase with more deeply lobed leaves, seems not to be separable, except mechanically, from *P. nivea.*

Potentilla norvegica L. Sp. Pl. 499. 1753.
  Fragaria norvegica Crantz, Inst. 2:179. 1766. *P. monspeliensis* var. *norvegica* Farw. Asa Gray Bull. 3:7. 1895 (corrected reprint). *Tridophyllum norvegicum* Greene, Leafl. 1:189. 1905. ("Habitat in Norvegiae, Sueciae, Borussiae, Canadae agris")
  P. monspeliensis L. Sp. Pl. 499. 1753. *Fragaria monspeliensis* Crantz, Inst. 2:179. 1766. *Tridophyllum monspeliense* Greene, Leafl. 1:189. 1905. *P. norvegica* ssp. *monspeliensis* Aschers. & Graebn. Syn. Mitteleur. Fl. 6[1]:748. 1905. ("Habitat Monspelii")
  P. hirsuta Michx. Fl. Bor. Am. 1:303. 1803. *P. norvegica* var. *hirsuta* Lehm. Stirp. Pug. 9:75. 1851. *P. norvegica* ssp. *hirsuta* Hylander, Uppsala Univ. Årssk. 1945[7]:203. 1945. ("Hab. in Canada, a Quebec ad ostium fluminis s. Laurentii")
  P. leurocarpa Rydb. N. Am. Fl. 22[4]:307. 1908. *(Suksdorf 2011,* near Bingen, Klickitat Co., Wash., in 1891) The uncommon phase with smooth achenes.
  Taprooted, erect to ascending annual or biennial; stem (1) 3-6 (8) dm. tall, simple to branching and often floriferous for much of the length with leafy-bracteate, long-pedunculate (but rather compact) cymes, strongly hirsute below to subtomentose above, eglandular, the larger hairs slightly pustulose; leaves mainly cauline; stipules well developed, ovate, entire to more commonly strongly toothed; leaflets usually 3 (very occasionally 5), from broadly ovate to obovate below to narrowly oblong above, (1) 3-6 (8) cm. long, strongly crenate-serrate, spreading- to appressed-hirsute; calyx strigose to hirsute, eglandular to weakly glandular-puberulent, at anthesis about 7-11 mm. broad, the broadly lanceolate, erect sepals and oblong-elliptic to lanceolate calyx bracteoles subequal in length or the bracteoles the longer, considerably longer than the hypanthium, strongly accrescent in fruit; petals yellow, broadly obovate and often retuse, from 3/4 as long to almost as long as the sepals; stamens usually 20 (15), often more or less abortive; pistils numerous; style terminal, thickened basally; achenes light brown, 1-1.3 mm. long, ovate, flattened, usually strongly undulate-corrugate longitudinally (rarely smooth). N=35.
  Widespread in the Northern Hemisphere, more common in e. N. Am. and c. U.S., to Mex., but occasional throughout our area, usually on moist soil, often along irrigating ditches or in waste places, from Alas. to Calif., on both sides of the Cascades, and eastward. Late May-Aug.
  The species is variable and questionably native, at least in our area.

Potentilla ovina Macoun, Can. Rec. Sci. 6:464. (for Oct. 1895) pub. in early 1896.
  P. diversifolia var. *pinnatisecta* Wats. Bot. King Exp. 87. 1871. *P. pinnatisecta* A. Nels. Bull. Wyo. Exp. Sta. 28:104. (May) 1896. *(Watson 331,* which was cited first, has been annotated at the Gray Herb. as the type; it came from the "Uintas," in Aug., 1869, as did *Watson 332,* the other number cited)
  P. wyomingensis A. Nels. Bull. Torrey Club 27:32. 1900. *(A. Nelson 5781,* Druid Peak, Yellowstone Nat. Park, July 12, 1899, the second of 2 collections cited, was said to be typical of the species)
  P. monidensis A. Nels. Bull. Torrey Club 27:266. 1900. *(A. Nelson 5414,* near Monida, Mont., June 16, 1899)
  P. klamathensis Rydb. N. Am. Fl. 22[4]:343. 1908. *(Leiberg 660,* near Ft. Klamath, Oreg., in 1894) See discussion on page 151.
  P. versicolor Rydb. N. Am. Fl. 22[4]:344. 1908. *(Coville & Leiberg 307,* Grayhart Buttes, Lake Co., Oreg., in 1896) See discussion on page 151.
  P. nelsoniana Rydb. N. Am. Fl. 22[4]:344. 1908. *(A. Nelson 1819,* near La Plata Mines, Wyo., in 1895)
  Perennial with a strong crown and often with short thick rhizomes, mostly sericeous-hirsute but sometimes with a slight mixture of curled or crisped hairs, usually grayish, especially on the lower surface of the leaves; stems spreading to erect, commonly only 0.5-1.5 (but up to 2.5) dm. tall; basal leaves numerous, tufted, short-petiolate, pinnate, the leaflets (7) 9-21, crowded to rather distant, 5-12 mm. long, more or less pedately divided nearly to the base into 3-5 (7) linear segments usually 0.75-1.5 mm. broad; cymes mostly 3- to 7-flowered, open, the branches spreading to ascending; calyx bowl-shaped, rarely over 1 cm. broad even in fruit, the lobes lanceolate, acute to acuminate, 3.5-5 mm. long; petals yellow, obcordate, 1-2 mm. longer than the sepals; stamens usually 20; pis-

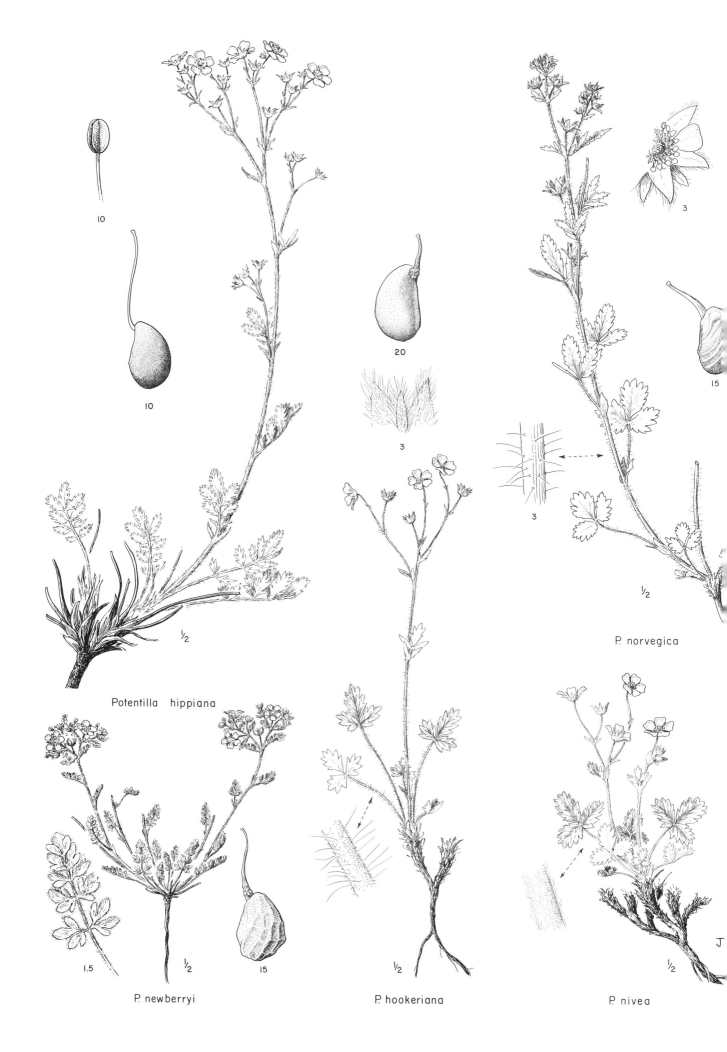

10

10

20

3

3

3

15

1/2

P. norvegica

Potentilla hippiana

1.5

1/2

15

1/2

P. newberryi

P. hookeriana

J

1/2

P. nivea

tils numerous; style slender, subapically inserted on the ovary and somewhat longer than the achene which at maturity is smooth and 1.5-2 mm. long.

Moist meadows to (more commonly) open ridges and barren slopes, from middle altitudes to well above timber line, chiefly in the Rocky Mts., B.C. to Sask., southward (and common) in Ida., Mont., and Wyo., to Utah, Colo., and N.M., also in e. Oreg. and n.e. Calif., and possibly in e. Nev. June-Aug.

Both numbers cited for var. *pinnatisecta* by Watson *(331* and *332)* actually came from the Uinta Mts. Although the range was listed as "East Humboldt Mountains, Nevada, and on the Wahsatch and Uintas," no Nevadan material has been seen that is referable here. *Watson 332* is good average material for this taxon, but *Watson 331* is slightly intermediate toward *P. plattensis,* tending to be sericeous-strigose.

The name, *P. ovina,* is used for this taxon since Macoun's specific epithet almost surely antedates *P. pinnatisecta* (Wats.) A. Nels. The species is one of several very closely related taxa that range over most of w. N.Am., s. of Alas. and Yukon. They tend to replace one another geographically or ecologically, and might with some reason be considered as races of one large specific complex. Several of the taxa concerned are not known to enter our area, but have often been listed for it. Of these, *P. plattensis* Rydb. is a distinctly strigose (never at all sericeous or grayish) species; common to the e. of our range from Alta. to Man., and s. in e. Mont. and the Dakotas to Wyo., Colo., N.M., and Ariz. *P. millefolia* Rydb. is a similar species which differs mainly in being more or less glandular and usually taller, and in having slightly broader leaf segments; it is chiefly Sierran, but extends into Nev. and s.e. Oreg. and possibly across s. Utah and n. Ariz. to Colo. It seems to approach *P. ovina* in s.e. Oreg., from whence two other taxa, *P. versicolor* and *P. klamathensis,* have been described, although neither is known from within our area. The latter taxa are somewhat intermediate phases between *P. ovina, P. millefolia,* and the Cascadian-Sierran, usually tomentose-leaved *P. breweri,* with which *P. ovina* tends to blend in the Steens Mts., s.e. Oreg.

In the greater part of its range (in our area), *P. ovina* is allopatric with its closest relatives, so that its recognition is not too difficult, even though it undergoes considerable variation in the number and degree of dissection of the leaflets and in general pubescence (plants with grayish-woolly to hirsute-sericeous and greenish leaves often growing together).

Potentilla pacifica Howell, Fl. N.W. Am. 179. 1898.

   P. anserina var. *grandis* T. & G. Fl. N.Am. 1:444. 1840. *Argentina anserina* var. *grandis* Rydb.
      Monog. Potent. 2:161. 1898. *Argentina pacifica* Rydb. N.Am. Fl. 22⁴:353. 1908. *(Dr. Scouler,*
      Oregon)
   Argentina occidentalis Rydb. N.Am. Fl. 22⁴:354. 1908. *P. occidentalis* Fedde, Just Bot. Jahresb.
      36²:488. 1910. *(C. F. Baker 3217,* Suisun, Solano Co., Calif., in 1903)
   Argentina subarctica Rydb. N.Am. Fl. 22⁴:354. 1908. *P. subarctica* Fedde, Just Bot. Jahresb. 36²:
      488. 1910, but not *P. subarctica* Rydb. N.Am. Fl. 22⁴:347. 1908. *P. yukonensis* Hultén, Fl. Alas.
      6:1033. 1945. *(Arthur Hollick,* Yukon R. near Palisades, Alas., in 1903)

Similar to *P. anserina* in leaf and floral character, differing mainly in pubescence; stems, petioles, and upper surface of the leaflets green, glabrous or only sparsely appressed-silky, the lower surface of the leaflets closely tomentose and moderately appressed-silky at least along the veins; larger leaves up to 4 dm. long; flowers perhaps slightly larger than in *P. anserina,* the calyx lanate as well as sericeous, the petals broadly oblong, mostly 8-13 mm. long; stamens mostly 25-30; mature achenes about 2 mm. long, rounded and very obscurely wrinkled to smooth on the back.

Coastal dunes, beaches, sand flats, and marsh edges and stream banks; Alas. to s. Calif., perhaps very occasionally inland. May-Aug.

The species is closely related to *P. anserina* and (in the absence of mature achenes) not always separable from it, a flowering specimen from Deschutes Co., Oreg. *(J. W. Thompson 30),* being exactly intermediate between the two.

Potentilla palustris (L.) Scop. Fl. Carn. 2nd ed. 1:359. 1772.

   Comarum palustre L. Sp. Pl. 502. 1753. *Fragaria palustris* Crantz, Stirp. Austr. 2:11. 1766. (Europe)
   Comarum palustre var. *villosum* Pers. Syn. 2:58. 1807. *P. palustris* var. *villosa* Lehm. Stirp. Pug.
      9:44. 1851. (Typification recondite)

Strongly rhizomatous perennial with floating or prostrate to ascending (and often nodally rooting), usually reddish stems up to 1 m. long, glabrous below but becoming more or less hairy and purplish-glandular above; leaves pinnate; leaflets usually (3) 5-7, approximate, glabrous to sparsely strigose (strigillose) and light green on the upper surface, paler, glaucous, and sparsely to copiously whitish-strigose (strigillose) to appressed-silky on the lower surface, obovate to oblong or narrowly elliptic-oblong, (2) 3-6 (10) cm. long, shallowly crenate-serrate to deeply serrate; flowers usually several to numerous in open, often unilateral cymes, deep wine red to purple except for the more greenish-purple, glandular and hirsute-pilose calyx; hypanthium shallowly bowl-shaped, 5-8 mm. broad; sepals ovate-lanceolate to lanceolate, acuminate, spreading and mostly 7-11 mm. long at anthesis, accrescent, ascending, and up to 20 mm. long in fruit; petals ovate-lanceolate to elliptic or spatulate, acute or acuminate, about as long as the calyx bracteoles; stamens about 25; pistils many; style reddish, smooth, slender and nearly uniform in diameter, inserted median-laterally on the ovary, 1.5-2 mm. long; mature achenes broadly ovate in outline, brownish-purple, plump, smooth, about 1.5 mm. long. 2N=28, 42-64.

Bogs, wet meadows, creek banks and lake margins, from sea level to subalpine; Alas. southward along the coast to n. Calif., e. to Lab. and Greenl., Ohio, Ia., and Wyo. June-Aug.

Our plants show great variation in many features but are not differentiated into geographic races even though some of them could be covered by certain infraspecific names proposed for the plants of n.c. N.Am.

Potentilla paradoxa Nutt. in T. & G. Fl. N.Am. 1:437. 1840.

  P. supina of Michx. Fl. Bor. Am. 1:304. 1803, but not of L. in 1753. *Tridophyllum paradoxum* Greene, Leafl. 1:189. 1905. (*Nuttall*, "Banks of the great western rivers, the Ohio! Mississippi! Missouri! &c to Oregon")

  P. supina var. *nicolletii* Wats. Proc. Am. Acad. 8:553. 1873. *P. nicolletii* Sheld. Minn. Bot. Stud. 1:16. 1894. *Tridophyllum nicollettii* Greene, Leafl. 1:189. 1905. (*Nicollet*, Devil's Lake, N.D.)

Annual or more probably a biennial or short-lived perennial with a taproot and simple to branched caudex, the stem usually 4-7 (9) dm. long, from glabrous below to strongly hirsute above, leafy throughout and often floriferous for over half the length, the inflorescence diffusely cymose, leafy-bracteate; leaves pinnate, the lower ones with 2 to 4 (5) pairs of elliptic to oblong, crenate-serrate leaflets 1-3 cm. long, the upper ones sometimes ternate; stipules well developed; calyx hirsute, 5-9 mm. broad at anthesis, the lobes ovate-triangular, 3-4 mm. long, erect; petals yellow, obovate, subequal to the sepals; stamens usually (10-15) 20, the anthers often abortive; pistils numerous; style terminal, somewhat thickened basally, about equaling the ovary; mature achenes about 1.2 mm. long, obscurely undulate-ridged longitudinally, bearing on their adaxial edge a wedge-shaped thickening nearly (or quite) as large as the rest of the fruit.

Sandy stream banks, lake shores, and moist flats; over much of N.Am. especially in the Mississippi R. Valley and e., s. to Mex., not common in our area but known from a few collections, as at Bingen, Klickitat Co., Wash.; perhaps also in Asia. June-July.

Potentilla pensylvanica L. Mant. 76. 1767. ("Canada")

  P. pensylvanica var. *strigosa* Pursh, Fl. Am. Sept. 356. 1814. *P. strigosa* Pall. ex Tratt. Ros. Monog. 4:31. 1824. (*Lewis*, on the Missouri)

  P. bipinnatifida Dougl. ex Hook. Fl. Bor. Am. 1:188. 1833. *P. pennsylvanica* var. *bipinnatifida* T. & G. Fl. N.Am. 1:438. 1840. ("Plains of the Saskatchawan and Red Rivers, *Douglas, Drummond*")

  P. missourica Hornem. ex Lindl. Bot. Reg. 17:pl. 1412. 1831. *P. pennsylvanica* var. *communis* T. & G. Fl. N.Am. 1:438. 1840. *P. pennsylvanica* var. *missourica* Lehm. Stirp. Pug. 9:41. 1851. (Garden specimens, from seed collected by Richardson in "Arctic America")

  P. sericea var. *glabrata* Hook. Fl. Bor. Am. 1:189. 1833. *P. pennsylvanica* var. *glabrata* Wats. Proc. Am. Acad. 8:554. 1873. *P. glabrella* Rydb. Monog. Potent. 94. 1898. (*Drummond*, "Rocky Mountains, between lat. 52° and 56°")

  P. pennsylvanica var. *conferta* Gray, Pl. Fendl. 42. 1849, not *P. conferta* Bunge in 1830. *P. pennsylvanica* var. *arachnoidea* Lehm. Stirp. Pug. 9:41. 1851. *P. arachnoidea* Dougl. in Lehm. Stirp. Pug. 9:41. 1851, in synonymy; ex Rydb. N.Am. Fl. 22[4]:350. 1908. (*Fendler 202*, "Valley of Santa Fe Creek, in the mountains")

  P. bipinnatifida platyloba Rydb. Monog. Potent. 100. 1898. *P. platyloba* Rydb. Bull. Torrey Club 33: 143. 1906. (*Smith & Pound 85*, Nebraska, in 1892)

  P. virgulata A. Nels. Bull. Torrey Club 27:265. 1900. (*A. Nelson 6011*, near Mammoth Hot Springs, Yellowstone N. Park, July 20, 1899)

  P. lasiodonta Rydb. N.Am. Fl. 22[4]:351. 1908. (*J. Macoun 16716*, near Calgary, Alta., in 1897)

Perennial from a usually branched caudex, the 1-several flowering stems (1) 2-5 (7) dm. tall, de-

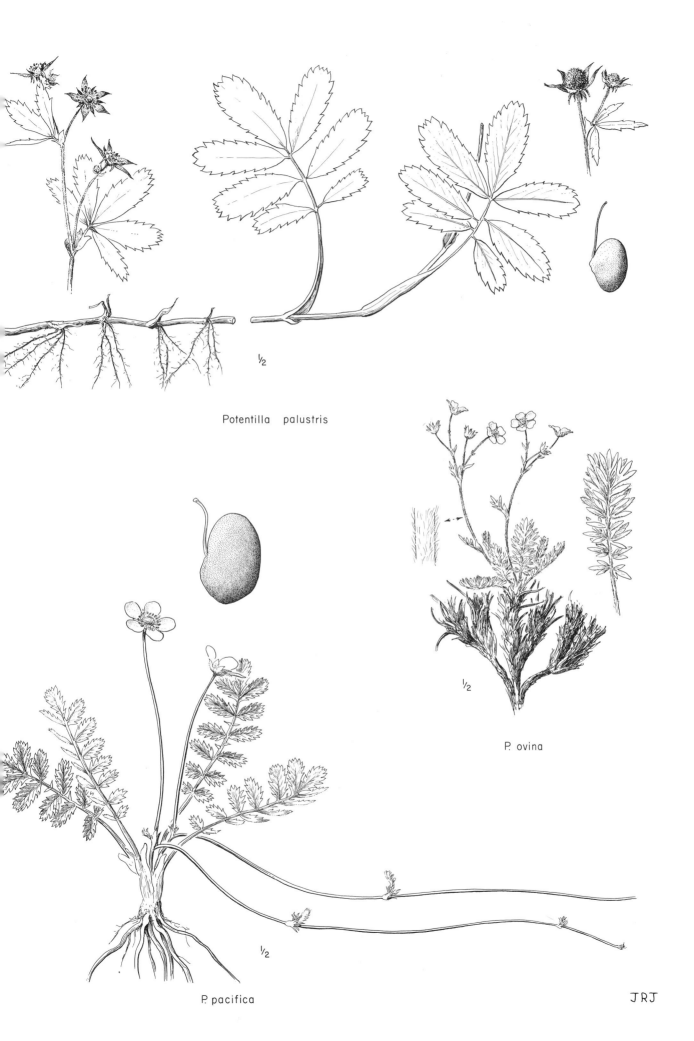

Potentilla   palustris

½

P. ovina

½

P. pacifica

½

JRJ

cumbent to erect, sparsely to rather thickly pubescent and usually tomentose; leaves pinnate, heavily pubescent, usually greenish above but nearly white on the lower surface; leaflets 5-9 (11), the lower ones often reduced, the upper 3 the largest, oblong to oblanceolate-obovate, (1) 1.5-3 (4) cm. long, laciniately lobed 1/2-4/5 the way to the midrib into linear, antrorse segments; stipules of cauline leaves laciniately cleft; cymes several-flowered, narrow, the bracts much reduced upward; calyx grayish-strigose-tomentose and more or less glandular, the sepals triangular-lanceolate, about 5 mm. long, usually slightly exceeding the lanceolate bracteoles and subequal to the yellow petals; stamens usually 20; pistils numerous; style subapical, about 1 mm. long, considerably thickened and glandular-roughened at base, tapered to the tip, little if any longer than the nearly smooth, often granular mature achene.

Grassland and sagebrush plains to montane ridges; Alas. to Greenl., s. to B.C., Nev., N.M., Kans., Minn., and N.H.; perhaps in Asia. June-Aug.

Potentilla pensylvanica is often separated into several intergradient and more or less sympatric taxa on the basis of variation in leaf lobation and pubescence. Our material, as described above, is largely referable to the var. *strigosa* Pursh, although it is not felt that recognition of infraspecific races is necessarily called for.

Potentilla quinquefolia Rydb. Monog. Potent. 76. 1898.
> *P. nivea* var. *pentaphylla* Lehm. Delect. Sem. Hort. Hamb. 1850:12. 1850, not *P. pentaphylla* Richt. in 1815. *P. nivea* var. *quinquefolia* Rydb. Bull. Torrey Club 23:302. 1896. (Typification recondite)

Perennial with a rather small, branched crown and short, ascending rootstocks, grayish-lanate on the lower surface of the leaves and usually also on the stems, the upper surface of the leaves greenish but finely hirsute-sericeous; stems ascending to erect, 1-2 dm. long; basal leaves numerous, mostly ternate but in part often with 5 leaflets and then usually digitate or sometimes more nearly pinnate, the leaflets more or less oblong, (0.5) 1-2 (3) cm. long, dentate-cleft for most of the length, the teeth oblong-rounded, extending about halfway to the midvein; cauline leaves usually 1 or 2, not much reduced, often glandular as well as sericeous to semilanate; cymes rather closely several-flowered, conspicuously bracteate; calyx bowl shaped, sericeous-villous, barely 1 cm. wide even in fruit, the lobes lanceolate, 3-4 mm. long; petals yellow, obcordate, slightly longer than the sepals; stamens usually 20; pistils numerous; style glandular-thickened and warty-papillose at base, slender, subapical on the ovary, and longer than the smooth, 1.5 mm. achene.

Alpine and subalpine gravelly meadows and river bars; B.C. to Sask., southward in the Rocky Mts. to Colo. and Utah, also in s.c. Oreg. June-July.

Potentilla quinquefolia is more or less intermediate in nature between *P. nivea* and *P. concinna*, but has more erect stems and smaller flowers than the latter and more floriferous and more leafy cymes and slightly smaller flowers than the former, with which it seems to be very closely related. In the Canadian Rockies a more glandular, less lanate phase is occasional.

Potentilla recta L. Sp. Pl. 497. 1753.
> *Hypargyrium rectum* Fourr. Ann. Soc. Linn. Lyon II, 16:372. 1868. ("Habitat in Italia, Narbona")

Perennial with an erect, simple to branched caudex but without rootstocks, sparsely to rather copiously hirsute to semihispid and with more abundant, finer, shorter, spreading, sometimes glandular pubescence, greenish throughout; stems one to few, erect, branched above, very leafy, (2) 3-8 dm. tall; leaves digitately 5- to 7-foliolate; leaflets oblanceolate, strongly veined, 3-8 cm. long, serrate-lobate about halfway to the midvein, the teeth divergent-antrorse; stipules lanceolate to ovate, 1-2 cm. long, laciniate above; cymes large and many-flowered, leafy-bracteate at the lower nodes, the branches ascending, the whole flat-topped; calyx cupuliform, considerably accrescent and up to 12 mm. broad, the lobes strongly veined, more or less acuminate, 5-9 mm. long; petals yellow, obovate, emarginate, equaling or up to as much as 3 mm. longer than the sepals; stamens usually 25 (reportedly 30); pistils numerous; style somewhat thickened and glandular-warty near the base, subapically attached; achenes strongly reticulate-veiny, slightly keeled dorsally, brownish-purple, 1 mm. long. 2N=28, 42.

This Eurasian species is well established as a ruderal in e. N.Am., and is known in our area in e. Wash. and n.w. Mont. June-July.

Potentilla rivalis Nutt. in T. & G. Fl. N.Am. 1:437. 1840.
> *Tridophyllum rivale* Greene, Leafl. 1:189. 1905. (*Nuttall*, "In alluvial soil along the Lewis River")

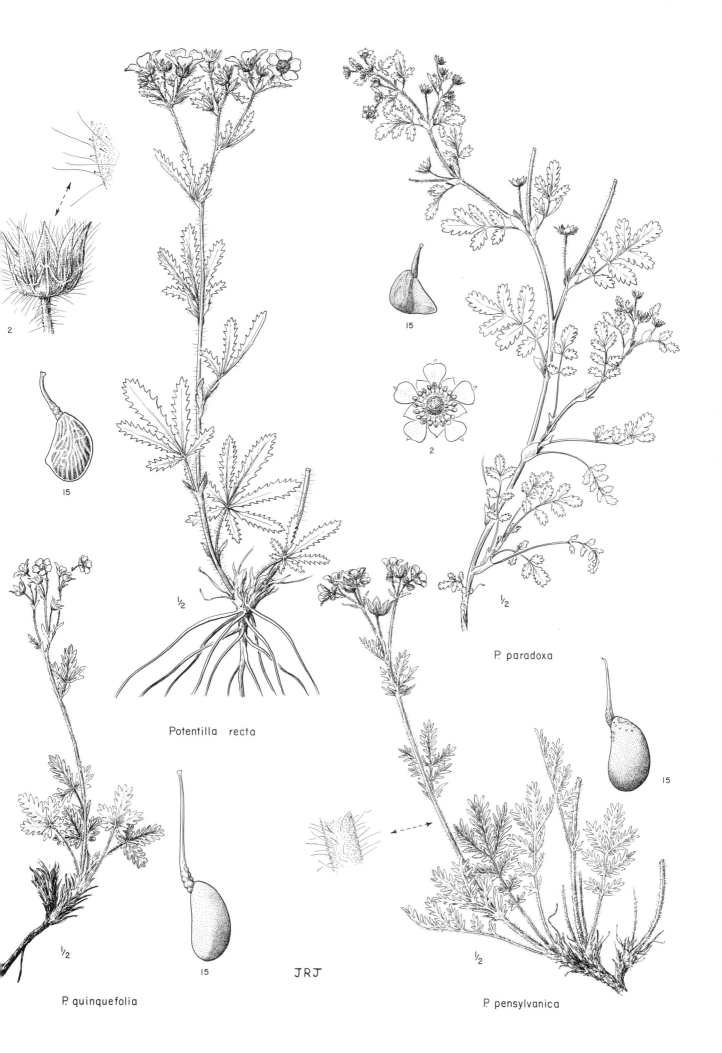

2

15

2

15

Potentilla recta

P. paradoxa

15

P. quinquefolia

1/2

15

JRJ

1/2

1/2

1/2

P. pensylvanica

P. millegrana Engelm. ex. Lehm. Delect. Sem. Hort. Hamb. 1849:11. 1849. *P. rivalis* var. *mille-grana* Wats. Proc. Am. Acad. 8:553. 1873. (North America)

P. leucocarpa Rydb. in Britt. & Brown, Ill. Fl. 2:212. 1897. ("Missouri to Minnesota, west to California and Washington," no collections cited)

Spreading to erect annual or biennial with a strong taproot and a usually simple (branched) caudex, strongly pubescent with fine, spreading to appressed and often semitomentose, eglandular hairs; stem usually freely branched and floriferous most of the length, the inflorescence leafy-bracteate, many-flowered but rather diffuse, long-pedunculate; leaflets oval, broadly obovate to oblong-oblanceolate, 1-4 (5) cm. long, coarsely crenate-serrate, often 5 and crowded-pinnate on the lower cauline and basal leaves, those of the upper cauline leaves always 3 and mostly proportionately more slender; calyx cup-shaped, 5-10 mm. broad at anthesis, much accrescent in fruit, the ovate-triangular lobes erect, longer than the hypanthium but usually somewhat shorter than the elliptic-lanceolate bracteoles; petals yellow, cuneate-obovate to oblanceolate, mostly rounded, about half as long as the sepals; stamens mostly 10 but sometimes 15 (even on the same plant); pistils numerous; style apical, thickened at the base; achenes yellow, ovoid-reniform, barely 0.8 mm. long, very lightly corrugate or smooth.

Damp soil, especially along rivers and around lakes, ponds, and swamps; B. C. southward, chiefly on the e. side of the Cascades in B. C. and Wash., down the Columbia R. Gorge and through the Willamette Valley, to s. w. Oreg. and to coastal s. Calif., e. to Sask., Minn., Ill., Mo., N. M., and n. Mex. May-Sept.

The species is highly variable but clearly distinguishable because of the fine, usually somewhat tomentose pubescence, smooth achenes, and small petals. The occasional plant with 5 leaflets on the lower cauline or basal leaves represents the var. *rivalis,* whereas the more common phase, with only 3 leaflets, has been called var. *millegrana* (Engelm.) Wats. The distinction seems arbitrary and unreal, the two taxa being completely sympatric.

Potentilla uniflora Ledeb. Mém. Acad. St. Pétersb. 5:543. 1812.

P. macrantha var. *uniflora* G. Don, Gen. Hist. Pl. 2:550. 1832. *P. villosa* var. *uniflora* Ledeb. Fl. Ross. 2:58. 1844. *P. nivea* var. *uniflora* Rydb. Bull. Torrey Club 23: 303. 1896. *(Eschscholtz,* "Hab. in terra Tschuktschorum ad sin. St. Lawrentii")

Perennial with a branched crown and short rootstocks, 5-15 (25) cm. tall, grayish-hairy throughout, the petioles and often the lower stems pilose with undulate to nearly straight, scarcely tangled hairs, the stems tomentose above; leaves grayish-tomentose beneath, hirsute and somewhat greenish above, the basal leaves ternate, the leaflets cuneate-obovate to obovate-rhombic, (0. 5) 1-2 cm. long, dentate-lobate about 1/2 the way to the midvein into (5) 7-11 more or less ovate-lanceolate teeth; cauline leaves usually only 1 or 2, greatly reduced above; cymes only 1- to 2 (3)-flowered, prominently bracteate; calyx saucer shaped, sericeous to partially tomentose, 8-12 mm. broad, the bracteoles linear-elliptic to narrowly oblong, subequal to, but much narrower than, the lanceolate, 3-4 mm. sepals; petals yellow, 4-5 mm. long; stamens usually 20; pistils numerous; style subapically attached and about equaling the mature achene, the lower portion swollen and papillate.

River bars to alpine ridges and rock crevices; Alas. southward in the Rocky Mts. to s. B. C., s. w. Alta., Mont., and Colo., reported (but not believed to occur) in the Wallowa Mts., Oreg.; Siberia. June-July.

Although this taxon is easily mistaken for depauperate *P. villosa* or *P. nivea,* it apparently is distinguishable from the former by the narrow calyx bracteoles and smaller flowers and from the latter because of the lack of tomentum on the petioles. The papillate style bases are similar to those of *P. hookeriana* and careful monographic study of this group of Siberian-American taxa is necessary to determine the true relationship between them.

Potentilla villosa Pall. ex Pursh, Fl. Am. Sept. 353. 1814.

P. fragiformis var. *villosa* Regel & Tiling, Fl. Ajan. 85. 1858. *P. grandiflora* var. *villosa* Kurtz in Engl. Bot. Jahrb. 19:374. 1894. ("On the north-west coast")

POTENTILLA VILLOSA var. PARVIFLORA C. L. Hitchc. hoc loc. A var. *villosa* similis sed floribus parvioribus, petalis 5-8 mm. longis. (Type: *J. W. Thompson 5580,* Mt. Angeles, Clallam Co., Wash., Aug. 2, 1930, in U. of Wash. Herb.)

Perennial with a branched crown and short thick ascending rootstocks, usually tufted, grayish-villous or somewhat sericeous throughout and usually lanate on the stems, calyx, and (especially) on the lower surface of the leaves; stems 3-15 (20) cm. tall; leaves ternate, the leaflets rather thick and coriaceous, strongly veined, cuneate-obovate to flabelliform, (0. 5) 1-2 cm. long, coarsely crenate-dentate up to 1/3 the way to the midvein, the teeth rounded; cauline leaves usually 2, subsessile but not much reduced; cymes prominently bracteate, (1) 2- to 5-flowered, rather open; calyx mostly 7-11 mm. long, the bracteoles elliptic to oval-elliptic, mostly 2-4 mm. broad and rarely more than twice as long as broad, subequal (or equal) to the triangular, 3.5-4.5 (6) mm. lobes; petals yellow, obcordate, exceeding the sepals, mostly 5-8 mm. long; stamens usually 20; pistils numerous; style basally thickened and tapered upward, slightly verrucose, subterminally attached and about equal to the smooth or eventually reticulate, 1 mm. achene. 2N=42, 49.

Coastal bluffs and arctic tundra from Alas. southward, at increasing elevations, to the Cascade and Olympic mts. of Wash., where alpine on ridges and talus slopes and in rock crevices, also in the Rocky Mts. to s. B.C. and s.e. Alta.; Aleutian Is.; n.e. Asia. July-Sept., in our area.

Our material, as described above, differs from that from n. Alas. as follows:

1 Calyx 7-11 mm. broad; petals 5-8 mm. long; stems rarely over 15 cm. tall; coastal Alas. from about Juneau, s. to the Olympic and Cascade mts. of Wash., and in the s. Rockies; gradually transitional n. to var. *villosa*   var. parviflora C. L. Hitchc.
1 Calyx mostly 9-14 mm. broad; petals mostly 8-12 mm. long; stems often well over 15 cm. tall; Alas. w. to e. Asia.   var. villosa

Prunus L. Plum; Cherry

Flowers in terminal racemes, corymbs, or umbels, or solitary on short spur shoots, complete, perigynous; calyx turbinate to campanulate, the hypanthium disc-lined, about equaling the 5 lobes, deciduous near the base shortly after anthesis; petals white to pink or red; stamens 20-30, the filaments slender, exserted; pistil 1, simple, the ovary borne at the base of the hypanthium, containing 2 pendulous ovules, the style terminal, the stigma discoid; fruit a 1- (rarely 2-) seeded drupe with usually a fleshy and juicy (ours) mesocarp and a much-hardened endocarp; trees or shrubs with alternate, deciduous (ours) or evergreen, crenate-serrulate to serrate and often gland-toothed leaves, usually bearing conspicuous glands either along the margin of the blade near the petiole or on the petiole itself.

Perhaps 200 species of temperate regions of the Northern Hemisphere and the Andes of S.Am. (The ancient Latin name for the plum.)

Reference:

Wight, W. F. Native American species of Prunus. U.S. Dept. Agr. Bull. 179:1-75. 1915.

Most of the species have edible fruit and all are apt to be animal- (especially bird-) disseminated.

1 Flowers numerous in elongate racemes
  2 Leaves evergreen, thick and leathery; the Laurel cherry, a very common ornamental shrub w. of the Cascades, often persisting in waste areas and occasionally bird-disseminated, but rarely becoming established   *P. laurocerasus* L.
  2 Leaves deciduous, fairly firm but not thick and leathery   P. VIRGINIANA
1 Flowers 1-several in umbels or corymbs
  3 Flowers corymbose, borne on a common axis developing the current season, the rachis often leafy-bracteate; flowers 10-15 mm. broad; drupe mostly 6-10 mm. broad, not glaucous
    4 Corymbs leafy-bracteate at base; leaves oval to broadly elliptic-ovate, abruptly acute, usually glabrous; fruit nearly black, 6-8 mm. thick; plant usually arborescent, a very occasional escape   P. MAHALEB
    4 Corymbs generally not leafy-bracteate at base; leaves various, but usually either rounded or obtuse, or gradually acuminate, often pubescent; fruit either red or else averaging more than 8 mm. in thickness; trees or shrubs
      5 Leaves gradually acuminate, finely serrate; fruit red, 4-7 mm. long; mostly e. of the continental divide   P. PENSYLVANICA
      5 Leaves rounded to acute, crenulate to serrate; fruit red to black, 8-12 mm. long; mostly w. of the continental divide   P. EMARGINATA
  3 Flowers single or umbellate, borne (1 or more per axil) at or near the tip of a shoot developed the previous year; flowers often well over 15 mm. broad; drupe usually either over 10 mm. broad or else strongly glaucous
    6 Calyx copiously hairy externally; leaves rounded to obtuse, often somewhat cordate at base; w. and s. Oreg.   P. SUBCORDATA
    6 Calyx glabrous externally or if at all pubescent then the leaves acute to acuminate

7 Pit of the fruit subglobose (not flattened), smooth; flowers 2.5-3.5 cm. broad; pedicels
usually at least 2 cm. long; plants generally arborescent; leaves 5-15 cm. long; fruit
not glaucous—the cultivated cherries, not uncommon (especially in w. Ida.) as occasional escapes or persistent from old plantations
   8 Leaves permanently hairy on the lower surface, at least along the main veins, mostly 8-15 cm. long; petals obovate; the sweet cherry   *P. avium* L.
   8 Leaves glabrous or glabrate on the lower surface, mostly 5-8 cm. long; petals suborbicular; the sour cherry   *P. cerasus* L.
7 Pit of the fruit either strongly flattened or, if turgid, then usually evidently rugose;
flowers often less than 2.5 cm. broad; pedicels usually less than 2 cm. long; leaves
often less than 5 cm. long or else not coarsely serrate; fruit usually glaucous
   9 Pit of the fruit turgid, rugose-pitted; flowers usually solitary from each node of the
spur branches; fruit 10-15 mm. broad, bluish-black; flowers 10-15 mm. broad   P. SPINOSA
   9 Pit of the fruit strongly flattened, smooth; flowers usually 2-several at each node of
the spur branches; fruit usually over 15 mm. broad, yellow to bluish-black; flowers
15-30 mm. broad
     10 Flowers 2-3 per node on the spur shoots; leaves convolute in the bud; fruit usual-
ly yellow, green, or blue, glaucous; the cultivated plums, which not uncommonly
escape and persist (often as thickets), but which are especially frequent along the
Clearwater R., Ida.; *P. domestica* L. is the more commonly encountered plant,
but about 30 mi. e. of Elk City, Idaho Co., there grows a colony of numerous
rounded (seemingly native) plum trees that bear fruits resembling the green gage
plum in color, flavor, and size, and seem to be referable to *P. institia* L. (bullace
or damson plum) but in many ways appear to be intermediate between that taxon and
P. domestica L. (of which *P. institia* is often considered varietal). The two taxa
are distinguished chiefly on minor differences in pubescence as well as in fruit
shape (see discussion under *P. spinosa*).
     10 Flowers usually 2-5 per node on the spur shoots; leaves conduplicate in the bud;
fruit usually red, slightly or not at all glaucous   P. AMERICANA

Prunus americana Marsh. Arbust. 111. 1785.
   P. domestica var. *americana* Castiglioni, Viag. Negli Stati Uniti 2:339. 1790. (Eastern U.S., no
specimens cited)
   From shrubby and only 1 m. tall to arborescent and up to 10 m. in height, often thicket forming,
usually with some of the branches more or less sharp pointed (spinose), the bark deep brownish-purple; leaves with rather stout, pubescent petioles (3) 5-12 mm. long, the blades lanceolate to elliptic
or oblong-oblanceolate, acuminate, acute (rounded) at base, serrate, somewhat hairy (at least beneath) to glabrous, 4-10 cm. long; flowers (1) 2-4, umbellate, the pedicels slender; calyx usually
reddish tinged, the hypanthium glabrous exteriorly (ours) to pubescent, the lobes about equaling the
hypanthium, pubescent on the upper surface, 2.5-3.5 mm. long, oblong-lanceolate, erose-serrulate;
petals elliptic-oblong to oblong-oblanceolate, 7-9 mm. long; stamens about 25; ovary glabrous; drupe
(yellow) orange to red (purplish), scarcely glaucous, the flesh yellow, 1.8-2.5 cm. long, the pit flattened, smooth. N=8. Wild plum.
   Along watercourses, and on open to wooded, moist to dry areas from the plains into the lower
mountains; over much of n.e. Can. and the U.S., (reportedly) w. to Ravalli and Sanders cos., Mont.,
and to Utah and Ariz., chiefly e. or s. of our range. Apr.-May.

Prunus emarginata (Dougl.) Walpers Rep. 2:9. 1843.
   Cerasus emarginata Dougl. ex Hook. Fl. Bor. Am. 1:169. 1833. *(Douglas,* "On the upper part of the
Columbia River, especially about the Kettle Falls")
   Cerasus mollis Dougl. ex Hook. Fl. Bor. Am. 1:169. 1833. *P. mollis* Walpers Rep. 2:9. 1843,
but not of Torr. in 1824. *P. emarginata* var. *mollis* Brew. in Brew. & Wats. Bot. Calif. 1:167.
1876. *Cerasus prunifolia* Greene, Proc. Biol. Soc. Wash. 18:57. 1905. *P. prunifolia* Shafer in
Britt. & Shafer, N. A. Trees 500, fig. 461. 1908. *(Douglas,* "North-West coast of America near
the mouth of the Columbia, and on subalpine hills, near the source of that river," is the first citation)
   Cerasus erecta Presl, Epim. Bot. 194. 1847. *P. emarginata villosa* Sudw. U.S. Div. Forest. Bull.

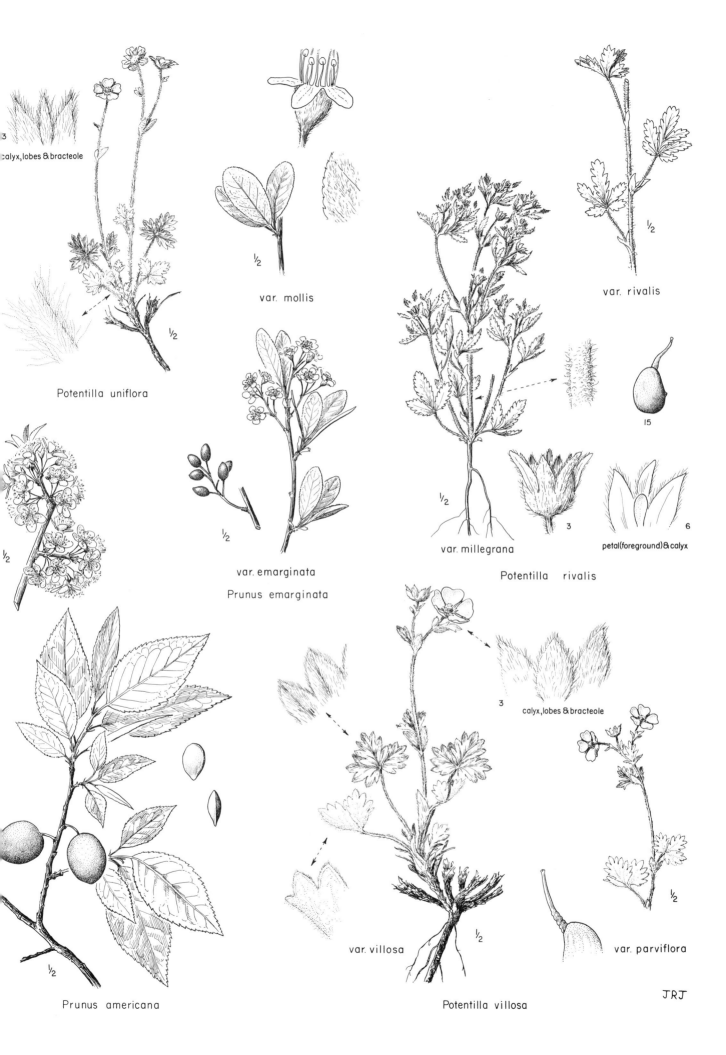

3
calyx,lobes & bracteole

var. mollis

var. rivalis

Potentilla uniflora

var. emarginata

Prunus emarginata

var. millegrana

15

6
petal(foreground)&calyx

Potentilla rivalis

3
calyx,lobes & bracteole

var. villosa

var. parviflora

Potentilla villosa

Prunus americana

JRJ

14:240. 1897. *P. emarginata* ssp. *erecta* Piper in Piper & Beattie, Fl. N.W. Coast 199. 1915. (*Haenke*, "Nutka-Sund") = var. *mollis*.

P. corymbulosa Rydb. Mem. N.Y. Bot. Gard. 1:226. 1900. (*Rydberg & Bessey 4437*, Bridger Mts., Mont., June 18, 1897, is the first collection cited) = var. *emarginata*.

Cerasus trichopetala Greene, Proc. Biol. Soc. Wash. 18:60. 1905. *P. trichopetala* Blank. Mont. Agr. Coll. Stud. Bot. 1:70. 1905. (*R. S. Williams*, Columbia Falls, Mont., May 24, 1894) = var. *emarginata* or possibly *P. pensylvanica*.

Straggly shrubs from 2 m. tall to erect, spreading trees as much as 15 m. tall, glabrous to densely pubescent throughout, the bark deep reddish-purple on the young twigs; leaves with petioles 5-12 mm. long, the blades elliptic or oblong to oblong-obovate, obovate, or oblanceolate, rounded to obtuse or (less commonly) acute, rarely at all acuminate, crenulate to serrulate, 3-8 cm. long; flowers mostly (3) 5-8 (10), corymbose, the peduncles sometimes leafy-bracteate; calyx glabrous to coarsely hairy, the hypanthium 2.5-3.5 mm. long, equaling or slightly exceeding the oblong-lanceolate, usually entire lobes; petals obovate to obovate-oblanceolate, 5-7 mm. long, often pubescent on the lower (dorsal) surface; stamens about 20; ovary glabrous; drupe red to almost black, 8-12 mm. long, very bitter. Bitter cherry.

Moist woods or along watercourses in grassland or sagebrush desert, into the mountains, at medium elevations; B.C. southward, from the coast inland, through Wash. and Oreg. to s. Calif., e. to Mont., Wyo., Utah, and Ariz. Apr.-June.

The plants are extremely variable in height, size of the leaves, and general pubescence, but most of those from w. of the Cascades differ from those from farther e. as follows:

1 Plant more arborescent than shrubby, up to 15 m. tall and with a main trunk as much
    as 2.5 dm. thick, heavily pubescent, especially on the lower surface of the leaves and
    on the calyx; mainly w. of the Cascades and in the lowlands, from B.C. to s. Oreg.
                                                   var. mollis (Dougl.) Brew.
1 Plant more shrubby than arborescent, generally several-stemmed, 1-4 (8) m. tall,
    from glabrous to rather heavily pubescent on the leaves, the petals frequently hairy
    on the back; chiefly in and e. of the Cascades in B.C., Wash., and Oreg., and s. and
    e., but also in the Olympic Mts.   var. emarginata

Prunus mahaleb L. Sp. Pl. 474. 1753. ("Habitat in Helvetia")

Spreading tree as much as 10 m. tall, the twigs copiously pubescent, the bark grayish-red, tardily glabrate; leaves with slender, generally glandless, puberulent petioles mostly 8-15 mm. long, the blades oval or broadly elliptic-ovate, 2-5 (6) cm. long, abruptly acute, finely crenulate and with tiny gland-teeth, usually glabrous and rather pale green on both surfaces; flowers 4-12 in short, leafy-bracteate, corymbose racemes, the pedicels up to 2 cm. long; calyx glabrous, turbinate-campanulate, greenish-white, the lobes entire, oblong-lanceolate, about 3 mm. long and about equal to the hypanthium; petals oblong-elliptic to oblanceolate, (6) 7-9 mm. long; stamens about 20, exserted; ovary glabrous; drupe ovate, 6-8 mm. long, nearly or quite black. N=8. Mahaleb cherry.

The species is often used as budding stock for cherries, and plants sometimes escape and become established, as near Lake Waha, Nez Perce Co., Ida. (*W. H. Baker 8898*), and w. of Spokane, Wash. May-June.

Prunus pensylvanica L. f. Suppl. 252. 1781.

Cerasus pensylvanica Lois. Nouv. Duham. 5:9. 1812. (North America)

Shrub to small tree 1-5 (12) m. tall, glabrous or subglabrous, the bark reddish; petioles slender, 5-15 mm. long; leaf blades lanceolate or ovate-lanceolate to elliptic or elliptic-obovate, 3-8 (12) cm. long, acuminate, acute to rounded at base, finely serrate; flowers 4-12, in leafless corymbs; calyx glabrous, the hypanthium turbinate-campanulate, 2-2.5 mm. long, about equal to the oblong-lanceolate, entire (erose) lobes; petals ovate to obovate, 5-7 mm. long, usually hairy on the outer (dorsal) surface; stamens about 30; drupe nearly globose, red, 4-7 mm. long. Bird or pin cherry.

Dry to moist woods, clearings, fence rows, sand dunes, and canyons; e. B.C. and c. and e. Mont. to Colo., e. to Lab. and Newf., Va., Ind., and the Dakotas, almost entirely e. of the continental divide, but within our range in s.e. B.C. and along the base of the Rockies in Mont. Apr.-June.

Prunus spinosa L. Sp. Pl. 475. 1753. (Europe)

Rigid, freely branched, often thorny shrub 1-3 (4) m. tall, sometimes thicket-forming; young branches finely puberulent to glabrous; leaves convolute in bud, elliptic-ovate to obovate, 2-4 cm.

long, pubescent on both surfaces when young, ultimately nearly or quite glabrous, rounded; flowers developing slightly before the leaves, clustered, but usually single (2) at each of the numerous nodes on the short lateral spur shoots (which were developed the previous year); pedicels generally glabrous (puberulent), 10-15 mm. long; calyx glabrous externally, the lobes glabrous to pubescent on the upper surface; petals 5-7 mm. long, white, oblong-obovate; fruit globose, deep bluish-purple, glaucous, 10-15 mm. broad, very astringent; pit subglobose, rugose-pitted. N=16. Blackthorn, sloe.

Occasional in moist draws and on hillsides; s.e. Wash. and adj. Ida., perhaps also in w. Oreg.; Europe. Apr.

This plant has been cultivated in certain localities within our area, as at Pullman, Wash., where it was grown as early as 1892. It has become well established in that general region, and perhaps also in Oreg., as near Sutherlin, Douglas Co. It is closely allied to *P. domestica* L. and *P. institia* L., and is said to be the ultimate source of both of those taxa, through hybridization involving *P. cerasifera* Ehrh., the cherry plum. The plants from s.e. Wash. are variable, some having glabrous, others pubescent juvenile branches. Whereas *P. spinosa* is usually said to have glabrous sepals, European material is not constant in this respect, some plants having pubescence on the upper surface of the sepals, as do our representatives. It would appear that the three taxa, *P. spinosa, P. domestica,* and *P. institia,* as they occur in s.e. Wash. and adj. Ida., are perhaps more or less modified by interhybridization.

Prunus subcordata Benth. Pl. Hartw. 308. 1849. *(Hartweg,* "in montibus Sacramento," Calif.)

Rigid and often spinose-branched, rounded, thicket-forming shrub 1-3 (8) m. tall, usually puberulent throughout; bark glabrate, grayish-purple; leaves with usually glandless petioles up to 12 mm. long, the blades ovate to obovate, 2-4.5 (6) cm. long, serrulate, acute to semicordate and usually with a few glands at the base, rounded to obtuse; flowers 1-4 from the branch tips, umbellate; pedicels 6-17 mm. long; calyx puberulent, the lobes oblong-lanceolate, spreading, 3.5-5 mm. long, about equaling the narrowly campanulate hypanthium; petals obovate, 6-10 mm. long; stamens about 30, the anthers yellow; ovary glabrous to strongly hirsute; drupe dark red to reddish-purple, rarely yellow, glabrous, about 2 cm. long, edible. Wild plum.

Stream banks to dry hillsides and open pine woodland; very occasional in the Willamette Valley, from Marion Co. southward, and more common in s. Oreg. (from Douglas to Lake cos.) and in the Coast Range and Sierra Nevada of Calif., to Monterey and Tulare cos. Apr.-May.

In s. Oreg. and in Calif. 2 or 3 varieties are usually recognized, but the plant which is occasional in the Willamette Valley is referable to var. *subcordata.*

Prunus virginiana L. Sp. Pl. 473. 1753. (Virginia)

Cerasus demissa Nutt. in T. & G. Fl. N.Am. 1:411. 1840. *Prunus demissa* Walpers Rep. 2:10. 1843. *Padus demissa* Roem. Syn. Rosifl. 3:87. 1847. *Prunus virginiana* var. *demissa* Torr. Bot. Wilkes Exp. 284. 1874. *Padus virginiana* var. *demissa* Schneider, Ill. Handb. Laubh. 1:642. 1906. (*Nuttall,* "Plains of the Oregon towards the sea, and at the mouth of the Wahlamet")

Cerasus demissa var. *melanocarpa* A. Nels. Bot. Gaz. 34:25. 1902. *Prunus melanocarpa* Rydb. Bull. Torrey Club 33:143. 1906. *Padus melanocarpa* Shafer in Britt. & Shafer, N.Am. Trees 504. 1908. *Prunus demissa* var. *melanocarpa* A. Nels. Mitt. Deutsch. Dendr. Ges. 1911:231. 1911. *Prunus virginiana* var. *melanocarpa* Sarg. Journ. Arn. Arb. 2:117. 1920. (Rocky Mountains, no specimens cited)

Prunus demissa var. *melanocarpa* f. *trichodisca* Koehne, Mitt. Deutsch. Dendr. Ges. 1911:234. 1911. (Specimens cited from Wash., Calif., Colo., Ida., and N.M.) = var. *melanocarpa.*

Prunus demissa var. *nuttallii* f. *howellii* Koehne, Mitt. Deutsch. Dendr. Ges. 1911:234. 1911. (*Howell,* rocky hillsides and gulches [Oregon?] in 1880)

Prunus demissa var. *melanocarpa* f. *leiodisca* Koehne, Mitt. Deutsch. Dendr. Ges. 1911:234. 1911. (Specimens cited from Mont., Utah, Colo., N.M., and Neb.)

Prunus pinetorum Suksd. Werdenda 1:24. 1927. (*Suksdorf 10611,* Bingen, Wash., Aug. 20, 1920 and May 14, 1921) = var. *melanocarpa.*

Ours erect shrubs or small trees 1-4 (6) m. tall with purplish-gray bark; petioles 5-15 mm. long, bearing 1-2 distal purplish-red glands; leaf blades elliptic to ovate-oblong or oblong-obovate, finely serrate, (3) 4-10 cm. long, usually abruptly acute to somewhat acuminate, bright green and glabrous on the upper surface, paler and glabrous to pubescent beneath; flowers numerous in terminal racemes, the pedicels of rather uniform length, mostly 4-8 mm. long; calyx glabrous, the lobes spreading to recurved, oval, finely glandular-erose, 1-1.5 mm. long, 1/2-1/3 as long as the gland-lined hypanthium; petals suborbicular, 4-6 mm. long; stamens about 25; drupe red to purple or black, ovoid, 8-11 mm. long, sweet but astringent. Chokecherry.

British Columbia, eastward in Can. to Newf., s. to Calif. and N. M., the Dakotas, Kans., Mo., Tenn., and N. C., in many habitats, and in our area from coastal bluffs to desert grassland and sagebrush, where chiefly along watercourses, upward in the lower mountains to ponderosa pine forest. May-early July.

The chokecherries are mostly worthful shrubs. Although the flowers are fairly attractive, the species are valued chiefly because of the fruits, all of which are alluring to birds. Those species with red fruits are most valuable as ornamentals.

Although our plants are often considered to be specifically distinct from *P. virginiana,* they appear more realistically to represent only fairly clearly marked geographic races of that wide-ranging species, separable as follows:

1 Leaves generally glabrous on the lower surface, or pubescent only in the axils of the
    veins (very rarely more generally hairy on plants to the e. of our range); shrubs to small
    trees up to 15 m. tall
    2 Plants large shrubs or small trees up to 15 m. tall; leaves thin; fruit crimson to deep
       red; Sask. to Newf., s. to Kans., Mo., Tenn., and N.C.,   var. virginiana
    2 Plants small to medium-sized shrubs, rarely over 4 (6) m. tall; leaves rather thick
       in texture; drupes deep bluish-purple to nearly black; along the e. base of the Cas-
       cades from B. C. to n. e. Oreg., but w. to the coast in s. Oreg. and in Calif., e. in
       B. C. to Alta. and the Dakotas, s. in the Rockies to N. M.   var. melanocarpa (A. Nels.) Sarg.
1 Leaves usually pubescent over most or all of the lower surface; plants small to medium-
    sized shrubs 2-4 (6) m. tall; w. of the Cascades, from B. C. to n.w. Oreg.
                                           var. demissa (Nutt.) Torr.

Purshia DC. ex Poir.

Flowers complete, perigynous, solitary and terminal on short, lateral, leafy spurs; calyx turbinate-funnelform, persistent, the 5 lobes more or less spreading; petals 5, yellow, deciduous; stamens usually 25, inserted in one series with the petals at the top of the gland-lined hypanthium; pistils 1 (2), borne on a very short stipe at the base of the hypanthium, the ovary with a single erect ovule, tapered gradually into a short thick style and stigmatic almost the full length; fruit an ellipsoid-fusiform achene; shrubs with alternate, deciduous, stipulate, apically 3-cleft, usually glandular leaves closely aggregated and semifasciculate on short lateral spur shoots.

Only 1 other species, that in Calif. and Nev. (Named for F. T. Pursh, 1774-1820, author of Flora Americae Septentrionalis, one of the earliest floras of North America.)

Purshia tridentata (Pursh) DC. Trans. Linn. Soc. 12:158. 1818.
    Tigarea tridentata Pursh, Fl. Am. Sept. 333. 1814. *Kunzia tridentata* Spreng. Syst. Veg. 2:475.
    1825. *(Lewis,* "prairies of the Rocky-mountains and on the Columbia river")
Freely and rigidly branched shrubs 1-2 (4) m. tall; leaves cuneate, (5) 10-20 (25) mm. long, deeply 3-toothed at the tip, pubescent to arachnoid but greenish on the upper surface, grayish-tomentose beneath, usually more or less revolute; calyx strongly stipitate-glandular as well as tomentose or arachnoid, (5) 6-8 mm. long, the ovate-oblong, entire lobes about equaling or slightly shorter than the funnelform-turbinate hypanthium; petals obovate-oblong to spatulate, (5) 6-9 mm. long; stamens well exserted; achene rather cartilaginous, about 1. 5 cm. long, finely puberulent; seed black, pyriform, 6-8 mm. long. N=9. Antelope bush or brush, bitter brush.

Grassland and sagebrush desert to juniper woodland or ponderosa pine forest; B. C. southward along the e. side of the Cascades in Wash. and Oreg. (and in the Columbia Gorge), to Trinity Co. and the Sierra Nevada (mostly on the e. side) as far s. as Inyo Co., Calif., e. to (mostly) w. Mont., Wyo., Colo., and N. M. Apr. -June.

This is considered to be one of the best browse plants in w. U.S. and the species should be rated as ornamentally choice for any area e. of the Cascades. It is best grown from seeds.

Pyrus L. Pear; Apple

Flowers rather showy, complete, epigynous, borne in umbellate or corymbose clusters on spur shoots; calyx urceolate, ebracteolate, adnate to the ovary below, with an expanded free hypanthium

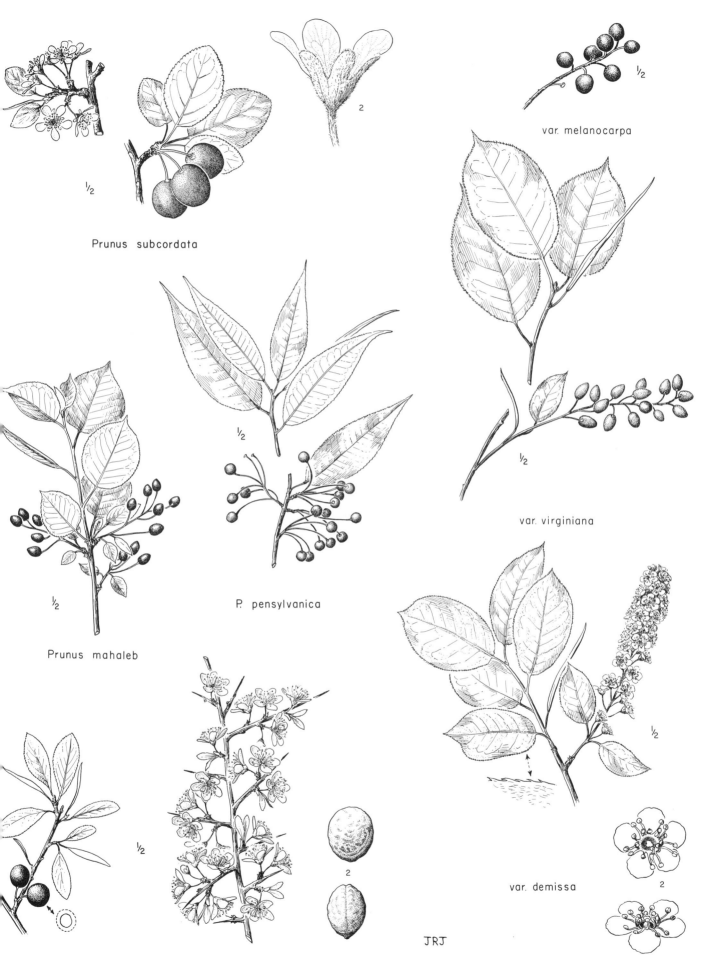

var. melanocarpa

Prunus subcordata

1/2

2

P. pensylvanica

1/2

var. virginiana

Prunus mahaleb

1/2

P. spinosa

1/2

2

var. demissa

2

JRJ

P. virginiana

above, the lobes 5; petals white or pink, usually short-clawed; stamens few (about 15) to many, inserted with the petals at the top of the disc-lined hypanthium; pistil 1, the ovary inferior, 2- to 5-carpellary; styles 2-5, free or connate basally; fruit a more or less fleshy pome, the carpels imbedded in the floral tube, the endocarp and septa cartilaginous; seeds (1) 2 per carpel; shrubs or trees with alternate, simple, toothed to lobed leaves.

About 50 species of temperate N. Am. and Eurasia. (Classical Latin name for the pear.)

The pears, apples, and mountain ashes are by some workers treated under the separate genera *Pyrus, Malus,* and *Sorbus,* respectively; by others they are combined in the one genus *Pyrus.* The fruit of *Pyrus* usually has a distinctive "pear" shape and numerous stone cells in the flesh, that of *Malus* an "apple" shape and no stone cells. *Sorbus* sometimes contains stone cells and does not always have pinnate leaves, but at least in our area it is habitally and morphologically sufficiently distinctive to be maintained as a separate genus.

Besides our one native species, the cultivated pear and apple both frequently escape and many plantings persist after evidence of human care is lost. Both species are therefore included in the following key but not described formally:

1 Fruit pyriform, containing grit cells; leaves ovate to elliptic, abruptly acuminate, crenate;
    flowers white; the cultivated pear   *P. communis* L.
1 Fruit applelike to obovoid, the flesh not containing grit cells; leaves more or less oblong-
    ovate, serrate to lobed; flowers white or pinkish
    2 Carpels (and styles) usually 5; fruit well over 2 cm. in diameter; leaves not lobed; the
        cultivated apple   *P. malus* L.
    2 Carpels (and styles) usually 3 (4); fruit less than 2 cm. in diameter; leaves sometimes
        lobed   P. FUSCA

Pyrus fusca Raf. Med. Fl. 2:254. 1830.

   Malus fusca Schneider, Ill. Handb. Laubh. 1:723. 1906. (No type given)
   P. diversifolia Bong. Mém. Acad. St. Pétersb. VI, 2:133. 1833. *Malus diversifolia* Roem. Fam.
      Nat. Syn. 3:215. 1847. *Malus rivularis* var. *diversifolia* Koehne, Deutsche Dendr. 262. 1893.
      P. fusca var. *diversifolia* Bailey, Cycl. Am. Hort. 5:2876. 1916. *(Mertens,* Sitka)
   P. rivularis Dougl. ex Hook. Fl. Bor. Am. 1:203, pl. 68. 1833. *Malus rivularis* Roem. Fam. Nat.
      Syn. 3:215. 1847. *(Menzies,* "Nootka Sound," is the first of 3 collections cited)
   P. rivularis var. *levipes* Nutt. N. Am. Sylva 2:24. 1846. *Malus rivularis* var. *levipes* Koehne,
      Deutsche Dendr. 262. 1893. *Malus fusca* var. *levipes* Schneider, Ill. Handb. Laubh. 1:724. 1906.
      P. fusca var. *levipes* Bailey, Cycl. Am. Hort. 5:2876. 1916. (No type specified)

Several-stemmed shrub to small tree 3-12 m. tall, the young twigs crisp-pubescent; leaf blades lanceolate to ovate-lanceolate or ovate-oblong, (3) 4-10 cm. long, gradually acute to acuminate, serrate, occasionally with a more or less prominent lobe on one or both margins, deep green and glabrous or pubescent on the upper surface, paler and somewhat crisp-pubescent to lanate beneath; flowers mostly 5-12, corymbose; calyx glabrous to tomentose externally, the lanceolate lobes recurved, about 5 mm. long, deciduous; petals white (pink), oblong-oval to broadly obovate, 9-14 mm. long; stamens about 20, somewhat shorter than the petals; styles usually 3(4), about equaling the stamens, connate 1/3-2/3 their length; fruit 3 (4)-chambered, oblong-ovoid to obovoid, not umbilicate, 10-16 mm. long, yellow to purplish-red, of no economic importance. N=17. Western crab apple.

Moist woods, stream banks, swamps, and bogs to open canyons and into the foothills of the mountains; s. Alas. southward to Sonoma Co., Calif., from the coast to the w. slope of the Cascades up to about 2500 ft. elevation. Late Apr.-early July.

Rosa L. Rose

Flowers large, complete, strongly perigynous, borne singly or in small (seldom large) cymes at the ends of lateral branches or main shoots; sepals mostly 5; petals mostly 5 (or the flowers especially in cultivated forms often double, with numerous petals), light pink to deep rose, less often yellow or white; stamens numerous; pistils more or less numerous (seldom less than 10), the slender styles more or less exserted from the usually narrow mouth of a globose to ellipsoid or pyriform hypanthium, the capitate stigmas commonly (in our spp.) forming a cluster closing the mouth of the hypanthium; ovaries each with a single ovule; achenes bony, usually long-hairy on at least part of the surface, enclosed within the hypanthium, which ripens into a more or less fleshy, colored (commonly reddish or purplish) berrylike structure called a hip; more or less prickly shrubs or woody vines, seldom nearly unarmed, often with a pair of prickles (the infrastipular prickles) at or just below each node; leaves alternate, odd-pinnately compound with 3-11 (rarely more) leaflets, these usually more or less toothed; stipules well developed, usually green and leafy in texture, adnate to the petiole. X=7.

More than a hundred species, native to the Northern Hemisphere. (The classical Latin name.)

The roses are a taxonomically difficult genus. The difficulty is due partly to infraspecific (perhaps Mendelian) variation in some conspicuous characters, such as pubescence, which are often taxonomically useful in other genera, partly to the frequent interspecific hybridization, and partly to other factors such as polyploidy and sometimes apomixis. Each of our native species will hybridize with any of the others, given the opportunity, but they are withal reasonably distinct. Once one is acquainted with them there is seldom any difficulty in recognizing the various species in the field.

Rydberg recognized 115 species of *Rosa* native to North America in 1918 (N.Am. Fl. 22[6]:483-533), and a number of others have been described since that time. Beginning with Mrs. Erlanson in 1934 (cited below) nearly all recent students have taken the point of view that the number of real species is much more limited, and that many of the conspicuous morphological differences, such as in pubescence, hip shape, and armature, which impressed Rydberg and others as being of specific value are actually without taxonomic significance in many cases. The present author fully concurs with this more recent approach. We are trying to recognize the self-perpetuating natural populations, not cataloging morphological variability.

The roses in general are attractive both in flower and fruit, *R. gymnocarpa* being perhaps the least important, whereas *R. nutkana* (especially that element often called *R. spaldingii)* is to be rated as choice.

References:

Cole, Donald. A revision of the Rosa californica complex. Am. Midl. Nat. 55:211-224. 1956.

Erlanson, Eileen. Experimental data for a revision of the North American wild roses. Bot. Gaz. 96:197-259. 1934.

Jones, George N. The Washington species and varieties of Rosa. Madroño 3:120-135. 1935.

Lewis, Walter H. A monograph of the genus Rosa in North America. I. R. acicularis. Brittonia 11: 1-24. 1959.

1 Introduced species; prickles mostly stout and strongly curved; sepals tending to be re-
    flexed after anthesis and usually sooner or later deciduous, some of them usually with
    conspicuous lateral lobes
  2 Lower surface of the leaflets stipitate-glandular   R. EGLANTERIA
  2 Lower surface of the leaflets glabrous or nearly so   R. CANINA
1 Native species; prickles stout or weak but seldom much curved; sepals generally as-
    cending or erect after anthesis, persistent except in *R. gymnocarpa,* seldom with
    lateral lobes
  3 Sepals, top of the hypanthium, and styles deciduous together as the fruit matures;
      achenes relatively few, up to about a dozen; sepals 5-12 mm. long   R. GYMNOCARPA
  3 Sepals and styles long-persistent, the sepals often well over 12 mm. long; achenes
      more numerous, mostly 15-30 or more
    4 Stems more or less bristly with slender prickles, the infrastipular prickles, if
        present, not differentiated from the others
      5 Low shrub of forested regions, producing solitary (seldom 2) flowers on lateral
          branches of the season; leaflets mostly 5-7 (9); boreal species, common in w.
          Mont.   R. ACICULARIS
      5 Semishrub of the plains and prairie region, encroaching into our range in Mont.;
          flowers cymose at the end of the main shoot of the season, often also on lateral
          shoots; leaflets (7) 9-11   R. ARKANSANA
    4 Stems mostly with well-defined infrastipular prickles, or sometimes nearly unarmed
      6 Flowers relatively small and commonly clustered, the sepals mostly 1-2 cm. long
        and 2-3.5 mm. wide at the base, the petals 1.2-2.5 cm. long, the hypanthium
        3-5 mm. thick at anthesis, commonly about 1 (1.5) cm. thick in fruit; plants
        mostly of the lowlands and hills
        7 Plants occurring west of the Cascade summits; sepals nearly always with gland-
          tipped bristles on the back; leaflets rather finely toothed, the inner margin of
          the tooth mostly 0.5-1.0 mm. long   R. PISOCARPA
        7 Plants occurring east of the Cascade summits; sepals nearly always without

bristles and seldom at all glandular on the back; leaves (especially in var. *ultra-montana)* averaging more coarsely toothed, the inner margin of the tooth often well over 1.0 mm. long   R. WOODSII

6 Flowers larger and commonly solitary, the sepals mostly 1.5-4 cm. long and 3-6 mm. wide at the base, the petals mostly 2.5-4 cm. long, the hypanthium mostly 5-9 mm. thick at anthesis, (1) 1.2-2 cm. thick at maturity; plants mostly of wooded or moist country, often at higher elevations than the two preceding species   R. NUTKANA

Rosa acicularis Lindl. Ros. Monog. 44. 1820. (Cultivated specimens, from material originally collected by Bell in Siberia)

R. sayi Schwein. in Keating, Narr. Exp. Long. 2:388. 1824. *R. acicularis* var. *sayi* Rehd. in L. H. Bailey, Cycl. Am. Hort. 1555. 1902. *R. acicularis* ssp. *sayi* W. H. Lewis, Brittonia 11:19. 1959. *(Say,* mouth of the St. Peter [Minnesota] River, Minn.)

R. acicularis var. *bourgeauiana* Crepin, Bull. Soc. Bot. Belg. 15:29. 1876. *R. bourgeauiana* Rydb. Fl. Rocky Mts. 442. 1917. *(Bourgeau,* Saskatchewan basin)

R. engelmanni S. Wats. Gard. & For. 2:376. 1889. *R. acicularis* var. *engelmanni* Crepin ex L. H. Bailey, Cycl. Am. Hort. 1555. 1902. *R. suavis* var. *engelmanni* Hara, Journ. Fac. Sci. Univ. Tokyo, sect. 3, Bot. 6:72. 1952. (Cultivated specimens, from seed collected by Engelmann near Empire City, Colo.)

R. butleri Rydb. N.Am. Fl. 226:506. 1918. *(Butler 796,* Helena, Mont.)

Slender lax shrub 2-12 dm. tall; twigs more or less densely (seldom sparsely) bristly with straight, slender, unequal prickles, the infrastipular prickles, if present, not clearly differentiated from the others; leaf rachis usually villous-puberulent and rather finely glandular, sometimes merely glandular or even subglabrous; leaflets 5-7 (9), elliptic to ovate or obovate, mostly 1.5-4.5 (8) cm. long and 0.7-2.5 (4.5) cm. wide, or the lowest ones smaller, serrate or more often doubly serrate, the teeth very often gland-tipped, the upper surface green and glabrous or nearly so, the lower pale and commonly somewhat villous or villous-puberulent, often glandular as well, varying to subglabrous; flowers mostly solitary (2) on lateral branches of the season, the slender pedicel glabrous or rarely glandular, not bristly; hypanthium glabrous, mostly 3-4.5 mm. thick at anthesis; sepals 1.5-3 cm. long, 2-4 mm. wide at the base, tapering gradually to the slender central part and generally somewhat dilated distally, seldom evidently stipitate-glandular on the back, persistent and erect or connivent in fruit; petals pink to deep rose, (1.5) 2-3 cm. long; hips globose to ellipsoid or pyriform, 1-2 cm. long and wide, dark blue or purplish; achenes commonly 15-25, about 4 mm. long, stiffly long-hairy along one side or toward the tip. In the American plants, 2N typically = 42, as contrasted to 56 in the Eurasian ones.

Woods, and sometimes in open places in wooded regions or in the mountains; circumpolar in the boreal coniferous forest region, extending s. in the cordillera to B. C., Ida. (rarely), Mont., Wyo., Colo., and n. N.M. June-Aug.

Our plants as here described belong to the American var. *bourgeauiana* Crepin. Var. *acicularis,* with glandular pedicels, is primarily Eurasian, extending also into Alas.

Rosa collaris Rydb. (Fl. Rocky Mts. 441. 1917), based upon *Macbride 1676,* Pinehurst, Blaine Co., is perhaps a hybrid, resembling *R. gymnocarpa* in its short sepals and small hips with few achenes, and resembling *R. acicularis* in the persistence of the sepals and type of pubescence of the leaves.

Rosa arkansana Porter in Port. & Coult. Syn. Fl. Colo. 38. 1874.

R. blanda var. *arkansana* Best, Bull. Torrey Club 17:145. 1890. *R. virginiana* var. *arkansana* Mac-Millan, Metasp. Minn. Valley 304. 1892. *(Brandegee 664,* banks of the Arkansas River near Cañon City, Colo.)

R. blanda var. *setigera* Crepin, Bull. Soc. Bot. Belg. 15:33. 1876. (Typification recondite)

R. suffulta Greene, Pitt. 4:12. 1899. *R. arkansana* (var.) *suffulta* Cockerell, Bull. Torrey Club 27:88. 1900. *(Vasey,* meadows of the Rio Grande at Las Vegas, N.M.) The common form with hairy leaves.

R. alcea Greene, Leafl. 2:63. 1910. *(Spreadborough,* Moose Jaw, Sask.; Canadian Survey no. *10624)* A form more strongly glandular than the *suffulta* form.

Plants mostly suffrutescent from a rhizome or running root, 1-5 dm. tall and dying back part way or wholly to the ground each year, or in protected places becoming truly shrubby and up to 10 or even 15 dm. tall; stems sparsely to more often densely beset with slender, unequal, straight prickles, the infrastipular ones, if present, not clearly differentiated from the others; leaves glabrous (in the type) to much more often villous-puberulent especially beneath, the rachis often subtomentosely so and sometimes also glandular; leaflets (7) 9-11, rather crowded, firm, 1-4 cm. long, 0.5-2.5 cm. wide, elliptic to obovate, serrate, the teeth often callous-tipped, but seldom at all glandular; flowers generally several in a corymbiform cyme terminating the main shoot of the season, often also terminating lateral shoots; pedicels stout, glabrous to glandular or loosely villous; hypanthium glabrous or with a few stipitate glands, 4-5 mm. thick at anthesis; sepals (1) 1.5-2 (3) cm. long, 3-5 mm. wide at the base, coarsely stipitate-glandular on the back (sometimes very sparsely so), persistent and more or less spreading in fruit; petals 1.5-2.5 (3) cm. long, pink or sometimes white to reputedly deep rose; hips globose to pyriform, 8-15 mm. thick, purplish; achenes commonly 15-30, about 4 mm. long or a little less, stiffly long-hairy especially along one side. 2N=28.

Prairies and high plains of c. N.Am., sometimes extending into open woods; Sask. and Man. to Mont., N.M., Tex., Mo., and Minn., barely encroaching onto the e. edge of our range. June-Aug.

Rosa canina L. Sp. Pl. 491. 1753. (Europe)

Coarse shrub 1-3 m. tall, provided with well-developed, flattened, more or less strongly curved or hooked prickles; leaflets 5-7, elliptic to ovate or obovate, acute or acuminate, 1.5-4 cm. long and 1-2.5 cm. wide, sharply serrate (sometimes doubly so, and the teeth then often gland-tipped), glandless and usually glabrous on both sides, or with a few deciduous glands along the main veins beneath; flowers in small cymes or solitary; sepals 1-2 cm. long, generally some of them with long, slender, lateral lobes, all reflexed soon after anthesis and deciduous before the fruit is mature; petals 2-2.5 cm. long, pink or white; hips 1.5-2 cm. long, globose to ellipsoid or ovoid, bright red, glabrous. Dog rose. 2N=35.

Native of Eurasia, naturalized in thickets and along roadsides at low elevations w. of the Cascade Mts. and in n. Ida., and in e. U.S. May-June.

Rosa eglanteria L. Sp. Pl. 491. 1753. (Habitat in Helvetia, anglia)

R. rubiginosa L. Mant. 2:564. 1771. (Habitat in Europa australi)

Coarse shrub 1-2 m. tall, with well-developed, flattened, unequal, strongly curved or hooked prickles; foliage sweetly aromatic; leaflets 5-7, firm, broadly elliptic to suborbicular or seldom ovate, 1-2.5 cm. long, doubly serrate with gland-tipped teeth, the lower surface coarsely stipitate-glandular and also somewhat hairy; flowers in small cymes or occasionally solitary, on short, stout, glandular-bristly pedicels; sepals 1-2 cm. long, strongly stipitate-glandular, usually some of them with some prominent slender lateral lobes, more or less spreading or reflexed after anthesis, and tending to fall off when the fruit is mature; petals 1.5-2 cm. long, bright pink; styles more or less densely short-hairy; fruit subglobose or ovoid, 1-1.5 cm. long, bright red, generally glabrous. 2N=35. Sweetbrier, eglantine.

Native of Europe, naturalized along roadsides, in pastures, and less often in more natural habitats; w. of the Cascade Mts., in our range; also in e. U.S. and occasionally elsewhere. June-July.

The closely related European plant, *Rosa micrantha* Sm., differing from *R. eglanteria* in its subglabrous styles and in some other minor features, has also been reported from w. Wash., but its presence as a regular member of our flora remains to be demonstrated.

Rosa gymnocarpa Nutt. in T. & G. Fl. N.Am. 1:461. 1840. *(Nuttall,* "Oregon, in shady woods")

R. prionota Greene, Leafl. 2:256. 1912. *(Heller 5858,* foothills s. of Mt. Sanhedrin, Lake Co., Calif.)

R. leucopsis Greene, Leafl. 2:258. 1912. *(H. E. Brown 99,* sage plains of s.e. Oreg., near Wagontire Mt., Lake Co.)

R. apiculata Greene, Leafl. 2:259. 1912. *(DeAlton Saunders,* Whidby I., near Coupeville, Puget Sound, Wash., July, 1899)

R. helleri Greene, Leafl. 2:259. 1912. *(Heller 3317,* Lake Waha, Nez Perce Co., Ida.)

R. dasypoda Greene, Leafl. 2:260. 1912. *(Sheldon 8815,* Bear Creek, Wallowa Co., Oreg.)

Slender, lax shrub 3-12 dm. tall; twigs varying from bristly with slender prickles as in *R. acicularis* to nearly unarmed, the infrastipular prickles, if present, seldom obviously different from the others; leaf rachis coarsely and rather sparsely stipitate-glandular, sometimes also with a few small prickles; leaflets 5-9, elliptic to elliptic-obovate or sometimes ovate, rounded to less often acute at the tip, 1-4 cm. long, 0.5-3 cm. wide, mostly doubly serrate with the teeth commonly gland-tipped, otherwise generally glabrous except sometimes for a few stipitate glands along the midrib beneath; flowers small, scattered at the end of the branches, mostly borne singly, seldom 2-4 together, the

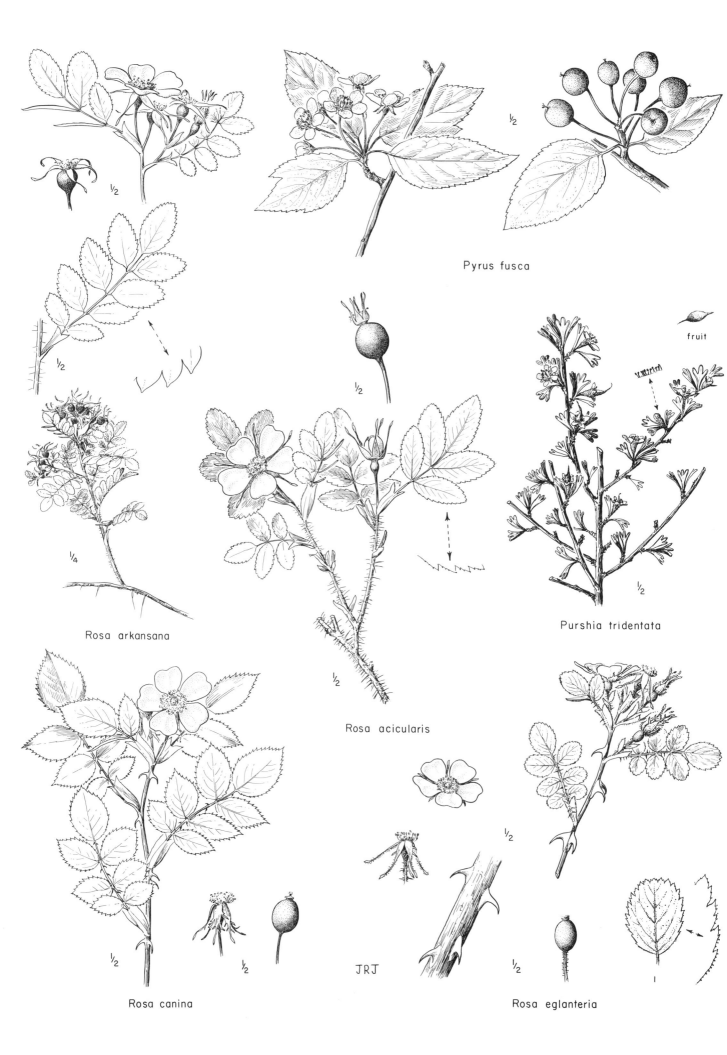

Pyrus fusca

fruit

Purshia tridentata

Rosa arkansana

Rosa acicularis

Rosa canina

JRJ

Rosa eglanteria

slender pedicel coarsely stipitate-glandular or seldom glabrous; sepals 5-12 mm. long, the broad, ovate to broadly lanceolate or triangular base contracted to a slender, commonly caudate tip; petals 1-1.5 cm. long, light pink to deep rose; hypanthium glabrous, only 1.5-3 mm. thick at anthesis; sepals and styles deciduous from the maturing fruit, the hypanthium circumscissile just below the tip; mature hips scarlet or vermilion, about 1 cm. long or a little less, subglobose to pyriform or narrowly ellipsoid, depending largely on the number of achenes maturing; achenes few, up to about a dozen, 3-5 mm. long, glabrous except commonly for a few long stout hairs at the stylar end. 2N=14.

Moist or dry woods, sometimes in open places in wooded regions, from near sea level to about 6000 ft.; s. B.C. and n.w. Mont. (wholly w. of the continental divide), s. along the w. edge of Ida. nearly to Boise, and s. in and w. of the Cascade Mts. to the Sierra Nevada of Calif.; also in the Wallowa Mts. and less commonly in the Ochoco-Blue Mt. region of e. Oreg. (May) June-July (Aug.).

This is the most sharply marked of our native roses, but even so it hybridizes occasionally with *R. acicularis* and perhaps with *R. nutkana*. In flowering condition it can be recognized by its very short sepals without an expanded tip, very narrow hypanthium, stipitate-glandular pedicels, and doubly gland-toothed leaves with glabrous surfaces.

Rosa nutkana Presl, Epim. Bot. 203. 1851. *(Haenke,* Nootka Sound)

R. durandii Crepin, Bull. Soc. Bot. Fr. 22:19. 1875. *(Hall 146,* Oregon) = var. *nutkana.*

R. spaldingii Crepin, Bull. Soc. Bot. Belg. 15:42. 1876, nom. provis. Date of first proper validation not ascertained. *(Spalding,* Clearwater, "Oregon") = var. *hispida.*

R. nutkana var. *hispida* Fern. Bot. Gaz. 19:335. 1894. *R. spaldingii* var. *hispida* G. N. Jones, Madroño 3:130. 1935. *(Watson 124,* Rock Creek, Mont., and *Piper 1540,* Pullman, Wash.)

R. macdougali Holz. Bot. Gaz. 21:36. 1896. *R. nutkana* (ssp.) *macdougali* Piper, Contr. U.S. Nat. Herb. 11:335. 1906. *R. nutkana* var. *macdougali* M. E. Jones, Bull. U. Mont. Biol. 15:35. 1910. *(Sandberg, MacDougal, & Heller 572,* canyons near Farmington Landing, south end of Lake Coeur d'Alene, Ida.) = var. *hispida.*

R. muriculata Greene, Leafl. 2:263. 1912. *R. nutkana* var. *muriculata* G. N. Jones, Madroño.3:128. 1925. *(Coville,* near Woodland, Cowlitz Co., Wash., July 15, 1898) = var. *nutkana.*

R. columbiana Rydb. N.Am. Fl. 22⁶:514. 1918. *(Sandberg, MacDougal, & Heller 381,* valley of the Little Potlatch R., Latah Co., Ida.) = var. *hispida.*

R. nutkana var. *pallida* Suksd. Werdenda 1:23. 1927. *(Suksdorf 10244,* Falcon Valley, Klickitat Co., Wash.) = var. *hispida.*

R. nutkana var. *alta* Suksd. Werdenda 1:23. 1927. *R. spaldingii* var. *alta* G. N. Jones, Madroño 3: 132. 1935. *(Suksdorf 10821,* Bingen, Klickitat Co., Wash.)

R. megalantha G. N. Jones, Proc. Biol. Soc. Wash. 41:194. 1928. *(Jones 614,* Spokane, Wash.) = var. *hispida.*

R. nutkana var. *setosa* G. N. Jones, Madroño 3:129. 1935. *(Jones 4908,* Deception Pass, north end of Whidbey I., Puget Sound, Wash.) = var. *nutkana.*

R. spaldingii var. *chelanensis* G. N. Jones, Madroño 3:133. 1935. *(Jones 1402,* Wenatchee R., near Cashmere, Chelan Co., Wash.) = var. *hispida.*

R. anatonensis St. John, Fl. S.E. Wash. 206. 1937. *(St. John & Palmer 9555,* near Anatone, Asotin Co., Wash.) = var. *hispida.*

R. caeruleomontana St. John, Fl. S.E. Wash. 207. 1937. *(G. N. Jones 1892,* Blue Mts., Asotin Co., Wash.) = var. *hispida,* or perhaps a hybrid with *R. woodsii* var. *ultramontana.*

R. jonesii St. John, Fl. S.E. Wash. 207. 1937. *(St. John & Jones 9621,* summit of Moscow Mt., Latah Co., Ida.) = var. *hispida.*

R. spaldingii var. *parkeri* St. John, Fl. S.E. Wash. 210. 1937. *(Parker 503,* Grizzly Camp, Latah Co., Ida.) = var. *hispida.*

R. rainierensis G. N. Jones, U. Wash. Pub. Biol. 7:174. 1938. *(Allen 292,* Goat Mts., Wash.) = var. *hispida.*

Shrub (0.5) 1-2 (4) m. tall, more or less prickly or nearly unarmed, the infrastipular prickles mostly larger than and well differentiated from any others which may be present, the stem in any case not indiscriminately bristly as in *R. acicularis;* leaflets mostly 5-7 (9), elliptic or ovate, serrate or doubly serrate, 1-7 cm. long and 0.7-4.5 cm. wide; flowers large, typically solitary at the ends of the lateral branches of the season, occasionally 2 or more; hypanthium glabrous or less often covered with gland-tipped bristles, 5-9 mm. wide at anthesis; sepals mostly 1.5-4 cm. long and 3-6 mm. wide near the base, generally constricted in the middle and expanded distally, the back glabrous or occasionally glandular-bristly; petals light pink to deep rose, 2.5-4 cm. long; hips purplish, globose or depressed-globose to pyriform, (1) 1.2-2 cm. long and thick; achenes rather numerous, 4-6 (8) mm. long, stiffly long-hairy along one side. 2N=42, perhaps also 28.

Mostly in wooded regions, or in open places at moderate elevations in the mountains; Alas. (n. as far as Unalaska) and B. C. , s. in and w. of the Cascade Mts. to n. Calif. , in e. Oreg. to the Blue Mt. region, and in the Rocky Mt. region to Colo. and Utah. May-July.

In areas where both *R. nutkana* and *R. woodsii* occur, *R. nutkana* is typically at higher elevations, in habitats (often wooded) where it is not subjected to such intense transpiration. They hybridize to a limited extent, and no one character provides a sharp separation, but there is seldom any difficulty in the field.

The species consists of two wholly intergradient varieties, as indicated in the following key. The var. *hispida,* as here recognized, has a much broader definition than originally intended by Fernald, who applied the name to the sporadic phase of the species that has gland-tipped bristles on the calyx and hypanthium. This character is now believed to be without taxonomic significance in this instance, since it is not obviously correlated with anything else, and plants with the bristles frequently grow intermingled among those without bristles. Under the rules of nomenclature, which permit priority to operate only *within* a taxonomic category, the epithet *hispida* must be taken up when the population of the interior is recognized in varietal status, inasmuch as it is the oldest varietal epithet whose type comes from this population.

1 Leaflets doubly serrate with glandular teeth; infrastipular prickles becoming much enlarged
   and conspicuously flattened toward the base; leaflets evidently glandular beneath; leaf rachis evidently stipitate-glandular; plants occurring chiefly or wholly w. of the Cascade
   summits, but introgressing into var. *hispida* and influencing the character of that var.
   over much of the northern portion of its range   var. nutkana
1 Leaflets singly or seldom doubly serrate, the teeth usually not gland-tipped; prickles
   seldom becoming large and flattened toward the base; leaflets and leaf rachis glandular or not, otherwise glabrous or puberulent; plants occurring chiefly e. of the Cascade summits   var. hispida Fern.

Rosa pisocarpa Gray, Proc. Am. Acad. 8:382. 1872. *(Hall 145,* Oregon)
  R. anacantha Greene, Leafl. 2:265. 1912. *(Greene,* Tacoma, Wash. , Aug. 24, 1889)
  R. pringlei Rydb. Bull. Torrey Club 44:79. 1917. *(Pringle s.n.,* Siskiyou Co. , Calif. , in 1882)
   Shrub 1-2 (3) m. tall, more or less prickly or nearly unarmed, the infrastipular prickles straight or sometimes wanting, the stem in any case not indiscriminately bristly as in *R. acicularis;* leaves puberulent on the lower surface and along the rachis, not glandular; leaflets 5-9, elliptic or ovate, mostly 1.5-4 cm. long and 0.7-2 cm. wide, rather finely and closely serrate, the inner margin of each tooth 0.5-1.0 mm. long; flowers rather small, mostly in small corymbiform cymes terminating the branches of the season, seldom solitary; hypanthium glabrous and bluish-glaucous, 3-5 mm. thick at anthesis; sepals 1-1.5 (2) cm. long and 2.5-3.5 mm. wide at the base, constricted near the middle and generally somewhat expanded distally, almost always coarsely stipitate-glandular on the back; petals 1.2-2 cm. long; hips purplish, globose to ellipsoid or pyriform, 6-12 (15) mm. long and thick, the rather numerous (commonly more than 20) achenes about 3 mm. long, stiffly long-hairy along one side. 2N=14.

Thickets, stream banks, and swampy places at lower elevations; s. B. C. to n. Calif. , w. of the Cascade summits. May-July.

This species is much less variable than our other native roses. Its stability is reflected in its short and simple synonymy. Occasional specimens which lack the gland-tipped bristles from the back of the sepals are difficult to distinguish from some forms of *R. woodsii* var. *ultramontana,* however.

Rosa woodsii Lindl. Ros. Monog. 21. 1820. (Cultivated specimens, thought to have come from near the Missouri River)
  R. fendleri Crepin, Bull. Soc. Bot. Belg. 15:91. 1876. *R. woodsii* (var.) *fendleri* Rydb. Fl. Nebr. 21:22. 1895. *(Fendler 210,* New Mexico) = var. *woodsii,* the form with glandular-toothed leaflets.
  R. californica var. *ultramontana* S. Wats. in Brew. & Wats. Fl. Calif. 1:187. 1876. *R. ultramontana* Heller, Muhl. 1:107. 1904. *R. woodsii* var. *ultramontana* Jeps. Fl. Calif. 2:210. 1936. *R. pisocarpa* var. *transmontana* (lapsus for *ultramontana)* Peck, Man. High. Pl. Oreg. 404. 1941. *(Watson 349,* Utah and Nev.)

R. macounii Greene, Pitt. 4:10. 1899. *(Macoun 10532* and *10533,* Assiniboia, Can., and *Greene s.n.,*
    near Cheyenne, Wyo., are cited) = var. *woodsii.*

R. grosseserrata E. Nels. Bot. Gaz. 30:119. 1900. *(Aven Nelson 6787,* Madison River, Yellowstone
    Nat. Park) = var. *ultramontana.*

R. fimbriatula Greene, Leafl. 2:135. 1911. *(Ward,* Missouri River, 15 miles below Round Butte,
    Mont.) = var. *woodsii.*

R. sandbergii Greene, Leafl. 2:136. 1911. *(Sandberg, MacDougal, & Heller 1099* [cited by Greene
    as *1009],* Colgate, Dawson Co., Mont.) = var. *woodsii.*

R. chrysocarpa Rydb. Bull. Torrey Club 44:74. 1917. *(Rydberg & Garrett 9302,* Allen Canyon, s.w.
    of the Abajo Mts., Utah) = var. *ultramontana.*

R. salictorum Rydb. Bull. Torrey Club 44:77. 1917. *(Nelson & Macbride 2113,* Gold Creek, Nev.)
    = var. *ultramontana.*

R. pyrifera Rydb. Fl. Rocky Mts. 445. 1917. *(Sandberg, MacDougal, & Heller 871,* Lake Pend
    d'Oreille, Ida.) = var. *ultramontana.*

R. puberulenta Rydb. Fl. Rocky Mts. 443. 1917. *(Rydberg & Garrett 9705,* Montezuma Canyon,
    east of Monticello, Utah) = var. *ultramontana.*

R. lapwaiensis St. John, Fl. S. E. Wash. 208. 1937. *(St. John et al. 9538,* Lapwai Creek, Nez
    Perce Co., Ida.) = var. *ultramontana.*

R. woodsii f. *hispida* W. H. Lewis, Rhodora 60:240. 1958. *(Kelsey s.n.,* near Helena, Mont., in
    June, 1892) = var. *woodsii.*

Strongly armed to nearly unarmed shrub, usually with well-developed, straight or somewhat curved
infrastipular prickles, and often with other stout or weak prickles as well; leaves variously puberulent
or glandular beneath to glabrous; leaflets 5-9, elliptic to ovate or obovate, often more distinctly cuneate toward the base than those of *R. pisocarpa,* up to 5 cm. long and 2.5 cm. wide, singly or sometimes partly doubly serrate, the teeth (at least in var. *ultramontana)* fewer and coarser than those of
R. pisocarpa, the inner margin of the tooth often well over 1 mm. long, the teeth not gland-tipped except sometimes in var. *woodsii;* flowers mostly in corymbiform cymes terminating the lateral
branches of the season, seldom solitary, relatively small, the glabrous hypanthium 3-5 mm. thick at
anthesis; pedicels glabrous, or in var. *woodsii* occasionally stipitate-glandular; sepals 1-2 cm. long,
2-3.5 mm. wide at the base, the tip usually a little expanded above the median constriction, the back
glabrous to puberulent or occasionally inconspicuously glandular, rarely coarsely stipitate-glandular
as in *R. pisocarpa;* sepals persistent, erect or spreading in fruit; petals light pink to deep rose, 1.5-
2.5 cm. long; hips globose to ellipsoid or pyriform, 6-12 (15) mm. long and wide, or sometimes up
to 2 cm. long when relatively slender; achenes mostly 15-30, 3-4 mm. long, stiffly long-hairy along
one side. 2N=14.

    From s. Mack. to e. Wash., e. Oreg., and s. Calif., e. to Minn., Wis., Mo., and Tex., wholly
e. of the Cascade summits in our range. May-July.

    The species consists of two well-marked varieties (with perhaps a third in Calif.), as follows:
1 Plains and prairie ecotype, encroaching onto the edge of our range in Mont., up to about
        1 m. tall, appearing stiff and with crowded leaves, the leaflets generally small, mostly
        1-2 cm. long, averaging less coarsely toothed than in var. *ultramontana,* and the teeth
        sometimes gland-tipped   var. woodsii
1 Cordilleran ecotype, occurring in moist places (as along stream banks and around ponds)
        in otherwise rather dry habitats, usually in the foothills and lowlands, taller than the
        var. *woodsii,* commonly 1-2 (3) m. tall and laxer, the leaflets usually larger, up to 5
        cm. long and 2.5 cm. wide, the teeth not gland-tipped   var. ultramontana (Wats.) Jeps.

    There is some question that the name *R. woodsii* properly applies to the species for which it is
here used, especially inasmuch as the geographic source of the original material is uncertain. No other application of the name has ever gained much acceptance, however.

Rubus L. Bramble; Blackberry; Raspberry

    Flowers mostly rather large, often showy, perigynous, regular, complete or sometimes imperfect,
single (terminal or axillary) to several and racemose, paniculate, umbellate, or cymose; calyx ebracteolate, the usually 5 (6, 7) lobes valvate, often more or less caudate, spreading to reflexed, persistent in fruit, the hypanthium saucerlike to shallowly campanulate, disc lined within; petals white to red,
erect to spreading, deciduous, equal in number to the sepals; stamens (15) 40-100 or more, the filaments slender to distinctly flattened, inserted with the petals at the outer edge of the hypanthium; pis-

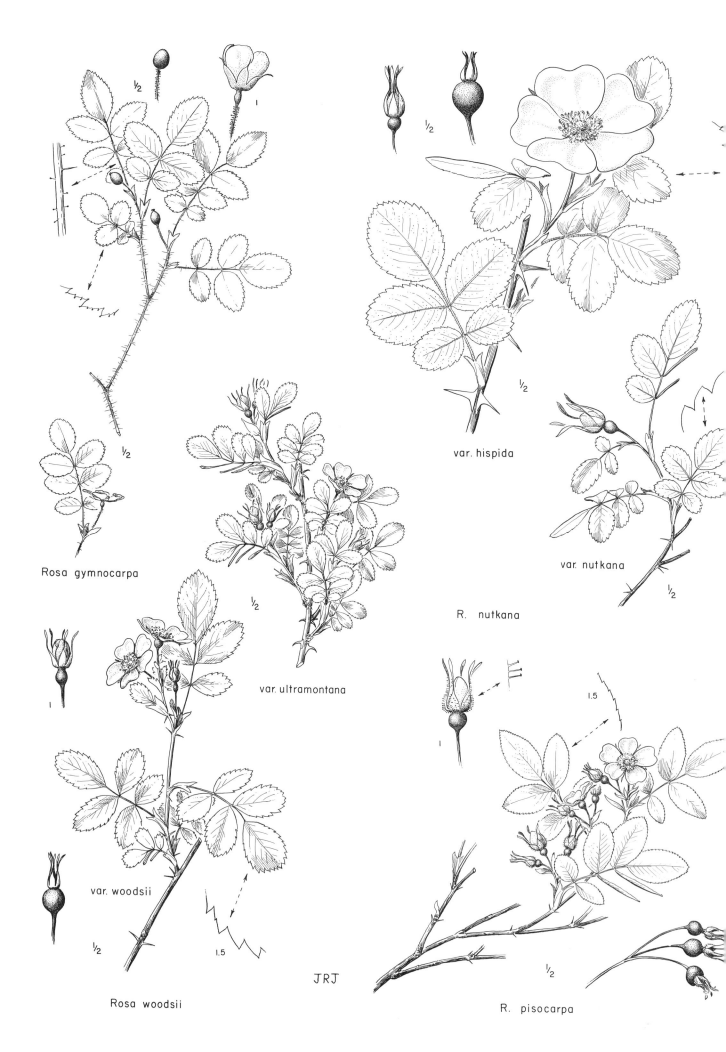

1/2

1

1/2

Rosa gymnocarpa

1/2

var. hispida

var. nutkana

R. nutkana

1/2

var. ultramontana

var. woodsii

1.5

Rosa woodsii

JRJ

1.5

R. pisocarpa

tils several to many, borne on a rounded to conic-oblong, often fleshy receptacle, the ovary 2-ovulate (1 ovule ultimately abortive), the style filiform to clavate; mature fruit an aggregation of rather weakly coherent drupelets separately or collectively deciduous, often closely associated with the fleshy receptacle, the "pit" (putamen) usually wrinkled or pitted; perennial shrubs (a few species semiherbaceous) with trailing to erect stems, often armed with weak to strong epidermal bristles or prickles, the leaves alternate, deciduous or persistent, mostly with well-developed stipules and simple to ternate or pinnate (rarely) or somewhat decompound blades, the stems often biennial, leaf-bearing the first year (primocanes) and floriferous the second (floricanes).

Probably several hundred species, nearly cosmopolitan but most common in the temperate regions of the Northern Hemisphere. (The Roman name for the plant.)

Taxonomically, *Rubus* is a notably difficult genus. Not only are many of the taxa interfertile in varying degree but most of them are polymorphic; many are known to have a varying chromosome number and not a few have been shown to be apomictic. These conditions are most pronounced in the blackberries, of which we probably have only one native species, but our other species, including the brambles and (especially) the raspberries, are variable and of such a nature that several alternatives confront the systematist. It will be seen that the course followed here is perhaps more than ordinarily conservative.

References:

Bailey, L. H. Species Batorum. The genus Rubus in North America. Gentes Herb. 5:1-932. 1941.

Brown, Spencer W. The origin and nature of variability in the Pacific Coast blackberries (Rubus ursinus Cham. & Schlecht. and R. lemurum sp. nov.) Am. Journ. Bot. 30:686-697. 1943.

Fernald, M. L. Rubus idaeus and some of its variations in North America. Rhodora 21:89-98. 1919.

1 Plants unarmed
  2 Stems erect, woody, rarely less than 0.5 m. tall
    3 Petals red; leaves trifoliolate   R. SPECTABILIS
    3 Petals white; leaves palmately lobed *(Anoplobatos)*
      4 Flowers solitary; leaf blades rarely as much as 5 cm. long; styles hairy their full length   R. BARTONIANUS
      4 Flowers (2) 3-9 in loose cymose clusters; leaf blades mostly over 5 cm. long; styles glabrous above   R. PARVIFLORUS
  2 Stems mostly trailing and more herbaceous than woody, seldom if ever as much as 0.5 m. tall
    5 Plant dioecious; leaves broadly cordate-reniform to cordate-orbicular, 4-10 cm. broad, shallowly (3) 5-lobed; flowers single and terminal on erect leafy branches; circumboreal, in N. Am. s. to the n. end of Vancouver I., Me., and N.Y., but not believed to reach our area, although the range in B.C. is uncertain *(Chamaemorus)*   *R. chamaemorus* L.
    5 Plant perfect flowered; leaves (except in *R. pedatus*) compound *(Cylactis)*
      6 Petals more or less reddish-tinged, (8) 10-16 mm. long; plant not stoloniferous   R. ACAULIS
      6 Petals white, mostly less than 8 mm. long; plant usually stoloniferous
        7 Leaflets 5 or occasionally only 3 but the lower 2 again divided nearly to the base   R. PEDATUS
        7 Leaflets 3 or blades simple but deeply 3-lobed
          8 Leaves usually (at least in part) simple and merely deeply lobed; ovaries densely pubescent; filaments very slender   R. LASIOCOCCUS
          8 Leaves compound (3-foliolate); ovaries glabrous; filaments flattened, narrowed abruptly (and often more or less bidentate) just below the anthers   R. PUBESCENS
1 Plants armed with bristlelike to broad-based and often curved prickles
  9 Leaves (at least in part) simple and cordate, evergreen; stipules ovate-lanceolate, slenderly acuminate; stems trailing; petals usually pink to purple *(Chamaebatus)*   R. NIVALIS
  9 Leaves nearly always compound, mostly deciduous; stipules various but usually slender or adnate to the petiole; petals mostly white
    10 Receptacle fleshy, forming part of the (blackberrylike) ripe fruit; stems mainly trailing or clambering, the plants strongly armed, at least some of the prickles usually flattened or hooked; petals white or pale pink *(Eubatus)*
      11 Plants more or less completely dioecious, the pistillate flowers with distinctly

rudimentary stamens, the staminate with small, nonfunctional pistils; stems slender, trailing, armed with rather slender and scarcely flattened prickles; leaves (simple) trifoliolate, deciduous   R. URSINUS
  11 Plants perfect flowered; stems thick, frequently clambering to erect, armed with large, often flattened prickles; leaves various
    12 Leaves evergreen, the leaflets laciniate to dissected   R. LACINIATUS
    12 Leaves deciduous or more or less evergreen, the leaflets merely toothed
      13 Leaves grayish- or white-tomentose beneath
        14 Inflorescence (especially the pedicels) stipitate-glandular   R. VESTITUS
        14 Inflorescence (including the pedicels) eglandular   R. PROCERUS
      13 Leaves soft-pubescent but not at all tomentose beneath   R. MACROPHYLLUS
10 Receptacle dry or only slightly fleshy, usually not forming part of the ripened (raspberrylike) fruit, or if partially succulent the petals deep pink; stems mainly erect or arched, rarely trailing, the prickles (except in *R. leucodermis* and *R. nigerrimus)* neither flattened nor hooked *(Idaeobatus)*
  15 Petals pink to red, usually well over 1.5 cm. long; fruit salmon-colored to red; stems erect, not at all vinelike, often armed only (or chiefly) near the base, the leaves not prickly   R. SPECTABILIS
  15 Petals white, usually less than 1.5 cm. long; fruit often black; the leaves frequently prickly on the back
    16 Main prickles flattened and often retrorsely hooked; fruit dark reddish-blue to black; flowers several in an umbellate to flat-topped cluster
      17 Leaves greenish and glabrous or glabrate beneath; local in the Snake R. Canyon in s.e. Wash.   R. NIGERRIMUS
      17 Leaves grayish-tomentose beneath; widespread   R. LEUCODERMIS
    16 Main prickles usually neither flattened nor retrorsely hooked, but sometimes slightly flattened; fruit (yellowish to) red; flowers usually few, in an open but not flat-topped raceme or thyrse   R. IDAEUS

Rubus acaulis Michx. Fl. Bor. Am. 1:298. 1803.
  Manteia acaulis Raf. Sylva Tellur. 161. 1838. ("Hab. in sphagnosis sinui Hudsonis adjacentibus")
  Strongly rhizomatous perennial with erect, herbaceous, more or less finely pilose annual flowering stems 2-15 cm. tall; leaves 2-5 per stem, with conspicuous ovate-lanceolate to oblong-obovate, entire stipules, the blades trifoliolate; leaflets (often obliquely) ovate to obovate, (1) 1.5-3 (5) cm. long, rounded to acute, serrate to biserrate; flowers usually single and terminal, rarely 2; calyx finely pubescent-pilose, the lobes narrowly lanceolate, reflexed, 8-11 mm. long; petals nearly erect, uniformly pink to crimson or rose or sometimes yellowish toward the base, narrowly rhombic-obovate to spatulate-obovate, (8) 10-16 mm. long, with a clawlike base; stamens 30-40; filaments rather broad, narrowed abruptly at the tip; pistils 20-30, glabrous; style slender but enlarged slightly upward to the rather large capitate-discoid stigma; drupelets coherent, forming a globular, reddish aggregate about 1 cm. broad; putamen 2-2.5 mm. long, slightly wrinkled.
  Tundra to mountain meadows, bogs, and woods; Alas. across N.Am. to Newf. and Lab., s. to Vancouver I. and s. B.C., in the Rocky Mts. of Mont., Wyo., and Colo., and in Minn. June-July.

Rubus bartonianus Peck, Rhodora 36:267. 1934. *(Peck 17611,* Snake R. Canyon, Wallowa Co., Oreg., opposite Hell's Canyon, Ida., July 12, 1933)
  Slender shrubs (1) 2-3 m. tall with more or less zigzag, finely puberulent but glabrate, often reddish branches, the old bark straw-colored and shredding; leaf blades cordate, 2.5-5 cm. long and usually somewhat broader, with 5-7 shallow, palmate, acute, doubly serrate lobes, more or less puberulent to finely sericeous; flowers single and terminal on leafy branches; calyx sericeous to lanate and stipitate-glandular, the lobes lanceolate, caudate, spreading to reflexed, 8-12 mm. long; petals 5 (to 8), white, oblong to oblong-oblanceolate or rhombic-elliptic, 14-20 mm. long; stamens and pistils numerous; ovary glabrous; style clavate, hairy the full length, about 1 (1.3) mm. long; stigma large, 2-lobed; fruits coherent into a hemispheric, deep red or brownish, raspberrylike cap 1-1.5 cm. broad; putamen about 2.5 mm. long, prominently reticulate-pitted.
  Canyon sides, usually where protected, forming thickets along streams; Snake R. Canyon, Wallowa Co., Oreg., and Idaho Co., Ida. Apr.-May.

Rubus idaeus L. Sp. Pl. 492. 1753. ("Habitat in Europae lapidosis")

R. strigosus Michx. Fl. Bor. Am. 1:297. 1803. *R. idaeus* var. *strigosus* Maxim. Bull. Acad. St. Pétersb. 17:161. 1872. *Batidaea strigosa* Greene, Leafl. 1:238. 1906. ("Hab. in montibus Pensylvaniae et in Canada") = var. *sachalinensis.*

R. idaeus ssp. *R. melanotrachys* Focke, Abh. Nat. Ver. Bremen 13:472. 1896, the name illegitimate acc. Art.24 of the Int. Rules of Bot. Nomen. *R. idaeus* var. *melanotrachys* Fern. Rhodora 21:97. 1919. *(Diecke,* northwestern America) = var. *sachalinensis*-a sporadic eglandular phase of w. N. Am. that has its origin from the glandular var. *sachalinensis.* The plants are native in our area, although so similar to the Eurasian var. *idaeus* that distinction between the two from herbarium specimens is not always possible, even though there is perhaps a tendency for the American plants to have less hairy inflorescences. The variant meets our concept of the category of "forma," but such a combination is not hereby formally proposed.

R. melanolasius Dieck, Neubert's Gart.-Mag. 47:177. 1894, not described. *R. idaeus* ssp. *R. melanolasius* Focke, Abh. Ver. Bremen 13:473. 1896, the name illegitimate (see above). *R. idaeus* ssp. *melanolasius* Focke, Bibl. Bot. 17, Heft 72^2:209. 1911. *R. idaeus* var. *melanolasius* Davis, Madroño 11:144. 1951. (Cultivated plant from n.w. N. Am.). =var. *sachalinensis,* or perhaps a more distinctive phase from n. and e. of our range.

Batidaea acalyphacea Greene, Leafl. 1:240. 1906. *R. acalyphaceus* Rydb. N. Am. Fl. 22^5:448. 1913. *R. idaeus* var. *acalyphaceus* Fern. Rhodora 21:98. 1919. *R. strigosus* var. *acalyphaceus* Bailey, Gentes Herb. 5: 872. 1945. *(E. A. Mearns 2353,* Yellowstone Park, July 28, 1902) =var. *sachalinensis*-the epithet "acalyphacea" is usually applied to a dwarfed, densely setose plant common in (but not restricted to) the Rocky Mts.

Batidaea subcordata Greene, Leafl. 1:240. 1906. *(E. A. Mearns 2553,* Yellowstone Park, July, 1902, is the first of 3 collections cited) = var. *sachalinensis.*

Batidaea unicolor Greene, Leafl. 1:241. 1906. *(J. N. Rose 52,* near Red Lodge, Mont., July 27, 1893) = var. *peramoenus?*

Batidaea peramoena Greene, loc. cit. 1906. *R. peramoenus* Rydb. N. Am. Fl. 22^5:446. 1913. *R. idaeus* var. *peramoenus* Fern. Rhodora 21:98. 1919. *(Leiberg 1105,* St. Mary's River, Ida., June, 1805 [sic])

Batidaea cataphracta Greene, Leafl. 1:241. 1906. *(Henderson 3598,* head of Pettit Lake, Ida., in 1895) = var. *peramoenus?*

Batidaea sandbergii Greene, Leafl. 1:242. 1906. *(Sandberg, McDougal, & Heller 859,* Packsaddle Peak, Ida., Aug., 1892) = var. *sachalinensis.*

Batidaea filipendula Greene, loc. cit. *(Henderson 4039,* Lost River Mts., Ida.) = var. *peramoenus.*

Batidaea viburnifolia Greene, Leafl. 1:242. 1906. *R. viburnifolius* Rydb. N. Am. Fl. 22^5:446. 1913, but not of Franch. in 1895. *R. idaeus* var. *viburnifolius* Berger, Rep. N.Y. State Agric. Exp. Sta. 1925 (pt. 2):51. 1925. *R. greeneanus* Bailey, Gentes Herb. 5:879. 1945. *(C. H. Shaw 472,* Selkirk Mts., B. C., Aug., 1904) = var. *peramoenus.*

R. sachalinensis Levl. in Fedde, Rep. Sp. Nov. 6:332. 1909. *R. idaeus* ssp. *sachalinensis* Focke, Bibl. Bot. 17, Heft 72^2:210. 1911. *(Korsakof 565,* Sagalien, Japan, July 30, Sept. 30, 1908)

R. idaeus var. *gracilipes* M. E. Jones, Bull. U. Mont. Biol. 15:35. 1910. ("Alta northward, and throughout the west," no specimens cited) = var. *sachalinensis?*

Strong perennial with sparsely to copiously bristly and prickly but otherwise glabrous to strongly pubescent branches (0.5) 1-2 (3) m. tall, the bark mostly yellow to cinnamon-brown, exfoliating; floricanes and the inflorescence eglandular to more or less strongly stipitate- or bristly-glandular and otherwise with or without pubescence; leaves 3- to 5-foliolate; stipules linear-subulate, 4-10 mm. long, often caducous; leaflets ovate to ovate-oblong or broadly lanceolate, (3) 4-10 cm. long, usually rather long-acuminate, irregularly biserrate to lobulate-biserrate, greenish and glabrous to more or less hirsute on the upper surface, usually grayish-lanate but sometimes tardily glabrate and greenish, or even semiglabrous beneath; flowers mostly several, 1-4 per axil, forming a racemose to thyrsoid, leafy inflorescence; calyx usually more or less lanate and eglandular to stipitate- or bristly-glandular, the lobes reflexed, lanceolate to lanceolate-caudate, 4-8 (12) mm. long; petals white, erect or ascending, narrowly oblong to oblong-spatulate, 4-6 mm. long; stamens 75-100, the filaments somewhat flattened, glabrous; pistils rather numerous; style slender, glabrous; ovary tomentulose; drupelets crumbling apart or weakly coherent to form a (yellowish to) red, finely tomentulose raspberry; putamen strongly reticulate-pitted. 2N=14, 21, 28. Red raspberry.

Stream banks and open, moist or dry woods to rocky montane slopes, often on talus; Alas. eastward across Can. to Newf., s. through B. C., Wash., and Oreg., to Calif., and to Ariz., N. M., n. Mex., Ia., Minn., Tenn., and N. C.; much of middle and n. Eurasia. May-July.

Several workers (among them Focke, Fernald, and more recently, Hultén) have considered the native red raspberries of N. Am. to comprise part of a circumboreal species complex, but have differed as to the number and nature of the infraspecific taxa to be recognized. In general, the plants of Europe and of much of Asia lack glands in the inflorescence, whereas those of the more eastern part of Asia and nearly all of N. Am. are glandular. On this basis, the species is divisible into two subspecies, the earliest legitimate epithet existent in subspecific rank for the glandular phase being ssp. *sachalinensis*. There are many variants in the N. Am. material, but in our area, only two are sufficiently distinctive that their recognition need not be considered nomenclatural flattery.

1 Plants eglandular; mainly Eurasian, from the British Isles to Siberia, s. to the Mediterranean and the Caucasus, occasionally escaped from cultivation in N. Am. ; closely
simulated by sporadic eglandular mutants of n. w. N. Am.   ssp. idaeus
1 Plants glandular (very rarely eglandular); Asia (where freely intergradient to ssp. *idaeus)*
and N. Am.   ssp. sachalinensis (Levl.) Focke
   2 Leaves glabrous or subglabrous and greenish on the lower surface; occasional in n. c.
Wash. , n. Ida. , and n. w. Mont. , n. to c. Alas. and Yukon (acc. Hultén); freely
intergradient to the next   var. peramoenus (Greene) Fern.
   2 Leaves permanently grayish-lanate on the lower surface; mainly N. Am. from Alas.
to Newf. , s. through B. C. , Wash. , and Oreg. to Calif. , and e. to Ariz. , n. Mex. ,
Ia. , Minn. , and N. C. ; Asia to the Ural Mts.   var. sachalinensis

Rubus laciniatus Willd. Hort. Berol. pl. 82. 1807.
   R. vulgaris var. *laciniatus* Dippel. Handb. Laubh. 3:529. 1893. (A horticultural specimen)
Strongly armed perennial with ascending or arched but usually trailing to clambering stems up to 10 m. long, the prickles flattened, recurved; leaves persistent, evergreen, primarily (3) 5-foliolate, the leaflets from laciniately lobed to divided into secondary leaflets in turn irregularly incised to jagged-lobed or deeply and coarsely serrate-dentate, greenish and sparsely hirsute to glabrous above, pilose-hirsute to copiously soft-pubescent (but greenish) beneath; flowers rather numerous in simple or (more often) partially compound and more or less flat-topped and somewhat leafy racemes, the pedicels and rachis strongly armed and copiously soft-pubescent but eglandular; calyx lanate, usually prickly, the lobes lanceolate and usually strongly caudate, reflexed, 8-15 mm. long; petals generally pinkish (white), 9-14 mm. long, cuneate-obovate and usually deeply trilobed; stamens numerous (at least 75); pistils rather few; style slender, glabrous; drupelets coherent and with the fleshy receptacle forming a globular to ovoid blackberry 1-1. 5 cm. thick, palatable; putamen strongly reticulate-pitted. Evergreen blackberry.

Very commonly escaped on the w. side of the Cascade Mts. from s. B. C. to Calif. , and occasional e. to Ida. , also on the Atlantic coast. ; a cultigen believed to be of European origin. June-early Aug.

Rubus lasiococcus Gray, Proc. Am. Acad. 17:201. 1882.
   Comarobatia lasiococca Greene, Leafl. 1:245. 1906. *(Hall 140,* in part, near Mt. Hood, Oreg. , in
1871, is the first of 2 collections cited)
Unarmed, more or less pilose to crisp-puberulent, trailing perennial with stolonous, freely rooting, herbaceous stems up to 2 m. long, and short, erect, 1- to 3-foliate flowering stems up to 1 dm. long; leaf blades broadly cordate-reniform, (2) 3-6 cm. broad, shallowly to deeply 3-lobed or sometimes divided into rounded or obtuse (rarely acute) lobes or segments, doubly serrate; flowers 1 or 2; calyx thickly puberulent, the lobes ovate-lanceolate to lanceolate, 4-7 (8) mm. long, acuminate, entire to toothed, reflexed; petals white, 5-8 mm. long; stamens numerous, the filaments very slender; pistils 7-15, the ovary grayish-puberulent-tomentose; style slender, glabrous; drupelets densely puberulent, juicy, semicoherent, the aggregate scarcely 1 cm. broad, red; putamen smooth.

Moist to dry soil, thickets, or sparse to dense woods in the mountains, at from perhaps 1500 to 5500 ft. elevation in our area, from B. C. southward in the Cascades to n. w. Calif. , also in the Olympic Mts. , Wash. June-Aug.

The plant makes a fine ground cover that is easily established.

Rubus leucodermis Dougl. ex T. & G. Fl. N. Am. 1:454. 1840.
   R. occidentalis var. *leucodermis* Focke, Abh. Nat. Ver. Bremen 4:147. 1875. *Melanobatus leucoder-*
mis Greene, Leafl. 1:243. 1906. *(Douglas,* Oregon)
Well-armed, deciduous perennial with erect but usually arching and sometimes apically rooting, glaucous branches (primocanes) 1-3 m. long, the prickles numerous, stoutish, somewhat flattened and distinctly hooked, up to 6 mm. long; stipules filiform-linear, up to 5 mm. long, often caducous;

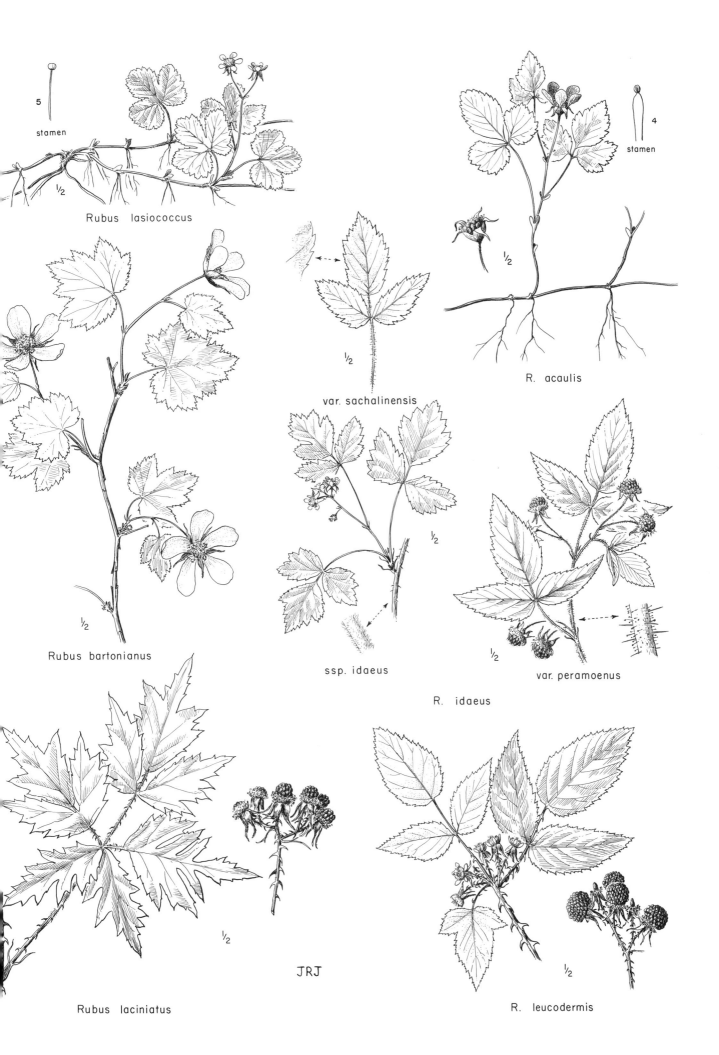

5
stamen
½
Rubus lasiococcus

½
Rubus bartonianus

½
var. sachalinensis

½
ssp. idaeus

R. idaeus

4
stamen
½
R. acaulis

½
var. peramoenus

½
Rubus laciniatus

JRJ

½
R. leucodermis

leaf blades 3 (5)-foliolate, greenish and glabrous to subglabrous above, white-tomentose beneath, the leaflets ovate to ovate-lanceolate, 1.5-8 cm. long, acute to acuminate, from shallowly lobed to lobulate and doubly serrate or merely irregularly doubly serrate; flowers mostly 2-7 (10), usually approximate in an umbellate to corymbose raceme, the rachis and pedicels tomentose, generally prickly but not glandular; calyx tomentose, often glandular, the segments narrowly lanceolate and more or less acuminate-caudate, reflexed, 6-12 mm. long; petals white, broadly spatulate-clawed, shorter than the sepals; stamens 70-100, the filaments slightly flattened, glabrous; pistils numerous; style glabrous; ovary tomentose; drupelets coherent and forming a reddish-purple to black, finely tomentulose raspberry up to 12 mm. broad; putamen strongly reticulate-pitted. 2N=14. Black raspberry, blackcap.

Fields and open to wooded hills; B.C. southward to s. Calif., from the coast inland and e. to Mont., Utah, and Nev. Late Apr.-early July.

Rubus macrophyllus Weihe & Nees, Rub. Germ. 35, pl. 12. 1822. (Near Mennighüffen, Germany)
   R. macrophyllus var. *amplificatus* Bab. Synop. Brit. Rubi. 20. 1846. *R. amplificatus* Lees in Steele, Handb. Field Bot. 58. 1847. (England)
Well-armed perennial with upright and arching stems 1-3 m. long, the prickles up to 7 mm. long, flattened, straight to slightly curved; leaves semipersistent, the stipules lanceolate, up to 1 cm. long, the blades 3- to 5-foliolate; leaflets ovate to cuneate-obovate, gradually to abruptly acuminate, 3-9 cm. long, rather shallowly serrate to biserrate-dentate, greenish and glabrous to subglabrous on the upper surface, paler green and rather coarsely pubescent or pilose (but not tomentose) beneath; inflorescence axillary, few-flowered, racemose or paniculate, leafy, flat-topped, coarsely pubescent and rather strongly stipitate-glandular; calyx semitomentose, the lobes reflexed, lanceolate, less than 1 cm. long; petals white to pink, about 1 cm. long; stamens at least 75; pistils rather numerous; style glabrous; drupelets coherent and with the fleshy receptacle forming a globose blackberry 1-1.5 cm. thick; putamen strongly wrinkled-pitted.

A native of Europe, *R. macrophyllus* has apparently escaped and become established in our area, at least in Pacific Co. and Mason Co. *(G. N. Jones 8542)*, Wash. May-June.

This blackberry is readily separable from both *R. procerus* and *R. vestitus* because of the more leafy inflorescence, smaller prickles, and nontomentose leaves.

Rubus nigerrimus (Greene) Rydb. N.Am. Fl. 22[5]:445. 1913.
   R. hesperius Piper, Erythea 5:103. 1897, but not of Rogers in 1896. *Melanobatus nigerrimus* Greene, Leafl. 1:244. 1906. *R. leucodermis* var. *nigerrimus* St. John, Fl. S.E. Wash. 213. 1937. (Snake R. Canyon at Wawawai and Almota, Whitman Co., Wash., no collection mentioned)
Strong perennial, the primocanes erect to trailing or clambering, greenish but somewhat glaucous, glabrous, the prickles flattened and mainly straight; flowering stems up to 5 m. long, trailing, armed with numerous flattened and more or less hooked prickles, greenish to bluish-brown or purplish-brown; leaves deciduous, greenish and glabrous on both surfaces, only slightly paler beneath, 3-foliolate on the flowering stems but often 5-foliolate (3 leaflets petiolulate and 2 sessile) on the primocanes, the leaflets ovate-lanceolate, 2 or 3 times shallowly lobed-serrate; flowers 1-5 (6) in small, loose clusters; pedicels and calyx very sparsely glandular-puberulent and sometimes very lightly pubescent (glabrous); sepals spreading, 5-8 mm. long; petals white, flared, elliptic, often more than 5 (up to 12) in number, 4-7 mm. long; stamens 75-100; pistils 25-40, the ovaries ripening into scarcely coalescent, blackish, strongly rugose, only moderately succulent, glabrous (very obscurely puberulent) drupelets that are free of the receptacle, the whole somewhat raspberrylike, but rather dry and tending to crumble.

Known only from moist hillsides (usually along streams) along the Snake R. in Whitman Co., Wash. May-June.

The species is apparently derived from *R. leucodermis,* although it has many distinct peculiarities. Information from residents of the area in which the plant grows is that the ripe fruits are not particularly succulent or palatable but that they rarely are seen as they are usually eaten in a semi-ripe state by birds and other animals.

Rubus nivalis Dougl. ex Hook. Fl. Bor. Am. 1:181. 1832.
   Cardiobatus nivalis Greene, Leafl. 1:244. 1906. *(Douglas,* "On the high snowy ridges of the Rocky Mountains," probably in Oreg., acc. Bailey, Gentes Herb. 5:44. 1941)

R. pacificus Macoun, Ott. Nat. 16:213. 1903, not of Hance in 1874. *(W. B. Anderson,* Comox, Vancouver I., B.C., in 1899)

Trailing perennial with freely rooting, slender, pubescent and retrorsely-prickly stems up to 2 m. long; stipules ovate to lanceolate, acuminate, entire, 4-10 mm. long; petioles armed with retrorse, curved prickles; leaf blades 3-6 cm. long, usually glabrous, bright green and shining, especially on the upper surface, mostly simple, cordate-ovate, and nonlobate, to prominently 3-lobed, or even in part 3-foliolate, usually prickly on the undersurface along the veins, the margins strongly dentate-serrate; flowers single or in pairs in the axils; calyx somewhat pubescent-pilose and sometimes weakly bristly, the lobes ovate-lanceolate to lanceolate, acute to acuminate, 6-9 mm. long, reflexed, usually more or less purplish; petals inconspicuous, (white) pink to dull purple, narrowly elliptic-lanceolate, tapered to each end, equaling (to half again as long as) the sepals; stamens about 15; filaments slender, pilose at base, usually purplish; pistils 4-9, the ovaries strongly pubescent, the styles slender, about 5 mm. long; drupelets large, red, the putamen strongly alveolate-wrinkled, 3-4 mm. long.

Open to deeply shaded, usually moist areas in the mountains up to about 5000 ft. elevation; B. C. southward to s. w. Oreg., in the Cascade, Olympic, and coastal mts., e. to Ida. June-July.

This is a plant that makes a charming ground cover and one that takes readily to cultivation.

Rubus parviflorus Nutt. Gen. Pl. 1:308. 1818.

Rubus nutkanus var. *nuttallii* T. & G. Fl. N. Am. 1:450. 1840. *Rubacer parviflorum* Rydb. Bull. Torrey Club 30:274. 1903. *Bossekia parviflora* Greene, Leafl. 1:211. 1906. *Rubus nutkanus* var. *parviflorus* Focke, Bibl. Bot. 17, Heft 72:124. 1911. *Rubus parviflorus* f. *nuttallii* Fassett, Ann. Mo. Bot. Gard. 28:323. 1941. *(Nuttall,* "On the island of Michilimackinak, Lake Huron")

Rubus nutkanus Moc. ex Ser. in DC. Prodr. 2:566. 1825. *Rubus parviflorus* var. *grandiflora* Farw. Am. Midl. Nat. 11:263. 1929. ("Moc. pl. Nutk. icon." [See Fern. Rhodora 37:275. 1935]) = var. *parviflorus.*

Rubus velutinus H. & A. Bot. Beechey Voy. 140. 1832, not of Vest in 1823. *Rubus nutkanus* var. *velutinus* Brew. Bot. Calif. 1:172. 1876. *Rubus parviflorus* var. *velutinus* Greene, Bull. Torrey Club 17:14. 1890. *Rubacer tomentosum* Rydb. Bull. Torrey Club 30:274. 1903. *Rubacer velutinum* Heller, Muhl. 1:106. 1904. *Rubus parviflorus* var. *grandiflora* subvar. *velutinus* Farw. Am. Midl. Nat. 11:263. 1929. (San Francisco)

Rubus nutkanus f. *lacera* Kuntze, Meth. Sp. 102. 1879. *Rubus parviflorus* var. *bifarius* f. *lacera* Fern. Rhodora 37:281. 1935. *(Lyall,* Vancouver I., in 1858) = var. *parviflorus.*

Rubus nutkanus var. *scopulorum* Greene ex Focke, Bibl. Bot. 17, Heft 72:124. 1911. *Rubus parviflorus* var. *scopulorum* Fern. Rhodora 37:283. 1935. *Rubus parviflorus* f. *scopulorum* Fassett, Ann. Mo. Bot. Gard. 28:323. 1941. (Rocky Mts. of s. Colo., no specimens cited) = var. *parviflorus.*

Rubus parviflorus var. *fraserianus* J. K. Henry, Torreya 18:54, fig. 1. 1918. *Rubus parviflorus* var. *bifarius* f. *fraserianus* Fern. Rhodora 37:281. 1935. *(Henry,* Ucluelet, Vancouver I., B.C., June 19, 1917) = var. *parviflorus.*

Rubus parviflorus var. *hypomalacus* Fern. Rhodora 37:277. 1935. *Rubus parviflorus* f. *hypomalacus* Fassett, Ann. Mo. Bot. Gard. 28:323. 1941. *(J. M. Grant 211,* Olympic Mts., Wash.) = var. *parviflorus.*

Rubus parviflorus var. *heteradenius* Fern. Rhodora 37:279. 1935. *Rubus parviflorus* f. *heteradenius* Fassett, Ann. Mo. Bot. Gard. 28:323. 1941. *(Suksdorf 1758,* Falcon Valley, Klickitat Co., Wash., June 26, 1893) = var. *parviflorus.*

Rubus parviflorus var. *bifarius* Fern. Rhodora 37:280. 1935. *Rubus parviflorus* f. *bifarius* Fassett, Ann. Mo. Bot. Gard. 28:323. 1941. *(W. R. Carter,* near Cameron Lake, Vancouver I., B.C., July 14, 1917) = var. *parviflorus.*

Rubus parviflorus f. *acephalus* Fassett, Ann. Mo. Bot. Gard. 28:325. 1941. *(Doris K. Gillespie 15399,* Port Orford, Oreg., Aug. 12, 1938) = var. *parviflorus.*

Upright unarmed shrub 0.5-2 (3) m. tall, puberulent and stipitate-glandular, eventually glabrate and with gray, flaking bark; leaves with membranous, lanceolate stipules and stipitate-glandular petioles, the blades deeply cordate based, palmately (3) 5 (7)-lobed, (5) 6-15 (25) cm. long and somewhat broader, doubly dentate-serrate, glabrous to somewhat hairy; flowers (2) 3-7 (9) in terminal corymbs or flat-topped panicles; calyx pubescent to villous, often stipitate-glandular, the 5 (6-7) lobes spreading, oblong-ovate, 10-18 mm. long, with a slender apical appendage about half the total length; petals usually 5 (6-7), white (sometimes pinkish-tinged), oblong-obovate to obovate, (1) 1.5-2.5 cm. long; stamens numerous; pistils numerous; ovary pubescent above but the style glabrous, 1-1.5 mm. long;

drupelets coherent as a thimblelike aggregate fruit; putamen about 3 mm. long, strongly wrinkled-pitted. Thimbleberry.

Open to wooded, moist to dry places from near sea level to subalpine mountain slopes; Alas. southward to s. Calif., from the coast e. to the Great Lakes region, the Dakotas, Wyo., Colo., N.M., and n. Mex. May-July.

This is an extremely variable species. Following a recent treatment (Fassett, Norman C. Rubus odoratus and R. parviflorus. Ann. Mo. Bot. Gard. 28:299-374. 1941), our plants are all referred to var. *parviflorus*, differing from the much more hairy phase, the var. *velutinus* (H. & A.) Greene, of coastal Calif., that has sometimes been reported for our area. For a discussion of the involved synonymy of the species and an explanation of the inappropriate specific name, see the older treatment of Fernald (Rhodora 37:273-284. 1935).

Although the species is sometimes listed as a worthful ornamental (especially by eastern or English authorities), most gardeners in the Pacific Northwest have reason to regard it as a pest. The fruit is considered palatable.

Rubus pedatus J. E. Smith, Plant. Ic. Ined. pl. 63. 1790.

Dalibardia pedata Stephan. in Mém. Soc. Nat. Mosc. 1:129. 1806. *Comaropsis pedata* DC. Prodr. 2:555. 1825. *Ametron pedatum* Raf. Sylva Tellur. 161. 1838. *Psychrobatia pedata* Greene, Leafl. 1:245. 1906. ("In Americae borealis tractu occidentali")

Mat-forming, unarmed, subglabrous to more or less sericeous-pilose perennial; stems herbaceous, strongly stolonous and nodally rooting and producing short, erect stems (rarely over 2 cm. long) which bear 1-3 leaves and a single-flowered filiform peduncle 2-6 (8) cm. long; leaves with prominent, brownish-membranous, entire stipules, the blades digitately 5-foliolate, or 3-foliolate with the lateral leaflets divided nearly to the base; leaflets obovate to deltoid-obovate, 1-3 cm. long, doubly serrate-dentate or incised-dentate; calyx glabrous or sparsely pilose, the lobes narrowly oblong-oblanceolate, (4) 5-11 mm. long, erose to irregularly few-toothed near the tip, reflexed; petals white, oblong, equaling (to slightly longer than) the sepals; stamens numerous, the filaments filiform; pistils 3-6, glabrous; style filiform, the stigma not enlarged; drupelets red, juicy, more or less coherent, palatable; putamen about 4 mm. long, with only a few prominent wrinkles.

Mossy banks to open or dense, usually rather moist woods, from near sea level upward nearly to timber line; Alas. southward to s. Oreg., from the coast inward to w. Alta., w. Mont., and n. Ida. May-early July.

The species is easily grown and makes one of the choicest of ground covers for n.w. gardens, especially in shady areas.

Rubus procerus Muell. in Boulay Ronces Vosgrennes 6:7. 1864. (Type not known)

R. thyrsanthus sensu Peck, Man. High. Pl. Oreg. 409. 1941, but not of Focke in 1877.

Strong, more or less evergreen perennial with ascending to nearly erect but eventually usually clambering to sprawling stems up to 10 m. long, armed with strong, flattened prickles; leaves of the primocanes mostly 5-foliolate, the leaflets broadly ovate to oblong or oblong-obovate, usually abruptly short-acuminate, 6-12 cm. long, bright green and generally nearly glabrous above, grayish-tomentose beneath, sharply serrate-dentate; inflorescence paniculate-corymbose, 5- to 20-flowered, lanate, eglandular; calyx tomentose, the segments reflexed, lanceolate, 8-12 mm. long; petals white or occasionally reddish tinged, oval-obovate, mostly 10-15 mm. long; stamens 100 or more; pistils many, the ovary usually sparsely pilose, the style slender, glabrous; drupelets coherent and with the fleshy receptacle forming a nearly globose blackberry about 1.5 cm. thick, palatable; putamen reticulate-pitted. Himalayan blackberry.

This species of the Old World has escaped and become established from s. B.C. to Calif., mostly on the w. side of the Cascades, where it is especially common along roadsides and railways, but it is also abundant along the Snake R. in s.e. Wash. June-Aug.

The plant often is a serious pest in which domestic animals (especially sheep) may become entangled. It is the most common wild blackberry harvested in w. Wash. and Oreg., although the fruit is inferior in flavor to that of *R. ursinus*.

The use of the name, *R. procerus*, for this widely cultivated plant follows Bailey's treatment, but it is probably more commonly called *R. fruticosus* L.

Rubus pubescens Raf. in Med. Rep. III, 2:333. 1811.

R. saxatilis var. *canadensis* Michx. Fl. Bor. Am. 1:298. 1803. *R. saxatilis* var. *americana* Pers.

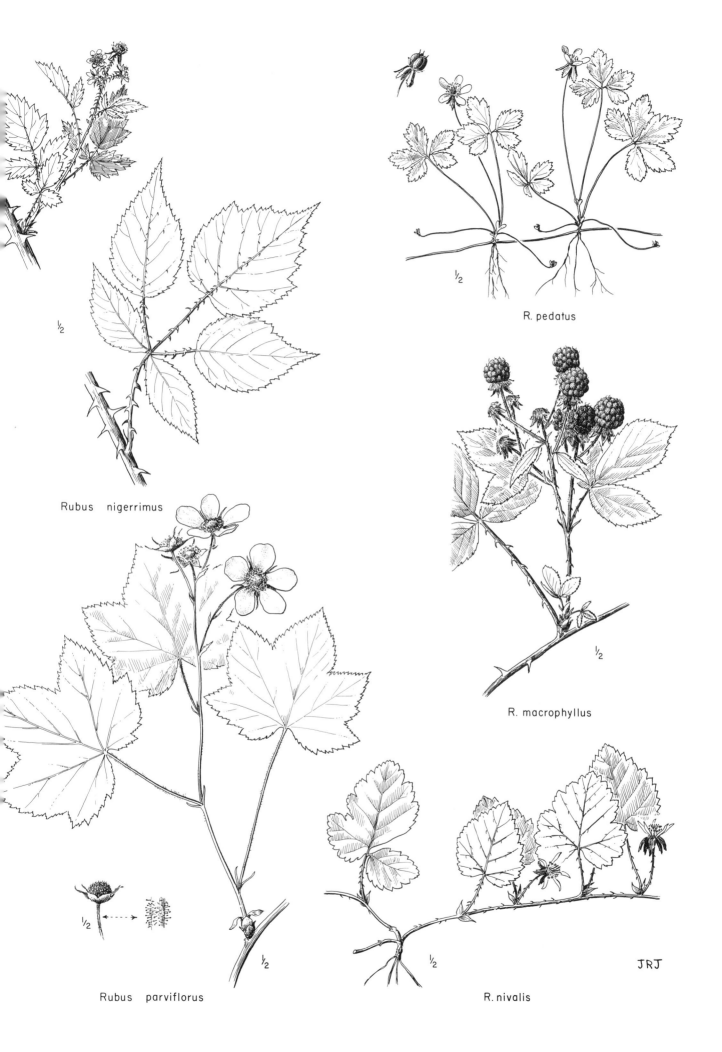

½

Rubus   nigerrimus

½   R. pedatus

Rubus   parviflorus   R. macrophyllus

½

½

½   R. nivalis

JRJ

Syn. 2:52. 1807. *R. triflorus* Richards. App. Frankl. Journ. 2nd ed. 19. 1823. *R. americanus*
Britt. Mem. Torrey Club 5:185. 1894. (Hudson Bay)

R. transmontanus Focke, Bibl. Bot. 17, Heft 72:27. 1910. (Columbia River near Revelstoke, Wash.
[really in B. C.], is the only locality definitely cited) Although applied in a rather vague way, this
name apparently falls to synonymy here.

Unarmed, more or less pilose perennial with rootstocks and herbaceous, leafy flowering stems up
to 1 m. long, these sometimes erect at first but ultimately reclining and often nodally rooting; leaves
with oblanceolate, entire stipules up to 1 cm. long, the blades trifoliolate; leaflets short-petiolulate,
(obliquely) ovate to deltoid, 2-6 cm. long, doubly serrate-dentate; flowers 1-2 (rarely more) on short,
erect, leafy shoots, the inflorescence usually stipitate-glandular; calyx pilose-pubescent and often
glandular, the lobes lanceolate, 3-5 (6) mm. long, entire, reflexed; petals white or greenish-white,
oblong-lanceolate to obovate-spatulate, usually erect, (4) 5-8 mm. long; stamens numerous, the filaments broad and flattened, narrowed abruptly and with a nearly square shoulder (or even bidentate)
near the tip; pistils 20-30, glabrous; style slender; stigma discoid; drupelets deep red, weakly coherent into a blackberrylike aggregate up to 1 cm. broad; putamen about 2 mm. long, smooth.

Stream banks and deep moist woods or on open burns and clearings where moderately dry; n. B.C.
eastward through all the Canadian provinces, to Newf. and Lab., southward, e. of the Cascades, to
Stevens and Ferry cos., Wash., in the Rocky Mts. to n. Colo., and in the central states to Iowa,
Wisc., and Ind.; to be expected in Ida. Late May-early July.

Rubus spectabilis Pursh, Fl. Am. Sept. 348, pl. 16. 1814.

Parmena spectabilis Greene, Leafl. 1:244. 1906. *(Lewis,* "On the banks of the Columbia" is the first
of 2 collections cited)

Strongly rhizomatous, thicket-forming perennial usually 1-3 (5) m. tall, the stem erect to arching,
usually strongly bristly below, less so (or even unarmed) above, the prickles acicular, the brownish
bark eventually shredding; leaves pinnately 3 (5)-foliolate; leaflets ovate, acute to acuminate, glabrous
or very sparsely appressed-pubescent above, usually pubescent along the veins beneath, doubly serrate to lobulate-serrate, the terminal one (3) 4-9 (11) cm. long, the lateral pair smaller, often unequally lobed or divided; flowers 1-2 (4) on short leafy branches; calyx pubescent, the lobes spreading,
ovate-lanceolate, acute to acuminate, 9-15 (up to 20) mm. long; petals red to reddish-purple, showy,
obovate-elliptic, up to half again as long as the sepals; stamens 75-100; pistils numerous; style glabrous, slender, about 2 mm. long; drupelets imperfectly coherent, separating from the semifleshy
receptacle and forming a somewhat raspberrylike, yellow to reddish, rather insipid fruit; putamen
strongly wrinkled-pitted, about 3 mm. long. Salmonberry.

Lowland moist woods, swamps, and stream banks to mountain slopes, at medium elevations; Alas.
southward to n. w. Calif., from the coast to the Cascades (rarely e. of the crest). Mar.-June.

The salmon berry is considered (at least by those who do not have to contend with its extremely aggressive natural spread) to be an attractive shrub because of the showy flowers and fruits.

Rubus ursinus Cham. & Schlecht. Linnaea 2:11. 1827.

R. vitifolius ssp. *ursinus* Abrams, Ill. Fl. Pac. St. 2:458. 1944. (California)

R. macropetalus Dougl. ex Hook. Fl. Bor. Am. 1:178. 1833. *R. ursinus* var. *macropetalus* Brown,
Am. Journ. Bot. 30:696. 1943. *(Douglas; Dr. Scouler,* "in the Valley of the Columbia, North-
West America")

R. helleri Rydb. N. Am. Fl. 22⁵:460. 1913. *(Heller & Heller 3990,* near Montesano, Chehalis [Gray's
Harbor] Co., Wash., June 30, 1898)

R. vitifolius sensu Abrams, Ill. Fl. Pac. St. 2:457. 1944.

Dioecious perennial with pubescent but ultimately glabrate, slender, trailing and terminally rooting
primocanes up to 5-6 m. long, abundantly armed with rather slender, slightly retrorsely hooked, almost terete prickles; floral branches (produced the second year) numerous, erect, mostly 1-3 (5)
dm. long, bearing several leaves and 1-several (2) 4- to 10-flowered corymbs; leaves trifoliolate; stipules slightly adnate to the petioles, the free tip linear, up to 15 mm. long; lateral leaflets ovate-lanceolate, 3-7 cm. long, doubly serrate-dentate to somewhat lobulate, the terminal leaflet larger, commonly up to 10 cm. long, frequently deeply 3-lobed and occasionally divided, the leaflets then 5; inflorescence more or less purplish stipitate-glandular; calyx villous-tomentose, usually stipitate-glandular, and sometimes weakly prickly, the lobes lanceolate-acuminate, 7-11 mm. long; staminate
flowers with narrowly elliptic to elliptic-spatulate, white petals mostly 12-18 mm. long, and with 75-
100 stamens; pistillate flowers with broader (sometimes oval) petals mostly 8-12 mm. long, and with

rudimentary but usually abortive-anthered stamens, the pistils numerous, borne on an elongate receptacle, the styles glabrous; drupes coherent and with the fleshy receptacle forming a globose to usually elongate blackberry up to 2.5 cm. long and about 1 cm. thick; putamen 2.5-3 mm. long, strongly reticulate-pitted. 2N=42, 56, 84. Douglas berry, Pacific blackberry, dewberry.

Prairies, clearings, and open to fairly dense woodlands, especially common on logged or burned forest land; B.C. to n. Calif. from the coast to middle elevations in the mountains, e. to c. Ida. Apr. -early Aug.

Our material, as described above, is all referable to var. *macropetalus* (Dougl.) Brown. Other varieties, including var. *ursinus* (to which *R. vitifolius* Cham. & Schlecht. is referable), are to be found in various parts of Calif. and s. Oreg.; most of these have nonglandular inflorescences.

This, our only native blackberry, has a fruit of excellent flavor; it is the ultimate source of several horticultural varieties, having been used in the production of the mammoth berry and the loganberry and its derivatives, the youngberry, boysenberry, and other strains. It is a decided pest in the woodland garden where it is extremely difficult to eradicate.

Rubus vestitus Weihe & Nees, Rub. Germ. 81:pl. 33. 1822. (N. Germany)

R. fruticosus sensu G. N. Jones, U. Wash. Pub. Biol. 5:175. 1936, not of L.

Strong perennial with arching to trailing branches up to 3 m. long, well armed with straight, flattened prickles up to 7 mm. long; leaves partially evergreen, 3- to 5-foliolate; stipules linear-lanceolate to narrowly ovate, 5-10 mm. long; leaflets ovate to rotund-ovate, abruptly acuminate, 5-9 (10) cm. long, doubly serrate-dentate, greenish and glabrous to sparsely strigose above, paler and pubescent (somewhat lanate) beneath; flowers rather numerous in corymbiform panicles, abundantly hairy and stipitate-glandular; calyx more or less tomentose, the lobes reflexed, lanceolate-acuminate, less than 1 cm. long; petals white (pinkish-tinged), 10-15 mm. long; stamens at least 75; pistils rather numerous; style glabrous; drupelets coherent and with the receptacle forming a globular blackberry about 1.5 cm. thick; putamen heavily reticulate-alveolate.

A European species that has been collected at Salem, Oreg., by J. C. Nelson, and along the Humptulips R., Grays Harbor Co., Wash. *(G. N. Jones 4593)*. Apr.-June.

There seems to be little doubt that *R. vestitus*, which differs from *R. procerus* chiefly in the straight spines and stipitate-glandular inflorescence, is well established in certain localities within our range.

Sanguisorba L. Burnet

Flowers sessile and closely packed in headlike to elongate spikes, each subtended by a more or less papery bract and 2 lateral bractlets, perfect or imperfect (the plants sometimes dioecious), apetalous, perigynous, white or greenish to red or purple; calyx 4-merous, the hypanthium urceolate, much contracted and nearly closed by a prominent disc at the mouth, the sepals broad, spreading, petaloid, imbricate in bud; stamens usually 4, but sometimes rather numerous or only 2, inserted at the top of the hypanthium, the filaments filiform to flattened, sometimes clavate; pistils 1 or 2 (3), the ovary with a single suspended ovule, the style terminal, the stigma papillate-discoid and brushlike; fruit a 1-seeded achene, enclosed in the usually hardened, 4-angled and more or less winged, ovoid to fusiform, smooth, reticulate, or strongly verrucose-rugose hypanthium; annual, biennial, or perennial, often rhizomatous herbs with alternate, pinnately compound leaves, the leaflets toothed to pectinately dissected, the stipules adnate or free and often dissected.

Fifteen to 20 species of the Northern Hemisphere in temperate to subarctic regions. (From the Latin *sanguis*, blood, and *sorbere*, to drink or absorb; some species having alleged astringent properties were used as styptics.)

1 Leaflets pectinately pinnatifid; plant annual or biennial with very leafy stems; stamens
    usually 2   S. OCCIDENTALIS
1 Leaflets merely toothed; plant perennial, often rhizomatous, the stems usually sparingly leafy; stamens 4 or about 12
    2 Stamens about 12, the filaments filiform; leaflets mostly 1-2 cm. long; flowers largely
        imperfect, the staminate usually below the pistillate in the same spike; fruiting hypanthium very strongly verrucose-papillate   S. MINOR
    2 Stamens 4, the filaments often flattened and clavate; leaflets mostly over 2 cm. long;
        flowers perfect; fruiting hypanthium smooth to wrinkled
        3 Calyx greenish or only lightly pinkish- or rose-tinged; spikes mostly 3-8 cm. long,

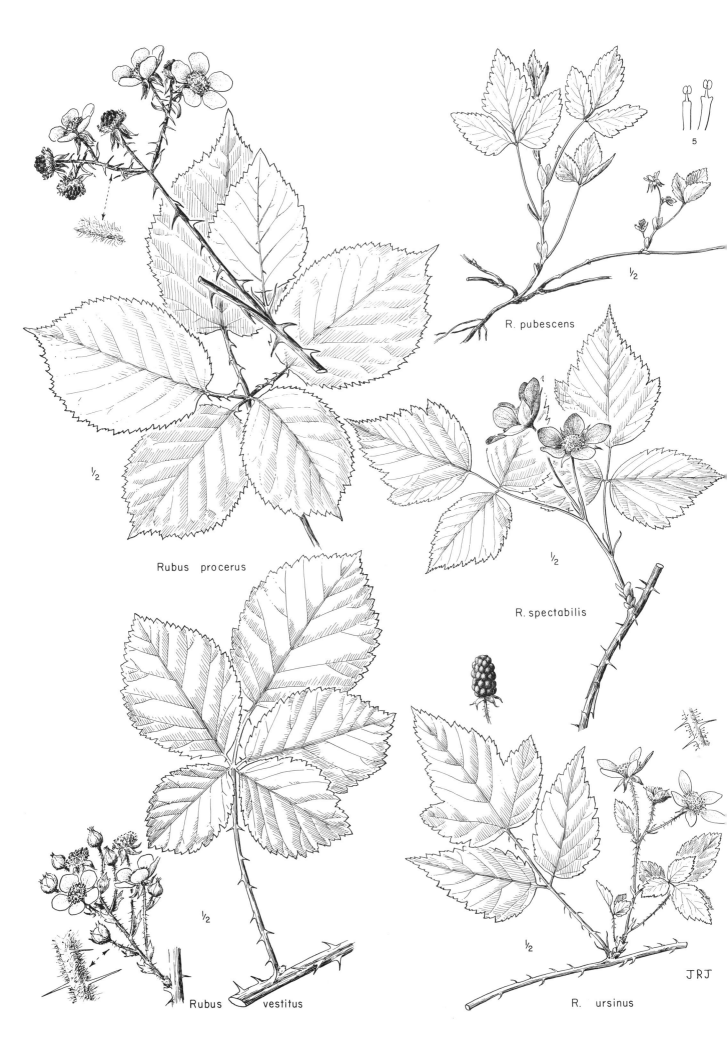

R. pubescens

½

Rubus procerus

½

R. spectabilis

½

Rubus vestitus

½

R. ursinus

½

5

JRJ

strongly tapered upward during anthesis; filaments strongly flattened and clavate, usually at least 3 times as long as the sepals   S. SITCHENSIS
3 Calyx reddish to purple; spikes mostly 1-3 (but up to 7) cm. long, not strongly tapered upward; filaments terete to clavate and flattened, mostly 1-2 times as long as the sepals
   4 Filaments flattened and clavate upward, about twice as long as the sepals   S. MENZIESII
   4 Filaments terete, not at all clavate, about equal to the sepals   S. OFFICINALIS

Sanguisorba menziesii Rydb. N. Am. Fl. 22^4:387. 1908.
   S. media sensu Hook. Fl. Bor. Am. 1:197. 1833, but not of L. in 1762. *(Thomas Howell 1620,* mts.
   back of Short Bay, Alas., in 1895)
   Very similar to *S. officinalis,* but more freely rhizomatous, the basal leaves usually larger, the
blades up to 2.5 dm. long, the leaflets 9-15, ovate to oblong, mostly 2-5 cm. long, coarsely serrate;
spikes 1.5-7 cm. long, 10-13 mm. thick; flowers reddish-purple; sepals 2.5-3 mm. long; stamens 4,
the filaments flattened and rather strongly clavate, usually about twice as long as the sepals or slightly longer; fruiting hypanthium wing-margined, pubescent.
   Occasional in coastal bogs and marshlands from Alas. to the Olympic Peninsula, Wash. July-Aug.
   Although some workers have regarded this plant as merely a purplish-flowered phase of *S. sitchensis,* in so many respects it is intermediate between that species and *S. officinalis* that its hybrid origin seems probable. Whether extant collections are representative of an established amphiploid or
merely the occasional intermediate hybrid is not known.
   This species, and also *S. sitchensis* and *S. officinalis,* take readily (but not aggressively) to cultivation,
their chief requirement being a fairly moist soil. They are desirable chiefly for their fern-like foliage.

Sanguisorba minor Scop. Fl. Carn. 2nd ed. 1:110. 1772.
   Poterium sanguisorba L. Sp. Pl. 994. 1753. *Poterium minus* J. E. Gray, Nat. Arr. Brit. Pl. 2:575.
   1821. *S. sanguisorba* Britt. Mem. Torrey Club 5:189. 1894. (Southern Europe)
   Perennial from a usually branched caudex, often somewhat rhizomatous, the flowering stems generally with several scarcely reduced leaves, simple or branched above, 2-6 (7) dm. tall, mostly
sparsely pilose with multicellular (more or less moniliform) hairs; basal leaves several and somewhat rosulate, with adnate stipules; cauline leaves several, reduced upwards but the stipules becoming
free and leafletlike above; leaflets mostly 9-17, oval to ovate-oblong, 1-2 cm. long, coarsely serrate;
spikes globose to ovoid, 8-20 mm. long, about 10 mm. thick; bractlets ovate, ciliate; flowers mostly
imperfect, the lower ones staminate, the upper ones pistillate, a few often perfect; calyx greenish to
rose-tinged, in fruit the hypanthium urceolate, 4-5 mm. long, woody, very prominently papillate-war-
ty between as well as along the rather prominent ridges; stamens about 12, the filaments filiform;
pistils and achenes 2. N=14, 28.
   Native to Europe and sometimes introduced in N. Am., usually in waste places; fairly common in e.
U.S. but only occasional in the west and known in our area from Lummi I., Whatcom Co., and Bingen,
Klickitat Co., Wash., but probably occurring elsewhere. June-Aug.

Sanguisorba occidentalis Nutt. in T. & G. Fl. N. Am. 1:429. 1840.
   Poteridium occidentale Rydb. N. Am. Fl. 22^4:388. 1908. ("Grand Rapids of the Oregon to the Wahlamet," *Douglas, Nuttall)*
   S. myriophylla Braun & Bouché, Ind. Sem. Hort. Berol. 1867:app. 10. 1868. (Described from cultivated plants from seeds collected by Geyer, "Spokan highlands")
   Glabrous annual or biennial with a strong taproot and freely branched, very leafy stem 2-6 (9) dm.
tall; leaves mostly 3-8 cm. long, pinnate-pinnatifid, the leaflets up to 17, the ultimate segments narrowly linear; stipules from simple (on the basal leaves) to pectinately dissected above; spikes from
subglobose to cylindric and up to 3 cm. long, 5-7 mm. thick; bracteoles papery, equaling the hypanthium; flowers perfect, about 2-3 mm. long; sepals longer than the urceolate hypanthium at anthesis,
broadly ovate-triangular, greenish with a lighter margin; stamens usually 2, the filaments filiform,
about equaling the sepals; fruiting calyx about 3 mm. long, ellipsoid, 4-sided with thick, rounded but
scarcely winged angles, the faces slightly reticulate.
   Grassy flats (most common on semiwaste land or where the soil is moist in the early spring), sagebrush, and woodland; e. of the Cascades from s. B.C. through Wash., s. through the Willamette Valley to s.w. Oreg. and s. Calif., e. to Ida. and w. Mont. June-July.
   The plant is very similar in general appearance to *S. annua* Nutt. of c. U.S., which differs in having
(mostly) 4 stamens and a much more prominently winged calyx in fruit.

Sanguisorba officinalis L. Sp. Pl. 116. 1753. (Europe)

S. microcephala Presl, Epim. Bot. 202. 1849. *(Haenke,* "Nutka-Sund")

Glabrous perennial with thick short rootstocks, mainly basal leaves, and erect, simple to sparingly branched, nearly leafless flowering stems up to 6 dm. tall; basal leaves with adnate stipules forming membranous margins to the petioles, the blades mostly 8-15 cm. long, the leaflets (7) 9-13 (15), ovate to oblong-lanceolate, round based to cordate, crenate to serrate, (1) 1.5-3 (4) cm. long; cauline leaves 1-2, much reduced, the stipules more or less leafletlike; spikes broadly cylindric, 1-2.5 (3) cm. long, about 1 cm. broad; flowers maroon or dull violet to deep maroon-purple, about 5 mm. long; bracteoles membranous, lanceolate to ovate, equaling or longer than the calyx tube, brownish-purple, ciliate and usually sparsely pubescent; calyx more or less pubescent, the lobes about 2-2.5 mm. long; stamens 4, the filaments from slightly shorter to slightly longer than the sepals, thick and nearly terete; fruiting calyx ellipsoid, narrowly winged on the 4 angles, pubescent. N=14, 21, 28.

Muskeg, swamps, and bogs, almost circumboreal; Iceland to much of n. Europe and Asia, in N. Am. from Alas. and Yukon, chiefly along the coast, to n. w. Calif., inland to Mt. Hood, Oreg., where up to 5000 ft. elevation. July-Aug.

Our plants have been treated as comprising a species, *S. microcephala,* closely related to (but separate from) the Eurasian *S. officinalis,* but there is no basis for such distinction, at least as evinced by herbarium material.

Sanguisorba sitchensis C. A. Meyer in Trautv. & C. A. Meyer, Fl. Ochot. 34. 1856.

Poterium sitchense Wats. Bibl. Bot. 1:303. 1878. *S. canadensis* var. *sitchensis* Koidzumi, Bot. Mag. Tokyo 31:137. 1917. *(Mertens,* Sitka Island)

S. canadensis var. *latifolia* Hook. Fl. Bor. Am. 1:198. 1833. *S. latifolia* Cov. Contr. U.S. Nat. Herb. 3:339. 1896. *S. stipulata* var. *latifolia* Hara, Journ. Jap. Bot. 23:31. 1949. *(Dr. Scouler,* "Observatory Inlet, North-West coast of America," is the first of 2 collections cited)

Glabrous, freely rhizomatous perennial with mostly basal leaves, the flowering stems simple or branched above, 2.5-10 dm. tall; basal leaves with completely adnate, membranous stipules, the blades (5) 10-30 cm. long, the 9-17 leaflets ovate to ovate-oblong, (1) 2-6 (7) cm. long, coarsely serrate, mostly cordate based; leaves of the flowering stems 1-3, reduced and with free, often leafletlike stipules; spikes (2) 3-8 cm. long, strongly tapered upward until fully flowered, 10-18 mm. thick; bractlets brownish-pubescent; calyx greenish-white or ochroleucous to slightly pinkish- or purplish-tinged, somewhat pubescent, the oval sepals about 2.5 mm. long; stamens 4, the filaments flattened and strongly clavate, usually at least 3 times as long as the sepals, white to mauve; fruiting hypanthium lightly winged on the 4 angles.

Bogs, swamps, and stream banks from Alas. southward to Oreg., in our area montane, from the Cascades e. to c. Ida. and n. e. Oreg.; Asia. July-Aug.

Sanguisorba sitchensis is very similar to, and perhaps as properly treated as a geographic race of, *S. canadensis* L. of e. N.Am., which has relatively more slender leaflets and thicker sepals.

Sibbaldia L.

Flowers inconspicuous, complete, perigynous, borne in small, axillary and terminal, leafy-bracteate cymes on short flowering stems; calyx 5-merous, 5-bracteolate, the hypanthium shallow, gland lined; petals 5, pale yellow; stamens 5 in ours (4-10); pistils 5-15 (20), the style lateral; ovule single, ascending; fruit an achene; dwarf, perennial, usually tufted herbs with long-petioled, stipulate, ternate leaves.

Six or 7 arctic to montane species, all except ours Asiatic. (Named for Sir Robert Sibbald, 1641-1722, Professor of Medicine at Edinburgh.)

Sibbaldia procumbens L. Sp. Pl. 284. 1753.

Potentilla procumbens Clairv. Man. 166. 1811, not of Sibth. in 1794. *Potentilla sibbaldii* Hall, f. in Ser. Mus. Helv. 1:51. 1818. ("Habitat in Alpibus Lapponiae, Helvetiae, Scothiae")

Rather strongly rhizomatous, mat-forming, strigillose perennial; leaves mostly from the horizontal stems, slender-petiolate, the 3 leaflets cuneate-obovate, 1-2 (3) cm. long, semitruncate and apically 3- to 5-toothed; flowering stems 4-8 (15) cm. tall, usually leafless below the axillary, pedunculate, (1) 2- to 15-flowered cymes; calyx lobes 2.5-3.5 (5) mm. long, ascending, longer than the narrower bracteoles and the internally hirsute hypanthium; petals oval to spatulate, about half the length of the

5

3

Sanguisorba
occidentalis

S. officinalis

½

mature calyx

5

8

Sanguisorba minor

½

JRJ

S. menziesii

3

½

sepals; stamens alternate with, and shorter than, the petals; style about 1 mm. long; mature achenes pyriform-ovoid, smooth or very lightly reticulate, supported on a strongly flattened, hirsutulous stipe from only slightly to much shorter than the approximately 1.5 mm. body. N=7.

Circumpolar, extending southward in N. Am., on open, dry to moist alpine slopes or meadows, to s. Calif., Utah, Colo., Que., and N.H. June-early Aug.

Sorbus L. Mountain Ash

Flowers complete, regular, borne in large compound corymbs; calyx ebracteolate, turbinate-obconic, adnate almost to the top of the ovary, with a short, disc-lined, free hypanthium, the 5 lobes triangular, persistent; petals 5, white (cream), ovate, orbicular, or obovate to elliptic, usually very short-clawed, ours sparsely pilose on the upper surface near the base, deciduous; stamens 15-20, the filaments broadened slightly at the base, usually shorter than the petals; carpels 2-5, the ovary compound, 1/2-4/5 inferior; styles 2-5, distinct; fruit fleshy, pomaceous, small, 2- to 5-celled, each loculus 1- or 2-seeded; seeds somewhat flattened; trees or shrubs with alternate, deciduous, pinnate (ours) to simple stipulate leaves.

Possibly 50 species of temperate to subarctic N. Am. and Eurasia. (Ancient Latin name for one of the species.)

All of our species are highly regarded as ornamentals. Although the flowers and foliage are attractive the plants are valued mainly because of the fruits, the orange-fruited *S. scopulina* var. *cascadensis* being perhaps best. The plants are rather readily transplanted from the wild, and are fairly quickly established through seedage.

Several hybrids have been noted between *Sorbus* and other genera, one of which was named *Amelasorbus jackii* Rehd. (Journ. Arn. Arb. 6:154. 1925), based on *Jack 1329*, from Clearwater Co., Ida. It was supposedly a hybrid between *Sorbus scopulina* and *Amelanchier florida* (= *A. alnifolia* var. *semiintegrifolia*).

Reference:

Jones, G. N. A synopsis of the North American species of Sorbus. Journ. Arn. Arb. 20:1-43. 1939.

1 Leaves (at least in part) with over 13 leaflets; winter buds grayish strigose-villous; inflorescence grayish-lanate at anthesis; styles 2-3; plant arborescent, introduced and sometimes escaped in our area, but persisting chiefly near present or past habitations

S. AUCUPARIA

1 Leaves rarely if ever with more than 13 leaflets; winter buds mostly either rufous-pubescent or glutinous and very sparsely hairy to glabrous; styles 3-5; plants fruticose, all native (chiefly montane) species

2 Winter buds and (less conspicuously) the young growth and the inflorescence rufous-hairy; calyx usually glabrous externally; leaflets 7-11, semitruncate to rounded (subacute) at the tip, rarely serrate more than 3/4 of the length, usually at least 1/3 as broad as long; styles 4-5; fruits red but glaucous and with a slight bluish cast   S. SITCHENSIS

2 Winter buds somewhat glutinous and only sparsely whitish-hairy, the young growth and the inflorescence grayish-strigillose-pilose; calyx usually somewhat pubescent externally; leaflets (7) 9-13, acute or more often shortly acuminate, finely serrate nearly the full length, mostly less than 1/3 as broad as long; styles 3-4; fruits orange to scarlet but glossy rather than glaucous   S. SCOPULINA

Sorbus aucuparia L. Sp. Pl. 477. 1753.

Pyrus aucuparia Gaertn. Fruct. 2:45. 1791. (Europe)

Small, gray-barked, spreading-branched tree 5-12 (20) m. tall, the young growth (especially the inflorescence) grayish-pilose to downy or somewhat lanate, tardily almost or quite glabrous; winter buds grayish-downy; stipules lanceolate, persistent most or all of the season; leaflets 11-15 (17), oblong or oblong-lanceolate, (2) 3-6 cm. long, rather abruptly acute, sharply but rather finely serrate nearly the full length; corymbs large, flat-topped, usually with at least 75 flowers; calyx hairy; petals white, semiorbicular, about 4 mm. long; stamens about equaling the petals; carpels 3-4, the styles 2-3 mm. long; fruit globose, bright red, pendulous, 10 (12) mm. broad. N=17. Rowan tree, mountain ash.

This European species is very commonly planted as an ornamental and is widely bird-disseminated, mostly near habitations, where it may seem to be native. May-June.

The flowers have a particularly strong, rather unpleasant odor.

Sorbus scopulina Greene, Pitt. 4:130. 1900.

Pyrus scopulina Longyear, Trees & Shrubs Rocky Mt. Reg. 152. 1927. *(C. F. Baker,* near Pagosa
  Peak, Colo., Aug. 10, 1899, and *Heller 3711,* Santa Fe Canyon, N. M., June, 1897)

S. angustifolia Rydb. Fl. Rocky Mts. 448, 1062. 1917. *(Piper,* Cedar Mt., Ida., July, 1898) = var.
  scopulina.

SORBUS SCOPULINA var. CASCADENSIS (G. N. Jones) C. L. Hitchc. hoc loc. *S. cascadensis* G.
  N. Jones, U. Wash. Pub. Biol. 7:174. 1938. *(Allen 291,* in part [the flowering specimen], Bear
  Prairie, Mt. Rainier, Wash., June, 1897)

Erect, several-stemmed shrub 1-4 (6) m. tall; winter buds glutinous and from glabrous to sparsely
white- or grayish-strigillose, the young growth sparsely to rather heavily grayish strigillose-pilose,
the older bark yellowish to reddish-purple, eventually grayish-red; stipules green but somewhat membranous, linear, deciduous to persistent; leaflets (7) 9-13, narrowly oblong or oblong-elliptic to oblong-lanceolate, (2) 3-7 (8) cm. long, usually less than 1/3 as broad, shortly cuneate-acute at base,
acute or more commonly shortly acuminate at the tip, finely and sharply serrate almost the full length,
glabrous and dark green above, much paler and glabrous to very sparsely pilose-villous beneath along
the midrib; corymbs nearly flat-topped, usually from 70- to over 200-flowered, sparsely to fairly
densely pubescent with whitish, more or less appressed hairs; calyx whitish-pubescent; petals oval,
(4) 5-6 mm. long; carpels 3-4, the styles about 2 mm. long; fruit (yellow) orange to scarlet, subglobose, glossy (not glaucous), about 1 cm. long.

Alaska southward to n. Calif., e. to w. Alta., the Dakotas, Wyo., Colo., and N. M., in our area
from the foothills to subalpine or even alpine. May-early July.

Sorbus scopulina contrasts strongly with *S. sitchensis* because of its acuminate and more finely
toothed leaves, glutinous winter buds, and white (rather than rufous) pubescence. There are 2 phases
of the species, each of which has been accorded specific rank, but because of their complete intergradation, treatment at the varietal level seems the more natural. Their only observable differences
are emphasized by the following key:

1 Leaflets of at least some of the leaves usually 13; stipules mostly caducous before the
    end of anthesis; widespread, from Alas. to w. Alta. and the Dakotas, s. to N. M.,
    Utah, Ida., and (mainly) on the e. side of the Cascade Mts. through Wash. and Oreg.
    to n. Calif.   var. scopulina
1 Leaflets rarely if ever over 11; stipules usually persistent until after anthesis; B.C.
    southward, mostly on the w. slope of the Cascades, and in the Olympic Mts. of Wash.,
    to n. and Sierran Calif., also with and often intergradient to var. *scopulina* on the
    e. side of the Cascades and occasional farther e. as in n.e. Wash., and in the Wal-
    lowa Mts., Baker Co., Oreg. *(Thompson 13414);* the 11-leafleted plants from e. of
    the Cascades apparently were referred to *S. scopulina* by Jones on the basis of the
    range   var. cascadensis (G. N. Jones) C. L. Hitchc.

Sorbus sitchensis Roemer, Fam. Nat. Syn. 3:139. 1847.

Pyrus sitchensis Piper, Mazama 2:107. 1901. *S. americana sitchensis* Sudw. Check List For. Trees
  133. 1927. *(Mertens,* Sitka Island)

S. pumilus Raf. Med. Fl. 2:265. 1830. *Pyrus sambucifolia* var. *pumila* Sargent, Silva N.Am. 4:82.
  1890. *S. sambucifolia* var. *pumila* Koehne, Deuts. Dendrol. 247. 1893. (Nomen dubium, no collections cited and name published with no description) = var. *grayi.*

SORBUS SITCHENSIS var. GRAYI (Wenzig) C. L. Hitchc. hoc loc. *S. sambucifolia* var. *grayi* Wenzig, Bot. Zentralb. 35:342. 1888. *(Lyall,* Cascade Mts., in 1859, lectotype, the first of 2 collections cited)

Pyrus occidentalis Wats. Proc. Am. Acad. 23:263. 1888. *S. occidentalis* Greene, Pitt. 4:131. 1900.
  (Lyall, Cascade Mts., lat. 49°, in 1859, presumably the same collection as cited by Wenzig) = var.
  grayi.

S. tilingii Gandg. in Bull. Soc. Bot. France 65:25. 1918. *(Tiling,* Sitka) = var. *sitchensis,* in part,
  as to type only.

Erect, several-stemmed shrub 1-4 m. tall; winter buds and young growth rather heavily rufous-
strigillose-villous, the bark reddish-purple, tardily glabrous, eventually grayish-red; stipules persistent or tardily dehiscent, brownish-membranous and usually rufous-pubescent; leaflets 7-11, rather
thick, glabrous and dark green above, paler beneath and often persistent-pubescent with rufous hairs
along the midvein, (oval) oblong to oblong-obovate, (1.5) 2-5 (7) cm. long, 1/3-2/5 as broad, rounded
(rarely subacute) to semitruncate at the tip, rather coarsely serrate from about 3/4 their length to on-

ly near the tip; corymbs rounded, 15- to 80-flowered; calyx glabrous (sparsely pubescent) without; petals rhombic to oval (3) 4-5 mm. long; carpels 4-5, the styles (1. 5) 2-3 mm. long, the top of the ovary hirsute; fruit ellipsoid to subglobose, red but glaucous and with a slight bluish cast, about 1 cm. long.

Alaska and Yukon southward through the Cascade and Olympic mts. of Wash., and the Cascades of Oreg., to n. Calif., eastward to e. B. C., n. Ida., and n. w. Mont., montane in our area at from 2000-10,000 ft. elevation. June-July.

There are 2 geographic races of this species which have by some authors been considered identical, by others as specifically distinct. They differ in only one morphological feature, the serration of the leaflets, and there is considerable intermediacy even in this respect, but they are largely separable as follows:

1 Leaflets toothed primarily only above midlength, sometimes nearly entire; Cascade Mts.
    from s. B. C. s. through Wash. (where also in the Olympic Mts.) to the Sisters in Oreg.,
    perhaps intermittently s. as far as n. Calif. where reported from the Siskiyou Mts.
                                                                                    var. grayi (Wenzig) C. L. Hitchc.
1 Leaflets toothed usually at least 1/2-3/4 of the length; Alas. and s. Yukon s. to s. w.
    Alta., n. w. Mont., and n. Ida., and to w. B. C., not uncommon in the Cascades of
    Wash. (Skamania Co., *G. N. Jones 6059* and *Suksdorf*; Kittitas Co., *J. W. Thompson
    14848* and *B. O. Mulligan),* where often intermediate to var. *grayi,* and not seldom
    (especially in the area between Mt. Rainier and Mt. Adams) transitional to *S. scopulina*
    (east side Mt. Adams, Yakima Co., *Suksdorf 7096;* Mt. Rainier, *G. N. Jones 9716*
    and *F. A. Warren 1783)*   var. sitchensis

Spiraea L.

Flowers complete, perigynous, borne in dense, terminal, compound corymbs or panicles, small but showy en masse; calyx persistent, turbinate to hemispheric, ebracteolate but usually the top of the pedicel with a tiny linear bracteole which may be partially adnate to the calyx; free hypanthium about equaling the 5 usually triangular, erect to reflexed lobes; petals 5, white to pink, rose, or purplish, deciduous; stamens 25-50 (ours), the filaments slender, inserted with the petals at the edge of the disc-lined hypanthium; pistils usually 5 (3-7), distinct, the style terminal, slender, the stigma capitate-discoid; ovules 2-several, pendulous; fruit a small, somewhat leathery, ventrally dehiscent, 2- to several-seeded follicle; seeds fusiform, tapered at each end; endosperm lacking; deciduous shrubs with simple, toothed (rarely shallowly lobed) leaves.

Possibly 70 species of the (chiefly temperate) Northern Hemisphere. (From the Greek *speira*, coil, spire, or wreath, the meaning not obvious; the Greeks are said to have applied the name *speiraia* to some plant used for garlands.)

Several Eurasian members of the genus are considered valuable ornamentals, but our species are mostly not particularly attractive, although worth a place in the native garden. One exception is *S. densiflora,* which because of the habit, foliage, and flower color, must be rated as choice.

1 Flowers in rather flat-topped corymbs; plants generally almost or quite glabrous, but some-
    times puberulent with tiny, straight, erect hairs
    2 Petals pink to red; inflorescence and the lower surface of the leaves often puberulent
                                                                                    S. DENSIFLORA
    2 Petals white or with only a pale pinkish or lavender tinge; plant glabrous or merely
        ciliolate on the leaves and bracts   S. BETULIFOLIA
1 Flowers in rounded to much elongate corymbs; plants with pubescence at least in the
    inflorescence, the hairs mostly crisp
    3 Petals pale pink or whitish; inflorescence rounded to obconic, 1-2 times as long as
        broad   S. PYRAMIDATA
    3 Petals dark pink to rose; inflorescence narrowly conic to much elongate, usually
        several times as long as broad   S. DOUGLASII

Spiraea betulifolia Pall. Fl. Ross. 1:33, pl. 16. 1784. (Asia)
    SPIRAEA BETULIFOLIA var. LUCIDA (Dougl.) C. L. Hitchc. hoc loc. *S. lucida* Dougl. ex Hook.
    Fl. Bor. Am. 1:172. 1833, in synonymy; ex Greene, Pitt. 2:221. 1892. *S. corymbosa* var. *lucida*
    Zobel, Handb. Laubh. Deuts. Dendr. Ges. 157. 1903. *(Douglas,* "In the subalpine regions of Mount
    Hood, and in the Blue Mountains, near Lewis and Clarke's River, North-West America")
    Strongly rhizomatous, glabrous or subglabrous shrubs mostly (1. 5) 2. 5-6 (10) dm. tall; leaves ovate-

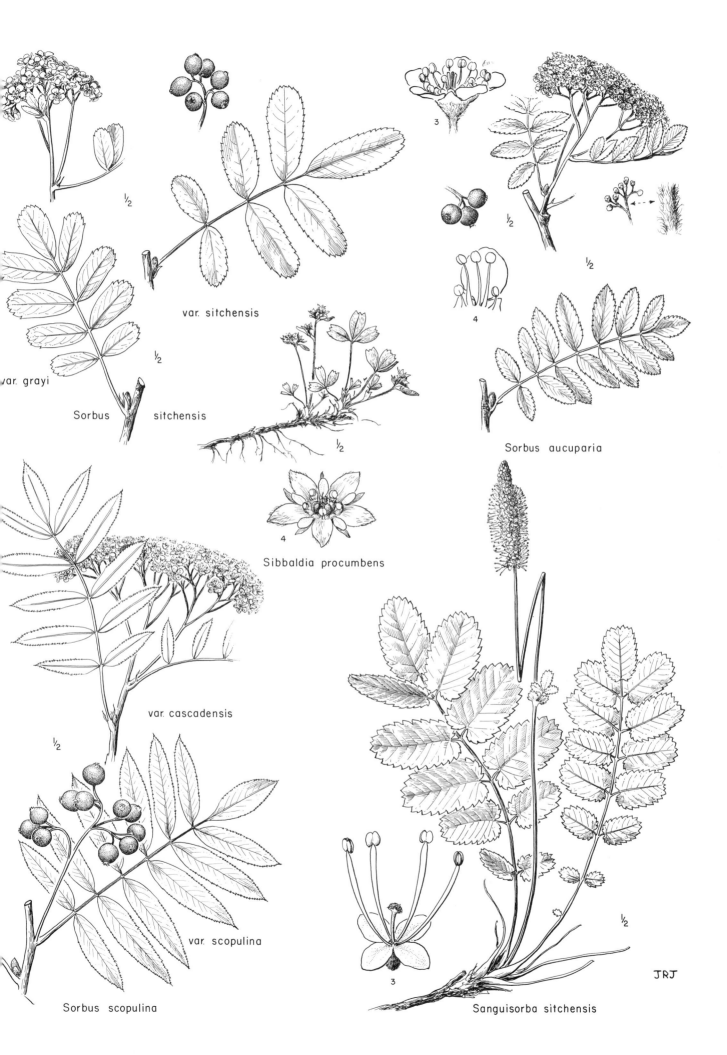

var. sitchensis

½

var. grayi

½

Sorbus sitchensis

3

½

4

½

Sorbus aucuparia

Sibbaldia procumbens

4

var. cascadensis

½

var. scopulina

Sorbus scopulina

3

½

Sanguisorba sitchensis

JRJ

oblong or obovate, (1.5) 2-7 (9) cm. long, rather coarsely (usually doubly) serrate, to sometimes shallowly lobulate, dark green on the upper surface, pale on the lower; corymbs commonly 3-8 cm. broad, nearly flat-topped; calyx glabrous externally, the hypanthium hemispheric, 1-1.5 mm. wide, slightly pubescent (glabrous) within; petals dull white, often with a pale pinkish or lavender tinge, about 2 mm. long; styles 1.5-2 mm. long; follicles about 3 mm. long, glabrous or sparsely ciliolate along the suture.

Stream banks and lake margins and open to wooded valleys and hillsides, often in rock slides; B.C. southward to n.c. Oreg., from near sea level to about 4000 ft. elevation in the Cascades, e. and up to 10,000-11,000 ft. elevation, from Sask. to S.D. and Wyo.; Asia. June-July.

Our material, as described above, is referable to the var. *lucida* (Dougl.) C.L. Hitchc., and differs from the Asiatic var. *betulifolia* in almost intangible ways, having slightly coarser-toothed leaves and glabrous or subglabrous (rather than pubescent) fruits. A third element of the species, var. *corymbosa* (Raf.) Wats., occurs in e. U.S.; it also has glabrous fruits, but is otherwise more pubescent than the var. *lucida*.

Spiraea densiflora Nutt. ex T. & G. Fl. N.Am. 1:414. 1840. (*Nuttall*, "Oregon," is the only specimen mentioned; acc. Rydberg, it was collected in the Blue Mts.)
    S. *betulaefolia* var. *rosea* Gray, Proc. Am. Acad. 8:381. 1872. S. *lucida* var. *rosea* Greene, Pitt. 2:221. 1892. S. *rosea* Koehne, Deuts. Dendrol. 218. 1893, not of Raf. in 1838. S. *arbuscula* Greene, Erythea 3:63. 1895. (*Elihu Hall*, Oregon, summer of 1871) = var. *densiflora*.
    SPIRAEA DENSIFLORA var. SPLENDENS (Baumann) C.L. Hitchc. hoc loc. S. *splendens* Baumann ex K. Koch in Monats. Ver. Bef. Gart. Preuss. 18:294. 1875. S. *densiflora* ssp. *splendens* Abrams, Ill. Fl. Pac. States 2:411. 1944. (Cultivated plants, from seed collected in Calif.)
    Low, spreading to erect, freely branched shrub (2) 5-10 dm. tall, with strong rootstocks, from glabrous to sparsely or rather thickly puberulent on the young stems, inflorescence, and lower surface of the leaves; bark reddish- or purplish-brown; leaves ovate-oval to oblong-elliptic, (1.5) 2-4 cm. long, finely serrate at least half their length, bright green on the upper surface, paler and strongly veined beneath; flowers in rather small, nearly flat-topped (to rounded) corymbs mostly 2-4 cm. broad; calyx glabrous (puberulent) externally, the hypanthium hemispheric, 1-1.5 mm. wide, pubescent within, the sepals triangular, scarcely 1 mm. long, erect to spreading; petals oval to narrowly obovate, 1.5-2 (2.5) mm. long, pink to rose; filaments pink, exceeding the petals; styles 1-1.5 mm. long; follicles 4-5, glabrous or ciliate along the suture, 2.5-3 mm. long.

Mountains of w. N.Am. at from 2000-11,000 ft. elevation, along streams and lakes or on wooded to open rocky slopes; B.C. southward, through the Cascade and Olympic mts. of Wash., to n.w. and Sierran Calif., e. to s.e. B.C., Mont., Ida., and e. Oreg. Late June-Aug.

Spiraea densiflora is often considered to be merely the red-flowered phase of S. *lucida* (= S. *betulifolia* var. *lucida*), but the two taxa have distinctive (although largely sympatric) ranges; they do not appear to intergrade and the pattern of variation in the pubescence of this species is not paralleled in the other.

1 Plant nearly or quite glabrous throughout except within the hypanthium, or the leaves and
    inflorescence bractlets merely ciliate; B.C. s. to Oreg., in both the Olympic and Cascade
    mts., e. in B.C. and n. Wash. to n.w. Mont. and n. Ida., largely transitional to the var.
    splendens in c. Ida.   var. densiflora
1 Plant finely puberulent with tiny, straight, erect hairs in the inflorescence (where most
    noticeable) and often on the lower surface of the leaves; the common phase in Calif., n.
    to s. Oreg., where transitional to var. *densiflora,* and to n.e. Oreg., c. Ida., and w.
    Mont., as far n. as Glacier Nat. Park   var. splendens (Baumann) C.L. Hitchc.

Spiraea douglasii Hook. Fl. Bor. Am. 1:172. 1833. (*Douglas*, "North-West coast of America, about the Columbia and the Straits of de Fuca")
    S. *menziesii* Hook. Fl. Bor. Am. 1:173. 1833. S. *douglasii* var. *menziesii* Presl, Epim. Bot. 195. 1852. S. *douglasii* f. *menziesii* Voss in Vilmorin, Blumengartn. 3rd ed. 1:246. 1894. (*Menzies*, "North-West coast of America")
    S. *cuneifolia* Raf. New Fl. N.Am. 3:67. 1836. ("Origon and New Albion")
    SPIRAEA DOUGLASII var. ROSEATA (Rydb.) C.L. Hitchc. hoc loc. S. *roseata* Rydb. N.Am. Fl. 22³:250. 1908, erroneously listed as S. *roseola* on p. 246. (*B. W. Everman 304*, Petitt Lake, Ida., Aug. 13, 1895)

S. subvillosa Rydb. N.Am. Fl. 22³:251. 1908. *(Thomas Howell*, Cascades of the Columbia, Oreg.,
in 1886) = var. *menziesii.*

S. idahoensis A. Nels. Bot. Gaz. 52:264. 1911. *(Macbride 630*, Trinity, Elmore Co., Ida., Aug.
23, 1910) = var. *roseata.*

Erect, freely branched shrub (0.5) 1-2 m. tall, the young growth usually semilanate with fine, soft,
crisped hairs, less commonly sparsely hairy to glabrous, the older bark brown, eventually glabrate;
leaf blades mainly oblong-elliptic or more nearly ovate-oblong or oblong-obovate, (2.5) 4-10 cm. long,
dark green and usually glabrous (to tomentulose) above, much paler beneath, glabrous to densely felty-
tomentose, rather remotely serrate mostly no more than half the length; inflorescence paniculate, (4)
6-20 (30) cm. long, usually oblong to conic and several times as long as broad, glabrous to (commonly) finely tomentose; calyx finely lanate to subglabrous externally, the triangular lobes about 1 mm.
long, reflexed, about equaling the hypanthium; petals pink to deep rose, orbicular-elliptic to obovate,
1.5-2 mm. long; filaments pink to rose; follicles shining, glabrous or sparsely ciliate along the suture, 2.5-3 mm. long. N=18.

Stream banks, swamps, bogs, lake margins, and damp meadows, from sea level to subalpine; s.
Alas. southward, along the coast and inland, to n. Calif., e. to s.e. B.C., n. and c. Ida., and n.e.
Oreg. June-Aug.

The plants vary from glabrous to tomentose on the leaves and inflorescence, but the following three
phases are rather well defined geographically as well as morphologically:

1 Leaves grayish-tomentose on the lower surface; inflorescence and calyces finely
tomentose; s. B.C. to n.w. Calif., from (mostly) the w. side of the Cascades to
the coast   var. douglasii
1 Leaves glabrous to pubescent but not grayish-tomentose beneath; inflorescence and
calyces tomentulose to glabrous
  2 Inflorescence and calyces tomentulose; leaves glabrous to pubescent beneath; s.
    Alas. to n.w. Oreg., through the Cascades, e. to n. Ida. and n.e. Oreg., freely
    transitional to var. *douglasii* especially in n. Oreg. and the Columbia R. Gorge,
    where one of the intermediate forms was designated *S. subvillosa* by Rydberg
                                                    var. menziesii (Hook.) Presl
  2 Inflorescence nearly or quite glabrous; leaves glabrous beneath; c. Ida., mainly
    in Valley, Elmore, Custer, and Boise cos., closely approached by occasional
    plants from the Columbia R. Gorge   var. roseata (Rydb.) C. L. Hitchc.

Spiraea pyramidata Greene, Pitt. 2:221. 1892.

S. menziesii ssp. *pyramidata* Piper in Piper & Beattie, Fl. S.E. Wash. 135. 1914. *(Greene,* Lower
Yakima River, near "Clealum," Wash., in 1889)

S. tomentulosa Rydb. N.Am. Fl. 22³:251. 1908. *(Suksdorf 5,* Falcon Valley, Klickitat Co., Wash.,
in 1881)

Rhizomatous, spreading to erect shrub mostly 5-10 (12) dm. tall, usually finely crisp-pubescent
above and in the inflorescence, the leaves from glabrous to (more commonly) slightly to moderately
crisp-puberulent at least beneath, ovate-lanceolate to oblong-elliptic or oblong-lanceolate, 2-7 (9) cm.
long, from subentire to coarsely once or twice serrate chiefly above the middle; panicle usually large,
from rounded to obconic, (2) 5-10 cm. broad, and 1-2 times as long; calyx usually sparsely hairy
without, the triangular lobes scarcely 1 mm. long, reflexed, about equaling the conic-hemispheric
hypanthium; petals white but with a distinct pinkish or lavender tinge (at least in the bud), (1.5) 2-2.5
mm. long; carpels glabrous to somewhat pubescent, 2.5-3 mm. long.

Valleys, often along water, to dry canyon slopes; e. side of the Cascade Mts., from s. B.C. to n.
Oreg., e. to c. Ida. June-Aug.

Spiraea pyramidata and *S. tomentulosa* seem not significantly different, although the type of the latter was considerably the more hairy. Both plants have been regarded as hybrids between *S. lucida* and
S. douglasii because of their suspiciously intermediate character. However there is no evidence of
backcrossing with either of the postulated parents. The plants appear to have normal flowers and
fruits, and in view of their abundance and natural range, to constitute a single self-perpetuating taxon,
maintainable at the specific level. Certain plants of c. Ida. (Custer, Elmore, and Valley cos.) are
subglabrous in the inflorescence but otherwise seem to belong here rather than with *S. lucida.*

Waldsteinia Willd.

Flowers complete, perigynous; calyx free of the pistils, ebracteolate (ours) or bracteolate, the hypanthium obconic, the sepals 5, triangular-lanceolate; petals white or yellow, deciduous; stamens numerous, the filaments slender, erect, borne with the petals at the outer margin of the disc-lined hypanthium, persistent in fruit; pistils 2-6, the ovaries very hairy, 1-ovulate, the style slender, usually equaling the stamens, soon deciduous; fruit an elongate achene with an erect seed; perennial, rhizomatous herbs with long-petiolate, 3 (5)-lobed or -foliolate leaves clustered in a basal tuft and naked or bracteate peduncles bearing (1) 2-several flowers in loose, bracteate, racemelike cymes.

Six species of the N. Temp. Zone; four in N. Am. (In honor of Count Franz Adam Waldstein-Wartenburg, 1759-1823, an Austrian botanist.)

Waldsteinia idahoensis Piper, Bull. Torrey Club 30:180. 1903. *(Piper,* Lochsa R. at the mouth of Lempke's Creek, Bitterroot Forest Reserve, Ida., July 31, 1902)

Strongly rhizomatous, sparsely hirsute perennial herb; leaves basal, the stipules membranous, adnate and forming a broad base to the slender, 5-12 cm. petioles, the blades cordate-suborbicular, shallowly 3- to 5-lobed and coarsely bicrenate-dentate, (2) 3-5 (6) cm. broad; peduncles with 1 or 2 ovate to lanceolate, simple to 3-lobed bracts (up to 1 cm. long) near the (1) 2- to 7-flowered, lanceolate-bracted, loose cymes, sparsely glandular-pubescent above; hypanthium narrowly obconic, 2-3 mm. long, the lobes spreading, triangular-lanceolate, about 4 mm. long; petals cream or yellowish, suborbicular, 4-5 mm. long; stamens about 70, equaling the petals; pistils 2-4 (possibly more), the ovary and achene canescent, the style puberulent near base, the stigma small.

In meadows along streams in w.c. Ida. June.

LEGUMINOSAE Pea Family

Flowers hypogynous to slightly perigynous, usually perfect, mostly in pedunculate racemes, spikes, or heads (or single); sepals usually five and at least partially connate; corolla (ours) characteristically "papilionaceous," consisting of one (usually the largest) upper petal (banner) and 2 lateral horizontal petals (wings) that usually are stuck to the two lower ventral, more or less connate petals (the keel) which enclose the stamens and pistil, sometimes the petals regular and 5, or reduced to 3 or only 1; stamens 10 and united into 1 group (monadelphous) or more commonly in two groups of 9 and 1 (diadelphous), but occasionally all distinct, less frequently reduced to 5, or (none of ours) numerous and distinct and then the corolla usually of 5 similar petals; pistil one, 1-carpellary, the ovary with parietal placentation and usually dehiscent on 2 sutures (legume) or less frequently indehiscent, sometimes breaking crosswise into 1-seeded segments (loment), very rarely spiny and burlike; seeds 1-several, endosperm usually lacking; annual to perennial herbs, shrubs, or trees, usually with alternate, stipulate, pinnately to palmately compound (rarely simple), often tendril-bearing leaves.

One of the three largest families of flowering plants, with between 500 and 600 genera; found in nearly all parts of the world and in widely varied habitats. Our members all have essentially papilionaceous flowers. More typically subtropical and tropical are the many legume-bearing members with nearly regular flowers and 5 to 10 stamens that are recognized either as the subfamily Caesalpinioideae or the family Caesalpiniaceae, and those with regular flowers of 5 (3-6) petals and few to numerous distinct stamens, also recognized at both levels, as the Mimosoideae or Mimosaceae.

The Leguminosae are a family of great importance, yielding many products used by man, including dyes, resins, valuable wood, oils, etc., and many food or forage plants, notably *Trifolium* (clover), *Medicago* (alfalfa), *Phaseolus* (bean), *Pisum* (pea), *Glycine* (soybean), and *Arachis* (peanut). Among the many ornamentals, *Lathyrus* (sweet pea), *Lupinus* (lupine), and *Cytisus* (scotch broom) are familiar examples. Three genera, *Lupinus, Astragalus,* and *Oxytropis,* contain members that are regarded as poisonous to livestock.

Alhegi camelorum Fisch., camelthorn, has become established in Grant Co., Wash. In the following key it will run to *Ulex,* from which it differs in its glabrosity, red flowers, and strongly constricted pods.

1 Plants introduced trees or shrubs with woody stems; branches either prickly or thorny or greenish and strongly grooved and the leaves reduced and inconspicuous
  2 Stipules modified into thorns; trees or large shrubs; branches neither spine-tipped nor greenish and prominently ridged; flowers white, in drooping racemes   ROBINIA
  2 Stipules not modified into thorns; shrubs; branches sometimes spine-tipped, generally greenish and often prominently grooved; flowers usually yellowish, not drooping
    3 Branches spine-tipped   ULEX
    3 Branches not spine-tipped   CYTISUS
1 Plants mostly native herbs or somewhat shrubby but with much of the stem dying back

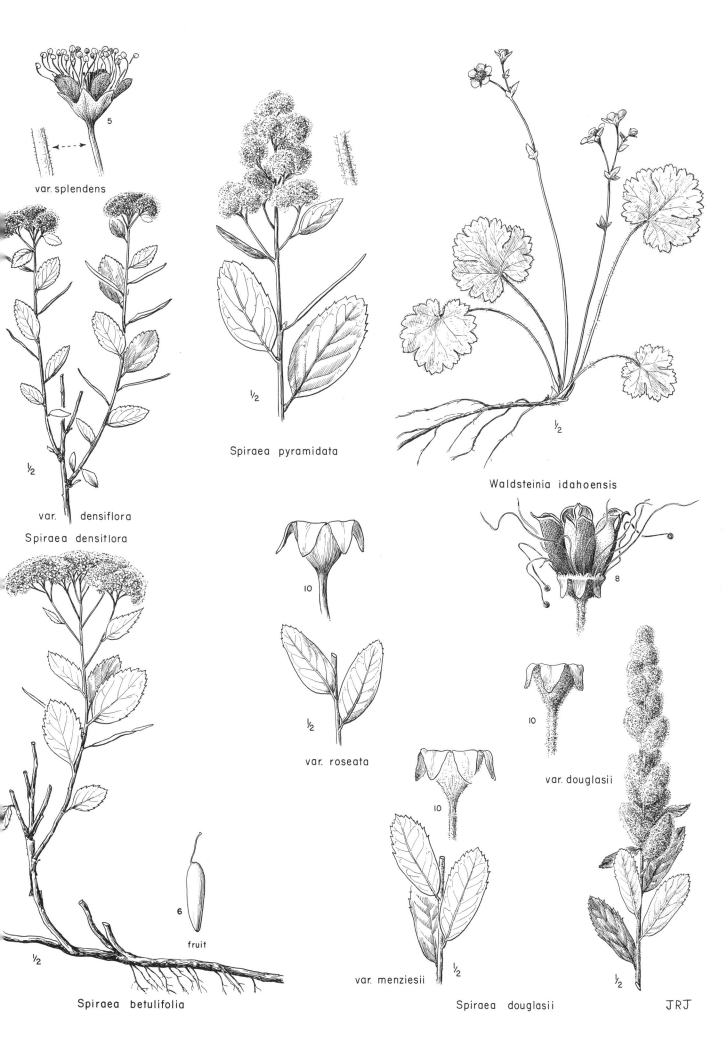

var. splendens

var. densiflora

Spiraea densiflora

5

½

Spiraea pyramidata

½

Waldsteinia idahoensis

½

10

var. roseata

½

8

10

var. douglasii

10

var. menziesii

½

Spiraea douglasii

½

Spiraea betulifolia

½

6

fruit

JRJ

each year, rarely at all prickly or spinose; branches seldom greenish and strongly
grooved; leaves usually conspicuous
4 Fertile stamens 5, alternate with 4 petaloid staminodia; flowers small, not truly
     papilionaceous, borne in dense spikes   PETALOSTEMON
4 Fertile stamens nearly always 10, if fewer then not alternate with colored stami-
     nodia; flowers papilionaceous
     5 Stamens distinct; leaves trifoliolate; flowers yellow, racemose; pod several times
          as long as the calyx   THERMOPSIS
     5 Stamens more or less connate; leaves various but if trifoliolate then the flowers
          other than yellow and racemose or the pods barely exceeding the calyx
          6 Herbage thickly dotted with small glands; leaflets either 3 (less commonly
               5) or the fruits covered with hooked spines
               7 Leaves pinnately many-foliolate; fruits cockleburlike, covered with hooked
                    spines   GLYCYRRHIZA
               7 Leaves 3 (5)-foliolate; fruits not spiny   PSORALEA
          6 Herbage seldom at all glandular-punctate but if so then the fruits neither bur-
               like nor the leaves trifoliolate
               8 Leaves (at least the upper ones) simple, the blades oblong, serrate; flow-
                    ers axillary, purplish or reddish; plants conspicuously pilose-villous   ONONIS
               8 Leaves usually all compound, if (as rarely) simple, then not serrate or
                    flowers other than red or plant not pilose-villous
                    9 Leaves trifoliolate, usually shallowly toothed; mature fruits either
                         spirally coiled or more or less completely enclosed within the calyx
                         10 Pod falcate to coiled, very heavily veined, sometimes prickly; flow-
                               ers either few and borne in axillary, pedunculate, small heads, or
                               more numerous and racemose, yellow or purplish   MEDICAGO
                         10 Pod neither falcate to coiled nor spiny; flowers mostly numerous in
                               heads or racemes, often other than yellow or purplish
                               11 Flowers in long narrow racemes, white or yellow; plants usual-
                                    ly 0.5-3 m. tall   MELILOTUS
                               11 Flowers in heads or short spikes, mostly other than yellow but
                                    if yellow then the plants usually decumbent or at least much less
                                    than 0.5 m. tall   TRIFOLIUM
                    9 Leaves seldom trifoliolate but if so then the leaflets entire or the pods elon-
                         gate beyond the calyx and not coiled
                         12 Fruits with short spiny teeth, 1- or 2-seeded, not much longer than
                               broad; flowers pinkish-lavender, very noticeably carmine- or purple-
                               lined; wings scarcely half as long as the keel; calyx lobes linear-lan-
                               ceolate, much longer than the tube; stipules membranous, brown   ONOBRYCHIS
                         12 Fruits usually not spiny but if prickly then several-seeded and much
                               longer than broad, or plants otherwise dissimilar to *Onobrychis*
                               13 Leaves palmately compound; leaflets (4) 5-17
                                    14 Flowers in elongate racemes; stamens monadelphous   LUPINUS
                                    14 Flowers in heads or headlike spikes or racemes; stamens
                                         diadelphous   TRIFOLIUM
                               13 Leaves not palmately compound
                                    15 Pods much constricted between the seeds, breaking cross-
                                         wise into 1-seeded segments; keel sometimes equaling or con-
                                         siderably exceeding the wings
                                         16 Flowers umbellate; petals all strongly clawed; loment lin-
                                               ear, 4-angled   CORONILLA
                                         16 Flowers racemose; petals not strongly clawed; loment flat-
                                               tened   HEDYSARUM
                                    15 Pods not conspicuously constricted between the seeds, dehis-
                                         cing lengthwise rather than breaking crosswise; keel usually
                                         shorter than the wings
                                         17 Leaves even-pinnate, the rachis prolonged as a slender
                                               bristle or a simple or branched tendril; plants often scandent

18 Style filiform, more or less equally hairy on all sides for about 1 mm. below
    the tip   VICIA
18 Style flattened, hairy only on one (the ventral) side   LATHYRUS
17 Leaves odd-pinnate or at least with a terminal leaflet, the rachis not prolonged as
    a bristle or tendril (although the terminal leaflet sometimes confluent with the
    rachis); plants not scandent
    19 Flowers axillary, solitary or in small pedunculate heads or umbels; free por-
        tion of the filaments (sometimes only every other one) dilated and usually
        broader than the anthers   LOTUS
    19 Flowers usually spicate or racemose; none of the filaments dilated
      20 Leaves trifoliolate; pods pubescent; flowers yellow   MEDICAGO
      20 Leaves pinnate; leaflets usually at least 5; pods and flowers various
        21 Calyx with 2 tiny, readily deciduous bracteoles at the base; style
            sparsely pubescent nearly the full length; flowers in loose axil-
            lary racemes; pedicels slender, 3-5 mm. long; plants mostly over
            5 dm. tall   SWAINSONA
        21 Calyx usually without bractlets but if bracteolate then the style gla-
            brous or flowers numerous in terminal or long axillary racemes, the
            pedicels often less than 3 mm. long; plant often less than 5 dm. tall
          22 Keel of the corolla abruptly narrowed to a beaklike point; plants
              usually without leafy stems   OXYTROPIS
          22 Keel of the corolla not abruptly beaked; plants mostly with leafy
              stems   ASTRAGALUS

Astragalus L. Locoweed; Milk vetch; Rattlepod

Flowers usually several (rarely single) in pedunculate, axillary racemes, papilionaceous, general-ly showy, white or yellowish to reddish or purple; calyx campanulate to tubular; banner usually well reflexed from the wings; wings generally short-auriculate at the base of the blade, mostly exceeding the blunt or rounded to acutish or barely beaked keel, and although generally shorter than the banner, sometimes equaling or exceeding it; stamens 10, diadelphous; pods sessile to conspicuously stipitate, globose to linear, straight to coiled, membranous to cartilaginous or fleshy and with a woody or bony texture when dried, often conspicuously inflated, dehiscent or indehiscent, from laterally compressed (the width then always greater than the thickness) to markedly obcompressed (the width then less than the thickness), one or both sutures (or perhaps more properly, the valves adjacent to the sutures) of-ten either very conspicuous or sulcate and the dorsal (lower) often inflexed and intruded to form a par-tial to complete longitudinal septum, the pod then often more or less 3-angled or triquetrous; annual to perennial herbs with odd-pinnate (imparipinnate) leaves, the leaflets usually jointed to the rachis, but sometimes at least the terminal one confluent with (and often similar to) it; stipules free or connate opposite the petioles; pubescence various, but usually appressed, the hairs mostly simple but some-times attached subterminally or near the middle and thus 2-branched (referred to as dolabriform or malpighiaceous).

Over 2000 species, circumboreal at high latitudes (except Greenl.), and highly developed in Medi-terranean, steppe, desert and cold montane climates of the Northern Hemisphere, most numerous in Eurasia, about 550 species in N.Am., 100 in S.Am. (principally Andean), and 1 in Natal. (An ancient Greek name for some leguminous plant, possibly from *astragalos*, ankle bone, in reference to the pod-or leaf-shape.)

Several species are known to be poisonous to livestock, especially horses, one of the symptoms of the poisoning being a crazed condition (hence the common name "locoweed"). Another disease pro-duced by some species of *Astragalus*, as well as by species of several other genera, is commonly re-ferred to as "alkali disease," and "blind staggers." In this instance the poisoning is due to the sele-nium content of the plant and therefore the disease is encountered only where selenium-bearing soils (usually derived from shales) occur.

Our members of the genus are seldom cultivated, but there are several species that have consid-erable merit, at least for e. of the Cascades, among them *A. purshii*, *A. kentrophyta*, and *A. gilvi-florus* for the rock garden, and almost any of the larger flowered or mottled-fruited species for more general use.

References:

Barneby, R. C. Pugillus Astragalorum I, Leafl. West. Bot. 3:97-114. 1944; II, Proc. Calif. Acad.
ser. 4, 25:147-170. 1944; III-VI, Leafl. West. Bot. 4:49-63, 65-147, 228-238. 1944-1946; 5:1-9.
1947; VII, Am. Midl. Nat. 37:421-516. 1947; VIII-IX, Leafl. West. Bot. 5:25-35, 82-89. 1947-
1948; X, Am. Midl. Nat. 41:496-502. 1949; XI, Leafl. West. Bot. 5:193-197. 1949; XII, El Aliso
2:203-215. 1950; XIII-XVII, Leafl. West. Bot. 6:89-101, 172-176; 7:31-37, 192-195; 8:14-23.
1951-1956; XVIII-XIX, Am. Midl. Nat. 55:477-503, 504-507. 1956; XX, Leafl. West Bot. 8:120-
125. 1957.

Jones, M. E. Revision of North American species of Astragalus. 330 pp. Salt Lake City. 1923.

Rydberg, P. A. Galegeae, subtribe Astragalanae. N. Am. Fl. 24[5-7]:251-462. 1929.

The genus *Astragalus* is one of the largest in the Leguminosae and one with great range in geographical and ecological distribution. It has been divided, largely on the basis of highly variable pod characters, into supraspecific taxa, variously called sections, subgenera, or genera (Rydberg recognized 28 such segregate genera).

There is a marked tendency toward speciation, many of our taxa being very local and only infrequently collected. The species themselves, unlike those of such genera as *Lupinus, Oxytropis,* and *Lathyrus,* are notably constant and comparatively well isolated from one another, genetically.

It is a pleasure to acknowledge my indebtedness, for help with *Astragalus,* to Mr. R. C. Barneby, whose knowledge of the genus is unexcelled. The preliminary study of the various taxa was based largely upon material annotated by him. He has been good enough to check the manuscript and to correct or add many details of morphology and range of the species as well as several descriptions of taxa omitted. For the sake of the record it should be noted that all new taxa proposed under Mr. Barneby's name were described by him; the Latin diagnoses are his. As can readily be seen, he has also proposed several new combinations. In order to adapt the treatment of *Astragalus* to the general taxonomic philosophy governing the preparation of this flora, his delineation of taxa has not been followed strictly in a very few cases. Of course Mr. Barneby is not to blame for such errors as may materialize in the process of this adaptation.

Although most of the more important differences between species are based upon fruit peculiarities, it often happens that material at hand does not include the pods; the first of the two keys that follow, based almost entirely upon floral and foliar characters, was supplied by Mr. Barneby in essentially its present form.

KEY USABLE ON FLOWERING MATERIAL*

1 Leaflets 5-11, linear-elliptic, all continuous with the rachis and mucronate or spinulose
at the tip; raceme 1- to 3-flowered, short-pedunculate or subsessile, the small flowers
(banner 4-8 mm. long) usually appearing axillary; pod less than 1 cm. long, 1-locular
  2 Pubescence of the herbage dolabriform; banner oblanceolate or narrowly obovate, whitish; ovules 2-4; plants of sandy deserts and badlands at low elevations   A. KENTROPHYTA
  2 Pubescence basifixed; banner broadly obovate-cuneate or suborbicular, commonly
pinkish- or bluish-purple; ovules 5-8; mostly high-montane, descending on talus to
middle elevations   A. TEGETARIUS
1 Leaflets not as above, if all continuous with the rachis, then the racemes several-flowered and long-pedunculate and the flowers larger
  3 Plants acaulescent and tufted, with 3-foliolate leaves, long narrow flowers that are
sessile among the leaves, and very large, hyaline stipules; banner 16-28 mm. long;
pubescence dolabriform   A. GILVIFLORUS
  3 Plants caulescent or if acaulescent then either the leaves pinnate or the flowers
racemose and the banner less than 16 mm. long
    4 Leaves all reduced to simple oblanceolate blades; acaulescent plants with dolabriform pubescence and small-flowered, pedunculate racemes   A. SPATULATUS
    4 Leaves pinnate or (if some reduced to a simple rachis or phyllode) the plant
caulescent and the pubescence basifixed
      5 Pubescence of the herbage dolabriform   GROUP I (page 199)
      5 Pubescence basifixed
        6 Terminal leaflet, at least of some of the upper leaves (and sometimes of all),
confluent with the rachis, or the leaves (or some of them) reduced to a

---

*For key usable on plants with fruits see page 207

naked rachis or to a simple grasslike blade (phyllode)   GROUP II (below)
6 All leaflets jointed to the rachis, the leaves imparipinnate
  7 Stipules at the lowest nodes (and sometimes at all) fully amplexicaul and connate
    opposite the petiole, or (if the lowest nodes bladeless) the stipules united around
    the stem as a low collar or sheath
    8 Banner (measured along the curvature of the midvein) over 15 mm. long GROUP III (page 200)
    8 Banner not over 15 mm. long
      9 Banner 10-15 mm. long   GROUP IV (page 200)
      9 Banner less than 10 mm. long   GROUP V (page 201)
  7 Stipules petiolar or petiolar-cauline, variably decurrent-amplexicaul but not united
    opposite the petiole
    10 Plant weedy, rhizomatous; stems fistulose, 3-7 dm. long; keel longer and broad-
      er than the wings; pods pendulous, stipitate, 2-celled   A. CHINENSIS
    10 Plant seldom either weedy or rhizomatous; stems mostly not fistulose; flowers
      and pods various
      11 Banner over 15 mm. long   GROUP VI (page 202)
      11 Banner not over 15 mm. long
        12 Banner 10-15 mm. long   GROUP VII (page 205)
        12 Banner less than 10 mm. long   GROUP VIII (page 206)

GROUP I
Plants with dolabriform pubescence

1 Stipules, at least those at the lower nodes (and sometimes at all), connate opposite the
  petiole
  2 Banner 13-19.5 mm. long; keel obtuse, 9-15 mm. long; pod bilocular
    3 Flowers declined and retrorsely imbricated; stems arising singly or few together
      from oblique or creeping rhizomes   A. CANADENSIS
    3 Flowers erect or ascending at a narrow angle; stems arising together from a
      superficial root crown or shortly forking caudex   A. STRIATUS
  2 Banner 6-13 mm. long; keel 6-9.5 mm. long, triangular and acutish at the tip   A. MISER
1 Stipules variably decurrent or amplexicaul but not connate opposite the petiole
  4 Plants acaulescent; wing petals deeply cleft or toothed at apex; pod bilocular   A. CALYCOSUS
  4 Plants caulescent or, if acaulescent, then the wings entire and the pod 1-loc-
    ular
    5 Banner 12-22 mm. long; keel 9-18 mm. long
      6 Flowers white (except for the purplish-tipped keel), drying ochroleucous,
        nodding at full anthesis; racemes 10- to 30-flowered; pod bilocular; s.w.
        Mont. and Ida.   A. TERMINALIS
      6 Flowers purple, ascending; racemes 5- to 15-flowered; pod 1-locular; e. of
        the Missouri and Madison rivers, Mont.   A. MISSOURIENSIS
    5 Banner not over 11 mm. long; keel 4-8 mm. long
      7 Plants low, tufted, silvery-cinereous, subacaulescent (in our range), the pe-
        duncles subscapose and often very short and the pods subradical and 1-
        locular   A. LOTIFLORUS
      7 Plants tall, strongly caulescent, green (introduced in s.e. Wash.); pods rac-
        emose on axillary peduncles, reflexed, bilocular   A. FALCATUS

GROUP II
Pubescence basifixed; terminal leaflet, at least of some of the upper leaves
and often of all, confluent with the rachis, or the leaves (or some of them)
reduced to a naked rachis or a simple grasslike blade (phyllode)

1 Stipules free   A. ATRATUS
1 Stipules connate at the lower nodes, at least
  2 Banner 15-24 mm. long; keel 11-16 mm. long; plants coarse, malodorous (selenium-
    scented); pods woody
    3 Petals white; leaves all pinnate, the linear, rather stiff leaflets 2-6 cm. long; e.
      of the continental divide in Mont.   A. PECTINATUS
    3 Petals pink-purple; upper leaves reduced to the rachis, the leaflets when present
      only 3-30 mm. long; Snake R. plains   A. TOANUS

2 Banner 4-14 mm. long (up to 16.5 mm. long in *A. cusickii* but the pod then bladdery
   and the range otherwise); keel 4-11 mm. long; plants scentless; pods papery
    4 Stems arising singly or few together from creeping rhizomes or from subterranean
      caudex branches; pod bladdery-inflated, 1-locular
      5 Stems from true rhizomes, these not marked by stipular sheaths; calyx teeth less
        than 1 mm. long; s.w. Ida., near (but not known from within) our range   *A. sterilis* Barneby
      5 Stems from buried caudex branches, these beset at intervals with stipular sheaths;
        calyx teeth at least 1 mm. long; s.e. Ida.   A. CERAMICUS
    4 Stems arising together from the root crown or from an aerial caudex; if plants at all
      rhizomatous then the pod not bladdery-inflated and 1-locular
      6 Ovary and the immature pod stipitate
        7 Lateral leaflets usually 9-12 pairs in most leaves; banner 6-8 mm. long; keel
          about 5 mm. long; pod triquetrous, bilocular   A. MULFORDIAE
        7 Lateral leaflets generally 4-9 pairs; banner 10-15.5 mm. long; keel 7-9.5 mm.
          long; pod 1-locular
          8 Pod bladdery-inflated; plants mostly of the Snake and Salmon R. canyons
            along the Oreg.-Ida. boundary and adj. s.e. Wash.   A. CUSICKII
          8 Pod linear-oblong, not inflated, 2-sided; plants widespread, but not within
            the canyons of the Snake and Salmon rivers   A. FILIPES
      6 Ovary and the immature pod quite sessile
        9 Plants tufted, densely leafy, with a suffruticulose caudex beset by rigid persistent
          petioles; pod less than 1 cm. long, 2-locular; Cascade foothills   A. PECKII
        9 Plants slender, caulescent, dying back annually to the root crown and without a
          caudex; leaves flaccid; pod at least 1 cm. long, 1-locular; far e. of the Cascades
          10 Pod narrowly oblong, 3-4 mm. wide; leaflets (or most of them) flat, thin, and
            grasslike; stems slender, flexuous, prostrate; plants of moist soil
                                  A. DIVERSIFOLIUS
          10 Pod linear to linear-oblanceolate, about 2 mm. wide; leaflets involute and
            very narrow; stems rigid or wiry, rushlike, ascending or erect; plants of
            dry soil   A. CONVALLARIUS

GROUP III

Pubescence basifixed; stipules connate; leaves imparipinnate; flowers large,
the banner over 15 mm. long

1 Stems and herbage villous-hirsute with horizontally spreading hairs up to 1-2 mm. long;
  pod pendulous, stipitate, glabrous, 2-locular   A. DRUMMONDII
1 Stems and herbage pilose or strigillose, the hairs either appressed or, if loose, then less
  than 1 mm. long; pod either erect or 1-locular
  2 Flowers erect, crowded into ovoid heads; stems arising from a buried root crown; pod
    small, hirsute   A. DASYGLOTTIS
  2 Flowers variously oriented, but if erect then loosely racemose; stems from a super-
    ficial root crown or caudex; pod glabrous or strigillose
    3 Calyx gibbous-saccate behind the pedicel; pod pendulous, stipitate, papery; eastern,
      extending w. to Fremont Co., Ida.   A. BISULCATUS
    3 Calyx somewhat oblique at base but not gibbous-saccate; pod erect, fleshy, be-
      coming leathery; Wash., Oreg., and extreme s.w. Ida.
      4 Pod sessile or subsessile, the stipe, if any, less than 2 mm. long   A. REVENTUS
      4 Pod stipitate, the stipe at least 3 mm. long
        5 Stipe stout, 3-5 (8) mm. long; pod erect, the body grooved dorsally, partly or
          wholly bilocular; local in e. Wash.   A. LEIBERGII
        5 Stipe slender, (6) 9-16 mm. long; pod pendulous, the body compressed, 1-
          locular; widespread
                     A. FILIPES

GROUP IV

Pubescence basifixed; stipules connate; leaves imparipinnate; flowers of
medium size, the banner 10-15 mm. long

1 Plants of the Olympic Mts.   A. COTTONII
1 Plants of the e. slope of the Cascades and eastward
  2 Wings deeply bidentate at the apex; pod stipitate, laterally compressed   A. ABORIGINUM
  2 Wings entire or at most shallowly truncate-emarginate

3 Flowers erect, subsessile, closely crowded in ovoid heads; caudex subterranean, with
    few to numerous partially buried branches; calyx tube 4-7 mm. long; pods hirsute
                                                                 A. DASYGLOTTIS
3 Flowers variously oriented, but if erect then loosely racemose or plant otherwise not as
    in *A. dasyglottis*
    4 Stems arising singly or few together from slender, widely creeping and adventitiously
        rooting, subterranean caudex branches; plants of cool or moist soils in the mts.
        5 Keel broad, purple tipped, equaling the purple-margined banner, longer and wider
            than the narrow, white wings; pod usually black-pilose   A. ALPINUS
        5 Keel narrow, lilac tipped, shorter than the white wings and banner; pod minutely
            strigillose   A. LEPTALEUS
    4 Stems arising together from the root crown or caudex; if the root crown subterra-
      nean the caudex branches not rooting and the plants not montane
        6 Calyx tube gibbous-saccate at base behind the pedicel; pod pendulous, stipitate,
            2-grooved ventrally   A. BISULCATUS
        6 Calyx tube more or less oblique, but not gibbous, at base
          7 Root crown subterranean; plants of the plains, extending w. in Mont. to the Mis-
            souri and Yellowstone rivers   A. FLEXUOSUS
          7 Root crown superficial or, if somewhat buried, then plants from far w. of the
            Missouri River
            8 Immature pod erect and fleshy, often stipitate
              9 Pods plainly stipitate; calyx tube 3-5.5 mm. long   A. LEIBERGII
              9 Pods sessile or subsessile; calyx tube usually over 5.5 mm. long   A. REVENTUS
            8 Immature pod pendulous, thin textured
              10 Ovary and pod pubescent with (at least some) black hairs
                11 Leaflets 11-19, mostly elliptic and acute; pod laterally compressed, the
                   sutures both prominent; ovules 2-6; plants of the Rocky Mts. (w. only
                   into Ida.)   A. BOURGOVII
                11 Leaflets usually 9-13 (to 19), oval or oblong and obtuse, and ovules
                   7-9 (at least on those plants within the range of *A. bourgovii);* pod
                   flattened dorsally, obtusely triquetrous   A. ROBBINSII
             10 Ovary and pod glabrous or white-strigillose
                12 Ovary and pod stipitate; stems all leafless at base
                  13 Pod linear-oblong, laterally flattened   A. FILIPES
                  13 Pod obovoid or half-ellipsoid, greatly inflated, bladdery
                    14 Stems mostly less than 3 dm. long; pod symmetrically obovoid,
                     balloon shaped, broadest just below the beakless apex; leaflets
                     mostly broader than linear   A. WHITNEYI
                  14 Stems 3-7 dm. long; pod variable in outline, ovoid to half-ellip-
                     soid, but contracted at the apex into a definite, deltoid beak   A. CUSICKII
            12 Ovary and pod sessile or nearly so; some short, sterile, leafy stems
              present, forming a basal leafy tuft   A. MISER

GROUP V

Pubescence basifixed; stipules connate; leaves imparipinnate; flower small,
the banner not over 10 mm. long

1 Keel petals 2.5-6 mm. long
    2 Herbage gray-villosulous; stems erect or ascending, tomentose at base; pod 2-locular
                                                  A. LYALLII
    2 Herbage variably pubescent, but if villosulous the stems prostrate and matted and the
      pod 1-locular or nearly so
      3 Ovules 2 (rarely 3); extremely slender, prostrate plants of e.c. Oreg., the leaves
        with 5-11 small obovate-cuneate to oblanceolate leaflets   A. TEGETARIOIDES
      3 Ovules 4-11; plants mostly from e. of Oreg.
        4 Stems arising from a buried root crown, subterranean for a space of 1-7 cm.; pod
          small, fleshy but becoming leathery and transversely rugulose; leaflets retuse; a
         plains species extending w. in Mont. to the Missouri and Yellowstone rivers
                                               A. GRACILIS
        4 Stems from a superficial root crown or caudex; pod papery or papery-membranous,

not rugulose; plants usually more western but if from e. of the Missouri R. then the
leaflets only exceptionally retuse
  5 Immature pod nearly always bearing some black or fuscous hairs   A. EUCOSMUS
  5 Immature pod strigillose with white hairs exclusively, or glabrous
     6 Pod inflated, obovoid or subglobose, sessile; plants from w. of the continental
         divide, from upper Deer Lodge Valley, Mont., n.w. to e. Wash.; stipules
         mostly turning black on drying   A. MICROCYSTIS
     6 Pod laterally compressed and oblong or lentiform in profile, not inflated, some-
         times stipitate; plants either from e. and s. of the range given or with stipules
         that do not blacken on drying
        7 Stems mostly erect; flowers white; pod stipitate; stipules blackening   A. TENELLUS
        7 Stems diffuse or prostrate, often matted, if erect the petals purplish; pod ses-
           sile; stipules not blackening   A. VEXILLIFLEXUS
1 Keel petals over 6 mm. long
  8 Wings deeply bidentate at apex   A. ABORIGINUM
  8 Wings entire or obscurely truncate-emarginate
     9 Stems arising singly or few together from slender, widely creeping and adventitious-
        ly rooting, subterranean caudex branches; plants of moist or cool soils in the mts.
      10 Petals irregularly graduated, the broad purple-tipped keel about equaling the
          purple-margined banner and both longer and wider than the narrow, white wings;
          pod usually black-pilose   A. ALPINUS
      10 Petals regularly graduated, the lilac-tipped keel shorter than the white wings and
          banner; pod minutely strigillose   A. LEPTALEUS
     9 Stems arising together from the root crown or caudex; if the root crown subterra-
        nean then the plants of dry foothills and prairies or alpine scree and ridges
      11 Keel about as long as the banner, its lower edge abruptly incurved distally through
         a right angle, appearing broadly truncate at the apex; pod sessile, bladdery-
         inflated, 2-locular
        12 Stems 1-4 dm. long, diffuse or trailing, freely branched, the axillary peduncles
          numerous; root crown subterranean   A. AMBLYTROPIS
        12 Stems less than 1 dm. long, often almost lacking, tufted on the superficial
          root crown or shortly forking caudex, the few peduncles subscapose
                  A. PLATYTROPIS
      11 Keel much shorter than the banner or, if nearly as long, then triangular and
        acutish at the apex; pod 1-locular or nearly so, either not inflated or if inflated
        then stipitate (at least within the calyx)
        13 Ovary and pod pubescent with (at least some) black hairs
          14 Leaflets 11-19, mostly elliptic and acute; pod laterally compressed, the
            sutures both prominent; ovules 2-6; plants of the Rocky Mts. (w. only
            into Ida.)   A. BOURGOVII
          14 Leaflets (of plants within the range of *A. bourgovii*) 9-13, oval or ob-
            long, obtuse; ovules 7-9 (but leaflets up to 17 and the ovules only 3-5 in
            var. *alpiniformis* of the Wallowa Mts., Oreg.); pod flattened dorsally, ob-
            tusely triquetrous   A. ROBBINSII
        13 Ovary and pod glabrous or white-strigillose
          15 Keel tip bluntly rounded at the apex; pod linear-oblanceolate, terete or a
            trifle dorsiventrally compressed; a plains species extending w., in Mont.,
            to the Missouri and Yellowstone rivers   A. FLEXUOSUS
          15 Keel tip triangular, acute or acutish; pod either greatly inflated or lat-
            erally flattened; mostly montane in the Rocky Mts. and westward
            16 Stems erect, 3-9 dm. tall; pods stipitate, strongly compressed   A. FILIPES
            16 Stems often prostrate, 1-5 dm. long; pods not at once stipitate and
              strongly compressed
              17 Pod greatly inflated, balloon-shaped, stipitate   A. WHITNEYI
              17 Pod linear-oblanceolate, compressed, nearly or quite sessile   A. MISER

GROUP VI

Pubescence basifixed; leaves imparipinnate; stipules free; flowers large,
the banner over 15 mm. long

1 Stems arising singly or few together from widely creeping rhizomes; flowers purple; pod
   coiled in a ring; just s. of our range in s.w. Ida.   *A. camptopus* Barneby
1 Stems arising together from a determinate root crown or caudex
   2 Wing petals 1-4 mm. longer than the banner, dilated upward and emarginate at or be-
      low the apex; plant hirsute; pod glabrous, 2-locular   A. SUCCUMBENS
   2 Wing petals subequal to or shorter than the banner; plant and pod various
      3 Ovary and pod glabrous
         4 Stems and leaves hirsute, the longest hairs 1-2 mm. long   A. DRUMMONDII
         4 Stems and leaves mostly strigillose or glabrous, if loosely pubescent then the
            hairs less than 1 mm. long
            5 Ovary and pod stipitate (in *A. beckwithii* elevated on a stipelike gynophore)
               6 Calyx glabrous except for a few hairs on the teeth, the teeth (3.5) 4-7 mm.
                  long   A. BECKWITHII
               6 Calyx pubescent on the tube (even though thinly so), if glabrescent the teeth
                  much less than 3.5 mm. long
                  7 Calyx gibbous-saccate dorsally behind the pedicel; pod 1-locular
                     8 Flowers ascending; pod erect or ascending   A. TWEEDYI
                     8 Flowers nodding or declined; pod pendulous
                        9 Herbage green, nearly glabrous; reticulations of the ripe pod sunken
                           below the general surface   A. COLLINUS
                        9 Herbage gray-pubescent; reticulations of the pod prominent, ele-
                           vated above the general surface   A. CURVICARPUS
                  7 Calyx more or less oblique at base but not gibbous-saccate; pod various
                     10 Stems diffuse or decumbent; peduncles 2-6 cm. long; pod 10-15 mm.
                        thick; local in the Snake R. Canyon below Weiser, Ida.   A. VALLARIS
                     10 Stems erect or ascending; peduncles mostly over 8 cm. long (some-
                        times shorter in *A. eremiticus)*; pod mostly not over 8 (rarely to
                        10) mm. thick
                        11 Stipe of the pod 6-15 mm. long, widely incurved-ascending, the
                           body erect but distant from the axis of the raceme
                           12 Body of the pod 6-10 mm. thick, truncate at base, sulcate along
                              both sutures; along both slopes of the Bitterroot Mts., in Lem-
                              hi Co., Ida., and Beaverhead Co., Mont.   A. SCAPHOIDES
                           12 Body of the pod 3.5-7 mm. thick, cuneately tapering into the
                              stipe, sulcate only dorsally; s.w. Ida. and southward A. EREMITICUS
                        11 Stipe of the pod 2.5-6 mm. long, straight, erect in the same plane
                           as the body, the pod subappressed to the axis of the raceme
                                                                                A. STENOPHYLLUS
            5 Ovary and pod sessile
               13 Peduncles 1-2 dm. long, erect; stems erect or incurved-ascending; pod
                  1-locular
                  14 Stems 2-4.5 dm. long, exceeding the longest (usually the lowest) pe-
                     duncle and raceme together; s.w. Ida.   A. ADANUS
                  14 Stems less than 2 dm. long, shorter than the longest inflorescence; Blue
                     Mts. of Wash. and Oreg.   A. REVENTUS
               13 Peduncles 1-7 cm. long, mostly humistrate or weakly incurved-ascending;
                  pod 2-locular
                  15 Pod subglobose, merely cuspidate at the apex, not beaked, the valves
                     thickly fleshy, at least 2 mm. thick in section; Mont. eastward
                                                                                A. CRASSICARPUS
                  15 Pod subglobose to lance-ellipsoid, but always terminating in a well-de-
                     fined, laterally flattened, triangular or lance-acuminate, usually in-
                     curved beak; pod leathery or thinly fleshy, the valves much less than
                     1 mm. thick in section   A. LENTIGINOSUS
      3 Ovary and pod pubescent
         16 Calyx gibbous-saccate dorsally behind the pedicel
            17 Pod usually straight or nearly so; reticulations of the valves immersed, sunken
               below the general surface; n.e. Oreg. to w.c. Ida. and northward   A. COLLINUS

17 Pod lunately to hamately incurved; reticulations of the valves prominent, elevated
   above the general surface; upper Deschutes Valley, Oreg., s. and s.e. to Ida.
                                                                        A. CURVICARPUS
16 Calyx not gibbous, though often oblique at base
   18 Racemes not over 10-flowered; leaflets uniformly and densely silvery-strigose or
      softly villous-tomentose on both sides
      19 Plants strictly acaulescent, the leaflets silvery with straight, subappressed hairs;
         pod densely hirsute and tomentose; entering our range from the s. in Crook Co.,
         Oreg.   A. NEWBERRYI
      19 Plants caulescent or if nearly stemless then either the leaflets villous-tomentose
         with spreading or curly and tangled hairs, or the pod strigillose-villosulous
         (and then not w. of Ida.)
         20 Herbage softly villous with extremely fine, curly or sinuous, entangled hairs;
            pod hirsute, or both hirsute and tomentose
            21 Pod thinly hirsute, the vesture not concealing the surface of the valves;
               w. end of Snake R. plains and vicinity   A. NUDISILIQUUS
            21 Pod densely tomentose and hirsute, the vesture concealing the surface of
               the valves; widespread   A. PURSHII
         20 Herbage pubescent with straight or merely incumbent hairs; pod strigillose
                                                                        A. ARGOPHYLLUS
   18 Racemes usually over 10-flowered, or if rarely less than 10-flowered, then the
      herbage green and the leaflets glabrous or medially glabrescent above
      22 Herbage and pod either hirsute or both softly hirsute and tomentose, some of
         the hairs at least 1.3 mm. long
         23 Flowers ascending; pod 1-locular, both hirsute and tomentose; n.e. Oreg.
            northward and eastward   A. INFLEXUS
         23 Flowers declined; pod 2-locular, only hirsute; just reaching our s. border
            in Ida. and Oreg.   A. MALACUS
      22 Herbage and pod strigillose or shortly villosulous, but the hairs all less than 1
         mm. long
         24 Flowers ascending at full anthesis (sometimes declined after fertilization)
            25 Banner bright pink-purple; pod both deflexed and sessile; local (in our range)
               near the Big Bend of the Columbia R.   A. CASEI
            25 Banner usually whitish, ochroleucous, or tinged with lavender, but if at
               all purplish the pod either ascending or stipitate
               26 Pod long-stipitate, laterally compressed, pendulous; at low elevations
                  in the Columbia Basin, extreme n. Oreg. northward
                  27 Pubescence of the calyx and leaflets mostly straight and appressed;
                     leaflets linear to linear-oblong; body of the pod falcately or lunately
                     incurved, 6-9 mm. thick; ovules 30-36   A. SCLEROCARPUS
                  27 Pubescence of the calyx and leaflets incumbent or curly; leaflets
                     narrowly oblong to obovate-cuneate; body of the pod less than 6
                     mm. thick; ovules 20-30
                     28 Pod coiled into a ring   A. SPEIROCARPUS
                     28 Pod merely falcate   A. SINUATUS
               26 Pod sessile or subsessile, the stipe, if any, not over 1.5 mm. long
                  29 Pod fully bilocular, quite sessile; leaflets glabrous beneath (oc-
                     casionally with a few hairs along the midrib)   A. LENTIGINOSUS
                  29 Pod 1-locular or nearly so, obscurely stipitate; leaflets pubescent
                     beneath   A. CIBARIUS
         24 Flowers nodding or declined at full anthesis
            30 Pod erect; n.e. Oreg. and eastward
               31 Leaflets 19-27; calyx teeth 1-2 mm. long; stipe of pod 3-6 mm. long
                                                                        A. STENOPHYLLUS
               31 Leaflets 23-39; calyx teeth 2-6 mm. long; stipe of the pod not over
                  1.5 mm. long   A. REVENTUS
            30 Pod pendulous; n.c. Oreg.   A. HOWELLII

GROUP VII

Pubescence basifixed; leaves imparipinnate; stipules free; flowers medium
sized, the banner 10-15 mm. long

1 Stems arising from widely creeping rhizomes; flowers purple; pod coiled into a ring; s.w.
  Ida., just s. of our range   *A. camptopus* Barneby
1 Stems arising together from a determinate root crown or caudex
  2 Plants of the Oregon Coast Range; flowers white; stems nearly glabrous; pod glabrous,
    triquetrous-compressed, bilocular   A. UMBRATICUS
  2 Plants of the interior, e. of the Cascade crest
    3 Cauline stipules very large, foliaceous, becoming papery, several-nerved, de-
      flexed; leaflets 9-15, ample and thin-textured, 1.5-6 cm. long; pod pendulous,
      stipitate, bladdery, membranous; plants of stream banks and mountain woods,
      mesophytic   A. AMERICANUS
    3 Cauline stipules not foliaceous; leaflets mostly less than 1.5 cm. long; pod sessile
      if inflated; plants xerophytic
      4 Leaflets 7-13, broadly ovate-oblong to oblong-elliptic, emarginate to retuse, thin
        in texture, greenish, 3-18 mm. long; calyx 4-6 mm. long, the teeth about equal-
        ing the tube; banner and wings about equal, the keel only 0.5-1 mm. shorter; pods
        sessile   A. AMNIS-AMISSI
      4 Leaflets more numerous, or plants otherwise not as above
        5 Banner (usually because of its erectness) shorter than the wings (sometimes only
          slightly so)
          6 Calyx gibbous-saccate behind the pedicel; flowers nodding; pod pendulous, stip-
            itate, 1-locular
            7 Plants of n.e. Oreg. (Morrow to Wallowa Co.), w.c. Ida., and n. to B.C.;
              reticulation of the pod sunken below the general surface   A. COLLINUS
            7 Plants of n.c. Oreg. (Gilliam to Wasco Co.), s. and s.e. to the Snake
              River plains; reticulation of the pod prominent, raised above the gen-
              eral surface   A. CURVICARPUS
          6 Calyx somewhat oblique at the base, but not gibbous
            8 Leaflets 21-33; petals ochroleucous; pods erect; s.e. Wash. and adj. Ida.
             9 Ovary and pod sessile, glabrous   A. RIPARIUS
             9 Ovary and pod stipitate, pubescent   A. PALOUSENSIS
            8 Leaflets 9-19; petals whitish and more or less purple tinged; pods (as-
             cending) declined or deflexed; Blue Mts., Oreg., and southeastward
             10 Racemes 2- to 8-flowered; pod 4.5-6.5 mm. in thickness, semi-
              bilocular; e. Oreg. and the s.w. corner of Ida.   A. SALMONIS
             10 Racemes 6- to 15-flowered; pod 3-4.2 mm. in thickness, 1-locular;
              Camas Co., Ida.   A. ATRATUS
        5 Banner, as measured along the curved dorsal axis, longer than the wings (some-
          times only slightly so)
          11 Plants densely villous-tomentose with fine entangled hairs; ovary and pod silky-
            hirsute; flowers pink or purple; stems very short, the plants tufted or matted;
            around the s. edge of our area   A. PURSHII
         11 Plants mostly strigillose to nearly glabrous, but if villous-tomentose then of
            more northern range and the flowers ochroleucous or whitish and merely
            tinged with lavender, the stems well developed
            12 Calyx villous; flowers subsessile in heads or oblong spikes; ovules 4-10
                                        A. SPALDINGII
            12 Calyx strigillose; flowers loosely racemose; ovules more numerous
             13 Banner and keel of nearly equal length, the latter narrowly triangular
              and acute at the apex; pod erect, sessile, bilocular   A. OBSCURUS
             13 Banner evidently (often greatly) exceeding the keel, the latter round-
              ed and obtuse at the apex
              14 Ovary and pod quite sessile
                15 Petals pink-purple; leaflets linear, distant   A. CASEI
                15 Petals whitish except for the keel tip (lilac in *A. iodanthus*

which occurs near our s. limit in s. w. Ida.); leaflets broader than linear, mostly
oval or obovate
  16 Pod 1-locular, bladdery-inflated; local along the Salmon R. in Custer and Lemhi
     cos., Ida.   A. WOOTONII
  16 Pod either bilocular (if inflated) or 1-locular and not inflated
    17 Pod 2-locular; banner white or cream-colored   A. LENTIGINOSUS
    17 Pod 1-locular; banner lilac; probably not quite reaching our area, in Ida.
                                                  A. iodanthus Wats.
14 Ovary and pod more or less stipitate
  18 Pod erect or incurved to spreading
    19 Stipe 6-15 mm. long, widely incurved-ascending, the body of the pod erect but
      distant from the axis of the raceme   A. EREMITICUS
    19 Stipe 2.5-6 mm. long, straight, erect in the same plane as the body, the pod sub-
      contiguous to the axis of the raceme   A. STENOPHYLLUS
  18 Pod pendulous
    20 Flowers nodding; pod more or less triquetrous, bilocular
      21 Leaflets glabrous above; body of the pod 3-4.5 cm. long, about 10 times longer
        than wide   A. ARTHURI
      21 Leaflets pubescent above; body of the pod about 2-3 cm. long, 5-7 times longer
        than wide   A. HOWELLII
    20 Flowers ascending; pod laterally compressed, bicarinate by the sutures, 1-locular
      22 Leaflets oblong to cuneate-obcordate, loosely pubescent with incumbent hairs;
        body of the pod 3.5-5.5 mm. wide, coiled into a ring or contorted through
        1-2.5 turns; ovules 20-30   A. SPEIROCARPUS
      22 Leaflets linear to linear-oblong, strigillose; body of the pod 6-9 mm. wide,
        lunate or falcate; ovules 30-36   A. SCLEROCARPUS

GROUP VIII

Pubescence basifixed; leaves imparipinnate; stipules free; flowers small,
the banner not over 10 mm. long

1 Plants annuals or biennials with a very slender taproot; leaflets 3-13
  2 Leaflets linear-oblong, the terminal one longer than the uppermost pair; pod lunately
    half ellipsoid, inflated; Snake R. plains   A. GEYERI
  2 Leaflets oval or obovate, the terminal one no longer than the rest; pod various; John
    Day and Columbia rivers
    3 Pod linear-oblong to obliquely elliptic in profile, slightly to strongly incurved, 3-6
      mm. wide, only slightly if at all inflated, more or less compressed, partially to
      fully 2-locular; ovules 6-14   A. DIAPHANUS
    3 Pod half-ovoid, distinctly inflated and somewhat bladdery, 1-locular; ovules about
      22   A. DIURNUS
1 Plants either (mostly) perennials, or if flowering the first year or characteristically biennial, then the leaflets more than 13
  4 Peduncles paired in at least some of the upper axils   A. LEMMONII
  4 Peduncles solitary
    5 Leaflets 5-9, linear, remote; racemes very loosely and remotely 7- to 30-flowered;
      pod deflexed, long-stipitate, laterally compressed, 1-locular; near our s. border
      in s. e. Oreg.   *A. solitarius* Peck
    5 Leaflets either more numerous or broader and closely set
      6 Leaflets 5-7 (or in var. *lagopinus* as many as 11), densely pannose-tomentose;
        pod thickly hirsute-tomentose, 1-celled   A. PURSHII
      6 Leaflets more than 7; if plant tomentose then the ovary nearly or quite 2-celled
        7 Herbage densely gray-villous or -tomentose; racemes 10- to 40-flowered; pod
          4-8 mm. long, 2-locular
          8 Calyx 5-8 mm. long; banner at least 7 mm. long; racemes densely spicate in
            flower, interrupted only in fruit; pods spreading or ascending   A. SPALDINGII
         8 Calyx usually less than 5 (rarely to 5.5) mm. long; banner rarely over 6.5 (to
            7.5) mm. long; racemes loosely flowered from the first; pods declined   A. LYALLII
        7 Herbage strigillose or thinly villosulous, if the pubescence loose then the

racemes only 3- to 12-flowered or the pod either over 8 mm. long or 1-locular (or both)
9 Leaflets (11) 17-27; pod pendulous, stipitate, triquetrous, 2-locular
  10 Racemes 4- to 12-flowered; flowers veined with lilac, 6-7.5 mm. long; leaflets 17-
    27, elliptic to oval or oblong, 1-7 mm. long; locally abundant in the foothills of the
    Sawtooth Mts. in Blaine Co., Ida.   A. ONICIFORMIS
  10 Racemes 10- to 25-flowered; flowers whitish-yellow, 6-12 mm. long; leaflets 11-
    23, linear to linear-lanceolate or narrowly oblong, 5-10 mm. long; c. Wash. and
    Oreg.   A. HOWELLII
9 Leaflets 7-19 (23), if over 17 on any one leaf then the ovary and pod sessile or 1-
  locular
  11 Wings slightly longer than the banner, the keel much shorter than either
    12 Racemes 2- to 8-flowered; pod 4.5-6.5 mm. thick, semibilocular; e. Oreg.
      and the s.w. corner of Ida.   A. SALMONIS
    12 Racemes 6- to 15-flowered; pod 3-4.2 mm. thick, 1-locular; Camas Co.,
      Ida.   A. ATRATUS
  11 Wings shorter than the banner or, if nearly as long, then the keel also nearly as
    long
    13 Petals lilac or purplish, often drying bluish; Rocky Mts., Mont. and e. Ida.
      14 Stems prostrate; pod 2-3.5 cm. long, arcuate; near our s. border in s.w.
        Ida.   *A. iodanthus* Wats.
      14 Stems erect; pod 8-12 mm. long, not arcuate   A. EUCOSMUS
    13 Petals white or ochroleucous, often drying yellowish, only the tip of the keel
      lilac (if the banner faintly lilac- or pink-veined then plant often more western)
      15 Keel 3.5-5.5 mm. long; ovules 4-10
        16 Pod sessile, inflated, 1-locular; plants of the Cascade Range, villosulous
          A. PULSIFERAE
        16 Pod shortly to prominently stipitate, more or less lunate and nearly or
          quite 2-locular; plants from e. of the Cascades, more nearly strig-
          illose
          17 Plant green and nearly glabrous; stems 20-45 cm. long; leaflets 5-20
            mm. long; c. Ida. in our range   A. PAYSONII
          17 Plant grayish-strigillose; stems 5-15 (20) cm. long; leaflets 5-10
            (14) mm. long; c. Wash. and Oreg.   A. HOWELLII
      15 Keel over 6 mm. long; ovules 14-40
        18 Leaflets 7-13, broadly ovate-oblong to oblong-elliptic, retuse or emar-
          ginate, thin and greenish, 3-18 mm. long; calyx mostly 4-6 mm. long,
          the tube and teeth about equal; pods sessile, strigillose   A. AMNIS-AMISSI
        18 Leaflets, calyx, and pod never at once as above
          19 Peduncles 1-6.5 cm. long, shorter than the leaves; either the tip of
            the keel obtusely rounded or the flowers declined; pod inflated, neith-
            er linear-oblong nor strictly erect
            20 Calyx tube broadly campanulate, 3.5-4 mm. long, 2.5-3.3 mm.
              wide, the teeth about 3-4 mm. long; pod 1-locular   A. WOOTONII
            20 Calyx tube narrowly campanulate, about 3-4 mm. long but only
              1.5-2.5 mm. wide, the teeth less than 2.5 mm. long; pod fully
              2-locular   A. LENTIGINOSUS
          19 Peduncles 3-15 cm. long, mostly longer than the leaves; tip of the keel
            narrowly triangular, acutish; pod strictly erect, linear-oblong, 2-
            locular   A. OBSCURUS

KEY USABLE ON PLANTS IN FRUIT

1 Plant prostrate, densely matted; leaflets linear-filiform, 3-10 mm. long, silvery-strigose,
  acerose; flowers 1 or 2 (3) in each raceme; pods 1- to 4-seeded, indehiscent, about 5 mm.
  long *(Kentrophyta)*
  2 Pubescence of the herbage rather loose, basifixed; corolla usually purple (whitish); high-
    montane   A. TEGETARIUS
  2 Pubescence of the herbage strictly appressed, dolabriform; corolla whitish; lowland
      A. KENTROPHYTA

1 Plant not at once prostrate and matted and with linear-filiform, acerose leaflets, only 1 or
    2 flowers per raceme, and 1- to 4-seeded, indehiscent pods
    3 Inflorescence, even in fruit, a headlike spike nearly as broad as long; calyx often black-
        hairy; corolla purplish, about 15 mm. long; pod erect, about 1 cm. long, ovoid, grayish-
        or blackish-pilose, cordate in cross section due to the complete intrusion of the lower
        suture; plant rhizomatous   A. DASYGLOTTIS
    3 Inflorescence seldom capitate or as broad as long, but if so, plants then not otherwise
        like A. dasyglottis
        4 Pods woolly or villous with long white or grayish hairs that almost or quite conceal
            the surface, usually woody or coriaceous in texture with the lower suture intruded
            to divide the cavity more or less completely into 2 cells (see, also, A. nudisiliquus)
            5 Corolla 6-14 mm. long, white with purplish pencilling or spotting; pod 4-10 mm.
                long, villous but not woolly, cordate in section, the lower suture completely in-
                truded
                6 Racemes dense; flowers closely crowded, 10-14 mm. long, spreading to erect;
                    pod ovoid, included in, or not much longer than, the calyx   A. SPALDINGII
                6 Racemes lax; flowers not crowded, reflexed, 6-9 mm. long; pod twice as long
                    as the calyx   A. LYALLII
            5 Corolla 15-30 mm. long, usually deep reddish or purplish; pod 15-30 mm. long
                (rarely shorter but then woolly), often completely 2-celled
                7 Herbage hirsute; pod compressed, slightly falcate, 2-celled   A. MALACUS
                7 Herbage nearer to being woolly or silky; pod obcompressed, incompletely
                    2-celled
                    8 Stems usually over 1.5 dm. long; calyx teeth linear, 4-7 mm. long, 1/2
                        the length of the tube or longer; flowers rose-purple   A. INFLEXUS
                    8 Stems seldom over 1 dm. long; calyx teeth linear-lanceolate to triangular,
                        usually less than 4 mm. long, and less than 1/2 the length of the tube; flow-
                        ers purple to white
                        9 Pubescence of the leaves nearly straight, mostly appressed; stems essen-
                            tially lacking   A. NEWBERRYI
                        9 Pubescence of the leaves more nearly tomentose or villous, not appressed;
                            stems often obvious
                            10 Leaflets mostly acutish, or fewer than 11; pods tomentose; plants usually
                                subacaulescent   A. PURSHII
                            10 Leaflets obtuse or rounded, 11 or more on at least some of the leaves;
                                pods villous; plants usually with evident stems   A. NUDISILIQUUS
        4 Pods neither woolly nor with the surface concealed by the often dense but short, or
            less abundant longer pubescence, the texture of the pod various, the lower suture
            often not intruded
            11 Pod fleshy, nearly globose, 2-3 cm. long, 2-celled, glabrous; calyx with 2 (3)
                membranous scales at base (Geoprumnon)   A. CRASSICARPUS
            11 Pod usually not fleshy, or if fleshy then not otherwise as above; calyx usually
                without scales at base
                12 Plant annual or biennial
                    13 Pod greatly inflated, 1-celled   A. GEYERI
                    13 Pod not inflated, 2-celled   A. DIAPHANUS
                12 Plant perennial
                    14 Flowers 4-8 mm. long; pod 5-6 mm. long, nearly completely 2-celled by
                        the intrusion of the lower suture; leaflets not over 1 cm. long
                        15 Terminal leaflet confluent with, and not noticeably different from, the
                            rachis; plant grayish-strigillose   A. PECKII
                        15 Terminal leaflet unlike (usually much broader than) the rachis, with
                            which it is jointed; plant greenish; n.e. Calif., n. nearly to Bend,
                            Oreg., and probably into our area   A. LEMMONII
                    14 Flowers over 8 mm. long; pod either over 6 mm. long or not 2-celled
                        or leaflets more than 1 cm. long
                        16 Plant essentially acaulescent, silvery; pubescence dense, usually tight-
                            ly appressed, dolabriform; leaflets often lacking or no more than 3

17 Leaflets (at least on many leaves) 5 or more
  18 Pod nearly 2-celled by the intrusion of the lower suture, obcordate in section, ob-
    long, arcuate, not at all inflated; leaflets commonly 3-5 (7)   A. CALYCOSUS
  18 Pod 1-celled, usually obcompressed, more or less conspicuously inflated, not
    arcuate; leaflets commonly 7-15
    19 Racemes long-pedunculate, usually equaling the leaves; flowers 15-20 mm.
      long; pod sparsely hairy   A. MISSOURIENSIS
    19 Racemes mostly very short-pedunculate, the flowers about 10 mm. long, ap-
      parently nearly sessile in the leaf axils; pod fairly heavily pubescent with
      long crisp hairs   A. LOTIFLORUS
17 Leaflets 3 or fewer
  20 Leaves all (or nearly all) simple; flowers several in long-pedunculate racemes,
    borne above the leaves   A. SPATULATUS
  20 Leaves trifoliolate; flowers in very short-pedunculate, mostly 1- or 2-flowered
    racemes, apparently sessile in the leaf axils   A. GILVIFLORUS
16 Plant usually caulescent, or if subacaulescent then with basifixed pubescence or not at all
  silvery; leaflets mostly numerous
  21 Pod inflated, thin and papery, and not at all woody, ovoid, globose, or obovoid,
    usually neither conspicuously compressed nor obcompressed   GROUP A (below)
  21 Pod not inflated, or if inflated then tough and often woody, usually elliptic to linear or
    oblong in outline, often strongly compressed or obcompressed
    22 Pod not visibly stipitate, the stalk (if any) shorter than the calyx and concealed
      by it   GROUP B (page 210)
    22 Pod stipitate, the body narrowed to a stalk at the mouth of the calyx tube or
      outside the calyx, this stipe at least as long as the calyx tube and visible at
      maturity of the fruit   GROUP C (page 214)

GROUP A
Pod inflated, thin and papery, ovoid, globose, or obovoid, usually
not conspicuously compressed

1 Pod stipitate, narrowed to a stalklike base at least as long as the calyx
  2 Leaflets 2-5 cm. long, 7-15 mm. broad; stipules leafletlike, often reflexed; flowers
    ochroleucous; calyx lobes scarcely 1 mm. long; plant glabrous or subglabrous
        A. AMERICANUS
  2 Leaflets usually less than 2 cm. long, either less than 7 mm. broad, or the plants
    otherwise not as above
    3 Leaflets often lacking on many of the leaves, linear, usually about 1 mm. broad,
      the rachis 6-20 cm. long; calyx lobes 0.5-2 mm. long
      4 Stems several from a branched caudex (sometimes the branches more or less
        rhizomatous at base), with stipular sheaths below; calyx teeth mostly 1-2 mm.
        long
        5 Flowers usually over 1 cm. long, not purple-tipped; calyx lobes scarcely 1 mm.
          long; pod not mottled; lateral leaflets 10-16   A. CUSICKII
        5 Flowers no more than 1 cm. long, usually purple-tipped; calyx lobes linear, 1-
          2 mm. long; pod mottled; lateral leaflets often lacking, if present then fewer
          than 10   A. CERAMICUS
      4 Stems single or few together, from creeping rhizomes that lack stipular sheaths;
        calyx teeth about 0.5 mm. long; s.w. Ida., near our area, but not known from
        within it   *A. sterilis* Barneby
    3 Leaflets present on all leaves, seldom less than 2 mm. broad, if narrower, the
      rachis much less than 6 cm. long, or calyx lobes well over 2 mm. long
      6 Flowers ochroleucous, 15-20 mm. long; stipules not connate   A. COLLINUS
      6 Flowers usually purple or at least purplish-tinged, 6-11 mm. long; basal stip-
        ules (at least) usually somewhat connate
        7 Body of pod less than 3 cm. long, glabrous, not mottled, somewhat compressed,
          elliptic-oblong in outline, the lower suture intruded   A. COTTONII
        7 Body of pod usually at least 3 cm. long, usually strigillose, often mottled, not
          compressed, elliptic-obovate in outline, the lower suture not intruded   A. WHITNEYI
1 Pod not visibly stipitate, the stalk, if any, concealed by the calyx

8 Legume about 1 cm. long, broadest at or above the middle, hairy, not particularly oblique;
   stipules connate; calyx teeth linear, subequal to the tube   A. MICROCYSTIS
8 Legume usually well over 1 cm. long, if only that long then usually very oblique; stipules
   often free; calyx teeth often much shorter than the tube
    9 Pod nearly or quite 2-celled
      10 Fruits only slightly inflated, 7-8 mm. in diameter, partially bilocular, the septum
         only about 1.5 mm. wide; stipules not connate   A. AMNIS-AMISSI
      10 Fruits more strongly inflated, usually well over 8 mm. in diameter, more com-
         pletely bilocular, the septum much more than 1.5 mm. wide; stipules often con-
         nate
        11 Leaves grayish-hairy; leaflets oblong-obovate, 5-10 mm. long; pod densely strig-
          illose; keel nearly or quite as long as the banner; plant sometimes almost acau-
          lescent; stipules connate
          12 Plant strongly caulescent; pod not mottled; Salmon R. Canyon, Custer and
            Lemhi cos., Ida., on shale or basalt   A. AMBLYTROPIS
          12 Plant essentially acaulescent; pod usually mottled; mostly on alpine to sub-
            alpine ridges, usually on limestone   A. PLATYTROPIS
        11 Leaves greenish, although sometimes rather hairy; larger leaflets over 10 mm.
          long; pod glabrous or only very sparsely hairy; keel much shorter than the ban-
          ner; plant strongly caulescent; stipules not connate   A. LENTIGINOSUS
    9 Pod strictly 1-celled, the sutures only slightly or not at all intruded
      13 Flowers 14-18 mm. long; calyx tube 4-7 mm. long   A. COLLINUS
      13 Flowers less than 14 mm. long; calyx tube usually less than 4 mm. long
        14 Stems wiry, 1 to few together from creeping rhizomes that lack stipular sheaths,
         up to 15 cm. tall; just to the south of our area in s.w. Ida.   *A. sterilis* Barneby
        14 Stems usually several from a branched caudex, often over 15 cm. tall, if the
         branches at all rhizomatous then with stipular sheaths
          15 Pod 8-20 mm. long; leaves 1.5-6 cm. long; flowers less than 10 mm. long;
           leaflets always present
           16 Legume usually over 15 mm. long, not strongly compressed; plants of
            John Day Valley, Oreg.   A. DIURNUS
           16 Legume not over 15 mm. long, often compressed; not known from the
            John Day Valley
            17 Stipules connate; pod rather symmetrical, broadest at or above the
              middle, the lower suture not at all sulcate; corolla white to deep
              magenta-purple   A. MICROCYSTIS
            17 Stipules not connate; pod oblique, the lower suture slightly sulcate;
              corolla ochroleucous or very pale purple   A. PULSIFERAE
        15 Pod 25-40 mm. long; some of the leaves usually over 6 cm. long; flowers
         about 10 mm. long, if shorter then the leaves in part without leaflets
         18 Leaflets present on all leaves; pod not mottled   A. WOOTONII
         18 Leaflets lacking on many leaves; pod mottled   A. CERAMICUS

GROUP B

Pod not inflated, usually elliptic to linear or oblong in outline, often strongly
compressed or obcompressed, not visibly stipitate, the stipe (if any) shorter than, and concealed by, the calyx

1 Legumes 1-celled with the lower suture very slightly (if at all) intruded, either round in
   section or, more commonly, compressed
    2 Calyx and pod both very conspicuously pubescent with appressed black (or at least dark)
      hairs; plant not at all rhizomatous
      3 Racemes usually at least 20-flowered; stipules free, or the lowest ones slightly con-
        nate; pod usually over 15 mm. long   A. ROBBINSII
      3 Racemes mostly 5- to 10-flowered; stipules strongly connate; pod about 15 mm. long
                                               A. BOURGOVII
    2 Calyx and pod both not black-hairy, or the plant strongly rhizomatous
      4 Pod 4-10 (12) mm. long, elliptic-oblong, acute; leaflets 1-4 mm. broad, grayish-
        hairy; flowers 4-8 (10) mm. long
        5 Flowers 4-6 mm. long, whitish, the banner purple-pencilled; leaflets obovate to

obcordate, 3-7 mm. long; pod 4-6 mm. long; ovules 2 (3); seeds usually single
A. TEGETARIOIDES

5 Flowers 5-8 (10) mm. long, generally more or less purplish; leaflets linear- to
   oblong-elliptic, 5-12 mm. long; pod 7-11 mm. long; ovules 4-7; seeds usually 2
   or more   A. VEXILLIFLEXUS
4 Pod over 10 mm. long and (usually) linear or oblong, or leaflets either over 4 mm.
   broad or not grayish-hairy, or flowers over 10 mm. long
   6 Plant delicate, diffuse, the stems arising at intervals from a subterranean rootstock;
      lower stipules connate; pod somewhat obcompressed   A. LEPTALEUS
   6 Plant usually ascending to erect, the stems mostly from a woody, branched caudex;
      stipules often all free; pod usually compressed
      7 Plant often rushlike or broomlike, the leaves with comparatively few, narrow, much
         elongate leaflets, the terminal leaflet (if any) confluent with the rachis; pod linear
         to narrowly oblong in outline, 2-5 mm. broad, strongly compressed
         8 Corolla 13-20 mm. long; plant erect; stems many, erect, stiff and broomlike;
            leaves few; leaflets 0-9, very narrow, 5-30 mm. long; along the Snake R. in
            s.w. Ida., and just at the edge of our range   A. TOANUS
         8 Corolla not over 13 mm. long; plant often decumbent
            9 Rachis of the leaves 1-1.5 mm. wide, attenuate toward the tip; lateral leaflets
               2, 4, or more (sometimes lacking), linear, 1-3 cm. long, often involute and
               scarcely as broad as the rachis; pod linear, about 2 mm. wide; plant erect,
               occurring chiefly on dry bench land   A. CONVALLARIUS
            9 Rachis of the leaves 1-2 mm. broad, usually widened into a terminal leaflet,
               or if not widened, the lateral leaflets 2-5 mm. broad; pod 3-4 mm. broad;
               plant prostrate, occurring in alkaline meadows   A. DIVERSIFOLIUS
      7 Plant not rushlike, the leaves usually with many leaflets, the terminal leaflet gen-
         erally jointed to the rachis; pod various
         10 Pod terete or slightly compressed, 12-21 mm. long, 2-3 mm. thick, partially
             filled between the seeds with criss-crossing fibrous strands   A. FLEXUOSUS
         10 Pod usually either compressed or obcompressed, if terete then over 3 mm.
             thick or not filled with fibrous material
             11 Pod strongly obcompressed, about twice as thick as broad, very deeply fur-
                 rowed on either side of the prominent ventral (upper) suture, strictly 1-celled;
                 wings and keel subequal; racemes usually over 30-flowered   A. BISULCATUS
             11 Pod not as above, although sometimes strongly obcompressed
                 12 Plant fleshy, malodorous; leaves with the upper leaflet confluent with the
                     rachis; flowers about 2 cm. long, ochroleucous; pod fleshy, nearly solid,
                     drying to woody, nearly terete, 5-7 mm. thick   A. PECTINATUS
                 12 Plant not fleshy and malodorous; leaves usually with the upper leaflet not
                     confluent with the rachis; flowers often less than 2 cm. long, or purplish;
                     pod not as above
                     13 Calyx tube 8-12 mm. long; corolla usually purple
                         14 Leaves appressed-villous or sericeous; stems usually less than 10
                             cm. long; pod hirsute-strigose, 1.5-2.5 cm. long A. ARGOPHYLLUS
                         14 Leaves villous to tomentose; stems 5-25 cm. long; pod thinly vil-
                             lous, 2-3.5 cm. long   A. NUDISILIQUUS
                     13 Calyx tube usually less than 7 (never as much as 8) mm. long; corolla
                         often ochroleucous
                         15 Corolla, or at least the banner, pinkish-purple to purple; pods 2-
                             5.5 cm. long
                             16 Banner 13.5-18 mm. long; wing tips white; leaflets narrowly
                                 elliptic-oblong to linear; pod 2-5.5 cm. long; Walla Walla
                                 region, Wash.   A. CASEI
                             16 Banner 8.5-15 mm. long; wing tips purple; leaflets oval to
                                 obovate; Owyhee Co., Ida., probably always s. of our area
A. iodanthus Wats.
                         15 Corolla mostly white or ochroleucous, the keel sometimes pur-
                             plish; if banner purplish then the pods not over 2.5 cm. long

17 Pod nearly solid when green, very tough and leathery when ripe and dry, more or less cordate in section, both sutures raised, the lower one also sulcate; leaflets oblong-elliptic; plant usually over 3 dm. tall   A. ADANUS

17 Pod neither solid when green nor very tough and leathery and with raised sutures when ripe; leaflets usually linear to linear-elliptic; plant mostly less than 3 dm. tall

  18 Pod compressed, both sutures acute; stipules sometimes connate

    19 Legume obliquely elliptic in outline, 8-12 mm. long   A. EUCOSMUS

    19 Legume linear or narrowly oblong in outline, usually well over 12 mm. long

      20 Stipules connate opposite the petiole; corolla 8-12 mm. long; calyx tube usually less than 3 mm. long   A. MISER

      20 Stipules not connate opposite the petiole; corolla mostly over 12 mm. long; calyx tube 4-7 mm. long   A. COLLINUS

  18 Pod obcompressed, sulcate dorsally; stipules free

    21 Pedicels 3-6 mm. long; pod nearly straight, about 3.5 (3-4.2) mm. wide and thick, more membranous than woody, 1-celled, the lower suture sulcate but not intruded; racemes 6- to 15-flowered   A. ATRATUS

    21 Pedicels 1-4 mm. long; pod about 5.5 (4.5-6.5) mm. in width and thickness, oblique or curved, woody (fleshy when immature), deeply sulcate on the lower suture, nearly 2-celled; racemes 2- to 8-flowered   A. SALMONIS

1 Legumes either partially to completely 2-celled (the lower suture at least somewhat intruded) or more or less cordate, reniform, or obcompressed in section

  22 Pod membranous, often somewhat transparent, strigillose, strongly arcuate; root annual or biennial   A. DIAPHANUS

  22 Pod thick and opaque, or if thin, then copiously dark-hairy or root strongly perennial

    23 Plant rhizomatous, the stems low and stout, not over 1.5 dm. long; leaflets more or less cordate; stipules connate; pod 2-celled; c. Wyo., possibly into our area in e. Ida.   *A. oreganus* Nutt. ex T. & G.

    23 Plant either not rhizomatous or rhizomatous but the stems taller, the leaflets not cordate, the stipules free, or the pod 1-celled

      24 Pod inflated to turgid, ovoid to ovoid-lanceolate, usually terete or more or less didymous in section, tapered to a distinct and prominent beak that is laterally compressed to filiform and 1-celled; plant usually greenish   A. LENTIGINOSUS

      24 Pod not as above, usually more nearly linear to oblong in outline, seldom beaked as above; plant often grayish-hairy

        25 Flowers (15) 30-150, borne in spikelike congested racemes, 12-18 mm. long; pedicels about 1 mm. long; pubescence of the calyx and stems appressed, dolabriform; stipules membranous, mostly connate opposite the petioles; pod 8-20 mm. long, seldom over twice as long as the calyx, the lower suture deeply sulcate

          26 Plant strongly rhizomatous; pod 8-20 mm. long, 4-5 mm. broad   A. CANADENSIS

          26 Plant nonrhizomatous; pod 8-12 mm. long, 3-4 mm. broad   A. STRIATUS

        25 Flowers seldom borne in spikelike racemes; pubescence usually basifixed; stipules seldom connate; pod mostly over 1 cm. long

          27 Leaflets 3-10 mm. long, often filiform or linear, distant, the terminal leaflet often confluent with the rachis; stems either filiform and more or less decumbent, or short (scarcely 5 cm. long); flowers 7-10 mm. long; pod contracted to a short (0.5-2 mm. long) stipe within the calyx, the body 10-22 mm. long, cordate in section, 3-7 mm. thick

            28 Pedicels 3-6 mm. long; pod nearly straight, about 3 mm. wide and thick, more membranous than woody, 1-celled, the lower suture sulcate but not intruded   A. ATRATUS

            28 Pedicels 1-4 mm. long; pod (4) 5-7 mm. in width and thickness, oblique or curved, woody, deeply sulcate on the lower suture and nearly 2-celled   A. SALMONIS

          27 Leaflets either over 10 mm. long, or plants (especially the pods) not as above

            29 Legume strictly 1-celled, somewhat obcompressed, the valves

papery-membranous, not inflexed, strigillose with black or white hairs; stipules connate; branches rhizomelike at base   A. LEPTALEUS
29 Legume either more or less 2-celled due to the intrusion of the valves along the dorsal
     (lower) suture, or plant otherwise not as above
     30 Pod completely 2-celled to falsely 2-celled due to the intrusion of the lower suture, the
         partition extending at least 3/4 of the way across the cavity
         31 Flowers 4.5-8 (9) mm. long, white or purplish-tinged; pods 5-9 mm. long, re-
           flexed; leaflets 5-20 mm. long, linear, grayish-hairy   A. LYALLII
         31 Flowers either over 8 mm. long, or not whitish, or the pods either over 10 mm.
           long or ascending to erect
           32 Stems villous to hirsute
             33 Pod glabrous, shining, not mottled   A. SUCCUMBENS
             33 Pod hirsute, dull, often mottled   A. MALACUS
           32 Stems glabrous or strigose
             34 Calyx about 4 (5) mm. long, the blunt teeth scarcely 1/4 as long as the tube;
               flowers about 15 mm. long, white but drying ochroleucous, the keel purplish-
               tipped; leaflets usually oblong-oblanceolate to obovate; pod erect, glabrous,
               15-20 mm. long, about 4 mm. broad, slightly obcompressed, broadly cor-
               date in section, the lower suture almost completely intruded   A. TERMINALIS
             34 Calyx teeth 1/3-4/5 the length of the tube, or plants otherwise not as above
               35 Pubescence dolabriform; calyx about 5 mm. long; pod sessile   A. FALCATUS
               35 Pubescence basifixed; calyx either less than 4.5 mm. or well over 5
                 mm. in length; pod sometimes stipitate
                 36 Calyx 7.5-15 mm. long   A. REVENTUS
                 36 Calyx rarely as much as 7 mm. long
                   37 Plant silvery-strigillose; pod sessile   A. OBSCURUS
                   37 Plant pale green, nearly glabrous; pod with a stipe 1-1.5 mm.
                     long
                     38 Banner about 7 mm. long; body of the pod 10-17 mm. long;
                       leaflets 7-15   A. PAYSONII
                     38 Banner 10-14 mm. long; body of the pod 14-24 mm. long;
                       leaflets 15-23   A. UMBRATICUS
     30 Pod imperfectly 2-celled, the partition (or the intruded lower suture) less than 3/4
         complete
         39 Legume gray- and black-hairy, less than 13 mm. long, pendulous; flowers 6-9
           mm. long   A. EUCOSMUS
         39 Legume not at once black-hairy, less than 13 mm. long, and pendulous, when
           the flowers are less than 10 mm. long
           40 Flowers less than 10 mm. long; pod 6-9 mm. long, obcompressed, grayish-
             hairy, usually 2-seeded; prairies and low hills on the e. slopes of the Rockies,
             apparently not known w. of the continental divide in our area   A. GRACILIS
           40 Flowers or pods either well over 10 mm. long, or the pods not grayish-hairy
             41 Pod at least 2 cm. long (usually considerably longer), distinctly arcuate,
               7-10 mm. thick, usually glabrous; leaves greenish, oblong-elliptic to
               obovate, glabrous on the upper surface   A. CIBARIUS
             41 Pod either less than 2 cm. long, or straight, or the leaflets not oblong-
               elliptic to obovate and glabrous on the upper surface
               42 Leaflets 9-15, linear, 2-6 cm. long, at least the terminal one confluent
                 with the rachis; fruits reflexed; calyx tube over 6 mm. long, usually
                 black-hairy; e. of the continental divide in our area   A. PECTINATUS
               42 Leaflets either not linear and so much as 2 cm. long, or the fruits
                 erect, or the calyx tube less than 6 mm. long
                 43 Plant greenish; stems 3-7 dm. tall; leaflets 10-25 mm. long, ob-
                   long-elliptic, some, at least, usually over 5 mm. broad; pod fleshy
                   and nearly solid when green, hardened and filled with fibers when
                   dried, 10-15 mm. long, the lower suture intruded only slightly;
                   calyx teeth 1.5-2 mm. long   A. ADANUS

43 Plant not at once greenish and with stem, leaflets, pods, and calyx as above
    44 Calyx teeth broadly triangular, about 1 mm. long; pod partially septate; leaflets
        oblong-oblanceolate to obovate, (2) 3-5 mm. broad   A. TERMINALIS
    44 Calyx teeth at least 2 mm. long; pod not septate; leaflets usually linear, mostly less
        than 3 mm. broad
        45 Wing petals 1-4 mm. longer than the banner; pod glabrous   A. RIPARIUS
        45 Wing petals shorter than the banner; pod often pubescent   A. REVENTUS

GROUP C
Pod not inflated and papery, usually elliptic to linear or oblong in outline,
often strongly compressed or obcompressed, stipitate, the stipe at least
as long as the calyx tube and visible at maturity of the fruit

1 Pod conspicuously curved, often making a complete turn, occasionally curved less than 1/2
  turn but then the fruits always over 5 mm. broad and with the two sutures very prominent
    2 Pubescence of the pod spreading, stiff; plant rhizomatous; known only from Owyhee Co.,
      Ida., just to the south of our range   *A. camptopus* Barneby
    2 Pubescence of the pod appressed or lacking
      3 Pod curved less than 1/2 turn, but averaging at least 5 mm. broad, the sutures very
        prominent   A. SCLEROCARPUS
      3 Pod usually curved at least 1/2 turn, rarely as much as 5 mm. broad, the sutures
        not prominent
        4 Fruit grayish-hairy, at least when young, finally becoming strongly reticulate; stipe
          about equal to the calyx   A. SPEIROCARPUS
        4 Fruit not grayish-hairy even when young, never strongly reticulate; stipe about
          twice as long as the calyx   A. CURVICARPUS
1 Pod nearly or quite straight, never curved into as much as a semicircle, if strongly arcuate then (except in *A. beckwithii* and *A. vallaris*) not over 5 mm. broad
    5 Calyx at least 8 mm. long; pod either over 5 mm. in width or thickness, or stipe 1.5 cm.
      long or evidently jointed to the body of the pod
      6 Flowers 11.5-16.5 mm. long   A. LEIBERGII
      6 Flowers usually about 20 mm. long
        7 Pod 1-celled, the body jointed to the stipe   A. BECKWITHII
        7 Pod nearly or quite 2-celled, with a broad (at least 1/2 complete) partition, the
          body continuous with the stipe
          8 Pod and stem erect   A. SCAPHOIDES
          8 Pod pendulous; stems spreading   A. VALLARIS
    5 Calyx less than 8 mm. long or pod much less than 5 mm. broad or with the stipe much
      less than 1.5 cm. long and not jointed to the body
      9 Pod from strongly compressed to round in cross section, strictly 1-celled, the lower
        suture rarely protruding into the cavity, the partition (if any) rudimentary and less
        than 1/4 the height of the cavity
        10 Calyx tube 5-9 mm. long; pod oval to round in cross section
          11 Pod glabrous, erect; flowers spreading to erect   A. TWEEDYI
          11 Pod hairy, pendulous; flowers sometimes reflexed
            12 Calyx gibbous at base; flowers reflexed   A. COLLINUS
            12 Calyx not gibbous-based; flowers ascending   A. SINUATUS
        10 Calyx tube either less than 5 mm. long or pod strongly compressed
          13 Pod black-hairy
            14 Leaflets usually less than 15 (very rarely to 20) mm. long; pod 10-15
              (16) mm. long   A. BOURGOVII
            14 Leaflets usually over 15 mm. long; pod usually over 15 mm. long   A. ROBBINSII
          13 Pod not black-hairy
            15 Rachis of leaves seldom as much as 5 cm. long; leaflets usually 2-5 mm.
              broad   A. ABORIGINUM
           15 Rachis of leaves usually well over 5 cm. long; leaflets often less than 2
             mm. broad
             16 Body of the pod 7-15 mm. long; stipe scarcely twice as long as the ca-
                lyx; leaflets (1.5) 2-5 mm. broad   A. TENELLUS

16 Body of the pod mostly over 15 mm. long; stipe usually over twice as long as the calyx;
    leaflets 1-2 mm. broad
   17 Flowers less then 9 mm. long; calyx 3-4 mm. long; pod acute at each end; a little-
      known species from Malheur Co. , Oreg. , just to the s. of our range; related to
      A. filipes but apparently distinct from it   *A. solitarius* Peck
   17 Flowers 9-13 mm. long; calyx 4-6 mm. long; pod usually rounded at the ends   A. FILIPES
9 Pod round, cordate, or triangular in section, or obcompressed, usually with the lower
  suture rather deeply sulcate, if pod round or compressed then with a partition extending at least 1/3 across the cavity
   18 Legume woody, ovoid or subglobose, the body 1-2 cm. long, 2-locular; plant rhi-
      zomatous; keel exceeding the narrower wings   A. CHINENSIS
   18 Legume not at once woody, ovoid or subglobose, 1-2 cm. long, and 2-locular if the
      plant is also rhizomatous and the keel is longer than the wings
     19 Pod and calyx black-hairy; pod pendulous, the body 8-12 mm. long; plant rhizo-
        matous   A. ALPINUS
     19 Pod and calyx not at once black-hairy when the plant is rhizomatous and has
        reflexed pods with the body of the pod less than 15 mm. long
      20 Flowers purple (occasionally lavender or nearly white); pod very much obcom-
          pressed and deeply 2-grooved on the upper (ventral) surface   A. BISULCATUS
      20 Flowers either whitish or yellowish if pod strongly obcompressed and deeply
         2-grooved on the upper surface
        21 Leaflets not over 10 mm. long; flowers not over 9 mm. long
          22 Pod at least 4 mm. broad and thick, mottled   A. SALMONIS
          22 Pod either less than 4 mm. broad and thick, or not mottled
            23 Pod obliquely elliptic-oblong in outline, not markedly arcuate;
               leaflets linear, (1) 3-8 mm. long; calyx at least partially brownish-
               or blackish-hairy
              24 Lower stipules connate   A. MULFORDIAE
              24 Lower stipules, as the upper, more or less amplexicaul but not
                 connate   A. ONICIFORMIS
           23 Pod obliquely and narrowly oblong, often strongly arcuate; leaflets
               mostly narrowly lanceolate to narrowly oblong, 7-10 mm. long;
               calyx white-hairy   A. HOWELLII
       21 Leaflets well over 10 mm. long or flowers at least 10 mm. long
        25 Body of the pod linear, 3.5-5 cm. long, 3-4 mm. broad, strigillose;
          leaflets 6-12 (14) mm. long   A. ARTHURI
        25 Body of the pod not at once linear, 3.5-5 cm. long, 3-4 mm. broad, and
          strigillose if the leaflets are only 6-12 mm. long
          26 Pod pendulous; plant grayish-villous with long soft hairs, (3) 4-8 dm.
           tall   A. DRUMMONDII
          26 Pod erect or plant either not grayish-villous or less than 3 dm. tall
           27 Stipe 8-12 mm. long, about twice the length of the calyx
             28 Pod compressed, nearly completely 2-celled, usually pendent,
               softly short-strigose-lanate   A. HOWELLII
             28 Pod obcompressed, imperfectly 2-celled, usually erect, gla-
               brous to strigillose   A. EREMITICUS
           27 Stipe (2.5) 4-7 (8) mm. long, usually only slightly exceeding the
            calyx
             29 Flowers ascending to erect; banner shorter than the wings; stip-
               ules free opposite the petioles   A. PALOUSENSIS
             29 Flowers nodding; banner longer than the wings; stipules some-
               times connate opposite the petioles
               30 Stipules, at least the lower ones, connate opposite the
                 petioles   A. LEIBERGII
               30 Stipules all free opposite the petioles   A. STENOPHYLLUS

Astragalus aboriginum Richards. in Frankl. Journ. 746. 1823.
   Phaca aboriginorum Hook. Fl. Bor. Am. 1:143. 1831. *Tragacantha aboriginum* Kuntze, Rev. Gen.

2:942. 1891. *Homalobus aboriginorum* Rydb. Mem. N.Y. Bot. Gard. 1:246. 1900. *Homalobus abo-riginum* Rydb. ex Britt. Man. 554. 1901. *Atelophragma aboriginum* Rydb. Bull. Torrey Club 32: 660. 1906. *(Dr. Richardson,* Lake Winnipeg to the Rocky Mts.)

Astragalus vaginatus sensu Hook. Fl. Bor. Am. 1:149. 1831, not Pall. 1800. *Astragalus richard-sonii* Sheld. Minn. Bot. Stud. 1:126. 1894. *(Dr. Richardson,* "between 54° and 64° north")

Phaca glabriuscula Hook. Fl. Bor. Am. 1:144. 1831. *Astragalus glabriusculus* Gray, Proc. Am. Acad. 6:204. 1864. *Tragacantha glabriuscula* Kuntze, Rev. Gen. 2:945. 1891. *Astragalus abori-ginum* var. *glabriusculus* Rydb. Contr. U.S. Nat. Herb. 3:492. 1896. *Homalobus glabriusculus* Rydb. Mem. N.Y. Bot. Gard. 1:246. 1900. *Atelophragma glabriusculum* Rydb. Bull. Torrey Club 32:660. 1905. *(Drummond,* "Vallies of the Rocky Mountains")

Astragalus forwoodii Wats. Proc. Am. Acad. 25:129. 1890. *Atelophragma forwoodii* Rydb. Bull. Torrey Club 40:51. 1913. *(Forwood,* Black Hills, S.D.) = var. *aboriginum.*

Astragalus aboriginum var. *fastigiorum* M. E. Jones, Rev. Astrag. 135. 1923. *(M. E. Jones,* Mt. Haggin, Mont.) = var. *aboriginum.*

Atelophragma wallowense Rydb. Bull. Torrey Club 55:122. 1928. *Astragalus forwoodii* var. *wal-lowensis* Peck, Man. High. Pl. Oreg. 447. 1941. *(Cusick 2267,* Wallowa Mts., Oreg., July 28, 1899) = var. *aboriginum.*

Caespitose, rather sparsely to densely sericeous or crisp strigillose-villous perennial from a woody taproot and a much-branched crown; stems 5-40 cm. long, prostrate to ascending; leaves 2-5 cm. long, very shortly petiolate to sessile; stipules oblong, rounded, 2-5 mm. long, connate below but free on the upper portion of the stem; leaflets 7-15, oblong-elliptic to linear, 10-20 mm. long, 1.5-6 mm. broad, mostly acute; peduncles usually well exceeding the subtending leaves; racemes closely 10- to 20-flowered, elongated and open in fruit; pedicels 1-2.5 mm. long; flowers 8-13 mm. long, spreading, whitish (but with a purple keel) to purplish overall; calyx 4-8 mm. long, usually black-hairy, the teeth linear, from half as long to nearly as long as the tube; banner erect; wings shorter than the banner, deeply bidentate at the tip, about 1 mm. longer than the keel; pod usually pendulous and somewhat falcate-lunate, with a slender stipe about as long as the calyx, the body mem-branous, glabrous or crisp-villous, 2-3 cm. long, 4-5 mm. broad, slightly compressed and oval in section, 1-celled, both sutures fairly prominent, the dorsal one often intruded as a hyaline partition up to 0.5 mm. wide. N=8.

Alpine and subalpine knolls and scree, descending in open places to bluffs and riverbanks in the foothills and plains; Can., through the Rockies to Colo., w. to Nev. and the Wallowa Mts., Oreg., often on limestone. June-Aug.

The species shows much variation in pubescence, flower size, degree of union of the stipules, and lobing of the petals. Subglabrous to heavily villous plants frequently grow intermingled. Therefore, the more glabrous plants that may or may not have narrower leaflets are not considered very significant and may perhaps needlessly be distinguished as the var. *glabriusculus* (Hook.) Rydb.

Astragalus adanus A. Nels. Bot. Gaz. 53:222. 1912.

Ctenophyllum adanum Rydb. Fl. Rocky Mts. 502, 1063. 1917. *Cnemidophacos adanus* Rydb. N.Am. Fl. 24⁵:285. 1929. *(J. F. Macbride 260,* Boise hills, Ida., June 18, 1910)

Minutely strigillose to somewhat villous, greenish perennial with a thick taproot and branched cau-dex; stems many, erect, 3-7 dm. tall, striate; leaves 7-12 cm. long, short-petiolate; stipules 3-6 mm. long, lanceolate, not connate; leaflets 15-25, oblong-elliptic, obtuse to retuse, 10-25 mm. long, glabrous on the upper surface; racemes loosely 10- to 30-flowered, the peduncles erect, mostly con-siderably exceeding the leaves; flowers 15-20 mm. long, ochroleucous; pedicels stout, 3-4 mm. long; calyx 5-7 mm. long, the teeth lanceolate, 1/4-1/3 the length of the campanulate tube; pod sessile, glabrous or glabrate, erect, fleshy and nearly solid, becoming much hardened and partially filled with fibers when mature, 10-15 mm. long, 4-5 mm. wide, nearly straight, 1-celled, cordate in section with the lower of the prominent sutures sulcate, distal portion of the body forming a distinct sterile beak.

Dry hillsides, often in sagebrush; along the n. edge of the Snake R. plains from Camas and Custer to Elmore and Canyon cos., and also in the Goose Creek Mts., Cassia Co., Ida. May-June.

Astragalus alpinus L. Sp. Pl. 760. 1753.

Tragacantha alpina Kuntze, Rev. Gen. 2:942. 1891. *Tium alpinum* Rydb. Bull. Torrey Club 32:659. 1906. *Phaca alpina* Piper, Contr. U.S. Nat. Herb. 11:371. 1906. *Atelophragma alpina* Rydb. Bull. Torrey Club 55:130. 1928. ("Habitat in Alpibus Lapponicis, Helveticis")

Phaca astragalina DC. Astrag. 64. 1802. *Astragalus astragalinus* Sheld. Minn. Bot. Stud. 1:65.
   1894. (Type not ascertained)
Phaca andina Nutt. ex T. & G. Fl. N. Am. 1:345. 1838, in synonymy. *Astragalus andinus* M. E.
   Jones, Rev. Astrag. 137. 1923. *(Nuttall,* Thornburg's Pass, lat. 43°, Rocky Mts.)
Astragalus alpinus var. *giganteus* sensu Sheld. Minn. Bot. Stud. 1:65. 1894, not of Pall. Astrag.
   42. 1800. *Astragalus giganteus* Sheld. loc. cit., not of Wats. in 1882.
Astragalus alpinus ssp. *alaskanus* Hultén, Fl. Alas. 7:1082. 1946. *(Anderson,* Five Finger Rapids,
   Yukon District, June 18, 1898)
   Sparsely to densely strigillose to silky (but usually greenish) perennial with widespread rootstocks;
stems slender, ascending to erect, 5-20 cm. long; leaves 5-15 cm. long; stipules 1-3 mm. long, deltoid, all except the uppermost connate; leaflets 13-23, ovate to oblong-elliptic, often retuse, 5-15
mm. long, rarely as much as 10 mm. broad; peduncles equaling or exceeding the leaves; racemes
closely 10- to 30-flowered, elongate, lax, often secund in fruit; pedicels about 1 mm. long; flowers
7-12 mm. long, pale lilac to purplish, the keel usually darkest in color (about equaling the erect banner, both longer than the wings); calyx black-hairy, 3-4.5 mm. long, the teeth about half the length
of the tube; pod usually pendulous (spreading), with a slender stipe about equal to the calyx teeth, the
body black-hairy, membranous, narrowly ellipsoid, 8-12 mm. long, cordate-triangular in section,
the lower suture deeply sulcate and intruded to form a nearly complete partition. N=about 28.
   Circumpolar, arctic to subalpine regions, s. to Okanogan Co., Wash., the Wallowa Mts. of Oreg.,
and through the Rocky Mts., Mont. and c. Ida., to n.e. Nev. and N. M. June-Aug.

Astragalus amblytropis Barneby, Am. Midl. Nat. 41:501. 1949. *(Hitchcock & Muhlick 14115,* 10 mi.
   w. of Clayton, Custer Co., Ida., June 28, 1946)
   Grayish-strigillose perennial with a deep taproot and branched crown; stems several, prostrate to
ascending, usually partially buried in talus and therefore rhizomelike at the base, 1-3 dm. long;
leaves 2-5 cm. long; stipules deltoid, 1-2 mm. long, purplish, the lower ones connate; leaflets (7)
9-13, oblong-obovate, usually emarginate, 5-10 mm. long; racemes 3- to 11-flowered, the peduncles
much shorter than the leaves; pedicels 0.5-2.5 mm. long; calyx finely whitish-strigillose, about 1/3
the length of the corolla, the linear teeth subequal to the tube; corolla 6-8 mm. long, ochroleucous to
yellowish, usually tinged with purple, the wings rather narrow, about equaling the much broader, very
conspicuous keel; pod sessile, inflated, membranous, ellipsoid-ovoid, 2-3.5 cm. long, abruptly contracted to a short, acute, beaklike tip, completely 2-celled by the intrusion of the lower suture,
broadly oval in section, with both sutures somewhat sulcate, strigillose, not mottled.
   Shale and volcanic cliffs and talus along the Salmon R., for about 30 mi. up and downstream from
Challis, Custer and Lemhi cos., Ida. June.
   Similar in flower and fruit to *A. platytropis* Gray, but always caulescent and with larger leaflets and
an unmottled pod.

Astragalus americanus (Hook.) M. E. Jones, Contr. West. Bot. 8:8. 1898.
   Phaca frigida var. *americana* Hook. Fl. Bor. Am. 1:140. 1831. *A. frigidus* var. *americanus* Wats.
   Bibl. Ind. 193. 1878. *A. alpinus* var. *americanus* Sheld. Minn. Bot. Stud. 1:133. 1894. *Phaca*
   americana Rydb. in Britt. & Brown, Ill. Fl. 2:304. 1897. ("Woody regions of the Rocky Moun-
   tains in lat. 52° to 56°, to Slave Lake in lat. 61°," *Dr. Richardson, Drummond)*
   A. americanus f. *glabrescens* Rouss. Contr. Lab. Bot. U. Montreal 24:50. 1933. *(J. M. Macoun,*
   Kicking Horse Lake, B. C., Aug. 11, 1890) The rather uncommon form with puberulent pods.
   Glabrous or very sparsely pilose, greenish perennial from a woody caudex and often (always?)
somewhat rhizomatous; stems more or less fistulose, 3-10 dm. tall; stipules 1-3 cm. long, oblong,
membranous to semiherbaceous, usually deflexed, not connate; leaflets 9-17, ovate to oblong or elliptic-oblanceolate, 2-5 cm. long, 7-15 mm. broad; racemes loosely 15- to 40-flowered; peduncles
usually shorter than the leaves; pedicels slender, 3-10 mm. long; flowers usually reflexed, ochroleucous, 10-15 mm. long; calyx campanulate, 4-5 mm. long, nearly glabrous except for the cilia of
the broadly triangular, scarcely 1 mm. long teeth; wings narrow, the blade shorter than the claw,
auriculate, not nearly concealing the broader and often longer keel; pod with a slender stipe about
twice as long as the calyx, the body membranous, inflated, ellipsoid, acute at each end, 2-3 cm. long,
glabrous or rarely puberulent, nearly terete in section, the lower suture not at all intruded. N=8.
   Stream banks and meadows; Alas. southward in the Rocky Mts. through e. B. C. and Mont. to n.
Colo., e. to Ont. and the Black Hills, S.D., also in Que. June-Aug.

Astragalus aboriginum

A. americanus

stipules

A. alpinus

A. amblytropis

A. adanus

JF

Our plant is frequently and perhaps more properly treated as var. *americanus* (Hook.) Wats. of the arctic *A. frigidus* (L.) Gray.

ASTRAGALUS AMNIS-AMISSI Barneby, hoc loc. (Type: *Hitchcock & Muhlick 21331,* along base of cliffs and in rock crevices, Pass Creek Gorge, 10 miles n. of Leslie, Custer Co., Ida., July 2, 1957, holotype in U. of Wash. Herb.)
  Perennial, with a taproot and superficial root crown or shortly forking caudex, thinly strigillose with subappressed, basifixed hairs, the herbage green, the leaflets nearly glabrous to moderately pubescent above; stems weakly ascending, 1-2.5 dm. long, simple, floriferous from near or below the middle; stipules 1.5-5 mm. long, the small lower ones papery, the rest herbaceous, strongly amplexicaul-decurrent but free; leaves (3) 4-9.5 cm. long, with (7) 9-13 broadly ovate-oblong to oblong-elliptic, retuse or emarginate, thin-textured leaflets 3-15 (18) mm. long; peduncles slender, (2) 3-8 cm. long, divaricate or recurved in fruit; racemes loosely 5- to 12-flowered, the flowers ultimately spreading or declined, the axis 1-3 cm. long in fruit; calyx 4-6 mm. long, black- or partly white-strigillose, the campanulate tube 2-3 mm. long; the teeth subulate or lance-subulate, 1.8-3 mm. long; petals whitish suffused with purplish-blue, the keel deeply maculate; banner ovate-cuneate, 9-10.5 mm. long; wings slightly longer or shorter, the oblong-oblanceolate, straight blades 7-8.5 mm. long; keel only 0.5-1 mm. shorter than the banner, the broadly half-oblong-obovate blades 5.3-6.5 mm. long, abruptly incurved through a right angle distally, appearing truncate; pod ascending or loosely spreading, humistrate, sessile, subsymmetrically ellipsoid, moderately inflated, about 15-17 mm. long, 7-8 mm. in diameter, a little obcompressed, shallowly sulcate ventrally, the thin, green or purplish, strigillose valves becoming papery, inflexed as a partial septum about 1.5 mm. wide; ovules 16-20.
  Affinis *A. amblytropidi* Barneby, sed caules breviores simplices basi haud subterranei, folia parce strigulosa viridia, flores paullo majores, et praesertim legumin subdimidio minus, vix vesicarium, semibiloculare, ovula minus numerosa pariens.
  Known only from the type locality (other material seen: *Ripley & Barneby 8799,* at 6300-6600 ft. elevation, June 21, 1947). June-July.

Astragalus argophyllus Nutt. in T. & G. Fl. N.Am. 1:331. 1838.
  Xylophacos argophyllus Rydb. Bull. Torrey Club 40:49. 1913. *(Nuttall,* "Valleys of the Rocky Mountains, near the sources of the Platte")
  A. uintensis M. E. Jones, Proc. Calif. Acad. Sci. II, 5:670. 1895, nom. provis. *Xylophacos uintensis* Rydb. Bull. Torrey Club 32:662. 1906. *(Jones,* locality not stated in the original description) = var. *argophyllus.*
  A. argophyllus var. *martini* M. E. Jones, Rev. Astrag. 207. 1923. *(Rev. Geo. W. Martin,* Soda Spgs., Ida., June 19, 1901) = var. *argophyllus.*
  A. argophyllus var. *cnicensis* M. E. Jones, Rev. Astrag. 208. 1923. *(M. E. Jones,* Thistle, Utah, in 1901) = var. *argophyllus.*
  A. argophyllus var. *pephragmenoides* Barneby, Am. Midl. Nat. 37:460. 1947. *(Ripley & Barneby 5437,* n. slope of Grand Mesa, Mesa Co., Colo.)
  Low, caespitose, silvery-sericeous or appressed-villous perennial from a branched and woody caudex, with short, spreading or decumbent stems 2-10 cm. long; leaves 3-15 cm. long; stipules membranous, lanceolate-acuminate, 3-7 mm. long, not connate; leaflets 13-19, elliptic, acute, 5-15 mm. long; racemes short, loosely 2- to 8-flowered; peduncles shorter than the leaves; flowers 2-2.5 cm. long, light to deep purplish; pedicels 1-3 mm. long; calyx tubular, about half the length of the corolla, the subulate teeth about 1/3 the length of the tube or less; pods sessile, hirsute-strigose, woody, oblong-ovoid, slightly falcate, 15-25 mm. long, about 8 mm. thick, somewhat obcompressed, lightly corrugate, short-beaked, 1-celled but the lower suture very slightly intruded.
  Open sagebrush hills and along meadows and streams; Snake R. drainage of Ida., s. to Nev. and e. Calif., e. to Mont., Wyo., and Colo. May-Aug.
  There are two intergradient phases of the species in our area, separable as follows:
1 Flowers deep purple; keel 17-19 mm. long; pubescence appressed; mostly along streams and
    in alkaline meadows; Snake R., Ida., to Mont., s. to Wyo., Utah, and Calif.   var. argophyllus
1 Flowers pale pinkish; keel mostly 13-17 mm. long; pubescence spreading; mostly in sage-
    brush or juniper woodland; Snake R., Ida., and adjacent Mont. and Wyo., s. to Colo.
    and Ariz.   var. pephragmenoides Barneby

Astragalus arthuri M. E. Jones, Contr. West. Bot. 8:20. 1898.

Atelophragma arthuri Rydb. Bull. Torrey Club 40:51. 1913. *Tium arthuri* Rydb. N. Am. Fl. 24⁷:388. 1929. *(Heller & Heller,* Lake Waha, Nez Perce Co., Ida., May 19, 1896; this was collection no. *3259)*

Strigillose, greenish perennial from a stout taproot and branched crown; stems numerous, ascending or erect, 2-4 dm. long; leaves 7-12 cm. long; stipules lanceolate, 2-4 mm. long, not connate; leaflets 17-29, oblong-elliptic, 6-12 (14) mm. long, 2-4 mm. broad, glabrous on the upper surface, those of the upper leaves linear and 1-2 mm. broad; peduncles from about equal to the subtending leaves to nearly twice their length; racemes loosely 5- to 20-flowered; pedicels 2-5 mm. long; flowers white, spreading or drooping, 10-12 mm. long; calyx 5-7 mm. long, the oblong-triangular teeth about 1/2 the length of the tube; banner erect, the wings about 1 mm. longer than the rounded keel; pod spreading to reflexed, strigillose, with a stipe about twice the length of the calyx, the body 3.5-5 cm. long, 3-4 mm. broad, compressed, triangular-cordate in section, the lower suture deeply sulcate and intruded as a nearly complete partition.

Dry grassy hills and stony meadows, on basalt; n. foothills of the Blue and Wallowa mts., Umatilla and Wallowa cos., Oreg., Asotin Co., Wash., and Idaho and Nez Perce cos., Ida. May-June.

Astragalus atratus Wats. Bot. King Exp. 69, pl. 11. 1871.

Hamosa atrata Rydb. Bull. Torrey Club 34:48. 1907. *Tium atratum* Rydb. N. Am. Fl. 24⁷:394. 1929. *(Watson 265,* Pah-Ute, Havallah and Toyabe ranges, Nev.)

A. owyheensis Nels. & Macbr. Bot. Gaz. 55:375. 1913. *A. atratus* var. *owyheensis* M. E. Jones, Rev. Astrag. 182. 1923. *Tium owyheense* Rydb. N. Am. Fl. 24⁷:395. 1929. *(Nelson & Macbride 1887,* above the "Hot Hole" of the East Bruneau, Owyhee Co., Ida., July 2, 1912)

ASTRAGALUS ATRATUS var. INSEPTUS Barneby, hoc loc. A var. *atrato* legumine 1-loculari, alisque integris, a var. *owyheensi* foliolis omnibus articulatis, legumine crassiori subcoriaceo, habituque robustiori absimilis. (Type: *Ripley & Barneby 10673,* Camas Prairie, 10 miles s.e. of Fairfield, Camas Co., Ida., June 10, 1951, in Herb. Calif. Acad. Sci.; isotypes at NY, POM, IDS, RM, US, WTU)

Grayish-strigillose perennial with a branched caudex and many very slender, often prostrate and creeping (to erect) stems (0) 1-2 (3) dm. long; leaves 4-15 cm. long, the rachis almost filiform; stipules membranous, 1-2 mm. long, not connate; leaflets 5-17, remote, filiform or linear, 3-10 mm. long, less than 0.5 mm. broad, the terminal one sometimes continuous with the rachis; racemes 6- to 15-flowered, very lax, subequal to the long filiform peduncles; pedicels slender, 3-6 mm. long; flowers 8-9 mm. long, whitish, purplish-lined or -tinged, the wings usually equaling or slightly exceeding the banner; calyx slightly less than half the length of the corolla, the teeth about 1/3 the length of the tube; pod pendulous, with a short stipe 1-2 mm. long which is concealed by the calyx, the body membranous to slightly leathery, purplish-mottled, narrowly oblong, 15-22 mm. long, about 3 mm. wide, obcompressed and about 4 (3-4.2) mm. thick, 1-celled, slightly sulcate on the lower suture, strigillose. N=12.

Sagebrush slopes; Baker Co., Oreg., s. to Nev., e. to adj. Ida. May-June.

Although our plants, which usually are called *A. owyheensis,* appear to be amply distinct, they grade gradually into *A. atratus* var. *atratus* to the south of our range in n.e. Nev. Our material is not uniform and is referable to two varieties, as follows:

1 Leaflets all very small, narrow and remote, the terminal one continuous with the rachis or represented by a small dilation of the rachis; pod of papery texture; sagebrush slopes, Baker and Malheur cos., Oreg., s.e. through Owyhee Co., Ida., to n.e. Nev.

var. owyheensis (Nels. & Macbr.) M. E. Jones

1 Leaflets more ample and less scattered, the terminal one jointed to the rachis; pod of leathery texture; stony flats where moist in the spring, n. edge of Snake R. plains in Camas Co., Ida.

var. inseptus Barneby

Astragalus beckwithii T. & G. in Pac. R. R. Rep. 2:120. 1855.

Tragacantha beckwithii Kuntze, Rev. Gen. 2:943. 1891. *Phaca beckwithii* Piper, Contr. U.S. Nat. Herb. 11:371. 1906. *Phacomene beckwithii* Rydb. Am. Journ. Bot. 16:205. 1929. *(J. A. Snyder,* "on the Cedar Mountains, west of Lone Rock, and south of Great Salt Lake," May, 1854)

A. beckwithii var. *weiserensis* M. E. Jones, Zoë 5:47. 1900. *Phacomene weiserensis* Rydb. N. Am. Fl. 24⁷:383. 1929. *A. weiserensis* Abrams, Ill. Fl. Pac. St. 2:590. 1944. (No type cited, but presumed to be *M. E. Jones,* Weiser, Ida.)

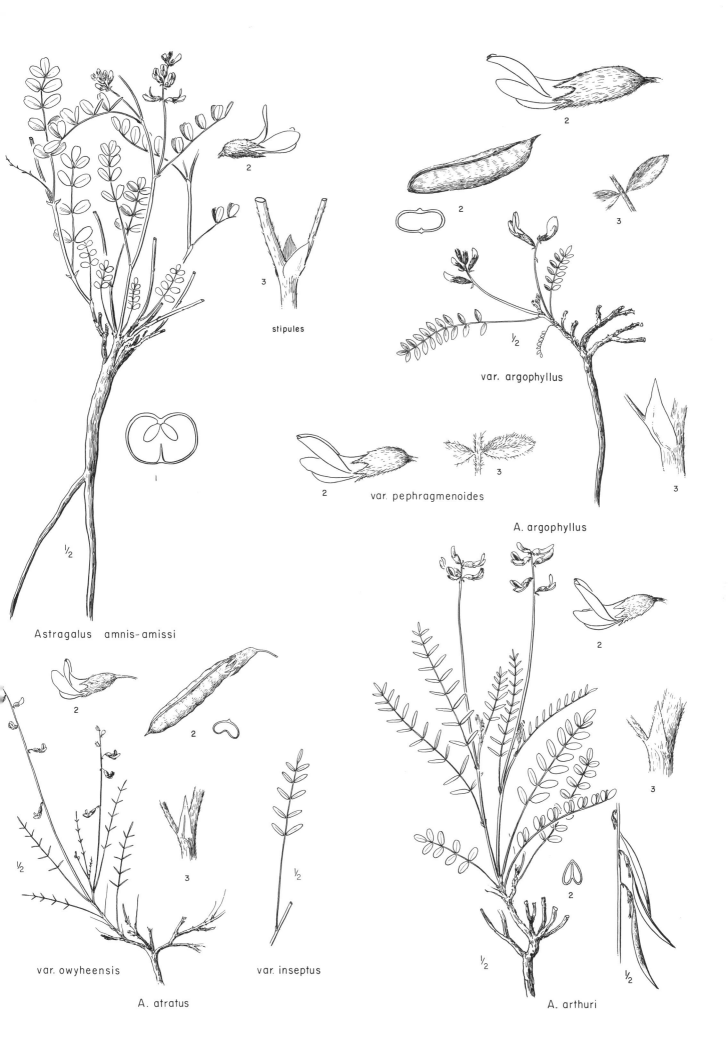

2

3

stipules

1

½

Astragalus   amnis-amissi

2

2

3

var. argophyllus

½

2

var. pephragmenoides

3

A. argophyllus

3

2

2

2

½

3

var. owyheensis

½

var. inseptus

A. atratus

2

½

2

½

A. arthuri

Phacomene pontina Rydb. N.Am. Fl. 24⁷:383. 1929. *(Macoun 450,* Spence's Bridge, B.C., May 18, 1875)

Caespitose, glabrous to sparsely strigillose-puberulent, pea-greenish and rather succulent perennial from a taproot; stems several, 2-6 dm. long, usually prostrate (ascending); leaves 7-15 cm. long; stipules ovate-lanceolate, 2-5 mm. long, not connate; leaflets 15-25, 10-20 mm. long, 4-10 mm. broad, oblong-elliptic to obovate, rounded to retuse; peduncles usually considerably shorter than the leaves; racemes 5- to 15-flowered; pedicels 1-3 mm. long; flowers creamy white to dirty yellow, about 2 cm. long, spreading; calyx glabrous or very sparsely blackish-strigillose, 9-14 mm. long, the teeth linear-lanceolate, about equal to the broadly campanulate tube; banner erect, the margins well reflexed; wings slender, 2-4 mm. longer than the rounded keel; pod glabrous, erect, stipitate; stipe stout, about equal to the calyx, jointed to the fruit; body of the pod lunate, 1.5-3 cm. long, somewhat inflated, sometimes more or less purplish-mottled, coriaceous, 1-celled, obcompressed and nearly or fully 1 cm. thick, with the upper suture very prominent, the lower slightly intruded but the false septum less than 2 mm. high. N=11.

Mostly in the lower foothills, on grassy, rocky, usually heavy clay soils; considerably interrupted in distribution, s. B.C. and n.c. Wash., southward, e. of the Cascades, to Nev. and Utah, e. through the Blue Mts., Oreg., to s.w. and c. Ida. Apr.-June.

The plants of our range might appropriately be referred to var. *weiserensis* M. E. Jones, differing from the more southern var. *beckwithii* (of Utah and e. Nev.) in the fewer (11-17) leaflets, longer calyx teeth (mostly 5-7 mm. long), and slightly larger flowers. In the species as a whole, the body of the pod is nearly always contracted at the base (just above the joint with the stipe) into an obconic, sterile neck. Plants from a small area along the upper Salmon R. in Lemhi Co., Ida., considerably isolated from the main range of the species, at elevations of 3600-4500 ft., have the pod broadly rounded at base and more deeply sulcate dorsally. They may be varietally distinct.

Astragalus bisulcatus (Hook.) Gray, Pac. R. R. Rep. 12:42. 1860.

Phaca bisulcata Hook. Fl. Bor. Am. 1:145. 1831. *Tragacantha bisulcata* Kuntze, Rev. Gen. 2:943. 1891. *Diholcos bisulcatus* Rydb. Bull. Torrey Club 32:664. 1906. *(Drummond,* "plains of the Saskatchawan")

Sparsely soft-strigillose, greenish, caespitose perennial from a thick branched crown; stems several, thick, spreading to erect, 4-10 dm. tall; leaves 4-8 (10) cm. long, short-petiolate; stipules lanceolate, 3-6 mm. long, the basal ones connate; leaflets 11-27, lanceolate to elliptic or oblong-elliptic, mostly 15-25 mm. long, 3-8 mm. broad; peduncles stout, from about as long to twice as long as the leaves; racemes closely 30- to 150-flowered, lengthening and more open in fruit; pedicels 1-2 mm. long; flowers purplish to (occasionally) pale lavender or nearly white, 11-15 mm. long; calyx 6-9 mm. long, the teeth linear-lanceolate, subequal to the strongly gibbous based, grayish- or blackish-strigillose tube; banner erect; wings and keel subequal; pod pendulous, stipitate, the stipe about equal to the calyx tube, the body linear-oblong, nearly straight, membranous, glabrous, 15-22 mm. long, strongly obcompressed, 1.5-2 mm. wide and nearly twice as thick, sulcate on either side of the upper (ventral) suture, 1-celled, neither suture intruded. N=12.

Sagebrush desert and grassland, especially where somewhat alkaline; mostly on the e. slope of the Rocky Mts., from Alta. to N.M., e. to Nebr., w. to c. Beaverhead Co., Mont., and to Clark Co., Ida. June-July.

Astragalus bourgovii Gray, Proc. Am. Acad. 6:227. 1864.

Tragacantha bourgovii Kuntze, Rev. Gen. 2:943. 1891. *Homalobus bourgovii* Rydb. Mem. N.Y. Bot. Gard. 1:247. 1900. *(Bourgeau,* "in Palliser's expedition, Rocky Mountains on the British Boundary")

Very finely strigillose, caespitose perennial from a stout taproot and branched crown, the stems numerous, from erect and only slightly decumbent at base to nearly prostrate, 1-3 dm. long; leaves 3-8 cm. long; stipules 1-3 mm. long, connate nearly half their length; leaflets 11-19, elliptic-lanceolate to elliptic-oblanceolate, 5-10 (20) mm. long; racemes loosely 5- to 10-flowered, on peduncles usually several times the length of the leaves; flowers bluish-purple, 7-10 mm. long; calyx blackish-strigillose, mostly slightly more than half the length of the corolla, the teeth linear-lanceolate, about half the length of the tube; keel about equal to the wings; pod narrowed to a stipe shorter than (and concealed by) the calyx, the body blackish-strigillose, membranous, compressed, not inflated, obliquely oblong-elliptic in outline, 12-16 mm. long, about 3 mm. wide, 1-celled, the lower suture not sulcate or at all intruded.

Gravel bars and stream banks to alpine slopes, mostly on limestone; Rocky Mts., Alta. and Mont., w. to B.C. and the Coeur d'Alene Mts., Ida. June-Aug.

Astragalus calycosus Torr. ex Wats. Bot. King Exp. 66, pl. 10. 1871.
  Tragacantha calycosa Kuntze, Rev. Gen. 2:943. 1891. *Hamosa calycosa* Rydb. Bull. Torrey Club
    40:50. 1913. *(Watson,* West Humboldt Mts., Nev.)
  A. brevicaulis A. Nels. Bull. Torrey Club 26:9. 1899. *(A. Nelson 4601,* near Fort Bridger, Wyo.,
    June 9, 1898)
Caespitose, subacaulescent perennial, silvery-strigose with dolabriform hairs; leaves many, tufted, 3-6 cm. long; stipules ovate-deltoid, 2-4 mm. long, not connate; leaflets (3) 5 (7), mostly oblanceolate, 5-15 mm. long; racemes (1) 2- to 5-flowered, peduncles usually about equaling the leaves; flowers 10-16 mm. long, white to purplish-tipped; calyx 6-9 mm. long, the teeth narrowly lanceolate, 1/2-3/4 the length of the tube; wings about equaling the well-reflexed banner, strongly 2-lobed at the tip and often with a slender tooth in the sinus, 2-4 mm. longer than the keel; pod sessile, straight or arcuate, 12-16 mm. long, cartilaginous, narrowly obcordate in section, nearly completely 2-celled by the intrusion of the lower suture, the partition sometimes contacting but not grown to the upper (ventral) suture.
  Grassland and sagebrush desert into the drier open slopes of the lower mts.; c. Ida. to Wyo., s. to Calif., Ariz., and n.w. N.M. May-July.

Astragalus canadensis L. Sp. Pl. 757. 1753. ("Habitat in Virginia, Canada")
  A. mortoni Nutt. Journ. Acad. Phila. 7:19. 1834. *A. canadensis* var. *mortoni* Wats. Bot. King Exp.
    68. 1871. *Tragacantha mortoni* Kuntze, Rev. Gen. 2:946. 1891. *Phaca mortoni* Piper, Contr. U.S.
    Nat. Herb. 11:372. 1906. ("Sources of the Missouri in Montana")
  A. spicatus Nutt. in T. & G. Fl. N.Am. 1:336. 1838, not of Pallas in 1753. *A. pachystachys* Rydb.
    N.Am. Fl. 24[7]:448. 1929, not of Bunge in 1869. *A. mortoni* f. *rydbergii* Gandg. Bull. Soc. Bot.
    Fr. 48:16. 1902. *(Nuttall,* "plains near streams, in the Rocky Mountain range") = var. *brevidens*.
  A. tristis Nutt. in T. & G. Fl. N.Am. 1:336. 1838. *(Nuttall,* "Rocky Mountains, toward the sources
    of the Platte") = var. *mortonii*.
  A. mortoni f. *brevidens* Gandg. Bull. Soc. Bot. Fr. 48:16. 1902 *A. canadensis* var. *brevidens*
    Barneby, Leafl. West. Bot. 4:238. 1946. *(Nelson,* Evanston, Wyo.)
  A. brevidens Rydb. N.Am. Fl. 24[7]:450. 1929. *(Peck 10360,* 3 mi. n. of Whitney, Baker Co., Oreg.,
    Feb. 22, 1921) = var. *brevidens*.
Erect to decumbent perennial with extensive rootstocks, 3-8 dm. tall, glabrate and greenish to grayish-strigillose with dolabriform hairs; stipules lanceolate-acuminate, 6-15 mm. long, some of them more or less connate or all free; leaves mostly 10-20 cm. long; leaflets 13-29, ovate-lanceolate to oblong, 1.5-4 cm. long, (3) 5-18 mm. broad, truncate to retuse, usually glabrous on the upper surface; peduncles 5-20 cm. long; racemes spikelike, 5-15 cm. long, closely 30- to 150-flowered; pedicels stout, 1-2 mm. long, with 2 tiny linear bracteoles at the top just below the calyx; flowers ochroleucous to white, spreading-drooping, 12-18 mm. long; calyx 6-9 mm. long, finely strigillose, the base asymmetrical, the tube about twice the length of the linear-lanceolate teeth; wing petals very narrow, about equal to the banner, 1-2 mm. longer than the blunt, usually purplish-tipped keel; pod erect, sessile, 8-20 mm. long, 4-5 mm. broad, cartilaginous-woody, cordate-terete in section, completely 2-celled by the intrusion of the lower suture, abruptly sharp pointed, rather generally strigillose.
  British Columbia to Hudson's Bay and the Atlantic coast, s. to Calif., Colo., and on the Great Plains to Tex. June-July.
  Astragalus canadensis var. *canadensis* occurs e. of the Rocky Mts., from Alta. s. to n. Tex. and e. to Sask., Hudson's Bay, and the Atlantic coast. In our area, the following two phases (which are more or less sympatric and freely intergradient) occur:
1 Calyx pubescent with both white and black hairs, the teeth subequal, 1.5-3 mm. long; plant
    greenish, not silvery-pubescent; chiefly in ponderosa pine and lower montane forest, less
    commonly along watercourses in sagebrush; Okanogan Co., Wash., e. to Ida. and Mont.
    N=8.   var. mortonii (Nutt.) Wats.
1 Calyx white-hairy, the teeth 1-2 mm. long, the upper ones much broader and shorter than
    the lower three; plants usually more or less silvery-pubescent; mostly along water-
    courses or in somewhat alkaline flats, less commonly in open forest; B.C. to Calif., on
    the e. side of the Cascades, e. to the Dakotas and Colo.   var. brevidens (Gandg.) Barneby

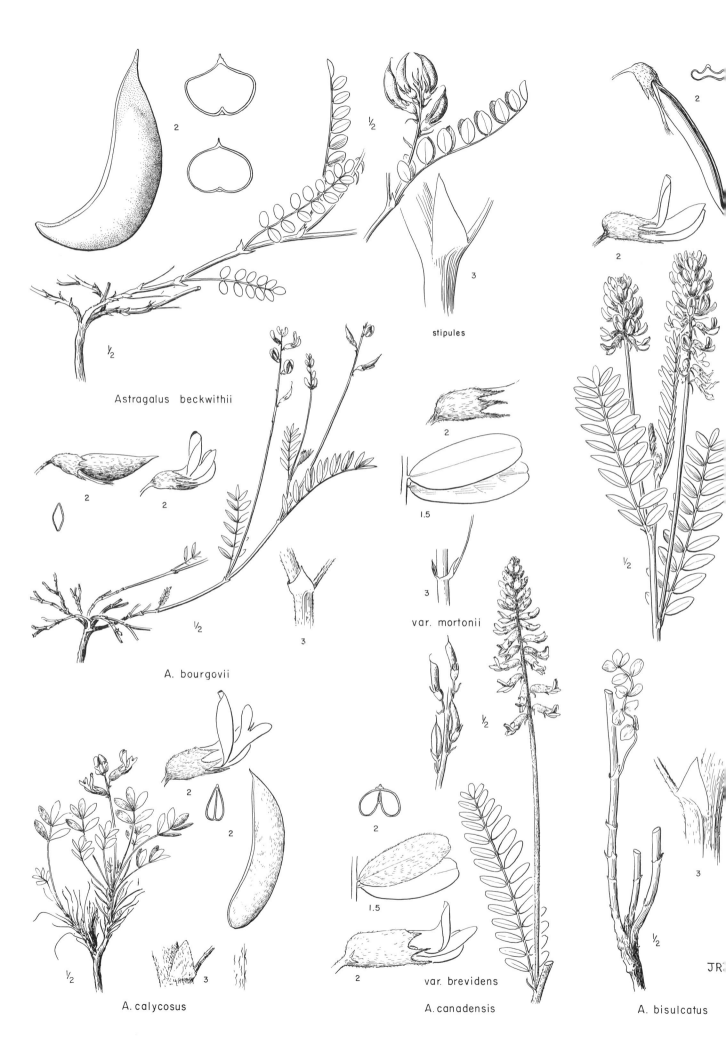

Astragalus beckwithii

A. bourgovii

stipules

var. mortonii

var. brevidens

A. calycosus

A. canadensis

A. bisulcatus

JR

Astragalus casei Gray in Brew. & Wats. Bot. Calif. 1:154. 1876.

Tragacantha casei Kuntze, Rev. Gen. 2:943. 1891. *Xylophacos casei* Rydb. Bull. Torrey Club 52: 147. 1925. *(Lemmon & Case,* near Pyramid Lake, Nev.)

Wiry, sparsely leaved perennial from a taproot and buried root crown, strigillose throughout, the leaflets commonly more densely pubescent above than beneath; stems usually diffuse, 1-4 dm. long; leaves 3-10 cm. long; stipules 2-5 mm. long, the papery lower ones strongly amplexicaul but free, the upper ones herbaceous, deflexed; leaflets 9-15, distant, narrowly elliptic-oblong to linear, obtuse or retuse, 5-25 mm. long; peduncles 3-9 cm. long; racemes mostly 8- to 20-flowered, the flowers ascending or declined in age, the axis elongating, at length 3-12 cm. long; flowers pink-purple (the wing tips white), the banner 13.5-18 mm. long; calyx 6-9 mm. long, blackish-strigillose, the cylindric-campanulate tube 4.5-7.5 mm. long, the teeth about 1.5 mm. long; pod declined, sessile, lance- or oblong-ellipsoid, 20-55 mm. long, 5-10 mm. wide, obcompressed below the triangular-acuminate, laterally flattened beak, either straight, gently incurved, or sigmoidally arcuate (decurved proximally and incurved distally), 1-locular, the fleshy, red-mottled valves becoming brownish and stiffly leathery, white-strigillose.

Dry hillsides and valley floors, in sandy or gravelly alkaline soils; locally plentiful in w. Nev. and adj. Calif. and collected many years ago by T. S. Brandegee in the Walla Walla region, Wash., although not since seen in our area. May-July.

Astragalus ceramicus Sheld. Minn. Bot. Stud. 1:19. 1894.

Phaca picta Gray, Pl. Fendl. 37. 1849. *A. pictus* Gray, Proc. Am. Acad. 6:214. 1864, but not of Steud. in 1840. *(Fendler,* Rio del Norte)

Psoralea longifolia Pursh, Fl. Am. Sept. 741. 1814. *Orobus longifolius* Nutt. Gen. Pl. 2:95. 1818. *Phaca longifolia* Nutt. ex T. & G. Fl. N.Am. 1:346. 1838. *A. filifolius* Gray, Pac. R. R. Rep. 12:42. 1860, but not of Clos, in 1846. *A. pictus* var. *filifolius* Gray, Proc. Am. Acad. 6:215. 1864. *A. ceramicus* var. *imperfectus* Sheld. Minn. Bot. Stud. 1:19. 1894. *A. longifolius* Rydb. Fl. Nebr. 21:47. 1895, but not of Lam. in 1783. *A. mitophyllus* Kearney, Leafl. West. Bot. 4:216. 1946. *A. ceramicus* var. *filifolius* Hermann, Journ. Wash. Acad. Sci. 38:237. 1948. *(Bradbury,* "in upper Louisiana")

Finely strigillose, usually canescent perennial with a deep root and long slender filiform rootstocks, easily mistaken for an annual; stems 1-3 dm. tall, flexuous and often zigzag; stipules linear-lanceolate, 3-8 mm. long, the lower ones usually very shortly connate, the upper ones free; leaflets mostly lacking, the rachis elongate and flexuous, 5-13 cm. long, occasionally with 1-several linear leaflets 1-3.5 cm. long; racemes (1) 3- to 10-flowered; peduncles much shorter than the leaves; pedicels 1-3 mm. long; flowers 6-9 mm. long; calyx about 4 mm. long, strigillose, often purplish mottled, the linear-lanceolate teeth about equal to the tube; corolla ochroleucous, the banner erect, the wings with short auricles, little if any longer than the (usually) purplish-tipped keel; pod ordinarily stipitate but in our range usually sessile, the body membranous, much inflated, purplish-mottled, oblong-ellipsoid, 2.5-3.5 cm. long, glabrous, 1-celled, the lower suture often prominent but only very slightly if at all intruded.

Sand dunes and prairie lands; c. Ida. to e. Mont., the Dakotas, and Neb., s. to Kans., w. Okla., and Ariz., just at the southern limit of our range. May-July.

Our plants are confined to a small area of dunes and sandy plains in Butte, Jefferson, Madison, and Bonneville cos., Ida., and are widely separated from the nearest stations of var. *filifolius* (Gray) Hermann in e. Wyo., e. Mont., and Colo. They resemble that variety in every way, except that the pod is nearly or quite sessile; they may deserve varietal status. The var. *ceramicus,* characterized by oval to elliptic leaflets on the lower leaves, ranges from the Dakotas to Okla. and Ariz.

Astragalus chinensis Linn. f. Decad. 1:5, pl. 3. 1762. (China)

Strigillose, greenish perennial with short rootstocks; stems thick, fistulose, 3-7 dm. long; leaves 5-25 cm. long; stipules not connate; leaflets 13-19, oblong, 1-2.5 cm. long; peduncles shorter than the leaves; racemes 8- to 12-flowered; flowers spreading, ochroleucous, 13-17 mm. long; calyx short, tubular, about 5 mm. long, the teeth slightly shorter than the tube; keel very broad, conspicuously longer and broader than the wings; pod pendulous, with a stipe nearly twice as long as the calyx, the body woody, subglobose, 1-2 cm. long, strongly reticulate, the lower suture sulcate and intruded to completely partition the loculus.

An introduced weed in Clark Co., Ida., on waste or cultivated ground. June-Aug.

Astragalus cibarius Sheld. Minn. Bot. Stud. 1:149. 1894.

Xylophacos cibarius Rydb. Bull. Torrey Club 40:48. 1913. *A. webberi* var. *cibarius* M. E. Jones, Contr. West. Bot. 10:87. 1902. *(M. E. Jones,* Utah Valley, Utah, May, 1880)

A. arietinus M. E. Jones, Proc. Calif. Acad. Sci. II, 5:653. 1895. *(Jones 55540,* Fairview, Utah)

A rather fleshy-leaved, greenish, caespitose, sparsely strigillose perennial with decumbent to ascending stems 1-3 dm. tall; stipules membranous, hyaline, prominently veined, oblong-rounded, 3-5 mm. long, all but the very basal ones not connate; leaflets 13-19, oblong-elliptic to obovate, 5-15 mm. long, sparsely strigillose on the lower surface, usually glabrous on the upper; peduncles mostly shorter than the leaves; racemes closely 4- to 15-flowered, more or less headlike; pedicels scarcely 1 mm. long; flowers violet-purple to nearly white, 15-20 mm. long, ascending; calyx 7-10 mm. long, usually darkish-hairy, the teeth linear-lanceolate, 1/2-3/4 the length of the tube; banner narrow, well reflexed, wings scarcely equal to the banner, narrower (but 2-4 mm. longer) than the purplish, blunt keel; pod erect, sessile or subsessile, glabrous to sparsely grayish-hairy, usually arcuate, fleshy and drying to cartilaginous, 2.5-3.5 cm. long, obcompressed, usually thicker (7-10 mm.) than broad, narrowed to a conspicuous beak, 1-celled, the lower suture usually sulcate but not intruded. N=11.

Mostly in sagebrush; c. Ida. to Mont., s. to n. e. Nev., Utah, and n. w. Colo. May-June.

Astragalus collinus Dougl. ex Hook. Fl. Bor. Am. 1:141. 1830, as a synonym; ex G. Don, Gen. Syst. 2:256. 1832.

Phaca collina Hook. loc. cit. *Homalobus collinus* Rydb. Bull. Torrey Club 40:53. 1913. *Tragacantha collina* Kuntze, Rev. Gen. 2:944. 1891. *(Douglas,* "Native of N. America, on the banks of the Columbia River")

A. cyrtoides Gray, Proc. Am. Acad. 6:201. 1864. *(Spalding,* hillsides, Clear Water River, Apr. 14) = var. *collinus*.

Homalobus laurentii Rydb. Bull. Torrey Club 51:15. 1924. *A. laurentii* Peck, Man. High. Pl. Oreg. 443. 1941. *A. collinus* var. *laurentii* Barneby, Am. Midl. Nat. 55:487. 1956. *(W. E. Lawrence 774,* e. of Heppner, Oreg.)

Finely crisp-puberulent or -strigillose perennial with a taproot and several spreading to erect stems 1-5 dm. tall; leaves 3-8 (10) cm. long; stipules linear-lanceolate to oblong, 2-5 mm. long, not connate; leaflets 11-21 (25), linear to oblong or oblanceolate, 8-20 mm. long, 1-4 (4.5) mm. broad, often retuse; peduncles usually considerably longer than the leaves, as much as 15 cm. long; racemes closely 15- to 40-flowered, elongated and rather open in fruit; pedicels 2-3 mm. long; flowers reflexed, about 1.5 (1.3-1.7) cm. long, creamy white to yellowish or greenish-tinged; calyx slightly gibbous at base, 7-10 mm. long, crisply white (blackish) -puberulent, the teeth 1-3 mm. long; banner "stubby" and upturned only slightly, the wings considerably exceeding the banner but scarcely exceeding the blunt keel; pod pendulous, stipitate, the stipe 5-15 mm. long, the body linear to lance-oblong or more rarely lunately lance-oblong in outline, straight or gently incurved, 7-25 mm. long, 2.5-4.2 mm. broad, cuneate at each end, laterally compressed and bicarinate by the prominent sutures, the somewhat fleshy, villosulous or loosely strigillose (exceptionally glabrous) valves becoming leathery and impressed-reticulate, not inflexed. N=12.

Basaltic grassland and sagebrush desert from w. B.C. to Oreg., eastward to w. c. Ida., along the Snake and Clearwater rivers. May-June.

The species is represented by the following well-marked phases:

1 Pod slenderly linear-oblong, 7-25 mm. long, 2.5-3.5 mm. broad, straight or rarely a little incurved, crisply villosulous with hairs up to 0.25-0.5 mm. long, 12- to 18-ovulate; basaltic grassland and sagebrush desert, n. e. Oreg. (Umatilla and Wallowa cos.), n. through the Columbia Basin to Grand Coulee and e. to the lower Salmon and Clearwater rivers in w. c. Ida.; also in s. B.C., where somewhat isolated at the head of the Okanogan and on the middle Fraser rivers   var. collinus

1 Pod plumper and averaging shorter, obliquely ovate-oblong, lunately incurved, 8-15 mm. long, 3.3-4.2 mm. broad, either villosulous with hairs up to 0.5-1 mm. long or glabrous, 8- to 12-ovulate; similar habitats, but known only from Morrow Co., Oreg.   var. laurentii (Rydb.) Barneby

Astragalus convallarius Greene, Erythea 1:207. 1893.

Homalobus campestris Nutt. in T. & G. Fl. N. Am. 1:351. 1838. *A. campestris* Gray, Proc. Am. Acad. 6:229. 1864, but not of L. in 1753. *Tragacantha campestris* Kuntze, Rev. Gen. 2:943. 1891. *A. serotinus* var. *campestris* M. E. Jones, Proc. Calif. Acad. Sci. II, 5:668. 1895. *A. decum-*

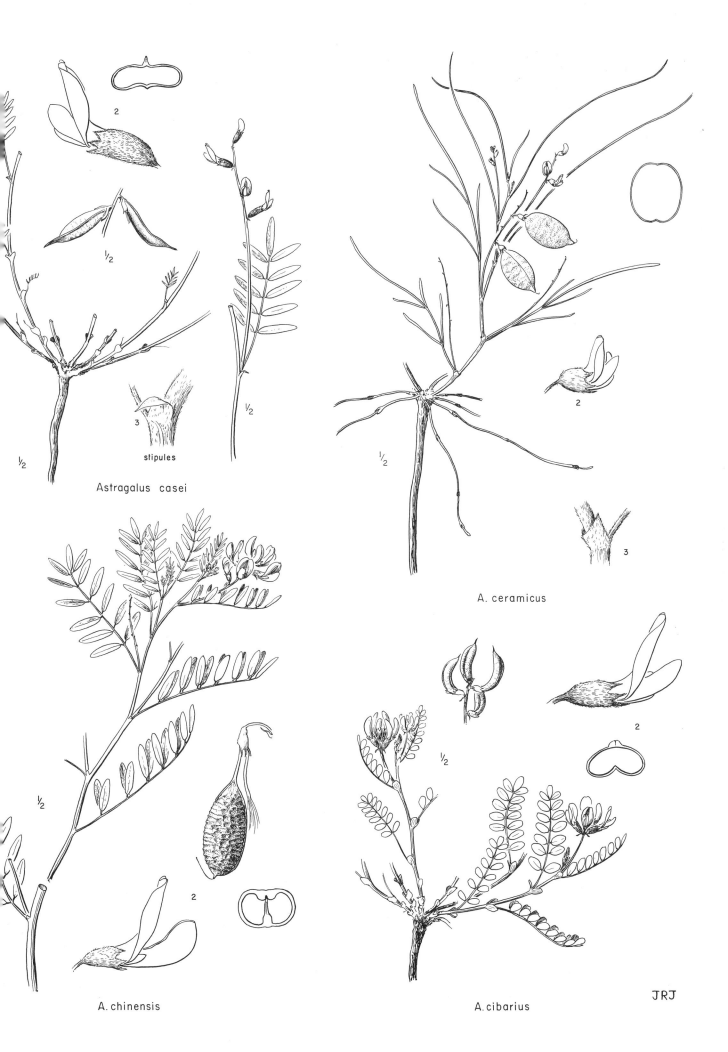

2

½

3
stipules

½

½

Astragalus casei

A. ceramicus

3

2

½

½

½

2

A. chinensis

A. cibarius

JRJ

bens var. *convallarius* M. E. Jones, Contr. West. Bot. 10:58. 1902. *Phaca convallaria* Piper, Contr. U. S. Nat. Herb. 11:373. 1906. *(Nuttall,* "sandy prairies of the Colorado of the west near the sources of the Platte")

Homalobus junceus Nutt. in T. & G. Fl. N. Am. 1:351. 1838. *A. junceus* Gray, Proc. Am. Acad. 6:230. 1864, but not of Ledeb. in 1826. *Tragacantha juncea* Kuntze, Rev. Gen. 2:945. 1891. *A. diversifolius* var. *junceus* M. E. Jones, Contr. West. Bot. 8:13. 1898. *A. diversifolius* var. *róborum* M. E. Jones, Contr. West. Bot. 10:61. 1902. *(Nuttall,* "with the preceding" [i. e. , *H. campestris]* "and in sandy places in the Rocky Mountain range towards the Oregon")

Minutely strigillose, rushlike perennial, often with rootstocks; stems erect, 3-6 dm. tall; stipules short, triangular, the lowermost connate; leaves 4-12 cm. long, rachis broadened and 1-1.5 mm. wide, leaflets usually lacking on many of the leaves, or as many as 7-9, linear and scarcely as broad as the rachis, 1-3 cm. long, the terminal leaflet confluent with and not distinguishable from the rachis; stipules 3 mm. long, slightly connate below, usually free above; racemes loosely 8- to 20-flowered; peduncles from shorter to longer than the leaves; flowers 8-12 mm. long, ochroleucous, shading to purplish; pedicels 1-2 mm. long; calyx about half the length of the corolla, the teeth scarcely 1/3 as long as the tube; wings and keel subequal to the strongly reflexed banner; pod pendulous, sessile, linear, 20-40 mm. long, 2-2.5 mm. broad, compressed, strigillose, 1-celled, neither suture sulcate or intruded.

Grasslands and sagebrush desert; Snake R. drainage, s. e. Ida. to s. c. Mont. , s. to Wyo. and Utah, more or less at the edge of our range. June-July.

There is reason to question that this plant is more than an extreme phase of *A. diversifolius* which it resembles in floral characters.

Astragalus cottonii M. E. Jones, Rev. Astrag. 135. 1923.

Astragalus olympicus Cotton, Bull. Torrey Club 29:573. 1902, not of Pall. in 1800. *Atelophragma cottoni* Rydb. Bull. Torrey Club 55:121. 1928. *(Elmer,* Olympic Mts. , Clallam Co. , Wash. , July, 1900)

Grayish villous-hirsutulous perennial with a well-branched crown and deep, stout taproot; stems rather thick, 1-2.5 dm. long, the internodes very short, the closely crowded, subsessile leaves 2-4 (6) cm. long; stipules ovate to oblong-lanceolate, rounded, 3-6 mm. long, only the lowest ever connate; leaflets 11-17, linear-elliptic to oblong or oblong-obovate, 8-16 mm. long; racemes rather loosely 7- to 20-flowered, peduncles about equaling to slightly exceeding the leaves; pedicels 0.5-2 mm. long; flowers 7-10 mm. long; calyx about 2/3 the length of the corolla, the linear-lanceolate lobes from half as long to nearly as long as the villous tube; corolla ochroleucous but at least the keel purplish tinged, the banner erect; pod with a slender stipe somewhat longer than the calyx, the body strongly inflated, membranous, glabrous, not mottled, oblong-ellipsoid, about 2.5 cm. long, 5-10 mm. broad, slightly compressed, not sulcate but the lower suture intruded to form a septum somewhat less than 1 mm. high.

Alpine talus slopes of the Olympic Mts. , Clallam Co. , Wash. June-Aug.

Astragalus crassicarpus Nutt. in Fraser's Cat. 1. 1813.

Geoprumnon crassicarpum Rydb. ex Small, Fl. S. E. U. S. 615, 1332. 1903. *(Nuttall,* Platte River, Neb.)

A. carnosus Pursh, Fl. Am. Sept. 740. 1814. *(Bradbury,* "in upper Louisiana")

A. caryocarpus Ker, Bot. Reg. 2:pl. 176. 1816. *Tragacantha caryocarpa* Kuntze, Rev. Gen. 2:943. 1891. (Cultivated plants from seeds sent by Nuttall from "Upper Louisiana")

A. succulentus Richards. Frankl. Journ. 746. 1823. *Geoprumnon succulentum* Rydb. Bull. Torrey Club 32:658. 1906. *(Richardson,* sandy plains in the neighborhood of Carlton, Sask.)

A. prunifer Rydb. Mem. N. Y. Bot. Gard. 1:239. 1900. *(A. Nelson 3143,* Medicine Bow, Wyo. , in 1897)

A. succulentus var. *paysoni* Kelso, Rhodora 39:151. 1937. *A. crassicarpus* var. *paysoni* Barneby, Am. Midl. Nat. 55:497. 1956. *(Payson & Payson 2514,* near Encampment, Carbon Co. , Wyo. , June 30, 1922)

Minutely strigillose, greenish, fleshy perennial with a thick taproot, branched crown, and prostrate to ascending stems 1-5 dm. long; stipules lanceolate, 3-8 mm. long, not connate; leaflets 13-21, oblong-oblanceolate to elliptic, 5-20 mm. long, rather fleshy, greenish, strigillose on the lower surface, usually glabrous above; racemes 5- to 20-flowered, usually shorter than the leaves; pedicels 1-3 mm. long; flowers 22-30 mm. long, spreading; calyx closely subtended by (usually) 2 membranous

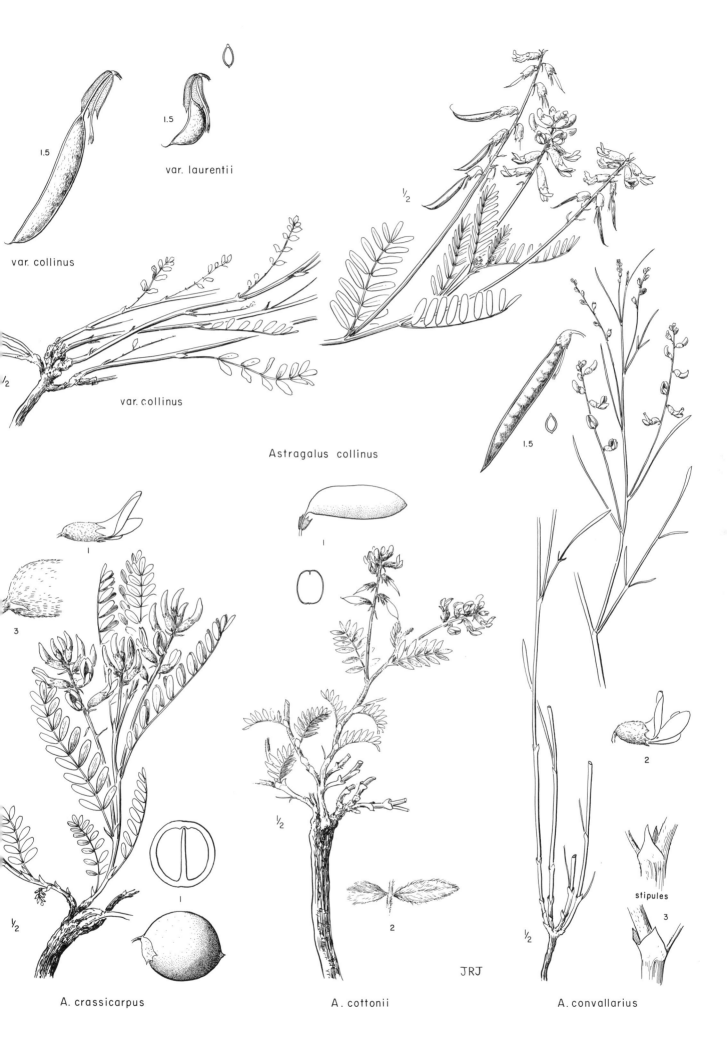

1.5

1.5

var. laurentii

1/2

var. collinus

1/2

var. collinus

Astragalus collinus

1.5

1

3

1

1/2

A. crassicarpus

1

1/2

A. cottonii

2

JRJ

1/2

stipules

2

3

A. convallarius

bractlets, more or less blackish-strigillose, tubular, 1/2-2/3 as long as the corolla, the teeth lanceolate, 1/4-1/3 the length of the tube; banner oblanceolate, retuse, usually clear white, varying to purplish; wings about equaling the banner, the blade auriculate, whitish but usually purplish-tinged to purple; keel much shorter than the wings, rounded, purplish-tinged to purple; pods sessile, subglobose, 2-2.5 cm. long, glabrous, very fleshy (hard and bony when dry), indehiscent, nearly terete in section, almost completely 2-celled, but the partition not grown to the upper suture.

Prairies and foothills; mostly e. of the Rocky Mts., from N. M. to Alta. and Sask., e. to Mo., Tenn., and Tex.; entering our range in Mont., and extending w. as far as Flathead, Granite, and Beaverhead cos. June-July.

There is considerable variation in the color of the corolla, but there seems to be little basis for the recognition of infraspecific taxa on this feature, even though the more western, whitish-flowered plants are sometimes maintained either as a distinct species, *A. succulentus* Richards., or as the var. *paysoni* (Kelso) Barneby, in contrast to the purplish-flowered *A. crassicarpus*.

Astragalus curvicarpus (Sheld.) Macbr. Contr. Gray Herb. n.s. 65:38. 1922.
　　A. speirocarpus var. *falciformis* Gray, Bot. Calif. 1:152. 1876. *A. speirocarpus* var. *curvicarpus* Sheld. Minn. Bot. Stud. 1:125. 1894. *A. gibbsii* var. *curvicarpus* M. E. Jones, Contr. West. Bot. 10:62. 1902. *Homalobus curvicarpus* Heller, Muhl. 2:86. 1905. (*Lemmon*, Sierra Co., Calif.)
　　Homalobus subglaber Rydb. Bull. Torrey Club 51:17. 1924. *A. subglaber* Peck, Man. High. Pl. Oreg. 444. 1940. *A. curvicarpus* var. *subglaber* Barneby, Am. Midl. Nat. 55:487. 1956. (*Thomas Howell*, John Day River, Oreg., May 11, 1882)
　　A. whitedii var. *brachycodon* Barneby, Am. Midl. Nat. 41:496. 1949. *A. curvicarpus* var. *brachycodon* Barneby, Am. Midl. Nat. 55:487. 1956. (*Peck 21491*, 3 mi. n. of Redmond, Deschutes Co., Oreg., July 2, 1942)

Loosely caespitose, minutely puberulent-strigillose and often somewhat grayish (or sometimes deep green and nearly glabrous) perennial with several spreading to erect stems 2-5 dm. long; leaves 4-10 cm. long; stipules 1-3 mm. long, not connate; leaflets 9-17, oblong to oblong-obovate, 5-17 mm. long, 2-7 mm. broad, truncate to retuse or obcordate; peduncles usually about equaling, but at anthesis up to twice as long as, the leaves; racemes rather closely to loosely 10- to 40-flowered; pedicels 1-3 mm. long; flowers ochroleucous, reflexed, 13-19 mm. long; calyx gibbous based, (6) 8-10 (13.5) mm. long, the teeth scarcely 1 mm. long; banner well reflexed, the wings 2-7 mm. longer than the rounded keel; pod reflexed, slender-stipitate, the stipe usually at least twice as long as the calyx, the body 1-celled, strongly compressed, glabrous or more commonly pubescent, 2.5-3.5 cm. long, 3-5 mm. broad, coiled to form 1/2 to a full turn, the ventral suture very prominent, but neither suture at all intruded.

Sagebrush deserts, from Nev. and n. Calif. to s. Ida. and c. Wash. May-July.

Two rather local variants merit recognition in addition to the more widespread var. *curvicarpus*:
1 Ovary and pod pubescent; banner 16.5-20 mm. long, with a well-developed blade; keel
　　12-15 mm. long; leaflets pubescent (although sometimes sparsely so) on the upper surface; sagebrush plains and foothills, sometimes on dunes, s.w. foothills of the Blue Mt. system, Oreg., s.e. to the Snake R. plains, s. Ida., and s. to Nev. and e. Calif.
　　　　　　　　　　　　　　　　　　　　　　　　　　　　　　var. curvicarpus
1 Ovary and pod glabrous or plant otherwise not as above, the flowers and leaflets various
　　2 Calyx mostly 6-8.5 mm. long; banner mostly 13.5-15.5 mm. long; keel 9.5-11.5 mm.
　　　　long; leaflets pubescent on the upper surface, even if only thinly so; dry hillsides and plains, on basalt or in pumice sand, locally plentiful on the upper forks of the Deschutes R., Crook, Deschutes, and Jefferson cos., Oreg.; intergrading with the other vars. to the n. and s.　　　　　　　　　　　　　　　　　var. brachycodon Barneby
　　2 Calyx 9-13.5 mm. long; banner (with poorly developed blade) 13.5-19.5 mm. long; keel
　　　　11.5-14.5 mm. long; leaflets nearly glabrous, sparsely strigillose on the lower surface along the midvein; dry rocky hillsides and sagebrush flats, in the John Day and lower Deschutes drainage systems, Grant to Wasco cos., s. to n. Jefferson Co., e.c. Oreg.　　　　　　　　　　　　　　　　　var. subglaber (Rydb.) Barneby

Astragalus cusickii Gray, Proc. Am. Acad. 13:370. 1878.
　　Phaca cusickii Rydb. Bull. Torrey Club 40:47. 1913. (*Cusick 68*, Union Co., Oreg.)
　　A. cusickii var. *flexilipes* Barneby, Am. Midl. Nat. 55:485. 1956. (*Ripley & Barneby 10709*, n. of Pollock, Idaho Co., Ida.)

Greenish, glabrous to sparsely strigillose perennial from a woody taproot and branched caudex; stems many, 3-6 dm. tall, angled and grooved; leaves (5) 6-10 cm. long; stipules lanceolate, 2-5 mm. long, the lowermost sometimes connate; leaflets 7-19, remote, linear, 5-20 mm. long, about 1 mm. broad; racemes loosely 10- to 30-flowered, usually long-pedunculate and borne above the leaves; flowers (white) ochroleucous, with at most only a light purplish tinge except sometimes the tip of the keel, spreading to reflexed, (9) 11-16.5 mm. long; pedicels 1-4 mm. long; calyx campanulate, about 1/3 as long as the corolla, sparsely and finely blackish (white)-strigillose, the teeth triangular, not quite 1 mm. long; wing petals with a narrow auriculate blade 2-4 mm. longer than the rounded keel; pod with a slender stipe once to twice as long as the calyx, the body membranous and thin, more or less translucent, glabrous, ellipsoid-obovoid, 2.5-4.5 cm. long, nearly terete in section, 6-22 mm. broad, 1-celled, the lower suture not at all intruded. N=11.

From sagebrush plains to grassy or rocky slopes, expecially on talus; e. Oreg. to Custer Co., Ida. May-July.

The species is represented by two phases, separable as follows:

1 Leaflets mostly 7-11 (13), nearly always strigillose on the upper surface; calyx 5-6.5 mm.
    long; petals white, without any trace of purple; pod obovoid, 12-22 mm. broad when pressed;
    Snake R. Canyon and tributaries from the west, from Weiser downstream to the mouth of the
    Grande Ronde, e. Oreg., immediately adj. Ida., and s.e. Wash.   var. cusickii
1 Leaflets 9-15 (19), generally glabrous on the upper surface; calyx 4-5 mm. long; petals
    pale purple or white except for the purple-tipped keel; pod half obovoid or half ellipsoid,
    the ventral suture straight or a little concave, 6-12 mm. broad when pressed; canyons
    of the lower Salmon and Little Salmon rivers, Idaho and n. Adams cos., Ida.
                                                                        var. flexilipes Barneby
There is evidence of intergradation between these two ordinarily distinct forms along the Snake R. below its confluence with the Salmon, where intermediate plants combining the larger flower of var. *cusickii* and the half ellipsoid pod of var. *flexilipes* occur. The slight difference in corolla color is in general dependably distinctive.

Astragalus dasyglottis Fisch. ex DC. Prodr. 2:282. 1825.

A. hypoglottis var. *dasyglottis* Ledeb. Fl. Alt. 3:293. 1831. ("In Sibiria Altaica")

A. agrestis Dougl. ex Hook. Fl. Bor. Am. 1:148. 1831, as a synonym; ex G. Don, Gen. Syst. 2:258. 1832. *Phaca agrestis* Piper, Contr. U.S. Nat. Herb. 11:372. 1906. *(Douglas,* "on the fertile plains of the Red River, and in the south, towards Pembina;" Don, however, cited the type as *Douglas,* "Native of North America, near the Columbia River, in fields")

A. goniatus Nutt. in T. & G. Fl. N.Am. 1:330. 1838. *(Nuttall,* "Rocky Mts., near the sources of the Platte")

A. hypoglottis var. *polyspermus* T. & G. Fl. N.Am. 1:328. 1838. *A. agrestis* var. *polyspermus* M. E. Jones, Contr. West. Bot. 10:65. 1902. *(Nuttall,* "on the Platte and near the sources of the Canadian")

A. tarletonis Rydb. Bull. N.Y. Bot. Gard. 2:175. 1901. *(Tarleton 78,* Five Finger Rapids, Yukon)

Low, appressed-pubescent perennial with long rootstocks from a buried crown; stems numerous, slender, decumbent to erect, 1-3 dm. tall; stipules linear to ovate, usually basally connate; leaves 4-10 cm. long; leaflets 11-19, linear-lanceolate to oblong-lanceolate, mostly retuse, 1-2 cm. long; racemes headlike, axillary, 7- to 20-flowered, 1.5-2.5 cm. broad and not much longer, even in fruit; peduncles shorter to longer than the leaves; flowers erect, about 17 (13-19) mm. long; pedicels thick, scarcely 0.5 mm. long; calyx tubular-campanulate, about half the length of the corolla, grayish- to blackish-strigose, the teeth linear, somewhat shorter than the tube; corolla usually purplish, or the wings whitish, rarely more nearly uniformly whitish and merely purplish-tinged, the banner narrow, longer than the slender wings which are about 4 mm. longer than the slightly acutish keel; pod sessile, erect, about 1 cm. long, grayish- to blackish-hirsute, ovoid, cordate in section, deeply sulcate, 2-celled by the complete intrusion of the lower suture. N=8.

Moist spots in sagebrush plains and montane meadows to alpine slopes; Yukon southward, on the e. side of the Cascades, from B.C. to n. Calif. and in the Rocky Mts. to N.M., e. to Man., Minn., Ia., and Kans.; e. Asia. June-Aug.

Plants with black-hairy calyces are sometimes treated as distinct from those in which the calyx is white-hairy. Since there is complete transition between the two extreme conditions, with most plants having a mixed pubescence, there appears to be no natural basis for the maintenance of more than the one taxon.

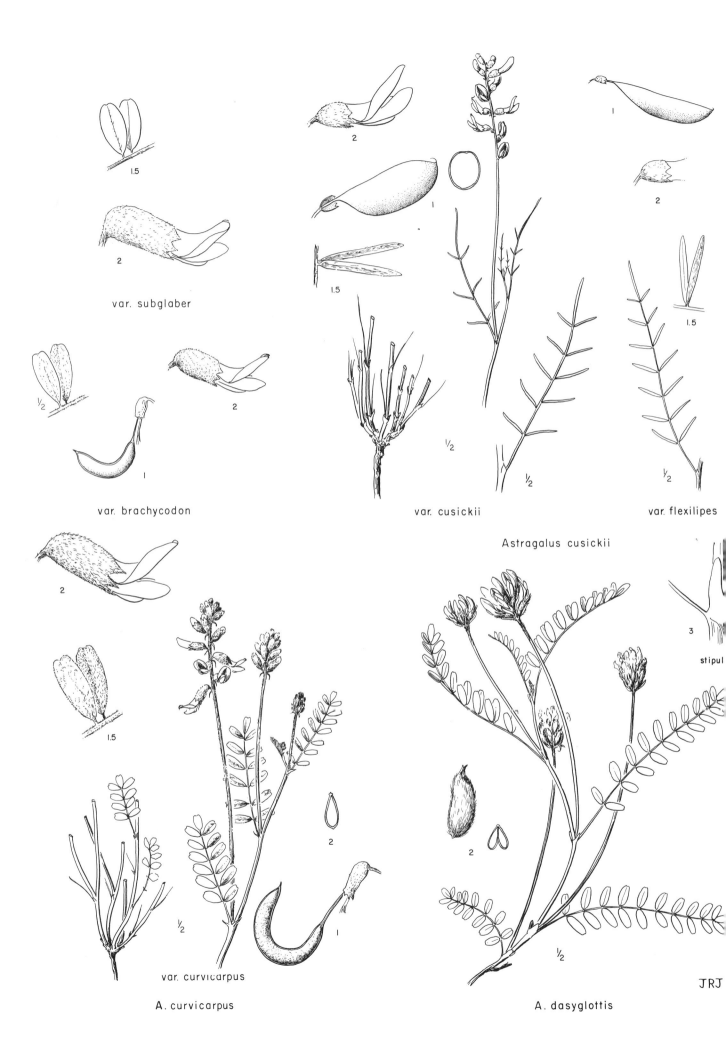

var. subglaber

var. brachycodon

var. cusickii

var. flexilipes

Astragalus cusickii

stipul

var. curvicarpus

A. curvicarpus

A. dasyglottis

JRJ

Astragalus diaphanus Dougl. ex Hook. Fl. Bor. Am. 1:151. 1831.
   A. lentiginosus var. *diaphanus* M. E. Jones, Proc. Calif. Acad. Sci. II, 5:675. 1895. *(Douglas,*
      "Abundant on sandy soil near the Great Falls of the Columbia")
   A. drepanolobus Gray, Proc. Am. Acad. 19:75. 1883. *Hamosa drepanoloba* Rydb. Bull. Torrey Club
      54:21. 1927. *(T. Howell,* gravel bars near mouth of John Day R. , Oreg.)
   Prostrate annual or biennial (definitely not a long-lived perennial) from a stout taproot, strigillose
to subglabrate; stems numerous, 1-4 dm. long; leaves 2-4 cm. long; stipules tiny, 1-2 mm. long, not
connate; leaflets 9-13, obovate, 4-12 mm. long, usually glabrous on the upper surface; racemes compactly 5- to 20-flowered; peduncles equaling or shorter than the leaves; flowers 4-8 mm. long, white
but the banner pinkish- or purplish-veined and the keel more or less purplish-tipped; calyx 2-3 mm.
long, often purplish, the teeth linear, somewhat shorter than the tube; pods sessile, linear-oblong in
outline, strongly arcuate, membranous, 1.5-2 cm. long, 3-4 mm. thick, cordate and 2-celled in section due to the complete intrusion of the lower suture. N=14.
   Gravel bars, alluvial slopes, and in thin gravelly soil overlying basaltic rocks along the Columbia
R. from the mouth of the John Day R. , in Wasco Co. , Oreg. , to Klickitat Co. , Wash. May-June.

Astragalus diurnus Wats. Proc. Am. Acad. 21:450. 1886.
   Phaca diurna Rydb. N.Am. Fl. 24^6:353. 1929. *(Thomas Howell,* "at Dayville, on John Day River,"
      Oreg. , but an isotype at WSU, so specified by Howell, is labeled "Trout Creek, Blue Mts. , May
      18, 1885")
   A. craigi M. E. Jones, Contr. West. Bot. 9:42. 1900. *(Howell,* "John Day's River," Oreg., May, 1885)
   Strigillose, rather grayish biennial or possibly a short-lived perennial from a taproot and branched
crown; stems several, ascending or decumbent, 1-2 dm. long; leaves 3-5 cm. long; stipules membranous, 1-3 mm. long, not connate; leaflets 9-15, oblong-obovate to oblong-cordate, 5-11 mm. long,
glabrous on the upper surface; peduncles shorter than the leaves; racemes small, 3- to 7-flowered;
pedicels 1-2 mm. long; flowers cream-colored, 5-7 mm. long; calyx about 3.5 mm. long, the slender
teeth subequal to the tube; banner well reflexed from the shorter wings and keel; pod sessile, spreading, membranous, finely whitish-strigillose, inflated, ovoid-reniform in outline, 15-20 mm. long and
over half as broad, 1-celled but the lower suture slightly intruded and forming a thin partial partition
of varying degree of development.
   Lower John Day Valley, Oreg. May.
   This is a very poorly understood species which apparently has not been collected more than once or
possibly twice; perhaps it is only an aberrant state of *A. diaphanus* which it resembles in all but the
character of the fruit. The pods of *A. diaphanus* are known to vary greatly in outline and curvature
and some that are less strongly incurved and more inflated closely approach those of the type of *A.
diurnus,* except for the more complete intrusion of the lower suture.

Astragalus diversifolius Gray, Proc. Am. Acad. 6:230. 1864.
   Homalobus orthocarpus Nutt. in T. & G. Fl. N.Am. 1:351. 1838. *A. campestris* var. *diversifolius*
      Macbr. Contr. Gray Herb. n.s. 65:35. 1922. *A. convallarius diversifolius* Tidestr. Proc. Biol.
      Soc. Wash. 50:20. 1937. *(Nuttall,* "with the preceding," i.e., *Homalobus junceus)* Not *A. ortho-
      carpus* of Boiss. in 1842.
   A. reclinatus Cronq. Madroño 7:79. 1943. *(Cronquist 3086,* 2 mi. s. of Dickey, Custer Co. , Ida. ,
      July 14, 1941)
   Strigillose perennial with trailing stems 3-6 dm. long that often are rhizomelike at the base; leaves
4-9 cm. long, the rachis 1-2 mm. broad; lateral leaflets usually 1-2 pairs, linear to linear-elliptic,
1-3 cm. long, mostly 2-5 mm. broad; terminal leaflet confluent with the rachis, 3-10 mm. broad;
stipules 1-2 mm. long, lanceolate, only the lowest ones connate; racemes loosely 5- to 12-flowered,
with peduncles equaling or longer than the leaves; pedicels 1-2 mm. long; flowers 10-12 mm. long,
yellowish or purplish-tinged; calyx about half as long as the corolla, the teeth linear-lanceolate, about
1/2 the length of the whitish- to blackish-strigillose tube; keel and wings subequal to the banner; pods
strigillose, sessile, membranous, compressed, 15-30 mm. long, 3-4 mm. broad, 1-celled, the
sutures neither sulcate nor intruded.
   Chiefly in moist alkaline soil; c. Ida. to Wyo. and Utah. June-Aug.

Astragalus drummondii Hook. Fl. Bor. Am. 1:153. 1831.
   Tragacantha drummondii Kuntze, Rev. Gen. 2:944. 1891. *Tium drummondii* Rydb. Bull. Torrey Club
      32:659. 1906. *(Douglas,* "Eagle and Red-Deer Hills of the Saskatchawan")

Grayish-villous-hirsute perennial from a heavy root and branched crown, usually with several erect stems (3) 4-7 (8) dm. tall; leaves 6-14 cm. long; stipules submembranous, 3-12 mm. long, the lowest either fully amplexicaul and connate into a sheath or decurrent around no more than 2/3 of the stem and free, the upper ones lance-acuminate, always free; leaflets 13-31, ovate to oblong or linear-elliptic, 2-3.5 cm. long, 2-10 mm. broad, villous on the lower surface, glabrous above; peduncles about equaling the leaves, racemes fairly closely 20- to 50-flowered, considerably elongate and usually exceeding the peduncles in fruit; pedicels 1-3 mm. long; flowers white, sometimes with a purplish-tipped keel, 18-25 mm. long; calyx black-hairy, 8-11 mm. long, the teeth about half as long as the tube; banner not much reflexed from the narrow wings which are 4-6 mm. longer than the keel; pod glabrous, pendulous, with a stipe about equal to the calyx, the body 2-4 cm. long, 4-5 mm. broad, somewhat obcompressed, broadly cordate in section, the lower suture deeply sulcate and intruded to form a nearly complete partition. N=11.

Mostly in the plains region on the e. side of the Rockies from Sask. and Alta. to N.M., w. in Mont. to Deer Lodge, Granite, and Beaverhead cos., and in Ida. to Clark Co., s. to Utah. June-July.

This is one of the few species in which the stipular attachment varies in different populations.

Astragalus eremiticus Sheld. Minn. Bot. Stud. 1:161. 1894.
   A. arrectus var. eremiticus M. E. Jones, Proc. Calif. Acad. Sci. II, 5:665. 1895. Tium eremiticum Rydb. Bull. Torrey Club 40:49. 1913. (C. C. Parry, Beaverdam Mts., Utah, in 1874)
   A. cusickii Rydb. Bull. Torrey Club 26:541. 1899, not of Gray in 1878. A. malheurensis Heller, Cat. N.Am. Pl. 2nd ed. 7. 1900. Tium malheurense Rydb. N.Am. Fl. 24⁷:391. 1929. A. eremiticus var. malheurensis Barneby, Am. Midl. Nat. 41:501. 1949. (Cusick 1238, Malheur, Oreg., in 1885)
   A. eremiticus var. spencianus M. E. Jones, Contr. West. Bot. 10:60. 1902. (M. E. Jones, Spencemont, Nev., in 1891)
   A. boiseanus A. Nels. Bot. Gaz. 53:223. 1912. Cystium boiseanum Rydb. Bull. Torrey Club 40:50. 1913. (Macbride 112, Big Willow, Ida., May 27, 1910) = var. spencianus.
Sparsely to densely strigillose, greenish to cinereous perennial from a heavy root and branched crown, with several erect, stiff stems 2-5 dm. tall; leaves 6-15 cm. long; stipules membranous, 2-6 mm. long, not connate; leaflets 13-27, linear to narrowly oblong, 9-25 mm. long, mostly 1-3 mm. broad, glabrous on the upper surface; peduncles usually not exceeding the leaves; racemes laxly 5- to 15-flowered; pedicels stout, 1-2 mm. long; flowers 14-18 mm. long, white to purplish; calyx black-hairy, 5-7 mm. long, the teeth 1/3-1/4 as long as the tube; wings 2-3 mm. longer than the rounded keel; pod strigillose to glabrous, ascending-erect but not appressed to the rachis of the raceme, with a stipe 8-12 mm. long, the body coriaceous, mottled, 18-25 mm. long, obcompressed and broadly cordate in section, 6-7 mm. thick, the lower suture deeply sulcate and intruded and incompletely partitioning the loculus. N=12 (var. malheurensis).

A species primarily of the Great Basin. May-July.

Our material, as described above, from s.c. Ida. to Malheur Co., Oreg., s. to Nev., is referable to the var. spencianus M. E. Jones. The var. eremiticus is a more southern plant of the Colorado R. drainage, and the purple-flowered var. malheurensis (Heller) Barneby is just out of our range in s.w. Ida. and adj. Oreg.

Astragalus eucosmus Robins. Rhodora 10:33. 1908.
   Phaca elegans Hook. Fl. Bor. Am. 1:144. 1831. Astragalus oroboides var. americanus Gray, Proc. Am. Acad. 6:205. 1863. Astragalus elegans Sheld. Minn. Bot. Stud. 1:154. 1894, not of Bunge in 1869. Astragalus minor M. E. Jones, Contr. West. Bot. 10:64. 1902, but not of Clos in 1846. Atelophragma elegans Rydb. Bull. Torrey Club 32:660. 1906. (Drummond, "prairies in the Rocky Mountains")
   Phaca parviflora Nutt. in T. & G. Fl. N.Am. 1:348. 1838, not Astragalus parviflorus of Lam. in 1783. Astragalus elegans var. curtiflorus Rydb. Mem. N.Y. Bot. Gard. 1:242. 1900. Astragalus curtiflorus M. E. Jones, Contr. West. Bot. 10:64. 1902, corrected in the appendix on page 87, which was published at the same time, to A. minor var. curtiflorus. (Nuttall, Valleys of the Rocky Mountains)
Sparsely to rather densely strigillose perennial with branched crown and sometimes short rootstocks, 3-7 dm. tall; stipules lanceolate, 4-9 mm. long, none (or only a few of the lowermost) at all connate; leaves 3-8 cm. long; leaflets 11-19, thin, greenish and usually glabrous on the upper surface, linear-elliptic to oblong or oblong-elliptic, 1-2.5 cm. long; peduncles usually about twice the length

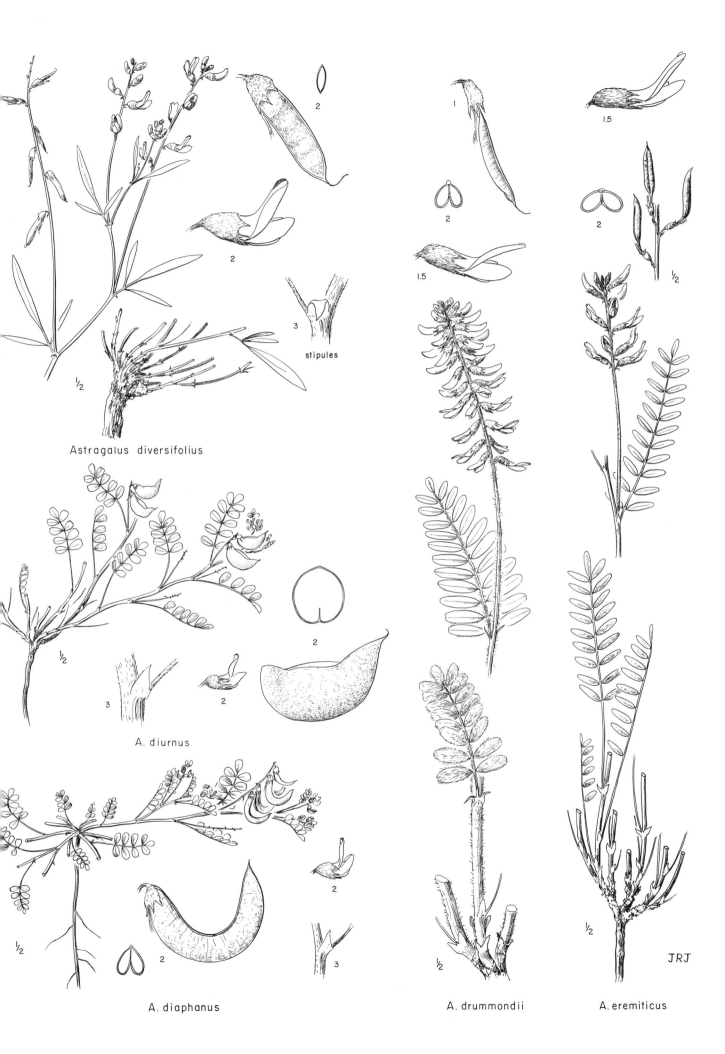

Astragalus diversifolius

stipules

A. diurnus

A. diaphanus

A. drummondii

A. eremiticus

JRJ

of the leaves; racemes (in fruit) at least as long as the peduncles, very loosely 10- to 30-flowered; pedicels 1-3 mm. long; flowers 6-9 mm. long, purplish; calyx 2.5-4 mm. long, grayish-hairy, the linear-lanceolate teeth less than 1/2 as long as the tube; banner only slightly upturned; wings much narrower and only slightly longer than the keel; pod gray- and blackish-hairy, pendulous, subsessile, 8-12 mm. long, obliquely elliptic (the upper suture more rounded), acute, membranous-cartilaginous, strongly compressed, the lower suture intruded to form a partial partition 0.5-1 mm. high. N=16.

Moist montane forest and meadows; Rocky Mts., from Colo. n. to Mont. and B.C. June-Aug.

Astragalus falcatus Lam. Encyc. 1:310. 1783. (Cultivated plants, derived from specimens sent from Russia by Demidow)

Greenish perennial 4-8 dm. tall, sparsely strigillose with white and black dolabriform hairs; leaves 7-12 cm. long; stipules membranous, lanceolate, 7-12 mm. long, not connate; leaflets 23-31, thin, oblong-elliptic, 15-25 mm. long, 2.5-5 mm. broad; peduncles equaling or shorter than the subtending leaves; racemes 20- to 50-flowered, congested, spikelike; flowers pendulous, 7-10 mm. long, ochroleucous; calyx about 5 mm. long, the teeth 1/3-1/2 as long as the tube; banner stubby, scarcely reflexed, equaled or exceeded by the narrow wings and about equaled by the keel; pod pendulous, sessile, leathery, 2-3 cm. long, 3-4 mm. thick and broad, strongly falcate, cordate-triangular in section, 2-celled, the lower suture completely intruded.

An introduced Caucasian plant, often escaping, as near Pullman and Anatone, Wash. July.

Astragalus filipes Torr. ex Gray, Proc. Am. Acad. 6:226. 1864.
  Tragacantha filipes Kuntze, Rev. Gen. 2:944. 1891. *Homalobus filipes* Heller, Muhl. 9:67. 1913.
  A. stenophyllus filipes Tidestr. Proc. Biol. Soc. Wash. 50:20. 1937. *(Dr. Pickering,* "near Fort Okanogan, interior of Washington Territory")
Strigillose, greenish perennial with a taproot and branched crown and numerous erect, slender stems 3-9 dm. tall; leaves 5-10 cm. long; stipules triangular, 1-3 mm. long, connate below, free above; leaflets 9-25, linear to linear-oblanceolate, 7-20 mm. long, 1-2 mm. broad; peduncles usually exceeding the leaves, as much as 2 dm. long; racemes laxly 10- to 30-flowered, much elongate in fruit and usually nearly equaling the peduncles; pedicels slender, 2-6 mm. long; flowers cream, 9-13 mm. long, spreading to pendent; calyx blackish (or partially white)-strigillose, 4-6 mm. long, the teeth scarcely 1 mm. long; banner erect; wings rounded, shorter than the banner but 1-4 mm. longer than the slightly beaked keel; pod spreading to pendulous, with a slender stipe 10-15 mm. long, the body membranous, greenish, not mottled, glabrous or strigose, linear to narrowly oblong-elliptic in outline, acute to more nearly rounded at either or both ends, 15-35 mm. long, (3) 4-6 mm. broad, strongly compressed, elliptic in section, 1-celled, both sutures prominent but neither sulcate.

Sagebrush plains and lower foothills; B.C. to n.e. Calif. and c. Nev., and e. to c. and s. Ida.; also disjunctly in s. Calif. and n. Baja Calif. May-June.

Two phases of the plant are common, one with glabrous, the other with strigose pods. The latter is almost confined to the Columbia R. drainage in Wash., s. B.C., and n.c. Oreg., but occurs sporadically southward in the range of the glabrous form. The northern plant tends to be more slender and diffuse in growth habit, with leaflets averaging narrower and often fewer and more scattered, and with racemes tending to be looser and longer; however none of these characters is stable. If desired, the phase with glabrous pod may be distinguished as var. *residuus* Jeps. (Man. Fl. Pl. Calif. 571. 1925, based on *Jepson 1469*, Coahuilla Valley, s. Calif.), the one with strigose pods as var. *filipes*.

Astragalus flexuosus (Dougl.) G. Don, Gen. Hist. Pl. 2:256. 1832.
  A. flexuosus Dougl. ex Hook. Fl. Bor. Am. 1:141. 1831, cited in synonymy. *Phaca flexuosa* Hook. loc. cit. *Tragacantha flexuosa* Kuntze, Rev. Gen. 2:945. 1891. *Homalobus flexuosus* Rydb. Bull. Torrey Club 32:666. 1906. *Pisophaca flexuosa* Rydb. N. Am. Fl. 24⁶:324. 1929. *(Douglas,* "Red River and Assinaboin, lat. 50°"; as first noted by Hooker, the type locality was incorrectly cited by Don as "North America, near the Columbia River")
  Phaca elongata Hook. Fl. Bor. Am. 1:140. 1831. *A. flexuosus* var. *elongatus* M. E. Jones, Contr. West. Bot. 10:58. 1902. *Pisophaca elongata* Rydb. N. Am. Fl. 24⁶:325. 1929. *(Dr. Richardson,* "Plains of the Saskatchawan")
Canescent-strigillose perennial with a branched crown and many slender, erect to decumbent-based stems 4-7 dm. long; leaves 4-12 cm. long, very short-petioled; lower stipules 2-4 mm. long, ovate-lanceolate, and slightly connate, the upper ones more narrowly lanceolate, 3-5 mm. long, and not connate; leaflets 15-21, linear-oblanceolate, narrowly oblong to oblanceolate-obovate, 5-22 mm. long,

acute and often petiolulate at base, rounded to emarginate, strigillose on the lower surface but usually glabrous on the upper, the rachis often somewhat flexuous; flowers 10-30 in elongate loose racemes which in fruit are usually at least equal to the 5-10 cm. peduncles; pedicels 1-3 mm. long; calyx grayish- to blackish-strigose-hirsute, 4-5 mm. long, the slender teeth 1/4-1/3 as long as the tube; corolla from nearly white and merely purplish-tinged, to light lavender-purple, 6-10 (11) mm. long, the rounded keel 5-7 mm. long, shorter than the wings and banner; pods very shortly stipitate, the stipe scarcely 1 mm. long and concealed by the calyx tube, the body narrowly oblong, 12-21 mm. long, 2-3.5 mm. wide, acute at each end, straight or slightly arcuate, nearly terete, somewhat leathery, 1-celled, neither suture intruded, the loculus partially filled with spongy filaments which dry to fibrous material attached to the walls. N=11.

A plant primarily of the Great Plains region, often on rather strongly alkaline soil; along the e. slope of the Rocky Mts., from Alta. to N.M., e. to Sask., Minn., and Kans. June-July.

The species varies in flower size, type of pubescence on the calyx, length of calyx lobes, and (more especially) in the size and degree of curvature of the pods; those plants with smaller, curved pods and less deeply colored flowers are usually distinguished as var. *elongatus* (Hook.) M. E. Jones, a distinction that hardly seems merited in view of the extreme variability of the species as a whole.

Astragalus geyeri Gray, Proc. Am. Acad. 6:214. 1864.

Phaca annua Geyer ex Hook. Lond. Journ. Bot. 6:213. 1847, not *A. annuus* of DC. in 1802. *Tragacantha geyeri* Kuntze, Rev. Gen. 2:945. 1891. *(Geyer,* "drift sand plains of the Upper Platte")

Grayish-strigose to greenish annual with spreading to erect stems 5-20 cm. long; leaflets 5-13, linear to oblong, obtuse to retuse, 5-15 mm. long; racemes 2- to 7-flowered, shorter than the leaves; flowers 6-8 mm. long, ochroleucous to pale lavender; calyx scarcely half as long as the corolla, the short linear-lanceolate teeth about 1/3 as long as the tube; pods sessile, about 2 cm. long, membranous, greatly inflated, slightly compressed, oblique, from slightly arcuate to somewhat lunate, 1-celled, the lower suture not at all intruded.

Sandy desert, especially on dunes; s.e. Oreg. to Calif. and Nev., eastward, through the Snake R. drainage of s. Ida. (where barely reaching our range), to Wyo. and Utah. June-July.

Astragalus gilviflorus Sheld. Minn. Bot. Stud. 1:21. 1894.

A. triphyllus Pursh, Fl. Am. Sept. 740. 1814, not of Pall. in 1800. *Phaca triphylla* Eat. & Wright, N.Am. Bot. 351. 1840. *Tragacantha triphylla* Kuntze, Rev. Gen. 2:948. 1891. *(Bradbury,* in upper Louisiana)

Phaca caespitosa Nutt. Gen. Pl. 2:98. 1818. *Orophaca caespitosa* Britt. in Britt. & Brown, Ill. Fl. 2:306. 1897. ("in arid gravelly hills near the confluence of Sawanee river and the Missouri") Not *A. caespitosus* of Pall. in 1800, or of Gray in 1864.

Caespitose, pulvinate, subacaulescent perennial from a multicipital woody caudex, silvery-strigose with dolabriform hairs; stems 1-3 cm. long, densely cushioned with old stipules and petioles; leaves 2-6 (10) cm. long; stipules membranous, acute, 6-15 mm. long; leaflets 3, elliptic to oblanceolate or obovate, 8-20 (35) mm. long; racemes mostly 1- or 2-flowered, subsessile in the leaf axils; flowers (15) 20-30 mm. long; calyx silvery-sericeous, 2/3-3/4 the length of the corolla, the subulate teeth about 1/3 as long as the tube; corolla ochroleucous to yellow, sometimes purplish-tipped; pods sessile, leathery, ovoid, 7-10 mm. long, scarcely equaling the calyx, sericeous, indehiscent. N=12.

Plains region, along the e. base of the Rocky Mts., from Mont. to Colo., e. to Sask. and Neb. June-July.

Astragalus gracilis Nutt. Gen. Pl. 2:100. 1818.

Microphacos gracilis Rydb. Bull. Torrey Club 32:663. 1906. ("From White river to the Mountains, on the plains of the Missouri")

A. microlobus Gray, Proc. Am. Acad. 6:203. 1864. *Tragacantha microloba* Kuntze, Rev. Gen. 2:946. 1891. *Microphacos microlobus* Rydb. Bull. Torrey Club 32:663. 1906. *A. parviflorus* var. *microlobus* M. E. Jones, Rev. Astrag. 193. 1923. ("Plains of Nebraska etc., to the Rocky Mountains," *Gray* and *Hall & Harbour)*

Grayish-strigillose perennial with a taproot and multicipital crown, the numerous stems freely branched, slender, erect or somewhat ascending, often buried and rhizomelike at base, 2-5 dm. tall; leaves numerous, 2-7 cm. long; stipules triangular to oblong-ovate, 1-3 mm. long, the lower ones slightly connate, the upper free; leaflets 9-15, linear to narrowly oblanceolate, 5-13 mm. long, 1-3 mm. broad, rounded to (commonly) retuse, glabrous on the upper surface except next to the margin,

2

2.5

2

½

stipules

3

½

½

2

2

3

½

2

2

½

2

½

2

2

½

2

3

A. geyeri

Astragalus eucosmus   A. filipes   A. flexuosus   A. falcatus

grayish-strigillose beneath; peduncles from shorter than the leaves to twice their length; racemes rather laxly 10- to 20-flowered, considerably elongate in fruit and then usually exceeding the peduncles; pedicels 1-2 mm. long; flowers 7-8 mm. long, pale lavender to purple, spreading; calyx 2-2.5 mm. long, the teeth triangular, 1/4-1/3 the length of the tube; banner strongly reflexed; wings conspicuously arched, considerably exceeding the rounded keel but shorter than the banner; pod reflexed, sessile, grayish-strigillose, arcuate-ovate, strongly obcompressed, 6-9 mm. long, 3.5-4 mm. thick, about 2 mm. wide, rather fleshy, when dry filled with fibrous material, 1-celled, but sulcate on the lower suture, and more or less reniform in section.

A plant of the Great Plains, from the e. foothills of the Rocky Mts., in Mont., to N.M., e. to the Dakotas and Okla. June-July.

Astragalus howellii Gray, Proc. Am. Acad. 15:46. 1879.
   Tium howellii Rydb. N.Am. Fl. 24⁷:389. 1929. *(J. & T. J. Howell*, near The Dalles, Oreg.)
   A. misellus Wats. Proc. Am. Acad. 21:449. 1886. *Phaca misella* Piper, Contr. U.S. Nat. Herb. 11:371. 1906. *A. howellii* var. *misellus* M. E. Jones, Rev. Astrag. 262. 1923. *Tium misellum* Rydb. N.Am. Fl. 24⁷:389. 1929. *(Howell*, Mitchell, Oreg.) = var. *aberrans*.
   ASTRAGALUS HOWELLII var. ABERRANS (M. E. Jones) C. L. Hitchc. hoc loc. *A. drepanolobus* var. *aberrans* M. E. Jones, Contr. West. Bot. 10:64. 1902. ("Columbia Basin," apparently collected by M. E. Jones)
   Grayish, soft- and somewhat crisp-strigillose to -villous perennial from a taproot and branched crown, with several prostrate to erect stems 5-20 cm. long; leaves 3-14 cm. long; stipules deltoid-lanceolate, 2-5 mm. long, not connate; leaflets 11-17, linear to linear-lanceolate or narrowly oblong, 5-14 mm. long, 1-2.5 mm. broad, from soft-hairy on both surfaces to glabrous above; peduncles from shorter to conspicuously longer than the leaves, as much as 18 cm. in length; racemes loosely 5- to 25-flowered, pedicels slender, 2-4 mm. long; flowers whitish-yellow, 6-15.5 (16.5) mm. long, spreading to reflexed; calyx 3-7.5 mm. long, copiously pubescent with soft, crisp, white and brown hairs, the linear-lanceolate teeth from about half as long to nearly as long as the tube; banner erect; wings narrow, 2-3 mm. longer than the very slightly beaked keel; pod spreading to pendulous, with a slender stipe once to twice the length of the calyx, the body obliquely linear-oblong, somewhat falcate, acute at each end, 15-30 mm. long, 2.5-4.5 (5) mm. broad, rather densely short-strigose-lanate, cordate-triquetrous in section, the lower suture almost completely intruded.

Sagebrush plains from c. Wash. to Oreg. Apr.-June.

The species is none too well understood, having been collected but few times. It is usual to recognize two species rather than what are here treated as two varieties, but the characters generally considered distinctive, even with the meager collections available, prove variant and intergradient.

1 Leaves 5.5-14 cm. long, with mostly 21-27 leaflets, these pubescent above; peduncles
   mostly 8-18 cm. long; racemes 10- to 25-flowered; calyx 5-7.5 mm. long; banner 10-   .
   15.5 (16.5) mm. long; stipe of the pod 7-14 mm. long; Wasco and Sherman cos., Oreg.
                                                                                            var. howellii
1 Leaves 3-7 cm. long, with 11-21 leaflets, these glabrous or subglabrous on the upper
   surface; peduncles 2.5-10 cm. long; racemes 5- to 15-flowered; calyx 3.5-5 mm.
   long; banner 6.5-9.5 (10) mm. long; stipe of the pod 2.5-5 mm. long; Wheeler, Grant,
   Crook, and Deschutes cos., Oreg., to Kittitas and Franklin cos., Wash.
                                                                      var. aberrans (M. E. Jones) C. L. Hitchc.

Astragalus inflexus Dougl. ex Hook. Fl. Bor. Am. 1:151. 1831.
   Tragacantha inflexa Kuntze, Rev. Gen. 2:945. 1891. *Phaca inflexa* Piper, Contr. U.S. Nat. Herb. 11:369. 1906. *Xylophacos inflexus* Rydb. Bull. Torrey Club 40:49. 1913. *(Douglas*, "barren sandy grounds of the Columbia, from the juncture of the Lewis and Clarke's River to the mountains")
   Perennial, grayish-villous-lanate throughout; stems several from a taproot and branched crown, prostrate to ascending, often matted, 1-5 dm. long; stipules ovate-lanceolate, acuminate, 1-3 cm. long, not connate; leaflets (13) 15-29, oval to elliptic-lanceolate, mostly acute, 7-18 mm. long; racemes rather closely 6- to 19-flowered, 1-3 cm. long at anthesis, as much as 6-8 cm. long in fruit; peduncles scarcely exceeding the leaves; pedicels 1-3 mm. long; flowers rose-purplish, 20-30 mm. long, spreading to erect, the bracts linear-acuminate, 5-15 mm. long; calyx tubular, 2/3-3/4 the length of the corolla, woolly-villous, the teeth linear, from about equal to the tube to only about half as long; banner arched, hairy on the back; wings considerably shorter than the banner but well surpassing the keel; pods spreading, sessile, densely silky-villous, 1.5-3 cm. long, chartaceous, ar-

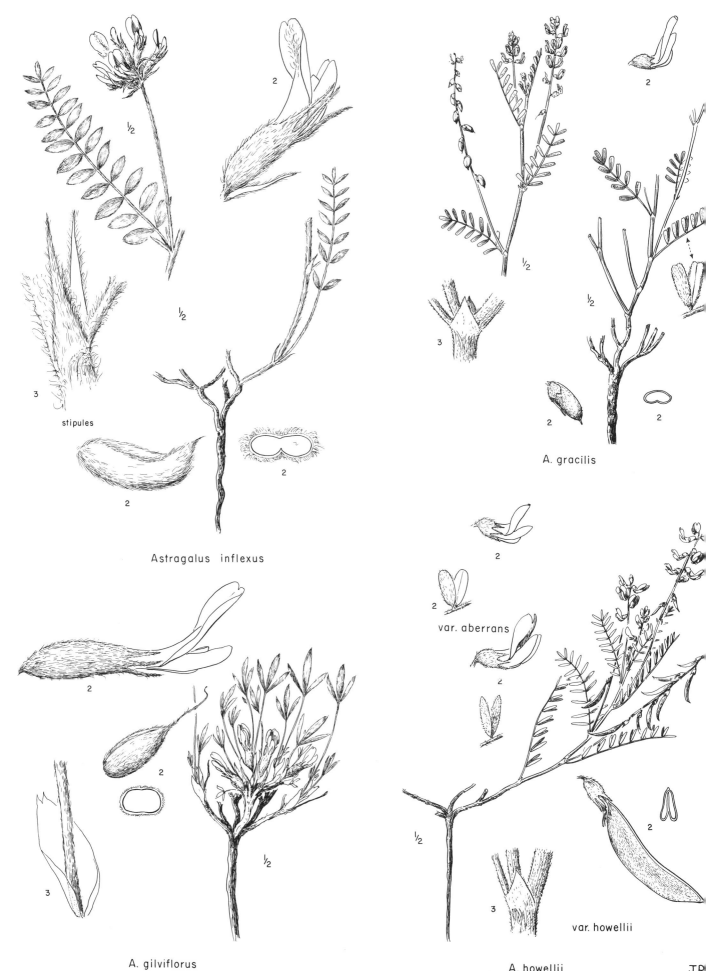

stipules

Astragalus inflexus

A. gilviflorus

A. gracilis

var. aberrans

var. howellii

A. howellii

JR

cuate, strongly obcompressed, the lower suture intruded but not completely dividing the ovary. N=11.
Sagebrush desert and dry hillsides from c. Wash. to w. Mont., s. to n. Oreg. Apr.-July.

On the edge of our range in Ida. there is a closely related species, *A. utahensis* (Torr.) T. & G., which differs in having fewer (11-17), obovate to suborbicular, obtuse leaflets. It is densely tomentose, with the pods completely concealed by the long silky hairs; it ranges s. to Nev. and e. to Wyo.

Astragalus kentrophyta Gray, Proc. Acad. Phila. 1863:60. 1863.

Kentrophyta montana Nutt. in T. & G. Fl. N.Am. 1:353. 1838. *Tragacantha montana* Kuntze, Rev. Gen. 2:941. 1891. *Homalobus montanus* Britt. in Britt. & Brown, Ill. Fl. 2:306. 1897. *A. centrophytus* Clements, Rocky Mt. Fls. 173. 1914. *A. montanus* M. E. Jones, Rev. Astrag. 80. 1923, not of L. in 1753. *(Nuttall,* hills of the Platte)

A. kentrophyta var. *elatus* Wats. Bot. King Exp. 77. 1871. *A. viridis* var. *impensus* ·Sheld. Minn. Bot. Stud. 1:118. 1894. *A. viridis* var. *elatus* Cockerell, Bot. Gaz. 26:437. 1898. *A. kentrophyta* var. *impensus* M. E. Jones, Contr. West. Bot. 10:63. 1902. *Kentrophyta impensa* Rydb. Bull. Torrey Club 32:665. 1905. *A. impensus* Woot. & Standl. Contr. U.S. Nat. Herb. 19:369. 1915. *(Watson 291,* Holmes Creek Valley, Nev.)

A. jessiae Peck, Leafl. West. Bot. 4:180. 1945. *A. tegetarius* var. *jessiae* Barneby, Leafl. West. Bot. 6:96. 1951. *A. kentrophyta* var. *jessiae* Barneby, Leafl. West. Bot. 6:154. 1951. *(Peck 21220,* 10 mi. s. of Adrian, Malheur Co., Oreg., June 13, 1942) = var..*kentrophyta.*

Very similar to *A. tegetarius* but the stems longer; plants often 5-10 cm. tall; pubescence appressed, dolabriform; stipules connate, strongly acerose; leaflets rigid, acerose, 5-12 mm. long; flowers usually ochroleucous but often purplish-tinged, 4-7 mm. long, usually 2-3 per raceme, the peduncles shorter than the leaves; ovules 2-4; pod sessile, (3) 4-6 (7) mm. long, ovoid to lance-ovoid, compressed, strictly 1-celled, grayish-sericeous, indehiscent, the tip shortly acuminate.

Lower foothills and plains, from s.e. Oreg. to Wyo., s. to Ariz.; mostly just coming to the edge of our range on the south. July-Sept.

The species is represented in our area by the following 2 varieties:

1 Plants low, tufted or matted, the internodes short, mostly less than 1.5 cm. long; pod
subsymmetrically ellipsoid or ovoid, nearly or quite beakless, 3-4.5 mm. long; s.e.
Oreg. to Wyo. and s.w. Sask.   var. kentrophyta
1 Plants taller, bushy-branched, rarely trailing, the internodes mostly well developed
and up to 1.5-4.5 cm. long; pod narrowly ovoid-acuminate, gently incurved into the
beak, 4-7 mm. long; sandy knolls, badlands, and talus under cliffs, common nearly
throughout Utah, to w. Colo., n.w. N.M., n. Ariz., and e. Nev., and apparently
isolated near the Great Bend of the Columbia near Walla Walla, Wash., and perhaps
in adj. Oreg.   var. elatus Wats.

Astragalus leibergii M. E. Jones, Proc. Calif. Acad. Sci. II, 5:663. 1895.

A. arrectus var. *leibergi* M. E. Jones, Contr. West. Bot. 10:68. 1902. *Phaca arrecta* var. *leibergi* Piper, Contr. U.S. Nat. Herb. 11:372. 1906. *Tium leibergi* Rydb. N.Am. Fl. 24[7]:392. 1929. *(Sandberg & Leiberg 354,* Egbert Spring, Douglas Co., Wash.)

Tufted, shortly caulescent or subacaulescent perennial with a woody taproot and multicipital caudex, silky-cinereous, gray-strigillose, or greenish, the leaflets pubescent or medially glabrescent above; stems mostly 1-10 (rarely up to 25) cm. long, composed of only a few developed internodes, nearly always shorter than the leaves and peduncles; leaves subradical or partly cauline, 8-28 cm. long; stipules scarious, pallid, at least the lowest (and commonly all) fully amplexicaul and connate into a bidentate sheath, the uppermost ones sometimes free; leaflets 15-31, linear, lance-elliptic or rarely lance-oblong, obtuse or acutish, (2) 4-10 (up to 30) mm. long; peduncles stout, erect, 10-28 cm. long; racemes loosely 7- to 20-flowered, the flowers nodding at full anthesis, the axis 4-17 cm. long in fruit; flowers whitish, 11.5-16.5 mm. long, the banner longer than the wings; calyx 5.5-9.5 mm. long, black-pilosulous, the campanulate tube 3-5.5 mm. long, the teeth 1.5-4 mm. long; pod erect, stipitate, the stipe 3-8 mm. long, the narrowly oblong-ellipsoid, straight or gently incurved body 16-27 mm. long, 4-7.5 mm. broad, obcompressed-triquetrous, keeled ventrally by the prominent suture, openly sulcate dorsally, the fleshy green (but at length leathery or almost woody) valves glabrous or minutely strigillose with black or white hairs, inflexed as a narrow, sometimes subobsolete septum 0.5-1.3 mm. wide.

Dry hillsides and plains, commonly in sagebrush scabland on basalt, more rarely in open pine

forest of foothill canyons on serpentine or granite; Douglas, Chelan, and Kittitas cos. of c. transmontane Wash. Mid Apr. -June.

A remarkable form of *A. leibergii,* perhaps varietally distinct, has been collected several times in
Swakane Canyon, Chelan Co., sometimes associated with the endemic *Trifolium thompsonii* or with
Douglasia nivalis. It differs from the ordinary form of the sagebrush desert in its strongly developed
stems, ample green herbage, and apparently free stipules, which are, however, connate in vernation
but ruptured by the growth of the exceptionally stout stems.

Astragalus lemmonii Gray, Proc. Am. Acad. 8:627. 1873.
   Tragacantha lemmoni Kuntze, Rev. Gen. 2:946. 1891. *(Lemmon,* Sierra Valley, Calif.)
   Slender, prostrate, thinly strigillose perennial, the herbage green; stems 1-4 dm. long; leaves 1-4.5
cm. long; stipules 2-5 mm. long, free; leaflets 7-15, elliptic, acute or acutish, 2-10 mm. long; racemes 2- to 13-flowered, borne in pairs in the leaf axils on peduncles 6-17 (30) mm. long, one raceme
of each pair usually developing much sooner than the other; flowers whitish or lilac-tinged, the banner
5-6 mm. long; calyx 3-4 mm. long; pod spreading or declined, sessile, oblong-ellipsoid, nearly or
quite straight, 4-7 mm. long, 1.5-2.5 mm. broad, bilocular, the papery valves strigillose, compressed-triquetrous, the lateral faces flat, the dorsal face narrower and sulcate.

Moist, or summer-dry meadows and rushy flats bordering streams and lake shores along the e.
foot of the Sierra-Cascade axis, from e.c. Calif. to the head of the Deschutes R. in s. Deschutes Co.,
Oreg., perhaps not quite reaching our range. June.

Astragalus lentiginosus Dougl. ex Hook. Fl. Bor. Am. 1:151. 1831.
   Tragacantha lentiginosa Kuntze, Rev. Gen. 2:946. 1891. *Phaca lentiginosa* Piper, Contr. U.S. Nat.
      Herb. 11:368. 1906. *Cystium lentiginosum* Rydb. Bull. Torrey Club 40:50. 1913. *(Douglas,* Blue
      Mts., Oreg.)
   A. salinus Howell, Erythea 1:111. 1893. *Cystium salinum* Rydb. Fl. Rocky Mts. 492, 1063. 1917.
      A. lentiginosus var. *salinus* Barneby, Leafl. West. Bot. 4:86. 1945. *(T. Howell,* southeastern
      Oreg.)
   Cystium heliophilum Rydb. Fl. Rocky Mts. 491, 1063. 1917. *A. heliophilus* Tidestr. Contr. U.S.
      Nat. Herb. 25:325. 1925. *(C. L. Shear 3430,* Lima, Mont., July 1, 1895) = var. *salinus.*
   A. lentiginosus var. *carinatus* M. E. Jones, Rev. Astrag. 125. 1923. *(M. E. Jones,* on flats, Bak-
      er City, Oreg., in 1902)
   A. lentiginosus var. *scorpionis* M. E. Jones, Rev. Astrag. 124. 1923. *(Purpus 6365,* Morey Peak,
      Nev., is the first of 3 specimens cited) = var. *lentiginosus.*
   Cystium platyphyllidium Rydb. N.Am. Fl. 24[7]:410. 1929. *A. lentiginosus* var. *platyphyllidium* Peck,
      Man. High. Pl. Oreg. 449. 1941. *(Leiberg 171,* Pine Creek, Gilliam Co., Oreg., June 7, 1891)
   Cystium merrillii Rydb. N.Am. Fl. 24[7]:410. 1929. *A. merrillii* Tidestr. Proc. Biol. Soc. Wash.
      50:21. 1937. *(Merrill & Wilcox 680,* Leucite Hills, Wyo., June 18, 1901) = var. *platyphyllidius.*
   Cystium cornutum Rydb. N.Am. Fl. 24[7]:412. 1929. *(T. Howell,* Muddy Station, John Day Valley,
      Oreg., May 13, 1885) = var. *platyphyllidius.*
   A. araneosus sensu Abrams, Ill. Fl. Pac. St. 2:598. 1944, not of Sheld.
   Greenish and often more or less succulent, glabrous to strigose perennial with a stout taproot and
multicipital crown and decumbent to erect stems 1-4 dm. long; stipules lanceolate, not connate, 2-5
(or the upper sometimes as much as 15) mm. long; leaflets 11-19, often glabrous (at least on the upper surface), oblanceolate to obovate, 8-15 mm. long; racemes closely 10- to 30-flowered, borne on
peduncles usually shorter than the leaves; flowers spreading to erect, 8-18 mm. long, white to pinkish or pale purplish-tipped; calyx strigillose to hirsute, 1/3-1/2 the length of the corolla, the teeth
linear-lanceolate, about 1/2 as long as the tube; banner considerably exceeding the wing and keel; pod
sessile, inflated or turgid, membranous to cartilaginous, glabrous to strigillose, usually somewhat
arcuate, 1-3 (4) cm. long, not at all compressed, 2-celled for most of the length by the intrusion of
the lower suture, but produced into a sterile, unilocular, compressed and flattened beak toward the
tip, sulcate on both sutures.

Throughout much of w. U.S., from desert or salt flat to barren subalpine slopes. May-July.

Astragalus lentiginosus is a species that has segregated into a great many more or less local, fairly
distinctive, but freely intergradient ecological or geographical races, of which there are four in our
area.

1 Flowers 13-18 mm. long; pods mostly 1.5-3 cm. long; n.c. Oreg. to n.e. Calif., e.
      to Wyo. and Utah   var. platyphyllidius (Rydb.) Peck

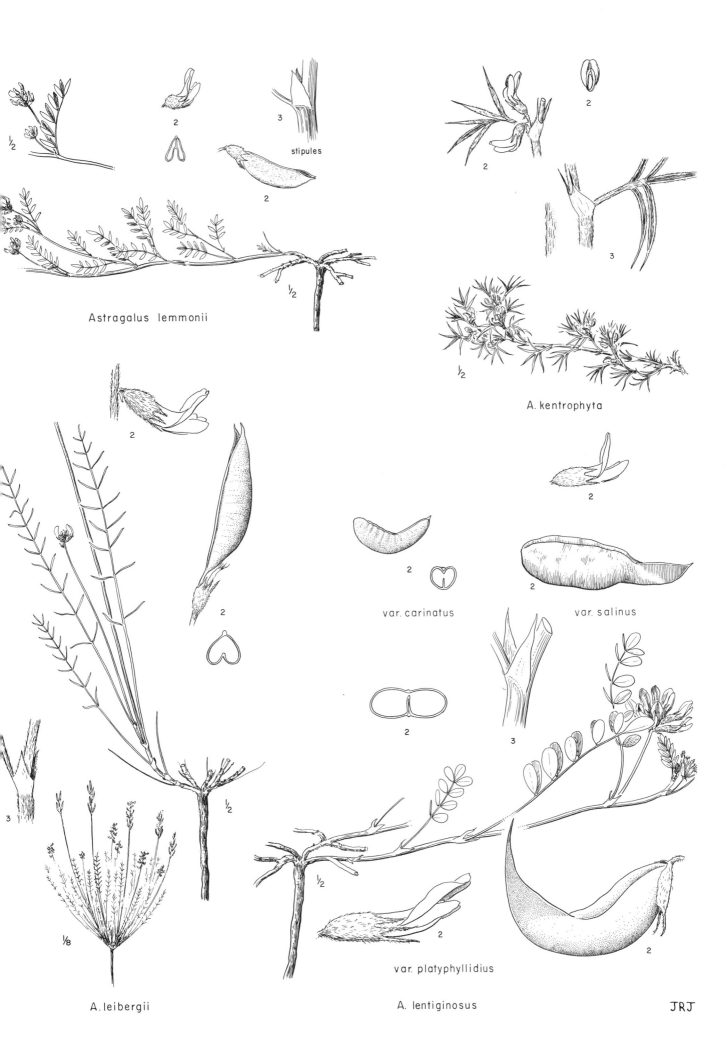

Astragalus lemmonii

stipules

A. kentrophyta

var. carinatus

var. salinus

A. leibergii

var. platyphyllidius

A. lentiginosus

JRJ

1 Flowers usually less than 13 mm. long, if longer then the pods only 1-2 cm. long
  2 Pods glabrous, shining, 1.5-2.5 cm. long; alkaline flats, c. Oreg. to n.e. Calif., e.
    to s.w. Mont. and Utah   var. salinus (Howell) Barneby
  2 Pods hairy
    3 Legumes ovoid to lanceolate, broadest below the middle, the ventral suture sulcate;
      mostly in sagebrush, c. Wash. to Oreg. and Ida.; N=11   var. lentiginosus
    3 Legumes broadest at or above the middle, the ventral suture acute; sagebrush des-
      ert and ponderosa pine forest, e. Oreg. to Calif.   var. carinatus M. E. Jones

Astragalus leptaleus Gray, Proc. Am. Acad. 6:220. 1864.
  Phaca pauciflora Nutt. ex T. & G. Fl. N.Am. 1:348. 1838, not of Persoon in 1808. *A. pauciflorus*
    Gray, Proc. Acad. Phila. 1863:60. 1863, not of Pall. in 1800, or of Hook. in 1831. *Tragacantha*
    leptalea Kuntze, Rev. Gen. 2:946. 1891. *Phaca leptalea* Rydb. Bull. Torrey Club 40:48. 1913.
    (Nuttall, "Plains of the Rocky Mountains, near streams")
  A delicate, diffuse perennial from a deeply buried taproot and extensively creeping subterranean
caudex, thinly strigillose, bright green; stems 5-20 cm. long, floriferous from near or well below
the middle; leaves 2.5-10 cm. long; stipules 2-5 mm. long, connate; leaflets 15-27, mostly lance-
elliptic and acute, often ovate and obtuse on some of the lower leaves, 3-15 mm. long, glabrous on
the upper surface; peduncles filiform, 2-5.5 cm. long; racemes loosely 1- to 5-flowered, the flowers
ascending or declined in age, the axis of the inflorescence only about 1 cm. long; flowers white, the
tip of the keel purplish; banner 8.5-12 mm. long; calyx 4-5.5 mm. long, the campanulate tube about
2.5-3.5 mm. long, the teeth 1-2.5 mm. long; pod pendulous, subsessile or shortly stipitate, the stipe
up to 1.5 mm. long, the body oblong-ellipsoid, slightly arched downward, 8-14 mm. long, 2.5-4 mm.
broad, somewhat obcompressed, flattened or openly sulcate dorsally, the papery-membranous valves
thinly strigillose with black or white hairs, not inflexed.
  Moist sedgy meadows, swales, and turfy hummocks at the edge of mountain brooks; local, w. Mont.
(Flathead Lake and Monida Pass) to c. Ida. (Custer Co.), s. in the Rocky Mts. to Colo., and (reportedly) n. to Alta. June-July.

Astragalus lotiflorus Hook. Fl. Bor. Am. 1:152. 1831.
  Phaca lotiflora T. & G. Fl. N.Am. 1:349. 1838. *Tragacantha lotiflora* Kuntze, Rev. Gen. 2:946.
    1891. *Cystopora lotiflora* Lunell, Am. Midl. Nat. 4:428. 1916. *(Drummond,* "About Carlton-House
    on the Saskatchawan")
  Phaca cretacea Buckl. Proc. Acad. Phila. 1861:452. 1861. *Batidophaca cretacea* Rydb. N.Am. Fl.
    24⁶:322. 1929. *(S. B. Buckley,* northern Texas)
  A. lotiflorus var. *brachypus* Gray, Proc. Am. Acad. 6:209. 1864, not *A. brachypus* of Schrenk. in
    1841. *A. elatiocarpus* Sheld. Minn. Bot. Stud. 1:20. 1894. *A. lotiflorus* var. *elatiocarpus* Rydb.
    Mem. N.Y. Bot. Gard. 1:254. 1900. *Phaca elatiocarpa* Rydb. Bull. Torrey Club 32:665. 1906.
    Cystopora elatiocarpa Lunell, Am. Midl. Nat. 4:428. 1916. *(Hall & Harbour 131,* in part, Colo.)
  Caespitose, short-lived perennial with a taproot and branched caudex, more or less silvery with
appressed to spreading, dolabriform hairs, nearly acaulescent, the stems only 1-6 cm. long; leaves
5-10 cm. long; stipules lanceolate, 2-5 mm. long, not connate; leaflets (3) 5-13, oblong-elliptic, 10-
20 mm. long; racemes 2- to 12-flowered, mostly subsessile in the axils, the peduncles mostly 0.5-2
(up to 8) cm. long; pedicels about 1 mm. long; flowers 7-10 mm. long (often cleistogamous and the
banner then only 4.5-7 mm. long), yellowish or ochroleucous to somewhat purplish-tinged; calyx usually a little more than half the length of the corolla, the linear lobes subequal to the tube; pod sessile,
15-25 mm. long, inflated, 6-8 mm. wide, short-beaked, somewhat lunate, rather leathery, slightly
obcompressed, 1-celled, strigillose-lanate.
  Plains area, from the e. base of the Rocky Mts., Alta. and Mont., to N.M., e. to Sask., Minn.,
Ia., and Tex. May-June.

Astragalus lyallii Gray, Proc. Am. Acad. 6:195. 1864.
  Tragacantha lyallii Kuntze, Rev. Gen. 2:946. 1891. *Phaca lyallii* Piper, Contr. U.S. Nat. Herb. 11:
    370. 1906. *(Dr. Lyall 8,* Upper Yakima River, on the boundary between British Columbia and
    Washington Territory)
  A. lyallii var. *caricinus* M. E. Jones, Rev. Astrag. 174. 1923. *A. caricinus* Barneby, Am. Midl.
    Nat. 55:502. 1956. *(M. E. Jones,* Glenn's Ferry, Ida., June 17, 1911)
  Grayish-strigose to villous-lanate or sericeous perennial with a long woody taproot and knotty,

branched crown, the stems several, 1.5-4 dm. tall, often floriferous from below the middle; leaves 3-11.5 cm. long; stipules membranous, linear-tipped, 3-8 mm. long, the lower ones often amplexicaul and connate, the upper ones free, or all free and more or less lanceolate-acuminate; leaflets 11-21, linear to linear-elliptic or narrowly oblong, (5) 10-15 (20) mm. long; racemes spikelike, axillary, loosely 10- to 35-flowered, rachis 5-15 cm. long; peduncles usually shorter than the leaves, 1-6.5 cm. long; pedicels stout, less than 1.5 mm. long; flowers 4.5-9 mm. long, white or purplish-tinged or -veined, drying to yellowish, spreading to reflexed, at least with age; calyx campanulate, 3.5-5.5 mm. long, grayish- to black-hairy, the teeth linear and from about once to nearly twice as long as the tube; banner narrow, well reflexed, 4.5-7.5 mm. long; keel rounded, scarcely exceeding the calyx; pod 5-9 mm. long, sessile or subsessile, reflexed, glabrate to hirsutulous or villous, triquetrous, broadly cordate in section, with flat or low-concave to convex lateral faces, 2-celled by the complete (or nearly complete) intrusion of the lower suture, the papery valves greenish to canescently tomentulose.

Sagebrush and desert, especially on sand dunes, from Kittitas Co., Wash., s.e. along the Snake R. to s.e. Ida. and s.e. Oreg. May-June.

Although there is considerable variation in the amount of hair on the pods, much of the material from Ida. differs in several respects from that of most of Wash., the species thus being represented by the following two well-marked phases:

1 Stipules, at least the lower ones, amplexicaul and connate into a pallid scarious sheath, the upper ones nearly or quite free; pod 6-8.5 mm. long, 2-3 mm. broad, with flat or slightly concave lateral faces, sessile, fully bilocular; sandy bluffs, hillsides, and fallow fields, on shale or basalt, common locally along the Snake R., at the edge of our range, from the mouth of the Raft R., Cassia Co., Ida., downstream to Owyhee Co., Ida., an' e. Malheur Co., Oreg., less commonly, and perhaps disjunctly, along the lower Yakima and Columbia rivers in Yakima, Benton, and Grant cos., Wash.
var. caricinus M. E. Jones

1 Stipules usually all free; pod 5-8 mm. long, 2-3.5 mm. broad, with slightly convex lateral faces, very slightly stipitate (the stipe not over 0.5 mm. long) or sessile, sometimes not fully bilocular; low hills and rolling plains, in sandy soil derived from basalt, locally plentiful over the s. half of the Columbia Basin in e. Wash., from the Yakima R. and Grand Coulee s.e. to the Palouse country and the lower Walla Walla R.
var. lyallii

Astragalus malacus Gray, Proc. Am. Acad. 7:336. 1868.
Tragacantha malaca Kuntze, Rev. Gen. 2:946. 1891. *Hamosa malaca* Rydb. Fl. Rocky Mts. 496. 1917. *(Anderson,* near Carson City, Nev.)
A. *obfalcatus* A. Nels. Bot. Gaz. 54:411. 1912. A. *malacus* var. *obfalcatus* M. E. Jones, Rev. Astrag. 227. 1923. *(Macbride 1023,* Reynolds Cr., Owyhee Co., Ida., July 3, 1911)

Grayish-hirsute to -pilose perennial with a woody taproot and 1-several erect to decumbent stems 5-10 (15) cm. long; leaves crowded and more or less imbricated at base, (5) 8-12 cm. long; stipules membranous, 10-15 mm. long, ovate-lanceolate to acuminate, not connate; leaflets (9) 11-19, oblong to oblong-obovate, (5) 10-20 mm. long; peduncles usually equaling or exceeding the leaves; racemes spikelike, 10- to 40-flowered, congested, considerably elongate in fruit; pedicels stout, 1-2.5 mm. long; flowers 15-20 mm. long, usually deep magenta, sometimes more nearly ochroleucous but with a purplish banner and keel; calyx tubular, somewhat gibbous-based, 8-12 mm. long, the lanceolate teeth 1/3-1/2 as long as the tube; wings shorter than the banner, nearly equaled by the keel; pod nearly or quite sessile, 2-3 cm. long, spreading to erect, membranous-cartilaginous, densely pilose-hirsute, often mottled, terete-cordate in section, the lower suture deeply intruded and almost to quite partitioning the cavity; stipe, if any, stout and not over 2 mm. long.

Desert sagebrush to forest land; s. Calif. and Nev. n. to s.e. Oreg., and to Owyhee Co. and the Snake R. in Elmore and Gooding cos., Ida.; although reported as far n. as the Blue Mts., Oreg., not actually known at present from within our area. Apr.-June.

Astragalus microcystis Gray, Proc. Am. Acad. 6:220. 1864.
Phaca microcystis Rydb. Mem. N.Y. Bot. Gard. 1:245. 1900. *(Dr. Lyall,* "Interior of Washington Territory, Fort Colville to the Rocky Mts.")
A. *miser* sensu M. E. Jones, Rev. Astrag. 98. 1923, but not of Dougl. ex Hook. in 1834.

Grayish-strigillose or -hirsutulous to subglabrate and greenish, caespitose perennial with a branched crown and long taproot; stems many, decumbent to ascending, 1-5 dm. long; leaves 3-6 cm.

long; stipules 3-5 mm. long, connate over half their length; leaflets 9-15, linear-elliptic to oblong, oblong-lanceolate, or oblanceolate, 5-15 mm. long; racemes loosely 5- to 12-flowered, the peduncles slender, usually exceeding the subtending leaves; flowers spreading, 6-8 (9) mm. long, pink or pale lavender to deep magenta-purplish; calyx 1/3-1/2 the length of the corolla, the linear-lanceolate teeth subequal to the grayish- to blackish-strigillose or -hirsutulous tube; banner deeply emarginate, semi-erect; pods sessile, usually strigillose, 8-12 mm. long, inflated, papery, ellipsoid-obovoid, slightly compressed, 1-celled, the lower suture not intruded.

Open prairies and foothills to ponderosa pine forest; e. of the Cascades, B.C. and Wash. e. to Mont. May-July.

The species is very diverse in habit, but plants of the open usually are much dwarfed in every respect.

Astragalus miser Dougl. ex Hook. Fl. Bor. Am. 1:153. 1831.

Phaca miser Piper, Contr. U.S. Nat. Herb. 11:373. 1906. *Homalobus miser* Rydb. Bull. Torrey Club 40:52. 1913. *Tium miserum* Rydb. N. Am. Fl. 24⁷:394. 1929. *(Douglas,* "On low hills of Spokan River, sixty miles from its confluence with the Columbia")

Homalobus decumbens Nutt. in T. & G. Fl. N.Am. 1:352. 1838. *A. decumbens* Gray, Proc. Am. Acad. 6:229. 1864. *Phaca decumbens* Piper, Contr. U.S. Nat. Herb. 11:373. 1906. *A. campestris* var. *decumbens* M. E. Jones, Rev. Astrag. 74. 1923. *A. miser* var. *decumbens* Cronq. Leafl. West. Bot. 7:18. 1953. *(Nuttall,* "Sandy plains of the Colorado of the West, near the sources of the Platte")

Homalobus tenuifolius Nutt. in T. & G. Fl. N.Am. 1:352. 1838. *A. miser* var. *tenuifolius* Barneby, Leafl. West. Bot. 7:195. 1954. *(Nuttall,* "Hills of the Rocky Mountains") Occurring to the s. and s.e. of our range.

A. serotinus Gray, Pac. R.R. Rep. 12:51. 1860. *Homalobus serotinus* Rydb. Mem. N.Y. Bot. Gard. 1:248. 1900. *A. decumbens* var. *serotinus* M. E. Jones, Contr. West. Bot. 10:58. 1902. *Phaca serotina* Piper, Contr. U.S. Nat. Herb. 11:374. 1906. *A. campestris* var. *serotinus* M. E. Jones, Rev. Astrag. 75. 1923. *A. miser* var. *serotinus* Barneby, Am. Midl. Nat. 55:481. 1956. *(Cooper,* near the Columbia R., Wash.)

A. palliseri Gray, Proc. Am. Acad. 6:227. 1864. *A. serotinus* var. *palliseri* Macbr. Contr. Gray Herb. n.s. 65:37. 1922. *(Bourgeau,* in Palliser's expedition, Rocky Mountains on the British Boundary) = var. *serotinus.*

A. strigosus Coult. & Fisher, Bot. Gaz. 18:299. 1893. *A. griseopubescens* Sheld. Minn. Bot. Stud. 1:24. 1894. *Homalobus strigosus* Rydb. Bull. Torrey Club 40:53. 1913. *A. serotinus* var. *strigosus* Mach. Contr. Gray Herb. n.s. 65:37. 1922. *(Kelsey,* Basin, Mont., July, 1892) = var. *miser.*

Homalobus hylophilus Rydb. Mem. N.Y. Bot. Gard. 1:247. 1900. *A. hylophilus* Nels. in Coult. & Nels. New Man. Bot. Rocky Mts. 291. 1909. *A. campestris* var. *hylophilus* M. E. Jones, Rev. Astrag. 75. 1923. *A. convallarius hylophilus* Tidestr. Proc. Biol. Soc. Wash. 50:20. 1937. *A. miser* var. *hylophilus* Barneby, Am. Midl. Nat. 55:482. 1956. *(Rydberg & Bessey 4490,* Bridger Mts., Mont., June 17, 1897)

A. divergens Blank. Mont. Agr. Coll. Stud. Bot. 1:73. 1905. *Homalobus divergens* Rydb. Bull. Torrey Club 34:417. 1907. *(Blankinship,* Big Coulee Creek, about 30 mi. n.e. of Big Timber, Sweet Grass Co., Mont., June 15, 1902) = var. *decumbens.*

Homalobus camporum Rydb. Bull. Torrey Club 32:666. 1906. *(A. Nelson 7085,* Bush Ranch, Sweetwater Co., Wyo., June 10, 1900) = var. *decumbens.*

Homalobus oblongifolius Rydb. Bull. Torrey Club 34:50. 1907. *A. hylophilus* var. *oblongifolius* Macbr. Contr. Gray Herb. n.s. 65:37. 1922. *A. decumbens* var. *oblongifolius* Cronq. Leafl. West. Bot. 3:253. 1943. *A. miser* var. *oblongifolius* Cronq. Leafl. West. Bot. 7:18. 1953. *(C. F. Baker 409,* Cerro Summit, Colo., in 1901) Occurring to the south of our range, in s. Wyo., Colo., Utah, and e. Nev.

A. campestris var. *crispatus* M. E. Jones, Rev. Astrag. 75. 1923. *A. decumbens* var. *crispatus* Cronq. & Barneby, Leafl. West. Bot. 5:34. 1947. *A. miser* var. *crispatus* Cronq. Leafl. West Bot. 7:18. 1953. *(M. E. Jones,* "Alta, Montana in pine woods," acc. Barneby on the e. slope of the Bitterroot Mts., probably in Beaverhead Co.)

A. miser var. *praeteritus* Barneby, Am. Midl. Nat. 55:483. 1956. *(C. L. Hitchcock 16944,* Ruby River, s. of Vigilante Experiment Sta., Madison Co., Mont., July 28, 1947)

Grayish-strigillose or -villosulous (with basifixed to dolabriform hairs) to greenish perennial, often with extensive rootstocks, the stems prostrate to erect, 1-5 dm. long; stipules lanceolate, 2-4 mm.

stipules

3

2

2

2

½

2

3

Astragalus microcystis

3

A. lotiflorus

2

2

½

2

½

2

A. leptaleus

3

2.5

A. malacus

var. caricinus

½

2

2

½

3

½

var. lyallii

A. lyallii

2

JRJ

long, connate; leaves 3-16 cm. long; leaflets 9-17, linear to lanceolate, oblong or oval, 5-30 mm. long, racemes 3- to 10-flowered; peduncles from shorter to longer than the leaves; flowers spreading to erect, 8-12 mm. long, mostly white but with the banner and wings bluish-pencilled, sometimes very light pinkish-purple, the acuminate keel usually purplish-tipped, subequal to the wings; pedicels 1-3 mm. long; calyx mostly 3-4 (to 6) mm. long, the teeth triangular, usually less than 1 (2) mm. long; pod sessile, pendulous, glabrous to hairy, linear or linear-oblanceolate, 2-3 cm. long, 3-4 mm. broad, compressed, 1-celled, with neither suture at all sulcate or intruded.

Grasslands and foothills to above timber line in the mountains, in moist meadowland to open dry ridges; e. of the Cascades, from s. B. C. and n. Wash. e. to Alta. and S. D. , s. to Nev. , and through the Rocky Mts. to Colo. and Utah. May-July.

This is a widespread, variable species that is divisible, in our area, into several fairly well-marked but completely intergradient races.

1 Pubescence dolabriform; c. Ida. , s. and e.
  2 Herbage strigillose with straight, appressed or narrowly ascending hairs; pod strigil-
     lose
    3 Leaflets narrowly linear to linear-elliptic; petals, except for the tip of the keel,
       whitish or straw colored; ovules 7-11; s. w. Mont. , on the upper forks of the Mis-
       souri R. , e. to c. Ida. (Custer, Lemhi, Butte, Clark, and Fremont cos.), s. to
       Yellowstone Nat. Park   var. praeteritus Barneby
    3 Leaflets broader (on the average), elliptic or oblanceolate; petals nearly always
       purplish or bright purple, rarely whitish; ovules 12-18; s. Mont. , in the drainage
       of the Yellowstone R. (Carbon, Stillwater, and Sweet Grass cos.), southward, e.
       of the divide, to s. c. Wyo.   var. decumbens (Nutt.) Cronq.
  2 Herbage villosulous with loose twisted hairs; pod minutely crisp-villosulous; Bit-
     terroot Mts. , Lemhi Co. , Ida. , and Beaverhead Co. , Mont.
                              var. crispatus (M. E. Jones) Cronq.
1 Pubescence basifixed; nearly throughout our range e. of the Cascades
  4 Leaflets equally pubescent on both faces, the herbage silvery or cinereous; calyx
    4. 5-6 mm. long; keel (8) 8. 5-10. 5 mm. long; pod consistently pubescent; upper
    Columbia and Spokane rivers in n. e. Wash. (Ferry, Stevens, and Spokane cos.)
    and immediately adjacent B. C. , e. to Flathead Lake and the Blackfoot and Deer
    Lodge valleys in n. w. Mont. , and extending just into w. Jefferson Co. , Mont.   var. miser
  4 Leaflets commonly glabrous or glabrescent above but (if pubescent above) then the
    flowers much smaller; if the flowers as large then the petals whitish (except for
    the tip of the keel) and the pod glabrous
    5 Stems less than 15 cm. long; keel usually 8-10 mm. long; leaflets mostly of
      broad outline, elliptic or oval; pod glabrous; the common form throughout the
      Rocky Mts. in Mont. , w. just into Fremont and Lemhi cos. , Ida. , and s. to
      n. w. Wyo.   var. hylophilus (Rydb.) Barneby
    5 Stems (when fully developed) 15 cm. long or more; keel 6-8 mm. long; pod either
      glabrous or puberulent; Rocky Mts. of s. Alta. , southward, through s. B. C. and
      along the e. slope of the Cascades, to Kittitas Co. , Wash.   var. serotinus (Gray) Barneby

Astragalus missouriensis Nutt. Gen. Pl. 2:99. 1818.
  Tragacantha missouriensis Kuntze, Rev. Gen. 2:946. 1891. *Xylophacos missouriensis* Rydb. in
  Small, Fl. S. E. U. S. 620, 1332. 1903. *(Nuttall, "Upper Louisiana")*
  Low, caespitose perennial with a thick root and much-branched crown, silvery with appressed, dolabriform hairs; stems very short, 1-4 cm. long; leaves (3) 4-10 cm. long; stipules lanceolate, 2-6 mm. long, not connate; leaflets (7) 9-17 (19), elliptic to obovate-elliptic, 7-15 mm. long; racemes closely 3- to 9-flowered, borne on peduncles usually equaling or exceeding the leaves; pedicels stout, 1-2 mm. long; flowers 15-25 mm. long; calyx about half as long as the corolla, the tube cylindric, somewhat gibbous at base, grayish- and often blackish-strigillose, 2-3 times as long as the linear teeth; corolla rose-purple but sometimes the petals yellowish-based; pod sessile, leathery, inflated, oblong-ovoid, 20-25 mm. long, 6-8 mm. wide, narrowed abruptly to a short sharp beak, strigillose, cross-corrugated, somewhat obcompressed, 1-celled. N=11.

Great Plains region, e. foothills of the Rocky Mts. , from Alta. to N. M. , e. to Sask. , Minn. , and Tex. June-July.

the teeth about 1/3 the length of the tube; banner erect; keel acute, 6-7 mm. long, semibeaked, about equaling the wings; pod erect, sessile, 1.5-2.5 cm. long, 3-4 mm. broad, almost straight, terete-cordate in section, nearly completely 2-celled by the intrusion of the lower suture, membranous-cartilaginous, minutely strigillose.

Sagebrush slopes (chiefly) from Malheur Co., Oreg., and s.w. Ida., to Nev. and Calif. May-June.

Astragalus oniciformis Barneby, Leafl. West. Bot. 8:122. 1957. *(Ripley & Barneby 8795,* Picabo, Blaine Co., Ida.)

Slender, diffuse, loosely strigillose perennial with suffruticulose caudex, greenish-cinereous, the leaflets medially glabrescent above; stems 5-25 cm. long, floriferous upward from below the middle; leaves 2.5-7.5 cm. long; stipules 1.5-4 mm. long, the lowest strongly amplexicaul but free; leaflets 17-27, elliptic to oval or oblong, obtuse or retuse, 1-7 mm. long; peduncles 5-25 mm. long, much shorter than the leaves; racemes loosely 4- to 12-flowered; flowers ochroleucous, veined with dull lilac, the strongly recurved banner about 5.5-7 mm. long; calyx 3-4 mm. long, the campanulate tube about 2 mm. long; pod pendulous, stipitate, the stipe 1.5-4 mm. long, the body 7-12 mm. long, 2-3.5 mm. broad, slightly incurved, triquetrous with flat lateral and openly sulcate dorsal faces, almost fully bilocular, the thin, green, densely strigillose valves becoming papery.

Sandy flats, on basalt; plentiful at the foot of the Sawtooth Mts., in Blaine Co., Ida. May-June.

This species is very closely related to *A. mulfordiae,* differing chiefly in having free (rather than connate) lower stipules.

Astragalus palousensis Piper, Bot. Gaz. 22:489. 1896.

A. arrectus var. *palousensis* M. E. Jones, Contr. West. Bot. 10:68. 1902. *(Piper 1493,* Pullman, Wash.)

Erect perennial with a woody taproot and shortly forking caudex, strigillose or pilosulous, the herbage green or cinereous, the leaflets glabrous above; stems numerous, in clumps, 2-4 dm. long, the main axis abruptly inhibited above the second or third flowering axil and forming less than half the height of the plant; leaves 9-22 cm. long; stipules 4-8 mm. long, becoming papery, more or less decurrent-amplexicaul but free; leaflets mostly (17) 21-31 (33), linear-oblong, lanceolate or rarely oval, obtuse or deeply notched, 8-22 mm. long; peduncles strict, 1-2.5 dm. long, much surpassing the leaves; racemes 15- to 35-flowered, early-elongating, 8-22 cm. long in fruit; flowers ascending to erect, ochroleucous but drying yellowish, about 12-13 mm. long; calyx 5-6.5 mm. long, the black-strigillose, campanulate tube 3.5-4.5 mm. long, the teeth about 1.5 mm. long; petals irregularly graduated, the wings about 2 mm. longer than the abruptly recurved banner, the keel 10-10.5 mm. long; pod erect, stipitate, the stipe 2.5-6 mm. long, the narrowly oblong-ellipsoid body 15-23 mm. long, 2.5-6.5 mm. in thickness, nearly or quite straight, obcompressed, the low, convex, ventral face keeled by the prominent thick suture, the dorsal face openly but deeply sulcate, the green, somewhat fleshy valves becoming leathery, loosely strigillose with black, white, or mixed black and white hairs, inflexed as a narrow but nearly complete septum.

Grassy hillsides, sagebrush flats, and river bluffs, sometimes in open pine forest, locally plentiful in the Palouse country, Whitman Co., Wash., and adj. w.c. Ida.; also (apparently disjunctly) along the lower Spokane and adj. Columbia rivers in e.c. Wash. May-June.

In recent correspondence Mr. Barneby has expressed the opinion that the proper epithet for this species will prove to be *A. arrectus* Gray.

Astragalus paysonii (Rydb.) Barneby, Leafl. West. Bot. 4:60. 1944.

Hamosa paysonii Rydb. Bull. Torrey Club 54:22. 1927. *(Payson & Payson 2748,* Horse Cr., 7 mi. w. of Merna, Sublette Co., Wyo.)

Perennial from a taproot and short caudex, pale green and nearly glabrous, the leaflets thinly strigillose beneath; stems ascending, 2-4.5 dm. long; stipules 2-5 mm. long, free; leaves 4-9 cm. long; leaflets 7-15, ovate-oblong to obovate-cuneate, mostly retuse, 5-20 mm. long; peduncles 3-8 cm. long; racemes 5- to 16-flowered, 1-4 cm. long in fruit; flowers white, the banner about 7 mm. long; calyx about 3.5 mm. long, at least partly black-strigillose; pod declined or deflexed, short-stipitate, the stipe 1-1.5 mm. long, the body lunately linear-ellipsoid, 10-17 mm. long, 2.5-3.5 mm. broad, cuspidate at the apex, compressed-triquetrous, narrowly grooved dorsally, fully 2-locular, the green, puberulent valves becoming papery and straw-colored.

Open slopes and ridges in the timber belt, rare and local, known only from the type locality on the

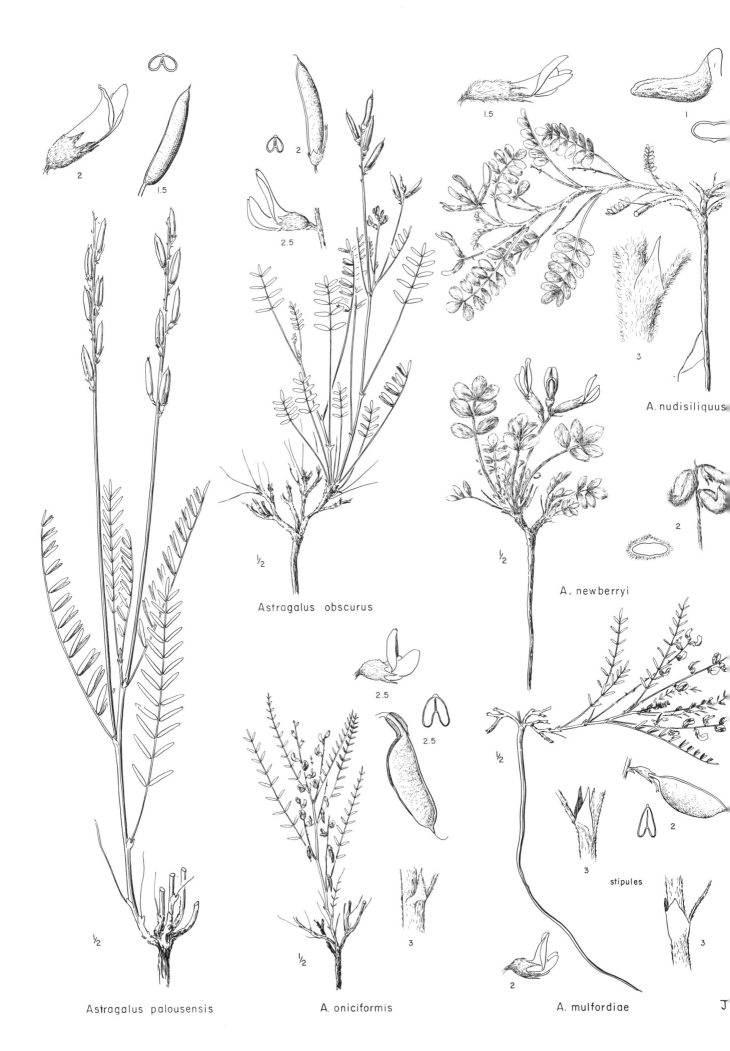

2

1.5

2

2.5

1.5

1

3

A. nudisiliquus

A. newberryi

Astragalus obscurus

2

2.5

2.5

A. oniciformis

½

3

½

½

2

3

stipules

3

Astragalus palousensis

A. oniciformis

A. mulfordiae

J

Astragalus pulsiferae Gray, Proc. Am. Acad. 10:69. 1874.

Phaca pulsiferae Rydb. N.Am. Fl. 24⁶:357. 1929. *(Mrs. Pulsifer Ames,* Sierra and Plumas cos., Calif.)

A. suksdorfii Howell, Erythea 1:111. 1893. *Phaca suksdorfii* Piper, Contr. U.S. Nat. Herb. 11: 369. 1906. *A. pulsiferae* var. *suksdorfii* Barneby, El Aliso 4:131. 1958. *(Suksdorf,* near base of Mt. Adams, Wash.)

Grayish-pubescent perennial with a deep taproot and multicipital crown, the stems slender and naked at base, 5-20 cm. long, prostrate to ascending; stipules 1-3 mm. long, ovate, not connate; leaves 1.5-4 cm. long; leaflets 7-15, linear-oblanceolate to obovate, 3-10 mm. long; racemes very compactly 4- to 10-flowered, much shorter than the leaves; flowers spreading, ochroleucous, purplish-tinged, 5-7 mm. long; calyx hairy, about half as long as the corolla, the teeth linear-lanceolate, from half as long to as long as the tube; banner erect; pod sessile, 1-1.5 cm. long, obliquely ellipsoid, villous, membranous and inflated, nearly terete in section but sulcate on both sutures, 1-celled, the lower suture not intruded.

Sandy and gravelly flats, in sagebrush and open pine forest on basaltic formations; n. Sierra Nevada and s. end of the Cascades Mts., in Calif., apparently absent or at least not collected in the Oregon Cascades, but reappearing in Wash. on Mt. Adams, and reported from Falcon Valley, Klickitat Co. July.

Our material, as described above, is referable to the var. *suksdorfii* (Howell) Barneby; the var. *pulsiferae,* a more villous plant, occurs in n.e. Calif. and n.w. Nev.

Astragalus purshii Dougl. ex Hook. Fl. Bor. Am. 1:152. 1831.

Tragacantha purshii Kuntze, Rev. Gen. 2:947. 1891. *Phaca purshii* Piper, Contr. U.S. Nat. Herb. 11:369. 1906. *Xylophacos purshii* Rydb. Bull. Torrey Club 32:662. 1906. *(Douglas,* "on the low hills of the Spokan River, North-West America")

A. glareosus Dougl. ex Hook. Fl. Bor. Am. 1:152. 1831. *Tragacantha glareosa* Kuntze, Rev. Gen. 2:945. 1891. *A. inflexus* var. *glareosus* M. E. Jones, Contr. West. Bot. 10:62. 1902. *Phaca glareosa* Piper, Contr. U.S. Nat. Herb. 11:369. 1906. *Xylophacos glareosus* Rydb. Fl. Rocky Mts. 506, 1063. 1917. *A. purshii* var. *glareosus* Barneby, Am. Midl. Nat. 37:503. 1947. *(Douglas,* "dry gravelly banks of rivers, from the confluence of Lewis and Clarke's River with the Columbia and the mountains")

Phaca mollissima Nutt. in T. & G. Fl. N.Am. 1:350. 1838, not *A. mollissimus* Torr. in 1828. *(Nuttall,* "plains of the Rocky Mts. and on the hills of 'Hamm's Fork' of the Colorado") = var. *purshii.*

A. purshii var. *tinctus* M. E. Jones, Zoë 4:269. 1893. *Phaca purshii* var. *tincta* Piper, Contr. U.S. Nat. Herb. 11:369. 1906. *(M. E. Jones,* Soda Spgs., Nevada Co., Calif., in 1881)

A. allanaris Sheld. Minn. Bot. Stud. 1:141. 1894. *(Suksdorf,* Rattlesnake Mt., Yakima Co., Wash., June, 1884) = var. *glareosus.*

A. candelarius Sheld. Minn. Bot. Stud. 1:142. 1894. *Xylophacos candelarius* Rydb. Bull. Torrey Club 52:370. 1925. *(W. H. Shockley,* Candelaria, Esmeralda Co., Nev., Apr. and May, 1888) = var. *tinctus.*

A. lanocarpus Sheld. Minn. Bot. Stud. 1:144. 1894. *(T. J. Howell,* "Klikitat Prairie," Wash., June, 1880) = var. *glareosus.*

A. booneanus A. Nels. Bot. Gaz. 53:223. 1912. *(W. J. Boone 2,* Caldwell, Ida.) = var. *glareosus.*

A. purshii var. *interior* M. E. Jones, Rev. Astrag. 222. 1923. *(M. E. Jones,* Aurum, Nev., in 1893, lectotype by Barneby) Probably not reaching our area.

A. purshii var. *lectulus* sensu M. E. Jones, Rev. Astrag. 223. 1923, not *A. lectulus* Wats. in 1887. = var. *lagopinus.*

Xylophacos incurvus Rydb. Bull. Torrey Club 52:366. 1925. *A. purshii* var. *incurvus* Jeps. Fl. Calif. 2:360. 1936. *A. incurvus* Abrams, Ill. Fl. Pac. St. 2:577. 1944, not *A. incurvus* Desf. in 1800. *(Lemmon 76,* in Calif., in 1875) = var. *purshii.*

Xylophacos ventosus Suksd. ex Rydb. Bull. Torrey Club 52:370. 1925. *A. ventosus* Abrams, Ill. Fl. Pac. St. 2:578. 1944. *(Suksdorf 10662,* "several kilos east of Bingen," Wash., Nov. 10, 1920) = var. *tinctus.*

Xylophacos lagopinus Rydb. Bull. Torrey Club 52:372. 1925. *A. lagopinus* Peck, Madroño 6:134. 1941. *A. purshii* var. *lagopinus* Barneby, Am. Midl. Nat. 37:511. 1947. *(J. B. Leiberg 326,* between Preneville [= Prineville?] and Bear Buttes, Crook Co., Oreg., June 25, 1894)

A. viarius Eastw. Leafl. West. Bot. 1:178. 1935. *(J. T. Howell 7171,* 7 mi. n. of Bend, Deschutes
   Co., Oreg., July 3, 1931) = var. *lagopinus.*
A. purshii var. *concinnus* Barneby, Leafl. West. Bot. 4:231. 1946. *(Macbride & Payson 3224,* Chal-
   lis, Custer Co., Ida.)
ASTRAGALUS PURSHII var. OPHIOGENES Barneby, hoc loc. *A. ophiogenes* Barneby, Leafl. West.
   Bot. 4:232. 1946. *(Ripley & Barneby 6478,* e. of King Hill, Ida., May 31, 1945)
   Caespitose, grayish-woolly perennial from a taproot and branched crown, matted, often pulvinate,
the stems prostrate, 5-10 cm. long; leaves closely crowded, 2-15 cm. long, white-tomentose; stipules ovate to oblong-lanceolate, acuminate, 3-15 cm. long, usually exceeding the internodes, not connate at base; leaflets (3) 7-19, elliptic or oblong to obovate or suborbicular, rounded to acute, 5-20
mm. long; racemes 3- to 10 (11)-flowered, from nearly sessile to fairly long-pedunculate and exceeding the leaves, subcapitate at anthesis, more elongate in fruit; flowers ochroleucous or yellowish to
deep reddish-purple, 1-3 cm. long; calyx tubular, 1/2-2/3 as long as the corolla, somewhat black-
hairy to grayish-lanate, the linear-lanceolate teeth only 1/3-1/2 as long as the tube; pod sessile, 1-
2.5 cm. long, coriaceous, grayish-woolly, dorsiventrally flattened, more or less reniform in section, from nearly straight to somewhat lunate or arcuate, 1-celled, the lower suture usually not intruded but sometimes intruded and the two sutures nearly contiguous.
   Prairies and sagebrush desert to the foothills and lower mountains; e. of the Cascades from s.
B. C. to n. Calif., e. to Alta., the Dakotas, and Mont., and s. to N. M., Utah, and Nev. Apr.-June.
   This is an extremely variable species with several well-marked but completely intergradient geographic races, six of which occur in our area.
1 Flowers ochroleucous or very pale lavender, the keel purple-tipped, 20-30 mm. long;
      leaflets usually 9-13, acute; pods 2-2.5 cm. long, mostly nearly straight; sagebrush
      valleys to lower hills, more or less general e. of the Cascades from B.C. to n.e. Calif.,
      e. to Alta., the Dakotas, and Colo.; N=11   var. purshii
1 Flowers reddish-purple, 10-25 mm. long; pods usually arcuate
   2 Pods less than 2 cm. long; flowers 10-16 mm. long; mostly s. of our range, but n. in
         Oreg. to Deschutes and Crook cos.
      3 Racemes mostly 3- to 5 (rarely 7)-flowered; leaflets 5-7 (11); hairs of the pod up to
         2-3 mm. long; Deschutes and Harney cos., Oreg., southward
                                                            var. lagopinus (Rydb.) Barneby
      3 Racemes mostly 5- to 11-flowered; leaflets 9-15 (17); hairs of the pod up to 3-5
         mm. long; w. Snake R. plains, Elmore Co., Ida., to s.e. Malheur Co., Oreg.,
         probably beyond our range   var. ophiogenes Barneby
   2 Pods and flowers usually at least 2 cm. long
      4 Pod very strongly sulcate and nearly 2-celled toward the base; leaflets chiefly ob-
         lanceolate-obovate, acute to rounded; usually in sagebrush, e. Wash. and Oreg. to
         Nev. and s.w. Ida.; N=11   var. glareosus (Dougl.) Barneby
      4 Pod not so strongly sulcate, the two sutures not contiguous within; leaflets various
         5 Leaflets (5) 7-9, obovate to suborbicular; calyx slightly inflated, usually at least
            5 mm. broad in pressed specimens; c. Ida. to w.c. Mont.   var. concinnus Barneby
         5 Leaflets (5) 9-13 (17), elliptic to obovate; calyx not inflated, less than 4 mm.
            broad when pressed; lower Columbia Valley, Wash., southward, e. of the
            Cascades, to Nev. and e. Calif.   var. tinctus M. E. Jones

Astragalus reventus Gray, Proc. Am. Acad. 15:46. 1879.
   Phaca reventa Piper, Contr. U.S. Nat. Herb. 11:372. 1906. *Cnemidophacos reventus* Rydb. Bull.
      Torrey Club 40:52. 1913. *(Douglas,* "interior of Oregon," and *Cusick,* "Grand Round Valley and
      Blue Mts. in e. Oreg.")
   A. conjunctus Wats. Proc. Am. Acad. 17:371. 1882. *A. reventus* var. *conjunctus* M. E. Jones,
      Contr. West. Bot. 10:61. 1902. *Phaca conjuncta* Piper, Contr. U.S. Nat. Herb. 11:373. 1906.
      Tium conjunctum Rydb. N.Am. Fl. 24⁷:393. 1929. *(J. Howell,* John Day Valley, Oreg., May,
      1880)
   A. hoodianus Howell, Erythea 1:111. 1893. *A. conjunctus* var. *hoodianus* M. E. Jones, Contr. West.
      Bot. 8:9. 1898. *Phaca hoodiana* Piper, Contr. U.S. Nat. Herb. 11:373. 1906. *Cnemidophacos hood-
      ianus* Rydb. N.Am. Fl. 24⁵:285. 1929. *A. reventus* var. *hoodianus* Peck, Man. High. Pl. Oreg.
      444. 1941. (Wasco Co., Oreg., "from Hood River to a point a few miles east of the Dalles, and also
      on the opposite side of the river;" presumably collected by Howell) = var. *oxytropidoides.*

ASTRAGALUS REVENTUS var. OXYTROPIDOIDES (M. E. Jones) C. L. Hitchc. hoc loc. *A. conjunctus* var. *oxytropidoides* M. E. Jones, Proc. Calif. Acad. Sci. II, 5:665. 1895. *Tium oxytropoides* Rydb. N. Am. Fl. 24[7]:393. 1929. *(Howell 798,* near The Dalles, Oreg. , May 8, 1885)

A. reventus var. *canbyi* M. E. Jones, Contr. West. Bot. 8:11. 1898. *(Brandegee 36,* Yakima region, Wash. , in 1882)

Cnemidophacos reventiformis Rydb. N.Am. Fl. 24[5]:284. 1929. *A. reventiformis* Barneby, Am. Midl. Nat. 55:492. 1956. *(Suksdorf,* Klickitat, Wash. , Apr. 26, 1882) = var. *canbyi.*

Cnemidophacos knowlesianus Rydb. N.Am. Fl. 24[5]:284. 1929. *(Suksdorf 482,* w. Klickitat Co. , Wash. , Apr. 23 and June, 1886) = var. *oxytropidoides.*

ASTRAGALUS REVENTUS var. SHELDONII (Rydb.) C. L. Hitchc. hoc loc. *Tium sheldoni* Rydb. N.Am. Fl. 24[7]:393. 1929. *A. conjunctus* var. *sheldoni* Peck, Man. High. Pl. Oreg. 448. 1941. *A. sheldoni* Barneby, Am. Midl. Nat. 55:489. 1956. *(Sheldon 8032,* Horse Creek Canyon, Wallowa Co. , Oreg. , May 14, 1897)

Robust perennial with a large woody taproot and a short, branched crown with many stems 2-40 cm. long, greenish or commonly more or less grayish-strigillose to -strigillose-villosulous or -silky; leaves 7-30 cm. long, numerous, closely crowded on the short stems, the internodes often not much longer than the stipules, the stipules triangular, membranous, 3-7 mm. long, from amplexicaul and connate to merely decurrent below but nearly or quite free upward toward the tips of the stems, the petiole often equaling the blade; leaflets 13-41, linear to oblong-oblanceolate, 3-25 mm. long, 1-4 (occasionally to 6) mm. broad, glabrous above or strigillose on both surfaces; peduncles usually exceeding the leaves, mostly 1-3 dm. long; racemes closely to laxly 7- to 35-flowered; pedicels about 1 (2-3) mm. long, lengthening to as much as 6 mm. in fruit; flowers drooping or spreading to erect, ochroleucous or the keel purplish, or lilac-purplish overall, (11) 13-23 mm. long; calyx 7. 5-15 mm. long, whitish- to blackish-villous-strigose, the teeth 2-7. 5 mm. long and 1/3-4/5 as long as the tube; banner well reflexed; wings slender, 2-4 mm. longer than the rounded to slightly beaked keel; pod erect, nearly or quite sessile (stipe up to 1. 5 mm. long but completely concealed by the calyx), glabrous to grayish- or blackish-strigose-villous, fleshy when green, coriaceous-cartilaginous when dried, obcompressed, straight to somewhat arcuate, 15-30 mm. long, 3-6 mm. wide and at least twice as thick, abruptly acute, often with a laterally flattened, cuspidate beak, 1-celled, but the lower suture sulcate and partially to nearly completely dividing the cavity.

Sagebrush desert and scabland or stony flats and hilltops to grassy hillsides and ponderosa pine forest, often on basalt; e. side of the Cascades from near Ellensburg, Kittitas Co. , Wash. , s. to c. Oreg. , e. to w.c. and s.w. Ida. Apr. -June.

This taxon includes a group of fairly clearly marked geographic races that have been (and possibly more correctly should be) recognized at the specific level. However, they are obviously very closely related, they tend to intergrade at least to some degree, and in line with the treatment of similar complexes in this flora, they are here recognized as well-marked varieties of one species.

1 Lower stipules connate (except sometimes in var. *oxytropidoides,* which see), the upper
   ones connate to amplexicaul or even free; leaflets 13-37
    2 Calyx tube cylindric or subcylindric, slightly less than once to a little more than twice
      as long as thick; ovary glabrous; leaflets 13-31; corolla often distally suffused with
      bluish-purple; locally abundant in n.c. Oreg. , from the e. Cascade foothills to the
      Blue Mts. , and s.e. along the Malheur R. to the Steens Mts. and s. w. Ida. , mostly
      above 2000 ft. elev. ; N=12   var. conjunctus (Wats.) M. E. Jones
    2 Calyx tube campanulate, about 1. 5 times as long as thick; ovary and pod either gla-
      brous or pubescent; leaflets 17-37; flowers mostly white, the keel rarely pale li-
      lac at the tip; mostly n. of Oreg. , except occasional along the Columbia R. , where
      occurring below 2000 ft.
       3 Ovary and pod sessile, pubescent; calyx 11-15 mm. long, the teeth 4. 5-7. 5 mm.
         long; ovules 18-22; stems 4-30 cm. tall; Columbia R. gap, Wasco and Hood River
         cos. , Oreg. , and Klickitat Co. , Wash.   var. oxytropidoides (M. E. Jones) C. L. Hitchc.
       3 Ovary glabrous or (if pubescent) the pod stipitate and the calyx less than 1 cm.
         long; calyx teeth 2. 5-5 mm. long; ovules 22-32; stems 2-12 (20) cm. tall; com-
         mon along the e. slope and foothills of the Cascades in the Yakima R. drainage,
         from near Ellensburg s. to the Horse Heaven and Klickitat hills, Wash. , and at
         one station on the s. bank of the Columbia R. in Sherman Co. , Oreg. , ascend-
         ing to 5000 ft. elev.   var. canbyi M. E. Jones

1 Lower stipules, like the upper ones, not at all connate, although sometimes decurrent
   over halfway around the stem; leaflets 23-41
   4 Ovary and pod glabrous, ultimately 7-10 mm. thick, the lower suture intruded only
      slightly, the septum less than 1 mm. wide; stems 5-18 cm. long; dry rocky slopes
      and grassy openings in pine forest, known only from the Blue Mts., from the head-
      waters of the Umatilla and Grande Ronde rivers, Oreg., n.e. into extreme s.e.
      Wash.   var. reventus
   4 Ovary and pod usually strigillose with black or white hairs (rarely glabrous), be-
      coming 4-6.5 mm. thick, the lower suture intruded to form a septum 1-1.5 mm.
      wide; stems 5-38 cm. long; common, usually on basalt, in the hill country about
      the lower Salmon and adjacent Snake rivers in Wallowa Co., Oreg., Idaho, Lewis,
      and Nez Perce cos., Ida., and s. Asotin Co., Wash.; N=12
                                                              var. sheldonii (Rydb.) C. L. Hitchc.

Astragalus riparius Barneby, Am. Midl. Nat. 55:490. 1956. *(Piper 4133,* Wawawai, Wash.)

Erect perennial from a woody taproot and knotty caudex, gray-strigillose or greenish, the leaflets glabrous above; stems tufted, 4-15 cm. long, shorter than the peduncles; leaves 1-2 dm. long; stipules 2.5-7 mm. long, soon becoming papery, more or less decurrent-amplexicaul but free; leaflets 21-33, linear-oblong or narrowly oblanceolate, obtuse or acutish, 4-21 mm. long; peduncles stout, 14-30 cm. long, usually surpassing the leaves; racemes mostly 10- to 20-flowered, the flowers ascending, the axis 4-12 cm. long in fruit; pedicels ascending, in fruit clavately thickened and 2-5 mm. long; flowers ochroleucous or greenish-white but drying yellowish; calyx 7-10.5 mm. long, black-pilosulous, the campanulate tube 4.5-6 mm. long, the teeth 2-5 mm. long; petals irregularly graduated, the wings 1-4 mm. longer than the banner, the banner 11.5-14 mm. long, abruptly recurved distally, the margins erose or undulate, becoming lacerate, the keel 9.5-13 mm. long; pod erect, sessile, oblong- or narrowly ovoid-ellipsoid, 1.5-2.5 cm. long, 6-10 mm. thick, nearly straight, broadly rounded at base, contracted into a triangular-acuminate, rigidly cuspidate beak, a little obcompressed, keeled by the thick ventral and commonly undulate dorsal sutures, flattened or shallowly sulcate dorsally, the green, fleshy, glabrous valves becoming leathery or almost woody, slightly or not at all inflexed. N=12.

Dry bluffs and canyon banks; locally plentiful along the lower Snake R. and affluent creeks between the mouth of the Clearwater and that of the Tucannon R., Whitman and Columbia cos., Wash., and Nez Perce Co., Ida. May-June.

This species combines the pod of *A. reventus* with the uniquely modified flower of *A. palousensis* and may have originated as a hybrid population, but it is now an established and self-perpetuating entity.

Astragalus robbinsii (Oakes) Gray, Man. 2nd ed. 98. 1856.

Phaca robbinsii Oakes, Hovey's Hort. Mag. 7:181. 1841. *Atelophragma robbinsii* Rydb. Bull. Torrey Club 55:124. 1928. *Astragalus labradoricus* var. *robbinsii* M. E. Jones, Rev. Astrag. 134. 1923. *(Dr. James W. Robbins,* banks of the Onion River, Burlington, Vt., in 1829)

Astragalus robbinsii var. *occidentalis* Wats. Bot. King Exp. 70. 1871. *Astragalus occidentalis* M. E. Jones, Contr. West. Bot. 8:17. 1898. *Astragalus labradoricus* var. *occidentalis* M. E. Jones, Rev. Astrag. 134. 1923. *Atelophragma occidentale* Rydb. Bull. Torrey Club 55:128. 1928. *(Watson,* East Humboldt Mts., Nev.)

Astragalus macounii Rydb. Mem. N.Y. Bot. Gard. 1:243. 1900. *Atelophragma macounii* Rydb. Bull. Torrey Club 32:660. 1906. *(John Macoun 25,* Deer Park, Lower Arrow Lake, B.C., in 1890) = var. *occidentalis.*

ASTRAGALUS ROBBINSII var. ALPINIFORMIS (Rydb.) Barneby, hoc loc. *Atelophragma alpiniforme* Rydb. Bull. Torrey Club 55:129. 1928. *Astragalus alpinus* var. *alpiniformis* Peck, Man. High. Pl. Oreg. 477. 1940. *(Cusick 2103,* Hurricane Creek, Wallowa Mts., Oreg., Aug. 25, 1898)

Sparsely strigillose to glabrate, greenish perennial with a taproot and multicipital caudex and prostrate or spreading to erect but decumbent-based stems 2-6 dm. long; leaves 6-12 cm. long; stipules oblong-lanceolate, 4-7 mm. long, the lower ones connate; leaflets 7-19, ovate-lanceolate to oblong, 1-2.5 cm. long, thin and succulent, paler on the lower surface; racemes loosely 10- to 60-flowered, more or less secund, in fruit often equal to the peduncles which much exceed the leaves; flowers 6-10 mm. long, purplish (sometimes pale); pedicels slender, 3-4 mm. long; calyx 1/2-2/3 the length of the corolla, usually blackish-puberulent to -strigillose, the teeth linear-lanceolate, 1/4-1/2 the length of the tube; pod spreading to reflexed, with a slender stipe about equal to the calyx, the body mem-

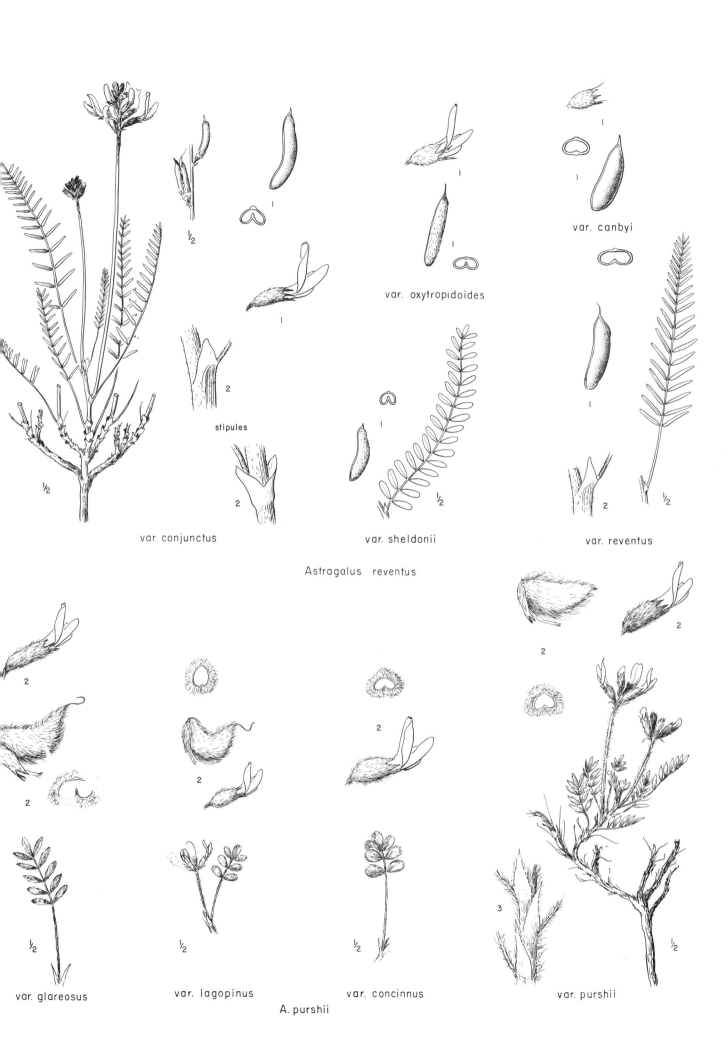

var. canbyi

var. oxytropidoides

½

2

stipules

2

var. conjunctus

var. sheldonii

var. reventus

2

Astragalus reventus

2

2

2

2

2

2

3

½

½

½

½

var. glareosus

var. lagopinus

var. concinnus

var. purshii

A. purshii

branous, black-hairy, 8-25 mm. long, 4-5 mm. wide, compressed, elliptic in outline, acute at both ends, neither suture sulcate but the dorsal (which is usually uppermost in this species due to the twisting of the pedicel) usually intruded very slightly as a thin partial partition about 1 mm. high.

Stream banks to alpine slopes, Alas. and B.C. southward (on the e. side of the Cascades) to Okanogan Co., Wash., e. to Great Bear Lake and Alta., and southward through Mont. and Ida. to Colo. and e. Nev. June-Aug.

This is a species of wide but interrupted range in N.Am, composed of 5 or perhaps 6 vars., 3 in n.e. U.S. and adj. Can., the rest (as described above) cordilleran. Our races are separable as follows:

1 Pod (ours) 15-25 mm. long, 7- to 9-ovulate; widespread, the range as given for the species in our area   var. occidentalis Wats.
1 Pod 8-11 mm. long, 3- to 5-ovulate; local in the Wallowa Mts., Oreg.
                                                                                 var. alpiniformis (Rydb.) Barneby

The stipe varies from longer than the whole calyx to much shorter than the calyx tube and the pods may or may not have a partial partition. Therefore *A. macounii,* based on these variant characters, is not believed to be significant. Under a less conservative treatment, however, var. *occidentalis* might be considered to be a very local species restricted to the East Humboldt and Ruby mts. in n.e. Nev. It has a short, subsessile, 7- to 9-ovulate pod 8-15 mm. long, in contrast with the common Rocky Mt. form (usually called *A. macounii)* which has a stipe varying from 1-3 mm. in length, and a 7- to 9-ovulate pod 15-25 mm. long.

Astragalus salmonis M. E. Jones, Contr. West. Bot. 8:9. 1898.
    Tium salmonis Rydb. N.Am. Fl. 24[7]:396. 1929. *(Howell,* Trout Creek, Blue Mts., Oreg., May 25, 1885)

Sparsely strigillose, caespitose, more or less prostrate perennial, the stems only 1-5 cm. tall; internodes very short, scarcely equaling the 3-4 mm., lanceolate, free stipules; leaves 3-8 cm. long; leaflets 9-17, linear or linear-elliptic to oblong-elliptic, usually folded or rolled, 3-8 mm. long; racemes loosely 2- to 8-flowered, borne on peduncles that usually about equal the leaves; pedicels stout, 1-4 mm. long; flowers (8) 10-12 (14) mm. long, whitish or cream with a purplish distal tinge, the keel usually dull purplish; calyx about half the length of the corolla, the teeth linear-lanceolate, about 1/3 as long as the tube; pod subsessile, the stipe from longer than the calyx to less than half as long, the body 10-20 mm. long, more or less mottled, fleshy but drying to woody and rather fibrous within, strigillose, 1-celled, somewhat arcuate, strongly obcompressed, 4.5-6.5 mm. thick and not quite so broad, cordate-reniform in section, the lower suture strongly sulcate.

Dry sagebrush-covered hills and valleys; Blue Mts., Oreg., s. and e. to Owyhee Co., Ida. May-June.

Astragalus scaphoides M. E. Jones, Contr. West. Bot. 10:69. 1902.
    A. arrectus var. *scaphoides* M. E. Jones, Proc. Calif. Acad. Sci. II, 5:664. 1895. *A. scophioides* Rydb. Mem. N.Y. Bot. Gard. 1:241. 1900. *Phacopsis scaphoides* Rydb. Bull. Torrey Club 40:52. 1913. *Hesperonix scaphoides* Rydb. N.Am. Fl. 24[7]:439. 1929. ("hills west of Clark's Canyon, Beaverhead Co., Mont.," July, 1888, collector not specified, but probably Tweedy)

Sparsely strigillose perennial with a taproot and branched crown; stems several, stout, ascending to erect, 2-6 dm. tall; leaves 10-25 cm. long; stipules lanceolate, 1-4 mm. long, not connate; leaflets 15-21, lance-oblong to elliptic-oblong, 1.5-3.5 cm. long, as much as 13 mm. broad, glabrous on the upper surface at least; peduncles mostly 10-15 cm. long; racemes closely 15- to 30-flowered but elongating and open in fruit; pedicels 2-5 mm. long; flowers spreading to slightly reflexed, white to ochroleucous, about 2 cm. long; calyx usually blackish-hairy, 8-10 mm. long, the narrowly lanceolate lower teeth about 2 mm. long; banner erect; wings 2-4 mm. longer than the keel; pod erect, with a stout upward-arching stipe about twice as long as the calyx, the body 1.5-2 cm. long, cartilaginous, glabrous, slightly mottled, corrugate-wrinkled, oblong-ovoid, inflated and slightly obcompressed, 4-6 mm. broad, 6-10 mm. thick, with both sutures sulcate, the lower intruded to form a 3/4 complete partition.

Sagebrush plains and lower slopes; a local species, from Beaverhead Co., Mont., to Lemhi Co., Ida. Late May-early July.

Astragalus sclerocarpus Gray, Proc. Am. Acad. 6:225. 1864.
    Phaca podocarpa Hook. Fl. Bor. Am. 1:142. 1831, not *A. podocarpus* Meyer in 1831. *Tragacantha*

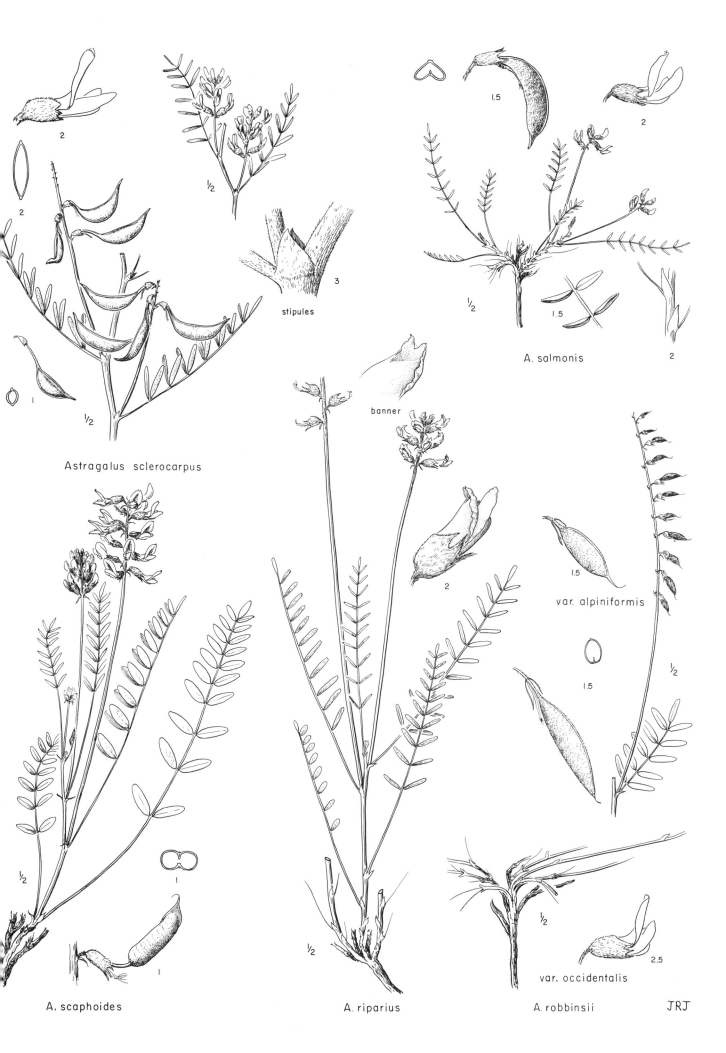

stipules

banner

1.5

2

A. salmonis

Astragalus sclerocarpus

var. alpiniformis

1.5

var. occidentalis

A. scaphoides

A. riparius

A. robbinsii

JRJ

sclerocarpa Kuntze, Rev. Gen. 2:948. 1891. *Homalobus podocarpus* Rydb. Bull. Torrey Club 51:
18. 1924. *(Douglas,* Great Falls of the Columbia)
Sparsely to densely silvery-strigillose perennial from a branched crown; stems several, 2-5 dm.
tall, usually sand-buried at base and the subterranean portion with only the rusty stipules at the nodes;
leaves 3-12 cm. long; stipules triangular, 2-4 mm. long, those above ground not connate, the subterranean ones mostly decurrent but sometimes connate; leaflets 15-21, linear to oblong-lanceolate,
(10) 15-30 mm. long, (1) 2-4 mm. broad; peduncles about equal to the leaves; racemes closely to laxly 10- to 30-flowered; pedicels 1-2 mm. long; flowers spreading-ascending, 10-14 mm. long; calyx
5-7 mm. long, grayish- and blackish-strigillose, the teeth about 1/4 as long as the tube, the upper 2
noticeably the broadest; banner from nearly pure white to greenish-white, erect; wings from almost
white to pale purplish, exceeding the rounded, usually purplish-tipped keel by about 1 (2) mm.; pod
stipitate, pendulous, grayish-strigillose to nearly glabrous, fleshy, filled with stringy twisted fibers,
drying to leathery-cartilaginous, nearly straight when young, but curving to nearly a half circle at
maturity and often becoming purplish mottled, the body 2-3 cm. long, 7-9 mm. broad, strongly compressed, elliptic in section, 1-celled, both sutures very prominent but neither intruded; stipe slender,
1.5-2 cm. long. N=11.
    Dunes and sandy barrens; along both sides of the Columbia R., from The Dalles upstream to the
Great Bend, thence n. through interior Wash. to Kettle Falls and, following the Okanogan R., n. just
into s. B.C. Apr.-June.

Astragalus sinuatus Piper, Bull. Torrey Club 28:40. 1901.
    Phaca sinuata Piper, Contr. U.S. Nat. Herb. 11:370. 1906. *Homalobus sinuatus* Rydb. Bull. Torrey
    Club 51:16. 1924. *(Brandegee 739,* e. Wash., in 1883)
    A. whitedii Piper, Bull. Torrey Club 29:224. 1902. *Homalobus whitedii* Rydb. Bull. Torrey Club
    51:16. 1924, as to name but not as to description. *(Whited 1353* and *1042,* Colockum Creek, 20
    miles s.w. of Wenatchee, Wash., in 1899)
Robust perennial from a woody taproot and knotty root crown, villous-cinereous throughout with
incumbent or curly hairs; stems decumbent, 2-4.5 dm. long; leaves 2-7 cm. long; stipules 2-4.5 mm.
long, more or less decurrent, free; leaflets 11-19, obovate-cuneate or oblong-oblanceolate, truncate-
retuse, 6-16 mm. long; peduncles 4.5-12 cm. long; racemes 8- to 16-flowered, the flowers ascending, the axis up to 4.5 cm. long in fruit; flowers whitish, the banner 16.5-20 mm. long; calyx 9-11.5
mm. long, black- or partly white-villosulous, the deeply campanulate tube 7-8.5 mm. long, the teeth
1.5-3 mm. long; pod pendulous or loosely spreading, stipitate, the stipe 5-7 mm. long, the body
lunately ellipsoid, cuneate at both ends, 24-27 mm. long, 4-5 mm. in diameter, compressed but the
valves somewhat transversely dilated, bicarinate by the sutures, 1-locular, the fleshy valves becoming leathery, gray-villosulous.
    Rocky hillsides, among sagebrush; known with certainty only from Colockum Creek, a tributary of
the Columbia R., in Chelan Co., Wash. Apr.-May.
    The epithet "whitedii" has for several decades been erroneously applied to *A. curvicarpus* Macbr.,
a species widespread around the s. periphery of our area and southward, but unknown in the Columbia
Basin. *A. sinuatus* differs from *A. curvicarpus* in the ascending flowers and in the calyx, which is
obliquely obconic at base and not gibbous-saccate behind the insertion of the pedicel. The pod of *A.
sinuatus* is similar to that of *A. curvicarpus,* but a little broader, and the species is apparently more
nearly related to *A. sclerocarpus.*

Astragalus spaldingii Gray, Proc. Am. Acad. 6:524. 1865.
    A. chaetodon Torr. ex Gray, op. cit. 194, but not of Bunge in 1851. *Tragacantha spaldingii* Kuntze,
    Rev. Gen. 2:948. 1891. *Phaca spaldingii* Piper, Contr. U.S. Nat. Herb. 11:370. 1906. *(Spalding,*
    "plains on the Kooskooskie River, interior of Washington Territory")
    ASTRAGALUS SPALDINGII var. TYGHENSIS (Peck) C. L. Hitchc. hoc loc. *A. tyghensis* Peck, Proc.
    Biol. Soc. Wash. 49:110. 1936. *(M. E. Peck 17367,* dry slope near Tygh Valley, Wasco Co., Oreg.)
Grayish villous-woolly perennial with a taproot and several erect to decumbent-based stems 1-4 (6)
dm. tall, from a branching caudex; stipules linear, not connate, 4-8 mm. long; leaves 5-12 cm. long;
leaflets 15-27, linear-elliptic or linear-lanceolate to lanceolate-oblong, 5-18 mm. long; peduncles
usually exceeding the leaves; racemes 15- to 60-flowered, headlike or spicate, 1.5-2.5 cm. broad,
often not quite so long at anthesis but growing to several times this length in fruit; flowers spreading
to erect, (8) 10-14 mm. long, nearly sessile, the pedicels stout, less than 1 mm. long; calyx tubular-
campanulate, densely villous-lanate, about 2/3 as long as the corolla, the linear teeth subequal to the

1/2

1/2

var. tyghensis

2

2

var. spaldingii

3

A. spaldingii

1/2

1/8

2

3

Astragalus sinuatus

2.5

2

1/2

stipules

3

A. spatulatus

JRJ

2

2

3

1/2

A. speirocarpus

tube; banner obovate, glabrous or hairy on the back, white but bluish-lined; wings white, slightly exceeding the rounded, purplish-tipped keel, with a glabrous or hairy, obovate blade about as long as the claw; pod sessile, very densely strigose-villous, ovoid, 4-7 mm. long, coriaceous, 2-celled by the complete intrusion of the lower suture, 1- or 2-seeded. N=12.

Sagebrush and grasslands of the valleys and foothills e. of the Cascades, from n. e. Oreg. to c. Wash., and e. to w. Ida. May-June.

In Wasco Co., Oreg., there is a very striking variant of this species that, in contrast to the more widespread var. *spaldingii,* is ranker in growth; it is also characterized by slightly broader leaflets and by coarse hairs on the back of the somewhat more yellowish banner and on the outer surface of the blades of the wing- and keel-petals. This is the var. *tyghensis* (Peck) C. L. Hitchc.

Astragalus spatulatus Sheld. Minn. Bot. Stud. 1:22. 1894.
   Homalobus caespitosus Nutt. in T. & G. Fl. N.Am. 1:352. 1838. *A. caespitosus* Gray, Proc. Am. Acad. 6:230. 1864, not of Pall. in 1800. *A. simplicifolius* var. *caespitosus* M. E. Jones, Proc. Calif. Acad. Sci. II, 5:647. 1895. *A. simplicifolius* var. *spatulatus* M. E. Jones, Contr. West. Bot. 10:65. 1902. *(Nuttall,* "dry and lofty hills of the Platte towards the Rocky Mountains")
   Homalobus canescens Nutt. in T. & G. Fl. N.Am. 1:352. 1838, not *A. canescens* DC. in 1802. *(Nuttall,* "on the dry chalky hills of the Platte towards the Rocky Mountains")
   Homalobus brachycarpus Nutt. in T. & G. Fl. N.Am. 1:352. 1838, not *A. brachycarpus* Bieb. in 1809. *(Nuttall,* "dry and lofty hills of the Platte towards the Rocky Mountains")
   Caespitose, more or less pulvinate, acaulescent perennial with a thick root and freely branched crown, silvery-strigillose with dolabriform hairs; leaves tufted, 1-4 cm. long, almost all simple, linear to oblanceolate; stipules connate; racemes loosely 3- to 15-flowered, borne on peduncles that usually exceed the leaves; pedicels about 1 mm. long; flowers purple, 7-9 mm. long; calyx about half as long as the corolla, the tube and linear teeth subequal; pod sessile, erect or ascending, strongly compressed, oblong, 7-12 mm. long, 2-3 mm. broad, acute, 1-celled, with both of the sutures prominent. N=12.
   Plains area, from the e. base of the Rocky Mts., Mont. to Colo., e. to Sask. and the Dakotas, w. into Utah. June-July.

Astragalus speirocarpus Gray, Proc. Am. Acad. 6:225. 1864.
   Tragacantha speirocarpa Kuntze, Rev. Gen. 2:948. 1891. *Phaca speirocarpa* Piper, Contr. U.S. Nat. Herb. 11:370. 1906. *Homalobus speirocarpus* Rydb. Bull. Torrey Club 51:18. 1924. *(Dr. Lyall,* "Wenass in the valley of the Upper Columbia River")
   Caespitose, grayish-strigillose perennial with a heavy root and freely branched crown, the stems numerous, ascending to erect, 1-4 dm. long, usually somewhat decumbent and partially buried at the base; leaves 3-6 cm. long; stipules triangular, 1-2 mm. long, not connate; leaflets (3-7) 9-17, oblong and rounded to obovate and retuse or even oblong-obcordate, 2-5 mm. broad; peduncles equaling or shorter than the leaves; racemes closely to rather loosely 10- to 40-flowered; pedicels 1-2 mm. long; flowers spreading to erect, white to lavender- or bluish-tinged (especially the keel), 14-20 mm. long; calyx grayish (or grayish and blackish) -hairy, 6-9 mm. long, the short, triangular teeth scarcely 1/4 as long as the tube; wings 2-4 mm. longer than the keel; pod pendulous, the stipe equaling to somewhat longer than the calyx, the body strongly compressed, 1-celled, 3-5.5 mm. broad, usually coiled about 1.5 (1-2.5) turns, strigillose and generally strongly reticulate.
   Sagebrush desert, especially near the Columbia R., probably restricted to Yakima, Kittitas, Klickitat, and Benton cos., Wash. Apr.-May.

Astragalus stenophyllus T. & G. Fl. N.Am. 1:329. 1838.
   A. leptophyllus Nutt. Journ. Acad. Phila. 7:18. 1834, not of Desf. in 1800. *Tragacantha stenophylla* Kuntze, Rev. Gen. 2:948. 1891. *Homalobus stenophyllus* Rydb. Mem. N.Y. Bot. Gard. 1:249. 1900, as to name but not as to concept. *Phaca stenophylla* Piper, Contr. U.S. Nat. Herb. 11:371. 1906. Not *A. stenophyllus* of most recent authors, which is *A. filipes.* *(Wyeth,* headwaters of the Missouri, perhaps on the upper Bighole R. in Beaverhead Co., Mont., in 1883)
   A. leucophyllus sensu Hook. Lond. Journ. Bot. 6:211. 1847, not of T. & G. in 1838. *A. arrectus* Gray, Proc. Am. Acad. 8:289. 1870. *Phaca arrecta* Piper, Contr. U.S. Nat. Herb. 11:371. 1906. *Tium arrectum* Rydb. Bull. Torrey Club 40:49. 1913. *(Geyer,* Kooskooskie River)
   A. atropubescens Coult. & Fisher, Bot. Gaz. 18:300. 1893. *A. kelseyi* Rydb. Mem. N.Y. Bot. Gard. 1:241. 1900, a typonym. *Tium atropubescens* Rydb. Bull. Torrey Club 40:49. 1913. *A. arrectus*

var. *kelseyi* M. E. Jones, Rev. Astrag. 161. 1923, as to name only. *(F. D. Kelsey,* Deer Lodge, Mont., June, 1892.)

Slender (but sometimes tall) perennial from a woody taproot and shortly forking caudex, thinly strigillose, the leaflets glabrous above; stems clumped, erect and ascending, 1.5-4 dm. long; leaves 5-21 cm. long; stipules 2.5-6 mm. long, the lowest strongly amplexicaul but not connate, the upper ones narrower; leaflets 19-29, oblong-oblanceolate to linear, truncate or retuse or on the upper leaves mucronate, 5-25 mm. long; peduncles erect, 8-25 cm. long; racemes loosely 10- to 25-flowered, the flowers nodding at full anthesis, the axis elongating, 5-27 cm. long in fruit; pedicels at first spreading but ultimately becoming erect, thickened, and 2.5-4 mm. long; flowers white or cream-colored, the banner 13-17.5 mm. long, the wings and keel successively shorter; calyx 5-6.5 mm. long, black-strigillose, the campanulate tube about 4-5 mm. long, the teeth 1-2 mm. long; pod erect, stipitate, the stipe 2.5-6 mm. long, the body linear-oblong or narrowly oblong-ellipsoid, nearly or quite straight, 14-28 mm. long, 2.5-4.5 mm. broad, obtusely triquetrous with convex lateral and shallowly grooved dorsal faces, the somewhat fleshy greenish-stramineous valves becoming thinly leathery or papery, glabrous or minutely black-strigillose, keeled ventrally by the suture, and with a partial or complete septum about 1-1.5 mm. wide.

Gravelly or sandy flats and hillsides, mostly among sagebrush; s.w. Mont. and e.c. Ida., also locally on steep grassy banks in the canyons of the lower Salmon and Snake rivers in Idaho Co., Ida. May-July.

Over most of its range *A. stenophyllus* has a glabrous pod. Plants with black-puberulent legumes are known to occur only along the Salmon and Pahsimeroi rivers in Lemhi Co., Ida., where they replace the more ordinary form. This phase has been confused with *A. palousensis,* which is, however, readily distinguished by its erect flowers with irregularly graduated petals (the banner shorter than the wings).

Astragalus striatus Nutt. in T. & G. Fl. N.Am. 1:330. 1838.

A. laxmanni sensu Nutt. Gen. Pl. 2:99. 1818, but not of Jacq. in 1776. *(Nuttall,* "on the hills of the Missouri")

A. adsurgens Hook. Fl. Bor. Am. 1:149. 1831, not of Pall. in 1800. *A. nitidus* Dougl. ex Hook. Fl. Bor. Am. 1:149. 1831, in synonymy, and sensu M. E. Jones, Proc. Calif. Acad. Sci. II, 5:646. 1895. *Phaca adsurgens* Piper, Contr. U.S. Nat. Herb. 11:372. 1906. *(Dr. Richardson,* "Plains of the Assinaboin and Saskatchawan Rivers, as far as the mountains," is the first collection cited)

A. adsurgens var. *robustior* Hook. Fl. Bor. Am. 1:149. 1831. *A. nitidus* var. *robustior* M. E. Jones, Rev. Astrag. 170. 1923. *(Douglas,* "from the Kettle Falls to the sources of the Columbia, on the West side of the Rocky Mountains")

A. hypoglottis var. *robustus* Hook. Lond. Journ. Bot. 6:210. 1847. *(Geyer 126,* "Upper Platte near Laramie's Fork; also on the Upper Missouri, Teton River")

A. adsurgens var. *albifolius* Blank. Mont. Agr. Coll. Stud. Bot. 1:71. 1905. *(E. N. Brandegee,* 7-mile road, Helena, Mont., July 19, 1898)

A. adsurgens var. *pauperculus* Blank. Mont. Agr. Coll. Stud. Bot. 1:72. 1905. *(Blankinship,* Billings, Mont., July 7, 1902)

A. chandonnetii Lunell, Am. Midl. Nat. 2:127. 1911. *(Z. L. Chandonnet,* McHugh, near Detroit, Minn., June 16, 1911)

Caespitose perennial 1-4 dm. tall, greenish- to rather grayish-strigillose with dolabriform hairs; leaves 4-12 cm. long; stipules membranous-scarious, connate, 5-15 mm. long; leaflets 9-23, narrowly oblong to oblong-obovate, 1-2.5 cm. long, (2) 3-10 mm. broad; peduncles about equaling or up to twice as long as the leaves; racemes oblong, congested, spikelike, 15- to 80-flowered; pedicels about 1 mm. long; flowers erect, white or very pale purplish to fairly dark purple, 14-18 mm. long; calyx 5-9 mm. long, the teeth from about 1/3 as long to nearly as long as the tube, whitish- or blackish-strigillose; banner only slightly reflexed, 1-3 mm. longer than the wings; wings 2-3 mm. longer than the keel; pod sessile, erect, membranous, 8-12 mm. long, 3-4 mm. thick and broad, strigillose with basifixed hairs, cordate in section, the lower suture broadly sulcate, intruded as a narrow membrane 2/3-3/4 the width of the pod. N=16.

Prairie grassland to rocky foothills; c. Wash. to Alta. and Minn., s. to N.M. June-Aug.

Astragalus succumbens Dougl. ex Hook. Fl. Bor. Am. 1:151. 1831.

Tragacantha succumbens Kuntze, Rev. Gen. 2:948. 1891. *Phaca succumbens* Piper, Contr. U.S.

Nat. Herb. 11:370. 1906. *Hamosa succumbens* Rydb. Bull. Torrey Club 54:14. 1927. *(Douglas,* "on the barren grounds of the Columbia and near the Wallawallah River")

A. dorycnioides Dougl. ex G. Don, Gen. Syst. 2:258. 1832. *(Douglas,* near the Columbia River)

Grayish-hirsute-pilose perennial from a deep taproot; stems 1-5 dm. long, erect to decumbent but the plant usually with one central, freely branched, ascending main stem; leaves rather fleshy, 3-10 cm. long; stipules ovate-lanceolate, 3-15 mm. long, not connate; leaflets (9) 13-19, oblong to oblong-obovate, 5-15 mm. long; peduncles mostly considerably shorter than the leaves; racemes densely 10- to 60-flowered; pedicels stout, 1-2 mm. long; flowers 18-26 mm. long, pinkish, with the banner more deeply pencilled or tinged and the keel purplish-tipped; calyx tubular, 9-15 mm. long, the teeth linear-lanceolate, 1/2-2/3 as long as the tube; margins of the banner scarcely reflexed; wings oblong and nearly straight, the tips more or less erose, from about as long as the banner to as much as 4 mm. longer; keel 3-6 mm. shorter than the wings, rounded; pod erect, sessile, cartilaginous, 3-4 cm. long, 5-8 mm. wide, glabrous, from slightly arcuate to straight, strongly compressed and narrowly cordate in section, 2-celled by the complete intrusion of the lower suture. N=12.

Sagebrush desert, sandy barrens, and lower foothills, from Klickitat and Grant cos., c. Wash., to Umatilla and Gilliam cos., Oreg. Apr.-June.

Astragalus tegetarioides M. E. Jones, Contr. West. Bot. 10:66. 1902.

Homalobus tegetarioides Rydb. N. Am. Fl. 24⁵:265. 1929. *(Cusick 2619,* Buck range, southern Blue Mts., Oreg., June 28, 1901)

Caespitose, grayish-strigillose perennial forming mats 1-4 dm. broad, with numerous freely branched stems 2-5 cm. tall; leaves 2-4 cm. long, the petiole nearly equal to the blade; stipules membranous, the lower ones shortly connate, 3-5 mm. long; leaflets (5) 7-11, obovate to obcordate-cuneate, 3-6 (7) mm. long, 2-4 mm. broad, distinctly petiolulate; flowers 3-7, about 4.5 (4-6) mm. long, borne in loose pedunculate racemes about equaling (or shorter than) the leaves; calyx grayish-pubescent, about half as long as the corolla, the linear-lanceolate teeth about equal to the campanulate tube; corolla whitish, the strongly arched banner purplish-lined, the purple-tipped keel rounded and conspicuously shorter than the oblique wings; pod 4-5 (6?) mm. long, 2-3 mm. wide, laterally flattened, 1-celled, 2 (3)-ovuled but usually 1-seeded, grayish-strigillose, very inconspicuously stipitate, the stipe much shorter than the calyx, the legume rupturing the calyx as it matures.

Chiefly in ponderosa pine forests; c. Oreg., in n. Harney, probably adj. Grant, and perhaps Deschutes cos. July-Aug.

Astragalus tegetarius Wats. Bot. King Exp. 76. 1871.

Homalobus tegetarius Rydb. Bull. Torrey Club 31:563. 1904. *Kentrophyta tegetaria* Rydb. Bull. Torrey Club 34:421. 1907. *A. montanus* var. *tegetarius* M. E. Jones, Rev. Astrag. 81. 1923. *(Watson 286,* East Humboldt Mts., Nev.)

A. tegetarius var. (?) *implexus* Canby ex Port. & Coult. Fl. Colo. Add. 1874. *A. kentrophyta* var. *implexus* Barneby, Leafl. West. Bot. 6:154. 1951. *(Canby,* South Park, Colo.)

A. aculeatus A. Nels. Bull. Torrey Club 26:10. 1899. *Homalobus aculeatus* Rydb. Mem. N.Y. Bot. Gard. 1:249. 1900. *Kentrophyta aculeata* Rydb. Bull. Torrey Club 32:665. 1906. *(A. Nelson 2445,* near Dome Lake, Big Horn Mts., Wyo., July 18, 1896)

Kentrophyta minima Rydb. Bull. Torrey Club 34:420. 1907. *(Tweedy 83,* Yellowstone Nat. Park, Aug., 1884)

Low, matted, caespitose, usually grayish-strigose perennial 1-4 dm. broad but less than 3 cm. high; pubescence basifixed and usually appressed; leaflets 5-9, mostly 3-7 mm. long, scarcely 1 mm. broad, spinulose tipped; stipules connate and often somewhat acerose; flowers single or in 2's or 3's, 6-9 mm. long; peduncles 3-10 mm. long; calyx campanulate, about 1/3 as long as the corolla, grayish- to blackish-hairy, the teeth linear-lanceolate, about 1/2 the length of the tube; corolla more or less uniformly purplish; ovules 5-8 but the seeds only 1-3; pod sessile, ovoid, compressed, 1-celled, 3-5 mm. long, indehiscent.

Alpine and subalpine peaks; Wallowa Mts., Oreg., n.e. to the Rocky Mts. in Mont., s. to Calif., Utah, and n. N.M. June-Sept.

The species is not always sharply distinct from *A. kentrophyta.*

Astragalus tenellus Pursh, Fl. Am. Sept. 473. 1814.

Tragacantha tenella Kuntze, Rev. Gen. 2:942. 1891. *Homalobus tenellus* Britt. & Brown, Ill. Fl. 2: 305. 1897. *(Lewis,* "on the banks of the Missouri")

Astragalus striatus

A. succumbens

Astragalus stenophyllus

A. tegetarioides

stipules

A. tegetarius

JRJ

Ervum multiflorum Pursh, Fl. Am. Sept. 739. 1814. *Homalobus multiflorus* T. & G. Fl. N. Am. 1:
351. 1838. *A. multiflorus* Gray, Proc. Am. Acad. 6:226. 1864. *(M. Lewis; Bradbury,* "In Upper
Louisiana")

Orobus dispar Nutt. Gen. Pl. 2:95. 1818. *Homalobus dispar* Nutt. in T. & G. Fl. N. Am. 1:350.
1838. *(Nuttall,* "On arid soil near Ft. Mandan")

Phaca nigrescens Hook. Fl. Bor. Am. 1:143. 1831. *A. nigrescens* Gray, Am. Journ. Sci. II, 33:
410. 1862. *(Dr. Richardson; Drummond,* "On the Saskatchawan, to the Rocky Mountains, and as
far north as Fort-Franklin, on the Mackenzie River, in lat. 65°"')

Homalobus stipitatus Rydb. Bull. Torrey Club 34:419. 1907. *(Geyer,* "Upper Missouri," said to be
on the hills between Fort Pierre and Devil's Lake, in 1839)

Homalobus strigulosus Rydb. Bull. Torrey Club 34:420. 1907. *A. tenellus* f. *strigulosus* Macbr.
Contr. Gray Herb. n. s. 65:34. 1922. *A. tenellus* var. *strigulosus* Hermann, Journ. Wash. Acad.
Sci. 38:237. 1948. *(Watson 285,* East Humboldt Mts., Nev., in 1860)

Sparsely strigillose, greenish perennial with a heavy root and freely branched crown, the numerous
slender, nearly erect stems 2-6 dm. tall; leaves (3) 5-8 (10) cm. long; stipules connate opposite the
petioles, 3-6 mm. long; leaflets 11-25 (up to at least 31), linear-lanceolate to linear-oblanceolate,
(5) 10-20 mm. long, 1.5-5 mm. broad; peduncles occasionally 2 per node, mostly shorter than the
leaves and usually much shorter than the very loose, 7- to 20-flowered racemes; flowers 6-9 mm.
long, mostly ochroleucous but sometimes pinkish-tinged and the keel usually somewhat purplish-tipped;
pedicels 1-3 mm. long; calyx 2.5-4 mm. long, the linear-lanceolate teeth about 2/3 as long as the
tube; banner erect; wings 1-2 mm. longer than the keel; pod pendulous, with a stipe as long to nearly
twice as long as the calyx, the body glabrous (ours), often finely mottled, membranous, linear-elliptic, 7-15 mm. long, 3-5 mm. broad, strongly compressed, 1-celled, with neither suture sulcate or
intruded. N=12.

Prairies and foothills to lower mountains; Yukon s. through e. B. C. to Minn. and N. M., w. to
Nev. and c. Ida. June-July.

Our material, as described above, is all referable to the var. *tenellus,* although the var. *strigulosus*
(Rydb.) Hermann, a small-flowered plant (corolla 6-7 mm. long) with strigillose pods, approaches
(and may reach) our range in the Snake R. Canyon of s. Ida., and ranges to c. Nev.

Astragalus terminalis Wats. Proc. Am. Acad. 9:370. 1882.

Tium terminale Rydb. N. Am. Fl. 24[7]:393. 1929. *(Watson,* Red Rock Creek, southern Mont., July,
1880)

A. reventoides M. E. Jones, Proc. Calif. Acad. Sci. II, 5:661. 1895. *A. terminalis* var. *reventoides*
M. E. Jones, Rev. Astrag. 167. 1923. *(Tweedy 7,* Grasshopper Creek, Beaverhead Co., Mont.)

Caespitose perennial with a taproot, the stems 0.5-3 dm. tall, ashy-strigillose with dolabriform
hairs; leaves 5-20 cm. long; stipules lanceolate, 3-5 mm. long, not connate; leaflets 13-21, oblong-
oblanceolate to obovate, 6-15 mm. long, rounded to retuse; peduncles 10-15 cm. long, usually surpassing the leaves; racemes rather compactly 10- to 30-flowered, elongating and more open in fruit;
pedicels 1-3 mm. long; flowers spreading-reflexed, 12-16 mm. long, almost pure white but drying to
ochroleucous except for the lilac-tipped keel; calyx white- or black-hairy, 4-5 mm. long, the teeth
triangular, less than 1/4 the length of the tube; banner erect, the blunt keel 3-4 mm. shorter than the
wings; pod glabrous, erect, essentially sessile, 15-20 mm. long, about 4 mm. broad, coriaceous-
woody, nearly straight, obcompressed and rather broadly cordate in section, the lower suture intruded and partially to completely partitioning the cavity.

Sagebrush benchland to montane ridges, especially on limestone; c. Ida. to Beaverhead Co., Mont.,
e. to the Gros Ventre R., Teton Co., Wyo. May-June.

Astragalus toanus M. E. Jones, Zoë 3:296. 1893.

Cnemidophacos toanus Rydb. N. Am. Fl. 24[5]:287. 1923. ("Toano Range, Eastern Nevada," probably
M. E. Jones)

Coarse, sparsely leafy, somewhat rushlike, strigillose-cinereous to nearly glabrous perennial;
stems clumped, ascending to erect, 1.5-5 dm. long; leaves 2.5-10 cm. long; stipules 1.5-6.5 mm.
long, the lowest connate into a scarious sheath, the upper ones free; leaflets 9 or fewer, linear-
filiform, scattered and distant, 3-30 mm. long, wanting in the upper (sometimes all) leaves, the
terminal one continuous with the rachis; peduncles stout, erect, 6-25 cm. long; racemes 7- to 35-
flowered, the flowers ascending, the axis 3-30 cm. long in fruit; flowers pink-purple, the banner 15-
20 mm. long; calyx tube 4-6 mm. long, the teeth 0.5-2 mm. long; pod erect, sessile, oblong-ellip-

soid, 13-25 mm. long, 4-5.5 mm. broad, straight, stiffly cuspidate, compressed, bicarinate by the prominent cordlike sutures, the fleshy, green- or purple-speckled, glabrous or strigillose valves becoming woody, not inflexed.

Gullied bluffs and knolls, sometimes on gumbo-clay flats with *Sarcobatus,* in selenium-rich soils, at low elevations along the Snake R., from Fremont to Owyhee Co., Ida. (where barely within our range), to n.w. Utah and Nev. May-June.

Astragalus tweedyi Canby, Bot. Gaz. 15:150. 1890.

Phaca tweedyi Piper, Contr. U.S. Nat. Herb. 11:371. 1906. *Homalobus tweedyi* Rydb. Bull. Torrey Club 51:14. 1924. *(Tweedy 613,* "hills along the Columbia river," Yakima Co., Wash., is the second of three collections cited)

More or less grayish, strigillose or crisply puberulent-strigillose perennial from a taproot, usually with several erect stems 2-6 dm. tall; leaves 4-10 cm. long; stipules lanceolate, 2-5 mm. long, not connate; leaflets 13-21, linear to narrowly oblong, 8-22 mm. long, 1-2 mm. broad; peduncles equaling or exceeding the leaves, as much as 15 cm. long; racemes closely 10- to 50-flowered, elongate in fruit and usually subequal to the peduncles; pedicels stout, 1.5-2.5 mm. long; flowers spreading or ascending, ochroleucous or white, 15-18 mm. long; calyx 7-9 mm. long, crisp-puberulent, slightly gibbous at base, the teeth 1-2 mm. long; wings acute, 1-2 mm. longer than the rounded keel; pod erect, glabrous, with a stipe about half again as long as the calyx, the body coriaceous, about 15 mm. long, 4-5 mm. broad, nearly terete in section, slightly arcuate, 1-celled, acute at each end, both sutures prominent but neither sulcate.

Sagebrush plains and foothills; Yakima Co., Wash., to n.c. Oreg., near the Columbia and lower Deschutes rivers. May-June.

Astragalus umbraticus Sheld. Minn. Bot. Stud. 1:23. 1894.

A. sylvaticus Wats. Proc. Am. Acad. 23:262. 1888, not of Willd. in 1803. *Hamosa umbratica* Rydb. Bull. Torrey Club 54:19. 1927. *(L. F. Henderson;* and *T. Howell* in 1887, Glendale, s. Oreg.)

Perennial from the crown of a taproot or short caudex, nearly glabrous, the inflorescence thinly black-strigillose; stems ascending, 2-5 dm. long; leaves 4-12 cm. long; stipules 3.5-9.5 mm. long, more or less amplexicaul-decurrent but free; leaflets 15-23, oblong-obovate and retuse or obcordate, 4-20 mm. long; peduncles 5-12 cm. long; racemes loosely 10- to 25-flowered, 1.5-5 cm. long in fruit; flowers white, the banner 10-14 mm. long; calyx 5-7 mm. long, the teeth 2-3 mm. long; pod spreading or declined, shortly stipitate, the stipe (concealed by the calyx) 1-2 mm. long, the body lunately or falcately linear or linear-lanceolate in profile, 14-24 mm. long, 2.5-3.5 mm. wide, compressed-triquetrous, narrowly but deeply grooved dorsally, fully 2-locular, the thin, green, glabrous valves becoming straw-colored or ultimately blackish.

Dry, open, oak and pine woods, uncommon; coast ranges of n.w. Calif. and s.w. Oreg., extending n., apparently interruptedly, into our range in Yamhill Co., Oreg. May-June.

Astragalus vallaris M. E. Jones, Contr. West. Bot. 10:59. 1902.

Hesperonix vallaris Rydb. N.Am. Fl. 24⁷:439. 1929. *(Cusick,* Snake R. Canyon, near Ballard's Landing, Oreg.)

Caespitose, sparsely strigillose, greenish perennial from a heavy root and branched crown, with several decumbent stems (1) 2-4 dm. long; leaves 5-10 cm. long; stipules 2-3 mm. long, greenish, ovate, not connate; leaflets 13-21, oblong to elliptic-obovate, usually slightly retuse, (4) 10-20 mm. long; peduncles generally not exceeding the leaves; racemes 5- to 10-flowered, compact; pedicels 2-3 mm. long, stout; flowers about 2 cm. long, white, ascending; calyx sparsely blackish-strigose, 8-10 mm. long, usually minutely lanceolate-bracteolate at base, the teeth linear, about 1/2-2/3 as long as the tube; wings narrow, 2-3 mm. longer than the rounded, usually purplish-tipped keel; pod pendulous, glabrous, with a stipe as long to nearly twice as long as the calyx tube, the body 2.5-4 cm. long, obcompressed, slightly falcate and rather broadly lunate in outline, coriaceous, reticulate-corrugate, 6-8 mm. wide and considerably thicker, both sutures prominent and somewhat sulcate, the lower one intruded and nearly completely partitioning the cavity.

This is a local species of the rocky slopes of the Snake R. Canyon in Washington and Adams cos., Ida., and adj. Oreg. May-June.

Astragalus vexilliflexus Sheld. Minn. Bot. Stud. 1:21. 1894.

A. pauciflorus Hook. Fl. Bor. Am. 1:149. 1831, not of Pall. in 1800. *Tragacantha pauciflora*

Astragalus toanus   A. tenellus   A. tweedyi   A. terminalis

stipules

Kuntze, Rev. Gen. 2:947. 1891. *Homalobus vexilliflexus* Rydb. Mem. N.Y. Bot. Gard. 1:249. 1900. *(Drummond,* "among rocks in the more elevated regions of the Rocky Mountains")

A. *amphidoxus* Blank. Mont. Agr. Coll. Stud. Bot. 1:72. 1905. *Homalobus amphidoxus* Rydb. N. Am. Fl. 24⁵:264. 1929. *(E. N. Brandegee,* Sky High, Unionville, Mont., July 10, 1898) = var. *vexilliflexus.*

A. *miser* sensu Gray, Proc. Am. Acad. 6:228. 1864, but not of Dougl. in 1831 = var. *vexilliflexus.*

A. *vexilliflexus* var. *nubilus* Barneby, Am. Midl. Nat. 55:484. 1956. *(Hitchcock & Muhlick 10857,* e. of Castle Peak, White Cloud Range, Custer Co., Ida., Aug. 8, 1944)

Grayish-strigillose, caespitose, often matted perennial from a heavy taproot and branched crown; stems 1-3 dm. long, prostrate to ascending; leaves 1.5-3 (5) cm. long; stipules lanceolate, 2-4 mm. long, connate half their length; leaflets 7-13, linear-elliptic to oblong-elliptic, acute, 5-12 mm. long; racemes loosely 5- to 10-flowered, the peduncles equaling or shorter than the leaves; pedicels 1-2 mm. long; flowers 5-8 (10) mm. long, from ochroleucous (but with a purplish keel) to rather deep lavender-purplish; calyx about half the length of the corolla, the teeth linear-lanceolate, about 2/3 as long as the tube; wings and keel much shorter than the erect banner; pod membranous, sessile or subsessile, narrowly oblong-elliptic in outline, strongly compressed (until rounded by the seeds), 7-11 mm. long, 2.5-3 mm. broad, strigillose, 1-celled, with neither suture sulcate or intruded.

Stream banks and open forest to sagebrush plains; B.C. to Sask. and southward, on both sides of the Rocky Mts. in Mont., to Wyo. and c. Ida. June-July.

The species is variable in flower-size and -color as well as in the type of pubescence, the hairs being appressed to somewhat spreading or even curly; two geographic races are notable:

1 Plants prostrate, densely matted, with the aspect of A. *tegetarius;* herbage loosely silky-villosulous; banner about 5-6 mm. long; open stony crests near timber line, Custer Co., Ida.   var. nubilus Barneby

1 Plants of various habit, if matted then the flowers larger (the banner 7-9 mm. long), if the flowers small then the stems well developed, decumbent or assurgent, and bushy-branched; rocky slopes and knolls in the foothills, ascending in open places through forests to timber line, along the Rocky Mts., mostly e. of the divide, from s. Alta. to s.w. Mont. and w. Wyo., e. to the Cypress Hills, s.w. Sask.; variable in stature and flower size, "A. *amphidoxus*" representing the taller, small-flowered phase prevalent at low elevations southward   var. vexilliflexus

Astragalus whitneyi Gray, Proc. Am. Acad. 6:526. 1865. *(Brewer,* mountains near Sonora Pass, Calif.)

Phaca hookeriana T. & G. Fl. N.Am. 1:693. 1838. A. *hookerianus* Gray, Proc. Am. Acad. 6:215. 1864, not of D. Dietr. in 1850. A. *sonneanus* Greene, Pitt. 3:186. 1897. A. *whitneyi* var. *sonneanus* Jeps. Fl. Calif. 2:347. 1936. *(Douglas,* "interior of Oregon, probably near the Rocky Mountains")

Very finely strigillose, grayish to green perennial with a woody, much-branched caudex, heavy taproot, and usually somewhat decumbent stems 5-30 cm. long, the internodes very short; leaves more or less tufted, 2-5 (8) cm. long; stipules 2-4 mm. long, usually purplish, all but the uppermost connate; leaflets 11-17, linear to obovate, 5-15 (20) mm. long; peduncles usually exceeding the leaves; racemes compactly 5- to 20-flowered; flowers spreading to erect, ochroleucous, tinged with lavender or purple, about 10 mm. long; banner longer than the acute wings and keel (as measured along the curved adaxial surface) but usually exceeded by both due to its erect position; calyx about half as long as the corolla, strigillose, the teeth linear-lanceolate, about half as long as the often purplish tube; pod pendulous, stipitate, the stipe from subequal to the calyx to twice as long, the body inflated, membranous, purplish-mottled, 2-6 cm. long, ellipsoid-obovoid, nearly terete in section, finely strigillose (glabrous), 1-celled, the lower suture not intruded.

Open stony slopes and mountain crests, commonly but not exclusively on serpentine; e. slope of the Cascades, Wash. to n.e. Calif. and n.w. Nev., e. to s.w. Ida. May-Aug.

Our material, as described above, is referable to the var. *sonneanus* (Greene) Jeps., which occurs in Chelan, Yakima, and Kittitas cos., and in the Blue Mts., in Columbia and Garfield cos., Wash., and in the mountains of Crook, Grant, and Wheeler cos., and in the Steens Mts., Harney Co., Oreg. The glabrous fruited var. *whitneyi* is limited to a few ranges in Nev. and Calif. To the south of our range, in n.e. Calif., s. Oreg., n.e. Nev., and s.w. Ida., the species is represented by the poorly defined, larger flowered, pubescent ovaried var. *confusus* Barneby (El Aliso 2:206. 1950). Plants from the s. slope and foothills of the Sawtooth Mts., in Blaine and Butte cos., Ida., seem nearest the latter, but tend to have glabrous pods, and perhaps merit recognition as a second variety in our area.

Astragalus wootonii Sheld. Minn. Bot. Stud. 1:138. 1894. *(E. O. Wooton,* near Las Cruces, N. M.,
    May, 1892)
    A. wootoni var. *aquilonius* Barneby, Am. Midl. Nat. 41:499. 1949. *(Hitchcock & Muhlick 9218,* 4
        mi. s. of Lemhi, Lemhi Co., Ida., June 23, 1944)
    Strigillose to villous, short-lived, greenish perennial with a branched crown from a taproot and nu-
merous usually decumbent based or trailing stems about 1.5 dm. long; stipules 1-4 mm. long, ovate
to lanceolate, not connate; leaves 4-10 cm. long, with 9-23 oblong to oblanceolate, rounded to retuse
leaflets 5-16 mm. long; racemes loosely 4- to 15-flowered; peduncles mostly shorter than the leaves;
flowers spreading, about 1 cm. long, greenish-white, the keel often purplish-tipped; pedicels slender,
2-5 mm. long; calyx 1/2-2/3 the length of the corolla, the linear-lanceolate teeth from somewhat
shorter to slightly longer than the tube; pod sessile, inflated, membranous, ellipsoid, not mottled,
glabrous to puberulent, 3-4 cm. long, 1-celled, the lower suture slightly sulcate but not intruded.
    Shale and gravel banks; Custer and Lemhi cos., Ida. May-June.
    Our material, as described above, is referable to the var. *aquilonius* Barneby. The var. *wootonii*
ranges from s.e. Calif. to N.M. and w. Tex.

Coronilla L.

Flowers papilionaceous, borne in axillary, long-pedunculate umbels; calyx campanulate, 5-toothed,
more or less bilabiate; petals subequal, all clawed; banner nearly round; wings almost completely concealing the acute keel; stamens 10, diadelphous; fruit an elongate, slender, terete to angled, several-
jointed loment; perennial herbs (ours) to shrubs with odd-pinnate leaves.
    A genus of about 25 species of n. Africa and Eurasia, several of which are considered horticultural-
ly desirable. (From the Latin *corona,* a crown, in reference to the inflorescence, the flowers often
spreading in a ring and suggestive of a crown.)

Coronilla varia L. Sp. Pl. 743. 1753. ("Habitat in Lusatia, Bohemia, Dania, Gallia")
    Glabrous, herbaceous perennial with spreading to diffuse stems up to 6 dm. long; leaflets (9) 11-15
(21), oblong or elliptic to obovate, mostly 1-2 cm. long, acute to truncate or retuse; peduncles usually exceeding the leaves; umbels (10) 14- to 20-flowered; calyx about 2 mm. long; corolla mostly 10-12
mm. long, white to pink, the banner with an arched claw longer than the calyx; wings and keel strongly auriculate, the latter purple-tipped; loment linear, 4-angled, usually arcuate, up to 5 cm. long and
with as many as 10 seeds. N=12. Crown vetch.
    A native of the Old World, often used as an ornamental, and reported as established in parts of w.
Oreg. May-July.

Cytisus L. Broom

Flowers papilionaceous, axillary or in terminal elongate or umbelliform racemes, white, yellow,
or purplish-tinged; calyx tubular to campanulate, bilabiate, the upper lip 2-lobed, the lower 3-lobed;
stamens 10, monadelphous, 4 usually considerably longer than the others; style curved, broadened and
concave near the tip; legumes flattened, several-seeded; seeds with a wartlike strophiole; deciduous
or evergreen shrubs with strongly angled green stems and alternate (ours) mostly 3-foliolate leaves,
the leaflets mostly obovate and entire (ours) but frequently the leaves themselves entire and minute
and the stems essentially naked; stipules very small, thickened.
    Nearly 50 species in the warmer and drier regions of the Old World. (Derivation uncertain, perhaps
from Cythrus, the place where the plant grew, or from the Greek *kutisus,* the name of a supposedly
leguminous plant.)
    Cytisus, a well-known genus in cultivation, is closely related to *Genista,* differing because of the
strophiolate seeds, although most species of *Genista* have entire calyx lips. Certain of the brooms are
prone to escape and sometimes form dense brushy stands that are practically impenetrable. The
brooms can be grown from cuttings, although they are usually propagated by seed.

1 Flowers white   C. MULTIFLORUS
1 Flowers yellow or purplish
    2 Flowers axillary; pods glabrous except on the margins   C. SCOPARIUS
    2 Flowers racemose; pods hairy on the margins as well as the valves   C. MONSPESSULANUS

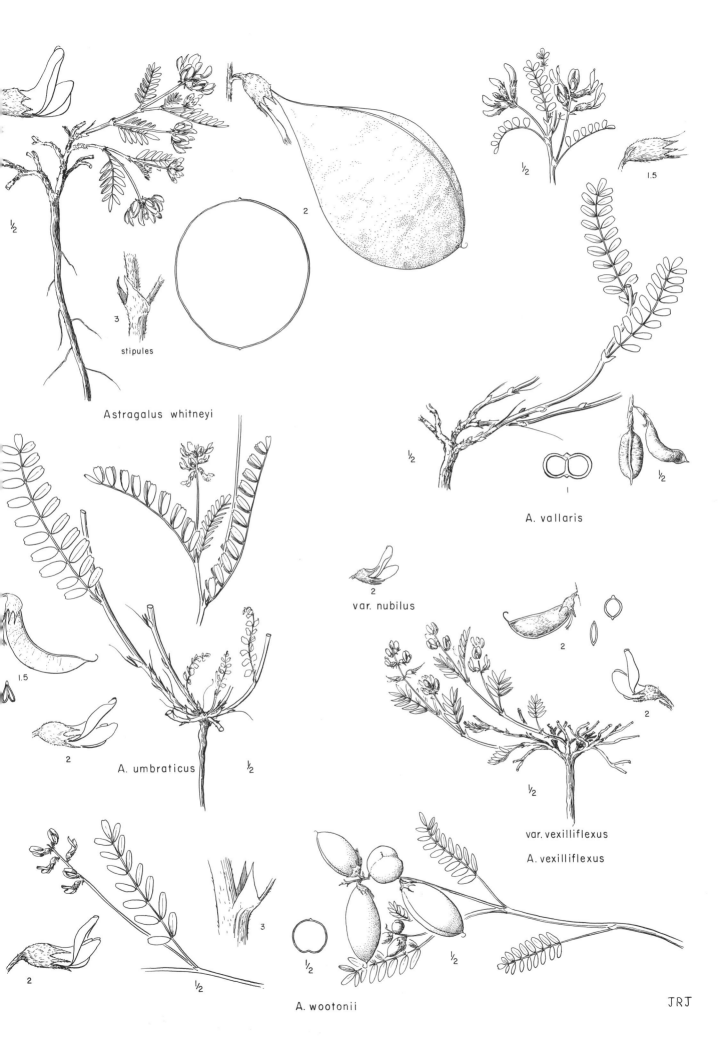

1/2

2

3
stipules

Astragalus whitneyi

1/2

1.5

A. vallaris

1

1/2

var. nubilus

2

A. umbraticus   1/2

1.5

2

var. vexilliflexus

A. vexilliflexus

2

2

2

3

2

1/2

1/2

A. wootonii

1/2

JRJ

Cytisus monspessulanus L. Sp. Pl. 740. 1753. ("Habitat Monspelii," Montpellier, France)
  Genista candicans L. Cent. 1:22. 1756. *C. candicans* Lam. Fl. Fr. 2:623. 1778, merely a nomen-
    clatural substitute for *C. monspessulanus.* ("Habitat Monspelii and in Italia")
  Shrub up to 3. 5 m. tall, villous-pubescent or glabrate, the branches leafy; leaves 3-foliolate; flow-
ers 3-10 in compact racemes on short lateral shoots, light yellow, about 1 cm. long; calyx campanulate,
hairy, the upper lip much more deeply lobed than the lower one; pods reddish-hairy. N=about 23.
  Occasionally escaped from gardens w. of the Cascades, from B. C. to n. w. Calif. June-July.

Cytisus multiflorus (Ait.) Sweet, Hort. Brit. 112. 1827.
  Genista alba Lam. Encyc. 2:623. 1788, not *C. albus* Hacquet in 1790. *Spartium multiflorum* Ait.
    Hort. Kew. 3:21. 1789. *Genista multiflora* Spach, Ann. Sci. Nat. Bot. III, 3:155. 1845. (Portugal)
  A low, deciduous shrub 0. 5-4 m. tall, soft-pubescent to glabrate; leaves on the lower part of the
branches 3-foliolate, the upper ones often simple; leaflets oblong to oblong-obovate, sericeous; flow-
ers 1 or 2 (3) in the leaf axils, white or pinkish tinged, about 1 cm. long; styles shorter than the keel,
only slightly curved; pods appressed-pubescent. N=12. White Spanish broom.
  The species, a native of the Mediterranean region, has occasionally escaped in w. Oreg. and Wash.
May-June.
  A second white-flowered species, *C. alba* Hacquet, is also cultivated, but is not known as an escape.
It has tubular calyces considerably longer than broad.

Cytisus scoparius (L.) Link, Enum. Hort. Berol. 2:241. 1822.
  Spartium scoparium L. Sp. Pl. 709. 1753. *Sarothamnus scoparius* Koch, Syn. Fl. Germ. 152. 1837.
    (S. Europe)
  Deciduous shrub up to 3 m. tall, pubescent to glabrous, the branches very strongly angled; leaves
at the base of the branches 3-foliolate, becoming simple above; flowers usually solitary but some-
times 2 or 3, about 2 cm. long, yellow or purplish-tinged; styles very strongly curved, longer than
the keel; pods glabrous except along the villous margins. N=12, 23, 24. Common broom, Scots broom,
Scotch broom.
  Scots broom is very commonly established and is spreading rapidly in many areas from Calif. to
B. C., on the w. side of the Cascades. In cultivation there are many horticultural varieties, including
dwarf, prostrate or creeping, and double- or varicolored-flowered plants. The var. *andreanus,* a form
with purplish wing petals, is rather frequently planted but seldom escapes to become established.

Glycyrrhiza L. Licorice; Licorice root

  Flowers papilionaceous, dirty white or yellowish to bluish, borne in dense, axillary, bracteate, pe-
duncled spikes; calyx tubular-campanulate, with 5 teeth, the upper two partially fused; banner only
slightly reflexed from the narrow wings and the acuminate keel; stamens 10, diadelphous; ovary covered with glands or hooked bristles, the style nearly straight; stigma capitate; legume burlike (ours),
indehiscent, 1-celled, few-seeded; more or less glandular perennial herbs with extensive aromatic
and sweetish rootstocks, odd-pinnate leaves, and small, linear, membranous, deciduous stipules.
  About 12 species, very widely distributed, but only one in N. Am. (Greek *glykys,* sweet, and *rhiza,*
root; most species have a sweetish licorice-flavored root.)
  Licorice of commerce is obtained from *G. glabra L.* which is grown chiefly in s. Europe.

Glycyrrhiza lepidota Pursh, Fl. Am. Sept. 480. 1814. ("On the banks of the Missouri")
  G. glutinosa Nutt. in T. & G. Fl. N. Am. 1:298. 1838. *G. lepidota* var. *glutinosa* Wats. Bot. Calif.
    1:144. 1876. *(Nuttall,* "Banks of Lewis's River")
  Stems 3-12 dm. tall, arising from deep, extensive woody rhizomes, viscid with stalked or sessile
glands; leaflets 7-15, lanceolate to oblong, 2-4 cm. long; racemes dense; flowers numerous, 10-15
mm. long, ochroleucous; calyx teeth subulate; pod 10-15 mm. long, densely hooked-prickly and cock-
leburlike. N=8.
  A plant of waste places and low ground, especially common along streams; B. C. to Ont. and Minn.,
s. to Calif., Ariz., n. Mex., Ark., and Tex. May-Aug.
  Within our range the species is represented by 2 well-marked varieties, separable as follows:
1 Plant with stalked glands throughout the inflorescence and often also on the petioles, leaf
    rachises, and main stem; e. of the Cascades, from B. C. to Calif., eastward to Yellow-
      stone Nat. Park
                                                                    var. glutinosa (Nutt.) Wats.

1 Plant with stalked glands only on the calyx; common from s. Calif. northeastward to
   Ont., and in most of Mont., but only occasional in Ida. and Wash.   var. lepidota

Hedysarum L.

Flowers numerous in axillary, pedunculate, bracteate, often secund racemes, yellowish-white,
pink, or purplish; calyx campanulate, with (usually) 2 subulate bracteoles at the base, the teeth subequal; corolla papilionaceous, the keel considerably surpassing the (commonly) auriculate-based
wings, usually longer than the banner; stamens 10, diadelphous; style slender, curved; fruit flat, constricted into more or less oval, indehiscent segments, usually breaking transversely; herbaceous perennials from woody taproots, the several stems generally hairy; leaves odd-pinnate, the leaflets minutely glandular dotted, at least above; stipules membranous, usually united and sheathlike, but sometimes free.

About 100 species, mostly in the northern part of the N. Temp. Zone, chiefly Eurasian. (From an
early Greek name, *hedusaron,* of uncertain application.)

Our species have very definite value as ornamentals and merit a trial in almost any company, *H.
occidentale* being perhaps at least as attractive as the others and more easily grown.

Reference:

Rollins, Reed C. Studies in the Genus Hedysarum in North America. Rhodora 42:217-239. 1940.

1 Flowers yellowish to nearly white   H. SULPHURESCENS
1 Flowers pink or carmine to purplish
  2 Upper calyx lobes broader but considerably shorter than the lower 3; wing petals with a
     slender, basal, auricular lobe nearly as long as the claw, the 2 petals weakly joined
     (by these lobes) above the ovary; pods not noticeably cross-corrugated, the reticula-
     tions rather regular
     3 Loment 7-12 mm. broad, the wing-margins 1-2 mm. broad; flowers 16-22 (seldom
        less than 18) mm. long   H. OCCIDENTALE
     3 Loment 3.5-6 mm. broad, the wing-margins less than 1 mm. broad; flowers 11-18
        (usually less than 16) mm. long   H. ALPINUM
  2 Upper calyx lobes slender, subequal to the lower ones and to the calyx tube; basal lobe
     of the wing petals broad, much shorter than the claws, not joined over the ovary; pods
     plainly cross-corrugated, the reticulations laterally elongated   H. BOREALE

Hedysarum alpinum L. Sp. Pl. 750. 1753. ("Habitat in Sibiria")
  H. alpinum var. *americanum* Michx. Fl. Bor. Am. 2:74. 1803. *H. americanum* Britt. Mem. Torrey
     Club 5:201. 1894. ("In borealibus Canadae, et in cataractis montium Alleghanis")
  H. alpinum var. *grandiflorum* Rollins, Rhodora 42:223. 1940. *(Fernald et al. 28625,* Pistolet Bay,
     Newf. , Aug. 11, 1925)

Much like *H. occidentale* but with the flowers smaller, ours mostly 11-15 mm. long; calyx teeth
more nearly subequal, narrowly triangular; loments consisting of 1-5 segments, 3.5-6 mm. broad,
with narrow wing-margins scarcely 0.5 mm. broad. N=7 (var. *americanum).*

A more or less circumpolar species. Late May-July.

Hedysarum alpinum is represented in N.Am. by several phases, only one of which, the var. *americanum* Michx., is known definitely to reach our area. It has flowers 11-15 mm. long and is found from
Alta. to B.C. and s. to n.c. Mont. The var. *grandiflorum* Rollins, which differs only in having flowers 14-18 mm. long, has almost the same range as var. *americanum* but does not quite reach our region on the north. To the e. and s.e. of our range var. *philoscia* (Nels.) Rollins occurs in Wyo., S.D.,
and Alta. It is characterized by flowers 11-15 mm. long and (in contrast to the other two varieties) by
pubescent fruits.

Hedysarum boreale Nutt. Gen. Pl. 2:110. 1818. ("Fort Mandan, on the banks of the Missouri," col-
   lector not mentioned, but surely Nuttall)
  HEDYSARUM BOREALE var. MACKENZII (Richards.) C. L. Hitchc. hoc loc. *H. mackenzii* Richards.
     Frankl. Journ. App. 745. 1823. *H. americanum* var. *mackenzii* Britt. Mem. Torrey Club 5:202.
     1894. *(Richardson,* Barren grounds, Point Lake to the Arctic Sea)
  H. canescens of Nutt. in T. & G. Fl. N.Am. 1:357. 1838, but not of L. in 1753. *H. cinerascens*
     Rydb. Mem. N.Y. Bot. Gard. 1:257. 1900. *H. macquenzii* f. *canescens* Fedtsch. in Acta Hort.
     Petrop. 19:272. 1902. *H. macquenzii* var. *canescens* Fedtsch. op. cit. 362. *H. boreale* var. *ci-*

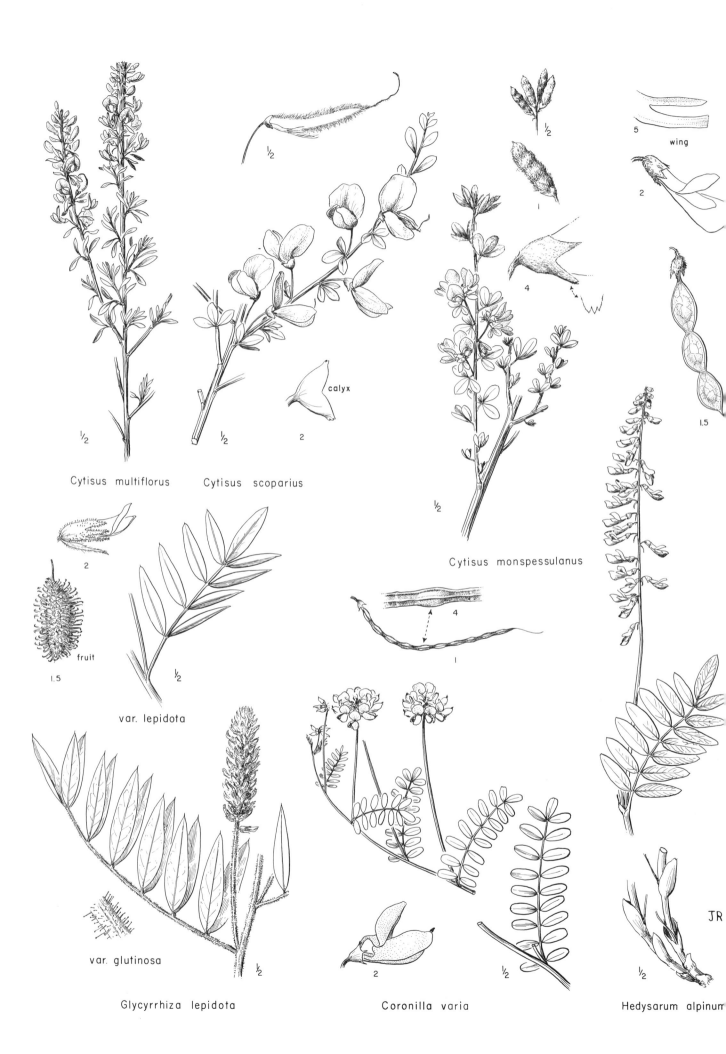

5

wing

2

Cytisus multiflorus   Cytisus scoparius

1/2   1/2   2   calyx

1/2   1/2   2

4

1

1.5

var. lepidota

2

fruit

1.5

1/2

Cytisus monspessulanus

1/2

4

1

var. glutinosa

Glycyrrhiza lepidota   Coronilla varia

2

1/2

1/2   1/2

JR

Hedysarum alpinum

nerascens Rollins, Rhodora 42:234. 1940. *(Nuttall,* "plains of the Rocky Mountains, particularly near Lewis's River")

H. carnosulum Greene, Pitt. 3:212. 1897. *(Greene?,* "about the mouth of the Cañon of the Arkansas, in southern Colorado") = var. *boreale.*

H. pabulare A. Nels. Proc. Biol. Soc. Wash. 15:185. 1902. *H. mackenzii* var. *pabulare* Kearney & Peebles, Journ. Wash. Acad. Sci. 29:485. 1939. *(A. Nelson 752,* Wind R. , Wyo. , in 1894) = var. *boreale.*

H. pabulare var. *rivulare* Williams, Ann. Mo. Bot. Gard. 21:344. 1934. *(Williams 975,* along the Snake R. , Teton Co. , Wyo. , July 31, 1932) = var. *boreale.*

Stems many from a very heavy crown, usually branched above, (1. 5) 2-6 dm. tall, sparsely to copiously grayish-strigillose; stipules small but the lower ones as much as 1 cm. long, somewhat leathery, tan or brownish-mottled; racemes 5- to 50-flowered, from compact to elongated and several times as long as broad; flowers erect or spreading, 11-22 mm. long, pink or pinkish-purple to reddish-purple; calyx 5-8 mm. long, the lobes all linear-lanceolate, subequal, at least the lower ones usually longer than the tube; wing petals with a blunt basal lobe scarcely 1/3 as long as the claw, the lobes of the two petals not joined over the ovary; loment 5-7 mm. broad, the (1) 2-6 segments nearly round, 1-4 times as broad as some of the connections, their margins very narrow and thickened but not at all winged, more or less corrugate because of the very prominent, transversely elongated reticulations.

Yukon southward to n. e. Oreg. , e. to Newf. , and s. in the Rocky Mts. to N. M. and Ariz. May-Aug.

Hedysarum boreale is a highly variable species that has been divided, largely on the basis of the quantity of pubescence and on flower size, into several varieties, 2 (possibly 3) of which occur in our area. Variety *mackenzii* is usually maintained as an admittedly closely related species, because of the shorter, more compact, fewer-flowered racemes, darker colored and larger flowers, and more-segmented loments. Plants from the Wallowa Mts. , Oreg. , and from the Fraser R. Canyon, B. C. , are intermediate in many characters between var. *mackenzii* and var. *borealis,* and weaken any case for the maintenance of both at the specific level. In general the differences between the two are little more than tendencies, rather than consistent quantitative or qualitative distinctions.

1 Racemes usually compact, 2-4 (6) cm. long; flowers 5-15 (20), purple, mostly 18-20 (17-21) mm. long; plants generally unbranched, greenish, the upper surface of the leaves usually glabrous to sparsely strigillose; loments with 3-6 (8) segments; Yukon to Newf. , s. to B. C. , Alta. , and Man. , just reaching the n. limit of our area; 2N=16, 17

var. mackenzii (Richards.) C. L. Hitchc.

1 Racemes often as much as 15 cm. long; flowers mostly 12-20 (5-25), carmine, magenta, or purple, 11-17 mm. long; plants usually somewhat grayish-strigillose and branched; loments with 2-5 (6) segments

  2 Plants usually conspicuously grayish-hairy, the leaves pubescent on both surfaces; little more than one extreme in an essentially continuous series grading to var. *boreale,* occurring chiefly on dry banks, in Ida. and Mont. , but also in Wyo. and Alta. ; 2N=16

var. cinerascens (Rydb.) Rollins

  2 Plant usually greenish, the leaves only sparsely hairy, often glabrous on the upper surface; grading into var. *cinerascens* throughout much of that taxon's range, and to the n. merging with the var. *mackenzii;* B. C. to Alta. , s. through Ida. to S. D. and to N. M. and Ariz. , also in n. e. Oreg.

var. boreale

Hedysarum occidentale Greene, Pitt. 3:19. 1896. *(C. V. Piper,* Olympic Mts. , Wash. , in 1890)

H. lancifolium Rydb. Mem. N. Y. Bot. Gard. 1:256. 1900. *(Canby 93,* headwater of Jocko R. , Mont., in 1883)

H. uintahense A. Nels. Proc. Biol. Soc. Wash. 15:186. 1902. *(A. Nelson 7198,* Evanston, Uinta Co. , Wyo. , June, 1900)

Stems several from a woody crown, 4-8 dm. tall, usually branched above, sparsely strigillose-pubescent, greenish; stipules membranous, brownish, the lower ones connate and as much as 3 cm. long, the upper ones mostly free; leaflets 9-21, 1-3 cm. long, minutely brownish-glandular-punctate on the upper surface; flowers 20-80, reddish to purplish, 16-22 mm. long, pendent; calyx 4-5 mm. long, strigillose, the lobes shorter than the tube, the lower 3 more slender and slightly longer than the upper ones; auricles of the wing petals slender, about equal to the claws, weakly joined over the ovary; segments of the loment 1-4, ovate-elliptic, 7-12 mm. broad, irregularly and rather lightly reticulate except on the somewhat erose-winged margins which are 1-2 mm. broad.

At higher elevations in the mountains; Ida. and Mont., s. to Wyo. and Colo., also in the Cascade (rare) and Olympic mts. of Wash. June-Sept.

Hedysarum sulphurescens Rydb. Bull. Torrey Club 24:251. 1897.
  H. flavescens of Coult. & Fisher, Bot. Gaz. 18:300. 1893, but not of Regel in 1882. *H. boreale* var. *leucanthum* M. E. Jones, Proc. Calif. Acad. Sci. II, 5:677. 1895. *H. boreale* var. *flavescens* Fedtsch. Bull. Herb. Boiss. 7:256. 1899. *(F. D. Kelsey,* near Helena, Mont., May, 1892)
  H. boreale var. *albiflorum* Macoun, Cat. Can. Pl. 1³:510. 1886. *H. albiflorum* Fedtsch. Acta Hort. Petrop. 19:252. 1902. *(Macoun,* Columbia Valley at Donald, Lat. 51°)
Stems several from a thick crown, 3-6 dm. tall, usually branched above, sparsely strigillose-pubescent, greenish; stipules brownish-membranous, connate or free, 10-15 mm. long; flowers 20-100, pale yellow to nearly white, 14-18 mm. long, pendent; calyx 3-4 mm. long, the teeth shorter than the tube, the upper two broader and shorter than the lower three; wings with a slender auriculate lobe subequal to the claw, those of the two petals joined by their tips over the ovary; segments of the loment 1-4, more or less obovate, 6-10 mm. broad, lightly and irregularly reticulate except along the narrowly winged margins.
Open forested areas on the e. slope of the Cascade Mts. from B.C. to Okanogan Co., Wash., e. to Alta., Mont., and Wyo. June-Aug.

Lathyrus L. Sweet Pea

Flowers either single (and pedunculate) in the axils or 2-many and racemose; calyx obliquely campanulate, the teeth subequal or the upper two much the shortest, linear to narrowly ovate-lanceolate or deltoid, shorter to longer than the tube; banner generally differentiated into a distinct blade and claw, the blade usually reflexed at about a right angle to the wings and keel; keel not at all beaked; stamens diadelphous, the tube of the 9 united filaments truncate at the apex; style usually sharply upturned from the ovary at almost a right angle, flattened but not grooved, hairy on the ventral surface for approximately half the length, persistent on the fruit; legumes 1-celled, several-seeded, usually somewhat flattened; cotyledons hypogaeous; annual or perennial, often rhizomatous herbs with erect to twining and angled to distinctly winged stems and pinnately compound leaves of 2-8 (0) leaflets, the stipules hastate to obliquely sagittate-lobed, entire to toothed; tendrils usually well developed, branched or simple and sometimes reduced to only a short bristle.
About 150 species, nearly all of the N. Temp. Zone in both hemispheres, a few in S. Am. (Ancient Greek name for this or some other member of the pea family.)
Lathyrus approaches *Vicia* in general characters, but is fairly consistently distinguished therefrom by the styles and by the proportionately broader flowers.
Like *Lupinus* and *Oxytropis*, *Lathyrus* undergoes a great deal of variation of a fortuitous nature. Among the more striking variations are leaflet width, general pubescence, degree of development of the tendrils, and flower size and color. The habit of the plant itself depends largely upon the immediate surroundings, particularly upon whether other plants are available to serve as a support. Several of the species apparently intercross fairly freely where their ranges overlap and for this reason, too, many plants cannot be referred to taxon with complete confidence.
References:
Bradshaw, R. V. Pacific Coast species of Lathyrus. Bot. Gaz. 80:233-261. 1925.
Hitchcock, C. Leo. A revision of the North American species of Lathyrus. U. Wash. Pub. Biol. 15:1-104. 1952.
Senn, Harold A. Experimental data for a revision of the genus Lathyrus. Am. Journ. Bot. 25:67-78. 1938.
White, T. G. A preliminary revision of the genus Lathyrus in North and Central America. Bull. Torrey Club 21:444-458. 1894.
The gardener east of the Cascade Mountains can grow most of our species easily, one of which, *L. rigidus,* is believed to have great potential as a garden subject; surely it should be growable in sunny, well-drained areas.
Pisum sativum L., the common pea, has many horticultural forms and is extensively cultivated in e. Oreg. and Wash., where it tends to persist on waste land; it may be mistaken for *Lathyrus.* It is an annual, with large stipules and 2-6 leaflets, whereas our annual species of *Lathyrus* have no more than 2 leaflets.

1 Leaflets lacking; stipules ovate-triangular and hastate; flowers yellowish; calyx lobes at least
    twice as long as the tube   L. APHACA
1 Leaflets 2 or more; stipules seldom ovate-triangular or hastate; flowers variously colored;
    calyx lobes seldom as much as twice as long as the tube
    2 Leaflets 2; annuals or weedy perennials
        3 Corollas 25-30 mm. long; plant an annual
            4 Pods glabrous, 7-10 cm. long, 7-9 mm. broad; corollas rose-purple   L. TINGITANUS
            4 Pods hairy, 3-6 cm. long, 4-7 mm. broad; corollas variously colored   L. ODORATUS
        3 Corollas much less than 25 mm. long; plant either annual or perennial
            5 Flowers yellow; plant perennial   L. PRATENSIS
            5 Flowers never yellow; plant annual or perennial
                6 Plant annual
                    7 Peduncles 1-flowered, the rachis extending beyond the pedicel of the single
                        flower as a straight to curved bristle; plant (including the pods) glabrous
                                                                                      L. SPHAERICUS
                    7 Peduncles usually 2-flowered, if only 1-flowered then either the rachis not
                        prolonged as a bristle or plants definitely hairy
                        8 Plant hirsute; calyx lobes about equal to the tube; pods hairy, 5-8 mm.
                            broad   L. HIRSUTUS
                        8 Plant usually sparsely pubescent; calyx teeth about twice as long as the
                            tube; pods glabrous, 2-4 mm. broad; introduced from s. U.S. in Douglas
                            Co., Oreg., and perhaps also in our area   L. *pusillus* Elliott
                6 Plant a perennial
                    9 Stems wingless; flowers 12-16 mm. long; leaflets obovate to elliptic-oblan-
                        ceolate, 2-4 cm. long, obtuse to rounded   L. TUBEROSUS
                    9 Stems winged; flowers 14-20 mm. long; leaflets usually acute, mostly over
                        4 cm. long
                        10 Pod 6-10 cm. long, 7-10 mm. broad; stipules 3-5 cm. long, ovate to
                            broadly lanceolate; flowers 15-20 mm. long, white to red   L. LATIFOLIUS
                        10 Pod 4-6 cm. long, 4-6 mm. broad; stipules 1-3 cm. long, linear-lan-
                            ceolate; flowers about 15 mm. long, red   L. SYLVESTRIS
    2 Leaflets at least 4 on some of the leaves; native perennials
        11 Tendrils lacking, usually represented only by a terminal, simple, nonprehensile
            bristle
            12 Flowers 1-2 per raceme, 8-13 mm. long; lower calyx teeth half again as long
                as the tube; leaflets elliptic to ovate or oval, apiculate   L. TORREYI
            12 Flowers either more than 2 per raceme, or plants otherwise not as above
                13 Plant densely villous, occurring on coastal sand dunes; rachis of the leaves
                    flattened, only 1.5-3 cm. long   L. LITTORALIS
                13 Plant scarcely villous, if copiously hairy then definitely not coastal in range;
                    rachis of the leaves usually over 3 cm. long and not flattened
                    14 Flowers 8-13 (16) mm. long
                        15 Leaflets 2-4 (very occasionally 6)   L. BIJUGATUS
                        15 Leaflets very rarely less than 6   L. LANSZWERTII
                    14 Flowers averaging at least 18 mm. long
                        16 Plants many-stemmed from a thick taproot, not rhizomatous; rachis
                            of the leaves 2-5 cm. long, somewhat winged; leaflets very heavily
                            veined; plants of sagebrush valleys and hillsides; flowers white   L. RIGIDUS
                        16 Plants rhizomatous; rachis of the leaves mostly over 5 cm. long, not
                            winged; leaflets not heavily veined; plants mostly of woodlands;
                            flowers white to blue   L. NEVADENSIS
        11 Tendrils present, usually well developed, either branched or prehensile
            17 Plant usually found along the coast in brackish sloughs or on sand dunes; stems
                either winged or else the stipules obliquely hastate or sagittate-ovate and
                from slightly shorter to longer than the leaflets
                18 Stems winged; leaflets mostly 6; stipules 1/6-1/2 as long as the leaflets
                                                                                      L. PALUSTRIS

18 Stems not winged; leaflets 6-12; stipules from only slightly shorter to longer than the
    leaflets   L. JAPONICUS
17 Plant ordinarily not from along the coast, if semimaritime then neither with winged stems
   nor with stipules nearly as large as the leaflets
    19 Corolla white
       20 Blade of the banner slightly shorter than the claw; lateral calyx lobes linear to lan-
          ceolate, not broadened above the base; leaflets (6) 8-12; flowers 13-17 mm. long;
          Willamette and Umpqua river valleys, Oreg.   L. HOLOCHLORUS
       20 Blade of the banner equaling or longer than the claw; lateral calyx lobes often
          broadened above the base and more or less ovate; leaflets and flowers various
          21 Lateral calyx lobes oblong-ovate to lanceolate but broadened noticeably just
             above the base
               22 Leaflets usually 6; east of the Cascades from Okanogan Co., Wash., n.
                  and e. to n. Ida. and Mont.   L. OCHROLEUCUS
               22 Leaflets 8-12; w. of the Cascades from King Co., Wash., to s.w. Oreg.
                                                  L. VESTITUS
          21 Lateral calyx lobes lanceolate, not broadened above the base; e. of the Cas-
             cade Mts., from Chelan to Kittitas Co., Wash., e. to s.e. Wash. and adj.
             Oreg. and Ida.   L. NEVADENSIS
    19 Corolla blue to red but sometimes pale and fading to brownish
       23 Leaflets 10-16, scattered; stipules sagittate-ovate, mostly at least half as long as
          the leaflets; plant erect, scarcely at all clambering, glabrous except on the ca-
          lyx, the lower half or third of the stem usually leafless at anthesis; racemes 5-
          to 13-flowered, secund; flowers 16-20 mm. long   L. POLYPHYLLUS
       23 Leaflets mostly averaging no more than 10, if (occasionally) more, then the stip-
          ules either smaller and narrower or the plant pubescent or otherwise not as above
          24 Keel 2-4 mm. shorter than the wing petals; calyx glabrous, or the teeth ciliate,
             the lowest tooth usually longer than the tube; stipules mostly broadly lance-
             olate to ovate and at least half as long as the leaflets (but sometimes much
             smaller)   L. PAUCIFLORUS
          24 Keel usually subequal to the wing petals but if conspicuously shorter then the
             calyx often hairy; lowest calyx tooth always shorter than the tube; stipules
             various
               25 Flowers pale lavender tinged to pinkish-violet or pinkish-orchid, 8-16 mm.
                 long; leaflets 4-12, linear to oblong-elliptic, mostly 3-8 times as long as
                 broad, very rarely (if ever) over 1 cm. broad   L. LANSZWERTII
               25 Flowers more nearly rose, blue, or purple, 13-27 mm. long; leaflets 4-10
                 but usually 6 or 8, often elliptic, usually over 1 cm. broad and not more
                 than 3 times as long as broad   L. NEVADENSIS

Lathyrus aphaca L. Sp. Pl. 729. 1753. ("Habitat in Italia, Gallia, Anglia")
    Glabrous annual 2-6 dm. tall; stems slender and reclining or twining to erect, angled but not winged;
stipules 1-4 cm. long, lanceolate to ovate or triangular, hastate-sagittate at the base; leaflets lacking, the rachis stiff, but tapering to a prehensile unbranched tendril; flowers 1 (2), lemon-yellow,
10-14 mm. long; calyx 7-10 mm. long, its lobes linear to narrowly oblong-lanceolate, about twice as
long as the tube; banner about half again as long as the calyx, the claw about 1/4 as long as the blade;
wings and keel subequal to the banner, with claws about 1/3 the length of the blades; pod 2-4 cm. long,
3-5 mm. broad. N=7.
    Fairly well established in several localities in the Willamette and Umpqua valleys of Oreg., also in
Calif. and a few other states; native to Europe. May-July.

Lathyrus bijugatus White, Bull. Torrey Club 21:457. 1894. (*J. H. Sandberg*, "copses," Latah Co.,
    Ida., June, 1892)
    L. bijugatus var. *sandbergi* White, Bull. Torrey Club 21:457. 1894. (*J. H. Sandberg*, Latah Co.,
    Ida., in 1892)
    Perennial with slender rhizomes, glabrous to sparsely crisp-puberulent; stems 1-3 (4) dm. tall,
slender, erect, not winged; stipules narrow, 1/8-1/3 the length of the leaflets; leaflets 2 or 4 (6), usually paired but sometimes scattered, from linear to oblong-elliptic or oblong-obovate, 2-15 cm. long

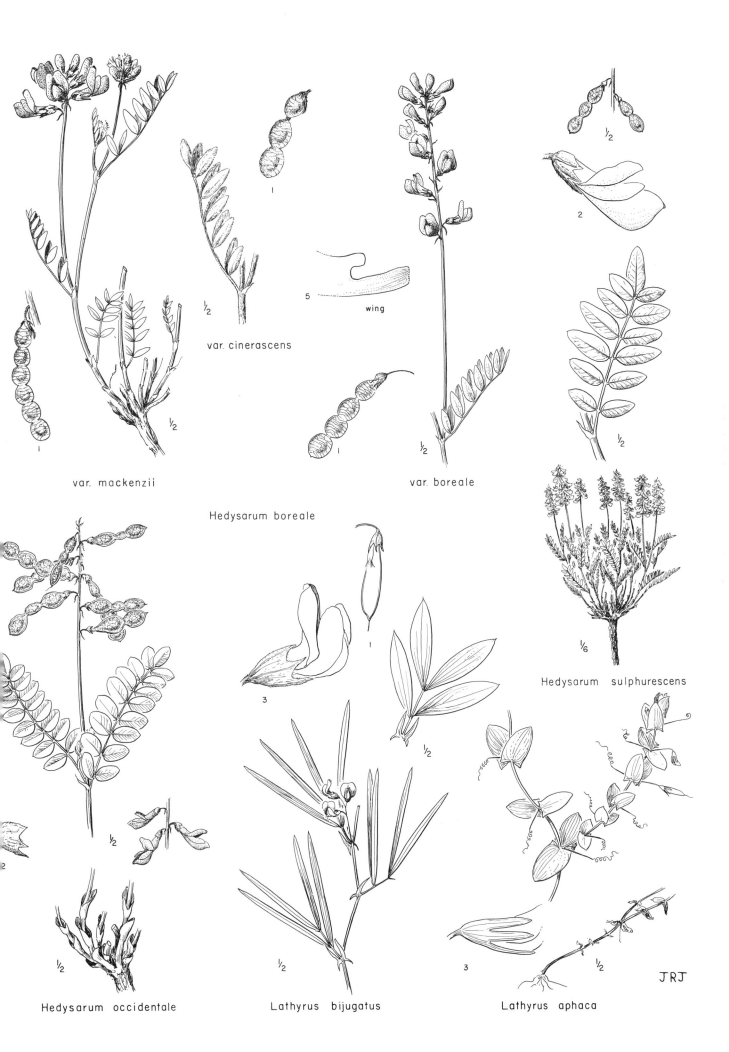

var. cinerascens

wing

var. mackenzii

var. boreale

Hedysarum boreale

Hedysarum sulphurescens

Hedysarum occidentale

Lathyrus bijugatus

Lathyrus aphaca

JRJ

and (1) 2-14 mm. broad, firm and fairly prominently veined but not coriaceous; tendrils bristlelike, 1-3 mm. long; flowers 2-3, pink to bluish, (8) 10-13 mm. long; calyx (4) 6-8 mm. long, the teeth narrowly lanceolate, subequal or the upper two not over 2/3 as long as the lower three, the lowest one only slightly shorter than the tube; banner very broadly obcordate, (9) 12-14 mm. long, about as broad; wings about equal to the banner, the blade slightly longer than the claw, 3/5 as broad as long; keel shorter than the wings, the tip slightly recurved; pod (2) 3-4 cm. long, 4-7 mm. broad, glabrous, 7- to 12-seeded.

Lower foothills, in open parks or under trees; extreme e. Wash. and adj. Ida., reported from Mont. but the record questionable. May-July

Lathyrus hirsutus L. Sp. Pl. 732. 1753. ("Habitat inter Angliae, Galliae")
Sparingly hirsute annual, 2-10 dm. tall; stems generally clambering, narrowly to broadly winged; stipules linear-lanceolate, usually entire; leaflets 2, linear-lanceolate to elliptic, 3-8 cm. long; tendrils well developed; flowers 1-2 (4), blue to red, 9-12 mm. long; peduncles usually longer than the leaves, the rachis often projecting past the base of the uppermost pedicel as a small bristle; calyx 5-7 mm. long, the teeth lanceolate to narrowly ovate-lanceolate, about equaling or slightly exceeding the tube; banner broadly obcordate, the blade twice as long as the claw; wings and keel with claw somewhat shorter than the blade; pod 2.5-4 cm. long, 5-8 mm. broad, conspicuously hirsute with pustular hairs, 4- to 10-seeded. N=7.

Escaped and established in the Willamette Valley, Oreg., and very occasional in a few other places in the U.S.; Europe. May-July.

The species is easily recognized because of the winged stems and hairy pods.

Lathyrus holochlorus (Piper) C. L. Hitchc. U. Wash. Pub. Biol. 15:31. 1952.
  L. ochropetalus ssp. holochlorus Piper, Proc. Biol. Soc. Wash. 31:190. 1918. (Gilbert 115, hills s. of Corvallis, Oreg.)
Rhizomatous perennial, from subglabrous to sparsely hairy on the calyces, lower surface of the leaves, and stipules; stems strongly angled but not winged, 3-10 dm. long, scandent; stipules ovate to ovate-lanceolate, mostly 1/5-1/2 the length of the leaflets, sometimes constricted into 2 lobes but more usually not, the margin coarsely undulate to dentate or dentate-lobed; leaflets (6) 8-12, ovate or oblong-ovate to elliptic, 2-5 cm. long, 0.7-3 cm. broad; tendrils fairly well developed but occasionally reduced to a mere bristle; flowers 5-15, whitish, aging to buff, 13-17 mm. long; calyx 9-12 mm. long, the teeth ciliate, the upper two deltoid-lanceolate and about 1/2 the length of the lateral pair, these linear to lanceolate and somewhat broader but shorter than the lowest one, which is about equal to the tube; banner pale greenish-cream, faintly pencilled with purplish-rose, 14-17 mm. long, the claw slightly longer but narrower than the reflexed blade; wings pale lemon, about equal in over-all length to the banner, the claw nearly equaling the blade; keel nearly white, slightly shorter than the wings, the tip strongly recurved; pod 3-5 cm. long, 4-7 mm. broad.

Fence rows and partially cleared land, Willamette Valley, s. to Roseburg, Oreg. May-July.

Lathyrus japonicus Willd. Sp. Pl. 3:1092. 1802.
  Orobus japonicus Alefeld, Bonplandia 9:143. 1861. (Asia)
  Pisum maritimum L. Sp. Pl. 727. 1753. L. maritimus Fries, Fl. Scand. 106. 1835. Orobus maritimus Reichenb. Fl. Germ. Excurs. 538. 1832, not L. maritimus Bigel. Fl. Bost. 2nd ed. 268.
    1824, which was (perhaps erroneously) stated not to be the same as Pisum maritimum L. (Europe)
  Pisum maritimum var. glabrum Ser. ex DC. Prodr. 2:368. 1825. L. maritimus var. glaber Eames, Rhodora 11:95. 1909. L. japonicus var. glaber Fern. Rhodora 34:181. 1932. (Canada)
Rhizomatous, glabrous to rather densely short-pubescent perennial; stems 1-15 dm. long, decumbent to clambering, strongly angled but not winged; stipules obliquely hastate or sagittate-ovate, usually subequal to (but often longer than) the leaflets; leaflets 6-12, paired or scattered, green and rather fleshy, strongly nerved when dry, 1-7 cm. long, 0.5-4 cm. broad; tendrils well developed; flowers 2-8, reddish-purple to light blue, 17-30 mm. long; calyx 10-15 mm. long, the teeth lanceolate, the upper two about half as long as the lateral pair which are 1-2 mm. shorter than the lowest one, all three of the lower teeth usually longer than the calyx tube; banner 16-20 mm. long, the claw about equal to the blade; wings and keel somewhat shorter than the banner, the tip of the keel slightly recurved; pod 3-7 cm. long, about 1 cm. broad, usually pubescent. N=7.

Sandy coasts; Alas. to n. Calif., Atlantic coast to Minn., Mich., and Ohio; Eurasia. May-Sept.

Our plants are the glabrous form of the species, the var. glaber (Ser.) Fern.

Lathyrus lanszwertii Kell. Proc. Calif. Acad. Sci. 2:150, fig. 44. 1861. *(Kennedy 1624,* Dinsmore
Camp, Hunter Creek Canyon, Washoe Co., Nev., June 20-25, 1907, lectotype)
  L. coriaceus White, Bull. Torrey Club 21:452. 1894. *(Watson 297,* Wahsatch Mts., Utah) = var.
    lanszwertii.
  L. oregonensis White, Bull. Torrey Club 21:456. 1894. *(Cusick 1372,* Oreg., is the first of 2 col-
    lections cited) = var. *lanszwertii.*
  L. coriaceus ssp. *aridus* Piper, Proc. Biol. Soc. Wash. 31:190. 1918. *L. lanszwertii* ssp. *aridus*
    Bradshaw, Bot. Gaz. 80:247. 1925. *L. lanszwertii* var. *aridus* Jeps. Fl. Calif. 2:389. 1936.
    (Cusick 2814, Black Butte, Crook [or Jefferson?] Co., Oreg.)
  Rhizomatous perennial 1.5-8 dm. tall, usually moderately crisp-puberulent (glabrous), the stems
angled but not winged, clambering, trailing, or (sometimes) erect; stipules mostly linear to linear-
lanceolate, 1/4-3/4 as long as the leaflets, usually 2-lobed but otherwise entire; leaflets 4-10 (12),
usually paired, linear to oblong-elliptic, 3-10 cm. long, 1.5-10 mm. broad, often coriaceous and
heavily veined; tendrils bristlelike or well developed and prehensile; flowers 2-8, 8-16 mm. long;
calyx 5-8 mm. long, glabrous or pubescent, the teeth lanceolate, all shorter than the tube; banner
pale lavender tinged to pinkish-violet or orchid, penciled with reddish-purple, the claw at least as
long as the blade; wings more nearly lavender and usually paler than the banner (to nearly pure white),
the blade longer than the claw; keel white or very pale lavender, the tip well recurved; pod glabrous,
4-6 cm. long, 3-6 mm. broad.
  Sagebrush-ponderosa pine woodland to montane; Wash. to the c. Sierra Nevada, e. to Ida. and Utah.
May-June.
  There are two rather well-marked races of the species, both of which occur in our area:
1 Tendrils lacking; flowers 8-12 mm. long, nearly white; e. side of the Cascade Mts. from
    Wash. to c. Calif., e. to extreme w. Ida.; N=7   var. aridus (Piper) Jeps.
1 Tendrils usually well developed; flowers 13-16 mm. long, usually pinkish to lavender;
    with the var. *aridus* on the e. side of the Cascade Mts. from Wash. to c. Calif., but
    farther e. in Ida., to Utah; N=7, 14   var. lanszwertii

Lathyrus latifolius L. Sp. Pl. 733. 1753. (Europe)
  Glabrous, rhizomatous perennial 8-20 dm. tall; stems very broadly winged, climbing; stipules
broadly lanceolate to ovate, 3-5 cm. long, usually entire, the upper lobe 2-3 times as long as the
lower; leaflets 2, lanceolate-elliptic to obovate-lanceolate, as much as 14 cm. long and 5 cm. broad;
tendrils well developed; flowers 5-15, pinkish-red (white, red, or striped), 15-20 mm. long; calyx
8-12 mm. long, the linear-lanceolate teeth subequal to the tube; banner well reflexed, nearly as broad
as long, short-clawed like the wings and keel; pod 6-10 cm. long, 7-10 mm. broad, 10- to 25-seeded.
N=7.
  This is our most common weedy sweet pea; native to Europe, it is established in many parts of the
U.S., especially along railroads and highways. May-July.

Lathyrus littoralis (Nutt.) Endl. in Walpers Rep. 1:722. 1842.
  Astrophia littoralis Nutt. in T. & G. Fl. N. Am. 1:278. 1838. *Orobus littoralis* Gray, Pac. R. R.
    Rep. 4:54. 1856. *(Nuttall,* "Sand hills near the estuary of the Oregon")
  Strongly rhizomatous perennial, gray-villous throughout; stems 1-6 dm. long, prostrate to erect,
neither scandent nor winged; stipules obliquely ovate or ovate-hastate to lanceolate, sometimes constricted into upper and lower lobes, usually equaling the leaflets in length and somewhat exceeding
them in width; leaflets 4-8, oblong-oblanceolate, 1-2 cm. long, the rachis only 1.5-3 cm. long, considerably flattened and prolonged as a broad, not at all tendril-like bristle; flowers 2-6 (10), 12-18 mm.
long; calyx 8-11 mm. long, the teeth lanceolate, subequal and about as long as the tube; banner 14-18
mm. long, (white) pink to red or purple, the claw about 2/3 as long as the blade; wings and keel light
colored, usually white, each with a claw nearly equal to the blade, the tip of the keel recurved slightly;
pod about 3 cm. long, 1 cm. broad, 1- to 5-seeded, hairy.
  On sand dunes along the Pacific coast from Vancouver I. to Monterey Co., Calif. May-July.

Lathyrus nevadensis Wats. Proc. Am. Acad. 11:133. 1876. *(Bigelow,* Mammoth Grove, Calif.)
  Vicia nana Kell. Proc. Calif. Acad. Sci. 7:89. 1876. *(Eisen,* near Fresno, Calif.)
  L. cusickii Wats. Proc. Am. Acad. 17:371. 1882. *L. nevadensis* ssp. *cusickii* C. L. Hitchc. U.
    Wash. Pub. Biol. 15:44. 1952. *(Cusick,* dry mountain slopes, Union Co., Oreg.)
  L. nuttalli Wats. Proc. Am. Acad. 21:450. 1886. *L. nevadensis* ssp. *lanceolatus* var. *nuttallii* C.

var. aridus

var. lanszwertii

Lathyrus lanszwertii

3

L. holochlorus

½

3

L. latifolius

½

2

½

½

L. japonicus

JRJ

3

½

L. hirsutus

L. Hitchc. U. Wash. Pub. Biol. 15:45. 1952. *(Nuttall,* U. Calif.) = var. *pilosellus.*
L. lanceolatus Howell, Fl. N.W. Am. 158. 1898. *L. nuttallii* ssp. *lanceolatus* Piper, Proc. Biol.
Soc. Wash. 31:191. 1918. *L. nevadensis* ssp. *lanceolatus* C. L. Hitchc. U. Wash. Pub. Biol. 15:
45. 1952. *(Howell,* in forests at Glendale, s. Oreg.)
LATHYRUS NEVADENSIS ssp. LANCEOLATUS var. PILOSELLUS (Peck) C. L. Hitchc. hoc loc. L.
rigidus var. *pilosellus* Peck, Torreya 28:55. 1928. *(Peck 7869,* summit of Horse Mt., 11 mi. s.e.
of McKenzie Bridge, Lane Co., Oreg., July 1, 1914)
L. pedunculatus St. John, Fl. S.E. Wash. 223. 1937. *(St. John et al. 4281,* Turner Creek, Lake
Coeur d'Alene, Ida.) = var. *pilosellus.*
L. parkeri St. John, Fl. S. E. Wash. 223. 1937. *L. nevadensis* ssp. *lanceolatus* var. *parkeri* C. L.
Hitchc. U. Wash. Pub. Biol. 15:45. 1952. *(Parker 511,* Grizzly Camp, Latah Co., Ida., July 2,
1922)
L. nevadensis ssp. *lanceolatus* var. *puniceus* C.L. Hitchc. U. Wash. Pub. Biol. 15:46. 1952.
(C. L. Hitchcock 18973, Tumwater Mt., 10 mi. n.w. of Leavenworth, Chelan Co., Wash., May
21, 1949)
Rhizomatous perennial, sparsely to rather densely soft-pubescent, especially on the lower surface
of the leaves and on the calyces; stems 1.5-8 dm. tall, angled but not winged, erect to clambering;
stipules lanceolate to linear-lanceolate, 2-lobed, usually considerably less than half as long as the
leaflets; leaflets 4-10, paired or scattered, linear to ovate-lanceolate or lanceolate-elliptic, 2-12 cm.
long, 0.2-3 cm. broad; tendrils from well developed to reduced and bristlelike; flowers 2-7 (10), 13-
27 mm. long, blue, bluish-purple, or mauve-red to white and fading to tan or brownish, the banner
usually darker in color than the wings or keel; calyx 6-12 mm. long, generally somewhat hairy (sometimes merely ciliate), the teeth lanceolate or linear-lanceolate, subequal, 1/3-3/4 as long as the tube,
the sinuses open; banner 15-27 mm. long, narrowly to broadly obcordate, the claw about equal to the
blade; wings subequal to the banner, the claw somewhat shorter than the blade; keel 1-3 mm. shorter
than the wings, the tip slightly recurved; pod 3-7 cm. long, 4-9 mm. broad, 4- to 12-seeded.
   Widespread, B.C. to Calif., on both sides of the Cascades, e. to Ida., usually in woodland. May-
July.
   The species is differentiated into several distinctive populations, all represented in our area.
1 Flowers mostly over 17 mm. long, white to blue; tendrils bristlelike or, if larger,
   mostly unbranched; leaflets 4-7 (8)
   2 Corolla pinkish to fairly dark blue or reddish-purple (very rarely white); Lane and
      Coos cos., Oreg., s. to Humboldt and Fresno cos., Calif.   ssp. nevadensis
   2 Corolla white, pinkish-veined; Blue and Wallowa mts., Oreg., and adj. Ida.
                                                  ssp. cusickii (Wats.) C. L. Hitchc.
1 Flowers mostly less than 17 (13-20) mm. long, variously colored; tendrils usually
   present, often well developed, the stems mostly scandent; leaflets mostly 8-14
                                                  ssp. lanceolatus (Howell) C. L. Hitchc.
   3 Flowers bluish to reddish-purple; mostly w. of the Cascades, from B.C. to Calif.;
      N=14   var. *pilosellus* (Peck) C. L. Hitchc.
   3 Flowers white or pinkish; plants from e. of the Cascade Mts.
      4 Banner white, at most only pinkish-lined; n.c. Ida., from Kootenai to Idaho cos.;
         N=14   var. parkeri (St. John) C. L. Hitchc.
      4 Banner definitely pinkish-tinged; Wenatchee Mts., Kittitas and Chelan cos., Wash.
                                                  var. puniceus C. L. Hitchc.

Lathyrus ochroleucus Hook. Fl. Bor. Am. 1:59. 1834.
   Orobus ochroleucus Brown, Ind. Sem. Hort. Berol. 1853. *(R. Wright,* Hudson's Bay)
   L. obovatus var. *stipulaceus* White, Bull. Torrey Club 21:455. 1894. *(Wilkes' Expedition 592,* Colville to Spokane, Wash.)
   Glabrous, rhizomatous perennial, the stems 3-8 dm. tall, inconspicuously angled, not winged; stipules 1/3-2/3 the length of the leaflets, ovate to ovate-lanceolate, not constricted into 2 lobes, the
margins entire to dentate; leaflets mostly 6 (4 or 8), ovate to lance-elliptic, 3-7 cm. long, 1-4 cm.
broad; tendrils well developed, usually branched; flowers 6-14, white or cream to ochroleucous, 12-
16 mm. long; calyx 8-10 mm. long, glabrous but the teeth ciliate, the upper two teeth deltoid-ovate,
1/3-1/2 as long as the oblong-lanceolate to narrowly ovate-oblong lower lateral pair, the latter broader but somewhat shorter than the lowest tooth which is nearly linear-lanceolate and usually slightly
longer than the tube; banner 14-17 mm. long, the claw not quite equaling the blade; wings about equal-

ing the keel, the tip of which is slightly recurved; pod 4-7 cm. long, 4-7 mm. broad, 5- to 12-seeded. N=7.

Transcontinental in Can., s. to n. e. Wash., Ida., Mont., and the central states, mostly in moist woods and at the edge of thickets. May-July.

Lathyrus odoratus L. Sp. Pl. 732. 1753. (Origin not specified)

Annual, rather strongly crisp-hairy; stems winged, 8-30 dm. tall, clambering; stipules linear-lanceolate, 2-lobed, 1/4-1/2 the length of the single pair of elliptic to lanceolate or oval leaflets; tendrils well developed, pinnately branched; peduncles 2- to 5-flowered, usually exceeding the leaves; flowers 25-30 mm. long, fragrant, variously colored; pod 3-6 cm. long, 4-7 mm. broad, rough-hairy, 4- to 10-seeded. N=7, 14. Sweet pea.

Often found as an escape from cultivation but probably never really becoming established. May-July.

Lathyrus palustris L. Sp. Pl. 733. 1753. (Northern Europe)

    L. *pilosus* Cham. Linnaea 6:548. 1831. L. *palustris* var. *pilosus* Ledeb. Fl. Ross. 1:686. 1842. L. *palustris* ssp. *pilosus* Hultén, Fl. Aleut. Is. 236. 1937. (*Chamisso*, "ad portum Petro-Pauli Kamtschatcae")

    L. *myrtifolius* var. *macranthus* White, Bull. Torrey Club 21:448. 1894. L. *macranthus* Rydb. Brittonia 1:92. 1931. (*Oakes*, Quoddy Head, Maine) = var. *pilosus*.

Glabrous to densely pubescent, rhizomatous perennial, the stems 3-10 dm. tall, scandent, usually narrowly winged; stipules 1/6-1/2 the length of the leaflets, sagittate-ovate to narrowly lanceolate, usually constricted into 2 lobes, the margins generally dentate; leaflets (4) 6 (8), paired or scattered, linear to ovate, usually elliptic-lanceolate, 2-7 cm. long, 0.3-2 cm. broad; tendrils well developed; flowers 2-5 (8), (whitish) pink to bluish-purple, 12-20 mm. long; calyx 8-12 mm. long, glabrous to uniformly hairy, the upper teeth broadly triangular-oblong, 1/3-2/3 as long as the lower lateral pair, the latter usually lanceolate and slightly shorter but broader than the lowest lobe which is subequal to the tube; banner 15-22 mm. long, the claw considerably shorter and narrower than the blade; wings about equal to the banner, the claw not so long as the blade; keel nearly equaling the wings, the tip slightly recurved; pod 4-6 cm. long, 4-6 mm. broad, 5- to 8-seeded, strigillose. N=7, 21.

Along both coasts of N. Am. and in the Great Lakes region; on the Pacific coast from Alas. to n. Calif. Apr.-July.

The species is found only along the coast in our area, chiefly on tidelands; it undergoes the usual variation in leaflet width and pubescence. Conventionally the pubescent plant is recognized as var. *pilosus* (Cham.) Ledeb. and the glabrous one as var. *palustris,* a distinction which is not particularly significant since the two types of plants commonly occur together.

Lathyrus pauciflorus Fern. Bot. Gaz. 19:335. 1894. (*Thomas Howell 677,* Roseburg, Oreg.)

    L. *utahensis* M. E. Jones, Proc. Calif. Acad. Sci. II, 5:678. 1895. L. *pauciflorus* ssp. *utahensis* Piper, Proc. Biol. Soc. Wash. 31:194. 1918. L. *pauciflorus* var. *utahensis* Peck, Man. High. Pl. Oreg. 457. 1941. (*Jones 54411,* Ireland's Ranch, Salina Canyon, Utah)

    L. *parvifolius* var. *tenuior* Piper in Piper & Beattie, Fl. Palouse Region 108. 1901. L. *pauciflorus* ssp. *tenuior* Piper, Proc. Biol. Soc. Wash. 31:194. 1918. L. *pauciflorus* var. *tenuior* Peck, Man. High. Pl. Oreg. 457. 1941. (*Elmer 52,* Snake R. Bluffs near Almota, Wash.) = var. *pauciflorus*.

    L. *brownii* Eastw. Bull. Torrey Club 30:491. 1903. L. *pauciflorus* ssp. *brownii* Piper, Proc. Biol. Soc. Wash. 31:195. 1918. L. *lanszwertii* var. *brownii* Jeps. Fl. Calif. 2:389. 1936. (*Brown 391,* Mt. Shasta, Calif., in 1890)

    L. *bradfieldianus* A. Nels. Bot. Gaz. 54:411. 1912. (*Macbride 927,* Silver City, Owyhee Co., Ida., June 19, 1911) = var. *utahensis*.

Perennial with a strong taproot and slender, fairly short rootstocks, glabrous except for the ciliate calyx teeth or some occasional pubescence on the stipules; stems strongly angled but not winged, thick, erect, scarcely at all scandent, 2-6 (8) dm. long; stipules usually obliquely ovate-lanceolate and dentate but not lobed, mostly 1/3-2/3 the length of the leaflets (but when the leaflets are linear, the stipules are much narrower, often but 1/6-1/5 as long as the leaflets, and frequently constricted into two lobes either of which may be dentate to cleft); leaflets rather thick and fleshy but not particularly coriaceous, mostly 8 or 10 (6-13), paired or scattered, from linear or linear-lanceolate and as much as 8 cm. long and only 2-4 mm. broad to ovate or ovate-elliptic, 4-5 cm. long, and over half that broad; tendrils well developed; flowers 4-7 (10) per raceme, 13-27 mm. long, orchid or pinkish-lavender to violet-purple, aging to bluish; calyx 8-18 mm. long, glabrous or the teeth ciliate; upper pair of calyx

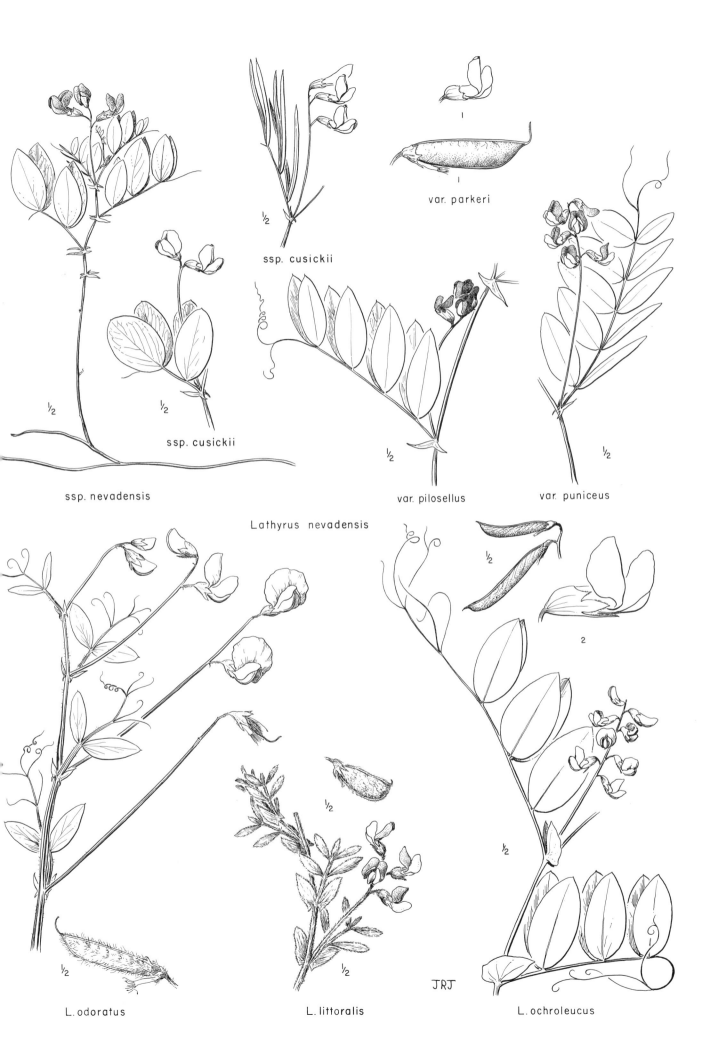

ssp. cusickii

var. parkeri

ssp. cusickii

½

½

ssp. nevadensis

var. pilosellus

var. puniceus

Lathyrus nevadensis

½

2

JRJ

L. odoratus

L. littoralis

L. ochroleucus

teeth triangular to linear-lanceolate, about 1/2 the length of the lower three, these lanceolate to linear and (at least the lowest one) usually longer than the calyx tube, the sinuses open; banner orchid to lilac, usually purplish on the back, the blade and claw subequal; wings from nearly white to pale orchid or blue, about equal to the banner, 2-4 mm. longer than the keel, the claw considerably shorter than the blade; keel nearly white to bluish, the tip strongly recurved; pod 3-5 cm. long, 3-6 mm. broad, glabrous.

Grassland and sagebrush slopes to ponderosa pine or montane forest on the e. side of the Cascade Mts., from Chelan Co., Wash., southward to s. Calif., e. to Ida., Utah, s.w. Colo., and n.e. Ariz. Apr.-June.

There are three variants of the species that are fairly distinctive:
1 Flowers usually at least 18 mm. long; leaflets generally over 3 cm. long, linear to ovate
                                                                                           ssp. pauciflorus
    2 Leaflets linear to oblong-elliptic, usually over twice as long as broad; e. base of the
         Cascade Mts., from Chelan Co., Wash., to c. Oreg., e. to w. Ida.   var. pauciflorus
    2 Leaflets ovate to ovate-lanceolate, usually at least half as broad as long; w. side of
         the Wasatch Mts., Utah, to s.w. Colo. and n.e. Ariz., occasional in Ida., Oreg.,
         and Wash.   var. utahensis (M. E. Jones) Peck
1 Flowers usually less than 17 mm. long; leaflets either linear, or (more commonly)
         obovate and not over 3 cm. long; s.c. Oreg., s. to Calif.; closely approached by
         plants from Wasco Co., Oreg.   ssp. brownii (Eastw.) Piper

Lathyrus polyphyllus Nutt. ex T. & G. Fl. N.Am. 1:274. 1838. *(Nuttall,* "Forests of the Oregon, to the sea")

Rhizomatous perennial, glabrous or subglabrous (usually pubescent only on the calyx teeth); stems 4-10 dm. tall, erect or somewhat scandent, terete or angled but not at all winged, at anthesis generally without leaves on most of the lower half; stipules sagittate-ovate, usually well over half as long as the leaflets, not constricted into 2 lobes, rounded to acute at base, the tip acuminate, the margins erose-dentate; leaflets 10-16, scattered, mostly 2.5-6 cm. long, 1-2 (4) cm. broad, ovate-lanceolate to lanceolate-elliptic; tendrils present but usually small, often simple or only forked; flowers 5-13, bluish-purple, aging to bluish, 16-20 mm. long; calyx 7-13 mm. long, the upper teeth deltoid to lanceolate, 1/3-2/3 as long as the linear-lanceolate lateral pair, the lowest tooth linear or linear-lanceolate, shorter to longer than the tube and 1-2 mm. longer than the lateral pair, the sinuses all open; banner purplish-blue to purplish-red, more reddish on the back, narrowly cordate, the claw about equal to the very shallowly emarginate blade; wings lighter and more nearly orchid-bluish or bluish-violet to nearly white, 1-3 mm. shorter than the banner; keel lighter and very noticeably shorter than the wings, the tip slightly recurved; pod 4-7 cm. long, 4-9 mm. broad, glabrous, many-seeded. N=7.

Coastal mts. and prairies, from the Puget Trough s. to Lake Co., Calif. May-July.

Lathyrus pratensis L. Sp. Pl. 733. 1753. (Europe)

Rhizomatous perennial, subglabrous or (more commonly) rather densely villous-hirsute; stems wingless, decumbent to erect, 4-10 dm. long, usually freely branched; stipules lanceolate to ovate-lanceolate, the lobes entire, the lower one scarcely 1/2 as long as the upper; leaflets 2, linear-elliptic, 2-5 cm. long, often equaled by the stipules; tendrils well developed; flowers 3-12, crowded, 12-17 mm. long, yellow; calyx 6-8 mm. long, strongly nerved, the main nerves hairy, the linear-lanceolate teeth scarcely equaling the tube; claw of all petals shorter than the blade; pod 2-3 cm. long, 4-7 mm. broad, 4- to 8-seeded. N=7, 14, 21.

An occasional introduction, apparently established in a few places on both sides of the mts. in Wash.; our only yellow-flowered perennial; native to Europe. May-July.

Lathyrus rigidus White, Bull. Torrey Club 21:455. 1894.

L. albus Wats. Bot. Calif. 2:442. 1880, but not of Garcke in 1849. *(Cusick,* Union Co., Oreg.)

Perennial, glabrous throughout, the stems many from a thick crown and a heavy taproot, erect and not at all clambering, 1.5-4 dm. long, angled but not winged; stipules usually nearly as long as the leaflets, frequently toothed, the upper lobe broadly lanceolate, 2-3 times as long as the lower one; leaflets (4) 6-10, oblanceolate or oblong-oblanceolate, 1.5-3 cm. long, 1/4-2/5 as broad, abruptly mucronulate, paired, very heavily veined; rachis 2-4 (5) cm. long, somewhat winged, prolonged into a slender, bristlelike, nonprehensile tip 2-5 mm. long; flowers 2-5, white or pinkish (rarely bluish),

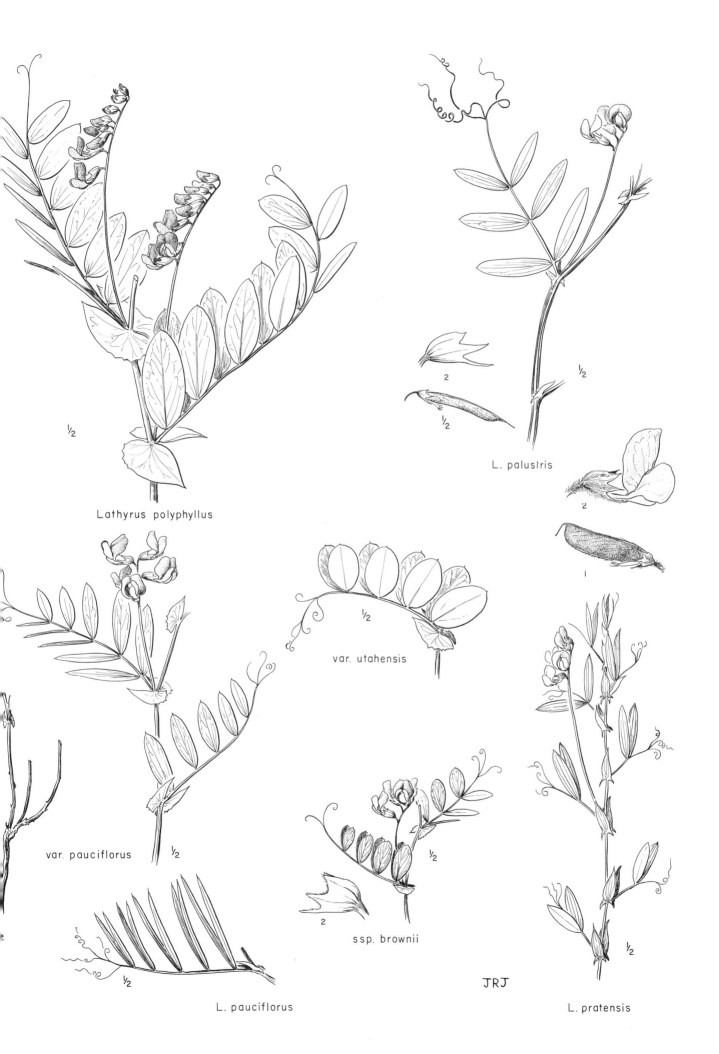

Lathyrus polyphyllus

L. palustris

var. utahensis

var. pauciflorus

L. pauciflorus

ssp. brownii

L. pratensis

JRJ

20-27 mm. long; calyx 7-9 mm. long, the teeth ciliate, narrowly to broadly lanceolate, about 2/3 as long as the glabrous tube; banner broadly obcordate, white but pinkish-veined to very faintly pinkish-tinged, nearly as broad as long, the claw a little over half as long as the blade and not clearly demarked therefrom; keel slightly shorter than the wings, the tip slightly recurved; pod 3-5 cm. long, 6-9 mm. broad, 5- to 12-seeded. N=7.

Sagebrush flats to juniper and pine woodland of the foothills, from Union Co., Oreg., and Adams Co., Ida., s. to Modoc Co., Calif. May-June.

Lathyrus sphaericus Retz. Obs. Bot. 3:39. 1785. (Europe)

Glabrous annual; stems several, slender, usually erect, not twining, narrowly winged to wingless, 2-5 dm. long; stipules bilobed, linear or linear-lanceolate, the upper lobe nearly twice as long as the lower, the whole scarcely 1/3-1/4 the length of the leaflets; leaflets 2, linear or linear-lanceolate to narrowly elliptic, 3-6 cm. long; lower leaves merely bristle-tipped, the upper ones with well-developed, simple or branched tendrils; peduncles single-flowered but the rachis extending beyond the pedicel as a straight, curved, or tendrilar bristle; flowers bluish, about 1 cm. long; calyx 5-8 mm. long, the teeth linear-lanceolate, equaling to nearly twice as long as the tube; banner not greatly reflexed, the claw about equal to the blade; wings and keel with a claw subequal to the blade; pod 3-6 cm. long, 3-5 mm. broad, glabrous. N=7.

A Eurasian species that occasionally escapes; fairly well established at several places in the Willamette and Umpqua valleys, Oreg. May-June.

Lathyrus sylvestris L. Sp. Pl. 733. 1753. (Europe)

Glabrous, rhizomatous perennial; stems 6-20 dm. long, broadly winged, climbing; stipules mostly linear-lanceolate, 2-lobed, 1-3 cm. long; leaflets 2, lanceolate to elliptic-lanceolate, 5-12 cm. long; tendrils well developed, branched; flowers 4-9, about 15 mm. long, red; calyx 7-9 mm. long, the teeth lanceolate, considerably shorter than the tube; banner broadly obcordate, the claw shorter than the blade; wings and keel with a claw about half the length of the blade; pod 4-6 cm. long, 4-6 mm. broad, 10- to 20-seeded. N=7.

Frequently escaping, and fairly well established in waste places and along roadsides in many parts of N.Am.; in our area known from Ida., Wash., and Oreg.; native of Europe. May-July.

Lathyrus tingitanus L. Sp. Pl. 732. 1753. ("Habitat in Mauritania")

Glabrous annual, the stems rather narrowly winged, 8-20 dm. long, clambering; stipules linear-lanceolate to ovate-lanceolate, entire or toothed between the upper and lower lobe, 1/2-2/3 the length of the leaflets; leaflets 2, usually not paired, 4-10 cm. long, linear to elliptic-lanceolate; tendrils well developed, pinnately branched; flowers 1-3, rose-purple, 25-30 mm. long; calyx about 10 mm. long, the teeth triangular-lanceolate, shorter than the tube; petals with a claw much shorter than the blade; pod glabrous, rather coriaceous, 7-10 cm. long, 7-9 mm. broad, 6- to 10-seeded. N=7.

Often grown as an ornamental, sometimes escaping and becoming established as in several localities in Oreg. and Calif. May-July.

The plant is similar to *L. odoratus*, differing in its glabrosity, nonodorous flowers, and larger, more leathery pods.

Lathyrus torreyi Gray, Proc. Am. Acad. 7:337. 1867. *(Bolander 6506, thickets near the coast, Shelter Cove, Humboldt Co., Calif.)*

L. torreyi var. *tenellus* Wieg. Bull. Torrey Club 26:135. 1899. *(Flett 276, Tacoma, Wash.)*

Rhizomatous perennial, usually moderately villous throughout; stems 0.5-4 dm. long, erect to decumbent, not scandent, slender, angled but not winged; stipules lanceolate, constricted into an upper lobe usually several times as long as the lower, the two scarcely half as long as the leaflets; leaflets 10-16, paired to scattered, elliptic to ovate, oval, or obovate-elliptic, 5-25 (30) mm. long, apiculate, the rachis prolonged as a short unbranched bristle (true tendrils lacking); flowers one or two, 8-13 mm. long; calyx 6-10 mm. long, the teeth linear-lanceolate, the lower three considerably longer than the calyx tube and nearly half again as long as the upper two; banner pale lilac to bluish-lilac or blue, 10-15 mm. long, the claw subequal to, but considerably narrower than, the well-reflexed blade; wings and keel about equal to the banner, each with a claw nearly equal to the blade, the wings pale bluish-lilac to nearly white, the keel white, with a well-recurved tip; pod about 2 cm. long, 4-5 mm. broad, 4- to 7-seeded. N=7.

Open prairies and clearings in the woods, mostly near the coast; Pierce Co., Wash., to Santa Cruz Co., Calif. May-July.

Lathyrus tuberosus L. Sp. Pl. 732. 1753. ("Habitat inter Belgii, Genevae, Tatariae")

Glabrous perennial with rootstocks and small tubers; stems 2-6 dm. long, not winged, usually erect from a decumbent base; stipules linear-lanceolate to lanceolate or obovate, the upper lobe much the larger, 1/2-2/3 as long as the leaflets; leaflets 2, oblanceolate to elliptic-oblanceolate, 2-4 cm. long; tendrils well developed; flowers 4-10, red, 12-16 mm. long; calyx 5-7 mm. long, the teeth ovate-lanceolate, subequal to the tube; banner well reflexed, very broadly obcordate, the claw (as of the wings and keel) about half as long as the blade; pod 2-4 cm. long, 3-4 mm. broad, coriaceous, 1- to 5-seeded. N=7.

Occasionally escaped along roadsides and in waste places; known from Okanogan Co., Wash., in our area; Eurasia. May-July.

Lathyrus vestitus Nutt. in T. & G. Fl. N.Am. 1:276. 1838.
  Orobus vestitus Alefeld, Bonplandia 9:145. 1861. *(Nuttall,* Columbia Plains)
  L. ochropetalus Piper, Proc. Biol. Soc. Wash. 31:189. 1918. *L. vestitus* ssp. *ochropetalus* C. L.
    Hitchc. U. Wash. Pub. Biol. 15:19. 1952. *(Piper 482,* Seattle, Wash., June, 1891)

Rhizomatous perennial, glabrous or with only a few cilia on the calyx teeth; stems 3-10 dm. long, suberect to climbing, angled but not winged; stipules 1/4-2/3 as long as the leaflets, usually lanceolate and not constricted into 2 lobes, the margins undulate-dentate; leaflets (8) 10 (12), imperfectly paired to scattered, linear or lanceolate-elliptic to ovate, (2) 3-6 cm. long, 5-30 mm. broad; tendrils well developed; flowers 5-20, 15-20 mm. long; calyx 12-15 mm. long, the upper teeth deltoid-lanceolate, less than 1/2 as long as the lateral pair, the latter lanceolate but noticeably widened just above the base, the sinus between them and the lowest lobe thus closed, the lowest tooth linear-lanceolate, 1-2 mm. longer than the lateral pair and usually equaling to considerably exceeding the tube; banner cream to very pale tannish-purple, pencilled with pinkish-purple, 17-22 mm. long, narrowly obcordate, the claw subequal (and broadened gradually) to the blade; wings slightly shorter than the banner, white to deep cream, often tinged with light tan, the claw subequal to the blade; keel about equaling the wings and of the same color, the tip scarcely at all recurved; pod 4-6 cm. long, 4-7 mm. broad, glabrous or pubescent. N=7.

Open to wooded areas; King Co., Wash., s. to the Willamette Valley and (mostly) along the coast from s. Oreg. to s. Calif. Apr.-June.

This widespread species is differentiated into several distinctive populations; our plants, as described above, are referable to the ssp. *ochropetalus* (Piper) C. L. Hitchc., which ranges from Seattle to s. Oreg., on the w. side of the Cascades. Other phases of the species occur, mostly in Calif., all of which have pink to lavender or purplish flowers; these taxa involve such names as *L. peckii* Piper and *L. bolanderi* Wats., both of which sometimes are erroneously listed for our area.

Lotus L. Lotus

Flowers papilionaceous, axillary and sessile to pedunculate in 1- to many-flowered umbels, white or yellow but often tinged with reddish-purple; peduncles ebracteate or with a simple to 3- to 5 (7)-foliolate bract between midlength and the tip; calyx tubular, the teeth deltoid to linear, from longer to much shorter than the tube; stamens 10, diadelphous, the free portion of alternate filaments considerably dilated; legumes 1- to many-seeded, usually laterally flattened and dehiscent, but sometimes much shortened and curved and indehiscent; glabrous or hairy annual or perennial herbs with spreading to erect, firm to fistulose stems, pinnately 3- to many-foliolate leaves, and glandlike, membranous, or expanded and leafletlike stipules.

A large genus of nearly 150 species, rather cosmopolitan but chiefly in the temperate zone. (The Greek name.)

Our native perennials, *L. pinnatus, L. crassifolius,* and *L. formosissimus,* should all be considered desirable garden subjects, probably as listed in order of importance. They are easily transplanted or grown from seed and do not become invasive, in spite of their rhizomatous habit.

References:
  Ottley, Alice M. A revision of the Californian species of Lotus. U. Calif. Pub. Bot. 10:189-305.
    1923.

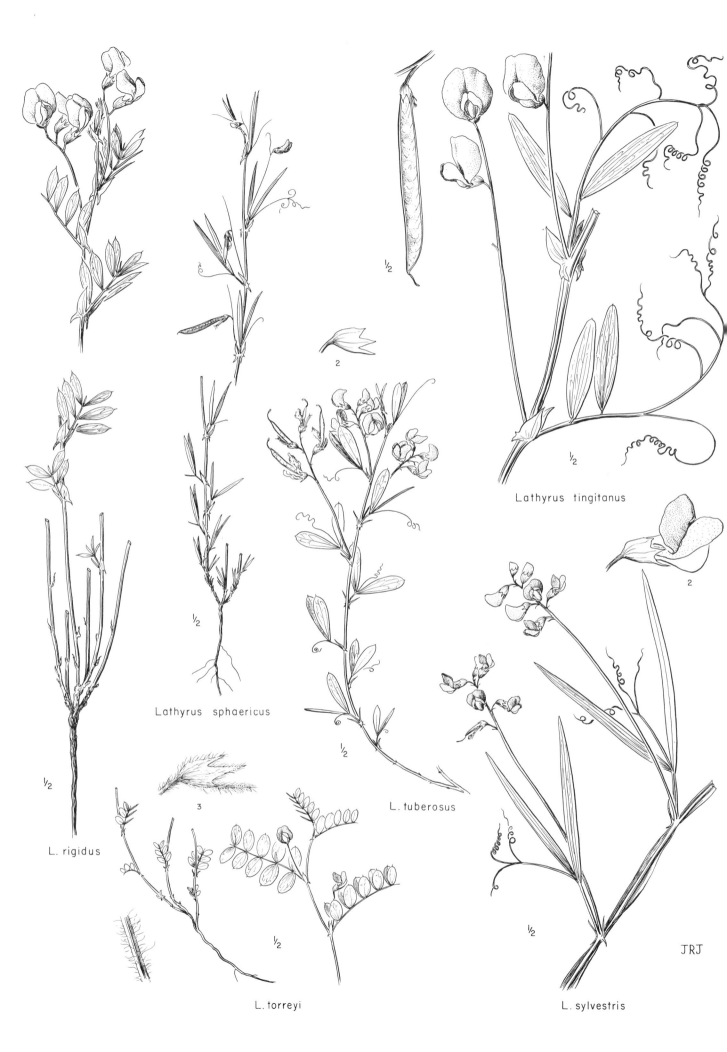

1/2

2

Lathyrus tingitanus

Lathyrus sphaericus

L. rigidus

3

L. tuberosus

L. torreyi

L. sylvestris

JRJ

————. The American Loti with special consideration of a proposed new section, Simpeteria.
  Brittonia 5:81-123. 1944.
1 Plant an annual; stipules reduced to blackish glands; flowers 1-3, sessile or short-
    pedunculate in the leaf axils
  2 Flowers subsessile in the leaf axils; rachis of the leaves flattened; pods 8-15 mm. long,
      3-4 mm. broad, not constricted between the seeds   L. DENTICULATUS
  2 Flowers pedunculate; rachis of the leaves not flattened; pods mostly well over 15 mm.
      long, usually less than 3 mm. broad, often noticeably constricted between the seeds
    3 Leaflets mostly 3, usually 3-10 (15) mm. broad; calyx teeth mostly well over 2 mm.
        long, exceeding the tube; pods only slightly or not at all constricted between the
        seeds; general over w. U.S.   L. PURSHIANUS
    3 Leaflets mostly 5, 1-4 mm. broad; calyx teeth usually less than 1 mm. long and
        considerably shorter than the tube; pods usually markedly constricted between
        the seeds   L. MICRANTHUS
1 Plant a perennial; stipules often well developed; flowers in axillary pedunculate umbels,
    usually at least 3 per umbel
  4 Leaves apparently sessile; leaflets 5, the lowest pair basal, petiolulate, similar to
      the other three, by their position resembling stipules, but true stipules represent-
      ed by tiny glands; flowers 3-15 in compact umbels   L. CORNICULATUS
  4 Leaves petiolate; leaflets 3-15, the lowest pair not basal on the petiole; stipules
      glandlike, or, if foliaceous, then dissimilar to the leaflets and not petiolulate
    5 Stipules glandlike, blackish; leaflets (3-4) 5; pods falcate, indehiscent, 1- to
        3-seeded   L. NEVADENSIS
    5 Stipules membranous, not glandlike or leafletlike; leaflets often 5-15; pods elon-
        gate, almost or quite straight, dehiscent, mostly 5- to 20-seeded
      6 Bract on the peduncle borne considerably below the umbel, usually compound
          (simple); pedicels 1-4 mm. long; pods usually over 3 mm. broad   L. CRASSIFOLIUS
      6 Bract of the peduncle (if any) just below the umbel, often simple; pedicels
          scarcely 1 mm. long; pods less than 3 mm. broad
        7 Bract usually 3 (1, 5, 7)-foliolate; banner yellow; wings pink to rose; keel
            purplish-tipped   L. FORMOSISSIMUS
        7 Bract (absent or) usually a mere tooth but rarely consisting of one leaflet; all
            petals essentially cream to yellowish, with little if any pink L. PINNATUS

Lotus corniculatus L. Sp. Pl. 775. 1753. (Europe)
  L. tenuis Waldst. & Kit. ex Willd. Enum. Hort. Berol. 797. 1809. (Hungary)
  L. macbridei A. Nels. Bot. Gaz. 53:221. 1912. (*Macbride 227*, near Falk's Store, Canyon Co., Ida.,
    June 7, 1910)
Perennial with numerous prostrate to ascending, usually trailing and often nodally rooting, glabrate
to pilose stems; stipules glandlike, the nearly sessile lowest pair of leaflets basal, distinctly short-
petiolulate, and nearly equal, in size and shape, to the 3 terminal leaflets which are elliptic to obovate, 5-17 mm. long, 2-7 mm. broad, and usually minutely serrulate and pilose-ciliate; leaf rachis
2-5 mm. long, rather definitely flattened; peduncles 3-12 cm. long, generally with a 3-foliolate (simple) bract just below the headlike umbel of 3-15 flowers; flowers 8-15 mm. long, yellow but mostly
tinged with red; calyx 5-8 mm. long, the narrowly linear teeth about equal to the tube; pod 20-40
mm. long, 2-3 mm. broad; seeds 10-25, brownish-black, about 1.5 mm. long. N=12. Bird's foot tre-
foil.
  A European escape, in bottom lands or wet places (including lawns), known from Ida. and from w.
Wash. and Oreg., and apparently slowly spreading; more common in e. U.S. May-Sept.

Lotus crassifolius (Benth.) Greene, Pitt. 2:147. 1890.
  Hosackia crassifolia Benth. Trans. Linn. Soc. 17:365. 1837. *Hosackia stolonifera* Lindl. Bot. Reg.
    23:pl. 1977. 1837. (*Douglas*, California)
  Hosackia platycarpa Nutt. in T. & G. Fl. N.Am. 1:323. 1838. (*Douglas*, '"mountain woods,' prob-
    ably the Blue Mountains of the Oregon") = var. *crassifolius*.
  Hosackia stolonifera var. *pubescens* Torr. Pac. R. R. Rep. 4:79. 1857. (*Bigelow*, Corte Madera,
    Marin Co., Calif.) = var. *crassifolius*.
LOTUS CRASSIFOLIUS var. SUBGLABER (Ottley) C. L. Hitchc. hoc loc. *Hosackia rosea* Eastw.

Proc. Calif. Acad. Sci. II, 6:424. 1896, not *L. roseus* Forskål in 1775. *L. stipularis* var. *sub-glaber* Ottley, U. Calif. Pub. Bot. 10:200. 1923. *L. aboriginus* Jeps. Fl. Calif. 2:315. 1936. *(Eastwood,* along rd. to Glen Blair, Ft. Bragg, Calif.)

Rhizomatous perennial; stems 2-10 dm. long, erect to spreading, fistulose, usually glabrous, sometimes very sparsely hairy above; leaves 5-15 cm. long, distinctly petiolate; stipules triangular, membranous, 4-10 mm. long; leaflets (7) 9-15, oblong to oblong-obovate, 1-3 (5) cm. long, bright green and glabrous on the upper surface, somewhat paler to distinctly glaucous and often strigillose on the lower surface; peduncles from shorter to considerably longer than the leaves, usually bracteate at (or somewhat above) midlength, the bract 3- to 5-foliolate or (less commonly) simple or even reduced to a mere membranous scale; umbels 7- to 20-flowered; pedicels 1-4 mm. long; flowers 8-13 mm. long, greenish-yellow or whitish but tinged with purple or reddish-purple; calyx 4-5 mm. long, the teeth triangular-lanceolate, about 1/4 the length of the tube; legume 2.5-4.5 cm. long, 3-4 mm. broad; seeds 4-10, dark brown, about 3.5 mm. long.

West of the Cascade Mts., except in the Columbia R. Gorge, from n.w. Wash. to s. Calif., from sea level to well up in the mountains, mostly in moist woods or along streams. May-July.

It is more or less conventional, following Ottley, to recognize a second species for our area, *L. stipularis* var. *subglaber* Ottley *(L. aboriginus* Jeps., *Hosackia rosea* Eastw.), almost solely on the basis that it has a whitish corolla tinged with red or purple, whereas *L. crassifolius* is said to have a greenish-yellow corolla that is often deeply purple-spotted. As there is no sharp contrast in our plants but rather transition from one phase to the other, it seems much more appropriate to recognize the two as varieties of *L. crassifolius,* as follows:

1 Corolla whitish, tinged with red or purple; n.w. Wash. southward, on the w. side of
  the Cascades, to s. Calif.   var. subglaber (Ottley) C. L. Hitchc.
1 Corolla greenish-yellow, often spotted with deep purple; s.w. Wash. to e. Wasco Co.,
  Oreg., southward to s. Calif.   var. crassifolius

Lotus denticulatus (Drew) Greene, Pitt. 2:139. 1890.
  Hosackia denticulata Drew, Bull. Torrey Club 16:151. 1889. *Anisolotus denticulatus* Heller, Muhl.
  7:139. 1912. *(Chestnut & Drew,* near Jarnigan's, along Mad R., Humboldt Co., Calif., July 11,
  1888)

Annual, mostly erect and sparingly branched, 1-5 dm. tall, finely appressed-pubescent or puberulent, but greenish; stipules glandlike; leaflets (2) 3 or 4, two borne at the tip of the distinctly flattened rachis, 1 or 2 lateral, elliptic to obovate, (5) 10-18 mm. long, 3-8 mm. broad; flowers single and subsessile in the leaf axils, 4-8 mm. long, cream, the banner purplish-tipped; calyx 4-5 mm. long, subglabrous to soft-pilose, the linear teeth nearly twice the length of the usually brownish-mottled tube; pod (8) 10-15 mm. long, 3-4 mm. broad, sparsely pubescent with soft appressed hairs, not constricted between the 2-4 seeds which are slate-colored, flattened, and 2.5-4 mm. long.

Southern B.C. to n. Calif., from the coast to the e. slopes of the Cascades, usually in sandy soil in the open. May-July.

Lotus humistratus Greene *(Hosackia brachycarpa* of Benth. but not of Hochst. ex Steud.) is a more prostrate, shorter, hairier plant with pods less than 1 cm. long, that occurs in s.w. Oreg. and has sometimes been mistaken for *L. denticulatus.*

Lotus formosissimus Greene, Pitt. 2:147. 1890.
  Hosackia gracilis Benth. Trans. Linn. Soc. 17:365. 1837. *L. gracilis* of Frye & Rigg, N.W. Flora
  234. 1912, but not of Salisb. in 1796. *(Douglas,* California)

Very similar to *L. pinnatus* but perhaps less lush in growth, the stems thinner and 2-5 dm. long, the plant more sprawling and essentially glabrous; leaflets commonly 5 (3-6); peduncles with a (usually) 3 (1-5, or even 7)-foliolate bract at the tip; flowers similar to those of *L. pinnatus* but the upper two calyx teeth usually not connate for more than half their length and the corolla much more conspicuously pinkish- to purplish-tinged, the keel purple-tipped.

Usually on moist soil, from near sea level into the lower mountains, from s.w. Wash. southward, on the w. side of the Cascades and in the coast ranges and along the coast, to Monterey Co., Calif. May-July.

Lotus micranthus Benth. Trans. Linn. Soc. 17:367. 1837. *(Douglas,* n.w. coast of America)
  Hosackia parviflora Benth. Bot. Reg. 15:pl. 1257. 1829. *Anisolotus parviflorus* Heller, Muhl. 3:
  100. 1907, not *L. parviflorus* Desf. *(Douglas,* N.W. coast of North America)

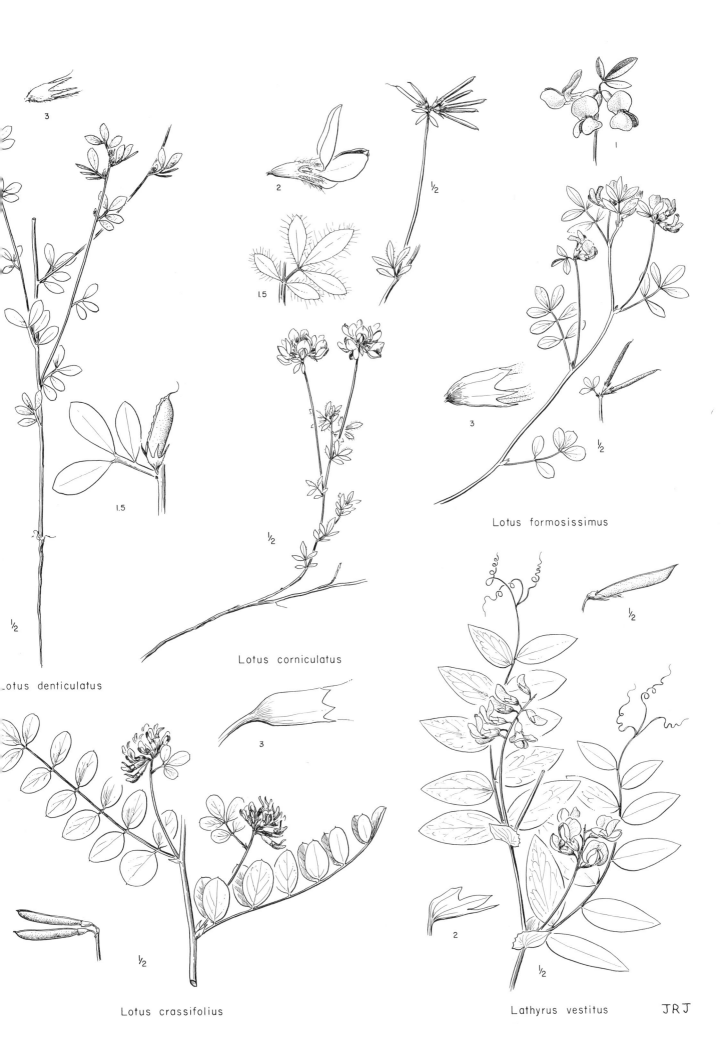

3

2

1.5

1/2

1.5

1/2

1/2

Lotus denticulatus

Lotus corniculatus

1/2

Lotus crassifolius

1

3

1/2

Lotus formosissimus

1/2

2

3

1/2

Lathyrus vestitus

JRJ

Hosackia microphylla Nutt. in T. & G. Fl. N.Am. 1:326. 1838, not *L. microphyllus* Hook. in Curtis'
   Bot. Mag. 55:pl. 2808. 1828. *(Nuttall,* Oregon)
   Glabrous to sparsely short-pubescent annual with prostrate to erect stems 1-3 dm. long; stipules
glandlike; leaflets 3-6, oblong to oblong-obovate, (2) 5-12 mm. long, (1) 2-3 (4) mm. broad; peduncles
axillary, 3-20 mm. long, with a single bract of (1) 2-3 leaflets just below the single flower; flowers
4-5 mm. long, pale yellow, usually reddish-tinged; calyx about 2 mm. long, the linear-lanceolate
lobes usually shorter than the tube; pod 15-30 mm. long, 1.5-2.5 mm. broad, noticeably constricted
between the 4-8 seeds.
   On open slopes and sandy flats, from the seashore into the mountains; w. of the Cascades from
B.C. to Calif. Apr.-Sept.

Lotus nevadensis (Wats.) Greene, Pitt. 2:149. 1890.
   Hosackia decumbens var. *nevadensis* Wats. Bot. Calif. 1:138. 1876. *(Lemmon,* Sierra Nevada from
      the Yosemite to Sierra Co.)
   Hosackia decumbens Benth. in Lindl. Bot. Reg. 15:pl. 1257. 1829, not *Lotus decumbens* Poir. in
      1813. *L. douglasii* Greene, Pitt. 2:149. 1890. *Anisolotus decumbens* Armst. & Thornb. West.
      Wild Fls. 244. 1915. *L. nevadensis* var. *douglasii* Ottley, Brittonia 5:81. 1944. *(Douglas,* N.W.
      coast of America)
   Hirsute-strigose to villous, usually grayish, matted perennial; stems many, slender, prostrate to
ascending, 1.5-5 dm. long; petioles mostly 2-4 mm. long; stipules glandlike, blackish; leaflets usually 5 (4 or 3), broadly elliptic to obovate, 5-15 mm. long, 2-7 mm. broad; rachis scarcely at all flattened, 5-10 mm. long; peduncles 5-30 mm. long, from shorter than the leaves to nearly twice their
length, naked or more commonly with a 3 (2 or 1)-foliolate bract at the base of the 3- to 12-flowered,
capitate umbel; flowers subsessile, 8-10 mm. long, yellow but often orange- or reddish-tinged; calyx 4-6 mm. long, the lobes linear, about equaling the tube; pod scarcely twice as long as the calyx,
falcate, indehiscent, 1- to 3-seeded.
   Very common, especially on rocky or sandy soil, from w. B.C. southward, on both sides of the
Cascades, to Calif., e. to Ida. and Nev. May-Sept.
   Our material, as described above, is referable to the var. *douglasii* (Greene) Ottley; it merges with
the var. *nevadensis* (which has flowers mostly only 4-8 mm. long and calyx teeth not much over half
the length of the tube) in n. Calif.

Lotus pinnatus Hook. in Curtis' Bot. Mag. 56:pl. 2913. 1829. *(Douglas,* Columbia River, between
   Fort Vancouver and Grand Rapids)
   Hosackia bicolor Dougl. ex Benth. in Lindl. Bot. Reg. 15:pl. 1257. 1829. *L. bicolor* Frye & Rigg,
      N.W. Flora 234. 1912. *(Douglas,* N.W. North America)
   Glabrous or very sparsely strigose perennial with a thick taproot and short rootstock; stems fistulose, (1) 1.5-6 dm. long, sprawling to erect; leaves 4-8 cm. long, petiolate; stipules membranous,
rounded to acuminate, 3-10 (15) mm. long; leaflets 5-9, elliptic or oblong to obovate, mostly 1-2
(2.5) cm. long; peduncles from shorter than the leaves to several times their length, ebracteate or
with a simple (usually lanceolate) membranous bract near the tip; umbels headlike, 3- to 12-flowered,
the pedicels scarcely 1 mm. long; flowers 10-15 mm. long; calyx tubular, 4-8 mm. long, the teeth
triangular-lanceolate, 1/5-1/3 as long as the tube, the upper two joined 3/4-4/5 their length and
merely notched between, the lower three acute-subulate; banner yellow; wings and keel with claw considerably exceeding the calyx, the wings cream, the keel yellow; pod 3-6 cm. long, 1.5-2 mm. broad;
seeds 8-20, about 1.5 mm. long.
   Moist soil, from n.w. Wash. to c. Calif., eastward, chiefly along streams and lakes, to Ida. May-
July.

Lotus purshianus (Benth.) Clements & Clements, Rocky Mt. Fls. 183. 1914.
   Hosackia purshiana Benth. in Lindl. Bot. Reg. 15:pl. 1257. 1829. *(Douglas,* N.W. coast of North
      America)
   L. sericeus of Pursh, Fl. Am. Sept. 489. 1814, but not of DC. in 1813. *Acmispon sericeum* Raf.
      New Fl. N. Am. 1:53. 1836. *Hosackia sericea* Trelease ex Bran. & Cov. in Geol. Surv. Ark. 4:
      174. 1888. ("On the banks of the Missouri")
   Trigonella americana Nutt. Gen. Pl. 2:120. 1818. *L. americanus* of Bisch. Linnaea 14 (App.):132.
      1840, but not of Vell. in 1825. *Hosackia americana* Piper, Contr. U.S. Nat. Herb. 11:366. 1906.
      (Nuttall, "On the dry and open alluvial soils of the Missouri, from the River Platte to the Mountains")

Hosackia unifoliolata Hook. Fl. Bor. Am. 1:135. 1833. *L. unifoliolatus* Benth. Trans. Linn. Soc.
17:368. 1837. *(Scouler,* Columbia River)
Hosackia elata Nutt. in T. & G. Fl. N.Am. 1:327. 1838. *Acmispon elatus* Rydb. Bull. Torrey Club
40:46. 1913. *(Nuttall,* "Gravelly bars of the Wahlamet & Oregon")
Hosackia elata var. *glabra* Nutt. in T. & G. Fl. N.Am. 1:327. 1838. *L. americanus* var. *minutiflorum* Ottley, U. Calif. Pub. Bot. 10:220. 1923. (Presumably collected by Nuttall, with *H. elata)*
Hosackia floribunda Nutt. in T. & G. Fl. N.Am. 1:327. 1838. *(Nuttall,* "Plains of the Rocky Mt.
range toward the Oregon")
Hosackia pilosa Nutt. in T. & G. Fl. N.Am. 1:327. 1838. *Acmispon pilosus* Heller, Muhl. 9:64.
1913. *(Nuttall,* "with the preceding," i.e. *H. floribunda)*
Hosackia mollis Nutt. in T. & G. Fl. N.Am. 1:327. 1838. *Acmispon mollis* Heller, Muhl. 9:62.
1913. *(Nuttall,* "Gravel bars and sandy shores of the Wahlamet, near the Falls")

Decumbent to erect, sparsely to strongly villous-pubescent annual; stems 1-6 dm. long; stipules
glandlike; petioles almost lacking; leaflets 3 (rarely 4 or 5, or on the upper leaves often only 1),
ovate-lanceolate to oblong-obovate, 6-30 mm. long, 3-10 (15) mm. broad; peduncles axillary, 5-30
mm. long, generally with a small, simple bract just below the single (rarely 2) flower; flowers 4-8
mm. long, pale yellow or cream, usually reddish-tinged; calyx 3-6 mm. long, the linear teeth from
only slightly longer than the tube to over twice its length; pod 15-35 mm. long, 2-2.5 mm. broad,
slightly constricted between the 4-8 seeds; seeds about 2 mm. long, only slightly flattened, brownish,
black-mottled. N=7.

General, chiefly in sandy or rocky, exposed to wooded areas on both sides of the Cascades, from
B.C. to Calif., e., through Mont., to the central states, s. to Mex. Apr.-Sept.

Although the species undergoes great variation in pubescence, in flower size, and in the number,
size, and shape of the leaflets, no distinctive populations meriting special nomenclatural status are
detectable.

Lupinus L. Lupine

Flowers racemose, verticillate to scattered, papilionaceous, the corolla white, yellowish, or pink
to blue or violet; calyx usually strongly bilabiate, the tube short, asymmetrical, the upper side from
only slightly bulged to definitely saccate or short-spurred, often with small, linear, usually deciduo-
ous bractlets below (less commonly above) the lateral sinuses; banner usually medianly grooved, the
sides generally turned back, from sharply to only slightly reflexed from the wings (purely for taxonomic convenience an arbitrary unit, the "banner index" is used in the following key: this index is found
by multiplying one unit, taken from the length of the wing exposed above the point where the banner
is reflexed, as measured in millimeters, by a second factor derived from the distance, also as measured in millimeters, between the tips of the banner and wing petals); wings usually connivent toward
their distal ends and more or less completely enclosing the keel; keel falcate, glabrous or more or
less ciliate along the upper margins; stamens 10, dimorphic, monadelphous; pods flattened, hairy,
(1) 2- to 12-seeded, commonly constricted between the smoothish, often mottled seeds; annual herbs
or herbaceous to suffrutescent perennials with palmately compound leaves of (3) 5-17 leaflets.

Probably 100 or more species, mostly N.Am., but represented on all continents except Australia.
(Latin *lupus,* wolf, the implication uncertain.)

Lupinus is a genus of economic importance, some forms being attractive cultivated perennials.
Among our native species *L. polyphyllus* (from which, by crosses with *L. arboreus,* our showy garden hybrids have been derived) is fairly easily grown in moist areas and desirable as a garden subject. Perhaps the choicest of the genus, at least for the rock gardener, are several of the varieties
of *L. lepidus,* especially the var. *lobbii,* which can be grown more successfully e. of the Cascades.
The seeds are readily eaten by many kinds of animals, although certain species are reputedly poisonous (especially the pods). One of our species has been introduced as a sandbinder along the coast, and
several are highly regarded as leguminous cover crops.

Taxonomically, the genus is probably in a more chaotic state than any other to be found in our area,
nearly 600 taxa having been described for the U.S. There are several reasons for this. For one thing
the species are extremely plastic, varying in a bewildering manner, and there seems to be little
doubt that specific boundaries are obscured by rather free interbreeding. Since lupines are mostly
rather showy, they attracted the attention of early collectors who obtained specimens of most of the
species, which were usually sent to Europe. These collections thus became the types from which the
earliest (usually inadequate) specific descriptions were drawn.

In this country, four or five taxonomists have been active in the description of novelties in *Lupinus*. The herbaria of two of the more prolific of these workers have not been readily available through loan, adding to the difficulty of study of the genus. C. P. Smith was, for nearly 50 years, regarded as the foremost specialist on the lupines. He obviously considered that the following characters (among others) were reliable for the foundation of species: 1) pubescence, whether appressed or spreading, and whether present or absent on the upper surface of the leaves, on the back of the banner, or on the wings; 2) the presence or absence, and the position, of cilia on the keel; 3) whether or not the stems branch; 4) the presence or absence of basal leaves and the relative length of the petioles; 5) the length of the racemes and the length of the pedicels; 6) the size of the spur or sac of the calyx; 7) the size of the flowers; 8) the amount of reflexion of the banner. Some of these features are known to behave as Mendelian characters in other genera, and they vary so greatly in *Lupinus* that it is almost certain that they cannot, except in combination, be used for the recognition of taxa. The possible variations of these and other similar characters are reflected in the extremely large number of taxa which have been described. It is believed that the number of species recognized in the present treatment is approximately all that can be maintained by a species concept comparable to that used for other genera of the Leguminosae. That the correct name has been selected in every case is doubtful, but from the proposed synonymy it can at least be seen with what the taxa correspond in other floras.

Three workers recently have made notable attempts to clarify the taxonomy of *Lupinus* and the treatment that follows has been based largely on their papers.

References:

Detling, L. E. The caespitose lupines of western North America. Am. Midl. Nat. 45:474-499. 1951.

Dunn, David B. Leguminosae of Nevada, II—Lupinus. Contributions toward a flora of Nevada, no. 39, Bureau of Plant Industry, U.S.D.A., Wash., D.C. 1956.

————. A revision of the Lupinus arbustus complex of the Laxiflori. El Aliso 14:54-73. 1957.

————. Lupinus pusillus and its relationship. Am. Midl. Nat. 62:500-510. 1959.

Phillips, Lyle L. A revision of the perennial species of Lupinus of North America. Res. Stud. State Coll. Wash. 23:161-201. 1955.

1 Plant an annual
  2 Seeds and ovules usually 2 but sometimes only 1
    3 Keel ciliate on the lower half; racemes verticillate, their peduncles about equaling the petioles of the leaves; flowers about 1.5 cm. long   L. MICROCARPUS
    3 Keel glabrous; racemes not verticillate, their peduncles shorter than the petioles; flowers about 1 cm. long   L. PUSILLUS
  2 Seeds and ovules usually 4 or more, rarely only 3
    4 Flowers 5-7 mm. long, the banner subequal to the wings but scarcely at all reflexed from them; keel blunt and not much upturned; pedicels mostly only 1-2 mm. long and rather stout   L. MICRANTHUS
    4 Flowers usually over 7 mm. long, but sometimes shorter; banner often much shorter than the wings and markedly reflexed from them; keel slender-tipped and usually rather sharply upturned; pedicels 2-5 (6) mm. long
      5 Flower (9) 10-13 mm. long; leaflets broadly oblanceolate; banner usually as broad as long, not markedly shorter than the keel   L. AFFINIS
      5 Flower (6) 7-9 (10) mm. long; leaflets narrowly oblanceolate to linear-oblanceolate; banner often longer than broad, frequently shorter than the keel   L. BICOLOR
1 Plant a perennial
  6 Either the banner conspicuously hairy over much of the abaxial (back) surface, or the calyx spurred (rarely the banner glabrous in *L. leucophyllus*, a species with much-contracted, almost spicate racemes)
    7 Calyx not spurred; either the pubescence of the banner extending to the upper third, or the flowers less than 9 mm. long, or the wing petals not pubescent on the upper half
      8 Flowers 10-18 mm. long, borne in rather loose racemes; banner pubescent to the upper third, well reflexed, the index (see generic description) averaging at least 15; pedicels 4-11 mm. long
        9 Flowers yellow (sometimes tinged with purple); banner very sparsely hairy; leaflets mostly over 6 cm. long and over 1 cm. broad   L. SABINII
        9 Flowers usually blue, but if yellow or yellowish the banner copiously hairy, or leaflets averaging less than 6 cm. long or less than 1 cm. broad

10 Keel slender, very strongly curved, narrowed rather gradually to the claw, the
    tip usually exposed; racemes very loose, usually long-pedunculate, the lowest
    flower well above all the leaves; flowers 12-16 mm. long, the banner usually
    yellow-centered; Willamette Valley, Oreg., southward   L. ALBIFRONS
10 Keel broader and not so strongly curved, abruptly contracted to the claw, the tip
    seldom exposed; racemes not especially loose, mostly short-pedunculate, the
    lowest flowers commonly equaled by some of the leaves; flowers generally 10-
    12 mm. long, the banner usually white-centered; e. of the Cascades   L. SERICEUS
8 Flowers either less than 10 mm. long or borne in compact racemes, in either case
   the banner not much reflexed (the index averaging 10 or less) and often not pubescent on the upper half; pedicels 1-4 mm. long
    11 Banner pubescent to the upper third; racemes densely flowered, mostly 15-20
       cm. long; flowers (7) 8-12 mm. long   L. LEUCOPHYLLUS
    11 Banner usually not pubescent to the upper third; racemes interrupted at least
       below, 5-15 cm. long; flowers 5-7 (8) mm. long   L. HOLOSERICEUS
7 Calyx spurred; pubescence of the banner usually not extending to the upper third; wing
   petal often hairy on the upper half; flowers at least 9 mm. long
   12 Wings pubescent near the upper tip; upper calyx lip 1/5-1/4 as long as the wings
              L. LAXIFLORUS
   12 Wings glabrous; upper calyx lip 1/3-3/4 as long as the wings   L. CAUDATUS
6 Neither the banner conspicuously hairy on the back nor the calyx spurred; pubescence of
  the banner, if any, largely concealed by the calyx, sometimes extending above in a
  line (in *L. sulphureus*)
  13 Banner slightly or moderately reflexed from the wings and keel, consequently (or
    because of the small-sized flowers) the banner index only 2-10; plants caulescent,
    the stems rarely less than 1 (and usually over 2) dm. long, exceeding the lower
    petioles
    14 Basal leaves usually present at flowering time, their petioles 3-5 times as long
      as the leaflets, conspicuously longer than the petioles of the cauline leaves
              L. SULPHUREUS
    14 Basal leaves usually absent at flowering time, if present their petioles, as those
      of the cauline leaves, less than 3 times as long as the leaflets   L. ARGENTEUS
  13 Banner mostly so well reflexed (or the flowers so large) that the banner index aver-
    ages well over 10 (usually at least 15), sometimes the index less than 15, but then
    the plants subacaulescent, with stems rarely over 1 dm. long and usually shorter
    than the longer petioles
    15 Flowers usually yellow (varying to bluish or purplish, or very rarely white),
      13-18 mm. long; plants either shrubby, mostly 1-2 m. tall, and from along the
      coast (especially on dunes), or plants of the Blue Mts. of s.e. Wash. and n.e.
      Oreg.
      16 Plants shrubby, coastal   L. ARBOREUS
      16 Plants herbaceous, local in the Blue Mts. of s.e. Wash. and n.e. Oreg.   L. SABINII
    15 Flowers almost never yellow, often less than 15 mm. long; plants not shrubby,
      when (as rarely) on coastal dunes then usually prostrate
      17 Plants of the immediate coast, prostrate; stems and petioles usually hirsute
        with rust-colored hairs 2-5 mm. long   L. LITTORALIS
      17 Plants not strictly coastal; stems and petioles seldom hirsute with rust-colored
       hairs as much as 2-4 mm. long
        18 Stems usually branched, the branches ending in racemes (lateral racemes
          often rudimentary at the time the terminal raceme starts to flower); leaves
          of the main stem usually averaging (3-7) 8 or more; plants (except some of
          those on the n. side of the Columbia R.) greenish, glabrous to moderately
          pubescent, the leaflets often glabrous on the upper surface; pedicels usually
          at least 5 mm. long; rarely e. of the Cascade Mts., except in the Columbia
          R. Gorge
         19 Upper calyx lip similar to the lower although shorter, usually at least 3
            times as long as the calyx tube; plants not villous; lowermost leaflets
            deciduous by the time the plant flowers

20 Keel ciliate, scarcely exposed; wings broad, nearly completely covering the keel;
     margins of the banner conspicuously turned back   L. RIVULARIS
20 Keel nonciliate, the tip exposed; wings slender, not completely covering the keel;
     margins of the banner turned back only slightly   L. ALBICAULIS
  19 Upper calyx lip strikingly broader than the lower, usually not over 2.5 times as long
       as the calyx tube; plants sometimes villous; lowermost leaves sometimes remaining
       until anthesis   L. LATIFOLIUS
18 Stems unbranched or if (as very rarely) branched, then the lateral stems not producing
     racemes; leaves of the main stem. usually less than 8; plants often grayish; pedicels
     sometimes much less than 5 mm. long; more general, often e. of the Cascade Mts.
  21 Plants low and spreading, usually less than 3 dm. tall, the leafy stems generally
       shorter than the scapes and racemes; upper calyx lip bidentate for 1/2 of the length
       or more   L. LEPIDUS
  21 Plants usually well over 3 dm. tall, mostly with leafy stems longer than the pe-
       duncles and equal to the longest basal petioles; upper calyx lip subentire to deep-
       ly bidentate
    22 Upper lip of the calyx cleft at least 1/3 of its length; leaflets oblanceolate, round-
         ed to abruptly acute, 1.5-3 (4) cm. long; racemes mostly 5-10 cm. long at an-
         thesis (to 15 cm. in fruit); plants usually brownish-villous to strigose on the
         stems and leaves   L. SAXOSUS
    22 Upper lip of the calyx cleft less than 1/3 of its length; leaflets various but usu-
         ally either acute (to acuminate) or over 4 cm. long; racemes mostly over 10
         cm. long, even at anthesis; plant generally not brownish-villous
      23 Leaflets pubescent (usually strigose) on the upper surface, mostly abruptly
           acuminate to apiculate, from linear-oblanceolate to elliptic-oblanceolate,
           2-4 (7) cm. long, 2-6 (14) mm. broad; stems only slightly or not at all
           fistulose   L. WYETHII
      23 Leaflets glabrous on the upper surface, mostly rounded or acute, elliptic-
           oblanceolate, (3) 4-10 (15) cm. long, 10-25 mm. broad; stems often strong-
           ly fistulose   L. POLYPHYLLUS

Lupinus affinis Agardh, Syn. Gen. Lup. 20. 1835. *(Douglas,* California)
  L. carnosulus Greene, Bull. Calif. Acad. Sci. II, 6:144. 1886. *L. affinis* var. *carnosulus* Jeps. Fl.
  W. Middle Calif. 371. 1901. *L. nanus* var. *carnosulus* C. P. Smith, Bull. Torrey Club 50:165.
  1923. *(Greene,* near Olema, Marin Co., Calif.)
  Erect annual, simple or branched from the base, 1-3 (5) dm. tall, strongly hirsute to subvillous or
appressed-hirsutulous; leaflets 5-8, oblanceolate, 1.5-2.5 (5) cm. long, usually more or less appressed-hirsute on both surfaces, 1/3-4/5 as long as the petioles; racemes long-pedunculate and exceeding the leaves; flowers in 2-5 (7) definite, well-separated verticils; pedicels slender, (2) 4-5 (6)
mm. long; calyx deeply bilabiate, the upper lip much the broader, deeply bilobed, subequal to the narrower, entire to lightly toothed lower lip; corolla 9-13 mm. long, bluish; banner suborbicular to
broader than long, reflexed at nearly a right angle from the keel, white- to purplish- centered; keel ciliate much of the length, and with a short tooth about midlength of the upper margin, the slender,
sharply upcurved tip bluish to black; pods 2.5-3.5 cm. long, 4-6 mm. broad; seeds 5-10, about 5 mm.
long, mottled.
  Interior valleys of Oreg., w. of the Cascades, from the s. part of the Willamette Valley southward
to Calif. Apr.-June.

Lupinus albicaulis Dougl. ex Hook. Fl. Bor. Am. 1:165. 1833. *(Douglas,* "About Fort Vancouver, on
  the Columbia")
  L. andersoni Wats. Bot. King Exp. 58. 1871. *(Anderson 9,* Carson City, Nev.)
  L. lignipes Heller, Muhl. 8:66. 1912. *(Heller 10042,* near Eugene, Lane Co., Oreg., May 18, 1910)
  Very similar to *L. rivularis,* the usually several stems ascending, generally tardily branching, up
to 10 dm. long, mostly strigillose; racemes generally 1-1.5 (up to 2.5) dm. long; flowers scattered
to subverticillate, not crowded, bluish to varicolored, fading to purplish-brown; pedicels (4) 5-8 (9)
mm. long; calyx thickly pubescent with more or less spreading hairs, strongly bilabiate, the upper
lip subequal to the (entire) lower one, usually at least 3 times as long as the tube, shallowly bidentate;
calyx tube asymmetrical at base but not saccate or spurred; banner glabrous, shorter than the keel

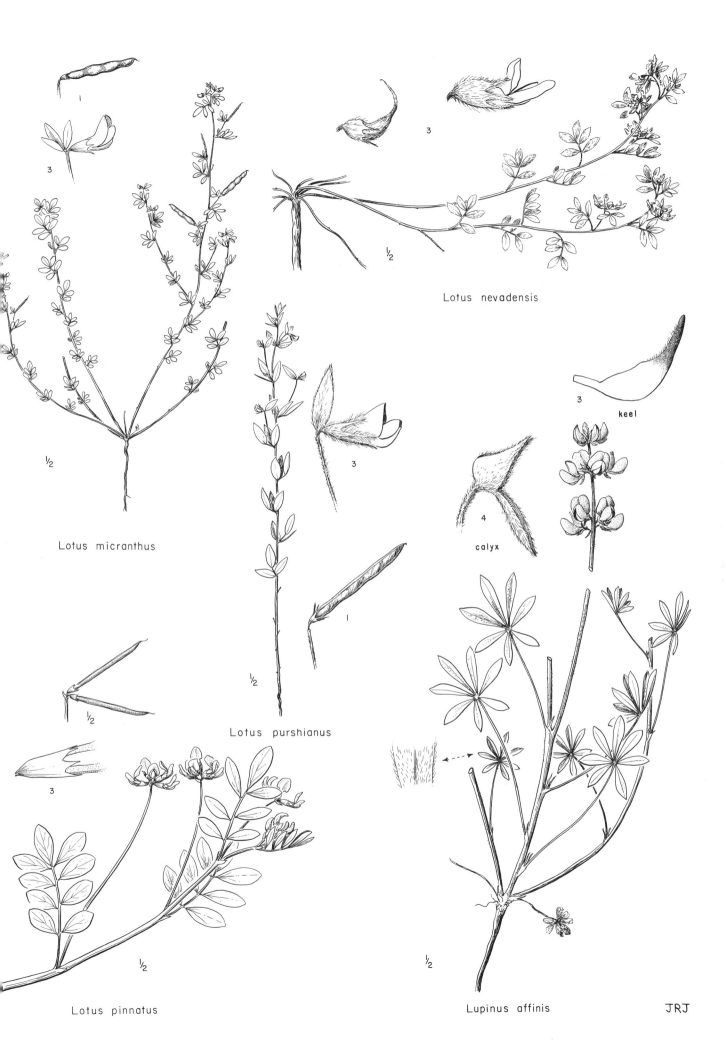

3

1

3

Lotus nevadensis

1/2

3

keel

4

calyx

1/2

Lotus micranthus

3

1

1/2

Lotus purshianus

1/2

3

1/2

Lotus pinnatus

1/2

Lupinus affinis

JRJ

and not very much reflexed from it (index averaging about 20), the edges not strongly reflexed; wings glabrous, narrow, usually not covering either the bottom edge or the tip of the strongly curved, non-ciliate keel; pods 3-4 cm. long, 8-10 mm. thick, hairy; seeds 6-8, about 4 mm. long, grayish, dark mottled. N=24.

Lowlands of the Puget Trough in w. Wash., southward, on the w. side of the Cascades (and at increasing elevation in the mountains), to Calif. and w. Nev., where the species is much more variable than in our area. June-July.

Lupinus albicaulis, *L. rivularis*, and *L. latifolius* are fairly closely related, somewhat intergradient taxa that might be treated as elements of one specific complex. However, they are as distinctive as others of our species and no more intergradient than some. *L. albicaulis* is distinguished chiefly by the largely exposed, glabrous keel.

Lupinus albifrons Benth. ex Lindl. Bot. Reg. 19:pl. 1642. 1833. (Described from a cultivated plant derived from seed collected in "California" by Douglas)

    L. eminens Greene, Erythea 1:125. 1893. *L. albifrons* var. *eminens* C. P. Smith in Jeps. Man. Fl. Pl. Calif. 531. 1925. *(G. W. Dunn,* Santa Inez Mts., Santa Barbara Co., Calif., June, 1891)

    L. albifrons var. *flumineus* C. P. Smith, Bull. Torrey Club 51:306. 1924. *(J. C. Nelson 741,* Willamette R., above Salem, Marion Co., Oreg., June 24, 1916)

    ? *L. johannis-howellii* C. P. Smith, Sp. Lup. 147. 1940. *(J. T. Howell 7157,* Elk Lake rd., 20 mi. w. of Bend, Deschutes Co., Oreg., July 2, 1931) Identity uncertain, possibly referable to *L. sericeus* rather than to *L. albifrons*.

Stems usually several from a slightly woody perennial base, 2-5 dm. tall, generally rather freely branched, the plant sericeous; foliage nearly all cauline; petioles of the lower leaves 3-4 times as long as the blades, becoming much shortened upward; leaflets 6-9, somewhat narrowly oblanceolate, 1.5-3 cm. long, abruptly apiculate, rather copiously grayish-sericeous on both surfaces; peduncles 5-15 cm. long, the racemes raised well above the leaves, 1-2 dm. long, very loose; flowers 12-16 mm. long, mostly verticillate, usually considerably shorter than the internodes; pedicels 3-8 mm. long; calyx sericeous, not at all saccate or spurred, the upper lip bidentate, the lower one entire; banner bluish, the center whitish or yellowish, darkening later to violet, rather copiously pubescent on the back, well reflexed from the keel (index 30-50); wings glabrous, bluish or violet; keel ciliate, conspicuously curved, slender, narrowed gradually to a claw, the purplish tip usually exposed in pressed specimens; pods 3-5 cm. long; seeds 4-9, about 4 mm. long, darkly mottled.

Southern Willamette Valley, Oreg,, southward; common throughout the foothills of Calif. Apr.-June.

In California, the species undergoes considerable variation, but our material, as described above, is all referable to var. *flumineus* C. P. Smith.

Lupinus arboreus Sims in Curtis' Bot. Mag. 17:682. 1803. (Described from a garden plant at Kensington, England, from seed of unknown origin, but surely from Calif.)

Diffusely branched perennial, suffrutescent to shrubby, mostly 1-2 m. tall, from deep heavy roots, pubescent to sericeous; leaflets 5-11, oblanceolate, 3-6 (8) cm. long, puberulent to sericeous; racemes usually numerous, loose, 1-2.5 dm. long; flowers imperfectly verticillate, 14-18 mm. long, yellow (white or bluish, the banner often somewhat purplish); pedicels 4-9 (12) mm. long; calyx heavily puberulent, the upper lip shallowly bidentate, the lower entire; banner well reflexed (index 50-100), glabrous; keel falcate, ciliate most of the length; pods (3) 4-6 (8) cm. long, about 1 cm. broad, hairy; seeds 8-12, about 5 mm. long, dark brown. N=20, 24.

Naturally occurring along the coast of Calif., but introduced in our range (probably usually as a sand binder) and now established, at least along the waterfront, near Port Townsend, at Blyn, and on Whidbey I., in Wash., and here and there along the coast of Oreg. May-Sept.

Lupinus argenteus Pursh, Fl. Am. Sept. 468. 1814. *(Lewis,* "on the banks of the Kooskoosky")

    L. decumbens Torr. Ann. Lyc. N.Y. 2:191. 1828. *L. argenteus* var. *decumbens* Wats. Proc. Am. Acad. 8:532. 1873. *(Torrey,* "on the southern branches of the Arkansas") = var. *argenteus*.

    L. tenellus Dougl. ex G. Don, Gen. Syst. 2:367. 1835. *L. laxiflorus* var. *tenellus* T. & G. Fl. N. Am. 1:377. 1840. *L. argenteus* ssp. *argenteus* var. *tenellus* Dunn, Leafl. West. Bot. 7:254. 1955. *(Douglas,* "Oregon from Fort Vancouver to the Rocky Mountains") = var. *argenteus*.

    L. laxiflorus var. *foliosus* Nutt. ex T. & G. Fl. N.Am. 1:377. 1840. *(Nuttall,* "Oregon plains") = var. *argenteus*.

    LUPINUS ARGENTEUS var. PARVIFLORUS (Nutt.) C. L. Hitchc. hoc loc. *L. parviflorus* Nutt. ex

calyx

3

2.5

keel

2.5

3

3

3

Lupinus arboreus

½

½

½

½

banner

2

½

L. albifrons

L. albicaulis

JRJ

Hook. & Arn. Bot. Beechey Voy. 336. 1840. *L. argenteus* ssp. *parviflorus* Phillips, Res. Stud. State Coll. Wash. 23:190. 1955. *(Nuttall,* "plains of the Rocky Mountains towards the Oregon")

L. foliosus var. *stenophyllus* Nutt. ex T. & G. Fl. N.Am. 1:377. 1840, as a synonym. *L. stenophyllus* Rydb. Bull. Torrey Club 34:42. 1907. *L. argenteus* var. *stenophyllus* incorrectly attributed to Rydb. by Davis, Fl. Ida. 439. 1952. *(Nuttall,* "Oregon plains")

L. floribundus Greene, Proc. Am. Acad. 28:364. 1893. *(Greene,* Middle and Upper Bear Cr., w. of Denver, Colo.) = var. *parviflorus.*

L. alpestris A. Nels. Bull. Torrey Club 26:127. 1899. *(E. Nelson 5070,* La Plata Mines, Wyo., Aug. 22, 1898) = var. *argenteus.*

L. myrianthus Greene, Pitt. 4:134. 1900. *(Greene,* Gunnison, Gunnison Co., Colo., Sept. 1, 1896) = var. *argenteus.*

L. alsophilus Greene, Pitt. 4:135. 1900. *(Greene,* Cimarron, Colo., Aug. 30, 1896) = var. *parviflorus.*

L. monticola Rydb. Mem. N.Y. Bot. Gard. 1:232. 1900. *(Rydberg & Bessey 4447,* Indian Cr., Mont., July 22, 1897) = var. *argenteus.*

L. rubricaulis Greene, Pl. Baker. 3:35. 1901. *L. caudatus* var. *rubricaulis* C. P. Smith, Contr. Dudley Herb. 1:29. 1927. *(C. F. Baker 342,* Crested Butte, Colo., July 6, 1901) = var. *argenteus.*

L. leptostachyus Greene, Pl. Baker. 3:36. 1901. *(Greene 182,* Sapinero, Gunnison Co., Colo.) = var. *stenophyllus.*

L. spathulatus Rydb. Bull. Torrey Club 29:244. 1902. *(S. Watson 225,* Wasatch Mts., in 1869) = var. *argenteus.*

L. lucidulus Rydb. Bull. Torrey Club 29:245. 1902. *(F. Tweedy 271,* Spread Cr., Wyo., in 1897) = var. *argenteus.*

LUPINUS ARGENTEUS var. DEPRESSUS (Rydb.) C. L. Hitchc. hoc loc. *L. depressus* Rydb. Bull. Torrey Club 30:255. 1903. *(J. B. Leiberg 1201,* divide between St. Joe and Clearwater rivers, Ida., in 1895)

L. evermannii Rydb. Bull. Torrey Club 30:255. 1903. *(B. W. Evermann 533,* near Sawtooth, Ida., in 1896) = var. *depressus.*

L. adscendens Rydb. Bull. Torrey Club 30:256. 1903. *(F. Tweedy 129,* headwaters of Tongue R., Wyo., in 1898) = var. *argenteus.*

L. maculatus Rydb. Bull. Torrey Club 30:257. 1903. *(M. E. Jones,* P. V. Junction, Wasatch Mts., Utah, in 1883) = var. *argenteus.*

L. laxus Rydb. Bull. Torrey Club 30:258. 1903. *(Rydberg & Bessey 4442,* Forks of the Madison, Mont., in 1897) = var. *argenteus.*

L. pulcherrimus Rydb. Bull. Torrey Club 30:258. 1903. *(F. Tweedy 4215,* Battle, Continental Divide, Wyo., in 1901) = var. *argenteus.*

L. jonesii Blank. Mont. Agr. Coll. Stud. Bot. 1:79. 1905. *(Blankinship,* Monida, Mont., June 26, 1902) = var. *argenteus.*

L. corymbosus Heller, Muhl. 2:69. 1905. *L. laxiflorus* var. *corymbosus* Jeps. Fl. Calif. 2:264. 1936. *(Heller 8015,* Montague, Siskiyou Co., Calif., June 9, 1905) = var. *argenteus.*

L. macounii Rydb. Bull. Torrey Club 34:42. 1907. *L. argenteus* var. *macounii* Davis, Madroño 11:144. 1951. *(J. Macoun 4070,* Cypress Hills, Sask., Can., in 1894) = var. *argenteus.*

L. roseolus Rydb. Bull. Torrey Club 34:44. 1907. *(F. Tweedy 270,* Continental Divide, Buffalo Fork, Wyo., Aug., 1897) = var. *depressus.*

L. lanatocarinus C. P. Smith, Sp. Lup. 317. 1942. *(Davis 137-35,* East of Fort Hull, Bingham Co., Ida., July 5, 1935) = var. *argenteus* toward *L. caudatus.*

L. alturasensis C. P. Smith, Sp. Lup. 318. 1942. *(R. J. Davis 464,* Alturas Lake, n.w. Blaine Co., Ida., June 29, 1938) = var. *depressus.*

L. davisianus C. P. Smith, Sp. Lup. 319. 1942. *(R. J. Davis 91-36,* Railroad Bridge, Custer Co., Ida., July 23, 1938) = var. *argenteus.*

L. fremontensis C. P. Smith, Sp. Lup. 320. 1942. *(Davis 326,* Sandhills, Fremont Co., Ida., June 18, 1938) = var. *argenteus.*

L. sparhawkianus C. P. Smith, Sp. Lup. 563. 1946. *(Martineau & Sparhawk 79,* Payette Forest, Valley Co., Ida., Aug. 11, 1912) = var. *depressus.*

L. christianus C. P. Smith, Sp. Lup. 564. 1946. *(J. H. Christ 2614,* Indian Lake, Ida., Oct. 11, 1933) = var. *depressus.*

L. minearanus C. P. Smith, Sp. Lup. 565. 1946. *(J. E. Minear 19,* Sawtooth Forest, Camas Co., Ida., July 7, 1929) = var. *argenteus.*

L. varneranus C. P. Smith, Sp. Lup. 566. 1946. *(I. M. Varner 82,* Boulder Lake, Idaho Forest,
Idaho Co., Ida., July 21, 1914) = var. *stenophyllus.*
L. lacuum-trinitatum C. P. Smith, Sp. Lup. 567. 1946. *(Hitchcock & Muhlick 10332,* Trinity Lakes,
10 mi. w. of Featherville, Elmore Co., Ida., July 25, 1944) = var. *depressus.*
L. serradentum C. P. Smith, Sp. Lup. 568. 1946. *(L. F. Henderson 3576,* 6 mi. above Sawtooth Di-
vide, Custer Co., Ida., July 26, 1895) = var. *depressus.*
L. capitis-amnicoli C. P. Smith, Sp. Lup. 569. 1946. *(J. W. Thompson 14141,* in part, head of
Boulder Cr., Sawtooth Mts., Blaine Co., Ida., Aug. 6, 1937) = var. *argenteus.*
L. equi-coeli C. P. Smith, Sp. Lup. 570. 1946. *(A. & R. Nelson 2983,* Horse Heaven, Seven Devils
Mts., Idaho Co., Ida., Aug. 10-15, 1938) = var. *argenteus.*
L. seclusus C. P. Smith, Sp. Lup. 570. 1946. *(H. J. Rust 885,* Burnside ranch, Clark Co., Ida.,
Aug. 1, 1916) = var. *argenteus.*
L. perplexus C. P. Smith, Sp. Lup. 571. 1946. *(J. W. Thompson 14141,* in part, Boulder Canyon,
Sawtooth Mts., Blaine Co., Ida., Aug. 6, 1937) = var. *argenteus.*
L. montis-liberatatis C. P. Smith, Sp. Lup. 571. 1946. *(J. H. Christ 14486,* Independence Mt.,
Minidoka Co., Ida., July 16, 1944) = var. *argenteus.*
L. summae C. P. Smith, Sp. Lup. 572. 1946. *(Christ & Ward,* Silver City rd. summit, Owyhee Co.,
Ida., June 28, 1937) = var. *argenteus.*
L. edward-palmeri C. P. Smith, Sp. Lup. 572. 1946. *(Edw. Palmer 558,* Big Butte Station, Bing-
ham Co., Ida., June 23, 1893) = var. *argenteus.*
L. hullianus C. P. Smith, Sp. Lup. 573. 1946. *(A. C. Hull 235,* Clark Co., Ida.) = var. *argenteus.*
L. cariciformes C. P. Smith, Sp. Lup. 573. 1946. *(Christ & Ward 14899,* 4 mi. s. of Macks Inn,
Fremont Co., Ida.) = var. *argenteus.*
L. montis-cookii C. P. Smith, Sp. Lup. 726. 1952. *(Sutton 70,* Cook Mt., Clearwater Forest, Ida.)
= var. *argenteus.*
L. achilleaphilus C. P. Smith, Sp. Lup. 730. 1952. *(J. W. West,* Paddy Flat Ranger Station, Idaho
Forest, Valley Co., Ida., July 23, 1930) = var. *stenophyllus.*
L. alicanescens C. P. Smith, Sp. Lup. 731. 1952. *(I. M. Varner 44,* Idaho Forest, Valley Co., Ida.,
July 11, 1914) = var. *stenophyllus.*
L. aliumbellatus C. P. Smith, Sp. Lup. 731. 1952. *(I. M. Varner 66,* Idaho Forest, Valley Co.,
Ida., July 20, 1914) = var. *argenteus.*
L. spathulatus var. *boreus* C. P. Smith, Sp. Lup. 746. 1952. ("*Herbarium 70,*"collector not known,
Targhee Forest, Fremont Co., Ida., July 19, 1913) = var. *argenteus.*

Perennial from a well-branched crown, (1) 1.5-4 (6) dm. tall; stems several, simple to freely
branched, more or less densely strigillose to subsilky, often grayish; leaves nearly all cauline, the
petioles from not quite so long to (the uppermost) as much as twice as long as the blades; leaflets (5)
6-9 (10), narrowly lanceolate to narrowly or rather broadly oblanceolate, mostly 2-3 (4) cm. long,
from glabrous on the upper surface and moderately strigillose on the lower, to almost equally strigillose or sericeous on both surfaces; racemes mostly 1-1.5 dm. long, rather slender, loose to tight,
the flowers subverticillate, (4) 5-10 (12) mm. long, from nearly white with bluish shading to (usually)
light or dark blue; pedicels (1) 3-6 (7) mm. long; calyx sericeous, the base sometimes subsaccate but
not spurred, the upper lip bidentate, the lower entire; banner usually whitish-centered and glabrous
(very rarely pubescent) on the back, fairly well reflexed, but the index only (2) 4-9 (10); wings glabrous; keel usually ciliate, less commonly glabrous, only slightly or not at all exserted; pods 1.5-2.5
cm. long, very hairy; seeds 2-5, about 4 mm. long, pinkish-gray to light brown. N=24.

From ponderosa pine forest to subalpine ridges; e.c. Oreg. southward to n.e. Calif., e. to Alta.,
Mont., and S.D., and s. to N.M., common in the Great Basin. May-July.

Ordinarily *L. argenteus* is fairly clearly separable from *L. caudatus* because of its glabrous banner
and only subsaccate (rather than spurred) calyx, but transition between the two is complete, especially in s. Ida. Some specimens have hairy banners and only slightly saccate calyces, some glabrous or
subglabrous banners and saccate calyces (both keyable to *L. argenteus),* whereas others have enough
pubescence on the banner and a sufficiently saccate calyx that they will key to *L. caudatus.*

The species varies greatly in height, pubescence, and flower size. The stature of the plant appears
to depend largely on elevation and other ecologic factors; variation in pubescence may also depend
upon ecological conditions, but appears to be largely fortuitous, and the same seems to be true of
flower size. There are, however, a few fairly clearly marked tendencies in the species as it occurs
in our range. Occasional plants are much less pubescent than others, those that have the upper surface of the leaves glabrous are often smaller flowered and are largely referable either to the var.

stenophyllus (Rydb.) Davis or to the much broader leafleted var. *parviflorus* (Nutt.) C. L. Hitchc.,
which ranges mostly from s.e. Ida. to w. Wyo. and southward. Variety *argenteus* undergoes much
variation in the pubescence of the leaves but has flowers (7) 9-10 (11) mm. long. In c. Ida. and in
Mont. and Wyo. there is a subalpine, usually grayish-hairy form of the plant, averaging perhaps 1.5-
2.5 dm. tall, with congested racemes, the flowers 7-10 mm. long. Although completely intergradient
with var. *argenteus*, it seems to constitute an ecologic race, var. *depressus* (Rydb.) C. L. Hitchc.
A somewhat larger flowered, less pubescent plant of e. Oreg. and southward *(L. corymbosus* Heller)
seems to parallel var. *stenophyllus* but to be less distinctive and therefore not maintainable.

Our varieties are imperfectly separable as follows:

1 Flowers mostly (4) 5-7 (9) mm. long; leaves usually glabrous on the upper surface
  2 Leaflets of basal leaves mostly oblanceolate to almost obovate, rounded to obtuse;
      flowers (4) 5-6 mm. long; s.e. Ida. to Wyo. and s. to Colo. and Utah, in our area
      only along the Wyo. border of Ida. and possibly in s. Mont.
                                        var. parviflorus (Nutt.) C. L. Hitchc.
  2 Leaflets of all leaves narrowly lanceolate, acute or acuminate; flowers (5) 6-7 (9)
     mm. long; c. Ida. to s. Mont., s. to N.M. and Utah   var. stenophyllus (Rydb.) Davis
1 Flowers mostly (7) 9-11 (12) mm. long; leaves various
  3 Plants subalpine, from c. Ida. to c. Mont. and Wyo., grayish-hairy, 1.5-2.5 dm.
     tall; racemes congested   var. depressus (Rydb.) C. L. Hitchc.
  3 Plants of various habitats, from Oreg. n.e. to Alta., Mont., and S.D., usually
     not grayish-hairy and often well over 2.5 dm. tall; racemes not notably con-
     gested   var. argenteus

Lupinus bicolor Lindl. Bot. Reg. 13:pl. 1109. 1827.
    L. micranthus var. *bicolor* Wats. Proc. Am. Acad. 8:536. 1873. *(Douglas,* "In the interior of the
      country about the Columbia River, from Fort Vancouver to the branches of the Lewis and Clarke's
      River")
    L. micranthus var. *microphyllus* Wats. Proc. Am. Acad. 8:535. 1873. *L. bicolor* var. *microphyllus*
      C. P. Smith, Bull. Torrey Club 50:382. 1923. *(Nuttall,* San Diego, Calif.)
    L. apricus Greene, Leafl. 2:67. 1910. *L. vallicola* var. *apricus* C. P. Smith, Muhl. 6:135. 1911.
      L. nanus var. *apricus* C. P. Smith, Bull. Torrey Club 50:170. 1923. *L. vallicola* ssp. *apricus*
      Dunn, El Aliso 3:166. 1955. *(C. F. Baker 610,* Stanford University, Calif., Apr., 1902.)
    L. hirsutulus Greene, Leafl. 2:152. 1911. *(John Macoun,* Beacon Hill, Vancouver I., B.C., June
      15, 1908)
    L. strigulosus Gandg. Bull. Soc. Bot. France 60:461. 1913. *(Suksdorf 5928,* Bingen, Wash.)
    L. bicolor var. *tridentatus* Eastw. ex C. P. Smith, Bull. Torrey Club 50:377. 1923. *L. bicolor*
      ssp. *tridentatus* Dunn, El Aliso 3:155. 1955. *(Eastwood 10369,* Santa Rosa, Sonoma Co., Calif.,
      Apr., 1921)

Annual, resembling *L. micranthus* in nearly all vegetative characters, the racemes perhaps with
shorter peduncles, but equally small, the few flowers more scattered than verticillate, (6) 7-9 (10)
mm. long; calyx similar, the upper lip bifid, the lower entire to 3-toothed; petals bluish, the banner
well reflexed from the wings, from broader than long and much shorter than the keel, to longer than
broad and subequal to the keel; keel ciliate on the upper half, the slender acute tip curved upward;
pods mostly 1.5-2 cm. long, 3-5 mm. broad; seeds 4-8, grayish and usually mottled, 2-3 mm. broad.

British Columbia southward, mainly w. of the Cascades, to s. Calif., and up the Columbia R. Gorge
to Wasco Co., Oreg., and Klickitat Co., Wash. Apr.-June.

Lupinus bicolor is very similar to *L. micranthus,* and perhaps might as consistently be combined
with that taxon. Watson followed this procedure, reducing *L. bicolor* to varietal status, as *L. micranthus* var. *bicolor* (Lindl.) Wats. However, *L. bicolor* was published 2 years earlier than *L. micranthus* and cannot properly be reduced to an infraspecific element of the latter. Rather than tamper ad-
ditionally with the nomenclature, both taxa are maintained at the specific level.

Lupinus bicolor is variable in the size of the flowers and in the degree of reflexion of the banner,
as well as in most other characters, but in general is not differentiated into geographic races within
our area. Although Smith and others have maintained the forms with broader and shorter banners as
infraspecific taxa under *L. nanus* or *L. vallicola,* the separation of these extreme forms from the
narrower and longer bannered plants is arbitrary. Smith seemed to realize this fact when he stated
(Bull. Torrey Club 50:171. 1923), "Several additional varieties could be indicated but not, as I now see
it, to any good purpose, especially if based primarily upon pubescence. Thus *L. hirsutulus* Greene

calyx

3

2

keel 2

2

var. parviflorus

var. argenteus

Lupinus argenteus

var. stenophyllus

2

1/2

1/2

1/2

1/2

2

var. depressus

JRJ

has at no time gained my respect and I am now unable to separate it from var. *apricus,* although it does seem to lean strongly toward *L. bicolor,* as the usual determinations indicate."

Lupinus caudatus Kell. Proc. Calif. Acad. Sci. 2:197. 1862. (No locality given)
   L. argentinus Rydb. Bull. Torrey Club 30:257. 1903. *(S. G. Stokes,* near Reno, Utah, in 1900)
   L. lupinus Rydb. Bull. Torrey Club 40:44. 1913. *L. utahensis* Moldenke, Torreya 34:9. 1934.
      (Rydberg & Garrett 9363, Western Bear's Ear, Elk Mts. , Utah, Aug. 2, 1911)
   L. hendersonii Eastw. Leafl. West. Bot. 2:266. 1940. *(Henderson 8119,* Alvord Ranch, Steen's Mts.,
      Harney Co. , Oreg.)
   L. gayophytophilus C. P. Smith, Sp. Lup. 734. 1952. *(Kenneth Pearse 229,* Headquarters Gulch,
      Intermountain Forest & Range Exp. Sta., Boise Forest, Elmore Co. , Ida., June 7, 1937)
   L. amniculi-vulpum C. P. Smith, Sp. Lup. 740. 1952. *(R. E. Allen A-35,* Summit Fox Creek,
      Challis Forest, Custer Co. , Ida., July 11, 1929)
   More or less silvery-strigillose perennial, usually with several simple or branched stems 2-3.5 (6) dm. tall; leaves mainly cauline, but basal leaves fairly common, often with a petiole 2-2.5 times as long as the blade, leaves of the upper part of the stem usually with petioles shorter than the blades; leaflets 7-9, narrowly to broadly oblanceolate, mostly 3-5 (6) cm. long, generally acute, almost equally strigillose (to silky) on both surfaces; racemes from nearly sessile to fairly long-pedunculate, 5-30 cm. long, the numerous flowers 10-12 mm. long, closely approximate to widely separated, scattered to subverticillate; pedicels 2-3 (5) mm. long; calyx (2.5) 3-4 mm. long, the base saccate to conspicuously spurred (spur to 2 mm. long), the upper lip bidentate and 1/3-3/4 as long as the wings, the lower lip entire; corolla from light blue to deep violet, the banner usually lighter at the center, fairly well reflexed (index 5-10), usually hairy on the back, at least on the lower half; wings glabrous or somewhat hairy near the base of the claw; keel ciliate most of the length; pods 2-3 cm. long, more or less silky; seeds 4-6, about 5 mm. long, pinkish-brown. N=24.
   Sagebrush and ponderosa pine forest; e. Oreg. s. to Calif. and e. to Mont. and Colo. May-July.
   Superficially this taxon resembles *L. leucophyllus* because of the dense racemes, and it is not always clearly delimitable from *L. argenteus.* However, it is most closely related to *L. laxiflorus,* with which it freely intergrades and from which it cannot always be separated by means of a key. The taxonomic picture is in no way clarified by the recognition of extreme variants in the species, even at the infraspecific level, as well differentiated geographic races do not exist in our area.

Lupinus holosericeus Nutt. in T. & G. Fl. N.Am. 1:380. 1840. *(Nuttall,* "Islands and gravelly banks
   of the Wahlamet")
   L. multicincinnis C. P. Smith, Sp. Lup. 735. 1952. *(R. B. Johnson 252,* Moores Flat, Sawtooth
      Forest, Elmore Co. , Ida. , July 16, 1931)
   Perennial with a branched crown and numerous stems 1.5-3 (4) dm. tall, simple to sparingly branched above, silvery-strigillose almost throughout, or the hairs in part somewhat spreading; leaves nearly all cauline, the petioles of the lower ones varying from about as long to nearly three times as long as the leaflets, the upper ones with the petiole rarely as long as the blade; leaflets 5-9, narrowly oblanceolate, mostly (1.5) 2.5-4 cm. long, about equally silky on both surfaces; peduncles very short; racemes 5-10 (15) cm. long, often not exceeding the leaves, from only slightly to strongly verticillate, crowded but interrupted at least below; flowers 5-7 (8) mm. long, pale violet-blue to deep blue; pedicels 2-3 mm. long; calyx silvery-strigillose, the upper lip bidentate, the lower entire to tridentate; banner only slightly reflexed from the keel (the index about 5), copiously pubescent on the back, the hairs mostly on the lower half but sometimes extending to the tip; wings glabrous or hairy near the base of the blade; keel ciliate most of the length or on the upper half only; pods 1.5-2.5 cm. long, silky; seeds 2-4, about 3.5 mm. long, pinkish, not mottled.
   Pine woodland and sagebrush slopes; mountains of s.c. Ida. (Adams to Blaine and Boise cos. and s. to Owyhee Co.) and adj. Oreg. June-July.
   This rather distinctive species is one of our more limited in range (it is the common belief that Nuttall's type probably was not really collected in the Willamette Valley). It both resembles and not uncommonly intergrades with at least three more-widespread taxa, namely *L. argenteus, L. leucophyllus,* and *L. sericeus.*

Lupinus latifolius Agardh, Syn. Gen. Lup. 18. 1835.
   L. rivularis var. *latifolius* Wats. Proc. Am. Acad. 8:525. 1873. *(Douglas,* California)
   L. cytisoides Agardh, Syn. Gen. Lup. 18. 1835. *(Douglas,* California)

banner

keel

3

Lupinus caudatus   L. holosericeus   L. bicolor   JRJ

L. arcticus Wats. Proc. Am. Acad. 8:526. 1873. *L. polyphyllus* ssp. *arcticus* Phillips, Res. Stud. State Coll. Wash. 181. 1955. *(Lyall,* "Washington Territory") = var. *subalpinus.*

L. ligulatus Greene, Pitt. 1:215. 1888. *L. latifolius* var. *ligulatus* C. P. Smith, Contr. Dudley Herb. 1:50. 1927. *(R. M. Austin,* Crooked Cr., Malheur Co., Oreg., July, 1886) = var. *latifolius.*

L. volcanicus Greene, Pitt. 3:308. 1898. *(C. V. Piper,* Mt. Rainier, Wash., Aug., 1895) = var. *subalpinus.*

L. viridifolius Heller, Muhl. 2:64. 1905. *L. latifolius* var. *viridifolius* C. P. Smith, Contr. Dudley Herb. 1:50. 1927. *L. rivularis* var. *viridifolius* Jeps. Fl. Calif. 2:260. 1936. *(Heller 7928,* Dunsmuir, Siskiyou Co., Calif., June 1, 1905) = var. *latifolius.*

L. suksdorfii Robins. ex Piper, Contr. U.S. Nat. Herb. 11:355. 1906. *(Suksdorf 110,* Columbia R., w. Klickitat Co., Wash., May 3, June, 1883, is the first of several collections cited) = var. *thompsonianus.*

L. alpicola Henderson ex Piper, Contr. U.S. Nat. Herb. 11:355. 1906. *(Henderson 1387,* Mt. Adams, Wash., is the first of 3 collections cited) Identity uncertain, perhaps *L. sericeus.*

L. subalpinus Piper & Robins. Contr. U.S. Nat. Herb. 11:356. 1906. *L. latifolius* var. *subalpinus* C. P. Smith, Bull. Torrey Club 51:308. 1924. *L. arcticus* var. *subalpinus* C. P. Smith in Abrams, Ill. Fl. Pac. St. 2:518. 1944. *(Lyall 1860,* Cascade Mts. to Ft. Colville)

L. confusus Heller, Muhl. 8:63. 1912, but not of Rose in 1905. *L. columbianus* Heller, Muhl. 8:84. 1912. *L. latifolius* var. *columbianus* C. P. Smith, Bull. Torrey Club 51:307. 1924. *(Heller 10107,* near Hood R., Wasco Co., Oreg., May 26, 1910) = var. *latifolius;* however, most material that was identified by Smith as "var. columbianus (Heller) C. P. Smith" = *L. polyphyllus* var. *burkei.*

L. latifolius var. *canadensis* C. P. Smith, Bull. Torrey Club 51:307. 1924. *(C. G. Newcombe,* Vancouver I., B.C., June 24, 1915) = var. *latifolius.*

LUPINUS LATIFOLIUS var. THOMPSONIANUS (C. P. Smith) C. L. Hitchc. hoc loc. *L. sericeus* var. *thompsonianus* C. P. Smith, Sp. Lup. 105. 1939. *(J. W. Thompson 10380,* near Rowena, Wasco Co., Oreg., Apr. 14, 1934)

L. glacialis C. P. Smith, Sp. Lup. 236. 1940. *(J. W. Thompson 14705,* Big Four Mountain, Snohomish Co., Wash., June 15, 1940) = var. *subalpinus.*

L. volcanicus var. *rupestricola* C. P. Smith, Sp. Lup. 236. 1940. *(J. W. Thompson 15210,* Goat Rocks, Lewis Co., Wash., Aug. 6, 1940) = var. *subalpinus?*

Much like *L. rivularis* vegetatively, the stems up to more than 1 m. tall, usually rather sparsely strigillose but sometimes rusty-villous, mostly with 6-8 leaves, the lowermost with petiole 2-3 times as long as the blade, often persistent at flowering time, the upper petioles 1-2 times as long as the blades; leaflets (6) 7-9 (10), elliptic to obovate, usually (2.5) 3-6 (9) cm. long, (5) 10-20 mm. broad, strigillose (villous) on the lower surface, the upper surface less densely strigillose (villous) to glabrate or glabrous; racemes 1-2 dm. long, the pedicels slender, 4-9 mm. long; flowers pale blue to bluish or lavender, 12-15 mm. long; calyx sericeous to somewhat shaggy-pubescent, not saccate or spurred, the upper lip shallowly bidentate, conspicuously broader than the lower entire lip and mostly only 2-2.5 times as long as the calyx tube; banner glabrous, fairly well reflexed from the keel (index 15-30), its edges moderately reflexed; wings glabrous, rather completely covering the curved, ciliate or glabrous keel; pods 2-3 cm. long, about 8 mm. broad; seeds grayish, darker mottled. N=24, 48.

Cascade Mts., from B.C. to Calif., w. to the coastal mountains, on open subalpine ridges to wooded slopes, occasionally on the lowland "prairies." June-Aug.

In our area *L. latifolius* is more variable than the other two species, *L. rivularis* and *L. albicaulis,* to which it is most similar. Some of the trends are sufficiently well established to justify the maintenance of infraspecific taxa. Much of the montane material from the Cascades of Wash. and B.C. has particularly long "shaggy" hairs on the calyx and is dwarfed, the plants seldom being over 2.5 dm. tall. Such plants have been called *L. volcanicus;* presumably it was also a specimen of this nature that was designated *L. arcticus,* and our more northern plants, including many from Alas., have usually been treated under that name. This dwarfed montane phase blends with the taller plants of the lowlands and appears to be no more than varietally distinct from them; at the varietal level, the earliest name available is "subalpinus." Unfortunately, it also merges with *L. polyphyllus* var. *burkei,* and there is no basis, other than the indefinite distinction of branched vs. unbranched stems, by which many intermediates can be referred to one or the other taxon.

In the Columbia R. Gorge var. *latifolius* is gradually transitional to a villous phase that is usually recognized at the specific level as *L. suksdorfii.* The pubescence is strongly suggestive of *L. saxosus* and also similar to that of much of the montane var. *subalpinus,* especially those plants that have been called *L. volcanicus.*

The three races are separable, in major part, as follows:

1 Plant usually rather heavily (whitish- to) rufous-villous, at least on the calyx and inflorescence, and usually also on the leaves and stems
   2 Keel tending to be nonciliate; plants mostly well over 2.5 dm. tall; Columbia R. Gorge of Wash. and Oreg., transitional (toward the west, and with elevation) to var. *latifolius*   var. thompsonianus (C. P. Smith) C. L. Hitchc.
   2 Keel tending to be ciliate; plants mostly 1-2.5 dm. tall; Alas. s., on alpine or subalpine ridges and meadows, in the Cascade Mts., to the region of Mt. Rainier, where an extreme phase with long yellow villosity *(L. volcanicus* Greene) occurs; completely transitional, in both habit and pubescence, to the taller, strigillose to subglabrous plant of the lowlands   var. subalpinus (Piper & Robins.) C. P. Smith
1 Plant strigillose (rarely villous even in the inflorescence), usually over 2.5 dm. tall; Cascade Mts. and w. to the coast, from the lowland prairies to wooded or open montane slopes, from s. B.C. to the Sierra Nevada of Calif.   var. latifolius

Lupinus laxiflorus Dougl. ex Lindl. Bot. Reg. 14:pl. 1140. 1828.
   L. arbustus ssp. *neolaxiflorus* Dunn, Leafl. West. Bot. 7:254. 1955. *(Douglas,* "Great rapids of the River Columbia")
   L. arbustus Dougl. ex Lindl. Bot. Reg. 15:pl. 1230. 1829. *L. laxiflorus* var. *arbustus* M. E. Jones, Contr. West. Bot. 14:33. 1912. *(Douglas,* "Growing only in gravelly soils in North California, common near Fort Vancouver") = var. *laxiflorus*.
   L. calcaratus Kell. Proc. Calif. Acad. Sci. 2:194. 1862. *L. laxiflorus* var. *calcaratus* C. P. Smith, Bull. Torrey Club 51:304. 1924. *L. arbustus* ssp. *calcaratus* Dunn, Leafl. West. Bot. 7:255. 1955. (Locality unknown)
   L. laxiflorus var. *montanus* Howell, Erythea 3:33. 1895. *L. arbustus* ssp. *arbustus* var. *montanus* Dunn, Leafl. West. Bot. 7:254. 1955. *(Howell,* Mt. Hood, Oreg.) = var. *laxiflorus*.
   L. pseudoparviflorus Rydb. Mem. N.Y. Bot. Gard. 1:232. 1900. *L. laxiflorus* var. *pseudoparviflorus* Smith & St. John, Fl. S.E. Wash. 227. 1937. *L. arbustus* ssp. *pseudoparviflorus* Dunn, Leafl. West. Bot. 7:255. 1955. *(Rydberg & Bessey 4441,* Bridger Mts., Mont., June 17, 1897)
   L. scheuberae Rydb. Bull. Torrey Club 29:244. 1902. *(E. W. Sheuber 135,* "Garnet county," Mont., in 1901) = var. *pseudoparviflorus*.
   L. laxiflorus f. *theiochrous* Robins. ex Piper, Contr. U.S. Nat. Herb. 11:358. 1906. *(J. S. Cotton,* Rattlesnake Mts., Yakima Co., Wash., July 16, 1900) = var. *laxiflorus*.
   L. laxispicatus Rydb. Bull. Torrey Club 34:42. 1907. *(Sandberg,* Kootenai Co., Ida., July, 1887) = var. *pseudoparviflorus*.
   L. silvicola Heller, Muhl. 6:81. 1910. *L. laxiflorus* var. *silvicola* C. P. Smith ex Jeps. Man. Fl. Pl. Calif. 527. 1925. *L. arbustus* ssp. *silvicola* Dunn, Leafl. West. Bot. 7:255. 1955. *(Heller 9857,* Summit, Placer Co., Calif., July 16, 1909) = var. *laxiflorus*.
   L. multitinctus A. Nels. Bot. Gaz. 53:221. 1912. *(Nelson & Macbride 114,* Big Willow, near Falk's Store, Canyon Co., Ida.) = var. *calcaratus*.
   L. variegatus Heller, Muhl. 8:89. 1912. *(Heller 10551,* Ruby Mts., near Deeth, Elko Co., Nev., July 8, 1912) = var. *calcaratus*.
   L. inyoensis var. *demissus* C. P. Smith, Bull. Torrey Club 51:304. 1924. *(Peck 5329,* Wallowa Mts., Baker Co., Oreg., Sept. 3, 1915) = var. *laxiflorus*.
   L. caudatus var. *subtenellus* C. P. Smith, Bull. Torrey Club 51:304. 1924. *(Leiberg 591,* Paulina Lake, Deschutes Co., Oreg., July 30, 1894) = var. *laxiflorus*.
   L. laxiflorus var. *cognatus* C. P. Smith ex Jeps. Man. Fl. Pl. Calif. 527. 1925. *(Cusick 3187,* Wallowa Mts., Wallowa Co., Oreg.) = var. *laxiflorus*.
   L. laxiflorus var. *durabilis* C. P. Smith, Am. Journ. Bot. 13:529. 1925. *(Leiberg 2731,* Priest River Range, Ida.) = var. *laxiflorus*.
   L. laxispicatus var. *whithamii* C. P. Smith ex St. John, Fl. S.E. Wash. 227. 1937. *L. sulphureus* ssp. *whithamii* Phillips, Res. Stud. State Coll. Wash. 23:193. 1955. *(Smith, St. John, & Whitham 4170,* junction of Divide and Kings Lake trails, Kaniksu National Forest, Pend Oreille Co., Ida., July 8, 1927) = var. *pseudoparviflorus* (see discussion under *L. sulphureus*).
   L. laxiflorus var. *lyleianus* C. P. Smith, Sp. Lup. 105. 1939. *(Eldon W. Lyle,* head of Sheep Creek, 7 mi. s.w. of Lick R. S., Wallowa Nat. For. & County, Oreg., Aug., 1930) = var. *laxiflorus*.
   L. caudatus var. *submanens* C. P. Smith, Sp. Lup. 106. 1939. *(Eldon W. Lyle,* 2 mi. e. of Anthony Lake, Blue Mts., Oreg., Aug., 1930) = var. *laxiflorus*.

L. laxiflorus var. *elmerianus* C. P. Smith, Sp. Lup. 106. 1939. *(Applegate 6483,* Paradise, Wallowa Co., Oreg., July, 1930) = var. *pseudoparviflorus.*

L. lyleianus C. P. Smith, Sp. Lup. 107. 1939. *(Eldon W. Lyle,* 7 mi. e. of Pearson R. S., Umatilla N.F., Oreg., July, 1930) = var. *laxiflorus.*

L. mucronulatus var. *umatillensis* C. P. Smith, Sp. Lup. 108. 1939. *(Applegate 6483,* Table Rock, Umatilla Co., Oreg., July, 1930) = var. *pseudoparviflorus.*

L. lutescens C. P. Smith, Sp. Lup. 235. 1940. *(J. W. Thompson 14626,* Badger Mt., Douglas Co., Wash., June, 1940) = var. *laxiflorus.*

L. yakimensis C. P. Smith, Sp. Lup. 238. 1940. *(J. W. Thompson 14572,* Cleman Mt., 25 mi. n.w. of Yakima, Yakima Co., Wash.) = var. *laxiflorus.*

L. wenatchensis Eastw. Leafl. West. Bot. 3:174. 1942. *(J. W. Thompson 14242,* Wenatchee Mt., Kittitas Co., Wash., Sept. 13, 1937) = var. *laxiflorus.*

L. henrysmithii C. P. Smith, Sp. Lup. 566. 1946. *(H. L. Smith 119,* Minidoka Nat. For., Ida., July, 1922) = var. *laxiflorus.*

L. fieldianus C. P. Smith, Sp. Lup. 567. 1946. *(R. C. Fields 224,* Thorn Cr., Idaho Nat. For., Ida.) = var. *pseudoparviflorus.*

L. lacus-payetti C. P. Smith, Sp. Lup. 574. 1946. *(M. E. Jones 6251,* Payette Lake, Ida.) = var. *pseudoparviflorus.*

L. merrillanus C. P. Smith, Sp. Lup. 574. 1946. *(Merrill & Wilcox 783,* Idaho Falls, Bonneville Co., Ida., July 4, 1901) = var. *laxiflorus.*

L. amniculi-putorii C. P. Smith, Sp. Lup. 575. 1946. *(Basil Crane,* Mink Cr., Bannock Co., Ida., July 30, 1935) = var. *laxiflorus.*

L. mackeyi C. P. Smith, Sp. Lup. 725. 1952. *(Mackey 65,* Clifty Block Mt. Range, Kaniksu Nat. For., Ida.) = var. *laxiflorus.*

L. geraniophilus C. P. Smith, Sp. Lup. 727. 1952. *(H. J. Helm 30,* Johnson Cr. R. S., Weiser Nat. For., Ida.) = var. *calcaratus.*

L. varneranus C. P. Smith, Sp. Lup. 730. 1952. *(I. M. Varner 82,* Boulder L., Idaho Nat. For., Ida.) = var. *calcaratus,* intermediate to *L. argenteus.*

L. augusti C. P. Smith, Sp. Lup. 733. 1952. *(Kenneth Pearce 23,* Head of Slater Cr., Boise Nat.-For., Elmore Co., Ida.) = var. *laxiflorus* with some doubt, possibly *L. caudatus*

L. stipaphilus C. P. Smith, Sp. Lup. 733. 1952. *(Kenneth Pearce 165b,* North Star Lake, Boise Nat. For., Elmore Co., Ida.) = var. *laxiflorus.*

L. multitinctus var. *grandjeana* C. P. Smith, Sp. Lup. 735. 1952. *(E. Grandjean 460,* Boise Nat. For., Elmore Co., Ida.) = var. *calcaratus.*

L. perconfertus C. P. Smith, Sp. Lup. 738. 1952. *(O. F. Cusick 59,* Horse Heaven Pass, Lemhi Forest, Lemhi or Custer Co., Ida., July 12, 1928) = var. ?

L. festucasocius C. P. Smith, Sp. Lup. 738. 1952. *(Johnson 20,* Copper Basin Potholes, Lemhi Nat. For., Ida.) = var. *laxiflorus.*

L. graciliflorus C. P. Smith, Sp. Lup. 739. 1952. *(G. A. Miller M-86,* Fairview R. S., Lemhi Nat. For., Ida.) = var. *calcaratus.*

L. stockii C. P. Smith, Sp. Lup. 743. 1952. *(Stock 186,* Bostetter R. S., Minidoka Nat. For., Ida.) = var. *laxiflorus.*

L. standingi C. P. Smith, Sp. Lup. 749. 1952. *(Standing 18,* Deep Cr., near Malad, Cache Nat. For., Ida.) = var. *laxiflorus.*

Multicipital perennial with (usually) numerous erect to spreading, simple or sparingly branched stems 2-5 (7) dm. tall, finely strigillose and usually greenish- to somewhat grayish-sericeous; leaves mostly cauline, the lower petioles 2-4 times (but the upper ones often scarcely half) as long as the blades; leaflets 7-11, oblanceolate to narrowly oblong-oblanceolate, acute or abruptly pointed to (occasionally) obtuse or rounded, mostly 3-5 cm. long, from equally pubescent on both surfaces to glabrous above; racemes rather short-pedunculate, (5) 7-20 cm. long; flowers many, mostly scattered but often rather closely crowded, (8) 9-14 mm. long; pedicels 3-7 mm. long; calyx with a conspicuous sac or spur (0.5) 1-3 mm. long, bilabiate, the upper lip bidentate, usually less than 1/3 as long as the wings, the lower lip entire; petals from white or cream over-all or with a tinge of bluish, pinkish, or violet, to rose or purple, often varying greatly on the same plant; banner not greatly reflexed (index 4-9), mostly rather copiously pubescent on the back to above the center; wings usually conspicuously hairy on the upper side above the middle; keel ciliate most of its length; pods 2-3.5 cm. long, more or less silky; seeds pinkish-brown. N=24, 48.

Mostly in sagebrush and ponderosa pine country; e. of the Cascades, from Wash. to Calif., spilling

w. through the Columbia and Rogue river drainages, e. to Ida., Mont., Utah, and Nev. May-July.

The use of the name *L. laxiflorus* for this taxon was considered correct procedure by Phillips (Res. Stud. State Coll. Wash. 23:196. 1955), but declared to be erroneous by Dunn (Madroño 14:54-75. 1957). The two agreed that the original description and the illustration (Lindl. Bot. Reg. 14:pl. 1140. 1828) of the taxon do not really apply to the entity with which we are dealing. Both have seen the supposed type, *Douglas,* "Great rapids of the River Columbia," and believe that it fits the above conception of *L. laxiflorus.* The specimen in question bears the notation "L. laxiflorus" and a reference to the Botanical Register, in what Phillips believes was Lindley's handwriting. Phillips therefore contends that there is no good reason to question the authenticity of this plant as the type of *L. laxiflorus* and cites Chap. III, Sect. II, Art. 18, of the 1952 ed. of the International Rules as his basis for giving pre-eminence to what he believes to be the type, rather than to the description and illustration, in interpreting the species. Dunn believes that Lindley's original description of *L. laxiflorus* actually matches a *Douglas* specimen labeled "L. tenellus" (which both he and Phillips refer to *L. argenteus*). Dunn is convinced that the specimen labeled "L. laxiflorus" was annotated not by Lindley, but by Agardh, and feels that the latter misinterpreted Lindley's work and mixed the two specimens on which *L. laxiflorus* and *L. tenellus* were based. He refers *L. laxiflorus* (on the basis of the description and illustration) to varietal status (as var. *tenellus*) under *L. argenteus* Pursh.

With the name "laxiflorus" removed from this taxon, Dunn took up the next oldest legitimate specific epithet for it, *L. arbustus* Dougl. ex Lindl., and proposed a new name, ssp. *neolaxiflorus* Dunn, to include the bulk of the plants which have more commonly been called *L. laxiflorus,* including the controversial *Douglas* specimen labelled "L. laxiflorus."

Although it seems not improbable that Dunn's interpretation of the situation is the correct one, the more conventional nomenclature is preferable for our use, especially since varietal combinations, already in existence under *L. laxiflorus,* would have to be proposed under the specific name *L. arbustus.*

The species is variable in many characters, especially flower color, the pubescence of the banner and wings, and the degree of development of the spur. Most of this variation does not appear to follow any constant or geographically established pattern, although several species have been proposed on striking color variation and others on the degree of development of the spur. Three infraspecific taxa are recognizable in our area: var. *calcaratus* (Kell.) C. P. Smith is very weakly tenable, being primarily the yellowish color-variant which crops up throughout the range of the species, but most commonly in Ida.; var. *pseudoparviflorus* (Rydb.) Smith & St. John, the less pubescent phase, occurring chiefly from the Blue Mts. to w. Mont., seems to be maintainable from the more heavily pubescent and more widespread var. *laxiflorus.* The varieties are separable as follows:

1 Flowers mostly light-colored, from pure white or cream to slightly or heavily tinged with
    pinkish, bluish, or violet, generally large, rarely less than 10 mm. long; primarily from
    n.e. Oreg. to Sierran Calif., e. to Ida., w. Mont., and Utah, but not uncommon else-
    where, especially in Wash.   var. calcaratus (Kell.) C. P. Smith
1 Flowers mostly blue to violet or purple, variable in size and from 8-14 mm. long
    2 Leaflets commonly glabrous or subglabrous on the upper surface, often obtuse or
        rounded; flowers mostly 8-11 mm. long; n.e. Wash. to n.e. Oreg., e. to Mont.
        (where very common); occasional elsewhere, especially along the e. base of the
        Cascades in Wash.; N=24   var. pseudoparviflorus (Rydb.) Smith & St. John
    2 Leaflets pubescent on the upper surface, often as hairy as beneath, usually acute to
        obtuse; flowers 9-14 mm. long; throughout the range of the species, from Wash. to
        Calif., e. to Mont. (where largely replaced by var. *pseudoparviflorus*); N=24, 48
                                                        var. laxiflorus

Although the species is fairly clearly marked, indicating that a comparatively small amount of interbreeding with other species occurs, there is definite intergradation with *L. caudatus,* and not a few specimens appear to be intermediate to *L. leucophyllus, L. argenteus,* and *L. sulphureus* (which see for discussion).

Lupinus lepidus Dougl. ex Lindl. Bot. Reg. 14:pl. 1149. 1828. *(Douglas,* "From Fort Vancouver to
    the Great Falls of the Columbia")
    L. aridus Dougl. ex Lindl. Bot. Reg. 15:pl. 1242. 1829. *L. lepidus* var. *aridus* Jeps. Fl. Calif.
      2:268. 1936. *L. lepidus* ssp. *aridus* Detling, Am. Midl. Nat. 45:491. 1951. *(Douglas,* "From the
      Great Falls of the Columbia in North America to the sources of the Missouri among the Rocky
      Mountains")
    L. minimus Dougl. ex Hook. Fl. Bor. Am. 1:163. 1833. *(Douglas,* "Mountain vallies in North-West

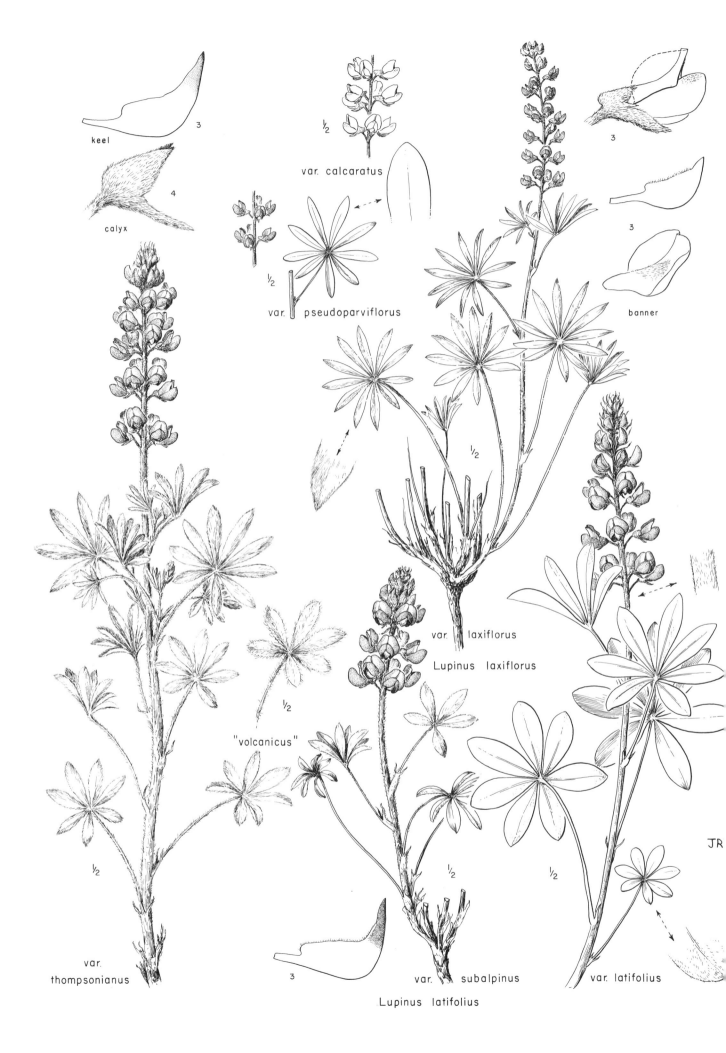

keel

3

calyx

4

var. calcaratus

½

var. pseudoparviflorus

½

banner

3

3

"volcanicus"

½

var. laxiflorus

Lupinus laxiflorus

½

var. thompsonianus

½

3

var. subalpinus

½

Lupinus latifolius

var. latifolius

½

JR

America near the Kettle falls and very abundant towards the Rocky Mountains, along the course of the Columbia") = var. *lepidus.*

L. caespitosus Nutt. ex T. & G. Fl. N.Am. 1:379. 1840. *L. lepidus* ssp. *caespitosus* Detling, Am. Midl. Nat. 45:494. 1951. *(Nuttall,* "In the grassy valleys of the Rocky Mountains, on the Sweet Waters of the Platte and the Colorado of the West") = var. *utahensis.*

L. lyallii Gray, Proc. Am. Acad. 7:334. 1868. *L. lepidus* ssp. *lyallii* Detling, Am. Midl. Nat. 45: 490. 1951. *(Lyall,* "Summit of the Cascade Mts., latitude 49°") = var. *lobbii.*

L. danaus Gray, Proc. Am. Acad. 7:335. 1868. *L. lyallii* var. *danaus* Wats. Proc. Am. Acad. 8: 534. 1873. *(Bolander,* Mt. Dana at 12,500 ft., Calif.) = var. *lobbii.*

LUPINUS LEPIDUS var. LOBBII (Gray) C. L. Hitchc. hoc loc. *L. lobbii* Gray ex Wats. Proc. Am. Acad. 7:334. 1868. *L. aridus* var. *lobbii* Wats. Proc. Am. Acad. 8:533. 1873. *L. lyallii* var. *lobbii* C. P. Smith in Jeps. Man. Fl. Pl. Calif. 525. 1925. *(Lobb,* High Sierras of California)

L. torreyi Gray ex Wats. Bot. King Exp. 58. 1871. *L. aridus* var. *torreyi* C. P. Smith, Bull. Torrey Club 51:303. 1924. *(Torrey 82,* near Washoe Lake, Nev., is the first of 3 collections cited) Identity uncertain, possibly the same as var. *aridus,* and if so, representing the earliest varietal name.

LUPINUS LEPIDUS var. UTAHENSIS (Wats.) C. L. Hitchc. hoc loc. *L. aridus* var. *utahensis* Wats. Proc. Am. Acad. 8:534. 1873. *L. watsonii* Heller, Muhl. 1:114. 1905. *(Watson,* Parley's Park in the Wasatch, Utah)

LUPINUS LEPIDUS var. CUSICKII (Wats.) C. L. Hitchc. hoc loc. *L. cusickii* Wats. Proc. Am. Acad. 22:469. 1887. *L. aridus* var. *cusickii* C. P. Smith, Bull. Torrey Club 51:303. 1924. *L. lepidus* ssp. *cusickii* Detling, Am. Midl. Nat. 45:493. 1951. *(Cusick 1361,* Forks of Upper Burnt River, Baker Co., Oreg.)

L. hellerae Heller, Bull. Torrey Club 25:265. 1898. *L. minimus* var. *hellerae* Smith & St. John, Fl. S.E. Wash. 2:228. 1937. *(Heller & Heller 3080,* 4 mi. e. of Lewiston, Nez Perce Co., Ida., May 18, 1896) = var. *aridus.*

L. brachypodus Piper, Bull. Torrey Club 29:642. 1902. *(Cusick 2561,* Barren Valley, Harney Co., Oreg., June 12, 1901) = var. *cusickii.*

L. piperi Robins. ex Piper, Contr. U.S. Nat. Herb. 11:353. 1906. *(Piper 2730,* Spokane, Wash.) = var. *aridus.*

L. piperi ssp. *imberbis* Robins. ex Piper, Contr. U.S. Nat. Herb. 11:353. 1906. *(K. Whited 121,* Wenatchee, Chelan Co., Wash., June, 1896) = var. *aridus.*

L. paulinus Greene, Leafl. 2:234. 1912. *(Leiberg 550,* Paulina Lake, Oreg., July 28, 1894) = var. *lobbii.*

L. fruticulosus Greene, Muhl. 8:117. 1912. *L. lyallii* var. *fruticulosus* C. P. Smith, Bull. Torrey Club 51:303. 1924. *(Coville & Applegate,* Klamath Co., Oreg., July 31, 1897, is the first of 2 collections cited) = var. *lobbii.*

L. perditorum Greene, Muhl. 8:117. 1912. *(Applegate,* along Rogue R., Jackson Co., Oreg., July 8, 1898) = var. *lobbii.*

L. abortivus Greene, Muhl. 8:117. 1912. *L. aridus* var. *abortivus* C. P. Smith, Bull. Torrey Club 51:303. 1924. *(Leiberg,* Stinking Water, Malheur Co., Oreg., June 21, 1896) = var. *cusickii.*

L. volutans Greene, Muhl. 8:118. 1912. *(Leiberg,* Malheur Valley, Oreg., June 8, 1896) = var. *cusickii.*

L. minutifolius Eastw. Leafl. West. Bot. 2:267. 1940. *(Henderson 8132,* above Fish Lake, Steens Mts., Harney Co., Oreg., July 20, 1927) = var. *aridus?*

L. alcis-temporis C. P. Smith, Sp. Lup. 558. 1946. *(J. H. Christ 14634,* Crystal Cr. Grade, 11 mi. e. of Pierce, Clearwater Co., Ida., Oct. 15, 1944) = var. *lobbii.*

L. markleanus C. P. Smith, Sp. Lup. 559. 1946. *(M. G. Markle 31,* Head of Casino Cr., Challis Forest, Custer Co., Ida., July 27, 1929) = var. *utahensis.*

L. lenorensis C. P. Smith, Sp. Lup. 560. 1946. *(J. H. Christ 7331,* along Clearwater R., 1 mi. w. of Lenore, Nez Perce Co., Ida., May 29, 1937) = var. *aridus.*

L. longivallis C. P. Smith, Sp. Lup. 561. 1946. *(L. F. Henderson 3089,* Long Valley, Valley Co., Ida., July 27, 1899) = var. *aridus.*

L. amnicoli-cervi C. P. Smith, Sp. Lup. 561. 1946. *(Nelson & Macbride 1844,* ·Deer Creek, Owyhee Co., Ida., July 1, 1912) = var. *utahensis.*

L. sinus-meyersi C. P. Smith, Sp. Lup. 562. 1946. *(Hitchcock & Muhlick 9468,* above Silver Cr., 8 mi. n.e. of Meyers Cove, Lemhi Co., Ida., July 1, 1944) = var. *utahensis.*

L. lepidus ssp. *medius* Detling, Am. Midl. Nat. 45:488. 1951. *(Detling 6546,* 3 mi. s. of Lapine, Deschutes Co., Oreg., Aug. 17, 1949) = var. *aridus* in greater part.

A low and spreading and somewhat matted, to erect and multicipital perennial 1-3.5 (rarely up to 5, but generally less than 2) dm. tall, grayish- to rusty-strigillose or -sericeous; stems usually very short, or if elongated then largely prostrate or decumbent, the erect stems seldom as long as the petioles of the lowermost leaves; leaves largely basal, the petiole usually 2-5 times as long as the blade; leaflets 5-9, oblanceolate, copiously hairy (sericeous to more or less villous) on both surfaces, 1-4 cm. long; racemes closely flowered, 4-15 cm. long, from subsessile to raised well above the leaves on peduncles as much as 15 cm. long; flowers 8-13 mm. long, crowded, verticillate, usually longer than even the lower internodes of the raceme, (white) bluish, the banner often lighter or darker in color; pedicels 2-5 mm. long; calyx sericeous to somewhat spreading-hairy, not saccate or spurred, the upper lip bidentate, the lower entire; banner glabrous, rather sharply reflexed from the keel (index 10-16); wings glabrous; keel ciliate; pods 10-20 mm. long, hairy; seeds 2-4, pinkish-gray or tan. N=24.

British Columbia s. to Calif., on both sides of the Cascades, e. to Mont., Wyo., and Colo. June-Aug.

This is a rather easily distinguished, variable complex in which a great many taxa have been proposed. The variation follows a more constant pattern than in most species of *Lupinus,* and (largely following Detling) 5 geographic or ecological (freely intergradient) races are delimitable in our area:

1 Racemes sessile or very short-pedunculate, surpassed and largely concealed by the
     leaves at flowering time
   2 Wing petals slender, 7-8 mm. long, the width about 1/3 of the length; banner slender,
        the width averaging less than 3/5 of the length; c. Oreg. through c. Ida. and s.e.
        Oreg. to w. Mont. and Wyo., s. to Colo. and Utah   var. utahensis (Wats.) C. L. Hitchc.
   2 Wing petals broader, usually over 8 mm. long, the width nearly 1/2 of the length;
        banner broader, the width averaging over 3/5 of the length; Blue Mts., Oreg.
        (and Wash.?), and in Okanogan Co., Wash.   var. cusickii (Wats.) C. L. Hitchc.
1 Racemes pedunculate, at least partially surpassing and not concealed by the leaves at
     flowering time
   3 Plants tending to be prostrate and matted; leaflets seldom over 15 mm. long; racemes
        mostly less than 5 cm. long at anthesis; subalpine, B.C. to s. Calif., in both the
        Cascade and Olympic mts., e. to w. Ida. and Nev.   var. lobbii (Gray) C. L. Hitchc.
   3 Plants not matted; leaflets usually over 15 mm. long; racemes mostly at least 5 cm.
        long; plants generally not high-montane
      4 Racemes usually partially concealed by (or at least the lower flowers equaled by)
           the longer leaves; flowers mostly 9-11 mm. long; s.c. Wash. to n. Calif. and
           Nev., e. of the Cascades, e. to w.c. Ida.   var. aridus (Dougl.) Jeps.
      4 Racemes usually exserted well beyond the longest leaves; flowers mostly 11-13
           mm. long; s. B.C. to n.w. Oreg., in the lowlands w. of the Cascade Mts.   var. lepidus

Lupinus leucophyllus Dougl. ex Lindl. Bot. Reg. 13:pl. 1124. 1827. *(Douglas,* "Great Falls of the River Columbia in North America to the source of the Missouri among the Rocky Mountains")
  L. plumosus Dougl. ex Lindl. Bot. Reg. 15:pl. 1217. 1829. *L. leucophyllus* ssp. *plumosus* Robins. ex Piper, Contr. U.S. Nat. Herb. 11:354. 1906. *(Douglas,* "common in northern California, in 45° North") = var. *leucophyllus.*
  L. canescens Howell, Erythea 1:110. 1893. *L. leucophyllus* var. *canescens* C. P. Smith, Bull. Torrey Club 51:306. 1924. *(T. Howell 787,* Buck's Mt., Blue Mts., Harney Co., Oreg., June, 1885) = var. *leucophyllus.*
  L. erectus Henderson, Bull. Torrey Club 27:343. 1900. *(Henderson,* Long Valley, Boise Co., Ida., July 4, 1895) = var. *leucophyllus.*
  L. retrorsus Henderson, Bull. Torrey Club 27:344. 1900. *L. leucophyllus* var. *retrorsus* Smith ex St. John, Fl. S.E. Wash. 228. 1937. *(Henderson 4608,* Coeur d'Alene Lake, near Harrison, Kootenai Co., Ida., Aug. 7, 1898) = var. *leucophyllus.*
  L. cyaneus Rydb. Bull. Torrey Club 28:35. 1901. *(E. V. Wilcox 446,* in Mont., in 1900) = var. *leucophyllus.*
  L. canescens ssp. *amblyophyllus* Robins. ex Piper, Contr. U.S. Nat. Herb. 11:354. 1906. *L. holosericeus* var. *amblyophyllus* C. P. Smith, Bull. Torrey Club 51:304. 1924. *(Sandberg & Leiberg 402,* Egbert Spgs., Douglas Co., Wash., July 5, 1893) = var. *leucophyllus.*
  L. macrostachys Rydb. Bull. Torrey Club 34:44. 1907. *(D. T. MacDougal 253,* Jocko Cr., Missoula Co., Mont., in 1901) = var. *leucophyllus.*

var. lobbii

banner

keel

var. utahensis

var. cusickii

var. lepidus

var. aridus

Lupinus lepidus

JRJ

L. tenuispicus A. Nels. Bot. Gaz. 54:410. 1912. *L. leucophyllus* var. *tenuispicus* C. P. Smith, Bull. Torrey Club 51:306. 1924. *(J. Clark 203,* Washington Co., Ida., Aug. 8, 1911)

L. leucophyllus var. *belliae* C. P. Smith, Bull. Torrey Club 51:305. 1924. *(M. B. Zundel,* Crystal Cr., Power Co., Ida., Sept., 1914) = var. *leucophyllus.*

L. sulphureus var. *echlerianus* C. P. Smith in St. John, Fl. S. E. Wash. 229. 1937. *(Smith & St. John 4152,* Echler Mt., Columbia Co., Wash., July 5, 1927) = var. *leucophyllus.*

? *L. pureriae* C. P. Smith, Sp. Lup. 192. 1940. *(E. A. Purer 7748,* near Coles Corner, Chelan Co., Wash., Aug. 9, 1938) Identity problematical.

L. agropyrophilus C. P. Smith, Sp. Lup. 728. 1952. *(W. A. McDowell 1321,* Peck Mt., Weiser Forest, Adams Co., Ida., Aug., 1928) = var. *leucophyllus.*

L. lysichitophilus C. P. Smith, Sp. Lup. 729. 1952. *(V. M. Brewer 78,* Smith Mt., Weiser Forest, Adams Co., Ida., Aug. 17, 1929) = var. *leucophyllus.*

L. salicisocius C. P. Smith, Sp. Lup. 747. 1952. *(M. Anderson 119,* Trail Station, Blackfoot R., Caribou Forest, Caribou Co., Ida., Sept. 7, 1913) = var. *leucophyllus.*

L. andersonianus C. P. Smith, Sp. Lup. 748. 1952. *(M. Anderson 119a,* Diamond Creek, Caribou Forest, Caribou Co., Ida., July 20, 1913) = var. *leucophyllus.*

L. falsoerectus C. P. Smith, Sp. Lup. 749. 1952. *(C. F. Korstian 44,* Mink Creek Planting Area, Cache Forest, Franklin Co., Ida., Aug. 4, 1917) = var. *leucophyllus.*

Perennial, with a branched crown and several stems 3-7 (9) dm. tall, simple to freely branched, usually grayish to rust-colored with a dense, mixed vesture of shorter tomentum and longer, spreading to appressed, straight hairs (occasionally the pubescence all appressed); leaves mainly cauline, with petioles 1-4 times as long as the blades; leaflets 7-10, oblanceolate to narrowly oblong, mostly 3-5 cm. long, almost equally silky or sericeous-pilose on both surfaces; racemes very short-pedunculate, (1) 1.5-2 dm. long; flowers (7) 8-10 (12) mm. long, from nearly white with the banner and wings merely pale lavender, to lilac, closely crowded in indistinct whorls, the racemes scarcely interrupted; pedicels 1-3 (4) mm. long; calyx shaggy, pubescent to silky, not spurred although often slightly saccate, the upper lip bidentate, the lower entire; banner only slightly reflexed (index about 8), thickly pubescent over much of the back, the hairs extending to the upper third or sometimes to the tip (very rarely glabrous); wings glabrous or sparsely hairy near the claw; keel ciliate much of the length (glabrous); pods 1.5-2.5 cm. long, 7-8 mm. broad, shaggy with rusty hairs; seeds smooth, pinkish-brown. N=24, 48.

Central Wash. to n. e. Calif., from the foothills of the Cascades e. to Mont., n. w. Wyo., Ida., and w. Nev. June-Aug.

Because of the rather small and nongaping flowers, dense spikelike racemes, and abundant pubescence, *L. leucophyllus* generally is fairly easily recognizable. Although these several features are variable, especially the size of the flowers and the amount and type of pubescence, the variation seems to be largely fortuitous, and the delimitation of infraspecific taxa, with one or two possible exceptions, is not believed to be merited. In the sagebrush and ponderosa pine area, from c. Wash. to c. and s. e. Oreg. and adj. Ida., there is what appears to be a weakly defined race separable from the greater bulk of the species as follows:

1 Flowers mostly 8-10 (12) mm. long, the pressed racemes usually well over 20 mm.
   broad; leaflets mostly over 7 mm. broad; plants generally with long spreading pubescence (except the subglabrous phase discussed below); with the range of the species,
   but replaced to some extent by var. *tenuispicus* throughout that taxon's range   var. leucophyllus
1 Flowers mostly 7-8 mm. long, the pressed racemes mostly 15-18 (20) mm. broad;
   leaflets usually less than 7 mm. broad; plants sometimes merely strigose, but varying to spreading-hairy; chiefly in the sagebrush and ponderosa pine region, from
   Lincoln Co., Wash., s. to c. and s. e. Oreg. and w. c. and s. w. Ida.
                                                                        var. tenuispicus (A. Nels.) C. P. Smith

Lupinus leucophyllus intergrades to *L. holosericeus,* largely through the var. *tenuispicus.* It also tends'to merge with *L. laxiflorus,* intermediate plants being not uncommon (e. g. *Cronquist 6663,* from Wheeler Co., Oreg.). It was a collection of this general nature which was named *L. canescens* by Howell. Very occasional individuals are only weakly strigillose, and glabrous or subglabrous on the upper surface of the leaves; a plant of this nature from Mont. was described as *L. cyaneus* by Rydberg. Such specimens (e. g. *Hitchcock & Muhlick 12469,* Gallatin Co., Mont.) are intermediate in nature to *L. polyphyllus,* and it is considered likely that they are hybrid in origin rather than representative of a distinctive genetic race. For a discussion of the intergradation with *L. sulphureus* see under that species.

Lupinus littoralis Dougl. ex Lindl. Bot. Reg. 14:pl. 1198. 1928. *(Douglas,* "Seashore, from Cape Mendocino to Puget Sound")

Prostrate perennial, usually matted; stems freely branching, as much as 6 dm. long, strigillose
but also conspicuously hirsute with spreading, straight, light-brownish hairs 2-5 mm. long, the same
type of pubescence usually also on the petioles and in the inflorescence; stipules conspicuous; petioles
1-2 times as long as the blades; leaflets 5-7 (8-9), oblanceolate, usually apiculate, mostly 2-3 cm.
long, from unequally strigillose on the two surfaces to glabrous above; racemes numerous, 5-15 cm.
long, loosely flowered; bracts very conspicuous in the bud but deciduous by anthesis; flowers mostly
verticillate, the lower verticils well separated, sometimes more or less scattered, 7-13 mm. long,
bluish to purplish; pedicels slender, 3-6 mm. long; calyx sericeous, neither spurred nor saccate,
the lips subequal, the upper one shallowly bidentate, the lower entire; banner well reflexed (index 20-
30), glabrous, shorter than the glabrous wings; keel ciliate most of the length, not exserted; pods
1.5-3.5 cm. long, 6-8 mm. broad, hairy; seeds 5-8, grayish but very conspicuously dark-mottled.
N=24.

A very easily recognized species, occurring along the immediate coast from B.C. to Mendocino
Co., Calif. May-Aug.

Lupinus micranthus Dougl. ex Lindl. Bot. Reg. 15:pl. 1251. 1829. *(Douglas,* "upon the gravelly banks
of the southern tributaries of the Columbia and on barren ground in the interior of California")

Simple to rather freely branched annual 1-4 dm. tall, sparsely to copiously brownish-villous and
with a more abundant, shorter pubescence; leaflets 5-8, mostly linear-oblanceolate, 1.5-3 cm. long,
appressed-villous on the lower surface, glabrous on the upper, 2/5-4/5 as long as the petiole; racemes small, pedunculate and usually raised above the leaves, the rather few flowers scattered to imperfectly verticillate, 5-7 (8) mm. long, on short pedicels 1-2 mm. long; calyx finely and closely
short-villous, the lips subequal, about 3 mm. long; petals bluish, the banner subequal to the wings
and only slightly reflexed from them, often white-centered, with or without violet blotches, the margins scarcely at all reflexed; keel ciliate on the upper half, blunt or very short-beaked, the tip upturned slightly; pods 2-3 cm. long, brownish with appressed hairs; seeds 5-7, about 2.5 mm. long,
smooth, grayish with fine, darker mottling. N=24.

Mostly w. of the Cascade Mts., in gravelly areas and drier "prairies," B.C. to Calif., and up the
Columbia R. Gorge to n.e. Oreg. Mar.-June.

Often growing with *L. bicolor* and intergrading with it to the extent that the specific boundary between the two is considerably blurred.

Lupinus microcarpus Sims, Curtis' Bot. Mag. 50:pl. 2413. 1823. (Described from a cultivated specimen; the seed supposedly from Chile)

L. ruber Heller, Muhl. 2:73. 1905. *L. microcarpus* var. *ruber* C. P. Smith, Bull. Torrey Club 45:
10. 1918. *(Heller 7827,* Tehachapi, Calif., May 5, 1905) = var. *microcarpus.*

L. subvexus C. P. Smith, Bull. Torrey Club 44:405. 1917. *(Heller & Brown 5415,* Madison, Yolo
Co., Calif., Apr. 29, 1902) = var. *microcarpus?*

L. subvexus var. *fluviatilis* C. P. Smith, Bull. Torrey Club 45:14. 1918. *(G. R. Vasey 259,* Wash.,
in 1889) = var. *microcarpus.*

L. subvexus var. *transmontanus* C. P. Smith, Bull. Torrey Club 45:15. 1918. *(T. Howell,* Wasco
Co., Oreg., May, 1885) = var. *microcarpus.*

L. subvexus var. *leibergii* C. P. Smith, Bull. Torrey Club 45:16. 1918. *(Leiberg 317,* near Prineville, Crook Co., Oreg., in 1894) = var. *microcarpus.*

L. densiflorus var. *scopulorum* C. P. Smith, Bull. Torrey Club 45:201. 1918. *L. microcarpus* var.
scopulorum C. P. Smith, Bull. Torrey Club 51:100. 1924. *(J. Macoun 21,* Beacon Hill, Vancouver I., B.C., July 4, 1889)

Annual, usually well branched, the stems spreading to erect, 0.5-4 dm. long, generally fistulose
at least at the base, sparsely to copiously long brownish-pilose, the pilosity more abundant on the
petioles, lower surface of the leaflets, and in the inflorescence, including the calyx; leaflets 6-9, oblanceolate, 2.5-4 cm. long, glabrous on the upper surface; petioles several times as long as the
leaflets; racemes from shorter to much longer than the leaves; flowers spreading to ascending, 1-1.5
cm. long, distinctly verticillate, the whorls 3-7, approximate or the internodes 2 or 3 times as long
as the flowers; calyx deeply bilobed, the upper lip very short, membranous, only slightly notched,
the lower lip greenish, 6-11 mm. long, 3-toothed at the tip; corolla pale yellow or pinkish to dark red

keel

½

Lupinus littoralis

3

banner

3

3

var. leucophyllus

var. tenuispicus

Lupinus leucophyllus

JRJ

or reddish-purple, the banner well reflexed; keel ciliate on the lower half; pods 1-2 cm. long, ovate, densely to sparsely long-pilose; seeds 2, wrinkled, 3-5 mm. long, light brown to olive.

Dry to moist soil; Vancouver I. and coastal Puget Sound southward, on the e. side of the Cascade Mts., to Baja Calif.; Chile. May-July.

This, the most showy of our annual species, has an involved taxonomic history, and the specific name used, based as it is on a cultivated plant of somewhat dubious ultimate origin, is open to question. The stature of the plant and the length of the racemes (characters on which, in the main, several segregates have been based) vary greatly in response to the habitat. Although the plants of Vancouver I. and coastal Wash. appear to differ slightly from those that are transmontane in the Northwest, they surely are members of one species and represent a type of geographic distribution that is paralleled by several other semidesert plants of our area. Assuming that our species is the same as that originally described from Chile, our two varieties may be delimited as follows:

1 Flowers pale yellowish; Vancouver I. and adj. islands of Wash.   var. scopulorum C. P. Smith
1 Flowers pink to reddish-purple; Yakima Valley s. to Baja Calif.; S. Am.   var. microcarpus

Lupinus polyphyllus Lindl. Bot. Reg. 13:pl. 1096. 1827. (Douglas, "northwest North America")
    L. grandifolius Lindl. ex Agardh, Syn. Gen. Lup. 18. 1835. L. polyphyllus var. grandifolius T. & G. Fl. N.Am. 1:375. 1840. (Douglas, "California") = var. polyphyllus.
    LUPINUS POLYPHYLLUS var. BURKEI (Wats.) C. L. Hitchc. hoc loc. L. burkei Wats. Proc. Am. Acad. 8:525. 1873. (Burke, "Snake Country")
    L. prunophilus M. E. Jones, Contr. West. Bot. 13:7. 1910. L. wyethii var. prunophilus C. P. Smith in St. John Fl. S. E. Wash. 2:229. 1937. L. arcticus var. prunophilus C. P. Smith, Sp. Lup. 235. 1940. L. polyphyllus ssp. polyphyllus var. prunophilus Phillips, Res. Stud. State Coll. Wash. 23:180. 1955. (M. E. Jones, no locality specified, but [acc. C. P. Smith in Abrams' Flora] from Mammoth Hills, Juab Co., Utah)
    L. apodotropis Heller, Muhl. 7:14. 1911. (Cusick 3359a, West Eagle Creek, Wallowa Mts., Baker Co., Oreg., Aug. 5, 1909) = var. burkei.
    L. pallidipes Heller, Muhl. 7:91. 1911. L. polyphyllus var. pallidipes C. P. Smith, Contr. Dudley Herb. 1:47. 1927. (Heller 10041, Eugene, Oreg., May 18, 1910)
    L. cottoni C. P. Smith, Bull. Torrey Club 51:309. 1924. L. arcticus var. cottonii C. P. Smith in Abrams, Ill. Fl. Pac. St. 2:518. 1944. (J. S. Cotton 1518, Hell Roaring R., Mt. Rainier Nat. For., Yakima Co., Wash., Sept. 6, 1903) = nearest var. polyphyllus, approaching var. burkei.
    L. superbus var. bernardinus Abrams ex C. P. Smith, as reported for Oreg. in Peck, Man. High. Pl. Oreg. 426. 1941 and in Abrams, Ill. Fl. Pac. St. 2:515. 1944 = var. burkei.
    L. garfieldensis C. P. Smith, Sp. Lup. 665. 1949. (Christ & Smith 15413, 7 mi. e. of Pomeroy, Garfield Co., Wash., July 11, 1946) = var. prunophilus. This is a puzzling plant which has been collected several times in Garfield Co.; it is definitely transitional to L. saxosus. Smith, op. cit., cited (apparently inadvertently) Hitchcock & Muhlick 8234, from 6 mi. s. of Pomeroy, both as L. garfieldensis (p. 665) and as L. saxosus but suggestive of L. prunophilus (p. 667).
    L. haudcytisoides C. P. Smith, Sp. Lup. 727. 1952. (C. S. Crocker 42, O'Hara Ranger Station, Idaho Co., Ida., July 31, 1930) Identity not certain, the pedicels are said to be only 1-1.5 mm. long, otherwise the plant seems to belong with var. burkei.
    L. valdepallidus C. P. Smith, Sp. Lup. 732. 1952. (L. O. Miles 155, S. Fk. Salmon R., Garden Valley Ranger Station, Payette Forest, Valley Co., Ida., July 21, 1941) = var. burkei.
Stems 1 to many from a perennial, usually branched crown, simple, commonly fistulose at least at the base, (3) 5-10 (15) dm. tall, nearly glabrous, or more or less strigillose to sparsely or thickly hirsute, usually with rufous hairs; basal and lower cauline leaves with petioles 3-6 times as long as the blades, the petioles of the 3-7 cauline leaves reduced upward, the upper ones sometimes shorter than the blades, with the same pubescence as the stems but often more densely hairy; leaflets mostly (5) 9-13 (18), elliptic-oblanceolate, usually rounded to acute or acuminate, (3) 4-10 (15) cm. long, 1-2.5 cm. broad, mostly glabrous or glabrate (very occasionally somewhat strigillose) on the upper surface, sparsely to densely strigillose or pilose-hirsute on the lower surface; racemes usually dense, mostly 1.5-4 dm. long, the flowers not verticillate, 10-15 mm. long, bluish to violet; pedicels (3) 4-8 mm. long, spreading or ascending but none erect after anthesis; calyx strigillose to sericeous, slightly if at all saccate, never spurred, the upper lip shallowly bidentate, the lower entire; banner glabrous, well reflexed (index 25-30); wings glabrous; keel very slightly exposed, (ciliate) nonciliate; pods 3-5 cm. long, hairy; seeds 4-8, grayish, mottled with dark brown. N=24.

Stream banks, meadows, and moist forest; B.C. to Calif., from the coast into the mountains, e. to

½

calyx

3

keel

3

3

3

3

var. scopulorum

½

½

var. microcarpus

Lupinus micranthus

L. microcarpus

J

Alta. and s. to Colo.; the most lush and rank-growing of our native species, characteristically occurring along streams or in fairly moist soil in our mountains. June-Aug.

In the greater part of its range, and especially w. of the Cascades, *L. polyphyllus* is a very distinctive species, notable for its fistulose stems, broad leaflets, sparse pubescence and elongate, closely flowered racemes that are usually considerably contracted below that particular portion in actual blossom. East of the Cascades, the species is usually considerably dwarfed, but otherwise not appreciably different from the (mostly) coastal var. *polyphyllus*. In two areas especially, however, it undergoes much range in the quantity and type of pubescence. In the region of the Wenatchee Mts., Wash., the plants are often much more strongly strigillose. This variation is considered to result from hybridization and introgression with *L. wyethii*. No useful purpose can be served by giving infraspecific names to these extremely diverse individuals which completely bridge the morphological gap between the two species. In s.e. Wash. and adj. Oreg. and Ida. there is a gradient series from the (nearly) glabrous or strigillose plants to a strongly rufous-hirsute race that closely resembles *L. saxosus*, but is usually separable from it on calyx character.

The species not uncommonly crosses with others, and in the region of the Blue Mts., s.e. Wash., frequently produces hybrids with *L. sulphureus (Hitchcock & Muhlick 21839)*.

The following key will distinguish (perhaps somewhat deceptively) the several races in our area:

1 Plants strongly fistulose stemmed, mostly over 6 dm. tall; chiefly in and w. of the
  Cascades Mts.
  2 Stems sparsely hirsute with long, white or rufous, stiff hairs; Willamette Valley,
    Oreg., n. to Thurston Co., Wash.   var. pallidipes (Heller) C. P. Smith
  2 Stems glabrous or strigillose; B. C. to Calif., from the coast into (and sometimes
    e. of) the Cascade summit, occasional in n.e. Oreg. and Ida.   var. polyphyllus
1 Plants usually only slightly fistulose stemmed, averaging less than 6 dm. tall
  3 Stems, petioles, and lower surface of the leaflets glabrous to strigillose, rarely
    with short, whitish, spreading pubescence; mainly from the e. side of the Cascades, from B. C. to Calif., e. to Ida., Mont., and southward, with much variation in the Wenatchee Mts., Wash.   var. burkei (Wats.) C. L. Hitchc.
  3 Stems, petioles, and lower surface of the leaflets strongly rufous-hirsute; chiefly
    in c. and e. Wash. to adj. Ida. and n.e. Oreg., approached by occasional plants
    from other areas in Ida., Nev., and Utah   var. prunophilus (M. E. Jones) Phillips

Lupinus pusillus Pursh, Fl. Am. Sept. 468. 1814. *(Lewis, "On the banks of the Missouri")*
  L. intermontanus Heller, Muhl. 8:87. 1912. *L. pusillus* var. *intermontanus* C. P. Smith, Bull.
    Torrey Club 46:408. 1919. *L. pusillus* ssp. *intermontanus* Dunn, Leafl. West. Bot. 7:255. 1955.
    (Heller 9599, Wadsworth, Churchill Co., Nev., May 7, 1909)
  L. allimicranthus C. P. Smith, Sp. Lup. 319. 1942. *(R. J. Davis*, Warm R. Dugway, Fremont Co.,
    Ida., June, 1930) = var. *intermontanus*—with some question, as the flowers are said to be only 5-7
    mm. long.

Low, villous annual 0.5-2 dm. tall, usually freely branched from the base; leaflets 5-7, oblanceolate to nearly elliptic, glabrous above, but more or less appressed-villous on the lower surface, usually over half as long as the petiole; racemes short-pedunculate, from considerably shorter to only slightly longer than the leaves; pedicels 1.5-2.5 mm. long; flowers scattered, not verticillate but usually rather closely crowded, 7-12 mm. long; calyx bilabiate, the upper lip about 1/3 as long as the lower; corolla light blue to deep bluish-purple, the banner rather well reflexed; keel not ciliate; pods 1-2 cm. long, oblong, shaggy-villous; seeds 2, light brown, wrinkled. N=24.

Sandy desert soil, often on dunes; c. Wash. southward to Calif. and Ariz., e. to Alta., the Dakotas, and Neb. May-June.

There are two phases of the plant in our area:

1 Flowers mostly at least 10 mm. long; chiefly e. of the Rocky Mts., from Alta. to
  Colo., but w. along the Snake R. into c. Ida.   var. pusillus
1 Flowers mostly 7-10 mm. long; e. Wash. to Calif. and Ariz., e. to Ida., Nev.,
  Wyo., Utah, and Ariz.   var. intermontanus (Heller) C. P. Smith

Lupinus rivularis Dougl. ex Lindl. Bot. Reg. 19:pl. 1595. 1833. *(Douglas, "America boreali-occidentalis," surely from Wash. or Oreg.)*
  L. amphibium Suksd. Werdenda 1:13. 1923. *(Suksdorf 7154*, on rocky island in Columbia River at
    Bingen, Klickitat Co., Wash., Sept. 19, 1910). The flowers with the keel only slightly ciliate.

2
keel

2

var. pallidipes

var. pallidipes

var. burkei

var. prunophilus

var.
polyphyllus

Lupinus polyphyllus

JRJ

½ ½ ½ ½

Stems usually several from a rather woody, often somewhat decumbent, perennial base, mostly slightly fistulose below, rather sparsely strigillose, (3) 4-10 dm. tall, typically with 8-20 nodes, the upper ones (at least) generally with reduced axillary racemes that are mostly incompletely developed at the time the terminal inflorescence starts to blossom; leaves rather short-petiolate, the leaflets of the lower ones generally deciduous before anthesis, the upper ones with thinly strigillose petioles from scarcely as long to as much as twice as long as the blades; leaflets mostly 6-10, oblanceolate, greenish, rather fleshy, 2.5-4 cm. long, 5-10 mm. broad, usually slightly apiculate, from equally strigillose on both surfaces to considerably less hairy on the upper; terminal racemes mostly 1-1.5 (2) dm. long, generally with peduncles about equaling the uppermost leaves; flowers 12-16 mm. long, borne in somewhat distant verticils, pale to rather deep blue, often more or less varicolored, fading to purplish-brown; pedicels 4-8 mm. long; calyx sericeous to strigillose, neither at all saccate nor spurred, the upper lip (4) 5-7 mm. long, shallowly bidentate, mostly (2.5) 3-5 times as long as the calyx tube, the lower lip entire to inconspicuously toothed near the tip, about as broad as the upper lip but slightly longer; banner well reflexed from the keel (index 20-30), glabrous on the back; wings glabrous; keel ciliate, arcuate but seldom exposed; pods 3-4.5 cm. long, about 8 mm. broad; seeds 6-10, grayish and strongly darker mottled. N=24.

Gravelly prairies, riverbanks, open woods, etc., always at low elevations; B.C. to n. Calif., on the w. side of the Cascades. Apr.-June (Sept.).

Although the type collection of *L. amphibium* is probably referable to this species on the basis of its branched stems, large flowers, glabrous banner, and nearly or quite glabrous keel (the pubescence of the keel varying in the same raceme), the plant is also suggestive of *L. sericeus* and *L. latifolius*. It does not very closely resemble any other specimen seen, and seems to be an aberrant individual rather than representative of a natural population.

Lupinus sabinii Dougl. ex Hook. Fl. Bor. Am. 1:166. 1833.

> L. *sericeus* ssp. *sabinii* Phillips, Res. Stud. State Coll. Wash. 23:168. 1955. *(Douglas,* "On the Blue Mountains of North-West America, and on the Dividing Ridge of the Rocky Mountains, near the confines of perpetual snow")

Strong perennial from a branched crown, with several stems (5) 7-12 dm. tall, more or less strigose-sericeous, the hairs often yellowish; leaves mostly cauline (at anthesis), the lower with petioles up to twice the length of the blades, the upper with petioles and blades about equal, leaflets about 9 (8-11), elliptic-oblanceolate to oblanceolate and acute, (4) 6-12 (15) cm. long, (8) 10-15 (25) mm. broad, strigose on both surfaces; peduncles usually about equaling the leaves; racemes 1.5-3 dm. long; flowers subverticillate, numerous (but not closely crowded), (13) 15-18 mm. long; pedicels 8-11 mm. long; calyx sericeous, slightly oblique but not truly saccate or spurred, the upper lip shallowly notched, the lower entire; petals usually bright yellow, although sometimes purplish-tinged; banner glabrous or with very few hairs on the back, the index 15-30; wings glabrous; keel ciliate; pods 3-4.5 cm. long, 11-15 mm. broad, yellowish-sericeous; seeds 4-7, pinkish-brown. N=24.

In the Blue Mts. of s.e. Wash. and n.e. Oreg., mostly in open ponderosa pine forest. Late May-early June.

In many respects these plants are suggestive of *L. sericeus*, but they have larger leaflets and a nearly or quite glabrous banner in addition to other minor but distinctive features.

Lupinus saxosus Howell, Erythea 1:110. 1893. *(Th. Howell,* "On high stony ridges, from near the Dalles eastward, in Oregon and Washington")

> L. *subsericeus* Robins. ex Piper, Contr. U.S. Nat. Herb. 11:354. 1906. *L. saxosus* var. *subsericeus* C.P. Smith in Abrams, Ill. Fl. Pac. St. 2:518. 1944. *(Whited 602,* Ellensburg, Wash., May 5, 1898)

Perennial with a branched crown; stems 1.5-2.5 (3) dm. tall, seldom branched, usually villous with rusty hairs, sometimes the pubescence nearly all appressed; leaves mostly basal; petioles of the basal leaves 2-4 (5) times as long as the blades, those of the upper leaves subequal to the blades; leaflets 7-12, rather broadly oblanceolate, 1.5-3 (4) cm. long, rounded, nearly always rubiginose-villous on the lower surface, usually glabrous above; racemes generally produced well above the leaves, closely flowered, 5-15 cm. long; pedicels 4-8 mm. long; flowers 10-14 mm. long, deep violet, the banner usually yellowish-centered; calyx villous to sericeous, not at all saccate or spurred, the upper lip deeply lobed nearly half the length, the lower lip entire; banner glabrous, well reflexed (index 25-30); wings glabrous; keel ciliate; pods 1.5-2 (2.5) cm. long; seeds 2-4. N=48.

Chiefly on dry rimrock, usually in areas of sagebrush; c. Wash. through e. Oreg. to n. Calif.,

2.5

1/2

1/2

2.5

2.5
keel

2.5
banner

1/2

L. rivularis

1/2

1/2

1/2

L. sabinii

3

3

3

calyx

1/2

var. pusillus

var. intermontanus

Lupinus rivularis

L. pusillus

JRJ

entirely e. of the Cascade crests. Apr.-May.

Lupinus saxosus is in general a clearly marked taxon, with short, rounded, rufous-villous leaflets that are unlike any other of our species. Although *L. polyphyllus* var. *prunophilus* might be confused with it, since the pubescence is similar, that plant has much longer, more pointed leaflets, and a distinctly dissimilar calyx.

The var. *subsericeus* (Robins.) C. P. Smith was described on the basis of the appressed pubescence, but it does not appear to constitute either a geographical or an ecological race, since the species as a whole varies from spreading- to appressed-pubescent.

Lupinus sericeus Pursh, Fl. Am. Sept. 468. 1814. *(Lewis,* "On the banks of the Kooskoosky," Ida.)
> L. *ornatus* Dougl. ex Lindl. Bot. Reg. 14:pl. 1216. 1828. *(Douglas,* "On the banks of the Spokane River, near Kettle Falls on the Columbia River") = var. *sericeus.*
> L. *leucopsis* Agardh, Syn. Gen. Lup. 29. 1835. *(Douglas,* "Hab. in America Boreali-occidentali") = var. *sericeus,* or as probably = *L. sulphureus.*
> L. *flexuosus* Lindl. ex Agardh, Syn. Gen. Lup. 34. 1835. L. *sericeus* var. *flexuosus* C. P. Smith, Bull. Torrey Club 51:307. 1924. *(Douglas,* "America Boreali-occidentali") = var. *sericeus.*
> L. *ornatus* ssp. *bracteatus* Robins. ex Piper, Contr. U.S. Nat. Herb. 11:355. 1906. *(Henderson 2338,* Spokane, Wash.) = var. *sericeus.*
> L. *alpicola* Henderson ex Piper, Contr. U.S. Nat. Herb. 11:355. 1906. *(Henderson 1387,* Mt. Adams, Wash.) = var. *sericeus,* or possibly = *L. latifolius* var. *thompsonianus.*
> L. *subulatus* Rydb. Bull. Torrey Club 34:43. 1907. *(R. S. Williams 3120,* Columbia Falls, Mont., June 3, 1897) = var. *sericeus.*
> L. *obtusilobus* Heller, Muhl. 8:115. 1912. L. *ornatus* var. *obtusilobus* C. P. Smith, Bull. Torrey Club 51:307. 1924. *(G. B. Grant 730,* Mt. Shasta, Calif., Sept., 1902) = var. *sericeus.*
> L. *sericeus* var. *egglestonianus* C. P. Smith, Sp. Lup. 104. 1939. L. *egglestonianus* C. P. Smith, Sp. Lup. 664. 1949. *(C. P. Smith 35127,* 2 mi. s. of Grass Valley, Sherman Co., Oreg.) = var. *sericeus.* Referred here with some doubt, possibly more properly belonging under *L. leucophyllus.*
> L. *sericeus* var. *wallowensis* C. P. Smith, Sp. Lup. 105. 1939. *(E. W. Lyle,* 5 mi. s. of Chico R. S., Enterprise Rd., Wallowa Mts., Oreg., July, 1930) = var. *sericeus.*

LUPINUS SERICEUS var. FIKERANUS (C. P. Smith) C. L. Hitchc. hoc loc. L. *fikeranus* C. P. Smith, Sp. Lup. 236. 1940. *(C. B. Fiker 887,* Scotch Creek basin, Okanogan Co., Wash., June 21, 1932)
> L. *tuckeranus* C. P. Smith, Sp. Lup. 318. 1942. *(H. M. Tucker 1193,* Moore Cr., Boise Co., Ida., July, 1938) = var. *sericeus.*
> L. *enodatus* C. P. Smith, Sp. Lup. 689. 1951. *(L. Benson 1646,* n. of Lewiston, overlooking the Clearwater R., probably in Nez Perce Co., Ida., June 16, 1929) = var. *sericeus.*
> L. *lyman-bensoni* C. P. Smith, Sp. Lup. 690. 1951. *(L. Benson 1635,* State Line between Spokane Co., Wash., and Post Falls, Kootenai Co., Ida., June 15, 1929) = var. *sericeus.*
> L. *diaboli-septem* C. P. Smith, Sp. Lup. 691. 1951. *(M. E. Jones 26583,* Seven Devil Mts., Washington Co., Ida., Aug. 5, 1899) = var. *sericeus.*
> L. *falsocomatus* C. P. Smith, Sp. Lup. 736. 1952. *(Kenneth Pearse 38,* Fall Creek Enclosure, Boise Forest, Elmore Co., Ida., Aug. 29, 1930) = var. *sericeus.*
> L. *herman-workii* C. P. Smith, Sp. Lup. 741. 1952. *(Herman Work 421,* head of Swamp Cr., Challis Forest, Custer Co., Ida., Aug. 22, 1914) = var. *sericeus.*
> L. *buckinghami* C. P. Smith, Sp. Lup. 741. 1952. *(Arthur Buckingham 103,* near Loon Creek Ranger Station, Challis Forest, Custer Co., Ida., July 8, 1932) = var. *sericeus.*
> L. *equi-collis* C. P. Smith, Sp. Lup. 742. 1952. *(Chas. Shaw 1207,* Horse-Hill-Bonanza, Challis Forest, Custer Co., Ida., July 25, 1916) Immature specimen, referred here because of the hairy banner.
> L. *forslingi* C. P. Smith, Sp. Lup. 743. 1952. *(C. L. Forsling S-27,* U.S. Sheep Exp. Sta., Clark Co., Ida., Aug. 4, 1923) = var. *sericeus.*
> L. *hiulcoflorus* C. P. Smith, Sp. Lup. 745. 1952. *(Ray Pickett 35,* Targhee Forest, Clark Co., Ida., July 24, 1925) = var. *sericeus.*
> L. *spiraeaphilus* C. P. Smith, Sp. Lup. 746. 1952. *(R. H. Hall,* Ranger Station Pump House, Targhee Forest, Fremont Co., Ida., July 10, 1936) = var. *sericeus.*

LUPINUS SERICEUS var. ASOTINENSIS (Phillips) C. L. Hitchc. hoc loc. L. *sericeus* ssp. *asotinensis* Phillips, Res. Stud. State Coll. Wash. 23:170. 1955. *(Phillips 181,* 10 mi. s.w. of Clarkston on highway 3K, Asotin Co., Wash.)

Perennial, generally with a branching crown and (1 to) several stems 2-5 (8) dm. tall, simple or sparingly branched above to (rarely) freely branched from the base, usually more or less silvery to somewhat rust-colored from the spreading to appressed pubescence which generally is of two distinct lengths; leaves at flowering time almost entirely cauline, the petioles of the lower ones as much as 3 times as long as the blades, reduced upward, the petioles of the upper leaves subequal to the blades; leaflets (6) 7-9, oblanceolate, 3-6 (9) cm. long, mostly 3-6 (9) mm. broad, about equally sericeous-silky on the two surfaces; racemes rather loose, (5) 10-15 cm. long, usually several times as long as the peduncles; pedicels mostly 4-7 mm. long; flowers scattered to imperfectly verticillate, (9) 10-12 (13) mm. long, variable in color, mostly lavender or blue, but sometimes yellowish or whitish; calyx silky, not spurred although sometimes slightly saccate, deeply bilabiate, the lower lip entire, the upper bidentate; banner well reflexed from the keel (index from 14 to 25), silky-hairy on the back for over 2/3 of the length, usually whitish- or yellow-centered; wings usually glabrous, occasionally with a few hairs near the base of the blade; keel ciliate (sometimes for almost the full length), the tip upturned; pods silky, 2-3 cm. long, nearly 1 cm. broad; seeds 3-5, light pinkish-brown. N=24.

British Columbia southward, e. of the Cascade Mts., to Calif. and Ariz., e. to Alta., and through the Rocky Mts. to N. M. May-Aug.

Lupinus sericeus varies even more than the majority of our other lupines, in both floral and vegetative characters, the pubescence ranging from appressed (and then usually short) to longer and spreading, and from very abundant to rather sparse. The flowers show considerable range in size and in the amount of pubescence on the back of the banner. Although many segregate species have been described, three (*L. flexuosus,* with longer floral bracts and spreading pubescence; *L. ornatus,* with abundant appressed pubescence; and *L. alpicola,* with sparse appressed pubescence) are more frequently recognized than the others, although they appear to represent little more than extreme phases in the pattern of variation, without distinctive natural ranges.

Albino plants occur sporadically throughout the range, and are particularly common in Okanogan Co., Wash., but mostly do not seem to represent true geographic races. However, in s.e. Wash., from the Snake R. southward, there occurs an albino phase that completely replaces the blue-flowered one in most of upland, nonforested Asotin Co. These plants tend to have a simple to sparingly branched crown and to have rather moderately hairy banners, but on the whole are remarkable chiefly because of their uniformity. Where they were seen in company with the var. *sericeus* (at the edge of the range to the e.) they did not appear to be intergrading with it.

Our races of the species are separable as follows:

1 Plant usually with 1-3 (4) flowering stems; flowers white, banner only moderately
   pubescent (mostly in a median line) on the back; nonforested Asotin Co., Wash.,
    from the Snake R. s., also n. of the Snake R. into Whitman Co., where tending
    more constantly to be multistemmed   var. asotinensis (Phillips) C. L. Hitchc.
1 Plant usually several-stemmed; flowers mostly bluish or lavender; banner often
   abundantly pubescent on the back
   2 Banner only sparsely hairy on the back; mostly along the e. base of the Cascade
    Mts., from Okanogan Co., Wash., to Hood River Co., Oreg.
                          var. fikeranus (C. P. Smith) C. L. Hitchc.
   2 Banner rather copiously pubescent over much of the back; with the range of the
    species   var. sericeus

The plants here called var. *fikeranus* are those referred to by Piper (Contr. U.S. Nat. Herb. 11:355. 1906) in the last paragraph under *L. sericeus.* They are most abundant in the general area where *L. wyethii* and *L. polyphyllus* are freely intergradient, although not entirely limited to that region, since they are occasional e. to Spokane and s. to Wasco Co., Oreg., and are not greatly different from the plants of Mt. Adams that were called *"L. alpicola* Henderson" by Piper.

Lupinus sulphureus Dougl. ex Hook. Fl. Bor. Am. 1:166. 1833. (*Douglas,* "Blue Mountains of North-
   West America, on elevated grounds near the source of the Clarke's River")
   L. leucopsis Agardh, Syn. Gen. Lup. 29. 1835. (*Douglas,* "Hab. in America Boreali-occidentali")
    = var. *sulphureus,* or possibly more properly referred to *L. sericeus.*
   L. oreganus Heller, Muhl. 7:89. 1911. (*Heller 10044,* Eugene, Lane Co., Oreg., May 18, 1910) =
    var. *kincaidii.*
   L. mollis Heller, Muhl. 8:105. 1912. *L. leucopsis* var. *mollis* C. P. Smith, Sp. Lup. 112. 1939.
    (*Heller 10080,* The Dalles, Wasco Co., Oreg., May 23, 1910) = var. *subsaccatus.*
   L. amabilis Heller, Muhl. 8:114. 1912. (*Heller 10043,* Eugene, Oreg., May 18, 1910) = var. *kincaidii.*

3

keel   2

2

2

banner   Lupinus saxosus

½

var. asotinensis   ½   var. sericeus   ½   var. fikeranus

½

2   2   2

Lupinus sericeus

JRJ

L. bingenensis Suksd. Werdenda 1:12. 1923. *L. leucopsis* var. *bingenensis* C. P. Smith, Sp. Lup. 111. 1939. *(Suksdorf 5036,* near Bingen, Klickitat Co. , Wash. , Apr. 24, and June 12, 1905) = var. *subsaccatus.*

LUPINUS SULPHUREUS var. SUBSACCATUS (Suksd.) C. L. Hitchc. hoc loc. *L. bingenensis* var. *subsaccatus* Suksd. Werdenda 1:13. 1923. *L. sulphureus* ssp. *subsaccatus* Phillips, Res. Stud. State Coll. Wash. 23:193. 1955. *(Suksdorf 11405,* Bingen Mt. , near Bingen, Klickitat Co. , Wash. , May, 1923)

LUPINUS SULPHUREUS var. KINCAIDII (C. P. Smith) C. L. Hitchc. hoc loc. *L. oreganus* var. *kincaidi* C. P. Smith, Bull. Torrey Club 51:305. 1924. *L. sulphureus* ssp. *kincaidii* Phillips, Res. Stud. State Coll. Wash. 23:193. 1955. *(T. Kincaid,* Corvallis, Benton Co., Oreg., June 8, 1898)

L. bingenensis var. *dubius* C. P. Smith, Bull. Torrey Club 51:305. 1924. *L. leucopsis* var. *dubius* C. P. Smith, Sp. Lup. 111. 1939. *(J. S. Cotton 1106,* Prosser, Benton Co. , Wash. , May 27, 1903) = var. *subsaccatus.*

L. bingenensis vars. *albus* and *roseus* Suksd. Werdenda 1:24. 1937. *(Suksdorf 10191* and *11855,* respectively, Bingen, Klickitat Co. , Wash.) = var. *subsaccatus.*

L. leucopsis var. *shermanensis* C. P. Smith, Sp. Lup. 111. 1939. *(W. H. Baker 21,* Columbia R. Hiway near Biggs Junction, Oreg., Apr., 1938) = var. *subsaccatus*—a particularly puzzling collection combining characters of *L. caudatus* (saccate calyx) with the pubescence and branching of *L. latifolius* var. *thompsonianus,* to which it is as similar as to var. *subsaccatus.*

L. leucopsis var. *hendersonianus* C. P. Smith, Sp. Lup. 111. 1939. *(Henderson 14418,* Eugene, Lane Co. , Oreg. , May, 1932) = var. *kincaidii.*

L. ostiofluminis C. P. Smith, Sp. Lup. 239. 1940. *(H. T. Rogers 390,* confluence of Spokane R. with the Columbia, Lincoln Co. , Wash.) = var. *subsaccatus.*

Perennial with a branched crown, the usually numerous stems mostly simple, (3) 4-8 (10) dm. tall, strigillose to somewhat whitish- or brownish-sericeous; basal leaves usually present and persistent until after flowering, the lowermost petioles (2) 3-5 times as long as the blades, the upper cauline leaves with petioles sometimes shorter than the blades; leaflets (7) 9-11 (12), rather narrowly oblanceolate, usually acutish, (2) 2.5-4 (5) cm. long, from nearly equally strigillose or sericeous on both surfaces to glabrous above and sparsely to copiously hairy beneath; racemes short-pedunculate, 6-15 (18) cm. long; flowers rather numerous but not closely crowded, scattered to imperfectly verticillate, 9-12 mm. long, from yellowish to bluish or purple; pedicels (2) 4-10 mm. long; calyx silky, often noticeably asymmetrical but not truly saccate or spurred, the upper lip bidentate, the lower entire; banner not much reflexed from the wings and keel (index 4-10), glabrous or very sparsely hairy on the back; wings glabrous; keel usually ciliate most of the length but sometimes glabrous; pods 2-3 cm. long; seeds 4-5, pinkish-brown.

From B.C. southward, e. of the Cascade Mts. , to Calif. , spreading through the Columbia R. Gorge to s.w. Wash. and the Willamette Valley, e. to s.e. Wash. , n.e. Oreg. , and probably to adj. Ida. Apr.-June.

Lupinus sulphureus is a rather distinctive species, characterized by the fairly small flowers with only slightly reflexed and glabrous or very slightly pubescent banner, usually somewhat gibbous calyx, and strigillose to sericeous, mostly narrow leaflets. However, in certain areas it loses much of its specific sharpness, apparently through crossing with other species. In the Columbia R. Gorge it shows some intermediacy with *L. sericeus,* whereas in the area slightly s. of Pomeroy, Garfield Co. , Wash. , it apparently is involved in the widespread hybridization that more closely involves *L. saxosus* and *L. polyphyllus.* Although occasional specimens of *L. sulphureus* are intermediate to *L. leucophyllus,* they probably will most often be keyed to the latter species because of the usual fairly pubescent banner, but this is not true of a few of these plants, especially some from Chelan Co. , Wash. , (e.g. *Kruckeberg 3258),* which, because of their glabrous banners, will key to *L. sulphureus.* Such individuals are intermediate in other respects, having pubescence in general more like that of *L. sulphureus,* but with the tight, scarcely interrupted, semispicate racemes so characteristic of *L. leucophyllus,* to which species they might arbitrarily be referred, in spite of their intermediacy.

The calyx varies from only slightly asymmetrical to definitely gibbous in the var. *sulphureus,* and hence tends to approach *L. laxiflorus.* Plants (chiefly) from n. e. Wash. , Ida. , and n. Mont. , that have the calyx only slightly more gibbous than in var. *sulphureus,* were recently included under *L. sulphureus* as ssp. *whithamii* (Smith & St. John) Phillips. Such specimens are also characterized by the glabrous or subglabrous upper surface of the leaflets, and by more abundant pubescence on the back of the banner than normal for the plants here included under *L. sulphureus;* they are therefore referred to *L. laxiflorus* var. *pseudoparviflorus* (Rydb.) Smith & St. John.

There are three recognizable geographic races of the species in our area, mostly readily separable as follows:

1 Flowers usually yellow; plants of the Blue Mts. of s.e. Wash. and n.e. Oreg.; not such
  a notable variant as usually considered, since the corollas undergo a wide range in
  color, sometimes being more nearly white, and often varying to bluish; furthermore,
  yellowish-flowered (rather than true albino) plants occur more or less sporadically
  elsewhere, but are especially common in Okanogan Co., Wash.; N=24   var. sulphureus
1 Flowers generally blue or purplish, but if white then mostly mere albino variants
  that do not usually comprise self-perpetuating populations
  2 Leaflets pubescent on both surfaces; racemes mostly less than 10 cm. long; B.C.
    southward to c. Oreg., mostly along the e. base of the Cascade Mts. and in the
    e. end of the Columbia R. Gorge, but fairly common in the sagebrush and ponderosa
    pine area farther e., although not at present known from Ida.; N=24, 48
                                                          var. subsaccatus (Suksd.) C. L. Hitchc.
  2 Leaflets glabrous on the upper surface; racemes mostly 10-18 cm. long; Willamette
    Valley s. to Douglas Co., Oreg.; N=24   var. kincaidii (C. P. Smith) C. L. Hitchc.

Lupinus wyethii Wats. Proc. Am. Acad. 8:525. 1873. *(Wyeth,* Flathead River)
  L. humicola A. Nels. Bull. Torrey Club 25:204. 1898. *L. arcticus* var. *humicola* C. P. Smith in
    Abrams, Ill. Fl. Pac. St. 2:518. 1944. *(Nelson 151,* Laramie Hills, Albany Co., Wyo., June 2,
    1894, is the first of two collections cited, both apparently collected by E. Nelson)
  L. humicola var. *tetonensis* E. Nels. Bot. Gaz. 30:120. 1900. *L. arcticus* var. *tetonensis* C. P.
    Smith in Abrams, Ill. Fl. Pac. St. 2:518. 1944. *(A. & E. Nelson 6341,* Teton Mts., Uinta Co.,
    Wyo., Aug. 16, 1899—the printed labels of what is believed to be this collection bear the number
    6541)
  L. candicans Rydb. Bull. Torrey Club 28:35. 1901. *(E. V. Wilcox 451,* Mont., in 1900)
  L. flavescens Rydb. Bull. Torrey Club 29:245. 1902. *(Wyeth,* Medicine Clay Prairies, Ida. or
    Mont.)
  L. comatus Rydb. Bull. Torrey Club 30:257. 1903. *(F. N. Pease,* Lake City, Colo., in 1878)
  L. rydbergii Blank. Mont. Agr. Coll. Stud. Bot. 1:78. 1904. *(Blankinship,* Big Coulee Cr., Sweet
    Grass Co., Mont., June 15, 1902, is the first of several collections cited) Possibly this is *L. lepi-
    dus.*
  L. diversalpicola C. P. Smith, Sp. Lup. 237. 1940. *(J. W. Thompson 15057,* Nelson Ridge, near
    Mt. Aix, Yakima Co., Wash., July, 1940)
  L. amniculi-salicis C. P. Smith, Sp. Lup. 737. 1952. *(R. B. Johnson 251,* Willow Creek, Sawtooth
    Forest, Elmore Co., Ida., July 20, 1931) Referred here with some doubt, possibly more properly
    placed under *L. lepidus.*
Multicipital perennial (2) 4-5 (6) dm. tall, the stems not at all (or only slightly) fistulose, simple,
strigose (as also the petioles) or with yellowish villosity and with shorter, finer hairs; basal leaves
numerous, their petioles 3-6 times as long as the blades; cauline leaves with petioles much reduced
upward and often shorter than the blades; leaflets (8) 9-11 (13), usually narrow, from linear-oblanceolate to elliptic-oblanceolate, rounded or more commonly abruptly acuminate to apiculate, mostly 2-
4 (up to 6 or even 7) cm. long, 2-6 (14) mm. broad, yellowish (grayish)-strigillose on both surfaces
(very occasionally glabrous above); racemes borne well above the leaves, mostly 1.5-2.5 dm. long,
many-flowered but eventually fairly open, usually tapered from the base; pedicels 5-10 mm. long;
flowers 11-16 mm. long, subverticillate to scattered, deep violet or purple, the banner often reddish,
yellowish, or white-centered; calyx sericeous to (more commonly) rather hirsutulous, the upper lip
shallowly bidentate, the lower entire; banner glabrous, well reflexed (index 25-40); wings glabrous;
keel usually (always?) ciliate; pods 2.5-4 cm. long, hairy; seeds 4-8, grayish, mottled. N=24.
  Sagebrush plains and valleys to montane or subalpine forests and open ridges; B.C. to Calif., e. to
Alta., and s. to Colo. May-July.
  Lupinus wyethii is most abundant in Ida. and Mont. and southward, where it is relatively uniform in
nature, the chief variation being an occasional plant with spreading, rather than appressed, pubescence. There is no indication that it interbreeds with *L. polyphyllus* in that area. It reappears at (or
possibly extends to) the eastern edge of the Cascade Mts., from Chelan to Yakima cos., Wash., but in
an extremely inconstant phase that seems to result from free interbreeding with *L. polyphyllus.* Such
plants (called *L. diversalpicola* by Smith) might be considered as varietally distinct from *L. wyethii*
as represented in the Rocky Mts., but no change in nomenclature is here proposed. The assignment

banner

2.5

2.5

2.5

2.5

var. subsaccatus

keel

½

2

2

Lupinus wyethii

½

½

calyx

3

var. sulphureus

Lupinus sulphureus

var. kincaidii

½

JRJ

of many specimens to one or the other of these species must of necessity be based arbitrarily upon the presence or absence of pubescence on the upper surface of the leaves; as might be expected, there are some plants that have such pubescence but which otherwise appear more similar to *L. polyphyllus* than to *L. wyethii*. Aside from the usually dependable difference in pubescence between the two taxa, *L. wyethii* tends to have narrower, more pointed leaflets, somewhat thicker and more open racemes, and shorter stature.

In some regions *L. wyethii* tends to approach, and may sometimes even be mistaken for, *L. lepidus,* but it is in general the taller species, with larger flowers, and a less deeply cleft upper calyx lip.

Medicago L.

Flowers papilionaceous, 2-many in short to fairly elongate spikelike racemes or heads on short axillary peduncles, yellow or bluish-purple (white); calyx teeth equal or subequal, lanceolate-acuminate, usually longer than the calyx tube; banner erect, considerably longer than the wings and keel; stamens 10, diadelphous; legumes 1- to several-seeded, strongly reticulate, from falcate to (usually) spirally coiled when mature, often armed with 2 rows of prickles on the exposed, keeled margins and forming a small prickly bur when mature, indehiscent; annual or perennial herbs with trifoliolate leaves, the leaflets apically dentate to weakly acicular-dentate. N=7, 8.

About 50 species of Eurasia and Africa, our species all introduced as weeds or escaped from cultivation. (Greek name, *medice,* apparently applied to alfalfa which the Greeks introduced from Media.)

Reference:

Shinners, Lloyd H. Authorship and nomenclature of Bur Clovers (Medicago) found wild in the United States. Rhodora 58:1-13. 1956.

1 Plant perennial, deep-rooted; flowers 6-10 mm. long; fruits unarmed
  2 Flowers blue (rarely pink or white); pods spirally coiled; leaflets mostly 2-4 cm. long
           M. SATIVA
  2 Flowers yellow (violet); pods merely slightly falcate; leaflets mostly 1-2 cm. long
           M. FALCATA
1 Plant annual, rather shallow-rooted; flowers less than 6 mm. long; fruits often armed
  3 Pod 1-seeded, unarmed, reniform, curved to less than 1 spiral, only the style spirally
    coiled; flowers 2-3 mm. long   M. LUPULINA
  3 Pod several-seeded, often prickly, spirally coiled as is the style; flowers 4-5 mm. long
    4 Leaflets mostly over 1.5 cm. long, with a central dark spot   M. ARABICA
    4 Leaflets mostly less than 1.5 cm. long, not spotted   M. HISPIDA

Medicago arabica (L.) Hudson, Fl. Ang. 288. 1762.
  M. polymorpha var. *arabica* L. Sp. Pl. 780. 1753. (Europe)
  M. cordata Desr. Lam. Encycl. 3:636. 1789. (Europe)
  M. maculata Sibth. Fl. Oxon. 232. 1794. (Europe)
  M. arabica ssp. *inermis* Ricker, U.S. B.P.I. Bull. 267:33, pl. 12, fig. 1. 1913. (Described from cultivated material)

Annual, more or less pilose with long multicellular hairs; stems prostrate or decumbent, 3-8 dm. long; stipules dentate-lobate about half their length; leaflets obovate-cuneate to obcordate, conspicuously purplish-spotted near the center, mostly 1.5-3 cm. long, rather prominently erose-dentate, sparsely strigillose to glabrate; flowers few (1-5), yellow, about 5 mm. long, borne in much reduced racemes on peduncles 1-3 cm. long; calyx about 3 mm. long, with lanceolate, acuminate teeth somewhat longer than the tube; mature pod several-seeded, spirally coiled, 5-8 mm. broad, the outer margin with 2 rows of grooved prickles 2.5-3.5 mm. long. N=8. Spotted medick.

Introduced and established from Wash. to Calif., mostly in waste places near the coast, occasional throughout much of c. and e. U.S.; native of Europe. Apr.-July.

The occasional plant with unarmed fruit, which has been called ssp. *inermis* Ricker, is not considered of sufficient significance to be accorded special nomenclatural status.

Medicago falcata L. Sp. Pl. 779. 1753. (Europe)
  Perennial with prostrate to erect stems 4-12 dm. long, usually rather thickly strigillose; stipules entire; leaflets narrowly oblanceolate, 1-2 cm. long, 2-4 mm. broad, toothed at the tip only; flowers pale yellow (violet), 6-8 mm. long, 10-50 in short, dense, pedunculate racemes 1-2 cm. long; pod only slightly falcate, pubescent, not spiny, 3-4 mm. long. N=8, 16. Yellow lucerne.

Eurasian, rarely introduced in N. Am., but well established on sagebrush slopes near Midway, B. C. *(J. W. Thompson 14403)*. June-Aug.

Medicago hispida Gaertn. Fruct. 2:349. 1791. (S. Europe)
  Medicago polymorpha L. Sp. Pl. 779. 1753. (Europe)
  Medicago polymorpha var. *nigra* L. Mant. Pl. 2:454. 1771. *Medicago nigra* Willd. Sp. Pl. 4th ed. 3:1418. 1802. (S. Europe)
  Medicago denticulata Willd. Sp. Pl. 4th ed. 3:1414. 1802. *Medica denticulata* Greene, Man. Bay Reg. 102. 1894. (Europe)
  Medicago apiculata Willd. Sp. Pl. 4th ed. 3:1414. 1802. *Medicago hispida* var. *apiculata* Burnat, Fl. Alpes Mar. 2:106. 1896. *Medicago polymorpha* var. *ciliaris* f. *apiculata* Shinners, Rhodora 58:11. 1956. (Europe) The form with short prickles on the fruits.
  Medicago denticulata var. *ciliaris* Ser. in DC. Prodr. 2:176. 1825. *Medicago polymorpha* var. *ciliaris* Shinners, Rhodora 58:9. 1956. (Europe) The more common form of the plant, with conspicuous prickles on the fruits.
  Medicago polycarpa var. *tuberculata* Gren. & Godr. Fl. Fr. 1:390. 1848. *Medicago polymorpha* var. *ciliaris* f. *tuberculata* Shinners, Rhodora 56:11. 1956. (Europe) The form with unarmed fruits.
    Glabrous or sparsely strigillose annual with prostrate to nearly erect stems 1-4 dm. long; stipules deeply lacerate-dentate for 1/2-3/4 their width; leaflets cuneate to obovate or obcordate, 1-2 (but mostly less than 1.5) cm. long, not spotted, denticulate to dentate at the apex; flowers 2-5, yellow, 4-5 mm. long; calyx about 3 mm. long, the teeth slightly longer than the tube; pod several-seeded, spirally coiled, the margins narrowly keeled, spineless or with 2 rows of slightly curved divergent prickles as much as 3 mm. long. N=7, 8. Bur clover.
    Widespread in w. U.S., especially abundant w. of the Cascade Mts. and Sierra Nevada, but found almost throughout the U.S.; native of Europe. Mar.-June.
    Medicago hispida is one of the most important of western introduced European "weeds." In the interior valleys of Calif., especially, it furnishes excellent forage in the early spring for cattle and sheep, the agents that have disseminated the plant so widely.
    Shinners (Rhodora 58:1-13. 1956) maintained that the name, *M. polymorpha* L., should be used for this taxon, and enumerated the several difficulties in typifying the epithets. He suggested that the name should be based upon the later-published *M. polymorpha* var. *nigra*. In the belief that this is an erroneous interpretation of the situation, the name, *M. polymorpha* L., is here rejected in favor of *M. hispida* Gaertn.
    Usually the plants bear prickles on the pods (for which reason they have been called "Bur clover"), but occasional specimens lack prickles. Since there is a continuous series from the condition of unarmed pods to the more common state where the pods have prickles 1.5-3 mm. long, there seems to be no necessity to accord nomenclatural recognition to the plants with unarmed fruits.
    Medicago minima (L.) Bartalini, which differs from *M. hispida* in having conspicuously soft-hairy stems and essentially entire stipules, has been collected in Jackson Co. and near Portland, Oreg., and near Bingen, Wash., and will probably be found elsewhere in our area occasionally.
    Medicago turbinata Willd., another European species, is occasionally found in the U.S., and was collected by Suksdorf at Linnton, near Portland, Oreg. It has a tightly 5-6 times spiralled pod, the margins of which are thickened but not spiny. In *M. hispida,* with which *M. turbinata* is otherwise similar, the pods are not so tightly (3 or 4 times) coiled.

Medicago lupulina L. Sp. Pl. 779. 1753. (Europe)
    Finely strigillose to villous annual with decumbent to erect stems 1-4 dm. long; stipules entire to shallowly denticulate-toothed; leaflets mostly elliptic-obovate, 0.5-2 cm. long; flowers 10-40, yellow, 2-3 mm. long, crowded in spikelike racemes 0.5-1 cm. long; calyx nearly as long as the corolla; pod 1-seeded, 2-3 mm. long, heavily reticulate, unarmed, glabrous to (glandular-) pilose, turning black at maturity, reniform in outline and curved to not quite one spiral, the style curved into a second spiral. N=8, 16. Black medick, hop clover.
    Waste places and sandy or gravelly soil throughout most of the U.S.; introduced from Europe. Apr.-Aug.

Medicago sativa L. Sp. Pl. 778. 1753. ("Habitat in Hispaniae, Galliae")
  Medica legitima Clus ex Greene, Man. Bay Reg. 101. 1894. (Europe)
  Medicago sativa f. *alba* Benke, Am. Midl. Nat. 16:424. 1935. *(H. C. Benke 5665,* McHenry, Ill.)

Long-taprooted, finely strigillose to glabrous perennial with more or less erect stems 3-10 dm. tall; stipules entire; leaflets elliptic-oblanceolate, 2-4 cm. long; flowers 20-100, bluish-purple (pink or white), about 1 cm. long, borne in short, pedunculate racemes 1-3 cm. long; calyx nearly as long as the corolla; pod many-seeded, coiled to 2-3 spirals, strongly reticulate but not armed, (2) 3-4 mm. long. N=16. Alfalfa, lucerne.

Alfalfa, native to the Old World, is to be found as an escape from cultivation in many temperate regions, in our area chiefly along roadsides or railroads or near cultivated fields. June-Oct.

Melilotus Mill. Sweet Clover; Melilot

Flowers very small, papilionaceous, white or yellow, borne in numerous, axillary, elongated and slender, spikelike, linear-bracteate racemes; calyx turbinate-campanulate, the teeth subequal; corolla deciduous, free of the 10 diadelphous stamens; pods ovoid, slightly longer than the calyx, reticulate, 1 (2-4)-seeded, usually indehiscent; annual or biennial, taprooted, erect and freely branched, glabrous or sparsely pubescent, characteristically sweetish-odored herbs with trifoliolate leaves, serrulate leaflets, and linear stipules partially adnate to the petiole.

Perhaps 20 species, native to Eurasia and Africa, with several well established throughout much of N. Am. (Greek *meli*, honey, and *lotos*, name for some cloverlike plant.)

1 Corolla white   M. ALBA
1 Corolla yellow
   2 Flowers 4-6 mm. long; calyx teeth narrowly lanceolate-subulate; plants usually over 1
     m. tall   M. OFFICINALIS
   2 Flowers 2-3 mm. long; calyx teeth oblong-lanceolate, obtuse; plants usually less than
     1 m. tall   M. INDICA

Melilotus alba Desr. in Lam. Encycl. 4:63. 1797.
   M. officinalis var. *alba* Nutt. Gen. Pl. 2:104. 1818. (Siberia)
Erect annual (biennial) 0.5-3 m. tall, freely branched above, sparsely strigillose or puberulent to glabrous; leaflets narrowly elliptic-oblanceolate to oblong, 1.5-3 cm. long, serrate-dentate almost to the base; racemes 4-12 cm. long; bractlets about 2 mm. long; flowers white, about 5 mm. long; calyx teeth narrowly lanceolate-subulate; pod about 4 mm. long. N=8, 12, 16. White sweet clover.

Native of Eurasia; widely distributed as a weed over most of the U.S. and s. Can. May-Oct.

Melilotus indica All. Fl. Pedem. 1:308. 1785. (Algeria)
   M. occidentalis Nutt. in T. & G. Fl. N.Am. 1:321. 1838. *(Nuttall,* "sides of naked hills near the sea, California")
   M. parviflora Desf. Fl. Atlan. 2:192. 1800. (Algeria)
Annual or biennial, up to 1 m. tall, branched above, glabrous or very sparsely pubescent; leaflets mostly oblanceolate, serrulate on the upper half only; racemes 2-8 cm. long; bractlets linear, about 0.5 mm. long; flowers yellow, 2-3 mm. long; calyx teeth oblong-lanceolate, obtuse; pod 2-2.5 mm. long. N=8.

This Mediterranean plant is sparingly introduced in the Pacific Northwest, mostly in w. Oreg. and Wash. Although it is more common in Calif., and fairly widely distributed in the U.S., it is seldom as abundant as the other 2 species. May-July.

Melilotus officinalis (L.) Lam. Fl. Fr. 2:594. 1778.
   Trifolium officinalis L. Sp. Pl. 765. 1753. (Europe)
Biennial (annual) herb from a strong taproot, 0.5-3 m. tall, freely branched above, glabrous to sparsely pubescent; leaflets linear to elliptic, oblanceolate, or obovate, serrulate nearly the entire length; racemes 3-10 cm. long; bractlets about 2 mm. long; flowers yellow, 4-6 mm. long; calyx teeth narrowly lanceolate, subulate; pod about 3 mm. long. N=8.

A native of the Mediterranean region, *M. officinalis* is common as a weed throughout the Rocky Mts., less frequent further w., and not at all common w. of the Cascades. May-Sept.

Onobrychis Adans.

Flowers numerous in bracteate, spikelike, axillary and terminal racemes, papilionaceous, white or pink to purple; calyx campanulate; wing petals much smaller than the keel or banner; stamens 10,

Medicago arabica

Medicago lupulina

Medicago falcata

Medicago hispida

Medicago sativa

Melilotus alba

Melilotus officinalis

Melilotus indica

monadelphous, but the upper filament free at the base and also for the upper third of its length; fruits indehiscent, 1- or 2-seeded, usually somewhat prickly; perennial herbs with membranous stipules and odd-pinnate leaves.

Nearly 100 species, largely Mediterranean. (Ancient Greek name for the plant.)

Onobrychis viciaefolia Scop. Fl. Carn. 2nd ed. 2:76. 1772.

Hedysarum onobrychis L. Sp. Pl. 751. 1753. *O. onobrychis* Rydb. Mem. N.Y. Bot. Gard. 1:256. 1900. (Europe)

O. sativa Lam. Fl. Fr. 2:652. 1778. (Europe)

Sparsely strigose perennial herb 2-4 dm. tall; stipules lanceolate, 5-12 mm. long, reddish-brown; leaflets 11-17 (up to 27), narrowly elliptic to oblanceolate, 1-2 cm. long, apiculate; racemes 10- to 50-flowered, brownish-bracteate; flowers 10-13 mm. long; calyx campanulate, the teeth linear-lanceolate, subequal, about twice the length of the tube; corollas pink to lavender, very prominently lined with reddish-purple, the wings not over half as long as the keel; fruits ovate in outline, 6-8 mm. long, strongly pubescent and prominently rugose, 1-seeded. N=7, 14. Saintfoin, sandfain, holy clover.

A native of Europe, where it is grown for stock feed, sandfain occasionally has been introduced in B.C. and Wash. (Whidbey I. and Yakima Co.), and in Mont., where it is now well established in Powell Co. June-Aug.

Ononis L.

Flowers usually axillary, papilionaceous; calyx campanulate, 5-lobed; stamens 10, monadelphous; keel of corolla short-beaked; pods dehiscent, few-seeded; prostrate to erect, usually hairy annual, biennial, or perennial herbs or shrubs with simple or 3-foliolate leaves, the leaflets serrulate to dentate, stipules very large, adnate to the petioles.

Nearly 100 species, largely from the Mediterranean region. (Ancient Greek name for one of the plants.)

Ononis repens L. Sp. Pl. 717. 1753. ("Habitat in Angliae littoribus maris")

Prostrate, copiously long-villous and glandular perennial herb; stems 2-5 dm. long; leaves 3-foliolate below, simple above (at least among the flowers); stipules membranous, large, serrulate; leaflets oblong-obovate, (5) 10-20 mm. long, serrulate at the apex; flowers single in the axils, purplish, 15-20 mm. long; calyx tube campanulate, 1/2-1/3 as long as the lanceolate, subequal lobes; pods (0.5) 1-2 cm. long, villous. N=15, 16, 30.

Several species of *Ononis* are cultivated as ornamentals. This one is not particularly attractive and although it is reported to be established at Bingen, Wash. *(G. N. Jones 2476),* probably it will not be introduced often. Flowering in Aug.

Oxytropis DC. Nom. Conserv. Stemless Loco

Flowers spicate or racemose, papilionaceous, white or cream to reddish or purple; calyx 5-toothed, usually accrescent but generally finally ruptured although occasionally enclosing the mature pod; banner commonly erect; wings usually exceeding the keel; keel prolonged into a point or tooth or generally into a straight or curved beak; stamens 10, diadelphous; legume sessile or stipitate, often inflated, membranous to leathery or considerably hardened, from 1-celled to nearly completely 2-celled by the intrusion of the upper (ventral) suture, several-seeded; caespitose perennials, acaulescent or with short leafy stems; leaves odd-pinnate, the leaflets scattered or subopposite to verticillate; stipules more or less adnate to the petiole and often connate around the stem.

About 300 species, chiefly in w. N.Am. and Siberia. (Greek *oxys*, sharp, and *tropis*, keel, referring to the beaked keel.)

Oxytropis as a whole is sometimes merged with *Astragalus,* but our species, at least, can readily be distinguished from that genus because of the beaked keel and more consistently acaulescent habit. Many of the species are extremely variable and not at all easily classified, in part at least due to interspecific hybridization.

For the gardener e. of the Cascades several of the species of *Oxytropis* have considerable potential, such as *O. besseyi, O. lagopus,* and *O. splendens* (See Barr, Bull. Am. Rock Gard. Soc. 18:65-67. 1960).

References:

Barneby, R. C. A revision of the North American species of Oxytropis. Proc. Calif. Acad. Sci.

IV, 27:177-312. 1952 (from which the following treatment is taken with very few changes).

Fernald, M. A. Oxytropis in Northeastern America. Rhodora 30:137-155. 1928.

Gray, A. Revision of the North American species of Oxytropis. Proc. Am. Acad. Sci. 20:1-7. 1884.

Nelson, Aven. The western species of Aragallus. Erythea 7:57-64. 1899.

1 Stipules foliaceous, adnate to the base of the petiole for only 1-3 mm., the free portion
    lanceolate, 6-12 mm. long; pods pendulous; plants often with short leafy stems   O. DEFLEXA
1 Stipules adnate to the base of the petiole for half their length or more; pods spreading
    or (more commonly) erect; plants acaulescent
    2 Leaflets of all the leaves fasciculate and appearing verticillate, 2 or more attached
        at a point on either side of the rachis, the herbage not glandular   O. SPLENDENS
    2 Leaflets subopposite or scattered, or occasionally a few paired on one side of the
        rachis, but then the herbage usually glandular
        3 Plant more or less glandular-verrucose and often viscid, the warts especially
            evident on the calyx teeth (sometimes lacking elsewhere); flowers cream to
            purple   O. VISCIDA
        3 Plant neither viscid nor glandular-verrucose
            4 Racemes 1- to 3-flowered; flowers purplish; plants alpine; leaves usually less than
                5 cm. long
                5 Pods stipitate, bladdery-inflated, usually over 8 mm. thick; leaflets linear;
                    flowers generally over 1 cm. long; Rocky Mts. of Mont., in our area O. PODOCARPA
                5 Pods sessile, oblong-ovoid, not greatly inflated, not over 8 mm. broad; leaf-
                    lets narrowly oblong; flowers usually less than 1 cm. long; found only in
                    Ida., in our area   O. PARRYI
            4 Racemes 4- to many-flowered, if with fewer flowers (in depauperate specimens)
              then the corolla ochroleucous or the plants not alpine or the leaves well over
              5 cm. long
              6 Flowers purple or reddish-purple
                7 Calyx usually accrescent and not ruptured by the growing (generally sessile)
                    pod, densely silky-villous and mostly concealed by the hairs; scapes usu-
                    ally less than 10 cm. tall; bracts membranous, shaggy-pilose on the back;
                    leaflets mostly fewer than 15 and seldom over 15 mm. long   O. LAGOPUS
                7 Calyx not accrescent, usually ruptured by the maturing short-stipitate pod,
                    hispid-hirsute, the surface not concealed by the hairs; scapes commonly
                    at least 10 cm. tall; bracts herbaceous, sparsely appressed-pilose; leaf-
                    lets often more than 15 or over 15 mm. in length   O. BESSEYI
              6 Flowers white to yellowish, the keel sometimes purple-tinged
                8 Pod fleshy before maturity, then hardened and bony, the dried wall nearly
                    1 mm. thick; flowers mostly well over 15 mm. long; plants of the Rocky
                    Mts., westward in our area only to c. Ida.   O. SERICEA
              8 Pod more membranous than fleshy, the dried wall scarcely 0.5 mm. thick;
                  flowers frequently less than 15 mm. long; transcontinental in n. N.Am.,
                  s. in the Rocky Mts. to Colo., w. to the Olympic Mts. in Wash.   O. CAMPESTRIS

Oxytropis besseyi (Rydb.) Blank. Mont. Agr. Coll. Stud. Bot. 1:80. 1905.
    Aragallus besseyi Rydb. Mem. N.Y. Bot. Gard. 1:250. 1900. *(Rydberg & Bessey 4501*, Spanish
        Basin, Madison Co., Mont., June 23, 1897)
    Aragallus argophyllus Rydb. Mem. N.Y. Bot. Gard. 1:255. 1900. O. *besseyi* var. *argophylla* Bar-
        neby, Leafl. West. Bot. 6:111. 1951. *(J. G. Cooper*, Little Blackfoot R., Mont., in 1860)
    O. *besseyi* var. *salmonensis* Barneby, Proc. Calif. Acad. Sci. IV, 27:234. 1952. *(Ripley & Barne-*
        by 8829, Salmon R. Canyon, 12 mi. below Clayton, Custer Co., Ida., June 22, 1947)
Caespitose perennial, silvery with appressed hairs throughout or (as generally) with spreading
hairs on the calyx and pod; leaves 2-11 cm. long, the stipules membranous, adnate to the petiole over
half their length, only slightly or not at all connate; leaflets (5) 7-21, not verticillate, 5-20 mm. long;
racemes closely 5- to 30-flowered, the peduncles erect, (3) 8-20 cm. tall; bracts herbaceous, sparsely appressed-pilose; flowers deep reddish-purple, fading to bluish, 18-24 mm. long; calyx 10-15 mm.
long, sparsely to rather densely hispid-hirsute, usually investing the pod until near maturity (but not
accrescent) and finally ruptured, the linear-lanceolate teeth 2/3 as long as the tube; banner erect;

wings usually bifid; beak of the keel 1-1.5 mm. long; pod 1-2 cm. long, the upper suture intruded about halfway, the beak about 5 mm. long.

Gravel benches, prairies, riverbanks, and lower foothills; n.w. Colo. to Mont. and c. Ida. June-July.

In our area the species is differentiated into 3 fairly clearly marked varieties:
1 Pubescence of the calyx distinctly appressed; local in Custer Co., Ida.   var. salmonensis Barneby
1 Pubescence of the calyx spreading
  2 Scapes 1-2 dm. tall; racemes loosely 8- to 20-flowered; e. slope of the Rocky
    Mts. from c. Mont. to n.e. Wyo. and s. Sask.   var. besseyi
  2 Scapes 2-9 cm. tall; racemes closely 3- to 10-flowered; s.w. Mont. and adj. Ida.
                                         var. argophylla (Rydb.) Barneby

Oxytropis campestris (L.) DC. Astrag. 59. 1802.
  Astragalus campestris L. Sp. Pl. 761. 1753. ("Habitat in Oelandia, Germania, Helvetia")
  O. monticola Gray, Proc. Am. Acad. 20:6. 1884, not *Astragalus monticola* Phil. in 1864-65. *Spiesia monticola* Kuntze, Rev. Gen. 1:206. 1891. *Aragallus monticola* Greene, Pitt. 3:212. 1897. *Astragalus grayanus* Tidestr. ex Tidestr. & Kitt. Fl. Ariz. & N.M. 216. 1941. (*Canby 91*, Jocko R., Lake Co., Mont., lectotype by Barneby) = var. *gracilis*.
  Aragallus gracilis A. Nels. Erythea 7:60. 1899. *O. gracilis* K. Schum. Just Bot. Jahresb. 27[1]: 496. 1901. *O. campestris* var. *gracilis* Barneby, Leafl. West. Bot. 6:111. 1951. (*A. Nelson 2545*, Limestone Range, Black Hills, Weston Co., Wyo., July 30, 1896)
  O. cusickii Greenm. Erythea 7:116. 1899. *O. campestris* var. *cusickii* Barneby, Leafl. West. Bot. 6:111. (Jan.) 1951. (*Cusick 1365*, Wallowa Mts., Oreg., in 1891, is the first of 2 collections cited)
  Aragallus alpicola Rydb. Mem. N.Y. Bot. Gard. 1:252. 1900. *O. alpicola* M. E. Jones, Mont. Bot. Notes 37. 1910, but not of Turcz. in 1842. *O. rydbergii* A. Nels. U. Wyo. Pub. Bot. 1:117. 1926. *Astragalus alpicola* Tidestr. Proc. Biol. Soc. Wash. 50:19. 1937. *O. campestris* var. *rydbergii* Davis, Madroño 11:144. (Aug.) 1951. (*Rydberg & Bessey 4503*, Old Hollowtop, near Pony, Mont., July 9, 1897, is the first of 3 collections cited) = var. *cusickii*.
  Aragallus villosus Rydb. Bull. Torrey Club 28:36. 1901. *O. villosa* K. Schum. Just Bot. Jahresb. 29[1]:543. 1903. *Astragalus rydbergianus* Tidestr. Proc. Biol. Soc. Wash. 50:19. 1937, not *Astragalus villosus* of Michx. in 1803. (*Wilcox 378*, Craig, Lewis & Clark Co., Mont., in 1900) = var. *gracilis*.
  Aragallus albertinus Greene, Proc. Biol. Soc. Wash. 18:15. 1905. *O. albertina* Rydb. Fl. Prair. & Pl. 484. 1932. *Astragalus albertinus* Tidestr. Proc. Biol. Soc. Wash. 50:19. 1937. (*Macoun 12535*, Prince Albert, Sask., July, 1896) = var. *gracilis*.
  Aragallus cervinus Greene, Proc. Biol. Soc. Wash. 18:16. 1905. (*Macoun 5358*, Deer Park, Lower Arrow Lake, B.C., June 8, 1890) = var. *gracilis*.
  Aragallus macounii Greene, Proc. Biol. Soc. Wash. 18:16. 1905. *O. macounii* Dayton, Proc. Biol. Soc. Wash. 40:120. 1927. (*Macoun 18516*, about Calgary, Alta., June and July, 1897, is the first of two numbers cited) = var. *gracilis* in part, but see also *O. sericea*.
  Aragallus luteolus Greene, Proc. Biol. Soc. Wash. 18:17. 1905. *O. luteola* Piper & Beattie, Fl. N.W. Coast 227. 1915. (*Elmer 2532*, Olympic Mts., Clallam Co., Wash., July, 1900) = var. *gracilis*.
  O. paysoniana A. Nels. U. Wyo. Pub. Bot. 1:119. 1926. (*Payson & Payson 2700*, summit of Piney Mt., Sublette Co., Wyo.) = var. *cusickii*.
  O. columbiana St. John, Proc. Biol. Soc. Wash. 41:100. 1928. *O. campestris* var. *columbiana* Barneby, Leafl. West. Bot. 6:111. 1931. (*St. John 6482*, Columbia R., Marcus, Stevens Co., Wash., June 27, 1924)
  O. mazama St. John, Proc. Biol. Soc. Wash. 41:101. 1928. *Astragalus mazama* G. N. Jones, U. Wash. Pub. Biol. 7:175. 1939. (*O. D. Allen 245*, Goat Mts., Pierce Co., Wash., July 6, and Sept. 30, 1896) = var. *gracilis*.
  O. okanoganea St. John, Proc. Biol. Soc. Wash. 41:102. 1928. (*St. John 7728*, n.w. of Riverside, Okanogan Co., Wash., July 2, 1923) = var. *gracilis*.
  O. olympica St. John, Proc. Biol. Soc. Wash. 41:103. 1928. (*Flett 134*, Olympic Mts., Jefferson Co., Wash., July 20, 1897) = var. *gracilis*.
  O. cascadensis St. John, Proc. Biol. Soc. Wash. 41:105. 1928. (*St. John 5113*, Grouse Ridge, Mt. Baker, Whatcom Co., Wash., Aug. 8, 1923) = var. *gracilis*.
Caespitose perennial, 5-30 cm. tall, from densely strigose-hirsute and grayish to sparsely stri-

gose-pilose and greenish; leaves 3-20 cm. long, the stipules membranous, 5-15 mm. long, adnate at least half their length, long-ciliate, usually hairy on the lower surface; leaflets 5-41, not verticillate, 5-30 mm. long; racemes spikelike, 5- to 40-flowered; peduncles from slightly shorter to considerably longer than the leaves; flowers 10-20 mm. long, white to yellowish, the keel sometimes purple; calyx loosely grayish- to black-hairy, about 1/2 the length of the corolla, the teeth more or less linear-ob-long, 1-4 mm. long, the sinuses rounded; pod 1-2.5 cm. long, sessile or very short-stipitate, almost 2-celled by the intrusion of the upper suture, the wall more membranous than fleshy, scarcely 0.5 mm. thick, beak about 5 mm. long.

More or less circumboreal; montane to submontane in our area. May-July.

Oxytropis campestris was originally described from the Old World, but is represented in N.Am. by several geographic races, the following three from our area being imperfectly distinguishable as follows:

1 Corolla white, the keel strongly purple-spotted; leaflets mostly no more than 17 (up to 23);
　　n.e. Wash. and the Flathead Lake region, Mont.　　　　　　var. columbiana (St. John) Barneby
1 Corolla ochroleucous, the keel not spotted; leaflets often more than 17
　　2 Stipules strongly hairy; scapes mostly over 15 cm. tall; leaflets usually more than 17;
　　　　w. Wash. to Alta. and N.D., southward to Colo.; N=16　　var. gracilis (A. Nels.) Barneby
　　2 Stipules glabrous or glabrate; scapes rarely over 15 cm. tall; leaflets seldom over 17;
　　　　range of var. *gracilis* to a large extent, but not on the w. side of the Cascades al-
　　　　though in n.e. Oreg. and c. Ida.　　　　　　　　　　var. cusickii (Greenm.) Barneby

The species is extremely variable and other local populations might be distinguished with scarcely less basis than there is for the maintenance of the above taxa.

Oxytropis deflexa (Pall.) DC. Astrag. 96. 1802.

　　Astragalus deflexus Pall. Acta Acad. Petrop. 2:268, pl. 15. 1779. *Aragallus deflexus* Heller, Cat.
　　　N.Am. Pl. 4. 1898. (Asia)
　　Astragalus retroflexus Pall. Astrag. 33, pl. 27. 1800. (Asia) = var. *deflexa*.
　　O. foliolosa Hook. Fl. Bor. Am. 1:146. 1831. *O. foliosa* Hook. ex T. & G. Fl. N.Am. 1:339. 1838.
　　　Aragallus foliolosus Macoun (incorrectly attr. to Hook.) Ott. Nat. 13:163. 1899. *Astragalus de-
　　　flexus* var. *foliolosus* Tidestr. Proc. Biol. Soc. Wash. 50:19. 1937. *O. deflexa* var. *foliolosa*
　　　Barneby, Leafl. West. Bot. 5:111. 1951. ("From Carlton House to the Rocky Mountains, in lat.
　　　54°")
　　O. deflexa var. *sericea* T. & G. Fl. N.Am. 1:342. 1838. *O. retrorsa* var. *sericea* Fern. Rhodora
　　　30:140. 1928. *(Nuttall,* Rocky Mountains)
　　O. retrorsa Fern. Rhodora 30:140. 1928. *(Crandall & Cowen 152,* Como, Park Co., Colo., Aug. 3,
　　　1895) = var. *sericea*.
　　O. deflexa var. *culminis* Jeps. Fl. Calif. 2:381. 1936. *(Victor Duran 1650,* White Mts., Inyo Co.,
　　　Calif.) = var. *sericea*.

Caespitose perennial from a strong taproot, pilose-villous and greenish to grayish, acaulescent or with stems 3-20 (30) cm. long; stipules foliaceous, lanceolate, 6-12 mm. long, not connate, adnate to the petiole for only 1-3 mm.; leaflets 15-30, ovate to lanceolate-elliptic, 3-25 mm. long; racemes spikelike, compactly 2- to 40-flowered, often much elongate and from 2-15 cm. long in fruit; pedun-cles 5-20 cm. long; flowers white to bluish-purple, 6-10 mm. long; calyx 2/3-3/4 the length of the corolla, darkish-strigillose to strigose-pilose, the teeth linear to lanceolate, subequal to the tube; banner scarcely at all reflexed; beak of the keel short, blunt; pod reflexed, pendulous, with a short stipe 1-2 mm. long, the body grayish- to black-strigose, 10-18 mm. long, 2-4 mm. broad, terete to dorsi-ventrally flattened, one-celled but the upper suture somewhat intruded. N=16.

Siberia and N.Am., in meadows and along streams, from the foothills to well up in the mountains. June-July.

Our material is referable to two varieties which are often treated as distinct species (*O. foliolosa* and *O. retrorsa*):

1 Plant nearly or quite acaulescent, greenish in color, with sparsely pilose, subcapitate,
　　7- to 10-flowered racemes; flowers bluish-purple, about 1 cm. long, the sinuses be-
　　tween the calyx teeth broad; Rocky Mts., from Alta. to Colo., and e. across Can.
　　　　　　　　　　　　　　　　　　　　　　　　　　　　var. foliolosa (Hook.) Barneby
1 Plant with stems 3-20 cm. tall, often strongly hairy and grayish; racemes elongate
　　(up to 15 cm. in fruit), 10- to 40-flowered, the flowers pale blue, 6-8 mm. long;
　　calyx sinuses narrow and acute; B.C. southward, on the e. side of the Cascades,

1.5

stipules

1/2

1/2

var. columbiana

var. cusickii

var. gracilis

Oxytropis campestris

1.5

calyx

var. salmonensis

3

1/2

3

2

1.5

Onobrychis viciaefolia

calyx 3

2

3

1/2

var. besseyi

JRJ

1/2

var. argophylla

Ononis repens

Oxytropis besseyi

to Wash. and Oreg. , e. across n. Can. , s. in the Rocky Mts. to N. M.   var. sericea T. & G.

Oxytropis lagopus Nutt. Journ. Acad. Phila. 7:17. 1834.

   Aragallus lagopus Greene, Pitt. 3:212. 1897. *Astragalus lagopus* Tidestr. Proc. Biol. Soc. Wash.
      50:19. 1937. *(Wyeth,* "Sources of the Missouri")
   Aragallus blankinshipii A. Nels. Erythea 7:58. 1899. *O. blankinshipii* K. Schum. Just Bot.
      Jahresb. 27[1]:496. 1901. *Astragalus blankinshipii* Tidestr. Proc. Biol. Soc. Wash. 50:18. 1937.
      (J. W. Blankinship, Middle Creek, 15 mi. s. w. of Bozeman, Mont. , in 1898) = var. *lagopus.*
   Aragallus atropurpureus Rydb. Bull. Torrey Club 34:424. 1907. *O. lagopus* var. *atropurpurea* Bar-
      neby, Leafl. West. Bot. 6:111. 1951. *(F. Tweedy 125,* Big Horn Mts. , Wyo. , in 1898)
   O. lagopus var. *conjugens* Barneby, Proc. Calif. Acad. Sci. IV, 27:227. 1952. *(E. O. Wooton,*
      near Helena, Mont. , June, 1921)
Sericeous-silky, caespitose and usually pulvinate perennial; leaves 3-7 (9) cm. long; stipules membranous, connate and adnate to the petiole 2/3 of their length; leaflets mostly 7-15, not verticillate,
lanceolate to elliptic, 7-15 mm. long; racemes spikelike, 5- to 20-flowered, 2-4 cm. long, peduncles often not exceeding the leaves, rarely as much as 9 cm. long; flowers reddish-purple, 15-20 mm.
long; calyx thickly hirsute with white (or some admixture of blackish), spreading hairs, about 2/3 as
long as the corolla, usually much accrescent and not split by the maturing pod, the teeth linear-lanceolate, about 2/3 as long as the tube; keel with a slender beak 1. 5-2 mm. long; pod ovoid-oblong,
about 1. 5 cm. long, 4-7 mm. broad, with long spreading hairs, broadly terete-cordate in section,
about half partitioned by the intrusion of the upper suture; beak distinct, 3-5 mm. long.
   Sagebrush plains to lower mountains; Rocky Mts. from Wyo. to w. Mont. and Ida. May-June (July).
   Oxytropis lagopus is very similar to *O. besseyi* when in flower, but it blossoms considerably earlier and is usually recognizable in fruit because of the much more inflated calyx. We have 2 (possibly
3) intergrading phases of the species in our area, separable as follows:
1 Leaflets 5-9, about equal to the rachis of the leaf; calyx usually persistent until after the
      pod dehisces; c. Mont. , mostly e. of the Rockies   var. conjugens Barneby
1 Leaflets mostly (9) 11 or more, rarely as long as the rachis; calyx usually deciduous
      with the enclosed pod before seed dispersal
   2 Corolla mostly pinkish-purple; Ida. and Mont.   var. lagopus
   2 Corolla rather deep purple; Wyo. and extreme s. Mont. , possibly not reaching our
         area   var. atropurpurea (Rydb.) Barneby

Oxytropis parryi Gray, Proc. Am. Acad. 20:4. 1884.

   Spiesia parryi Kuntze, Rev. Gen. 1:207. 1891. *Aragallus parryi* Greene, Pitt. 3:211. 1897. *Astra-
      galus parryanus* Tidestr. Proc. Biol. Soc. Wash. 50:19. 1937, not *Astragalus parryi* Gray. *(Par-
      ry,* Rocky Mts. of n. N. M. and Colo. , lectotype by Barneby)
Grayish silky-strigose, pulvinate, acaulescent perennial; leaves 3-5 cm. long, stipules membranous, connate, adnate to the petioles for 2/3 of their length; leaflets mostly 9-19, not verticillate,
narrowly oblong, 3-10 mm. long; racemes usually 2 (1-3)-flowered, peduncles only slightly if at all
longer than the leaves; flowers purplish, 7-10 mm. long; keel with a very short beak; calyx black-
hairy, about 3/4 as long as the corolla, the teeth lanceolate, 2/3-3/4 as long as the campanulate tube;
banner only slightly upturned; pod sessile, black-hairy, oblong-ovoid, 15-23 mm. long, 4-8 mm.
broad, nearly completely 2-celled by the intrusion of the upper suture, short-beaked.
   Alpine slopes and ridges; Great Basin area, N. M. to Wyo. , w. to the mountains of c. Ida. , Nev. ,
and Calif. July-Aug.

Oxytropis podocarpa Gray, Proc. Am. Acad. 6:234. 1864.

   Spiesia podocarpa Kuntze, Rev. Gen. 1:207. 1891. *Aragallus podocarpus* Nels. in Coult. & Nels. New
      Man. Bot. Rocky Mts. 294. 1909. *(Bourgeau,* Labrador, and *Schwenitz,* Alta.)
   O. arctica var. *inflata* Hook. Fl. Bor. Am. 1:146. 1831. *Spiesia inflata* Britt. Mem. Torrey Club
      5:201. 1894. *Aragallus inflatus* A. Nels. Erythea 7:59. 1899. *(Drummond,* "highest summits of the
      Rocky Mountains")
   O. hallii Bunge, Gen. Oxytr. 162. 1874. *Aragallus hallii* Rydb. Bull. Torrey Club 33:144. 1906.
      (Hall & Harbour, Rocky Mts. , Colo. , in 1862)
Pulvinate, more or less silvery-hirsute perennial; leaves 2-5 cm. long; stipules 4-10 mm. long,
membranous, connate, adnate to the petiole for 2/3 of their length; leaflets 9-27, not verticillate, linear, 5-10 mm. long; racemes 1- or 2-flowered; peduncles subequal to the leaves; flowers 12-17 mm.

long, purplish; calyx purple, 2/3 the length of the corolla, the lobes lanceolate, about 1/3 the length of the tube; corolla nearly erect; pod with a stipe 2-5 mm. long, usually concealed by the calyx, the body much inflated, rather papery, ovoid, 15-25 mm. long, usually at least 8 mm. thick, short-beaked.

Alpine ridges and slopes; Rocky Mts., from Alta. to Colo., e. to the Atlantic coast. July-Sept.

Oxytropis sericea Nutt. in T. & G. Fl. N. Am. 1:339. 1838.

O. lamberti var. *sericea* Gray, Proc. Am. Acad. 20:7. 1884. *Spiesia lamberti* var. *sericea* Rydb. Bot. Surv. Neb. 3:32. 1894. *Aragallus sericeus* Greene, Pitt. 3:212. 1897. *Aragallus lamberti* var. *sericeus* A. Nels. Erythea 7:62. 1899. *(Nuttall,* Rocky Mountains)

O. campestris var. *spicata* Hook. Fl. Bor. Am. 1:147. 1831. *Aragallus spicatus* Rydb. Mem. N. Y. Bot. Gard. 1:251. 1900. *O. spicata* Standl. Contr. U. S. Nat. Herb. 22:373. 1921. *O. sericea* var. *spicata* Barneby, Leafl. West. Bot. 6:111. 1951. *(Douglas,* "sources of the Columbia")

O. lambertii var. *ochroleuca* A. Nels. Bull. Wyo. Exp. Sta. 28:98. 1896. *Aragallus albiflorus* A. Nels. Erythea 7:62. 1899. *Aragallus saximontanus* A. Nels. Erythea 7:190. 1900. *Astragalus albiflorus* Gandg. Bull. Soc. Bot. France 48:14. 1901. *O. albiflora* of K. Schum. Just Bot. Jahresb. 27^1:496. 1901, but not of Bunge in 1874. *O. saximontanus* A. Nels. U. Wyo. Pub. Bot. 1: 113. 1926. *Astragalus saximontanus* Tidestr. in Tidestr. & Kitt. Fl. Ariz. & N. M. 216. 1941. *(A. Nelson 119,* Albany Co., Wyo.) = var. *sericea.*

Aragallus pinetorum Heller, Bull. Torrey Club 26:548. 1899. *O. pinetorum* K. Schum. Just Bot. Jahresb. 27^1:496. 1901. *(Heller & Heller 3751,* near Santa Fe, N. M., June 23, 1897) = var. *sericea.*

? *Aragallus collinus* A. Nels. Erythea 7:57. 1899. *O. collina* K. Schum. Just Bot. Jahresb. 27^1: 496. 1901. *(E. Nelson 4925,* Seminole Mts., Wyo., July, 1898) = var. *sericea.*

Aragallus albiflorus var. *condensatus* A. Nels. Erythea 7:63. 1899. *Aragallus saximontanus* var. *condensatus* A. Nels. Erythea 7:190. 1900. *O. condensatus* A. Nels. U. Wyo. Pub. Bot. 1:115. 1926. *(A. Nelson 4773,* Point of Rocks, Wyo., July 16, 1898; acc. Barneby, isotypes of this variety were collected at Bitter Creek, Sweetwater Co.) = var. *sericea.*

Aragallus majusculus Greene, Proc. Biol. Soc. Wash. 18:12. 1905. *(M. E. Jones 5674,* Mt. Ellen, Henry Mts., Utah, July, 1894) = var. *sericea.*

Aragallus·melanodontus Greene, Proc. Biol. Soc. Wash. 18:15. 1905. *(Macoun 18513,* Elbow R., Alta., in 1897) = var. *spicata.*

Aragallus macounii Greene, Proc. Biol. Soc. Wash. 18:16. 1905. *O. macounii* Rydb. Fl. Prair. & Pl. 485. 1932, in part. *(Macoun 18516,* about Calgary, Alta., June and July, 1897, is the first number mentioned, but Barneby cites *Macoun 18517,* Elbow R., Alta., as the type) = var. *spicata,* in part, see also *O. campestris.*

Caespitose, densely grayish villous-strigose to pilose perennial; leaves 5-30 cm. long; stipules membranous, 10-30 mm. long, adnate to the petiole for 3/4 of their length; leaflets 7 (9-21), not verticillate, 1-3.5 cm. long; racemes spikelike, compactly (5) 10- to 35-flowered, usually considerably elongated in fruit; peduncles 10-25 cm. long, erect; flowers 15-27 mm. long, white to yellowish, often pinkish-tinged, the keel frequently purplish-tipped; calyx 9-13 mm. long, the teeth linear-lanceolate to oblong, 1.5-5 mm. long; banner erect; wings greatly dilated, retuse; pod 1.5-2.5 cm. long, short-beaked, 5-7 mm. in diameter, fleshy when green, the wall considerably hardened and nearly 1 mm. thick when dried, the upper suture about halfway intruded.

Prairie land to subalpine meadows and slopes; N. M. to n. Can., on both slopes of the Rocky Mts. June-July.

The species is variable in flower color, height of plant, general pubescence, and length of the calyx teeth. Although many variants have been described, few of them can be delimited satisfactorily, the two more conspicuous being:

1 Corolla cream to white, often more or less tinged with pink, the keel frequently purple-tipped; w. Mont. to n. e. Nev., N. M., and Tex.   var. sericea
1 Corolla lemon to sulfur yellow, the keel usually not purple-tipped; n. B. C. to c. Ida. and n. Wyo.; N=24   var. spicata (Hook.) Barneby

Oxytropis splendens Dougl. ex Hook. Fl. Bor. Am. 1:147. 1831.

O. splendens var. *vestita* Hook. op. cit. 148. *Spiesia splendens* Kuntze, Rev. Gen. 1:207. 1891. *Aragallus splendens* Greene, Pitt. 3:211. 1897. *Astragalus splendens* Tidestr. Proc. Biol. Soc. Wash. 50:18. 1937. *(Douglas,* "On limestone rocks of the Red River, and south towards Pembina")

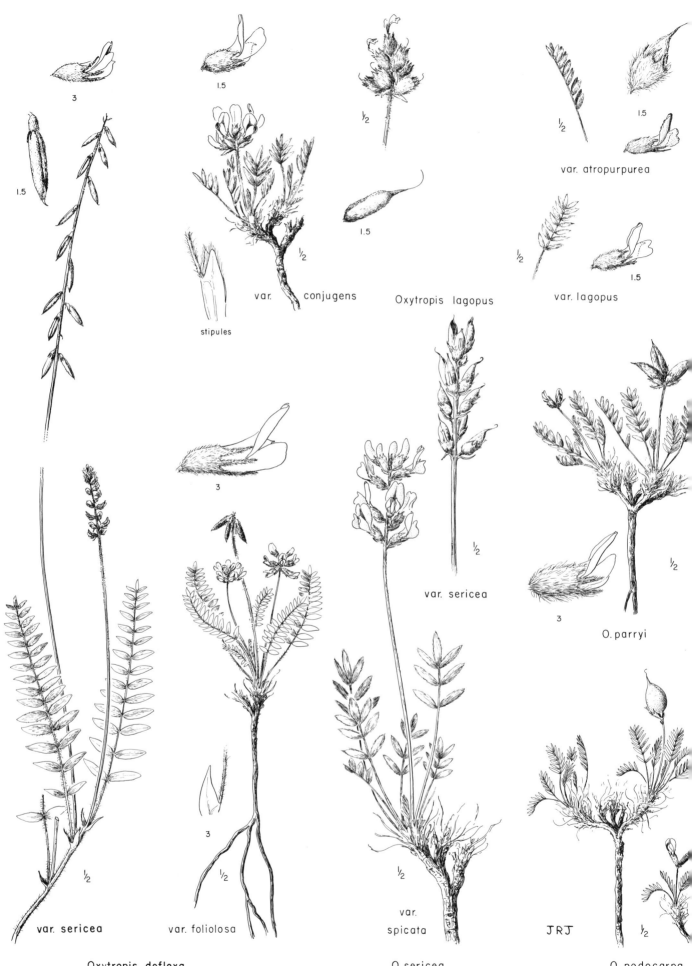

3

1.5

1.5

var. atropurpurea

1.5

1.5

var. conjugens

stipules

Oxytropis lagopus

var. lagopus

3

1/2

1/2

var. sericea

O. parryi

3

var. sericea

var. foliolosa

3

var. spicata

JRJ

Oxytropis deflexa

O. sericea

O. podocarpa

O. splendens var. *richardsoni* Hook. Fl. Bor. Am. 1:148. 1831. *Aragallus richardsoni* Greene, Pitt. 4:69. 1899. *O. richardsoni* K. Schum. Just Bot. Jahresb. 27[1]:496. 1901. *Astragallus splendens* var. *richardsonii* Tidestr. Proc. Biol. Soc. Wash. 50:18. 1937. *(D. Richardson,* "From Cumberland-House on the Saskatchawan, north to Fort Franklin and the Bear Lake, and west to the dry Prairies of the Rocky Mountains")

Aragallus caudatus Greene, Pitt. 4:69. 1899. *O. caudatus* K. Schum. Just Bot. Jahresb. 27[1]:496. 1901. *(Macoun 13957,* Moose Jaw, Assiniboia, June 26, 1896)

Aragallus galioides Greene, Proc. Biol. Soc. Wash. 18:16. 1905. *(McCalla,* Bow River near Banff, Alta., in 1899)

A nearly or quite acaulescent, densely silky perennial with a branched caudex; leaves 10-25 cm. long; stipules membranous, 10-15 mm. long, adnate to the petiole for 2/3 of their length and tubular-connate; leaflets mostly in verticils of 3-6, lanceolate to elliptic, 0.5-2 cm. long; peduncles usually slightly exceeding the leaves; racemes spikelike, densely 20- to 80-flowered, usually 5-10 cm. long but sometimes more elongate, especially in fruit; flowers reddish-purple, 12-15 mm. long; calyx 1/2-2/3 as long as the corolla, the lobes linear-lanceolate, obtuse, 1/3-1/4 the length of the tube; beak of the keel straight to slightly curved, slender, up to 1 mm. long; pod 10-15 mm. long, 3-4 mm. broad, cordate in section because of the intrusion of the upper suture, narrowed to a distinct beak about 3 mm. long. N=8.

Gravel bars and meadowland; foothills and lower levels of the Rocky Mts. (mostly on the e. slope), from Alas. to N.M., e. in Can. to Ont., and s. to n. Minn. June-Aug.

Oxytropis viscida Nutt. in T. & G. Fl. N.Am. 1:341. 1838.

O. campestris var. *viscida* Wats. Bot. King Exp. 55. 1871. *Spiesia viscida* Kuntze, Rev. Gen. 1: 207. 1891. *Aragallus viscidus* Greene, Pitt. 3:211. 1897. *Astragalus viscidus* Tidestr. Proc. Biol. Soc. Wash. 50:19. 1937. *(Nuttall,* "Rocky Mountains, near the sources of the Oregon")

Aragallus viscidulus Rydb. Mem. N.Y. Bot. Gard. 1:253. 1900. *O. viscidula* Tidestr. Contr. U.S. Nat. Herb. 25: 332. 1925. *(Rydberg 2716,* Melrose, Mont., in 1895, is the first of 5 collections cited)

Aragallus viscidulus var. *depressus* Rydb. Mem. N.Y. Bot. Gard. 1:253. 1900. *(Tweedy 120,* Haystack Peak, Park Co., Mont., in 1887)

Caespitose perennial with a branched caudex, nearly or quite acaulescent, pilose-hirsute and viscid more or less throughout (and especially on the calyx) with wartlike glands, but greenish; leaves 3-20 cm. long; stipules 6-12 mm. long, membranous, adnate to the petiole over half their length, connate opposite the petiole; leaflets 15-45, not verticillate, linear-lanceolate to narrowly oblong, 5-25 mm. long; racemes 2-7 cm. long, spikelike, mostly 7- to 30-flowered, becoming elongate in fruit, the peduncles usually slightly exceeding the leaves; flowers cream-colored to reddish-purplish, 10-15 mm. long; calyx about 2/3 as long as the corolla, grayish- to black-hairy, the teeth linear, 1/3-2/3 as long as the tube; banner slightly upturned; wings broadened and often retuse at the apex; keel with a short straight beak; pod 10-15 mm. long, grayish- to black-hairy, 4-6 mm. broad, with a distinct beak 3-6 mm. long, the upper suture deeply intruded and nearly half partitioning the cavity.

Alaska to Que., southward, in alpine or subalpine habitats, chiefly in the Rockies, to Colo., w. to the Sierra Nevada of Calif., the Wallowa Mts. of Oreg., and the Olympics of Wash., scattered eastward to Minn. and Ont. June-Aug.

The species shows an unusual amount of variability, with many more or less distinctive localized populations. In our area there are two color forms; plants of the Wallowa Mts. are usually purplish-flowered, whereas those of the Olympic Mts., restricted to Mt. Olympus and Hurricane Ridge, have cream colored flowers with little or no purplish tinge. The population of the Olympics seems to intergrade with *O. campestris,* but is sufficiently unique that it would be given distinctive infraspecific appellation were a published name available for it. Other workers have referred it either to *O. luteola* or to *O. olympica,* both of which are here included under *O. campestris* var. *gracilis.*

Petalostemon Michx. Nom. Conserv. Prairie Clover

Flowers many in dense, terminal and axillary, bracteate spikes, white, pink, or red; calyx campanulate to tubular, usually prominently nerved, the 5 teeth lanceolate, subequal; corolla irregular but not truly papilionaceous, apparently of 5 petals, one (probably the only true petal) larger, broader-clawed, and adnate at base to the calyx, the other 4 (due to their position almost surely staminodia) narrow-clawed, adnate to the short staminal tube and alternate with the 5 fertile stamens; legume 1-

or 2-seeded, indehiscent, usually contained in the calyx; glandular-punctate, herbaceous perennials (all of ours) or less commonly annuals, with small, odd-pinnate leaves and small, linear, setaceous stipules.

About 36 species of N. Am., especially common in the plains states and southeastward. (Greek *petalon*, petal, and *stemon*, stamen, in reference either to the union of the petals and stamens or to the presence of petaloid stamens.)

Although not choice ornamentals, all three of our species are worth a place in the arid garden e. of the Cascades, even though a little too large to be usable for the ordinary rockery.

1 Flowers white; calyx tube glabrous   P. CANDIDUM
1 Flowers pink to purplish (rarely whitish); calyx tube usually conspicuously hairy
  2 Leaflets elliptic to obovate, at least some 3-8 mm. broad; e. Wash. and Oreg.
    to s. w. Ida.   P. ORNATUM
  2 Leaflets linear, 1-2 mm. broad; e. slope of the Rocky Mts. and eastward   P. PURPUREUM

Petalostemon candidum Michx. Fl. Bor. Am. 2:49, pl. 37, fig. 1. 1803.
    Kuhniastera candida Kuntze, Rev. Gen. 1:192. 1891. *Dalea candida* Shinners, Field & Lab. 17:83.
      1949. ("Hab. in Tennassée et in regiòne Illinoensi")

Multistemmed perennial from a thick woody root, 2-6 dm. tall, essentially glabrous throughout; leaflets 5-9, narrowly oblong to oblong-oblanceolate, truncate or slightly retuse to acute or apiculate, 5-30 mm. long, 2-5 mm. broad; spikes mostly 2-5 cm. long, about 12 mm. thick; bracts acuminate, exceeding the calyx; calyx tube usually glabrous, about 2.5 mm. long, strongly 10-ribbed, the lobes lanate-ciliate; corolla white. N=7.

Dry open plains; e. slope of the Rocky Mts., from Mont. to Sask., and s. to most of c. and s. U.S. June-Aug.

Petalostemon ornatum Dougl. ex Hook. Fl. Bor. Am. 1:138. 1831.
    Dalea ornata Eat. & Wright, N. Am. Bot. 219. 1840. *Kuhniastera ornata* Kuntze, Rev. Gen. 1:192.
      1891. *(Douglas,* "in the arid prairies near the Blue Mountains of Lewis River, North-West Amer-
      ica")

Multistemmed perennial from a woody crown, 2-6 (7) dm. tall, glabrous except for the calyces and sometimes the stipules; leaflets 5-7, elliptic to obovate, 1-3 cm. long, (1) 3-8 mm. broad; spikes 2-7 cm. long, about 1.5 cm. thick; calyx and bracts tawny-villous to glabrous, calyx tube about 3 mm. long, the teeth somewhat shorter; corolla pink to rose, blades of the "petals" 3-4 mm. long, the claw nearly as long.

In rocky or sandy soils, often in sagebrush; e. of the Cascade Mts., from Yakima Co., Wash., s. into Oreg. and e. to w. Ida. May-July.

Several species of *Petalostemon* are represented by both a villous and a glabrous phase, and although most plants of *P. ornatum* have villous calyces and fruits, occasional specimens *(St. John et al. 4970)* have only a few long cilia on the calyx.

Petalostemon purpureum (Vent.) Rydb. Mem. N.Y. Bot. Gard. 1:238. 1900.
    Dalea purpurea Vent. Hort. Cels. pl. 40. 1800. (Garden specimens, from plants originally taken in
      Illinois by Michaux)
    P. violaceum Michx. Fl. Bor. Am. 2:50, pl. 37, fig. 2. 1803. *Kuhniastera purpurea* MacM. Metasp.
      Minn. Valley 329. 1892. ("Hab. in regione Illinoensi")
    P. molle Rydb. Mem. N.Y. Bot. Gard. 1:238. 1900. *P. purpureum* var. *molle* Nels. in Coult. &
      Nels. New Man. Bot. Rocky Mts. 299. 1909. *(Canby,* Snowy Mts., Mont., in 1882)

Multistemmed perennial 3-6 dm. tall, with thick woody roots, from subglabrous (except the inflorescence) to grayish-villous-woolly throughout; leaflets 3-7, linear, mostly 10-20 mm. long, 1-2 mm. broad; spikes numerous, (1) 2-7 cm. long, about 1.5 cm. thick; bracts long-acuminate, slightly exceeding the buds; calyx densely villous-hirsute and grayish to rusty, the tube 2.5-3 mm. long, concealed by the pubescence; corolla purple, sometimes very pale. N=7.

Dry plains and foothills; e. slope of the Rocky Mts., in Mont., to Sask. and Man., s. to Colo., Tex., and Ala. July-Aug.

Often, especially in the Rocky Mt. states, two distinct phases of the plant are to be found, a more copiously hairy form, which has been called var. *molle* (Rydb.) A. Nels., and a greenish, sparsely hairy to subglabrate form, the var. *purpureum.* They intergrade so completely that there is little reason to maintain them at higher than the varietal level.

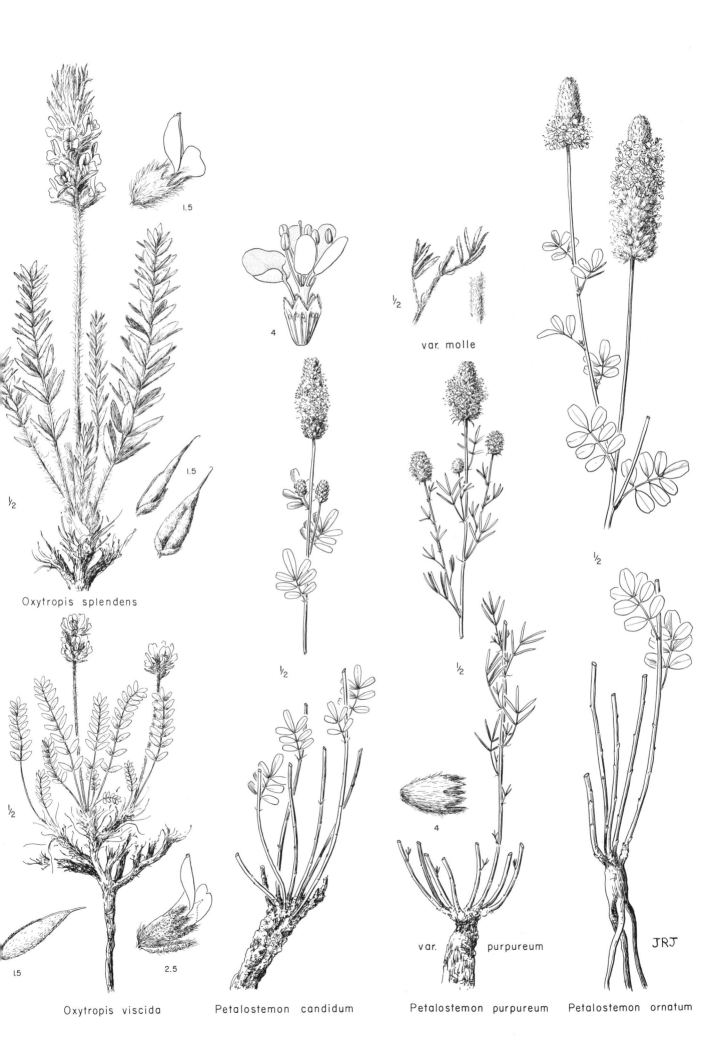

1.5

4

1/2

var. molle

1/2

1.5

1/2

Oxytropis splendens

1/2

1/2

4

1/2

1.5   2.5   var.   purpureum   JRJ

Oxytropis viscida   Petalostemon candidum   Petalostemon purpureum   Petalostemon ornatum

Psoralea L.

Flowers borne in small, axillary, pedunculate, spikelike to elongate and interrupted racemes, papilionaceous; calyx tubular to campanulate, the lobes subequal; corolla whitish to purple, the wings and keel only slightly shorter than the banner; stamens (9) 10, monadelphous (diadelphous); legumes 1-seeded, indehiscent (circumscissile or irregularly rupturing), more or less globose and often beaked, included in the calyx or only slightly longer than it; glandular-dotted, herbaceous perennials with 3- to 5-foliolate leaves and stipules free of the petioles.

A rather large genus with over 100 species, widespread but mostly in drier warmer regions. (Greek *psora,* itch or mange, hence rough or scablike, referring to the small scablike glands.)

1 Roots enlarged and tuberlike; flowers mostly over 1 cm. long, borne in dense spikes; calyx
    not black-hairy; Rocky Mts. and e.
   2 Plant long-hirsute   P. ESCULENTA
   2 Plant appressed-strigose   P. HYPOGAEA
1 Roots not enlarged and tuberlike; flowers mostly less than 1 cm. long, but if that long then
    the calyx black-hairy and plants from chiefly w. of the Cascades
   3 Flowers 9-12 mm. long; calyx black-hairy, markedly accrescent and as much as 12 mm.
     long in fruit; leaflets ovate   P. PHYSODES
   3 Flowers usually less than 9 (7-10) mm. long; calyx not black-hairy and not markedly
     accrescent; leaflets usually linear to lanceolate, oblanceolate, or oblong
     4 Leaves silvery-hairy, the lower surface obscured by the pubescence; probably en-
      tirely e. of the Rocky Mts.   P. ARGOPHYLLA
     4 Leaves greenish, often the upper surface glabrous
      5 Flowers blue, borne in long, loose, interrupted racemes that much exceed the
       leaves; Rocky Mts. eastward   P. TENUIFLORA
      5 Flowers white or the keel blue, borne in short, congested racemes often exceeded
       by the leaves; more general through much of our area   P. LANCEOLATA

Psoralea argophylla Pursh, Fl. Am. Sept. 475. 1814.
   Lotodes argophyllum Kuntze, Rev. Gen. 1:194. 1891. *Psoralidium argophyllum* Rydb. N.Am. Fl.
   24[1]:16. 1919. ("On the banks of the Missouri")
   Psoralea incana Nutt. Gen. Pl. 2:102. 1818. (*Nuttall,* "On the open plains of the Missouri")
   Psoralea argophylla robustior Bates, Am. Bot. 20:16. 1914. (Nebraska)
   Rhizomatous, white-sericeous perennial 3-6 dm. tall; stems usually freely branched above; leaves 3- to 5-foliolate; stipules linear-lanceolate, 3-8 mm. long; petioles from slightly longer to shorter than the blades, the leaflets oblong-elliptic to oblong-obovate, apiculate, 2-5 cm. long, 5-15 (20) mm. broad; flowers 7-10 mm. long, deep blue, subsessile and 2-4 at each node in interrupted spikes 2-5 cm. long; calyx grayish-hairy, the tube 2-3 mm. long, the upper four teeth narrowly lanceolate, about 3 mm. long, the lowest about twice as long, the whole elongating to as much as 1 cm. in fruit; legume about 8 mm. long, ovoid, with a short straight beak. N=11.
   Great Plains, to the e. slope of the Rocky Mts., Mont. to Sask., Mo., and N.M. June-July.

Psoralea esculenta Pursh, Fl. Am. Sept. 475. 1814.
   Pediomelum esculentum Rydb. N.Am. Fl. 24[1]:20. 1919. (*Lewis,* "On the banks of the Missouri")
   Perennial from fleshy tuberous roots, 1-3 dm. tall, hirsute throughout except on the glabrous upper surface of the leaves; leaflets (3) 5, oblong to oblanceolate, 2-6 cm. long; flowers numerous in dense spikelike racemes 2-10 cm. long; calyx tube about 5 mm. long, the lanceolate teeth somewhat longer; corolla ochroleucous to pale blue, about 1.5 cm. long; legume ovoid, the body about half as long as the slender beak. N=11. Breadroot.
   Prairies and lower foothills, from the e. side of the Rocky Mts. to Wis., and from Alta. to Okla., just on the border of our range. June-July.

Psoralea hypogaea Nutt. in T. & G. Fl. N.Am. 1:302. 1838.
   Lotodes hypogaeum Kuntze, Rev. Gen. 1:194. 1891. *Pediomelum hypogaeum* Rydb. N.Am. Fl. 24[1]:
   21. 1919. (*Nuttall,* "Plains of the Platte")
   Strigose perennial, acaulescent above ground, arising from fairly deep fusiform roots 2-6 cm. long and 5-15 mm. thick, with a slender, erect, bracteate, subterranean connecting-stem; stipules ovate-lanceolate, membranous, 1-2.5 cm. long; petioles 3-10 cm. long, usually much exceeding the ra-

cemes; leaflets (3) 5, digitate, linear-lanceolate to narrowly oblanceolate, 2-6 cm. long, 4-9 mm. broad, the upper surface much greener than the lower; racemes spikelike, 1-3 cm. long; peduncles 0.5-5 cm. long; flowers closely crowded; calyx (and bracts) sericeous to hirsute, the tube 3-4 mm. long, 4 of the lobes subequal and narrowly lanceolate, slightly longer than the tube, the fifth (the lowest) lanceolate or broader and 7-9 mm. long; corolla 10-13 mm. long, purplish; legume ovoid, about 5 mm. long exclusive of the flattish, 5-8 mm. beak. N=11.

Great Plains region, from the e. foothills of the Rocky Mts. e. to Neb., s. to N.M. and Okla., and in Utah. Apr.-July.

Psoralea lanceolata Pursh, Fl. Am. Sept. 475. 1814.
Psoralidium lanceolatum Rydb. N.Am. Fl. 24[1]:13. 1919. *(Lewis,* "On the banks of the Missouri")
Psoralea elliptica Pursh, Fl. Am. Sept. 741. 1814. *(Bradbury,* "In Upper Louisiana")
Psoralea scabra Nutt. in T. & G. Fl. N.Am. 1:300. 1838. *Psoralea lanceolata* ssp. *scabra* Piper, Contr. U.S. Nat. Herb. 11:364. 1906. *(John Townsend,* "on the Walla Wallah")
Lotodes ellipticum var. *latifolium* Kuntze, Rev. Gen. 1:193. 1891, not *Psoralea latifolia* Torr.
Psoralea purshii Vail, Bull. Torrey Club 21:94. 1894. *Psoralea lanceolata* var. *purshii* Piper in Piper & Beattie, Fl. Palouse Reg. 106. 1901. *(M. E. Jones,* Empire City, Nev.)

Freely branching perennial 3-6 dm. tall, glabrate to conspicuously strigose, strongly glandular; leaflets 3, greenish, narrowly elliptic-oblanceolate to narrowly obovate, 2-3 (4) cm. long; lower stipules as much as 2 cm. long, the upper ones much reduced; racemes rather closely 10- to 40-flowered; peduncles usually shorter than the leaves; flowers from white or with the keel bluish-tipped to leadblue, 4-7 mm. long; calyx campanulate, 2-2.5 mm. long, the teeth equal, very short, ovate-triangular; banner slightly exceeding the oblong wings; legume elliptic-globose, 4-5 mm. long, very glandular-warty, sparsely strigose to grayish-villous. N=11.

Sandy soil, often with sagebrush; e. side of the Cascade Mts., from Wash. to Calif. and Nev., e. to Neb. May-Sept.

In general, the more western material, which has been called both ssp. *scabra* (Nutt.) Piper and var. *purshii* (Vail) Piper, tends to have more abundant, longer hair on the pods.

Psoralea physodes Dougl. ex Hook. Fl. Bor. Am. 1:136. 1838.
Lotodes physodes Kuntze, Rev. Gen. 1:194. 1891. *Hoita physodes* Rydb. N.Am. Fl. 24[1]:8. 1919.
(Douglas, "from the Great Falls of the Columbia to the Rocky Mountains")

Sparingly branched, somewhat strigose perennial 5-7 dm. tall; leaflets 3, ovate-lanceolate, (2) 2.5-4 (5) cm. long; stipules linear; racemes short, congested, almost capitate, 15- to 20-flowered; flowers 9-12 mm. long; calyx black-hairy, tubular, 6-8 mm. long at anthesis, accrescent and as much as 12 mm. long in fruit, the teeth lanceolate, subequal; banner yellowish-green, about equaling the wings; keel purplish-tipped; pod about 5 mm. long, globose, sparsely dark-hairy, otherwise smooth. California tea.

West of the Cascades, commonly on logged-off land, from Wash. to Calif., also reported from Troy, Ida. May-June.

Psoralea tenuiflora Pursh, Fl. Am. Sept. 475. 1814.
Psoralidium tenuiflorum Rydb. N.Am. Fl. 24[1]:15. 1919. *(Lewis,* "On the banks of the Missouri")

Diffusely branched, strigose, greenish perennial 4-10 dm. tall; leaflets 3 or 5, entire, oblanceolate, 1-4 cm. long, strigose beneath, glabrous on the upper surface; stipules linear, 2-4 mm. long; racemes 5-15 cm. long, interrupted, the nodal groups of 2-4 flowers widely spaced; calyx 2-3 mm. long, whitish-strigose; corolla light to dark bluish-purple, 4-6 mm. long; legume ovoid, 7-8 mm. long.

Plains and foothills, possibly always on the e. side of the Rocky Mts.; Mont. to the Dakotas, s. to Tex. June-Sept.

Robinia L.

Flowers papilionaceous, showy, borne in large, drooping, axillary racemes; calyx campanulate, the teeth very broad; stamens 10, diadelphous; legumes flattened, 2-valved, several-seeded; trees or shrubs with thick bark, odd-pinnate leaves, and small stipules often modified into large thorns.

About 20 species of the U.S. and Mex. (Named for Jean and Vespasian Robin, the latter of whom is believed to have introduced the tree into cultivation in Europe, in the 16th century.)

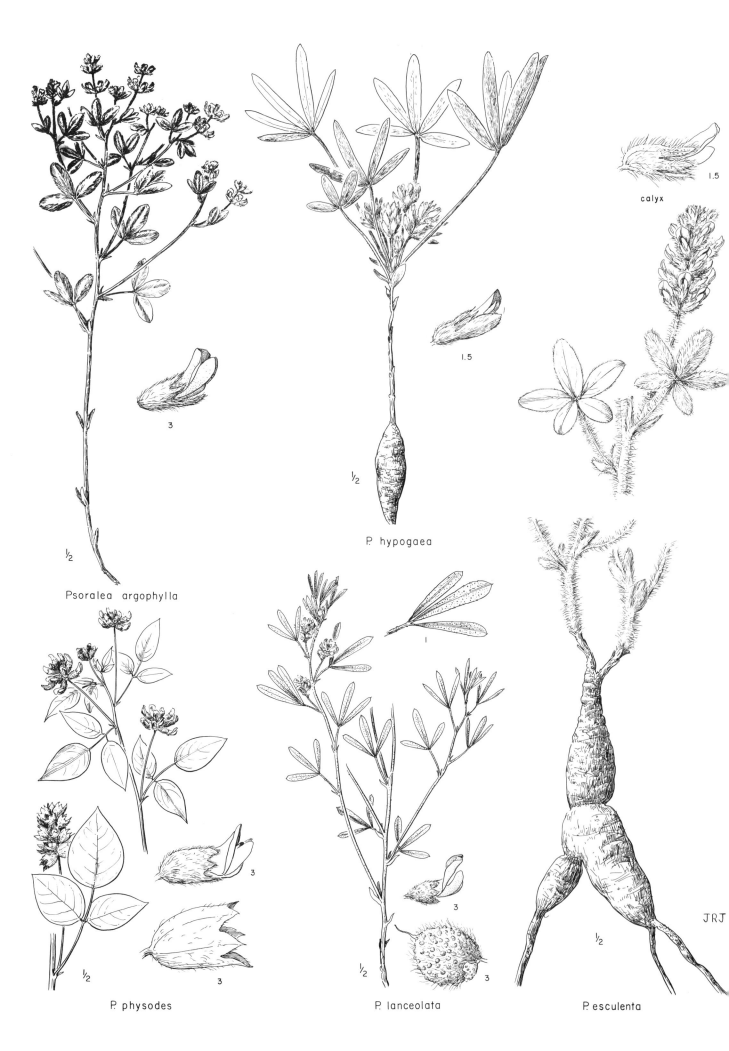

calyx

1.5

3

1/2

Psoralea argophylla

P. hypogaea

1.5

1/2

1

3

3

1/2

P. physodes

1/2

3

3

P. lanceolata

JRJ

1/2

P. esculenta

Robinia pseudo-acacia L. Sp. Pl. 722. 1753. (Virginia)

Well-armed tree to 25 m. tall; leaflets 11-21, lanceolate to elliptic-oval or oblanceolate, 2-4 cm. long, bright green on the upper surface, paler beneath; racemes 10-14 cm. long, 30- to 70-flowered; flowers fragrant, white, 14-20 mm. long; calyx 5-6 mm. long, broadly campanulate, thickly pubescent; banner well reflexed, usually yellow-blotched at base; legume 6-10 cm. long. N=10. Black or yellow locust, false acacia.

The wood of the black locust, which is native in e. U.S., is hard, tough, and resistant to decay, hence valuable around the farm for fence posts, firewood, etc. Therefore the tree has been widely introduced in other parts of N. Am., including the Pacific states, especially in rural areas where it is planted both as a shade tree and for its wood. Since it has a tendency to spread by suckers as well as by seed, it is often found so extensively established as to appear to be native; mostly around former habitations or along riverbanks. May-June.

Swainsona Salisb.

Flowers papilionaceous, loosely racemose; calyx 5-toothed, the tube campanulate, with 2 small lanceolate bracteoles at the base; stamens 10, diadelphous; style sparsely hairy nearly the full length on the upper side; legume stipitate, inflated, 1-celled, several-seeded; perennial herbs with odd-pinnate leaves.

About 50 species, one in New Zealand, the rest in Australia (mostly) and Eurasia. (Named for Isaac Swainson, English botanist, 1746-1812.)

Swainsona salsula (Pall.) Taub. in Engl. & Prantl, Nat. Pflanzenf. 3^3:281. 1894.

Phaca salsula Pall. Reise 3:747. 1776. *Sphaerophysa salsula* DC. Prodr. 2:271. 1825. ("Lacum siccum Tarei Dauuriae," Siberia)

Astragalus violaceus St. John, Res. Stud. State Coll. Wash. 1:98. 1929, but not of Basil. in 1922.

Astragalus iochrous Barneby, Leafl. West. Bot. 4:55. 1944. *(P. A. Ruppert,* Sunnyside, Yakima Co., Wash., July 23, 1928 and Aug. 6, 1928)

Perennial with extensive woody rootstocks; stems arising singly or several together, sparsely strigillose, at least above, 4-9 dm. tall; leaflets 9-25, elliptic-oblanceolate or oblong-elliptic, 1-2 cm. long, pale, glabrous on the upper surface, sparsely strigillose beneath; stipules linear, 2-7 mm. long, basally united to the petiole; racemes axillary, loosely 5- to 12-flowered, bracteate, about equaling the subtending leaves; calyx campanulate, about 5 mm. long, the teeth subequal, the 2 basal subtending bracteoles scarcely 1 mm. long; corolla brick red or pinkish-brown, sometimes drying more violet or purplish; petals 10-15 mm. long, subequal; ovary white-strigillose; legume membranous, inflated, 1.5-3 cm. long, 1-1.5 cm. broad, many-seeded, with a stipe once to twice as long as the calyx.

An Asiatic plant, now well established and rapidly spreading on alkaline soil in several w. states; Yakima Valley to e. Wash., and in e. Oreg., within our area. May-July.

Thermopsis R. Br. Buck bean; Yellow or Golden Pea

Flowers numerous to few in bracteate, subterminal axillary racemes, papilionaceous; floral bracts conspicuous; calyx campanulate, bilabiate, the lips toothed; corolla yellow; stamens 10, distinct; legumes coriaceous, elongate, many-seeded, not inflated; perennial herbs with thick woody rhizomes, trifoliolate leaves, and large leafletlike stipules that are adnate to the petiole.

A rather small genus of perhaps eight species of the U.S. and Asia. (Greek *thermos,* the lupine, and *opsis,* resemblance or likeness.)

Thermopsis montana is easily grown and well deserving of a place in the wild garden, having excellent foliage and showy, attractive flowers. Although transplantable with difficulty, the plant is best propagated from seeds.

References:

Larisey, M. M. A revision of the North American species of Thermopsis. Ann. Mo. Bot. Gard. 27:245-258. 1940.

St. John, Harold. Notes on North American Thermopsis. Torreya 41:112-115. 1941.

1 Pods erect or ascending to rarely somewhat spreading, nearly or quite straight, never recurved; plants mostly 5-9 dm. tall; Pacific coast to the Rocky Mts.   T. MONTANA

1 Pods spreading and usually recurved to annular; plants mostly 2-4 dm. tall; Rocky Mts. eastward   T. RHOMBIFOLIA

Thermopsis montana Nutt. in T. & G. Fl. N. Am. 1:388. 1840. *(Nuttall,* "High vallies of the Rocky
Mountains, in bushy places by streams, near the line of Upper California")
  T. fabacea sensu Hook. Fl. Bor. Am. 1:128. 1838, not of (Pall.) DC.
  T. gracilis Howell, Erythea 1:109. 1893. *(Howell,* mountains of southwestern Oregon) = var. *venosa.*
  T. angustata Greene, Pl. Baker. 3:34. 1901. *(Greene,* Star Valley, foothills of Ruby Mts., Nev.,
    July 20, 1896) = var. *montana.*
  T. stricta Greene, Pl. Baker. 3:34. 1901. *(C. F. Baker 173,* Sapinero, Colo., June 19, 1901, is
    the first of 2 collections cited) = var. *montana.*
  T. venosa Eastw. Bull. Torrey Club 32:198. 1905. *T. gracilis* var. *venosa* Jeps. Man. Fl. Pl.
    Calif. 515. 1925. *T. montana* var. *venosa* Jeps. Fl. Calif. 2:245. 1936. *(Eastwood,* Lewiston
    Trail, Trinity-Shasta Co. line, Calif., July 3, 1901)
  T. montana ssp. *ovata* Robins. ex Piper, Contr. U.S. Nat. Herb. 11:349. 1906. *T. ovata* Rydb.
    Bull. Torrey Club 40:43. 1913. *T. montana* var. *ovata* St. John, Torreya 41:112. 1941. *(Piper
    1489,* Cedar Mt., Latah Co., Ida.)
  T. xylorhiza A. Nels. Bot. Gaz. 52:265. 1911. *(Macbride 99,* Falk's Store, Canyon Co., Ida., May
    24, 1910) = var. *ovata.*
  T. subglabra Henderson, Rhodora 32:26. 1930. *(Rollo Patterson,* also *Henderson 9959,* Culp Creek,
    near base of Bohemia Mt. Divide, between Lane and Douglas cos., Oreg.) = var. *venosa.*
  More or less fistulose-stemmed perennial (4) 5-10 dm. tall, densely short villous-pubescent to
sparsely appressed-pubescent or glabrous below; stipules cordate-ovate to oblong or linear-lanceolate, often considerably wider than the leaflets, the latter broadly ovate-elliptic to oblanceolate or
linear-elliptic and as much as 10 cm. long; flowers 5-60, erect, 20-25 mm. long, borne in dense to
interrupted racemes 1-3 dm. long; pod erect to somewhat spreading, straight, villous, 2- to 5-seed-
ed. N=9.
  On sandy, well-drained soil to wet meadowland, B.C. to Calif., e. to the Rocky Mts. from Mont.
to Colo. May-Aug.
  The species shows much variation in the shape of the leaflets and stipules and especially in the
amount of pubescence present. On the whole, however, the variations appear to be largely fortuitous
and most of the segregates that have been described are not considered significant natural populations,
with the following exceptions:

1 Leaflets and stipules narrow, 4-6 times as long as broad; whole plant densely pubes-
    cent; legumes erect; racemes 10- to 60-flowered; chiefly Rocky Mts., from Mont.
    to Colo., but occasional as far w. as e. Oreg.   var. montana
1 Leaflets and stipules usually less than 4 times as long as broad; pubescence variable
    in amount; legumes erect to spreading; racemes sometimes with less than 10 flowers;
    chiefly coastal, but occasional to the Rocky Mts.
  2 Plant, or at least the leaflets, nearly or quite glabrous; racemes mostly 5- to 15-
      flowered; legumes sometimes spreading; Lane Co., Oreg., s. to Calif., occasional
      plants from Ida. and Wash. are so nearly identical as to be included here
                                                                        var. venosa (Eastw.) Jeps.
  2 Plant usually rather densely pubescent; racemes mostly 10- to 60-flowered; legumes
      erect; from the coast to the Rocky Mts., and from Wash. to Calif.; extremely var-
      iable, but a taller, conspicuously hairy plant from n.w. Oreg. strongly suggests
      T. macrophylla of s.w. Oreg. and Calif.   var. ovata (Robins.) St. John

Thermopsis rhombifolia Nutt. ex Richards. in Frankl. Journ. App. 737. 1823.
  Cytisus rhombifolius Nutt. in Fraser's Cat. 1813, nom. nud.; ex Pursh, Fl. Am. Sept. 742. 1814.
    Thermia rhombifolia Nutt. Gen. Pl. 1:282. 1818. *(Bradbury,* Upper Louisiana)
  Thermopsis arenosa A. Nels. Bot. Gaz. 25:276, pl. 18, fig. 4. 1898. *Thermopsis rhombifolia* var.
    arenosa Larisey, Ann. Mo. Bot. Gard. 27:251. 1940. *(A. Nelson 3182a,* Laramie Hills, Wyo.,
    July 17, 1897)
  Glabrate to conspicuously appressed-pubescent perennial 1.5-4 dm. tall; stems mostly simple, not
fistulose; stipules ovate-cordate, from narrower to broader than the leaflets, the latter mostly oblanceolate, 1.5-3 cm. long, 1-2 cm. broad; flowers about 2 cm. long, 10-30 per inflorescence, the
racemes short, crowded, scarcely 1 dm. long; legume usually silky, more or less lomentlike, spreading and also usually from recurved to annular, 4-8 cm. long. N=9.
  Alberta to N.D., s. to Colo., mostly on the grasslands e. of the Rocky Mts., but occasional w. of
the continental divide in Mont. May-June.

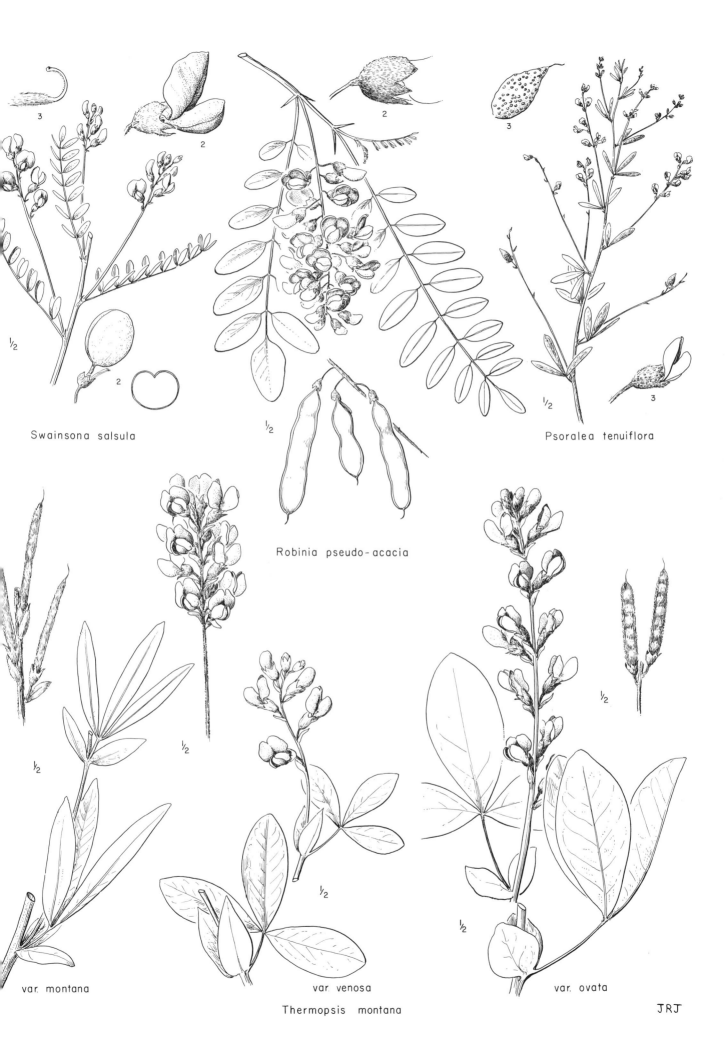

1/2

3

2

2

1/2

2

Swainsona salsula

2

1/2

Robinia pseudo-acacia

3

3

1/2

3

Psoralea tenuiflora

1/2

1/2

1/2

var. montana

1/2

1/2

var. venosa

Thermopsis montana

1/2

1/2

var. ovata

JRJ

Trifolium L. Clover

Flowers papilionaceous, white, yellow, or pink to red or purple, pedicellate to sessile and erect to reflexed, in (usually) pedunculate, terminal or axillary, often basally involucrate, more or less capitate spikes or racemes; calyx persistent, 5-toothed, the teeth sometimes bi- or trifid; corolla withering persistent, the banner usually only slightly reflexed from the wings and keel; stamens 10, diadelphous; legume globose to elongated, usually shorter than (and included within) the calyx, indehiscent, 1- to several-seeded; seeds globose to more or less reniform; cotyledons epigaeous; annual or perennial, taprooted or rhizomatous herbs, with usually palmately to semipinnately 3-foliolate (palmately 4- to 9-foliolate) leaves and membranous to foliaceous (often connate) stipules.

Nearly 300 species, mostly of the N. Temp. Zone, especially in w. N.Am., less abundant in S.Am. and Africa. (Latin, referring to the trifoli[ol]ate leaves.)

Trifolium subterraneum L. has recently been reported by Howell (Leafl. West. Bot. 11:115. 1960) near Corvallis, Oreg. It will probably be found to key with *T. gracilentum* and *T. bifidum,* from both of which it differs markedly because of the development of numerous sterile involucre-like sterile flowers and subterraneanly developed fruits; the flowers are cream colored and the tube of the calyx is 3-4 mm. long and subequal to the acicular teeth. Although our native clovers are seldom grown as ornamentals, this is somewhat unfortunate, as many are attractive.

References:

Martin, J. S. A revision of the native clovers of the United States. Ph.D. thesis. U. of Wash. Library. 1943.

McDermott, L. F. An illustrated key to the North American species of Trifolium. San Francisco. 325 pp. 1910.

The following treatment is taken without essential change from the Martin manuscript.

1 Plant an annual
  2 Heads involucrate
    3 Involucre long-villous, usually cup-shaped; calyx teeth simple
      4 Calyx usually hairy, the lower teeth at least as long as the tube and often surpassing the corolla; involucral lobes usually almost or quite entire   T. MICROCEPHALUM
      4 Calyx nearly or quite glabrous, the teeth all shorter than the tube and noticeably shorter than the corolla; involucral lobes several-toothed   T. MICRODON
    3 Involucre glabrous or subglabrous, often flared and saucer-shaped; calyx teeth various
      5 Lower calyx teeth usually bi- or trifid; involucre shallowly round-lobate, the lobes finely erose-dentate, with teeth less than 2 mm. long   T. CYATHIFERUM
      5 Lower calyx teeth entire or the involucre with teeth more than 2 mm. long
        6 Corolla strongly inflated, narrowed toward the tip; teeth of the involucre and calyx not stiffly spinulose; calyx tube 5- to 10-nerved
          7 Flowers and stipules both usually well over 10 mm. long; stems mostly fistulose; native to s. Oreg. and southward and thus out of our range, but collected at least once in Seattle, where probably introduced   *T. fucatum* Lindl.
          7 Flowers mostly less than 10 mm. long; stems not fistulose; stipules 5-10 mm. long   T. DEPAUPERATUM
        6 Corolla not strongly inflated; teeth of the involucre and calyx usually stiffly spinulose; calyx tube 10- to 25-nerved
          8 Sinuses of the calyx equally cleft; calyx lobes narrowly lanceolate-subulate, entire; heads 1-2 cm. broad; involucre irregularly lobed and cleft for (usually) at least half its length   T. VARIEGATUM
          8 Sinuses of the calyx not equally cleft, the upper much the deepest; calyx lobes often 3-toothed; either the heads less than 1 cm. broad or the involucre usually lobed less than half its length
            9 Heads 3-12 mm. broad, the 2-8 (15) flowers 4-8 mm. long   T. OLIGANTHUM
            9 Heads 10-30 mm. broad, the 6-60 flowers (8) 10-17 mm. long   T. TRIDENTATUM
  2 Heads not involucrate
    10 Calyx densely pilose-villous to strigose overall
      11 Flowers 12-15 mm. long; heads mostly at least 2.5 cm. long; stipules sheathing at base, the free portion not acuminate or subulate   T. INCARNATUM
      11 Flowers 5-8 (12) mm. long; heads usually less than 2.5 cm. long; stipules either not sheathing, or sheathing but the free portion acuminate to subulate

12 Calyx teeth shorter than the tube; calyx tube very prominently 10-striate and
    ovoid in fruit; known from a collection in Linn Co., Oreg., but probably not es-
    tablished in our area   *T. striatum* L.
12 Calyx teeth mostly 2-4 times as long as the tube; calyx tube not inflated and prom-
    inently striate in fruit
    13 Corolla white or pink, about 2/3 the length of the calyx; calyx teeth about twice
        as long as the tube; leaflets linear-oblanceolate   T. ARVENSE
    13 Corolla purplish, often subequal to the calyx; calyx teeth usually more than
        twice as long as the tube; leaflets oblanceolate to obcordate   T. MACRAEI
10 Calyx glabrous or pubescent only on the teeth
    14 Calyx teeth strongly fimbriate-denticulate, the 2 upper ones much larger than
        the lower 3   T. CILIOLATUM
    14 Calyx teeth nearly or quite entire, the upper ones often smaller than the lower 3
        15 Flowers white to pink or purplish; calyx 10-nerved
            16 Calyx teeth lanceolate-acuminate, glabrous; calyx tube over 1.5 mm. long
                on the upper side   T. GRACILENTUM
            16 Calyx teeth slender, acicular, sparsely villous-pilose; calyx tube not over
                1.5 mm. long on the upper side   T. BIFIDUM
        15 Flowers yellowish, fading to brown; calyx 5-nerved
            17 Pressed heads less than 8 mm. thick, usually with less than 30 flowers;
                corolla 3-3.5 mm. long   T. DUBIUM
            17 Pressed heads mostly over 8 mm. thick, usually with more than 30 flowers;
                corolla at least 4 mm. long
                18 Length of the stalk of the terminal leaflet twice that of the lateral leaf-
                    lets; heads mostly 8-11 mm. broad; stipules ovate, united to the pet-
                    iole for about half their length, the width of the adnate portion nearly
                    or quite equal to the length of the free tip   T. PROCUMBENS
                18 Length of the stalk of all leaflets about the same; heads usually well
                    over 1 cm. broad; stipules linear, often united to the petiole for 2/3
                    of their length, the width of the adnate portion much less than the
                    length of the free tips   T. AGRARIUM
1 Plant a perennial
   19 Flowers subtended by a true involucre
       20 Heads 1- to 4-flowered; plant 1-3 cm. tall   T. NANUM
       20 Heads several-flowered; plant usually well over 3 cm. tall
           21 Plants plainly pubescent, at least on the calyx
               22 Calyx becoming inflated and bladdery at maturity; plants sparingly to mod-
                   erately pubescent; weedy perennial in waste places   T. FRAGIFERUM
               22 Calyx not becoming greatly enlarged at maturity; plants rather heavily
                   pubescent; high montane in the Rocky Mts.   T. DASYPHYLLUM
           21 Plants nearly or quite glabrous, at least on the calyx
               23 Plant caespitose, taprooted, less than 1 dm. tall; heads borne on scape-
                   like peduncles; alpine or subalpine in the Rocky Mts.   T. PARRYI
               23 Plant leafy stemmed, rhizomatous, usually well over 1 dm. tall; general
                   from the Rocky Mts. w. to near the coast   T. WORMSKJOLDII
   19 Flowers not subtended by a true involucre, although sometimes the stipules of the up-
       per leaves somewhat involucral
       24 Leaflets commonly more than 3; flowers usually 30-100 per head, mostly over 18 mm.
           long; rachis of the head not prolonged above the upper flowers
           25 Leaflets linear to lanceolate, acute; calyx pubescent but not plumose   T. THOMPSONII
           25 Leaflets oblanceolate to obcordate; calyx teeth plumose   T. MACROCEPHALUM
       24 Leaflets generally 3, if more than 3 then flowers usually fewer than 30 per head
           and less than 15 mm. long or the rachis of the head extended well above the up-
           per flowers
           26 Calyx glabrous or only very sparsely pubescent with scattered hairs
               27 Heads 3-5 cm. long, but usually not so thick; calyx tube with about 20 prom-
                   inent nerves, the upper teeth curved downward; plants mostly over 5 dm.
                   tall   T. DOUGLASII

27 Heads usually as thick as long; calyx about (5) 10-nerved; plants mostly less than
    5 dm. tall
    28 Flowers 5-9 mm. long; heads axillary
        29 Corolla white or slightly pinkish-tinged; plant stoloniferous; calyx glabrous;
            leaflets usually retuse to obcordate; stipules 3-10 mm. long   T. REPENS
        29 Corolla usually pink to reddish (white); plant usually not stoloniferous, the
            stems ascending to erect; calyx with a few hairs at the base of the teeth;
            leaflets frequently rounded; stipules 5-20 mm. long   T. HYBRIDUM
    28 Flowers at least 10 mm. long; heads often terminal and solitary
        30 Leaflets (2) 2.5-5 cm. broad   T. HOWELLII
        30 Leaflets averaging less than 2 cm. broad
            31 Plant glabrous, caespitose or nearly so, the leafy stems scarcely 5 cm.
                tall; high montane in the Rocky Mts.   T. HAYDENII
            31 Plant either pubescent or with leafy stems over 5 cm. tall
                32 Rachis of the inflorescence prolonged and usually forked above the
                    strongly reflexed flowers; calyx scarcely 1/3 as long as the co-
                    rolla   T. PRODUCTUM
                32 Rachis not prolonged, flowers not reflexed, or the calyx over 1/3
                    as long as the corolla
                    33 Calyx usually less than 1/2 as long as the corolla, the teeth usu-
                        ally about equal in length to the tube   T. BECKWITHII
                    33 Calyx usually at least half as long as the corolla, the teeth usually
                        more than twice as long as the tube   T. LONGIPES
26 Calyx strongly pubescent to villous or plumose
    34 Heads sessile or very short-pedunculate, 50- to 200-flowered; peduncles shorter than
    the subtending leaves; stipules of the upper leaves forming a false involucre T. PRATENSE
    34 Heads pedunculate, sometimes few-flowered; peduncle either longer than the leaves,
    or the upper stipules not involucral
        35 Peduncles mostly shorter than the leaves; leaflets thick and leathery, usually not
            over 2.5 cm. long and with less than 30 serrations; plant seldom over 1.5 dm.
            tall; heads usually less than 15-flowered   T. GYMNOCARPON
        35 Peduncles usually longer than the leaves; leaflets either thin or more than 2.5 cm.
            long or with more than 30 serrulations; plants mostly over 1.5 dm. tall; heads
            usually more than 15-flowered
            36 Heads cylindric, usually nearly twice as long as thick; calyx densely villous-
                plumose, the tube 20- to 25-nerved, the teeth attenuate-aristate; leaflets
                linear to linear-elliptic, usually over 6 cm. long   T. PLUMOSUM
            36 Heads globose to hemispheric; calyx often merely ciliate or short-pubescent,
                the tube 5- to 10-nerved; leaflets seldom over 6 cm. long
                37 Calyx plumose-villous; flowers all strongly reflexed, the curvature chiefly
                  in the base of the calyx, the calyx tube therefore somewhat gibbous
                                        T. ERIOCEPHALUM
                37 Calyx strigose to short-villous; flowers erect to reflexed, the curvature all
                  in the pedicels, the calyx tube not gibbous at base
                  38 Calyx lobes subplumose; leaflets seldom as much as 3 times as long as
                    broad; pedicels often as much as 2 mm. long   T. LATIFOLIUM
                  38 Calyx lobes short-pubescent; leaflets usually more than 3 times as long
                    as broad; pedicels mostly less than 2 mm. long
                      39 Wings and keel abruptly acuminate-apiculate; flowers purplish, the
                      lower ones reflexed; leaflets usually less than 2 cm. long
                                    T. MULTIPEDUNCULATUM
                  39 Wings (and usually the keel) not acuminate-apiculate; flowers usu-
                    ally either ochroleucous or not reflexed; leaflets usually over 2
                    cm. long   T. LONGIPES

Trifolium agrarium L. Sp. Pl. 772. 1753.
  Chrysaspis agraria Greene, Pitt. 3:205. 1897. (Europe)
  Pubescent to glabrous annual with usually several erect or ascending stems 2-5 dm. tall; stipules

10-18 mm. long, rather narrow, adnate to the petiole for 1/2-2/3 of their length, the free portion acuminate; petioles usually shorter than the stipules; leaflets 3, oblong-elliptic to oblanceolate, 1-3 cm. long, all subsessile, serrulate about 2/3 of their length; heads noninvolucrate, 30- to 100-flowered, axillary on peduncles longer than the leaves, 10-16 mm. thick; flowers 5-7 mm. long, yellow, pendulous at anthesis; calyx glabrous (or the lower 3 teeth sparsely tufted-hairy), about 1/2 the length of the corolla, the teeth entire, the upper 2 scarcely half as long as the lower 3, the latter usually somewhat longer than the calyx tube; banner somewhat flared and spreading (not folded), considerably longer than the wings and keel; style subequal to the mature, usually 1-seeded legume. N=7. Hop clover.

A native of Europe, fairly common in or near waste places, roadsides, railways, etc., in Wash., Ida., and B.C., more common on the Atlantic coast. June-Sept.

Trifolium arvense L. Sp. Pl. 769. 1753. ("Habitat in Europa, America septentrionali")

Erect, pubescent annual with 1-several freely branched stems 1-4 dm. tall; stipules united to the petioles, narrow-margined but extending into linear-subulate free tips 5-10 mm. long; leaflets 3, linear-oblanceolate, 8-25 mm. long; heads many-flowered, noninvolucrate, 5-25 mm. long, 10-13 mm. thick, axillary as well as terminal on the numerous branches; flowers 5-6 mm. long; calyx strongly strigose-plumose, the teeth subulate-aristate, about twice as long as the tube, considerably surpassing the white or pale pink corolla; legume 1-seeded. N=7.

A European plant, occasionally escaped in old fields or other waste places; chiefly w. of the Cascade Mts., in B.C., Wash., and Oreg.; more common in e. U.S. May-July.

Trifolium beckwithii Brew. ex Wats. Proc. Am. Acad. 11:128. 1876. *(James A. Snyder,* "on the Sierra Nevada, California")

Glabrous perennial with a strong taproot and a branched crown but without true rhizomes, the usually several stems simple, 1.5-5 dm. tall; stipules 1-2.5 cm. long, ovate-lanceolate; leaflets 3, narrowly elliptic to ovate-lanceolate or oblong-obovate, 2-6 cm. long, finely serrulate; heads terminal, 2.5-4 cm. broad and about as long, long-peduncled from the axils of the upper 2-several somewhat involucral stipules, but without a true subtending involucre, the rachis usually not prolonged above the head; flowers 12-17 mm. long, only the lower ones reflexed; calyx glabrous, 1/3 (1/2) the length of the corolla, the lanceolate-subulate teeth about equal to the 5 (10)-veined tube; corollas reddish or pale purplish; legume 1- to 3-seeded.

Mostly in meadows in the mountains, from s.e. Oreg. e. to Deer Lodge Co., Mont., and s. to the middle Sierra Nevada of Calif. May-Aug.

This species is closely related to *T. productum;* it differs chiefly because the flowers are less reflexed and the rachis of the inflorescence is not prolonged above the head.

Trifolium bifidum Gray, Proc. Calif. Acad. Sci. 3:102. 1864. *(Brewer 1184,* near Mt. Diablo, Calif.)
    T. bifidum var. *decipiens* Greene, Fl. Fran. 24. 1891, not *T. decipiens* Hornem. in 1815. *T. greenei* House, Bot. Gaz. 41:334. 1906. *(Greene,* San Francisco Bay Region)
    T. hallii Howell, Fl. N.W. Am. 135. 1903. *(Hall,* Silverton, Oreg., in 1871)

Sparsely villous to glabrous annual, the stems simple or sparingly branched, erect to decumbent, 1-4 dm. long; stipules 1-2 cm. long, ovate, acuminate, the margins nearly or quite entire; leaflets 3, oblanceolate to obcordate, retuse to deeply bifid, 1-2.5 cm. long, rather coarsely serrulate; heads 5- to 30-flowered, 8-15 mm. broad, noninvolucrate, axillary as well as terminal; flowers pinkish, 5-7 mm. long, soon reflexed on pedicels 1-3 mm. long; calyx about equaling the corolla, the tube 10-veined and glabrous, the teeth sparsely villous, acicular, subequal, 2-4 times as long as the tube; legume 1- or 2-seeded.

Klickitat Co., Wash., southward, chiefly w. of the Cascades and Sierra Nevada, to s. Calif. Apr.-June.

Our plants have usually been recognized as the var. *decipiens* Greene, the leaves tending to be retuse or rounded rather than truncate or deeply bifid as in the Californian var. *bifidum.*

Trifolium ciliolatum Benth. Pl. Hartw. 304. 1848. *(Hartweg 1697,* Sacramento Valley, Calif.)
    T. ciliatum Nutt. Journ. Acad. Phila. II, 1:152. 1848, but not of Clarke in 1813-16. *(Gambel,* Pueblo de los Angeles, Calif.)

Glabrous to very sparingly ciliate annual with several decumbent to erect stems 1.5-5 dm. long; stipules 1-2.5 cm. long, usually narrow-margined, prolonged into acuminate entire tips; leaflets 3, oblong-elliptic to oblanceolate, obovate, or obcordate, 1-3.5 cm. long, finely acerose-denticulate the full length; heads axillary as well as terminal, 10- to 50-flowered, noninvolucrate, 1-2 cm. long, subglobose; flowers white to purplish, 6-12 mm. long, erect at anthesis but gradually becoming reflexed on the elongating, eventually 2-6 mm. pedicels; calyx glabrous, from shorter to longer than the corolla, the tube 10-veined, the teeth strongly denticulate-fimbriate, 2-3 times as long as the tube, the upper 2 much the longest; legume 1-seeded.

Southern Wash. to Baja Calif., w. of the Cascades and Sierra Nevada, at elevations up to several thousand feet in the south. Apr.-June.

Trifolium cyathiferum Lindl. Bot. Reg. 12:pl. 1070. 1827. *(Douglas*, Columbia River)

Glabrous annual with ascending to erect stems 1-5 dm. long; stipules ovate to lanceolate, 1/4-1/2 the length of the leaflets; leaflets 3, oblanceolate to obcordate, 5-25 mm. long, setaceous-serrulate most of their length; heads usually 5- to 30-flowered, 5-15 mm. long and about as broad; involucre flared to deeply cup-shaped, shallowly 6- to 14-lobed, finely erose-dentate, generally setulose; calyx glabrous, the tube membranous, 2-5 mm. long, 13- to 20-nerved, the teeth setaceous, usually exceeding the tube, the lower 3 generally slenderly bi- or trifid; corolla white or ochroleucous to pink, about equaling the calyx; legume usually 2-seeded, often short-stipitate.

Wet meadows to fairly dry sandy soil; e. of the Cascades, from B.C. to Calif., e. to Ida.; probably introduced in a few localities along the coast from Juneau to Los Angeles. May-July.

Trifolium dasyphyllum T. & G. Fl. N.Am. 1:315. 1838. *(James*, Rocky Mts., Colo.)

T. uintense Rydb. Bull. Torrey Club 34:47. 1907. *T. dasyphyllum* f. *uintense* McDerm. N.Am. Sp. Trif. 16. 1910. *(Watson 241*, Uinta Mts., Utah)

Matted, caespitose, sparsely appressed-pubescent to canescent perennial from a large woody taproot; leafy stems spreading, usually only 1-3 cm. tall, covered with old, scarious stipules; leaflets 3, linear to narrowly oblanceolate, acute to rounded, 7-30 mm. long, entire; petioles 5-30 mm. long; peduncles 2-10 cm. long; heads 5- to 30-flowered, subglobose, 20-35 mm. thick; involucres 3-10 mm. high, the bracts 7-15, distinct or basally connate, narrowly lanceolate and often setose; flowers 1-2 cm. long, spreading to erect, with pedicels 0.5-2 mm. long; calyx hairy, 6-11 mm. long, 1/2-3/4 the length of the banner, the tube 10- to 12-veined, somewhat shorter than the subequal, very narrow to subulate teeth; banner ochroleucous to yellowish but purple-tipped, the wings and keel purplish-tipped to uniformly purplish; legume 1- to 4-seeded.

Alpine and subalpine slopes from Big Horn, Carbon, Stillwater, Beaverhead, and Madison cos., Mont., to Colo., and in the mountains of e. Utah. July-Sept.

Trifolium depauperatum Desv. Journ. Bot. 4:69. 1814. (Western coasts of both N. and S.Am.)

T. amplectens T. & G. Fl. Fl. N.Am. 1:319. 1838. *T. depauperatum* var. *amplectens* Wats. Bot. Calif. 2:441. 1880. *(Douglas*, California)

Glabrous or very sparsely glandular-hairy annual usually with numerous spreading to ascending stems 0.5-4 dm. long; stipules acuminate, 5-10 mm. long; leaflets 3, linear to obcordate, 5-30 mm. long; heads axillary as well as terminal, long-pedunculate, 2- to 15-flowered, 4-15 mm. broad; involucres glabrous, reduced, often entire but sometimes acutely or bluntly 5- to 7-lobed; calyx membranous, broadly funnelform, 1/4-1/2 as long as the corolla, the tube 5- or 6-nerved, the teeth lanceolate, subequal to the tube; corolla 3-11 mm. long, white to pinkish but fading to brown, the banner much inflated and sometimes nearly as broad as long; legume 1- to several-seeded, sessile to stipitate.

Central Oreg. to s. Calif., sporadic n. (and possibly always introduced) to Wash. and Vancouver I. Apr.-June.

The plant undergoes great morphological variation to the south of our area, where a number of segregate taxa have been proposed. In our region the species is known only through a few collections from Vancouver I. and one from Pt. Ludlow, Wash., in 1890, all referable to var. *depauperatum*.

Trifolium douglasii House, Bot. Gaz. 41:335. 1906.

T. altissimum Dougl. ex Hook. Fl. Bor. Am. 1:130. 1831, but not of Lois. in 1807. *(Douglas*, "North-West America, between the Spokan River and Kettle Falls of the Columbia")

Perennial from a thick taproot, nonrhizomatous, usually glabrous; stems generally several, erect, simple or subsimple, 4-8 dm. tall; stipules oblong-lanceolate, 2-7 cm. long, adnate to the petiole most of their length, the margins serrulate; leaflets 3, linear to oblong-elliptic, 4-10 cm. long, the

½

½

5

½

Thermopsis rhombifolia

Trifolium cyathiferum

3

1.5

involucre

Trifolium beckwithii

8

4

1.5

4
calyx

4

1.5

½

½

½

½

folium agrarium   Trifolium bifidum   Trifolium arvense   Trifolium ciliolatum   JRJ

margins very finely serrulate-spinulose; petioles usually shorter than the stipules; heads noninvolucrate, axillary as well as terminal and long-pedunculate, globose to ovoid-cylindric, about 3 cm.
thick, as long to nearly twice as long, closely 50- to 200-flowered; flowers erect, spreading, or the
lowest reflexed, 14-20 mm. long, reddish-purple; calyx 1/2-3/5 as long as the corolla, glabrous, the
tube strongly 20 (17-25)-nerved, oblique, 1/3-3/4 as long as the subulate teeth; upper pair of calyx
teeth broader than the lower three and usually conspicuously curved downward; sinuses between the
lateral teeth deeper than that between the upper pair; legume usually 1-seeded.

In meadows and along streams; Spokane Co., Wash., to Baker Co., Oreg., e. to adj. Ida. June-
July.

Trifolium dubium Sibth. Fl. Oxon. 231. 1794.
   Chrysaspis dubia Greene, Pitt. 3:206. 1897. (Europe)
Annual, usually sparsely pubescent; stems 1-several, prostrate or decumbent to erect, 1-5 dm.
long; stipules 3-5 mm. long, acuminate, the margins entire to remotely denticulate; petioles usually
shorter than the blade; leaflets 3, cuneate-obovate to obcordate, 5-20 mm. long, shallowly denticulate above midlength; heads noninvolucrate, axillary, rather short-pedunculate, 3- to 20 (30)-flowered,
6-7 (8) mm. broad; flowers yellow, reflexed at anthesis, 3-3.5 mm. long; calyx glabrous, about half
as long as the corolla, 5-nerved, the teeth narrowly lanceolate, entire, the upper ones broader than,
but scarcely half as long as, the lower three; banner rather hoodlike, folded, nearly hiding the slightly shorter wings and keel; style shorter than the 1-seeded legume. N=7, 8, 14.

Native in Europe, this weedy plant is common throughout much of the U.S., in waste places, along
roads, and (in our area) on recently cleared land; a bad weed in lawns; B.C. to Calif. Apr.-Sept.

Trifolium eriocephalum Nutt. in T. & G. Fl. N.Am. 1:313. 1838. *(Nuttall,* "Prairies of the Wahlamet,
   and near Fort Vancouver")
   T. scorpioides Blasdale, Erythea 4:187. 1896. *(Blasdale,* Mad R., Humboldt Co., Calif.) = var.
      eriocephalum.
   T. harneyense Howell, Fl. N.W. Am. 134. 1898. *T. eriocephalum* var. *harneyense* Piper ex Mc-
      Derm. N.Am. Sp. Trif. 243. 1910. *(T. Howell,* Harney Valley, Oreg., in 1897) = var. *cusickii.*
   T. arcuatum Piper, Bull. Torrey Club 28:39. 1901. *T. eriocephalum* f. *arcuatum* McDerm. N.Am.
      Sp. Trif. 242. 1910. *(Suksdorf 270,* Simcoe Mts., Wash., June 6, 1884) = var. *eriocephalum.*
   T. arcuatum var. *cusickii* Piper, Bull. Torrey Club 29:642. 1902. *T. eriocephalum* var. *cusickii*
      Martin, Madroño 8:156. 1946. *(Cusick 2628,* Camp Creek, Maurey's Mt., e. Oreg., July 2, 1901,
      is the first of 2 collections cited)
   T. tropicum A. Nels. Bot. Gaz. 54:409. 1912. *(Macbride 967,* Jordan Valley, Owyhee Co., Ida.,
      June 22, 1911) = var. *cusickii.*
   T. eriocephalum var. *butleri* Jeps. Fl. Calif. 2:302. 1936. *(Butler 384,* Log Lake, Shackelford
      Creek, Siskiyou Co., Calif.) = var. *eriocephalum.*
   T. eriocephalum var. *piperi* Martin, Madroño 8:154. 1946. *(Cusick 2405,* Paradise, Wallowa Co.,
      Oreg., June 12, 1900)
Sparsely to densely villous-pubescent perennial from a thick taproot with a branched crown and 1-
several stems 2-6 dm. tall; stipules ovate to lanceolate, 2-5 cm. long, nearly entire; leaflets 3, elliptic or ovate to lanceolate or almost linear, 2-7 cm. long, entire to serrulate; heads occasionally
axillary as well as terminal, noninvolucrate, subglobose, 2.5-3 cm. broad, borne on peduncles usually considerably exceeding the subtending leaves; flowers 25-80, pinkish to red, 12-17 mm. long,
strongly reflexed, but the pedicels barely 1 mm. long, the curvature mostly in the calyx and corolla;
calyx plumose-villous, half the length of the corolla or longer, the tube slightly gibbous on the upper
side, the teeth filiform, 2-4 times as long as the 10-nerved tube; legume 1- to 4-seeded.

On both sides of the Cascades, from s.c. Wash. to n. Calif. and Nev., e. to w. Ida. and s.c. Utah.
May-July.

This well-marked species of drier meadows and woods is differentiated into several distinctive geographic races, 3 of which occur in our area:
1 Ovary with 2 (very rarely 3) ovules; Klickitat Co., Wash., to n. Calif.   var. eriocephalum
1 Ovary with 4 ovules
   2 Leaflets linear, acuminate; Union Co., Oreg., s. to s.e. Oreg. and s.w. Ida.
                                                                            var. cusickii (Piper) Martin
   2 Leaflets (at least the basal) elliptic to oblong, rounded to retuse; Blue Mts. of
      s.e. Wash. to Harney Co., Oreg., and adj. Ida.   var. piperi Martin

Trifolium fragiferum L. Sp. Pl. 772. 1753. ("Habitat in Suecia, Gallia, Anglia")

Rhizomatous perennial, sparingly to moderately pubescent, especially on the calyx; stems decumbent to creeping and rooting at the nodes, 5-30 cm. tall; stipules mostly 15-20 mm. long, narrow, the free portion acuminate; leaflets 3, obovate, 1-2.5 cm. long, serrulate, retuse, and apiculate; heads many-flowered, 10-12 mm. in diameter at anthesis but increasing to 15-20 mm. in fruit; flowers 4-6 mm. long, purplish, subtended by narrow bracts, the basal (outer) bracts connate and involucral; calyx from 2/3 as long to as long as the corolla, the tube gradually inflating greatly and becoming strongly reticulate-veined, causing the head to resemble a mass of veined bladders. N=8.

Introduced from Europe and becoming established in waste places in Wash., Oreg., and Ida., as well as in e. U.S. Apr.-July.

Trifolium gracilentum T. & G. Fl. N.Am. 1:316. 1838. (*Douglas,* California)

Glabrous or subglabrous annual with many usually decumbent or ascending stems 1-6 dm. long; stipules from entire to serrulate, acuminate, 1-2 cm. long; leaflets 3, obovate to obcordate, finely serrulate their entire length, 1-2 cm. long; heads axillary and terminal, long-pedunculate, noninvolucrate, rather loosely 5- to 50-flowered, usually as broad as long; flowers 5-9 mm. long, whitish to purple, reflexed after anthesis on pedicels 0.5-4 mm. long; rachis of the inflorescence often prolonged above the flowers; calyx glabrous, about 2/3 as long as the corolla, the teeth lanceolate-acuminate, 2-3 times as long as the usually 10-nerved tube, the upper 2 usually considerably the largest; legume 1- or 2-seeded.

Mainly Calif., but occasional on grassy knolls, etc., w. of the Cascades, from Wash. to Baja Calif. Apr.-June.

Trifolium gymnocarpon Nutt. in T. & G. Fl. N.Am. 1:320. 1838. (*Nuttall,* "Dry hills of the Rocky Mountain range, near the sources of the Sweetwater of the Platte")

T. *subcaulescens* Gray, Ives Rep. 10. 1860. T. *gymnocarpon* var. *subcaulescens* Nels. in Coult. & Nels. New Man. Bot. Rocky Mts. 279. 1909. (*Newberry,* near Ft. Defiance, Ariz.) = var. *gymnocarpon.*

T. *plummerae* Wats. Bot. Calif. 2:440. 1880. T. *gymnocarpon* f. *plummerae* McDerm. N.Am. Sp. Trif. 192. 1910. T. *gymnocarpon* var. *plummerae* Martin, Bull. Torrey Club 73:368. 1946. (*J. G. Lemmon;* and *Sara A. Plummer,* peaks west of Pyramid Lake, Nev.)

T. *nemorale* Greene, Pitt. 4:136. 1900. (*C. F. Baker,* Los Pinos, Colo., May 17, 1899) = var. *gymnocarpon.*

Strigose, caespitose perennial from a stout taproot; stems usually many and less than 1.5 dm. tall; leaves crowded, overlapping, the scarious stipules 5-20 mm. long, persistent as shreds on the old crown; petioles 3-10 cm. long; leaflets 3 (occasionally 4 or 5), 5-20 (30) mm. long, ovate to obovate, sharply serrate-dentate with from 5 to 15 teeth on each side; heads terminal, on peduncles usually shorter than the leaves, loosely 3- to 15-flowered, noninvolucrate, 1-2.5 cm. broad; pedicels 1-4 mm. long; flowers ochroleucous to flesh-colored, 8-14 mm. long; calyx 1/3-1/2 as long as the corolla, rather thickly villous-pubescent, the triangular-subulate teeth equaling or somewhat longer than the tube; legume generally 1-seeded.

Usually in dry soil of sagebrush desert to ponderosa pine forest; n.e. Oreg. to n.e. Calif. and n. Ariz., e. to Mont. and N.M. May-June.

The plants of our area, var. *plummerae* (Wats.) Martin, are moderately to heavily pubescent, with the leaves hairy on both surfaces. Variety *gymnocarpon* is the common phase of the southern Rocky Mts., with leaves glabrous on the upper surface.

Trifolium leibergii Nels. & Macbr. (Bot. Gaz. 65:58. 1918), known only from the type collection (*Leiberg 2344,* near "Dewey" [probably Drewsey], Harney Co., Oreg.), may prove to be a rare species that extends into our area. It differs from T. *gymnocarpon* in being appressed-villous and in having no more than three shorter and more nearly rotund leaflets.

Trifolium haydenii Porter, Ann. Rep. U.S. Geol. Surv. Mont. 1871:480. 1872. (*Allen & Adams,* ·s. of Virginia City, Mont., June 25-30, 1871)

Glabrous, more or less caespitose, matted perennial from a heavy taproot; stems short, 2-5 cm. tall, covered with the marcescent, usually scarious, lanceolate stipules; leaflets 3, broadly ovate to subrotund, prominently veined, 4-20 mm. long, rounded to acute or apiculate, denticulate; peduncles axillary, exceeding the leaves; heads loosely 5- to 20-flowered, noninvolucrate, 1.5-3 cm. thick, with the rachis usually prolonged; flowers mostly with pedicels 1-2 mm. long, soon becoming reflexed,

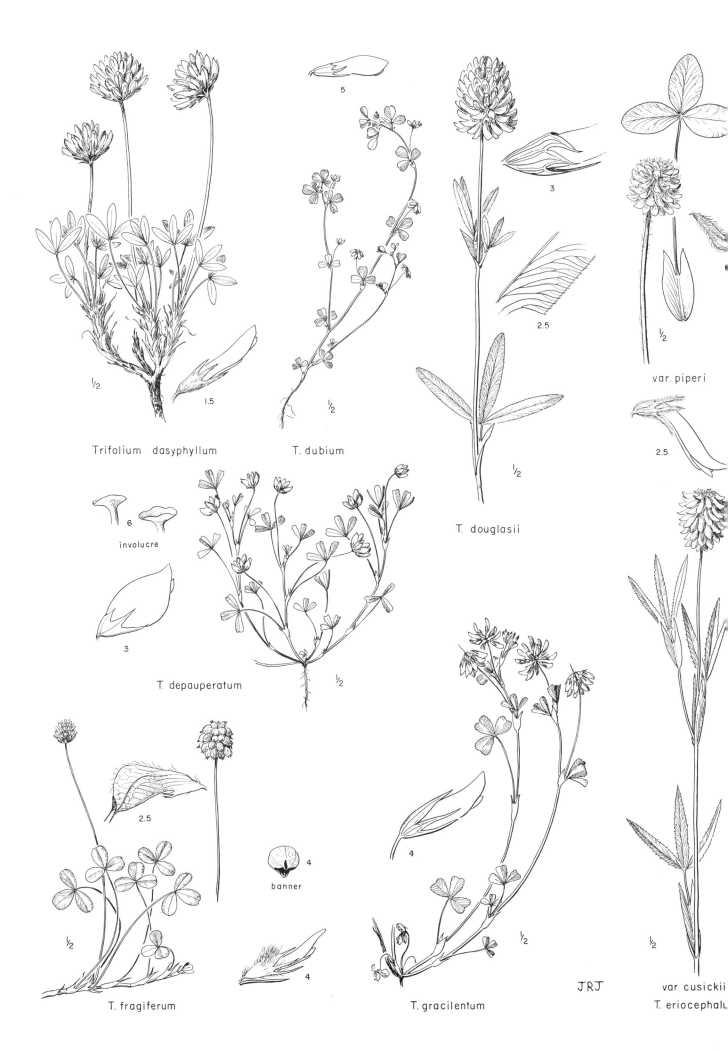

5

Trifolium dasyphyllum

1.5

½

T. dubium

½

3

2.5

½

var. piperi

2.5

T. douglasii

6

involucre

3

T. depauperatum

½

2.5

banner

4

4

4

T. fragiferum

½

½

T. gracilentum

½

JRJ

var. cusickii

T. eriocephalum

ochroleucous (but pinkish-tinged) to salmon, 13-17 mm. long; calyx glabrous, 1/2-2/3 as long as the corolla, the tube 10-veined, usually slightly shorter than the subequal, triangular-acuminate teeth; legume 1 (2)-seeded.

Alpine or subalpine slopes and ridges; Rocky Mts. of Carbon, Gallatin, and Park cos., Mont., and adj. Wyo. July-Aug.

Trifolium howellii Wats. Proc. Am. Acad. 23:262. 1888. *(T. Howell,* Siskiyou Mts., Oreg., July 11, 1887)

Glabrous perennial with stout rootstocks; stems simple, 3-8 dm. tall, fistulose; stipules ovate, 1.5-3 cm. long, the margins erose-serrulate; leaflets 3, thin, ovate to obovate, (2) 2.5-5 cm. broad, 4-9 cm. long, finely denticulate; heads axillary, fairly long-pedunculate but often surpassed by the leaves, noninvolucrate, about 2 cm. broad and as long, rather loosely 20- to 80-flowered; flowers ochroleucous or somewhat pinkish-tinged, 11-15 mm. long, at least the lower ones reflexed to pendulous; calyx glabrous, about half as long as the corolla, the tube thin and membranous, 10-nerved, from only slightly to considerably shorter than the subulate teeth, slightly gibbous on the upper side; legume 1- or 2-seeded.

West side of the Cascade Mts. from Lane Co., Oreg., to n.w. Calif., in wet or shady places. June-Aug.

Trifolium hybridum L. Sp. Pl. 766. 1753. (Europe)

Sparsely pubescent perennial; stems usually several, ascending to erect but sometimes stolonous; stipules 5-20 mm. long, mostly long-attenuate; leaflets 3, ovate to obovate but sometimes retuse to obcordate, 1-3 cm. long, serrulate; heads axillary, noninvolucrate, 1.5-2.5 cm. broad and nearly as long; peduncles from shorter than the leaves to several times their length; flowers white to reddish, 5-9 mm. long, soon reflexed, the pedicels 1-5 mm. long; calyx 1/2-1/3 as long as the corolla, sparsely hairy at the sinuses between the teeth, otherwise glabrous, the teeth subulate, about equal to the tube; banner nearly erect; legume 1- to 3-seeded. N=8. Alsike clover.

General in w. U.S. Apr.-Sept.

Trifolium hybridum has been introduced from Europe; it is very similar to *T. repens,* but differs in several minor details, as follows: the stipules tend to be longer, there is somewhat more pubescence on the vegetative parts, the calyx has tufts of hairs toward the top of the tube, the plant is less prone to creep, and the flowers average a deeper pink in color.

Trifolium incarnatum L. Sp. Pl. 769. 1753. (Italy)

Rank-growing, villous-pubescent annual with 1-several nearly simple, erect stems 2-8 dm. tall; stipules sheathing at base, 1-2 cm. long, entire to erose-dentate, usually reddish- or purple-margined; leaflets 3, broadly obovate to obcordate, 1-3 cm. long, denticulate on the upper half; heads terminal, (2) 2.5-6 cm. long, about 1.5 cm. thick, noninvolucrate, many-flowered; flowers sessile, ascending, crimson, 12-15 mm. long; calyx about 3/4 as long as the corolla, densely strigose-pilose, the slender aristate teeth slightly longer than the tube; banner considerably longer than the wings and keel, acute; seeds usually 1. N=7.

A native of Europe, occasionally established on waste land in w. Wash. and Oreg. May-June.

Trifolium latifolium (Hook.) Greene, Pitt. 3:223. 1897.

T. *longipes* var. *latifolium* Hook. Lond. Journ. Bot. 6:209. 1847. *T. howellii* var. *latifolium* McDerm. N.Am. Sp. Trif. 267. 1910. ("Open pine woods on the undulating ridges of the Coeur d'Aleine Mountains near St. Joseph's," Ida.)

T. *aitonii* Rydb. Bull. Torrey Club 34:46. 1907. *(G. B. Aiton 65,* "Palouse County," Ida., in 1892)

T. *orbiculatum* Kenn. & McDerm. Muhl. 3:8. 1907. *(Blankinship 6,* Thompson Falls, Mont., June 7, 1902)

Moderately appressed-pubescent, rhizomatous perennial from a thick taproot; stems usually several, 1-4 dm. tall; stipules ovate to lanceolate, 5-15 mm. long, erose-lacerate to entire; leaflets 3, ovate to obovate, 5-40 mm. long, finely serrulate; heads axillary and terminal, rather short-pedunculate (often exceeded by the leaves), noninvolucrate, 2-3 cm. broad and about as long; flowers 30-60, yellow to purple, 10-18 mm. long, erect to reflexed but not curved, the reflection all in the (1-4 mm. long) pedicels; calyx finely pubescent, 1/3-1/2 as long as the corolla, the teeth acicular, subequal, about twice the length of the tube; legume 1 (2)-seeded.

Moist meadows to rocky ridges; Wallowa Co., Oreg., to Missoula Co., Mont. June-Aug.

This species is more closely related to *T. longipes* than to *T. howellii,* but distinctive from both, chiefly because of its broader leaves, reflexed flowers on longer pedicels, and shorter peduncles.

Trifolium longipes Nutt. in T. & G. Fl. N.Am. 1:314. 1838. *(Nuttall,* "Valleys of the central chain of the Rocky Mountain range, and on the moist plains of the Oregon as low as the Wahlamet")

   T. longipes var. *reflexum* A. Nels. 1st Rep. Fl. Wyo. 94. 1896. *T. rydbergii* Greene, Pitt. 3:222. 1897. *T. oreganum* var. *rydbergii* McDerm. N.Am. Sp. Trif. 261. 1910. *(A. Nelson 918,* Wind R., Wyo.)

   T. caurinum Piper, Erythea 6:29. 1898. *T. longipes* f. *caurinum* McDerm. N.Am. Sp. Trif. 250. 1910. *(Lamb 1395,* Big Creek Prairie, Chehalis Co., Wash., Aug. 1, 1897) = var. *reflexum.*

   T. pedunculatum Rydb. Bull. Torrey Club 30:254. 1903. *(Henderson 3096,* Long Valley, Ida., in 1895) = var. *longipes.*

   T. brachypus sensu Blank. Mont. Agr. Coll. Stud. Bot. 1:81. 1905, not *T. longipes* var. *brachypus* Wats. = var. *reflexum.*

   T. shastense House, Bot. Gaz. 41:336. 1906. *T. longipes* var. *shastense* Jeps. Fl. Calif. 2:203. 1936. *(H. E. Brown 362,* Mt. Shasta, Calif., June 11-16, 1897)

   T. covillei House, Bot. Gaz. 41:337. 1906. *(Coville 1180,* Wenatchcc Mts., Kittitas Co., Wash., Sept. 4, 1901) = var. *longipes.*

   T. oreganum var. *multiovulatum* Henderson, Madroño 3:231. 1936. *(J. R. Patterson,* Saddle Mt., Clatsop Co., Oreg., June 10, 1928) = var. *shastense.*

Sparsely to rather copiously pubescent perennial with a taproot and a branched crown; stems 5-30 cm. tall, usually decumbent-based and often trailing and stolonous or rhizomatous for some distance; stipules ovate-lanceolate, 1-3 cm. long, erose-lacerate to entire; leaflets 3, mostly elliptic-lanceolate to oblanceolate (ovate to obovate), (1) 2-6 cm. long, serrulate to nearly entire; heads terminal, 20- to 70-flowered, noninvolucrate, 1.5-3.5 cm. broad, subglobose; peduncles usually exceeding the leaves; flowers ochroleucous to purplish, 11-18 mm. long, erect to reflexed; pedicels 1-3 mm. long; calyx glabrous to rather densely hairy, about 1/2 as long as the corolla, the teeth 2-4 times as long as the tube, subulate, the upper 2 often considerably reduced; banner acute; wings narrow, acute; keel acute to rounded; legume 1- to 4-seeded.

Widely distributed in w. N.Am., from B.C. southward, on both sides of the Cascade Mts., to s. Calif., e. to the Rocky Mts., from Mont. to Colo. and Utah. Late May-Aug.

This is a variable taxon consisting of several races that have often been accorded rank as separate species but which do not appear to be sufficiently distinctive or constant in character to merit more than infraspecific status. The three most conspicuous of these variants in our region are the following:

1 Flowers spreading to erect but usually not reflexed; on both sides of the Cascades, from
    Wash. to s. Calif., e. to Ida. and Utah   var. longipes
1 Flowers becoming strongly reflexed
  2 Plant few-stemmed from a taproot, not at all, or only slightly rhizomatous; mt. mead-
     ows, Clatsop Co. to s.w. Oreg. and n.w. Calif.   var. shastense (House) Jeps.
  2 Plant usually strongly rhizomatous; moist meadows from s.e. Wash. and n.e. Oreg.
     e. to Mont., s. to Colo. and Ariz.   var. reflexum A. Nels.

Trifolium macraei H. & A. in Hook. Bot. Misc. 3:179. 1833. *(Macrae,* "baths of collima," Chile)

   T. albopurpureum T. & G. Fl. N.Am. 1:313. 1838. *T. macraei* var. *albopurpureum* Greene, Fl. Fran. 26. 1891. *(Douglas,* California)

   T. dichotomum H. & A. Bot. Beechey Voy. 330. 1838. *T. macraei* var. *dichotomum* Brew. ex Wats. Proc. Am. Acad. 11:129. 1876. *(Douglas,* California)

   T. neolagopus Lojac. Nuov. Giorn. Bot. Ital. 15:194. 1883. *T. albopurpureum* var. *neolagopus* McDerm. N.Am. Sp. Trif. 209. 1910. *(Mrs. Elwood Cooper,* near Santa Inez, Calif., June, 1879)

Villous-pubescent annual with 1-several simple, erect or ascending stems 1-3 (5) dm. tall; stipules 5-10 mm. long, denticulate, prolonged to an acuminate tip; leaflets 3, oblanceolate to obcordate, 1-2 cm. long, coarsely serrulate their full length; heads axillary and terminal, noninvolucrate, 10- to 60-flowered, 1-2 cm. long and about as thick; calyx villous, 5-8 (12) mm. long, the tube 10- to 30-veined, the teeth subequal, acicular, plumose-villous-pilose, (1) 2-4 (6) times as long as the tube, from slightly shorter to much longer than the purplish (pink to white) corolla; legume 1-seeded.

Throughout much of Calif. and n. Baja Calif., from the foothills of the Sierra Nevada to the coast, n. to Vancouver I., B.C.; also in Chile. Apr.-June.

The synonymy involved in this species is complex. Although the plant shows great diversity, there

Trifolium haydenii

T. hybridum

T. incarnatum

T. gymnocarpon

T. latifolium

T. howellii

var. reflexum

var. shastense

var. longipes

T. longipes

JRJ

are not many truly significant variants, at least in the Pacific Northwest, where two sparsely represented taxa occur. The var. *macraei,* of coastal s. Oreg. and Calif. (and Chile), has subsessile or sessile heads, whereas our vars. have pedunculate heads.

1 Calyx usually no longer than the corolla, the teeth 2-3 times as long as the tube; w. side
   of the Cascade Mts. , in Wash. and on Vancouver I. , common from s. Oreg. to c. Calif.
                                                                        var. dichotomum (H. & A.) Brew. & Wats.
1 Calyx usually exceeding the corolla, the teeth 3-6 times as long as the tube; w. of the
   Cascades, in Wash. and Oreg. , s. to Baja Calif.   var. albopurpureum (T. & G.) Greene

Trifolium macrocephalum (Pursh) Poiret, Lam. Encycl. Meth. 5:336. 1817.
   Lupinaster macrocephalus Pursh, Fl. Am. Sept. 479. 1814. *T. megacephalum* Nutt. Gen. Pl. 2:105.
      1818. *(Lewis,* "At the head-waters of the Missouri")
   T. macrocephalum var. *caeruleomontanum* St. John, Fl. S. E. Wash. 237. 1937. *(W. T. Shaw,*
      Blue Mts. , Wash. , June 19, 1922)
   Sparsely to densely pubescent perennial with a thick root and extensive rhizomes; stems erect, 1-3 dm. tall; stipules ovate to ovate-lanceolate, their margins lacerate; leaflets (3-4) 5-9, rather thick and leathery, oblanceolate to obcordate, 0. 5-2. 5 cm. long; heads mostly solitary and terminal, noninvolucrate, 3-5 cm. broad and about as long, the peduncles exceeding the leaves; flowers 22-28 mm. long, pinkish to rose-pink, spreading to erect, the pedicels very short; calyx about 2/3 as long as the corolla, the teeth subulate, plumose, many times as long as the tube; legume mostly 1-seeded.
   Sagebrush desert and ponderosa pine woodland from Douglas, Grant, and Kittitas cos. , Wash. , southward through e. Oreg. to Nev. , e. to w. Ida. Apr. -June.

Trifolium microcephalum Pursh, Fl. Am. Sept. 478. 1814. *(Lewis,* "On the banks of Clarck's river")
   Sparsely to densely villous annual with prostrate to erect stems 1-7 dm. long; stipules ovate to ovate-lanceolate, denticulate-serrulate, about 1/2 as long as the leaflets; leaflets 3, obovate-oblanceolate, 1. 5-2. 5 cm. long; heads involucrate, 5-10 mm. long, 10- to 60-flowered; involucres villous, sometimes exceeding the lower flowers, shallowly crateriform, with about 10 (6-12) nearly entire, acute, shallow lobes; flowers 4-5 (7) mm. long; calyx usually either hirsute or villous (rarely glabrous), the teeth simple, setaceous, longer than the tube; corolla white to pinkish; legume 2-3 mm. long, 1- or 2-seeded, usually rupturing the calyx by maturity. N=8.
   Moist meadows, sandy riverbanks, and drier hillsides; s. w. B. C. to Baja Calif. , e. to Mont. , Nev. , and Ariz. , from the coast and other lowlands to well up in the mountains. Apr. -July.

Trifolium microdon H. & A. Bot. Misc. 3:180. 1833. *(Cummings 747,* Valparaiso, Chile)
   T. microdon var. *pilosum* Eastw. Proc. Calif. Acad. Sci. III, 1:100. 1898. *(Trask,* San Nicolas I. ,
      Calif.)
   Erect to decumbent, sparsely puberulent to rather conspicuously villous or pilose-woolly annual; stems 1-5 dm. long; stipules ovate, about 1/2 as long as the leaflets; leaflets 3, narrowly to broadly obcordate, 5-15 mm. long; heads involucrate, 5-8 mm. long, from about as thick to nearly twice as thick; involucres villous, cup-shaped, with 10-12 lacerate-toothed lobes; flowers white to pale red, aging to dark pinkish-brown, 4-6 mm. long; calyx glabrous, the teeth triangular and scarcely at all setose, more or less serrulate-erose, much shorter than the tube; corolla considerably longer than the calyx; legume large, often rupturing the calyx, 1-seeded.
   In meadows or on sandy or rocky soil, from s. w. B. C. southward, on the w. side of the Cascades, to Calif. ; S. Am. Apr. -June.
   There is considerable variation in the amount of pubescence on the plant, but the more woolly-pilose individuals usually are recognized as the var. *pilosum* Eastw. They are freely intergradient to the inconspicuously pubescent plants characteristic of the var. *microdon,* the taxon to which our plants all seem to belong.

Trifolium multipedunculatum Kennedy, Muhl. 5:59. 1909. *(Cusick 3190,* China Cap, Wallowa Mts. ,
      Oreg. , July 30, 1907)
   Sparingly appressed-pubescent perennial with a thick taproot and branched crown, more or less caespitose, the stems often trailing and rhizomelike, seldom over 1 dm. tall; stipules ovate-lanceolate, 5-20 mm. long, somewhat lacerate; leaflets 3, lanceolate to obovate, rather thick and leathery, 5-15 (20) mm. long, fairly coarsely setose-serrulate; heads terminal, subglobose, 2-3 cm. broad, noninvolucrate; peduncles usually longer than the stems and considerably exceeding the leaves; flowers

25-60, red to purplish, 12-19 mm. long, the lower ones reflexed; pedicels 1-1.5 mm. long; calyx pubescent, about 1/2 as long as the corolla, the tube (5) 10-nerved, the teeth filiform, subequal, 2-4 times as long as the tube; banner acuminate, the wings and keel abruptly acuminate-apiculate; legume 2- to 4-seeded.

Open dry slopes and subalpine ridges; c. Wash. (Wenatchee Mts.) and e. Oreg. (Wallowa and Steens mts.). June-Aug.

This clover resembles *T. longipes,* with which it apparently is closely related but not seemingly intergradient. On the average it is a shorter plant, with smaller leaflets and darker flowers, also differing because of the characteristically acuminate petals. It usually grows in drier, more open habitats.

Trifolium nanum Torr. Ann. Lyc. N.Y. 1:35. 1824. *(James,* James' Peak, Colo.)

Densely matted-caespitose, glabrous to obscurely pubescent perennial with a large taproot; stems 1-3 cm. long, thickened with the persistent, lanceolate stipules; petioles 5-20 mm. long; leaflets 3, obovate or oblanceolate, 5-15 mm. long, rounded to acute, mucronulate, slightly serrulate beyond midlength; heads 1- to 4-flowered, with an involucre of 1-4 small, distinct to connate bracts 0.5-2 mm. long; peduncles from shorter to longer than the leaves; flowers 15-22 mm. long; pedicels 1-2 mm. long; calyx 5-7 mm. long, not over 1/3 as long as the corolla, the teeth triangular, not spinose, subequal to the tube, the upper two only slightly shorter than the lower three; corollas lilac-purple, aging brown; legume very large, 5-15 mm. long, 1- to 4-seeded.

Alpine or subalpine in the Rocky Mts., from s.w. Mont. to N.M., and in the e. ranges of Utah. July-Aug.

Trifolium oliganthum Steud. Nom. II, 2:707. 1841.
T. pauciflorum Nutt. in T. & G. Fl. N.Am. 1:319. 1838, not of Urv. in 1822. *T. variegatum* var. *pauciflorum* McDerm. N.Am. Sp. Trif. 67. 1910. *(Nuttall,* "Wet places on the higher plains of the Oregon, particularly abundant nearly the outlet of the Wahlamet")

Glabrous or very sparsely glandular annual with 1-many erect to decumbent stems 5-20 (50) cm. long; stipules ovate, lacerate, about 1/4 as long as the leaflets; leaflets 3, linear-elliptic to linear-oblanceolate or narrowly obcordate, 5-30 mm. long, from serrulate-setulose full length to subentire; heads involucrate, small, 3-12 mm. thick, 2- to 8 (15)-flowered; involucres flared, from saucer-shaped and deeply, irregularly, and acuminately many-toothed to below the middle, to greatly reduced and more or less annular; flowers 4-8 mm. long; calyx tube narrowly campanulate, usually 10-veined, the teeth triangular-acicular, subequal, somewhat shorter than the tube, sometimes broadened and 2-toothed about midlength, the upper sinus much deeper than the others; corolla lavender to purple, white-tipped; legume mostly 2-seeded.

Meadowland to dry rocky soil; w. of the Cascades, from s.w. B.C. to Calif. Mar.-July.

Trifolium parryi Gray, Am. Journ. Sci. 83:409. 1862. *(Parry 178,* headwater of Clear Creek, e. of Middle Park, Colo.)
T. montanense Rydb. Mem. N.Y. Bot. Gard. 1:236. 1900. *(Rydberg & Bessey 4461,* Old Hollowtop, Pony Mts., Mont., July 7 and 9, 1897)

Glabrous to brownish-pubescent, tufted perennial from a thick taproot; stems numerous, 1-5 cm. long, covered with thin, scarious, marcescent, entire to toothed, rounded to acutely pointed stipules; leaflets 3, broadly elliptic to obovate, entire to serrulate or denticulate, rounded to acute, 10-40 mm. long; petioles 1-6 cm. long; heads involucrate, subglobose, 10-35 mm. thick, 4- to 30-flowered; peduncles usually exceeding the leaves; involucral bracts 6-12, distinct, thin and scarious, entire-margined, purplish-brown, rounded to acute or bifid, usually about equaling the calyces; flowers 11-22 mm. long , spreading to erect, the pedicels 0.5-2 mm. long; calyx glabrous, scarious, about 1/2 the length of the corolla, the tube from 1/2 as long to as long as the subulate to triangular, subequal teeth; corolla dark reddish-purple, aging brown; legume 1- to 4-seeded.

Moist subalpine to alpine meadows and stream banks, from 8000 ft. upward; Rocky Mts., from Park, Sweet Grass, Madison, Stillwater, and Carbon cos., Mont., and adj. Ida., to n. N.M. and the mountains of e. Utah. July-Sept.

Trifolium plumosum Dougl. ex Hook. Fl. Bor. Am. 1:130. 1831. *(Douglas,* "Blue Mountains in North-West America")

16

sepal tip

5

2.5

½

3

Trifolium macraei

½

T. microdon

2

½

½

T. oliganthum

½

1.5

T. macrocephalum

2

½

T. multipedunculatum

2.5

involucre

5

1.5

2

½

2

IC

T. microcephalum

½

T. parryi

½

J

T. nanum

T. plumosum var. *amplifolium* Martin, Bull. Torrey Club 73:369. 1946. *(M. E. Jones 6254,* Salmon Meadows, Washington Co., Ida., June 22, 1899)

Strigillose to villous perennial with short rhizomes; stems simple, 2-5 dm. tall; stipules lanceolate, 1.5-3 cm. long; leaflets 3, linear to linear-elliptic, 5-9 cm. long, usually less than 1 cm. broad, often folded and more or less falcate, very finely denticulate to entire; heads 50- to 150-flowered, terminal, cylindric, 2-2.5 cm. broad, 3-5 cm. long, noninvolucrate; peduncles often shorter than the upper 1 or 2 leaves; flowers 14-20 mm. long, spreading to nearly erect, nearly sessile, whitish but with pink to reddish tips; calyx usually well over half as long as (often nearly equal to) the corolla, villous-plumose, the teeth acicular, the lowest one much the longest, the tube 20- to 25-veined, from nearly as long to only half as long as the teeth; keel slightly longer than the wings; legume usually 1-seeded.

Dry hillsides and meadows from the Blue Mts. of Wash. to Union Co., Oreg., e. to Washington Co., Ida. May-Aug.

The species is delimitable into two well-marked races:

1 Leaflets of the basal leaves 2-5 (7) mm. broad, acuminate; Blue Mts. of s.e. Wash., and the Blue and Wallowa mts. of n.e. Oreg.   var. plumosum

1 Leaflets of basal leaves (8) 9-16 mm. broad, acute; Washington Co., Ida.

var. amplifolium Martin

Trifolium pratense L. Sp. Pl. 768. 1753. (Europe)

Sparsely soft-hairy, short-lived, taprooted perennial; stems several, 3-10 dm. tall; stipules ovate-lanceolate, 1-3 cm. long, conspicuously greenish veined; leaflets 3, lanceolate to oblong-obovate, 2-6 cm. long, very inconspicuously serrulate; heads terminal, sessile or with peduncles shorter than the 2 subtending leaves, 50- to 200-flowered, globose-conic, 2.5-3.5 cm. broad and about as long, noninvolucrate but the stipules of the upper leaves often somewhat involucral; flowers spreading to erect, deep red, 13-20 mm. long; calyx 1/2-2/3 as long as the corolla, short villous-hirsute, the teeth acicular, pubescent with straight, somewhat pustulose hairs, the 2 upper ones about equal to the tube, the lower 3 nearly twice as long; legume 2-seeded. N=7, 14. Red clover.

A European species, often cultivated, and now fairly widely introduced throughout much of w. U.S. June-Aug.

Trifolium procumbens L. Sp. Pl. 772. 1753.

Chrysaspis procumbens Desv. Fl. Anjou 338. 1827. *Amarenus procumbens* Presl, Symb. Bot. 1:46. 1830. (Europe)

Annual, much like *T. agrarium* in general habit and flower structure, the stems usually procumbent to ascending and 1-3 dm. tall; stipules ovate, 4-9 mm. long, adnate for about half their length, the width of the adnate portion from slightly less to slightly more than equal to the length of the free tip; petioles usually considerably longer than the stipules; petiolule of the terminal leaflet at least twice as long as those of the lateral leaflets; heads noninvolucrate, usually not over 12 mm. thick; flowers yellow, mostly 4-6 mm. long; style much shorter than the mature, 1- or 2-seeded legume. N=8. Low hop clover.

An introduced plant, now widespread in much of the U.S., on wasteland, roadsides, etc. May-Aug.

Trifolium productum Greene, Erythea 2:181. 1894.

T. kingii var. *productum* Jeps. Fl. Calif. 2:304. 1936. ("From Mt. Shasta southward to Placer County")

Glabrous or sparsely puberulent, taprooted perennial; stems several, erect, 2-5 dm. tall; stipules 1-3 cm. long, ovate-lanceolate, attenuate, entire to lacerate; leaflets 3 (rarely 4-5), ovate-lanceolate or elliptic to obovate, 1-5 cm. long, spinulose-serrulate; heads (15) 20- to 50-flowered, 2-2.5 cm. broad and about as long, noninvolucrate, axillary as well as terminal, often 2 or 3 from the axils of involucrelike, opposite, reduced upper leaves; rachis of the inflorescence prolonged (and usually branched) above the upper flowers; flowers 11-18 mm. long, strongly reflexed even in bud, the pedicels not over 1 mm. long; calyx scarcely 1/4 the length of the corolla, glabrous, the teeth subulate, about as long as the tube; corolla whitish, purple-tipped; legume 1-seeded.

Stream banks to fairly dry forest soil or open ridges; Crook to Harney Co., Oreg., s. to the c. Sierra Nevada of Calif. June-Sept.

Trifolium repens L. Sp. Pl. 767. 1753. (Europe)

Glabrous or very sparsely pubescent perennial; stems creeping and stolonous to erect, 1-6 dm.

long; stipules 3-10 mm. long, connate most of their length, the free portion shortly acuminate; petioles from only slightly to many times longer than the leaflets; leaflets 3, obovate but usually somewhat retuse to obcordate, 1-2 cm. long, finely serrulate; heads axillary, often long-pedunculate, 1.5-2 cm. broad, nearly as long, noninvolucrate; flowers 5-9 mm. long, white or cream to pinkish-tinged, pendulous at anthesis on pedicels 1-5 mm. long; calyx glabrous, about half the length of the corolla, the teeth lanceolate-subulate, about equal to the tube; banner much more erect than in the native clovers; legume 1- to 3-seeded. N=16, 24. White clover.

An introduced species, established throughout w. N.Am., from the Aleutian Is. southward, in waste places or apparently native habitats such as mountain meadows. Apr.-Sept.

Trifolium thompsonii Morton, Journ. Wash. Acad. Sci. 23:270. 1933. *(J. W. Thompson 8467,* along Swakane Creek, Chelan Co., Wash., June 23, 1932)

Grayish, strigillose-villous, taprooted perennial; stems several, erect, not rhizomatous, 2-7 dm. tall; stipules 2-4.5 cm. long, oblong-lanceolate, adnate for about half their length, the margins entire or nearly so; leaflets (3-4) 5-8, linear or linear-elliptic to lanceolate, acute, 2-6 cm. long, rather sharply serrulate; heads many-flowered, noninvolucrate, terminal on peduncles exceeding the leaves, 3-4.5 cm. broad and nearly as long; flowers spreading to somewhat recurved, 18-22 mm. long, bright reddish-lavender to deep orchid; pedicels 1-1.5 mm. long; calyx 1/2-4/5 as long as the corolla, the teeth subulate, villous, about twice as long as the tube; legume 1-seeded.

Known only from the type locality, where it is rather abundant on well-drained grassy hillsides just below the edge of ponderosa pine woodland; Swakane Canyon, Chelan Co., Wash. May-June.

Trifolium tridentatum Lindl. Bot. Reg. 13:pl. 1070. 1827. *(Douglas,* Columbia River)
   T. obtusiflorum Hook. Ic. Pl. pl. 281. 1840. *T. tridentatum* var. *obtusiflorum* Wats. Proc. Am. Acad. 11:130. 1876. *(Douglas,* near Monterey,Calif.)

Glabrous annual with spreading to erect stems 1-7 (10) dm. long; stipules setose-lacerate, ovate, 1/3-1/2 as long as the leaflets; leaflets linear to elliptic, narrowly oblong or oblanceolate, 1-4 (6) cm. long, usually setose-serrulate their full length (entire); heads 6- to 60-flowered; 10-30 mm. broad, about as long, the involucre flared and saucerlike, irregularly lacerate (less than half the length) into many unequal teeth, but not regularly lobed; flowers (8) 10-13 (17) mm. long; calyx often purplish, the tube usually 15- to 25-nerved, the teeth about equal to the tube, lanceolate, shortly setaceous-tipped, usually 2-toothed just above the expanded basal half, sometimes these teeth also setaceous, sinuses unequal, the upper one much the deepest; corolla exceeding the calyx slightly, purplish, often either lighter or darker in color at the tip; legume usually 2-seeded.

Grassy hillsides and meadowland; B.C. to Calif., on the w. side of the Cascades and Sierra Nevada. Apr.-July.

Although several taxa have been proposed on minor variations detectable in the more abundant and variable Californian material, such as the size of the head, length of the lateral teeth of the calyx lobes, and general habit of the plant, no significant phyletic tendencies are discernible in the species within our area. All our known plants are glabrous and referable to var. *tridentatum,* although the var. *obtusiflorum* (Hook.) Wats., a more or less glandular-pubescent form with fistulose stems, possibly may reach our range in the south.

Trifolium variegatum Nutt. in T. & G. Fl. N.Am. 1:317. 1838. *(Nuttall,* "Springy places near the mouth of the Wahlamet")

TRIFOLIUM VARIEGATUM var. ROSTRATUM (Greene) C. L. Hitchc. hoc loc. *T. rostratum* Greene, Proc. Acad. Phila. 47:547. 1896. *T. appendiculatum* f. *rostratum* McDerm. N.Am. Sp. Trif. 92. 1910. *T. appendiculatum* var. *rostratum* Jeps. Man. Fl. Pl. Calif. 539. 1925. *(V. K. Chestnut,* Lake Merritt, Oakland, Calif., Apr., 1889)

   T. dianthum Greene, Pitt. 3:217. 1897. *(Macoun,* Vancouver I., B.C., in 1893) = var. *variegatum.*

Glabrous annual with 1-several prostrate to erect stems 1-6 (8) dm. long; stipules ovate, deeply lacerate; leaflets 3, obovate to elliptic-oblanceolate, 5-20 (30) mm. long, margins prominently acicular-serrate; heads 1-2 cm. broad, 3- to 40-flowered, the involucre flaring and saucer-shaped, irregularly lobed and lacerate about half its length; flowers 5-20 mm. long; calyx tube narrowly campanulate, 10- to 25-veined, the teeth considerably longer than the tube, narrow and setose-tipped, the sinuses about equally deep; corolla slightly longer than the calyx, purplish, often white-tipped, aging to purplish-brown; legume 1- or 2-seeded, usually rupturing the calyx. N=8.

inside of calyx tube

Trifolium plumosum

T. pratense

T. procumbens

T. repens

T. productum

T. thompsonii

T. tridentatum

involucre

JRJ

Dry sandy soil to moist meadows; B. C. to Calif., mostly w. of the Cascades and Sierra Nevada, but occasional e. to Mont. and Utah. Apr. -July.

Trifolium variegatum resembles *T. oliganthum* very closely, but differs in having relatively longer calyx lobes, longer leaflets with the teeth more acicular, and usually more-floriferous heads. Although the species undergoes considerable variation, especially in the more southern part of its range, our plants, with one exception, are sufficiently uniform to be maintained as one taxon, var. *variegatum*, with the keel petals beakless or with a short broad beak not over 0.3 mm. long. The var. *rostratum* (Greene) C. L. Hitchc., which has the keel tipped with a slender tooth 0.5 mm. long, has been collected on Vancouver I., but is otherwise known only from Calif.

Trifolium wormskjoldii Lehm. Ind. Sem. Hort. Hamb. 17. 1825. (Grown from Californian seed collected by Wormskjold)
  T. *involucratum* Ortega, Nov. Rar. Pl. Hort. Bot. Matrit. Desc. 33. 1797, but not of Lam. in 1778.
  T. *willdenovii* Spreng. Syst. 3:208. 1826. (From seeds collected in "Cuba" [Mexico?], by Sesse)
  T. *fimbriatum* Lindl. Bot. Reg. 13:pl. 1070. 1827. *T. involucratum* var. *fimbriatum* McDerm. N. Am. Sp. Trif. 52. 1910. *T. wormskjoldii* var. *fimbriatum* Jeps. Fl. Calif. 2:294. 1936. *T. willdenovii* var. *fimbriatum* Ewan, Leafl. West. Bot. 3:224. 1943. *(Douglas,* Columbia River)
  T. *spinulosum* Dougl. ex Hook. Fl. Bor. Am. 1:133. 1834. *(Douglas,* "In the vallies between Spokan and Kettle Falls, North-West America")
  T. *heterodon* T. & G. Fl. N. Am. 1:318. 1838. *T. involucratum* var. *heterodon* Wats. Proc. Am. Acad. 11:130. 1876. *(Nuttall,* "Borders of marshes near the mouth of the Oregon")
Glabrous, taprooted perennial with decumbent-based and often rhizomatous stems 1-8 dm. long; stipules 1-4 cm. long, lacerate margined and usually acuminate; leaflets 3, linear-elliptic to oblong-obovate, 1-3 cm. long, finely serrulate; heads involucrate, axillary, 2- to 60-flowered, 2-3 cm. broad; involucres flared, from as much as 2 cm. broad and lacerately 8- to 12-lobed, to shallowly lobed or toothed, the lobes entire; flowers 10-18 mm. long, erect or spreading, reddish to purple, often white-tipped; pedicels 0.5-2 mm. long; calyx glabrous, 2/3-3/4 as long as the corolla, the tube 10-veined, about equaled by the 5 subequal, narrowly lanceolate-acicular (occasionally bifid) teeth; legume 1- to 4-seeded. N=about 24.

Meadows and stream banks to coastal dunes from B. C. to Calif. and Mex., e., at elevations up to 8000 ft., to Ida., Utah, Colo., and N. M. May-Sept.

Ulex L. Furze; Gorse

Flowers papilionaceous, showy, yellow, axillary and clustered near the branch tips; calyx deeply 2-lobed, the lobes nearly entire; stamens 10, monadelphous; legumes hairy, dehiscent; spiny, green-branched, nearly leafless shrubs with scalelike or rigid and prickly leaves.

A fairly small genus of about 20 species, chiefly in the Mediterranean region. (Old Latin name for the plant.)

Ulex europaeus L. Sp. Pl. 741. 1753. ("Habitat in Anglia, Gallia, Brabantia")
Shrub, 1-3 m. tall, pubescent, the branches greenish, prominently angled, sharply spine-tipped; flowers borne along the short, stout, lateral spinose branches; pedicels 3-6 mm. long, 2-bracteolate near the tip; calyx 10-15 mm. long, hairy; corolla yellow, 15-20 mm. long; keel hairy along the lower margin, slightly shorter than the wings. N=32, 48.

A European introduction; frequently escaped on the w. side of the Cascades and becoming a very serious pest and a bad fire hazard in some areas. Apr. -Sept.

Vicia L. Vetch

Flowers axillary, solitary to unilaterally racemose; calyx more or less unequally 5-toothed; corolla papilionaceous, the wings usually longer than the keel and joined with it to near midlength; stamens 10, usually diadelphous but all the filaments sometimes slightly connate at base; style filiform, densely bearded just below the stigmatic tip, the bearding frequently densest on the side next to the keel; pods flat, 2- to several-seeded, sometimes cross-septate between the nearly globose and usually slightly arillate seeds; cotyledons hypogaeous; annual or perennial herbs with weak, usually angled, trailing to climbing stems and pinnate leaves ending in simple to branched tendrils (rarely nontendrilar), the leaflets few to many, the stipules entire to sagittate.

Nearly 200 species, widely distributed except for Australia. (Classical Latin name for the plant.)
Reference:

Hermann, F. J. Vetches in the United States—native, naturalized, and cultivated. U.S. D. A. Agr.
Handb. 168:1-84. 1960.

Many of the species are extensively cultivated and frequently escape where they may persist for several seasons if they do not become established permanently. In addition to the species treated formally, others might occasionally be regarded as elements of our flora, including *V. faba* L., horse bean (flowers at least 2 cm. long, the wings prominently spotted); *V. dasycarpa* Ten., woollypod vetch (plant fewer-flowered than *V. villosa,* but otherwise rather similar, probably keying to *V. pannonica,* but with 5- to 20-flowered racemes that are strongly pedunculate); and *V. ervilia* (L.) Willd., bitter vetch (the nontendrilar leaves with 15 or more leaflets).

1 Flowers 20-80 in dense, 1-sided, spikelike racemes, purplish, 10-18 mm. long
  2 Plant a perennial, glabrous to appressed-pubescent; flowers 10-15 mm. long; calyx
    slightly gibbous at the base of the tube, but the bulge not projecting backward past the
    point of insertion of the pedicel, the teeth scarcely 0.5 mm. long, the lower ones usu-
    ally no longer than the tube   V. CRACCA
  2 Plant an annual or a biennial, rather densely hirsute-villous; flowers mostly over 15
    mm. long; calyx markedly gibbous, the pedicel attached slightly under the bulge, up-
    per teeth over 0.5 mm. long, the lower ones often longer than the tube   V. VILLOSA
1 Flowers 1-20 in often loosely flowered racemes, sometimes only 1-2 in the axils, various
  in color and size
  3 Flowers 3-6 mm. long
    4 Pod hairy, 2 (3)-seeded; racemes 3- to 8-flowered   V. HIRSUTA
    4 Pod glabrous, 2- to 5-seeded; racemes often only 1- or 2 (to 5)-flowered
      5 Racemes 2- to 5-flowered; pods 10-16 mm. long, 5-8 mm. broad, 2-seeded; leaf-
        lets 8-10 pairs; to be expected, but not at present known from extant collections,
        in our area   *V. disperma* DC.
      5 Racemes 1- or 2 (3)-flowered; pods 10-15 mm. long, less than 5 mm. broad, (3)
        4- or 5-seeded; leaflets 4-6 pairs   V. TETRASPERMA
  3 Flowers 12-25 mm. long
    6 Leaflets 19-29; flowers white to orange; plant perennial, 1-2 m. tall; stems fistulose
                            V. GIGANTEA
    6 Leaflets 8-20; flowers various; plant annual or perennial, mostly less than 1 m.
      tall; stems not fistulose
      7 Flowers ochroleucous; plant annual   V. PANNONICA
      7 Flowers purplish; plant perennial
        8 Flowers 1-3 per leaf axil   V. SATIVA
        8 Flowers in 4- to 10 (20)-flowered, pedunculate racemes
          9 Calyx not gibbous-based; flowers 12-25 mm. long   V. AMERICANA
          9 Calyx gibbous-based; flowers 10-15 mm. long   V. CRACCA

Vicia americana Muhl. ex Willd. Sp. Pl. 3:1096. 1802. (Pennsylvania)
  V. sylvatica Nutt. Gen. Pl. 2:97. 1817. *(Nuttall,* "Banks of the Missouri as far as Ft. Mandan") =
  var. *minor.*
  V. americana var. *minor* Hook. Fl. Bor. Am. 1:157. 1831. *(Drummond,* "On the Saskatchawan
  about Carlton-House")
  V. oregana Nutt. in T. & G. Fl. N. Am. 1:270. 1838. *V. americana* var. *oregana* Nels. in Coult. &
  Nels. New Man. Bot. Rocky Mts. 301. 1909. *V. americana* ssp. *oregana* Abrams, Ill. Fl. Pac.
  St. 2:617. 1944. *(Nuttall,* "Plains of the Oregon") = var. *truncata.*
  V. truncata Nutt. in T. & G. Fl. N.Am. 1:270. 1838. *V. americana* var. *truncata* Brew. in Brew.
  & Wats. Bot. Calif. 1:158. 1876. *(Nuttall,* "Plains of the Oregon")
  V. sparsifolia Nutt. ex T. & G. Fl. N.Am. 1:270. 1838. *(Nuttall,* "Plains of the Oregon") = var.
  minor.
  Lathyrus linearis Nutt. in T. & G. Fl. N.Am. 1:276. 1838. *V. americana* var. *linearis* Wats. Proc.
  Am. Acad. 11:134. 1876. *V. linearis* Greene, Fl. Fran. 30. 1891. *(Nuttall,* "Plains of the Platte")
  = var. *minor* in large part.
  Lathyrus dissitifolius Nutt. in T. & G. Fl. N.Am. 1:277. 1838. *V. dissitifolia* Rydb. Bull. Torrey

Club 33:144. 1906. *(Nuttall,* "with the preceding" namely *Lathyrus linearis,* thus also "Plains of the Platte") = var. *minor.*

V. americana var. *angustifolia* Nees in Wied. Reise Inn. Nord.-Am. 2:434. 1841. ("Auf dem Missouri") = var. *minor.*

V. truncata var. *villosa* Kell. Proc. Calif. Acad. Sci. 1:57. 1855. *V. americana* var. *villosa* Hermann, U.S.D.A. Agr. Handb. 168:83. 1960. *(Kellogg,* Placerville, Calif.)

V. californica Greene, Fl. Fran. 3. 1891. *(Greene,* Calaveras Co., Calif., June, 1889) = var. *villosa.*

V. caespitosa A. Nels. Bull. Torrey Club 25:373. 1898. *V. linearis* var. *caespitosa* Nels. in Coult. & Nels. New Man. Bot. Rocky Mts. 301. 1909. *(A. Nelson 2949,* Laramie Plains, Wyo., June 12, 1897) = var. *minor.*

V. americana var. *pallida* Suksd. Deuts. Bot. Monats. 18:26. 1900. *(Suksdorf 2111,* Bingen, Klickitat Co., Wash., May 21, June, 1892) = var. *truncata.*

V. washingtonensis Suksd. W. Am. Sci. 15:59. 1906. *(Suksdorf 2643,* near Bingen, Klickitat Co., Wash., May 21, June, 1900)

V. callianthema Greene, Leafl. 2:269. 1912. *(A. Nelson,* Bitter Creek, Wyo., June 2, 1897) = var. *minor.*

V. vexillaris Greene, Leafl. 2:269. 1912. *(I. T. Worthley,* Big Horn Co., Wyo., in 1896) = var. *minor.*

V. trifida Dietr. sensu Rydb. Fl. Rocky Mts. 526. 1917. = var. *minor.*

Subglabrous to densely pubescent perennial, the hairs often recurved; stems 1.5-8 dm. tall; stipules usually more or less lunate, deeply lacerate or toothed, 3-8 mm. long; tendrils from simple to branched and prehensile; leaflets mostly 8-12, variable in size, shape, and texture, linear to oval, very thin to prominently veined and coriaceous, subglabrous to densely pubescent, acute to retuse or truncate and sometimes 3-toothed, 1-3 cm. long, apiculate; racemes loose, with 4-10 bluish-purple flowers 12-25 mm. long; calyx about 1/3 as long as the corolla, the teeth scarcely 1/2 the length of the tube; pod 2.5-3.5 cm. long. N=7.

Southern Alas. e. to Ont., s. to Calif., n. Mex., Mo., and W. Va. May-July.

This is an extremely variable plant, but with two or three exceptions it is not differentiated into genetically distinctive phases in our area. Many of the plants of the Great Plains, from Wyo. to Alta., are short and often erect. They have rather narrow leaflets (not always linear) that are coriaceous, heavily veined, and rather densely pubescent with short curved hairs; the tendrils are usually simple and only slightly prehensile. This is the phase treated as *V. trifida* by Rydberg, and covered in part by the descriptions of *V. caespitosa* A. Nels. and *V. callianthema* Greene; it is at best a rather poorly defined variety, the earliest name available being var. *minor* Hook. Variety *americana* is largely a more eastern plant with thin, elliptic to ovate-oblong leaflets usually at least 6 mm. broad. Variety *villosa* is a distinctive coastal phase that extends n. to Island Co., Wash. It is characterized by somewhat smaller flowers and more abundant, soft pubescence, but is none too sharply delimited from the more wide-ranging var. *truncata* in w. Wash. Our three vars. are imperfectly delimitable on the following basis:

1 Leaflets narrow, thick and coriaceous, densely pubescent with short curly hairs, the lateral veins prominent, essentially unbranched, leaving the midrib at a narrow angle; chiefly e. of the Rocky Mts., in Alta., Mont., and Wyo., w. to Ida., occasional elsewhere

                                                                       var. minor Hook.

1 Leaflets various, but more usually thin, or glabrous, or with more-divaricate lateral veins

  2 Herbage, calyx, and (often) pod rather copiously villous-pubescent; stems usually zigzag; flowers mostly not over 15 mm. long; Sierra Nevada to n.w. Calif., northward, on the w. side of the Cascades, to Island Co., Wash., freely intergradient with var. *truncata* in Wash., but much more distinctive in Oreg.   var. villosa (Kell.) Hermann

  2 Herbage and calyx glabrous to coarsely pubescent; pod usually glabrous; stems usually not zigzag; flowers tending to average more than 15 mm. in length; B.C. to Calif., on both sides of the Cascades in our area, e. to the Rocky Mts. and plains region   var. truncata (Nutt.) Brew.

Vicia cracca L. Sp. Pl. 735. 1753. (Europe)

? *V. semicinecta* (printer's error for *semicincta)* Greene, Erythea 3:17. 1895. *(Austin,* Crane Cr., s.e. Oreg.)

4

4

2.5

involucre

1/2

Trifolium variegatum

2

2

1

1/2

Ulex europaeus

2.5

1/2

2.5

Trifolium wormskjoldii

3

1/2

Vicia cracca

1.5

1/2

var. villosa

1/2

var. minor

1/2

1/2

var. truncata

Vicia americana

JRJ

Glabrous to more or less appressed-pubescent, clambering perennial 0.5-1 m. tall; leaves 6-12 cm. long; stipules variously toothed to entire, 7-15 mm. long, leaflets mostly 12-18 (20), linear, apiculate, 1.5-3 cm. long; tendrils well developed; raceme usually exceeding the peduncle, the two together somewhat longer than the leaves; flowers (10) 20-70, closely packed, 10-15 mm. long, pendulous, violet-purple; calyx scarcely half as long as the corolla, the tube 2-3 mm. long, oblique and slightly gibbous-based but not conspicuously bulged (the pedicel attached at the lowermost point), the upper lobes very short, triangular, scarcely 0.5 mm. long, the lower lobes much longer, but usually not equaling the tube; pod 1.5-2 cm. long, 6-10 mm. broad, several-seeded. N=6, 7, 14. Bird vetch.

Native of Eurasia, but widely naturalized in e. U.S., occasional in the Rocky Mts. and westward. May-July.

The species is much less common than *V. villosa* but very similar to it in floral and vegetative characteristics, and possibly intergradient with it as some plants from our area are intermediate between the two.

Vicia gigantea Hook. Fl. Bor. Am. 1:157. 1831. *(Dr. Scouler; Douglas,* "Open woods on the Columbia")

Succulent, pubescent to glabrate, climbing perennial 1-2 m. tall; stems fistulose, conspicuously ridged, 3-7 mm. thick; leaves 10-20 cm. long, the leaflets 19-29, lanceolate to oblong, (1.5) 2.5-4 cm. long, less than twice the length of the stipules; tendrils very well developed; peduncles from shorter to much longer than the closely 7- to 20-flowered racemes; flowers 14-18 mm. long, ochroleucous to orange, often tinged with purple; calyx about half the length of the corolla, the upper teeth scarcely 1 mm. long, the lowest one subequal to the tube; pod 3-4 cm. long, 1-1.5 cm. broad, darkening greatly when dried, several-seeded; seeds black. Giant vetch.

Along the coast, especially near streams or in forest clearings, from Alas. to Calif., inland to the Willamette Valley in Oreg. May-July.

Vicia hirsuta (L.) S. F. Gray, Nat. Arr. Brit. Pl. 2:614. 1821.
Ervum hirsutum L. Sp. Pl. 738. 1753. (Europe)

Glabrate to puberulent, slender, clambering annual 3-7 dm. tall; stipules 2-5 mm. long, linear, generally lobed at base; tendrils well developed; leaflets (10) 14-18, linear, 1-2 cm. long, usually retuse and apiculate; peduncles rather slender but considerably thicker than the tendrils, shorter than the leaves; racemes with 3-8 whitish or pale blue flowers 3-4 (5) mm. long; calyx teeth lanceolate, about equal to the tube; pod hairy, 2 (3)-seeded, 1 cm. long. Tiny or hairy vetch.

A native of Europe, now fairly commonly established in w. Wash. and Oreg. May-July.

Vicia pannonica Crantz, Stirp. Aust. 2nd ed. 393. 1769. (Europe)

Villous annual 2-7 dm. tall, erect or clambering; stipules entire or few-toothed, 3-6 mm. long; tendrils simple or branched; leaflets 10-20, linear to narrowly oblong, truncate or retuse, apiculate, 1.5-2.5 cm. long; racemes subsessile, the 2 to 8 flowers clustered, pale yellow, 16-22 mm. long; banner purplish-lined and sometimes bluish-tinged, densely strigillose-villous on the back; calyx strongly villous, 10-14 mm. long, the teeth linear-subulate, the lower one subequal to the tube. N=6. Hungarian vetch.

Native of Europe, this species frequently escapes from cultivation and is established in Pierce Co., Wash., and in the Willamette Valley, Oreg., and probably elsewhere; Eurasia. May-June.

Vicia sativa L. Sp. Pl. 736. 1753. (Europe)
V. angustifolia L. Amoen. Acad. 4:105. 1759. *V. sativa* var. *angustifolia* Wahlb. Fl. Carp. 218. 1814. (Europe)
V. segetalis Thuill. Fl. Par. 2nd ed. 367. 1799. *V. angustifolia* var. *segetalis* Koch, Syn. Fl. Germ. 197. 1837. (Near Paris, France) = var. *angustifolia*.
V. sativa var. *linearis* Lange in Meddel. Kjöb. And. Act. 7:184. 1865. (Europe) = the linear-leaved phase of var. *sativa*.

Glabrate to rather strongly strigillose-villous perennial 3-8 dm. tall; stipules 3-8 mm. long, usually deeply toothed, sometimes sagittate; tendrils well developed, branched; leaflets generally 10-14, linear to oblanceolate or obovate-oblanceolate, emarginate to rounded, apiculate, mostly 1.5-3 cm. long; flowers 1-3 in the leaf axils, subsessile to short-pedunculate, 15-25 mm. long; banner erect, orchid to purplish; wings often red; calyx usually over half the length of the corolla, the teeth linear-subulate, subequal, equaling or exceeding the tube; pod 3-7 cm. long; seeds globose to somewhat compressed. N=6,7. Common vetch, tare.

European plants, long in cultivation, now widely introduced in the U.S. May-July.

Vicia sativa is extremely variable in the nature of the tips, and in the relative width, of the leaflets, as well as in the size of the flowers. The broader leafleted, larger flowered plants are usually called *V. sativa,* and the narrow leafleted, smaller flowered plants, *V. angustifolia.* However, because of the transition between the two, they are not considered to represent separate specific entities, although they are rather consistently separable as follows:

1 Flowers usually at least 18 mm. long, the wings often red; leaflets usually oblanceolate
   or broader; w. Wash. and Oreg.   var. sativa
1 Flowers usually less than 18 mm. long, wings usually not red; leaflets variable, often
     linear, more common and of more general distribution than var. *sativa* in our area
                                                                     var. angustifolia (L.) Wahlb.

Vicia tetrasperma (L.) Moench, Meth. 148. 1794.

Ervum tetraspermum L. Sp. Pl. 738. 1753. (Europe)

Very slender-stemmed, glabrous to sparsely strigillose annual 3-7 dm. tall; stipules 2-4 mm. long, linear, usually with a basal tooth; tendrils well developed; leaflets 8-10 (12), linear to narrowly oblong, apiculate, 10-25 mm. long; racemes 2- or 3-flowered; peduncles filiform, not much thicker than the tendrils, about equaling (exceeding) the leaves; flowers 5-6 mm. long, bluish, the banner only slightly reflexed; calyx lobes linear-lanceolate, about as long as the tube; pod 1-1.5 cm. long, (3) 4- or 5-seeded, glabrous. N=7.

An occasional escape w. of the Cascades, as on San Juan I., Wash., and in the Willamette Valley, Oreg., s. to Calif.; Europe. May-Aug.

Vicia villosa Roth, Tent. Fl. Germ. 2:182. 1789. (Europe)

Rather copiously hirsute-villous annual (biennial) 0.5-2 m. tall, clambering; leaves 6-15 cm. long, the stipules entire to toothed, 5-12 mm. long; tendrils well developed; leaflets 10-20, linear-lanceolate to narrowly oblong, 1.5-2 cm. long; racemes densely 20- to 60-flowered, strongly secund, about equaling the peduncles; flowers 15-18 mm. long, reddish-purple to violet; calyx 1/3-1/2 the length of the corolla, the tube 2.5-4 mm. long, very strongly gibbous at base, the pedicel attached somewhat ventrally just under the bulge; upper 2 calyx teeth triangular-based but usually prolonged to a slender tip, (0.7) 1-1.5 mm. long, the lower 3 linear and usually equaling or slightly exceeding the tube; pod 2-3 cm. long, 6-10 mm. broad, several-seeded. N=7. Hairy, woolly, or winter vetch.

A European introduction, very commonly escaped, especially along railroads and highways and on waste areas, and more or less general throughout the U.S. May-Aug.

GERANIACEAE Geranium Family

Flowers complete, regular (ours) to irregular, hypogynous, cymose to umbellate; sepals 5 (4 or 8), distinct, imbricate in bud, persistent; petals 5 (2, 4, 8, or lacking); stamens usually in 1 to 3 series of 5 each, often basally connate, one or more whorls often antherless, the anthers 2-celled, longitudinally dehiscent; pistil 1, carpels 3-5 (8), weakly united, the ovary deeply 3- to 5-lobed and -loculed, the styles adnate to the much elongate receptacle until the fruit is mature; stigmas capitate or elongate; fruit capsular but often septicidally dehiscent into 3-5 (8) segments, each 1- or 2-seeded and tipped with the slender, often spirally coiled, persistent styles; seeds usually exalbuminous; annual or perennial herbs (ours) or occasionally suffrutescent, with alternate or opposite, stipulate, mostly palmately-lobed or -divided to pinnately compound (but occasionally simple) leaves.

About a dozen genera and perhaps 500 species, common in the temperate regions, especially in Africa, and well represented in the tropics. The cultivated geranium, *Pelargonium,* is much more showy than members of the genus *Geranium* which are found in our area both as native and as introduced (and mostly weedy) species.

1 Leaves pinnately compound; fertile stamens 5   ERODIUM
1 Leaves palmately-lobed or -divided; fertile stamens usually 10   GERANIUM

Erodium L'Her. Alfilaria; Filaree; Stork's-bill; Crane's-bill; Clocks

Flowers few, umbellate on axillary peduncles, rather small; sepals usually setose, persistent; petals pinkish, often slightly dimorphic in size; filaments 10, distinct, alternately long and short, only the longer ones with anthers; styles spirally twisting at maturity; carpels sharp-pointed at the base;

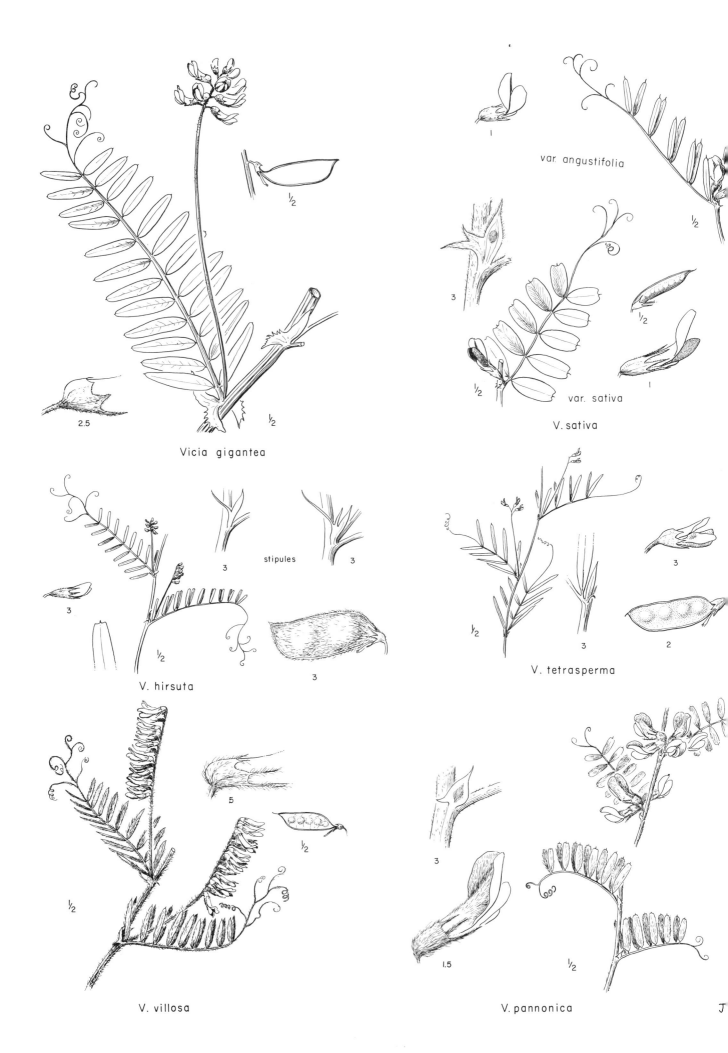

var. angustifolia

var. sativa

V. sativa

Vicia gigantea

2.5

stipules

V. hirsuta

V. tetrasperma

V. villosa

V. pannonica

seeds smooth; annual herbs (ours) to sub-shrubs with opposite, pinnate (simple) leaves and interpetiolar stipules.

About 50 species of s. Eurasia and Africa and Australia; adventive in many other parts of the world. (Greek *erodios*, a heron, referring to the long beak of the fruit.)

Erodium cicutarium (L.) L'Her. ex Ait. Hort. Kew. 2:414. 1789.
   Geranium cicutarium L. Sp. Pl. 680. 1753. (Mediterranean region)
   Annual, 0.3-3 (4) dm. tall; leaves mostly basal, rosulate, pinnate-pinnatifid to pinnately divided and the segments incised, the ultimate divisions very narrow; stipules acute; stems usually more or less reddish, the nodes swollen; flowers 10-15 mm. broad; sepals mucronate to aristate; petals pink, the claw ciliate; receptacle and styles 2.5-5 cm. long. N=10, 20.
   Native of the Old World, now widespread on drier plains and hillsides throughout w. U.S. Apr.-July.
   The species is palatable and nutritious and is valuable as stock feed, especially for sheep. It is also highly variable, and in Europe many infraspecific taxa have been described and often are maintained.
   Besides this species, 2 others have been reported for our region. *E. moschatum* (L.) L'Her. ex Ait. Hort. Kew. 2:414. 1789, is fairly common to the s. of our range, and may actually still occur in our area, but material confirming this speculation has not been seen, other than a *Suksdorf* collection from Albina, near Portland, Oreg. It is similar to *E. cicutarium* in most respects, but differs in having less-dissected leaves and in lacking the mucro to the calyx lobes. *E. aethiopicum* (Lam.) Brumh. & Thell. Mem. Soc. Nat. Sci. Cherb. IV, 38:352. 1911-12 (considered synonymous with *E. jacquinianum* F. & M., by Knuth, Pflanzenr. 4[129]:274. 1912) has been reported to be established in Lane Co., Oreg. It differs from *E. cicutarium* in having less finely divided leaves, mucronate but not aristate sepals, and longer (5-7 cm.) styles.

Geranium L. Crane's-bill; Wild Geranium

Flowers cymose (often only 2), usually somewhat showy; petals and sepals 5; stamens 10, generally all antheriferous, the filaments commonly more or less connate at the base; carpels often explosively separating and shedding the seeds, the styles usually recurving but not spirally coiling after the seeds are shed; annual or (more often) perennial herbs to semishrubs with alternate or opposite, palmately (rarely pinnately) -lobed or -divided leaves, and mostly swollen nodes.

About 200 species, widely distributed in temperate regions. (Greek *geranos*, crane, referring to the long beak of the fruit.)

Although not outstanding, *G. oreganum* is a rewarding species in cultivation, readily growing w. of the Cascades. Less valuable (but still worthwhile) are *G. richardsonii* and *G. viscosissimum*, both native and best used in gardens in the drier interior, the latter being the easier to grow.

References:
Fernald, M. L. Geranium carolinianum and allies of northeastern North America. Rhodora 37: 295-301. 1935.
Jones, G. Neville and Florence Freeman Jones. A revision of the perennial species of Geranium of the United States and Canada. Rhodora 45:5-26, 32-53. 1943.

1 Plants annual or biennial; petals less than 12 mm. long; seeds smooth or reticulate-pitted
  2 Petals (5) 7-11 mm. long; carpels glabrous; sepals conspicuously bristle-tipped
    3 Calyx eglandular, sparsely appressed-pubescent; herbage dull green   G. COLUMBINUM
    3 Calyx glandular-pilose; herbage light green; very sparingly introduced in both
      Wash. and Oreg., although usually a transient escape   *G. robertianum* L.
  2 Petals mostly less than 7 mm. long; carpels usually hairy, if glabrous then the sepals
    not bristle-tipped
    4 Ovaries glabrous, finely rugose; plants pilose-hirsute and somewhat glandular   G. MOLLE
    4 Ovaries usually hairy but not at all rugose; plants variously pubescent
      5 Fertile stamens 5; carpels pubescent with very fine appressed hairs; seeds smooth;
        sepals not bristle-tipped   G. PUSILLUM
      5 Fertile stamens 10; carpels pubescent with erect to spreading hairs; sepals bristle-
        tipped; seeds reticulate-pitted
        6 Beak of the stylar column (including the stigmas) 4-5 mm. long, the free stig-
          matic portion about 1 mm. long; fruiting pedicel much longer than the calyx
                        G. BICKNELLII

    6 Beak of the stylar column, including the stigmas, usually less than 3 mm. long,
         the stigmas less than 1 mm. long; fruiting pedicel mostly only slightly if at all longer than
         the calyx
        7 Pits of the seeds prominent, nearly isodiametric; inflorescence not particularly con-
           gested; carpels rather uniformly short-hirsute with spreading hairs   G. DISSECTUM
        7 Pits of the seeds not prominent, the alveolae elongate, not isodiametric; inflores-
           cence congested; carpels long-hirsute with ascending hairs and with shorter spread-
           ing pubescence   G. CAROLINIANUM
1 Plants perennial; petals at least 12 mm. long; seeds reticulate
    8 Petals white or pale pink, with pinkish or purplish veins, pilose about half their length
         on the inner face; inflorescence pilose with glandular, purplish-tipped hairs; beak of
         the stylar column no longer than the free lobes of the stigma   G. RICHARDSONII
    8 Petals pink to deep magenta-purple, rarely white, usually pilose on the inner face for
         not more than 1/3 their length; inflorescence eglandular or glandular with yellowish
         hairs; beak of the stylar column considerably longer than the stigmatic lobes
        9 Petals ciliate about 1/5 their length but not pilose on the inner surface; stigmas 2-
           2. 5 mm. long; w. of the Cascades   G. OREGANUM
        9 Petals pilose on the inner face; stigmas 4-5 mm. long; e. of the Cascades   G. VISCOSISSIMUM

Geranium bicknellii Britt. Bull. Torrey Club 24:92. 1897.
    G. nemorale var. *bicknellii* Fern. Rhodora 43:35. 1941. (No collections cited)
    G. carolinianum var. *longipes* Wats. Bot. King Exp. 50. 1871. *G. longipes* Goodding, Bot. Gaz. 37:
      56. 1904, not of DC. in 1822. *G. bicknellii* var. *longipes* Fern. Rhodora 37:297. 1935. *(Watson*
      206, locality not specified)
    G. nemorale Suksd. Deuts. Bot. Monats. 16:222. 1898. *(Suksdorf 2028,* Bingen, Klickitat Co.,
      Wash.) Weber (Rhodora 44:91. 1942) has pointed out that the date of publication for this name was
      1898, rather than 1892 as sometimes stated.
Retrorsely- to spreading-hirsute and more or less glandular annual (biennial) with erect to decumbent stems (0. 5) 1-6 dm. long; leaf blades cordate-rounded, 2-7 cm. broad, deeply divided into 5
cuneate segments, these variously cleft or toothed half their length or less; flowers mostly 2 per peduncle, the pedicels usually slender and in fruit 1. 5-3 times the length of the calyx; sepals 4-8 mm.
long, with a bristle-tip 1-2 mm. long, equaling or slightly longer than the retuse pinkish petals; fertile stamens 10, the filaments not connate; stylar column nearly 2 cm. long, the narrow beaklike tip
4-5 mm. long, including the stigmas; fruits hirsute-hispid, not wrinkled; seeds finely and shallowly
reticulate-pitted, the pits nearly half again as long as broad.
    Transcontinental in Can., southward to Wash., Mont., and Utah, in woodland or open fields; native.
May-Aug.
    There is considerable variation in the type and amount of pubescence of the plants, but in general
the western material matches that from e. U.S., so that varietal distinctions are lacking. Some specimens from the Olympic Mts. and from Klickitat Co., Wash., (including the type of *G. nemorale)* tend
to have more retrorse pubescence on the peduncles than average, but the variation is by no means
marked or constant and therefore not particularly significant.

Geranium carolinianum L. Sp. Pl. 682. 1753. (Carolina, Virginia)
    G. langloisii Greene, Pitt. 3:171. 1897. *(Rev. Father Langlois,* St. Martinsville, La.) The sube-
      glandular phase.
    G. thermale Rydb. Mem. N.Y. Bot. Gard. 1:478. 1900. *(Williams & Griffith,* Lo-Lo Hot Springs,
      Mont., in 1898) A dwarfed plant.
    G. sphaerospermum Fern. Rhodora 37:298. 1935. *(Fernald & Pease 3405,* Great Cloche I., Ont.,
      June 29, 1934) A phase with broad, 5-nerved sepals.
    G. carolinianum var. *confertiflorum* Fern. Rhodora 37:300. 1935. *(R. J. Webb 5263,* North Amherst,
      Lorain Co., Ohio, June 22, 1924) Proposed to include the villous-hirsute plants.
Annual, the pubescence spreading or retrorsely hirsute to pilose and usually somewhat glandular,
at least in the inflorescence; stems usually erect, 1. 5-5 (7) dm. tall; leaf blades cordate-rotund, 2-5
cm. broad, parted nearly to the base into (usually) 5 cuneate divisions, these cleft 1/3-1/2 their
length into linear segments; inflorescence congested, the peduncles often no longer than the pedicels
which usually do not exceed the calyx; sepals 4-8 mm. long, 3 (less commonly 5)-nerved, the bristle-
tip 1-1. 5 mm. long; petals subequal to the sepals, rounded to retuse, pink or rose; fertile stamens 10,

the filaments not connate at the base; stylar column about 1.5 cm. long, the beak, including the free stigmas, 2-2.5 mm. long; carpels pilose-hirsute and puberulent; seeds reticulate-alveolate, the alveolae elongate. N=26.

Throughout most of the U.S. and s. Can., in waste places and woodlands; like our other annual species, tending to be weedy. Apr.-July.

The species is rather variable in several respects, especially in the quantity and type of pubescence, some plants being almost eglandular, others more viscid. Although the inflorescence is more congested than in the other annuals, the degree of shortening of the peduncles and pedicels varies. The sepals of large vigorous plants sometimes enlarge noticeably as the fruit matures and then often are evidently 5-nerved. Such deviations are not considered to be of nomenclatural significance.

Geranium columbinum L. Sp. Pl. 682. 1753. ("Habitat in Gallia, Helvetia, Germania")

Strongly appressed-pubescent annual; stems decumbent to erect, 1.5-6 (8) dm. long; leaf blades much exceeded by the slender petioles, 2-7 cm. broad, 5- to 9-parted, the primary divisions deeply cleft into linear segments; peduncles very slender; pedicels spreading to reflexed; sepals 5-11 mm. long, with setose tips about 2 mm. long; petals purple, (5) 7-11 mm. long, rounded to very slightly retuse, barely exceeding the sepals; fertile stamens 10, the filaments not connate; stylar column about 1.5 cm. long, with a slender beak 3-5 mm. long; carpels glabrous, smooth; seeds finely reticulate-pitted. N=9.

Naturalized from Europe and not uncommon in waste places from the Dakotas to e. U.S.; known in the n.w. from Whatcom Co., Wash., and from along the Columbia R. in Oreg. June-Aug.

Geranium dissectum L. Cent. Pl. 1:21. 1755. (Europe)

G. *laxum* Hanks, N.Am. Fl. 25[1]:9. 1907. (*Frank H. Lamb 1263,* Oyhut, Chehalis Co., Wash., July 8, 1897)

Retrorsely- to spreading-hirsute annual, usually glandular above; stems spreading to erect, 1.5-6 (8) dm. long; leaf blades cordate-rotund, 2-6 cm. broad, parted nearly to the base, the 5-7 divisions from rather shallowly lobed (on the lower leaves) to incised-dissected into linear segments; pedicels commonly 2, subequal to the fruiting calyx and the peduncle; sepals 4-5 mm. long, about equaling the pink to purplish, retuse petals, elongating to 6-8 mm. in fruit, bristle-tipped; fertile stamens 10, the filaments not connate; stylar column about 1.5 cm. long, the beak 2-3 mm. long, including the (0.5 mm. long) free stigmas; fruits rather finely short-hirsute with spreading hairs; seeds prominently alveolate, the pits squarish-rounded. N=11.

An introduced weed from Europe, now established in scattered localities in c. and e. U.S. and along or near the coast from B.C. to Calif., mostly in waste places. Mar.-July.

Geranium molle L. Sp. Pl. 682. 1753. (Europe)

Pilose-hirsute and somewhat glandular, low, spreading annual 1-4 dm. tall; leaf blades (1.5) 2-5 cm. broad, reniform, about 2/3 divided into 5-7 broad lobes, these each cut about half their length into 3-5 segments; sepals soft-hairy, not bristle-tipped; petals pink, emarginate, 3-5 mm. long, slightly exceeding the sepals; fertile stamens 10, the inner filaments connate basally for about 1 mm.; stylar column 6-8 mm. long, with a filiform beak 1-2 mm. long; fruits glabrous, finely cross-rugose; seeds smooth. N=13.

A European weed now well established, usually on moist ground and mostly in waste places, especially in or at the edges of lawns; rather general throughout much of N.Am., but most common w. of the Cascades in our range. Apr.-Sept.

Geranium oreganum Howell, Fl. N.W. Am. 106. 1897. (*Howell,* Willamette Valley, Oreg.)

G. *albiflorum* var. *incisum* T. & G. Fl. N.Am. 1:206. 1838. G. *incisum* Nutt. ex T. & G. loc. cit. but not of Andrews in 1801. G. *hookerianum* var. *incisum* Walpers Rep. 1:450. 1842. (*Nuttall,* Oregon)

Perennial, (3) 4-8 dm. tall, glabrous to spreading- or retrorsely-pilose below, conspicuously glandular-pilose above; leaf blades rather densely hirsute on both surfaces, 6-14 cm. broad, divided nearly to the base into 5-7 cuneate, lobed segments; sepals 9-12 mm. long, the setose tips 2-3 mm. long; petals 18-24 mm. long, reddish-purple, the margins ciliate for 3-5 mm. from the base, the inner surface not pilose; filaments ciliate-based; stylar column 3.5-4.5 cm. long; stigmas 2-2.5 mm. long; fruit hirsute, each segment often 2-seeded.

Southern Wash. to n. Calif., w. of the Cascade Mts.; in meadows and woodlands. May-July.

Geranium dissectum

½
2

seed
9

Geranium bicknellii

½

Geranium columbinum

½

½
2
2
2
4
sepal

Erodium cicutarium

carpel
9

petal
1

Geranium carolinianum

½
1
seed
9

Geranium molle

½

Geranium oreganum

½

JRJ

Geranium pusillum Burm. Sp. Geran. 27. 1759. (Europe)

Hirsute to hirsute-strigillose and somewhat glandular-pubescent annual (biennial); stems prostrate to erect, 1-5 dm. long; leaf blades cordate-rotund, (1.5) 2-7 cm. broad, cleft 1/2-3/4 their length into 5-9 broad, irregularly incised and toothed segments; peduncles mostly 2-flowered, borne in the axils of nearly all of the upper leaves; sepals 2.5-4 mm. long, not bristle-tipped; petals obcordate, purple, about equaling the sepals; fertile stamens 5, alternating with shorter antherless filaments, not connate at the base; fruits 2 mm. long, soft-strigillose, not rugose; stylar column beakless, 6-9 mm. long; seeds smooth. N=13.

A European weed widely established throughout our area in moist or waste places. May-Aug.

Geranium richardsonii Fisch. & Trautv. Ind. Sem. Hort. Petrop. 4:37. 1837.

G. *albiflorum* Hook. Fl. Bor. Am. 1:116. 1831, but not of Ledeb. in 1829. G. *hookerianum* Walpers Rep. 1:450. 1842. *(Drummond,* "Vallies in the Rocky Mountains")

G. *loloense* St. John, Fl. S. E. Wash. 242. 1937. *(Piper 4027,* Musselshell, Lolo Trail, "Bitterroot Mts.," Clearwater Co., Ida., July 14, 1902)

Perennial, (2) 3-8 dm. tall, glabrous or sparsely strigillose below, purplish glandular-pilose in the inflorescence; leaf blades (4) 6-14 cm. broad, cleft 3/4 the length or more into 5-7 irregularly lobed main segments; sepals 6-11 mm. long, the setose tips 1.5-2 mm. long; petals 12-17 mm. long, white to pinkish, purplish-veined, pilose at least 1/2 their length on the inner surface; beak of the stylar column about equal to the free lobes of the stigma; carpels hirsute. N=26.

Throughout much of w. N.Am. from e. B.C. southward, on the e. side of the Cascades, to s. Calif. and N.M., from the foothills to well up in the mountains, mostly in partial shade. June-Aug.

Occasional light lavender-flowered specimens are found which are believed to result from hybridization with G. *viscosissimum.*

Geranium viscosissimum F. & M. Ind. Sem. Hort. Petrop. 11:suppl. 18. 1846. ("Hab. in America septentrionali occidentali")

GERANIUM VISCOSISSIMUM var. NERVOSUM (Rydb.) C. L. Hitchc. hoc loc. G. *nervosum* Rydb. Bull. Torrey Club 28:34. 1901. *(F. Tweedy 494,* Fish Cr., Teton For. Reserve, Wyo., in 1897)

G. *strigosum* of Rydb. Bull. Torrey Club 29:243. 1902, but not of Burm. f. in 1768. G. *strigosior* St. John, Fl. S. E. Wash. 243. 1937. *(F. Tweedy 4591,* Copperton, Wyo., in 1901) = var. *nervosum.*

G. *canum* Rydb. ex Hanks, N.Am. Fl. 25[1]:14. 1907, and Rydb. Fl. Rocky Mts. 532. 1917, in large part, but not as to type. = var. *nervosum.*

G. *incisum* sensu Hanks, N.Am. Fl. 25[1]:14. 1907, and Rydb. Fl. Rocky Mts. 532. 1917, in large part. = var. *nervosum.*

G. *viscosissimum* var. *album* Suksd. Werdenda 1:24. 1927. G. *viscosissimum* f. *album* St. John, Proc. Biol. Soc. Wash. 41:195. 1928. *(Suksdorf 8710,* Spangle, Spokane Co., Wash., June 29, 1916) = var. *viscosissimum.*

Perennial, (2.5) 4-9 dm. tall, the lower part of the stems (and petioles) spreading-hirsute to strigillose (rarely glabrous) and often glandular-pilose, pilose and glandular-villous above, especially in the inflorescence; leaf blades (4) 5-12 cm. broad, strigillose to hirsutulous and more or less glandular, parted more than 3/4 their length into 5 (7) obovate-cuneate, sharply toothed divisions; sepals 8-12 mm. long, the setose tips 1-3 (averaging about 2) mm. long; petals 14-20 mm. long, rounded to emarginate, pinkish-lavender to purplish (white), pilose at base usually for about 1/4 of the length but not uncommonly with pilosity past midlength; filaments ciliate-pilose; stylar column (2.5) 3-5 cm. long, the beak (including the 4-5 mm. stigmas) 10-14 mm. long; fruits glandular-hirsute.

From B.C. to n. Calif., e. of the Cascade Mts., e. to Sask., s. to w. S.D., Colo., Utah, and Nev., from the lowlands to well into the mountains. May-Aug.

The species is separable as follows into two rather clearly marked, freely intergradient varieties:

1 Lower petioles and stems hirsute and also glandular-puberulent, rather uniformly glandular-villous above; B.C. southward to n. Calif. and n. Nev., e. to Sask. and n. Wyo.

var. viscosissimum

1 Lower petioles and stems glabrous, strigillose, appressed- to spreading-puberulent, or hirsute, not glandular; with the var. *viscosissimum* throughout most of its range, but a little more southern, reaching Colo., Utah, and n. Calif.; N=26

var. nervosum (Rydb.) C. L. Hitchc.

The var. *nervosum* includes plants with two general types of pubescence, some being spreading-hirsute, others appressed-puberulent. Plants of the former type appear to be more northern in distribution (mostly to the n. of Oreg.).

OXALIDACEAE Oxalis or Wood-sorrel Family

Flowers nearly or quite regular, perfect, hypogynous, from singly pedunculate in the axils to cymose or racemose; sepals 5, distinct, imbricate; petals 5 (lacking in the often numerous cleistogamous flowers), distinct or basally connate, contorted in the bud; stamens basally connate, usually 10 (ours), of 2 unequal lengths, rarely only 5 fertile, the anthers 2-celled; pistil 5-carpellary, the ovary 5-loculed; styles 5; stigmas usually capitate; fruit a loculicidal, several- to many-seeded capsule (berry); seeds albuminous, sometimes arillate, the embryo straight; herbs (ours) or rarely shrubs or trees, often with watery, sour juice; leaves alternate, exstipulate or with very much reduced stipules, ternately (pinnately) compound, the leaflets usually obcordate.

About seven genera and perhaps 900 species, mostly tropical but extending well into the temperate zones.

Oxalis L. Oxalis; Wood sorrel*

Sepals 5, persistent; petals 5, showy, white or yellow to pinkish or violet, usually quickly withering and deciduous; stamens 10, dimorphic as to length, the filaments connate at base into a short tube; styles 5, slender, distinct; capsule elongate, 5-lobed and -loculed, membranous; seeds albuminous, pendulous, often cross-corrugated; acaulescent or caulescent herbs, often with corms or rhizomes, obcordate and strongly nyctitropic leaflets, and usually long (sometimes membranous-margined) petioles.

Perhaps 300 species, cosmopolitan, but mostly in the warmer temperate and subtropical regions. (Greek *oxys*, sharp [hence sour], because of the strongly acid juice which is more or less characteristic of the family as a whole.)

Numerous species of *Oxalis* are cultivated, and two of our native species are attractive. *O. suksdorfii* is seldom grown, but it has considerable charm and does not become weedy. Probably all gardeners who visit the coastal redwoods in California marvel at the beauty of the solid carpet of *O. oregana* under these giant trees. Many, in all innocence, will not rest until they get the plant established in their own gardens. Few will have much time for rest thereafter, unless they soon attempt to eradicate the aggressive pest—every last invasive rootstock of it. The gardener who feels he must grow the species should be sure he can spare it almost unlimited territory as it knows no compromise.

Reference:

Wiegand, K. M. Oxalis corniculata and relatives. Rhodora 27:113-124, 133-139. 1925.

1 Plants acaulescent; flowers white or pinkish
  2 Scapes 1-flowered; petals usually at least 13-20 mm. long; capsules ovoid, scarcely
    1 cm. long   O. OREGANA
  2 Scapes 2- to several-flowered; petals about 10 mm. long; capsules linear, 2-3 cm.
    long   O. TRILLIIFOLIA
1 Plants with leafy stems above ground; flowers yellow
  3 Petals 12-20 mm. long; peduncles 1 (2)-flowered   O. SUKSDORFII
  3 Petals 4-9 mm. long; peduncles (1) 2- to 7-flowered
    4 Hairs of the stems and petioles usually at least in part septate, their tips blunt, generally collapsing when dried or pressed; plants rhizomatous; stipules lacking   O. STRICTA
    4 Hairs of the stems and petioles nonseptate, their tips pointed, not collapsing when dried or pressed; plants not rhizomatous although often more or less stoloniferous; stipules usually present, although often greatly reduced
      5 Seeds with transverse whitish ridges; plant caespitose, the outer branches sometimes rooting at the base, the erect portion light green (often tan when dried), usually antrorsely strigose   O. DILLENII
      5 Seeds not white-ridged; plants with creeping and rooting main stems, the erect portion green to purple, glabrous or antrorsely to retrorsely strigose   O. CORNICULATA

---

*In the preparation of the treatment of the genus *Oxalis,* valuable suggestions have been received from George Eiten and are incorporated herein.

Oxalis corniculata L. Sp. Pl. 435. 1753.

    O. pusilla Salisb. Trans. Linn. Soc. 2:243. 1794. *Xanthoxalis corniculata* Small, Fl. S. E. U.S. 667. 1903. ("Habitat in Italia, Sicilia")

    O. repens Thunb. Oxalis (bot. dissert.) 14. 1781. *O. corniculata* var. *repens* Zucc. Mon. der amer. Oxalis-Arten, (Denkschr. Akad. Wiss. Muenchen) 54. 1831. *Xanthoxalis repens* Moldenke, Castanea 9:42. 1944. (Europe, S. Africa, Madagascar, Ceylon)

    O. corniculata var. *atropurpurea* Planch. in Van Houtte, Fl. Serres 12:47. 1857. *Xanthoxalis europaea* var. *atropurpurea* Moldenke, Phytologia 2:324. 1947. (In hort., probably originally from France). A variant with purple leaves.

    O. tropaeoloides Schlecht. ex Planch. in Van Houtte, Fl. Serres 12:47. 1857, in synonymy. *O. corniculata* var. *tropaeoloides* Cannarella, Bull. Soc. Bot. Ital. 78. 1909. *O. corniculata* var. *corniculata* f. *tropaeoloides* Knuth, Pflanzenr. IV, 130:149. 1930. (Type not ascertained)

    O. corniculata var. *purpurea* Parl. Fl. Ital. 5:271. 1872. *O. corniculata* var. *corniculata* f. *purpurea* Knuth, Pflanzenr. IV, 130:149. 1930. (Naples) A variant with purple leaves.

    O. corniculata var. *rubra* Nichols, Ill. Dict. Gard. 2:540. 1886. (No type specified)

    O. corniculata var. *viscidula* Wieg. Rhodora 27:121. 1925. *(Mrs. E. H. Terry,* Northampton, Mass., in 1902.

Nonrhizomatous perennial, but the stems trailing and freely rooting, glabrous or more or less antrorsely to retrorsely strigose, the erect portions usually less than 1 dm. tall, sometimes greenish but more frequently the herbage brownish, reddish, or purple; stipules mostly brownish, from broadened and 2-4 mm. long to linear and scarcely 1.5 mm. long; leaf blades 1-4 cm. long, leaflets obcordate; peduncles mostly 2- to 5-flowered; petals 4-8 mm. long, yellow, entire; filaments glabrous; capsule oblong, 1.5-2.5 cm. long, grayish-strigose; seeds brownish, transversely rugose but the crests not whitish. N=12, 24.

An introduced weed in our area, in waste places and gardens; widespread in N.Am. and in various other parts of the world; native to Europe. May-Oct.

Oxalis dillenii Jacq. Oxal. Monogr. 28. 1794.

    O. corniculata var. *dillenii* Trel. in Gray, Syn. Fl. 1¹:365. 1897. (Carolina)

    O. navieri Jord. in Schultz, Arch. Fl. Fr. et Allem. 1:310. 1854. *O. stricta* var. *navieri* Knuth, Pflanzenr. IV, 130:145. 1930. (Paris)

    O. stricta var. *viridiflora* Hus, Rep. Mo. Bot. Gard. 18:99. 1907. *O. stricta* f. *viridiflora* Fern. Rhodora 38:425. 1936. *Xanthoxalis stricta* f. *viridiflora* Moldenke, Boissiera 7:6. 1943. (St. Louis, Mo.)

    O. oneidica House, Bull. N.Y. State Mus. 243-244:43. 1921. *(House 6140,* N.Y., June 5, 1919)

    O. stricta var. *piletocarpa* Wieg. Rhodora 27:123. 1925. *O. corniculata* var. *dillenii* subvar. *piletocarpa* Farw. Am. Midl. Nat. 10:36. 1926. *Xanthoxalis stricta* var. *piletocarpa* Moldenke, Boissiera 7:6. 1943. *(E. F. Williams,* Alstead, N.H., in 1901)

    O. stricta sensu Am. authors, not of L.

A caespitose perennial without rhizomes but with the several branches spreading and sometimes rooting although not usually trailing, mostly antrorsely strigose with nonseptate, pointed hairs, the erect stems light greenish, 1-4 dm. tall; stipules greenish, from fairly broad and 2-3 mm. long to much narrower and shorter; leaf blades 1-4 cm. long, leaflets obcordate; peduncles (1) 2- to 7-flowered; petals 4-8 mm. long, yellow, entire; filaments glabrous; capsule oblong, 1.5-2.5 cm. long, grayish-strigose; seeds brownish, conspicuously cross-rugose, the ridges whitish.

Seldom found in our area, and then only as a weed, chiefly on the e. side of the Cascades, but transcontinental in Can. and widespread in e. U.S. and s. to Mex.; common as a weed in many other parts of the world. May-Oct.

This species is very similar to *O. corniculata* and is distinguishable therefrom chiefly because of its more erect habit, greener foliage, and whitish crests of the seeds.

Oxalis oregana Nutt. ex T. & G. Fl. N.Am. 1:211. 1838.

    O. acetosella var. *oregana* Trel. Mem. Bost. Soc. Nat. Hist. 4:90. 1888. *Oxys oregana* Greene, Man. Bay Reg. 71. 1894. *(Nuttall,* "Shady woods of the Oregon")

An acaulescent perennial from rhizomes that are scaly with the persistent stipules; petioles 3-20 cm. long, brownish-pilose, jointed to the stipules; leaflets as much as 4.5 cm. broad; scapes 5-15 cm. tall, 1-flowered, 2-bracteate near the flower; petals white to pinkish, more darkly pinkish- or

seed
18

petal
1.5

1/2

Oxalis corniculata

Geranium richardsonii

5

2

5

1/2

1/2

Oxalis dillenii

seed
18

petal
1.5

1/2

1/2

carpel
9

1/2

1/2

JRJ

Geranium viscosissimum   Geranium pusillum   Oxalis oregana

reddish-veined, 13-20 mm. long; staminal tube 1-2 mm. long; capsule ovoid-globose, 7-9 mm. long; seeds almond shaped, about 4 mm. long, with (mostly) longitudinal corrugations.

Usually in moist woods from the coast onto the e. slope of the Cascades, from the Olympic Peninsula, Wash., to Monterey Co., Calif. Apr.-Sept.

In some areas the flowers are nearly pure white, in others much more pinkish, but there is no apparent tendency toward ecological or geographical segregation of the two color phases.

Oxalis stricta L. Sp. Pl. 435. 1753.

   O. ambigua Salisb. Trans. Linn. Soc. 2:242. 1794, but not of Jacq. or of A. Rich. *O. corniculata* var. *stricta* Sav. in Lam. Encyc. Meth. 4:683. 1797. *Xanthoxalis stricta* Small, Fl. S.E. U.S. 667. 1903. *Ceratoxalis stricta* Lunell, Am. Midl. Nat. 4:468. 1916. (Virginia)

   O. europaea Jord. in Schultz, Arch. Fl. Fr. et Allem. 1:309,311. 1854. *O. stricta* var. *europaea* Knuth, Pflanzenr. IV, 130:144. 1930. *Xanthoxalis europaea* Moldenke, Castanea 7:125. 1942. (Europe)

   O. cymosa Small, Bull. Torrey Club 23:267. 1896. *Xanthoxalis cymosa* Small, Fl. S.E. U.S. 668. 1903. *Ceratoxalis cymosa* Lunell, Am. Midl. Nat. 4:468. 1916. *O. europaea* f. *cymosa* Wieg. Rhodora 27:135. 1925. *Xanthoxalis europaea* f. *cymosa* Moldenke, Boissiera 7:5. 1943. ("Ontario to the Lake Superior region and Nebraska south to the Gulf of Mexico", no specimens cited)

   O. bushii Small, Bull. Torrey Club 25:611. 1898. *Xanthoxalis bushii* Small, Fl. S.E. U.S. 667. 1903. *O. stricta* var. *bushii* Farw. Rep. Mich. Acad. Sci. 20:183. 1918. *O. europaea* var. *bushii* Wieg. Rhodora 27:135. 1925. *Xanthoxalis europaea* var. *bushii* Moldenke, Castanea 7:125. 1942. (*B. F. Bush 30,* Jackson Co., Mo., June 28, 1893)

   O. rufa Small in Britt. Man. 577. 1901. *Xanthoxalis rufa* Small, Fl. S.E. U.S. 668. 1903. *O. stricta* var. *rufa* Farw. Rep. Mich. Acad. Sci. 20:183. 1918. (Eastern N.A.)

   O. coloradensis Rydb. Bull. Torrey Club 29:243. 1902. *Xanthoxalis coloradensis* Rydb. Fl. Colo. 220. 1906. *Ceratoxalis coloradensis* Lunell, Am. Midl. Nat. 4:468. 1916. (*Rydberg & Vreeland 5920,* Sangre de Christo Cr., Colo., in 1900)

   Xanthoxalis interior Small, Fl. S.E. U.S. 668. 1903. *O. interior* Fedde, Just Bot. Jahresb. 32[1]: 410. 1905. (Mo. and Ark.)

   O. europaea f. *pilosella* Wieg. Rhodora 27:135. 1925. (*Bush 6701,* Greenwood, Mo.)

   O. europaea f. *villicaulis* Wieg. Rhodora 27:135. 1925. *O. stricta* var. *villicaulis* Farw. Am. Midl. Nat. 11:62. 1928. *Xanthoxalis europaea* f. *villicaulis* Moldenke, Castanea 9:42. 1944. (*C. K. Dodge 41,* Port Huron, Mich., in 1914)

   O. europaea var. *bushii* f. *subglabrata* Wieg. Rhodora 27:136. 1925. *Xanthoxalis europaea* var. *bushii* f. *subglabrata* Moldenke, Castanea 7:125. 1942. (*Pammel & Ball 4,* Ames, Ia.)

   O. europaea var. *bushii* f. *vestita* Wieg. Rhodora 27:136. 1925. *Xanthoxalis europaea* var. *bushii* f. *vestita* Moldenke, Castanea 7:125. 1942. (*I. W. Anderson,* Cambridge, Mass., in 1904)

Perennial with slender, widespread, fleshy rhizomes; stems from prostrate and matted to (more commonly) erect and as much as 5 dm. tall, glabrous to pubescent mostly with spreading, blunt-tipped, septate hairs but sometimes also somewhat strigose; stipules lacking; leaflets 1-5 cm. long, usually glabrous or only ciliate; peduncles generally exceeding the leaves, (1) 2- to 7-flowered; petals 4-9 mm. long, yellow; filaments glabrous; fruiting pedicels straight, erect to spreading; capsules sparsely hirsute with long septate hairs. N=12.

The plant is believed to be native to N.Am., where it is widespread and much more common in many places than in most of our area, although it is an extremely troublesome weed in w. Wash.; flowering almost throughout the year.

Several infraspecific taxa often are recognized largely on the basis of the variation in pubescence, but there is no good reason to apply more than one name to our plants.

Oxalis suksdorfii Trel. Mem. Bost. Soc. Nat. Hist. 4:89. 1888.

   O. pumila Nutt. ex T. & G. Fl. N.Am. 1:212. 1838, but not of d'Urv. in 1829. *Xanthoxalis suksdorfii* Small, N.Am. Fl. 25[1]:53. 1907. (*Nuttall,* "Forests of the Rocky Mountains and Oregon")

Rhizomatous, finely hirsute to pilose-villous perennial, most of the hairs unicellular, a few often septate; stems 3-15 cm. long, decumbent to erect; stipules very small or lacking; petioles 2-5 cm. long; leaf blades 1.5-3 cm. broad; peduncles exceeding the leaves, 1- to 3-flowered; petals yellow, 12-20 mm. long, rounded; filaments hairy, connate for 2-3 mm.; capsules oblong, 1-1.5 cm. long, hoary.

Usually in moist coastal woods but sometimes on rather dry open slopes, from s. Wash. southward, w. of the Cascade divide, to n.w. Calif. Apr.-Aug.

Oxalis trilliifolia Hook. Fl. Bor. Am. 1:118. 1831.

   O. macrophylla Dougl. ex Hook. loc. cit. as a synonym, not of Kunth. *Hesperoxalis trilliifolia*
Small, N. Am. Fl. 25¹:27. 1907. *(Douglas,* "North-West America, on the summits of high mountains near the 'Grand Rapids' of the river Columbia; and also in the vallies of the Rocky Mountains")
   Acaulescent perennial with stout scaly rhizomes; petioles 5-20 cm. long, usually somewhat brownish-pilose; leaf blades (3) 5-8 (12) cm. broad, the leaflets usually more nearly obovate-cuneate and
emarginate than obcordate; peduncles usually longer than the leaves, mostly brownish-pilose; scapes
(2) 3- to 7-flowered; petals about 1 cm. long, white or pinkish-tinged, emarginate; capsule linear, 2-
3 cm. long, slender-beaked; seeds brownish, shining, 3.5-4 mm. long, shallowly rugose-corrugate.
   In meadows and moist woods in the coastal mountains and up to 4000-6000 ft. on the w. slope of the
Cascades, from Wash. to n.c. and n.w. Calif. May-Aug.

LINACEAE Flax Family

   Flowers perfect, regular, hypogynous, polypetalous, borne in cymes or racemes; sepals 5 (4), imbricate, distinct or connate at the base; petals 5 (4), often showy but usally fugacious, usually short-
clawed, convolute in the bud; fertile stamens 5 (ours) to 10 or more, usually basally connate, often
alternate with small staminodia; pistil 1, with 5 (2-4) carpels and mostly 5 or 10 locules; styles as
many as the carpels, usually distinct; stigmas capitate or slightly elongate; fruit a capsule (ours) or
rarely drupelike, generally with twice as many locules as styles, and with 1 or 2 seeds per locule;
seeds albuminous to exalbuminous, the embryo straight; annual or perennial herbs (rarely suffruticose) with simple, usually entire, alternate (opposite), exstipulate or glandular-stipulate leaves.
   About 10 genera and nearly 200 species, widely distributed, but mainly in temperate regions.

Linum L. Flax

   Flowers racemose to cymose-paniculate; sepals 5; petals 5, usually showy, white, yellow, pinkish,
red, or blue, promptly deciduous; stamens 5, basally connate and forming a very short tube bearing
small toothlike staminodia (?) usually alternate with the filaments; pistil 2- to 5-carpellary, 2- to 5-
loculed, the styles distinct or basally connate; capsule septicidal, each locule more or less completely false-septate; seeds flattened to rounded, often mucilaginous when wetted.
   Nearly 100 species of temperate regions. (The original Latin name for flax.)
1 Petals yellow
   2 Plant a perennial with many spreading stems from the somewhat woody base; leaves
       numerous, densely crowded on the sterile branches; high mts. from s.e. Ida. to Wyo.
       and s. to Colo., possibly just reaching our range in Ida.   *L. kingii* Wats.
   2 Plant an annual or an erect short-lived perennial, usually simple at the base; leaves
       scattered, often deciduous on the lower part of the plant
       3 Leaves all alternate and entire; stigmas 4 or 5; plants of the Great Plains, from
           Alta. to Tex., possibly occasional in our area in Mont.   L. RIGIDUM
       3 Leaves (at least the lower cauline ones) opposite, the upper bracts serrulate; stig-
           mas 2; Wash. to Calif.   L. DIGYNUM
1 Petals white, pinkish, or blue
   4 Petals white to pink; stems usually pubescent   L. MICRANTHUM
   4 Petals blue; stems usually glabrous
       5 Plant an annual (sometimes perhaps a short-lived perennial); membranous margins
           of the inner sepals erose-serrulate or ciliate; leaves 3-nerved
           6 Petals less than 10 mm. long; leaves rarely as much as 2 cm. long   L. ANGUSTIFOLIUM
           6 Petals at least 10 mm. long; leaves often 2-2.5 cm. long   L. USITATISSIMUM
       5 Plant a perennial; margins of the sepals entire; leaves 1-nerved   L. PERENNE

Linum angustifolium Huds. Fl. Angl. 2nd ed. 134. 1778. (Europe)
   Slender, glabrous and glaucous annual (sometimes perennial?) 1-5 dm. tall, simple below but usually branched above; leaves sessile, narrowly linear-lanceolate, 1-2 cm. long, 1-2 mm. broad, distinctly 3-nerved; flowers on long slender flowering branches and filiform pedicels 1-3 cm. long; sepals about 5 mm. long, broadly ovate, cuspidate, with ciliate to erose, membranous margins; petals
blue, from only slightly longer than the sepals to as much as 7-8 mm. in length; staminodia represented by the merest teeth; stigma clavate; capsule 5-6 mm. long. N=15.

Along roadsides from Lane Co., Oreg., to coastal c. Calif.; Europe. May-Aug.

Linum digynum Gray, Proc. Am. Acad. 7:334. 1868.
Cathartolinum digynum Small, N.Am. Fl. 25¹:78. 1907. (*Bolander,* Yosemite Valley, Calif.)
Slender, glabrous and glaucous annual 0.5-4 dm. tall; stems simple or somewhat branched; lower leaves mostly opposite, 0.5-3 cm. long, sessile, 1-nerved, entire but the reduced upper floral bracts serrulate-glandular; pedicels about equaling the flowers; sepals 2-3 mm. long, very unequal, the outer ones serrate, all glandular-pectinate; petals pale yellow, from about as long to nearly twice as long as the shorter sepals; staminodia lacking; styles 2, basally connate; stigmas elongate; capsule 2-carpellary, falsely 4-celled, about equal to the calyx. N=8.
In meadows and prairies e. of the Cascades, from Wash. and Oreg. to the Sierra Nevada, Calif. May-July.

Linum micranthum Gray, Proc. Am. Acad. 7:333. 1868.
Hesperolinon micranthum Small, N.Am. Fl. 25¹:85. 1907. (*Bolander,* Mt. Bullion, Mariposa Co., Calif.)
Freely dichotomous-branched annual 1-4 dm. tall, the stems very slender and usually puberulent at the forks; leaves linear, 5-20 mm. long, 1-nerved; pedicels filiform, 1-2 cm. long; sepals ovate-lanceolate, 2-3 mm. long, their margins usually glandular-ciliate; petals white (pinkish), scarcely twice the length of the sepals; staminodia lacking; styles usually 3; stigmas capitate; capsule 2.5-3.5 mm. long.
On dry open ground, usually in timbered or brushy areas; Jefferson and Douglas cos., Oreg., to much of Calif. June-July.

Linum perenne L. Sp. Pl. 277. 1753. ("Siberia et Cantabrigiae")
L. lewisii Pursh, Fl. Am. Sept. 210. 1814. *L. sibiricum* var. *lewisii* Lindl. Bot. Reg. 14:pl. 1163. 1828. *L. perenne* var. *lewisii* Eat. & Wright, N.Am. Bot. 302. 1840. *L. perenne* ssp. *lewisii* Hultén, Fl. Alas. 7:1122. 1947. (*Lewis,* Rocky Mts.)
L. perenne f. *albiflorum* Cockerell, West Am. Sci. 3:217. 1887. *L. lewisii* f. *albiflorum* St. John, Fl. S.E. Wash. 244. 1937. (*Cockerell,* Montrose Co., Colo., Oct. 10, 1887)
Glabrous and somewhat glaucous, woody-crowned perennial 1-6 dm. tall; leaves alternate, linear, 1-3 cm. long, 1-nerved, acute to rounded; pedicels slender, about as long as the flowers, usually recurved in fruit; sepals 4-7 mm. long, the margins membranous, entire; petals blue, 10-23 mm. long; staminodia toothlike, about 1 mm. long; filaments slightly connate at the base; styles 5, considerably longer than the stamens; stigmas capitate; capsule 10-celled. N=9.
From prairies to alpine ridges, usually on dry, well-drained soil, throughout w. N.Am.; Eurasia. May-July.
This is an attractive plant that merits a place in the wild garden, or even in the rock garden, if taller plants are desired.
Although great variation occurs in our native plants, the various populations are uniform enough to merit their inclusion in one taxon. Early students of the American flora usually considered our native blue flax to be identical with the Eurasian *L. perenne,* whereas it is now customary to maintain it as a distinct species, even though there is only one significant difference between them. The plant of the Old World (var. *perenne),* which is occasional in our area as a garden escape, consistently produces flowers of two kinds (one with styles considerably longer than the stamens, the other with the stamens the longer), whereas the native N.Am. material, var. *lewisii* (Pursh) Eat. & Wright., as described above, does not exhibit this dimorphism.

Linum rigidum Pursh, Fl. Am. Sept. 210. 1814.
Cathartolinum rigidum Small, N.Am. Fl. 25¹:82. 1907. (*Nuttall,* "On the Missouri")
L. rigidum var. *tenerrimum* Blank. Mont. Agr. Coll. Stud. Bot. 1:85. 1905. (*Blankinship,* Big Horn R., Mont., June 15, 1890, is the first of several collections cited)
Glabrous to sparsely puberulent, olive-green annual; stems 1.5-5 dm. tall, usually simple below but diffusely and rather rigidly branched above, strongly striate-angled; leaves relatively few, alternate, ascending, linear to linear-lanceolate, 1-2.5 cm. long, acuminate, 1-nerved, entire to glandular-serrulate, the lower ones mostly deciduous by anthesis; flowers loosely cymose, the bracts scale-like; sepals narrowly lanceolate, acuminate, 4-8 mm. long, glandular-serrulate; petals yellow, 10-15 mm. long; stamens 5, the filaments very shortly connate at the base; style 4- to 5-branched near

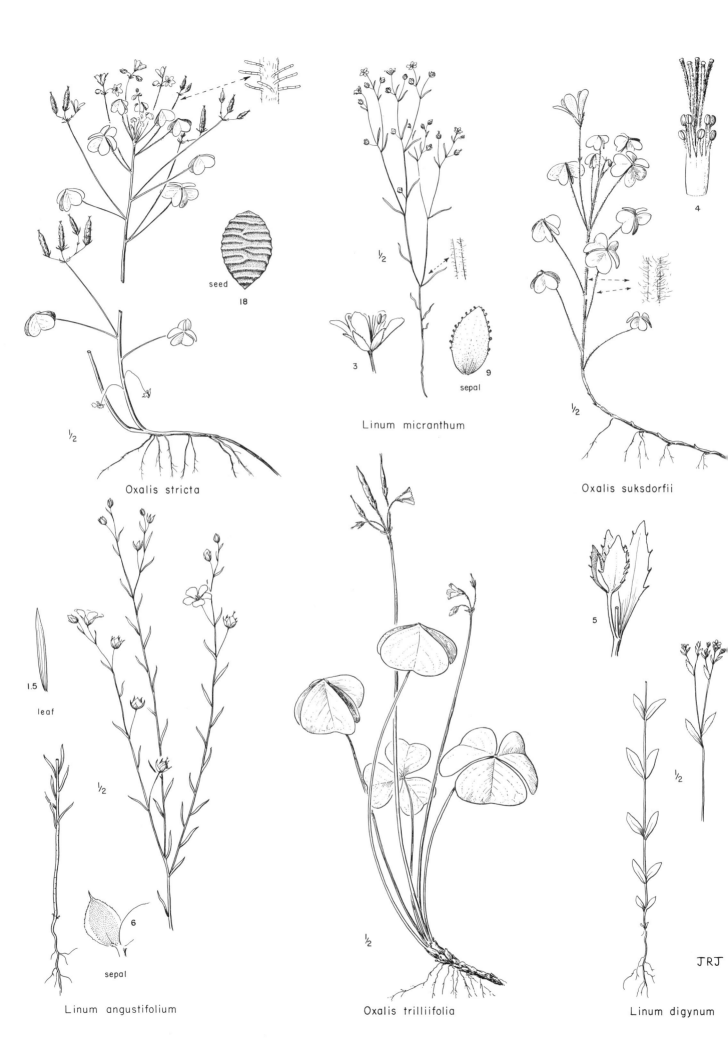

seed
18

Oxalis stricta

Linum micranthum

3

9
sepal

4

Oxalis suksdorfii

1.5
leaf

6

sepal

Linum angustifolium

5

Oxalis trilliifolia

Linum digynum

JRJ

the tip, the stigmas subglobose; capsule ovoid, 4-5 mm. long, 5 (4)-celled. N=15. Yellow flax.

From the prairies into the foothills; Man. to Alta., southward to Tex. and N.M., almost entirely e. of the Rocky Mts. but probably reaching our area in Mont. June-July.

Another species of the Great Plains region, from the e. side of the Rocky Mts., possibly may enter our range in Mont., namely *L. compactum* A. Nels. (Bull. Torrey Club 31:241. 1904). This plant, which seems to intergrade with *L. rigidum* and which is perhaps not specifically distinct from it, differs in its tendency to be more hairy, shorter, basally branched, and smaller flowered, with the petals usually less than 10 mm. long.

Linum usitatissimum L. Sp. Pl. 277. 1753. (Europe)
Simple or branched annual 3-5 dm. tall; leaves alternate, linear, 3-nerved, 1-2.5 cm. long; flowers on slender, elongate pedicels; sepals 5-7 mm. long, the inner ones broadest, minutely ciliate-serrulate; petals blue, 10-14 mm. long; staminodia minute, toothlike, alternating with the 5 fertile stamens; styles 5, exceeding the stamens slightly; stigma capitate; capsule about twice the length of the sepals, 10-celled. N=15, 16. Common flax, linseed.

This European species is cultivated for both the fiber (the source of linen) and the seed (common flaxseed, a source of linseed oil); in w. U.S. it is not rare as an escape, perhaps persisting only a few years at most. June-Nov.

ZYGOPHYLLACEAE Caltrop Family

Flowers regular, perfect, hypogynous, borne on axillary peduncles; sepals 5 (4), nearly or quite distinct; petals 5 (4), distinct (rarely lacking); stamens 10-15, distinct, the filaments often with a scalelike basal appendage; pistil 4- or 5-carpellary, usually with a single style and a lobed or unlobed stigma, but rarely the styles as many as the carpels; fruit usually a thin-walled to somewhat fleshy, 4- or 5-loculed capsule, but sometimes either more nearly baccate or a true schizocarp of 5 much-hardened carpels; seeds usually 2 to many per locule, with or without endosperm; annual or perennial herbs (ours) or shrubs or trees with opposite (rarely alternate), pinnatifid to pinnately compound leaves and persistent, often leathery or spinescent stipules.

About 25 genera and perhaps 150 species, primarily tropical and subtropical, a few in the Temp. Zone.

1 Leaves irregularly pinnatifid-dissected, the segments linear; stamens 12-15; fruit an ovoid
    capsule   PEGANUM
1 Leaves pinnately compound or bifoliolate, the leaflets not linear; stamens 10
  2 Leaflets 2; filaments each with a linear, 2-lobed scale adnate to the inner surface;
    fruit an oblong, unarmed capsule   ZYGOPHYLLUM
  2 Leaflets 8-14; filaments without scales; fruit a spiny schizocarp   TRIBULUS

Peganum L.

Flowers white or pale yellow, perfect, regular, borne singly on axillary peduncles; perianth 4- or 5-merous; stamens 12-15; pistil 2- or 3-carpellary, the stigma 1- to 3-winged; fruit a 3 (4)-celled, many-seeded capsule; annual or perennial herbs with alternate pinnatifid leaves.

Five or 6 species of the Mediterranean region and 1 in Mex. (Name used by Theophrastus for the plant.)

Peganum harmala L. Sp. Pl. 444. 1753. ("Habitat in arena Madritii, Alexandriae, Cappadociae, Galatia")
Glabrous and glaucous annual with decumbent to erect, freely branched stems 2-5 dm. long; leaflets irregularly pinnatifid into narrowly linear segments 1-3 cm. long; stipules setaceous, about 1 mm. long, quickly deciduous; flowers axillary on peduncles 1-2 cm. long; sepals linear, sometimes lobed, 1-2 cm. long; petals 14-18 mm. long, white; stamens 15; capsule membranous, nearly or quite globose, 1-1.5 cm. long. N=12.

A very occasional escape in w. U.S., collected at Ephrata, Wash.; Europe. June-Aug.

Tribulus L.

Flowers solitary on axillary peduncles; sepals 5; petals 5, yellowish or white, inserted at the outer edge of a 10-lobed, perigynous disc; stamens 10, borne within the disc, the filaments without scales; pistil 1, the ovary 5-celled; style 1; stigma 5-lobed; fruits depressed, deeply lobed, separating at maturity into 5 horny, 2-spined, nutlike, 2- to 4-seeded segments; freely branched, prostrate herbs with pinnately compound, opposite, stipulate leaves.

About a dozen species, native to desert and tropical regions, but introduced elsewhere. (Latin *tribulus*, a thorn or thistle.)

Tribulus terrestris L. Sp. Pl. 387. 1753. (Europe)

Strigose to hirsute, prostrate annual forming mats 3-10 dm. broad; leaflets 4-8 pairs, obliquely oblong-ovate, 5-15 mm. long, the stipules 1-2 mm. long; flowers short-pedunculate; petals yellow, 3-5 mm. long; fruit segments armed with 2 large and divergent spines 2-6 mm. long and with numerous smaller spines that form a broad, dorsal, longitudinal row. N=12, 24. Puncture vine.

Introduced from the Old World, and well established along highways and railroads in e. Wash. and Oreg., although more common from Calif. to N.M. May-Sept.

Puncture vine is one of the worst weeds in much of w. U.S. The burs are a source of painful injury to barefoot children, agricultural workers, and livestock, as well as the cause of innumerable flat tires, as implied by the common name. When a vehicle travels over the mature fruits, the burs are picked up by the tires but usually do not immediately puncture them. In travel, the spines soon break, the segments of the fruit being thrown off, thus insuring dissemination of the plant. The broken spines remaining in the tire gradually work inward through the casing, where eventually they are apt to produce exasperating slow leaks, at least if travel is over graveled roads.

Zygophyllum L.

Flowers perfect; stamens 10, filaments wing-margined by the basal scale adnate to the inner surface; ovary (4) 5-carpellary, the style and stigma 1; fruit a (4) 5-celled, usually more or less winged capsule; annual or perennial herbs with opposite, usually compound leaves and small interpetiolar stipules.

Nearly 100 species, mostly in subtropical Eurasia. (Greek *zygo*, yoke, and *phyllum*, leaf, referring to the bifoliolate leaves.)

Zygophyllum fabago L. Sp. Pl. 385. 1753. (Syria)

Glabrous, somewhat succulent annual; stems freely branched, decumbent to ascending, 2-5 dm. long; leaflets 2, obliquely oblong-obovate, 2-4 cm. long; stipules oblong, 1-3 cm. long, sometimes fused; peduncles about 1 cm. long; flowers axillary, about 1.5 cm. broad, the petals yellow, slightly longer than the sepals; stamens 10, their attached scales serrulate; capsule oblong, 2-3 cm. long. N=11.

Introduced at Ephrata, Wash., and at Minidoka, Ida., also occasional from Calif. to Colo.; native from w. Asia to Spain. June-Aug.

SIMAROUBACEAE Quassia Family

Flowers borne in various sorts of inflorescences, most commonly in a mixed panicle, perfect or (more often) imperfect, the plants then dioecious or subdioecious; calyx of 3-8 distinct or (commonly) partly connate sepals; petals as many as the sepals, distinct (occasionally lacking), seldom showy; stamens as many or (more often) twice as many as the sepals or petals; carpels usually 5, typically weakly united in flower (sometimes solely by the styles) and ripening into separate fruits, less often persistently united (but then with mostly free styles) or wholly distinct, borne on a short, broad gynophore; fruit drupaceous or samaroid, less often achenelike or a berry, with 1 or 2 ovules per locule; woody plants with alternate, pinnately compound or sometimes simple, exstipulate leaves and mostly very bitter bark.

Chiefly a tropical and subtropical family of wide distribution, with about 30 genera and nearly 200 species, some of which are cultivated in various parts of the U.S.

Reference:

Cronquist, A. Studies in the Simaroubaceae—IV. Resumé of the American genera. Brittonia 5:128-147. 1944.

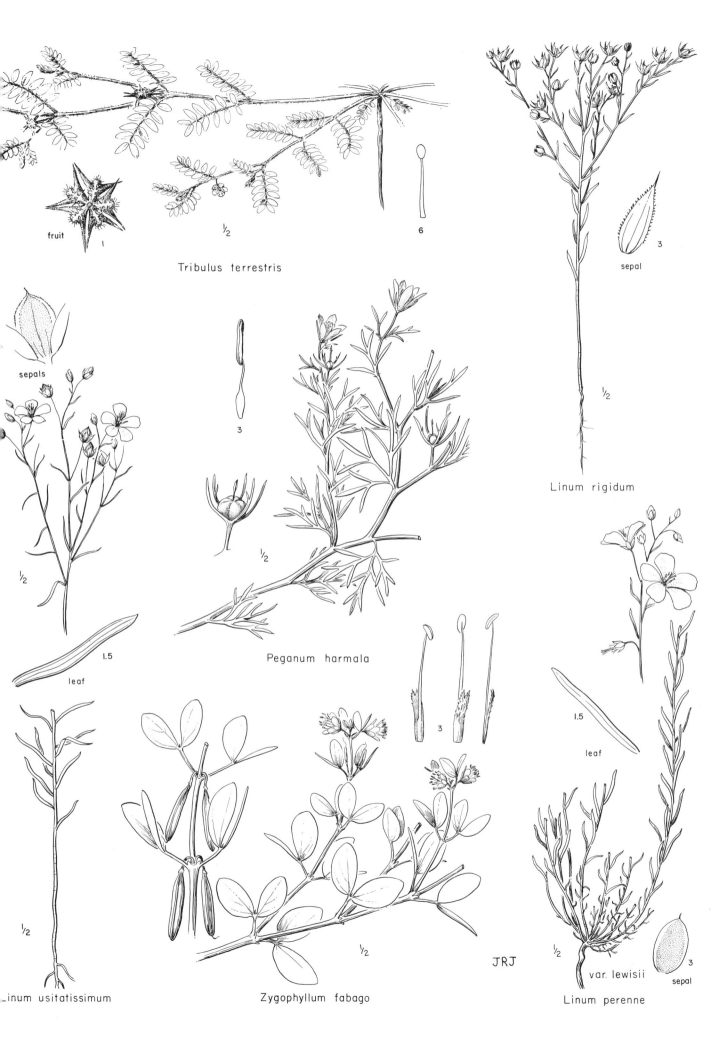

fruit

1

½

6

Tribulus terrestris

sepal

3

Linum rigidum

½

sepals

3

½

leaf

1.5

Peganum harmala

½

3

3

1.5

leaf

½

Linum usitatissimum

Zygophyllum fabago

½

JRJ

½

var. lewisii

sepal

3

Linum perenne

Ailanthus Desf. Nom. Conserv. Tree of Heaven

Flowers small, the numerous cymules arranged in a large open panicle; plants subdioecious, some trees staminate, others pistillate or with reduced but functional stamens; sepals 5 (6), persistent; petals 5 (6), deciduous; stamens mostly 10 (12); gynophore (disk) lobed; carpels 5 (6), mostly separate except for the short common style, ripening into distinct, 1-seeded samaras, often some of them abortive before maturity; stigmas distinct, divergent; trees with large, pinnately compound, deciduous leaves.

About 15 species, native to e. Asia, the East Indies, and n. e. Australia. (Supposedly an oriental name for the plant, meaning "tree of heaven.")

Ailanthus altissima (Mill.) Swingle, Journ. Wash. Acad. Sci. 6:495. 1916.
  Toxicodendron altissimum Mill. Gard. Dict. 8th ed. no. 10. 1768. (China)
  A. glandulosa Desf. Mém. Acad. Sci. Paris 1786:265. 1789. (China)
Rather large, rapidly growing, smooth-barked trees with very large, glabrous or puberulent leaves sometimes over 1 m. long; leaflets mostly 10-25 (the terminal one present or absent), lance-ovate, 4-15 cm. long, mostly with 1-3 coarse rounded teeth on each side near the base, each of these bearing a conspicuous gland on the lower side; inflorescences large, the flowers greenish, 6-8 mm. broad, ill-scented, the staminate and perfect with 10 stamens; samaras 2.5-5 cm. long, somewhat contorted or spirally twisted.

Often used as ornamental trees, especially along highways, and established in many parts of the U.S., as in the Snake R. Canyon, and along the Columbia R. near The Dalles; China. June-July.

EUPHORBIACEAE Spurge Family

Flowers imperfect, sometimes the staminate and pistillate borne separately and then one or both types with a calyx or (occasionally) with both a (5-merous) calyx and corolla and generally 10 stamens, but in most of ours the flowers greatly reduced, borne in cup-shaped perianthlike involucres, the staminate flowers usually several, mostly included in the involucre, each consisting of one stamen, the filament jointed with the pedicel, the pistillate flower single, terminal on the minute axis and usually exserted; pistil mostly 3 (1-4)-carpellary, the ovary superior, 3 (1)-celled; styles 3, partially or quite distinct, often deeply bifid; fruit capsular, usually elastically separating into three (very rarely two) 1-seeded segments, but sometimes 1-celled and 1-seeded; seeds with oily endosperm and commonly with a prominent caruncle; monoecious (mostly) to dioecious, sometimes fleshy herbs, shrubs, vines, or trees, usually with acrid, milky juice, the leaves commonly opposite (alternate); simple to compound, mostly stipulate, the stipules sometimes merely glandlike or spinelike.

A large family of nearly 300 genera and probably at least 7000 species, widely distributed but especially common in desert regions, best represented in the drier tropics. It is of considerable economic importance, yielding such products as rubber and castor oil, and including such popular ornamentals as the Poinsettia.

1 Plants sivery-hairy and somewhat hirsute-hispid; leaves ovate, 1-3 cm. broad   EREMOCARPUS
1 Plants not silvery-hairy; leaves variable   EUPHORBIA

Eremocarpus Benth. *

Flowers tiny, apetalous, the staminate borne in terminal clusters and consisting of 5-6 sepals and

---

*Piper, Contr. U.S. Nat. Herb. 11:382. 1906, proposed a substitute name for this genus, concerning which the following facts are cogent: Reichenbach, Handb. nat. Pflanzensystems 307. 1837, listed 2 genera, *Eremocarpus* Spach and *Drosanthe* Spach, under the group Drosantheae of the group Hypericeae of the family Hypericineae. Although the suprageneric groups are all characterized by Reichenbach, the genera are merely listed without description, a formal description is not given for *Eremocarpus,* and there is no way indicated by which *Eremocarpus* could be distinguished from *Drosanthe.* *Eremocarpus* Spach ex Reisenbach therefore was not validly published, whereas *Eremocarpus* Benth. Bot. Sulph. 53. 1844, was; consequently *Piscaria* Piper was superfluous and must be relegated to generic synonymy under *Eremocarpus* Benth.

5-9 stamens, the pistillate naked, mostly borne in the axils, with a single unbranched style and 1 stigma; fruit a 2-valved, 1-seeded capsule; monoecious, prostrate or spreading, grayish-stellate annual herbs with simple petiolate leaves and nonmilky juice.

One species in Pacific U.S. (Greek *eremos*, lonely, and *karpos*, fruit, in reference to the single carpel.)

Eremocarpus setigerus (Hook.) Benth. Bot. Sulph. 53, pl. 26. 1844.

Croton setigerum Hook. Fl. Bor. Am. 2:141. 1838. *Piscaria setigera* Piper, Contr. U.S. Nat.
  Herb. 11:382. 1906. *(Douglas, "*Menzies' Island, and on sandy banks of the Columbia upwards")
Stellate-hirsute and grayish-green, more or less dichotomously branched, matted, prostrate, musky-scented annual 2-10 cm. tall and 0.5-5 dm. broad; leaves forming rosettes at the branchlet ends and subtending the clusters of tiny flowers, the lower ones long-petiolate and mostly alternate, the blades 3-nerved, ovate or deltoid-ovate, 1-3 cm. long; staminate flowers in miniature racemes, 2-3 mm. broad; pistillate flowers either terminal on the racemes or more frequently axillary and solitary, naked except for 3-6 very tiny perianth vestiges; capsules 3-5 mm. long; seeds 3-4 mm. long, smooth, shining, grayish and mottled. N=10. Doveweed, turkey mullein.

From the Wash.-Oreg. border to Baja Calif., e. to Nev., chiefly under semidesert conditions, but on both sides of the Cascades in Oreg. and to the coast in much of Calif. June-Aug.

The seeds are a source of considerable forage to domesticated and wild fowls, hence the common names. The Indians reputedly used to mash the plant in large quantities and throw it into streams to stupefy the fish therein. There is conjecture as to why the plant was efficacious in this way, whether due to the clogging of the gills by the stellate hairs or to a narcotic effect.

Euphorbia L. Spurge

Flowers naked, borne in short, compact, involucrate clusters; involucres solitary in the axils or in terminal cymes, campanulate to obconic, perianthlike, usually bearing 4 glands alternate with inconspicuous toothlike lobes and sometimes also with entire and more or less crescent shaped or divergent-horned, often membranous- and whitish- or colored-bordered appendages, a fifth gland often apparently replaced by a small (often toothed) lobe usually concealed by the reflexed stalk of the pistil; staminate flowers numerous, included in the involucre, each represented by a single stamen, the filament jointed to the pedicel; pistillate flower single and terminal, often protruding from the involucre, 3-carpellary; fruit a capsule, usually becoming reflexed, separating into three 1-seeded segments; glabrous to sparsely hairy, annual or perennial herbs (ours) to shrubs, the leaves often stipulate, simple, the lower ones alternate or opposite, but the floral bracts opposite or whorled.

Chiefly tropical and subtropical, with probably well over 1000 species, many of which are reported to be poisonous. (Named in honor of Euphorbus, physician to King Juba II.)

Some workers have circumscribed the genus more narrowly than others, and several of our species will be found to have other generic names.

References:

Norton, J. B. S. North American species of Euphorbia, section Tithymalus. Ann. Rep. Mo. Bot.
  Gard. 11:85-144. 1900.
Wheeler, L. C. Euphorbia subgenus Chamaesyce in Canada and the United States exclusive of south-
  ern Florida. Rhodora 43:97-154, 168-205, 223-286. 1941.
In addition to the species keyed below, *E. marginata* Pursh (snow-on-the-mountain), an annual of c. U.S. (to e. Mont.) occasionally may be found in our area as an escape from gardens. It is grown because of the attractive foliage, the upper leaves and bracts being very broadly white-margined.

1 Plants mostly prostrate; leaves usually less than 1.5 (3) cm. long and less than 5 mm. broad,
    all opposite; glands of the involucre petaloid *(Chamaesyce, Anisophyllum)*
  2 Stems crisp-puberulent to pilose; leaves usually pilose on the lower surface, the larger
      ones sometimes over 15 mm. long
    3 Plants prostrate; capsules hairy; leaves 4-17 mm. long   E. SUPINA
    3 Plants erect or ascending; capsules glabrous; leaves 10-30 mm. long   E. MACULATA
  2 Stems glabrous; leaves usually glabrous, the larger ones seldom over 15 mm. long
    4 Seeds coarsely transcorrugated; leaves thick-margined, linear-oblong, entire to
        denticulate; involucres more or less turbinate   E. GLYPTOSPERMA
    4 Seeds smooth to wrinkled or pitted but not coarsely transcorrugated; leaves more

nearly obovate-oblong or ovate-oblong, the margins serrulate but not thickened; involucres approximately campanulate   E. SERPYLLIFOLIA
1 Plants erect; lower stem leaves (at least) often alternate, the larger ones mostly over
   1. 5 cm. long and more than 4 mm. broad; glands of the involucre usually not petaloid
   5 Leaves stipulate, all opposite, serrate; seeds less than 1. 5 mm. long, somewhat pris-
      matic, strongly mucilaginous when wet; caruncle absent or much reduced *(Lepadenia)*
                                                                           E. MACULATA
   5 Leaves not obviously stipulate, the lower ones alternate or if in part opposite then the
      seeds well over 1. 5 mm. long, not angled, and not mucilaginous when wet; caruncle
      often conspicuous *(Esula, Tithymalus)*
      6 Lower stem leaves opposite, oblong, 4-12 cm. long; seeds 4-6 mm. long   E. LATHYRUS
      6 Lower stem leaves alternate, usually less than 4 cm. long; seeds less than 4 mm.
         long
         7 Plants perennial; seeds smooth; blades of the lower leaves entire, linear to ob-
            long, usually at least 6 times as long as broad; rays of the umbel usually more
            than seven
            8 Lower leaves linear, seldom as much as 3 mm. broad, usually only 1-2 cm.
               long; plants usually not over 3 dm. tall   E. CYPARISSIAS
            8 Lower leaves mostly over 4 mm. broad, 3-6 cm. long; plants usually well over
               4 dm. tall   E. ESULA
         7 Plants annual, or if perennial then the lower leaves neither linear nor oblong and
            not over three times as long as broad; rays of the umbel often fewer than five
            9 Leaves crenulate or serrulate; seeds finely reticulate-pitted; glands of the
               involucre entire
               10 Capsules smooth; floral leaves usually more than 1 cm. broad   E. HELIOSCOPIA
               10 Capsules papillose-warty; floral leaves less than 1 cm. broad   E. SPATHULATA
            9 Leaves entire, or if (rarely) very minutely serrulate then the seeds more
               wrinkled than reticulate and the glands of the involucre horned
               11 Plants perennial; lower stem leaves sessile or nearly so   E. ROBUSTA
               11 Plants annual; lower stem leaves petiolate
                  12 Seeds with about 5 longitudinal rows of large deep pits; lower leaves
                     with slender (almost filiform) petioles, the blades ovate or rhombic-
                     ovate, entire   E. PEPLUS
                  12 Seeds finely and irregularly reticulate-pitted; lower leaves spatulate,
                     gradually narrowed to a petiolar base but not filiform-petiolate, often
                     finely serrulate   E. CRENULATA

Euphorbia crenulata Engelm. Bot. Mex. Bound. Surv. 192. 1859.
   E. leptocera var. *crenulata* Boiss. in DC. Prodr. 15^2:143. 1862. *Tithymalus crenulatus* Heller,
   Muhl. 1:55. 1904. *(Hartweg 1950,* Calif., is the first of 2 collections cited)
   E. leptocera Engelm. ex Torr. Pac. R. R. Rep. 4:135. 1857, nomen nudum. *Tithymalus leptocerus*
   Arthur, Torreya 22:30. 1922. *(Bigelow,* Grass Valley, Calif.)
   E. nortoniana A. Nels. Bot. Gaz. 47:437. 1909. *(Heller 6625,* San Francisco, Calif., Apr. 25, 1903,
   is the first collection cited)
Glabrous annual or biennial with several stems from the base, erect, 1. 5-4 dm. tall; leaves minutely serrulate, the lower ones alternate, petiolate, oblanceolate to spatulate, 1-3 cm. long; floral leaves
in 2's or 3's, sessile, very broadly ovate, the inflorescence mostly of axillary, 2- to 3 (5)-rayed
compound umbels; involucres 2-3 mm. long, fringed between the 4 greenish, horned, spreading
glands; seeds smooth, grayish, 1. 5-2 mm. long, irregularly rugose-pitted; caruncle prominent,
brownish.
   In the interior valleys and lower mountains; c. Oreg. (possibly entirely s. of our range) to Calif.,
and e. to Ariz. and Colo. May-June.
   Euphorbia crenulata is very similar to the eastern species, *E. commutata* Engelm. (in Gray, Man.
2nd ed. 389. 1856), which differs almost solely in having well-defined, rounded pits on the seeds.
Since the two taxa apparently do not overlap in range, there is no chance for their interbreeding and
they may consistently be maintained at the specific level. *E. commutata* ranges w. to Mont. but is not
believed to reach our area.

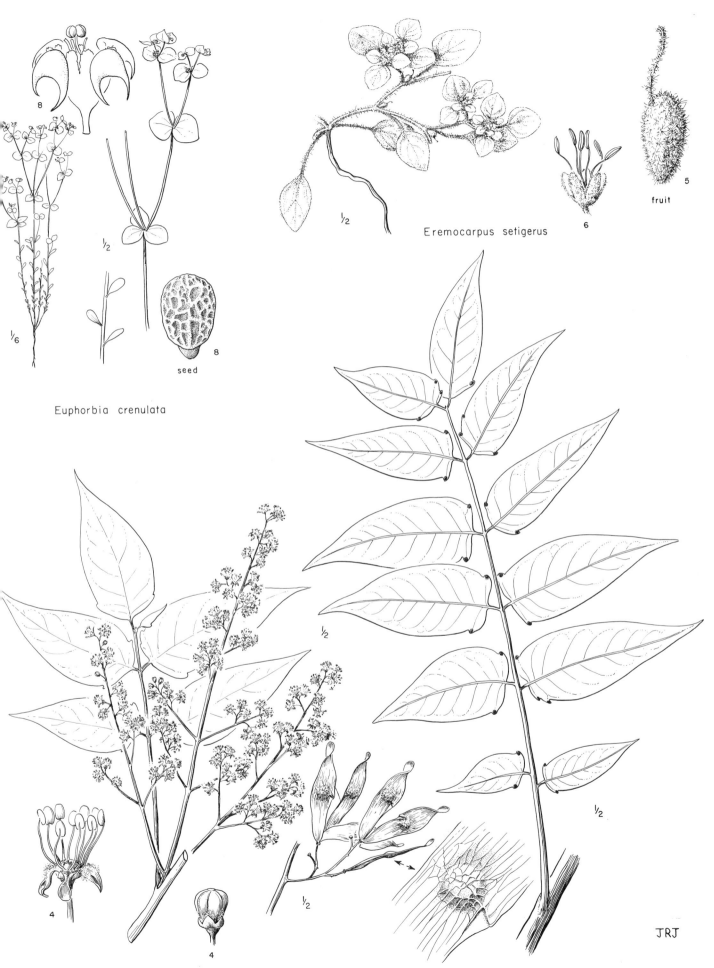

8

½

⅙

8
seed

Euphorbia crenulata

½

Eremocarpus setigerus

6

5
fruit

½

½

½

½

4

4

Ailanthus altissima

JRJ

Euphorbia cyparissias L. Sp. Pl. 461. 1753.
   Tithymalus cyparissias Hill, Hort. Kew. 172. 1768. *Esula cyparissias* Haw. Syn. Pl. Succ. 155.
      1812. *Keraselma cyparissias* Raf. Fl. Tellur. 4:116. 1836. *Galarrhoeus cyparissias* Small ex
      Rydb. Fl. Prair. & Pl. 520. 1932. ("Habitat in Misnia, Bohemia, Helvetia")
   Glabrous perennial 1.5-3 (4) dm. tall, the stems simple below but freely branched above; cauline
leaves alternate, the lower ones linear, 1-3 cm. long, 1-3 mm. broad, but those of the axillary upper branches more numerous and narrower; inflorescence a many-rayed umbel, the floral bracts
broadly ovate-cordate; involucres about 3 mm. long, the glands reddish-green, horned; capsules finely verrucose; seeds about 2 mm. long, pale yellow, smooth; caruncle rather smooth, flattened. N=10,
20.
   A Eurasian garden weed well established in n.e. U.S. and occasional in Ida., Wash., and Oreg.
May-Aug.

Euphorbia esula L. Sp. Pl. 461. 1753.
   Tithymalus esula Hill, Hort. Kew. 172. 1768. *Keraselma esula* Raf. Fl. Tellur. 4:116. 1836.
      Galarrhoeus esula Rydb. Brittonia 1:93. 1931. ("Habitat in Germania, Belgio, Gallia")
   E. virgata of Waldst. & Kit. Pl. Rar. Hung. 2:176, pl. 162. 1805, but not of Desf. in 1804.
   E. intercedens sensu Croizat, Am. Midl. Nat. 33:239. 1945.
   Perennial with heavy rootstocks, 2-9 dm. tall, glabrous and glaucous to sparsely pubescent above;
stems erect, simple below but freely and umbellately branched above; lowest leaves scalelike, the
main cauline leaves oblong to linear-oblanceolate, entire, 2-6 cm. long, 3-8 mm. broad, nearly or
quite sessile; floral leaves very broadly cordate-ovate, 12-16 mm. long; involucres 2-3 mm. long,
the glands 4 (5), brownish-green, with short, divergent horns; capsule inconspicuously warty to nearly
smooth; seeds mostly 1.5-2 mm. long, brownish, smooth; caruncle small. N=30, 32.
   A bad Eurasian weed now well established in many parts of the U.S.; in our area mostly on waste-
land in Mont. and Ida., but spreading in Wash. May-June.
   The correct specific name for this plant is open to clarification. Both *E. virgata* Waldst. & Kit. and
E. intercedens Podp., taxa closely related to (if not conspecific with) *E. esula*, have been claimed
more truly to encompass our plants.

Euphorbia glyptosperma Engelm. Bot. Mex. Bound. Surv. 187. 1859.
   Chamaesyce glyptosperma Small, Fl. S.E. U.S. 712, 1333. 1903. *(Engelmann*, Fort Kearney, Neb.)
   E. greenei Millsp. Pitt. 2:88. 1890. *Chamaesyce greenei* Rydb. Fl. Rocky Mts. 544. 1917. *(Greene,*
      Beaver Canyon, Ida., in 1889)
   Chamaesyce glyptosperma var. *integrata* Lunell, Am. Midl. Nat. 3:142. 1913. *Lunell,* Leeds,
      N.D., Aug. 20. 1906)
   Glabrous annual with prostrate, freely branched stems 0.5-4 dm. long; leaves all opposite, obliquely
lanceolate to oblong, 5-15 mm. long, crenulate-serrulate to entire; stipules linear, often laciniate,
about 1 mm. long; involucres axillary, more or less turbinate, 1 mm. long; glands 5, four of them
pinkish, depressed in the center, and with whitish-membranous appendages, the fifth represented by
a short fringed lobe; capsules smooth, about 1.5 mm. long; seeds about 1.2 mm. long, grayish, more
or less prismatic and coarsely wrinkled-corrugated transversely, very mucilaginous when wet; caruncle minute.
   On the e. side of the Cascades from B.C. to n. Calif., e. through the central states to N.B., Me.,
and N.Y., s. to Tex.; mostly on dry and rather sandy soil, from the plains to the lower mountains.
June-Sept.

Euphorbia helioscopia L. Sp. Pl. 459. 1753.
   Tithymalus helioscopius Hill, Hort. Kew. 172. 1768. *Galarhoeus helioscopius* Haw. Syn. Pl. Succ.
      152. 1812. ("Habitat in Europae cultis")
   Annual, 2-5 dm. tall, glabrous or sparsely pilose below to rather densely pilose above, the stems
somewhat fleshy, simple or branched below but umbellately branched above; lower cauline leaves
fleshy, alternate, oblanceolate to spatulate, narrowed gradually to the base, 1.5-3 cm. long, finely
serrate-dentate; floral leaves opposite to whorled, finely toothed; involucres about 2.5 mm. long;
glands 4, greenish-yellow, not horned; capsules smooth; seeds ovoid, about 2 mm. long, yellowish-
brown to brownish-black, finely reticulate-pitted, the ridges narrow, prominent; caruncle flattened.
N=21.

A fleshy, weedy annual found mostly in cultivated areas or recently abandoned gardens, chiefly on the w. side of the Cascades; more common in c. U.S. and Can.; Eurasia. Apr. -July.

Euphorbia lathyrus L. Sp. Pl. 475. 1753.
Tithymalus lathyris Hill, Hort. Kew. 172. 1768. *Galarhoeus lathyris* Haw. Syn. Pl. Succ. 143. 1812.
Epurga lathyris Fourr. Ann. Soc. Linn. Lyon 7:150. 1869. ("Habitat in Gallia, Italia ad agrorum margines")
Glabrous and glaucous, erect, rather fleshy, more or less dichotomously branched annual 3-10 dm. tall; leaves all opposite, the cauline ones from closely decussate near the stem base and narrowly ob-long, 6-12 cm. long, 7-15 mm. broad, and sessile, to broadened in shape upward on the stem, the upper ones distinctly ovate and subcordate; involucres about 3 mm. long, purplish, the glands with short rounded horns; capsules globose, coarsely wrinkled, 5-10 mm. long; seeds 4-6 mm. long, shal-lowly rugose-reticulate, brownish; caruncle conspicuous but rather flattened. N=10. Caper spurge.
Occasionally established in various parts of the U.S., especially e. of the Rocky Mts.; Eurasia. Apr. -May.
The plant is said to be moderately poisonous and it is sometimes (probably uselessly) planted as a mole and gopher repellent. In our area it appears to be no more than a temporary escape.

Euphorbia maculata L. Sp. Pl. 455. 1753.
Tithymalus maculatus Moench, Meth. 666. 1794. *Chamaesyce maculata* Small, Fl. S.E. U.S. 713, 1333. 1903. ("Habitat in America septentrionali")
E. nutans Lag. Gen. & Sp. Pl. 17. 1816. *Chamaesyce nutans* Small, Fl. S.E. U.S. 712, 1333. 1903. ("Nova Hispania")
E. preslii Guss. Prodr. Sic. 1:539. 1829. *Chamaesyce preslii* Arthur, Torreya 11:260. 1911. (North America)
An ascending to erect annual 1-5 dm. tall, the stems freely branched, tomentose to pilose below and glabrous or somewhat pilose above; leaves all opposite, very short-petiolate, oblong or oblong-lanceolate, more or less oblique, 1-3 cm. long, 4-9 mm. broad, sparsely to moderately long-pilose; stipules triangular, scarcely 1 mm. long; involucres obconic, about 1.5 mm. long; glands 4, the ap-pendages reniform, entire, white to pink, mostly 0.2-0.4 mm. long; capsules about 2 mm. long; seeds about 1.3 mm. long, golden-brown, more or less flattened and shallowly pitted-corrugated be-tween the 3 or 4 rounded longitudinal ridges, conspicuously mucilaginous when wet; caruncle lacking. N=14.
North Dakota to Que., s. to Fla. and Mex.; introduced in w. U.S. and known in our area from Skamania Co., Wash. *(Suksdorf 12327)*. June-Sept.

Euphorbia peplus L. Sp. Pl. 456. 1753.
Tithymalus peplus Hill, Hort. Kew. 172. 1768. *Esula peplus* Haw. Syn. Pl. Succ. 158. 1812.
Keraselma peplus Raf. Fl. Tellur. 4:116. 1836. *Galarrhoeus peplus* Rydb. Fl. Prair. & Pl. 520. 1932. ("Habitat in Europae cultis oleraceis")
Glabrous annual 1-3 (5) dm. tall, the stems usually freely branched and erect; lower leaves alter-nate, the blades 1-3 cm. long, rhombic-ovate to rhombic-obovate, entire, narrowed abruptly to very slender petioles 3-10 mm. long; floral leaves opposite, short-petiolate, broadly ovate, 10-25 mm. long; involucres 1-1.5 mm. long; glands 4, bearing long slender horns; capsules smooth, about 2.5 mm. long; seeds gray, 1-1.5 mm. long, oblong, conspicuously marked with 1-4 pits in each of 4-6 longitudinal rows; caruncle white, delicate, not very prominent. N=8.
A European weed now commonly introduced in much of N.Am., in our area found chiefly on waste-land w. of the Cascades from Wash. to Calif. May-Nov.

Euphorbia robusta (Engelm.) Small ex Britt. & Brown, Ill. Fl. 2:381. 1897.
E. montana var. *robusta* Engelm. Bot. Mex. Bound. Surv. 192. 1859. *(James, on the Upper Platte)*
Glabrous or puberulent perennial with a deep taproot and branched crown and with erect stems 1-3 dm. tall; leaves thick and rather fleshy, entire, the lower ones alternate, nearly sessile, ovate to ob-long, (0.5) 1-2 cm. long, the floral leaves about as long, cordate-ovate, in 2's or 3's; involucres about 3 mm. long, turbinate, hairy within and often toothed between the 4 reddish, short-horned glands; capsules depressed-globose, smooth, about 4 mm. long; seeds slightly over 2 mm. long, gray, finely reticulate-pitted; caruncle prominent, conical.

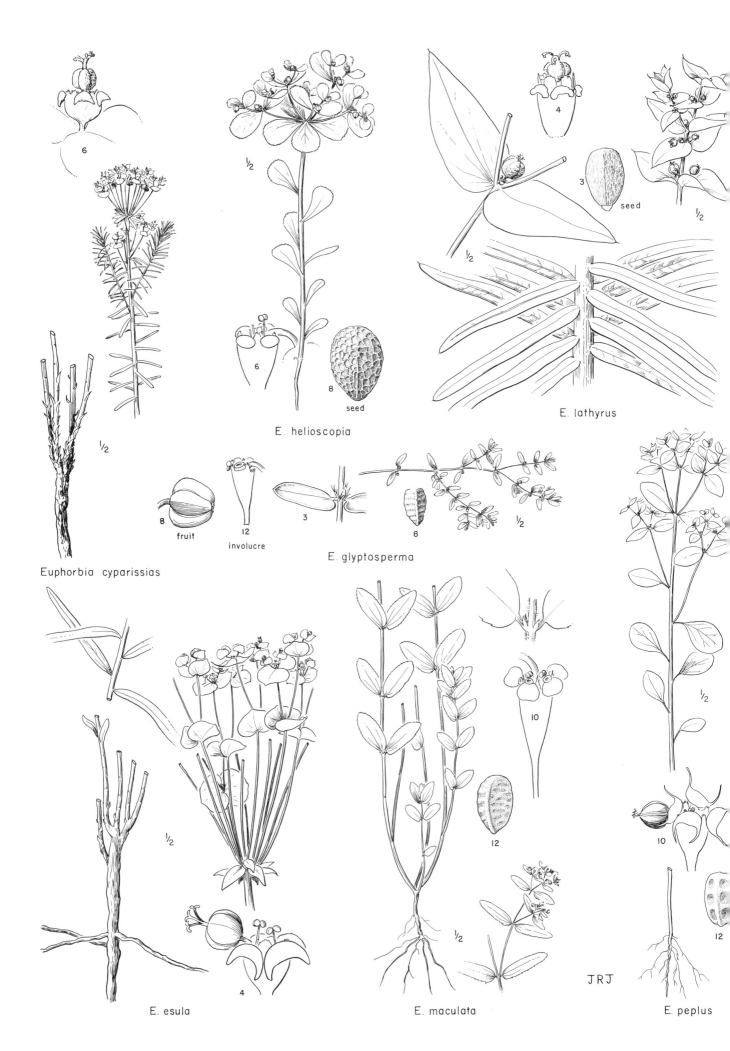

6

½

8
seed

E. helioscopia

½

3
seed

4

½

E. lathyrus

8
fruit

12
involucre

3

8

½

E. glyptosperma

½

Euphorbia cyparissias

½

½

10

12

½

10

12

½

4

E. esula

½

E. maculata

JRJ

E. peplus

In the foothills and lower mountains, mostly e. of the continental divide, from Mont. to S.D. and s. to N.M. and Ariz.; reported from Missoula, Mont. June-Aug.

Euphorbia serpyllifolia Pers. Syn. 2:14. 1807.
  Chamaesyce serpyllifolia Small, Fl. S.E. U.S. 712, 1333. 1903. ("Habitat in America calidiore")
  E. serpyllifolia var. *rugulosa* Engelm. ex Millsp. Pitt. 2:85, pl. 1, fig. 19. 1890. *E. rugulosa*
    Greene, Fl. Fran. 92. 1891. *Chamaesyce rugulosa* Rydb. Bull. Torrey Club 33:145. 1906. *(Parish
    & Parish,* San Bernardino Valley, Calif.)
  E. albicaulis Rydb. Mem. N.Y. Bot. Gard. 1:266. 1900. *Chamaesyce albicaulis* Rydb. Bull. Torrey
    Club 33:145. 1906. *(Rydberg 356,* Cheyenne Co., Neb., in 1891)
  Chamaesyce aequata Lunell, Am. Midl. Nat. 1:204. 1910. *(Lunell,* Leeds, N.D.)
  Chamaesyce aequata var. *claudicans* Lunell, Am. Midl. Nat. 1:205. 1910. *(Lunell,* Leeds, N.D.)
  Chamaesyce erecta Lunell, Am. Midl. Nat. 1:206. 1910. *(Lunell,* Leeds, N.D.)
  Glabrous annual with freely branched, usually prostrate stems 0.5-3 dm. long; leaves 5-15 mm. long, obliquely oblong-obovate or oblanceolate to more nearly oblong-ovate, serrulate near the tips; stipules lanceolate, 0.5-1.5 mm. long, usually lacerate; involucres about 1 mm. long, more or less campanulate, the 4 glands with sunken centers and with lobed or crenulate, whitish appendages; capsules smooth, 1.5-2 mm. long; seeds about 1.2 mm. long, grayish-brown, mucilaginous when wet, sharply prismatic, the sides sunken and pitted to transversely wrinkled or corrugated but the corrugations not extending across the angles; caruncle minute.
  On dry ground from the plains to the lower mountains e. of the Cascades, from B.C. southward, through Wash., and further w. in Oreg. and Calif., to Baja Calif., e. to Alta., Minn., Tex., and N.M. May-Sept.

Euphorbia spathulata Lam. Encyc. Meth. 2:428. 1788. (Uruguay)
  E. dictyosperma F. & M. Ind. Sem. Hort. Petrop. 2:37. 1835. *Tithymalus dictyospermus* Heller,
    Muhl. 1:56. 1904. (North America)
  E. arkansana var. *missouriensis* Nort. Rep. Mo. Bot. Gard. 11:103. 1900, at least as to our material.
  Glabrous annual 1-3 dm. tall, the stems erect and simple or branched below; leaves alternate, the lower ones 1-3 cm. long, obovate, spatulate, or narrowly oblanceolate, narrowed to a short basal petiole, crenulate-serrulate, rounded to acute; floral bracts ovate, 10-15 mm. long and about 2/3 as broad, crenulate to serrulate; involucres campanulate, 1-2 mm. long, the glands entire, yellow-green; capsules about 2.5 mm. long, prominently papillose-verrucose; seeds about 1.5 mm. long, ovoid, brownish-black, finely but distinctly reticulate-pitted.
  On dry hills and lower mountains e. of the Cascades, from Wash. to Calif., e. to Mont. May-July.

Euphorbia supina Raf. Am. Mag. 2:119. 1817.
  Chamaesyce supina Moldenke, Annot. Class. List Moldenke Coll. 135. 1939. (Long Island, New
    Jersey, Sand Hook, etc.)
  E. maculata sensu Am. auth. but not of L.
  An essentially prostrate annual with sparsely to densely strigose-hirsute or pilose stems 1-4 dm. long; leaves all opposite, obliquely oblong, 5-15 mm. long, about 1/3 as broad, frequently with an elongate central wine-colored spot, somewhat remotely serrulate, slightly to copiously hairy (glabrate) beneath, glabrous above; stipules linear, about 0.5 mm. long; involucres obconic, barely 1 mm. long, strigillose; glands 4, yellow, with conspicuous, reniform, white to deep pink, membranous appendages; capsules hairy, about 1.5 mm. long; seeds about 1 mm. long, pinkish, not reticulate but very lightly transcorrugate on the flattened surfaces between the 4 or 5 longitudinal ridges.
  Native in e. U.S. from Que. to Fla. and s. to the Dakotas and Tex.; introduced on the w. coast in Calif. and Oreg., but in our range known at present only from Walla Walla Co., Wash. June-Oct.

CALLITRICHACEAE Water Starwort Family

Flowers inconspicuous, borne singly or in 2's or 3's in the leaf axils, naked or subtended by 2 more or less inflated hyaline bracts, usually imperfect (and the plants polygamo-monoecious), each consisting of a single pistil or of 1 (rarely more) stamen, or of both; pistil 2-carpellary, the ovary deeply 4-lobed, separating at maturity into 4 one-seeded achenelike fruits; styles 2, distinct, slender; small aquatic herbs, either more or less submerged or emergent and rooting in the mud, the leaves exstipulate, either opposite (rarely whorled) and linear and 1-nerved or borne in tufts at the ends of the branches and then often broadened and 3-nerved.

Only the following genus.

Callitriche L.

Inconspicuous aquatics, widely distributed, with 24 species in the New World. (From the Greek *callos*, beautiful, and *trichos*, hair, in allusion to the slender stems.)

Reference:

Fassett, N. C. Callitriche in the New World. Rhodora 53:137-155, 161-182, 185-194, 209-222. 1951.

1 Pistillate flowers borne on distinct pedicels that are considerably longer than the fruits; styles sharply reflexed   C. MARGINATA

1 Pistillate flowers nearly or quite sessile, the pedicels (if any) less than 1/4 the length of the fruits

  2 Fruit encircled by a conspicuous winglike margin; leaf bases joined by tiny winged ridges   C. STAGNALIS

  2 Fruit not at all winged or with only a narrow wing toward the tip; leaf bases sometimes not joined by winged ridges

    3 Leaves all linear, 1-nerved, light green, their bases not joined by a ridge or a wing; floral bracts absent   C. HERMAPHRODITICA

    3 Leaves varied, often the upper ones broadened and several-nerved, their bases joined by a small winglike ridge; floral bracts usually present (the following three species are especially difficult to distinguish)

      4 Tiny pitlike markings of the carpel faces in rather regular vertical lines; fruit usually slightly wing-margined at the top, 1/5-1/3 longer than broad   C. VERNA

      4 Tiny pitlike markings of the carpel faces irregularly distributed rather than in definite rows; fruits scarcely at all winged, usually about as broad as long

        5 Fruits usually widest above the middle and therefore more or less obovate in outline; linear (generally submerged) leaves bidentate but with the midvein barely perceptibly thickened at the end; emergent leaves often well over 5 mm. broad; plants long-stemmed   C. HETEROPHYLLA

        5 Fruits usually more nearly round or oblong (rather than obovate) in outline; linear (submerged) leaves with the midnerve conspicuously thickened and slightly protuberant at the tip; none of the leaves over 5 mm. broad; plants with short and slender stems   C. ANCEPS

Callitriche anceps Fern. Rhodora 10:51. 1908. *(Fernald & Collins 234,* Table Top Mt., Gaspé Co., Que., Aug. 1, 1906)

Stems slender, usually less than 5 cm. long; leaves crowded, less than 10 mm. long, mostly linear, those that are emergent more oblanceolate; flowers ebracteate or very inconspicuously bracteate; fruits round or more or less oblong in outline, less than 1 mm. long, not winged, the tiny pits of the faces not in definite rows; styles usually early-deciduous; in general very similar to *C. heterophylla* but commonly much smaller.

Usually submerged, although rarely emergent as the water level lowers but even then not strongly dimorphic; fairly common in e. U.S. and Can.; on the Pacific coast known from Alas., Wash., and Utah. Mostly Aug.-Sept.

Callitriche hermaphroditica L. Cent. Pl. 1:31. (Feb.) 1755. (Type not known)

  C. palustris var. *bifida* L. Sp. Pl. 969. 1753. *C. bifida* Morong, Mem. Torrey Club 5:215. 1894. (Europe)

  C. autumnalis L. Fl. Suec. 2nd ed. 4. (Oct.) 1755. (Apparently a renaming of *C. hermaphroditica*)

  C. autumnalis var. *bicarpellaris* Fenley ex Jeps. Fl. Calif. 2:436. 1936. *(Mason 4445,* Clements, San Joaquin Valley, Calif.)

Stems slender, 5-15 cm. long, with rather widely spaced internodes; leaves all linear, 5-20 mm. long, 1-nerved, delicate, bright lustrous-green, inconspicuously white-margined, the tips bifid, the nerves slightly thickened at the end but not protruding, the bases usually broadened and more or less clasping but not forming either ridges or wings on the stem between their bases; bracts wanting; styles

1-2 mm. long, soon deciduous; carpels 1-2 mm. long, oval or rectangular-rounded in outline, the margins narrowly winged above, the faces irregularly and shallowly pitted. N=3.

In sloughs and streams, growing emersed; Greenl., Alas., and probably across n. N. Am., southward to c. Calif., n. N. M., the Great Lakes, and n. e. U. S.; Europe. July-Sept.

Callitriche heterophylla Pursh ex Darby, Bot. South. St. 311. 1841.
   C. austini Engelm. ex Gray, Man. 5th ed. 428. 1867. *C. deflexa* var. *austini* Hegelm. Verh. Bot. Ver. Brandenb. 9:15. 1867. *(Austin,* New Jersey)
   C. bolanderi Hegelm. Verh. Bot. Ver. Brandenb. 10:116. 1869. *C. palustris* var. *bolanderi* Jeps. Fl. Calif. 2:435. 1936. *C. heterophylla* var. *bolanderi* Fassett, Rhodora 53:177. 1951. *(Bolander 4528,* Auburn, Placer Co., Calif., Apr. 11, 1865)

Stems 0.5-4 dm. long, emersed to emergent and terrestrial, with small membranous ridges between the leaf bases, the emersed leaves linear, 1-nerved, 0.5-2.5 cm. long, the veins only slightly thickened and not projecting into the small apical notch, the floating leaves and those of emergent plants mostly broadly obovate, as much as 1 cm. wide, 3-nerved; floral bracts present; styles 1-5 mm. long, often persistent; carpels squarish-obovate in outline, 0.7-1.2 mm. long, about as broad at the widest point somewhat above the middle, wingless, irregularly pitted, the pits very small and shallow.

Greenland and N. and S. Am. Apr. -July (Sept.).

The species is represented in our area by two phases; the var. *heterophylla,* which has fruits mostly less than 0.9 mm. long, is abundant in the e. half of the U. S. but occasional from Wash. to Argentina; whereas the (scarcely significant) var. *bolanderi* (Hegelm.) Fassett, which has fruits mostly over 0.9 mm. long, is known from Greenl., but is found chiefly in Pacific coastal U. S. The distinction between the two varieties appears to be rather artificial.

Callitriche marginata Torr. Pac. R. R. Rep. 4:135. 1857. *(Bigelow,* Calif.)
Stems usually matted, very slender, 5-10 cm. long; leaves linear to linear-spatulate, 5-12 mm. long, the bases joined by a tiny winged ridge, those that are submerged narrowly 1-nerved, very slightly notched, the emergent leaves broader, incompletely 3-nerved; flowers bractless or with tiny, linear, brownish bracts, the staminate sessile, the pistillate on somewhat fleshy stipes or pedicels usually several times the length of the mature fruits; carpels more or less oval in outline, usually about 1 mm. long and slightly broader, conspicuously wing-margined, irregularly and finely reticulate-pitted; styles 1-2 mm. long, usually persistent.

Vernal pools in w. Calif.; reported from near The Dalles and at Grants Pass, Oreg. Mar. -May.

Callitriche stagnalis Scop. Fl. Carn. 2nd ed. 251. 1772. (Europe)
Stems 3-15 cm. long, aquatic to terrestrial; leaf bases joined by small winged ridges, the blades from linear and 1-nerved to broadly obovate-spatulate, 3- to 5-nerved, and nearly 1 cm. broad; floral bracts conspicuous, white; carpels from round to oblong-oval in outline, 1-1.5 mm. long, not quite so broad, conspicuously wing-margined all around, the pitting minute, not in rows; styles 1-3 mm. long, reflexed to erect, usually not persistent on the mature fruits. N=5, 10.

Chiefly European, where a great many forms have been described; in N. Am. mostly near the Atlantic seaboard but occasional in Wash., Oreg., and B. C. May-Aug.

Callitriche verna L. Fl. Suec. 2nd ed. 2:3. 1755.
   C. palustris var. *verna* Fenley ex Jeps. Fl. Calif. 2:435. 1936. (Europe)
   C. palustris L. sensu Morong, Mem. Torrey Club 5:215. 1894.

Stems slender, 5-20 cm. long, submerged, partially floating, or stranded on mud; leaf bases joined by small winged ridges, the submerged leaves linear, 5-20 mm. long, thin, 1-nerved, very narrowly margined, retuse, the vein thickened but not protruding, the floating and emergent leaves broadly spatulate, as much as 4 mm. broad, 3-nerved; floral bracts present; carpels oblong-obovate in outline, mostly about 1 mm. long, about 2/3 as broad, the tiny pitlike markings tending to lie in vertical lines; styles 1-2.5 mm. long, usually deciduous from the maturing fruits. N=10.

General in N. Am. except Mex. and s. and s. e. U. S.; Eurasia. June-Aug.

EMPETRACEAE Crowberry Family

Flowers axillary (ours) or in small, compact, terminal heads, regular, hypogynous, apetalous, perfect to imperfect and the plants monoecious to dioecious; perianth segments 2-6, the inner often

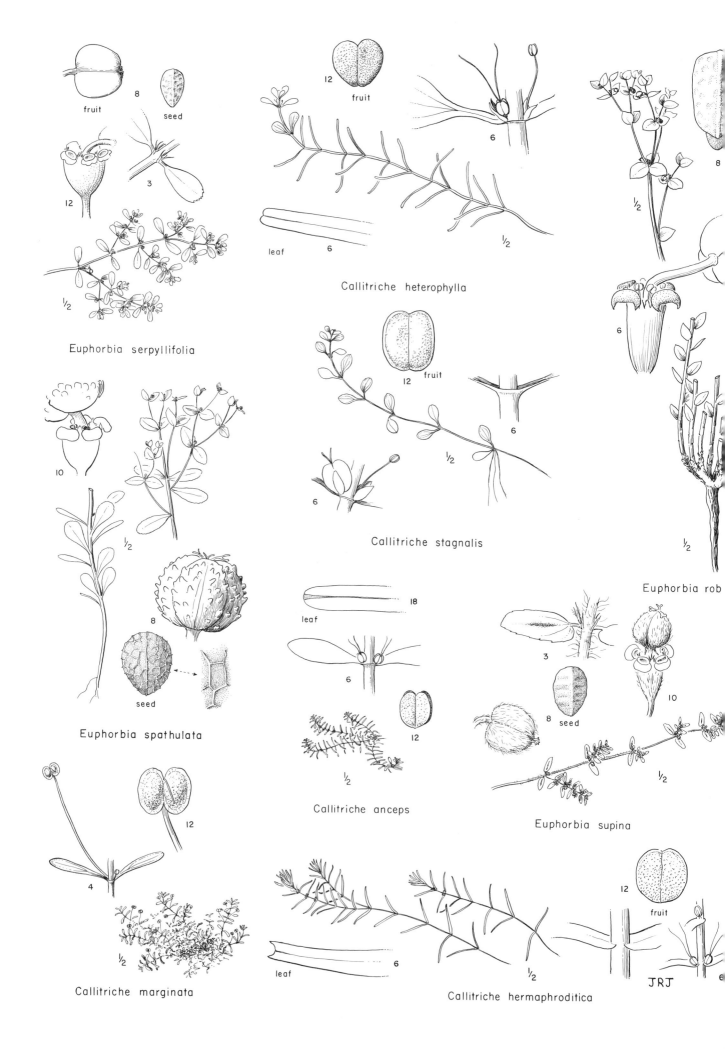

fruit   8
seed

12   3

½

Euphorbia serpyllifolia

10

½

8
seed

Euphorbia spathulata

12

4

½

Callitriche marginata

12
fruit

6
leaf   6

½

Callitriche heterophylla

12
fruit

6

6

½

Callitriche stagnalis

18
leaf

6

12

½

Callitriche anceps

3

8   seed

½

Euphorbia supina

½

8

6

½

Euphorbia rob

10

12
fruit

leaf   6

½

Callitriche hermaphroditica

JRJ   6

petaloid; stamens 2-4; pistil 1, the ovary superior, 2- to 9-carpellary and -loculed, each locule 1-seeded; style 1; stigmas more or less discoid with as many lobes as carpels; fruit baccate, fleshy or dry, with 2-9 hardened stonelike "seeds" (pyrenes); low evergreen shrubs with alternate, or whorled, entire, linear, heathlike leaves, the blades jointed with the petiole.

Three genera and 8 species of the cooler mountains and frigid zones of N. and S. Am. and Eurasia.

Empetrum L. Crowberry

Flowers inconspicuous, purplish, the perianth apparently of 9 segments, but 3 of these usually considered to be subtending bracts, the others sepals, although the inner 3 (which are more brightly colored) are perhaps as properly regarded as petals; fruit globular, drupelike, each of the 6-9 carpels separating when ripe into 1-seeded pyrenes; heatherlike, procumbent or depressed shrubs.

Two species, one in N.Am. and Eurasia, the other in S.Am. (The ancient Greek name, from *en*, upon, and *petros*, rock.)

Our species is a useful, low-growing, heatherlike shrub, perhaps more desirable to the collector than to the landscaper. It can be grown readily from cuttings and should not be dug in the wild.

Empetrum nigrum L. Sp. Pl. 1022. 1753. (Europe)

Low, spreading, heatherlike shrub up to 1.5 dm. tall; branches more or less lanate, as much as 3 dm. in length; leaves subterete, usually in part alternate, in part in whorls of 4, glandular-puberulent, 4-8 mm. long, revolute-margined, the lower surface grooved; flowers about 3 mm. long, immediately subtended by about 3 chaffy bracts not so large as, but otherwise similar to, the sepals; inner 3 sepals brownish-purple; perfect as well as staminate flowers generally with 3 stamens, the staminate usually with a rudimentary ovary; fruit purplish or black, globular, about 4-5 mm. long; stigmas peltate, with 6-9 short lobes. N=13, 26.

More or less circumpolar, extending southward, in the Cascade Mts. and along the coast, to Del Norte Co., Calif., mostly on exposed rocky bluffs but also in peat bogs. May-July.

Löve & Löve (Can. Journ. Gen. & Cytol. 1:34-38. 1959) reported N.Am. plants to be tetraploid (N=26), the dioecious European counterpart to be diploid. Doris Löve (Rhodora 62:265-292. 1960) used the name *E. eamesii* Fern. & Wieg. for most of the N.Am. material, and has called the black-fruited phase (which occurs in our area) *E. eamesii* ssp. *hermaphroditum* (Hagerup) D. Löve.

LIMNANTHACEAE Meadow-foam Family

Flowers complete, regular, borne singly on (often very long) axillary peduncles; perianth 3- to 5 (6)-merous; sepals free of the ovary, distinct except at the base; petals distinct, white or yellowish (ours) to pinkish-tinged, withering persistent; stamens equal in number to the petals or twice as many; pistil 2- to 5 (6)-carpellary, the ovary deeply divided into 2-5 globular segments which mature into 1-seeded nutletlike fruits; styles as many as the segments of the ovary, free except at the base; endosperm lacking, the embryo straight; annual, usually glabrous, juicy herbs with alternate, pinnate to pinnatifid, exstipulate leaves, generally to be found in moist places.

Two genera and perhaps 10 species of N.Am., nearly all in the Pacific states, mostly in Calif.

1 Perianth 4- to 5 (6)-merous; petals (3) 4-12 mm. long   LIMNANTHES
1 Perianth usually 3-merous; petals scarcely 2 mm. long   FLOERKEA

Floerkea Willd.

Flowers minute; perianth (2) 3 (4)-merous, the petals shorter than the sepals; stamens 3-6; carpels usually 2 or 3; fruits papillose-warty; small, succulent, annual herb with pinnate leaves.

One species of N.Am. (Named for the German botanist, Heinrich Gustav Floerke, 1764-1835.)

Floerkea proserpinacoides Willd. Neue Schrift. Ges. Nat. 3:449. 1801. (Pennsylvania)

F. occidentalis Rydb. Mem. N.Y. Bot. Gard. 1:268. 1900. (*Tweedy 525*, Swan Lake, Yellowstone Park, is the first of 3 collections cited)

Glabrous, weak, decumbent to erect, fleshy herb 2-10 cm. tall; petioles 1-4 cm. long; leaflets oval to elliptic or narrowly oblanceolate, 3-20 mm. long; sepals about 3 mm. long, free except at the base; petals white, spatulate, about half the length of the sepals, sparsely pilose at the base; stamens 3 (4), alternate with the petals and each adnate to a scalelike gland, or twice as many and those opposite the petals glandless; carpels usually 2, less commonly 3, free except for the common basal union of the styles; carpels separating at maturity, obovoid-globose, about 2.5 mm. long, indehiscent, the upper portion acutely to obtusely wrinkled, verrucose. False mermaid.

In wet places, especially under shrubs; rather general from B. C. to Calif. and through the Rocky Mt. states, eastward to the Atlantic coast, s. to Ga. Apr.-July.

The plant shows considerable variation in height, leaflet size, and length of the tubercles on the fruit, but our western material is essentially identical with the eastern, except that it is often (but by no means constantly) smaller, a condition which appears to be ecological rather than genetic.

Limnanthes R. Br. Nom. Conserv. Meadow-foam

Flowers solitary in the axils, the perianth 4- to 5 (6)-merous; petals white or yellowish; stamens 8 or 10; ovary of 5 (4) segments; low, glabrous, juicy annuals with alternate, pinnately dissected leaves.

Eight or 9 species of the Pacific coast. (Greek *limne*, marsh, and *anthos*, flower, referring to the habitat.)
Reference:
Mason, Charles T. A systematic study of the genus Limnanthes R. Br. U. Calif. Pub. Bot. 25: 455-512. 1952.

Nearly all the species of *Limnanthes* occur to the south of our range, although two were recorded for B. C. by Henry (Fl. So. B. C. 199. 1915). One of these was *L. douglasii* R. Br. in Lond. & Edinb. Phil. Mag. III, 2:70. 1833 (Type: *Douglas,* California), which was reported from Ucluelet, Vancouver I., (N=5). The other was *L. macounii* Trel. Mem. Bost. Soc. Nat. Hist. 4:85. 1887 (Type: *Macoun,* Vancouver I.). The former, which has showy petals 8-15 mm. long, is not otherwise known to occur n. of Douglas Co., Oreg., and if correctly identified, could well represent an escape, since the meadow-foams are not uncommonly cultivated. *L. macounii* is less easily accounted for. Characterized chiefly by the small, 4-merous flowers (the petals scarcely 5 mm. long), it has been collected only at the type locality, and not recently, suggesting the possibility that it was a very local species which no longer survives.

ANACARDIACEAE Sumac Family

Flowers usually in large panicles, mostly rather small, regular, perfect to imperfect and then the plants dioecious to polygamo-dioecious; calyx 5 (3-7)-parted; petals equal in number to the sepals, or lacking; stamens mostly equal to the petals or twice as many, but sometimes either more or less numerous than this, inserted at the outer edge of a fleshy, lobed, perigynous disc, usually in 2 series; pistil generally 3-carpellary, the ovary superior, 1-celled, 1-seeded; style mostly single, the stigmas 3; fruit usually drupaceous; shrubs (ours) or trees with alternate, pinnate (trifoliolate), exstipulate leaves and acrid or milky juice.

About 70 genera and over 500 species, widely distributed in tropic and temperate regions.

Rhus L. Sumac

Flowers in axillary or terminal panicles or thyrses, inconspicuous, regular, mostly imperfect and the plants polygamo-dioecious; fertile stamens 5; pistillate flowers with 5-10 staminal vestiges, the stamens separated from the ovary by a flat, lobed, fleshy disc; ovary 1-celled, style short, 3-parted just below the stigmas; fruit a berrylike drupe; shrubs or woody vines with pinnately 3- to many-foliolate leaves.

About 120 species, widely distributed outside of the Arctic and Antarctic; especially common in s. Africa. (The classical name for the plants.)
Reference:
Barkley, Fred A. A monographic study of the genus Rhus and its allies in North and Central America including the West Indies. Ann. Mo. Bot. Gard. 24:265-498. 1937.

1 Fruit abundantly reddish-hairy; petals pilose on the inner surface
  2 Leaflets 7-29 *(Sumac)*   R. GLABRA
  2 Leaflets 3-5 *(Schmaltzia)*   R. TRILOBATA
1 Fruit nearly or quite glabrous, not red; petals glabrous *(Toxicodendron)*
  3 Leaflets acute or acuminate; fruit about 4 mm. long; e. of the Cascade Mts.   R. RADICANS
  3 Leaflets rounded, obtuse, or very abruptly acutish; fruits about 5 mm. long; mainly
     w. of the Cascade Mts.   R. DIVERSILOBA

Rhus diversiloba T. & G. Fl. N.Am. 1:218. 1838.

R. toxicodendron ssp. *diversiloba* Engl. in DC. Monog. Phaner. 4:395. 1883. *Toxicodendron diver-silobum* Greene, Leafl. 1:119. 1905. *(Douglas, Oregon)*

Toxicodendron lobadioides Greene, Leafl. 1:119. 1905. *(Suksdorf, Columbia R., Klickitat Co., Wash., May 6 and July, 1885)*

R. diversiloba f. *radicans* McNair, Field Mus. Pub. Bot. 4:61. 1925, not *R. radicans* L. *(Mrs. Austin 780, Little Chico Creek, Butte Co., Calif., is the first of 2 collections cited)* A name for the more vining phase of the species.

Glabrous to pubescent shrubs or climbing vines 1-15 m. tall; leaflets 3 (5), ovate to obovate, sinuate to rather deeply lobed, rounded to shortly acute, 3-7 cm. long; flowers 1-2 mm. long, borne in rather loose, axillary, often reflexed panicles; sepals about half as long as the petals, pubescent to glabrate; staminate flowers with exserted stamens; fruits usually glabrous, about 5 mm. long, globose; seeds white, rather deeply many-grooved. Poison oak, poison ivy.

From the Puget Sound region to Mex., on the w. side of the Cascades and Sierra Nevada, inland along the Columbia R. to Klickitat Co., Wash. Apr.-July.

This is the common poison oak (also sometimes called poison ivy) of the coastal region. Many persons are allergic to the slightly volatile oil of the plant and break out in a burning or itching rash when it is contacted in any way.

Rhus glabra L. Sp. Pl. 265. 1753.

Toxicodendron glabrum Kuntze, Rev. Gen. 1:154. 1891. *Schmaltzia glabra* Small, Fl. S.E. U.S. 729, 1334. 1903. (North America)

R. glabra var. *occidentalis* Torr. Bot. Wilkes Exp. 257. 1874. *R. occidentalis* Blank. Mont. Agr. Coll. Stud. Bot. 1:86. 1905. *(Wilkes Exp., near Ft. Okanogan and Ft. Vancouver)*

Smooth-barked shrubs 1-3 m. tall, the branches glabrous or sparsely pubescent near the inflorescence; leaflets 7-29, oblong-lanceolate to elliptic, serrate, 5-12 cm. long; flowers in large, dense, compound panicles; petals 1-2 mm. long, acute, yellowish; fruits 4-5 mm. long, somewhat flattened, thickly reddish-hairy. Sumac.

From B.C. southward, on the e. side of the Cascade Mts., through Oreg. and Nev. to Mex., e. to N.H. and Ga. Apr.-July.

Although the plant spreads by greatly elongate, shallow roots and tends to form large thickets, it is often used as an ornamental because the leaves color to a deep red in the fall.

Rhus radicans L. Sp. Pl. 266. 1753.

Philostemon radicans Raf. Fl. Ludovic. 107. 1817. *R. toxicodendron* var. *radicans* Torr. Fl. N. & Mid. U.S. 324. 1824. *Toxicodendron radicans* Kuntze, Rev. Gen. 1:153. 1891. *R. toxicodendron radicans* Schelle in Schelle, Beiss. & Zabel, Handb. Laubh. Benen. 286. 1903. *R. toxicodendron* f. *radicans* McNair, Field Mus. Pub. Bot. 4:68. 1925. *R. toxicodendron* ssp. *radicans* R. T. Clausen, Cornell U. Agr. Exp. Sta. Mem. 291:8. 1949. (Virginia)

R. toxicodendron sensu auct., in part, not of L. in 1753.

R. rydbergii Small in Rydb. Mem. N.Y. Bot. Gard. 1:268. 1900. *Toxicodendron rydbergii* Greene, Leafl. 1:117. 1905. *R. toxicodendron* var. *rydbergii* Garrett, Spring Fl. Wasatch 2nd ed. 69. 1912. *R. radicans* var. *rydbergii* Rehd. Journ. Arn. Arb. 20:416. 1939. *(Williams 291, Great Falls, Mont., in 1885)*

Toxicodendron coriaceum Greene, Leafl. 1:117. 1905. *(Suksdorf, eastern Wash., in 1885)*

Toxicodendron hesperium Greene, Leafl. 1:118. 1905. *(Whited 241, Wenatchee, Wash., Aug. 16, 1896)*

A (sometimes scandent) shrub 0.5-2 m. tall, usually somewhat pubescent; leaflets 3, broadly ovate, acute or acuminate, entire or shallowly crenate to lobed, 5-15 cm. long; flowers 2-3 mm. long, borne in congested axillary panicles; fruit white with a greenish or yellowish cast, glabrous, about 4 mm. long; seeds white, longitudinally many-grooved. Poison ivy (poison oak).

Rather general on the plains and foothills and into the lower mountains; e. Wash. and Oreg., eastward to the Atlantic coast and s. to Mex. Apr.-July.

Rhus radicans is closely related to *R. diversiloba*, and, like it, causes dermatitis on many persons. Typical *R. radicans*, as it occurs in e. U.S., is ordinarily a vine, whereas our plants are usually slender shrubs. Although the more western phase has been segregated as the var. *rydbergii* (Small) Rehd., the distinction is somewhat dubious since similar plants occur in parts of e. U.S.

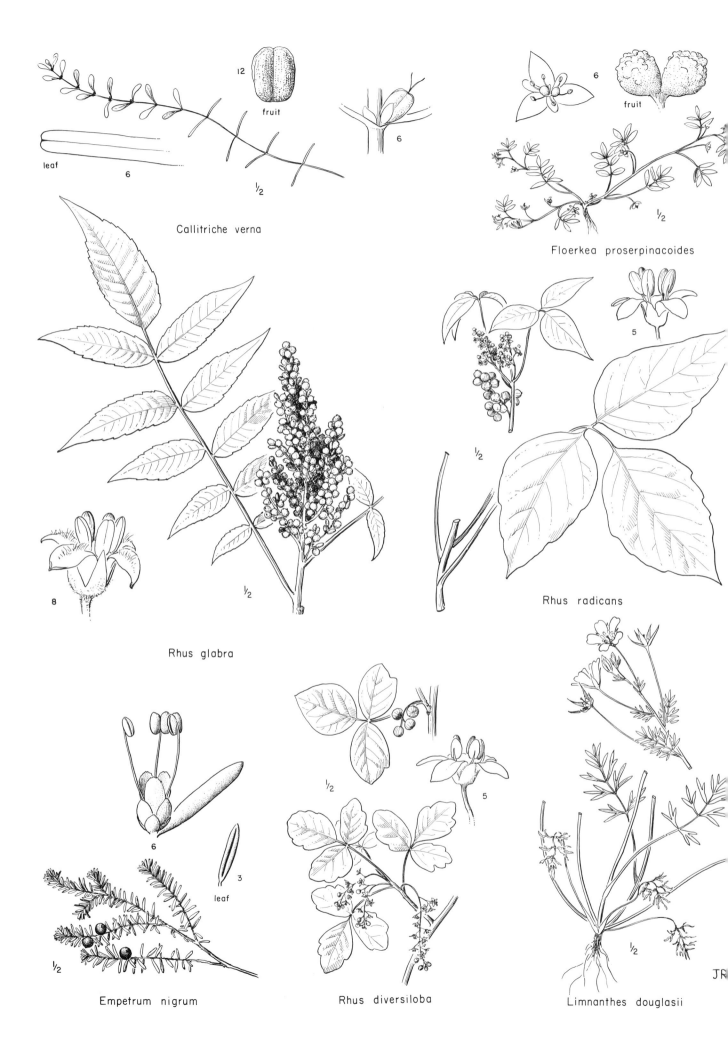

12
fruit

leaf
6

½

Callitriche verna

6
fruit

½

Floerkea proserpinacoides

5

½

Rhus radicans

8

½

Rhus glabra

½

5

6

3
leaf

½

Empetrum nigrum

½

Rhus diversiloba

½

Limnanthes douglasii

JR

Rhus trilobata Nutt. in T. & G. Fl. N. Am. 1:219. 1838.

R. aromatica var. *trilobata* Gray, Am. Journ. Sci. II, 33:408. 1861. *Toxicodendron trilobatum*
Kuntze, Rev. Gen. 1:154. 1891. *R. canadensis* var. *trilobata* Gray, Contr. U.S. Nat. Herb. 2:68.
1891. *Schmaltzia trilobata* Small, Fl. S. E. U.S. 728, 1334. 1903. *(Nuttall,* "In the central chain
of the Rocky Mountains")

R. trilobata var. *quinata* Jeps. Erythea 1:141. 1893. *R. canadensis* var. *quinata* Gray, Syn. Fl.
1:386. 1897. *Schmaltzia quinata* Greene, Leafl. 1:139. 1905. *(Jepson,* Napa Co., Calif.)

Schmaltzia oxyacanthoides Greene, Leafl. 1:134. 1905. *R. oxyacanthoides* Rydb. Fl. Rocky Mts.
551. 1917. *(Greene,* Grand Junction, Colo., Aug. 27, 1896)

Schmaltzia pubescens Osterh. Muhl. 7:11. 1911. *R. osterhoutii* Rydb. Fl. Rocky Mts. 551. 1917.
(Osterhout 7306, Rule Creek, Bent Co., Colo., June 9, 1910)

Rounded shrubs up to 2 m. tall, the stems puberulent, at least when young; leaves trifoliolate, 3-7
cm. long, leaflets puberulent on the lower surface at least, the terminal one fan-shaped, shallowly to
deeply incised-lobed and coarsely crenate, the lower 2 about half as large, shallowly crenate-lobed;
flowers in close clusters of spikes near the tips of the branches, developing before the leaves, yellowish-green, about 3 mm. long; sepals about half the length of the petals; disc orange; drupe reddish-
orange, 6-8 mm. long, minutely papillose-pubescent. Squawbush, skunkbush.

Along watercourses and on lower hills; Alta. to Ia., southward, chiefly on the e. side of the Rocky
Mts. (in Mont.), to Mex. and w. to Ida., s. e. Oreg., and Calif. May-July.

This is a widespread, variable species, consisting of several varieties, only one of which, var.
trilobata, as described above, occurs in our area.

CELASTRACEAE Staff-tree Family

Flowers mostly axillary, single or cymose, small, greenish to reddish, regular, perfect or imperfect and then the plants polygamo-monoecious or polygamo-dioecious; sepals 4-5 (6), connate at
base; petals as many as the sepals or rarely absent; stamens opposite the sepals, equal in number
thereto, or twice as many, borne at the outer edge of a flattened, lobed or cuplike perigynous disc;
pistil 2- to 5-carpellary, style 1; fruit a drupe, capsule, samara, or follicle; seeds usually with a
prominent, fleshy, often showy aril and with endosperm; trees and shrubs (often scandent) with (usually) opposite, simple, deciduous or evergreen leaves with or without small caducous stipules.

About 40 genera and nearly 500 species, widely distributed in temperate and tropical regions.

1 Leaves alternate, entire; petals whitish, clawed; branches spinose at the tips   GLOSSOPETALON
1 Leaves opposite, serrate; petals reddish, not clawed; branches not spiny
  2 Leaves persistent, 1-3 cm. long; flowers 4-merous   PACHISTIMA
  2 Leaves deciduous, usually well over 3 cm. long; flowers mostly 5-merous   EUONYMUS

Euonymus L.

Flowers 5 (4)-merous, perfect, regular; stamens 5, borne at the edge of a flattened, lobed disc;
ovary sunken in the disc; styles very short, almost lacking, the stigma globose to 4- or 5-lobed; fruit
a 4- or 5-celled, rather leathery capsule; seeds arillate; shrubs or small trees with opposite leaves
that (in ours) are deciduous.

A genus of about 150 species, mostly in the Northern Hemisphere, especially in Asia. (Greek *eu,*
good, and *onoma,* name.)

Euonymus occidentalis Nutt. ex Torr. Pac. R. R. Rep. 4:74. 1857.

E. atropurpureus Hook. Fl. Bor. Am. 1:119. 1831, but not of Jacq. in 1772. *E. atropurpureus* β?
T. & G. Fl. N. Am. 1:258. 1838. *E. occidentalis* Nutt. ex T. & G. loc. cit., in synonymy. *(Douglas,* "Banks of streams about the Columbia")

Straggling shrubs 2-5 m. tall, the branches glabrous, striate; leaves thin, oblong-lanceolate, serrate, acuminate, 5-10 cm. long; stipules minute or lacking; flowers in slender-peduncled, axillary,
mostly 3-flowered cymes; sepals rounded; petals about 5 mm. long, greenish- and purplish-mottled to
purplish-red; anthers transversely dehiscent, much longer than the filaments; capsules 3-lobed; seeds
covered by the reddish-orange aril. Western wahoo or burning bush.

In woods on the w. side of the Cascade Mts., from Lewis Co., Wash., to c. Calif. May-June.

This species is probably to be classed as a botanical collector's item, rather than a plant of much
horticultural merit.

Glossopetalon Gray

Flowers generally axillary on slender pedicels, mostly perfect, or imperfect and the plants (ours) polygamo-monoecious; sepals 4-6, persistent; petals 4-6, white; stamens equal to the sepals or twice as many, inserted under the outer edge of a small, perigynous, lobed disc; pistil 1, carpels 1-3; ovary 1-celled, superior, the stigma nearly sessile; fruit a 1 (2)-seeded, leathery follicle; aril of the seed white, rather small; small greenish-barked shrubs, with spinescent, angled and grooved stems and alternate, entire, deciduous, often stipulate leaves.

Five or 6 species of the more desert regions of N. Am., mostly n. of Mex. (Greek *glossa*, tongue, and *petalon*, petal, referring to the narrow petals.)

References:

Ensign, Margaret. A revision of the Celastraceous genus Forsellesia (Glossopetalon). Am. Midl. Nat. 27:501-511. 1942.

St. John, Harold. Nomenclatural changes in Glossopetalon. Proc. Biol. Soc. Wash. 55:109-112. 1942.

Glossopetalon nevadense Gray, Proc. Am. Acad. 11:73. 1876.

Forsellesia nevadensis Greene, Erythea 1:206. 1893. *(Lemmon, Pyramid Lake, Nev., in 1875)*
GLOSSOPETALON NEVADENSE var. STIPULIFERUM (St. John) C. L. Hitchc. hoc loc. *G. stipuliferum* St. John, Fl. S. E. Wash. 250. 1937. *Forsellesia stipulifera* Ensign, Am. Midl. Nat. 27: 507. 1942. *(Henderson 4855, near Lewiston, Ida., May 24, 1898)*

Erect to spreading, freely branched, glabrous shrubs 1-3 m. tall; branches spinose-tipped, conspicuously grooved; leaves entire, grayish-green, oblanceolate to oblong or oblong-lanceolate, 3-15 mm. long, the slender petioles articulate with the fleshy, cushionlike, often purplish base; stipules linear, about 1 mm. long, persistent; flowers mostly single (2-4) in the axils; sepals usually 5 (4-6), about 2 mm. long; petals slender-clawed, 4-9 mm. long, deciduous; stamens usually opposite and equal in number to the sepals in the staminate flowers, the staminal vestiges often twice as many as the sepals in the pistillate flowers; fruit about 3 mm. long, longitudinally grooved, usually 1-seeded; seeds dark brown, the aril fleshy and carunclelike.

On rocky canyon walls from w. c. Ida. to Calif., Ariz., and Utah. Apr.-June.

Our material, as described above, is referable to the var. *stipuliferum* (St. John) C. L. Hitchc., which some recent workers have treated as a distinct species on the basis of variable and minor characteristics. Variety *stipuliferum* ranges from Lewiston, Ida., to Calif., and e. to c. Ida., whereas the var. *nevadense*, which is somewhat more spinescent and has thicker, more yellowish branches and somewhat smaller stipules, is found in the desert canyons of Nev., Calif., Ariz., and Utah, although Ensign recorded it from Ida. on the basis of a specimen from Whitebird, on the Salmon R.

Pachistima Raf.

Flowers axillary, solitary or in small cymes, 4-merous except the pistil, perfect, regular; stamens borne at the outer edge of a flattened disc, the filaments equaling the longitudinally dehiscent anthers; ovary sunken in the disc, 2-celled; stigma 1; fruit a 1- or 2-seeded capsule; evergreen shrubs with leathery opposite leaves and tiny deciduous stipules.

Two species of N. Am. (Greek *pachus*, thick, and *stigma*, stigma.)

For a discussion of the spelling of this generic name, see Wheeler, Am. Midl. Nat. 29:792-795. 1943, where it was contended that "Paxistima" was the spelling to be used, instead of either "Pachistima" or "Pachystima." On the other hand, Rehder (Bibl. of Cult. Trees and Shrubs 410. 1949) maintained the spelling as used here, suggesting that "Paxistima" was probably an inadvertent misspelling.

Pachistima myrsinites (Pursh) Raf. Sylva Tell. 42. 1838.

Ilex? myrsinites Pursh, Fl. Am. Sept. 119. 1814. *Myginda myrtifolia* Nutt. Gen. Pl. 1:109. 1818. *Oreophila myrtifolia* Nutt. ex T. & G. Fl. N.Am. 1:259. 1838. *Paxistima myrsinites* Wheeler, Am. Midl. Nat. 29:793. 1943. *(Lewis, Hungry Creek, Ida., is the first of 2 collections cited)*
Myginda myrtifolia α minor Hook. Fl. Bor. Am. 1:120, pl. 41. 1830, cited as 41A, but drawings on plate 41, on the copy at the U. of Wash., are not labeled "A" and "B." (No specimens cited)
Myginda myrtifolia β major Hook. Fl. Bor. Am. 1:120. 1830. *Pachystima myrsinites* var. *major* Gray, Pl. Fendl. 29. 1849. *(Douglas,* "Valley of the Rocky Mountains, particularly abundant near the sources of the Columbia, in 52° N. lat. and 118° W. long.," is the first collection cited)

Pachystima macrophylla Farr, Contr. Bot. Lab. Univ. Pa. 2³:421. 1904. *(E. M. Farr,* Bear
   Creek, Selkirk Mts. , B.C. , Aug. 20, 1904)
Pachystima schaefferi Farr, Ott. Nat. 20:108. 1906. *(Mrs. Charles Schäffer,* Bear Creek Station,
   Selkirk Mts. , B.C. , May 25, 1905)
   Glabrous shrub 2-6 (10) dm. tall, the leaves glossy, oblong-lanceolate to lanceolate or oblanceolate,
serrate, 1-3 cm. lo ng; flowers 3-4 mm. broad, the petals maroon; capsules 3-4 mm. long; seeds dark
brown, about 2/3 covered by the thin, lacerate, whitish aril. Mt. box, mt. lover.
   British Columbia to Calif. , e. to the Rocky Mts. ; mostly at medium altitudes but downward to near
sea level in w. Wash. Apr. -June.
   This is surely one of the finest low-growing shrubs in the northwest, readily adaptable to shady or
open, well-drained sites, graceful when growing untouched, but readily shaped into a low hedge. Although the flowers are somewhat inconspicuous, nevertheless they have real beauty, but it is because
of the handsome, glossy, evergreen leaves that the plant is rated so highly.

ACERACEAE Maple Family

   Flowers corymbose to paniculate, usually regular, perfect or imperfect, generally with a lobed
disc between the stamens and the pistil, or the disc extrastaminal; sepals 4-5, from distinct to basally connate; petals 4-5 (or lacking), distinct; stamens 4-10, mostly 8, a rudimentary pistil usually
present in the staminate flowers; pistil 1, styles 1 or 2, stigmas usually 2, ovary superior, 2-celled,
rapidly developing an elongate divergent wing from each carpel; fruit a double samara, the two halves
tardily separating; seeds without endosperm; perfect-flowered to monoecious, dioecious, or polygamo-
dioecious trees or shrubs with opposite, exstipulate, palmately lobed to pinnately compound leaves.
   Two genera and perhaps 125 species, all but 2 in the genus *Acer;* chiefly in the N. Temp. Zone.

Acer L. Maple

   Petals usually smaller than the sepals, or lacking entirely; leaves generally palmately lobed, less
commonly 3- to 5-foliolate; flowers in axillary panicles, racemes, or corymbs.
   (Latin name for the maple tree.)
   A genus of great value as ornamentals and a source of valuable hardwoods. Two of our native species, *A. glabrum* and *A. circinatum,* are beautiful shrubs that assume deep autumn coloration.
   Reference:
   Keller, A. C. Acer glabrum and its varieties. Am. Midl. Nat. 27:491-500. 1942.
1 Leaves 3- to 5-foliolate, the terminal leaflet stalked; plants dioecious; petals lacking   A. NEGUNDO
1 Leaves simple, palmately lobed or very occasionally trifoliolate but then the leaflets
     not stalked; petals usually present
   2 Flowers 10-50 in elongate racemes or panicles; fruit bristly-hairy; larger leaves
        over 15 cm. broad; large trees   A. MACROPHYLLUM
   2 Flowers usually fewer than 10, corymbose or umbellate; fruit glabrous or sparse-
        ly pilose; larger leaves usually less than 15 cm. broad
      3 Leaves 3- to 5-lobed; stamens usually outside the disc; sepals green
         4 Sinuses of the leaves narrowly acute, the lobes ovate, sharply and finely bi-
              dentate; flowers corymbose; petals usually present   A. GLABRUM
         4 Sinuses of the leaves broad, the lobes oblong, coarsely and sparingly sinuate-
              toothed; flowers umbellate; petals lacking   A. GRANDIDENTATUM
      3 Leaves 7- to 9-lobed; stamens borne between the inner edge of the disc and the
           ovary; sepals usually red   A. CIRCINATUM

Acer circinatum Pursh, Fl. Am. Sept. 267. 1814. *(Lewis,* "On the great rapids of the Columbia
   river")
   Shrub or small tree 1-8 m. tall, often propagating vegetatively by layering and sometimes forming
dense thickets; stems purplish-red, brown with age, glabrous or sparsely pilose; leaves cordate-
flabellate, 7- to 9-lobed, serrate, 3-6 cm. long, usually considerably broader, more or less pilose
on the lower surface and often hairy at least along the veins on the upper surface; flowers few, 6-9
mm. broad, corymbose and terminal on short, lateral, mostly 2-leaved shoots; sepals 4 or 5, purplish, sparsely pilose near the tip; petals white, shorter than the sepals; stamens 8 (10), inserted on
the inner edge of a fleshy disc, those of the staminate flowers longer than the perianth, those of the

perfect flowers much shorter, the disc pilose at the base of the ovary; styles connate only at base; wings of the samaras widely spreading. N=13. Vine maple.

Alaska to coastal n. Calif., from the e. side of the Cascade Mts. westward (and there much more common) to the coast. Mar.-June.

Acer glabrum Torr. Ann. Lyc. N.Y. 2:172. 1827. *(Dr. James,* "in the Rocky Mts. about lat. 40°"')
    A. barbatum sensu Hook. Fl. Bor. Am. 1:113. 1831, but not of Michx. in 1803. *A. douglasii* Hook. Lond. Journ. Bot. 6:77, pl. 6. 1847. *A. glabrum* ssp. *douglasii* Wesmael, Bull. Soc. Bot. Belg. 29:46. 1890. *A. glabrum* var. *douglasii* Dippel, Handb. Laubh. 2:438. 1892. *(Douglas,* "on the west side of the Rocky Mountains, about the sources of the Columbia")
    A. subserratum Greene, Pitt. 5:1. 1902. *(Heller 3089,* Lewiston, Ida., May 20, 1896)

Shrub or small tree 1-10 m. tall; stems glabrous, the bark grayish to reddish-purple; leaves glabrous to sparsely glandular-puberulent, 2-14 cm. long and usually as broad, cordate in outline, palmately 3- to 5-lobed (trifoliolate), coarsely twice-serrate, paler on the lower surface; plants from dioecious to polygamo-monoecious, the staminate flowers with only tiny rudiments of the pistil but the pistil-bearing flowers with well-developed (if not always functional) stamens; flowers about 8 mm. broad, corymbose in the leaf axils; sepals 5 (4 or 6); petals equal to and somewhat smaller than the sepals, or lacking; stamens usually twice the number of the sepals, inserted at the outer edge of a lobed disc; styles 2; samaras 2.5-3 cm. long, their wings not greatly divergent, the angle between them usually less than 90°. Mt. maple.

Alaska to Calif., e. to the Rocky Mts. from s. Alta. to N.M., and to Neb. Apr.-June.

Our material is mostly referable to the var. *douglasii* (Hook.) Dippel, which usually has reddish stems and shallowly lobed leaves averaging over 6 cm. broad; it is found in the coastal lowlands and lower mountains from Alas. to s. Oreg., e. to Ida. and w. Mont., where it merges with var. *glabrum* which usually has grayish branches and more deeply lobed leaves seldom over 6 cm. in width; the latter ranges in the mountains from Mont. and Ida. s. to N.M., and e. to Neb.

Acer grandidentatum Nutt. in T. & G. Fl. N.Am. 1:247. 1838.
    A. barbatum var. *grandidentatum* Sarg. Gard. & For. 4:148. 1891. *A. saccharum* var. *grandidentatum* Sudw. U.S. Dept. Agr. Rep. 1892:325. 1893. *Saccharodendron grandidentatum* Nieuwl. Am. Midl. Nat. 3:183. 1914. *A. saccharum* ssp. *grandidentatum* Desmarais, Brittonia 7:383. 1952. *A. nigrum* var. *grandidentatum* Fosberg, Castanea 19:27. 1954. *(Nuttall,* "Rocky Mountains, on Bear River of Timpanagos")
    A. barbatum sensu Dougl. in Hook. Fl. Bor. Am. 1:113. 1833, not of Michx. *(Douglas,* "Rocky Mountains, about the sources of the Columbia")

Shrubby or somewhat arborescent, mostly 3-5 (6) m. tall; leaves (3) 5-8 cm. broad, deeply 3 (or imperfectly 5)-lobed over half way to the base, the lobes oblong, coarsely serrate-toothed; flowers umbellate, apetalous; calyx long-hairy; samaras about 3 cm. long, divergent at about 120°.

Eastern Ida., s.c. Mont., and w. Wyo. to Tex. and Ariz. Apr.-May.

Opinions differ as to how closely this taxon is related to the more eastern sugar maple, but for our purposes it can perhaps as well be treated at the specific level. It colors beautifully in the fall.

Acer macrophyllum Pursh, Fl. Am. Sept. 267. 1814. *(Lewis,* "Great Rapids of the Columbia")

Large spreading tree to 30 m. tall; branches greenish-barked; leaves puberulent on both surfaces, cordate, 10-30 cm. broad, deeply 5-lobed; flowers perfect or staminate, the two types usually borne together in large, many-flowered, axillary racemes or panicles, greenish-white, 10-15 mm. broad; perianth mostly 5 (4 or 6)-merous; filaments bearded at the base, borne toward the outer margin of a thick 10-lobed disc; styles 2; samaras golden-hirsute, the wings divergent at less than a 90° angle. Big-leaf maple, common maple.

Alaska to s. Calif., mostly from the w. slope of the Cascades and Sierra Nevada to the coast, but also in Idaho Co., Ida., where possibly introduced. Mar.-June.

Acer negundo L. Sp. Pl. 1056. 1753.
    Negundo aceroides Moench, Meth. 334. 1794. *Negundo negundo* Karst. Deutsch. Fl. 596. 1880-83. *Rulac negundo* A. S. Hitchc. Spr. Fl. Manhattan 25. 1894. (Virginia)
    A. fraxinifolium Nutt. Gen. Pl. 1:253. 1818, not *Negundium fraxinifolium* Raf. 1808. *A. negundo* var. *violaceum* Booth ex Kirchn. in Kirchn. & Petzola, Arb. Mosc. 190. 1864. *Rulac nuttallii* Nieuwl. Am. Midl. Nat. 2:137. 1911. *Negundo nuttallii* Rydb. Bull. Torrey Club 40:55. 1913.

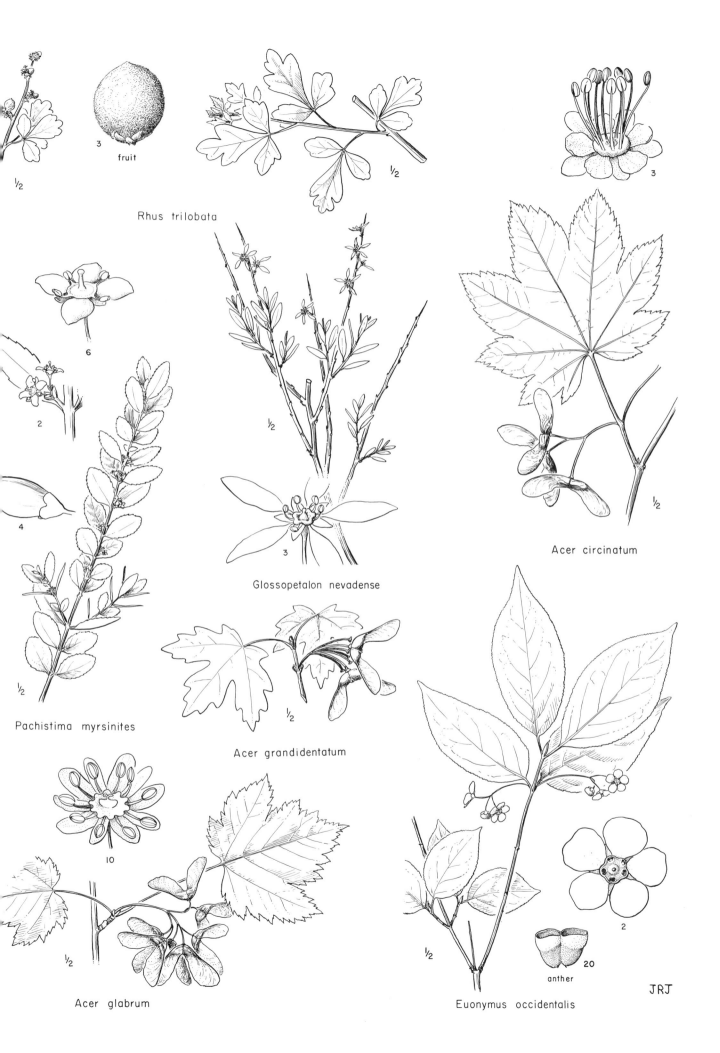

1/2

3
fruit

1/2

3

Rhus trilobata

6

2

4

1/2

Pachistima myrsinites

1/2

3

Glossopetalon nevadense

1/2

Acer grandidentatum

10

1/2

Acer glabrum

1/2

Acer circinatum

1/2

2

20
anther

JRJ

Euonymus occidentalis

A. nuttallii Lyon, Am. Midl. Nat. 12:39. 1930. *(Nuttall,* "North-westward on the banks of the Missouri to the mountains")

A. interior Britt. in Britt. & Shafer, N. Am. Trees 655, fig. 608. 1908. *Rulac interior* Nieuwl. Am. Midl. Nat. 2:139. 1911. *Negundo interius* Rydb. Bull. Torrey Club 40:56. 1913. *A. negundo* var. *interior* Sarg. Bot. Gaz. 67:239. 1919. *(Underwood & Selby 11,* s. e. of Ouray, Colo., Sept. 7, 1901)

Pale greenish, straggly to shapely dioecious tree as much as 20 m. tall, the young branches glabrous to finely puberulent; leaves trifoliolate, the leaflets oblong-lanceolate to oblanceolate, coarsely few-toothed, usually pubescent; staminate flowers in dense axillary clusters on slender pedicels 1-4 cm. long, sepals and stamens 4 or 5, petals and disc absent; pistillate flowers in axillary 6- to 15-flowered racemes, sepals 4 or 5, petals, disc, and staminal rudiments lacking; style 1, short, the stigmas elongate; samaras usually pubescent, the wings divergent less than 90°. N=13. Box elder.

A widespread species in N. Am., in our area mostly escaped from cultivation. Apr. -June.

Several geographic races are usually distinguished as varieties. Much of our material in w. Wash. and Oreg. is apparently the introduced var. *negundo,* of e. U. S., but in the Rocky Mt. area the var. *interior* (Britt.) Sarg. is native. This phase of the species, which ranges from much of Can. through the Rocky Mts. to Ariz., is characterized by pubescent branches, whereas var. *violaceum* Booth ex Kirchn., chiefly from the Rocky Mts. to the Mississippi Valley and n. e. U. S., but reported from Nez Perce Co., Ida., has glabrous, glaucous branches (var. *negundo* is neither pubescent nor particularly glaucous on the branches).

BALSAMINACEAE Touch-me-not or Balsam Family

Flowers one to several (and more or less umbellate or paniculate) in the leaf axils, perfect, or (some) cleistogamous and much reduced, zygomorphic, hypogynous; sepals 3 or (less commonly) 5, imbricate, petaloid in texture, one usually much enlarged, saccate, and often spurred; petals 5, distinct or the lateral pairs partially connate and apparently single and bilobed; stamens 5, often connate, filaments flattened, anthers 2-celled; pistil 5-carpellary, style single, short or quite lacking, the stigma 5-lobed; fruit generally an elastically dehiscent, explosive capsule (ours) with axile placentation, or rarely a berry; seeds numerous, without endosperm, the embryo straight; plants herbaceous and succulent (ours) to slightly woody at the base, with exstipulate, simple, alternate to whorled leaves.

Only 2 genera with about 400 species; fairly widespread, but most abundant in tropical Asia and Africa.

Impatiens L. Balsam; Touch-me-not; Jewelweed

Flowers showy, white to yellow, red, or purplish, 2-several in loose, open, leafy, axillary racemes; sepals apparently 4 and the petals only 2, the latter unequally bifurcate, but actually the sepals 3 (one saccate and often narrowed to a distinct, often recurved spur, the other 2 much smaller) and the petals 5, the upper one usually notched at the tip, the other (lateral) 4 united in pairs, one of each pair much smaller than the other and apparently only a lobe of it; filaments short, more or less geniculate and often somewhat connate, the 2-celled anthers frequently connate around the stigma; capsules 5-carpellary, 5-celled, or 1-celled by the decomposition of the filmy membranous dissepiments; annual or perennial, succulent herbs with alternate (opposite or whorled) simple leaves.

Nearly 400 species, mostly Indian, a few in temperate and tropical Am. (Latin, meaning impatient, in reference to the explosively dehiscent fruits.)

1 Saccate sepal not spurred   I. ECALCARATA
1 Saccate sepal spurred
  2 Leaves in part opposite or whorled, finely and closely serrate with 20 or more teeth
    on each side; spur usually less than 6 mm. long   I. GLANDULIFERA
  2 Leaves all alternate, coarsely crenate-serrate, with fewer than 20 teeth on a side; spur
    usually at least 6 mm. long   I. NOLI-TANGERE

Impatiens ecalcarata Blank. Mont. Agr. Coll. Stud. Bot. 1:84, pl. 1. 1905.
  I. biflora var. *ecalcarata* M. E. Jones, Bull. U. Mont. Biol. ser. 15, no. 61:39. 1910. *(Blankinship,* Plains, Missoula Co., Mont., Aug. 9, 1901)
  Stems 4-10 dm. tall, freely branched; leaves all alternate, the petioles very slender, 1-3 cm. long,

the blades ovate-elliptic to elliptic-lanceolate, 3-10 cm. long, coarsely serrate, the teeth 3-7 on each margin; flowers mostly paired (2-5), pale yellow to orange, 1-2 cm. long; saccate sepal not spurred.

In moist shady situations on both sides of the Cascades, from s. B.C. to n.w. Oreg., e. to Mont. Aug.-Sept.

The species is very similar to *I. noli-tangere* except in the lack of a spur, and is perhaps only a phase of that species.

Impatiens glandulifera Royle, Ill. Bot. Him. Mts. pl. 28, fig. 2, 1834, page 151. 1835, not of Arnott in Comp. Bot. Mag. 1:322. 1836.

    I. roylei Walpers Rep. 1:475. 1842. (India)

    I. roylei var. *pallidiflora* Hook. in Curtis' Bot. Mag. 125:pl. 7647. 1899. *I. glanduliflora* f. *pallidiflora* Weatherby, Rhodora 48:414. 1946. (Described from a garden plant, presumably from seeds collected in India)

Plant 6-13 dm. tall, the stems often purplish-tinged; leaves alternate to whorled, the petioles stout, 2-5 cm. long, the blades oblong-ovate to ovate-elliptic, 6-15 cm. long, sharply and closely serrate; flowers several on much-elongated axillary peduncles; lateral sepals purplish, the saccate sepal whitish to red, usually purplish-spotted, 2-3 cm. long, with a short recurved spur 4-5 mm. long; petals white to red.

This rather showy Asiatic ornamental occasionally escapes and becomes established in moist places, especially along streams, in w. Wash. and B.C. July-Sept.

Impatiens noli-tangere L. Sp. Pl. 938. 1753. (Europe)

    I. aurella Rydb. Bull. Torrey Club 28:34. 1901. *(D. T. MacDougal 20,* Priest R., Ida., in 1900)

    I. occidentalis Rydb. N.Am. Fl. 25^2:94. 1910. *(Suksdorf 960,* N. Fork Nooksack R., Wash., in 1890)

    I. aurella f. *badia* St. John, Res. Stud. St. Coll. Wash. 1:102. 1929. *(St. John 9210,* Indian Canyon, Spokane Co., Wash., Sept. 11, 1925)

    I. aurella f. *coccinea* St. John, loc. cit. *(St. John & Warren 6749,* Little Spokane R., Wash., Sept. 27, 1924)

Glabrous, succulent annual 2-6 dm. tall; stems freely branched; all leaves alternate, the petioles 2-4 cm. long, the blades elliptic-ovate, 3-12 cm. long, coarsely crenate-serrate, the teeth mucronate; flowers mostly in 2's, yellowish and often spotted or mottled with crimson or reddish-brown; saccate sepal 1-2 cm. long, scarcely 1/2 as broad at the orifice, narrowed gradually to the strongly recurved, 6-10 mm. spur. N=10.

In moist woods; Alas. to Oreg., from the coast inland to Ida.; Eurasia. July-Sept.

The very closely allied but more eastern species, *I. biflora* Walt. *(I. capensis* Meerb. ?), has been reported (although perhaps incorrectly) from Ida. It has the flowers typically brown-spotted on an orange background, and in comparison with *I. noli-tangere* seems not only to have more shallowly and remotely toothed leaves but also a slightly broader, more saccate sepal with often a more abruptly bent spur.

Our species differs from *I. pallida* Nutt., of e. U.S., in having a more slenderly saccate sepal that is narrowed gradually (rather than abruptly) to the spur. Formae *badia* and *coccinea* are mere sporadic color variants, not deemed to merit recognition.

RHAMNACEAE Buckthorn Family

Flowers in axillary, cymose to paniculate-corymbose or umbellate clusters, regular, perfect to imperfect and the plants sometimes dioecious; calyx 5 (4)-lobed, usually forming a short basal tube or free hypanthium; petals 5 (4), rarely wanting, generally clawed, the blade more or less hooded; stamens opposite and equal in number to the petals (alternate with the sepals), borne at the outer margin of a perigynous disc, the anthers 2-celled; pistil 1, the ovary superior or partially buried in the disc, 3 (2-4)-carpellary; fruit capsular, baccate, baccate-drupaceous, or sometimes samaroid, usually with 1 (2) seeds per locule; endosperm lacking; deciduous to evergreen shrubs (ours) or trees (a few herbaceous or suffrutescent) with simple, alternate or opposite, usually stipulate leaves.

Over 40 genera and 500 species, of very wide distribution but best represented in arid, warm, temperate to semitropical regions.

1 Fruit fleshy; flowers greenish; petals short-clawed or lacking   RHAMNUS
1 Fruit capsular; flowers white to blue; petals long-clawed   CEANOTHUS

3

½

Acer macrophyllum

½

Impatiens glandulifera

½

5

½

Impatiens noli-tangere

½

Impatiens ecalcarata

JRJ

½

Acer negundo

Ceanothus L. Buckbrush; Wild lilac; Buckthorn

Flowers in axillary to terminal umbels or panicles, 3-5 mm. broad, white or pinkish to deep blue or purplish, showy en masse; calyx 5-lobed, the lower portion adnate to the fruit, the lobes deciduous; petals 5, long-clawed, hooded; stamens 5, opposite the petals, separated from the pistil by a flat, lobed disc which also surrounds and often more or less embeds the 3-carpellary ovary; style 1, stigmas 3; fruit a hardened capsule, separating at maturity into the three 1-seeded, dehiscent carpels; prostrate to erect shrubs or small trees with simple, alternate to opposite, deciduous or evergreen leaves, the branches sometimes spinose at the tip.

About 40 species of N. Am., chiefly in Calif. (The meaning uncertain, but the word used by the Greeks as a plant name.)

Although the climate of the Northwest is a little unkind to most of the Californian species of *Ceanothus,* our native species do well in cultivation and all are well worth growing. Perhaps *C. integerrimus* is the best, although *C. prostratus* can be grown with charming results in a rockery. *C. velutinus* should be planted much more extensively than it is, and *C. sanguineus* can be used to good effect in the shrub border.

Reference:

Van Rensselaer, M. and H. E. McMinn. Ceanothus. 308 pp. Santa Barbara, Calif. 1942.

1 Leaves opposite or whorled, persistent
  2 Plant prostrate; branches neither rigid nor spinose; leaves toothed   C. PROSTRATUS
  2 Plant erect; branches rigid, spinose; leaves usually not toothed   C. CUNEATUS
1 Leaves alternate
  3 Leaves persistent, glutinous and shining on the upper surface; stipules about 1 mm.
      long, persistent; flowers white   C. VELUTINUS
  3 Leaves deciduous, usually neither glutinous nor shining on the upper surface; stipules
      linear, 3-8 mm. long, deciduous; flowers white to blue
    4 Capsules crestless; leaves glandular-serrulate; flowers white   C. SANGUINEUS
    4 Capsules crested on the back; leaves entire; flowers white or blue   C. INTEGERRIMUS

Ceanothus cuneatus (Hook.) Nutt. ex T. & G. Fl. N.Am. 1:267. 1838.
  Rhamnus cuneatus Hook. Fl. Bor. Am. 1:124. 1831. *(Douglas,* "North-west America. Abundant
  near the sources of the Multnomak River")

Freely branched, rigid, evergreen shrub 1-2 (4) m. tall, young branches puberulent (later glabrous), more or less purplish-barked, frequently spinose-tipped; leaves opposite and often more or less fasciculate, the stipules about 1 mm. long, knoblike, the petioles 1-2 mm. long, the blades oblong-obovate to cuneate, thick and fleshy, 5-20 mm. long, rounded to slightly retuse, sometimes somewhat toothed, sparsely to rather densely puberulent and grayish-green on the upper surface, more hairy and much more grayish on the lower surface; umbels rounded, 2-3 cm. broad, 20- to 60-flowered, borne on short lateral branches; pedicels 5-15 mm. long; petals white; capsules about 5 mm. long, inconspicuously crested just below the tip. Common buckbrush.

On dry, open flats and the lower foothills e. of the coastal mountains, from the Willamette Valley, Oreg. (formerly as far n. as Oswego), to Baja Calif. Apr.-June.

Ceanothus integerrimus H. & A. Bot. Beechey Voy. 329. 1838. *(Douglas,* California)
  C. thyrsiflorus var. *macrothyrsus* Torr. Bot. Wilkes Exp. 263. 1874. *C. macrothyrsus* Greene,
  Leafl. 1:68. 1904. *C. integerrimus* var. *macrothyrsus* Benson, Contr. Dudley Herb. 2:121. 1930.
  (Wilkes Expedition, Umpqua River, Oreg.)
  C. peduncularis Greene, Leafl. 1:67. 1904. *(H. D. Langille,* Mt. Hood, Oreg.) = var. *macrothyrsus.*

Erect to widely spreading, glabrous to pubescent, deciduous, alternate leaved shrub 1-4 m. tall; stipules brownish, linear-lanceolate, 3-7 mm. long, soon deciduous; leaf blades thin, entire, 1.5-6 cm. long, oblong to ovate, glabrous or (more commonly) sparsely to densely puberulent on the lower (or on both) surfaces, usually with 3 rather prominent veins; flowers white to blue, borne in rather large panicles terminal on young branches; capsules 5-7 mm. long, the carpels with elliptic, low crests about midlength. Deerbrush.

On the e. side of the Cascades from Wash. to Baja Calif., e. to N.M. May-July.

There is much variation in the size and shape of the leaves and in the general pubescence of the plant, and some of the variants, which are neither geographically nor morphologically well delimited, have been considered as distinctive. Our material, as described above, is at most a poorly marked variety of the species, var. *macrothyrsus* (Torr.) Benson, differing from var. *integerrimus,* which occurs s. of our range, because of its somewhat larger leaves and panicles.

Ceanothus prostratus Benth. Pl. Hartw. 302. 1848. *(Hartweg 284,* "in montibus Sacramento," Calif.)
  Prostrate grayish-green shrub forming mats 1-3 m. broad and 3-5 cm. tall, the stems usually puberulent, glabrous with age; leaves opposite, persistent, thick and fleshy, oblong to obovate, 1-2.5 cm. long, 3- to 7-toothed above the middle, usually pubescent on the lower surface, the stipules brownish, about 2 mm. long; flowers 10-30, about 4 mm. broad, borne in small corymbs terminal on short lateral branches, bluish to grayish-blue or nearly white, the pedicels slender, 1.5-2.5 cm. long; capsules 6-8 mm. long, with 3 short divergent horns. Mahala mat.
  On dry forest floor, along the eastern slope of the Cascades from Yakima Co., Wash., through Oreg. to the Sierra Nevada, also in w. Nev., ranging farther w. in s. Oreg. and in Calif. May-July.

Ceanothus sanguineus Pursh, Fl. Am. Sept. 167. 1814. *(Lewis,* "Rocky mountains on the banks of the Missouri")
  C. oreganus Nutt. ex T. & G. Fl. N.Am. 1:265. 1838. *(Nuttall,* "Woods of the Oregon, from the Blue Mountains to the Sea")
  Stems erect, usually glabrous, somewhat purplish, 1-3 m. tall; leaves alternate, deciduous, the stipules 3-6 mm. long, quickly deciduous, the petioles slender, 1-2 cm. long, the blades ovate to ovate-elliptic, 3-10 cm. long, thin, usually hairy at least along the veins on the lower surface, finely crenate-serrate and glandular; flowers white, paniculate on short lateral branches of the previous year's growth; capsules about 4 mm. long, deeply 3-lobed, not crested. Buckthorn, tea tree.
  On both sides of the Cascades, from B.C. to Calif., e. to Ida. and w. Mont. May-July.
  The species resembles *C. velutinus,* especially because it sometimes has glutinous leaves, but differs in that it is deciduous and has a crestless capsule and much longer, deciduous stipules.

Ceanothus velutinus Dougl. ex Hook. Fl. Bor. Am. 1:125, pl. 45. 1831. *(Douglas,* "Subalpine hills near the sources of the Columbia; and at the 'Kettle Falls'")
  C. laevigatus Hook. Fl. Bor. Am. 1:125. 1831. *C. velutinus* var. *laevigatus* T. & G. Fl. N.Am. 1: 686. 1840. *(Menzies,* Nootka)
  Spreading, evergreen, heavy-scented shrub 0.5-2 (4) m. tall, the young twigs minutely puberulent to glabrous; leaves alternate, the stipules about 1 mm. long, the petioles 1-2 cm. long, the blades ovate to ovate-elliptic, 5-10 cm. long, glabrous, glutinous-varnished and shining, often bronze-tinged on the upper surface, grayish and puberulent to glabrous beneath, strongly 3-veined from the base, closely and finely glandular-serrate; flowers white, borne in large thyrses; capsules 4-5 mm. long, deeply 3-lobed, crested slightly above the middle Mt. balm, sticky laurel, greasewood, tobacco brush.
  Coastal B.C. to Calif., e. (chiefly in the mountains) to S.D. and Colo. June-Aug.
  The species is represented by the two following varieties in our area:
1 Leaves glabrous on the lower surface (at least on the veins); w. side of the Cascades,
    from B.C. to n. Calif.   var. laevigatus (Hook.) T. & G.
1 Leaves finely puberulent on the entire lower surface; e. side of the Cascades, B.C. to
    Calif. and Nev., e. to Ida., Mont., S.D., and Colo., occasional w. of the Cascades
    in s. Oreg.   var. velutinus

Rhamnus L. Buckthorn

  Flowers in small axillary cymes, greenish-yellow, perfect to imperfect and the plants polygamo-dioecious; calyx 4- or 5-lobed, the hypanthium shallowly to deeply campanulate, lined internally with a perigynous disc, the upper portion soon deciduous-circumscissile about midlength of the hypanthium, the lower portion persistent and collarlike at the base of the fruit; petals lacking or equal to and alternate with (but much shorter than) the sepals, very short-clawed and only slightly hooded; stamens shorter than the sepals; ovary 2- to 4-celled, superior, free of the disc; fruit fleshy, baccate but containing 3 (2-4) small hardened seed-containing "pits"; shrubs or small trees with alternate, deciduous (ours) or persistent, simple, prominently pinnate-veined leaves.
  About 100 species on all continents, in temperate and subtropic regions. (The Greek name for the plant.)
  Reference:
Wolf, C.B. North American species of Rhamnus. Rancho Santa Ana Bot. Gard. Monog. 1:1-136. 1938.

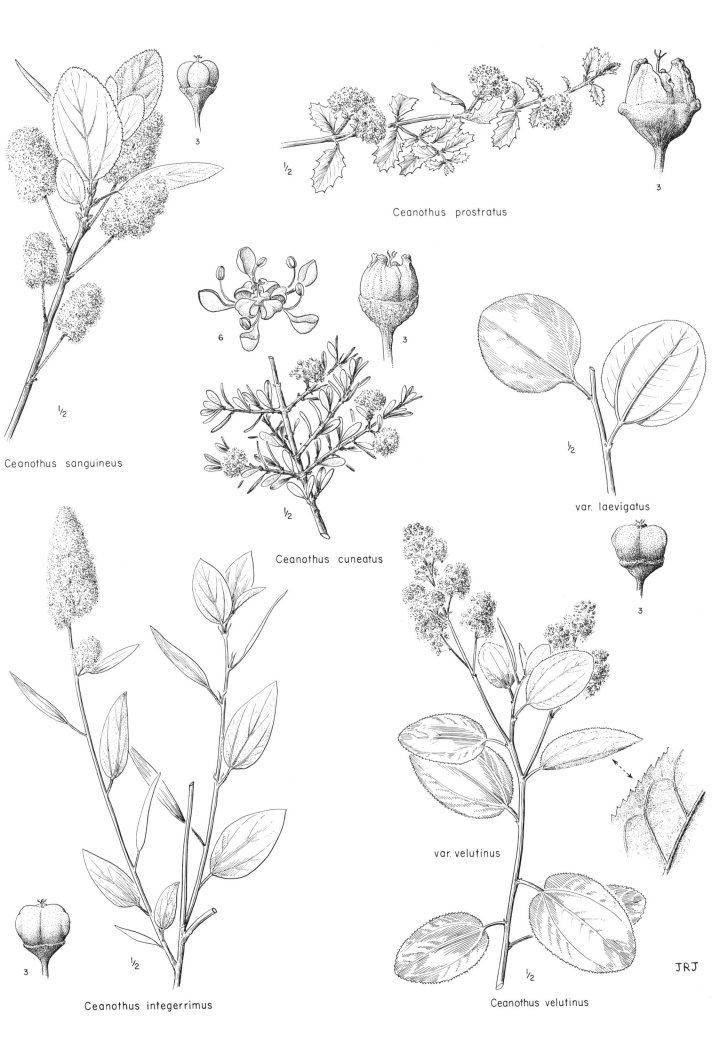

3

Ceanothus prostratus

1/2

6

3

Ceanothus sanguineus

1/2

1/2

Ceanothus cuneatus

var. laevigatus

1/2

3

var. velutinus

Ceanothus integerrimus

3

1/2

Ceanothus velutinus

1/2

JRJ

1 Flowers 8-40, in pedunculate umbels; hypanthium cup shaped; petals present; main
    lateral veins of the leaves usually more than 8 on each side   R. PURSHIANA
1 Flowers 2-5 in sessile umbels; hypanthium disc- or saucer-shaped; petals usually
    absent; main lateral veins of the leaves usually no more than 8 per side   R. ALNIFOLIA

Rhamnus alnifolia L'Her. Sert. Angl. 5. 1788. ("America septentrionale")
    Erect to rounded, deciduous shrub 0.5-1.5 m. tall, finely puberulent but the stems becoming gla-
brous; stipules oblong, 3-6 mm. long, soon deciduous; leaf blades thin, 6-11 cm. long, oblong-ovate
to oblong-elliptic, closely glandular-serrulate, the main lateral veins mostly 5-7; flowers 2-5 in
axillary, sessile umbels, developing with the leaves, about 5 mm. broad, functionally imperfect
(plants dioecious) but with vestiges of the other essential organs in both the pistillate and staminate
flowers; hypanthium shallowly campanulate, the sepals 5 (4); petals lacking; berry 6-8 mm. long, bluish-black, 3-seeded. Buckthorn.
    In the mountains, often on moist ground, especially along streams; B.C. to Que., Me., and Pa.,
southward, on the e. side of the Cascades, to the c. Sierra Nevada of Calif., and to Ida., Mont., and
Wyo. June-July.

Rhamnus purshiana DC. Prodr. 2:25. 1825.
    R. alnifolius Pursh, Fl. Am. Sept. 166. 1814, but not of L'Her. in 1788. *Frangula purshiana* Cooper,
        Pac. R. R. Rep. 12:29. 1860. (*Lewis*, Kooskoosky [Clearwater] River, Idaho)
    Yellow- to brownish-puberulent, deciduous shrub or small tree up to 10 m. tall; petioles stout, 5-
20 mm. long, leaf blades ovate-oblong to oblong-obovate, 6-13 cm. long, very finely serrulate, with
10-12 prominent, parallel, lateral veins on each side; flowers 8-50 in pedunculate axillary umbels,
5-merous, 3-4 mm. long, greenish, perfect or imperfect (if imperfect, the plant monoecious); calyx
campanulate, the hypanthium lined with a thin disc; petals small, hooded, scarcely exceeding the nearly sessile stamens; berries purplish-black, 6-9 mm. long. Cascara, chittam bark.
    British Columbia southward, chiefly on the w. side of the Cascades, to n. coastal and Sierran
Calif., e. through Ida. to w. Mont. Apr.-June.
    Cascara bark is much sought after because of its medicinal value. The plant is by no means unat-
tractive, but scarcely a valuable ornamental.

VITACEAE Grape Family

    Flowers cymose to paniculate, small, complete to imperfect, regular, perigynous (hypogynous);
calyx minutely (3) 4- to 5 (7)-lobed; petals as many as the sepals, valvate, white or greenish, free
or partially connate and calyptrate at anthesis; stamens as many as the petals and opposite them, usually inserted at the edge of the small, perigynous, entire to lobed disc; pistil one, 2 (3-6)-carpellary,
the ovary superior, 2 (3-6)-celled, each cell 2 (1)-ovuled; stigma capitate to peltate, sessile or on a
short style; fruit a berry; (arborescent) vining and tendril-bearing shrubs with alternate (opposite),
simple to compound, stipulate, deciduous or evergreen leaves, the nodes often swollen.
    Ten or 11 genera and perhaps 600 species, chiefly tropical and subtropical, with several genera of
horticultural value. Besides *Vitis, Parthenocissus* (Boston ivy and Virginia creeper) may sometimes
escape and persist for short periods, but is not known to become established. It differs from *Vitis* in
having cymose flowers with distinct petals, and usually disc-tipped tendrils.

Vitis L. Grape

    Flowers paniculate, small, greenish, perfect or imperfect, the petals connate above and caducous
as a whole at anthesis; fruit a fleshy berry with 2 to 4 pyriform seeds; deciduous (a few evergreen),
tendril-bearing vines with large, palmately veined, subentire to palmately lobed, toothed leaves.
    About 60 species, chiefly of the N. Temp. Zone. (The ancient Latin name for the grape.)
    Although there are no grapes known to be native to our area, at least 2 or 3 species have escaped
from cultivation and might well be mistaken for native elements of our flora.
1 Leaves mostly 5-lobed over half their length, tomentose beneath when young, floccose
    to glabrous with age   V. VINIFERA
1 Leaves unlobed or lobed less than half their length, often glabrous or merely pubescent
    2 Leaves usually floccose or lanate, crenate-serrate, the teeth much broader (basally)
        than long; native from s. Oreg. to Calif., not known from within our range   *V. californica* Benth.

2 Leaves usually glabrous to pubescent, strongly serrate-dentate, the teeth longer than
   broad   V. RIPARIA

Vitis riparia Michx. Fl. Bor. Am. 2:231. 1803.
   V. cordifolia var. *riparia* Brunet, Cat. Pl. Can. 34. 1865. ("Hab. ad ripas et in insulis fluviorum
     Ohio, Mississippi, etc. ")
   Vigorous, strongly climbing vine; leaf blades 6-15 cm. long, mostly shallowly and acuminately 3-
lobed and coarsely serrate-dentate, bright green and glabrous above, paler beneath and glabrous to
somewhat pubescent; flowers paniculate, fragrant; fruit glaucous, black, about 8 mm. long. 2N=38.
June or riverbank grape.
   From N.S. to Va., Mo., Tex., and N.M. May-July.
   This grape is sometimes grown for its handsome fall foliage, and tends to escape or to persist after
most signs of human habitation have been lost. It is known from our area in s.c. Mont. and from near
Portland, Oreg.

Vitis vinifera L. Sp. Pl. 202. 1753. (Europe)
   Strongly vining, the young growth usually tomentose or floccose, the pubescence tending to persist;
leaf blades 7-12 cm. long, deeply 3- to 5-lobed over half their length and coarsely once or twice serrate-dentate; flowers paniculate; fruit red to black (green), usually glaucous, about 1 cm. long. 2N=
38, 57, 76.
   Native to Eurasia, but long cultivated and consisting of many varieties and phases; in our area
known from the Snake R. Canyon in s.e. Wash., and apparently also from near Portland, Oreg. June-
July.

MALVACEAE Mallow Family

   Flowers single and axillary to clustered-cymose or racemose, regular or slightly irregular, polypetalous, hypogynous, perfect to functionally pistillate (rarely the plants dioecious); sepals 5, distinct
or more or less fused, sometimes alternately bracteolate at base; petals 5, white or yellow to pink or
lavender, short-clawed, inserted on the staminal tube somewhat above the receptacle; stamens numerous, monadelphous, forming a slender elongate tube, the anthers freed separately or in groups of
2-6 from the upper portion of the tube; anthers reniform, 1-locular, dehiscing all the way around;
pistil 2- to many-carpellary, the stigmas globose to linear and stigmatose full length; ovary superior,
2- to many-loculed, the segments usually in a weakly connate ring around a central axis; fruit a capsule or a schizocarp, rarely a berry or samara; seeds reniform, 1-several per locule, the embryo
curved, endosperm scanty; herbs or shrubs (ours) to trees, with alternate, stipulate, entire to palmately lobed leaves, usually stellate-pubescent, the sap often mucilaginous.
   A cosmopolitan family of perhaps 75 genera and 1200 species, largely of temperate or tropical (often truly desert) areas, including several ornamental plants such as *Hibiscus* and the hollyhock *(Al-
thaea)* and one plant of great economic importance, cotton *(Gossypium)*.
   Reference:
   Kearney, Thomas H. The American genera of Malvaceae. Am. Midl. Nat. 46:93-131. 1951.
   In addition to the following members that are well established in our area, two others may occasionally escape and be mistaken for part of our native flora: *Abutilon theophrasti* Medic. has bright yellow
petals, ebracteolate calyces, and 10-15 two-horned, several-seeded carpels; and *Modiola caroliniana*
(L.) G. Don has deep orange-reddish petals barely 5 mm. long, capitate stigmas, and carpels that
are partitioned into two 1-seeded cells.
1 Petals yellow; carpels 1-seeded   SIDA
1 Petals white, pink, or lavender to reddish; carpels often more than 1-seeded
   2 Stigmas terminal and capitate
     3 Petals mostly over 2 cm. long; leaves large and suggestive of those of the grape; car-
       pels smooth on the sides, dehiscent full length   ILIAMNA
     3 Petals mostly less than 2 cm. long; leaves small and often deeply lobed to incised;
       carpels strongly reticulate on the sides below the middle   SPHAERALCEA
   2 Stigmas extending the full length of the style branches (on the inner surface)
     4 Stamens freed from the top of the staminal tube in 2-4 series of connate groups of
       2-6; calyx mostly ebracteolate   SIDALCEA

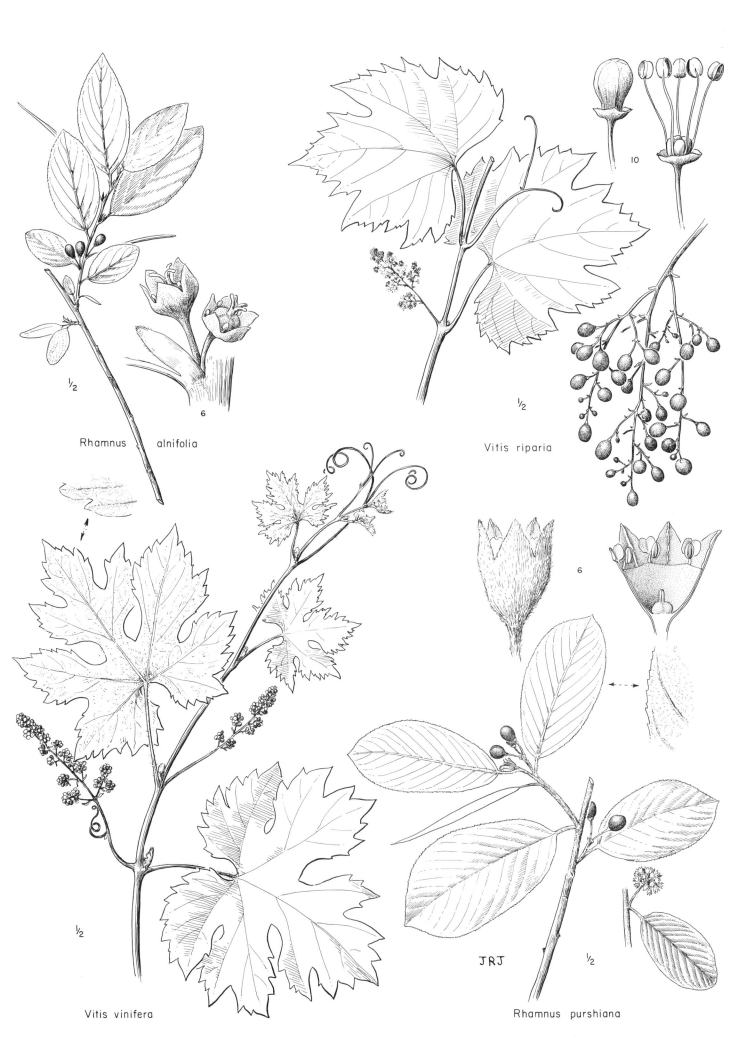

Rhamnus alnifolia

½

6

Vitis riparia

½

10

Vitis vinifera

½

Rhamnus purshiana

½

6

JRJ

4 Stamens freed singly or in pairs from the upper third of the staminal tube; calyx basally
   bracteolate   MALVA

Iliamna Greene

Flowers borne in loose racemes with usually several per node, pinkish to lavender; calyx tribracteolate; petals ciliate-clawed; filaments freed separately from the upper 3/4 of the staminal tube;
stigmas capitate; carpels many, dehiscent full length, prominently hairy with long stiff hairs and
small stellae, their sides smooth; seeds 2-4 per carpel; herbaceous stellate perennials with deciduous
stipules and (3) 5- to 9-lobed leaf blades.
   Seven N.Am. species, mostly along watercourses in the west. (Meaning unknown.)
   These plants are attractive perennials, perhaps too large for many gardens, but worth a place in
most.
   Reference:
   Wiggins, I. L. A resurrection and revision of the genus Iliamna. Contr. Dudley Herb. 1:213-229.
      1936.
1 Sepals less than 1 cm. long; pedicels stout, mostly less than 1 cm. long   I. RIVULARIS
1 Sepals usually at least 1 cm. long; pedicels slender, some of them usually over
   1 cm. long   I. LONGISEPALA

Iliamna longisepala (Torr.) Wiggins, Contr. Dudley Herb. 1:227. 1936.
   Sphaeralcea longisepala Torr. Bot. Wilkes Exp. 255. 1874. *Phymosia longisepala* Rydb. Bull. Tor-
      rey Club 40:61. 1913. *(Wilkes' Expedition,* Upper Columbia, Wash.)
   Stems puberulent and hirsute, 1-2 m. tall; stipules linear-lanceolate, about 1 cm. long; leaf blades
4-10 cm. long, broadly cordate, 5- to 7-lobed, the lobes triangular, coarsely crenate-serrate, finely
stellate; flowers rose-purplish, 1-several in the axils, on slender pedicels 1-5 cm. long; bracteoles
of the calyx linear to lanceolate; sepals about 1.5 cm. long, acuminate; petals approximately 2 cm.
long; carpels about 8 mm. long, the sides smooth, the back with long stiff brownish hairs and tiny soft
stellae; seeds finely echinulate.
   From sagebrush to ponderosa pine areas in the lower levels on the e. side of the Cascade Mts.,
Kittitas Co. to Chelan and Douglas cos., Wash. June-Aug.

Iliamna rivularis (Dougl.) Greene, Leafl. 1:206. 1906.
   Malva rivularis Dougl. ex Hook. Fl. Bor. Am. 1:107. 1831. *Sphaeralcea rivularis* Torr. ex Gray,
      Pl. Fendl. 23. 1849. *Phymosia rivularis* Rydb. Bull. Torrey Club 40:60. 1913. *(Douglas,* "Com-
      mon on the banks in North-West America, from the ocean to the Rocky Mountains")
   Sphaeralcea acerifolia Nutt. ex T. & G. Fl. N.Am. 1:228. 1838. *I. acerifolia* Greene, Leafl. 1:206.
      1906. *Phymosia acerifolia* Rydb. Bull. Torrey Club 40:60. 1913. *(Nuttall,* "rivulets east of Walla-
      wallah") = var. *rivularis.*
   ILIAMNA RIVULARIS var. DIVERSA (A. Nels.) C. L. Hitchc. hoc loc. *Sphaeralcea rivularis* var.
      diversa A. Nels. Bot. Gaz. 52:266. 1911. *Iliamna rivularis* ssp. *diversa* Wiggins, Contr. Dud-
      ley Herb. 1:222. 1936. *(Macbride 582,* Manyon Creek, Elmore Co., Ida., Aug. 11, 1910)
   Very similar to *I. longisepala* but the leaves more variable, 3- to 7-lobed, 5-15 cm. long; pedicels
shorter and stouter, mostly less than 1 cm. long; sepals commonly obtuse (acute), 3-5 mm. long;
carpels more rounded at the tip. N=33. Wild hollyhock.
   East side of the Cascades, from B.C. to Oreg., e. to Mont. and Colo. June-Aug.
   The species shows much fortuitous variation in leaf shape and in the quantity of the pubescence, although there seem to be no recognizable geographic variants with the exception of var. *diversa* (A.
Nels.) C. L. Hitchc. Variety *diversa,* occasional from c. Ida. to Yellowstone Nat. Park, is characterized by 3- to 5-lobed, truncate-based leaves; in comparison, the more common and widespread
var. *rivularis* has cordate, 5- to 7-lobed leaves.

Malva L.

Flowers in axillary clusters or subterminal panicles, inconspicuous to showy; calyx basally tribracteolate; petals whitish to lavender, usually retuse; stamens not connate in groups, but freed from the
staminal tube singly or in pairs; style branches 10-15, stigmatic most of their length, not terminally
enlarged; carpels equal in number to the styles, 1-seeded, smooth to corrugated, glabrous to hairy,

separating at maturity; annual to perennial herbs with long-petiolate, stipulate leaves, the blades from ovate to reniform and from very shallowly lobed to dissected, the pubescence simple to branched or stellate.

Perhaps 25 species of the Old World, ours weedy introductions. (Name from the Greek *malakos,* soft, referring to the supposed emollient property of the leaves.)

Children usually refer to these plants as "cheese-weeds" because of the resemblance of the segmented fruits to old-fashioned "wheel" cheeses.

1 Upper cauline leaves dissected into linear segments; flowers showy; petals 2-3 cm. long
                                                                      M. MOSCHATA
1 Upper cauline leaves shallowly lobed, but not dissected into linear segments; flowers usually not showy; petals commonly less than 2 cm. long
  2 Petals (1) 1.5-2 cm. long; calyx bracteoles ovate or oblong   M. SYLVESTRIS
  2 Petals usually less than 1.5 cm. long; bracteoles linear
    3 Carpels rounded and smooth (except for the puberulence) on the back; petals 2-4 times as long as the calyx   M. NEGLECTA
    3 Carpels flattened and wrinkled on the back; petals from shorter to only slightly longer than the calyx   M. PARVIFLORA

Malva moschata L. Sp. Pl. 690. 1753. ("Habitat in Italia, Gallia")

Papillose-strigose to glabrate perennial usually 3-6 dm. tall; blades of the basal leaves cordate-reniform and crenate to shallowly lobed, those of the upper cauline leaves cleft to the base into 5 (3) lobes and then dissected into linear segments; stipules oblong, 4-7 mm. long; flowers white to deep pinkish, averaging 4-5 cm. broad; petals cuneate to obcordate; carpels densely hairy on the back. N = 21.

A fairly common escape from gardens, mostly along roadsides and chiefly w. of the Cascades in Oreg. and Wash.; native to Europe. May-July.

Malva neglecta Wallr. in Syll. Ratisb. 1:140. 1824. (Europe)

M. rotundifolia as commonly considered in the U.S., but not of L.

Hirsutulous, spreading annual (biennial) mostly 2-6 dm. tall; petioles several times the length of the blades, the blades cordate-reniform, averaging 1.5-4 cm. long, crenate-dentate, very inconspicuously 5 (7)-lobed; flowers pale bluish-lavender to nearly white; calyx 1/2-1/4 the length of the corolla, very shallowly lobed; petals about 10 mm. long, emarginate; carpels rounded on the back, puberulent but otherwise nearly or quite smooth. N=21.

A fairly common European weed throughout the U.S. May-Sept.

Malva parviflora L. Amoen. Acad. 3rd ed. 416. 1787. (Europe)

M. borealis Wallm. in Liljebl. Svensk. Fl. 3rd ed. 374. 1816. (Europe)

M. pusilla sensu Rydb. Fl. Rocky Mts. 557. 1922.

Sparsely to densely hirsutulous-puberulent, prostrate to spreading annual or biennial, with stems 2-6 dm. long; leaf blades reniform-cordate, 2-5 cm. long, somewhat broader, finely crenate-dentate and shallowly 5- to 7-lobed, usually no more than half as long as the petioles; flowers from nearly sessile to long-pedicellate; petals white or pale lavender, about equaling the calyx, the claw glabrous; carpels flattened and strongly cross-corrugated on the back, glabrous to hirsutulous-puberulent. N=21.

A weed of European origin, more or less common in waste places in most of the U.S. Mar.-Aug.

Malva rotundifolia L. (Sp. Pl. 588. 1753), which differs from *M. parviflora* almost solely because of ciliate claws to the petals, is known from c. and e. N.Am., and is to be expected in our area, although not presently known to occur therein. It, too, is native to Europe.

Malva sylvestris L. Sp. Pl. 689. 1753. ("Habitat in Europae campestribus")

Strigillose or hirsute (glabrate) biennial or annual 2-5 dm. tall; stipules mostly 2-4 mm. long; leaf blades usually considerably shorter than the petioles, cordate-ovate, averaging 3-7 cm. long, 5 (7-9)-lobed about 1/3 of their length, the lobes crenate-dentate; flowers deep bluish-purple; bracteoles oblong or ovate; calyx lobes scarcely half as long as broad; petals (1) 1.5-2 cm. long, emarginate; carpels rugose and sparsely hairy. N=21.

A European weed common in waste places throughout much of N.Am.; in our area chiefly on the w. side of the Cascades in Wash. and Oreg. May-Sept.

Sida L.

Flowers axillary, perfect; calyx with 1-3 linear bracteoles (ours); petals yellow, stellate on the edge exposed in the bud; filaments freed individually from the staminal tube; stigmas capitate; carpels 5-12, very thin but strongly reticulate on the contacting faces, stellate on the back, 1-seeded, indehiscent (ours) or 2-valvate; densely stellate perennial herbs with linear, deciduous stipules.

A large genus of well over 100 species, general in the subtropics. (Greek, the meaning not known.)

Sida hederacea (Dougl.) Torr. ex Gray, Pl. Fendl. 23. 1849.
  Malva hederacea Dougl. ex Hook. Fl. Bor. Am. 1:107. 1831. *Disella hederacea* Greene, Leafl. 1:
    209. 1906. *(Douglas, "in the interior districts of the Columbia")*
  Malva plicata Nutt. ex T. & G. Fl. N. Am. 1:227. 1838. *(Nuttall, "On the Wallawallah, Oregon")*
  Sida? obliqua Nutt. ex T. & G. op. cit. 233. *(Nuttall, "On the Wallawallah River")*
  Prostrate to erect, grayish-stellate perennial herb with stems 1.5-4 dm. long; leaf blades reniform, 1.5-3 cm. long, crenate and very shallowly lobed; petals about 1 cm. long, yellow within, reddish-purple and stellate on the surface exposed in the bud; carpels 6-10. N=11. Alkali mallow.

In sandy (often saline) soil on the e. side of the Cascades, from Okanogan Co., Wash., to Calif., Ariz., Okla., Tex., and Mex. June-Sept.

Sidalcea Gray

Flowers white to deep pink or pinkish-lavender, borne in open or contracted and sometimes semispicate racemes, strongly dimorphic, those of the perfect-flowered plants the largest; petals conspicuously ciliate on the claws, usually erose to deeply emarginate; stamens 40-70, freed from the staminal tube in 2 or 3 series, the outer series commonly of 5 phalanges of 4-6 (2) stamens, the middle series usually of 5 groups of 2-4 stamens each (although often lacking), the inner series usually freed somewhat above the others and generally of 10 pairs of partially connate filaments; pistillate flowers mostly with the anthers sterile or rudimentary, sometimes merely with a crenate-topped staminal tube and no vestige of the anthers; styles equal in number to the segments of the ovary, elongate, stigmatic full length; carpels 5-10, tardily separating, 1-seeded, from nearly smooth all over to conspicuously reticulate-alveolate at least on the sides, sparsely glandular-puberulent to stellate at the top and on the back, tipped by a rather inconspicuous, persistent, short, solid beak; annual or (ours) perennial herbs with taproots or short rootstocks, usually stellate and often somewhat hirsute.

Perhaps 25 species of w. N. Am., almost all in the U.S. *(Sida,* plus *Alcea,* both names of other genera of the *Malvaceae.)*

Sidalcea is a notably variable genus, with a basic diploid chromosome number of 20; but several species have diploid, tetraploid, and hexaploid populations that are not visibly distinguishable from one another. Many of the various taxa are interfertile and often hybridize in the field.

Almost all the perennial species of *Sidalcea* are attractive (if not choice) plants, readily grown, and completely hardy on both sides of the Cascades, but the best are probably S. *cusickii* and S. *hendersonii.*

References:
Hitchcock, C. Leo. A Study of the perennial species of Sidalcea. U. Wash. Pub. Biol. 18:1-79. 1957.
Roush, E. M. F. A monograph of the genus Sidalcea. Ann. Mo. Bot. Gard. 18:117-244. 1931.
1 Plants from enlarged, rather fleshy, tap or fascicled roots, not at all rhizomatous,
    rather sparsely hirsute throughout, but especially so on the calyx; racemes elongate
    and loosely flowered; carpels nearly smooth; Baker Co., Oreg., southward   S. NEOMEXICANA
1 Plants from woody roots, often rhizomatous, usually rather conspicuously stellate, at
    least on the calyx; racemes often congested and spicate; carpels mostly strongly reticulate-alveolate
  2 Petals white to pale pink or pinkish-orchid; lower portion of the stems usually hirsute with simple (geminate) hairs; racemes elongate, loosely flowered; carpels about
      3.5 mm. long, rather prominently reticulate-pitted on the sides; plant with short
      rhizomes; from Portland s. through the Willamette Valley, Oreg.   S. CAMPESTRIS
  2 Petals usually fairly deep pink to pinkish-lavender; lower stems often stellate; racemes
      frequently closely flowered and spikelike; plant with or without rhizomes
    3 Rootstocks lacking, the plants frequently with decumbent branches but these not rooting; herbage often conspicuously glaucous; carpels from nearly smooth to reticulate,

        but not truly pitted; mostly e. of the Cascades in our area   S. OREGANA
3 Rootstocks present although often short and thick; stems frequently trailing and rather
      freely rooting; plants not particularly glaucous; carpels sometimes pitted; mostly w.
      of the Cascades
    4 Stems somewhat fistulose, 5-15 dm. tall, glabrous or only sparsely pubescent with
         short, simple to geminate hairs at the base; racemes compounded, closely flow-
         ered and spikelike; calyx nearly or quite glabrous (except for the ciliation), purplish,
         enlarging considerably in fruit; carpels about 4 mm. long, nearly or quite smooth,
         the beak usually at least 1 mm. long; coastal in distribution, mostly on or near tide-
         lands   S. HENDERSONII
    4 Stems seldom fistulose (except sometimes in *S. cusickii* and *S. hirtipes),* mostly rather
         densely pubescent throughout; racemes often open; calyx usually abundantly hairy; car-
         pels sometimes less than 3.5 mm. long, often prominently reticulate-alveolate, the
         beak mostly less than 1 mm. long
      5 Calyx 4-6 mm. long, subglabrous to fairly thickly pubescent with tiny stellae, usu-
           ally purplish; petals 5-15 mm. long; carpels lightly reticulate on the sides, the
           beak less than 0.5 mm. long; stems glabrous to sparsely hirsute with short, ap-
           pressed, simple hairs; n. Willamette Valley, Oreg.   S. NELSONIANA
      5 Calyx mostly over 6 mm. long, usually densely hairy; petals frequently over 15 mm.
           long; carpels often coarsely reticulate, the beak commonly at least 0.5 mm. long;
           stamens generally very hairy
        6 Stems hirsute at base with rather stiff, simple or geminate hairs 1-2.5 mm. long;
           racemes spikelike, congested, usually less than 8 cm. long; calyx 9-15 mm. long,
           enlarged considerably in fruit, finely stellate but also prominently hirsute with
           hairs 1-2 mm. long; carpels prominently reticulate-alveolate; near the coast in
           n. Oreg., and in Clark and Lewis cos., Wash.   S. HIRTIPES
        6 Stems usually stellate; racemes elongate; calyx less than 9 mm. long, or plant
           otherwise not as above
          7 Calyx lobes widened above the base and somewhat ovate-lanceolate, prominent-
             ly veined; rhizomes usually short and thick; stems (4) 5-18 dm. tall, often
             somewhat fistulose; racemes compounded, spikelike, closely flowered; petals
             often more than 5, truncate or erose, 10-18 mm. long; carpels about 3 mm.
             long, smooth or only slightly reticulate on the sides; barely reaching our area
             s. of Eugene, Oreg.   S. CUSICKII
          7 Calyx lobes not widened above the base, usually tapered rather uniformly to the
             tip; stems mostly less than 10 dm. tall, not fistulose, usually trailing and
             freely rooting at base, often truly rhizomatous; racemes commonly rather
             open; petals 5, rounded, retuse, 15-30 mm. long; carpels usually rather con-
             spicuously reticulate-alveolate; c. Willamette Valley, Oreg., southward   S. VIRGATA

Sidalcea campestris Greene, Bull. Calif. Acad. Sci. 1:76. 1885. *(Howell 614,* dry prairies, Oreg.)
    S. asplenifolia Greene, Pitt. 3:158. 1897. *(Piper,* Seattle, Wash., July, 1891)
    Perennial from a thick taproot and short thick rootstocks; stems 0.5-2 m. tall, usually hirsute, or
hirsute and stellate below; leaves 5-15 cm. broad, (5) 7- to 9-lobed; racemes elongate and open, the
pedicels 3-6 (20) mm. long; calyx from uniformly stellate to coarsely stellate and bristly-hirsute; petals from nearly white to pale pink or pinkish-orchid, 12-25 mm. long; carpels about 3.5 mm. long,
prominently reticulate-alveolate on the sides, the beak about 0.5 mm. long. N=30.
    In dry fields and along roadsides; Willamette Valley, Oreg., from Portland southward; introduced,
probably as a garden plant, in Seattle, Wash., but not persistent there. May-July.

Sidalcea cusickii Piper, Proc. Biol. Soc. Wash. 29:99. 1916.
    S. oregana var. *cusickii* Roush, Ann. Mo. Bot. Gard. 18:174. 1931. *(Cusick 4147,* in swales near
    Roseburg, Oreg., June, 1914)
    Perennial from a heavy taproot and with rootstocks varying from short and thick to slender and rath-
er elongate; stems 4-18 dm. tall, often fistulose, glabrous or very finely scabrid-pubescent with 2-
to many-rayed, appressed hairs below, but always finely stellate above; racemes usually compounded,
very tightly flowered and spikelike, the pedicels 1-2 (5) mm. long; calyx 6-10 mm. long, finely grayish-stellate to subglabrous and purplish, the lobes broadened above the base and somewhat ovate-lan-

8
carpel

1.5

5   Sida hederacea   1/2

Malva neglecta

6

1/2   Malva parviflora

8

1/2   Malva   sylvestris

1/2

1/16

Sidalcea campestris

2   carpel

2

Malva moschata   Iliamna rivularis   Iliamna longisepala

JRJ

1/2   1/2   1/2

ceolate, prominently veined; petals deep pink, 10-18 mm. long; carpels about 3 mm. long, nearly smooth on the back, lightly reticulate on the sides, the beak 0. 5-1 mm. long. N=10.

In open fields and along roadsides, mostly on rather heavy soil; valleys of the Coquille and Umpqua rivers, Oreg., barely reaching our area. May-June.

Sidalcea hendersonii Wats. Proc. Am. Acad. 23:262. 1888. *(Henderson,* Clatsop Bay, Oreg.)

Perennial from a heavy taproot and short, thick, ascending rootstocks, the stems 5-15 dm. tall, fistulose, glabrous or subglabrous at the base and usually purplish-tinged, as also the stipules, petioles, and calyces; racemes compound, congested and spikelike; calyx 8-12 mm. long, enlarging conspicuously in fruit, usually glabrous except for the ciliate lobes, sometimes more or less finely stellate all over; petals deep pink, carpels about 4 mm. long, smooth or slightly wrinkled on the margins, the beak 0. 8-1. 3 mm. long. N=10.

Along the coast from Vancouver I. to the mouth of the Umpqua R. in Douglas Co., Oreg., on (or adjacent to) tideland. June-Aug.

Sidalcea hirtipes C. L. Hitchc. U. Wash. Pub. Biol. 18:42. 1957. *(C. L. Hitchcock 19529,* 15 mi. s. of Tillamook, Tillamook Co., Oreg., July 1, 1951)

A large-clumped perennial with short thick rhizomes; stems 7-13 dm. tall, occasionally slightly fistulose, copiously hirsute with rather stiff, simple or geminate (to cruciate) hairs as much as 2. 5 mm. long; racemes usually compound, spikelike, closely many-flowered; pedicels 1-3 (5) mm. long; calyx 9-15 mm. long, enlarged considerably in fruit, finely stellate but conspicuously ciliate and hirsute with simple to stellate hairs 1-2 mm. long; petals pink; carpels 3. 5-4 mm. long, prominently reticulate-alveolate on the sides, the beak 0. 6-0. 8 mm. long. N=30.

Northern Lincoln Co. to Tillamook and Clatsop cos., Oreg., and Clark and Lewis cos., Wash., from coastal mountains to bluffs along the ocean, but not on tideflats. June-July.

Sidalcea nelsoniana Piper, Proc. Biol. Soc. Wash. 32:41. 1919. *(J. C. Nelson,* Salem, Oreg.)

Perennial with a stout taproot and short rootstocks, the stems 4-10 dm. tall, glabrous to hirsute with short, appressed, simple hairs; racemes somewhat spikelike but elongate and open, many-flowered, the pedicels usually about 3 mm. long; calyx 4-6 mm. long, usually purplish-tinged, subglabrous to fairly thickly pubescent with tiny stellae; petals 5-15 mm. long, pinkish-lavender; carpels about 3 mm. long, lightly reticulate on the sides, the beak less than 0. 5 mm. long. N=10.

On gravelly, well-drained soil; Willamette Valley from Salem to Portland, w. to e. Tillamook Co., Oreg. May-July.

Sidalcea neomexicana Gray, Pl. Fendl. 23. 1849. *(Fendler,* Santa Fe, N.M.)

SIDALCEA NEOMEXICANA var. CRENULATA (A. Nels.) C. L. Hitchc. hoc loc. *S. crenulata* A. Nels. Proc. Biol. Soc. Wash. 17:93. 1904. *S. neomexicana* ssp. *crenulata* C. L. Hitchc. U. Wash. Pub. Biol. 18:73. 1957. *(Goodding 1091,* Juab, Utah)

Perennial from an enlarged fleshy tap- or fascicled root, without rhizomes; stems 2-7 dm. tall, sparsely pubescent with short, appressed, partially geminate and partially simple hairs; racemes elongate, many-flowered, the pedicels slender, 5-8 (10) mm. long; calyx 5-8 mm. long, sparsely pubescent with short, simple to 4-rayed, pustulose hairs, usually without stellae; carpels about 2. 5 mm. long, lightly reticulate, the beak about 0. 6 mm. long. N=10.

On moist, usually strongly alkaline, heavy soil from e.c. Oreg. to Wyo., southward to s. Calif., N. M., and n. Mex. June-July.

Our plants, as described above, are referable to the var. *crenulata* (A. Nels.) C. L. Hitchc., which occurs from e. c. Oreg. to c. Utah and Nev. Other varieties of the species range from Wyo. to N. M., Mex., and s. Calif.

Sidalcea oregana (Nutt.) Gray, Pl. Fendl. 20. 1849.

Sida oregana Nutt. ex T. & G. Fl. N. Am. 1:234. 1838. *(Nuttall,* "West side of the Rocky Mountains") *Callirhoë spicata* Regel, Gartenfl. 21:291, pl. 737. 1872. *Sidalcea spicata* Greene, Bull. Calif. Acad. Sci. 1:76. 1885. *Sidalcea oregana* var. *spicata* Jeps. Fl. Calif. 2:492. 1936. *Sidalcea oregana* ssp. *spicata* C. L. Hitchc. U. Wash. Pub. Biol. 18:64. 1957. (Plate in Gartenflora, drawn from plants grown from seeds supposedly collected in the Sierra Nevada of Calif.) *Sidalcea nervata* A. Nels. Proc. Biol. Soc. Wash. 17:94. 1904. *(A. Nelson 4101,* Evanston, Wyo.) = var. *oregana.*

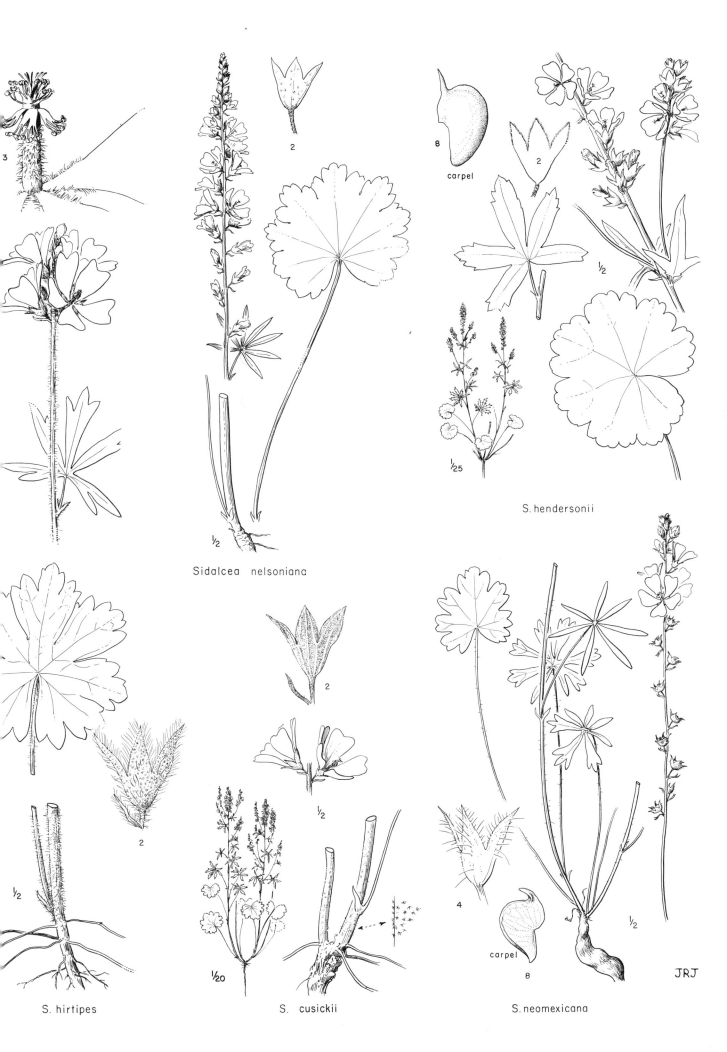

3

2

8
carpel

2

½

S.hendersonii

½

½

Sidalcea nelsoniana

2

2

½

½

½

½

carpel

4

½

carpel

8

JRJ

S. hirtipes

½

½

½

½

S. cusickii

S. neomexicana

Sidalcea maxima Peck, Madroño 6:13. 1941. *Sidalcea oregana* ssp. *oregana* var. *maxima* C. L.
Hitchc. U. Wash. Pub. Biol. 18:61. 1957. *(Peck 15435,* Dairy Cr. , 20 mi. n.w. of Lakeview,
Lake Co. , Oreg. , July 3, 1927)
Sidalcea oregana ssp. *oregana* var. *calva* C. L. Hitchc. U. Wash. Pub. Biol. 18:61. 1957. *(C. L.
Hitchcock 19427,* 12 mi. s. e. of Cashmere, Chelan Co. , Wash. , June 21, 1951)
Sidalcea oregana ssp. *oregana* var. *procera* C. L. Hitchc. U. Wash. Pub. Biol. 18:62. 1957. *(C. L.
Hitchcock 19432,* about 10 mi. e. of Coulee City, Grant Co. , Wash.)

Perennial from a stout taproot and branched crown, but without rootstocks; stems 2-15 dm. tall,
from glabrous to hirsute or stellate below, finely stellate above; racemes simple to compound, from
closely many-flowered and spikelike to open and lax, the pedicels 1-10 mm. long; calyx 3.5-9 mm.
long, from uniformly finely stellate to bristly with a mixture of longer, simple to 4-rayed, spreading
hairs sometimes as much as 2.5 mm. long; petals 1-2 cm. long, light pinkish to fairly deep water-
melon pink; carpels 2.5-3 mm. long, from nearly smooth to reticulate-alveolate on the sides and
back; beak 0.3-0.7 mm. long.

In sagebrush plains, meadowland, and ponderosa pine forest; c. Wash. to n. Calif. , almost entirely e. of the Cascades, eastward to Wyo. and Utah. Late May-July.

This is an extremely variable species, represented in our area by the following two subspecies and
several varieties:

1 Stems usually hirsute with simple hairs at the base; inflorescence densely crowded and
    spicate, the pedicels mostly 1-2 mm. long in flower; calyx averaging about 5 mm. long
    at anthesis; carpels nearly or quite smooth; c. Oreg. to the s. Sierra Nevada, Calif. ;
    N=10, 20   ssp. spicata (Regel) C. L. Hitchc.
1 Stems usually stellate near the base; inflorescence open (at least after flowering); ped-
    icels averaging 3 mm. in length; calyx usually over 5 mm. long at anthesis; carpels
    reticulate-rugose   ssp. oregana
  2 Stems pubescent near the base with coarse, spreading (not appressed), simple to
     stellate hairs; calyx usually well over 6 mm. long; extreme e. Wash. and adj.
     Ida. and n. Oreg. ; N=10, 20   var. procera C. L. Hitchc.
  2 Stems variously pubescent to glabrous, if stellate then the hairs usually appressed;
     calyx various
    3 Leaves nearly or quite glabrous, rather thick and fleshy; calyx sparsely stellate,
       the lobes subglabrous on the back but strongly ciliate with simple to stellate
       hairs 0.5-1 mm. long; stems glabrous at base or sparsely hairy with large, 4-
       rayed, appressed hairs; plants of the Wenatchee Mts. , Wash. ; N=30
                                           var. calva C. L. Hitchc.
    3 Leaves usually stellate; calyx uniformly stellate, but the lobes sometimes with
       cilia as much as 0.5 mm. long; stems variously pubescent to glabrous
      4 Stems glabrous or only sparsely stellate at base with 4- to 9-rayed, appressed
        hairs, often heavily glaucous; petals of the perfect flowers usually over 15 mm.
        long; s.c. Oreg. e. to Wyo. , s. to Utah and n. e. Calif.   var. oregana
      4 Stems usually moderately to strongly pubescent at base; if plants glaucous, then
        the petals of the perfect flowers seldom over 15 mm. long; c. and s. Oreg. to
        n. Calif. ; N=20   var. maxima (Peck) C. L. Hitchc.

Sidalcea virgata Howell, Fl. N.W. Am. 101. 1897.
  S. malvaeflora ssp. *virgata* C. L. Hitchc. U. Wash. Pub. Biol. 18:24. 1957. *(Howell,* Silverton,
  Oreg.)

Perennial from a strong taproot and short rootstocks or trailing, rooting branches, the stems 2-10
dm. tall, mostly uniformly pubescent with soft, several-rayed, fairly long hairs at the base, and becoming more finely stellate above; leaves 2-8 cm. broad, the lower blades often more or less reniform; racemes usually elongate and open, the pedicels 3-8 mm. long; calyx 6-12 mm. long, uniformly finely stellate; petals from pale to deep pinkish-rose, often conspicuously lighter-veined; carpels
3-4 mm. long, prominently reticulate-alveolate on the sides, sparsely glandular-puberulent to stellate
on the back, the beak about 0.5 mm. long. N=10, 20.

From fields and roadsides to grassy hillsides and lower mountains, especially in moist meadows;
Willamette Valley to s. Oreg. May-June.

The species is sympatric, and freely interfertile, with two or three subspecies of *S. malvaeflora*

(DC.) Gray ex Benth. in s. Oreg. Therefore it might just as consistently be considered as a variety or subspecies of that taxon.

Sphaeralcea St. Hil.

Flowers in simple to compound racemes; calyx 5-lobed, usually 3 (1-2)-bracteolate at base; petals 5, short-clawed, usually emarginate; filaments freed from the staminal tube individually; carpels 8-12, the styles equal in number to the carpels, the stigmas capitate; fruits stellate dorsally, rugose-reticulate and indehiscent on the lower 1/3-4/5, the upper portion dehiscent and smooth on the sides; seeds 1 or 2 per carpel, glabrous to puberulent; stellate-pubescent, perennial herbs (often somewhat frutescent at base) with deciduous stipules and thickish, simple to divided leaf blades.

Over 200 species of Africa and the Western Hemisphere. (Greek *sphaera*, sphere, and *alcea*, mallow, alluding to the globose fruits.)

These plants are good subjects for many gardens, perhaps best e. of the Cascades, but *S. coccinea*, the best of the 3 species, does well in sunny well-drained areas in the Puget Sound area.

Reference:

Kearney, T. H. The North American species of Sphaeralcea subgenus Eusphaeralcea. U. Calif. Pub. Bot. 19:1-128. 1935.

1 Leaves crenate to lobed somewhat less than halfway to the midvein; calyx bracteolate; seeds usually 1 per carpel   S. MUNROANA
1 Leaves divided nearly to the midvein; usually either the calyx ebracteolate or the seeds 2 per carpel
  2 Calyx usually ebracteolate; carpels 1-seeded, with over 2/3 of the lateral faces rugose-reticulate; flowers mostly in simple racemes   S. COCCINEA
  2 Calyx usually tribracteolate; carpels generally 2-seeded, with no more than the lower half of the lateral faces rugose-reticulate; inflorescence thyrsoid   S. GROSSULARIAEFOLIA

Sphaeralcea coccinea (Pursh) Rydb. Bull. Torrey Club 40:58. 1913.

Cristaria coccinea Pursh, Fl. Am. Sept. 453. 1814. *Malva coccinea* Nutt. in Fraser's Cat. 1813 (nomen), and in Gen. Pl. 2:81. 1818. *Sida coccinea* DC. Prodr. 1:465. 1824. *Malvastrum coccineum* Gray, Pl. Fendl. 21. 1849. *(Lewis*, "On the dry prairies and extensive plains of the Missouri")

Low, spreading, rhizomatous perennial, 1-2 dm. tall; leaf blades 2-5 cm. long, 3-parted, the divisions variously lobed, the upper surface yellowish-green, the lower surface more grayish, the stipules deciduous; flowers in short terminal bracteate racemes, the lower bracts leaflike, the upper ones reduced and deciduous; pedicels much shorter than the usually ebracteolate calyx; petals 1-2 cm. long, rusty-red, emarginate; carpels 1-seeded, about 3 mm. long, strongly pubescent on the back, the sides prominently reticulate, the apical, empty, nonreticulate portion usually not more than 1/10 the overall length of the carpel. N=5.

Primarily a species of the Great Plains which extends into w. Mont. as far as Granite and Powell cos. June-July.

Sphaeralcea grossulariaefolia (H. & A.) Rydb. Bull. Torrey Club 40:58. 1913.

Sida grossulariaefolia H. & A. Bot. Beechey Voy. 326. 1840. *Malvastrum grossulariaefolium* Gray, Pl. Fendl. 21. 1849. *Malvastrum coccineum* var. *grossulariaefolium* Torr. in Stansb. Exp. 384. 1852. *(Tolmie*, Snake River, Ida.)

Grayish-stellate, erect perennial 3-7 (10) dm. tall; leaf blades 2-5 cm. long, 3-parted, the lower divisions again deeply cleft; inflorescence interrupted, compound; calyx 5-10 mm. long, usually with 3 linear bracteoles at the base; petals reddish, 1.5-2 cm. long; carpels about 3 mm. long, generally 2-seeded, rugose-reticulate and indehiscent on the lower 1/3 to 1/2; seeds mostly puberulent. N=21.

Central Ida. to s.c. Wash., southward to Utah and Nev., from the open desert into the lower levels of the mountains. June-July.

Sphaeralcea munroana (Dougl.) Spach ex Gray, Proc. Am. Acad. 22:292. 1887.

Malva munroana Dougl. ex Lindl. Bot. Reg. 16:pl. 1306. 1830. *Nuttallia munroana* Nutt. Journ. Acad. Phila. 7:16. 1834. *Malvastrum munroanum* Gray, Pl. Fendl. 21. 1849. *Malveopsis munroana* Kuntze, Rev. Gen. 1:72. 1891. *(Douglas*, plains of the Columbia)

Thick-rooted, grayish-hairy to greenish, multistemmed perennial 2-8 dm. tall; leaf blades 2-6 cm.

8

2

4

carpel

3

Sphaeralcea coccinea

2

½

½

½

½

Sidalcea virgata

var. maxima

2

½

½

½

JRJ

8

2

ssp. spicata

var. procera

var. calva

var. oregana

Sidalcea oregana

long, reniform to ovate-deltoid, from rather shallowly 3- to 5-lobed to merely crenate; pedicels usually shorter than the calyx; calyx bracteoles usually 3, linear; petals 1-2 cm. long, apricot-pink to reddish; carpels about 3 mm. long, rugose-reticulate only on the lower (indehiscent) third, mostly 1-seeded.

From the open desert plains to lower mountain slopes; s. c. B. C. to w. Mont., Calif., and Utah. May-Aug.

HYPERICACEAE St. John's Wort Family

Flowers in leafy or bracteate, terminal or axillary cymes (rarely single), usually showy, white or yellow, complete, regular, hypogynous; sepals 5 (4), imbricate, sometimes partially connate; petals 5 (4), distinct, often short-clawed and glandular at the base; stamens 15-100, the filaments free or slightly connate basally into 3 (4-8) separate groups, the anthers 2-celled; pistil 3- to 5-carpellary, styles distinct, ovary 3 (4-5)-celled with axile placentation or 1-celled with 3 (5) parietal placentae; fruit a membranous capsule or a berry; seeds minute, cylindric, exalbuminous, the embryo straight or curved; annual or perennial herbs (shrubs or trees) with opposite or whorled, simple, sessile (ours) or petiolate, exstipulate, finely translucent-dotted or black-maculate leaves.

Seven or 8 genera and about 300 species, widely distributed but most abundant in the tropics and subtropics; only one genus in our area.

Hypericum L. St.-John's-Wort

Foliage and perianth usually blackish- or purplish-dotted along the margins; sepals 5; petals 5, yellow; stamens numerous, free or basally connate into groups; ovary 1-celled with 3 (5) parietal placentae, or 3- to 5-celled; fruit a septicidal capsule; annual or perennial, glabrous, sessile-leaved herbs.

Perhaps 250 species, cosmopolitan. (Greek, the meaning not certain.)

Many of the species are of horticultural value, and in addition to the following native or well-established weedy species, some of the large-flowered, leathery-leaved cultivated species, such as *H. moserianum* Andre., occasionally escape from gardens and persist for some time.

Of our native species *H. anagalloides* is easily introduced and in moist areas will form attractive mats, sometimes too readily. *H. formosum* is much larger-flowered and on the whole rather choice.

1 Petals usually less than 6 mm. long, scarcely exceeding the sepals, not black-dotted along the margins; stamens less than 50, only slightly if at all connate at base
  2 Stems erect, over 1 dm. tall; leaves mostly more than 1.5 cm. long   H. MAJUS
  2 Stems procumbent, matted; leaves usually less than 1.5 cm. long   H. ANAGALLOIDES
1 Petals either well over 6 mm. long and usually much longer than the sepals, or blackish-glandular along the margins; stamens usually over 50, basally connate into 3 (4-5) groups
  3 Sepals linear-lanceolate, 3-5 times as long as broad, mostly acute; leaves lanceolate to obovate-oblanceolate or narrowly spatulate-oblanceolate; seeds brownish, usually over 1 mm. long, distinctly pitted in longitudinal rows, not striate   H. PERFORATUM
  3 Sepals triangular to ovate-lanceolate, less than 3 times as long as broad, rounded to acute; leaves ovate-oblong to ovate; seeds yellowish, 1 mm. long or less, indistinctly reticulate-alveolate but not pitted, apparently longitudinally striate   H. FORMOSUM

Hypericum anagalloides C. & S. Linnaea 3:127. 1828. (*Chamisso*, San Francisco, Calif.)
  H. bryophytum Elmer, Bot. Gaz. 36:60. 1903. (*Elmer 2833*, Olympic Mts., Clallam Co., Wash., Aug., 1900)
  H. tapetoides A. Nels. Bot. Gaz. 52:266. 1911. (*Macbride 453*, Silver City, Owyhee Mts., Ida., July 22, 1910)

A low, diffuse perennial with numerous prostrate, freely rooting stems forming dense mats, the upright stems usually 5-15 cm. long, simple or branched above; leaves ovate or elliptic to obovate, 5-15 mm. long, 5- to 7-nerved, rounded, slightly clasping at the base; cymes leafy, few-flowered; sepals 2-3 mm. long, ovate-elliptic, rounded to acute, usually only slightly shorter than the salmon-yellow petals; stamens 15-25, longer than the sepals, the filaments rather stout, slightly or not at all connate at base; styles 3, slender, 1.5-2 mm. long; seeds yellow, about 0.5 mm. long, longitudinally striate and finely corrugate transversely.

On moist ground from along the coast to well up in the mountains; B. C. to Baja Calif., e. to Mont. June-Aug.

Hypericum formosum H. B. K. Nov. Gen. & Sp. 5:196. 1821. *(Humboldt & Bonpland,* Pazcuaro, Mex.)
   HYPERICUM FORMOSUM ssp. SCOULERI (Hook.) C. L. Hitchc. hoc loc. *H. scouleri* Hook. Fl.
      Bor. Am. 1:111. 1831. *H. formosum* var. *scouleri* Coult. Bot. Gaz. 11:108. 1886. *(Scouler,*
      "North-west coast of America, near the Columbia")
   HYPERICUM FORMOSUM var. NORTONIAE (M. E. Jones) C. L. Hitchc. hoc loc. *H. nortonae* M.
      E. Jones, Bull. U. Mont. Biol. 15:39, pl. 5. 1910. *(M. E. Jones,* McDonald Peak, Mont., is the
      first of several specimens rather vaguely cited)
   Perennial with rather widespread slender stolons and rhizomes, the erect stems usually many,
(0. 5) 1-8 dm. tall, simple to freely branched; leaves ovate-lanceolate to oblong-elliptic or obovate,
1-3 cm. long, usually purplish-black dotted (especially along the margins), the bases somewhat clasping; cymes few-flowered, leafy-bracteate; sepals triangular to ovate-lanceolate, 4-5 mm. long, more
or less purplish-black dotted, usually obtuse or rounded, sometimes denticulate; petals pale to bright
yellow, about twice the length of the sepals, conspicuously purplish-black dotted along the margins, or
blackish-denticulate; stamens 75-100, connate in 3 (4-5) distinct groups; styles slender, 3-5 mm.
long; capsule 6-9 mm. long, 3-celled; seeds about 0. 8 mm. long, yellow to brownish, finely reticulate-alveolate but apparently longitudinally striate, the alveolae very shallow.
   Mostly in moist places from along the coast to well up in the mountains; B. C. to Baja Calif., e. to
Mont., Wyo., and c. Mex. June-Sept.
   Our plants, as described above, belong to the ssp. *scouleri* (Hook.) C. L. Hitchc., from which ssp.
formosum of Mex. and s.w. U.S. (n. to Colo. and Utah) differs in its mostly narrower and sharper-
pointed sepals that are often more conspicuously black-striate or black-glandular on the margins. In
our area ssp. *scouleri* consists of two varieties: 1) var. *scouleri,* mostly over 2 dm. tall and frequently branched, occurring from the lowlands to moderate altitudes in the mountains, and 2) the subalpine ecotype, var. *nortoniae* (M. E. Jones) C. L. Hitchc., a more slender, mostly simple-stemmed
plant up to about 2 dm. tall, often with relatively broader leaves, occurring from c. Ida. (and n. e.
Oreg.) to Wyo., Mont., Alta., and B. C.

Hypericum majus (Gray) Britt. Mem. Torrey Club 5:225. 1894.
   H. canadense var. *majus* Gray, Man. 5th ed. 86. 1867. *(Robbins,* Lake Superior)
   Perennial with short leafy rhizomes, the upright stems 1-5 dm. tall, simple or branched above;
leaves 1-3. 5 cm. long, lanceolate to oblong, rounded, 5- to 7-nerved; cymes inconspicuously bracteate; flowers 4-7 mm. long, the petals about equal to the lanceolate sepals; stamens (10 ?) 15-35,
the filaments almost capillary, distinct; capsule 1-celled, blunt; styles 3, short; seeds yellow, about
0. 5 mm. long, longitudinally striate and finely transversely corrugate.
   On wet ground; B. C. to Que., southward to Pa., N. J., Ill., Ia., and Colo. ; collected in 1891 at
Green Lake, Seattle, Wash., but probably no longer growing in our area. July-Sept.

Hypericum perforatum L. Sp. Pl. 785. 1753. (Europe)
   Taprooted perennial but with short rhizomes and also sometimes stoloniferous, the erect stems 1-
several, usually rather freely branched, 3-8 dm. tall; leaves lanceolate to obovate-oblanceolate or
(mostly) narrowly spatulate-oblanceolate, 1-3 cm. long, more or less purplish-black dotted, narrowed and not clasping, but connected by narrow winglike structures at the base; cymes large and
compound, many-flowered, leafy-bracteate; sepals lanceolate, acute, 5-7 mm. long; petals about
twice as long as the sepals, their margins conspicuously purplish-black maculate; stamens 75-100,
connate at base into 3 (4-5) distinct groups, the anthers purplish-tipped; styles slender, 3-4 mm. long;
capsules 5-8 mm. long, 3-celled, acute; seeds brown, about 1.25 mm. long, conspicuously reticulate-
pitted, the prominent alveolae in longitudinal rows, the seeds not at all striate in appearance. N=16,
18. Klamath weed.
   A European weed, now a most serious pest on waste land and pastures throughout much of U. S., but
especially common from c. Calif. to near Tacoma, Wash., less abundant n. into B. C. and e. to Mont.
June-July.
   It is difficult to eradicate, and reputedly poisonous to livestock.

ELATINACEAE Waterwort Family

   Flowers solitary or in small cymes, axillary, inconspicuous, perfect, regular or irregular, hypogynous, many often cleistogamous; sepals and petals 2-5, distinct, or the persistent sepals somewhat
connate at base; stamens as many or twice as many as the petals, distinct; pistil 1, carpels 2-5, the

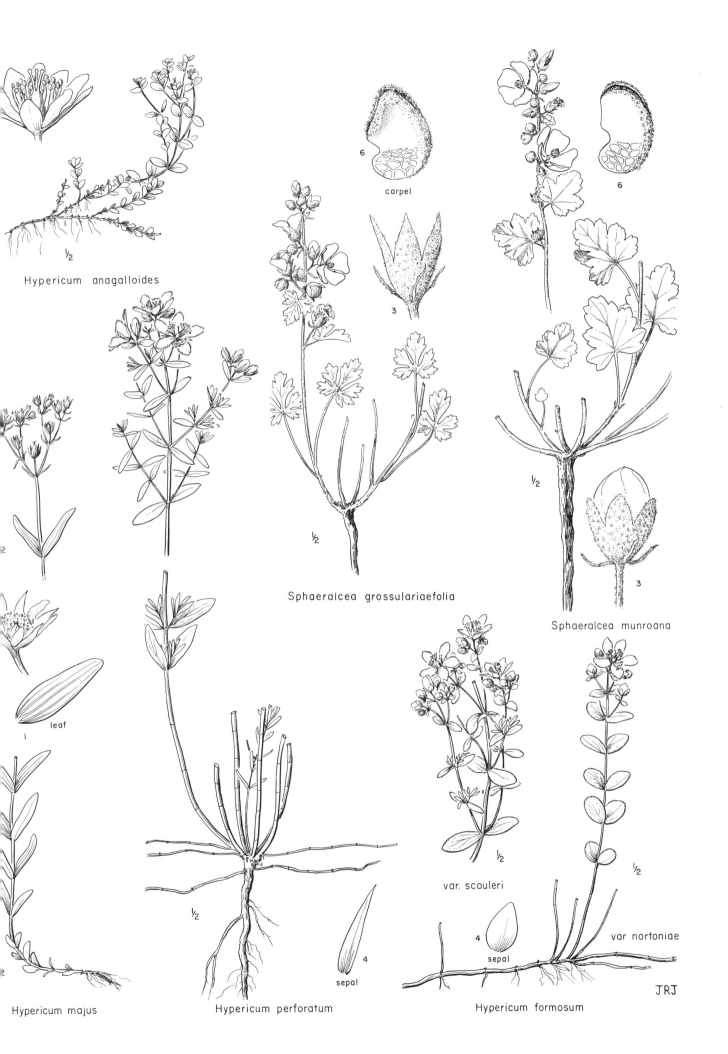

Hypericum anagalloides

½

carpel

6

3

Sphaeralcea grossulariaefolia

½

6

½

3

Sphaeralcea munroana

2

leaf

1

½

var. scouleri

½

½

var nortoniae

4
sepal

4
sepal

2

Hypericum majus

Hypericum perforatum

Hypericum formosum

JRJ

ovary superior, 2- to 5-loculed, stigmas nearly sessile, the short styles distinct; fruit a septicidal capsule with axile placentation; seeds numerous, reticulate or longitudinally striate, without endosperm; annual (perennial) aquatic or terrestrial herbs or subshrubs with opposite or whorled, membranous, stipulate, simple leaves.

Two genera and about 30 species, widely distributed.

1 Flowers 5-merous; sepals distinctly keeled; stems more or less woody at base, glandular-puberulent   BERGIA
1 Flowers 2- to 4-merous; sepals not keeled; stems entirely herbaceous, glabrous   ELATINE

Bergia L.

Flowers 1-several in the axils, pedicellate, 5-merous; sepals cuspidate, strongly ribbed; capsule globose; seeds lightly striate longitudinally; prostrate to erect herbs (ours somewhat woody at the base) with opposite leaves and whitish, glandular-pectinate stipules.

Fifteen to 20 species, widely distributed, but chiefly tropical or subtropical. (Named for Peter Jonas Bergius, Swedish Botanist, 1730-1790.)

Bergia texana (Hook.) Seubert, Walpers Rep. 1:285. 1842.

Merimea texana Hook. Ic. Pl. pl. 278. 1840. *Elatine texana* T. & G. Fl. N.Am. 1:678. 1840.

(*Drummond,* Texas)

Erect to decumbent, glandular-puberulent annual with freely branched stems 0.5-3 dm. long; leaves elliptic to oblanceolate, slender-petioled, glandular-denticulate, 2-4 cm. long; sepals about 3 mm. long, strongly carinate, coarsely glandular-denticulate and pubescent; petals white, shorter than the sepals; stamens 10 (5?), many of the flowers cleistogamous and with the anthers of the longer stamens closely appressed to the stigmas and usually adherent to them when the flower finally is forced open by the growth of the capsule.

On wet ground, often in vernal pools; known only from along the lower Columbia R. in our area; Calif. to Ill., s. to Tex. June-Nov.

Elatine L. Waterwort

Flowers minute, usually solitary in the axils, 2- to 4-merous; stigmas globose; capsules membranous, 2- to 4-celled; seeds regularly reticulate in longitudinal rows; tiny, glabrous, prostrate herbs with opposite, entire leaves and entire stipules, usually growing on mud flats or at the edge of ponds, freely rooting at the nodes.

About 15 species, widely distributed; blossoming from early summer until fall. (Supposedly from the Greek, meaning firlike.)

References:

Fassett, N. Elatine and other aquatics. Rhodora 41:367-377. 1939.

Fernald, M. L. Elatine in E. North America. Rhodora 19:10-15. 1917.

Mason, Herbert L. New species of Elatine in California. Rhodora 13:239-240. 1956.

Fernald was of the opinion that the (mainly) Eurasian *E. triandra* occurred, but not commonly, in e. U.S. Fassett was convinced that it was common in most of N.Am. and that it was separable into several infraspecific taxa, including vars. *triandra* and *brachysperma* (Gray) Fassett for our area. Mason has recently expressed the view that neither *E. americana* nor *E. triandra* occurs in Calif., and proposed the name, *E. gracilis* Mason, for the Californian plants usually called *E. triandra,* partially on the basis that *E. gracilis* has flowers with 2 sepals much larger than the third, whereas *E. triandra* has 3 equal or subequal sepals. Although the plants concerned from the Pacific Northwest seem more nearly to qualify as *E. triandra, E. gracilis* (if maintainable as a separate taxon) is to be expected also.

1 Capsules 4-celled, pedicellate; seeds curved to an approximate right angle   E. CALIFORNICA
1 Capsules 2- or 3-celled, sessile; seeds only slightly curved
  2 Pits of the seeds usually 10-15 per longitudinal row; leaves seldom at all notched at the tip   E. BRACHYSPERMA
  2 Pits of the seeds usually 18-27 per row; leaves often distinctly notched at the tip
    3 Leaves obovate, rounded at the tip, not notched   E. AMERICANA
    3 Leaves linear to narrowly oblanceolate, usually distinctly notched at the tip   E. TRIANDRA

Elatine americana (Pursh) Arnott. Edinb. Journ. Sci. 1:431. 1830.

Peplis americana Pursh, Fl. Am. Sept. 238. 1814. *E. triandra* var. *americana* Fassett, Rhodora
33:72. 1931. (Pennsylvania)

Plants matted or with some branches ascending and 2-4 cm. long; leaves spatulate to obovate, less
than 10 mm. long, the tips rounded and not at all emarginate; sepals and petals 3; stamens 3; seeds
somewhat curved, the pits mostly 18-26 in each of about 10 longitudinal rows.

Scarcely distinguishable from *E. triandra* and possibly not within our range; mostly along the Atlantic coast.

Elatine brachysperma Gray, Proc. Am. Acad. 13:361. 1878.

Alsinastrum brachyspermum Greene, Man. S. F. Bay Reg. 62. 1894. *E. triandra* var. *brachysperma*
Fassett, Rhodora 41:374. 1939. (Specimens from Ill., *E. Hall;* Texas, *E. Hall 37,* 1872; and Calif.,
Kellogg & Harford 257, were cited)

Plants submerged to terrestrial; leaves linear to narrowly oblong; flowers 2- or 3-merous; seeds
about 0.5 mm. long, brownish, the pits 10-15 in each of the 6-8 longitudinal rows.

Central Oreg. and s. Calif., also in Colo. and c. U.S.

Elatine californica Gray, Proc. Am. Acad. 13:361. 1878. *(J. G. Lemmon,* Sierra Valley, Calif.)

E. williamsii Rydb. Mem. N.Y. Bot. Gard. 1:260. 1900. *E. californica* var. *williamsii* Fassett,
Rhodora 41:375. 1939. *(R. S. Williams 855,* Big Belt Mts., Mont., Aug. 25, 1891)

Plants forming small mats; flowers short-pedicellate; sepals and petals 4; stamens 8; capsule 4-
celled; seeds rounded at one end, blunt at the other, curved to over a right angle, the pits 20-30 in
each of the usually 10 longitudinal rows.

Washington to s. Calif., e. to Mont.

Elatine triandra Schkuhr, Bot. Hand. 1:345, pl. 109b, fig. 2. 1791. (Europe)

Matted, but commonly with ascending to erect branches 2-15 cm. long; leaves 5-12 mm. long, linear to narrowly spatulate, the tips truncate to emarginate; flowers 2- or 3-merous; seeds straight or
curved, with 18-27 pits in each of the 8-10 longitudinal rows. 2N=about 40.

Rather general in w. U.S.; Eurasia.

TAMARICACEAE Tamarisk Family

Flowers small, perfect, regular, usually in long, slender, often compound, spikelike racemes;
sepals and petals 4 or 5, distinct; stamens as many or twice as many as the petals, and borne with
them on a fleshy perigynous disc, the filaments sometimes connate at base, the anthers 2-celled; pistil one, 3- to 5-carpellary; ovary superior, 1-celled or falsely septate, usually with 3-5 parietal placentae or with basal placentation; styles 3-5; fruit a capsule; seeds usually numerous, erect and
bearded at one end to pubescent overall; endosperm lacking; more or less ericoid shrubs or small
trees with alternate, scalelike, exstipulate leaves.

Four genera and nearly 100 species of the Old World, mostly Mediterranean or Asiatic; typically
xerophytic or halophytic.

Tamarix L. Tamarisk

A genus of nearly 70 species of Eurasia. (Named for the River Tamaris of Spain.)

Tamarix parviflora DC. Prodr. 3:97. 1828. (Europe)

Spreading shrub 2-4 m. tall, with numerous slender, arching to recurved branches and overlapping,
4-ranked, lanceolate, sessile, scalelike leaves 1-1.5 mm. long; racemes produced on wood of the
previous season; flowers pale pink, 2-3 mm. broad, 4-merous.

Established in many moist spots in the desert regions of w. U.S., as in Malheur Co., and along the
John Day R. in Gilliam Co., Oreg. May-June.

Our plant has more commonly been called either *T. gallica* L. or *T. pentandra* Pall., but both of
those species have 5-merous flowers.

Tamarisk is commonly used as a windbreak in desert regions, and most species are neat and attractive ornamental shrubs, valued as much for their foliage as for their flowers.

VIOLACEAE Violet Family

Flowers perfect, hypogynous to somewhat perigynous, 5-merous except for the pistil, irregular (ours) to nearly regular, usually nodding, often cleistogamous, solitary (ours) or clustered in diverse ways; sepals usually persistent, nearly or quite distinct; petals distinct, the lowermost largest and often spurred, the others of 2 dissimilar pairs; stamens usually connivent by their 2-celled anthers around the pistil, the filaments very short or lacking; pistil compound, the ovary superior, 1-celled, with 3 (to 5) parietal placentae; style 1; stigma more or less globose; fruit a 3-valved capsule (ours) or a berry; seeds rather large, albuminous, the embryo straight; perennial (annual) herbs (ours) or shrubs with alternate (opposite), simple or dissected, stipulate leaves.

A family of about 15 genera and several hundred species of cosmopolitan distribution; 2 genera native in the U.S., but only one in our area.

Viola L. Violet

Flowers mostly zygomorphic and showy but sometimes also in part cleistogamous and inconspicuous or even subterranean; sepals persistent, auriculate below the point of attachment; petals blue, violet, yellow, or white, the lowest one spurred or saccate at the base; stamens with short filaments but with broad connectives extending past the anther sacs, connivent around the pistil and closely investing the ovary; capsule explosively dehiscent; annual or perennial, acaulescent to caulescent herbs with axillary, 1-flowered peduncles, and simple to compound, stipulate leaves.

About 300 species of all continents, often in cool or damp places. (Old Latin name for the plant.)

Several of our violets have considerable value as ornamentals, among the best being *V. adunca*, *V. glabella* (apt to become invasive), *V. sempervirens* (with perhaps the nicest foliage of all), and *V. trinervata*, a *must* for gardeners e. of the Cascades.

References:

Baker, M. S. Studies in western violets. I-Madroño 3:51-56. 1935; II-Madroño 3:232-238. 1936; III-Madroño 5:218-231. 1940; IV-Leafl. West. Bot. 5:141-147. 1949; V-Leafl. West. Bot. 5:173-177. 1949; VI-Madroño 10:110-128. 1949; VII-Madroño 12:8-18. 1953; VIII-Brittonia 9:217-230. 1957. IX-Madroño 15:199-204. 1960.

Brainerd, E. Violets of North America. Bull. Vt. Agr. Exp. Sta. 224:1-205. 1921.

1 Plant an annual; stipules very large, laciniate into 5-9 linear segments and usually with
  one segment itself leafletlike and often nearly as large as the main blade of the leaf V. ARVENSIS
1 Plant a perennial; stipules usually small and not leaflike or leafletlike
  2 Leaves compound or deeply dissected into linear to oblong segments
    3 Leaf segments leathery, more or less prominently 3-nerved, tending to be elliptic,
      glabrous; petals with little if any yellow coloring   V. TRINERVATA
    3 Leaf segments not leathery, usually 1-nerved, often pubescent; lower petals often
      predominantly yellow in color
      4 Blade of the leaves considerably broader than long in outline, pedatifid; petals all
        yellowish on the inner surface   V. SHELTONII
      4 Blade of the leaves usually at least as long as broad in outline, 2-3 times pinnat-
        ifid, or the upper petals purplish on the inner surface
        5 Plant glabrous; lower petals usually pale yellowish; style head long-bearded;
          w. of the Cascades   V. HALLII
        5 Plant puberulent to conspicuously pubescent; lower petals mauve (white), yel-
          lowish only at the base; style head copiously short-bearded; e. of the Cas-
          cades   V. BECKWITHII
  2 Leaves neither compound nor deeply dissected
    6 Petals predominantly white, often with bluish or purplish shading, but not yellow
      7 Leaves narrowly elliptic to elliptic-lanceolate, 3-6 times as long as broad, not
        at all cordate at base   V. LANCEOLATA
      7 Leaves reniform to oval, usually somewhat cordate-reniform, not over twice
        so long as broad
        8 Plants producing annual, flower-bearing, elongate to short and tufted-leafy
          stems
          9 Lateral petals not violet-spotted, upper petals often not purple on the back
            10 Petals shaded with blue to purple on the back, not at all yellowish; w.

of the Cascades (and in the Columbia R. Gorge)   V. HOWELLII
10 Petals usually shaded with some yellow, at least basally; often e. of the Cascades   V. CANADENSIS
9 Lateral petals violet spotted at base, upper petals deep purplish on the back   V. OCELLATA
8 Plants not producing annual floriferous stems, the flowers pedunculate on the main
rhizome
11 Petals generally tinged with some violet or blue on the back, usually at least 8 mm.
long; stolons well developed; leaves mostly over 2.5 cm. broad, glabrous   V. PALUSTRIS
11 Petals seldom other than pure white (aside from the purplish pencilling), often less
than 8 mm. long; stolons present or absent; leaves often less than 2.5 cm. broad
12 Stolons lacking; leaves usually well over 3 cm. broad, reniform, often pilose on
the lower surface   V. RENIFOLIA
12 Stolons present; leaves usually less than 3 cm. broad, more cordate than reniform, glabrous   V. MACLOSKEYI
6 Petals predominantly blue, violet, or yellowish, rather than white
13 Corolla bluish to purple, definitely not yellow
14 Petals not uniform in color, the lateral pair usually conspicuously dark-spotted
basally   V. OCELLATA
14 Petals uniform in color, or at least the lateral pair not conspicuously spotted
15 Stolons present, very slender and elongate; erect leafy stems lacking, the
leaves arising from the rhizomes; plants of very moist or boggy ground
V. PALUSTRIS
15 Stolons absent or if present the plants with erect leafy stems
16 Aerial stems well developed, often longer than the leaves and peduncles,
floriferous on the upper 2/3; either the leaves more or less acuminate,
or the plants subalpine in the Olympic Mts.
17 Leaves reniform, mostly less than 4 cm. broad, purplish-green, not
acuminate; plants mainly less than 15 cm. tall; subalpine in the Olympic Mts., Wash., usually on talus   V. FLETTII
17 Leaves cordate, usually many of them over 4 cm. broad, acuminate,
bright green; plants mainly over 15 cm. tall; widespread, usually in
woodland   V. CANADENSIS
16 Aerial stems lacking or at least shorter than the peduncles which are
borne mainly on the rhizomes; rarely the stems longer than the peduncles, but then the plants neither native to subalpine regions in the Olympic Mts. nor with acuminate leaves
18 Petioles and peduncles sparsely to evidently hirsute; sepals ciliate;
spur much less than 1/2 the length of the blade of the lowest petal;
style not bearded   V. SEPTENTRIONALIS
18 Petioles and peduncles glabrous or the sepals not ciliate or the styles
bearded
19 Head of the style not bearded; plants glabrous, caulescent or acaulescent
20 Plants with rather conspicuous erect leafy stems; from near the
coast   V. LANGSDORFII
20 Plant essentially acaulescent, with scarcely any development of
erect leafy stems; from e. of the Cascades   V. NEPHROPHYLLA
19 Head of the style bearded; plants glabrous or hairy, usually caulescent
21 Spur broad and pouched, much less than half as long as the blade
of the lowest petal; Cascades and westward   V. HOWELLII
21 Spur slender, usually nearly half as long as the blade of the lowest petal; widespread   V. ADUNCA
13 Corolla partially or wholly yellow
22 Flowers from near the tip of erect leafy stems that are naked below; leaves mostly
large, thin, and cordate-based
23 Petals mostly bluish, yellow only at the base   V. FLETTII
23 Petals mostly yellowish or whitish, sometimes blue to brown on the back
24 Upper petals yellow on the back; leaves mostly abruptly acute   V. GLABELLA

24 Upper petals bluish-red on the back; leaves usually acuminate   V. CANADENSIS

22 Flowers not confined to the tip of erect branches, the stems leafy and floriferous below

  25 Leaves cordate based to reniform, usually as broad as long

    26 Leaves finely purplish-dotted, -flecked, or reticulately -mottled, rather firm and leathery, persistent, usually hairy; plants stoloniferous, w. of the Cascades

                                                      V. SEMPERVIRENS

    26 Leaves not purple-flecked, thin, usually withering during the winter, often glabrous; plants not stoloniferous, widely distributed   V. ORBICULATA

  25 Leaves neither truly cordate nor reniform, usually longer than broad

    27 Leaf blades coarsely veined, usually not over 4 cm. long, coarsely few-toothed or -lobed, not regularly serrate or dentate, often glaucous and more or less purplish, at least along the veins; upper petals deep purple on the back; capsules puberulent   V. PURPUREA

    27 Leaf blades not coarsely veined, from entire to finely or obscurely serrate or dentate, often well over 4 cm. long, usually not glaucous and not noticeably purplish, even along the veins; upper petals often not purple on the back; capsules glabrous or puberulent   V. NUTTALLII

Viola adunca Sm. in Rees, Cyclop. 37, no. 63. 1817.

  V. canina var. *adunca* Wats. Bot. Calif. 1:55. 1880. *(Menzies,* west coast of North America)

  V. longipes Nutt. in T. & G. Fl. N.Am. 1:140. 1838. *V. canina* var. *longipes* Wats. Bot. Calif. 1: 56. 1880. *V. adunca* var. *longipes* Rydb. Mem. N.Y. Bot. Gard. 1:263. 1900. *(Nuttall,* "borders of woods and in brushy plains near the Oregon, and in the Rocky Mountains") = var. *adunca.*

  V. canina var. *oxyceras* Wats. Bot. Calif. 1:56. 1880. *V. oxyceras* Greene, Pitt. 3:255. 1897. *V. adunca* ssp. *oxyceras* Piper, Contr. U.S. Nat. Herb. 11:395. 1906. *V. adunca* var *oxyceras* Jeps. Man. Fl. Pl. Calif. 647. 1925. *(Brewer,* Yosemite Valley, Calif.) = var. *adunca.*

  V. montanensis Rydb. Mem. N.Y. Bot. Gard. 1:263. 1900. *(Rydberg & Bessey 4532,* Jack Creek Canyon, Mont., July 15, 1897) = var. *adunca.*

  V. odontophora Rydb. Mem. N.Y. Bot. Gard. 1:264. 1900. *(Williams 114,* Grafton, Mont., in 1892) = var. *adunca.*

  V. monticola Rydb. Mem. N.Y. Bot. Gard. 1:264. 1900. *(Rydberg & Bessey 4529,* Cedar Mt., Mont., July 16, 1897) = var. *adunca.*

  V. retroscabra Greene, Pitt. 4:290. 1901. *(Greene,* Cimarron, Colo., in 1896, is the first of several collections cited) = var. *adunca.*

  V. bellidifolia Greene, Pitt. 4:292. 1901. *V. adunca* var. *bellidifolia* Harrington, Man. Pl. Colo. 377, 641. 1954. *(Baker, Earle, & Tracy,* Slide Rock Canyon, w. of Mt. Hesperus, Colo.)

  V. verbascula Greene, Leafl. 2:32. 1910. *(Sandberg & Leiberg 33,* Hangman Cr., Spokane Co., Wash., May 14, 1893) = var. *adunca.*

  V. drepanophora Greene, Leafl. 2:32. 1910. *(J. T. Jardine,* Wallowa National Forest, Oreg., in 1909) = var. *adunca.*

  V. mamillata Greene, Leafl. 2:33. 1910. *(Goodding 1202,* Dyer Mine, Uinta Mts., Utah, June 30, 1902) = var. *adunca.*

  V. oxysepala Greene, Leafl. 2:34. 1910. *(Tidestrom,* Wasatch Mts., July 15, 1909) = var. *adunca.*

VIOLA ADUNCA var. UNCINULATA (Greene) C. L. Hitchc. hoc loc. *V. uncinulata* Greene, Leafl. 2:97. 1910. *V. adunca* ssp. *uncinulata* Applegate, Am. Midl. Nat. 22:282. 1939. *(Applegate,* near Crater Lake, Oreg., Aug. 17, 1896)

  V. adunca var. *glabra* Brain. Rhodora 15:109. 1913. *V. adunca* f. *glabra* G. N. Jones, U. Wash. Pub. Biol. 5:194. 1936. *(Collins & Fernald 111,* Carlton, Bonaventure Co., Que., July 19, 1905, is the first of several collections cited) = var. *adunca.*

VIOLA ADUNCA var. CASCADENSIS (M.S. Baker) C.L. Hitchc. hoc loc. *V. cascadensis* M.S. Baker, Leafl. West. Bot. 5:173. 1949. *(W. H. Baker 5279,* Indian Ford Cr., 5 mi. n.w. of Sisters, Deschutes Co., Oreg.)

Perennial with short to elongate slender rhizomes, glabrous to densely puberulent, usually stemless in the early season but later with aerial stems as much as 10 cm. long; stipules linear-lanceolate, 3-10 mm. long, entire to remotely slender-toothed; leaf blades mostly 1-3 cm. wide, ovate-lanceolate to reniform but usually cordate-ovate, finely crenate, from considerably narrower to broader than long; peduncles from shorter to considerably longer than the leaves; flowers 5-15 mm. long; spur conspicuous, slender, usually over half the length of the lowest petal and somewhat hooked; petals

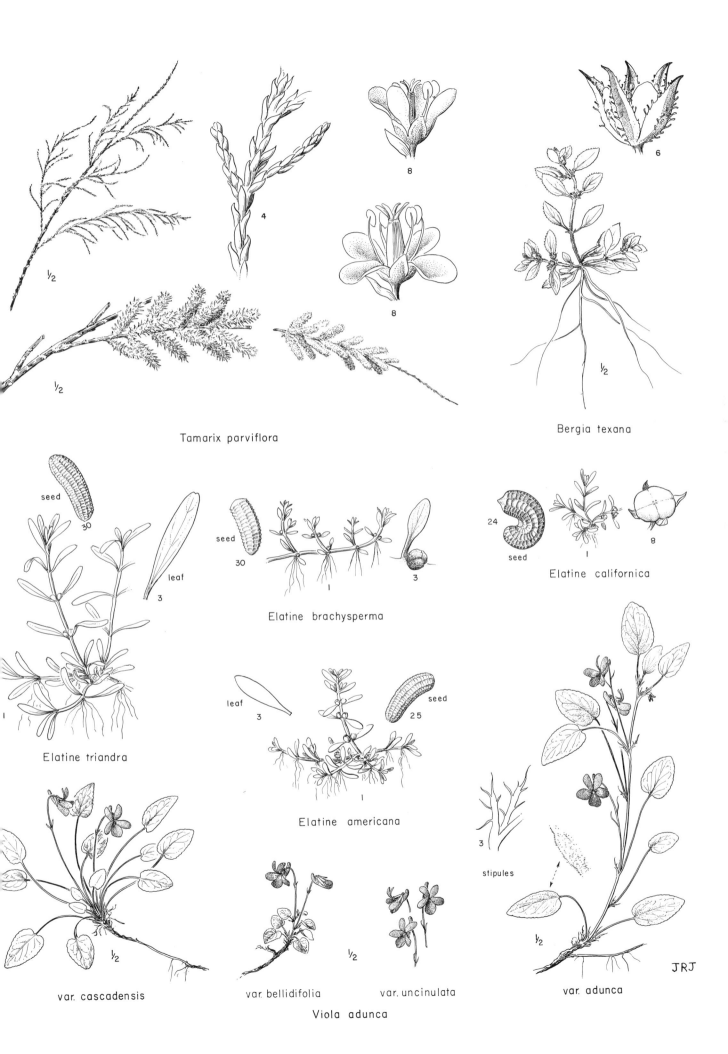

Tamarix parviflora

Bergia texana

Elatine triandra

Elatine brachysperma

Elatine californica

Elatine americana

var. cascadensis

var. bellidifolia

var. uncinulata

var. adunca

Viola adunca

JRJ

blue to deep violet, the lower three often whitish-based, pencilled with purplish-violet, the lateral pair white-bearded; style head bearded with thick, short to fairly long hairs. N=10.

In dry to moist meadows, woods, and open ground near timber line; throughout most of w. N. Am., eastward to the Atlantic coast. Apr. -Aug.

A very widespread taxon, including many variants that have been considered distinct species, but most of which appear to be fortuitous or inconsequential. Others, however, are much more significant, but as they resemble one another very closely and apparently intergrade more or less completely, they are here treated at the varietal level, as follows:

1 Plant stemless at the time the normal (petaliferous) flowers are produced, but with cleistogamous fertile flowers produced later from well-developed but nonpersistent stems; known only from one or two localities in Deschutes Co. , Oreg. , and Okanogan Co. , Wash. , but possibly a growth phase of *V. adunca* to be found in other localities
                                                                var. cascadensis (M. S. Baker) C. L. Hitchc.
1 Plant caulescent, the normal flowers produced on short aerial stems
    2 Leaves usually pubescent, mostly considerably longer than broad; plants usually over 5 cm. tall, with the range of the species   var. adunca
    2 Leaves glabrous; plants mostly dwarf, generally not over 5 cm. tall
        3 Petals usually whitish at the base, about 5 mm. long; chiefly in the Rocky Mts.
                                                                var. bellidifolia (Greene) Harrington
        3 Petals not whitish at the base, mostly well over 5 mm. long; local in s. Oreg. , especially near Crater Lake, and possibly not reaching our range
                                                                var. uncinulata (Greene) C. L. Hitchc.

Viola arvensis Murr. Prodr. Stirp. Gotting. 73. 1770. (Europe)

Freely branched annual 1-3 dm. tall; leaves puberulent, the blades ovate to lanceolate, coarsely crenate-serrate, 1-3 cm. long, about equaling the petioles, the stipules very large, laciniate into 5-9 linear segments and usually with 1 leaflike segment nearly as large as the main blade; flowers long-pedunculate, single in the axils, whitish or light yellow with a varied tinge of blue; sepals nearly or quite as long as the petals, lanceolate, the spur short; style head copiously short-hairy. N=17. Wild pansy.

The pansy is native to Europe but widely cultivated; it tends to escape, and is to be found in a more or less transient state in most of the U.S. and Can. , especially near cities. Mar. -June.

Viola beckwithii T. & G. Pac. R. R. Rep. 2:119, pl. 1. 1855. *(Snyder,* between Great Salt Lake and the Sierra Nevada)

Perennial from a deep-seated, erect rootstock, the aerial portion 5-12 cm. tall, puberulent (especially the leaves) with short thick hairs; subterranean stipules membranous, as much as 3 cm. long, those above ground linear and entire or adnate their full length to the petioles; leaf blades glaucous, ternately decompound into linear segments; peduncles from shorter to longer than the leaves; flowers about 1.5 cm. long; sepals obtuse to rounded, rarely acute; upper pair of petals reddish-purple, the lower ones mauve (white in plants from Elmore Co. , Ida. , and e.), purple-pencilled, yellowish at the base, the lateral pair with yellow bearding; style head densely short-bearded.

On sagebrush hills and plains and ponderosa pine woodland from n. e. Oreg. to Ida. and Utah, s. to Inyo Co. , Calif. Mar. -May.

Viola canadensis L. Sp. Pl. 936. 1753. *(Kalm,* n. e. North America)

VIOLA CANADENSIS var. RUGULOSA (Greene) C. L. Hitchc. hoc loc. *V. rugulosa* Greene, Pitt. 5: 26. 1902. *(Sandberg,* Hennepin Co. , Minn. , June, 1891)

V. rydbergii Greene, Pitt. 5:27. 1902. *V. canadensis* ssp. *rydbergii* House in Rydb. Fl. Colo. 233. 1906. *(Rydberg 2726,* Mont. , is the first of several collections cited)

V. geminiflora Greene, Pitt. 5:29. 1902. *(Heller 3281,* Lake Waha, Nez Perce Co. , Ida. , June, 1896)

Perennial with short thick rootstocks and often with slender stolons; stems 1-4 dm. tall, glabrous to puberulent; stipules lanceolate, 1-2 cm. long, entire, glabrous to ciliate; petioles as much as 3 dm. long; leaf blades cordate, acute to rather conspicuously acuminate, from (usually) puberulent on one or both surfaces to glabrous; flowers from the upper portion of the stem, the peduncles shorter than the leaves; sepals lanceolate, often ciliate or pubescent, the spur short; petals about 1.5 cm. long, white, yellow-based, the 3 lower ones purplish-pencilled, the lateral pair bearded, all (but especially the upper pair) more or less purplish-tinged on the outside and sometimes less conspicuously so on the inside; style head sparsely long-bearded; capsule granular to puberulent, the seeds brownish. N= 12.

In moist woodland and forest, usually on loamy soil; s. Alas. and B. C. throughout Wash. and Oreg., e. to the Atlantic coast and s. in the Rocky Mts. to N. M. and Ariz. May-July.

Most of our western material is referable to the var. *rugulosa,* distinguishable from var. *canadensis* in the following way:

1 Stolons present but often buried and not obvious; plants pubescent; leaves often wider than
　　long, ciliate-margined; Alas. to Oreg. and through the Rocky Mts. to Colo. , e. occa-
　　sionally to c. U. S. and the southern Appalachians　　　var. rugulosa (Greene) C. L. Hitchc.
1 Stolons lacking; plants glabrous to puberulent; leaves usually longer than broad, not
　　ciliate; c. and e. U. S. and Can. , extending w. , occasionally to the Rocky Mts. , from
　　Alta. to N. M. , also in Ariz.　　　　　　　　　　　　　　　var. canadensis

Viola flettii Piper, Erythea 6:69. 1898. *(Flett,* Mt. Constance, Wash.)

Perennial with short, rather thick, horizontal or ascending rootstocks, glabrous, caulescent, the stems 3-15 cm. long; leaves fleshy, purplish-green; stipules lanceolate, callous-serrulate, saliently toothed, leaf blades reniform, 1. 5-4 cm. broad, finely crenate-serrate; peduncles usually about equaling the leaves; flowers about 1. 5 cm. long, the spur very short, yellow; petals purplish-violet, yellow at the base, the lower 3 darker veined, the lateral pair yellow-bearded; style head well bearded.

In alpine rock crevices and on talus slopes of the n. Olympic Mts. , Wash. June-Aug.

This is one of the choicest of our violets but it is not at all easily grown, even in a good scree.

Viola glabella Nutt. in T. & G. Fl. N. Am. 1:142. 1838. *(Nuttall,* shady woods of the Oregon)

Perennial with widely spreading, scaly, fleshy rootstocks, the flowering stems 5-30 cm. tall, leafless except on the upper 1/3-1/5 of their length; stipules membranous, ovate to obovate, 5-10 mm. long, entire; leaf blades reniform to ovate-cordate, usually abruptly acute, crenate-serrulate, glabrous or (more commonly) puberulent, the basal leaves with petioles 10-20 cm. long; flowers 8-14 mm. long, borne chiefly on the upper part of the stem, on peduncles about as long as the leaves; spur very short; petals clear yellow on both surfaces, the three lower ones purplish-pencilled within, the lateral pair well bearded; style head copiously bearded; seeds brown. N=12.

Mostly in fairly moist woods or along streams; Alas. southward, through B. C. , Wash. , and Oreg. , to the coastal ranges and the Sierra Nevada of Calif. , e. to Mont. ; n. e. Asia. Mar. -July.

Viola hallii Gray, Proc. Am. Acad. 8:377. 1872. *(Elihu Hall,* Salem, Oreg.)

Perennial with deep-seated, short, erect rhizomes, glabrous or subglabrous throughout, the aerial portion 5-12 cm. tall; stipules of the subterranean parts of the plant membranous, entire, several cm. in length, those above ground greenish, 5-15 mm. long, usually sparingly toothed; leaf blades ternately decompound into linear segments; peduncles equaling or slightly exceeding the leaves; flowers 10-15 mm. long, the upper petals purplish-blue on both surfaces, the lower three pale yellow and purplish-pencilled, the lateral pair clavate-bearded; style head copiously long-bearded. N=30-36.

Mostly in open woodland or on light gravelly plains and slopes; Willamette Valley, Oreg. , to n. w. Calif. Mar. -May.

Viola howellii Gray, Proc. Am. Acad. 22:308. 1887. *(Thomas Howell,* near Portland, Oreg.)

Perennial with fairly widespread, thickish, scaly rootstocks; stems 2-10 cm. tall, sparsely pubescent; stipules 3-10 mm. long, salient-toothed; leaf blades cordate to reniform-cordate, crenate, puberulent to glabrous, the petioles of the many basal leaves often as much as 10-15 cm. long; peduncles usually exceeding the leaves; flowers 1. 5-2 cm. long; sepals obtuse, conspicuously auriculate at the base; spur very prominent, saccate, not at all hooked; petals from bluish-violet (but shading to white at the base) to nearly white, strongly veined with purple, the lateral pair white-bearded; style head sparsely bearded; seeds light brown. N=about 40.

In moist woods and on prairies on the w. side of the Cascade Mts. , from s. B. C. to Mendocino Co. , Calif. ; in s. Oreg. extending as far e. as Klamath Lake. Apr.-May.

Viola lanceolata L. Sp. Pl. 934. 1753. (Canada)

Perennial with very slender, widespread, creeping rhizomes, glabrous, 5-15 cm. tall, nearly or quite stemless; stipules linear-lanceolate, about 1 cm. long, slender-toothed; leaves with very long

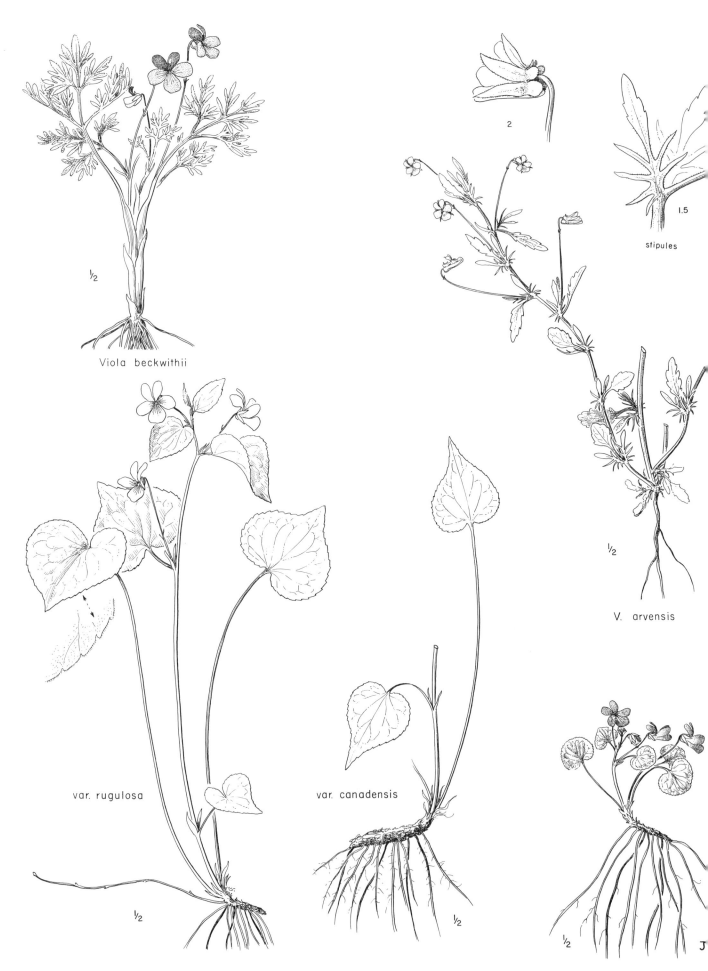

Viola beckwithii

2

1.5

stipules

V. arvensis

var. rugulosa

var. canadensis

V. flettii

Viola canadensis

J

and slender petioles, the blades narrowly elliptic-lanceolate, 0.5-1.5 (2) cm. broad, 3-6 times as long, very shallowly crenate-dentate; peduncles about equaling the leaves; flowers (6) 8-11 mm. long, white, the lower three petals purplish-pencilled, beardless or nearly so; style head beardless. N=12.

Bogs and moist meadows; rather general from Minn. eastward; in our area known only from Pierce and Pacific cos., Wash., where undoubtedly introduced. May-June.

Viola langsdorfii (Regel) Fisch. in DC. Prodr. 1:296. 1824.
V. mirabilis var. langsdorfii Regel, Bull. Soc. Imp. Mosc. 34:472, pl. 6, fig. 26, 1861. (Chamisso, Unalaska)
V. simulata M. S. Baker, Madroño 3:237. 1936. (J. H. Henry, near Shawnigan Lake, Vancouver I., B.C., May 9, 1915)
Perennial with widespread, rather thick, horizontal to ascending rootstocks, glabrous or sometimes sparsely pubescent, acaulescent or with well-developed, leafy, erect stems 2-10 cm. tall; stipules ovate to lanceolate, 3-10 mm. long, entire or remotely glandular-denticulate; leaf blades reniform to ovate-cordate, crenate to crenate-serrulate, 2-5 cm. broad; peduncles from shorter to considerably longer than the leaves; flowers 1.5-2 cm. long, the spur conspicuous, saccate; petals violet, the lower three whitish-based, the lateral pair white-bearded; style head glabrous.
In moist places, bogs, etc., from Alas. southward along the coast to s. Oreg.; n.e. Asia. Apr.-Aug.
Viola simulata appears to be no more than a very minor variant of this more widespread species, although it was said to differ in being less succulent and in having shorter, less erect stems, smaller stipules and spurs, and a peculiar style head. Until such time as it can be shown that the type collection is representative of a distinctive population, rather than a chance variant, it would seem best (at least for our purpose) to regard V. simulata as synonymous with V. langsdorfii.

Viola macloskeyi Lloyd, Erythea 3:74. 1895.
V. blanda var. macloskeyi Jeps. Man. Fl. Pl. Calif. 648. 1925. V. pallens ssp. macloskeyi M. S. Baker, Madroño 12:18. 1953. (F. E. Lloyd, s.e. of Mt. Hood, Oreg., in 1894)
VIOLA MACLOSKEYI var. PALLENS (Banks) C. L. Hitchc. hoc loc. V. rotundifolia var. pallens Banks ex DC. Prodr. 1:295. 1824. V. pallens Brain. Rhodora 7:247. 1905. V. macloskeyi ssp. pallens M. S. Baker, Madroño 12:60. 1953. (Labrador and Kamchatka, no specimens cited)
Perennial with slender rootstocks and usually with filiform stolons, mostly 3-6 cm. tall, without erect aerial flowering stems; stipules lanceolate, membranous, glandular-denticulate; petioles 2-4 cm. long; leaf blades ovate-cordate, 1-3 cm. long, very indistinctly to fairly prominently crenate; peduncles usually exceeding the leaves; flowers 5-10 mm. long, the spur fairly prominent; petals white, the lower three generally purple-pencilled, the lateral pair usually bearded; style head glabrous. N=12.
Plants of boggy or wet ground in the mountains throughout much of N. Am. May-Aug.
The species is represented in our area by the following strongly intergradient varieties:

1 Leaf blades inconspicuously crenate, usually less than 2.5 cm. broad; B.C. to Alta., s. in the mts. to s. Calif. var. macloskeyi
1 Leaf blades prominently crenate, often over 2.5 cm. broad; B.C. and Wash. to the Atlantic coast in Can., s. on the Atlantic coast to the Carolinas and in the Rocky Mts. to Colo.; not known from Ida. var. pallens (Banks) C. L. Hitchc.

Viola nephrophylla Greene, Pitt. 3:144. 1896. (Greene, valley of Cimarron R., Colo., Aug. 29, 1896)
VIOLA NEPHROPHYLLA var. COGNATA (Greene) C. L. Hitchc. hoc loc. V. cognata Greene, Pitt. 3:145. 1896. ("Northern Rocky Mountains," the type believed to be Greene, Dale Creek, Wyo.)
V. austinae Greene, Pitt. 5:30. 1902. (Austin, Butterfly Valley, Plumas Co., Calif., in 1876) = var. nephrophylla.
V. subjuncta Greene, Pitt. 5:31. 1902. (Piper, Rock Creek, Whitman Co., Wash.) = var. nephrophylla.
V. macabeiana M. S. Baker, Madroño 5:226. 1940. (McCabe 6149, s.e. corner of Columbia Lake, Kootenay District, B.C., May 21, 1938, is the first collection cited) = var. nephophylla.
Glabrous, acaulescent perennial with fairly widespread, shallow, fleshy, spreading rhizomes; stipules linear-lanceolate, entire; leaf blades deeply ovate-cordate to cordate-triangular, as much as 7 cm. in width, prominently crenate-dentate; petioles 5-25 cm. long; peduncles usually equaling or ex-

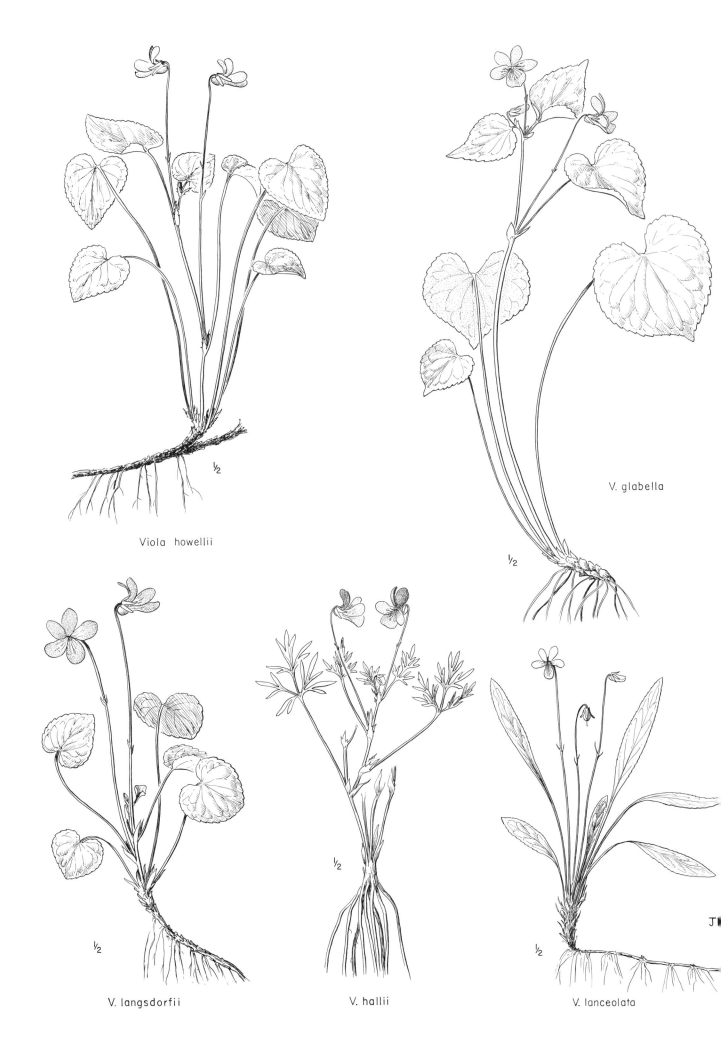

Viola howellii

V. glabella

½

½

V. langsdorfii

½

V. hallii

½

V. lanceolata

½

ceeding the leaves; flowers 10-20 mm. long; spur short, saccate; petals bluish-violet, the lower three more or less whitish at base, prominently bearded, the upper pair sometimes not bearded; style head glabrous.

Moist places, especially in meadows and along streams; B. C. to Newf., southward on the e. side of the Cascades to Calif., Ariz., N. M., Minn., and N. Y. May-July.

In the Rocky Mt. area, from Alta. to Colo., and w. to B. C. and Wash., a variant of *V. nephrophylla*, the var. *cognata* (Greene) C. L. Hitchc., is distinguishable because of its thicker, smoother, purplish-backed leaves, more slender rootstock, and the bearding on all 5 petals; var. *nephrophylla*, ranging from New England and e. Can. to B. C. and along the e. side of the Cascades in Wash., s. to Ariz., usually does not have purple-backed leaves and lacks the bearding on the upper petals.

Viola nuttallii Pursh, Fl. Am. Sept. 174. 1814.

Crocion nuttallii Nieuwl. & Lunell, Am. Midl. Nat. 4:478. 1916. *(Nuttall,* "On the banks of the Missouri")

V. praemorsa Dougl. ex Lindl. Bot. Reg. 15:pl. 1254. 1829. *V. nuttallii* var. *praemorsa* Wats. Bot. King Exp. 35. 1871. *V. nuttallii* ssp. *praemorsa* Piper, Contr. U. S. Nat. Herb. 11:393. 1906. *(Douglas,* banks of the lower Columbia)

V. nuttallii var. *major* Hook. Fl. Bor. Am. 1:79. 1830. *V. praemorsa* var. *major* Peck, Man. High. Pl. Oreg. 486. 1941. *V. praemorsa* ssp. *major* M.S. Baker, Madroño 10:128. 1949. *(Douglas,* "dry sandy plains of the Columbia")

V. linguaefolia Nutt. in T. & G. Fl. N. Am. 1:141. 1838. *V. nuttallii* ssp. *linguaefolia* Piper in Piper & Beattie, Fl. S.E. Wash. 166. 1914. *V. nuttallii* var. *linguaefolia* Jeps. Man. Fl. Pl. Calif. 645. 1925. *V. praemorsa* var. *linguifolia* Peck, Man. High. Pl. Oreg. 486. 1941. *V. praemorsa* ssp. *linguaefolia* M. S. Baker, Madroño 10:128. 1949. *(Wyeth,* "Kamas Prairie, near the sources of the Oregon") = var. *major.*

V. flavovirens Pollard, Bull. Torrey Club 24:405. 1897. *(Heller 3106,* Lake Waha, Nez Perce Co., Ida., June 3-4, 1896) = var. *major.*

VIOLA NUTTALLII var. BAKERI (Greene) C. L. Hitchc. hoc loc. *V. bakeri* Greene, Pitt. 3:307. 1898. *(M. S. Baker,* Bear Valley Mts., n. Calif., June, 1896)

V. vallicola A. Nels. Bull. Torrey Club 26:128. 1899. *Crocion vallicola* Nieuwl. & Lunell, Am. Midl. Nat. 4:478. 1916. *V. nuttallii* var. *vallicola* St. John, Fl. S. E. Wash. 262. 1937. (Four collections from Wyo., *A. Nelson 43, 4340, 4345,* and *4525,* are cited in that order)

V. erectifolia A. Nels. Bot. Gaz. 29:143. 1900. *(A. Nelson 5481,* Henry's Lake, Ida., June 22, 1899) = var. *major.*

V. gomphopetala Greene, Pl. Baker. 3:11. 1901. *(C. F. Baker 225,* Grand Mesa, Colo., June 23, 1901) = var. *major.*

V. physalodes Greene, Pl. Baker. 3:12. 1901. *(C. F. Baker 67,* Cimarron R., Colo., June 7, 1901) = var. *vallicola.*

V. praemorsa var. *altior* Blank. Mont. Agr. Coll. Stud. Bot. 1:83. 1905. *(Blankinship,* Bozeman, Mont., June 20, 1899, is the first collection cited) = var. *major.*

V. xylorrhiza Suksd. Werdenda 1:25. 1927. *(Suksdorf 10200,* e. of Husum, Klickitat Co., Wash.) = var. *major.*

V. subsagittifolia Suksd. Werdenda 1:25. 1927. *(Suksdorf 8530,* s. e. of Spangle, Spokane Co., Wash., Apr. and May, 1916) = var. *vallicola.*

V. praemorsa var. *oregana* Peck, Man. High. Pl. Oreg. 486. 1941. *V. praemorsa* ssp. *oregana* Baker & Clausen ex Peck, loc. cit. in syn., published without Latin diagnosis or designated type; M.S. Baker, Brittonia 9:228. 1957. *(M. S. Baker 8862,* 20 mi. from Klamath Falls, Oreg.) = nearest var. *praemorsa.*

V. praemorsa ssp. *arida* M. S. Baker, Brittonia 9:227. 1957. *(M. S. Baker 11462* and *12806,* Klamath Falls, Oreg.) = nearest var. *major.*

Perennial from short, erect rootstocks; stems largely subterranean but the aerial portions sometimes as much as 15 cm. long; stipules adnate much of their length, the free portion entire to few-toothed; petioles 3-15 cm. long, somewhat wing-margined; leaf blades glabrous to copiously puberulent, elliptic-lanceolate to ovate or lanceolate, cuneate to rounded or subcordate at base, entire to sinuate or sparingly crenate-dentate, 2-10 cm. long; peduncles equaling or shorter than the leaves; flowers 5-15 mm. long, the spur short; petals yellow, the upper ones usually brownish-backed, the lower three pencilled with brownish-purple, the lateral pair bearded; style head bearded, lobed to rounded; ovary hairy to glabrous. N=6, 12, 18, 24.

With the exception of one variety the species is found east of the Cascades and coastal mountains, usually in rather dry areas, often in open sagebrush prairie but more frequently in ponderosa pine forest or up to middle altitudes in the mountains, B. C. to Calif., e. to c. U.S. Apr.-July.

Viola nuttallii is a wide-ranging plant which undergoes much variation in leaf shape, general pubescence, and flower size, for which reason other taxa are usually recognized in the complex, sometimes at the specific level. These several taxa are more or less sympatric in range and intergrade with one another, possessing no distinctive gross morphological features by which they can be recognized consistently. Regardless of their cytological composition, therefore, it seems that for our purposes they can best be regarded as varieties, and ours are separable, in large part, as follows:

1 Leaf blades narrowly lanceolate or elliptic-lanceolate, usually at least 3 times as long as
    broad, narrowed to petioles nearly or quite as long; chiefly e. of the Rocky Mts., from
      Alta. to Ariz., e. to Mo., possibly in our range in Mont.; N=12   var. nuttallii
1 Leaf blades various, but usually less than three times as long as broad, often truncate
    or subcordate at the base; in the Rockies and westward
    2 Upper petals not brownish-backed; flowers 5-12 mm. long; leaf blades mostly 2-5 cm.
      long; capsules glabrous; Cascade Mts. from Mt. Adams, Wash., to Calif.; by far
        the most distinctive of the several varieties; N=24   var. bakeri (Greene) C. L. Hitchc.
    2 Upper petals usually brownish-backed; flowers 8-15 mm. long; leaf blades 3-10 cm.
      long; capsules glabrous or hairy
      3 Leaves glabrous to sparsely hairy, the blades ovate to ovate-lanceolate, usually
        more or less truncate or subcordate at the base, generally less than 5 cm. long;
        capsules glabrous; B.C. to Oreg., e. to the Rocky Mts., chiefly in sagebrush
        and on sagebrush-ponderosa pine benchland; N=12   var. vallicola (A. Nels.) St. John
      3 Leaves densely hairy to glabrous, the blades usually at least 5 cm. long, seldom
        at all cordate-based; capsules often hairy (the last 2 varieties are very poorly
        marked and represent little more than extremes in a continuous series of var-
        iation)
        4 Leaves usually conspicuously hairy, the blades thick and fleshy, ovate-lanceo-
          late; chiefly w. of the Cascades in valleys and "prairies," B.C. to n. Calif.;
          N=24   var. praemorsa (Dougl.) Wats.
        4 Leaves glabrous to moderately hairy, the blades variable but not noticeably
          fleshy; e. of the Cascades, mostly in ponderosa pine forest and in the lower
          mts. in general; N=18, 24   var. major Hook.

Viola ocellata T. & G. Fl. N.Am. 1:142. 1838. *(Douglas,* California)

Perennial with rather short, thick, scaly, horizontal to ascending rootstocks; flowering stems 5-30 cm. tall, puberulent; stipules lanceolate, 5-15 mm. long; petioles as much as 1 dm. long; leaf blades cordate-ovate to nearly triangular, 2-5 cm. long, serrate, puberulent on the lower surface, at least; peduncles mostly shorter than the leaves; flowers 10-15 mm. long; sepals narrowly lanceolate, acute; spur rather conspicuous; petals white, the three lower ones yellow at the base and purplish-pencilled, the lateral pair purple maculate and yellow bearded, all, but especially the upper two, purplish-red on the back; seeds dark brown. N=6. Eyed violet.

In woodland of the coast ranges from Monterey Co., Calif., to n. Douglas Co., Oreg., probably not quite reaching our area. Apr.-June.

Viola orbiculata Geyer ex Hook. Lond. Journ. Bot. 6:73. 1847.
    V. sarmentosa var. *orbiculata* Gray, Syn. Fl. 1¹:199. 1895. *V. sempervirens* var. *orbiculata*
    Henry, Fl. So. B.C. 208. 1915. *V. sempervirens* ssp. *orbiculata* M. S. Baker, Madroño 5:226
    (mimeographed insert in reprints). 1940. *(Geyer,* Coeur d'Alene Mts., Ida.)
    V. sempervirens ssp. *orbiculoides* M. S. Baker, Madroño 5:224. 1940. *(M. S. Baker 736,* Nisqual-
    ly Glacier, Mt. Rainier, Wash., July, 1924)

Very similar to *V. sempervirens,* but without stolons, the aerial stems mostly not over 5 cm. long; herbage usually glabrous, not at all purplish-mottled; leaf blades from ovate-cordate to nearly orbicular, 2-4 cm. broad, serrulate-crenulate, thin in texture, often persistent over the winter; flowers as in *V. sempervirens* but the stigmatic area considerably longer. N=12. Round-leaved violet.

On alpine and montane slopes of the Olympic and Cascade mts., from B.C. to n. Oreg., eastward, in coniferous forests, to Ida. and Mont. May-Aug.

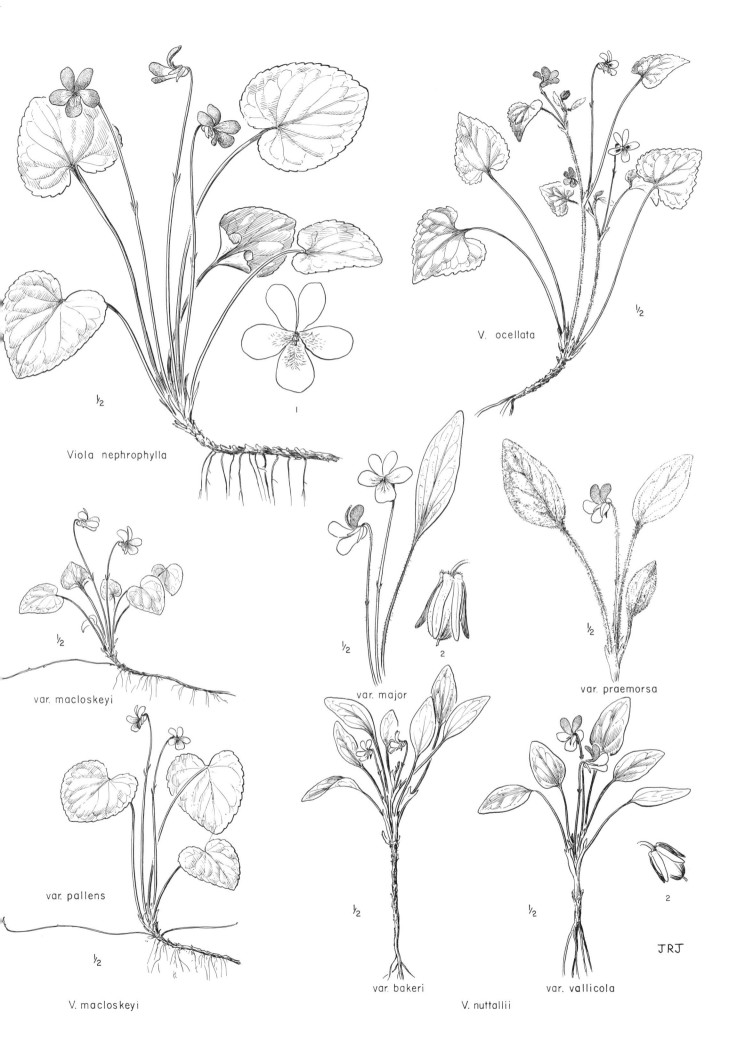

V. ocellata

½

½

Viola nephrophylla

I

var. macloskeyi

½

var. major

2

½

var. praemorsa

½

var. pallens

½

V. macloskeyi

var. bakeri

½

var. vallicola

½

2

V. nuttallii

JRJ

Viola palustris L. Sp. Pl. 934. 1753. (Europe)

    V. palustris ssp. *brevipes* M. S. Baker, Madroño 3:235. 1936. *V. palustris* var. *brevipes* Davis,
    Fl. Ida. 477. 1952. *(M. S. Baker 7629a* and *7629b,* transplants from Estes Park and from Moraine
    Park, Colo.)

    Glabrous perennial from slender widespread rhizomes and creeping stolons, without erect flowering
stems, the flowers and leaves both arising from the rhizomes; stipules scarious, lanceolate, entire;
petioles as much as 15 cm. long; leaf blades cordate to reniform, 2.5-3.5 cm. broad, crenate; peduncles shorter to longer than the leaves; flowers 10-13 mm. long; sepals auricled at the base; spur
conspicuous, saccate; petals white, the 3 lower with purple lines, or more or less deeply tinged with
lilac, the lateral pair sparsely bearded; style head glabrous, lobed; seeds dark brown. N=24. Marsh
violet.

    In moist meadows and along streams, etc.; B.C. to Calif., e. to the Rockies and more northerly
to Lab. and Me. May-July.

    The pure white-flowered plants have been called var. *brevipes* (M. S. Baker) Davis, but do not ap-
pear to constitute a geographic variant of taxonomic significance.

Viola purpurea Kell. Proc. Calif. Acad. Sci. 1:55. 1855. *(M. S. Baker 8655,* 2 mi. w. of Paynes
    Creek, Tehama Co., Calif., lectotype)

    V. nuttallii var. *venosa* Wats. Bot. King Exp. 35. 1871. *V. aurea* var. *venosa* Wats. in Brew. &
    Wats. Bot. Calif. 1:56. 1876. *V. praemorsa* var. *venosa* Gray, Syn. Fl. 1¹:200. 1895. *V. venosa*
    Rydb. Mem. N.Y. Bot. Gard. 1:262. 1900. *V. purpurea* var. *venosa* Brain. Vt. Agr. Exp. Sta.
    Bull. 224:111. 1921. *V. purpurea* ssp. *venosa* Baker & Clausen, Madroño 10:125. 1949. *(Watson*
    145, West Humboldt Mts., Nev.)

    V. pinetorum Greene, Pitt. 2:214. 1889. *V. purpurea* var. *pinetorum* Greene, Fl. Fran. 243. 1891.
    (Greene, s. of Tehachapi, Kern Co., Calif., June 25, 1889)

    V. atriplicifolia Greene, Pitt. 3:38. 1896. *V. purpurea* var. *atriplicifolia* Peck, Man. High. Pl.
    Oreg. 486. 1941. *V. purpurea* ssp. *atriplicifolia* Baker & Clausen, Madroño 10:126. 1949. *(Bur-*
    glehaus, near Mammoth Hot Spgs., Yellowstone Nat. Park, in 1893) = var. *venosa.*

    V. thorii A. Nels. Bot. Gaz. 30:193. 1900. *V. atriplicifolia* var. *thorii* Nels. in Coult. & Nels. New
    Man. Bot. Rocky Mts. 321. 1909. *(A. Nelson 5816,* Yellowstone Park, July 13, 1899) = var. *venosa.*

    V. purpurea ssp. *dimorpha* Baker & Clausen, Madroño 10:122. 1949. *(M. S. Baker 8100,* Child's
    Meadow, Plumas Co., Calif., June 26, 1935) = var. *purpurea.*

    V. purpurea ssp. *geophyta* Baker & Clausen, Madroño 10:124. 1949. *(Keck & Clausen 3707,* 20 mi.
    s. of Lapine, Klamath Co., Oreg., June 23, 1935) = var. *purpurea.*

    Perennial with shallow to deep-seated, rather slender, scaly rhizomes, the aerial stems 5-15
cm. tall; herbage puberulent and usually glaucous-green, often purplish, or at least purplish-veined;
stipules lanceolate, entire to few-toothed; petioles 2-6 cm. long; leaf blades rather thick and fleshy,
1-3 cm. broad, ovate or orbicular to lanceolate in outline, cuneate to cordate, sinuately and deeply
toothed to subentire; peduncles shorter to longer than the leaves; flowers 5-12 mm. long; spur very
short; petals yellow, brownish-pencilled, sometimes fading to light brownish-purple, the lateral pair
bearded; style head bearded; capsules puberulent. N=6.

    From the lowlands to high in the mountains; Chelan Co., Wash., to Calif. and Ariz., e. to Mont.,
Wyo., and Colo. May-Aug.

    This is a violet of wide geographical and ecological range that undergoes much variation. It has
been segregated into numerous specific or infraspecific taxa, most of which are impossible of deter-
mination and recognition, although two major variants are distinguishable as follows:

1 Seeds with a distinctly feathered and discoid caruncle usually nearly 1 mm. broad;
    Chelan Co., Wash., s. to the e. slopes of the Sierra Nevada, e. to Mont., Wyo.,
    and Colo.   var. venosa (Wats.) Brain.
1 Seeds with a rounded or amorphous caruncle less than 0.5 mm. broad; Cascade Mts.
    from Deschutes Co., Oreg., s. in the mts. of Calif. to Inyo Co.   var. purpurea

Viola renifolia Gray, Proc. Am. Acad. 8:288. 1870. *(Sherman,* Hanover, N.H.)

    V. brainerdii Greene, Pitt. 5:89. 1902. *V. renifolia* var. *brainerdii* Fern. Rhodora 14:88. 1912.
    (J. M. Macoun 18903, Beaver Meadows, near Ottawa, Ont., in 1898)

    Perennial from short ascending rootstocks but without horizontal rhizomes or stolons, acaulescent;
petioles 3-15 cm. long; stipules lanceolate, 3-10 mm. long, saliently toothed; leaf blades cordate-
orbicular to reniform, 2-6 cm. broad, crenulate-serrate, rounded to obtuse, sparsely to rather

heavily pubescent, at least beneath; peduncles usually shorter than the leaves; corolla 10-15 mm. long; petals white, the lower three purple-pencilled, all beardless; style head beardless, slightly bilobed; capsules purplish; seeds brown. N=12.

From lowland forest to subalpine slopes; n. e. U.S. , westward to B. C. and reputedly to extreme n. Wash. , southward in the Rocky Mts. to Colo. , but not known from Ida. June-Aug.

Two phases of the plant are often to be found, a glabrous or glabrate form (var. *renifolia)* and a more hairy one that has been called var. *brainerdii* (Greene) Fern. Certain large-leaved plants from Idaho *(Holmgren & Tillett 9578)* would seem to fit this species in all respects other than that they are strongly stoloniferous, for which reason they are referred to *V. macloskeyi.*

Viola sempervirens Greene, Pitt. 4:8. 1899.
  V. sarmentosa Dougl. ex Hook. Fl. Bor. Am. 1:80. 1830, not of Bieberstein in 1808. *(Douglas,* Ft. Vancouver)
Perennial with scaly rhizomes, the aerial stems slender, stolonous, often greatly elongated; herbage usually puberulent and rather conspicuously maculate with tiny purplish blotches that often form a reticulum; stipules brownish, membranous, lanceolate, entire or sparsely slender-toothed; petioles 2-10 cm. long; leaf blades cordate-lanceolate to cordate-ovate or nearly reniform, 1-3 cm. broad, serrulate-crenulate, thick and leathery, persisting through the winter; peduncles usually exceeding the leaves; flowers 5-15 mm. long, the spur short, saccate; petals lemon-yellow to gold, the lower three purplish-pencilled, the lateral pair yellow-bearded; style head short-bearded; the stigmatic area very short; capsules purplish-mottled; seeds brown. N=12, 24. Evergreen violet.

In moist woods from the w. slope of the Cascades to the coast, B. C. to Monterey Co. , Calif. Mar. -June.

Viola septentrionalis Greene, Pitt. 3:334. 1898. *(J. M. Macoun,* near Ottawa, Ont. , Can. , May 10, 1898)
Conspicuously pubescent, acaulescent perennial from rather thick, scaly, shallow rhizomes; petioles 3-20 cm. long; stipules membranous, oblong-lanceolate, 5-10 mm. long, glandular-ciliate; leaf blades deeply cordate-triangular, as much as 7 cm. broad, coarsely crenate-serrate; peduncles usually about equaling the leaves; flowers 12-20 mm. long, the spur rather prominent; sepals ciliate; petals deep bluish to violet, the lower three white-based and copiously white-bearded; style head glabrous. N=27.

In moist open woods; n. e. N.Am. , as far s. as Pa. , westward to Ont. , also at several spots near the U.S. border in B. C. Apr. -June.

Perhaps the species is more general in our area, but it is not easily detected as it flowers very early in the spring.

Viola sheltonii Torr. Pac. R. R. Rep. 4:67, pl. 2. 1857. *(Bigelow,* near Downieville, Calif.)
Perennial with slender, deep-seated rootstocks, glabrous to sparsely puberulent, the flowering stems 5-15 cm. tall; stipules membranous, small, usually more or less pectinate; leaf blades glaucous and often somewhat purplish on the lower surface, 2-5 cm. long, commonly at least as broad, deeply cleft into 3 main lobes and usually dissected into ultimate linear segments; peduncles generally exceeding the leaves; flowers about 12 mm. long, yellow, the upper pair of petals brownish-backed, the lower three purplish-pencilled, the lateral pair sparsely clavate-bearded; style head sparsely bearded.

In chaparral or forest, especially under ponderosa pines, from Baja Calif. northward intermittently to the e. side of the Cascades near Cle Elum, Kittitas Co. , Wash. (also in Colo. ?). Apr. -June.

Viola trinervata Howell, Bot. Gaz. 11:290. 1886.
  V. beckwithii var. *trinervata* Howell, Bot. Gaz. 8:207. 1883. *(Th. Howell,* Goldendale, Wash. , Apr. 1, 1878)
  V. chrysantha var. *glaberrima* Torr. Bot. Wilkes Exp. 238. 1874. *(Pickering,* "High, dry prairies between the Spipen River and the Columbia")
  V. trinervata f. *semialba* St. John, Am. Bot. 34:94. 1928. *(St. John, Pickett, & Warren 6860,* 10 mi. n.w. of Almira, Wash:)
Perennial with short, thick, very deep-seated rootstocks, glabrous and more or less glaucous; stems several, naked below, 5-15 cm. tall; basal leaves usually several, the blades pedately-lobed to -compound, the main segments again irregularly once or twice dissected, the ultimate segments most-

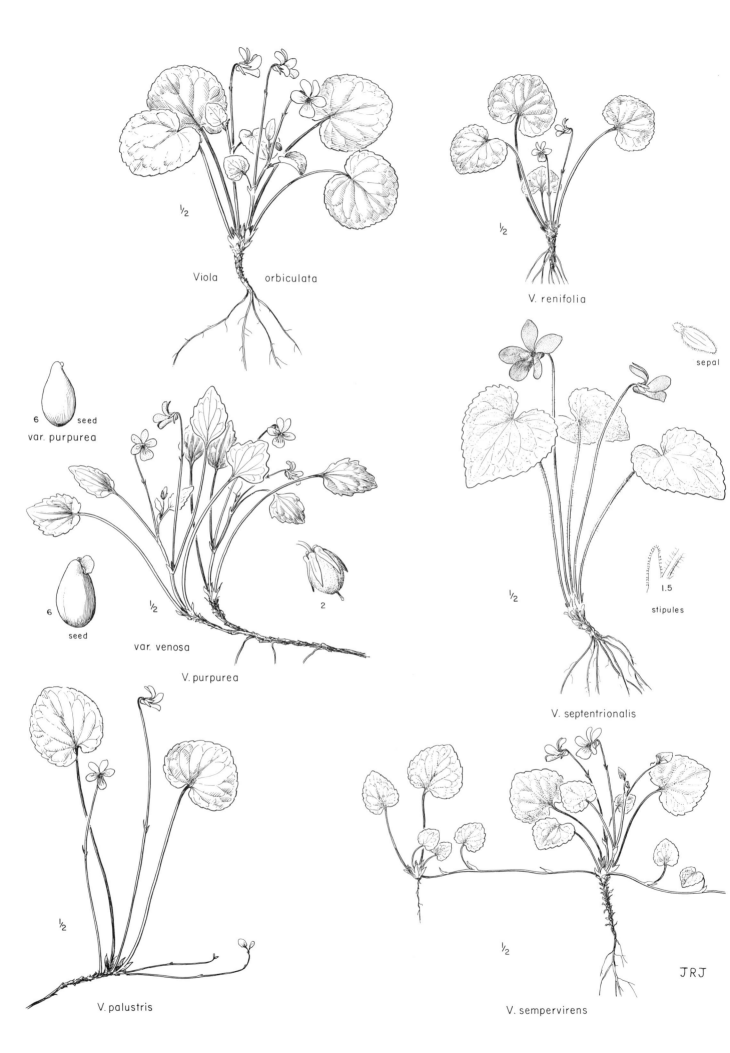

1/2

Viola   orbiculata

V. renifolia

sepal

6 seed
var. purpurea

6
seed

1/2

2

var. venosa

V. purpurea

1.5

stipules

V. septentrionalis

1/2

V. palustris

1/2

V. sempervirens

JRJ

ly narrowly elliptic, 1.5-4 cm. long, rather leathery, usually prominently 3 (5)-nerved on the lower surface; peduncles mostly exceeding the leaves, bractless or with 1-2 tiny bractlets; flowers about 1.5 cm. long, bicolored, the upper pair of petals deep reddish-violet, the lower three pale to fairly deep lilac, with yellowish or whitish base and purple pencilling or blotching; style head hairy. Sagebrush violet, desert pansy.

Sagebrush flats and rocky hillsides; Okanogan Co., Wash., to n.e. Sherman and Malheur cos., Oreg. Mar.-June.

LOASACEAE Blazing-Star Family

Flowers solitary or in open to condensed cymose clusters, regular, perigynous to epigynous; calyx (ours) adnate to the ovary, the free portion with a short to rather well-developed (often flared) hypanthium, the lobes 5 (4), persistent; petals 5 (4), often short-clawed, distinct (all ours) to connate at base; stamens usually numerous (all ours), distinct or basally connate into groups, some often expanded and forming petaloid staminodia alternating with the true petals; anthers mostly 2-celled; ovary inferior, 1 (3)-celled; style 1, simple (branched above); placentation parietal or (less commonly) axile; fruit a capsule; seeds with or without endosperm, the embryo straight; herbs or shrubs, often (all ours) scabrous or bristly with rough, barbed (glochidiate), sometimes stinging hairs; leaves alternate (as in most of ours) to opposite, without stipules.

About 15 genera and 250 species, chiefly in the New World; most abundant in tropical and semi-desert areas, especially in S.Am.

Mentzelia L. Blazing Star

Flowers solitary to cymose toward the ends of the branches, often showy; calyx with a short, flared, free hypanthium and 5 lobes; petals 5, yellowish to orange; stamens numerous, the filaments linear or (the outer five) expanded into petaloid staminodia; ovary inferior, 1-celled with 3 parietal placentae; capsule many-seeded, the seeds prismatic to flattened; annual or perennial herbs with alternate, scabrous, barbellate-pubescent, brittle leaves that readily attach to any foreign body.

About 50 species of the desert and drier tropics of N. and S.Am. (Named for the German botanist, C. Mentzel, 1622-1701.)

Two of our species, *M. decapetala* and *M. laevicaulis,* are showy, beautiful (but somewhat coarse) herbaceous perennials that do well in dry sunny areas, although they are best adaptable to gardens e. of the Cascades, in our area.

Reference:

Darlington, Josephine. A monograph of the genus Mentzelia. Ann. Mo. Bot. Gard. 21:103-226. 1934.

Besides the 4 species formally treated, three others are to be found adjacent to our range and may prove to be elements of our flora. *M. torreyi* Gray, a perennial, has petals less than 1 cm. long. The other two are annuals; *M. congesta* (Nutt.) T. & G. closely resembles *M. albicaulis* but has a rather congested inflorescence and broader floral leaves and bracts than described for *M. albicaulis*. It is a diploid plant (N=9). *M. gracilenta* T. & G. is also very similar to *M. albicaulis* but has larger flowers, the petals averaging 6 to 14 (20) mm. in length; it, too, is a diploid species.

1 Plants biennial to perennial; petals 1.5-8 cm. long; seeds flattened

  2 Petals apparently 10, the inner five (staminodia) not quite so broad as the outer (true) petals; floral bracts adherent to the ovary; seeds thin-margined but not winged

                                                                           M. DECAPETALA

  2 Petals 5, the five outer stamens sometimes flattened and somewhat petaloid, but much narrower than the true petals; floral bracts not adherent to the ovary; seeds distinctly wing-margined   M. LAEVICAULIS

1 Plants annual; petals less than 1.5 cm. long; seeds not flattened

  3 Floral bracts mostly ovate-lanceolate to ovate; inflorescence congested; capsules linear; seeds uniseriate throughout, prismatic, grooved on the vertical margins, so obscurely tuberculate-muricate as apparently to be smooth, even when viewed with a 10x lens   M. DISPERSA

  3 Floral bracts narrowly to broadly lanceolate; inflorescence not congested; capsules usually noticeably broadened upward; seeds obviously tuberculate-muricate when viewed with a 10x lens, and prismatic and often grooved on the vertical angles in

the lower portion of the capsules, those above arranged and shaped irregularly
and not grooved on the angles

M. ALBICAULIS

Mentzelia albicaulis Dougl. ex Hook. Fl. Bor. Am. 1:222. 1833.
  Bartonia albicaulis Dougl. ex Hook. loc. cit. *Acrolasia albicaulis* Rydb. Bull. Torrey Club 30:277.
    1903. *(Douglas,* "On arid plains of the river Columbia")
  M. parviflora Heller, Bull. Torrey Club 25:199. 1898. *Acrolasia parviflora* Heller, Muhl. 1:138.
    1906. *(Heller 3750,* 11 mi. s. e. of Santa Fe, N. M. , June 23, 1897)
  M. tenerrima Rydb. Mem. N. Y. Bot. Gard. 1:271. 1900. *Acrolasia tenerrima* Rydb. Bull. Torrey
    Club 30:277. 1903. *(Rydberg & Bessey 4542,* "Foot of Electric Peak," Mont. , Aug. 18, 1897)
  M. tweedyi Rydb. Mem. N. Y. Bot. Gard. 1:271. 1900. *Acrolasia tweedyi* Rydb. Bull. Torrey Club
    30:277. 1903. *(Tweedy 152,* Trail Creek, Park Co. , Mont. , in 1887)
  M. ctenophora Rydb. Bull. Torrey Club 28:33. 1901. *Acrolasia ctenophora* Rydb. Bull. Torrey
    Club 30:277. 1903. *M. albicaulis* var. *ctenophora* J. Darl. Ann. Mo. Bot. Gard. 21:184. 1934.
    (Rydberg & Vreeland 5769, Cucharas R. , below La Veta, Colo.)
  Acrolasia gracilis Rydb. Bull. Torrey Club 31:566. 1904. *M. albicaulis* var. *gracilis* J. Darl. Ann.
    Mo. Bot. Gard. 21:184. 1934. *M. gracilis* Thompson & Lewis, Madroño 13:103. 1955. *(J. H. Co-*
    wen, Larimer Co. , Colo., in 1895) This taxon has been reported for our area and is said to differ
    from *M. albicaulis* chiefly because of the irregularly divided (as compared with more shallowly
    scalloped) leaves; it has been found to be an octoploid (N=36). It is not feasible to separate the
    plants of our area into two taxa on the basis of leaf lobation.
  Simple to freely branched annual 1-4 dm. tall, glochidiate-pubescent throughout but the whitish
shining stems usually becoming glabrous, at least below; leaves various, 2-10 cm. long, the basal
usually linear and entire to shallowly few-lobed, narrowed to petioles, the cauline linear to lanceolate, from subentire to laciniate into linear lobes, sessile to petiolate, the floral leaves usually broader and entire to few-toothed; flowers usually rather few in modified, more or less open cymes; calyx
(including the adnate portion) 1-2 cm. long, the lobes 2-4 mm. long; petals yellow, 2-6 mm. long;
stamens 15-35, shorter than the petals; capsules linear-clavate, 1-2.5 cm. long, about 2 mm. thick;
seeds 15-30, in a single row below and often grooved on the angles but usually arranged and shaped
irregularly and not grooved on the angles in the upper part of the capsule, all fairly prominently
tuberculate to papillose-muriculate. N=27.
  On dry, usually sandy soil in desert valleys and foothills from B. C. to s. Calif. , mostly e. of the
Cascades and Sierra Nevada, e. to Mont. and N. M. May-July.
  The species has been divided into several varieties that are considered to be of insufficient significance to merit recognition, since they were based largely on apparently plastic characters.

Mentzelia decapetala (Pursh) Urb. & Gilg, Ber. Deuts. Bot. Ges. 10:263. 1892.
  Bartonia decapetala Pursh ex Sims, Curtis' Bot. Mag. 36:pl. 1487. 1812. *Bartonia ornata* Pursh, Fl.
    Am. Sept. 327. 1814. *Torreya ornata* Eaton, Man. Bot. 7th ed. 560. 1836. *M. ornata* T. & G. Fl.
    N. Am. 1:534. 1840. *Touterea ornata* Eat. & Wright, N. Am. Bot. 454. 1840. *Hesperaster decapetalus*
    Cockerell, Torreya 1:142. 1901. *Touterea decapetala* Rydb. Bull. Torrey Club 30:276. 1903. *Nut-*
    tallia decapetala Greene, Leafl. 1:210. 1906. *(Nuttall,* "on the banks of the Missouri, from the
    river Platt to the Andes"; but Pursh obviously intended that the species should have been based upon
    Lewis, "white bluffs near the Maha village")
  Perennial from a deep taproot, harshly scabrid-glochidiate-pubescent overall; stems usually single
(occasionally 2-3), 3-10 dm. tall, mostly simple below and branching above; leaves alternate, fleshy,
lanceolate, 4-15 cm. long, sharply sinuate-pinnatifid, the lower ones petiolate, the upper sessile and
often more or less clasping; flowers terminal at the ends of the various branches, vespertine, pleasantly odorous, subtended by several alternate, pinnatifid bracts usually adnate to the lower portion of
the ovary; calyx (including the adnate portion) 1-2 cm. long at anthesis, rapidly lengthening as the
fruit matures, the lobes 2-4 cm. long, lanceolate, persistent on the capsule; petals apparently 10,
cream to pale yellow, about twice the length of the calyx lobes, the inner five (staminodia) narrower;
stamens very numerous, the filaments about 3/4 the length of the petals; style usually equaling the
stamens; stigma 3-4 mm. long, not lobed, papillose on the 4-5 somewhat spiral angles; capsules 3-
4.5 cm. long; seeds many, borne horizontally, 3-3.5 mm. long, dark grayish-brown, strongly flattened and thin-margined but not truly winged. N=11.
  On the plains and into the lower mountains; e. Mont. , s. Alta. , and the Dakotas to Mex. , mostly
e. of the Rocky Mts. , but occasional in Mont. and Ida. July-Sept.

Mentzelia dispersa Wats. Proc. Am. Acad. 11:137. 1876.

Acrolasia dispersa Davids. Bull. So. Calif. Acad. Sci. 5:14. 1906. *(Vasey 195,* Colorado)

M. albicaulis var. *integrifolia* Wats. Bot. King Exp. 114. 1871. *M. integrifolia* Rydb. Mem. N.Y.
Bot. Gard. 1:271. 1900. *Acrolasia integrifolia* Rydb. Bull. Torrey Club 30:278. 1903. *Acrolasia
albicaulis* var. *integrifolia* Daniels, Fl. Boulder 174. 1911. *(Geyer 663* is the first of 3 collections
cited from Nev. and Utah)

M. compacta A. Nels. Bull. Torrey Club 25:275. 1898. *Acrolasia compacta* Rydb. Bull. Torrey
Club 30:278. 1903. *M. dispersa* var. *compacta* Macbr. Contr. Gray Herb. n.s. 56:26. 1918. *(A.
Nelson 2454,* Parkman, Sheridan Co., Wyo., July 22, 1896)

M. pinetorum Heller, Bull. So. Calif. Acad. Sci. 2:69. 1903. *Acrolasia pinetorum* Heller, Muhl.
2:99. 1905. *M. dispersa* var. *pinetorum* Jeps. Man. Fl. Pl. Calif. 651. 1925. *(Heller 5910,* Mt.
Sanhedrin, Lake Co., Calif.)

Acrolasia latifolia Rydb. Bull. Torrey Club 31:567. 1904. *M. latifolia* Nels. in Coult. & Nels. New
Man. Bot. Rocky Mts. 324. 1909. *M. dispersa* var. *latifolia* Macbr. Contr. Gray Herb. n.s. 56:
26. 1918. *(Tweedy 5149,* Mts. between Sunshine and Ward, Colo., in 1902)

Low annual 1-4 dm. tall, finely glochidiate-pubescent throughout, simple to freely branched or multistemmed; leaves lanceolate to ovate-lanceolate, entire to sinuate-lobed or even subpinnatifid, 3-10
cm. long, the lower ones short-petiolate and alternate, the upper ones sessile or subsessile, the
bracteate leaves often opposite; flowers irregularly cymose in fairly compact clusters; calyx lobes
about 2 mm. long; petals yellow, 3-6 mm. long; stamens 15-35, shorter than the petals; capsules 1-3
cm. long, linear, 1.5-2 mm. thick; seeds in one row throughout the capsule, mostly conspicuously
prismatic and grooved on the vertical edges, grayish, lightly tessellate-muriculate, the excrescences
scarcely visible even at 10x magnification.

On dry soil, from the plains into the canyons and slopes of the lower mountains, from Wash. (e. of
the Cascades) to s. Calif., e. to Mont., Wyo., and Colo. May-July.

The species shows considerable variation in the size and indenture of the leaves and is usually divided into 2 or more varieties which nearly completely overlap in range and appear to be little more
than growth forms, rather than geographic races. The three recognized by Darlington were a robust
phase, var. *latifolia* (Rydb.) Macbr., with petals 5-6 mm. long and leaves sometimes coarsely
toothed, and two phases with small flowers and fruits: one, the var. *dispersa,* usually simple to sparingly branched and with variable leaves that may be as much as 10 cm. long, the other the var. *compacta* (A. Nels.) Macbr., including the plants that are more freely branched and compact in habit,
with usually smaller, ovatish leaves less than 5 cm. long.

Mentzelia laevicaulis (Dougl.) T. & G. Fl. N.Am. 1:535. 1840.

Bartonia laevicaulis Dougl. in Hook. Fl. Bor. Am. 1:221. 1833. *Hesperaster laevicaulis* Cockerell,
Torreya 1:143. 1901. *Touterea laevicaulis* Rydb. Bull. Torrey Club 30:276. 1903. *Nuttallia laevicaulis* Greene, Leafl. 1:210. 1906. *(Douglas,* "On the gravelly islands and rocky shores of the Columbia, near the 'Great Falls'")

MENTZELIA LAEVICAULIS var. PARVIFLORA (Dougl.) C. L. Hitchc. hoc loc. *Bartonia parviflora*
Dougl. in Hook. Fl. Bor. Am. 1:221. 1833. *Touterea parviflora* Rydb. Bull. Torrey Club 30:276.
1903. *Nuttallia parviflora* Greene, Leafl. 1:210. 1906. *(Douglas,* "Abundant on calcareous rocky
situations and micaceous sandy banks of streams, in the interior parts of the Columbia")

M. brandegei Wats. Proc. Am. Acad. 20:367. 1885. *Touterea brandegei* Rydb. Bull. Torrey
Club 30:276. 1903. *Nuttallia brandegei* Greene, Leafl. 1:210. 1906. *(Brandegee,* near the Simcoe
Mountains, Washington Territory, on the mesa bordering Satus Creek, in 1883) = var. *parviflora.*

Nuttallia acuminata Rydb. Bull. Torrey Club 40:61. 1913. *M. laevicaulis* var. *acuminata* Nels. &
Macbr. Contr. Gray Herb. n.s. 65:40. 1922. *M. acuminata* Tidestr. Contr. U.S. Nat. Herb. 25:
362. 1925. *(Sandberg, MacDougal, & Heller 651,* Spokane R., Kootenai Co., Ida., in 1892) = var.
laevicaulis.

Biennial or short-lived perennial from a deep taproot, 3-10 dm. tall; stem usually one and branched
chiefly above but not uncommonly branched near the base and the plant apparently multistemmed,
harshly scabrid-glochidiate-pubescent throughout or nearly smooth near the base; leaves alternate,
the lower ones oblanceolate, deeply sinuate-pinnatifid and somewhat runcinate-lobed to the slender
petioles, as much as 15 cm. long, the upper leaves sessile, oblong to ovate-oblong, less deeply lobed;
flowers terminal on numerous branch ends and often in the top 1 or 2 leaf axils, diurnal, inodorous,
each subtended by 1-several linear, entire to few-toothed bracts that are free of the ovary; calyx (including the adnate portion) 1-2.5 cm. long at anthesis, accrescent as the fruit matures, the lobes

linear, 1.5-4 cm. long; petals lemon yellow, 2.5-8 cm. long, narrowly oblong or oblong-lanceolate; stamens very numerous, about 2/3 the length of the petals, the 5 outer ones often not antheriferous, considerably expanded and more or less petaloid, although much narrower than the petals with which they alternate; style usually slightly longer than the stamens; stigma 3-4 mm. long, angled but not lobed; capsules 1.5-3.5 cm. long; seeds numerous, borne horizontally, light grayish-green or brownish, flattened and distinctly winged. N=11.

In desert valleys and lower mountains; Mont. to B.C. and e. Wash., southward to Calif., Utah, and Wyo. July-Sept.

The species is represented in our area chiefly by the var. *laevicaulis* which has large flowers (the petals at least 4 cm. long) and fruits (mostly over 2 cm. long). In s. B.C. and along the valleys e. of the Cascades in Wash., var. *laevicaulis* is largely replaced by var. *parviflora* (Dougl.) C. L. Hitchc., a smaller-flowered plant with the fruit usually less than 2 cm. long and the petals 1.5-4 cm. long. There is no sharp line of demarcation between the two taxa.

CACTACEAE Cactus Family

Flowers showy, solitary, regular (ours), usually perfect; sepals and petals generally numerous, imbricate, these and the stamens basally coalescent and adnate, enclosing and more or less adnate to the ovary, forming a well-developed, free hypanthium; stamens numerous, borne in groups or spirally arranged; pistil compound, the style one, the stigma 3- to 10-lobed; ovary inferior, 1-celled, with several parietal placentae; fruit many-seeded, either dry or baccate and leathery to fleshy; fleshy, mostly herbaceous (ours) to shrubby or woody, spiny plants with globose, flattened, or cylindric and often fluted or ribbed and tuberculate, jointed stems, usually with special cushions (areoles) which may bear coarse woolly hair, coarse spines, soft sharp bristles, and flowers.

A family of approximately 1200 species, confined almost entirely to the desert and tropical regions of N. and S.Am. There is great difference of opinion as to proper generic limits in the Cactaceae, some workers recognizing about 25 genera, others up to 5 times that many. This may be due in part to the fact that the plants usually are so poorly preserved that herbarium studies may not be particularly fruitful.

Many of the cacti, such as the night-blooming cereus, are considered attractive ornamentals, but ours are largely of interest as plant oddities and as indicators of overgrazing.

Reference:

Britton, N. L. and J. N. Rose. The Cactaceae. 4 vols. Washington, D.C. 1919-1923.

Grateful acknowledgment is hereby made for many helpful suggestions contributed by Dr. Lyman Benson, who read, corrected, and augmented the original manuscript. He is, of course, in no way responsible for the ultimate interpretation of our taxa.

1 Stems jointed, the joints of ours more or less flattened, not tuberculate; areoles producing soft, sharp, barbed bristles (glochids) as well as long spines; leaves small, scalelike, quickly deciduous   OPUNTIA
1 Stems not jointed, ours more or less globose, conspicuously tuberculate; areoles usually bearing wool in addition to the main spines, but without spinose, barbed bristles; leaves lacking
  2 Stems with low, longitudinally spiralled ribs bearing the grooved tubercles; flower borne in a well-defined area at the side of the areole near the tip of the tubercle
                                                                                                    PEDIOCACTUS
  2 Stems not longitudinally ribbed, the tubercles conspicuous, more or less mammillate, grooved on the upper side; flower borne on the side of the tubercle at the base of a groove connecting with the spiniferous areole   CORYPHANTHA

Coryphantha (Engelm.) Lemaire

Flowers solitary at the base of the groove on the ventral side of the mature tubercles somewhat back from the stem tip, more or less funnelform, ours greenish-white to deep red or purplish; ovary not spinose; berry somewhat fleshy; plants fleshy, the stems globose, leafless; spines produced from areoles at the tip of spirally arranged, mammillate tubercles that are longitudinally grooved on the upper side.

About 35 species, all in N.Am. (Greek *koryphe*, a cluster, and *anthos*, flower.)

stipules

Viola sheltonii

Viola trinervata

Mentzelia laevicaulis

18

seed

Mentzelia albicaulis

15

seed

Mentzelia dispersa

seed

5

Mentzelia decapetala

JRJ

1 Flowers greenish-white, barely reddish-tinged; main spines 1 per areole; fruit less than
    1 cm. long, subglobose, reddish   C. MISSOURIENSIS
1 Flowers reddish-purple; main spines 3-5 per areole; fruit 1-2 cm. long, oblong,
    greenish   C. VIVIPARA

Coryphantha missouriensis (Sweet) Britt. & Rose in Britt. & Brown, Ill. Fl. 2nd ed. 2:570, fig. 2984.
    1913.
    Cactus mamillaris sensu Nutt. Gen. Pl. 1:295. 1818, not of L. *Mammillaria missouriensis* Sweet,
    Hort. Brit. 171. 1827. *Mammillaria nuttallii* Engelm. Pl. Fendl. 49. 1849. *Cactus missourien-*
    sis Kuntze, Rev. Gen. 1:259. 1891. *Neobesseya missouriensis* Britt. & Rose, Cactaceae 4:53,
    pl. 11. 1923. *Neomamillaria missouriensis* Britt. & Rose ex Rydb. Fl. Prair. & Pl. 560. 1932.
    ("On the high hills of the Missouri probably to the mountains," collector not specified, but prob-
    ably Nuttall)
    Stems usually 1, globose or subglobose, up to 5 cm. tall; tubercles nearly terete except for the 8
grooves; areoles with 1 main spine 9-12 mm. long and with 10-20 smaller, slender, marginal spines;
flowers about 2.5 cm. long, greenish-yellow or dull whitish to sometimes reddish-tinged; fruit subglobose, less than 1 cm. long, eventually somewhat reddish.
    Valleys and hills of the desert and grasslands; mainly e. of the Rocky Mts., from Man. to Kans.,
but w. to e. Mont. and Colo., and known in our area from c. Ida., along the Salmon R. in Custer Co.
June-July.
    The species is in general so similar to *C. vivipara,* except for flower color, that it is not illustrat-
ed separately.

Coryphantha vivipara (Nutt.) Britt. & Brown, Ill. Fl. 2nd ed. 2:571, fig. 2985. 1913.
    Cactus vivipara Nutt. Gen. Pl. 1:295. 1818. *Mammillaria vivipara* Haw. Syn. Pl. Succ. Suppl. 72.
    1819. *Escobaria vivipara* F. Buxb. Oest. Bot. Zeits. 98:78. 1951. *(Nuttall,* "With the above"
    [Cactus mamillaris] "on the summits of gravelly hills")
    Stems 1-several, subglobose, the base turbinate, 3-6 cm. tall and about as thick; tubercles terete
except for the groove; areoles with 3-5 main spines 8-11 mm. long and with 10-20 smaller slender
marginal spines; flowers 2.5-3 cm. long, bright reddish-purple; fruit greenish, 1-2 cm. long.
    Desert valleys and hills; Alta. to s.e. Oreg., Colo., and Kans. May-June.

Opuntia Mill. Cholla or Prickly-Pear Cactus

    Flowers borne in areoles of previous years' growth; perianth tube campanulate, more or less ro-
tate above, areolate to the summit of the ovary; sepals several-seriate, greenish without, often grading into the numerous yellowish to red or reddish-purple petals; stamens numerous, shorter than the
petals; stigma with 5-8 short lobes; berry more or less pear-shaped, truncate, dry (in ours) to succulent; seeds strongly flattened, usually disc-shaped; fleshy plants with flattened, broad, jointed (ours)
or terete and tuberculate stems and scalelike, spirally arranged, caducous leaves, the areolae axillary, bearing (usually) both larger rigid spines and slender sharp bristles.
    About 250 species of N. and S. Am. (Derivation uncertain.)
    The prickly pear cacti (with flattened stems) are often singed of spines and fed to stock in drought
years. The cholla cacti (not in our area) provide one of the most picturesque features of the deserts
of the Southwest, with their intricately branched stems and fearsome spines that are occasionally fatal
to grazing stock.
1 Joints of the stems not greatly flattened, often nearly as thick as broad, usually less than
    5 cm. long, readily detached from the plant; spines rather strongly barbed; areoles usu-
    ally noticeably white-woolly; flowers yellow   O. FRAGILIS
1 Joints of the stems conspicuously flattened, the larger ones 5-13 cm. long, not readily
    detached from the plant; spines only slightly barbed; areoles usually with rusty wool, if
    any; flowers yellow to reddish   O. POLYACANTHA

Opuntia fragilis (Nutt.) Haw. Syn. Pl. Succ. Suppl. 82. 1819.
    Cactus fragilis Nutt. Gen. Pl. 1:296. 1818. *Tunas fragilis* Nieuwl. & Lunell, Am. Midl. Nat. 4:479.
    1916. *(Nuttall,* "From the Mandans to the mountains")
    O. *polyacantha* var. *borealis* sensu Piper, Contr. U.S. Nat. Herb. 11:397. 1906, not of Coult. in
    1896.

Plants prostrate, forming low mats 0.5-2 dm. tall, the joints of the stem usually not greatly flattened, often nearly as thick as broad, subglobose to obovoid, mostly 2-5 (but the lower ones sometimes as much as 10) cm. long, the upper one, at least, usually easily dislodged; areoles commonly coarsely white-woolly and with a few yellowish spinose bristles as well as 2-7 straight, yellowish to brownish spines 1-3 cm. long; flowers yellow, 3-5 cm. long and broad; filaments reddish; fruit dry, 1.5-2 cm. long, somewhat spiny. Prickly pear.

Dry hillsides and open ground; B.C. southward, chiefly on the e. side of the Cascades, to n. Calif., e. to Tex. and Wisc., also on drier islands and part of the mainland in the Puget Trough. May-June.

Opuntia polyacantha Haw. Syn. Pl. Succ. Suppl. 82. 1819.

  Cactus ferox Nutt. Gen. Pl. 1:296. 1818, not of Willd. in 1813. *O. missouriensis* DC. Prodr. 3:472.
    1828. *Tunas polyacantha* Nieuwl. and Lunell, Am. Midl. Nat. 4:479. 1916. *(Nuttall, "In arid situations on the plains of the Missouri")*
  O. missouriensis platycarpa Engelm. Proc. Am. Acad. 3:300. 1856. *O. polyacantha platycarpa*
    Coult. Contr. U.S. Nat. Herb. 3:436. 1896. *(Hayden collections in 1853-54, from Neb. and Mont.)*
  O. missouriensis var. *microsperma* Engelm. & Bigel. Proc. Am. Acad. 3:300. 1856. *O. polyacantha* var. *borealis* Coult. Contr. U.S. Nat. Herb. 3:436. 1896. *("Fur Traders" collection,*
    Pierre, S.D., in 1847)
  O. columbiana Griffiths, Bull. Torrey Club 43:523. 1916. *(Griffiths 10041, near Pasco, Wash.)*

Plants forming rounded clumps 1-3 (4) dm. tall, often spreading into mats several meters broad; stem joints obovate to suborbicular in outline, strongly flattened, mostly 5-15 cm. long; areoles often with grayish-yellow wool as well as tawny spinose bristles and 5-11 straight spines 1-5 cm. long; flowers 5-7 cm. long, yellow, but the perianth sometimes reddish-tinged to reddish; fruit dry, about 2.5 cm. long, spinose.

From the plains into the foothills and lower mountains; B.C. to Alta., s. to e. Oreg., Ariz., Tex., and Mo. May-June.

Usually the flowers are yellow when they blossom, with a tendency to age to pinkish, but some plants (in our area only in s.c. Ida.) produce flowers that are reddish in color at anthesis. There is no published varietal name available for this phase. According to Dr. Benson (private correspondence), *O. polyacantha* tends to have terete or subterete spines, whereas *O. erinacea* Engelm. & Bigel. has the spines distinctly flattened. Our plants, especially those of the Snake R. drainage (which were called *O. columbiana* Griffiths), have slightly flattened to subterete spines. That there has been some modification of *O. polyacantha* in our area by ultimate crossing with *O. erinacea* is not impossible, but whatever the reason, our material is so variable that there is no clearly established basis for including it in more than one taxon, whether at the specific or infraspecific level.

Pediocactus Britt. & Rose   Hedgehog Cactus

Flowers solitary in a specialized area near the tip of the tubercles at the edge of the spiniferous areoles; free hypanthium short, campanulate; petals and stamens numerous; fruit globose, greenish; seeds tuberculate; small, leafless plants with 1-several globose to ovoid stems bearing low tubercles in rows on 8-13 spiral longitudinal ridges.

One species of w. N.Am. (Greek *pedion*, field, and cactus.)

Pediocactus simpsonii (Engelm.) Britt. & Rose in Britt. & Brown, Ill. Fl. 2nd ed. 2:570, fig. 2983.
  1913.
  Echinocactus simpsoni Engelm. Trans. St. Louis Acad. 2:197. 1863. *Mammillaria simpsonii* M. E.
    Jones, Zoë 3:302. 1893. *(H. Engelmann, Utah, in 1859)*
  Echinocactus simpsoni var. *minor* Engelm. Trans. St. Louis Acad. 2:197. 1863. *(Hall & Harbour,*
    Colo., in 1862)
  Echinocactus simpsoni var. *robustior* Coult. Contr. U.S. Nat. Herb. 3:377. 1896. *(Watson, Nevada,*
    in 1868)

Stems subglobose to depressed, single to clustered, 7-12 cm. thick; tubercles 12-25 mm. long, in 8-13 spiral rows; central spines 8-12, straight, yellowish to reddish-brown, 8-25 mm. long; marginal spines 10-30, smaller, whitish; flowers 1.5-2 cm. long, yellowish-green to purplish; fruits subglobose, 6-8 mm. long; seeds black, about 3 mm. long.

Desert valleys and low mountains; e. Wash. to Nev., e. to Wyo., Utah, and Colo. May-July.

Our material is referable to the var. *robustior* Coult., which ranges from e. Wash. to Nev. It differs from the more eastern var. *simpsonii* in its slightly larger tubercles (15-25 vs. 12-16 mm. long) and main spines (15-28 vs. 8-15 mm. long).

ELAEAGNACEAE Oleaster Family

Flowers mostly axillary, rather inconspicuous, perfect or imperfect, apetalous; calyx 4-lobed, tubular below and forming a well-developed hypanthium that closely invests, but is not adnate to, the ovary; stamens 4 or 8, borne near the top of the hypanthium, usually alternating with the lobes of a disc; pistil 1-carpellary, the style short; fruit a 1-seeded achene or much more hardened and nutlike, closely invested by the thickened, fleshy calyx and thus pseudobaccate or apparently drupaceous; deciduous (ours) or evergreen, perfect-flowered to dioecious trees or shrubs, conspicuously silvery- to rusty-scurfy with lepidote to stellate hairs and alternate or opposite, simple leaves, sometimes spiny.

Three genera and perhaps 25 species, widely distributed but largely in n. temperate, very arid regions.

Reference:

Nelson, Aven. The Elaeagnaceae—a mono-generic family. Am. Journ. Bot. 22:681-683. 1935.

1 Leaves alternate; plants perfect flowered; stamens 4   ELAEAGNUS

1 Leaves opposite; plants dioecious; stamens 8   SHEPHERDIA

Elaeagnus L.

Flowers axillary, perfect; calyx tubular-turbinate above the globose, investing base, lobed less than half the length, deciduous from the top of the developing fruit, without a glandular thickening at the top of the hypanthium; stamens 4, borne near the top of the hypanthium; ovary wall becoming much hardened and bony, longitudinally fluted, the investing calyx more mealy than juicy, the fruit drupaceous; alternate-leaved shrubs, often of very arid regions.

About 20 species of the N. Temp. Zone. (Greek *elaia*, olive, and *agnos*, the Grecian name of the chaste-tree.)

Elaeagnus commutata Bernh. in Allg. Thuer. Gartenf. 2:137. 1843, nomen nudum; ex Rydb. Fl. Rocky Mts. 582. 1917.

E. *argentea* Pursh, Fl. Am. Sept. 114. 1814, not of Calla in 1791. *(Lewis,* banks of the Missouri)

Spreading to erect, unarmed shrubs 1-4 m. tall, with extensive rootstocks; young branches brownish-scurfy, old branches dark grayish-red; leaves short-petiolate, the blades lanceolate to oblanceolate, 2-7 cm. long, acute to obtuse, silvery-scurfy on both surfaces or sometimes brownish-lepidote beneath; flowers 6-14 mm. long, 1-3 per axil or clustered at the base of new twigs; pedicels 1-2 mm. long; stamens 4, included; fruits obovoid, 9-12 mm. long, the investing hypanthium dry and mealy but the pericarp much hardened and bony. N=14. Silverberry.

On gravel benches and scabland, or more commonly along watercourses; B.C. and Yukon to Que., southward in the Rocky Mt. area, from Ida. and Mont. to Utah. June-July.

This plant, readily adaptable to dry soils, is sometimes used as a windbreak. It is related to the Russian olive, which is much more extensively used for that purpose. The latter, *Elaeagnus angustifolia* L., is more truly arborescent and has narrow leaves and juvenile branches that are silvery. It sometimes occurs as an escape or a remnant of older plantings.

Shepherdia Nutt. Nom. Conserv.

Flowers 1-several in the leaf axils, apetalous, appearing with or before the leaves; staminate flowers with a conspicuous 8-lobed disc in the throat of the hypanthium, the 4 calyx lobes usually spreading, the stamens 8, a pistil completely lacking; pistillate flowers with short, usually erect calyx lobes, but with no vestige of stamens, the hypanthium becoming very fleshy and forming a pseudobaccate fruit; pericarp not greatly hardened; dioecious shrubs or small trees with opposite entire leaves.

Three species of N. Am. (Named for John Shepherd, 1764-1836, an English botanist.)

Although our native species are scarcely choice, they have some horticultural value, being perhaps most attractive when in fruit.

1 Leaves green above, brownish-lepidote on the lower surface; branches unarmed   S. CANADENSIS

1 Leaves silvery on both surfaces; branches often spine-tipped   S. ARGENTEA

Shepherdia argentea (Pursh) Nutt. Gen. Pl. 2:240. 1818.

Hippophae argentea Pursh, Fl. Am. Sept. 115. 1814. *Lepargyraea argentea* Greene, Pitt. 2:122. 1890. *Elaeagnus utilis* A. Nels. Am. Journ. Bot. 22:682. 1935. *(Lewis,* on the banks of the Missouri)

Large, rigidly branched shrubs or small trees 2-6 m. tall, the young twigs silvery-scurfy, the older branches brownish, often spine-tipped; leaves with short, slender petioles, the blades mostly narrowly oblong to oblanceolate, 2-5 cm. long, 5-15 mm. broad, silvery-scurfy on both surfaces; flowers clustered on short branches, the staminate plants often blossoming before the leaves have started to expand; sepals brownish within, spreading to erect, 1-2 mm. long; fruit yellow to reddish, ellipsoid, 4-6 mm. long, acid. N=13. Buffalo berry, rabbitberry.

Chiefly along watercourses or other less xeric spots; B. C., s. e. Oreg., and e. and s. Calif., e. to Minn. and c. Can., not known from Wash. or from Ida., and in Mont. occurring chiefly e. of the Rocky Mts. May-July.

The fruits are sometimes used for making jelly.

Shepherdia canadensis (L.) Nutt. Gen. Pl. 2:241. 1818.

Hippophae canadensis L. Sp. Pl. 1024. 1753. *Lepargyraea canadensis* Greene, Pitt. 2:122. 1890. *Elaeagnus canadensis* A. Nels. Am. Journ. Bot. 22:682. 1935. *(Kalm,* Canada)

S. canadensis f. *xanthocarpa* Rehd. Mitt. Deuts. Dendrol. Ges. 1907:75. 1907. A phase with more yellowish fruits.

An unarmed, spreading shrub 1-4 m. tall; young twigs conspicuously brownish-scurfy, older branches brownish; leaves short-petiolate, the blades ovate to ovate-lanceolate, 1.5-6 cm. long, 1-3 cm. broad, mostly greenish on the upper surface, white-scurfy but conspicuously brownish-lepidote beneath; staminate flowers brownish, the sepals 1-2 mm. long, spreading to reflexed; fruit ellipsoid, yellowish-red, fleshy, bitter. N=11. Buffalo berry, soapberry.

On open to wooded areas, in various habitats; Alas. to Oreg., common e. through the Rocky Mt. states to the Atlantic coast. May-July.

LYTHRACEAE Loosestrife Family

Flowers perfect, regular (ours) or less commonly irregular, perigynous, 1-several in the leaf axils or in bracteate terminal racemes, spikes, or cymes; calyx gamosepalous and forming a campanulate to cylindric hypanthium free of the ovary, the 4-7 lobes usually alternating with toothlike appendages; petals distinct, generally equaling the sepals, borne at the top of the hypanthium, usually crumpled in the bud, often very quickly deciduous, sometimes lacking; stamens as many or (more commonly) up to twice as many as the petals (rarely fewer or numerous), borne on the hypanthium, the filaments generally of 2 lengths; anthers versatile; ovary (1) 2- to 6-celled, superior; style 1; stigma 2-lobed or discoid; fruit a capsule, usually dehiscent, the placentation commonly axile; seeds numerous, exalbuminous, the embryo straight; annual to perennial, often glabrous herbs (ours) to shrubs or trees, with entire, mostly opposite or verticillate (alternate), exstipulate leaves.

About 21 genera and nearly 500 species, chiefly tropical but common in temperate regions.

1 Hypanthium elongate-cylindric to conic, several times as long as thick, strongly nerved; petals 5-7, usually showy   LYTHRUM
1 Hypanthium campanulate to hemispheric, not strongly nerved; petals usually 4, not at all showy
  2 Style usually exserted from the calyx, at least 1 mm. long; capsules not dehiscing by regular sutures; leaves sessile, cordate-clasping at the base, bearing (usually) more than 1 flower per axil   AMMANNIA
  2 Style not exserted from the calyx, less than 1 mm. long; capsules dehiscent along regular sutures; leaves often petiolate, not at all cordate-based, usually bearing only 1 flower per axil   ROTALA

Ammannia L.

Flowers axillary, 4-merous (ours), borne in few-flowered cymes; hypanthium strongly 8-ribbed and 4-angled, the sinuses projecting as small teeth or horns; petals small or wanting; capsule thin-walled, 4-celled, rupturing irregularly, many-seeded; placentation axile; small, herbaceous, glabrous, opposite leaved annuals.

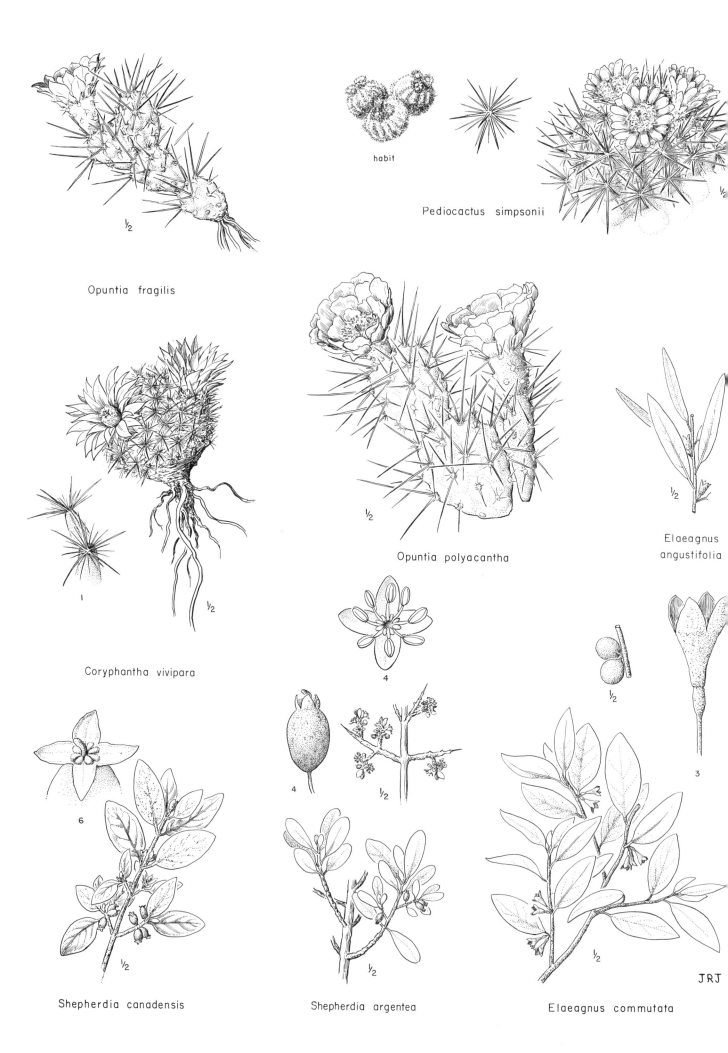

habit

Pediocactus simpsonii

Opuntia fragilis

Opuntia polyacantha

Elaeagnus
angustifolia

Coryphantha vivipara

Shepherdia canadensis

Shepherdia argentea

Elaeagnus commutata

JRJ

About 20 species, largely tropical. (Named after Paul Ammann, German botanist, 1634-1691.)

Ammannia coccinea Rottb. Pl. Hort. Havn. Descr. 7. 1773. (Europe)
 A. alcalina Blank. Mont. Agr. Coll. Stud. Bot. 1:87. 1905. *(Blankinship,* Lake Bowdoin, near
  Malta, Mont., Aug. 25, 1903)
   Simple to branched annual 5-40 cm. tall; leaves entire, oblong to oblong-lanceolate, 2-4 cm. long,
2-8 mm. broad, cordate-clasping at base; flowers 1-5 per axil, sessile or subsessile; hypanthium
globose to campanulate, 2-3 mm. long, becoming accrescent and up to 5 mm. in fruit; petals fugacious (sometimes wanting), purplish, the blade about 1 mm. long; stamens 4, projecting beyond the
hypanthium; style slightly exserted, 1-1.5 mm. long; stigma discoid; capsule membranous, nearly
globose, about 4 mm. long.
   Widely distributed in wet places, often where alkaline, from the tropics to much of N.Am.; rare in
the Pacific states but reported from Klickitat Co., Wash. May-June.

Lythrum L. Loosestrife

   Flowers perfect, often showy, clustered in the axils or in terminal spikes, often di- or trimorphic
as to the relative length of the stamens and the style; calyx forming a narrowly cylindric, longitudinally striate hypanthium, the 5-7 lobes alternating with linear, toothlike, erect to spreading processes;
petals 5-7, white to pink or purple, inserted at the top of the hypanthium; stamens inserted on the
lower portion of the hypanthium, from as many to twice as many as the petals; capsule elongate, usually 2-celled; annual or perennial herbs with opposite (or less commonly with some verticillate or alternate), simple, generally sessile and often cordate based leaves.
   About 30 species, widely distributed, but chiefly of semiswampy ground in the Northern Hemisphere.
(Greek *luthron,* blood, referring either to the staining properties of the plant or to the flower color,
the name used by Dioscorides.)
1 Plants annual, 1-4 dm. tall; lower leaves mostly alternate; petals less than 4 mm.
    long   L. HYSSOPIFOLIA
1 Plants perennial, mostly well over 4 dm. tall; lower leaves usually opposite; petals
    commonly over 4 mm. long
  2 Main leaves 2-5 cm. long, not cordate; petals purple, about 5 mm. long   L. ALATUM
  2 Main leaves 3-10 cm. long, more or less cordate-based; petals reddish-purple,
      7-10 mm. long   L. SALICARIA

Lythrum alatum Pursh, Fl. Am. Sept. 334. 1814. *(Enslen,* Lower Georgia)
   Glabrous and glaucous, rhizomatous perennial 4-10 dm. tall; stems 4-angled, erect, usually freely
branched above; lower leaves opposite, the upper ones more often alternate, narrowly lanceolate to
oblong-ovate, those of the main stems 1.5-5 cm. long, those of the lateral branches very much
smaller; flowers in terminal, many-flowered, bracteate spikes, dimorphic as to the relative length
of the stamens and the style; hypanthium purplish, about 5 mm. long, the linear appendages about 1
mm. long; petals purple, about 5 mm. long.
   On swampy to fairly dry soil, from S.D. to Tex., e. to Mass. and Ga.; possibly not in our area
although known from B.C. June-Aug.

Lythrum hyssopifolia L. Sp. Pl. 447. 1753. ("Hab. in Germaniae, Helvetiae, Angliae, Galliae")
   Glabrous, pale glaucous-green annual; stems simple to branched, erect to semiprostrate, 1-4 dm.
long, lightly angled; leaves alternate or partially opposite, especially the near-basal ones, linear to
oblong, 1-2.5 cm. long, 1-8 mm. broad, entire; flowers solitary and sessile in the upper axils; hypanthium 4-5 mm. long, the lobes about 0.5 mm. long, the linear appendages twice as long; petals
white to rose, 2-3.5 mm. long; stamens and style included in the hypanthium; seeds ovoid, about 0.7
mm. long. N=10?
   In moist (usually marshy) places, but chiefly coastal in our area, from Wash. to Calif., e. U.S.;
Europe. May-Sept.

Lythrum salicaria L. Sp. Pl. 446. 1753. (Europe)
   Rhizomatous perennial 0.5-2 m. tall; stems simple to branched above, angled; leaves mostly opposite but some often in 3's or alternate, 3-10 cm. long, lanceolate, acute, more or less cordate at
base, scabrid-puberulent to soft-pubescent; flowers in crowded, terminal, interrupted, elongate spikes,

usually several per bract, subsessile; hypanthium purplish, about 5 mm. long, the linear appendages 1.5-2 mm. long, twice as long as the calyx lobes; petals reddish-purple, 7-10 mm. long; stamens 8-10, from very slightly to conspicuously exserted, of three distinct types as to relative length as compared with the style. N=15, 25, 30.

A European species introduced in marshes in the Puget Sound area, Wash., and from Mich. to the Atlantic coast. Aug.-Sept.

The species is colorful growing en masse, and is worth a place in the damper spots of a wild garden, although it is not to be rated as choice.

Rotala L.

Flowers inconspicuous, axillary, perfect, ours 4-merous; calyx lobes minute; petals white, very small; capsules 4-celled, many-seeded, septicidally dehiscent; placentation axile; glabrous annual or perennial herbs with opposite leaves.

A widely distributed genus of about 40 species chiefly of the Old World tropics. (Latin *rotula*, a little wheel, supposedly referring to the whorled leaves of the type species.)

Rotala ramosior (L.) Koehne in Mart. Fl. Bras. 132:194. 1875.
    Ammannia ramosior L. Sp. Pl. 120. 1753. *(Gronovius,* Virginia)
    Ammannia humilis Michx. Fl. Bor. Am. 1:99. 1803. *Boykinia humilis* Raf. Aut. Bot. 9. 1840.
    (North Carolina)
Glabrous, erect to procumbent, simple or branched annual 5-15 cm. tall; stems angled; leaves 1.5-3 cm. long, the blades lanceolate to oblanceolate, usually attenuate to short petioles; flowers solitary (2-3) in the axils; hypanthium campanulate and 1.5-2.5 mm. long at anthesis, but becoming accrescent, nearly globose, and as much as 3-4 mm. in fruit; petals white, about 1 mm. long; stamens and style included in the hypanthium; capsule globose, about 3 mm. long. Toothcup.

In wet, usually swampy places; occasional in Wash., and s. to Mex. and the W. Indies, e. to the Atlantic coast. June-Aug.

ONAGRACEAE Evening Primrose Family

Flowers perfect, regular to (occasionally) irregular, polypetalous, epigynous, usually in simple (compound) leafy or bracteate spikes or racemes; petals and stamens borne on the calyx which is adnate to the ovary and usually projecting above the ovary as a free hypanthium, in some cases the hypanthium composed in part of the sterile upper portion of the ovary itself; sepals and petals 4 (2 in *Circaea*, rarely 3 or in *Jussiaea* 5, very occasionally lacking), inserted at the top of the hypanthium, the sepals distinct or more or less connate and often turned to one side, valvate; petals white, yellow, pink, red, or lavender to purplish, often with a clawed base; stamens usually as many or twice as many as the petals, often in 2 series; ovary inferior, usually 4 (1, 2, or 5)-celled; style elongate, with a globose, discoid, or 4-lobed stigma; fruit a many-seeded capsule (sometimes indehiscent, 1- or 2-seeded, and nutlike); seeds (1 to) many, small, smooth to angled and papillate or hairy, exalbuminous, the embryo straight; annual or perennial, often rhizomatous herbs with opposite, alternate, or basal, simple to pinnatifid, exstipulate or more or less glandular-stipulate leaves.

About 20 genera and 600 species, widely distributed but common in desert regions especially in N. Am.

1 Flowers 2-merous; fruits with hooked hairs, 1- or 2-seeded, indehiscent   CIRCAEA
1 Flowers usually 4 (5)-merous; fruits not bristly, usually several-seeded
    2 Petals 5, well developed; sepals persistent on the slender, elongate, eventually reflexed capsule; stamens usually 10   JUSSIAEA
    2 Petals usually 4, but rarely lacking; sepals either deciduous or the capsule short and thickened; stamens usually 8 or fewer
        3 Fruits not twice so long as broad, 4-angled and flattened between the angles, many-seeded; sepals persistent, the hypanthium prolonged only a little if at all beyond the ovary   LUDWIGIA
        3 Fruits usually several times as long as broad, seldom 4-angled or 4-sided; hypanthium usually prolonged as a tube above the ovary, but deciduous as the fruits mature
            4 Fruit nutlike, hardened, indehiscent, 1- to 4-seeded, 4-angled and flat-sided   GAURA

4 Fruit capsular, many-seeded, dehiscent, usually more or less terete
  5 Seeds with a conspicuous tuft of long hairs at the tip; plants often growing along
     streams or on moist soil; leaves often opposite   EPILOBIUM
  5 Seeds glabrous or very finely strigose-puberulent; plants mainly from dry hills or
     drying mud flats; leaves nearly always alternate
    6 Ovary 2-celled, slender; plants usually very freely branched; flowers small, white
      to pinkish, the petals up to 5 mm. long   GAYOPHYTUM
    6 Ovary 4-celled; habit and flowers diverse, the petals often well over 5 mm. long
      7 Flowers axillary, sessile or subsessile; petals 1-6 mm. long; calyx lobes erect;
        anthers basifixed; plants often villous-hairy   BOISDUVALIA
      7 Flowers often pedicellate, if sessile either the petals over 6 mm. long, the calyx
        lobes reflexed, or the anthers versatile
        8 Petals yellow or white, although sometimes aging reddish or purplish; anthers
          often versatile   OENOTHERA
        8 Petals pink to purple; anthers erect, attached near the base   CLARKIA

Boisduvalia Spach

Flowers small, axillary, sessile or subsessile, diurnal, 4-merous except for the stamens; hypanthium short-funnelform; petals small, bilobed 1/3-1/2 of their length; stamens 8, the epipetalous ones the shorter; stigma subcapitate to broadly and shortly 4-lobed; capsules elongate, 4-ribbed, straight to curved, 4-celled; seeds numerous, glabrous, in 1 row per cell; annual, usually rather copiously soft-hairy, simple to branched, spreading to erect herbs with alternate, simple, usually sessile leaves.

About 10 species of w. N. and S. Am., generally in places that are soggy with standing or slow-moving water during the winter and spring, less commonly along intermittent to permanent streams. (Named for Jean Alphonse Boisduval, 1801-79, a French naturalist, author of a flora of France.)

Reference:

Munz, P. A. A revision of the genus Boisduvalia. Darwiniana 5:124-152. 1941.

1 Petals (2.5) 3-8 (12) mm. long; capsules slenderly fusiform, very short-beaked, straight,
  the septa completely free of the valves, adherent entirely to the placentae which persist as a central core until after the shed of the usually less than 6 seeds of each locule;
  floral leaves crowded, ovate to ovate-lanceolate; plant usually densely strigose to pilose
                   B. DENSIFLORA
1 Petals 1.5-4 mm. long; capsules sometimes conspicuously beaked or curved; septa adherent to the valves, the placentae usually fragmenting as the usually 6 or more seeds
  of each locule are shed; floral leaves often linear and not crowded; plant sometimes
  glabrate
  2 Plant villous-pilose, usually canescent; inflorescence not crowded; floral bracts lin-
    ·ear to linear-lanceolate, entire; capsules 6-11 mm. long, short-beaked, usually
    somewhat torulose, rather strongly curved; seeds 6-8 per locule   B. STRICTA
  2 Plant greenish, strigose to glabrate; inflorescence usually crowded, the floral bracts
    lanceolate to ovate-lanceolate, denticulate to dentate; capsules averaging about 7 mm.
    long, nearly straight, pointed but not beaked; seeds often more than 8 per locule
                   B. GLABELLA

Boisduvalia densiflora (Lindl.) Wats. Bot. Calif. 1:233. 1876.
  Oenothera densiflora Lindl. Bot. Reg. 19:pl. 1593. 1833. *B. douglasii* Spach, Nouv. Ann. Mus. Par.
  III, 4:400. 1835. (From seeds collected by Douglas in "Northern California")
  Oenothera salicina Nutt. ex T. & G. Fl. N. Am. 1:505. 1840, in synonymy only. *B. salicina* Rydb.
  Bull. Torrey Club 40:62. 1913. *B. densiflora* var. *salicina* Munz, Darwiniana 5:139. 1941. *(Nuttall,*
  "on the Wahlamet and Wallawallah")
  B. densiflora var. *imbricata* Greene, Fl. Fran. 225. 1891. *B. imbricata* Heller, Muhl. 1:42. 1904.
  Oenothera densiflora var. *imbricata* Levl. Bull. Acad. Int. Geogr. Bot. 18:302. 1908. *B. densi-*
  flora f. *imbricata* Munz, Darwiniana 5:136. 1941. *(Bolander 6403,* Mt. Diablo, Calif.) = var.
  densiflora.
  B. bipartita Greene, Erythea 3:119. 1895. *B. densiflora* var. *bipartita* Jeps. Fl. W. Middle Calif.
  2nd ed. 276. 1911. *(Greene,* Arroyo del Valle, Calif., June 14, 1895) = var. *densiflora.*

B. densiflora var. *pallescens* Suksd. Deuts. Bot. Monats. 18:88. 1900. *(Suksdorf 2254,* east of Bingen, Klickitat Co. , Wash.) = var. *densiflora.*

B. densiflora var. *montanus* Jeps. Fl. W. Middle Calif. 330. 1901. *(Jepson 14390,* Howell Mt. , Napa Co. , Calif. , lectotype by Jepson, no type originally cited) = var. *densiflora.*

B. sparsiflora Heller, Muhl. 1:42. 1904. *(Heller 7021,* Donner Lake, Calif. , July 25, 1903) = var. *salicina.*

B. sparsifolia Nels. & Kenn. Muhl. 3:139. 1908. *(Kennedy 644,* Maggie Creek, Elko Co. , Nev. , Aug. 13, 1902) = var. *densiflora.*

Strigose-canescent to softly hirsute-pilose, glandular to eglandular and greenish, simple to branched annual 1. 5-10 dm. tall; leaves many, usually crowded, the lower ones lanceolate, mostly 1. 5-5 cm. long, nearly or quite sessile, entire to remotely denticulate; floral leaves usually crowded, mostly ovate to ovate-lanceolate, often acuminate, entire or remotely denticulate, as much as 15 mm. long and 12 mm. broad; flowers sessile in crowded terminal and lateral spikes that lengthen as the fruits mature; hypanthium 1. 5-2. 5 mm. long; petals pale pink to rose or purplish, (2. 5) 3-8 (12) mm. long, lobate 1/3-1/2 their length; capsule slenderly fusiform but usually thickest below mid-length, 6-10 mm. long, very short-beaked, straight and usually appressed to the stem; valves free of the septa at the time of dehiscence, the septa almost entirely adherent to the placentae; seeds 3-6 per locule, flattened, 1. 5-2 mm. long, brownish. N=10.

Throughout our area, ranging from Vancouver I. , B.C. , southward, on both sides of the Cascades, to Baja Calif. , and e. to w. Mont. , Ida. , and Nev. July-Sept.

The species varies widely in the quantity and type of pubescence, in the color and size of the flowers, and in the general habit. Although it is not felt that they constitute particularly significant taxa, two variants are sufficiently clearly marked to be recognizable. Plants with soft, spreading hairs frequently are somewhat glandular and usually have darker petals mostly at least 4 mm. long; they are referable to the var. *densiflora.* Other plants, the var. *salicina* (Rydb.) Munz, are usually canescent with short, appressed hairs, and have paler pinkish petals 3-4 mm. long; in our area they occur with, but are less common than the var. *densiflora.*

Boisduvalia glabella (Nutt.) Walpers Rep. 2:89. 1843.

Oenothera glabella Nutt. in T. & G. Fl. N. Am. 1:505. 1840. *(Nuttall,* "plains of the Oregon east of Wallawallah")

Sparsely strigose to subglabrous, pale-greenish annual, simple or more commonly branched from the base and spreading, 1-3 dm. tall; leaves numerous and rather uniformly spaced, lanceolate to slightly ovatish, 5-18 mm. long, mostly 3-6 mm. broad, denticulate, sessile; flowers crowded, the floral bracts similar to the leaves; petals pale pinkish to reddish-purple, 2-4 mm. long, deeply bilobed; capsule usually slightly curved, pointed but not beaked, averaging about 7 mm. long, 1-1.5 mm. thick; septa mostly adherent to the valves, the placentae fragmenting as the seeds are shed; seeds 6-14 per locule, brownish.

British Columbia and e. Wash. southward to s. Calif. and Nev. , e. to Sask. , the Dakotas, and Utah; S. Am. June-July.

Boisduvalia stricta (Gray) Greene, Fl. Fran. 225. 1891.

Gayophytum strictum Gray, Proc. Am. Acad. 7:340. 1867. *(Bolander 6535,* Cloverdale, California)

Oenothera densiflora var. *tenella* Gray, Proc. Am. Acad. 8:384. 1872. *(E. Hall 189,* Oregon)

Oenothera torreyi Wats. Proc. Am. Acad. 8:600. 1873. *B. torreyi* Wats. Bot. Calif. 1:233. 1876. *(Torrey 190,* New Almaden, Calif.)

B. diffusa Greene, Proc. Acad. Phila. 1895:547. 1896. (Upper Humboldt River near Deeth, Nevada, no collector mentioned, but probably *Greene,* Aug. 5, 1895)

B. parviflora Heller, Bull. Torrey Club 25:199. 1898. *(Heller 3411,* Lake Waha, Ida. , July 10, 1896)

Annual, 1-5 dm. tall, simple or more commonly branched from the base, the branches strict, spreading only slightly, grayish with dense, soft, straight, eglandular hairs; leaves numerous, uniformly spaced, mostly linear or very narrowly lanceolate, entire or remotely denticulate, 1-4 cm. long, 1-2. 5 (4) mm. broad; inflorescence loose to somewhat congested, the floral bracts similar to the leaves but more commonly entire and often smaller; petals pinkish to purple, mostly about 2 mm. long; capsule 6-11 mm. long, curved (especially near the short-beaked tip), usually torulose, the septa adherent to the valves, the placentae disintegrating with the shed of the (6) 7 (8) seeds of each locule. N=9.

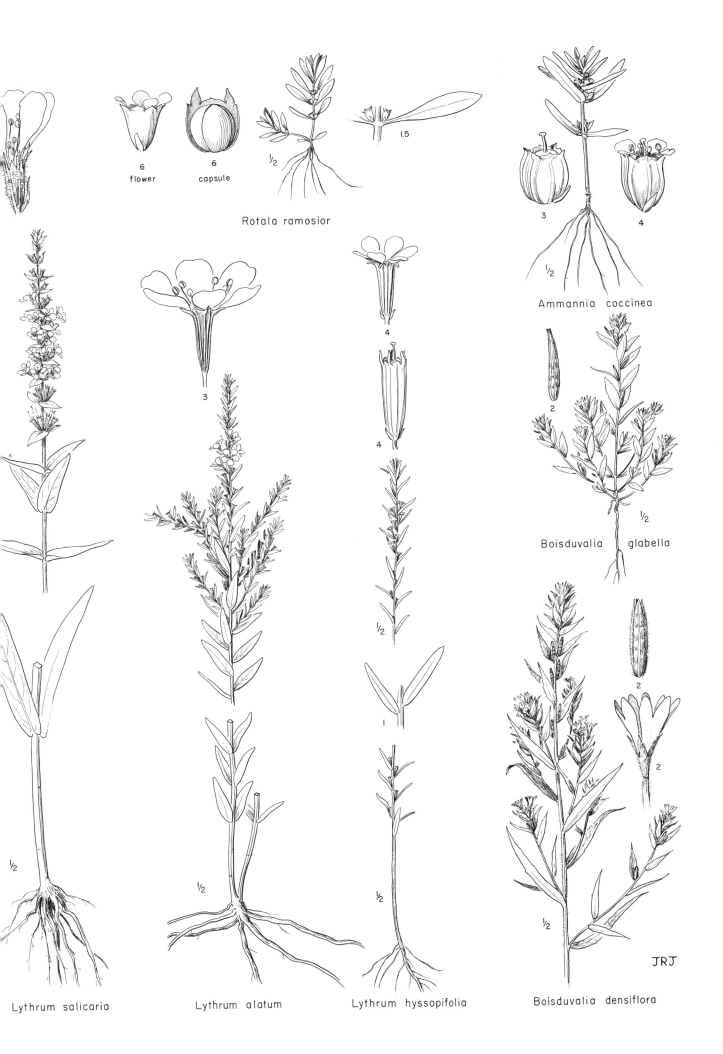

6
flower

6
capsule

1/2

1.5

Rotala ramosior

3

4

1/2

Ammannia coccinea

3

4

1/2

Boisduvalia glabella

2

2

Lythrum salicaria

3

1/2

1/2

Lythrum alatum

4

4

1/2

I

1/2

Lythrum hyssopifolia

1/2

2

2

1/2

JRJ

Boisduvalia densiflora

Eastern Wash., from Yakima Co. southward, on both sides of the Cascades in Oreg., to c. Calif., e. to Ida. and Nev. June-July.

Circaea L. Enchanter's Nightshade

Flowers small, white, 2-merous, borne in simple or (more commonly) in compound racemes; sepals reflexed; hypanthium very short, deciduous after flowering; petals notched; fruit 1 (2)-seeded, indehiscent, pear-shaped, covered with short, hooked hairs; seeds glabrous; small, opposite leaved, juicy, perennial herbs.

Perhaps seven species of N. Am. and Eurasia. (Named after the Greek goddess, Circe.)

Circaea alpina L. Sp. Pl. 9. 1753. (Europe)
C. *pacifica* Asch. & Magnus, Bot. Zeit. 29:392. 1871. *C. alpina* var. *pacifica* M. E. Jones, Bull. U. Mont. Biol. ser. 15, no. 61:39. 1910. *C. alpina* f. *pacifica* G. N. Jones, U. Wash. Pub. Biol. 5:195. 1936. *(Bolander,* near San Francisco [at Papermill Creek, Marin Co., acc. Jeps.])

Perennial from slender rootstocks, 1-5 dm. tall, simple to freely branched, clear green; stems usually glabrous on the lower half, often sparsely strigose to short-pilose above and in the inflorescence; leaf blades cordate-ovate to ovate, usually acuminate, subentire to saliently dentate, 2-6 cm. long, sparsely to rather thickly strigillose (especially on the lower surface), considerably exceeding the petioles; racemes often with 1 or 2 linear bracts at the base, the individual flowers with a minute, sometimes glandlike bractlet; pedicels spreading to erect in flower, becoming reflexed and equaling or slightly exceeding the fruit; petals and sepals 1-1.5 mm. long; fruits turbinate, about 2 mm. long. N=11.

Usually in cool, damp woods; Eurasia and N. Am., from Alas. to Newf., southward (mostly in the mountains) to s. Calif., Colo., and Ga. May-July.

This is a highly variable taxon often treated as two species, but not believed to be separable even into clearly marked geographic races, although much of our material (from B.C. to Ida. and Calif.) appears to be characterized by subcordate and subentire to denticulate leaves. This phase is somewhat unduly significantly distinguished as var. *pacifica* (Asch. & Magnus) M. E. Jones, in contrast to the more widespread var. *alpina* which has more evidently cordate and more strongly dentate leaves.

Lest the unwary be enchanted by the name or appearance of this plant, it should be pointed out that it easily rates as one of the worst weeds among our native plants.

Clarkia Pursh

Flowers long-pedicellate to subsessile, 4-merous except for the stamens, borne in simple to branched, loose to compact, leafy-bracted, drooping to erect spikes or racemes, open continuously or only during the day; buds nodding to erect; hypanthium well developed, often with a narrow ring of hairs internally somewhat above the base, abscissent at the top of the ovary as the fruit matures; calyx lobes distinct and sharply reflexed to spreading, or united in pairs, or all united and turned to one side, their tips often more or less free in the bud; corolla regular to slightly irregular, the petals pinkish to purple, often distinctly spotted, cuneate to obovate, from rounded to more or less lobate and from nearly uniformly tapered to the base to distinctly and conspicuously unguiculate; stamens 4 or (ours) 8, but the epipetalous the shorter and sometimes considerably reduced and nonfunctional, the anthers linear, basifixed; stigma with 4 ovate to linear, white to purple lobes; style from shorter to longer than the stamens; ovary elongate, slender or somewhat fusiform or clavate, terete or somewhat quadrangular, usually 4- or 8-sulcate; capsule slender, terete to quadrangular, often beaked, 4-valvate; seeds grayish to brown, more or less angular, not at all comose, but with a distinct crestlike border of elongated cells.

A genus of about 30 species largely of n.w. N. Am. (centered in Calif.); one species in S. Am. (Named for Captain William Clark of the Lewis and Clark Expedition.)

Plants usually growing in the open or in wooded, dry and usually grassy areas, blossoming as the grass dries and when most other flowers are past, hence the common name "Farewell-to-spring" for many of them. Several of the species are excellent, selfperpetuating ornamentals for the drier areas of the wild garden. These include, in order of excellence, *C. pulchella, C. amoena, C. viminea,* and *C. rhomboidea.* In our area it is probable that they behave as winter annuals, so it is suggested that seed be sown in the fall to establish the plants successfully.

The species included in *Clarkia* have been accorded varied generic status by different workers, but it has been perhaps conventional to assign the several taxa with clawed petals and more or less irregular flowers that do not close at night to *Clarkia,* and to refer to *Godetia* the bulk of the species which are hemeranthous and regular-flowered and which have unclawed petals. In our area such disposition of the species would be satisfactory on morphological grounds, but the experimental and cytological work done in recent years practically dictates the necessity of including the various species in one generic complex.

Reference:
Lewis, Harlan, and M. E. Lewis. The genus Clarkia. U. Calif. Pub. Bot. 20:241-392. 1955.

1 Petals distinctly clawed, usually more or less 3-lobed; flowers slightly irregular, not
    closing at night *(Clarkia)*
    2 Fertile stamens 4; petals very obviously 3-lobed   C. PULCHELLA
    2 Fertile stamens 8; petals not obviously 3-lobed, the lateral lobes, if any, very small
        and usually at the base of the blade   C. RHOMBOIDEA
1 Petals neither clawed nor lobed; flowers regular, closing at night *(Godetia)*
    3 Calyx lobes usually connate and turned to one side under the flower; ovaries 4-sulcate
        at anthesis; capsules terete and 8-nerved, elongate, slender, often beaked, usually
        with pedicels 2-10 mm. long; stigmas 1-7 mm. long
        4 Buds and the rachis of the inflorescence recurved, the latter straightening and be-
            coming erect as flowering proceeds; petals unspotted; plants erect   C. GRACILIS
        4 Buds and the rachis of the inflorescence erect; petals often spotted; plants spread-
            ing to erect   C. AMOENA
    3 Calyx lobes mostly distinct and closely reflexed; ovaries usually terete and not 4-sul-
        cate at anthesis; capsules often somewhat 4-angled at maturity, usually sessile, nev-
        er long-beaked; stigmas less than 2 mm. long
        5 Flowers usually closely crowded; capsules 3-5 mm. thick, generally noticeably en-
            larged near the center, largely concealed by the floral bracts; stigmas scarcely 1
            mm. long   C. PURPUREA
        5 Flowers not crowded; capsules 1.5-3 mm. thick and not conspicuously thicker near
            the middle, not concealed by the bracts; stigmas often over 1 mm. long
            6 Hypanthium 6-12 mm. long, slender at the base but flared at the top, enlarged con-
                siderably where joined to the ovary; petals mostly at least 15 mm. long; stigmas
                about 1.5 mm. long   C. VIMINEA
            6 Hypanthium 2-7 mm. long, more or less uniform in taper, not enlarged at the top
                of the ovary; petals 5-15 mm. long; stigmas about 1 mm. long   C. QUADRIVULNERA

Clarkia amoena (Lehm.) Nels. & Macbr. Bot. Gaz. 65:62. 1918.
  Oenothera amoena Lehm. Ind. Sem. Hort. Hamb. 8. 1821. *Godetia amoena* G. Don in Sweet, Hort.
    Brit. 3rd ed. 237. 1839. ("America Septentrionalis")
  CLARKIA AMOENA var. LINDLEYI (Dougl.)C. L. Hitchc. hoc loc. *Oenothera lindleyi* Dougl. in
    Curtis' Bot. Mag. 55:pl. 2832. 1828. *Godetia amoena* var. *lindleyi* Jeps. U. Calif. Pub. Bot. 2:
    329. 1907. *C. amoena* f. *lindleyi* Nels. & Macbr. Bot. Gaz. 65:62. 1918. *C. amoena* ssp. *lindleyi*
    Lewis & Lewis, U. Calif. Pub. Bot. 20:267. 1955. (Described from a cultivated plant from seed
    collected by Douglas, on "Northwest Coast of America")
  Godetia grandiflora Lindl. Bot. Reg. 27:misc. 61. 1841. *Oenothera grandiflora* Wats. Proc. Am.
    Acad. 8:596. 1873. *C. superba* Nels. & Macbr. Bot. Gaz. 65:60. 1918. (Garden specimens from
    seeds collected by Dyer on the Columbia River) = var. *lindleyi.*
  CLARKIA AMOENA var. CAURINA (Abrams) C. L. Hitchc. hoc loc. *Godetia caurina* Abrams ex
    Piper, Contr. U.S. Nat. Herb. 11:410. 1906. *C. caurina* Nels. & Macbr. Bot. Gaz. 65:62. 1918.
    C. amoena ssp. *caurina* Lewis & Lewis, U. Calif. Pub. Bot. 20:268. 1955. *(Elmer 2565,* Olympic
    Mts., Wash., June, 1900)
  Godetia romanzovii sensu Piper in Piper & Beattie, Fl. N.W. Coast 251. 1915. = var. *pacifica.*
  CLARKIA AMOENA var. PACIFICA (Peck) C. L. Hitchc. hoc loc. *Godetia pacifica* Peck, Proc.
    Biol. Soc. Wash. 47:187. 1934. *(Peck 16332,* open bluff above the sea, Otter Crest, Lincoln Co.,
    Oreg., May 23, 1931)
Simple to freely branched annual 1-10 dm. tall; leaves 2-7 cm. long, mostly 2-6 mm. broad; buds usually erect; hypanthium 2-10 mm. long; calyx lobes generally connate and turned to one side under the flower, occasionally partially free; flowers tending to close at night, the petals pale pinkish to rose-purple, usually carmine-maculate in the center, more or less obovate-cuneate, truncate or

slightly retuse to rounded or obtuse, not clawed, 1-4 cm. long; anthers glabrous or slightly hairy, straight or slightly curved after dehiscence; stigmas (1) 1.5-5 (7) mm. long, linear to oval, yellow; style usually exceeding the stamens; capsule (1) 1.5-4.5 cm. long, linear to somewhat clavate, straight to curved, beakless or with a beak several mm. long, 4-sulcate when young, later terete and 8-ribbed. N=7.

Occasional from Vancouver I. southward (and more common) to c. Calif.; w. of the Cascades in our area. June-Aug.

This is a variable species, especially as regards the size and coloration of the flowers and the length of the capsules and stigma lobes, but it is divisible into several poorly differentiated varieties largely separable as follows:

1 Stigmas linear, usually well over 2 mm. long; petals mostly over 2 cm. long; capsules
   not strongly curved; plants usually erect; occasional from Vancouver I. southward
                                              var. lindleyi (Dougl.) C. L. Hitchc.
1 Stigmas oval, usually not over 2 mm. long; petals mostly less than 2 cm. long; capsules often curved; plants often branched from the base and more or less decumbent
    2 Capsule curved; petals rhombic-ovate, striped with 2 lunate median bands; plant usu-
      ally basally branched and more or less decumbent; believed to be restricted to
      coastal bluffs of Lincoln Co., Oreg., but plants from that region are very closely
      matched by a collection *(Webster 1225)* from Port Angeles, Wash., which is un-
      doubtedly the plant referred to *G. romanzovii* by Piper   var. pacifica (Peck) C. L. Hitchc.
    2 Capsule straight; petals rounded to truncate, spotted to immaculate; plant erect;
      occasional from Vancouver I. to s. Oreg., w. of the Cascades *(G. epilobioides*
      sensu Howell, Fl. N.W. Am. 235. 1898)   var. caurina (Abrams) C. L. Hitchc.

Clarkia gracilis (Piper) Nels. & Macbr. Bot. Gaz. 65:63. 1918.
   Godetia gracilis Piper in Piper & Beattie, Fl. N.W. Coast 251. 1915. *Godetia amoena* var. *gracilis*
   C. L. Hitchc. Bot. Gaz. 89:342. 1930. *(Elihu Hall 192,* Silverton, Oreg.)

Very similar to *C. amoena,* the plants erect, 1.5-6 dm. tall; tip of the inflorescence and the buds reflexed, becoming erect only as flowering proceeds; hypanthium 1.5-3 mm. long; petals mostly 8-20 mm. long, pink to lavender, usually not spotted; stigma lobes 1-1.5 mm. long, cream; style shorter than the stamens; capsules straight, usually slightly enlarged above the middle, 3-5 cm. long, attenuate to a distinct slender beak. N=14.

Known from Klickitat and Walla Walla cos., Wash., southward in Oreg., on the w. side of the Cascades, to c. Calif. June-July.

In s.e. Oreg. and n.e. Calif. there grows a somewhat similar plant, *C. lassenensis* (Eastw.) Lewis & Lewis, which has been reported as far n. as Deschutes Co., Oreg. It differs from *C. gracilis* in having the ovary 8-sulcate and frequently bulged slightly at the middle. It has a chromosome complement of N=7.

Clarkia pulchella Pursh, Fl. Am. Sept. 260, pl. 11. 1814.
   Oenothera pulchella Levl. Monog. Onoth. 288. 1908. *(Lewis,* "On the Kooskoosky and Clarck's Rivers")

A simple to freely branched, finely strigillose annual 1-5 dm. tall; leaves linear-lanceolate to spatulate, entire to denticulate, 2-7 cm. long, 2-10 mm. broad; flowers few in short racemes, slightly irregular; buds nodding; hypanthium 1-3 mm. long, glabrous within, the calyx lobes usually connate and turned to one side; petals lavender to rose-purple, 3-lobed, the middle lobe the widest, the base narrowed to a fairly slender claw with a pair of opposite, short, divergent, blunt teeth near the base; fertile stamens 4, the anthers coiling after pollen-shed, the epipetalous ones mere vestiges without functional pollen; stigmas oval-oblong, 1-3 mm. long, white; capsule about 2 cm. long, straight or curved. N=12.

Along the Columbia R. in s. B.C., southward, on the e. side of the Cascades, to s.e. Oreg., e. to Ida. and w. Mont. May-June.

Clarkia purpurea (Curtis) Nels. & Macbr. Bot. Gaz. 65:64. 1918.
   Oenothera purpurea Curtis in Curtis' Bot. Mag. 10:pl. 352. 1796. *Godetia purpurea* G. Don in Sweet,
   Hort. Brit. 3rd ed. 237. 1839. (Described from plants grown from seed collected by Douglas on
   the west coast of North America)
   Oenothera decumbens Dougl. in Curtis' Bot. Mag. 56:pl. 2889. 1829. *Godetia decumbens* Spach, Hist.

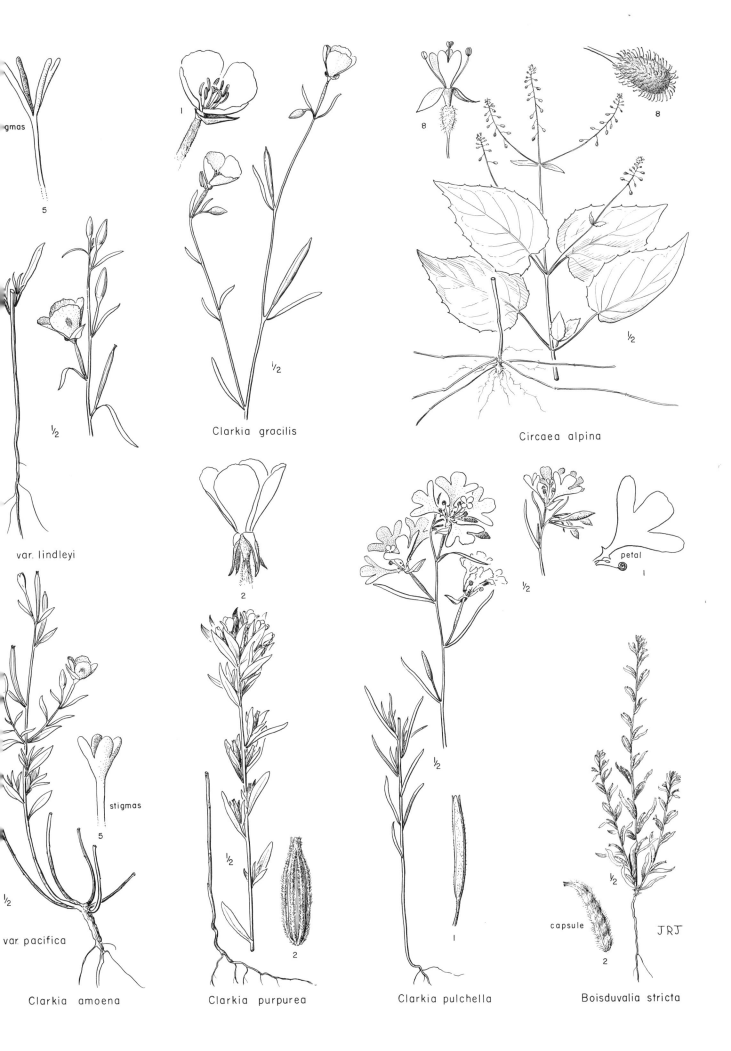

gmas

5

var. lindleyi

½

Clarkia gracilis

½

Circaea alpina

½

2

petal

1

stigmas

5

½

var. pacifica

Clarkia amoena

½

2

capsule

2

½

JRJ

1

Clarkia purpurea

Clarkia pulchella

Boisduvalia stricta

8

8

Veg. Phan. 4:288. 1835. *C. decumbens* Nels. & Macbr. Bot. Gaz. 65:64. 1918. (Described from
plants grown from seeds collected by Douglas in "northern California" in 1827)
Godetia lepida Lindl. Bot. Reg. 22:pl. 1849. 1836. *Oenothera lepida* H. & A. Bot. Beechey Voy.
342. 1838. (Garden plants, from seeds collected by Douglas in Calif.)
Oenothera arnottii T. & G. Fl. N. Am. 1:503. 1840. *Godetia arnottii* Walpers Rep. 2:88. 1843. *C.
arnottii* Nels. & Macbr. Bot. Gaz. 65:64. 1918. *(Douglas,* California)
Godetia albescens Lindl. Bot. Reg. 27:misc. 61. 1841. *Oenothera albescens* Wats. Proc. Am. Acad.
8:597. 1873. (Garden plants, from seeds collected by Dyer along the Columbia River)
Oenothera lepida var. *parviflora* Wats. Proc. Am. Acad. 8:597. 1873. *Godetia purpurea* var. *parviflora* C. L. Hitchc. Bot. Gaz. 89:335. 1930. (Based partly on *Oenothera decumbens* Dougl. and
perhaps to be considered strictly synonymous with it)
Annual, usually simple below but often with closely crowded branched above, less commonly
branched from the base and somewhat decumbent, 1-6 dm. tall, finely pubescent to subglabrous;
leaves 2-4 cm. long, 3-15 mm. broad; spikes closely flowered, the buds and immature fruits largely
concealed by the often expanded floral bracts; hypanthium 3-7 mm. long; calyx lobes usually distinct
and sharply reflexed but occasionally slightly connate; petals 5-20 mm. long, cuneate to obovate,
rounded, crimson to purple, spotted or immaculate; stigmas about 1 mm. long, oval, usually purplish; capsule mostly 15-25 mm. long, 3-5 mm. thick, generally obviously bulged near the center,
nearly or quite sessile, pointed but not beaked, 8-ribbed, terete but often drying to quadrilateral,
from glabrous to densely strigose or almost lanate. N=26.
Dry ground, usually in the open at the edge of vernal pools but rarely collected in recent years;
from Salem, Oreg. , to s. Calif. June-July.

Clarkia quadrivulnera (Dougl.) Nels. & Macbr. Bot. Gaz. 65:63. 1918.
Oenothera quadrivulnera Dougl. ex Lindl. Bot. Reg. 13:pl. 1119. 1828. *Godetia quadrivulnera* Spach,
Hist. Veg. Phan. 4:389. 1835. *Oenothera prismatica* var. *quadrivulnera* Levl. Monog. Onoth. 267.
1908. *C. purpurea* ssp. *quadrivulnera* Lewis & Lewis, U. Calif. Pub. Bot. 20:305. 1955. (Garden
plants grown from seeds collected by Douglas in "the northwest of North America")
Godetia tenella sensu Wats. Bot. Calif. 1:230. 1880, not of Cav. *Godetia bingensis* Suksd. Deuts.
Bot. Monats. 18:88. 1900. *(Suksdorf 86,* Bingen, Wash.)
Godetia brevistyla Piper, Proc. Biol. Soc. Wash. 37:92. 1924. *(Elmer 2567,* Olympic Mts. , Wash. ,
June, 1900)
Annual, usually erect, simple or branched from the base, 1-7 dm. tall; leaves 1-5 cm. long, 2-7
mm. broad; spikes loosely few-flowered, the floral bracts mostly narrower than the cauline leaves,
not concealing the flowers or capsules; hypanthium 2-7 mm. long; calyx lobes usually distinct and
sharply reflexed, but sometimes partially connate; petals 5-15 mm. long, cuneate, rounded to nearly
truncate, from pale pinkish-lavender and with or without a carmine or purplish central spot to deep
rose-purple; stigmas oval, about 1 mm. long, purplish; ovaries nearly or quite sessile, generally
conspicuously hairy; capsule 10-20 (25) mm. long, rather uniformly 2-3 mm. thick, usually rather
short-beaked, terete or quadrilateral. N=26.
Occasional, but rather rare, on "prairies" from the Olympic Peninsula to Tacoma, and up the Columbia R. Gorge to Klickitat Co. , Wash. , and in Hood River Co. and the Willamette Valley, Oreg. ,
s. (and more abundant) in the valleys and foothills to Baja Calif. May-July.

Clarkia rhomboidea Dougl. ex Hook. Fl. Bor. Am. 1:214. 1833.
Oenothera rhomboidea Levl. Monog. Onoth. 287. 1908. *Phaeostoma rhomboidea* A. Nels. Bot. Gaz.
52:267. 1911. *(Douglas,* "From the great falls of the Columbia to the Rocky Mountains")
C. gauroides Dougl. ex Sweet, Brit. Fl. Gard. II, 4:379. 1837. (Described from garden plants grown
from seeds collected by Douglas in California)
A usually simple, finely strigillose annual 1. 5-10 dm. tall; leaves few, mostly subopposite, the
petioles slender, 1-3 cm. long, the blades lanceolate to elliptic, 2-7 cm. long, 5-20 mm. broad, entire to denticulate; inflorescence a few-flowered, loose raceme; buds nodding; hypanthium 1-3 mm.
long, scaly and white-hairy at the base of the stamens; calyx lobes usually distinct; corolla slightly
irregular, rose-purple and often purple-dotted, the petals 5-10 mm. long, 3-6 mm. broad, with a
narrowly rhomboidal blade 2-4 times the length of the broad, often basally toothed claw; fertile stamens 8; stigma lobes oval, about 0. 5 mm. long, white to purple; capsule 1. 5-3. 3 cm. long, short-
beaked. N=12.

Southern B. C. southward to s. Calif., e. (spottily) to Ida., Utah, and Ariz., entirely e. of the Cascade Mts. in Wash., but extending to the west side in Oreg. June-July.

Clarkia viminea (Dougl.) Nels. & Macbr. Bot. Gaz. 65:64. 1918.

Oenothera viminea Dougl. ex Hook. in Curtis' Bot. Mag. 55:pl. 2873. 1828. *Godetia viminea* Spach, Hist. Veg. Phan. 4:388. 1835. *C. purpurea* ssp. *viminea* Lewis & Lewis, U. Calif. Pub. Bot. 20:503. 1955. (Described from plants grown from seeds collected by Douglas in "Northern California")

Annual, simple or more commonly branched from the middle or even from the base, 1.5-8 dm. tall; spikes usually much elongate and loosely flowered; leaves 2-5 cm. long, 2-5 mm. broad; buds distinctly pointed; hypanthium 6-12 mm. long, flared above, enlarged considerably where joined to the ovary, with an internal ring of hairs slightly below midlength; sepals usually sharply reflexed, free or partially connate (mostly) in pairs; petals 13-25 mm. long, cuneate, rounded, usually fairly deep rose-lavender to purplish, often with a large central oval or an apical triangular, wine-colored spot, sometimes yellowish toward the base; stigmas elliptic-oblong, about 1.5 mm. long; ovary at anthesis usually subequal to the hypanthium; immature fruits terete and 8-ribbed, showing a distinct annular thickening at the juncture with the floral tube, mature capsules mostly 15-25 mm. long, 2-3 mm. thick, terete or quadrilateral, not enlarged at the middle, short-beaked. N=26.

Very occasional from Multnomah to Wasco cos., Oreg., southward on the w. side of the Cascades to Calif., where much more common, from the coast to the Sierras. May-June.

Epilobium L. Willow herb

Flowers perfect, usually in terminal, simple or compound racemes, occasionally axillary to scarcely reduced leaves; hypanthium short or lacking; sepals 4; petals 4, white or yellowish to deep rose-purple, obcordate to shallowly notched; stamens 8; stigma 4-lobed or more or less oblong-clavate and not lobed; capsules usually pedicellate, linear to subclavate, or more nearly fusiform, 4-celled, loculicidal; seeds smooth to papillate, conspicuously comose at the tip; mostly annual or perennial herbs, but a few more or less suffruticose, the perennials mostly spreading by stolons or rhizomes and sometimes producing new rosettes of leaves at their tips, or forming bulblike offsets (turions); leaves usually opposite, sessile to petiolate, entire to dentate, often rather "willowlike."

Perhaps 150 species; cosmopolitan. (Greek *epi*, upon, and *lobos*, pod, referring to the inferior ovary.)

Reference:

Trelease, William. The species of Epilobium occurring north of Mexico. Rep. Mo. Bot. Gard. 2: 69-117. 1891.

The wise gardener will not be misled by the apparent desirability of most of these species as ornamentals. They spread all too rapidly, both vegetatively and by seed, perhaps the worst being *E. angustifolium*. On the other hand, *E. obcordatum* and *E. latifolium* have great potential, at least for rock gardens e. of the Cascades, but neither seems to have been grown with real success in the Puget Sound area.

1 Stigma 4-cleft (shallowly so in *E. obcordatum)*; petals either yellow or at least 1 cm. long (sometimes both yellow and over 1 cm. long)

  2 Petals yellow

    3 Plant woody based; stems 1-3 dm. tall; leaves entire, 1-3 cm. long; capsules fusiform, less than 3 cm. long; flowers slightly irregular; petals mostly 5-9 mm. long   E. SUFFRUTICOSUM

    3 Plant herbaceous throughout; stems erect, 2-7 dm. tall; leaves serrate-dentate, 3-8 cm. long; capsules linear, 4-8 cm. long; flowers regular; petals 10-18 mm. long   E. LUTEUM

  2 Petals pink to purple (white)

    4 Free hypanthium 1-3 mm. long; plants more or less prostrate, scarcely 1 dm. tall; ovaries with short, glandular puberulence; petals obcordate   E. OBCORDATUM

    4 Free hypanthium lacking, the calyx cleft to the top of the ovary; plants erect; ovary not at all glandular; petals usually not obcordate

      5 Leaves (in part at least) usually over 8 cm. long; plants mostly over 4 dm. tall, strongly "rhizomatous"; styles exceeding the stamens, hairy near the base;

floral bracts much reduced, linear; racemes elongate, usually at least 15-flowered
E. ANGUSTIFOLIUM

    5 Leaves less than 8 cm. long; plants usually less than 4 dm. tall, not rhizomatous;
styles shorter than the stamens, glabrous; floral bracts similar to the leaves although somewhat reduced; racemes usually not more than 15-flowered   E. LATIFOLIUM

1 Stigma usually entire or, if with short lobes, the petals never yellow (sometimes
creamy-white) and seldom as much as 1 (rarely to 1.3) cm. long

    6 Plant an annual, taprooted, usually growing on well-drained soil; epidermis on the
lower portion of the stem usually exfoliating

        7 Leaves mostly opposite, at least below; seeds 0.5-0.9 mm. long, very incon-
spicuously alveolate; stems soft-pubescent; plants mostly less than 4 dm. tall;
petals seldom over 4 mm. long   E. MINUTUM

        7 Leaves, except the lowermost, usually alternate; seeds at least 1 mm. long, very
minutely spinose-papillate; plants usually over 4 dm. tall; petals usually over 4
mm. long   E. PANICULATUM

    6 Plant a perennial, usually rhizomatous, generally growing in moist places; lower
epidermis mostly not exfoliating

        8 Plants usually grayish-strigillose, at least above, producing turions; leaves linear
to narrowly lanceolate, (1) 2-6 cm. long, 1-4 (8) mm. broad; petals white to
pink   E. PALUSTRE

        8 Plants not grayish-strigillose, or if so not producing turions or the leaves not lin-
ear or the petals not white

            9 Turions (see generic description) usually present; seeds often papillate
E. GLANDULOSUM

            9 Turions absent; seeds various

                10 Plants glabrous throughout or with minute pubescence in the inflorescence
or on the ovary, usually distinctly glaucous, often matted at the base; leaves
lanceolate to ovate, entire to denticulate   E. GLABERRIMUM

                10 Plants usually pubescent, not glaucous

                    11 Stems usually well over 3 dm. tall and freely branched above the mid-
dle; rhizomes short or lacking; leaves often well over 4 cm. long, usually denticulate to serrate; seeds conspicuously crested-papillate in numerous parallel longitudinal lines; coma white   E. WATSONII

                    11 Stems seldom over 3 dm. tall, simple or with a few basal branches;
rhizomes usually well developed and the plants matted; leaves seldom
over 4 cm. long, often entire; seeds smooth or the papillae very small
and not in distinct rows; coma usually dingy   E. ALPINUM

Epilobium alpinum L. Sp. Pl. 348. 1753. ("Alpibus Helveticis, Lapponicis")
   E. anagallidifolium Lam. Encyc. Meth. 2:376. 1788. *(Lamarck,* Mont-d'Or, France) = var. *alpinum.*
   E. nutans Hornem. Fl. Dan. pl. 1387. 1810, but not of F. W. Schmidt about 1795. *E. horneman-*
   nii Reichb. Icon. Bot. Crit. 2:73. 1824. *E. alpinum nutans* Hornem. Nom. Fl. Dan. 66. 1827. *E.*
   alpinum var. *nutans* Hook. Fl. Bor. Am. 1:205. 1832. (Norway)
   EPILOBIUM ALPINUM var. LACTIFLORUM (Hausskn.) C. L. Hitchc. hoc loc. *E. lactiflorum*
   Hausskn. Oest. Bot. Zeits. 29:89. 1879. *E. alpinum* f. *lactiflorum* Moore, Rhodora 11:147. 1909.
   (Indicated to be circumboreal; no specimen cited)
   E. oregonense Hausskn. Monog. Epilob. 276. 1884. *(Hall 179,* Oregon, in 1871) = var. *gracillimum.*
   EPILOBIUM ALPINUM var. GRACILLIMUM (Trel.) C. L. Hitchc. hoc loc. *E. oregonense* var.
   gracillimum Trel. Rep. Mo. Bot. Gard. 2:109. 1891. (Of the two collections cited, the second,
   Suksdorf 860, from Wash., more nearly fits the description)
   EPILOBIUM ALPINUM var. CLAVATUM (Trel.) C. L. Hitchc. hoc loc. *E. clavatum* Trel. Rep.
   Mo. Bot. Gard. 2:111, pl. 48. 1891. *(Macoun,* Kicking Horse River, British America, in 1890,
   is the first collection of many cited)
   E. treleasianum Levl. Rep. Nov. Sp. 5:8. 1908. *(Farr,* Rogers' Pass, Selkirk Mts., B.C., Dec.
   8, 1903) = var. *nutans.*
   E. alpinum sensu Rydb. Fl. Rocky Mts. 588. 1917, and many other Am. authors = var. *lactiflorum.*
   E. pulchrum Suksd. Werdenda 1:26. 1927. *(Suksdorf 5741,* Wodan Valley, Mt. Adams, Wash., July
   26, 1906) = var. *albiflorum.*

EPILOBIUM ALPINUM var. ALBIFLORUM (Suksd.) C. L. Hitchc. hoc loc. *E. pulchrum* var. *albiflorum* Suksd. Werdenda 1:27. 1927. (Garden specimen)

E. glareosum G. N. Jones, U. Wash. Pub. Biol. 7:175. 1938. *(Allen 252,* Goat Mts., Pierce Co., Wash.) = var. *albiflorum.*

Low, usually matted perennial, spreading by rhizomes and stolons but not producing turions; stems decumbent based to erect, usually simple but occasionally with a few basal branches, 0.5-3 (4) dm. tall, mostly glabrous below and crisp-puberulent in lines above, or occasionally glabrous up to (but usually glandular-puberulent in) the inflorescence; leaves generally opposite, occasionally alternate above, usually about equally spaced, sessile to short-petiolate, linear to ovate, (0.5) 1-5 cm. long, entire to serrulate; flowers few, nodding to erect; pedicels 0.5-5 cm. long; hypanthium 1-2 mm. long; sepals 1.5-6 mm. long; petals white to deep pink or lilac-rose, notched, 3-13 mm. long; stigma entire; capsule 2-7 cm. long, linear to subclavate; seeds about 1 mm. long, smooth to inconspicuously papillate, the coma dingy. N=18.

Moist banks and rocks, talus slopes, and mountain meadows, often above timber line; throughout the mountains of w. N.Am., southward as far as s. Calif. and Colo., e. to the n. Atlantic coast; Eurasia. June-Sept.

As here treated, the species includes at least four other taxa that are nearly always accorded specific rank, namely *E. clavatum, E. oregonense, E. lactiflorum,* and *E. hornemannii.* These names appear to distinguish little more than conspicuous phases of an almost continuous variation in which neither genetic nor geographic barriers have developed to isolate distinctive taxa. For this reason, keys and descriptions cannot be drawn that will serve to delimit the different variants except in a mechanical way such as the following:

1 Stems erect, glabrous or sparsely glandular above, not puberulent in lines; leaves sessile, linear to narrowly oblong-lanceolate, usually nearly erect, borne mostly on the lower portion of the stem, the upper ones spaced much more distantly than the lower; seeds smooth; B.C. to s. Calif., e. to Ida.   var. gracillimum (Trel.) C. L. Hitchc.
1 Stems usually decumbent based, generally crisp-puberulent in decurrent lines; leaves mostly short-petiolate, lanceolate to ovate or ovate-oblong, spreading, generally rather uniformly spaced on the stem; seeds smooth to papillate
  2 Petals white or cream-colored to pale pinkish, 2-5 mm. long; seeds smooth; plants 1-3 dm. tall, usually not matted; leaves 2-5 cm. long; Alas. to the Atlantic coast, s. in the mts. to Calif. and Colo.   var. lactiflorum (Hausskn.) C. L. Hitchc.
  2 Petals usually either pink or else more than 5 mm. long; seeds often papillate; plants usually matted and often not over 1.5 dm. tall; leaves often less than 2 cm. long
    3 Plant seldom as much as 1.5 dm. tall; petals 3-6 mm. long; leaves usually ovate, 1-2 cm. long; seeds smooth or capsules more or less clavate
      4 Capsules subclavate, 1.5-2 mm. thick; seeds papillate; alpine and subalpine, B.C. and Alta. to Calif. and Colo.   var. clavatum (Trel.) C. L. Hitchc.
      4 Capsules linear, about 1 mm. thick; seeds smooth; Alas. to Lab., southward, usually at or above timber line, to Calif. and Colo.; Eurasia   var. alpinum
    3 Plant mostly 1.5-3 dm. tall; petals 5-13 mm. long; leaves ovate to lanceolate, 1.5-5 cm. long; seeds papillate; capsules linear
      5 Petals 5-8 mm. long; Alas. to Lab., southward to Colo. and Calif.; Eurasia   var. nutans (Hornem.) Hook.
      5 Petals 8-13 mm. long; Cascade Mts. of Wash.   var. albiflorum (Suksd.) C. L. Hitchc.

Epilobium angustifolium L. Sp. Pl. 347. 1753.
  Chamaenerion angustifolium Scop. Fl. Carn. 2nd ed. 1:271. 1772. ("Habitat in Europa boreali")
  E. spicatum Lam. Fl. Fr. 3:482. 1778. *Chamaenerion spicatum* S.F. Gray, Nat. Arr. Brit. Pl. 2: 559. 1821. *(Lamarck,* near Paris, France)
  Chamaenerion angustifolium var. *platyphyllum* Daniels, Fl. Boulder 176. 1911. *E. angustifolium* var. *platyphyllum* Fern. Rhodora 20:5. 1918. *(Daniels 268,* Green Mt., near Boulder, Colo.)
  Chamaenerion exaltatum Rydb. Fl. Rocky Mts. 584. 1917. (No type cited)

Perennial from widespread rhizomelike roots that form adventitious buds freely; stems usually simple, 1-3 m. tall, glabrous except for fine puberulence in the inflorescence and (especially) on the ovaries; leaves alternate, narrowly lanceolate, subsessile, (5) 10-15 (20) cm. long; racemes terminal, many-flowered, greatly elongate, small-bracted; hypanthium practically lacking; sepals 8-12 mm.

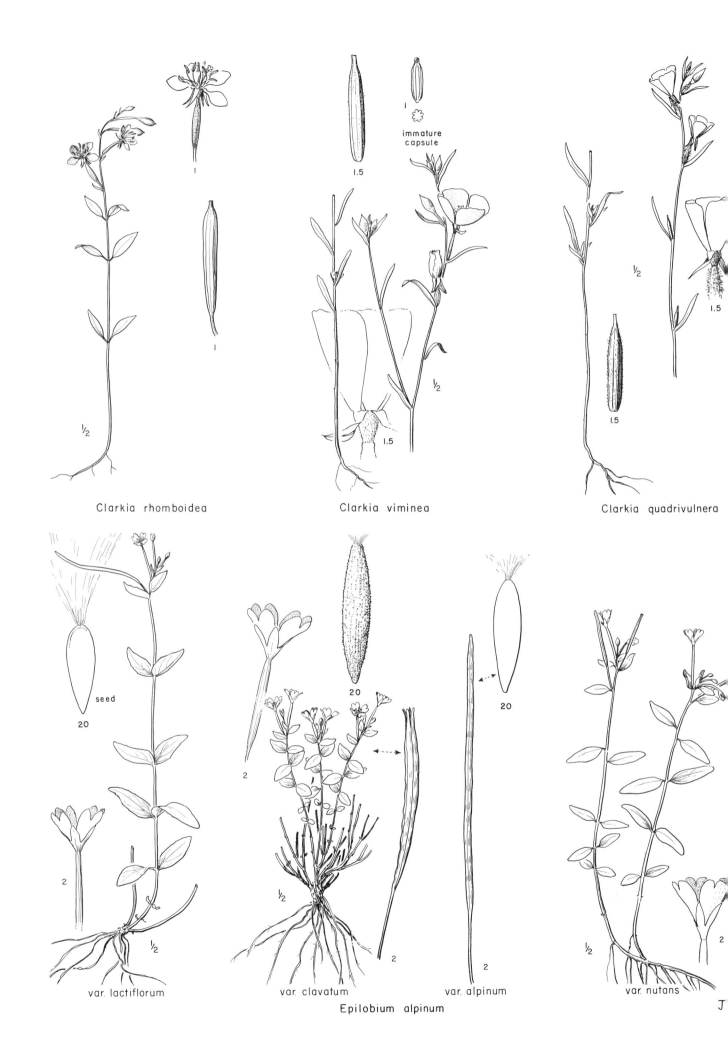

immature
capsule

1.5

Clarkia rhomboidea

Clarkia viminea

Clarkia quadrivulnera

seed

20

var. lactiflorum

var. clavatum

var. alpinum

var. nutans

Epilobium alpinum

J

long, short-clawed; petals 8-20 mm. long, rose to purple (white); style longer than the stamens, pilose on the lower portion; stigma 4-cleft; capsules 5-8 cm. long; coma dingy. N=18, 36. Fireweed.

Widespread, Alas. to Calif., eastward to the Atlantic coast; Eurasia. June-Sept.

The species is common in our area from the coast to well up into the mountains, and especially evident along highways and railroads and on old burns, hence the common name. It is sometimes grown as an ornamental, but is apt to become a bothersome weed.

Epilobium glaberrimum Barbey in Brew. & Wats. Bot. Calif. 1:220. 1876.

   E. fastigiatum var. *glaberrimum* Piper, Contr. U.S. Nat. Herb. 11:404. 1906. *(Bolander,* Yosemite Valley, is the first collection cited)

   E. affine var. *fastigiatum* Nutt. in T. & G. Fl. N.Am. 1:489. 1840. *E. glaberrimum* var. *fastigiatum* Trel. Rep. Mo. Bot. Gard. 2:105, pl. 39. 1891. *E. fastigiatum* Piper, Contr. U.S. Nat. Herb. 11:404. 1906. *(Nuttall,* plains of the Oregon)

   E. glaberrimum var. *latifolium* Barbey in Brew. & Wats. Bot. Calif. 1:220. 1876. *E. platyphyllum* Rydb. Bull. Torrey Club 40:63. 1913; not *E. latifolium* L. *(Lemmon,* Sierra Co., Calif., is the first collection cited) = var. *fastigiatum.*

   E. pruinosum Hausskn. Monog. Epilob. 252, pl. 15. 1884. *(Bridges 252* and *Lobb 232,* both from Calif., are cited in that order)

   E. atrichum Levl. Rep. Nov. Sp. 7:99. 1909. *(Cusick 3295,* Wallowa Mts., Oreg.) = var. *fastigiatum?*

Perennial from branching rootstocks, without turions, often matted at the base, the stems usually many, mostly simple or basally branched, only occasionally sparingly branched above, erect or slightly decumbent at the base, 1-5 dm. tall, entirely glabrous or slightly glandular-pubescent in the inflorescence, usually glaucous throughout and especially on the foliage; leaves opposite, sessile and often slightly clasping, numerous and usually overlapping, oblong-lanceolate to ovate, obtuse, entire to distinctly denticulate, 1.5-4.5 cm. long, very gradually reduced upward and through the inflorescence; pedicels 5-20 mm. long; hypanthium about 1.5 mm. long; sepals 2-4 mm. long, usually rose-to purplish-tinged; petals notched, deep rosy-purple to light pink, 4-8 mm. long; stigma entire; capsules linear, usually arcuate, 4-7 cm. long, glabrous or sparsely glandular-strigillose; seeds about 1 mm. long, plainly (though minutely) papillate, the coma white. N=18.

Wet places in the mountains; B.C. southward through the Olympic and Cascade mts. of Wash. to s. Calif., e. to Ida., w. Mont., and Utah. June-Aug.

There are two rather clearly marked varieties of the species:

1 Plant usually over 3 dm. tall, often branched above; leaf blades lanceolate to narrowly ovate, mostly 2.5-4.5 cm. long, barely overlapping; ovaries hairy; Wash. to Calif., e. to Ida.   var. glaberrimum

1 Plant mostly 1-3 dm. tall, simple; leaf blades mostly 1-3 cm. long, ovate to lanceolate,. closely crowded and overlapping; ovaries usually glabrous; more widespread and common, from w. Wash. to Calif., e. to Mont. and Utah   var. fastigiatum (Nutt.) Trel.

Epilobium glandulosum Lehm. Stirp. Pug. 2:14. 1830. (No specimen cited)

   E. brevistylum Barbey in Brew. & Wats. Bot. Calif. 1:220. 1876. *(Lemmon,* Sierra Co., Calif.) = var. *tenue.*

   E. saximontanum Hausskn. Oest. Bot. Zeits. 29:119. 1879. (Rocky Mts., *Bourgeau, Parry,* etc.) = var. *glandulosum.*

   E. leptocarpum Hausskn. Monog. Epilob. 258, pl. 14, fig. 67. 1884. *(Hall 188,* Oregon) = var. *macounii.*

   E. halleanum Hausskn. Monog. Epilob. 261. 1884. *(E. Hall,* Oregon) = var. *macounii.*

   E. drummondii Hausskn. op. cit. 271. *(Drummond,* Rocky Mts.) = var. *glandulosum.*

   E. pringleanum Hausskn. Mitt. Bot. Ver. Jena 7:5. 1888. *E. brevistylum* var. *pringleanum* Jeps. Man. Fl. Pl. Calif. 670. 1925. (Type not ascertained) = var. *tenue.*

   E. exaltatum Drew, Bull. Torrey Club 16:151. 1889. *E. californicum* var. *exaltatum* Jeps. Man. Fl. Pl. Calif. 670. 1925. *E. glandulosum* var. *exaltatum* Munz, Man. So. Calif. Bot. 333. 1935. *E. brevistylum* var. *exaltatum* Jeps. Fl. Calif. 2:570. 1936. *(Chestnut & Drew,* Grouse Creek, Humboldt Co., Calif., Aug. 1, 1888) = var. *glandulosum.*

   E. delicatum Trel. Rep. Mo. Bot. Gard. 2:98. 1891. *(Cusick 911,* Union Co., Oreg., in 1880, in part) = var. *tenue.*

EPILOBIUM GLANDULOSUM var. TENUE (Trel.) C. L. Hitchc. hoc loc. *E. delicatum* var. *tenue*

Trel. Rep. Mo. Bot. Gard. 2:99. 1891. *E. brevistylum* var. *tenue* Jeps. Man. Fl. Pl. Calif. 670. 1925. *E. pringleanum* var. *tenue* Munz, El Aliso 4:95. 1958. *(Cusick 911,* Union Co., Oreg., in 1880, in part)

E. ursinum Parish ex Trel. Rep. Mo. Bot. Gard. 2:100. 1891. *E. brevistylum* var. *ursinum* Jeps. Man. Fl. Pl. Calif. 670. 1925. *(Parish 1619,* San Bernardino Co., Calif., in 1882, is the first of 3 collections cited) = var. *tenue.*

E. ursinum var. *subfalcatum* Trel. Rep. Mo. Bot. Gard. 2:101. 1891. *E. brevistylum* var. *subfalcatum* Munz in Abrams, Ill. Fl. Pac. St. 3:176. 1951. *(Gray,* California, in 1872, is the first of several collections cited) = var. *tenue.*

EPILOBIUM GLANDULOSUM var. MACOUNII (Trel.) C. L. Hitchc. hoc loc. *E. leptocarpum* var. *macounii* Trel. Rep. Mo. Bot. Gard. 2:103, pl. 37. 1891. *(Macoun 692,* Lake Athabasca, in 1875, is the first of 2 collections cited)

E. ovatifolium Rydb. Bull. Torrey Club 31:567. 1904. *(Patterson 205,* near Empire City, Colo., in 1892) = var. *glandulosum.*

E. rubescens Rydb. Bull. Torrey Club 31:568. 1904. *(Baker,* Pagosa Spgs., Colo., in 1899) = var. *glandulosum.*

E. stramineum Rydb. loc. cit. *(Rydberg,* Idaho Spgs., Colo., in 1905) = var. *glandulosum.*

E. palmeri Rydb. Bull. Torrey Club 31:569. 1904. *(E. Palmer 156,* "South Utah") = var. *glandulosum.*

E. mirabile Trel. ex Piper, Contr. U.S. Nat. Herb. 11:404. 1906. *(Piper 2344,* Olympic Mts., Wash., Aug., 1895) = var. *macounii.*

E. paddoense Levl. Rep. Nov. Sp. 5:8. 1908. *(Suksdorf,* Mt. Paddo, Wash., Aug. 6, 1885) = var. *macounii.*

E. sandbergii Rydb. Bull. Torrey Club 40:64. 1913. *(Sandberg, MacDougal, & Heller 737,* Mud Lake, Kootenai Co., Ida., July 25, 1892) = var. *glandulosum.*

Perennial from slender to filiform rhizomes ending in globose turions, the scales of previous offsets usually persistent through the season; stems simple to branched at the base or above, erect, 0.5-9 dm. tall, glabrous to pilose, often crisp-puberulent in lines, frequently glandular-puberulent in the inflorescence; leaves opposite, from sessile and often somewhat clasping to stalked, with slender to broad petioles as much as 6 mm. long, the blades linear-lanceolate to ovate-lanceolate, 1-12 cm. long, entire to conspicuously denticulate, from crowded and overlapping to remotely spaced, gradually reduced to the floral bracts; racemes rather few-flowered; pedicels 5-30 mm. long; hypanthium 1.5-3 mm. long; sepals 2-5 mm. long; petals pale to dark pink or purplish, notched, 3-10 mm. long; stigma entire; capsules slender, 2-7 cm. long, from nearly glabrous to strigillose, glandular-puberulent, or glandular-pilulose; seeds 0.5-2 mm. long, from nearly smooth or lightly alveolate (the borders of the alveolae appearing in silhouette as papillae) to papillate-echinulate; coma white to tawny.

Moist ground in the mountains (in our area), usually at middle elevations; Alas., through Wash. and Oreg. to s. Calif., in the Rocky Mts. as far s. as Colo., e. to the n. Atlantic coast; Asia. June-Aug.

The species is an unusually variable one, showing much range in the size and color of the flowers, in the stature and pubescence of the plant, and in the size and shape of the leaves. Many of the variants have been described and several consistently have been recognized at the specific level; mostly they have overlapping ranges, where they blend more or less completely with one another, making their separation impossible by other than strictly arbitrary criteria.

1 Petals mostly over 6 (2-10) mm. long; leaves sessile; stems glabrous below but crisp-puberulent (usually in lines) above; capsules glandular-puberulent; occurring throughout the range of the species   var. glandulosum

1 Petals 3-5 mm. long; leaves sessile or petiolate; stems pubescent, the hairs sometimes not in lines; capsules glandular to eglandular

  2 Stems not pubescent in lines; capsules glandular-pilulose; leaves sessile or petiolate; Alas. to s. Calif., e. to Colo.   var. tenue (Trel.) C. L. Hitchc.

  2 Stems pubescent in lines; capsules strigillose, eglandular; leaves usually petiolate; Alas. to Oreg., e. to Ida.   var. macounii (Trel.) C. L. Hitchc.

Epilobium latifolium L. Sp. Pl. 347. 1753.

Chamaenerion latifolium Sweet, Hort. Brit. 2nd ed. 198. 1830. ("Habitat in Sibiria")

E. latifolium var. *tetrapetalum* Pallas ex Pursh, Fl. Am. Sept. 259. 1814. ("On the north-west coast")

Chamaenerion latifolium f. *albiflorum* Nath. Öfvers. Kgl. Sv. Vetensk.-Akad. Förh. 46. 1884, nomen nudum. *Chamaenerium latifolium* var. *albiflorum* (incorrectly attributed to Nath. by) Pors. Meddel. Groenl. 58:110. 1926. *E. latifolium* var. *albiflorum* Pors. Rhodora 41:264. 1939. (Greenland) A white form.

Chamaenerion latifolium var. *grandiflorum* Rydb. Mem. N.Y. Bot. Gard. 1:479. 1900. *(R. S. Williams 1011,* Stanton Lake, Mont., in 1894)

Chamaenerion latifolium var. *megalobum* Nieuwl. Am. Midl. Nat. 3:132. 1913. *(Heacock 455,* Cheops Draw, Selkirk Mts., B.C.) A near-albino.

Chamaenerion subdentatum Rydb. Fl. Rocky Mts. 585. 1917. (Alta. and B.C., no specimen cited)

E. latifolium ssp. *leucanthum* Ulke, Can. Field Nat. 49: 108. 1935. *E. latifolium* f. *leucanthum* Fern. Rhodora 46:252. 1944. *(Ulke 1267,* Horsethief Creek, Kootenay District, B.C.) A white form.

Perennial with a stout caudex, nonrhizomatous; stems usually many, simple to freely branched, erect, (0.5) 1-3 (4) dm. tall; leaves opposite below, lanceolate to ovate, entire to denticulate, (1) 2-6 cm. long, subsessile, glaucous, usually very finely puberulent-strigillose; racemes 3- to 12-flowered, congested to elongate, the lower floral leaves only slightly reduced; pedicels 3-40 mm. long; free hypanthium absent, the sepals joined directly to the top of the ovary, purplish, lanceolate, 10-15 (18) mm. long; petals obovate to broadly deltoid-obovate, short-clawed, 15-25 mm. long, rose-purple to pale purple or occasionally white; ovary canescent; style shorter than the stamens, glabrous; stigma deeply 4-cleft; capsules 3-8 cm. long; coma tawny. N=36.

River bars, along streams, and on drier subalpine to alpine slopes, often on talus; Alas. southward, in both the Olympic and Cascade mts. of Wash. and the Blue-Wallowa Mts. of Oreg., to the Sierra Nevada, e. to Ida. and Mont., and in the Rockies to Colo.; Eurasia. June-Sept.

Within the species there is considerable variation in the size and shape of the leaves, but on the whole there has not been sufficient differentiation to merit the recognition of geographic races. The albino form is fairly frequently encountered.

Epilobium luteum Pursh, Fl. Am. Sept. 259. 1814. *(Pallas,*"On the north-west coast")

E. luteum var. *lilacinum* Henderson, Mazama 10:50. 1928. *(Leach & Leach 2109,* Elwha Basin, Olympic Mts., Wash.)

Herbaceous perennial from widespread rhizomes; stems mostly simple, erect, 2-7 dm. tall, pubescent in decurrent lines; leaves mostly opposite, subsessile, ovate-lanceolate to lanceolate, (2) 3-8 cm. long, glandular-serrulate or -dentate, acuminate to rounded, glabrous except usually for the puberulent margins; flowers 2-10, axillary in the somewhat reduced upper leaves, glandular-puberulent on the ovary and usually on the 5-30 mm. pedicels; hypanthium cup-shaped, 1-2 mm. long; sepals linear-lanceolate, 8-13 mm. long; petals yellow, shallowly obcordate, (10) 14-18 mm. long; style usually conspicuously exceeding the petals; stigma lobes 2 mm. long, oblong-ovate; capsule linear, 4-8 cm. long; coma rust-colored.

Moist soil; Alas. southward, at middle and higher elevations, to the Cascade Mts., Oreg., w. to the Olympic Mts. and Vancouver I. July-Sept.

Epilobium luteum var. *lilacinum* was described from a collection (all of which is believed to have been taken from a single plant) that is intermediate in character between *E. luteum* and *E. glandulosum* in leaf character and in the size and color of the flower. The plant is completely sterile, with poorly developed anthers and ovaries shrivelled after anthesis, and it is therefore assumed that it was a hybrid.

Epilobium minutum Lindl. ex Hook. Fl. Bor. Am. 1:207. 1833.

Crossostigma lindleyi Spach, Ann. Mus. Paris II, 4:404. 1835. *(Menzies,* "North-West coast of America," is the first of several collections cited)

E. minutum var. *foliosum* T. & G. Fl. N.Am. 1:490. 1840. *E. foliosum* Suksd. Deuts. Bot. Monats. 18:87. 1900. *(Nuttall,* "Oregon and the Rocky Mts. of Calif.")

E. adscendens Suksd. Deuts. Bot. Monats. 18:87. 1900. (No specimens cited, but *Suksdorf 81,* Klickitat Co., Wash., is the type)

E. adscendens var. *canescens* Suksd. loc. cit. *(Suksdorf 2147,* Klickitat Co., Wash., June 27, 1892)

A simple or (more commonly) freely branched annual 0.3-4.5 dm. tall, finely strigillose-puberulent throughout except on the ovaries and the calyx, and sometimes also weakly glandular; leaves predominantly opposite, short-petiolate, the blades mostly oblong-lanceolate, 1-2.5 cm. long; flowers pedicellate in the axils of somewhat reduced, alternate leaves; free hypanthium cup-shaped, scarcely 1 mm. long; petals white to rose, 2-4 mm. long, bilobed; stigma entire; capsules slightly curved, linear-cla-

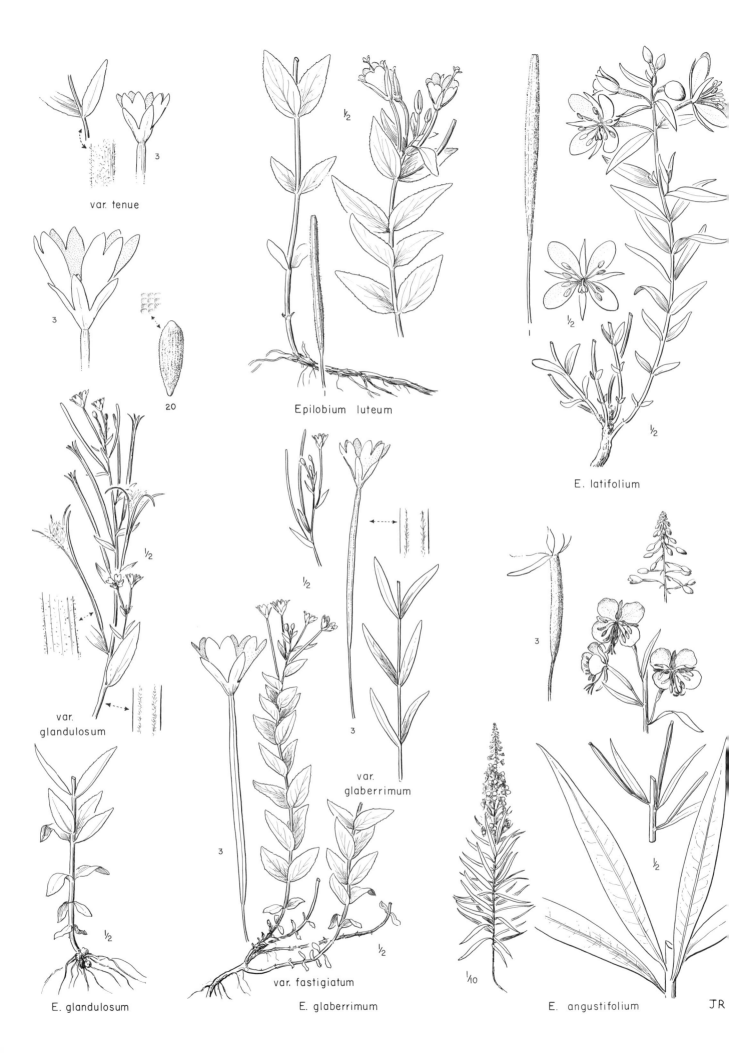

var. tenue

20

var.
glandulosum

E. glandulosum

Epilobium luteum

var.
glaberrimum

var. fastigiatum

E. glaberrimum

E. latifolium

E. angustifolium

JR

vate, 1.5-2.5 cm. long; seeds slightly less than 1 mm. long, finely but very indistinctly alveolate; coma pale yellowish.

Widespread from B.C. to Calif. and e. to Mont., from sea level to well up in the mountains, usually on gravelly or dry soil. Apr.-Aug.

Epilobium obcordatum Gray, Proc. Am. Acad. 6:532. 1865. *(Brewer,* "In the Sierra Nevada, at Squaw Valley and Ebbett's Passes")

Matted perennial from an eventual taproot, glabrous except in the inflorescence, the branches of the caudex slender, ebracteate, rhizomelike; stems many, 3-10 cm. tall; leaves opposite, crowded, ovate, entire to remotely denticulate, (0.5) 1-2 cm. long, subsessile, very glaucous; flowers 1-5 from the leaflike upper bracts, erect; pedicels, ovary, and calyx usually finely glandular-puberulent (in ours); sepals purplish, about 1 cm. long; free hypanthium 1-3 mm. long; petals rose-purple, 14-20 mm. long, obcordate; styles exceeding the stamens, glabrous; stigma deeply 4-lobed; capsule 2-4 cm. long; coma somewhat tawny.

High mountain meadows, talus slopes, and alpine ledges; Sawtooth Mts., Ida., from Blaine, Custer, and Elmore cos., and in the Blue Mts. and c. Cascades of Oreg., s. to Nev. and the Sierra Nevada of Calif. July-Aug.

Epilobium palustre L. Sp. Pl. 348. 1753. (Europe)

E. lineare Muhl. Cat. Pl. 39. 1813, nomen subnudum. *E. palustre* var. *lineare* Gray, Man. 2nd ed. 130. 1856. (Virginia, Carolina)

E. leptophyllum Raf. Précis Découv. 41. 1814. (Pennsylvania and Maryland)

E. densum Raf. Précis Decouv. 42. 1814. (No specimen cited)

E. davuricum Fisch. ex Hornem. Suppl. Hort. Bot. Hafn. 44. 1819. (Dauria, i.e. east of Lake Baikal in Siberia)

E. palustre var. *albiflora* Lehm. in Hook. Fl. Bor. Am. 1:207. 1833. *(Dr. Richardson, Drummond,* "Throughout Canada and as far north as lat. 64°, and among the Prairies of the Rocky Mountains")

E. wyomingense A. Nels. Bot. Gaz. 30:194. 1900. *(A. Nelson 5902,* Yancey's, Yellowstone Park, July 16, 1899, is the first of 2 collections cited)

Simple to branched perennial 1-4 (8) dm. tall, from slender rhizomes which often end in small turions, finely canescent-strigillose throughout or only sparsely so (subglabrate) below; leaves mainly opposite, sessile or subsessile, entire to slightly denticulate, obtuse, linear to lanceolate or narrowly oblong, (1) 2-6 cm. long, mostly 4 (8) mm. broad; inflorescence loosely racemose to paniculate; pedicels slender, 1-4 cm. long; free hypanthium 1-1.5 mm. long, the sepals about twice as long; petals white to pinkish, notched, 3-5 mm. long; styles shorter than the petals; stigma about 1 mm. long, 4-lobed, but the lobes usually completely coalescent; capsule linear, 3-6 cm. long, usually canescent; seeds minutely papillate, the coma white to tawny. N=18.

Wet soil, often in bogs; Alas. to the Cascades of c. Wash., e. to the Atlantic coast and s. in the Rockies to Colo.; Eurasia. June-Aug.

This taxon is often treated, chiefly on the basis of flower color and degree of pubescence, as three species: *E. densum* the densely hairy phase, *E. palustre* often with leaves glabrous but with the stems strigillose, and *E. davuricum* with even less pubescence, that of the stems occurring chiefly in decurrent lines. It has been reported that *E. davuricum* lacks the filiform rhizomes of the other two, but this apparently is not the case. It would seem that *E. densum* and *E. davuricum* could more consistently be recognizable as varieties of *E. palustre,* if their minor peculiarities were to be emphasized.

Epilobium paniculatum Nutt. ex T. & G. Fl. N.Am. 1:490. 1840. ("Oregon, near Fort Vancouver and Straits of Da Fuca," *Scouler,* is the first of 3 collections mentioned)

E. jucundum Gray, Proc. Am. Acad. 12:57. 1876. *E. paniculatum* var. *jucundum* Trel. Rep. Mo. Bot. Gard. 2:85. 1891. *(Greene,* Scott Valley, Siskiyou Co., Calif., Aug. 28, 1876)

E. paniculatum f. *bracteata* Hausskn. Monog. Epilob. 247. 1884. (Specimens cited from Oreg., Wash., and Colo.)

E. paniculatum f. *adenocladon* Hausskn. Monog. Epilob. 247. 1884. *E. adenocladon* Rydb. Bull. Torrey Club 33:146. 1906. (Specimens cited from Colo., Oreg., and Utah)

E. paniculatum f. *subulata* Hausskn. Monog. Epilob. 247. 1884. *E. subulatum* Rydb. Bull. Torrey Club 40:64. 1913.. *E. paniculatum* var. *subulatum* Fern. Rhodora 37:324. 1935. (Several specimens from w. U.S. and Can. are cited)

E. hammondi Howell, Fl. N.W. Am. 224. 1898. (Southwestern Oregon)

E. apricum Suksd. W. Am. Sci. 11:77. 1901. *(Suksdorf 2640,* near Bingen, Klickitat Co., Wash.)

E. fasciculatum Suksd. W. Am. Sci. 11:77. 1901. *(Suksdorf 2641,* Falcon Valley, Klickitat Co., Wash.)

E. tracyi Rydb. Bull. Torrey Club 40:63. 1913. *E. paniculatum* f. *tracyi* St. John, Fl. S. E. Wash. 275. 1937. *E. paniculatum* var. *tracyi* Munz, El Aliso 4:94. 1958. *(Tracy & Evans 547,* Ogden, Utah, July 31, 1887)

E. laevicaule Rydb. Bull. Torrey Club 40:64. 1913. *E. paniculatum* f. *laevicaule* St. John, Fl. S. E. Wash. 275. 1937. *E. paniculatum* var. *laevicaule* Munz, El Aliso 4:94. 1958. *(Rydberg 2728,* Manhattan, Mont., in 1895)

E. altissimum Suksd. Werdenda 1:28. 1927. *E. paniculatum* var. *laevicaule* f. *altissimum* Munz, El Aliso 4:94. 1958. *(Suksdorf 11907,* Bingen, Wash., Aug. 24, 1925, is the first of several collections cited)

E. jucundum var. *viridifolium* Suksd. Werdenda 1:29. 1927. *(Suksdorf 4342,* Bingen, Wash., Sept. 14, 1904)

E. paniculatum var. *tracyi* f. *fasciculatum* Munz, El Aliso 4:94. 1958, not the same as *E. fasciculatum* Suksd. *(Raven 5822,* Mono Creek, Fresno Co., Calif., July 25, 1953)

Annual, usually very freely branched, glabrous except sometimes in the inflorescence, 3-10 (20) dm. tall; leaves alternate almost throughout, petiolate, narrowly lanceolate to linear, entire to somewhat denticulate, those of the main stem 3-7 cm. long but those in the axillary fascicles and on the side branches usually much smaller; racemes terminating the numerous branches, lax, few-flowered; bracts linear; pedicels 5-20 mm. long, glabrous to glandular-puberulent like the ovaries; free hypanthium narrowly funnelform, 2-12 mm. long; petals rose to pale pinkish, obovate, bilobed, 3-12 mm. long; stigma entire or if lobed the lobes more or less coalescent; capsule 1.5-2.5 cm. long, linear, clavate; seeds minutely papillate, the coma pale yellowish. N=12.

Common in much of w. U.S.; usually on dry soil in open or wooded areas, especially in the ponderosa pine belt. July-Aug., mostly.

This species is extremely variable in nearly all respects, but especially so as concerns the size of the flowers and the amount and type of pubescence on the pedicels and ovaries. Several varieties or forms are usually distinguished, based upon various combinations of characters involving flower size and degree of pubescence. Since the taxa thus delimited are essentially sympatric, it would seem to serve no particularly useful purpose to name these variants even though it is not improbable that some of them may be sufficiently isolated genetically that free interbreeding does not occur. The largest-flowered plants, found chiefly from coastal Wash. to Ida. and s. to Calif., which usually have rose-colored petals 7-13 mm. long and the free hypanthium over 5 mm. long, vary from glabrous to glandular, but are maintainable as var. *jucundum* (Gray) Trel., whereas the nomenclaturally typical plants, with smaller and often much lighter-colored flowers, range from B.C. to the Dakotas, and s. to N.M. and s. Calif.

Epilobium suffruticosum Nutt. in T. & G. Fl. N. Am. 1:488. 1840.

Cordylophorum suffruticosum Rydb. Fl. Rocky Mts. 590. 1917. *(Nuttall,* "gravelly banks of streams, east of Wallawallah")

Suffruticose perennial forming spreading clumps, strigillose and often canescent almost throughout and especially on the ovaries, but not at all glandular; stems many, 1-3 dm. tall, more or less decumbent at base, usually rather freely branched; leaves mainly opposite, lanceolate to elliptic, entire, sessile, 1-3 cm. long; flowers in the axils of scarcely reduced upper leaves, slightly irregular, one petal usually noticeably the largest; style and stamens curved slightly to one side; pedicels 5-25 mm. long; free hypanthium broadly funnelform, about 3 mm. long; petals pale yellow, (5) 7-9 (10) mm. long, narrowly cuneate-obovate, with a narrow apical notch; stamens and style subequal and about equaling the petals; stigma 4-cleft; capsule 1.5-2.5 cm. long, narrowly fusiform, usually curved; coma cream-colored, deciduous as a unit.

Gravelly stream banks and moist rocky areas at medium elevations; c. Ida. to Mont. and Wyo. July-Aug.

Epilobium watsonii Barbey in Brew. & Wats. Bot. Calif. 1:219. 1876. ("Near the Russian settlement, Sonoma Co.; only the Russian collectors"—i.e. near Ft. Ross, Calif.)

E. franciscanum Barbey in Brew. & Wats. Bot. Calif. 1:220. 1876. *E. watsonii* var. *franciscanum* Jeps. Man. Fl. Pl. Calif. 670. 1925. *(Bigelow,* San Francisco, Calif.) = var. *watsonii,* the less pubescent to subglabrous phase.

E. adenocaulon Hausskn. Oest. Bot. Zeits. 29:119. 1879. *E. glandulosum* var. *adenocaulon* Fern. Rhodora 20:35. 1918. (Specimens cited from Ohio, Chile, Sask., and N.Y.) = var. *occidentale*.

E. americanum Hausskn. Oest. Bot. Zeits. 29:118. 1879. *(Bourgeau,* Saskatchewan River, Aug., 1857) = var. *occidentale*.

E. californicum Hausskn. Monog. Epilob. 260. 1884. *(Wrangell,* "Colonia Ross," Sonoma Co., Calif.) = var. *parishii*.

EPILOBIUM WATSONII var. PARISHII (Trel.) C. L. Hitchc. hoc loc. *E. parishii* Trel. Zoë 1:210. 1890. *E. californicum* var. *parishii* Jeps. Man. Fl. Pl. Calif. 670. 1925. *E. adenocaulon* var. *parishii* Munz, El Aliso 4:95. 1958. *(Parish 1889,* San Bernardino Co., Calif., is the first of several collections cited)

EPILOBIUM WATSONII var. OCCIDENTALE (Trel.) C. L. Hitchc. hoc loc. *E. adenocaulon* var. *occidentale* Trel. Rep. Mo. Bot. Gard. 2:95, pl. 23. 1891. *E. occidentale* Rydb. Mem. N.Y. Bot. Gard. 1:275. 1900. *E. glandulosum* var. *occidentale* Fern. Rhodora 20:35. 1918. *E. californicum* var. *occidentale* Jeps. Man. Fl. Pl. Calif. 670. 1925. ("Vancouver Island, and British Columbia to central California, and Nevada?" *Shockley 509)*

E. adenocaulon var. *perplexans* Trel. Rep. Mo. Bot. Gard. 2:96. 1891. *E. perplexans* Nels. in Coult. & Nels. New Man. Bot. Rocky Mts. 337. 1909. *E. glandulosum* var. *perplexans* Fern. Rhodora 20:35. 1918. *(Hall 176,* Oregon, is the first collection cited) = var. *occidentale*.

E. macdougalii Rydb. Fl. Rocky Mts. 587, 1064. 1917. *(MacDougal 793,* Flathead Lake, Mont., July 29, 1901) = var. *parishii*.

E. cinerascens Piper, Proc. Biol. Soc. Wash. 31:75. 1918. *(Peck 7817,* Sutherlin, Douglas Co., Oreg.) = var. *parishii*.

E. praecox Suksd. Werdenda 1:27. 1927. *(Suksdorf 10594,* Bingen, Wash., May-Nov., 1920-26, a garden specimen, is the first of several collections cited) = var. *occidentale*.

E. griseum Suksd. Werdenda 1:28. 1927. *(Suksdorf 10272,* Big White Salmon, Wash., July 3, 1919, is the first of several collections cited) = var. *parishii*.

Perennial but often blooming the first season, eventually spreading by short rootstocks that produce rosettes of leaves but no turions; stems (2) 3-10 dm. tall, often glandular above, especially in the inflorescence, weakly to densely pubescent, the hairs either in decurrent lines or more general, usually simple below but rather freely branched above, the inflorescence mostly compound; leaves opposite, short-petiolate to subsessile, narrowly lanceolate to rather broadly ovate-lanceolate, (2.5) 3-7 cm. long, generally obviously serrulate; sepals 2-5 mm. long, often purplish; petals 3-10 mm. long, notched, white or cream to deep purplish-red; stigma entire or if lobed the lobes usually coalescent; capsule linear, 4-8 cm. long, strigillose to glandular-puberulent; seeds 0.5-1.2 mm. long, distinctly crested-papillate in numerous parallel longitudinal lines, the coma white. N=18.

General over much of Alas. and the U.S., from sea level to the lower levels of the mountains, usually in wet places. June-Aug.

This is perhaps the most confusing complex in the genus in our area, not only because of the great variation in morphological characters, but also because of its complicated nomenclatural history. Even though intergradation between the various elements is notably extensive, it is more or less customary to call the greater part of the species *E. adenocaulon,* restricting the earlier name, *E. watsonii,* to a second of the several species usually recognized. Munz (Calif. Fl. 927-933. 1959) has treated the complex as consisting of three species, *E. adenocaulon, E. watsonii,* and *E. ciliatum* Raf. (Med. Repos. N.Y. II, 5:361. 1808), *E. ciliatum* including those plants with thin, flaccid, pale-green leaves tapered to definite slender petioles, the papillae of the seeds more or less conical, and the coma caducous. Rafinesque's name is the oldest epithet of these three and would be the one to use for the complex, as treated here, were there no doubt as to the exact nature of his species. Because there *is* such doubt, the name *E. ciliatum* is not being adopted for our plants.

Within the complex there are graded series in pubescence, the plants ranging from sparsely hairy to canescent and from eglandular to strongly glandular. The flowers vary from small to large and from white to deep rose-colored. There are all possible combinations of these several characters. Since there is little evidence of the development of geographic races in the species, it seems unfeasible to delimit more than the following three varieties:

1 Petals 6-10 mm. long, deep purplish; plants usually reddish on the lower leaves and stem
    and on the calyx; mostly near the coast from s. Wash. to c. Calif.   var. watsonii
1 Petals commonly not over 6 mm. long, often pale; plants various
  2 Inflorescence not glandular, often canescent with short, appressed to curved hairs; not

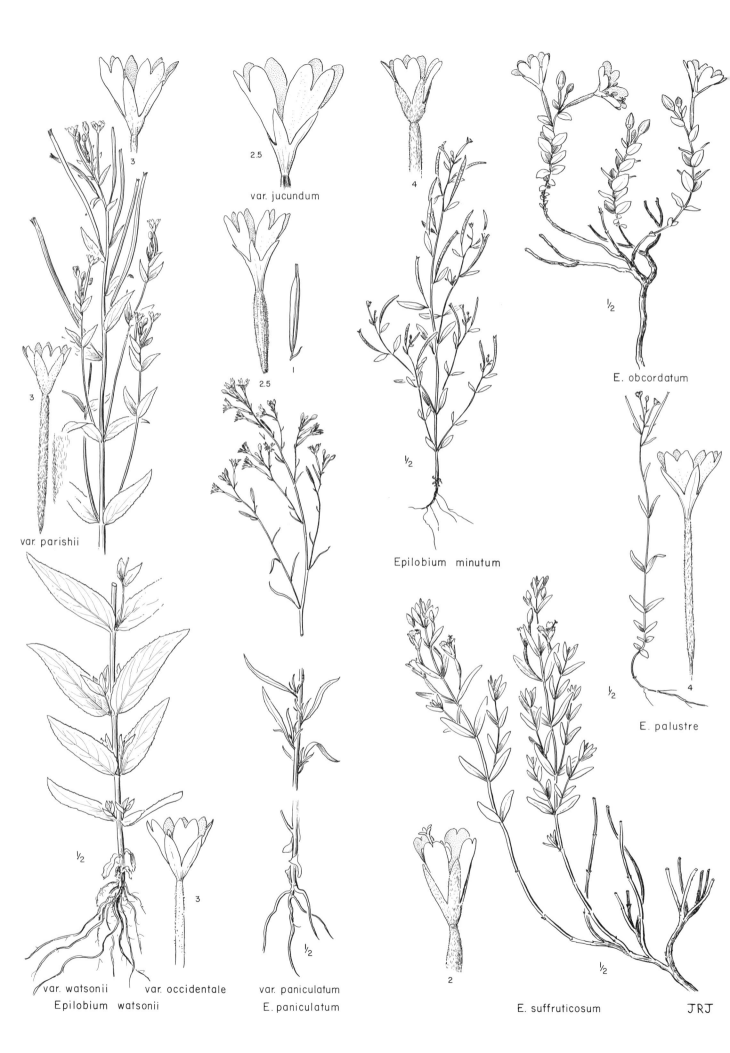

var. jucundum

var. parishii

2.5

E. obcordatum

Epilobium minutum

E. palustre

var. watsonii   var. occidentale

Epilobium watsonii

var. paniculatum

E. paniculatum

E. suffruticosum

JRJ

common, but occasional from the Rocky Mts. to Wash. and s. to s. Calif.

var. parishii (Trel.) C. L. Hitchc.

2 Inflorescence from sparsely to densely glandular; Alas. to the Atlantic coast, s. to Calif.
and Colo.   var. occidentale (Trel.) C. L. Hitchc.

Gaura L.

Flowers spicate or spicate-racemose, whitish to pink, 4-merous excepting the stamens, slightly irregular; free hypanthium narrowly cylindric-obconic, the sepals usually distinct and reflexed; stamens 8; stigma shortly 4-lobate from a cuplike base; ovary 4-celled or sometimes 1-celled; ovules 1 or 2 per locule; fruit hardened, indehiscent, more or less nutlike, usually fusiform, 1- to 4-seeded; annual or perennial, alternate leaved herbs.

About 18 species of tropical and temperate N. Am. (Greek *gauros*, superb or proud, because of the erect "proud" flowers.)

Reference:

Munz, P. A. A revision of the genus Gaura. Bull. Torrey Club 65:105-122, 211-228. 1938.

1 Plant annual, usually simple, villous with soft spreading hairs; larger leaves over 4 cm. long; petals less than 3 mm. long; floral bracts deciduous   G. PARVIFLORA

1 Plant perennial, usually several-stemmed from the base; leaves less than 4 cm. long; petals over 3 mm. long; floral bracts persistent   G. COCCINEA

Gaura coccinea (Nutt.) Pursh, Fl. Am. Sept. 733. 1814.
Malva coccinea Nutt. in Fraser's Cat. no. 51. 1813. *G. multicaulis* Raf. Atl. Journ. 146. 1832. *(Bradbury,* "in Upper Louisiana" on the Missouri River)
G. glabra Lehm. in Hook Fl. Bor. Am. 1:209. 1833. *G. coccinea* var. *glabra* T. & G. Fl. N. Am. 1:518. 1840. *(Drummond,* "about Carlton-House, on the Saskatchawan")

Glabrous or glabrate to strigose or hirsute perennial, usually with several decumbent-based, simple to freely branched stems 2-6 dm. long; leaves many, sessile, linear-lanceolate, oblong-lanceolate, or lanceolate, 1.5-3 (3.5) cm. long, reduced upward, usually with only a few shallow teeth; spikes 5-20 cm. long, closely many-flowered, the bracts 3-9 mm. long; free hypanthium slender, 5-9 mm. long; sepals distinct, sharply reflexed; petals pinkish or red to nearly white, 3-6 mm. long including the narrow claw, the blade oval to rhombic; filaments 3-5 mm. long; appendages with tiny linear scales at the base; anthers 3-5 mm. long; style slightly longer than the stamens; stigma lobes short, oval; fruit glabrous to hirsute, 5-9 mm. long, somewhat fusiform, sharply 4-angled and semiwinged on the upper half, 1-seeded. N=7.

Dry open slopes, chiefly in the sagebrush area in our region; Alta. to Mex., chiefly e. of the Rockies, but crossing into w. Mont.; reported (as an escape?) from Bingen, Wash. June-Aug.

There are two forms of the species, traditionally referred to as var. *coccinea,* the pubescent phase, and var. *glabra* (Lehm.) T. & G., which is glabrous or nearly so. They occur together and differ in no other respect.

Gaura parviflora Dougl. ex Hook. Fl. Bor. Am. 1:208. 1833. *(Douglas,* "Sandy banks of the Walla-wallah River, North-West coast of America")

Annual from a taproot and with a single stem that is simple below but usually branched above, 0.2-2 m. tall, hirsute to the inflorescence with soft spreading hairs 1-2 mm. long and (usually) more abundantly puberulent or puberulent-glandular; basal leaves spatulate, the lower cauline ones oblong, the others oblong- or ovate-lanceolate to lanceolate, 4-15 cm. long, sinuately dentate to entire, gradually reduced to the very small, linear, caducous floral bracts; spikes many-flowered, 1-3 dm. long; free hypanthium very slender, 2-3 mm. long; sepals 2-3 mm. long, reflexed, distinct; petals pinkish, short-clawed, about 2 mm. long; filaments without basal scales; anthers about 1 mm. long; fruit 6-10 mm. long, fusiform, distinctly 4-angled but not at all winged, glabrous, 1 (2)-seeded.

Sandy or rocky, often disturbed or waste land; e. Wash. to the c. U.S., s. to Mex. July-Sept.

Gayophytum Juss.

Flowers small, regular, leafy-bracteate, from racemose and with filiform pedicels, to subsessile or sessile in short spikes; perianth 4-merous; free hypanthium essentially lacking, the calyx divided nearly to the ovary, the segments distinct, usually reflexed; petals white to pinkish, short-clawed;

stamens 8, unequal, the set alternate with the petals reduced and often sterile; stigmas capitate; capsules linear to linear-clavate, usually somewhat torulose, 2-celled, 4-valved; seeds glabrous to puberulent, not comose, borne in 1 row per locule; subsimple to (usually) diffusely branched, slender-stemmed annuals with entire, linear to very narrowly lanceolate, alternate or in part (the lower ones) opposite leaves.

Perhaps six species, in drier, usually open soil of w. N.Am. to S.Am. (Named for G. Gay, author of a flora of Chile, and *phyton,* the Greek word for plant—"Gay's plant. ")

Reference:

Munz, P. A. The genus Gayophytum. Am. Journ. Bot. 19:768-778. 1932.

It would appear that there are at least four conspicuous characteristics that vary more or less independently in this genus: 1) flower size, the petals showing a continuous variation between about 0. 5 and 4 mm. in length; 2) seed vesture, the testa either glabrous or hairy; 3) foliage vesture, the leaves glabrous or pubescent; 4) general pubescence, mostly either short and appressed or spreading. Since plants with every conceivable combination of these several characters may be found (often many of them in the same area) and since such variants occur more or less throughout the range of the genus, it is felt that the separation of the complex into numerous taxa on these bases, at whatever level, will more nearly represent mechanical sorting than the delimitation of natural entities. Even the four species here admitted are not sharply delimited, although they seem to represent natural populations.

1 Capsules nearly or quite sessile, only slightly or not at all constricted between the seeds;
    plants basally branched, very leafy above, the leaves commonly much longer than the
    internodes
    2 Seeds about 1 mm. long, erect in the capsule   G. RACEMOSUM
    2 Seeds 0. 5-0. 75 mm. long, obliquely disposed to the axis of the capsule   G. HUMILE
1 Capsules pedicelled, usually constricted between the seeds; plants commonly branched
    above, the internodes often longer than the leaves
    3 Petals scarcely 1 mm. long; capsules seldom over 6 mm. long, shorter than the usu-
        ally sharply reflexed pedicels; plants glabrous   G. RAMOSISSIMUM
    3 Petals usually over 1 mm. long, or the capsules well over 6 mm. long and exceed-
        ing the erect to spreading (reflexed) pedicels; plants glabrous to pubescent   G. NUTTALLII

Gayophytum humile Juss. Ann. Sci. Nat. I, 25:18, pl. 4. 1832. (Chile)
    G. pumilum Wats. Proc. Am. Acad. 18:193. 1883. *(Torrey 97,* Lake Co., Calif.)
    G. humile var. *hirtellum* Munz, Am. Journ. Bot. 19:778. 1932. *(C. F. Baker 1373,* Snow Valley,
    Ormsby Co., Nev.)

Low, usually diffusely branched, glabrous to grayish-pubescent, very leafy annual 5-20 cm. tall; leaves linear to linear-spatulate, 5-30 mm. long, 1-2 mm. broad; flowers sessile or subsessile in crowded spikes, the floral bracts nearly as large as the cauline leaves, largely concealing the flowers and fruits; free hypanthium about 0. 2 mm. long; sepals and petals about 1 mm. long; capsule sessile or subsessile, linear and very slightly torulose to somewhat clavate and not at all torulose, 7-15 mm. long; seeds 15-20, obliquely ascending in the capsule, 0. 5-0. 75 mm. long, glabrous to puberulent. N=7.

From the foothills to medium elevations in the mountains on the e. side of the Cascades, Wash. to s. Calif., e. to Ida.; Chile. June-Aug.

Gayophytum nuttallii T. & G. Fl. N.Am. 1:514. 1840.
    Oenothera micrantha Nutt. ex T. & G. loc. cit. as a synonym. ("Rocky Mountains, *Nuttall! Douglas!")*
    G. diffusum T. & G. Fl. N.Am. 1:513. 1840. *(Nuttall,* "Rocky Mountains and plains of Oregon")
    G. ramosissimum var. *strictipes* Hook. Lond. Journ. Bot. 6:224. 1847. *(Gordon,* Upper Platte
    River)
    G. lasiospermum Greene, Pitt. 2:164. 1891. *(Greene,* near Julian, San Diego Co., Calif.)
    G. eriospermum Cov. Contr. U.S. Nat. Herb. 4:103. 1893. *G. lasiospermum* var. *eriospermum*
    Jeps. Man. Fl. Pl. Calif. 689. 1925. *(Coville,* East Fork of Kaweah River, Tulare Co., Calif.,
    July 28, 1891)
    G. intermedium Rydb. Bull. Torrey Club 31:569. 1904. *G. nuttallii* var. *intermedium* Munz, Am.
    Journ. Bot. 19:772. 1932. *(Underwood & Selby 193,* w. of Ouray, Colo., in 1901)
    G. nuttallii var. *abramsii* Munz, Am. Journ. Bot. 19:772. 1932. *(Abrams 2694,* San Antonio Mts.,
    Calif., July 13, 1920)

G. diffusum var. *villosum* Munz, Am. Journ. Bot. 19:773. 1932. *(Leiberg 435,* Farewell Bend, Crook Co., Oreg.)

Annual, 1.5-6 dm. tall, glabrous to strigillose or densely pubescent with soft spreading hairs, usually diffusely branched especially above and in the inflorescence, the ultimate branches rather filiform; leaves linear (the primary ones) to oblanceolate, 1.5-5 cm. long, 1-2 (7) mm. broad, gradually reduced upward to the linear bracts; pedicels filiform, 2-8 mm. long, usually erect to spreading, even in fruit; sepals 1-3 mm. long; petals white to pinkish, 1-5 mm. long; capsules pedicellate, mostly erect, glabrous to soft-pubescent, torulose, 4-12 mm. long; seeds glabrous to puberulent, 1-1.5 mm. long. N=7, 14.

Dry gravelly soil in the valleys, foothills, and mountains, sometimes to near timber line; throughout most of w. U.S. June-Aug.

Usually this complex taxon is divided into several species and varieties on the basis of the various combinations previously discussed. The following are the taxa most frequently recognized. They are practically sympatric and considered to be aspects of the one species and of no more nomenclatural significance than forms, were such to be recognized:

1 Petals not over 2 mm. in length
  2 Seeds pubescent   *G. lasiospermum* Greene
  2 Seeds glabrous
    3 Plants with spreading hairs   *G. nuttallii* var. *abramsii* Munz
    3 Plants strigillose
      4 Capsules erect   *G. nuttallii*
      4 Capsules reflexed   *G. intermedium* Rydb.
1 Petals 2-5 mm. long
  5 Seeds pubescent   *G. eriospermum* Cov.
  5 Seeds glabrous
    6 Plants strigillose   *G. diffusum* T. & G.
    6 Plants with short, soft, spreading hairs   *G. diffusum* var. *villosum* Munz

Gayophytum racemosum T. & G. Fl. N.Am. 1:514. 1840. *(Nuttall,* "Rocky Mountains near Black-Foot River")

G. caesium Nutt. ex T. & G. Fl. N.Am. 1:514. 1840. *G. racemosum* var. *caesium* Munz, Am. Journ. Bot. 19:776. 1932. *(Nuttall,* "Oregon on dry, open plains near Wallawallah") The form with spreading pubescence.

G. helleri Rydb. Bull. Torrey Club 40:65. 1913. *(Heller 3433,* "Forest, Nez Perces County," July 16, 1896) A phase with hairy seeds.

G. helleri var. *glabrum* Munz, Am. Journ. Bot. 19:777. 1932. *(Macbride 398,* Silver City, Ida.) A glabrous phase.

Very similar to *G. humile* in most respects, the plants low and diffusely branched, from nearly or quite glabrous to grayish puberulent; flowers sessile or subsessile, the petals barely 1 mm. long; capsules sessile (or with pedicels rarely up to 2 mm. long), 7-15 mm. long, linear and very slightly torulose to (commonly) slightly clavate and not torulose, the seeds rarely over 10, erect in the capsule, about 1 mm. long. N=14.

Open slopes, often where moist early in the season, from the foothills to well up in the mountains (often to timber line); chiefly e. of the Cascade summit, Wash., southward to s. Calif., e. to Mont. and Colo. June-Aug.

Gayophytum ramosissimum Nutt. ex T. & G. Fl. N.Am. 1:513. 1840. *(Nuttall,* "Rocky Mountains etc.")

G. ramosissimum var. *deflexum* Hook. Lond. Journ. Bot. 6:224. 1847. *(Geyer 547,* "Valley of Tshimakaine, abundant on the Upper Columbia")

Glabrous annual 1.5-4 dm. tall, freely and diffusely branched, especially above; leaves linear, 1-3.5 cm. long, 0.5-2.5 mm. broad, gradually but not greatly reduced to the linear floral bracts; pedicels filiform, 3-6 mm. long, often exceeding the capsules, spreading to erect in flower, but often rather sharply reflexed in fruit; sepals and petals about 0.5 mm. long; capsule pedicellate, 3-6 mm. long, distinctly torulose; seeds glabrous, about 0.5 mm. long. N=7.

Eastern Wash. to Calif., e. to Mont. and Colo.; in dry foothills and valleys to the lower mountains. June-July.

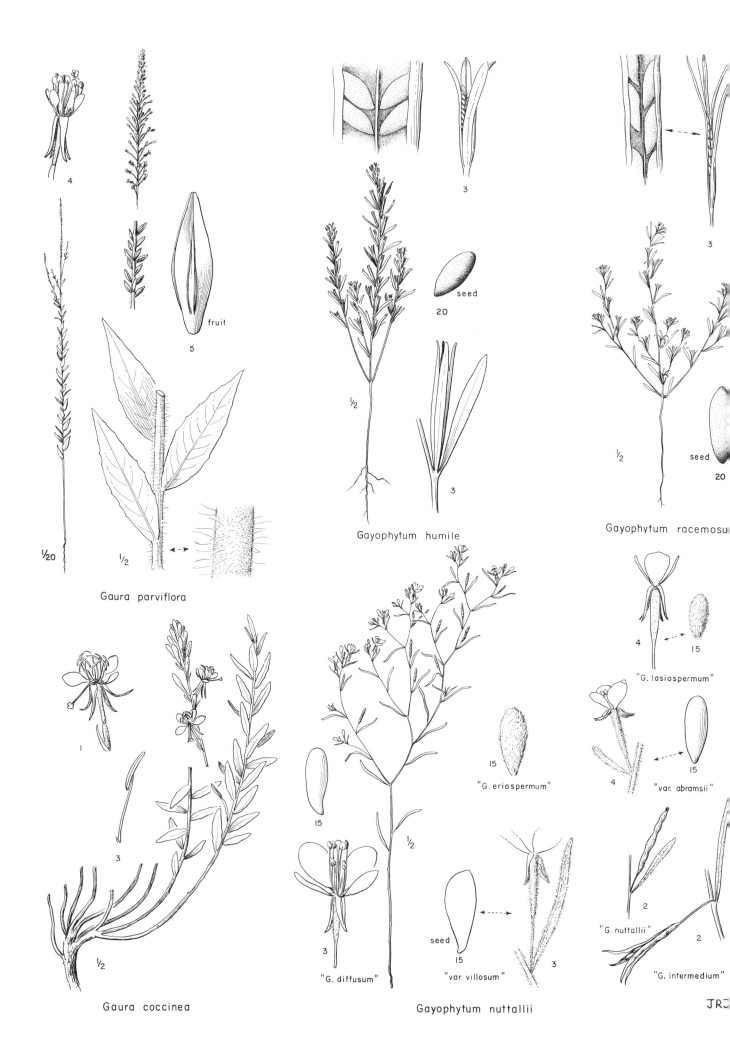

3

seed
20

Gayophytum humile

3

Gayophytum racemosu

1/20   1/2

Gaura parviflora

fruit

5

4

1/2

4

15

"G. lasiospermum"

4   15

"var. abramsii"

2

"G. nuttallii"

2

"G. intermedium"

1

3

1/2

Gaura coccinea

15

"G. eriospermum"

15

1/2

3

"G. diffusum"

seed

15

"var. villosum"

3

Gayophytum nuttallii

JRJ

Jussiaea L.

Flowers solitary in the leaf axils, pedicellate; calyx not prolonged above the ovary as a free hypanthium, the sepals usually 5 (ours), 4, or 6, persistent in fruit; petals 5 (ours), 4, or 6, usually yellow (ours) or white, deciduous; stamens twice as many as the petals, in two series; pistil 5 (4-6)-carpellary, the style simple; stigma capitate (ours) or somewhat 4- to 6-lobed; ovary inferior, 4- to 6-loculed, the placentation axile; capsule cylindrical to fusiform or obconic, many-seeded, loculicidal and septicidal; perennial herbs (ours) or semishrubs to somewhat arborescent, with alternate, simple, stipulate (ours), petiolate leaves.

Forty or more species, mostly of the tropics and subtropics of both hemispheres, but chiefly in the Americas. (In honor of Bernard de Jussieu, 1699-1777, one of the most famous of French taxonomists.)

Reference:

Munz, P. A. A revision of the New World species of Jussiaea. Darwiniana 4:179-284. 1942.

Jussiaea uruguayensis Camb. in St. Hilaire, Fl. Bras. Merid. 2:264. 1829.

Ludwigia uruguayensis Hara, Journ. Jap. Bot. 28:294. 1953. (Uruguay)

Glabrous to slightly hairy perennial with prostrate (often floating) to ascending, freely rooting, mostly reddish stems and widespread rhizomes, often forming mats up to 2 m. broad; leaves not greatly reduced upward, the blades entire, from nearly elliptic to oblong-spatulate, lanceolate, or linear-lanceolate, up to 12 cm. long, narrowed either gradually or rather abruptly to slender or winged petioles as much as 2-3 cm. long; flowers solitary and axillary; pedicels 1-3 cm. long, often bracteolate just below the flower; ovary slender, up to 1 cm. long, pubescent or glabrous; sepals usually 5, acute, 6-13 mm. long; petals clear deep yellow, oblong-ovate, 12-20 mm. long, slightly retuse and very shortly clawed; stamens 5-8 mm. long; style about 4 mm. long; stigma capitate; capsule more or less cylindric, 10-nerved, up to 25 mm. long and 3-4 mm. thick; seeds pendulous, 1.5 mm. long, about as broad.

Swamps, rivers, and lakes; native of S. Am., but introduced freely in s.e. U.S. and in our area along the Columbia R. below Portland, where reported to be a pest. July-Sept.

Our plant is the glabrous phase, var. *uruguayensis*.

Ludwigia L.

Flowers sessile in the upper axils; free hypanthium lacking, the 4 sepals persistent in fruit; petals lacking in ours; stamens 4, opposite the sepals, the filaments very short; stigma capitate; fruit a turbinate, 4-celled, many-seeded capsule; seeds glabrous; opposite, entire-leaved perennials with prostrate stems.

About 30 species, the majority in N. Am. (Named for C. G. Ludwig, 1709-1773, an outstanding German botanist.)

Reference:

Munz, P. A. The American species of Ludwigia. Bull. Torrey Club 71:152-165. 1944.

Ludwigia palustris (L.) Ell. Sk. Bot. S. C. & Ga. 1:211. 1821.

Isnardia palustris L. Sp. Pl. 120. 1753. (Europe)

Isnardia palustris var. *americana* DC. Prodr. 3:61. 1828. *L. palustris* var. *americana* Fern. & Griscom, Rhodora 37:176. 1935. (Canada to Georgia)

L. palustris var. *pacifica* Fern. & Griscom, loc. cit. *(W. R. Carter 128*, Vancouver I., B.C., July 14, 1914)

Glabrous, succulent perennial; leaves 2-6 cm. long, blades ovate-elliptic or rhombic-elliptic to obovate, tapered gradually to broad petioles nearly as long; flowers about 2 mm. long, greenish; sepals ovate; anthers considerably longer than the filaments; capsule 2.5-4 mm. long, 4-sided, turbinate-truncate above, with 2 tiny lateral, slightly adnate bracteoles about half as long.

Plants of marshes and bogs; very widespread in N. Am., Eurasia, and Africa. July-Sept.

Our plants are considered to be varietally distinct from the Old World plants and of two kinds: var. *americana* (DC) Fern. & Griscom, from the Cascades e. to the Atlantic coast, with leaf blades usually less than twice as long as broad; and var. *pacifica* Fern. & Griscom, to the w. of the Cascades and Sierras, with narrower leaves that usually are more than twice as long as broad.

Oenothera L. Evening Primrose

Flowers white or yellow, frequently aging to reddish or purplish, often nocturnal, sometimes fragrant, usually in leafy to bractless spikes or racemes, or sessile among the basal leaves; hypanthium often very conspicuous, usually deciduous after anthesis; sepals 4, mostly distinct, reflexed, sometimes connate and turned to one side in anthesis; petals 4, clawless, often retuse to emarginate; stamens 8, the anthers usually linear and versatile, sometimes sub-basifixed; stigma globose to deeply linear-lobate; capsules woody to membranous, straight to curved or partially coiled, dehiscent, 4-celled; seeds numerous, not comose; annual to perennial, acaulescent to strongly caulescent herbs with alternate or basal, simple to pinnatifid leaves.

About 90 species, all but one in N. or S. Am. (Greek name used by Theophrastus.)

References:

Munz, P. A. A revision of the subgenus Chylismia of the genus Oenothera. Am. Journ. Bot. 15: 223-240. 1928.

————. Revision of North American species of the subgenus Sphaerostigma, genus Oenothera. Bot. Gaz. 85:233-270. 1928.

————. Revision of the North American species of the subgenera Taraxia and Eulobus of the genus Oenothera. Am. Journ. Bot. 16:246-257. 1929.

————. The North American species of the subgenera Lavauxia and Megapterium of the genus Oenothera. Am. Journ. Bot. 17:358-370. 1930.

————. The subgenus Anogra of the genus Oenothera. Am. Journ. Bot. 18:309-327. 1931.

————. The subgenus Pachylophis of the genus Oenothera. Am. Journ. Bot. 18:728-738. 1931.

————. The Oenothera hookeri group. El Aliso 2:1-47. 1949.

The gardener e. of the Cascades easily may utilize several rather choice species, including especially *O. caespitosa* and *O. pallida,* the first, at least, excellent for rock gardens, even in the Puget Sound area. Species of less interest, more tractable in drier climates, include *O. tanacetifolia* and *O. flava* and the annuals *O. palmeri, O. deltoides,* and *O. clavaeformis.*

1 Plants mostly without leafy stems, the flowers borne among the rosettes of leaves, stems
  (if present) short and concealed by the leaves and flowers; hypanthium several times as
  long as the ovary; perennials except for *O. palmeri*
  2 Petals white or yellow, aging to pink or purple, 1-5 cm. long; stigma lobes linear, usu-
    ally at least 3 mm. long; hypanthium deciduous from the developing fruit
    3 Petals yellow, aging to purple, 1-2 cm. long; capsule distinctly wing-margined
      (Lavauxia)   O. FLAVA
    3 Petals white, aging to purple, well over 2 cm. long; capsules not wing-margined
      (Pachylophis)   O. CAESPITOSA
  2 Petals yellow, seldom aging to purple, less than 2 cm. long; stigmas globose or dis-
    coid; hypanthium filiform, persistent on the fertile portion of the ovary, flared only
    at the top *(Taraxia)*
    4 Plant annual; leaves less than 1 cm. broad; petals not over 5 mm. long   O. PALMERI
    4 Plant perennial; leaves at least 1 cm. broad; petals at least 6 mm. long
      5 Leaves entire to dentate or lobed; capsules glabrous   O. HETERANTHA
      5 Leaves pinnatifid; capsules hairy
        6 Hypanthium less than 2.5 cm. long; petals 6-8 (10) mm. long; capsules usually
          curved   O. BREVIFLORA
        6 Hypanthium usually over 2.5 cm. long; petals mostly over 10 mm. long; cap-
          sules straight   O. TANACETIFOLIA
1 Plants with leafy flowering stems; hypanthium often shorter than the mature capsules; annuals or perennials
  7 Petals white or pinkish and more than 1 cm. long *(Anogra)*
    8 Hypanthium and sepals glandular-pubescent; stems 4-10 dm. tall   O. NUTTALLII
    8 Hypanthium not glandular-pubescent; stems often only 1-2 dm. tall
      9 Basal leaf blades usually at least 1 cm. broad, rhombic to oblanceolate-ovate; plant
        an annual usually with several prostrate or decumbent basal branches   O. DELTOIDES
      9 Basal leaf blades mostly less than 1 cm. broad, linear to elliptic or linear-lanceo-
        late; plant a rhizomatous perennial with a simple or basally branched stem   O. PALLIDA
  7 Petals either yellow or less than 1 cm. long
    10 Hypanthium usually at least 2 cm. long; petals yellow; stigmas distinct, linear *(Onagra)*

 11 Petals at least 2.5 cm. long
    12 Flowers very pale yellow; calyx red; leaves crinkled, blades of the cauline leaves
        and floral bracts over 1/2 as broad as long; N=7   *O. erythrosepala**
    12 Flowers yellow to orange; calyx often not reddish; leaves not crinkled; blades usu-
        ally less than 1/3 as broad as long   O. HOOKERI
 11 Petals less than 2.5 cm. long   O. BIENNIS
10 Hypanthium much less than 2 cm. long; petals white (pink) or yellow; stigma from capitate to discoid or shallowly lobed
    13 Plant perennial; petals 8-25 mm. long *(Meriolix)*   O. SERRULATA
    13 Plant annual; petals often less than 8 mm. long
        14 Capsules slender-pedicellate, not attenuate; petals yellow *(Chylismia)*
            15 Petals not over 4 mm. long   O. SCAPOIDEA
            15 Petals mostly well over 4 mm. long   O. CLAVAEFORMIS
        14 Capsules sessile or subsessile (at least not slender-pedicellate), usually attenuate
          and often contorted; petals often white *(Sphaerostigma)*
            16 Capsule 4-15 mm. long, linear-fusiform to fusiform-lanceolate, usually not-
              iceably thickened at the base, straight or curved (but not conspicuously ar-
              cuate) in the upper half
              17 Petals yellow; capsules 4-8 mm. long   O. ANDINA
              17 Petals white; capsules 8-15 mm. long   O. BOOTHII
            16 Capsule usually over 15 mm. long, linear, scarcely at all thickened at the
              base, from conspicuously contorted or coiled to merely curved or arched
              18 Petals white, at least 3 mm. long; floral bracts much smaller than the
                  lower leaves   O. ALYSSOIDES
              18 Petals yellow or if whitish then less than 3 mm. long; floral leaves not
                greatly reduced in size
                  19 Petals 1-2 mm. long; capsules 12-25 mm. long   O. MINOR
                  19 Petals 2.5-4 mm. long; capsules 20-40 mm. long   O. CONTORTA

Oenothera alyssoides H. & A. Bot. Beechey Voy. 340. 1838.
  Sphaerostigma alyssoides Walpers Rep. 2:78. 1843. *(Douglas,* Pine Creek, Snake Country)
  Sphaerostigma implexa A. Nels. Bot. Gaz. 52:267. 1911. *(Macbride 27,* Falk's Store, Canyon Co.,
    Ida., May 17, 1910)
  Grayish-strigillose annual 5-30 (40) cm. tall, usually with several simple to sparingly branched,
leafy stems; leaves narrowly lanceolate to ovate-lanceolate or oblanceolate, (1) 2-8 cm. long, (2) 5-
20 mm. broad, narrowed to slender petioles, the upper ones nearly linear; flowers many, sessile in
short, crowded, secund spikes that elongate to as much as 10-15 cm. in fruit, leafy-bracteate below,
the bracts gradually reduced and linear above; free hypanthium 3-7 mm. long; sepals usually connate
and turned to one side; petals white, rhombic-ovate, (3.5) 4-5 mm. long; stamens conspicuously exserted, the anthers versatile, narrowly oblong, about 1.5 mm. long; stigma capitate; capsule linear,
slightly thickened at the base, usually from conspicuously contorted to slightly coiled, 12-20 mm. long.
N=7.
  Sagebrush plains and dry foothills, usually where very sandy; e. Oreg. to c. Ida., s. to Utah and s.
Calif. June.

Oenothera andina Nutt. in T. & G. Fl. N.Am. 1:512. 1840.
  Sphaerostigma andinum Walpers Rep. 2:79. 1843. *(Nuttall,* "Dry plains in the Rocky Mountains, near
    the Black-Foot River")
  O. hilgardi Greene, Bull. Torrey Club 10:41. 1883. *Sphaerostigma hilgardi* Small, Bull. Torrey Club
    23:188. 1896. *Sphaerostigma andina* var. *hilgardi* A. Nels. Bot. Gaz. 40:56. 1905. *O. andina* var.
    hilgardii Munz, Bot. Gaz. 85:251. 1928. *(E. W. Hilgard,* Klickitat Swale, Washington Territory,
    July, 1882)
  Finely strigillose annual, often broader than tall, usually with several to many stems, the branches

---

  *According to Munz *O. erythrosepala* Borb. (Magyar Bot. Lapok. 2:245. 1903) is an escape along
the w. coast from Wash. to Calif. It is not a natural species but a plant of hybrid origin which is frequently cultivated.

ascending, simple to freely forked, 3-15 cm. long; leaves alternate, linear to linear-spatulate, 5-25 mm. long, 0.5-2 mm. broad, mostly basal, the lower branches nearly leafless; flowers sessile in short, crowded, linear-bracteate spikes; free hypanthium 0.5-2 mm. long; sepals separately reflexed; petals lemon yellow, ovate, 1-4.5 mm. long; filaments short; anthers about 0.5 mm. long, oval; stigma capitate, the style usually exserted; capsules nearly straight, fusiform-lanceolate, thickened basally, 4-8 mm. long. N=14.

Dry fields and sagebrush scabland into the lower foothills; s. B.C. southward, e. of the Cascades, to Calif., e. to Wyo. and Utah. May-July.

Throughout nearly all the range the plants have small flowers with petals scarcely 1.5 mm. long, but in e. Wash., especially in the Yakima region, there occurs a larger flowered phase, with petals as much as 4.5 mm. long. This race is usually designated as var. *hilgardii* (Greene) Munz, in contrast to the smaller flowered var. *andina*. The two appear to interbreed, as individuals with flowers intermediate in size occur.

Oenothera biennis L. Sp. Pl. 346. 1753. (Europe)
  Oenothera depressa Greene, Pitt. 2:216. 1891. *Onagra depressa* Small, Bull. Torrey Club 23:170. 1896. (Cultivated at Berkeley, Calif., from seed collected by Blankinship, from Custer, Mont.)
  Onagra strigosa Rydb. Mem. N.Y. Bot. Gard. 1:278. 1900. *Onagra biennis* var. *strigosa* Piper in Piper & Beattie, Fl. Palouse Region 124. 1901. *Oenothera strigosa* Mack. & Bush, Man. Jackson Co. Mo. 139. 1902, but not *Oenothera strigosa* Willd. ex Spreng. in 1825. (*Rydberg & Bessey 4584*, Pony, Mont., July 8 and 12, 1897, is the first of several collections cited)
  Onagra strigosa var. *subulata* Rydb. Mem. N.Y. Bot. Gard. 1:279. 1900. *Onagra subulifera* Rydb. Bull. Torrey Club 40:66. 1913, not *Oenothera subulata* of R. & P. in 1794. *Oenothera rydbergii* House, N.Y. State Mus. Bull. no. 233-234:61. 1921. (*Rydberg & Bessey 4588*, Forks of the Madison, Mont., July 26, 1897)
  Oenothera cheradophila Bartlett, Bot. Gaz. 44:302. 1907. (*Suksdorf 5860*, Bingen, Wash., Aug. 20, 1906)
  Oenothera muricata sensu Rydb. Fl. Rocky Mts. 594. 1917.

Erect, simple or freely branched biennial or short-lived perennial 3-10 dm. tall, usually grayish-strigose and with a mixture of longer, spreading, reddish- and pustular-based hairs; lower leaves petiolate, the upper ones subsessile, lanceolate, the largest ones 1-2.5 cm. broad, entire to undulate-dentate; bracts from shorter to longer than the mature capsules; inflorescence spicate, the buds erect; flowers vespertine; free hypanthium 3-5 cm. long; sepals 10-15 mm. long, their free tips 1-3 mm. long in the bud; petals yellow, 1-2 cm. long; stamens subequal to the petals and style, the anthers 4-6 mm. long; stigma lobes linear, 4-7 mm. long; capsule 2.5-4 cm. long, linear-fusiform, terete. N=7.

Chiefly in meadows and along stream banks, from the plains to the lower mountains, throughout most of the N.W., to e. U.S. June-Aug.

The plants described above usually have been called either *O. strigosa* or *O. rydbergii*, and have been said to differ from *O. biennis* as follows:
1 Free tips of the sepals about 2 mm. long; plant grayish-strigose throughout   *O. rydbergii*
1 Free tips of the sepals about 3 mm. long; plants with mixed pubescence, the longer hairs
    with reddish-pustular bases   *O. biennis*
Our material does not separate on such a basis, as it practically all has pubescence that is reddish-pustular.

Oenothera boothii Dougl. ex Lehm. in Hook. Fl. Bor. Am. 1:213. 1833.
  Sphaerostigma boothii Walpers Rep. 2:77. 1843. (*Douglas*, "near the branches of Lewis and Clarke's River, lat. 46° north")
  O. pygmaea Dougl. ex Lehm. in Hook. Fl. Bor. Am. 1:213. 1833. *O. boothii* var. *pygmaea* T. & G. Fl. N.Am. 1:510. 1840. *Sphaerostigma boothii* var. *pygmaeum* Walpers Rep. 2:78. 1843. (*Douglas*, "interior of North-West America, near the Utalla River")
  Sphaerostigma senex A. Nels. Proc. Biol. Soc. Wash. 18:173. 1905. (*True 750*, Pyramid Lake, Calif., June 9, 1903)
  Sphaerostigma lemmonii A. Nels. Bot. Gaz. 40:61. 1905. (*J. G. Lemmon 103*, Sierra Nevada, Calif., in 1875)

Annual, finely glandular-puberulent to finely strigillose (especially above), simple or branched below and usually sparingly branched above, 1-4 dm. tall; leaves alternate, the blades lanceolate-ovate to oblanceolate, entire to dentate, (1) 2-6 (8) cm. long, the lower ones with petioles nearly as long, the upper ones reduced to linear-lanceolate, sessile bracts; flowers sessile in elongate, crowded, many-flowered spikes; free hypanthium 2-8 mm. long; sepals 1.5-7 mm. long, distinct; petals white, aging to pinkish, 1.5-8 mm. long; stamens about equaling the petals, the anthers subequal, versatile, 0.5-1 mm. long, slightly exceeded by the style; stigma globose; capsule linear-fusiform, 8-15 mm. long, somewhat curved in the upper half but not greatly contorted. N=7.

Sagebrush and lower foothills from the e. base of the Cascades, in Wash., to Calif. and Nev. June-July.

In the northern part of its range a smaller-flowered plant with the petals only 1.5-2.5 mm. long, the var. *pygmaea* (Dougl.) T. & G., is occasionally to be found along with the more wide ranging var. *boothii,* which is on the average a more glandular plant with petals usually over 3 mm. in length.

Oenothera breviflora T. & G. Fl. N.Am. 1:506. 1840.
Taraxia breviflora Small, Bull. Torrey Club 23:185. 1896. *(Nuttall,* "Plains of the Rocky Mountains")

Caespitose, acaulescent, more or less densely strigillose perennial from a long taproot; leaves (3) 5-15 cm. long, the blades lanceolate to oblanceolate, 5-15 mm. broad, deeply sinuate-pinnatifid, narrowed to slender petioles 1/2 to 2/3 as long; free hypanthium 10-25 mm. long, the flared upper portion 1.5-2 mm. long; petals yellow, not aging to purplish, 6-8 (10) mm. long; stamens unequal, the longer set slightly more than half the length of the petals; style about equaling the petals; stigma globose, very slightly lobed; capsule coriaceous, hairy, oblong-fusiform, 1-2.5 cm. long, usually curved above.

Drier meadowland and stream banks, from e. Oreg. to n.e. Calif., e. to Sask., Mont., and Wyo. May-July.

Oenothera caespitosa Nutt. in Fraser's Cat. no. 53. 1813.
Pachylophis caespitosa Raimann in Engl. & Prantl, Nat. Pflanzenf. 3^7:215. 1893. *(Nuttall,* banks of the Missouri, from White River to the Mandans)
O. *marginata* Nutt. in H. & A. Bot. Beechey Voy. 342. 1838. *Pachylophus marginatus* Rydb. Bull. Torrey Club 33:146. 1906. O. *caespitosa* var. *marginata* Munz, Am. Journ. Bot. 18:733. 1931. *(Nuttall,* Rocky Mts. in Upper California, about lat. 42°)
O. *montana* Nutt. in T. & G. Fl. N.Am. 1:500. 1840. O. *caespitosa* var. *montana* Durand, Bot. Gr. Salt Lake 164. 1859. *Pachylophus montanus* A. Nels. Bull. Torrey Club 26:128. 1899. *(Nuttall,* "Plains of the Platte in the Rocky Mountains")
O. *marginata* var. *purpurea* Wats. Bot. King Exp. 108. 1871. O. *caespitosa* var. *purpurea* Munz, Am. Journ. Bot. 18:730. 1931. *(Watson 412,* E. Humboldt Mts., Nev.)
O. *idahoensis* Mulford, Bot. Gaz. 19:117. 1894. *(Mulford,* near Boise, Ida., June, 1892) = var. *montana.*
Pachylophus glabra A. Nels. Bull. Torrey Club 31:242. 1904. *(A. Nelson 8340,* Platte River bottoms, near Badger, Wyo., June 1, 1901) = var. *caespitosa.*
Pachylophus hirsutus Rydb. Bull. Torrey Club 31:571. 1904. *(Rydberg,* Georgetown, Colo., in 1895) = var. *montana.*
Pachylophus canescens Piper, Contr. U.S. Nat. Herb. 11:409. 1906. *(Cotton 1345,* Sentinel Bluffs, Wash.) = var. *purpurea.*
Pachylophus cylindrocarpus A. Nels. Bot. Gaz. 47:429. 1909. *(Goodding 960a,* Carson's, Meadow Valley Wash, s. Nev., May 26, 1902) = var. *marginata.*
Pachylophus psammophilus Nels. & Macbr. Bot. Gaz. 61:32. 1916. O. *caespitosa* var. *psammophila* Munz, Am. Journ. Bot. 18:733. 1931. ("sand dunes in the vicinity of St. Anthony," Ida., no collections cited)

Caespitose, acaulescent to short-stemmed perennial rarely as much as 2.5 dm. tall, glabrous to densely pubescent or villous; leaves narrowly to broadly oblanceolate, (5) 10-25 cm. long, 1-2.5 cm. broad, the petioles slender, usually subequal to the runcinate or pinnatifid to remotely toothed or nearly entire blades; flowers sessile to pedicellate, vespertine; free hypanthium slender, somewhat flared above, (3) 5-12 cm. long, greenish or reddish; calyx lobes 2.5-3.5 cm. long, connate or free; petals white, aging to pinkish, obcordate, 2.5-4.5 cm. long; anthers 8-13 mm. long; stigma lobes 5-8 mm. long; capsule woody, oblong-ovoid, 1-4 cm. long, angled but not winged, often tuberculate, sessile or with a pedicel as much as 1 cm. long.

Talus slopes, roadcuts, and dry hills, over much of w. U.S. May-July.

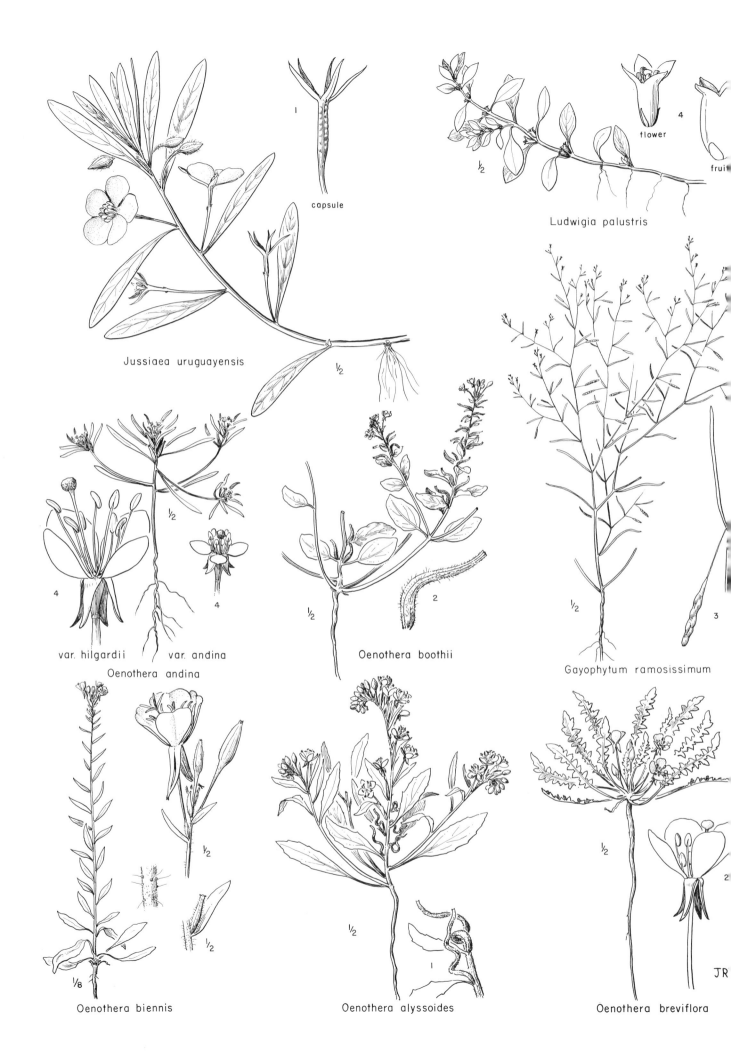

capsule

1

Jussiaea uruguayensis

1/2

Ludwigia palustris

flower

fruit

4

var. hilgardii   var. andina

Oenothera andina

4

4

1/2

Oenothera boothii

2

1/2

Gayophytum ramosissimum

1/2

3

Oenothera biennis

1/8

1/2

1/2

Oenothera alyssoides

1/2

1

Oenothera breviflora

1/2

2

JR

There are numerous variants in the species, some of which, including the following from our area, appear to be recognizable as varieties:
1 Capsules short-pedicellate, straight, 3-4 cm. long, tuberculate; plant usually caules-
  cent, villous-hirsute; from the Columbia R. , e. Wash. , s. to Calif. , e. to Ida. , and
  s.e. to Colo.   var. marginata (Nutt.) Munz
1 Capsules sessile, usually curved, mostly less than 3 cm. long, often smooth; plant
  often acaulescent
    2 Plant nearly or quite glabrous
        3 Capsules tuberculate on the angles, 1-2 cm. long; plant acaulescent; e. Oreg.
          eastward, through Mont. and Wyo. , to the Dakotas   var. caespitosa
        3 Capsules not at all tuberculate, well over 2 cm. long; plant caulescent; Fremont
          Co. , Ida.   var. psammophila (Nels. & Macbr.) Munz
    2 Plant pubescent, often villous
        4 Capsules tubercled, pubescent; plant canescent with appressed pubescence; e.
          Wash. s. to Nev.   var. purpurea (Wats.) Munz
        4 Capsules not tubercled, usually glabrous; leaves strigillose to villous, at least on
          the veins; e. Oreg. to Mont. and Ida.   var. montana (Nutt.) Durand

Oenothera clavaeformis Torr. & Frem. in Frem. Rep. 314. 1845.
  Chylisma clavaeformis Heller, Muhl. 2:105. 1906. *O. scapoidea* var. *clavaeformis* Wats. Bot. King
    Exp. 109. 1871. *Chylisma scapoidea clavaeformis* Small, Bull. Torrey Club 23:194. 1896. *(Fre-
    mont's Pac. R. R. Exp., 1843-44, Mojave Desert)*
  O. cruciformis Kell. Proc. Calif. Acad. Sci. 2:227, fig. 71. 1863. *Chylisma cruciformis* Howell,
    Fl. N.W. Am. 233. 1898. *Chylisma scapoidea cruciformis* Small, Bull. Torrey Club 23:193. 1896.
    (Steamboat Spgs. , Nev.)
Glabrous to canescent-puberulent annual 1-3 dm. tall; stems simple or with a few basal branches,
erect or ascending; leaves almost entirely basal, simple to lyrate-pinnatifid, the blade (or main lobe
of the blade) ovate to ovate-lanceolate or lanceolate-elliptic, (1) 2-6 cm. long, more or less coarsely
dentate, the petiole slender, usually exceeding the blade; racemes many-flowered, congested, elongating and somewhat open in fruit, the bracts greatly reduced, linear; pedicels 10-20 mm. long; hypanthium 3-5 mm. long; petals yellow, often flecked with purplish-red, 4-10 mm. long; stamens about
equaling the petals, the anthers linear, about 3 mm. long; style equaling or slightly exceeding the petals; stigma globose; capsule pedicellate, erect, linear, 10-20 mm. long, about 2 mm. thick; seeds
1.2 mm. long. N=7.
  Sandy desert soil; e. Oreg. to s. Calif. , e. to Ida. and Utah. May-July.
  Oenothera clavaeformis is very similar to *O. scapoidea* (differing chiefly in flower size) and sometimes is treated as a variety of it.

Oenothera contorta Dougl. ex. Hook. Fl. Bor. Am. 1:214. 1833.
  Sphaerostigma contortum Walpers Rep. 2:78. 1843. *(Douglas, "interior banks of the Columbia Riv-
    er")*
  O. parvula Nutt. in T. & G. Fl. N.Am. 1:511. 1840. *Sphaerostigma parvulum* Walpers Rep. 2:78.
    1843. *(Nuttall, "Plains of the Rocky Mountains toward Lewis' River")*
  Sphaerostigma filiforme A. Nels. Bot. Gaz. 40:57. 1905. ("New River [Reese's River], Utah, May
    28, 1889. Collector not known")
  Sphaerostigma contortum var. *flexuosum* A. Nels. Bot. Gaz. 40:58. 1905. *Sphaerostigma flexuosum*
    Rydb. Fl. Rocky Mts. 601. 1917. *O. contorta* var. *flexuosa* Munz, Bot. Gaz. 85:253. 1928. *(A.
    Nelson 4060,* Point of Rocks, Wyo. , June 16, 1898)
Low annual, simple to (more commonly) branched and often sprawling, strigillose or spreading-
pubescent to glandular-puberulent, 5-15 (40) cm. tall; leaves linear to linear-lanceolate or linear-
oblanceolate, entire to remotely denticulate, (5) 10-30 mm. long, the floral bracts similar or slightly reduced; flowers sessile or short-pedicellate; free hypanthium 1.5-2 mm. long; petals yellow, 2.5-
4 mm. long; stamens markedly unequal, the longer set nearly twice the length of the shorter; capsule
linear, scarcely 1 mm. thick, 2-4 cm. long, conspicuously arched to nearly coiled. N=7 (ours).
  Sandy soil along the coast; B. C. to Baja Calif. , inland along the rivers to the Rocky Mts. May-June.
  Our material is of two kinds, separable as follows:
1 Capsules sessile, beaked; widespread, with the range of the species as a whole   var. contorta
1 Capsules short-pedicellate, not beaked; with the var. *contorta* in much of the

range, but in our area not found along the coast; N=7   var. flexuosa (A. Nels.) Munz

Oenothera deltoides Torr. & Frem. in Frem. Rep. 315. 1845. *(Fremont,* no number or locality mentioned)

O. *deltoides* var. *piperi* Munz, Am. Journ. Bot. 18:314. 1931. *(Cusick 2566,* Man's Lake, e. Oreg., June 13, 1901)

More or less caespitose annual or (at most) biennial, often with many basal spreading branches 5-15 cm. long, strigillose to puberulent, long-hirsute, or sublanate; leaves nearly all basal or near the tips of the branches, the blades mostly oblanceolate-obovate, (2) 3-7 cm. long, subentire to (often) sinuate-pinnatifid, narrowed to slender petioles usually at least as long, the upper leaves often subsessile; buds nodding; flowers sessile, largely nocturnal; free hypanthium 2-5 cm. long, lanate to strigillose; sepals 10-30 mm. long, usually purplish, distinct; petals (1) 2-4 cm. long, rounded to obcordate, white, aging to pink or purplish; stamens 2/3 the length of the petals, the anthers 5-9 mm. long; style about equaling the stamens; stigma lobes linear, 4-6 mm. long; capsule woody, 2-7 cm. long, 3-5 mm. thick, tapered to the tip; seeds 1.5-2 mm. long. N=7.

A widespread desert plant of w. U.S., usually occurring on sandy soil from e. Oreg. to s. Calif., e. to Utah and Ariz. May-July.

The species is not common in our area, although one variety, var. *piperi* Munz, has been collected as far n. as Crook Co., Oreg.

Oenothera flava (A. Nels.) Garrett, Spring Fl. Wasatch Reg. 4th ed. 106. 1927.
Lavauxia flava A. Nels. Bull. Torrey Club 31:243. 1904. *(A. Nelson 219,* Laramie, Wyo., June, 1894)

Caespitose, acaulescent, rather sparsely strigillose to subglabrous perennial; leaves in tufts or rosettes, oblanceolate, 5-20 cm. long, up to 15 mm. broad, deeply runcinate or runcinate-pinnatifid on the lower third, the terminal lobe entire to undulate-dentate; flowers sessile among the leaves, vespertine; calyx usually purplish, the free hypanthium slender, 2-12 cm. long; calyx lobes free or slightly connate and turned to one side; petals yellow, aging to purplish, 10-20 mm. long; anthers linear, 4-8 mm. long; stigma lobes linear, about 3 mm. long; capsule woody, ovoid, 10-20 mm. long, conspicuously 4-winged.

Usually in hard-packed soil in swales or around vernal pools in the plains and lower foothills; Sask. to Mex., w. to Ida. and Calif., and along the Yakima R. in Wash. July-Aug.

Oenothera heterantha Nutt. ex T. & G. Fl. N.Am. 1:507. 1840.
Taraxia heterantha Small, Bull. Torrey Club 23:185. 1896. *(Wyeth,* "sources of the Oregon")
Jussieua subacaulis Pursh, Fl. Am. Sept. 304. 1814. *Taraxia subacaulis* Rydb. Mem. N.Y. Bot. Gard. 1:281. 1900. O. *subacaulis* Garrett, Spring Fl. Wasatch Reg. 64. 1911. *(Lewis,* "On the banks of the Missouri") Identity uncertain.
O. *heterantha* var. *taraxacifolia* Wats. Proc. Am. Acad. 8:589. 1873. *(Watson 418,* Austin, Nev.)

Caespitose, acaulescent perennial with a long taproot, from quite glabrous to pubescent on the margins and veins of the foliage; leaves 5-20 cm. long, 1-4 (5) cm. broad, the blades obovate to linear-elliptic or linear-oblanceolate, entire to sinuate or sinuate-dentate, often more deeply lobed to sublyrate-pinnatifid at the base; flowers sessile; free hypanthium (consisting in large part of the sterile upper portion of the ovary) filiform, 3-9 cm. long, persistent on the developing fruits, the flared summit 1-2.5 mm. long; sepals distinct; petals yellow, not aging to purple, 8-13 mm. long, slightly notched; stamens decidedly unequal, about 1/4 and 1/2 the length of the petals, the anthers 1-3 mm. long; style slightly exceeding the stamens; stigma globose-discoid; capsule 12-18 mm. long, coriaceous, 4-angled, glabrous, narrowed above to a slender point. N=7.

Meadows, benchland, and stream banks, from the sagebrush plains to medium elevations in the mountains, usually where dry by late summer; e. Wash. to the Sierras of Calif., e. to Mont. and Colo. Apr.-July.

Oenothera hookeri T. & G. Fl. N.Am. 1:493. 1840.
Onagra hookeri Small, Bull. Torrey Club 23:171. 1896. *(Douglas,* California)
Onagra macbrideae A. Nels. Bot. Gaz. 52:269. 1911. *Oenothera macbrideae* Heller, Muhl. 9:68. 1913. *(Macbride 473,* Twilight Gulch, Owyhee Mts., Ida., July 27, 1910) = var. *angustifolia.*
Onagra ornata A. Nels. Bot. Gaz. 52:268. 1911. *Oenothera ornata* Rydb. Bull. Torrey Club 40:66.

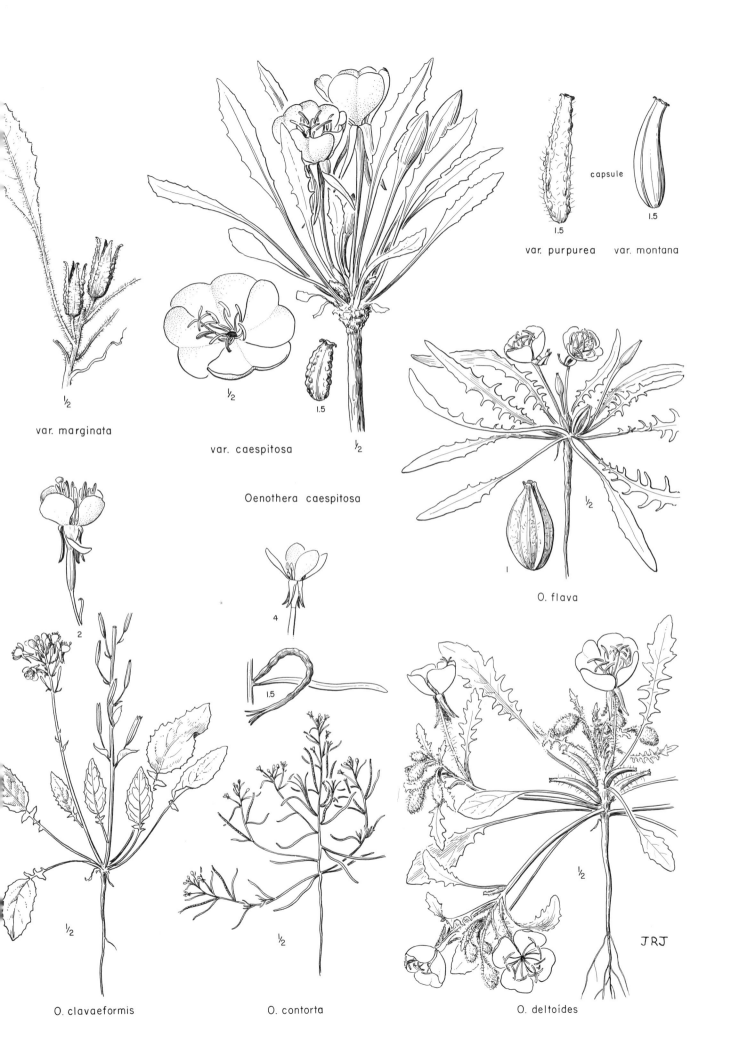

capsule

var. purpurea   var. montana

1.5   1.5

var. marginata

var. caespitosa

Oenothera caespitosa

O. flava

O. clavaeformis

O. contorta

O. deltoides

JRJ

1913. *Oenothera hookeri* ssp. *ornata* Munz, El Aliso 2:25. 1949. *Oenothera hookeri* var. *ornata*
Munz, loc. cit. *(Macbride 262*, near Boise, Ida., June 18, 1910)
Oenothera hookeri var. *angustifolia* Gates, Mut. Fact. in Evol. 10, 30. 1915. *Oenothera hookeri*
ssp. *angustifolia* Munz, El Aliso 2:26. 1949. *(M. E. Jones 5624*, Asphalt, Utah Co., Utah)

Erect, simple to sparingly branched biennial or perennial as much as 1 m. tall, usually spreading-
pubescent, the hairs often of 2 lengths, the longer ones basally reddish-pustulate; leaves numerous,
spatulate or oblanceolate, 5-20 cm. long, 1-2.5 cm. broad, entire to denticulate, the lower ones
narrowed to a short petiole, the upper ones subsessile; bracts from shorter to longer than the capsules; free hypanthium 3-4 cm. long; sepals 20-25 (30) mm. long, their free tips 2-5 mm. long in the
erect buds; petals yellow (usually aging to orange or reddish-purple), 2.5-4 cm. long; stamens subequal and over half the length of the petals, the anthers up to 1 cm. long; stigma lobes linear, about 5
mm. long; capsule nearly terete, 2-5 cm. long, linear-lanceolate in outline, usually shaggy-pubescent.

Widespread in w. U.S., from w. Ida. and adj. s.e. Wash. and n.e. Oreg. to Utah, Colo., N.M.,
s.w. Tex., n. Mex., much of Calif., and n. Baja Calif., also in the Columbia R. Gorge; from the
plains to middle elevations in the mountains, in varied habitats. June-Sept.

This is an extremely variable and complex species that has proven notably difficult to treat taxonomically; our material has been referred by Munz to two of many infraspecific variants:

1 Sepals greenish; corolla yellow to orange, not turning red with age; s.e. Wash. to c.
    and s.w. Ida., particularly common near Boise   var. ornata (A. Nels.) Munz
1 Sepals reddish; corolla yellow, but turning red with age; widespread in the lower mts.
    from s.e. Oreg. to e. Calif., e. to Nev., Utah, and Colo.   var. angustifolia Gates

Oenothera minor (A. Nels.) Munz, Bot. Gaz. 85:238. 1928.

Sphaerostigma minor A. Nels. Bull. Torrey Club 26:130. 1899. *Sphaerostigma nelsonii* Heller,
Muhl. 1:1. 1900. *(A. Nelson 3047*, cliffs bordering the Green and Platte rivers, Wyo., May 31,
1897)
O. alyssoides var. *minutiflora* Wats. Bot. King Exp. 111. 1871. *Sphaerostigma alyssoides* var.
minutiflorum Small, Bull. Torrey Club 23:192. 1896. *Sphaerostigma minutiflora* Rydb. Bull. Torrey Club 33:146. 1906, not *O. minutiflora* of D. Dietr. in 1840. *(Watson 421*, Monitor Valley, Nev.,
and Black Rock Point and Stansbury I., Salt Lake, Utah) = var. *minor*.
O. minor var. *cusickii* Munz, Bot. Gaz. 85:240. 1928. *(Cusick 2545*, Malheur River, Oreg., June
6, 1901)

Strigillose to spreading-pubescent and usually more or less glandular annual 5-25 cm. tall, the
main stem usually branched at the base but simple above; basal leaves long-petiolate, the blades oblong-lanceolate to oblanceolate, 1-3 (5) cm. long, 4-15 mm. broad, reduced gradually upward to linear sessile bracts in the upper part of the spikes; flowers sessile, the branches floriferous nearly to
the base; free hypanthium 0.5-2 mm. long; sepals 1-2 mm. long; petals obovate, pale yellowish or
off-white, usually aging to pink, 1-2 mm. long; filaments of 2 quite unequal lengths, the longer stamens about equaling the petals; anthers versatile, ovoid, 0.6-1 mm. long; stigma globose; style
slightly surpassing the stamens; capsule linear, scarcely enlarged at the base, 12-25 mm. long, usually very conspicuously contorted and often somewhat coiled.

Dry sandy or gravelly soil, often in sagebrush; e. Wash. to n.e. Calif. and to n. Nev., e. to Ida.,
Wyo., and Colo. June-July.

There are two phases of the species, separable as follows:

1 Petals about 1 mm. long; occasional with var. *minor* in Wash. and Oreg.   var. cusickii Munz
1 Petals about 2 mm. long; e. Wash. s. to Nev., e. to Wyo. and Colo.   var. minor

Oenothera nuttallii Sweet, Hort. Brit. 2nd ed. 199. 1830.

O. albicaulis var. *nuttallii* Engelm. Am. Journ. Sci. II, 34:334. 1862. *Anogra nuttallii* A. Nels.
Bot. Gaz. 34:368. 1902. *(Nuttall*, from the Missouri)

Rhizomatous, grayish-green perennial with 1-several simple or (usually) rather freely branched
stems 4-10 dm. tall, the epidermis glabrous, white, usually exfoliating; leaves strigillose, linear to
lanceolate, entire to repand-denticulate, up to 10 cm. long and 12 mm. broad, sessile or with short,
winged petioles, gradually reduced upward to the floral bracts, the lowest of which are longer than
the fruits, the upper ones usually shorter; inflorescence densely glandular-puberulent; buds nodding;
flowers sessile; free hypanthium about 3 cm. long; sepals 2-2.5 cm. long, purplish, connate and
turned to one side in flower; petals white, fading to pink, 1.5-2.5 cm. long; stamens 2/3 the length of
the petals, the anthers about 10 mm. long; style somewhat longer than the stamens, the stigma lobes

linear, 4-6 mm. long; capsule erect, 2-3 cm. long, linear, tapered slightly, 2.5-3 mm. thick; seeds about 2 mm. long.

Grasslands and hillsides; Rocky Mts. from Can. to Colo., e. to Minn., usually if not always on the e. side of the continental divide. July-Aug.

Oenothera pallida Lindl. Bot. Reg. 14:pl. 1142. 1828.
  Anogra pallida Britt. Mem. Torrey Club 5:234. 1894. *(Douglas,* northwestern North America)
  O. pallida var. *leptophylla* T. & G. Fl. N.Am. 1:495. 1840. *O. leptophylla* Wats. Proc. Am. Acad. 8:602. 1873. *Anogra leptophylla* Rydb. Bull. Torrey Club 40:65. 1913. *(Nuttall,* plains near the source of the Platte) = var. *pallida.*
  O. pallida var. *idahoensis* Munz, Am. Journ. Bot. 18:320. 1931. *(Merrill & Wilcox 869,* 12 mi. w. of St. Anthony, Ida., July 8, 1901)

Glabrous to grayish-strigillose, rhizomatous perennial with whitish, exfoliating bark, the stems from simple to freely branched basally, 1-5 dm. tall, very leafy; leaves linear to linear-lanceolate, 2-6 cm. long, mostly less than 5 (10) mm. broad, entire to serrulate or serrate or even with 1 or more prominent basal toothlike lobes, the blades narrowed gradually to petioles 5-15 mm. long; flowers in very leafy spikes, fragrant, vespertine; buds nodding; free hypanthium glabrous to strigose but not glandular, 1.5-3.5 cm. long, usually pink to purplish; sepals 1-2 cm. long, mostly connate and turned to one side in flower; petals 1.5-3 cm. long, white, aging pinkish, obovate, rounded to retuse; stamens about equaling the petals, the anthers 6-10 mm. long; style about equaling the petals; stigma lobes linear, 4-6 mm. long; capsule linear, 2-3 mm. thick at the base, tapered to the tip, usually arched upward, 1.5-3.5 cm. long. N=7.

Sandy or gravelly soil, mostly on dunes; e. Wash. and Oreg. to Ida., N.M., and Ariz. May-July.

The species is divisible, on the basis of pubescence, into the following two varieties:

1 Plant grayish-strigillose; usually less than 2 dm. tall; sand dunes near St. Anthony, Fremont Co., Ida.
　　　　　　　　　　　　　　　　　　　　　　　　　　　var. idahoensis Munz
1 Plant glabrous or subglabrous, usually at least 2 dm. tall; with the range of the species
　　　　　　　　　　　　　　　　　　　　　　　　　　　var. pallida

Oenothera palmeri Wats. Proc. Am. Acad. 12:251. 1877.
  Taraxia palmeri Small, Bull. Torrey Club 23:184. 1896. *(Palmer,* Arizona, in 1876)

Caespitose annual from a slender taproot, finely strigillose overall, 2-6 cm. tall; leaves linear-lanceolate to linear-oblanceolate, 2-6 cm. long, 2-7 mm. broad, entire to denticulate, narrowed to broad petioles that are flaring at the base; flowers sessile among the basal leaves; free hypanthium (largely sterile ovary) 8-20 mm. long, flared at the summit for 1-2 mm.; petals yellow, 3-5 mm. long, rounded; stamens distinctly unequal, about 1/4 and 1/2 the length of the petals; anthers about 0.5 mm. long; style about equaling the stamens, the stigma globose-discoid; capsule ovate, 5-7 mm. long, coriaceous, 4-angled below, winged above. N=7.

Sandy desert land, from s.e. Oreg. to the Mojave Desert of s. Calif., e. to Nev. and Ariz.; just entering our range in n. Malheur Co., Oreg. Mar.-June.

Oenothera scapoidea Nutt. ex T. & G. Fl. N.Am. 1:506. 1840.
  Chylisma scapoidea Small, Bull. Torrey Club 23:193. 1896. *O. brevipes* var. *scapoidea* Levl. Monog. Onoth. 146. 1905. *(Nuttall,* "Clay hills in the Rocky Mountains")
  Chylisma scapoidea var. *seorsa* A. Nels. Bot. Gaz. 54:140. 1912. *O. scapoidea* var. *seorsa* Munz, Am. Journ. Bot. 15:233. 1928. *(A. Nelson 4125,* Evanston, Wyo., July 27, 1897)

Glabrous to puberulent annual, often glandular in the inflorescence, the stems simple or more commonly several-branched from the base, erect or spreading, (0.5) 1-5 dm. tall; basal leaves mainly 4-8 cm. long, the blades ovate to ovate-lanceolate, 1-3 cm. broad, sinuate-denticulate, about equaled by the slender petioles, sometimes with 2-several broadish, short pinnules below the main blade; cauline leaves similar but smaller, abruptly reduced to the tiny linear bracts of the elongate, loosely flowered, simple or sometimes compound racemes; pedicels 5-20 mm. long; free hypanthium flared abruptly above the ovary, 2-3 mm. long; sepals distinct, reflexed; petals 2-4 mm. long, lemon to golden yellow, usually maculate with small, dark reddish-purple spots on the lower half; stamens slightly shorter than the petals; style about equaling the stamens; stigma discoid, very shallowly lobed; capsule pedicellate, usually erect, straight to slightly curved, 15-25 mm. long, 2-2.5 mm. thick; seeds about 1.7 mm. long.

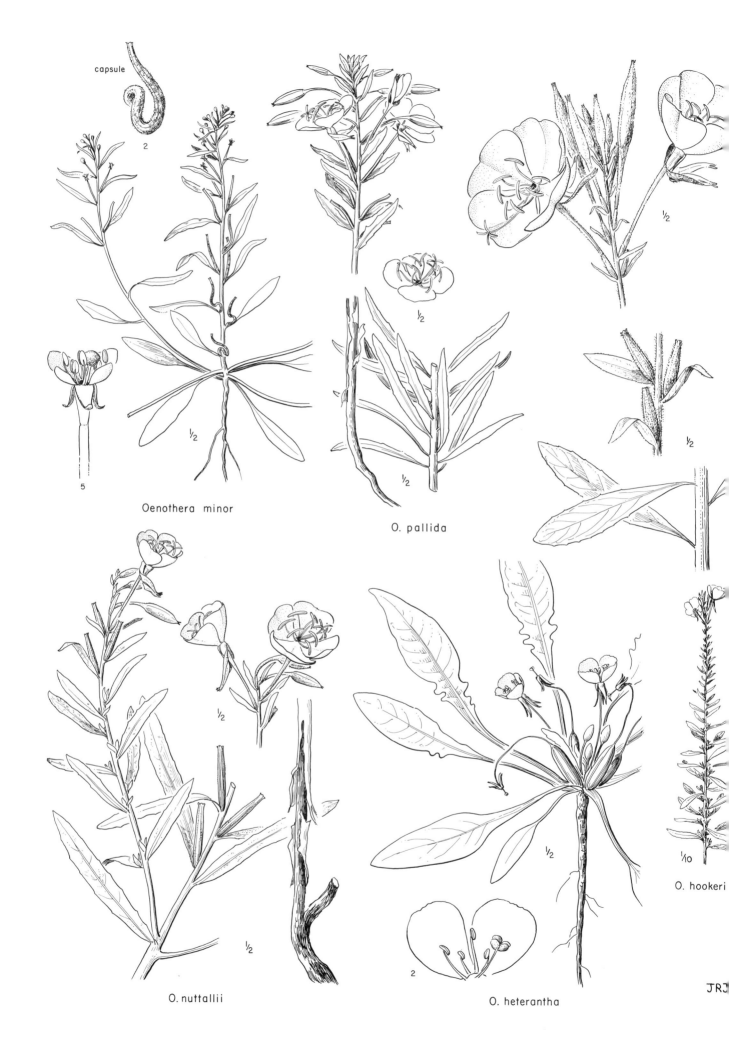

capsule

2

5

Oenothera minor

½

O. pallida

½

½

½

O. nuttallii

2

O. heterantha

½

⅒

O. hookeri

JRJ

Mostly in the sagebrush desert, especially on rocky or sandy soil in our area; e. Oreg. through s. Ida. to Wyo., s. to Colo. May-July.

Our plants, as described above, are usually referred to var. *seorsa* (A. Nels.) Munz.

Oenothera serrulata Nutt. Gen. Pl. 1:246. 1818.

Meriolix serrulata Walpers Rep. 2:79. 1843. *(Nuttall,* "From the river Platte to the mountains")

O. leucocarpa Lehm. in Hook. Fl. Bor. Am. 1:210. 1833. *O. serrulata* var. *douglasii* T. & G. Fl. N.Am. 1:502. 1840. *(Drummond,* "Dry banks of the Saskatchawan")

Canescent to sparsely strigillose, suffrutescent perennial 2-6 dm. tall, usually many-stemmed from the base; leaves linear to oblanceolate, (1) 2-5 cm. long, 3-10 mm. broad, remotely serrulate to serrate, usually with fascicles of smaller leaves in their axils; flowers sessile in the axils of scarcely reduced bracts; free hypanthium 6-15 mm. long, flared almost immediately above the ovary; sepals distinct, spreading to reflexed, keeled on the back; petals yellow, 8-25 mm. long; stamens scarcely half the length of the petals; style not exceeding the stamens; stigma discoid, very shallowly lobed; capsule linear, 15-30 mm. long. N=7.

Prairies, dry fields, and sand dunes; Man. to Alta., southward, almost entirely on the e. side of the Rockies, to Tex. and Ariz.; reported in our area from Bozeman, Mont. May-July.

Oenothera tanacetifolia T. & G. Pac. R. R. Rep. 2:121, pl. 4. 1855.

Taraxia tanacetifolia Piper, Contr. U.S. Nat. Herb. 11:405. 1906. *(Beckwith* [probably collected by J. A. Snyder], "On the higher parts of the Sierra Nevada, June 18, 1854)

O. nuttallii T. & G. Fl. N.Am. 1:506. 1840, but not of Sweet in 1830. *Taraxia longiflora* Nutt. ex Small, Bull. Torrey Club 23:185. 1896. *(Nuttall,* "Plains in the Rocky Mountains, near Blackfoot River")

Taraxia tikurana A. Nels. Bot. Gaz. 54:140. 1912. *(Nelson & Macbride 1302,* Tikura, Blaine Co., Ida.)

Caespitose, acaulescent, sparsely to canescently strigillose to hirsute perennial apparently from a taproot, but really from deepset, widely branching rootstocks, ultimately forming extensive patches; leaves 5-20 cm. long, 1-3.5 cm. broad, the blades narrowly to broadly lanceolate to oblanceolate, deeply sinuate-pinnatifid, narrowed to slender petioles from 1/2 to nearly as long; free hypanthium very slender, 2.5-9 cm. long, the flared upper portion 2-3.5 mm. long; petals yellow, sometimes aging to purplish, 10-16 mm. long, rounded; stamens unequal, the longer ones about half the length of the petals; style subequal to the petals; stigma globose; capsule 15-20 mm. long, coriaceous, oblong-ovoid, 4-sided and slightly winged, usually pubescent. N=21.

Moist soil (usually drying during the summer) of meadows, swales, and riverbanks in sagebrush plains to ponderosa pine forest; Wash. southward, on the e. side of the Cascades, to the Sierra Nevada of Calif., e. to Ida. and Mont. June-Aug.

HALORAGIDACEAE Water Milfoil Family

Flowers single in the axils (ours) to more or less paniculate, often on a fleshy axis, usually very small, regular, epigynous, mostly imperfect and the plants monoecious or polygamo-monoecious; free hypanthium poorly or not at all developed, the calyx segments 2 or 4 (lacking); petals lacking or 2 or 4 and small and soon deciduous; stamens commonly 4 or 8, the anthers basifixed; pistil usually 4-carpellary, the ovary 1- to 4-locular; ovules 1 per locule, pendulous; styles as many as the locules, short; stigmas often plumose; fruit drupaceous (nutletlike), the carpels usually separating as the fruit matures; ours aquatic perennial herbs with whorled, pinnately dissected leaves, other members varied, but usually aquatic or paludose herbs (somewhat shrubby) with alternate, opposite, or whorled and sometimes very large, mostly stipulate leaves.

A rather small but nearly cosmopolitan family of 8 genera and perhaps 100 species; often including the Hippuridaceae.

Myriophyllum L. Water Milfoil

Flowers mostly single in the axils of foliage or bracteal leaves, minute, imperfect, the staminate usually uppermost along the stems, each generally subtended by 2 tiny bracteoles; calyx adherent to the ovary, its 4 lobes small; petals 4, usually quickly deciduous (or lacking); stamens 4 or 8; pistil 4-carpellary, the stigmas plumose or nearly smooth, the carpels separating into 1-seeded achenes or nutletlike segments; perennial aquatic herbs with whorled (ours) leaves, the emersed leaves pectinately dissected to subentire and reduced, the submerged leaves filiformly pinnatifid and exstipulate to minutely stipulate.

Cosmopolitan, with between 30 and 40 species. (Greek *myrios*, thousand, and *phyllon*, leaf.)
Reference:

Patten, Bernard C., Jr. The status of some American species of Myriophyllum as revealed by the discovery of intergrade material between *M. exalbescens* Fern. and *M. spicatum* L. in New Jersey. Rhodora 56:213-225. 1954.

1 Leaves with 25 or more segments, those subtending flowers unmodified, all usually
    submersed   M. BRASILIENSE
1 Leaves usually with less than 25 segments, those subtending flowers modified, usually
    considerably reduced in size, and often emersed
    2 Bracteal leaves linear to narrowly oblong, 10-15 mm. long, serrate; petals tardily
        deciduous; stigmas very short, nearly smooth; stamens 4   M. HIPPUROIDES
    2 Bracteal leaves usually either more or less ovate or much less than 10 mm. long,
        sometimes dissected; petals quickly deciduous; stigmas plumose; stamens 8
        3 Inflorescence often forking; bracts oblong-ovate, 7-10 mm. long, pectinate to
            serrulate; bracteoles 1-2 mm. long, whitish, erose to pectinate   M. ELATINOIDES
        3 Inflorescence simple; bracts various, but usually considerably less than 7 mm.
            long; bracteoles 1-1.5 mm. long, usually greenish with whitish, entire to minutely
            erose margins   M. SPICATUM

Myriophyllum brasiliense Camb. in St. Hilaire, Fl. Bras. Merid. 2:252. 1829. (Brasil)
    M. proserpinacoides Gillies ex H. & A. in Bot. Misc. 3:313. 1833. (S.Am.)
    Stems sturdy, 2-4 mm. thick and 1-9 dm. long, usually completely submersed until the time of flowering; leaves in whorls of 4-6, all finely pinnatifid, 2.5-5 cm. long, the (21) 25-37 divisions filiform; stipules linear; flowers in the axils of submersed leaves along the main stem, with minute, linear to filiformly dissected bracteoles, imperfect, the staminate with 4 small petals, the pistillate apetalous; stamens 8; mature ovary 1.5-2 mm. long; stigmas plumose. Parrot's feather.
    A commonly grown S.Am. aquatic plant, introduced in scattered places in other parts of the world; in our area reported chiefly from Ida. May-July.

Myriophyllum elatinoides Gaud. Ann. Sci. Nat. 5:105. 1825.
    Stems 2-10 dm. long, simple or branched; leaves 1-2.5 cm. long, in whorls of 4 or 5, pectinately dissected into 13-21 filiform segments, very gradually transitional to the bracts; bracts conspicuous, ovate-oblong, 7-10 mm. long and from pectinate (the lower) to lightly serrulate; inflorescence simple to conspicuously forked, submersed to emergent; flowers subtended by small, whitish, erose to pectinate bracteoles 1-2 mm. long; sepals whitish, more or less serrate-pectinate; petals fugacious, about 2.5 mm. long; stamens 8, the anthers about 2 mm. long; stigmas plumose; mature carpels smooth, about 2 mm. long.
    A S.Am. and Australian species known in our area only from the Deschutes R., n. of Bend, Oreg. June-July.
    Peck (Man. High. Pl. Oreg. 508. 1941) described this taxon as having 4 stamens, but the plants I have seen have 8. Patten assigned our material (as exemplified by *Hitchcock & Martin 4910*) to *M. spicatum* ssp. *exalbescens* (Fern.) Hult., but the forking inflorescence, large main bracts and pectinate to serrate, whitish bracteoles are all noncharacteristic of that taxon, as was indirectly pointed out by Fernald (Rhodora 21:123. 1919).

Myriophyllum hippuroides Nutt. ex T. & G. Fl. N.Am. 1:530. 1840. *(Nuttall, "Oregon, in ponds of the Wahlamet")*
    Stems usually rather freely branched, 1-6 dm. long; leaves 4-5 per whorl, the submersed ones 1.5-3 cm. long, pinnately dissected into 13-23 filiform segments, and gradually transitional to the linear-oblong, 10-15 mm. long, merely serrate bracts; inflorescence terminal and from immersed to emersed; petals ochroleucous, tardily deciduous, about 1 mm. long; staminate flowers usually uppermost on the stem; stamens 4; stigmas very short, nearly smooth; mature ovary 1.5-2 mm. long, the carpels nearly or quite smooth.
    In ponds and slow-moving streams; Wash. and Oreg., where usually (always?) w. of the Cascades, scattered s. to Mex. and e. to N.Y. and to s.e. U.S. July-Oct.

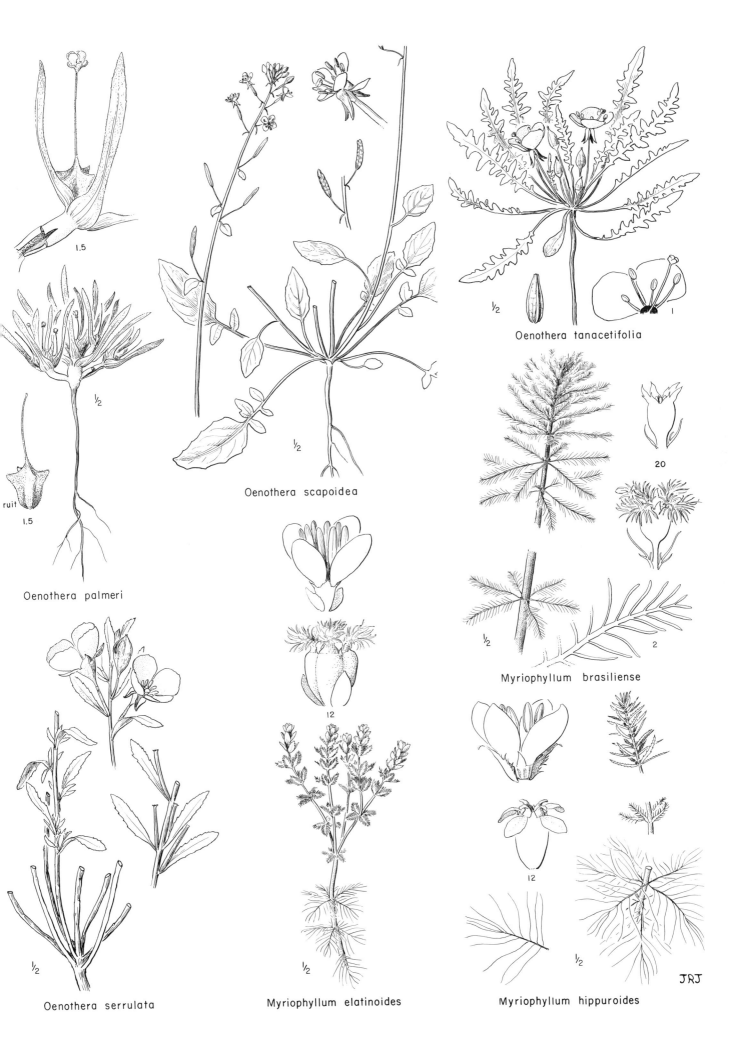

1.5

1/2

fruit

1.5

Oenothera palmeri

Oenothera scapoidea

Oenothera serrulata

1/2

12

Myriophyllum elatinoides

1/2

Oenothera tanacetifolia

1/2

20

Myriophyllum brasiliense

1/2

2

12

1/2

Myriophyllum hippuroides

JRJ

Myriophyllum spicatum L. Sp. Pl. 992. 1753. (Europe)

M. verticillatum L. loc. cit. (Europe)

M. exalbescens Fern. Rhodora 21:120. 1919. *M. spicatum* var. *exalbescens* Jeps. Man. Fl. Pl. Calif. 691. 1925. *M. spicatum* ssp. *exalbescens* Hultén, Fl. Alas. 7:1159. 1947. *(Williams, Collins, & Fernald,* York River, Que., July 29, 1905)

Stems simple or branched; leaves 3 or 4 per node, 1-3 cm. long, pinnately dissected into 13-23 filiform segments, abruptly reduced to the bracteal leaves of the usually emergent inflorescences, the bracts from shorter than the flowers and nearly entire, to considerably longer than the flowers and deeply pectinate; petals quickly deciduous, about 2.5 mm. long; stamens 8, anthers 1.5-2 mm. long; mature achenes about 2.5 mm. long, rounded. N=7, 14.

In quiet streams and ponds, often where strongly brackish; widespread in N.Am., from B.C. southward to Calif., e. to the Atlantic coast; Eurasia. June-Aug.

There has been diversity of opinion as to the status of the plants included in this species, and several workers have called those with the more reduced bracts *M. exalbescens,* and those with longer bracts *M. verticillatum.* On the basis of karyology, this procedure would seem to be the correct one. However, Patten contended that, because of their strong tendency to intergrade, *"spicatum"* and *"verticillatum"* should be regarded as elements of the one species, *M. spicatum.* The two taxa are separable, with due allowance for their intergradiency, as follows:

1 Bracteal leaves equaling or shorter than the fruits, serrulate to entire; N=7; our common form   var. exalbescens (Fern.) Jeps.

1 Bracteal leaves usually exceeding the fruits, pectinate to serrate; N=14; rarely collected in w. U.S., but more common eastward and in Eurasia   var. spicatum

HIPPURIDACEAE Mare's-tail Family

Flowers usually perfect but sometimes imperfect and the plants monoecious; stamen 1; ovary surrounded by the adnate calyx which is very obscurely (if at all) lobed at the tip, a perianth otherwise lacking; pistil 1-carpellary, the style slender, stigmatic the entire length, in perfect flowers usually lying in the groove between the lobes of the comparatively large anther; fruit nutlike, indehiscent, 1-celled and 1-seeded; perennial, glabrous, aquatic or amphibious herbs with creeping rhizomes and entire, linear, sessile, whorled leaves, those of the upper whorls each bearing a single sessile flower.

A small family generally considered to contain only the one genus; undoubtedly related to, and often included in, the Haloragidaceae, a family which differs in usually having toothed to pinnatifid leaves, a more evident perianth, unisexual flowers, and 4 to 8 stamens.

Hippuris L. Mare's-tail

Two or three species of fresh to brackish water in the cooler temperate regions. (Greek *hippos,* horse, and *oura,* tail.)

1 Leaves usually less than 1 mm. broad and 1 cm. long; stems less than 1 mm. thick; flowers mostly imperfect   H. MONTANUS

1 Leaves usually at least 1 mm. broad and over 1 cm. long; stems at least 1 mm. thick; flowers mostly perfect   H. VULGARIS

Hippuris montana Ledeb. in Reichenb. Icon. Bot. 1:71, pl. 86, fig. 181. 1823. (Unalaska)

Glabrous perennial from a slender creeping rhizome, the erect branches simple, scarcely 0.5 mm. thick, 1.5-10 cm. long; leaves 5-8 per whorl, 0.5-1 mm. broad, 5-10 mm. long; flowers nearly all imperfect, the staminate mostly in whorls below the pistillate, but often the two types intermixed, very occasionally a flower perfect; filaments from nearly lacking to several times as long as the 0.5 mm. anthers; mature ovary about 1 mm. long, the slender tapered stigma nearly as long.

Along streams and mossy banks, and in wet to boggy meadows; Olympic and Cascade mts. of Wash. n. to Alas. and e. in B.C. to the Selkirk Mts. July-Sept.

Hippuris vulgaris L. Sp. Pl. 4. 1753. (Europe)

Glabrous perennial with extensive creeping rhizomes, the erect branches mostly partly emersed, 1.5-2.5 mm. thick, 5-30 cm. long, usually simple but frequently with few to many short branches from the lower nodes; leaves (4) 6-12 per whorl, 1-2 mm. broad, (8) 10-35 (50) mm. long; flowers mostly perfect, the mature ovary about 2 mm. long, the filament about 1 mm. long, subequal to the anther. N=16.

Along streams and ponds or in shallow water of lakes, the plants usually at least partially emersed; more or less circumboreal, southward in N. Am. to Calif., N. M., and the states of c. and e. U. S.; Eurasia, and extreme S. Am. June-July.

The somewhat similar species, *H. tetraphylla* L., with 4 (3-6) shorter leaves at a whorl, is common in Alas. and has been reported from B. C., but apparently does not extend southward as far as our area.

ARALIACEAE Ginseng Family

Flowers small, usually more or less greenish, polypetalous, epigynous, perfect or imperfect, generally in globose to capitate umbels borne in corymbs, racemes, or panicles; calyx adnate to the ovary, the 5 lobes nearly obsolete; petals (3-10) 5, valvate to imbricate, spreading to reflexed, usually caducous and often somewhat connate and calyptrate; stamens 5 (ours) or rarely more numerous, the anthers versatile; pistil 5 (4-6)-carpellary, the ovary covered by a fleshy disc, mostly 5 (1-15)-locular with 1 pendulous ovule per locule; styles equal in number to the carpels or only one, sometimes lacking; fruit drupaceous-baccate; seeds with abundant endosperm, the embryo small; perennial, often armed, sometimes climbing, frequently polygamo-dioecious or dioecious herbs, shrubs, or trees with large alternate (ours), opposite, or whorled, exstipulate or stipulate, mostly palmately lobed (rarely subentire) to compound leaves, often stellate.

A large family of tropic (mainly) and temperate regions, with perhaps 60 genera and 800 species.

1 Plants armed with slender spines   OPLOPANAX
1 Plants unarmed
  2 Leaves large, compound, 1 per stem; plants not scandent   ARALIA
  2 Leaves simple, numerous on the scandent stems   HEDERA

Aralia L.

Flowers perfect, borne in paniculate to compound umbels, greenish-white, 5-merous; sepals minute, slightly spreading; petals imbricate in the bud, recurved, often caducous; styles distinct nearly to the base; stigma not at all or only slightly enlarged; ovary completely inferior, 5 (4-6)-celled, the ovules suspended, 1 per locule; fruit a 2- to 5-seeded berry; perennial herbs (ours) to shrubs or trees with large, alternate, pinnately or ternately compound to decompound leaves, often aromatic.

About 25 species of N. Am., Asia, and Australia. (Name, according to Fernald, from the French-Canadian word *"aralie"* under which name the plant, collected in Quebec, was originally sent to Tournefort, who first used the name *Aralia.)*

Besides the following species, *A. californica* Wats., a much taller plant (1-2.5 m. tall), approaches our area in s.w. Oreg.; it is frequently transplanted into woodland gardens in the northwest.

Aralia nudicaulis L. Sp. Pl. 274. 1753. (Hab. in Virginia)

Widely rhizomatous perennial herb with short, erect, woody stems barely reaching the soil surface; leaves generally single, 3-5 dm. long, from a membranous basal scale 1-2 cm. long, the petiole slender, mostly about equaling the usually ternate blade, the long-petiolulate pinnae from again ternate or (rarely) nearly biternate to (usually) more nearly pinnate, the ultimate segments somewhat oblique, ovate-oblong, serrate to doubly serrate, 5-12 cm. long; inflorescence shorter than the leaves, consisting of 3-7 globose, corymbose to paniculate umbels; pedicels slender, 5-10 mm. long; flowers about 5-6 mm. long; petals greenish-white, about 2.5 mm. long; berries dark purple, 6-8 mm. long. N=12. Wild sarsaparilla.

In moist shaded soil; e. B. C. and n. e. Wash. to Mont., s. to Colo., eastward, through the north-central states and Can., to much of the Atlantic coast. May-June.

Hedera L. Ivy

Flowers perfect, 5-merous; sepals minute; petals rather fleshy, greenish; ovary usually 5-celled but the style single; fruit a 2- to 5-seeded berry; evergreen, usually finely stellate, climbing shrubs with aerial supporting roots and alternate, leathery, long-petiolate leaves.

About 6 species of the Old World. (Ancient classical name for the Ivy.)

Hedera helix L. Sp. Pl. 202. 1753. (Europe)

A strongly scandent, stellate-puberulent, evergreen shrub as much as 30 m. tall; leaves leathery, broadly ovate to triangular, usually acuminate, entire to fairly deeply 3- to 5-lobed, 4-10 cm. long; flowers (if any) 5-7 mm. long, borne in inflorescences of (1) several racemose, globose umbels; pedicels 5-15 mm. long; berry deep bluish-black, 6-9 mm. long. N=24. English ivy.

This European plant is very commonly used as an ornamental vine and ground cover; it is freely disseminated by birds and has become more or less permanently established in many places in w. Oreg. and Wash. May-June.

Oplopanax (T. & G.) Miq.

Flowers small, greenish-white, mostly perfect; sepals 5, very small; petals usually 5, valvate in the bud; stamens 5, usually longer than the petals; ovary 2 (3)-celled, styles 2 (3), free, stigmatose nearly their full length; fruit a 2- or 3-seeded berry; thick-stemmed, deciduous, well-armed shrubs with large palmately lobed leaves.

The genus is usually considered to include a single Asiatic species in addition to our plant. (Greek *hoplon,* tool or weapon, and *Panax.*)

Oplopanax horridum (J. E. Smith) Miq. Ann. Mus. Bot. Lugd. Bat. 1:16. 1863.

Panax horridum J. E. Smith in Rees, Cycl. 26:no. 10. 1812. *Echinopanax horridum* Decne. & Planch. Rev. Hort. 4:105. 1854. *Fatsia horrida* Benth. & Hook. Gen. Pl. 1:939. 1865. *Ricinophyllum horridum* Nels. & Macbr. Bot. Gaz. 61:45. 1916. *(Menzies,* Nootka Sound)

Deciduous shrub 1-3 m. tall, with the thick and rather punky stems, petioles, and leaf veins densely armed with yellowish spines 5-10 mm. long; leaf blades (shallowly) palmately 7- to 9-lobed, 1-3.5 dm. broad, irregularly twice serrate, cordate-based, about equaled by the petioles; flowers 5-6 mm. long, subsessile in small capitate umbels borne in elongate racemes or panicles as much as 2.5 dm. long; berry bright red, considerably flattened, 5-8 mm. long. Devil's-club.

In moist woods, especially near streams, from Alas. southward, along the coast and on the w. side of the Cascades, to s. Oreg., e. in n. Wash. and B.C. to Ida. and Mont., also in Mich. and Ont. May-July.

Because it is such a fearsome plant in the uncleared path of the out-of-doorsman, the decided ornamental value of this plant is often overlooked. It merits a place in the garden, in a moist spot, where it can be seen but not necessarily encountered. The fruits are especially attractive.

UMBELLIFERAE Parsley Family

Flowers epigynous, polypetalous, mostly regular and perfect, occasionally some of them sterile or neutral, and then often with somewhat irregular corolla; calyx teeth 5, or obsolete; petals 5, usually inflexed at the tip, typically yellow or white, less often purple or other colors; stamens 5, inserted on an epigynous disk, alternate with the petals; ovary bicarpellate, 2-celled, each cell with a single anatropous ovule; styles 2, often swollen at the base to form a stylopodium; fruit a dry schizocarp, consisting of 2 halves (the mericarps) united by their faces (the commissure), each mericarp typically 5-nerved, usually with one or more vertical oil tubes in the intervals and on the commissure; mericarps separating at maturity, revealing a slender central carpophore to which they are apically attached, the carpophore entire to deeply bifid, or sometimes obsolete by adnation of the separate halves to the commissural faces of the mericarps; seeds with small embryo and copious firm endosperm; annual or perennial herbs, rarely woody at the base, often hollow-stemmed, often or usually aromatic; leaves alternate or rarely opposite, or sometimes all basal, usually with sheathing petiole and large, compound to variously cleft or dissected blade, simple and entire in a few genera; flowers typically in compound umbels, the umbel usually subtended by a few bracts, and the umbellets by bractlets, less often in simple or proliferating umbels, or in compound heads.

A large, cosmopolitan family of about 300 genera and 3000 species, most abundant in the drier parts of temperate zones. Mathias & Constance recognize 92 genera and 509 species in North America. The flowers are so uniform, and the vegetative features often so variable among closely allied species, that the genera must often be separated largely by technical characters of the mature fruit. Many of the genera are notoriously ill-defined, and no system yet devised avoids the wide separation of closely related genera which happen to differ in some convenient technical character of the fruit.

Although the family is not thought of as one of ornamental importance, many of our species have

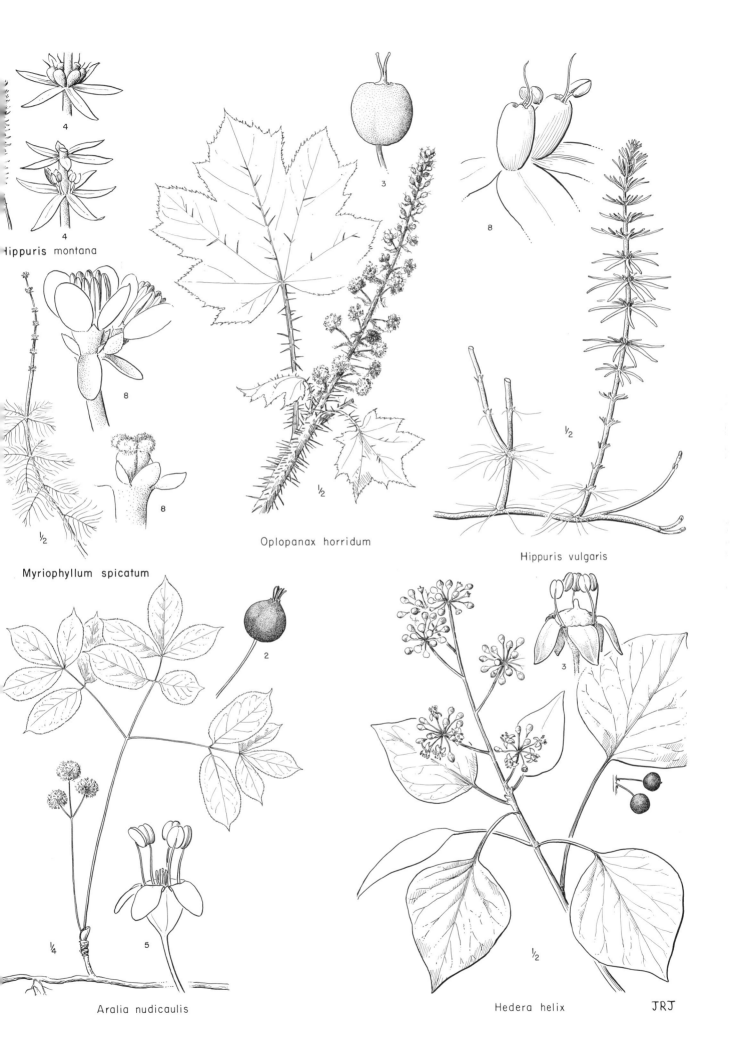

4

4

Hippuris montana

8

8

½

Myriophyllum spicatum

3

½

Oplopanax horridum

8

½

Hippuris vulgaris

2

¼

5

Aralia nudicaulis

3

½

Hedera helix

JRJ

considerable charm and merit the gardener's attention (especially members of the genus *Lomatium* in which excellent foliage, fairly showy flowers, and interesting fruits are often combined). *Heracleum* is usable with good effect in boggy soil, and even *Conium maculatum* is an attractive plant although not recommended because of its poisonous properties.

The family contains a number of species with edible roots, foliage, or seeds, as well as some that are highly poisonous. Species not definitely known to be edible should not be eaten, or should be tested only with the greatest of caution, inasmuch as very small quantities of some of the more violently poisonous species (e. g. *Cicuta* spp. , *Conium maculatum)* are quickly fatal.

Reference: Mathias, Mildred E. , and Lincoln Constance. Umbelliferae. N. Am. Fl. 28B:43-297. 1944-1945.

TECHNICAL KEY TO THE GENERA
Emphasizing Characters of the Fruit*

1 Fruit more or less strongly compressed dorsally (i. e. parallel to the commissure), unarmed, at least the lateral ribs generally winged, only occasionally wingless; carpophore bifid to the base

  2 Stylopodium obsolete or nearly so, though sometimes more or less evident in the fresh flower

    3 Umbellets capitate, the flowers sessile; maritime plants with definite broad leaflets; flowers white   GLEHNIA

    3 Umbellets with pedicellate flowers, the pedicels sometimes very short; plants not maritime; leaves usually more or less dissected, with small and often narrow ultimate segments, but sometimes with definite broad leaflets; flowers yellow to white or purple

      4 Dorsal ribs of the fruit generally more or less winged, rarely wingless; involucel dimidiate (asymmetrical, as if one half were wanting)   CYMOPTERUS

      4 Dorsal ribs of the fruit usually filiform and wingless, occasionally very narrowly winged; involucel dimidiate or more often not   LOMATIUM

  2 Stylopodium more or less well developed; leaves in most species with well-defined broad leaflets

    5 Umbellets capitate, the flowers sessile; fruit cuneate; dorsal as well as lateral ribs of the fruit generally winged; flowers white   SPHENOSCIADIUM

    5 Umbellets with pedicellate flowers; fruit not cuneate; dorsal ribs winged or wingless

      6 Leaves more or less dissected, without well-defined leaflets or with the leaflets more or less cleft; dorsal ribs of the fruit narrowly winged; maritime plants   CONIOSELINUM

      6 Leaves with more or less definite, entire or merely toothed leaflets, or some of the leaflets sometimes with a few irregular lobes; fruit various; plants not maritime except 2 spp. of *Angelica*

        7 Oil tubes reaching only part way from the apex toward the base of the fruit, readily visible through the pericarp; marginal flowers of the umbel enlarged and radiant, the outer corolla lobes commonly 2-cleft; leaves in our species trifoliolate, with very large leaflets mostly 1-4 dm. long and wide; flowers white   HERACLEUM

        7 Oil tubes extending all the way from the apex to the base of the fruit, conspicuous or inconspicuous; marginal flowers not enlarged, the corolla lobes not 2-cleft; leaves pinnately to ternately compound, with more than 3 leaflets, these generally less than 1 dm. wide

          8 Plant an introduced biennial weed with yellow flowers and merely once pinnate leaves; involucre and involucel usually wanting   PASTINACA

          8 Plants native perennials, not weedy, the leaves often 2-several times compound; flowers white (pink) or in one species yellowish; involucre and involucel often well developed   ANGELICA

1 Fruit subterete or more or less compressed laterally (i. e. at right angles to the commissure), dorsally somewhat compressed only in *Daucus,* which has armed fruits and entire or apically cleft carpophore; ribs of the fruit varying from obsolete to promi-

---

*For key to genera, emphasizing vegetative characters, see page 511.

nent and corky-thickened, but seldom (notable exception: *Ligusticum)* winged; carpophore entire or variously bifid, or wanting

9 Fruit with a conspicuous beak much longer than the body; introduced annual weed with
    dissected leaves   SCANDIX

9 Fruit beakless or with a short beak distinctly shorter than the body
  10 Fruit generally provided with conspicuous scales, or bristly hairs, or hooked or
     glochidiate or barbed prickles, or tubercles, glabrous and unarmed only in a few
     species (of *Osmorhiza* and *Anthriscus)* that have the fruit narrowed at the tip to a
     more or less distinct short beak
    11 Carpophore wanting; calyx teeth well developed and prominent; fruits sessile or
      nearly so, and the ribs usually obsolete; our species perennial
      12 Leaves simple, entire or merely toothed; inflorescence capitate, the flowers
        all alike and sessile; fruit in our species covered with scales   ERYNGIUM
      12 Leaves compound or dissected; flowers of two kinds, some staminate and
        generally pedicellate, others perfect and sessile or nearly so; fruit in
        our species with uncinate prickles   SANICULA
    11 Carpophore well developed, entire or bifid at the apex; calyx teeth evident to
      more often minute or obsolete; fruits pedicellate; ribs of the fruit evident to
      obsolete
      13 Plants annual or biennial, often weedy, the leaves in our species dissect-
        ed into small and narrow ultimate segments; native and introduced spe-
        cies
        14 Fruit shortly but distinctly beaked; ribs obsolete   ANTHRISCUS
        14 Fruit beakless; ribs evident
          15 Fruit dorsally somewhat compressed, provided with glochidiate or
            barbed prickles   DAUCUS
          15 Fruit laterally compressed, provided with uncinate prickles CAUCALIS
      13 Plants perennial, native, not weedy; leaves with well defined broad leaf-
        lets; fruit merely bristly-hispid or glabrous, beakless or short-
        beaked   OSMORHIZA
  10 Fruit unarmed and commonly glabrous, not beaked
    16 Leaves all simple, entire or toothed to sometimes palmately lobed
      17 Umbels simple (sometimes proliferous, or aggregated into a scarcely um-
        bellate inflorescence); carpophore wanting; plants fibrous-rooted, growing
        in water or in wet places
        18 Leaves reduced to narrow, hollow, septate phyllodes   LILAEOPSIS
        18 Leaves with a distinct normal petiole and broad blade   HYDROCOTYLE
      17 Umbels compound; carpophore present, bifid to the base; plants taprooted,
        growing in upland habitats   BUPLEURUM
    16 Leaves compound or dissected (basal leaves merely toothed in our sp. of *Zizia)*
      19 Stylopodium obsolete or nearly so; plants (except *Zizia)* low, scapose or
        subscapose, not more than about 3 dm. tall
        20 Plants perennial from a cluster of fleshy-fibrous roots, mostly growing
          in wet places; calyx teeth well developed; carpophore bifid about half-
          way to the base; flowers yellow; our species with the basal leaves
          simple, merely toothed   ZIZIA
        20 Plants perennial from a taproot (with or without a branching caudex), or
          from a thickened, cormlike root, mostly growing in dry places; calyx
          teeth, carpophore, and flower-color various; basal leaves compound or
          dissected
          21 Carpophore wanting
            22 Lateral ribs of the fruit developed into inflexed corky wings, a
              corky riblike projection also present along the length of the
              commissural face of each mericarp; leaf segments elongate
              and narrow; root sometimes short and tuberous-thickened,
              sometimes more elongate   OROGENIA
            22 Lateral and dorsal ribs of the fruit merely raised and thickened,
              with or without an irregular vestige of a wing; commissural

face of the mericarp without a projection; primary leaf segments 3-5, relatively broad and more or less confluent, merely toothed or lobed; root thickened and elongate   RHYSOPTERUS
21 Carpophore present, entire to deeply cleft
   23 Fruits linear, mostly 8-10 mm. long and 1-1.5 mm. wide; n. Ida. and adj.
      Wash. *(L. orogenioides)*   LOMATIUM
   23 Fruits linear-oblong or broader, 2-7 mm. long
      24 Calyx teeth in our species minute or obsolete; ribs of the fruit in our spe-
         cies filiform and inconspicuous, although the fruit may be grooved-sul-
         cate; our species occurring in Wash.   TAUSCHIA
      24 Calyx teeth well developed and more or less conspicuous; ribs of the fruit
         thickened and conspicuous; our species occurring in Ida. and Mont.   MUSINEON
19 Stylopodium more or less well developed; plants often taller and leafy-stemmed
   25 Carpophore entire; stem purple-spotted; oil tubes numerous, small, more or less
      confluent; robust taprooted biennial weed with dissected leaves   CONIUM
   25 Carpophore bifid to the base, or wanting; stem not purple-spotted; oil tubes and habit
      various
      26 Plants taprooted; carpophore bifid to the base, the halves distinct, not adnate to
         the mericarps
         27 Ribs of the fruit not winged; introduced, more or less weedy biennials or rath-
            er short-lived perennials; leaves, or some of them, dissected into small and
            narrow ultimate segments
            28 Lowermost leaves (in our species) merely once compound and with well-de-
               fined broad leaflets; flowers white; our species a rather short-lived per-
               ennial   PIMPINELLA
            28 Lowermost leaves, when well developed, dissected like the others, with-
               out well-defined broad leaflets
               29 Flowers yellow; stem glaucous; our species a robust, short-lived
                  perennial, 1-2 m. tall when well developed   FOENICULUM
               29 Flowers white (pink); stem not glaucous; our species biennial, up to
                  about 1 m. tall   CARUM
         27 Ribs of the fruit (except in one species) distinctly though sometimes rather
            narrowly winged; native perennials, not weedy; leaves in most species less
            dissected and with evident leaflets   LIGUSTICUM
      26 Plants with fibrous or tuberous-thickened roots, not taprooted; carpophore bifid
         to the base, or wanting, the halves sometimes adnate to the mericarps
         30 Ribs of the fruit inconspicuous; oil tubes numerous and contiguous; halves
            of the carpophore adnate to the mericarps, inconspicuous; involucre and
            involucel well developed; roots not thickened   BERULA
         30 Ribs of the fruit generally conspicuous and corky-thickened, except some-
            times in species of *Perideridia* which have tuberous-thickened roots and
            have the halves of the carpophore separate from the mericarps; oil tubes,
            involucre, and involucel various
             31 Calyx teeth well developed, forming a persistent conspicuous crown on
               the fruit; carpophore wanting   OENANTHE
            31 Calyx teeth small and inconspicuous, or obsolete; carpophore bifid,
              sometimes deciduous or the halves adnate to the mericarps
              32 Involucre of well-developed, subfoliaceous bracts; roots not tuber-
                 ous-thickened and plants not bulbiliferous; halves of the carpo-
                 phore (in our species) adnate to the mericarps   SIUM
              32 Involucre of inconspicuous, often scarious bracts, or wanting; some
                 of the roots generally tuberous-thickened, or if not so then the up-
                 per leaves bearing axillary bulbils; halves of the carpophore distinct,
                 not adnate to the mericarps, persistent or (in *Cicuta)* deciduous
                33 Plants with bulbils in the axils of at least the upper leaves   CICUTA
                33 Plants not bulbiliferous
                  34 Base of the stem thickened, hollow, and with well-developed
                    transverse partitions; leaves with well-defined broad leaflets CICUTA

34 Base of the stem normal in structure, not modified as in *Cicuta;* leaves compound or
    dissected with mostly narrow and elongate ultimate segments   PERIDERIDIA

ARTIFICIAL KEY TO THE GENERA
Emphasizing Vegetative Characters

1 Leaves all simple, entire or toothed (palmately lobed in *Hydrocotyle*)
  2 Inflorescence densely capitate, without rays, the flowers and fruits sessile; leaves
      often spiny-toothed   ERYNGIUM
  2 Inflorescence umbellate, the flowers and fruit more or less pedicellate; leaves not
      spiny-toothed
    3 Plants aquatic or semi-aquatic, fibrous-rooted, with simple umbels
      4 Leaves with a broad, mostly rotund-reniform, toothed and generally lobed blade
          well differentiated from the petiole   HYDROCOTYLE
      4 Leaves reduced to long, narrow phyllodes, without a differentiated blade   LILAEOPSIS
    3 Plants of uplands, with a branching caudex; leaves narrow and entire, but well de-
        veloped   BUPLEURUM
1 Leaves or most of them compound or very deeply cleft (basal leaves merely toothed in
    Zizia, but the cauline ones compound)
  5 Leaves or many of them with more or less well-defined leaflets, not dissected into
      small and narrow segments
    6 Basal leaves merely toothed   ZIZIA
    6 Basal leaves when well developed compound or deeply cleft
      7 Leaflets 3, very large, mostly 1-4 dm. long and wide   HERACLEUM
      7 Leaflets usually more than 3, seldom as much as 1 dm. wide
        8 Plants perennial from fibrous or fleshy-thickened, fascicled roots, without
            a taproot or well-developed caudex; not maritime
          9 Leaves palmately very deeply cleft or palmately once compound   SANICULA
          9 Leaves pinnately to ternately once to several times compound
            10 Base of the stem thickened, hollow, and with well-developed transverse
                partitions; some of the roots usually tuberous-thickened; primary lat-
                teral veins of the leaflets tending to be directed to the sinuses between
                the teeth   CICUTA
            10 Base of the stem without transverse partitions; roots not tuberous-thick-
                ened; veins not directed to the sinuses
              11 Ribs of the fruit inconspicuous; calyx teeth minute or obsolete; wide-
                    spread   BERULA
              11 Ribs of the fruit prominent, more or less corky-thickened
                12 Plants usually reclining or scrambling-ascending; primary lat-
                        eral veins of the leaflets tending to be directed to the teeth;
                        calyx teeth well developed; chiefly w. of the Cascades   OENANTHE
                12 Plants erect; primary lateral veins of the leaflets bearing no ob-
                        vious relation to the teeth or sinuses; calyx teeth minute or ob-
                        solete; widespread   SIUM
        8 Plants annual, biennial, or perennial from a taproot or stout caudex, or with
          fleshy-fibrous roots from a rhizome-caudex in one maritime species (of
          Conioselinum)
          13 Plants low, scapose or subscapose (or with a pseudoscape), up to 2 dm.
              tall; inland, nonmaritime species
            14 Fruit dorsally compressed and with evident lateral wings; leaflets
                  evidently toothed or cleft; flowers ochroleucous to yellow; Wash. and
                  Oreg., from the Cascades westward   LOMATIUM
            14 Fruit slightly flattened laterally, very slightly or not at all winged; leaf-
                  lets and flowers various
              15 Plants with a pseudoscape arising from the subterranean crown of a
                      taproot; leaflets toothed or lobed; flowers white; e. Oreg.   RHYSOPTERUS
              15 Plants scapose or subscapose from a taproot and slightly branched
                      caudex; leaflets mostly entire; flowers yellow; Cascades of Wash.
                                                                            TAUSCHIA

13 Plants (except some maritime species) taller and usually more or less leafy-
    stemmed
  16 Annual or biennial weeds
    17 Plant a biennial; flowers yellow; fruit unarmed   PASTINACA
    17 Plant an annual; flowers white or pink; fruit prickly   CAUCALIS
  16 Perennials, not (except *Pimpinella)* weedy
    18 Umbellets capitate, the flowers and fruits sessile
      19 Plants low, subacaulescent, less than 2 dm. tall, maritime   GLEHNIA
      19 Plants taller, mostly 5-18 dm. tall, leafy-stemmed, occurring inland
                                                     SPHENOSCIADIUM
    18 Umbellets not capitate, the flowers and fruits pedicellate
      20 Fruit dorsally flattened
        21 Stylopodium well developed; flowers (except one sp.) white or oc-
           casionally pinkish (see also *Conioselinum)*   ANGELICA
        21 Stylopodium obsolete or nearly so; flowers (except in *L. martindalei,*
           which might occasionally be sought here) yellow   LOMATIUM
      20 Fruit subterete or flattened laterally
        22 Fruit linear or linear-oblong to clavate, not at all winged, 8-22 mm.
           long; leaves always with well-defined leaflets   OSMORHIZA
        22 Fruit broader and often shorter, often some of the ribs winged; leaf-
           lets not always well defined
          23 Fruit 2-2.5 mm. long, the ribs filiform and rather obscure;
              Wash.   PIMPINELLA
          23 Fruit 3-7 mm. long, the ribs prominent and except in one spe-
              cies winged; widespread   LIGUSTICUM
5 Leaves more or less dissected into mostly rather small and narrow ultimate segments,
    without well-defined leaflets
  24 Stem purple-spotted; robust biennial weed of moist places, 0.5-3 m. tall, with
    white flowers   CONIUM
  24 Stem not purple-spotted; habit and habitat various
    25 Plant a robust, short-lived perennial weed, 1-2 m. tall when well developed,
      with glaucous stems, yellow flowers, and finely dissected leaves, the ultimate
      segments well under 1 mm. wide   FOENICULUM
    25 Plants distinctly otherwise, differing in one or usually more respects from the
      above
      26 Plants annual or biennial, taprooted, often weedy; native and introduced spe-
        cies; flowers white, seldom pink or yellowish, never yellow
        27 Fruit distinctly beaked
          28 Beak much longer than the body of the fruit; plants mostly 1-3 dm.
             tall   SCANDIX
          28 Beak shorter than the body of the fruit; plants mostly 4-15 dm. tall
                                              ANTHRISCUS
        27 Fruit beakless
          29 Fruit prickly
             30 Fruit dorsally somewhat compressed, provided with glochidiate
               or barbed prickles or bristles   DAUCUS
            30 Fruit laterally somewhat compressed, generally provided with un-
               cinate prickles   CAUCALIS
          29 Fruit unarmed, glabrous   CARUM
      26 Plants perennial, with or without a taproot; native species, not weedy
        31 Fruit armed with hooked prickles   SANICULA
        31 Fruit unarmed
         32 Plants maritime   CONIOSELINUM
         32 Plants not maritime
           33 Plants definitely taprooted, the taproot sometimes fleshy-thick-
             ened, but distinctly elongate; taproot often surmounted by a
             stout branching caudex
            34 Fruit essentially wingless, slightly compressed laterally

35 Fruit without a carpophore; Oreg. RHYSOPTERUS

35 Fruit with a definite carpophore; Mont. and Ida. MUSINEON

34 Fruit with at least the lateral ribs produced into evident wing-margins when fully mature, these inflexed in *Orogenia*

  36 Carpophore wanting; w. of the Cascade summits in Oreg. OROGENIA

  36 Carpophore present, bifid to the base; widespread

    37 Body of the fruit subterete or slightly compressed laterally LIGUSTICUM

    37 Body of the fruit distinctly compressed dorsally

      38 Dorsal ribs of the fruit wingless or nearly so, only the lateral ribs distinctly winged; involucel dimidiate or not LOMATIUM

      38 Dorsal ribs of the fruit generally winged, the wings sometimes narrower · than those of the lateral ribs; involucel dimidiate (all on one side of the umbellet) CYMOPTERUS

33 Plants with fibrous or fleshy-thickened roots; roots when fleshy-thickened either clustered or solitary, and when solitary always short, not over about an inch long

  39 Plants bearing bulbils in the axils of at least the upper leaves CICUTA

  39 Plants not bulbiliferous

    40 Roots strictly fibrous, not at all fleshy; plants growing in water or very wet places BERULA

    40 Roots or some of them more or less fleshy-thickened; plants of wet or more often dry places

      41 Fruit slightly to strongly flattened laterally, wingless or the mericarps with thickened, incurved lateral wings; flowers white or pink

        42 Fruit linear, mostly 8-10 mm. long; n. Ida. and adj. Wash. LOMATIUM

        42 Fruit linear-oblong or broader, 2-7 mm. long

          43 Plants low, up to about 2 dm. tall, with a solitary, globose, corm-like root

            44 Fruit linear-oblong, 5-7 mm. long, not winged; c. Wash. TAUSCHIA

            44 Fruit oblong-elliptic, 3-4 mm. long, the lateral ribs developed into inflexed corky wings; widespread OROGENIA

          43 Plants taller, 2-12 dm. tall, the thickened roots solitary or often clustered PERIDERIDIA

      41 Fruit more or less strongly flattened dorsally, the marginal ribs with thin spreading wings at maturity; flowers yellow or white LOMATIUM

Angelica L.

Inflorescence of one or usually several compound umbels, the terminal one generally the largest and blooming first; involucre and involucel of foliaceous or narrow and scarious bracts or bractlets, or wanting; flowers white, seldom pink or yellowish; calyx teeth minute or obsolete; stylopodium broadly conic; carpophore bifid to the base; fruit elliptic-oblong to orbicular, strongly compressed dorsally, glabrous to scabrous, hispidulous, or tomentose, the lateral and often also the dorsal ribs evidently winged, or the ribs all sometimes merely elevated and corky-thickened but scarcely winged; oil tubes few to numerous; stout perennials, usually or always single-stemmed from a stout taproot; leaves pinnately to ternately once to thrice compound, with broad, toothed or cleft leaflets; petioles of the lower leaves elongate, generally sheathing only at the base, those of the middle and upper leaves with progressively more prominent sheathing portion, the uppermost leaves often consisting merely of a bladeless petiolar sheath, or the leaves rarely all basal.

A circumboreal genus of about 50 species. (Name from the Latin *angelus,* angel, referring to the cordial and medicinal properties of some species.)

1 Plants maritime; involucre mostly wanting, but involucel present

  2 Leaves essentially glabrous; dorsal ribs or wings of the fruit similar to the lateral ones; oil tubes numerous, adhering to the seed, which is free within the pericarp at maturity (fruits unique among our species) A. LUCIDA

  2 Leaves tomentose or woolly beneath (unique among our species in this regard); lateral wings of the fruit much better developed than the dorsal; oil tubes few, the seed adhering to the pericarp A. HENDERSONII

1 Plants not maritime; involucre and involucel various

3 Flowers distinctly yellowish; inflorescence consisting of a single compound umbel (rarely
 2); involucral bracts conspicuous, foliaceous, often equaling the rays of the umbel, or
 rarely wanting; Alta. and B. C. to n. Mont. and n. Ida.   A. DAWSONII
3 Flowers white or pinkish; inflorescence usually of 2 or more compound umbels; involucre usually wanting, occasionally represented by a few small bracts, or by a single
 sheathing bract
  4 Rachis of the leaves abaxially geniculate at the point of insertion of the first pair of
   pinnae and commonly also at the points of insertion of successive pinnae, the primary pinnae generally deflexed; chiefly west of the Cascade summits   A. GENUFLEXA
  4 Rachis of the leaves not geniculate, and the pinnae not deflexed; east of the Cascade
   summits, except for the widespread A. arguta
    5 Leaves oblong to elliptic, pinnate to incompletely bipinnate; involucel wanting; s. w.
     Mont. and adj. Ida. to Utah and N. M.   A. PINNATA
    5 Leaves more deltoid, ternate-pinnate; involucel present or absent
     6 Ovaries and fruit scabrous-tuberculate or short-hairy; leaflets 1. 5-5. 5 cm. long
      7 Involucel generally wanting; dorsal ribs of the fruit very narrowly winged, not
       much thickened; c. Wash. to n. Oreg., east of the Cascade summits   A. CANBYI
      7 Involucel of evident narrow bractlets; dorsal wings of the fruit thickened and
       fairly well developed, though not so large as the lateral ones; w. Mont. and
       adj. Ida. to Colo. and Utah   A. ROSEANA
     6 Ovaries and fruit glabrous; leaflets mostly 4-14 cm. long; involucel usually
      wanting; widespread
                       A. ARGUTA

Angelica arguta Nutt. in T. & G. Fl. N. Am. 1:620. 1840. (Nuttall, Wappatoo Island, Oreg., and near
 Fort Vancouver, Wash.)
 A. lyallii Wats. Proc. Am. Acad. 17:374. 1882. (Lyall, in the Galton and Cascade mountains, near
 the Canadian boundary)
 A. piperi Rydb. Fl. Rocky Mts. 631, 1064. 1917. (Piper 2336, Blue Mts., Walla Walla Co., Wash.)
 Plants robust, 5-20 dm. tall; leaves large, ternate-pinnately about twice compound; leaflets glabrous or often hairy along the veins beneath, sessile or petiolulate, ovate to lanceolate, acute, mostly 4-14 cm. long and 1. 5-8 cm. wide, serrate or doubly serrate and sometimes irregularly few-cleft,
the primary lateral veins tending to be directed to some of the teeth, or sometimes scarcely reaching
the margin; compound umbels usually 2 or more; rays of the umbel mostly 18-45, often scabrous-
tuberculate, unequal, up to 8 cm. long, the outer ones often basally connate; involucre wanting; involucel wanting, or occasionally of a few narrow and inconspicuous bractlets; flowers white (reputedly
sometimes pinkish), the petals glabrous; ovaries and fruits glabrous; fruit broadly elliptic to orbicular or obovate, 4-7 cm. long and 4-5 mm. wide, the dorsal ribs narrowly winged, the lateral ones
broader and nearly or fully as wide as the body; oil tubes solitary in the intervals, several on the commissure.
 Stream banks, wet meadows, marshes, and bottomlands, from the foothills and valleys to moderate
elevations in the mountains; s. w. Alta. to Wyo. and Utah, w. to s. B. C., Wash., Oreg., and n. Calif.,
seldom found w. of the Cascades in Wash., but common enough in the Willamette Valley of Oreg.
June-Aug.
 Within the area of A. canbyi there are some specimens which suggest the possibility of hybridization and introgression.

Angelica canbyi Coult. & Rose, Rev. N. Am. Umbell. 40. 1888. (Suksdorf 763, Klickitat River, near
 Mt. Adams, Wash.)
 Plants 5-12 dm. tall, averaging less robust than A. arguta; leaves ternate-pinnately about twice
compound; leaflets sessile or petiolulate, lanceolate to elliptic or ovate, acute, mostly 2-5. 5 cm. long
and 0. 5-3. 5 cm. wide, serrate or sometimes doubly serrate or irregularly few-cleft, glabrous or
scabrous along the veins beneath, rather firm and often reticulate-veiny, the primary lateral veins
tending to be directed to some of the teeth; compound umbels usually 2 or more; rays of the umbel
mostly 8-34, often scabrous-tuberculate, unequal, the longer ones sometimes 7 cm. long; involucre
and involucel wanting, or rarely the involucel irregularly developed and consisting of a few scarious
bractlets; flowers white, the petals glabrous; ovaries evidently short-hairy, the fruit less markedly
so; fruit broadly elliptic or oblong to suborbicular, 4. 5-6 mm. long and 3-5 mm. wide, the dorsal

ribs very narrowly winged, the lateral ones broader, commonly half as wide as the body; oil tubes 1 or 2 in the intervals, several on the commissure.

Moist places, often along streams; Columbia plains of c. Wash. and adj. Oreg., extending onto the e. slope of the Cascade (and Wenatchee) Mts. July-Sept.

Angelica dawsonii Wats. Proc. Am. Acad. 20:369. 1885. *(Lyall,* Rocky Mountains near the Canadian boundary)

Thaspium aureum var. *involucratum* Coult. & Rose, Rev. N.Am. Umbell. 83. 1888. *(Leiberg,* Kootenai Co., Ida., in July, 1887)

Perennial from a stout, compactly branched caudex which surmounts a taproot, 3-12 dm. tall; basal leaves large and long-petiolate, ternately or ternate-pinnately about twice compound; leaflets petiolulate or sessile, thin, rather closely serrate, generally lanceolate or ovate, acute, often subcordate at the base, 2.5-9 cm. long and 1-4.5 cm. wide; cauline leaves few and short-petiolate, often more or less reduced, or none; inflorescence generally consisting of a single compound umbel (rarely 2), the rays 10-30, mostly 2-4 cm. long; involucre of several well-developed, toothed, leafy bracts sometimes as long as the rays, or rarely wanting; involucel of well-developed, entire or irregularly few-toothed bractlets, commonly as long as or longer than the pedicels; flowers reported to be pale greenish yellow in life, distinctly yellowish when dried, the petals glabrous; ovaries and fruit glabrous; fruit 4-7 mm. long, 3-4.5 mm. wide, the lateral wings from about half as broad to nearly as broad as the body, the dorsal wings well developed but narrower than the lateral ones; oil tubes solitary in the intervals, several on the commissure.

Along streams and on moist or wet slopes in the mountains, s.w. Alta. and s.e. B.C. to n. Ida. (s. to Clearwater Co.) and n.w. Mont. (s. to Ravalli Co.), in our range chiefly w. of the continental divide, but extending onto the e. slope in Glacier National Park as well as in Alta. A very distinct species. June-Aug.

Angelica genuflexa Nutt. in T. & G. Fl. N.Am. 1:620. 1840. *(Nuttall,* Wappatoo Island, Oreg., and near Fort Vancouver, Wash.)

Stout, leafy-stemmed perennial from an erect simple crown, possibly fibrous-rooted, often more than 1 m. tall, the stem commonly glaucous; leaves ternate-pinnately about twice compound, the rachis abaxially geniculate at the point of insertion of the first pair of pinnae and commonly also at the points of insertion of successive pinnae, the pinnae often deflexed (or seemingly spreading at right angles to the rachis in herbarium specimens, not directed obliquely forward as in other species of the genus); leaflets ovate or lanceolate to elliptic, serrate, mostly 4-10 cm. long and 1.5-5 cm. wide, glabrous, or hairy along the veins beneath; compound umbels generally several; rays mostly 22-45, unequal, up to 7 cm. long; involucre wanting; involucel of well-developed narrow bractlets; flowers white or pinkish, the petals glabrous; ovaries minutely hispidulous; fruits glabrous, suborbicular, 3-4 mm. long, the dorsal wings narrow, the lateral ones broader and about as wide as the body; oil tubes solitary in the intervals, 2 on the commissure.

Moist places; n. Calif. to Alas. and Siberia, apparently wholly w. of the Cascade summits in the U.S., but extending to the Selkirk Mts. of s. B.C. and to c. Alta. July-Aug.

Angelica hendersonii Coult. & Rose, Bot. Gaz. 13:80. 1888. *(Henderson 2158,* Long Beach, Ilwaco, Pacific Co., Wash.)

Stout perennial 3-15 dm. tall; leaves firm, tomentose or woolly beneath, green and subglabrous above, ternate-pinnately about twice compound, the leaflets sessile or petiolulate, mostly 4-8 cm. long and 2.5-6 cm. wide, irregularly serrate; compound umbels usually several, the rays mostly 30-45, up to 7 cm. long, tomentose or woolly; involucre wanting; involucel of narrow, tomentose bractlets; flowers white, the petals more or less tomentose on the back towards the base; ovaries strongly woolly-tomentose, the fruits less so or glabrate; fruit broadly elliptic or subrotund, 7-10 mm. long, the dorsal ribs not much elevated, the lateral ones produced into wings nearly or quite as wide as the body; oil tubes solitary in the intervals, 4 on the commissure.

Bluffs and sand dunes along the coast, from s. Wash. to c. Calif. July-Aug.

Angelica lucida L. Sp. Pl. 251. 1753.

Imperatoria lucida Spreng. Umbell. Prodr. 17. 1813. *Coelopleurum lucidum* Fern. Rhodora 21:146. 1919. (Canada)

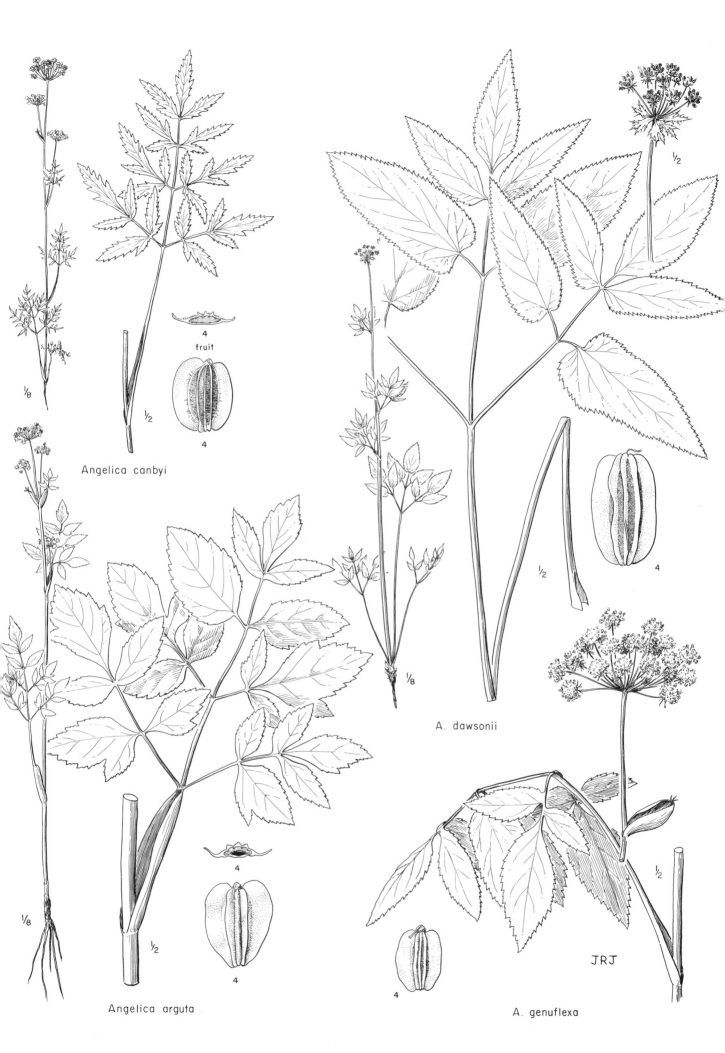

4
fruit

½

4

Angelica canbyi

½

⅛

1/2

4

⅛

½

4

Angelica arguta

A. dawsonii

½

½

4

JRJ

⅛

A. genuflexa

Archangelica gmelini DC. Prodr. 4:170. 1830. *Pleurospermum gmelini* Bong. Veg. Isl. Sitcha
    141. 1833. *Coelopleurum gmelini* Ledeb. Fl. Ross. 2:361. 1844. (Kamtschatka)
Coelopleurum maritimum Coult. & Rose, Bot. Gaz. 13:145. 1888. *(Henderson 384,* Long Beach, Il-
    waco, Pacific Co., Wash.)
Coelopleurum longipes Coult. & Rose, Contr. U.S. Nat. Herb. 7:142. 1900. *(Joseph Howell 735,*
    near Astoria, Oreg.)
    Stout perennial 4-14 dm. tall; compound umbels generally 2 or more, the primary one often over-
topped by one on an axillary branch; leaves glabrous, petiolate, the petiole with very broad, sheathing
base, or the upper wholly sheathlike; leaf blades ternate-pinnately about twice compound; leaflets
mostly 3-7 cm. long and 2-6 cm. wide, ovate to subrotund, serrate or crenate-serrate, sessile or
petiolulate; compound umbels usually several, with 20-45 rays up to 10 cm. long; involucre wanting,
or of a few small deciduous bracts; involucel of evident narrow bractlets; flowers white, the petals
glabrous; ovaries and fruit glabrous; fruit oblong-elliptic, 4-9 mm. long and 2-5 mm. wide, the ribs
all about alike, prominently corky-thickened, raised to a thin edge, but scarcely winged; oil tubes
small, numerous, continuous about the seed and adhering to it, the seed free within the pericarp at
maturity.
    Beaches and bluffs along the coast, apparently strictly maritime in our region, but often occurring
more or less inland elsewhere in its range; margin of the Pacific basin, from Siberia and the Kurile
Is. to n. Calif.; also from Lab. to N.Y. July-Aug.

Angelica pinnata Wats. Bot. King Exp. 126. 1871. *(Watson 458,* Wahsatch and Uinta mts., Utah)
    Plants 3-10 dm. tall, averaging less robust than *A. arguta;* leaf blades more or less oblong, pin-
nately or sub-bipinnately compound, the lower pinnae often with 3 crowded leaflets, the upper undivided; leaflets sessile or nearly so, lanceolate or lance-ovate to elliptic-oblong, serrate, mostly 2.5-9
cm. long and 0.5-3 cm. wide, glabrous or nearly so; compound umbels generally 2 or more, seldom
solitary; rays mostly 6-25, unequal, the longer ones up to about 7 cm. long; involucre wanting or
sometimes represented by a sheathing bract; involucel wanting; flowers white, or reputedly sometimes
pinkish, the petals glabrous; ovaries obscurely scabrous or hirtellous; fruit glabrous or nearly so,
3-6 mm. long, broadly elliptic-oblong to suborbicular, the dorsal wings well developed but narrower
than the lateral ones; oil tubes solitary (2) in the intervals, several on the commissure. N=11.
    Moist places, often along stream banks or in wet meadows, in the foothills or at moderate elevations
in the mountains; n.w. Wyo., barely extending into adj. Mont. (Park Co.) and Ida. (Fremont and
Clark cos.), s. to Colo., Utah, and N.M. July-Aug.

Angelica roseana Henderson, Contr. U.S. Nat. Herb. 5:201. 1899.
    Rompelia roseana K.-Pol. Bull. Soc. Nat. Mosc. II, 29:125. 1916. *(Henderson 4065,* foothills of the
    Lost River Mountains, near Salmon, Ida.)
    Stout perennial 1.5-8 dm. tall, the stem often glaucous; leaves ternate-pinnately 2-3 times com-
pound; leaflets firm, mostly sessile, sharply serrate or dentate, glabrous or scabrous, 1.5-3.5 cm.
long and 0.5-3 cm. wide; compound umbels usually several, the rays 15-35, unequal, up to 10 cm.
long; involucre wanting, or rarely of a few bracts; involucel of narrow bractlets; flowers white or
reputedly sometimes pink, the petals glabrous; ovaries and fruit scabrous-tuberculate; fruit 4-7 mm.
long and 3-4 mm. wide, the dorsal wings thickened and fairly well developed but still smaller than the
lateral ones; oil tubes solitary in the intervals, 4 on the commissure.
    Talus slopes and other rocky places in the mountains, often at high altitudes; Lake Co., Mont. to
Colo., w. to Lemhi Co. and e. Custer Co., Ida., and to Utah. July-Aug.

Anthriscus Hoffm. Nom. Conserv.

    Inflorescence of few-rayed compound umbels; involucre usually wanting; involucel of several narrow,
entire, usually reflexed bractlets; flowers white; calyx teeth obsolete; styles short, the stylopodium
conic; carpophore entire or bifid at the apex; fruit ovoid to linear, short-beaked at the tip, flattened
laterally and often constricted at the commissure, smooth or bristly, the ribs obsolete; oil tubes obscure or obsolete; branching, leafy-stemmed annuals or biennials, seldom perennial, with petiolate,
pinnately compound or dissected leaves.
    About 10 species of the Old World. (Name slightly modified from the ancient Greek name of some
umbellifer.)

1/6

fruit

4

4

1/2

1/4

1/2

1/2

4

Angelica hendersonii

Angelica lucida

JRJ

Anthriscus scandicina (Weber) Mansfeld in Fedde, Rep. Sp. Nov. 46:309. 1939.

Scandix anthriscus L. Sp. Pl. 257. 1753. *Chaerophyllum anthriscus* Crantz, Class. Umbell. 76.
1767. *Caucalis scandix* Scop. Fl. Carn. 2nd ed. 1:191. 1772. *Caucalis scandicina* Weber in Wig-
gers, Prim. Fl. Holsat. 23. 1780. *A. vulgaris* Pers. Syn. Pl. 1:320. 1805, not *A. vulgaris* Bernh.
1800. *Myrrhis anthriscus* Lag. Amen. Nat. 98. 1811. *A. scandix* Asch. Fl. Brand. 1:260. 1860,
not *A. scandix* Bieb. 1808. *A. anthriscus* Karst. Deuts. Fl. 857. 1882. *Myrrhodes anthriscus*
Kuntze, Rev. Gen. 1:268. 1891. *Cerefolium vulgare* Bubani, Fl. Pyren. 2:411. 1900. *Chaerefo-
lium anthriscus* Schinz. & Thell. Viert. Nat. Ges. Zürich 53:554. 1908. (Europe)

Taprooted, branching annual 4-9 dm. tall, the stem generally glabrous; leaves basal and cauline,
gradually reduced upwards, sparsely or moderately hirsute or hispidulous, petiolate, the blade pinnately dissected with small ultimate segments; peduncles slender, 2 cm. long or less, borne opposite
the upper leaves (i.e., the stem sympodial above); rays 3-6, 1-2.5 cm. long; involucel of a few small
lanceolate bractlets; umbellets 3- to 7-flowered, the pedicels short, each with a ring of short flattened
hairs at the summit; fruit ovoid, about 4 mm. long, the body covered with short thick uncinate prickles, the short stout beak unarmed.

Stream banks and moist open places, often in disturbed soil; native of Europe, occasionally introduced in e. U.S., and in the Pacific states from Wash. to Calif. May.

Anthriscus sylvestris (L.) Hoffm., a more robust biennial European weed, up to 15 dm. tall, with
short-beaked, smooth, ribless fruits 5-8 mm. long, has been collected as an introduction in Wash.
It does not have sympodially branched stems.

Berula Hoffm.

Inflorescence of compound umbels; involucre of evident, narrow, entire or toothed bracts; involucel of narrow bractlets; flowers white; calyx teeth minute or obsolete; styles short, the stylopodium
conic; carpophore bifid to the base, the halves inconspicuous, adnate to the mericarps; fruit elliptic
to orbicular, somewhat compressed laterally, glabrous, the ribs inconspicuous; oil tubes numerous
and contiguous; glabrous leafy-stemmed perennials from fibrous roots, often stoloniferous, the leaves
pinnately compound, with entire to toothed or cleft leaflets, or some leaves submerged and with filiform-dissected blade.

A single species of the north temperate regions. (A late Latin name of some umbellifer.)

Berula erecta (Huds.) Cov. Contr. U.S. Nat. Herb. 4:115. 1893.

Sium erectum Huds. Fl. Angl. 103. 1762. (England)

Sium angustifolium L. Sp. Pl. 2nd ed. 1672. 1763. *B. angustifolia* Mert. & Koch in Roehl. Deuts.
Fl. 3rd ed. 2:433. 1826. (southern Europe)

Sium pusillum Nutt. in T. & G. Fl. N.Am. 1:611. 1840, not of Poir. in 1810. *B. pusilla* Fern. Rhodora 44:189. 1942. *(Nuttall*, Wappatoo Island, Oreg.; this is Sauvie's I., at the mouth of the Willamette) = var. *incisa.*

BERULA ERECTA var. INCISA (Torr.) Cronq. hoc loc. *Sium? incisum* Torr. in Frem. Rep. 90.
1845. *(Fremont*, North Fork of the Platte)

Soft, fibrous-rooted perennial mostly 2-8 dm. tall, generally freely branched, often stoloniferous
at the base; submerged, filiform-dissected leaves sometimes present; aerial leaves more or less dimorphic, the lower mostly with 7-21 ovate or lance-ovate to elliptic, crenate or crenately lobulate to
occasionally serrate or laciniate leaflets up to 5 cm. long and 3 cm. wide, the upper with smaller and
relatively narrower, more sharply toothed or often irregularly incised or subpinnatifid leaflets; umbels usually several or many, the 6-15 rays mostly 1-2 (4) cm. long at maturity; fruit 1.5-2 mm.
long. N=6.

In wet places or in shallow water in the valleys and plains; Europe and the Mediterranean region; s.
B.C. to Baja Calif., e. to Ont., N.Y., and Fla. June-Sept.

The American plants, as here described, constitute the var. *incisa* (Torr.) Cronq. Plants of the Old
World do not have the markedly dimorphic leaves which characterize the American plants. The American plants have sometimes been treated as a distinct species, *B. pusilla* Fern., based on *Sium pusillum* Nutt., but *Sium pusillum* Nutt. is a later homonym of *S. pusillum* Poir., and Nuttall's epithet
therefore carries no priority under the Rules. The oldest available specific epithet for the American
plants is *incisum (Sium? incisum* Torr.), but this has never been transferred into *Berula* at the specific level.

Bupleurum L.

Inflorescence of compound umbels; involucre of conspicuous leafy bracts, or wanting; involucel of broad, mostly foliaceous, often connate bractlets; flowers yellow or sometimes purplish; calyx teeth obsolete; styles short, the stylopodium depressed-conic; carpophore bifid to the base; fruit oblong to elliptic or orbicular, slightly flattened laterally and constricted at the commissure, evidently ribbed, otherwise smooth to more or less roughened or tuberculate; oil tubes more or less numerous, or obscure or wanting; taprooted, glabrous and often glaucous, more or less leafy-stemmed annuals or perennials; leaves simple and entire, the basal ones mostly petiolate and more or less parallel-veined, the cauline ones often sessile and broader and clasping.

About 100 species, centering in the Mediterranean region, only one native to N. Am. (A nearly direct transliteration of an ancient Greek name, meaning ox rib, applied to some plant.)

Bupleurum americanum Coult. & Rose, Rev. N. Am. Umbell. 115. 1888. (*Lay & Collie*, Alaska)
   B. purpureum Blank. Mont. Agr. Coll. Stud. Bot. 1:89. 1905. (*Blankinship*, Mt. Hyalite, Gallatin
   Co., Mont., is the first of three specimens cited)
   Perennial from a taproot and branching caudex, 0.5-5 dm. tall, several-stemmed, glabrous and glaucous; leaves narrow and elongate with several principal nerves, entire, the basal ones up to 16 cm. long (including the ill-defined petiolar base) and 1 cm. wide, the cauline ones mostly sessile, usually several, progressively shorter but often nearly or quite as wide as the basal, the upper sometimes lanceolate or lance-ovate and clasping; involucre of 2-6 well-developed leafy bracts 5-15 mm. long; rays 4-14, up to 4 cm. long at maturity; involucel of well-developed foliaceous bractlets 3-5 mm. long; umbellets very compact in flower and fruit; flowers yellow or sometimes purple; fruit broadly oblong, 3-4 mm. long, glaucous, the ribs prominently raised, almost winged; oil tubes numerous and well developed.
   Rock outcrops and open grasslands or dry meadows, from the base of the mountains to above timber line; Alas. and Yukon to n. and w.c. Wyo., chiefly e. of the continental divide in our region, barely entering Ida. along the mountains of the divide from Lemhi to Fremont cos. July-Aug.

Carum L.

Inflorescence of compound umbels on terminal and lateral peduncles; involucre and involucel of a few narrow and inconspicuous bracts (or bractlets), or wanting; flowers mostly white; calyx teeth obsolete; styles short, spreading, the stylopodium low-conic; carpophore bifid to the base; fruit oblong to broadly elliptic-oblong, somewhat compressed laterally, glabrous, evidently ribbed; oil tubes solitary in the intervals, 2 on the commissure; glabrous, leafy-stemmed biennials or perennials from a taproot, the leaves pinnately dissected into small and narrow ultimate segments.

About 50 species of Eurasia. (Name slightly modified from the ancient Greek and Roman names of some umbellifer.)

Carum carvi L. Sp. Pl. 263. 1753. (Northern Europe)
   Biennial, 3-10 dm. tall, single-stemmed at the base, generally branched at least above; basal and lower cauline leaves well developed, petiolate, the blade mostly 8-17 cm. long and 3-10 cm. wide; middle and upper leaves few and more or less reduced, but still generally petiolate; flowers white, or reputedly rarely pink; rays of the umbel mostly 7-14, commonly 1-3 cm. long at anthesis, up to 5 cm. in fruit; fruit rather broadly oblong-elliptic, 3-4 mm. long. N=10. Caraway.
   Roadsides and meadows, seldom extending out into the sagebrush; native of Europe, now more or less established across the n. U.S., and found here and there in our range. June-July.

Caucalis L.

Inflorescence of compound umbels, the rays of the primary umbel few and well developed, the pedicels of the ultimate umbels short or obsolete; involucre a few entire or dissected often leafy bracts, or wanting; involucel a few entire to dissected, commonly somewhat scarious bractlets; flowers white; calyx teeth evident; petals cuneate or obovate, with a slender inflexed tip; styles short, the stylopodium thick and conic; carpophore entire or bifid at the apex; fruit oblong or ovoid, somewhat compressed laterally, beset with stout, spreading, uncinate prickles along alternate ribs, merely bristly-hairy on the others; oil tubes solitary under alternate ribs, and two on the commissure; taprooted an-

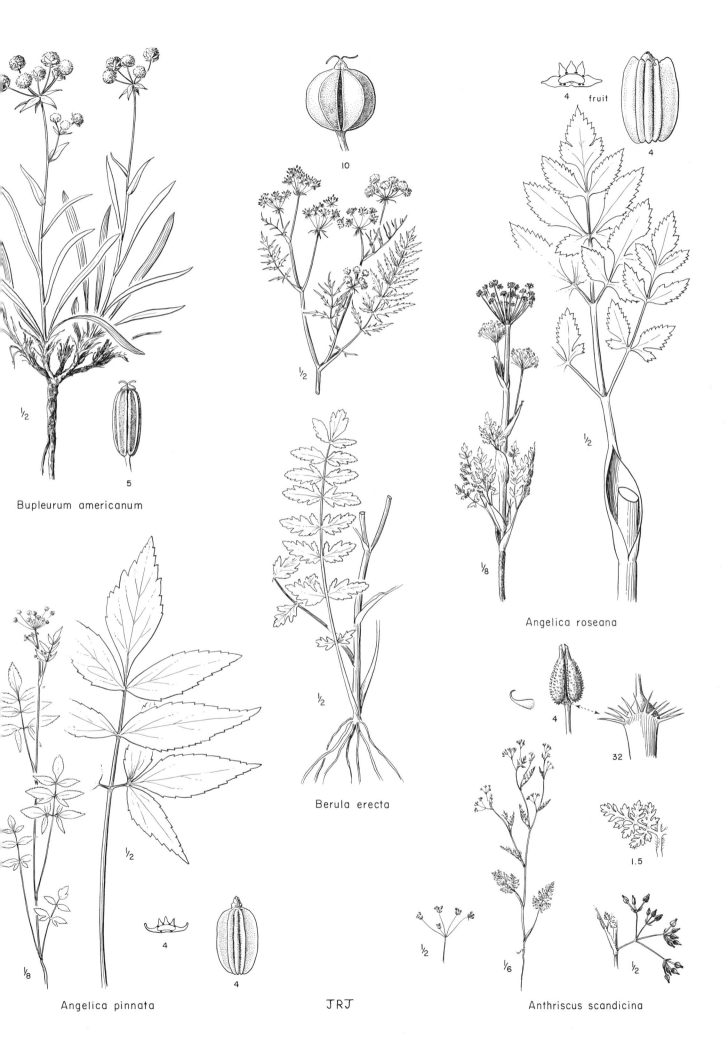

Bupleurum americanum

10

fruit

4

Angelica roseana

Angelica pinnata

Berula erecta

Anthriscus scandicina

JRJ

nuals with leafy, mostly branching stems and petiolate, pinnately compound or dissected leaves.

About 5 species, native to Asia, the Mediterranean region, and N. and C.Am. (The ancient Roman name.)

Several species of the related genus *Torilis* have been collected near or within the southern boundary of our range in Oreg., but do not appear to be established. All of them have the tubercles or glochidiate or uncinate prickles of the fruit generally distributed rather than in a few definite longitudinal rows. *T. nodosa* (L.) Gaertn. has sympodial stems with sessile or short-pedunculate simple capitate umbels opposite the leaves. *T. japonica* (Houtt.) DC. has compound umbels with one narrow bract to each ray. *T. arvensis* (Huds.) Link. and *T. scabra* (Thunb.) DC. have compound umbels without an involucre, or with a single involucral bract. *T. arvensis* has 2- to 10-rayed umbels with the fruits 3-5 mm. long; *T. scabra* has 2- or 3-rayed umbels with the fruit 5-7 mm. long.

Caucalis microcarpa H. & A. Bot. Beechey Voy. 348. 1838.
   Yabea microcarpa K.-Pol. Bull. Soc. Nat. Mosc. II, 28:202. 1915. *(Douglas,* Calif.)
   Slender annual 1-4 dm. tall, simple or branched especially above, more or less spreading-hirsute throughout; leaves chiefly cauline, the blade mostly 2-6 cm. long and 2-5 cm. wide, pinnately dissected into small, narrow ultimate segments; involucre of several fairly well-developed, scarcely modified leaves; rays of the umbel 1-9, ascending, 1-8 cm. long; involucel of several pinnatifid to entire bractlets; pedicels markedly unequal; fruit oblong, 3-7 mm. long, beset with uncinate prickles along the ribs, those of alternate ribs larger and tending to be confluent at the base.

   Along streams and on open, vernally moist slopes; s. B.C., southward through Wash. (both sides of the Cascades) and Ida. to Baja Calif. and Sonora, Mex. Apr.-June.

   Caucalis latifolia L., a stouter European species with less dissected leaves (mostly only once pinnate, with segments or leaflets 1-8 cm. long and 5-20 mm. wide) and larger fruits (mostly 10-12 mm. long), has been reported as a casual introduction in Oreg., but is not known to be established.

Cicuta L. Water Hemlock

Inflorescence of compound umbels; involucre wanting, or of a few inconspicuous narrow bracts; involucel of several narrow bractlets, or rarely wanting; flowers white or greenish; calyx teeth evident; styles short, spreading, the stylopodium depressed or low-conic; carpophore bifid to the base, deciduous; fruit glabrous, elliptic or ovate to orbicular, compressed laterally, the ribs usually prominent and corky; oil tubes solitary in the intervals, 2 on the commissure; glabrous, violently poisonous perennials, the tuberous-thickened base of the stem hollow, with well developed transverse partitions; roots clustered at the thickened stem-base, generally some of them tuberous-thickened; leaves once to thrice pinnate or ternate-pinnate, with well-developed leaflets or mere linear segments.

About 8 species, one Eurasian, the rest North American. Our two species are sharply distinct, but the remaining North American species are all closely allied to *C. douglasii.* (The ancient Latin name of some poisonous umbellifer, perhaps *Conium maculatum.)*

1 Plants bulbiliferous, the bulbils borne in the axils of at least the reduced upper leaves;
       segments of the principal leaves well under 5 mm. wide   C. BULBIFERA
1 Plants not bulbiliferous; leaflets of the principal leaves 5-35 mm. wide   C. DOUGLASII

Cicuta bulbifera L. Sp. Pl. 255. 1753.
   Cicutaria bulbifera Lam. Encyc. Meth. 2:3. 1786. *Keraskomion bulbiferum* Raf. New Fl. N.Am.
       4:21. 1836. *(Clayton,* Virginia)
   Plants generally single-stemmed, 3-10 dm. tall, mostly relatively slender, not much thickened at the base and sometimes without thickened roots; leaves all cauline, the middle and lower ones more or less dissected, with narrowly linear, entire or obscurely few-toothed segments mostly 0.5-1.5 mm. wide and 0.5-4 cm. long, the upper and rameal ones more or less reduced, with fewer segments, or undivided, many of them bearing one or more axillary bulbils; umbels frequently wanting, or present but not maturing fruit, the rays mostly 1-2.5 cm. long; fruit orbicular, 1.5-2 mm. long, constricted at the commissure, the ribs broader than the narrow intervals.

   In marshes, bogs, wet meadows, and shallow standing water, from the plains and lowlands to the mt. valleys; Newf. to Va., w. to n. Sask., n. Alta., B.C., s. Oreg., and Neb. Not common in our range. Aug.-Sept.

   The cordilleran plants, as here described, possibly constitute a distinct, as yet unnamed variety. The typical phase of the species, occurring in e. and c. U.S. and adj. Can., generally has more evidently toothed and somewhat broader, more definite leaflets up to about 5 mm. wide. Occasional eastern American specimens resemble our form of the species.

Cicuta douglasii (DC.) Coult. & Rose, Contr. U. S. Nat. Herb. 7:95. 1900.
  ? Sium ? douglasii DC. Prodr. 4:125. 1830. [Both question marks are by DC.] *(Douglas,* western North America)
  C. occidentalis Greene, Pitt. 2:7. 1889. *C. douglasii* var. *occidentalis* M. E. Jones, Bull. U. Mont. Biol. 15:42. 1910. *(Greene,* Trinidad, Colo. , July 17, 1889, and Bear Creek, 200 miles north of Trinidad, July 28, 1889)
  C. purpurata Greene, Pitt. 2:8. 1889. *(Greene,* "along the Yakima River near Clealum," Wash. , Aug. 14, 1889)
  C. vagans Greene, Pitt. 2:9. 1889. *(Greene,* Lake Pend d'Oreille, Ida. , Aug. 9, 1889)
  C. subfalcata Greene, Leafl. 2:237. 1912. (U. S. Forest Service, Gallatin National Forest, s. Mont. , Aug. 29, 1911)
  C. fimbriata Greene, Leafl. 2:240. 1912. *(Cooper,* "saline or brackish marshes of Washington near the sea, in 1854 or 1855)
  C. cinicola A. Nels. Bot. Gaz. 54:141. 1912. *(Nelson & Macbride 1315,* Rock Creek, near Twin Falls, Ida.)
  C. occidentalis f. *oregonensi-idahoensis* H. Wolff, Pflanzenr. IV, 228 (Heft 90):82. 1927. *(Cusick 2256,* Lower Wallowa Valley, Oreg.)
      Plants stout, 5-20 dm. tall, often glaucous, the stems solitary or few together from an erect or horizontal, tuberous-thickened and chambered base, several of the roots generally tuberous-thickened as well; leaves basal and cauline, 1-3 times ternate-pinnate, with well-defined, lance-linear to lance-ovate, serrate leaflets mostly 3-10 cm. long and 6-35 cm. wide, the primary lateral veins tending to be directed to the marginal sinuses; umbels several, the rays 2-6 cm. long at maturity; fruit rotund-ovate to orbicular, 2-4 mm. long, somewhat constricted at the commissure, the corky-thickened, subequal ribs wider than the often darker intervals. N=22.
      Marshes, edges of streams and ditches, and in other wet low places, from the plains and lowlands to the mt. valleys; Alas. to s. Calif. , e. to Alta. , Colo. , N. M. , and Chihuahua. June-Aug.
      Cicuta douglasii is highly poisonous to man as well as to livestock. It causes many deaths of cattle, especially in early spring when the shallow root is easily pulled up along with the young tops by cattle hungry for green feed. All parts of the plant are poisonous, but the tuberous-thickened roots and stem base are especially so. The basal parts of one plant are enough to kill a cow. The other species of the genus are also poisonous, but the only other species found in our range, *C. bulbifera,* is relatively uncommon, while *C. douglasii* is frequently found along streams in pastures.
      The pattern of leaf venation will serve to distinguish *C. douglasii* from most or all of the species in various other genera with which it has been confused. In *C. douglasii* the primary lateral veins tend to be directed to the sinuses rather than to the teeth of the leaflets. Often the venation is very regular, with one primary lateral vein for each sinus. Frequently the vein splits just before reaching the sinus, the upper or main branch continuing beyond the sinus to form the lower margin of one tooth, while the lower branch extends into the other tooth near its upper margin.

Conioselinum Hoffm.

      Inflorescence of one or more compound umbels; involucre of a few narrow or leafy bracts, or wanting; involucel of well-developed, narrow, often more or less scarious bractlets; flowers white; calyx teeth obsolete; stylopodium conic; carpophore bifid to the base or nearly so; fruit elliptic or elliptic-oblong, dorsally compressed, glabrous, the lateral ribs evidently thin-winged, the dorsal ribs more narrowly so or low and corky; oil tubes 1-2 in the intervals, 2-4 on the commissure, often not reaching the base of the fruit; perennial from a taproot or cluster of fleshy-fibrous roots, sometimes with a caudex, generally single-stemmed, the stem more or less leafy; leaves pinnately or ternate-pinnately decompound, without well-defined leaflets, or the leaflets more or less deeply cleft.
      About 10 species of N. Am. and Eurasia, mostly in boreal or mountainous regions. (Named for its resemblance to *Conium* and *Selinum.*)
      Closely related to *Angelica,* and apparently not distinguished from it by any one constant character. Forms of *C. pacificum* with a more or less definite taproot and with less than usually dissected leaves might well be mistaken for an *Angelica.*

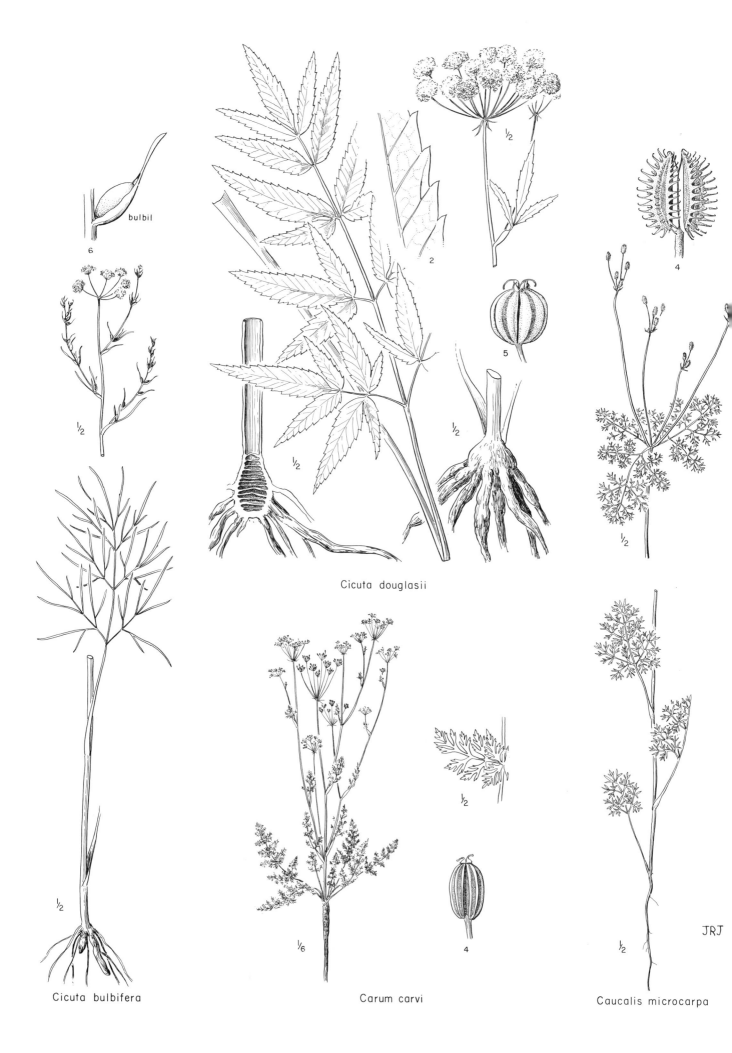

bulbil

6

½

½

½

½

2

5

½

4

½

Cicuta douglasii

½

Cicuta bulbifera

1/6

4

Carum carvi

½

½

JRJ

Caucalis microcarpa

Conioselinum pacificum (Wats.) Coult. & Rose, Contr. U.S. Nat. Herb. 7:152. 1900.

Selinum pacificum Wats. Proc. Am. Acad. 11:140. 1876. *(Kellogg & Harford 315,* Saucelito hills, near San Francisco)

C. gmelini (Cham. & Schlecht.) Coult. & Rose, 1900; not *C. gmelini* (Bray) Steud. 1840.

C. fischeri sensu Hook. and other authors, not Ledeb. 1844.

Selinum benthami Wats. Bibl. Ind. 432. 1878. *C. benthami* Fern. Rhodora 28:221. 1926. (Alas.; typification obscure)

Selinum hookeri Wats. ex Coult. & Rose, Rev. N. Am. Umbell. 45. 1888. (Alas. to Oreg.; the typification obscure)

Single-stemmed perennial 2-10 (or reputedly 15) dm. tall, typically from a short, stout rhizome-caudex with fleshy-fibrous roots, varying to more or less distinctly taprooted; plants essentially glabrous except for the scabrous roughening of the rays of the umbel and often also of the margins and main veins of the leaf segments, sometimes more or less glaucous; leaves all cauline, pinnately 2-4 times compound, typically giving the appearance of having deeply cleft leaflets, the leaflets mostly less definite than in *Angelica;* umbels 1-several, with 8-30 subequal rays mostly 1. 5-5 cm. long at maturity; involucre of a few narrow bracts, or more often wanting, seldom of leafy bracts; involucel of well-developed narrow bractlets commonly equaling or surpassing the pedicels, these 4-8 mm. long at maturity; fruit 5-8. 5 mm. long, the lateral wings well developed, the dorsal ones less so.

Bluffs and rocky or sandy beaches along the seashore; Alas. and the Aleutian Is. to Calif. July-Aug.

This species has often been confused with the wholly distinct *C. chinense* (L.) B. S. P. , an e. Am. species of cold swamps, marked by its smaller fruits that are mostly 2. 5-4. 5 mm. long. The misleading name *C. chinense* reflects a misunderstanding as to the source of the original material, which came from New York.

Conium L. Poison Hemlock

Inflorescence of compound umbels; involucre and involucel of several small, lanceolate to ovate bracts or bractlets; flowers white; calyx teeth obsolete; styles reflexed, the stylopodium depressed-conic; carpophore entire; fruit glabrous, broadly ovoid, laterally somewhat flattened, with prominent, raised, often wavy and somewhat crenate, almost winged ribs; oil tubes numerous and small, more or less confluent; glabrous biennials from a stout taproot with purple-spotted, freely branching hollow stem, large, pinnately or ternate-pinnately dissected leaves with rather small ultimate segments, and numerous terminal and axillary pedunculate compound umbels.

Two species, one Eurasian, the other South African. (Name derived from *koneion,* the ancient Greek name of *Conium maculatum.)*

Conium maculatum L. Sp. Pl. 243. 1753. (Europe)

Plants 0. 5-3 m. tall, coarse and freely branched, the blades of the larger leaves commonly 1. 5-3 dm. long; umbels numerous, the rays subequal, mostly 1-4 cm. long at maturity; fruit 2-2. 5 mm. long.

Roadside ditches and other moist disturbed sites; native to Eurasia, now established as a weed over most of N. Am. , and often common in waste places in cities especially w. of the Cascades. May-Aug.

This highly poisonous species is the "hemlock" of classical antiquity.

Cymopterus Raf.

Inflorescence of compound, loose to capitate umbels on terminal peduncles or scapes; involucre of scarious to herbaceous bracts, or wanting; involucel of large or small, ovate to filiform, herbaceous to scarious or subcoriaceous bractlets all on one side of the umbellet, often basally connate; flowers white, yellow, or purple; calyx teeth evident to obsolete; styles slender, without a stylopodium; carpophore bifid to the base; fruit ovoid to oblong, somewhat flattened dorsally, the lateral and usually one or more of the dorsal ribs conspicuously winged; oil tubes 1-many in the intervals, 2-many on the commissure, and sometimes one under each wing; low, taprooted perennials, caulescent or acaulescent, often with a pseudoscape, the leaves ternately or pedately to more often pinnately more or less dissected, with mostly small ultimate segments.

About 40 species of w. N. Am. (Name from the Greek *kyma,* wave, and *pteron,* wing.)

The genus has often been divided into several (sometimes as many as 9) smaller genera which lack substantial distinguishing characters. Mathias & Constance recognize two of the segregates *(Pteryxia* and *Pseudocymopterus)* in addition to *Cymopterus* in their treatment for the N. Am. Flora, but some of the species which they refer to *Cymopterus* are habitally more like *Pteryxia*. It is possible with some difficulty to divide *Cymopterus* in the broad sense into two groups on the basis of habit, as indicated in the following key to our species. If such a course were followed, 25 of the 32 species referred to *Cymopterus* in the N. Am. Flora would remain in *Cymopterus,* while the other 7 species, plus the species of *Pteryxia* and *Pseudocymopterus,* would form another group for which the oldest name is *Pseudocymopterus.* Mathias believes (personal communication) that this treatment would also cut across the natural relationships in some instances, so that it would be no more satisfactory than the treatment given in the N. Am. Flora. I am therefore constrained to expand the limits of *Cymopterus* to include *Pteryxia* and *Pseudocymopterus* as well as the other segregates which have been included by Mathias & Constance. As so limited *Cymopterus* is an essentially natural as well as morphologically recognizable group. Mathias believes that the available evidence at least permits the interpretation here adopted.

1 Plants with one or two (seldom more) short to elongate pseudoscapes arising from the
    simple or occasionally few-branched subterranean crown of the taproot; leaves tending
    to form a flat rosette
  2 Flowers white; fruiting umbel very dense and headlike; species primarily of the Great
      Plains and adj. westerly intermontane valleys, extending into the drier valleys of
      c. Ida. and thence through the Snake R. plains to Harney Co., Oreg.   C. ACAULIS
  2 Flowers yellow; fruiting umbel fairly compact, but not at all headlike; mts. of c. Ida.
      to n.w. Mont. (w. of the continental divide)   C. GLAUCUS
1 Plants caespitose, with numerous leaves and several or numerous peduncles or bas-
    ally leafy stems arising from the surficial branched caudex which surmounts the
    taproot
  3 Flowers white; calyx teeth minute and scarious, generally blunt; plants acaulescent
    4 Leaves subtripinnatisect, some of the secondary segments generally again cleft,
      the ultimate segments numerous and crowded; rays of the umbel, or some of
      them, often more than 5 (to 17) mm. long at maturity; foothills to high
      elevations in the mts. of w. Mont., c. Ida., and southward   C. BIPINNATUS
    4 Leaves sub-bipinnatisect, some of the distal primary segments generally entire,
      the other primary segments generally only once pinnatifid, the ultimate seg-
      ments fewer and less crowded than in *C. bipinnatus;* rays of the umbel less than
      5 mm. long; local at high elevations in the mts. of c. Ida. and n.e. Nev.   C. NIVALIS
  3 Flowers yellow; calyx teeth evident, usually rather narrow and pointed, mostly green
    5 Leaves appearing very open and skeletonlike, the segments remote; plants ordinarily
      caulescent; along the s. edge of our range, and southward   C. PETRAEUS
    5 Leaves with more dense, crowded segments
      6 Leaves more or less ovate in outline, ternate-pinnately dissected; plants caules-
        cent or sometimes acaulescent; widespread e. of the Cascades   C. TEREBINTHINUS
      6 Leaves oblong in outline, pinnately dissected; plants acaulescent; s.w. Mont.
        (Beartooth Mts.) to c. Ida. and southward   C. HENDERSONII

Cymopterus acaulis (Pursh) Raf. Herb. Raf. 40. 1838.
    Selinum acaule Pursh, Fl. Am. Sept. 732. 1814. *(Bradbury,* Missouri River, from the "river Nadu-
    et of the Mahas")
    C. leibergii Coult. & Rose, Contr. U.S. Nat. Herb. 7:182. 1900. *C. glomeratus* var. *leibergii* M.
    E. Jones, Contr. West. Bot. 12:25. 1908. *(Leiberg 2253,* Malheur Valley, near Harper Ranch,
    Oreg.)

Glabrous or obscurely viscid perennial with one or two slender pseudoscapes arising from the subterranean apex of an elongate, thickened, often upwardly tapering taproot, the pseudoscape not reaching much if at all above the ground level; leaves clustered on the pseudoscape, tending to form a rosette, petiolate, the blade 1.5-7 cm. long and 1-5 cm. wide, pinnately 2-3 times divided into rather small and narrow, scarcely crowded ultimate segments; peduncles usually several, 2-10 cm. long at maturity; inflorescence compact and subcapitate in flower, dense and strongly capitate in fruit, the rays seldom as much as 1 cm. long and the pedicels only about 1 mm. long; involucre wanting or rarely vestigial; involucel of prominent, narrow, partly connate, green or greenish, seldom scarious-margined bractlets; calyx teeth inconspicuous to evident, acutish to sharply acute, subscarious, sometimes fully 1 mm. long; flowers white; fruit ovoid to broadly oblong, 5-10 mm. long, 3-8 mm. wide, its prominent wings equaling or narrower than the body, often slightly wavy, but seldom corrugate;

dorsal wings 1-3, similar to the lateral ones, or narrower; oil tubes 3-17 in the intervals, 5-13 on the commissure, and sometimes one under each wing.

Dry flats and hillsides in the plains and valleys; c. Sask. and w. Minn. to Colo., Mont., and the drier intermontane valleys of c. Ida., thence through the Snake R. plains of Ida. to Harney Co., Oreg. Apr.-May.

Cymopterus bipinnatus Wats. Proc. Am. Acad. 20:368. 1885.
> *Pseudocymopterus bipinnatus* Coult. & Rose, Rev. N. Am. Umbell. 75. 1888. *Pseudoreoxis bipinnatus* Rydb. Bull. Torrey Club 40:73. 1913. *(Hayden 14, Rocky Mts. s. of Virginia City, Mont.)*
> *Cynomarathrum macbridei* A. Nels. Bot. Gaz. 54:142. 1912. *(Macbride 1502, Bear Canyon, Lemhi National Forest, Ida.)*

Glabrous or more or less scabrous-puberulent perennial from a stout taproot and branching caudex that tends to be clothed by the leaf bases from previous years; leaves numerous, all basal, glaucous, petiolate, the blade 1.5-7 cm. long and 0.5-2 cm. wide, subtripinnatisect, some of the secondary segments generally again cleft, the ultimate segments numerous and crowded, small and narrow; peduncles naked, arising from the caudex, 5-25 cm. long at maturity; umbel compact and sometimes headlike, the 3-5 fertile rays 1-17 mm. long; involucre wanting; involucel of linear to obovate, scarious bractlets with green or dark midvein; calyx teeth minute and scarious, generally blunt, up to about 0.5 mm. long; flowers white; fruit ovoid-oblong, 3-6 mm. long, the wings narrower than the body; dorsal wings 3, similar to the lateral ones, or narrower; oil tubes 1-7 in the intervals, 4-8 on the commissure.

Open, often rocky places, from the foothills to above timber line in the mountains; s.w. Mont. (n. to Helena) and w. Wyo. to c. Ida., and in the Steens Mts. of Oreg. and the mountains of n.e. Nev. and n. Utah. May-July.

Cymopterus glaucus Nutt. Journ. Acad. Phila. 7:28. 1834.
> *Aulospermum glaucum* Coult. & Rose, Contr. U.S. Nat. Herb. 7:176. 1900. *(Wyeth, on the borders of Flat-Head River, towards the sources of the Columbia)*

Glabrous and glaucous perennial from a stout taproot with a simple or few-branched subterranean crown, producing one short or elongate pseudoscape from the crown or each of its branches, the pseudoscape sometimes as much as 15 cm. long, and the upper part often distinctly aerial; leaves rather pale bluish-green, clustered on the pseudoscape, petiolate, the blade mostly 1.5-9 cm. long and 1-7 cm. wide, bipinnately or tripinnately cleft, with small, rounded or mucronate, crowded ultimate segments; peduncles 1-several on a pseudoscape, 5-15 cm. long at maturity; involucre of several narrow bracts; involucel of evident narrow bractlets; inflorescence fairly compact at anthesis but scarcely headlike, obviously umbellate at maturity, with unequal rays, the larger ones 1-4 cm. long; calyx teeth minute and more or less scarious, mostly well under 0.5 mm. long; flowers yellow; fruit oblong, 5-7 mm. long, the lateral wings equaling or narrower than the body, the dorsal ones often narrower than the lateral; oil tubes 3 or 4 in the intervals, about 6 on the commissure.

Open, usually gravelly or rocky slopes and flats from the foothills to above timber line in the mountains; c. Ida., from Idaho Co. to Boise Co., n. Camas and n. Elmore cos., and e. Custer Co., and w. of the continental divide in Mont. May-July.

CYMOPTERUS HENDERSONII (Coult. & Rose) Cronq. hoc loc.
> *Pseudocymopterus hendersoni* Coult. & Rose, Contr. U.S. Nat. Herb. 7:190. 1900. *Pseudopteryxia hendersonii* Rydb. Fl. Rocky Mts. 624, 1064. 1917. *Pteryxia hendersoni* Math. & Const. Bull. Torrey Club 69:248. 1942. *(Henderson 4068, summit of peak, source of Mill Creek, Ida.; this locality is w. of Challis, in Custer Co.)*

Glabrous, pleasantly aromatic perennial from a stout taproot and branching caudex that is clothed by the petioles and leaf bases from previous years; leaves numerous, all basal, petiolate, the blade mostly 2-13 cm. long and 1-3 cm. wide, oblong in outline, bi- or tripinnatisect, with small and narrow ultimate segments; peduncles naked, arising from the caudex, 5-30 cm. long at maturity; rays of the umbel unequal, the longer ones mostly 1-3 cm. long at maturity; involucre mostly wanting; involucel of prominent narrow green bractlets; calyx teeth well developed, rather narrow, 0.5-1 mm. long; flowers yellow; fruit ovoid-oblong, 4-7 mm. long, the wings narrower than the body; dorsal wings equaling or narrower than the lateral ones; oil tubes 1-5 in the intervals, 3-8 on the commissure.

Open, rocky places from the foothills to above timber line in the mountains; Beartooth Mts. of s.w. Mont., s. to n. N.M., w. to c. Ida., n.e. Nev., and s. Utah. Fruiting in mid- and late summer.

fruit

4

Conioselinum pacificum

1/2

1/2

Conium maculatum

1/8

1/2

5

4

1/2

4

Cymopterus glaucus

1/2

1/2

3

1/2

4

10

1/2

1/2

Cymopterus bipinnatus

1/2

1/2

4

umbel at anth

Cymopterus acaulis

JR

Cymopterus nivalis Wats. Bot. King Exp. 123. 1871.

Pseudoreoxis nivalis Rydb. Bull. Torrey Club 40:73. 1913. *Pseudocymopterus nivalis* Mathias, Ann. Mo. Bot. Gard. 17:327. 1930. *(Watson 448,* East Humboldt Mts., Nev.)

Resembling a dwarf form of *C. bipinnatus,* but with less-dissected leaves, differing as indicated in the key; mature fruit unknown.

Rocky places at high elevations in the mountains of Nev. (Nye Co., Elko Co.) and c. Ida.; rarely collected. July-Aug.

This is perhaps only a form of *C. bipinnatus.*

Cymopterus petraeus M. E. Jones, Contr. West. Bot. 8:32. 1898.

Pteryxia petraea Coult. & Rose, Contr. U.S. Nat. Herb. 7:172. 1900. *(Jones s.n.,* Palisade, Nev., June 14, 1882)

Glabrous, aromatic perennial from a stout taproot and branching caudex that is generally clothed by the leaf bases from previous years; leaves numerous, chiefly basal, petiolate, the blade mostly 4-17 cm. long, pinnately or subternate-pinnately 2-3 times dissected, with small ultimate segments, appearing very open and skeletonlike, the upper primary segments often much smaller than the lower; stems mostly 15-45 cm. high including the long terminal peduncle, commonly with 1 or more leaves below the middle, seldom strictly scapose; rays of the umbel markedly unequal, the longer ones mostly 1.5-5 cm. long at maturity; involucre wanting; involucel of inconspicuous, narrow, green or scarious-margined bractlets 1-3 mm. long; calyx teeth well developed, lance-linear to lance-ovate or triangular, barely to strongly acute, mostly 0.5-1 mm. long; flowers yellow; fruit ovoid-oblong, 4-7 mm. long, the wings equaling or narrower than the body; dorsal wings equaling or narrower than the lateral ones; oil tubes (1) 3 in the intervals, 5-15 on the commissure.

Open, rocky places in dry or desert regions, often in cliff crevices; Snake R. plains of Ida., to s. e. Oreg., s. to Inyo Co., Calif., and n. Ariz., barely reaching the southern edge of our range. May-June.

Cymopterus terebinthinus (Hook.) T. & G. Fl. N.Am. 1:624. 1840.

Selinum terebinthinum Hook. Fl. Bor. Am. 1:266. 1832. *Pteryxia terebinthina* Coult. & Rose, Contr. U.S. Nat. Herb. 7:171. 1900. *(Douglas,* sandy grounds of the Wallahwallah R.)

CYMOPTERUS TEREBINTHINUS var. FOENICULACEUS (T. & G.) Cronq. hoc loc. *C. foeniculaceus* T. & G. Fl. N.Am. 1:624. 1840. *Pteryxia foeniculacea* Nutt. ex Coult. & Rose, Contr. U.S. Nat. Herb. 7:171. 1900. *Pteryxia terebinthina* var. *foeniculacea* Mathias, Ann. Mo. Bot. Gard. 17:332. 1930. *(Nuttall,* Blue Mts. of Oreg.)

C. thapsoides T. & G. Fl. N.Am. 1:625. 1840. *Pteryxia thapsoides* Nutt. ex Coult. & Rose, Contr. U.S. Nat. Herb. 7:172. 1900. *(Nuttall,* Blue Mts. of Oreg.) = var. *foeniculaceus.*

? *C. albiflorus* T. & G. Fl. N.Am. 1:625. 1840. *Pteryxia albiflora* Nutt. ex Coult. & Rose, Contr. U.S. Nat. Herb. 7:173. 1900. *C. terebinthinus* var. *albiflorus* M. E. Jones, Contr. West. Bot. 10:56. 1902. *Pteryxia terebinthina* var. *albiflora* Mathias, Ann. Mo. Bot. Gard. 17:339. 1930. *(Nuttall,* hills of the Bear River in the Rocky Mountain Range) Identity uncertain.

CYMOPTERUS TEREBINTHINUS var. CALCAREUS (M. E. Jones) Cronq. hoc loc. *C. calcareus* M. E. Jones, Contr. West. Bot. 8:32. 1898. `Pteryxia calcarea` Coult. & Rose, Contr. U.S. Nat. Herb. 7:173. 1900. *Pteryxia terebinthina* var. *calcarea* Mathias, Ann. Mo. Bot. Gard. 17:334. 1930. *(Jones,* Green River, Wyo., June 23, 1896)

Pteryxia californica Coult. & Rose, Contr. U.S. Nat. Herb. 7:172. 1900. *C. californicus* M. E. Jones, Contr. West. Bot. 12:27. 1908. *C. terebinthinus* var. *californicus* Jeps. Man. Fl. Pl. Calif. 730. 1925. *Pteryxia terebinthina* var. *californica* Mathias, Ann. Mo. Bot. Gard. 17:337. 1930. *(Brown s.n.,* Sisson, Siskiyou Co., Calif., in 1897)

C. elrodi M. E. Jones, Bull. U. Mont. Biol. 15:41. 1910. *Pteryxia elrodi* Rydb. Fl. Rocky Mts. 621, 1064. 1917. *(Jones,* Alta, Ravalli Co., Mont., July 11, 1909) = var. *foeniculaceus.*

Glabrous, mildly and pleasantly aromatic perennial from a stout taproot and branching caudex that is clothed by the leaf bases from previous years; leaves numerous, chiefly basal or low-cauline, petiolate, the blade mostly 3-18 cm. long, broadly ovate in outline, from half as wide to fully as wide as long, ternate-pinnately dissected, with numerous small and crowded ultimate segments; stems mostly 10-60 cm. tall including the long terminal peduncle, commonly with one or more leaves below the middle, less often strictly scapose; rays of the umbel elongating unequally, the longer ones mostly 1.5-7 cm. long at maturity; involucre wanting; involucel of narrow green bractlets 2-6 mm. long; calyx teeth well developed, 0.5-1.5 mm. long, acutish to acuminate, generally greenish; flowers yellow; fruit ovoid to ovoid-oblong, 5-11 mm. long, the lateral wings about equaling or broader than the body, the dorsal ones similar to the lateral or more or less reduced; oil tubes 3-12 in the intervals, 6-20 on the commissure.

Dry, open, often rocky or sandy places, from the lowlands to moderate elevations in the mountains; Wash. to Calif., e. to w. Mont., Wyo., and n. Colo., wholly e. of the Cascades in our range. Apr.- May, or to June or July at higher elevations.

The species consists of 4 wholly confluent varieties, two of which occur in our range. Both of our varieties generally have bright green foliage. They differ as follows:

1 Wings of the fruit more or less strongly crisped, generally all about alike and equaling
    or often broader than the body; inflorescence at anthesis mostly 2-7 cm. wide; ultimate
    segments of the leaves usually relatively short and broad; Columbia Plateau of Wash.
    and n. Oreg., mostly at low elevations, often in sand dunes   var. terebinthinus
1 Wings of the fruit scarcely or not at all crisped, the dorsal ones often narrower than the
    lateral ones, which are often no broader than the body; inflorescence at anthesis most-
    ly 1.5-3 cm. wide; ultimate segments of the leaves relatively long and slender; foot-
    hills to moderate elevations in the mountains of c. Ida., from Idaho Co. to the border
    of the Snake R. plains, extending e. into Ravalli Co., Mont., and w. into the mts. and
    foothills of n.e. and c. Oreg. (and extreme s.e. Wash.) as far as the Maury Mts. and
    Ochoco Mts.   var. foeniculaceus (T. & G.) Cronq.

Two other varieties occur to the s. of our range. Both have rather firm, mostly gray-green foliage with relatively short and broad ultimate segments, and have the wings of the fruit essentially plane. The var. *calcareus* (T. & G.) Cronq., with initially small inflorescence as in var. *foenicula-ceus,* ranges from the c. and upper Snake R. plains (chiefly or wholly from Twin Falls Co. eastward) to Wyo., n. Colo., n. Utah, and n.e. Nev. The var. *californicus* (Coult. & Rose) Jeps., differing from the var. *calcareus* chiefly in being larger in all respects, occurs in the Sierra Nevada region of Calif. and adj. Nev., n. into the southern Cascades and coast ranges of n. Calif. The var. *terebin-thinus* is fairly well marked, especially by its fruits, but the other three varieties are so poorly de-fined that many specimens might be misidentified in the absence of geographic data.

Daucus L.

Inflorescence of compound umbels; involucre of numerous dissected or entire bracts, or wanting; involucel of numerous toothed or entire bractlets, or wanting; flowers white, the central flower of the umbel or of each umbellet often purple, or rarely all the flowers pink or yellow; outer flowers of the umbel or the umbellet often with slightly enlarged and irregular corolla; calyx teeth evident to obsolete; styles short, the stylopodium conic; carpophore entire or bifid at the apex; fruit oblong to ovoid, slightly compressed dorsally, evidently ribbed, beset with stout, spreading, glochidiate or barbed prickles along alternate ribs, merely bristly or hairy on the others; oil tubes solitary under alternate ribs, and two on the commissure; taprooted annuals or biennials, leafy-stemmed, the leaves pinnately dissected, with small and narrow ultimate segments.

About 25 species, widely distributed, only one native to the U.S. (The ancient Greek name of some umbellifer.)

1 Relatively coarse biennial; involucral bracts scarious margined below, the segments
    firm, elongate, filiform-subulate; umbellets mostly (10) 20- to many-flowered   D. CAROTA
1 Relatively slender annual; involucral bracts not scarious-margined, the segments lin-
    ear or lanceolate, scarcely elongate; umbellets mostly 5- to 12-flowered   D. PUSILLUS

Daucus carota L. Sp. Pl. 242. 1753.
    Caucalis carota Crantz, Class. Umbell. 113. 1767. *Caucalis daucus* Crantz, Stirp. Austr. 3:125.
    1767. *Carota sativa* Rupr. Fl. Ingr. 468. 1860. (Europe)
Relatively coarse biennial, 2-12 dm. tall, more or less hirsute throughout to subglabrous, single-stemmed from a well-developed bitter taproot; leaf blades mostly 5-15 cm. long and 2-7 cm. wide, the ultimate segments linear to lanceolate, 2-12 mm. long and 0.5-2 mm. wide, or those of the upper leaves more elongate; peduncles several or solitary, 2.5-6 dm. long; involucral bracts generally pin-natifid into firm, elongate, filiform-subulate segments, evidently scarious-margined from the base to the lowermost pair of segments, spreading or reflexed in fruit; bractlets of the involucel mostly lin-ear and entire, equaling or exceeding the flowers; inflorescence at anthesis showy, commonly 4-12 cm. wide, the numerous rays unequal; ultimate umbellets with mostly (10) 20-numerous flowers; flow-

ers white or yellowish, the central flower of the umbel commonly purple or pinkish, or rarely the flowers all pink; inflorescence commonly narrower in fruit than in flower, the outer rays longer than the others (to 7 cm.) and arching inwards; fruit ovoid, 3-4 mm. long, about 2 mm. broad, broadest at the middle. N=9. Wild carrot, Queen Anne's lace.

Roadsides, disturbed sites, and moist open places; native of Eurasia, now established as an occasional weed throughout most of N. Am. July-Aug.

This is the wild ancestor of the common cultivated carrot. The cultivated plants differ from the wild ones chiefly in the size and taste of the roots, and are regarded as being conspecific with them.

Daucus pusillus Michx. Fl. Bor. Am. 1:164. 1803. (Carolina)

D. microphyllus Presl ex DC. Prodr. 4:213. 1830. *D. pusillus* var. *microphyllus* T. & G. Fl. N. Am. 1:636. 1840. *(Haenke,* Nootka Sound)

Relatively slender annual, 0.5-6 (9) dm. tall, more or less hirsute throughout, single-stemmed from a slender taproot; leaf blades mostly 3-10 cm. long and 1.5-7 cm. wide, the ultimate segments linear, 1-5 mm. long, 0.5-1 mm. wide; peduncles solitary or several, up to 4 dm. long; involucral bracts pinnatifid into linear or lanceolate segments that are seldom as much as 8 mm. long, not scarious margined, closely appressed to the inflorescence in fruit; bractlets of the involucel mostly linear and entire, about equaling the pedicels; inflorescence at anthesis mostly 1-4 cm. wide, scarcely showy, slightly or scarcely larger (to 5 cm. wide) at maturity, the rays seldom as much as 4 cm. long; ultimate umbellets with mostly 5-12 flowers; flowers all white, or sometimes purplish; fruit oblong, 3-5 mm. long, about 2 mm. broad, usually broadest below the middle.

Dry, open places at lower elevations, often about rock outcrops; s. B.C. to Baja Calif., e. to Mo., S.C., and Fla., chiefly w. of the Cascades in our range. May-July.

Eryngium L.

Inflorescence of dense, bracteolate heads terminating the branches, the bracteoles representing the involucels of the 1-flowered, sessile umbellets; involucre of one or more series of entire or variously toothed or cleft bracts subtending the head; flowers sessile, white to blue or purple, the corolla lobes sometimes fimbriate; calyx lobes well developed and conspicuous, firm and sometimes spinescent; stylopodium and carpophore wanting; fruit globose to obovoid, slightly or scarcely flattened laterally, with a broad commissure, variously covered with scales or tubercles, the ribs obsolete; oil tubes mostly 5; taprooted or fibrous-rooted biennial or perennial herbs, caulescent or acaulescent, usually glabrous, the leaves commonly rather firm, entire or toothed to deeply cleft and often more or less spinose-toothed, the venation often subparallel; petioles often septate-nodose.

About 200 species, widely distributed in temperate and subtropical regions. (Name evidently Greek, but of uncertain origin and significance.)

1 Blades of the basal leaves almost equaling or longer than the petioles, rounded or sub-
    cordate at the base; heads blue or bluish; European weed, once collected at Salem,
    Oreg., but probably not fully established as a member of our flora   *E. planum* L.
1 Blades of the basal leaves much shorter than the petioles, tapering to the base, or the
    basal petioles bladeless; native species, not weedy
    2 Blades of the larger leaves up to about 1 cm. wide; heads green or greenish; w. of the
        Cascades, and extending to the e. end of the Columbia R. Gorge   E. PETIOLATUM
    2 Blades of the larger leaves mostly 1-3 cm. wide; e. of the Cascades, in our range
        3 Heads blue, the corolla blue and the bracteoles and calyx lobes bluish-tinged; plants
            mostly 3-10 dm. tall; n. Ida. and adj. Wash., irregularly s. to c. Calif.   E. ARTICULATUM
        3 Heads green or greenish, the corolla white, the bracteoles and calyx lobes scarcely
            or not at all bluish; plants mostly 1-3 dm. tall; n.e. Calif. and n. Nev. to n. Har-
            ney Co., Oreg., probably not quite reaching our range   *E. alismaefolium* Greene

Eryngium articulatum Hook. Lond. Journ. Bot. 6:232. 1847. *(Geyer 583,* stony edges of the Spokane River, and Skitsoe and Coeur d'Alene Lakes, Idaho)

Fibrous-rooted perennial from a short simple crown, 3-10 dm. tall, divaricately branched above; basal petioles elongate, hollow, septate-nodose, often bladeless; cauline leaves with progressively shorter petioles, the upper sessile; blades mostly 1-3 cm. wide, commonly more or less elliptic, spinulose-toothed or remotely spinulose-ciliate; heads mostly 1-1.5 cm. wide, the bracteoles and calyx lobes more or less prominently bluish-tinged and the corolla blue; calyx lobes 3-5 mm. long,

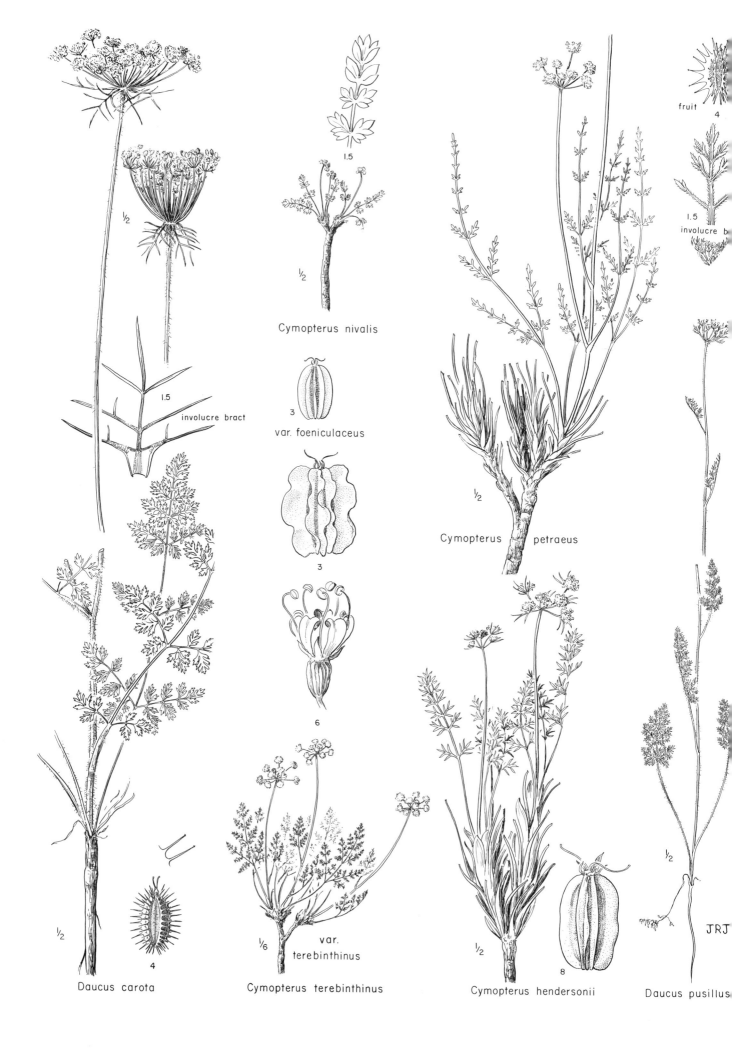

1.5

1/2

1/2

1.5

involucre bract

Cymopterus nivalis

3
var. foeniculaceus

3

6

var.
terebinthinus

Cymopterus terebinthinus

1/6

Daucus carota

1/2

1/2

4

Cymopterus petraeus

Cymopterus hendersonii

1/2

8

fruit

4

1.5

involucre b

Daucus pusillus

1/2

JRJ

acuminate; fruit ovoid, 2-3 mm. long, densely covered with minutely cellular-reticulate scales 0.3-1 mm. long, these with an inflated hollow base abruptly tipped by a slender point. N=16.

Low ground along streams and lakes, often where submerged in the spring; n. Ida. and adj. Wash., irregularly s. to c. Calif., wholly e. of the Cascades in our range. July-Sept.

Eryngium petiolatum Hook. Fl. Bor. Am. 1:259. 1832. *(Douglas,* plains of the Multnomah)
  E. petiolatum var. *juncifolium* Gray, Proc. Am. Acad. 8:385. 1872. *(E. Hall 200,* Oreg.)
  Fibrous-rooted perennial from a short simple crown, 1.5-5 dm. tall, divaricately branched, often from near the base; basal petioles more or less elongate, hollow, septate-nodose, commonly blade-less; cauline leaves with progressively shorter petioles, the blades narrowly elliptic, up to about 1 cm. wide, spinulose-toothed or subentire; heads mostly about 1 cm. wide or less; bracteoles often conspicuously surpassing the flowers and fruit, sometimes slightly bluish-tinted, but the heads not appearing bluish; calyx lobes about 3 mm. long, acuminate; corolla apparently white; fruit more or less ovoid, about 2 mm. long, covered with prominent gradually tapering scales that are basally hollow, slightly inflated, and mostly 1-1.5 mm. long.

Low ground, especially in places submerged in the spring and drier in the summer; Willamette Valley of Oreg., extending up the Columbia R. in Oreg. and Wash. to the e. end of the gorge. June-Aug.

Foeniculum Adans.

Inflorescence of compound umbels, the rays several or rather numerous; involucre and involucel wanting; flowers yellow; calyx teeth obsolete; styles very short, recurved, the stylopodium conic; carpophore 2-cleft to the base; fruit oblong, subterete or slightly flattened laterally, glabrous, with prominent ribs; oil tubes solitary in the intervals, 2 on the commissure; taprooted biennials or perennials with a strong anise odor, glabrous and glaucous, erect and leafy stemmed, the leaves petiolate, pinnately dissected with elongate, filiform ultimate segments; petioles broad and somewhat sheathing.

Four species of the Old World. (Name a diminutive of the Latin *foenum,* hay, referring to the odor.)

Foeniculum vulgare Mill. Gard. Dict. 8th ed. Foeniculum no. 1. 1768.
  Anethum foeniculum L. Sp. Pl. 263. 1753. *F. officinale* Allioni, Fl. Pedem. 2:25. 1785. *F. foeniculum*
    Karst. Deutsch. Fl. 837. 1882. (Europe)
  Stout short-lived perennial, 1-2 m. tall when well developed, the solitary stem commonly branched above; leaves (exclusive of the petioles) sometimes as much as 3 or 4 dm. long and broad, the ultimate segments filiform, 4-40 mm. long and well under 1 mm. wide; rays of the umbel 10-40, unequal, mostly 2-8 cm. long at maturity; fruit 3.5-4 mm. long. N=11. Sweet fennel.

Roadsides and waste places; native of the Mediterranean region, now widely introduced elsewhere, and found throughout much of the U.S., especially toward the south; established at least west of the Cascades in our range. July-Sept.

Glehnia Schmidt

Inflorescence of compound umbels on terminal peduncles, the rays of the primary umbel well developed; ultimate umbels capitate, the pedicels obsolete; involucre of a few narrow bracts, or wanting; involucel of several well-developed, lance-attenuate bractlets; flowers white; calyx teeth minute; styles short, without a stylopodium; carpophore bifid to the base; fruit ovate-oblong to subglobose, somewhat compressed dorsally, the ribs all broadly corky-winged; oil tubes several in the intervals, 2-6 on the commissure; stout, low, shortly or scarcely caulescent, taprooted perennials with once or twice ternate or ternate-pinnate leaves and evident, broad, toothed leaflets.

Two closely related species of North Pacific maritime distribution. (Presumably named for Peter von Glehn, 1835-1876, curator of the Botanic Garden at St. Petersburg.)

Glehnia leiocarpa Mathias, Ann. Mo. Bot. Gard. 15:95. 1928.
  Cymopterus? littoralis Gray, Pac. R.R. Rep. 12:62. 1860. *G. littoralis* ssp. *leiocarpa* Hultén,
    Fl. Alas. 7:1180. 1947. *(Cooper s.n.,* Shoalwater Bay, Wash., in 1854) Not *Glehnia littoralis*
    Schmidt, 1867, which is based directly on Asiatic material.
  Stout low perennial, nearly acaulescent, or with the stem and strongly sheathing petioles buried in the sand; leaves spreading, often prostrate, thick and firm, glabrous above, tomentose beneath, the

fruit

5

1/12

1/2

1/2

Foeniculum vulgare

15
scale

1/2

Eryngium petiolatum

2

fruit

15

scale

15
scale

2

fruit

Eryngium articulatum

1/2

2

Glehnia leiocarpa

J

leaflets crenate or crenate-serrate with callous teeth, 2-7 cm. long, broadly elliptic to obovate, or broader and deeply trilobed; peduncles stout, solitary or few, less than 1 dm. long; rays of the umbel mostly 5-13, up to 4. 5 cm. long at maturity, evidently woolly; fruit 6-13 mm. long, glabrous or with a few long hairs toward the tip.

Dunes and sandy beaches along the coast, from Kodiak I. , Alas. , to Mendocino Co. , Calif. June-July.

The closely allied but geographically apparently wholly disjunct *G. littoralis* Schmidt, of the Asiatic Pacific coast and islands, differs chiefly in its evidently hairy fruits.

Heracleum L. Cow Parsnip

Inflorescence of compound umbels on terminal and axillary peduncles, the rays unequal; involucre wanting, or of a few deciduous bracts; involucel of numerous slender bractlets, or rarely wanting; flowers white (yellow) or tinged with green or red; calyx teeth minute or obsolete; outer flowers of at least the marginal umbellets generally irregular, the outer corolla lobes enlarged and often bifid; stylopodium conic, the styles short, erect or recurved; carpophore bifid to the base; fruit orbicular to obovate or elliptic, strongly flattened dorsally, usually pubescent, the dorsal ribs narrow, the lateral ones broadly winged; oil tubes 2-4 on the commissure, solitary in the intervals, extending only part way from the stylopodium toward the base of the fruit, readily visible to the naked eye; large, coarse biennial or perennial herbs from a taproot or a fascicle of fibrous roots; leaves large, petiolate, ternately or pinnately compound, with broad, toothed or cleft leaflets; petioles sheathing and usually conspicuously expanded.

A circumboreal genus of about 60 species, only one native to N.Am. (Supposedly named for Hercules.)

Heracleum lanatum Michx. Fl. Bor. Am. 1:166. 1803.

Sphondylium lanatum Greene, Man. Bay Reg. 157. 1894. *Pastinaca lanata* K.-Pol. Bull. Soc. Nat. Mosc. II, 29:113. 1916. *(Michaux,* Canada)

H. douglasii DC. Prodr. 4:193. 1830. *(Douglas,* western North America)

Robust, single-stemmed perennial from a stout taproot or cluster of fibrous roots, 1-3 m. tall, thinly tomentose or woolly-villous at least on the lower surfaces of the leaves and toward the inflorescence, varying to sometimes nearly glabrous; leaves once ternate, with broad, distinctly petiolulate, coarsely toothed and palmately lobed leaflets mostly 1-3 (4) dm. long and wide, the lateral ones mostly narrower than the central one and often asymmetrical; involucre of 5-10 deciduous narrow bracts 0. 5-2 cm. long; involucel of bractlets similar to the bracts; rays mostly 15-30, up to 10 cm. long or even longer, the terminal umbel commonly 1-2 dm. wide; pedicels 8-20 mm. long; flowers white; fruit obovate to obcordate, 7-12 mm. long, 5-9 mm. wide, slightly hairy or glabrous. N=11.

Stream banks and moist low ground, from the lowlands to moderate elevations in the mountains; Alas. to Newf. , s. to Calif. , Ariz. , and Ga. ; also in Siberia and the Kurile Islands. June-Aug.

Hydrocotyle L. Water Pennywort

Inflorescence usually a simple umbel, sometimes proliferous or an interrupted spike, the pedicels ascending to reflexed; involucre inconspicuous or wanting; flowers white, greenish, or yellow; calyx teeth minute or obsolete; stylopodium conspicuously conic to depressed; fruit orbicular or ellipsoid, more or less flattened laterally, the dorsal surface rounded or acute, with narrow, acute ribs, or the ribs obsolete; low perennials with creeping or floating stems that root at the nodes, and with petiolate, not much divided, often peltate leaves.

About 75 species, widely distributed in both hemispheres, especially in tropical and warm-temperate regions; 17 spp. in N.Am. (Name from the Greek *hydor,* water, and *kotyle,* flat cup, referring to the peltate leaves of some species.)

Hydrocotyle ranunculoides L. f. Suppl. 177. 1781. *(Mutis,* Mexico)

Plants glabrous; stems floating or creeping, with well-developed internodes; petioles weak, elongate, mostly 0. 5-3.5 dm. long; leaf blades rotund-reniform, with an evident basal sinus, not peltate, mostly 1-6 cm. wide, 5- or 6-lobed to about the middle, the lobes with a few broad teeth or rounded lobules; peduncles axillary, much shorter than the petioles, mostly 1-5 cm. long, recurved in fruit; umbel simple, 5- to 10-flowered, the ascending or spreading pedicels only 1-3 mm. long; stylopodium

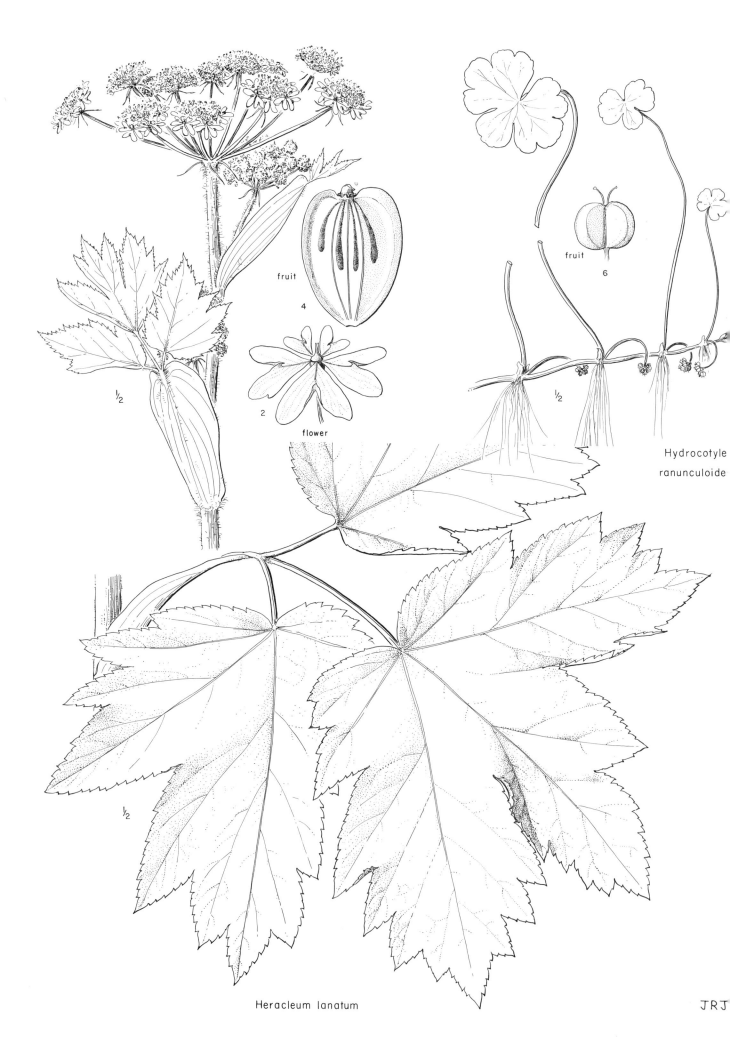

fruit

4

flower

2

fruit

6

Hydrocotyle
ranunculoide

½

½

½

Heracleum lanatum

JRJ

depressed; fruit suborbicular, 1-3 mm. long, 2-3 mm. wide, the dorsal surface rounded, the ribs obsolete.

Marshes, ponds, and wet ground; tropical Am., n. to Del., Ark., Calif., and w. of the Cascade Mts. to the Puget Sound region of Wash.

Hydrocotyle umbellata L., with merely crenate or lobulate, distinctly peltate leaves, extends n. to s. Oreg., but is not known to enter our range.

Ligusticum L.

Inflorescence of compound umbels; involucre and involucel wanting, or of a few inconspicuous narrow bracts or bractlets; flowers white or sometimes pinkish; calyx teeth evident or obscure; styles short, spreading, the stylopodium low-conic; carpophore bifid to the base; fruit oblong to ovate or suborbicular, subterete or slightly compressed laterally, glabrous, the ribs evident, often winged; oil tubes 1-6 in the intervals, 2-10 on the commissure; taprooted perennials, with basal and cauline or wholly basal leaves, the leaves ternately or ternate-pinnately compound or dissected, with or without well-defined broad leaflets.

About 20 species of both hemispheres, mostly in the north temperate regions. (Name from *Ligustikon*, the classical Greek name of an umbellifer from the region of the Ligurians, whose territory formerly included parts of what is now s.e. France as well as the present Italian province of Liguria.)

1 Leaves dissected into numerous more or less linear ultimate segments mostly 1-3
  mm. wide
  2 Plants relatively robust, generally 5-10 dm. tall, with one or more, more or less
    well-developed cauline leaves; basal leaves usually 10-25 cm. wide; rays of the
    main umbel mostly (10) 12-20; fruit 5-7 mm. long; Fremont Co., Ida., and Madison Co., Mont., s. through the mts. of w. Wyo. and e. Ida. to the Wasatch region of Utah   L. FILICINUM
  2 Plants relatively small and slender, 1-6 dm. tall, scapose or more often subscapose, the single cauline leaf much reduced or wanting; basal leaves seldom as much as 10 cm. wide; rays of the umbel mostly 5-13; fruit 3-5 mm. long; Wallowa and high Blue mts. of n.e. Oreg., e. across c. Ida. to s.w. Mont.; mts. of Colo. and the Uinta Mts. of Utah, but apparently not yet known from Wyo.   L. TENUIFOLIUM

1 Leaves less dissected, the ultimate segments broader, more or less toothed or
  cleft
  3 Ribs of the fruit narrowly winged; in and e. of the Cascade Mts.
    4 Leaflets relatively large and broad, mostly 3-8 cm. long and 2-5 cm. wide;
      robust leafy-stemmed plants mostly 1-2 m. tall; n. Ida. and adj. Mont.
                                       L. VERTICILLATUM
    4 Leaflets smaller, mostly 1-5 cm. long and 0.5-2 cm. wide; smaller, less
      leafy plants, up to about 12 dm. tall
      5 Rays of the terminal umbel 15-40; plants 5-12 dm. tall, with 1 or more,
        more or less well-developed cauline leaves; c. and n.c. Wash. to w. Mont.
        (w. of the continental divide), s. to Valley Co., Ida., and the Blue Mts. of
        n.e. Oreg.   L. CANBYI
      5 Rays of the terminal umbel 7-14; plants 2-6 dm. tall, scapose or with 1 or
        2 strongly reduced cauline leaves; Cascade Mts. of Wash., s. to the Sierra
        Nevada of Calif., e. to w. Valley Co., Elmore Co., and Cassia Co.,
        Ida., and n.e. Nev.   L. GRAYI
  3 Ribs of the fruit wingless; w. of the Cascade Mts.   L. APIIFOLIUM

Ligusticum apiifolium (Nutt.) Gray, Proc. Am. Acad. 7:347. 1868.
  Cynapium apiifolium Nutt. ex T. & G. Fl. N. Am. 1:641. 1840. *(Nuttall,* plains of the Oregon, near
    the confluence of the Wahlamet)
  Pimpinella apiodora Gray, Proc. Am. Acad. 7:345. 1868. *L. apiodorum* Coult. & Rose, Contr. U.S.
    Nat. Herb. 7:132. 1900. *(Bolander,* Mendocino Co., Calif., is the first of 3 collections cited)
  Plants 4-15 dm. tall, glabrous or slightly hairy, generally branched when well developed; leaves varying in the degree of dissection, sometimes with more or less evident ultimate leaflets up to 5 cm. long and 4 cm. wide, but these toothed or cleft, or so deeply cleft as to lose their individual identity;

cauline leaves mostly well developed, but progressively smaller and shorter-petiolate than the large, long-petiolate basal ones; rays of the umbel 12-30, mostly 2.5-6 cm. long at maturity; fruit broadly elliptic, 3-5.5 mm. long, the ribs raised but not winged.

Thickets, fence rows, and open or sparsely wooded slopes and prairies, at low elevations; Wash. to c. Calif., wholly w. of the Cascade Mts. May-July.

Ligusticum canbyi Coult. & Rose, Rev. N.Am. Umbell. 86. 1888. *(Canby 155,* near the headwaters of the Jocko River, Mont.)

L. leibergii Coult. & Rose, Contr. U.S. Nat. Herb. 7:134. 1900. *(Leiberg 614,* Traille River basin, Kootenai Co., Ida.)

Plants mostly 5-12 dm. tall, glabrous or scaberulous, generally branched above and with more than one umbel when well developed, the lateral umbels often a little smaller than the terminal one; leaves less dissected than in *L. filicinum* and *L. tenuifolium,* but more dissected than in *L. verticillatum,* commonly with toothed or cleft but more or less definite leaflets up to about 5 cm. long; cauline leaves 1-several, more or less reduced, but usually at least one still fairly well developed; rays of the terminal umbel 15-40, mostly 2.5-5 cm. long at maturity; fruit oblong or elliptic, 4-5 mm. long, the ribs narrowly winged.

Moist or wet meadows, stream banks, and boggy slopes in the mountains, seldom in drier soil; Cascade and Wenatchee mts. of c. and n. Wash., e. across n. Wash. to n. Ida. and adj. B.C., and w. of the continental divide in Mont.; s. in Ida. to Adams and Valley cos., and in the Wallowa Mts. of n.e. Oreg. May-Aug.

Forms of *L. canbyi* with the leaves more finely dissected than usual closely approach *L. filicinum* in appearance, but the ranges of these two taxa are not known to overlap. In the absence of mature fruit *L. canbyi* might easily be confused with *L. apiifolium,* but these two taxa occur in different habitats as well as different geographical areas. The even more similar *L. californicum* Coult. & Rose, of the coast ranges of n. Calif. and adj. Oreg., occurs in drier habitats than *L. canbyi* and is geographically well removed.

Ligusticum filicinum Wats. Proc. Am. Acad. 11:140. 1876. *(Watson 454,* Wahsatch and Uintah mts.)

Plants mostly 5-10 dm. tall, glabrous or slightly scaberulous, generally branched above and with more than one umbel when well developed; leaves dissected into numerous more or less linear ultimate segments mostly 1-3 mm. wide, the basal leaves large and long-petiolate, mostly 10-25 cm. wide; cauline leaves 1-several, reduced but still generally more or less well developed; rays of the main umbel mostly (10) 12-20, 2.5-6 cm. long at maturity; fruit oblong or oblong-elliptic, 5-7 mm. long, the ribs narrowly winged.

Open or wooded, moist or dry slopes and ridges in the mountains; Wasatch region of Utah, n. through the mountains of e. Ida. and w. Wyo. to n. Fremont Co., Ida., and Madison Co., Mont. July-Aug.

Ligusticum grayi Coult. & Rose, Rev. N.Am. Umbell. 88. 1888.

? *Pimpinella apiodora* var. *nudicaulis* Gray, Proc. Am. Acad. 8:385. 1872. *(E. Hall 206,* Oregon, in 1871)

L. apiifolium var. *minor* Gray ex Brew. & Wats. Bot. Calif. 1:264. 1876. *(Bolander 6341,* Ostrander's Meadows, Yosemite Valley, Calif.)

L. purpureum Coult. & Rose, Contr. U.S. Nat. Herb. 7:137. 1900. *(Allen 259,* Goat Mts. [Goat Rocks], Cascade Mts., Oreg.)

L. cusickii Coult. & Rose, Contr. U.S. Nat. Herb. 7:138. 1900. *(Cusick 1799,* higher mts. of e. Oreg.)

L. tenuifolium var. *dissimilis* A. Nels. Bot. Gaz. 53:224. 1912. *(Macbride 677,* Trinity Lake region, Elmore Co., Ida.) A form with the leaves more than usually dissected, approaching *L. tenuifolium.*

L. caeruleomontanum St. John, Fl. S.E. Wash. 297. 1937. *(St. John et al. 9654,* Table Rock, Columbia Co., Wash.)

Plants mostly 2-6 dm. tall, glabrous, scapose or with one or two strongly reduced cauline leaves; basal leaves well developed, less dissected than in *L. tenuifolium,* commonly with more or less distinct, toothed or cleft leaflets up to 3 cm. long; umbels 1-3, the rays 7-14, 2-3.5 cm. long at maturity; fruit elliptic-oblong, 4-6 mm. long, the ribs narrowly winged.

Moist or sometimes dry, open or wooded slopes and drier meadows in the mountains; Cascade Mts.

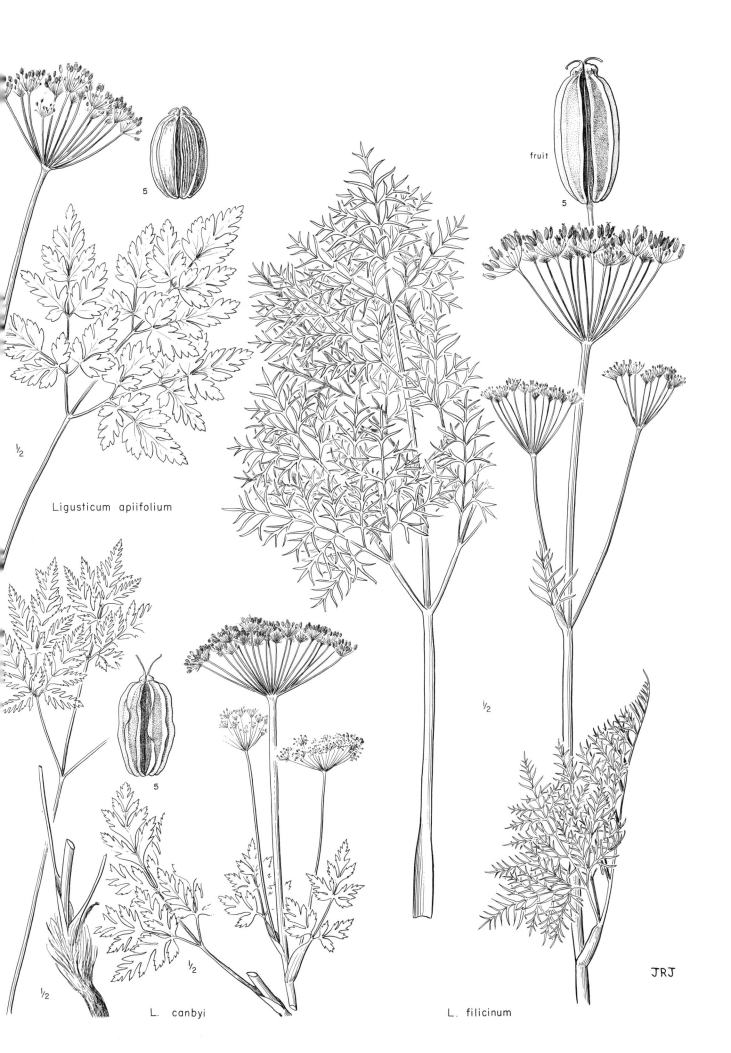

5

fruit

5

Ligusticum apiifolium

½

½

½

5

½

L. canbyi

L. filicinum

JRJ

of Wash. to the Sierra Nevada of Calif., e. to the Blue Mt. region of n.e. Oreg. and adj. Wash., w. Valley Co., Elmore Co., and Cassia Co., Ida., and Elko Co., Nev. July-Sept.

The species is reported from w. Mont., but the Montana specimens I have seen so labeled are *L. tenuifolium.* The more robust specimens of *L. grayi* approach the smaller specimens of *L. canbyi* in appearance, but the distinction can ordinarily be drawn on the number of rays of the umbel. Forms of *L. grayi* with the leaves more finely dissected than usual approach *L. tenuifolium,* and in areas where both species occur, such as in w.c. Ida., the separation is sometimes difficult.

Ligusticum tenuifolium Wats. Proc. Am. Acad. 14:293. 1879.
> *L. filicinum* var. *tenuifolium* Math. & Const. Bull. Torrey Club 68:123. 1941. *(Hall & Harbour 216* in part, mountains of Colorado)
> *L. oreganum* Coult. & Rose, Contr. U.S. Nat. Herb. 7:138. 1900. *(Cusick 1058,* Eagle Creek Mts. [= Wallowa Mts.], Union Co., Oreg.)

Plants slender, 1-6 dm. tall, scapose or more often subscapose, the single cauline leaf much reduced or sometimes wanting; basal leaves well developed but mostly less than 10 cm. wide, dissected into more or less linear ultimate segments 1-2 (3) mm. wide; umbel solitary or sometimes 2-3, with 5-13 rays, these 1.5-3 cm. long at maturity; fruit 3-5 mm. long, the ribs narrowly winged.

Meadows, marshes, stream banks, and moist slopes in the mountains; Wallowa and higher Blue mts. of n.e. Oreg., e. across c. Ida. (n. to c. Idaho Co.) to n. Fremont Co., and to Beaverhead Co., Mont.; mountains of Colo. and the Uinta Mts. of Utah; apparently not in Wyo. June-Aug.

Ligusticum tenuifolium has sometimes been treated as a variety of *L. filicinum,* and it must be admitted that no one of the several differences between the two shows real discontinuity. The two populations seem to be essentially distinct, however, and aside from the fine dissection of the leaves they actually have little in common beyond the generic characters. The differences between them are of the same order of magnitude as those separating other recognized American species of the genus.

Ligusticum verticillatum (Geyer) Coult. & Rose, Contr. U.S. Nat. Herb. 3:320. 1895.
> *Angelica ? verticillata* Geyer ex Hook. Lond. Journ. Bot. 6:233. 1847. *(Geyer 414,* high plains of the Nez Perces Indians)

Robust, soft-stemmed plants mostly 1-2 m. tall, glabrous or slightly scaberulous; leaves less dissected than in our other species, commonly with large, well-defined, toothed or lobed leaflets 3-8 cm. long and 2-5 cm. wide; cauline leaves several, well developed, differing from the large, long-petiolate basal leaves in their progressively shorter petioles and often fewer but still large leaflets; two uppermost more or less reduced leaves often opposite and bearing a verticil of long-pedunculate axillary umbels; rays of the terminal umbel mostly 15-30, 4-8 cm. long at maturity; fruit elliptic, 4-6 mm. long, the ribs narrowly winged.

Thickets, moist wooded slopes, stream banks, and swamps, in the foothills and valleys of n. Ida., from c. Idaho Co. (Clearwater drainage) to Bonner Co., and reputedly in w. Mont. May-Aug.

Lilaeopsis Greene

Inflorescence of simple, axillary, few-flowered pedunculate umbels; involucre of a few small bracts; flowers white; calyx teeth minute; styles short, the stylopodium depressed or obsolete; carpophore wanting; fruit globose or ovoid, terete or slightly flattened laterally, glabrous, the lateral ribs (those next to the commissure) corky-thickened, the dorsal ones narrow; oil tubes usually solitary in the intervals, two on the commissure; small rhizomatous perennials, the leaves reduced to elongate, narrow, entire, hollow, transversely septate phyllodes arising from the rhizome.

About a dozen species, of N. and S.Am. and the Australian region. (Named for its resemblance to the genus *Lilaea.)*

Lilaeopsis occidentalis Coult. & Rose, Bot. Gaz. 24:48. 1897.
> *Crantziola occidentalis* K.-Pol. Bull. Soc. Nat. Mosc. II, 29:125. 1916. *L. lineata* var. *occidentalis* Jeps. Madroño 1:139. 1923. *(Howell s.n.,* salt marshes of Tillamook Bay, Oreg., July 11, 1882)

Leaves tufted at intervals along the rhizome, 3-15 cm. long, 1-4 mm. wide, with 5-11 partitions; peduncles slender, 0.5-3 (4.5) cm. long, ascending or often recurved above, generally much shorter than the leaves, 5- to 12-flowered; fruit ovoid, about 2 mm. long, the lateral ribs pale and prominent, the dorsal ones inconspicuous.

Marshes, salt flats, and sandy or muddy beaches and shores along and near the coast, including the Puget Sound region; s. Vancouver I. to c. Calif. June-July.

Lomatium Raf. Desert Parsley; Biscuitroot

Inflorescence of compound umbels; involucre wanting or inconspicuous; involucel of evident to inconspicuous bractlets, or wanting; flowers mostly yellow or white, less often purple; calyx teeth minute or obsolete; stylopodium scarcely developed, sometimes visible in fresh flowers, but generally obscure or wanting in dried specimens or in fruit; carpophore bifid to the base; fruit linear to orbicular or obovate, glabrous to hairy or granular-roughened, dorsally flattened and with more or less well developed, thin or corky marginal wings, or the wings rarely obsolete and the narrow fruit scarcely flattened; dorsal ribs evident to obsolete, often raised, but only very narrowly if at all winged; oil tubes solitary to numerous in the intervals, 2-several on the commissure, sometimes obscure; taprooted perennials, the root sometimes short and tuberous-thickened, sometimes more elongate and woody, or moniliform; leaves usually mostly or all basal (with some notable exceptions), pinnately to ternately or in part quinately compound to dissected, the ultimate segments large and leafletlike to more often small and more or less confluent. X=11.

About 75 species of w. and c. N.Am. (Name from the Greek *loma,* a border, referring to the winged fruit.)

Most of the species of this notoriously difficult genus are fairly well defined (even though often highly variable), but the species present so many different combinations of characters that a satisfactory key is extraordinarily difficult to devise.

The thickened roots of a number of species were used for food by the Indians.

Reference:

Mathias, Mildred E. A revision of the genus Lomatium. Ann. Mo. Bot. Gard. 25:225-297. 1938.

1 Ultimate segments of the leaves relatively large, many or all of them 1 cm. long or
  longer
  2 Ultimate segments of the leaves forming more or less definite leaflets, these entire
    to deeply cleft, usually over 5 mm. wide
    3 Leaflets strongly toothed or cleft; flowers white or ochroleucous to sometimes yel-
      low; Cascade region and westward   L. MARTINDALEI
    3 Leaflets mostly entire or shallowly toothed; flowers yellow; widespread
      4 Herbage (or at least the stems and peduncles) more or less hirtellous-
        puberulent   L. TRITERNATUM
      4 Herbage essentially glabrous and often glaucous
        5 Leaflets mostly (4) 10-60 mm. wide; longer rays of the umbel mostly 6-20
          cm. long in fruit; widespread   L. NUDICAULE
        5 Leaflets up to about 8 mm. wide; longer rays of the umbel mostly 3-5 cm.
          long in fruit; c. Wash.   L. BRANDEGEI
  2 Ultimate segments of the leaves narrow and scarcely leafletlike, seldom as much
    as 5 mm. wide
    6 Fruits linear to narrowly oblong, mostly (2. 5) 3-8 times as long as wide, less
      than 4 mm. wide over-all, the wings up to 1/3 or rarely 1/2 as wide as the
      body, or obsolete
      7 Involucel generally wanting; pedicels elongate, mostly 4-13 mm. long at ma-
        turity; flowers yellow, or somewhat purplish in age
        8 Plants distinctly caulescent and generally few-branched above; root globose-
          thickened or moniliform to sometimes more slender and elongate; wide-
          spread e. of the Cascades   L. AMBIGUUM
        8 Plants acaulescent or nearly so, with 0-2 cauline leaves, the stems or scapes
          simple above the base; root elongate and not much thickened; mts. of c. Ida.
                                                                          L. IDAHOENSE
      7 Involucel present; pedicels short or elongate; flowers yellow or white
        9 Larger leaf segments usually over 2 mm. wide; stem and peduncle finely
          hirtellous-puberulent, rarely subglabrous; pedicels 2-12 mm. long at ma-
          turity; flowers yellow; widespread   L. TRITERNATUM
        9 Larger leaf segments mostly 1-2 mm. wide; stem and peduncle (or scape)
          glabrous to occasionally scaberulous or hirtellous; pedicels short, only

0. 5-3 mm. long at maturity; flowers yellow or white *(L. farinosum* and *L. hambleniae* might sometimes be sought here, except for the long pedicels)

    10 Leaf segments few, less than 15; flowers white; fruits wingless; plants acaulescent; local in n. Ida. and adj. Wash.   L. OROGENIOIDES

    10 Leaf segments relatively numerous; flowers yellow, rarely white; fruits very narrowly winged at maturity; caulescent or acaulescent; more widespread, from n. Ida. and s. e. Wash. to n. e. Calif.   L. LEPTOCARPUM

6 Fruits broader, either more than 4 mm. wide, or not more than about 2. 5 times as long as wide, the wings often well over 1/3 as wide as the body

    11 Bractlets of the involucel broadly oblanceolate to broadly obovate or elliptic; flowers yellow; Jefferson Co., Oreg., to s. e. Wash., Mont., Wyo., s. e. Oreg., and s. w. Ida.   L. COUS

    11 Bractlets of the involucel narrow, mostly linear or lanceolate, or wanting

        12 Flowers white or ochroleucous to occasionally purple; low, often scapose or subscapose plants up to about 3 (4) dm. tall, often with tuberous-thickened roots

            13 Plant, when well developed, generally with a branching caudex surmounting a taproot, the caudical branches often rather slender and elongate and frequently broken off in herbarium specimens at the point of attachment to the root; pedicels 1-4 (5) mm. long; moderate to high elevations in the mts. from n. e. Oreg. to s. w. Mont.   L. CUSICKII

            13 Plant with a globose-thickened or moniliform to more elongate root and a usually simple crown, without a branching caudex; foothills and lowlands

                14 Pedicels mostly 6-22 mm. long in fruit; n. Ida. and adj. Wash. and Mont.   L. FARINOSUM

                14 Pedicels short, seldom as much as 4 mm. long

                    15 Plants relatively small, mostly 1-1. 5 dm. tall at maturity, the fruits mostly 5-7 mm. long and the bractlets of the involucel often less than 2 mm. long; root subglobose; southward from Kittitas, Lincoln, and s. Spokane cos., Wash., and from Latah Co., Ida.   L. GORMANII

                    15 Plants larger, mostly 1. 5-4 dm. tall at maturity, the fruits mostly 7-12 mm. long and the bractlets of the involucel mostly 2-3 mm. long; root subglobose to more often elongate or moniliform; northward from Kittitas, Lincoln, and Spokane cos., Wash. (also on Cleman Mt., Yakima Co.), and from Kootenai Co., Ida.   L. GEYERI

        12 Flowers yellow or occasionally somewhat purplish, or wholly purple in some robust species mostly over 3 dm. tall

            16 Stems or scapes solitary or few from the simple or occasionally few-branched rootcrown

                17 Plants essentially acaulescent (or with a pseudoscape), mostly 1-2 dm. tall at maturity, glabrous, arising from a globose-thickened root; c. Wash.   L. HAMBLENIAE

                17 Plants either taller and often distinctly caulescent, or with a more elongate and not much thickened root, or often both; glabrous or hairy

                    18 Involucel usually absent; stems or scapes glabrous or merely scaberulous; moderate to high elevations in the mts. of c. Ida.   L. IDAHOENSE

                    18 Involucel present; stems or scapes hirtellous-puberulent to sometimes merely scaberulous; widespread, mostly in the foothills and lowlands, sometimes extending to moderate elevations in the mts.

                        19 Leaves irregularly and more or less distinctly pinnately dissected, the ultimate segments seldom much over 1 cm. long; canyons of the Snake and lower Salmon rivers in Ida. and Oreg.   L. ROLLINSII

                        19 Leaves ternately to more or less ternate-pinnately or in part quinately dissected, the ultimate segments usually though not always well over 1 cm. long; widespread   L. TRITERNATUM

            16 Stems several or numerous from a large, woody root that is often surmounted by a branching caudex

                20 Fruits with corky-thickened narrow wings; flowers yellow or purple

                    21 Plants robust, mostly 5-15 (20) dm. tall at maturity; foliage gen-

erally more or less scaberulous; fruit 8-18 mm. long; flowers yellow or purple;
    throughout our range   L. DISSECTUM
  21 Plants smaller, mostly 3-6 (8) dm. tall at maturity, wholly glabrous; fruit 16-
      28 mm. long; flowers purple; e. end of the Columbia R. Gorge, n. to Yakima
      Co., Wash.   L. COLUMBIANUM
20 Fruits with thin wings; flowers yellow (occasional specimens of the widespread spe-
    cies *L. triternatum* might be sought here)
    22 Fruit 6-12 mm. long; plants mostly 2-6 dm. tall at maturity
      23 Larger leaf segments mostly 2-8 mm. wide; fruiting pedicels short, 1-4 (5)
        mm. long; fruits usually more or less deflexed; involucel present; Kittitas
        Co. to Okanogan Co., Wash.   L. BRANDEGEI
      23 Larger leaf segments mostly 1-2 mm. wide; fruit not deflexed; involucel
        mostly wanting; e. end of the Columbia R. Gorge   L. LAEVIGATUM
    22 Fruit 15-32 mm. long; plants mostly 5-20 dm. tall at maturity
      24 Stem and peduncle glabrous; e. end of the Columbia R. Gorge   L. SUKSDORFII
      24 Stem and peduncle puberulent; Wenatchee region of c. Wash.   L. THOMPSONII
1 Ultimate segments of the leaves relatively small, rarely any of them as much as 1
  cm. long
  25 Bractlets of the involucel broadly oblanceolate to obovate, ovate, or subrotund,
    entire to sometimes ternately once or twice divided, sometimes more or less
    connate
    26 Wings of the fruit corky-thickened; bractlets of the involucel ternately or bi-
      ternately divided; Willamette Valley of Oreg.   L. BRADSHAWII
    26 Wings of the fruit thin, not corky-thickened; bractlets of the involucel entire
      to more or less cleft, but not deeply divided
      27 Leaves very finely dissected, with very narrow, thin and soft, acute or
        acuminate ultimate segments mostly well under 1 mm. wide and seldom
        over 5 (12) mm. long; plants usually distinctly caulescent; w. of the
        Cascades   L. UTRICULATUM
      27 Leaves less finely divided (the segments relatively and often actually broad-
        er and often longer, frequently also thicker and blunter), or the plants es-
        sentially acaulescent, or both; e. of the Cascades
        28 Fruiting pedicels more or less elongate, mostly 3-15 mm. long; taproot
          elongate and seldom much thickened
          29 Leaves glabrous or merely granular-scaberulous; fruit granular-
            roughened when young, not hairy; plants distinctly caulescent; in-
            volucel scarcely dimidiate; flowers yellow; Union, Grant, and Crook
            cos., Oreg., s. to Calif.   L. VAGINATUM
          29 Leaves evidently short-hairy; fruit usually short-hairy at least when
            young; plants acaulescent or nearly so; involucel dimidiate
            30 Flowers yellow; bractlets of the involucel puberulent or villous-
              puberulent; e. and s. from Mont., c. Ida., and c. Oreg.
                                        L. FOENICULACEUM
            30 Flowers white; bractlets of the involucel glabrous or only very fine-
              ly hirtellous; c. Oreg. and southward   L. NEVADENSE
        28 Fruiting pedicels very short, commonly 1-3 mm. long, or obsolete; tap-
          root usually short and tuberous-thickened, sometimes more slender and
          elongate
          31 Involucel dimidiate and the bracts usually connate below the middle or
            nearly to the tip; ovaries and young fruit finely puberulent or glabrous;
            s. Kittitas Co., Wash., to Jefferson Co., Oreg.   L. WATSONII
          31 Involucel not dimidiate and the bractlets distinct; fruit glabrous or
            granular-roughened, not hairy; Jefferson Co., Oreg., to s.e. Wash.,
            Mont., Wyo., s.e. Oreg., and s.w. Ida.   L. COUS
  25 Bractlets of the involucel narrow, mostly linear or lanceolate, distinct or merely
    connate at the base, or wholly wanting
    32 Plants dwarf alpine or subalpine perennials from a taproot and much-branched
      caudex, less than 1 dm. tall, with very compact inflorescence (the rays of the

umbel all less than 1 cm. long at maturity) and not much dissected leaves; mts. of
n. e. Oreg.

    33 Ultimate segments of the leaves glabrous or slightly scaberulous, mostly 3-8 mm.
long; ovaries and fruit glabrous   L. GREENMANII

    33 Ultimate segments of the leaves usually hirtellous-puberulent, mostly 1-3 (6) mm.
long; ovaries and fruit hirtellous-puberulent   L. OREGANUM

32 Plants distinctly otherwise, mostly larger and with looser mature inflorescence, or
if small then with a short, tuberous-thickened root

  34 Wings of the fruit more or less corky-thickened, usually narrow; flowers purple
or yellow (or salmon-colored)

    35 Foliage generally more or less scaberulous, rarely glabrous; robust species,
mostly 5-15 (20) dm. tall at maturity; throughout our range   L. DISSECTUM

    35 Foliage glabrous; smaller species, up to about 6 (8) dm. tall; more local in
distribution

      36 Flowers salmon-yellow; s. e. Wash. and adj. Oreg. and Ida.   L. SALMONIFLORUM

      36 Flowers purple (rarely yellow); more western or southern species

        37 Fruit large, mostly 16-28 mm. long; leaf segments 3-20 mm. long;
upper end of the Columbia R. Gorge, n. to Yakima Co., Wash.

                                                 L. COLUMBIANUM

        37 Fruit smaller, mostly 9-16 mm. long; leaf segments 1-8 mm. long;
not of the Columbia R. Gorge region

          38 Ultimate segments of the leaves rigidly cuspidate; root woody, ver-
tical, surmounted by a simple or usually branched caudex that is
clothed by the fibrous remnants of leaf bases from previous years;
Wenatchee region of c. Wash.   L. CUSPIDATUM

          38 Ultimate segments of the leaves rounded to acutish or barely apicu-
late; root somewhat tuberous-thickened and often horizontal, the
stem or scape arising from a simple crown that is not conspicuously
clothed by the leaf bases from previous years; more southern species

            39 Plant with very conspicuous large bladeless basal sheaths; Yakima
Co., Wash.   L. TUBEROSUM

            39 Plant without conspicuous bladeless basal sheaths; s. Morrow and
s. Wasco cos. to n. Malheur Co., Oreg.   L. MINUS

  34 Wings of the fruit thin, either narrow or broad; flowers yellow or white, occasionally somewhat purplish

    40 Leaves not much dissected, more nearly with toothed or cleft leaflets; flowers
white or ochroleucous to sometimes yellow; Cascade, Olympic, and Coast
ranges   L. MARTINDALEI

    40 Leaves more dissected, with small and narrow ultimate segments that do not
resemble leaflets

      41 Plants with a strongly tuberous-thickened, often subglobose or moniliform
root; low plants, acaulescent, not over about 3 (4) dm. tall at maturity

        42 Fruits narrow and elongate, 6-13 mm. long, 2-3 mm. wide, very nar-
rowly or scarcely winged   L. LEPTOCARPUM

        42 Fruits broader and often with well-developed wings

          43 Flowers yellow; wings of the fruit very narrow, only about 0.5 mm.
wide; fruiting peduncles recurved to the ground; Jefferson and
Wheeler cos., Oreg.   L. HENDERSONII

          43 Flowers white; wings of the fruit often well over 0.5 mm. wide;
fruiting peduncles ascending to suberect

            44 Pedicels very short, up to 3 mm. or occasionally 5 mm. long
at maturity

              45 Plants relatively small, mostly 1-1.5 dm. tall at maturity;
fruits mostly 5-7 mm. long; bractlets of the involucel often
less than 2 mm. long; main root subglobose; fruits glabrous
or granular-roughened; southward from Kittitas, Lincoln,
and s. Spokane cos., Wash., and from Latah Co., Ida. L. GORMANII

              45 Plants larger, mostly 1.5-3 (4) dm. tall at maturity, the

fruits mostly 7-12 mm. long and the bractlets of the involucel mostly 2-3
mm. long; root subglobose to more often elongate or moniliform; fruit glabrous; northward from Kittitas, Lincoln, and Spokane cos., Wash., and from
Kootenai Co., Ida. (also on Cleman Mt., Yakima Co.)   L. GEYERI
  44 Pedicels mostly 4-15 mm. long in fruit; fruit 6-10 mm. long; southward from Kittitas, Douglas, and Lincoln cos., Wash., and from Nez Perce Co., Ida.   L. CANBYI
41 Plants otherwise, either taller, or distinctly caulescent, or with the root elongate and
scarcely tuberous-thickened, often differing in more than one of these respects
  46 Ovaries and young (often also mature) fruits granular-scaberulous or more or less
hairy
    47 Herbage granular-scaberulous to subglabrous; fruit granular-scaberulous at
least when young; flowers yellow
      48 Pedicels 5-15 mm. long at maturity; fruit 8-12 mm. long; c. and n.e. Oreg.
to Calif.   L. VAGINATUM
      48 Pedicels 2-5 (7) mm. long at maturity; fruit 5-8 mm. long; n. Ida. to n.w.
Mont. and s.w. Alta.   L. SANDBERGII
    47 Herbage short-hairy (at least the scape or the stem and peduncle); fruit short-
hairy at least when young, not granular-scaberulous
      49 Flowers yellow
        50 Fruit large, mostly 16-28 mm. long; plants mostly 5-10 dm. tall at ma-
turity; c. Wash.   L. THOMPSONII
        50 Fruit small, mostly 5-10 mm. long; plants less than 5 dm. tall; not of
Wash.   L. FOENICULACEUM
      49 Flowers white or somewhat purplish, very rarely yellow in *L. macrocarpum*
        51 Bractlets of the involucel villous-puberulent, not markedly scarious-mar-
gined; fruit (7) 10-20 mm. long, (1.8) 2-5 times as long as wide; widespread e. of the Cascade summits   L. MACROCARPUM
        51 Bractlets of the involucel glabrous or very finely hirtellous, evidently
scarious-margined; fruit 6-10 mm. long, less than twice as long as
wide; c. Oreg. and southward   L. NEVADENSE
  46 Ovaries and fruit glabrous
    52 Flowers ordinarily white or somewhat purplish, very rarely yellow in *L. ma-
crocarpum*
      53 Herbage essentially glabrous; c. Wash. and n. Ida. to s. B.C. (extreme
specimens of *L. cusickii,* from n.e. Oreg. to s.w. Mont., might be sought
here)   L. GEYERI
      53 Herbage evidently puberulent, villous-puberulent, or hirtellous
        54 Bractlets of the involucel villous-puberulent, not markedly scarious-
margined; fruit (7) 10-20 mm. long, (1.8) 2-5 times as long as wide;
widespread e. of the Cascade summits   L. MACROCARPUM
        54 Bractlets of the involucel glabrous or very finely hirtellous, evidently
scarious-margined; fruit 6-10 mm. long, less than twice as long as
wide; c. Oreg. and southward   L. NEVADENSE
    52 Flowers yellow, rarely white in *L. leptocarpum*
      55 Taproot or tuberous-thickened root with a mostly simple, often subter-
ranean crown, the stems solitary or few
        56 Fruit narrowly oblong, 1.5-3 mm. wide, more than 2.5 times as long
as wide
          57 Involucel well developed; pedicels 1-3 mm. long   L. LEPTOCARPUM
          57 Involucel none; pedicels 4-13 mm. long   L. AMBIGUUM
        56 Fruit elliptic, often broadly so, less than 2.5 times as long as wide
          58 Herbage glabrous; leaves blue-glaucous, mostly or all clustered
at or near the base; c. and e. Oreg., not extending to the Snake
R. canyon   L. DONNELLII
          58 Herbage scaberulous or crisp-puberulent to subglabrous; leaves
cauline and basal, not glaucous; in and near the canyons of the
Snake and lower Salmon rivers in n. Oreg. and adj. Ida.   L. ROLLINSII
      55 Taproot usually surmounted by a branched caudex, the stems or scapes

several or numerous (taproot sometimes with a simple crown in *L. hallii,* but still with clustered stems)

    59 Leaves very finely dissected, usually with several hundred or more than a thousand very narrow and often subterete ultimate segments that lie in numerous different planes so that the leaf has "thickness"; leaves often evidently scaberulous; widespread e. of the Cascade summits   L. GRAYI

    59 Leaves less finely dissected, with not more than a few hundred flat, linear or oblong segments that tend to lie in nearly a single plane; leaves glabrous or very nearly so

        60 Plants acaulescent; oil tubes solitary in the intervals; Snake R. canyon region of e. Oreg. and adj. Ida. and Wash.   L. SERPENTINUM

        60 Plants more or less caulescent; oil tubes mostly 2-3 in the intervals; w. of the Cascade summits in Oreg.   L. HALLII

Lomatium ambiguum (Nutt.) Coult. & Rose, Contr. U.S. Nat. Herb. 7:212. 1900.

    Eulophus ambiguus Nutt. Journ. Acad. Phila. 7:27. 1834. *Peucedanum ambiguum* Nutt. ex T. & G. Fl. N. Am. 1:626. 1840. *Cogswellia ambigua* M. E. Jones, Contr. West. Bot. 12:33. 1908. *(Wyeth,* dry hillsides on the borders of the Flat-Head River)

    Peucedanum tenuissimum Hook. Lond. Journ. Bot. 6:235. 1847. *(Geyer 302,* Coeur d'Aleine country)

    Taproot short and tuberous-thickened to long and slender, sometimes moniliform; plant glabrous, 1-8 dm. tall, simple or more often branched; leaves basal and cauline or all cauline, ternately or ternate-pinnately 2-several times dissected into narrow and usually very unequal ultimate segments, the longer segments mostly 1-8 cm. long and up to 5 mm. wide, the smallest ones often only 1-2 mm. long; lower leaves often with larger ultimate segments than the upper; rays of the umbel commonly somewhat elongate even at anthesis, the longer ones mostly 3-10 cm. long at maturity; involucel none; flowers yellow; pedicels 4-13 mm. long at maturity; fruit glabrous, narrowly oblong, 5.5-12 mm. long, 1.5-3.3 mm. wide, the wings evident but narrow, 0.3-0.5 (0.7) mm. wide; oil tubes solitary in the intervals, 2 on the commissure.

    Open, often rocky slopes and flats, from the scablands, valleys, and foothills to moderate elevations in the mountains; Wash. and adj. B.C., e. (or barely w.) of the Cascade summits, and n.e. Oreg., e. across Ida. (n. of the Snake River plains) to w. Mont. (as far as Meagher and Park cos.) and w. Wyo., and apparently isolated in the Wasatch Mts. of Utah. May-July.

Lomatium bradshawii (Rose) Math. & Const. Bull. Torrey Club 69:246. 1942.

    Leptotaenia bradshawii Rose ex Mathias, Leafl. West. Bot. 1:101. 1934. *(Bradshaw 2047,* Eugene, Oreg.)

    Taproot long and slender; plants glabrous, acaulescent or nearly so, more or less erect, 2-6.5 dm. tall; leaves ternate-pinnately dissected into linear or filiform segments 3-10 mm. long and up to 1 mm. wide; bractlets of the involucel ternately or biternately divided; flowers light yellow; rays of the umbel elongating unequally, the longer ones 4-13 cm. long at maturity, generally only 2-5 fertile; pedicels 2-5 mm. long at maturity; fruit 8-13 mm. long, 5-7 mm. wide, glabrous, the corky-thickened wings only about half as wide as the body (or less) and nearly concolorous with it; dorsal ribs inconspicuous and only slightly raised; oil tubes obscure.

    Moist low ground in the Willamette Valley of Oreg. from near Salem to near Eugene. May.

Lomatium brandegei (Coult. & Rose) Macbr. Contr. Gray Herb. n.s. 56:35. 1918.

    Peucedanum brandegei Coult. & Rose, Bot. Gaz. 13:210. 1888. *Cynomarathrum brandegei* Coult. & Rose, Contr. U.S. Nat. Herb. 7:246. 1900. *Cogswellia brandegei* M. E. Jones, Contr. West. Bot. 12:32. 1908. *(Brandegee 799,* "Walla Walla region," Wash.; doubtless actually taken in c. Wash.)

    Perennial from a long stout taproot and compactly branched woody caudex; plants 2-6 dm. tall at maturity; herbage slightly glaucous, essentially glabrous except for the finely scaberulous-ciliolate margins of the leaf segments; leaves basal and cauline or nearly all basal, ternate-pinnately dissected into broadly linear to linear-elliptic or narrowly lance-elliptic ultimate segments mostly 1-5 cm. long, the better-developed segments mostly 2-8 mm. wide; flowers yellow; involucel of several evident narrow bractlets; rays of the umbel elongating unequally, the longer ones mostly 1.5-5 cm. long; pedicels short, 1-4 (5) mm. long at maturity; fruit glabrous, mostly deflexed, rather narrowly elliptic-oblong, 8-12 mm. long, the wings from half as wide to nearly as wide as the body; oil tubes (1) 3 (4) in the intervals, several on the commissure.

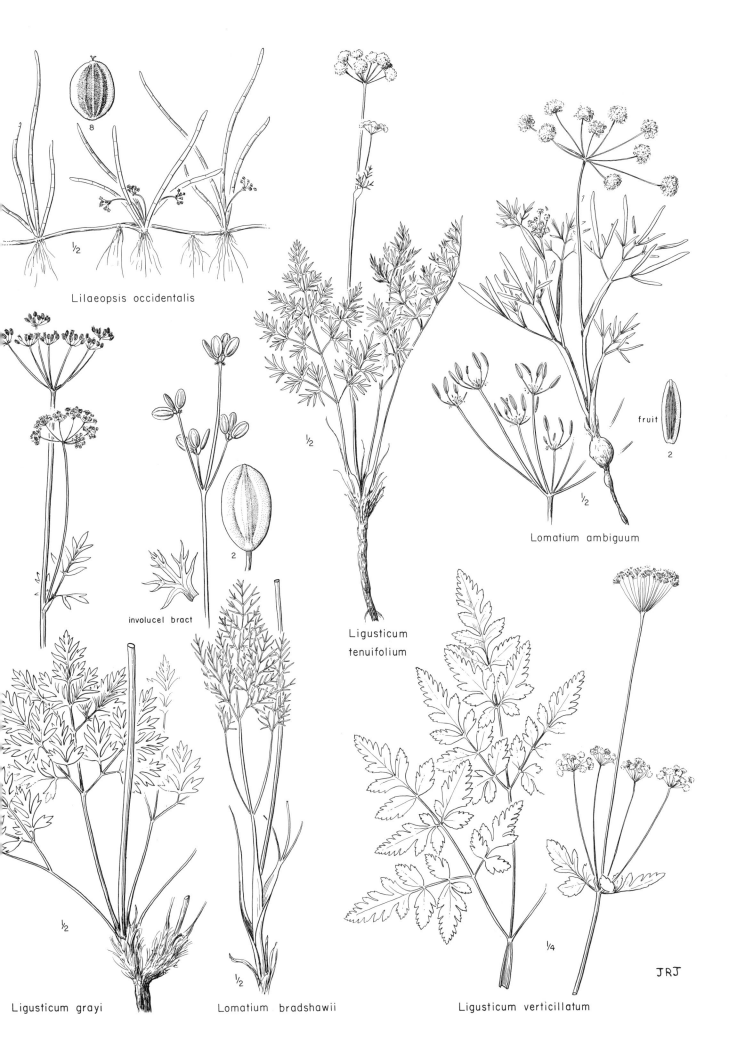

8

½

Lilaeopsis occidentalis

½

involucel bract

2

½

Ligusticum tenuifolium

½

fruit

2

Lomatium ambiguum

¼

JRJ

½

Ligusticum grayi

½

Lomatium bradshawii

Ligusticum verticillatum

Open or wooded slopes, from the foothills to fairly high elevations in the mountains; Wenatchee and Cascade regions from Kittitas Co. to Okanogan Co. , Wash. May-July.

Lomatium canbyi Coult. & Rose, Contr. U.S. Nat. Herb. 7:210. 1900.

Peucedanum canbyi Coult. & Rose, Bot. Gaz. 13:78. 1888. *Cogswellia canbyi* M. E. Jones, Contr. West. Bot. 12:33. 1908. *(Howell 67, high ridges, eastern Oreg.)*

Taproot with a globose-thickened base up to 3. 5 cm. thick surmounted by a short or rather elongate, more slender and cylindrical upper portion; herbage glabrous; leaves all essentially basal, pinnately or ternate-pinnately (more pinnate than ternate) dissected into numerous small, crowded, mostly blunt or rounded ultimate segments generally not over 5 (8) mm. long; scapes 1-several, ascending or suberect, 1-2 dm. long at maturity; bractlets of the involucel narrow and inconspicuous, mostly 1-3 mm. long; flowers white with purple anthers; rays of the umbel elongating unequally, the longer ones mostly 2. 5-7 cm. long; pedicels becoming elongate, mostly 4-15 mm. long in fruit; fruit glabrous, rather broadly elliptic, 6-10 mm. long, the wings about half as wide as the body; oil tubes 1 or 2 in the intervals, 2-4 on the commissure.

Open, rocky places at lower elevations, often with sagebrush; Kittitas, Douglas, and Lincoln cos. , Wash. , to Nez Perce Co. , Ida. , s. to n. Oreg. and reputedly to n. Calif. Mar. -Apr.

Lomatium columbianum Math. & Const. Bull. Torrey Club 69:246. 1942.

Ferula purpurea Wats. Proc. Am. Acad. 21:453. 1886. *Leptotaenia purpurea* Coult. & Rose, Rev. N. Am. Umbell. 52. 1888. *(Suksdorf 281, near the Columbia R. in Klickitat Co. , Wash.)* Not *Lomatium purpureum* A. Nels. in 1901.

Stout aromatic perennial from a very thick woody taproot and caudex or crown, glabrous and glaucous, mostly 3-6 (8) dm. tall at maturity, with several stems or scapes from the base; leaves all or mostly basal, ternate-pinnately dissected into numerous linear ultimate segments about 1 mm. wide or less and commonly about 1 (2) cm. long; bladeless basal sheaths usually well developed and conspicuous; flowers purple (rarely yellow), beginning to bloom before the leaves are fully expanded; rays of the umbel mostly 7-14, usually not very unequal, mostly 4-20 cm. long at maturity; involucel of well-developed narrow bractlets; pedicels mostly 10-30 mm. long at maturity; fruits elliptic to oblong-obovate, 16-28 mm. long, 8-15 mm. wide, the somewhat corky-thickened wings 1/5-1/3 as wide as the body; dorsal ribs barely raised, the oil tubes about 3 in each interval, readily visible through the pericarp.

Dry, rocky slopes along the Columbia R. near the east end of the gorge (Wasco and Hood River cos. , Oreg. , and Klickitat Co. , Wash.), n. to Yakima Co. , Wash. Mar. -Apr.

Lomatium cous (Wats.) Coult. & Rose, Contr. U.S. Nat. Herb. 7:214. 1900.

Peucedanum cous Wats. Proc. Am. Acad. 21:453. 1886. *Cogswellia cous* M. E. Jones, Contr. West. Bot. 12:33. 1908. *(Howell 270, John Day Valley, Oreg. , May, 1880)*

Peucedanum circumdatum Wats. Proc. Am. Acad. 22:474. 1887. *L. circumdatum* Coult. & Rose, Contr. U.S. Nat. Herb. 7:213. 1900. *Cogswellia circumdata* M. E. Jones, Contr. West. Bot. 12: 33. 1908. *(Cusick, hillsides in the Wallowa region of eastern Oreg. , June, 1886)*

L. montanum Coult. & Rose, Contr. U.S. Nat. Herb. 7:214. 1900. *Peucedanum montanum* Blank. Mont. Agr. Coll. Stud. Bot. 1:93. 1905. *Cogswellia montana* M. E. Jones, Contr. West. Bot. 12: 34. 1908. *(Rose 479, Yellowstone National Park)*

Root short and strongly tuberous-thickened to less often more slender and elongate; plants glabrous or short-hairy, acaulescent or less often with one or more fairly well-developed cauline leaves, sometimes with a more or less definite pseudoscape, 1-3. 5 dm. tall at maturity; leaves pinnately or ternate-pinnately dissected, highly variable, the segments often short, crowded, and relatively broad in smaller plants, longer, less crowded, and relatively narrower in larger plants, sometimes as much as 15 mm. long and 3 mm. wide; rays of the umbel mostly 5-20, elongating unequally, the longer ones 1. 5-10 cm. long at maturity; bractlets of the involucel well developed and persistent, broadly oblanceolate to broadly obovate or elliptic, 2-5 mm. long, entire or occasionally few-toothed, subherbaceous to more or less scarious, often purplish in part; flowers yellow; pedicels short, commonly only 1-3 mm. long at maturity; fruits oblong to rather broadly elliptic, 5-12 mm. long, 3-5 mm. wide, glabrous or often more or less evidently granular-roughened especially when young; lateral wings about equaling or often a little narrower than the body; dorsal ribs sometimes produced into very narrow wings; oil tubes inconspicuous, 1-4 in the intervals, 4-6 on the commissure. Cous (the Indian name).

Dry, open, often rocky slopes and flats, often with sagebrush, commonest in the foothills and low-lands, but sometimes extending to above timber line in the mountains; n. e. Oreg. (chiefly in the Blue Mt. region from Morrow and Grant cos. eastward, but extending as far w. as Jefferson Co.) and s. e. Wash. (Blue Mt. region, n. to Whitman and Franklin cos.), e. across c. Ida. (n. to Latah and Clear-water cos.) to Meagher and Carbon cos. , Mont. , and the mountains of n. Wyo. (e. to the Bighorn Mts.); s. to Harney Co. , Oreg. , and Owyhee Co. , Ida. , and reportedly to n. e. Nev. Apr. -July, de-pending on the elevation.

Lomatium cous is a highly variable but withal well-defined and readily recognizable species. It is customary to attempt to restrict the name *L. cous* to specimens with the young fruits granular-rough-ened, and to segregate the specimens with strictly glabrous ovaries and fruits under two other names: *L. montanum* for the common, acaulescent forms, and *L. circumdatum* for the less common, caules-cent forms. The cleavage on the basis of roughening of the young fruit is not at all sharp, and speci-mens authoritatively determined as *L. montanum* or *L. circumdatum* frequently have some trace of granular roughening. Most of the plants from the more eastern part of the range of the species do have the young fruits glabrous or nearly so, but specimens with evidently granular-roughened young fruits have been taken as far east as the Big Horn Mts. of Wyo. , at the eastern limit of range for the species as a whole. Specimens with the roots elongate and not much thickened are more common among the *L. montanum* element than among the remainder of the species, but the difference is purely statistical and not taxonomically useful.

The caulescent plants of both fruit types seem to be restricted to the western part of the range of the species. Among the plants with granular-roughened fruits, the caulescent forms have never been taxonomically segregated from the caulescent ones, and there is little reason to believe that this character is any more significant among the plants with strictly glabrous fruits.

Lomatium cusickii (Wats.) Coult. & Rose, Contr. U.S. Nat. Herb. 7:226. 1900.
Peucedanum cusickii Wats. Proc. Am. Acad. 21:453. 1886. *Cogswellia cusickii* M. E. Jones, Contr. West. Bot. 12:32. 1908. *(Cusick,* highest summits of the Eagle Creek [Wallowa] Mts. , Union Co. , Oreg.)
Cogswellia brecciarum M. E. Jones, Contr. West. Bot. 12:37. 1908. *Cynomarathrum brecciarum* Rydb. Fl. Rocky Mts. 630, 1064. 1917. *(Jones,* Mt. Haggin, Mont. , Aug. 3, 1905)
Cogswellia altensis M. E. Jones, Bull. U. Mont. Biol. 15:41. 1910. *(Jones,* Alta, Mont.)
Plants, when well developed, with a branching caudex surmounting a taproot, the caudical branches often rather slender and elongate, and frequently broken off in herbarium specimens at the point of attachment to the root; plants up to about 1. 5 (3) dm. tall at maturity; herbage glabrous or scaberulous; leaves chiefly basal or nearly so, but often one or two distinctly cauline, mostly ter-nately or ternate-pinnately (more ternate than pinnate) 2-3 times dissected into linear segments com-monly 1-6 cm. long and up to 2. 5 (4. 5) mm. wide; flowers white or ochroleucous to more or less purplish; rays of the umbel elongating unequally, the longer ones mostly 1-3 cm. long at maturity; in-volucel of several narrow bractlets that are generally broadened at the base and often connate into a narrow basal rim or collar, or the tips of the bractlets reduced so that the involucel is represented by a collar with one or more short teeth; pedicels short and stout, 1-4 (5) mm. long at maturity; fruit glabrous or nearly so, rather narrowly elliptic, 7-15 mm. long, the wings from about half as wide to fully as wide as the body; oil tubes 1-3 in the intervals, about 5 on the commissure.

Open or wooded, often rocky places at moderate to high elevations in the mountains; Wallowa Mts. (and probably the Blue Mts. near Meacham) of Oreg. , e. across c. Ida. (n. to Idaho Co.) to Beaver-head and Deer Lodge cos. , Mont. June-July.

Herbarium specimens which do not fully show the underground parts and in which the flower-color is obscure, are often confused with *L. idahoensis,* from which they are readily distinguished by the presence of an involucel.

Lomatium cuspidatum Math. & Const. Bull. Torrey Club 69:246. 1942.
Leptotaenia watsoni Coult. & Rose, Rev. N. Am. Umbell. 52. 1888. *(Brandegee 801,* Wenatchee region, Wash.) Not *Lomatium watsoni* Coult. & Rose, 1900.
Taproot woody, surmounted by a simple or branched caudex that is clothed with the fibrous rem-nants of the leaf bases of previous years; stems or scapes simple or few-branched, usually curved-ascending rather than strictly erect, up to about 5 dm. long; herbage glabrous and more or less glau-cous; leaves all borne at or near the base, ternate-pinnately dissected into numerous small, crowded, firm, sharp-pointed ultimate segments mostly 1-5 mm. long; flowers purple; rays of the umbel mostly

Lomatium brandegei

2

1/6

involucel bract
3

L. cous

1/2

L. canbyi

1/2

1/4

L. cusickii

1/2

L. columbianum

2

1/2

1/8

fruit
2

L. cuspidatum

1/2

3

JRJ

5-13, elongating unequally, the longer ones mostly 3. 5-10 cm. long at maturity; involucel of a few narrow and sometimes inconspicuous bractlets, tending to be dimidiate; fruiting pedicels mostly 5-20 mm. long; fruit elliptic, 9-13 mm. long, the lateral wings very narrow and somewhat corky-thickened, about 0. 5 mm. wide, the dorsal ribs inconspicuous and scarcely raised; oil tubes mostly 3 in the intervals, several on the commissure.

Open, rocky slopes, often on serpentine, from 2000 to 6500 feet altitude in the Wenatchee Mt. region of Chelan and Kittitas cos., Wash. May-July.

Lomatium dissectum (Nutt.) Math. & Const. Bull. Torrey Club 69:246. 1942.
   Leptotaenia dissecta Nutt. in T. & G. Fl. N. Am. 1:630. 1840. *Ferula dissecta* Gray, Proc. Am. Acad. 7:348. 1868. *(Nuttall,* plains of the Oregon, near the confluence of the Wahlamet)
   Leptotaenia multifida Nutt. in T. & G. Fl. N. Am. 1:630. 1840. *Ferula multifida* Gray, Proc. Am. Acad. 7:348. 1868. *Leptotaenia dissecta* var. *multifida* Jeps. Madroño 1:145. 1923. *Lomatium dissectum* var. *multifidum* Math. & Const. Bull. Torrey Club 69:246. 1942. *(Nuttall,* plains of the Oregon, east of Wallawallah, and in the Blue Mts.)
   Leptotaenia dissecta var. *foliosa* Hook. Lond. Journ. Bot. 6:236. 1847. *Leptotaenia foliosa* Coult. & Rose, Contr. U.S. Nat. Herb. 7:198. 1900. *(Geyer 517,* Nez Perce region, n. Ida.) Probably = var. *dissectum.*
   LOMATIUM DISSECTUM var. EATONII (Coult. & Rose) Cronq. hoc loc. *Leptotaenia eatoni* Coult. & Rose, Rev. N. Am. Umbell. 52. 1888. *Leptotaenia multifida* var. *eatoni* M. E. Jones, Contr. West. Bot. 12:40. 1908. *(Eaton 147,* Utah)
Robust perennial from an often very large, woody taproot which may be surmounted by a branching caudex, mostly 5-15 (20) dm. tall at maturity, the several glabrous stems usually ascending rather than strictly erect; leaves large, basal and cauline, the lower ones the largest, all generally more or less scaberulous, seldom glabrous, ternate-pinnately dissected into small and often narrow ultimate segments up to about 1 (2) cm. long; rays of the umbel mostly 10-30, equal or unequal, at least the longer ones mostly 4-10 cm. long at maturity; flowers yellow or purple, some of them always sterile; involucel of well-developed narrow bractlets; fruit elliptic, 8-17 mm. long and 4. 5-10 mm. wide, the lateral wings narrow and more or less corky-thickened, up to about 1 mm. wide, the dorsal ribs inconspicuous; oil tubes obscure.

Open, often rocky slopes and dry meadows, often on talus, from the foothills and valleys to moderate elevations in the mountains; s. Alta. and B.C. to Colo., Ariz., and s. Calif. Apr. -June.

The species consists of 3 geographic varieties. Two of these varieties, var. *eatonii* and var. *multifidum,* might perhaps properly be grouped into a subspecies, but no such combination is here proposed.

1 Fruits sessile or on very short pedicels shorter than the pedicels of the sterile flowers; leaves mostly of var. *eatonii;* chiefly w. of the e. base of the Cascades, from s. B.C. to n. Calif., but also in n. Ida.   var. dissectum
1 Fruits borne on well-developed pedicels mostly 4-20 mm. long; e. of the Cascade summits
  2 Leaves very finely dissected, the very numerous ultimate segments mostly linear and 0. 5-1. 5 (2) mm. wide; s. B.C. and Alta. to n. Wyo., c. Ida., c. Oreg., and occasionally n. Nev.   var. multifidum (Nutt.) Math. & Const.
  2 Leaves less finely dissected, the ultimate segments tending to be fewer, broader, and more confluent, the larger segments mostly (1. 5) 2-4 mm. wide; s. Wyo., s. Ida., and c. and n. e. Oreg. to Colo., Utah, Ariz., Nev., and s. Calif.   var. eatonii (Coult. & Rose) Cronq.

Lomatium donnellii Coult. & Rose, Contr. U.S. Nat. Herb. 7:231. 1900.
   Peucedanum donnellii Coult. & Rose, Bot. Gaz. 13:143. 1888. *Cogswellia donnellii* M. E. Jones, Contr. West. Bot. 12:34. 1908. *(Howell 829,* John Day Valley, Oreg.)
Taproot elongate and not strongly thickened, with a subterranean simple crown, the plant tending to develop a short pseudoscape; plants often branched at the ground level and producing several ascending peduncles, these becoming 1. 5-3 dm. long at maturity; leaves mostly or all clustered at or near the base, glabrous and strongly bluish-glaucous, ternate-pinnately dissected into more or less numerous ultimate segments, these broadly linear to elliptic-oblong, mostly (1. 5) 2-8 mm. long and (0. 7) 1-2 mm. wide; flowers yellow; rays of the umbel elongating unequally, the longer ones mostly 5-9 cm. long at maturity; involucel dimidiate, of several well-developed narrow bractlets; fruiting pedicels 4-12 mm. long; fruit 5-10 mm. long, broadly elliptic, the lateral wings 1/5-1/2 as wide as the body; oil tubes mostly 3-4 (6) in the intervals, visible through the pericarp.

Open, rocky or gravelly slopes and dry meadows in the foothills, valleys, and lower mountains of c. Oreg., from n. Jefferson (and doubtless s. Wasco) and s. Gilliam (and Union?) cos. to n. Harney and n. Malheur cos. Apr.-May.

Lomatium farinosum (Hook.) Coult. & Rose, Contr. U.S. Nat. Herb. 7:210. 1900.

Peucedanum farinosum Hook. Lond. Journ. Bot. 6:235. 1847. *Cogswellia farinosa* M. E. Jones, Contr. West. Bot. 12:33. 1908. *(Geyer 325,* Coeur d'Aleine Mts., Ida.)

Taproot prominently globose-thickened a little below the more slender top, commonly 1-2 cm. thick, sometimes shortly moniliform; herbage glabrous; leaves clustered at or near the base, often on a more or less evident pseudoscape, ternately or ternate-pinnately (more ternate than pinnate) 1-3 times cleft into long, narrow ultimate segments mostly 1-10 cm. long and 0.5-3 mm. wide; scapes or elongate peduncles usually several, curved-ascending, 1.5-5 dm. tall at maturity; flowers white, the inflorescence compact at anthesis; rays of the umbel elongating unequally, the longer ones mostly 3-7 cm. long at maturity; involucel tending to be dimidiate, the bractlets few, narrow, 1-3 mm. long at anthesis, up to 5 mm. in fruit, often connate at the base, or some of them bifid; pedicels slender and elongate, mostly 6-22 mm. long at maturity; fruit glabrous, elliptic, 5-6.5 mm. long, barely or scarcely half as wide, the narrow wings about 1/3 to 1/2 as wide as the body; oil tubes several in the intervals and on the commissure.

Rocky slopes and scablands in the valleys and foothills; Kootenai Co. to Nez Perce and Clearwater cos., Ida., extending e. to s.w. Missoula Co., Mont., and w. at least to Spokane, Whitman, and Adams cos., Wash. Apr.-May.

See comment under *L. hambleniae.*

Lomatium foeniculaceum (Nutt.) Coult. & Rose, Contr. U.S. Nat. Herb. 7:222. 1900.

Ferula foeniculacea Nutt. Gen. Pl. 1:183. 1818. *Pastinaca foeniculacea* Spreng. in Roem. & Schult. Syst. Veg. 6:587. 1820. *Peucedanum foeniculaceum* Nutt. ex T. & G. Fl. N. Am. 1:627. 1840. *Cogswellia foeniculacea* Coult. & Rose, Contr. U.S. Nat. Herb. 12:449. 1909. *(Nuttall,* on the high plains of the Missouri, commencing about the confluence of the river Jauke—probably the James R. in S.D.)

L. villosum Raf. Journ. de Phys. 89:101. 1819. *Pastinaca villosa* Spreng. in Roem. & Schult. Syst. Veg. 6:588. 1820. *Peucedanum villosum* Wats. Bot. King Exp. 131. 1871. *Cogswellia villosa* M. E. Jones, Contr. West. Bot. 12:34. 1908. *(Nuttall,* plains of the Platte) = var. *foeniculaceum.*

LOMATIUM FOENICULACEUM var. DAUCIFOLIUM (T. & G.) Cronq. hoc loc. *Peucedanum foeniculaceum* var. *daucifolium* T. & G. Fl. N. Am. 1:627. 1840. *Cogswellia daucifolia* M. E. Jones, Contr. West. Bot. 12:34. 1908. *(Nuttall,* on the Platte R., probably in Neb.)

LOMATIUM FOENICULACEUM var. MACDOUGALII (Coult. & Rose) Cronq. hoc loc. *L. macdougali* Coult. & Rose, Contr. U.S. Nat. Herb. 7:233. 1900. *Cogswellia macdougali* M. E. Jones, Contr. West. Bot. 12:34. 1908. *(MacDougal 84,* Mormon Lake, Ariz.)

L. jonesii Coult. & Rose, Contr. U.S. Nat. Herb. 7:233. 1900. *Cogswellia jonesii* M. E. Jones, Contr. West. Bot. 12:34. 1908. *(Jones 5435,* head of Salina Canyon, Utah) = var. *macdougalii.*

L. semisepultum Peck, Proc. Biol. Soc. Wash. 50:122. 1937. *(Peck 18913,* near Hampton, Deschutes Co., Oreg.) = var. *macdougalii.*

Perennial from a long woody or somewhat fleshy taproot which is sometimes surmounted by a branching caudex, acaulescent or sometimes with one or several short pseudoscapes; herbage sparsely to usually rather densely hirtellous-puberulent throughout; leaves ternate-pinnately dissected into numerous small and narrow ultimate segments that are mostly 1-3 mm. long; scapes or peduncles curved-ascending to prostrate, mostly 0.6-3 dm. long at maturity; flowers yellow; rays of the umbel elongating unequally, the longer ones mostly 2.5-6 cm. long at maturity; involucel dimidiate, the bractlets often connate below, tapering above; pedicels mostly 3-11 mm. long at maturity; ovaries and generally also the mature fruits in our forms distinctly hirtellous-puberulent; fruits elliptic to suborbicular, 5-10 mm. long, the wings 1/4-1/2 as wide as the body; oil tubes 1-several in the intervals, several on the commissure.

Dry, open slopes, from the valleys and plains to moderate or occasionally high elevations in the mountains; Man., Mo., and Tex., w. to Mont., the drier parts of c. and s. Ida., c. and s.e. Oreg., Nev., and Ariz. Apr.-Aug., depending on the elevation.

The species consists of three intergradient but geographically significant varieties. The var. *macdougalii* (Coult. & Rose) Cronq., occurring w. of the continental divide, has moderately to rather densely hirtellous herbage and fruits, and the bractlets of the involucel are seldom much connate.

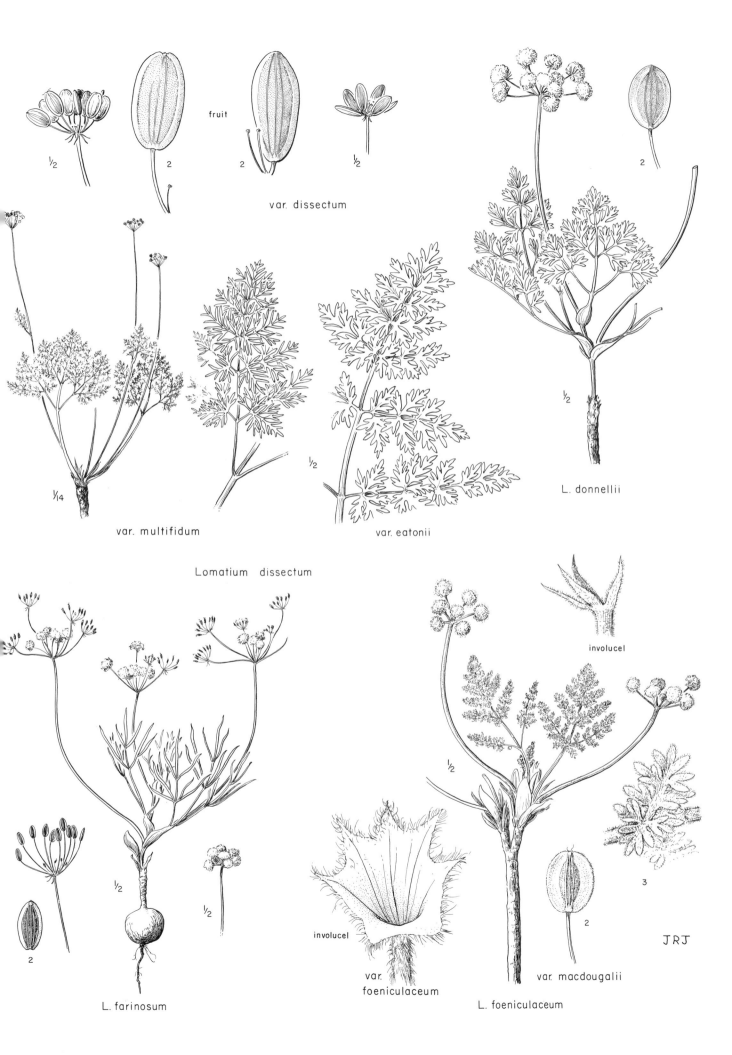

fruit

½   2   2   ½

var. dissectum

2

L. donnellii

½

1/14

var. multifidum

½

var. eatonii

Lomatium dissectum

½

½

½

2

L. farinosum

involucel

involucel

var.
foeniculaceum

½

2

3

involucel

var. macdougalii

L. foeniculaceum

JRJ

The var. *foeniculaceum*, occurring on the northern Great Plains and adjacent westerly intermontane valleys of Mont. and Wyo. (barely extending w. of the continental divide in Powell Co., Mont.), tends to be less densely hairy, though the hairs are often longer, and the bractlets of the involucel are generally more or less strongly connate. The ovaries of var. *foeniculaceum* are pubescent, but the fruit is sometimes glabrous at maturity. The var. *daucifolium* (Nutt.) Cronq., of the southern Great Plains region, from Neb., Kans., and Mo. to Okla. and Tex., has glabrous ovaries and fruits, scantily pubescent or glabrate herbage, and strongly connate bractlets.

Lomatium geyeri (Wats.) Coult. & Rose, Contr. U.S. Nat. Herb. 7:209. 1900.

   Peucedanum geyeri Wats. Bibl. Ind. 428. 1878. *Cogswellia geyeri* M. E. Jones, Contr. West. Bot. 12:33. 1908. *(Geyer 458,* Upper Columbia River)
   Orogenia fusiformis var. *leibergii* Coult. & Rose, Rev. N. Am. Umbell. 92. 1888. *Orogenia leibergii* Rydb. Fl. Rocky Mts. 611, 1064. 1917. *(Leiberg,* sandhills in the Bitterroot Mts., Ida., June, 1887)
   Peucedanum evittatum Coult. & Rose, Bot. Gaz. 14:277. 1889. *(Vasey,* Ellensburg, Wash., May, 1889)

   Much like *L. gormanii,* but taller and more robust, commonly 1.5-3 (4) dm. tall at maturity, glabrous; taproot subglobose to more often elongate or moniliform, up to 4 cm. thick; bractlets of the involucel mostly 2-3 mm. long; pedicels sometimes as much as 4 or even 5 mm. long at maturity; fruits mostly 7-12 mm. long, glabrous.

   Open slopes and flats from the foothills, valleys, and lowlands to moderate elevations in the mountains; Kittitas, Lincoln, and Spokane cos., Wash. (and on Cleman Mt., Yakima Co.), and Kootenai Co., Ida., n. to s. B.C., e. (or barely w.) of the Cascade summits. Early spring.

   It is possible that further study will show *L. geyeri* and *L. gormanii* to be geographical varieties of a single species, but the information currently available does not require such a treatment.

Lomatium gormanii (Howell) Coult. & Rose, Contr. U.S. Nat. Herb. 7:208. 1900.

   Peucedanum gormani Howell, Fl. N.W. Am. 252. (1 April) 1898. *Cogswellia gormani* M. E. Jones, Contr. West. Bot. 12:33. 1908. *(Howell,* high hills opposite the Dalles, presumably in Klickitat Co., Wash.)
   Peucedanum confusum Piper, Erythea 6:29. (10 April) 1898. *(Piper 73,* Pullman, Wash.)
   L. piperi Coult. & Rose, Contr. U.S. Nat. Herb. 7:211. 1900. *Cogswellia piperi* M. E. Jones, Contr. West. Bot. 12:33. 1908. *(Vasey,* Ellensburg, Wash., May, 1889) The form with strictly glabrous fruits.
   L. gormani f. *purpureum* St. John, Proc. Biol. Soc. Wash. 41:196. 1928. *Cogswellia gormani* f. *purpurea* St. John, Fl. S.E. Wash. 292. 1937. *(St. John & Pickett 3714,* Pullman, Wash.) A purple-flowered form.

   Taproot short and globose-thickened, up to 2 cm. thick; herbage glabrous or slightly puberulent; plants low, scapose or seemingly so, the well-developed leaves all attached at or below the ground level (sometimes the leaves all strictly basal on sheathing petioles, sometimes the plants tending to develop a pseudoscape); stems or peduncles 1-3 from the subterranean rootcrown, simple or with 1 or 2 branches, mostly ascending or erect, or (especially the branches) more or less prostrate, up to about 1.5 dm. tall at maturity; leaves dissected into small, narrow ultimate segments, these sometimes numerous, crowded, and less than 5 mm. long, sometimes fewer, longer, and less crowded, some of them often over 1 cm. long; inflorescence compact at anthesis, the flowers white with purple anthers (rarely wholly purple); rays of the umbel elongating unequally, the longer ones 1-4 cm. long at maturity; bractlets of the involucel narrow and inconspicuous, often less than 2 mm. long; pedicels very short, seldom as much as 3 mm. long at maturity; fruit broadly elliptic, glabrous or granular-roughened, mostly 5-7 mm. long, the wings well developed, mostly 1/4-1/2 as wide as the body; oil tubes 1-8 in the intervals, 2-6 on the commissure. Pepper and salt.

   Open slopes and scablands in the foothills, valleys, and plains, often with sagebrush; Kittitas, Lincoln, and s. Spokane cos., Wash., barely extending into adj. Ida., s. through Oreg. (and to Washington Co., Ida.), reputedly to n. Calif.; wholly e. of the Cascade summits in our range. Early spring.

   The difference between glabrous and papillate-roughened fruits appears to have no taxonomic value in this species, and may indeed prove to be merely Mendelian. Here, as elsewhere, individual morphologic characters cannot properly be assigned a particular taxonomic importance a priori. A character is only as important as it proves to be in each instance in helping to delimit a group which has been perceived on the basis of all the available evidence.

Lomatium grayi Coult. & Rose, Contr. U.S. Nat. Herb. 7:229. 1900.

Peucedanum millefolium Wats. Bot. King Exp. 129. 1981, not of Sonder in 1861-2. *Peucedanum grayi* Coult. & Rose, Bot. Gaz. 13:209. 1888. *Cogswellia millefolia* M. E. Jones, Contr. West. Bot. 12:35. 1908. *Cogswellia grayi* Coult. & Rose, Contr. U.S. Nat. Herb. 12:450. 1909. *L. millefolium* Macbr. Contr. Gray Herb. n. s. 53:15. 1918. *(Watson 466,* Antelope Island, Great Salt Lake, Utah)

Peucedanum grayi var. *aberrans* M. E. Jones, Contr. West. Bot. 10:55. 1902. (Several *Jones* collections from s. w. Ida. are cited)

Malodorous perennial from a long stout taproot that is usually surmounted by a branching caudex, the branches of the caudex often covered with the fibrous remnants of the leaves of previous years; plants 1. 5-5 dm. tall at maturity, the several glabrous stems or scapes ascending rather than strictly erect; foliage granular-scaberulous to glabrous, sometimes glaucous; leaves mostly or all borne at or near the base, ternate-pinnately dissected into very numerous (several hundred to a thousand or more) very narrow and often subterete ultimate segments up to about 6 mm. long, the segments disposed in many planes so that the leaf has "thickness"; flowers yellow; rays of the umbel elongating rather unequally, the longer ones mostly 3. 5-10 cm. long at maturity; involucel of several well-developed, very narrow bractlets, more or less dimidiate, seldom obsolete; pedicels 7-15 mm. long at maturity; fruit glabrous, elliptic (sometimes rather broadly so) 8-15 mm. long, the lateral wings mostly 1/3-2/3 as wide as the body; oil tubes mostly solitary in the intervals, 2-4 (6) on the commissure.

Dry, open, often rocky places from the foothills and lowlands to moderate elevations in the mountains; Wash. (e. of the Cascade summits) to n. Ida., s. through e. Oreg. and w. Ida. to n. e. Nev., e. irregularly to s. e. Ida., s. w. Wyo., Utah, and s. w. Colo. Apr. -May.

This species has sometimes been confused with *Cymopterus terebinthinus*, which may be distinguished by its pleasant odor, coarser leaf segments more nearly in a single plane, well-developed calyx teeth, and usually well-developed dorsal wings of the fruit.

Lomatium greenmanii Mathias, Ann. Mo. Bot. Gard. 25:274. 1938. *(Cusick 2458,* Wallowa Mts., head of Keystone Creek, 9000 feet, Oreg., Aug. 4, 1900)

Dwarf perennial from a taproot and much branched caudex, the stems or scapes slender, less than 1 dm. long at maturity, generally with a single more or less reduced leaf; leaves chiefly basal, pinnately or ternate-pinnately once or partly twice compound, the ultimate segments firm, pointed, mostly lanceolate or elliptic, 3-8 mm. long, up to 2. 5 mm. wide, scaberulous along the margins and sometimes on the midrib beneath, otherwise glabrous; flowers reported on the basis of dried specimens to be white; inflorescence small and compact, the rays few and only 1-3 mm. long when the fruit is submature; involucel of a few narrow bractlets; ovaries and young fruit glabrous; mature fruit reported to be ovate, 3. 5 mm. long, 2 mm. wide, with narrow wings, the oil tubes solitary in the intervals and 2 on the commissure.

Known only from the type collection, Wallowa Mts., Oreg.

Further collecting may well show *L. greenmanii* to be merely a glabrous form of *L. oreganum.* One does not expect to find two species as similar as these in the same habitat and local area.

Lomatium hallii (Wats.) Coult. & Rose, Contr. U.S. Nat. Herb. 7:224. 1900.

Peucedanum hallii Wats. Proc. Am. Acad. 11:141. 1876. *Cogswellia hallii* M. E. Jones, Contr. West. Bot. 12:35. 1908. *(Hall 211* in part, Silver Creek, Marion Co., Oreg.)

Perennial from a stout taproot which is sometimes surmounted by a very short branching caudex, 2-3. 5 dm. tall at maturity, glabrous throughout, or somewhat scaberulous in the inflorescence; stems clustered; leaves chiefly basal, but one or more low-cauline ones usually or always present as well, shining green, pinnately or ternate-pinnately dissected into more or less numerous (up to several hundred) small ultimate segments 1-6 mm. long which all tend to lie in nearly the same plane; flowers yellow; rays of the umbel elongating unequally, the longer ones mostly 2-5 cm. long at maturity; involucel of several well-developed narrow bractlets; fruiting pedicels 4-10 mm. long; fruit glabrous, elliptic, 5-9 mm. long, the wings about half as wide as the body; oil tubes 2-3 in the intervals, about 5 on the commissure.

Rocky crevices and bluffs in the foothills and valleys; Rogue and Umpqua region of s. w. Oreg., extending n. along the w. slopes of the Cascades to Marion Co. Apr.

Occasional specimens of *L. hallii* are suggestive of *L. martindalei,* indicating that there may be some small amount of introgression between the two species.

Lomatium hambleniae Math. & Const. Bull. Torrey Club 69:153. 1942. *(Hamblen s.n.*, Dry Falls,
  Grand Coulee, Wash., in 1941)
  Very similar to *L. farinosum* except for the yellow flowers.
  Scablands, often with sagebrush; c. Wash., from Grant Co. to Chelan and Yakima cos. Apr.
  Lomatium hambleniae may prove to be merely a color form of *L. farinosum,* and at best the two
are probably no more than varietally distinct. The apparent geographic segregation of the two color-
forms is based on a limited number of specimens with determinable flower color, although it is very
likely that the yellow-flowered plants do not extend appreciably east of the stated range. Further field
observations will be necessary before the proper taxonomic status of *L. hambleniae* can be determined, and in the meantime no nomenclatural change is here proposed.

Lomatium hendersonii Coult. & Rose, Contr. U.S. Nat. Herb. 7:209. 1900.
  Peucedanum hendersonii Coult. & Rose, Bot. Gaz. 13:210. 1888. *Cogswellia hendersoni* M. E.
    Jones, Contr. West. Bot. 12:33. 1908. *Leptotaenia hendersonii* Math. & Const. Bull. Torrey Club
    68:123. 1941. *(Howell,* John Day Valley, Oreg., May, 1882)
  Taproot short and strongly tuberous-thickened, napiform to subglobose; plants glabrous, acaulescent or tending to develop a pseudoscape, the leaves all attached at or below the ground level; leaves
ternate-pinnately dissected into small and crowded ultimate segments up to about 6 mm. long and 1 or
1.5 mm. wide; peduncles (scapes) short, in fruit about 1 dm. long and recurved to the ground; flowers
yellow (sometimes incorrectly reported to be white, the color soon fading in herbarium specimens);
rays of the umbel few, elongating unequally, the longer ones 1-3 cm. long at maturity; bractlets of the
involucel slender but fairly well developed, 1.5-4 mm. long; pedicels 2-5 (or reputedly 7) mm. long
at maturity; fruit rather broadly elliptic-ovate, 6-7 mm. long, becoming very narrowly winged at
maturity (wings about 0.5 mm. wide) and then promptly deciduous; oil tubes inconspicuous, solitary
in the intervals, 2 on the commissure.
  Dry, open slopes (especially in heavy clay soil) at lower elevations in Jefferson and Wheeler cos.,
Oreg., and n.w. Owyhee Co., Ida. Rarely collected. Apr.
  A similar but larger and more nearly erect plant, up to 2.5 dm. tall, from Malheur Co., Oreg.,
has been described as *Leptotaenia leibergii* Coult. & Rose, Contr. U.S. Nat. Herb. 7:202. 1900.
While it is possible that further collecting will demonstrate that *Leptotaenia leibergii* is a taxonomic
synonym of *Lomatium hendersonii,* as indicated by Mathias & Constance in the N. Am. Flora, it is at
least equally possible that *Leptotaenia leibergii* will need a new name in *Lomatium.*

Lomatium idahoense Math. & Const. Bull. Torrey Club 70:58. 1943. *(Cronquist 2856,* Beaver Creek,
  near Marsh Creek, 25 miles n.w. of Stanley, Custer Co., Ida.)
  Taproot elongate and not much thickened; plants low, mostly 1-2 (4) dm. tall, with one or a few
scapes arising from the root; herbage glabrous or rather sparsely scaberulous; leaves all basal or
one or two distinctly cauline, mostly short-petiolate, ternately or ternate-pinnately 1-3 times dissected (more ternately than pinnately), with narrow, obtuse or rounded to more or less acute ultimate
segments mostly 1-5 (10) cm. long; peduncles often scarcely surpassing the leaves; flowers yellow,
or more or less purplish in age; involucel wanting or rarely of a few inconspicuous narrow bractlets;
rays of the umbel elongating unequally, the longer ones mostly 3-8 cm. long at maturity; pedicels
mostly 5-11 mm. long at maturity; fruit rather narrowly elliptic, 6-11 mm. long, glabrous, the
wings less than half as wide as the body; oil tubes solitary in the intervals, 2 on the commissure.
  Open, often rocky slopes and dry meadows, at moderate to high elevations in the mountains of c.
Ida., in Custer, Elmore, Blaine, Boise, and Camas cos. June-July.

Lomatium laevigatum (Nutt.) Coult. & Rose, Contr. U.S. Nat. Herb. 7:225. 1900.
  Peucedanum laevigatum Nutt. in T. & G. Fl. N. Am. 1:627. 1840. *Cogswellia laevigata* M. E.
    Jones, Contr. West. Bot. 12:32. 1908. *(Nuttall,* "Blue Mts. of Oregon")
  Glabrous and slightly glaucous perennial from a stout, branching, woody caudex that surmounts a
taproot; plants 2.5-4 dm. tall at maturity; leaves chiefly or entirely basal, ternate-pinnately dissected into linear ultimate segments 1-2 mm. wide, the better-developed segments mostly 1-3 cm.
long; flowers yellow; involucel wanting, or occasionally of 1 or 2 inconspicuous setaceous bractlets;
rays of the umbel elongating unequally, the longer ones mostly 3-5 cm. long; fruiting pedicels mostly
4-10 mm. long; fruit glabrous, elliptic, 7-12 mm. long, the wings from about half as wide to nearly
as wide as the body; oil tubes solitary in the intervals, 2 on the commissure.
  Crevices in basalt cliffs along the Columbia R. in Klickitat Co., Wash., and adj. Oreg. Apr.

1/2

2

Lomatium geyeri

1.5

fruit

2

1/4

L. grayi

2

1/2

L. gormanii

1/2

L. greenmanii

1/2

2

L. hallii

2

L. hambleniae

2

1/2

L. hendersonii

JRJ

Lomatium leptocarpum (T. & G.) Coult. & Rose, Contr. U.S. Nat. Herb. 7:213. 1900.

    Peucedanum triternatum var. *leptocarpum* T. & G. Fl. N. Am. 1:626. 1840. *Peucedanum ambiguum*
      var. *leptocarpum* Coult. & Rose, Rev. N. Am. Umbell. 59. 1888. *Cogswellia leptocarpa*
      M. E. Jones, Contr. West. Bot. 12:33. 1908. *L. ambiguum* var. *leptocarpum* Jeps. Madroño 1:159.
      1923. *(Nuttall,* "plains of the Oregon near the confluence of the Wahlamet"; a specimen of the type
      collection at New York is labeled "near Fort Vancouver?"; the collection was doubtless made east
      of the Cascade Mts., rather than west of them)

    Peucedanum bicolor var. *gumbonis* M. E. Jones, Contr. West. Bot. 10:55. 1902. *(Jones,* Monroe
      Creek, Washington Co., Ida., Apr. 20, 1900, and Indian Valley, Adams Co., Ida., July 15, 1899)

Root short and cormose-thickened to more elongate and slender, sometimes with 2-several thickened parts connected by more slender segments, usually surmounted by a simple crown which gives
rise to only one or two stems, rarely surmounted by a more branching caudex which gives rise to a
cluster of stems; plants 1-5 dm. tall, simple or branched at the base, glabrous to occasionally scaberulous or hirtellous, essentially acaulescent (the underground part of the stem or peduncle commonly enclosed by the sheathing petioles), or larger specimens often caulescent with 1 or 2 fairly well-
developed leaves below the middle; leaves ternate-pinnately 2-several times dissected into mostly
linear and elongate, often very unequal ultimate segments mostly 0. 5-2 mm. wide, the longer segments mostly (0. 5) 1-5 cm. long; peduncles long, curved-ascending to erect; rays of the umbel elongating unequally, generally more or less elongate even at anthesis, sometimes up to 11 cm. at maturity; involucel of well-developed but mostly narrow and more or less linear-attenuate bractlets 2-7
(10) mm. long and seldom over 1 mm. wide; flowers yellow or occasionally white; pedicels very
short, only 1-2 (3) mm. long at maturity; fruits glabrous, crowded and usually more or less numerous in each umbellet, narrow and elongate, 6-13 mm. long, 2-3 mm. wide, very narrowly or scarcely
winged, the wings up to about 0. 5 mm. wide; oil tubes solitary in the intervals, 2-4 on the commissure.

    Open slopes, flats, meadows, and swales, especially in heavy clay soils, from the foothills and low-
lands to moderate elevations in the mountains; throughout most of the part of Oreg. that lies e. of the
Cascades, extending into extreme s. e. Wash., and from Clearwater Co., Ida., to e. Custer Co.,
Camas Co., and Owyhee Co., and in n. e. Calif.; apparently isolated in s. Wyo. and adj. Colo. May-
June.

    Plants from the Wasatch and Bear River Range region of Utah and adj. Ida. that have been referred to
L. leptocarpum have much more finely divided leaves, with numerous filiform segments; these represent a distinct species, *L. bicolor* (Wats.) Coult. & Rose.

Lomatium macrocarpum (Nutt.) Coult. & Rose, Contr. U.S. Nat. Herb. 7:217. 1900.

    Peucedanum macrocarpum Nutt. ex T. & G. Fl. N. Am. 1:627. 1840. *Cogswellia macrocarpa* M.
      E. Jones, Contr. West. Bot. 12:33. 1908. *(Nuttall,* barren hills on the Oregon)

    Peucedanum macrocarpum var. *? eurycarpum* Gray, Proc. Am. Acad. 8:385. 1872. *Peucedanum*
      eurycarpum Coult. & Rose, Rev. N. Am. Umbell. 61. 1888. *(Hall 210,* Oregon)

    L. macrocarpum var. *artemisiarum* Piper, Bull. Torrey Club 29:223. 1902. *L. artemisiarum*
      Piper, Contr. U.S. Nat. Herb. 11:423. 1906. *Cogswellia macrocarpa* var. *artemisiarum* St. John,
      Fl. S.E. Wash. 292. 1937. *(Piper 2976,* Pasco, Wash.)

    L. macrocarpum var. *semivittatum* Piper, Bull. Torrey Club 29:224. 1902. *L. macrocarpum* ssp.
      semivittatum Piper, Contr. U.S. Nat. Herb. 11:223. 1906. *(Henderson 397,* Hood River, Oreg.)

    L. flavum Suksd. Allg. Bot. Zeits. 12:6. 1906. *Cogswellia flava* Coult. & Rose, Contr. U.S. Nat.
      Herb. 12:449. 1909. *(Suksdorf 506,* Bingen, Wash.) A form reported to have yellow flowers; yel-
      low flowers are otherwise unknown in the species.

Taproot mostly elongate, strongly thickened throughout or with a tuberous or moniliform base and
more slender upper part, capped by a usually simple and often subterranean crown; plants usually
branched near the base and with several peduncles, shortly or scarcely caulescent and sometimes
with a short pseudoscape, the leaves clustered near the ground; herbage sparsely to rather densely
puberulent or villous-puberulent, the leaves commonly somewhat grayish (or anthocyanic) rather than
shining green; leaves more or less ternate-pinnately or merely pinnately dissected into rather small
ultimate segments up to about 9 mm. long and 2 mm. wide; peduncles spreading or ascending, mostly
1-2. 5 dm. long at maturity; flowers white or purplish-white, rarely yellow; rays of the umbel elongating unequally or subequally, the longer ones mostly 2-6 cm. long at maturity; involucel tending to be
dimidiate, the bractlets narrow, often irregularly connate, well developed and conspicuous, commonly equaling or surpassing the flowers, villous-puberulent, generally green or greenish, not markedly

scarious-margined; pedicels 1-11 mm. long at maturity; fruits mostly rather narrow, oblong or linear-oblong to more or less elliptic, (1. 8) 2-5 times as long as wide, (7) 10-20 (28) mm. long, glabrous or puberulent (as also the ovaries), the marginal wings narrow to fairly broad; oil tubes 1 (3) in the intervals, sometimes obscure, 2-6 on the commissure. N=11.

Open, rocky hills and plains, not extending much into the mountains; s. B. C. to c. Calif. , e. to Man. , N. D. , w. Wyo. , and n. c. Utah, wholly e. of the Cascade summits in our range. Late Mar. - May.

A common and highly variable species.

Lomatium martindalei Coult. & Rose, Contr. U.S. Nat. Herb. 7:225. 1900.
  Peucedanum martindalei Coult. & Rose, Bot. Gaz. 13:142. 1888. *Cogswellia martindalei* M. E. Jones, Contr. West. Bot. 12:34. 1908. *(Howell,* rocky places, Cascade Mts. , Oreg. , May-June, 1880)
  Peucedanum martindalei var. *angustatum* Coult. & Rose, Bot. Gaz. 13:143. 1888. *L. martindalei* var. *angustatum* Coult. & Rose, Contr. U.S. Nat. Herb. 7:225. 1900. *Cogswellia martindalei* var. *angustata* M. E. Jones, Contr. West. Bot. 12:34. 1908. *Cogswellia angustata* Coult. & Rose, Contr. U.S. Nat. Herb. 12:449. 1909. *L. angustatum* St. John, Mazama 11:83. 1929. *(Howell,* Cascade Mts. , Oreg.)
  LOMATIUM MARTINDALEI var. FLAVUM (G. N. Jones) Cronq. hoc loc. *L. angustatum* var. *flavum* G. N. Jones, U. Wash. Pub. Biol. 5:202. 1936. *(Piper 897,* Olympic Mts. , Wash. , is the first of several collections cited)

Taproot elongate and not much thickened, or with a deep-seated thickening, capped by a simple, often subterranean rootcrown, or occasionally by a more or less branching caudex, the plant sometimes developing a short pseudoscape; herbage glabrous or sometimes granular-scaberulous; leaves chiefly basal or nearly so, pinnately or ternate-pinnately once or twice compound, with a relatively small number (in any case less than 60) of toothed or cleft, more or less leafletlike ultimate segments; flowers white or ochroleucous, seldom yellow except in the Olympic Mts. ; rays of the umbel equal or unequal, the longer ones mostly (1) 1. 5-6 cm. long at maturity; involucel of a few inconspicuous narrow bractlets, or wanting; pedicels mostly 2-10 (15) mm. long at maturity; fruits oblong to broadly elliptic, mostly 8-16 mm. long, the wings equaling or narrower than the body; oil tubes solitary in the intervals, 2 on the commissure. N=11.

Rocky slopes, less often dry meadows, usually well up in the mountains; Cascade region from s. B. C. to s. Oreg. , extending w. nearly to the coast in s. Oreg. ; coast range of n. Oreg. (Tillamook Co.); Olympic Mts. of Wash. May-July (Sept.).

The species consists of 3 varieties. Var. *martindalei,* with relatively broad fruits mostly 1. 5-2 times as long as wide, occupies the more southern part of the range, extending n. to Mt. Hood, Oreg. , and occasionally to Skamania Co. , Wash. The var. *angustatum* Coult. & Rose, with narrower fruits mostly 2-3 times as long as wide, occurs in the Cascades of Wash. and s. B. C. , extending s. into Oreg. about to Mackenzie Pass. Both of these varieties typically have ochroleucous flowers, although there are occasional yellow-flowered individuals at least in some localities (e. g. Black Butte, Jefferson Co. , Oreg. , as noted on the label of *Cusick 2687).* The var. *flavum* (G. N. Jones) Cronq. , from the Olympic Mt. region, has consistently yellow flowers, along with narrow fruits like those of var. *angustatum.* The single collection of the species from the Coast Range of n. Oreg. has narrow fruits and may represent either var. *flavum* or var. *angustatum.*

Lomatium minus (Rose) Math. & Const. Bull. Torrey Club 69:246. 1942.
  Leptotaenia minor Rose ex Howell, Fl. N. W. Am. 251. 1898. *Cusickia minor* M. E. Jones, Contr. West. Bot. 12:40. 1908. *(Leiberg 98,* near Rock Creek, Morrow Co. , Oreg.)

Plants with a tuberous-thickened, vertical or horizontal root and a simple crown, the old leaf bases not prominent; stem or scape stout and more or less erect, 1-3 dm. tall, leafless or with a single reduced leaf; herbage glabrous and glaucous; leaves ternate-pinnately dissected into numerous small and crowded ultimate segments mostly 1-5 mm. long; flowers purple; rays of the umbel elongating unequally, the longer ones mostly 2-6 cm. long at maturity; involucel of well-developed narrow bractlets, more or less dimidiate; fruiting pedicels 5-8 mm. long; fruit elliptic, 12-16 mm. long, the lateral wings narrow and strongly thickened, about 1 mm. wide; dorsal ribs prominent and slightly raised at maturity, as wide as the intervals or wider; oil tubes mostly solitary in the intervals, 3 or 4 on the commissure.

In dry drainage channels that are covered with basaltic rocks, on the open plateaus at lower altitudes in the Blue Mt. region of Oregon. Apr.

This very distinct species is as yet insufficiently known.

Lomatium nevadense (Wats.) Coult. & Rose, Contr. U. S. Nat. Herb. 7:220. 1900.

Peucedanum nevadense Wats. Proc. Am. Acad. 11:143. 1876. *Cogswellia nevadensis* M. E. Jones, Contr. West. Bot. 12:33. 1908. *(Watson 469,* western Nevada, from the Washoe to the West Humboldt Mts.)

Taproot elongate and not much thickened, with a usually subterranean, simple or few-branched crown or caudex; plants often tending to develop a pseudoscape, otherwise subacaulescent, the leaves all or nearly all in a cluster at or near the ground level; scapes or peduncles 0. 5-3 dm. long at maturity, prostrate or ascending to suberect; herbage finely puberulent or hirtellous throughout; leaves pinnately 2-3 times divided into small and narrow ultimate segments, the basal pair of pinnae the largest, but the blade scarcely ternate, the leaf appearing more open than in some other species, and the ultimate segments sometimes appearing more as slender teeth on the penultimate ones; flowers white; rays of the umbel elongating unequally, the longer ones mostly 2-5 cm. long at maturity; involucel dimidiate, the bractlets glabrous or very finely hirtellous, linear to lanceolate or rather narrowly elliptic, distinctly scarious-margined, sometimes connate below; fruiting pedicels 4-14 mm. long; ovaries and young fruit hirtellous-puberulent or rarely glabrous; mature fruit 6-10 mm. long, broadly elliptic to suborbicular, often glabrate, the lateral wings 1/4-2/3 as wide as the body; oil tubes 2-9 in the intervals, 4-12 on the commissure.

Dry, open slopes and flats in the foothills and plains; c. Oreg. (Grant and Jefferson cos.) to s. Calif., n. w. Colo., s. Utah, w. Ariz., and n. Sonora. May.

The closely related *L. orientale* Coult. & Rose, of the Great Plains, has consistently glabrous ovaries and fruits that average a little narrower in proportion to their length than those of *L. nevadense,* but the two taxa are otherwise scarcely to be distinguished, and forms of *L. nevadense* with glabrous ovaries and fruits are not uncommon to the south of our range. *L. orientale* occurs in parts of Mont., but is not known to extend west into our range.

Lomatium nudicaule (Pursh) Coult. & Rose, Contr. U. S. Nat. Herb. 7:238. 1900.

Smyrnium nudicaule Pursh, Fl. Am. Sept. 196. 1814. *Ferula nudicaulis* Nutt. Gen. Pl. 1:183. 1818. *Pastinaca nudicaulis* Spreng. in Roem. & Schult. Syst. Veg. 6:587. 1820. *Ferula nuttallii* DC. Prodr. 4:174. 1830. *Peucedanum nudicaule* Nutt. ex T. & G. Fl. N. Am. 1:627. 1840. *Cogswellia nudicaulis* M. E. Jones, Contr. West. Bot. 12:31. 1908. *(Lewis,* on the Columbia River, probably at The Dalles)

Seseli leiocarpum Hook. Fl. Bor. Am. 1:263. 1832. *Peucedanum leiocarpum* Nutt. ex T. & G. Fl. N. Am. 1:626. 1840. *(Douglas,* near Fort Vancouver on the Columbia)

Peucedanum latifolium Nutt. ex T. & G. Fl. N. Am. 1:625. 1840, not of DC. in 1830. *Peucedanum nuttallii* Wats. Bot. King Exp. 128. 1871. *L. platyphyllum* Coult. & Rose, Contr. U. S. Nat. Herb. 7:238. 1900. *Cogswellia latifolia* M. E. Jones, Contr. West. Bot. 12:31. 1908. *Cogswellia platyphylla* Coult. & Rose, Contr. U. S. Nat. Herb. 12:450. 1909. *(Nuttall,* plains east of the Wallawalla River)

Peucedanum leiocarpum var. *campestre* Nutt. ex T. & G. Fl. N. Am. 1:626. 1840. *(Nuttall,* locality not stated)

Stems or scapes solitary or several from the simple or slightly branched crown of a stout taproot, the plant mostly 2-9 dm. tall at maturity; herbage glabrous and strongly blue-glaucous; leaves firm, ternately or ternate-pinnately 1-3 times compound, with 3-30 well-defined, veiny, often petiolulate ultimate leaflets, these lanceolate or oblong to ovate or subrotund, mostly 2-9 cm. long and (0. 4) 1-6 cm. wide, entire or often dentate toward the tip, sometimes also with one or more irregular lobes representing a partial division into additional leaflets; scapes or stems generally more or less erect, naked or leafy only at the base, often strongly fistulose just beneath the umbel; flowers yellow, the umbellets well separated from each other at anthesis and usually rather many-flowered; rays of the umbel elongating unequally, the longer ones mostly 6-20 cm. at maturity; fruiting pedicels 3-15 mm. long; fruit 7-15 mm. long, oblong or elliptic, sometimes narrowed to a short beaklike tip, the wings up to about half as wide as the body; oil tubes solitary in the dorsal intervals, 1-several in the lateral, several on the commissure.

Dry, open or sparsely wooded places from the lowlands to moderate elevations in the mountains,

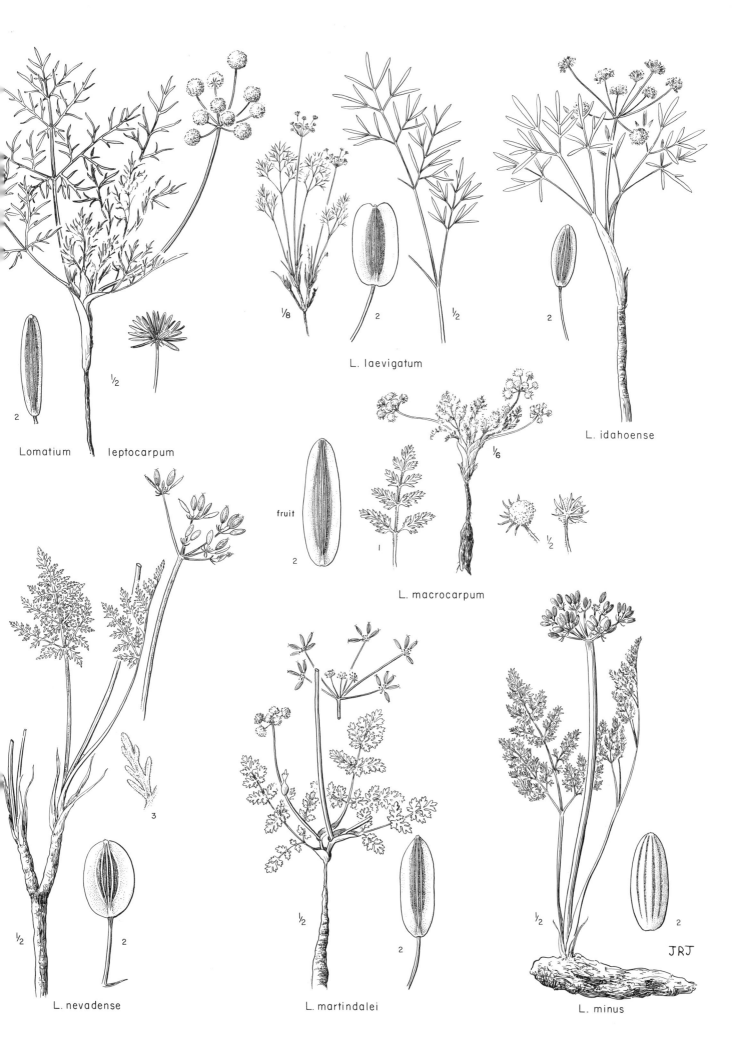

L. laevigatum

Lomatium   leptocarpum

L. idahoense

L. macrocarpum

fruit

L. nevadense

L. martindalei

L. minus

JRJ

often with sagebrush or ponderosa pine; both sides of the Cascades, from s. B. C. to c. Calif., e. to s. w. Alta., w. and s. Ida., and w. Utah. Apr. -June

Our most distinctive species of the genus.

Lomatium oreganum Coult. & Rose, Contr. U. S. Nat. Herb. 7:224. 1900.

Peucedanum oreganum Coult. & Rose, Rev. N. Am. Umbell. 64. 1888. *Cogswellia oregana* M. E. Jones, Contr. West. Bot. 12:35. 1908. *(Cusick 1390,* Blue and Eagle Creek [Wallowa] Mts., Oreg.)

Dwarf perennial from a taproot and much-branched caudex, acaulescent, the slender peduncles (scapes) only 2-6 cm. long at maturity; leaves, peduncles, inflorescence, and fruits hirtellous-puberulent; leaves all basal, pinnately or ternate-pinnately about twice compound, the ultimate segments 1-3 (6) mm. long; flowers yellow; inflorescence small and compact, the rays few and only 1-5 mm. long at maturity; involucel of a few narrow bractlets; pedicels about 1 mm. long; fruit elliptic-oblong, about 5 mm. long, 2. 5-3 mm. wide, the wings narrower than the body; oil tubes 2-3 in the intervals, 4 on the commissure.

Open rocky places at high altitudes in the Wallowa Mts. and the Elkhorn Ridge of the Blue Mts., n. e. Oreg. July.

Lomatium orogenioides (Coult. & Rose) Mathias, Ann. Mo. Bot. Gard. 25:242. 1938.

Leibergia orogenioides Coult. & Rose, Contr. U. S. Nat. Herb. 3:575. 1896. *Cogswellia orogenioides* M. E. Jones, Contr. West. Bot. 12:33. 1908. *(Leiberg 1027,* Santianne Creek bottoms, Coeur d'Alene Mts., Ida.)

Root subglobose, about 1 cm. thick or less; plants glabrous, acaulescent, blooming when only 3-10 cm. tall, but up to 3 dm. or more in fruit; leaves few, ternately or pinnately cleft into 3-13 narrowly linear or filiform segments up to 5 cm. long and 1. 5 mm. wide, the petiole very slender; flowers white; rays of the umbel mostly 4-8, elongating unequally, the longer ones 5-15 cm. long at maturity; bractlets of the involucel inconspicuous, linear or lanceolate, 1-3 mm. long; pedicels very short, 0. 5-3 mm. long; each umbellet commonly producing 1-6 fruits; fruits linear, 6-10 mm. long, 1-1. 5 mm. thick, quadrangular or subterete, the lateral wings obsolete; oil tubes small, solitary in the intervals, 2 on the commissure.

Meadows and moist bottomlands along streams; n. Ida. and adj. Wash. Apr. -May.

This species with wingless fruits is technically aberrant in *Lomatium,* but it appears to be allied to *L. leptocarpum,* in which the fruits are only very narrowly winged.

Lomatium rollinsii Math. & Const. Bull. Torrey Club 70:59. 1943. *(Constance, Rollins, & Dillon 1573,* near Deep Creek, Snake R. Canyon, Wallowa Co., Oreg.)

Taproot short, usually tuberous-thickened or moniliform; plants mostly 2-7 dm. tall at maturity, with one or two simple or sparingly branched stems from the base; herbage evidently scaberulous or crisp-puberulent to subglabrous; leaves cauline and basal, irregularly and more or less distinctly pinnately dissected into unequal narrow segments that are seldom much over 1 cm. long; peduncles elongate; involucel of inconspicuous narrow bractlets; flowers yellow; rays of the umbel elongating unequally, the longer ones mostly 3-7 cm. long at maturity; pedicels mostly 4-10 (15) mm. long at maturity; fruit glabrous, elliptic, 5-8 mm. long, the wings well developed but scarcely or barely half as wide as the body; oil tubes 1-2 in the intervals, 4 on the commissure.

Open slopes in and near the canyons of the Snake and lower Salmon rivers, e. Oreg. and w. Ida. Apr.

Lomatium salmoniflorum (Coult. & Rose) Math. & Const. Bull. Torrey Club 69:246. 1942.

Peucedanum salmoniflorum Coult. & Rose ex Holz. Contr. U. S. Nat. Herb. 3:228. 1895. *Leptotaenia salmoniflora* Coult. & Rose, Contr. U. S. Nat. Herb. 7:201. 1900. *(Sandberg 24,* near upper ferry, Clearwater R. above Lewiston, Ida.)

Taproot strongly thickened and more or less elongate, with a simple and sometimes deep-seated crown that usually is not conspicuously clothed by old leaf bases; stems or scapes solitary or several, often branched near the base, usually ascending rather than erect, often short at anthesis, mostly 2-6 dm. tall at maturity, usually with one or more reduced leaves below the middle; herbage glabrous and apparently glaucous; leaves chiefly basal or nearly so, ternate-pinnately dissected into very numerous small and narrow ultimate segments mostly 1. 5-5 mm. long; petiolar sheaths very prominent; flowers salmon-colored or salmon-yellow; rays of the umbel mostly 5-13, elongating unequally, the longer ones mostly 3-6 cm. long at maturity; involucel of a few inconspicuous narrow bractlets, tending to be dimidiate; fruit elliptic-oblong to broadly elliptic, 8-14 mm. long, the lateral wings corky-thickened, 0.5-1 mm. wide; dorsal ribs evident and slightly raised, narrower than the intervals; oil tubes mostly solitary in the intervals, 2 on the commissure.

Open rocky slopes near the Snake and Clearwater rivers in w. Idaho Co. and s. Latah and n. Nez Perce cos., Ida., and s. Whitman Co., w. to Palouse Falls, Wash.; probably also extending into e. Wallowa Co., Oreg. Mar.-Apr.

This species has often been confused with the superficially similar *L. grayi,* from which it differs in its simple rootcrown, consistently glabrous leaves, more or less salmon-colored flowers, and narrowly corky-winged fruit.

Lomatium sandbergii Coult. & Rose, Contr. U.S. Nat. Herb. 7:230. 1900.
  Peucedanum sandbergii Coult. & Rose, Bot. Gaz. 13:79. 1888. *Cogswellia sandbergii* M. E. Jones, Contr. West. Bot. 12:35. 1908. *(Sandberg 47,* bare mountain tops in Kootenai Co., Ida.)

Taproot more or less elongate and slightly thickened, with a simple crown; plants mostly 1-3 dm. tall at maturity, caulescent, the stem commonly branched at or near the base and producing several elongate peduncles; herbage granular-scaberulous to subglabrous; leaves basal and low-cauline, withering as the fruit ripens, small, the blade only 1.5-7 cm. long, ternate-pinnately dissected into small narrow segments mostly 1-4 (7) mm. long, these all tending to lie in nearly a single plane; flowers yellow; rays of the umbel elongating unequally, the longer ones mostly 2.5-10 cm. long at maturity; involucel of several slender bractlets; pedicels 2-5 (7) mm. long; fruit granular-scaberulous, elliptic, 5-8 mm. long, the wings only 1/5-1/3 as wide as the body; oil tubes several (mostly 4-5) in the intervals, mostly 6-8 on the commissure.

Open, rocky slopes and ridges, from moderate to high elevations in the mountains; n. Ida., n.w. Mont. (chiefly w. of the continental divide), and extreme s.w. Alta. May-July.

Lomatium serpentinum (M. E. Jones) Mathias, Ann. Mo. Bot. Gard. 25:271. 1938.
  Cogswellia serpentina M. E. Jones, Contr. West. Bot. 12:42. 1908. *(Cusick 3532 c,* rocky banks of the Snake R. near the mouth of McDougal Creek, Oreg.)
  Cogswellia fragrans St. John, Fl. S.E. Wash. 290. 1937. *(St. John 4193,* mouth of the Salmon R., Idaho Co., Ida.)

Pleasantly aromatic (parsley-scented) perennial from a stout taproot and branching caudex, essentially acaulescent, mostly 1.5-4 dm. tall at maturity; leaves bright green, glabrous or obscurely papillate-scaberulous, ternate-pinnately dissected into moderately numerous (commonly 100-300) small flat ultimate segments mostly 1-6 mm. long and up to 2.5 mm. wide, the segments all tending to lie in nearly the same plane; flowers yellow; rays of the umbel elongating unequally, the longer ones mostly 2-8 cm. long at maturity; involucel of several fairly well-developed narrow bractlets; fruiting pedicels 3-15 mm. long; fruit elliptic, 5.5-10 mm. long, glabrous, the wings from about half as wide to fully as wide as the body; oil tubes solitary in the intervals, 2 on the commissure.

Open, often rocky slopes in and near the Snake R. Canyon; w. Ida., e. Oreg., and extreme s.e. Wash. Apr.-July.

Lomatium suksdorfii (Wats.) Coult. & Rose, Contr. U.S. Nat. Herb. 7:239. 1900.
  Peucedanum suksdorfii Wats. Proc. Am. Acad. 20:369. 1885. *Cogswellia suksdorfii* M. E. Jones, Contr. West. Bot. 12:32. 1908. *(Suksdorf,* dry, rocky mountainsides, w. Klickitat Co., Wash.)

Stout, glabrous perennial from a taproot and branching caudex, mostly 5-20 dm. tall at maturity, caulescent, but the first (sterile) umbel of the season at least sometimes borne on a long scape; leaves large, ternate (or quinate)-pinnately dissected into long, narrow, acute ultimate segments mostly 1-5 cm. long and 1-5 mm. wide; flowers yellow; longer rays of the umbel 6-11 cm. long at maturity; involucel of well-developed narrow bractlets sometimes as much as 1 cm. long or more; fruiting pedicels 6-13 mm. long; fruit glabrous, very large, 15-32 mm. long, the lateral wings 1/3-1/2 as wide as the body, the dorsal ribs distinctly raised or very narrowly winged; oil tubes solitary (2-4) in the intervals, 2 on the commissure.

Dry, open slopes in w. Klickitat Co., Wash. May.

LOMATIUM THOMPSONII (Mathias) Cronq. hoc loc.
  L. suksdorfii var. *thompsonii* Mathias, Ann. Mo. Bot. Gard. 25:289. 1938. *(Sandberg & Leiberg 489,* Peshastin, Okanogan [now Chelan] Co., Wash.)

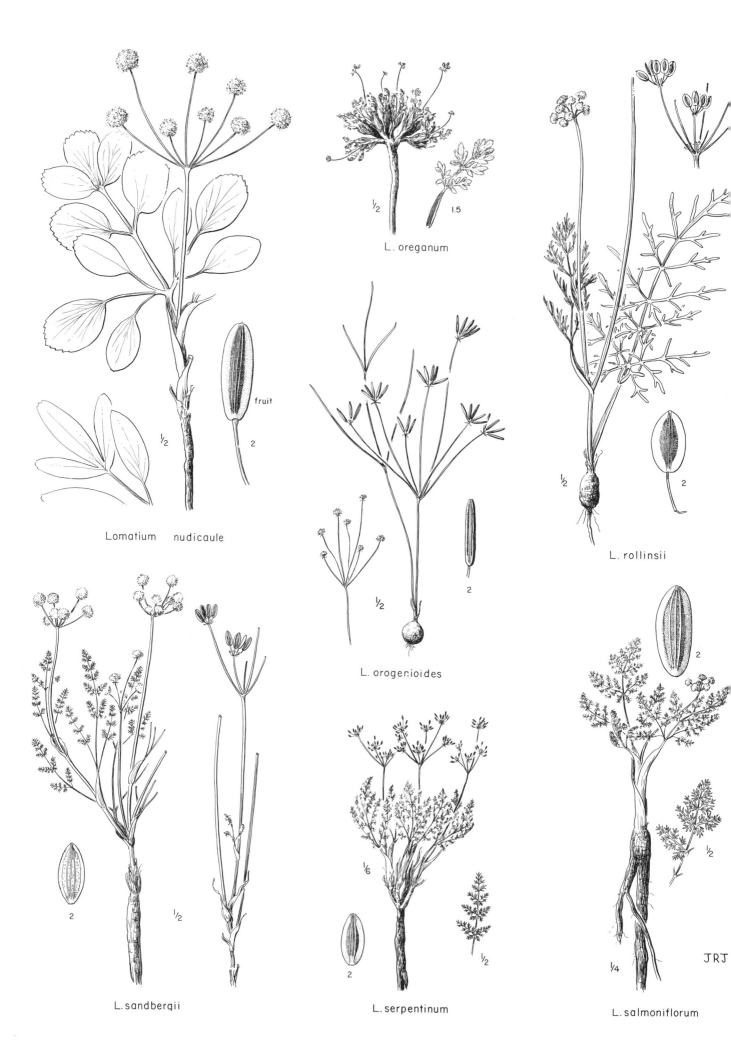

L. oreganum

fruit

2

½

Lomatium nudicaule

L. orogenioides

½

2

L. rollinsii

½

2

L. sandbergii

2

½

L. serpentinum

⅙

2

½

L. salmoniflorum

2

¼

½

JRJ

Coarse perennial from a large, woody taproot and compactly branched caudex; mostly 5-10 dm. tall at maturity; stem and peduncle hirtellous-puberulent, the foliage more sparsely so or often merely scaberulous to subglabrous; leaves basal and low-cauline, or all basal, large, ternate (or quinate)-pinnately dissected into rather small, linear ultimate segments mostly 1-2 (3) mm. wide and up to about 1 cm. or occasionally 2 cm. long; flowers yellow; involucel of a few well-developed bractlets; rays of the umbel 5-14, elongating unequally, the longer ones mostly 5-10 cm. long at maturity; fruiting pedicels mostly 8-15 mm. long; fruit puberulent when young, later glabrate, large, elliptic-oblong, 16-28 mm. long, the lateral wings well developed but scarcely half as wide as the body, the dorsal ribs narrowly raised.

Open or wooded slopes in the foothills of the Wenatchee region, Chelan Co., Wash. June.

Lomatium thompsonii differs from *L. suksdorfii* in its puberulent stems, shorter leaf segments, and more northern, disjunct distribution. There is no indication of any intergradation between the two taxa, and *L. thompsonii* seems to warrant specific status. *L. thompsonii* has also been confused with *L. dissectum,* which has glabrous stems, mostly more crowded leaf segments, more numerous rays in the umbel, and very different fruit.

Lomatium triternatum (Pursh) Coult. & Rose, Contr. U.S. Nat. Herb. 7:227. 1900.
> *Seseli triternatum* Pursh, Fl. Am. Sept. 197. 1814. *Eulophus triternatus* Nutt. Journ. Acad. Phila. 7:27. 1834. *Peucedanum triternatum* Nutt. ex T. & G. Fl. N. Am. 1:626. 1840. *Peucedanum nuttallii* Walpers Rep. 2:411. 1843. *Cogswellia triternata* M. E. Jones, Contr. West. Bot. 12:32. 1908. *(Lewis,* on the waters of the Columbia River; probably actually on the Clearwater R. near the mouth of Potlatch Creek, Ida.)
> *Peucedanum triternatum* var. *leptophyllum* Hook. Lond. Journ. Bot. 6:235. 1847. *Cogswellia leptophylla* Rydb. Bull. Torrey Club 40:74. 1913. *L. simplex* var. *leptophyllum* Mathias, Ann. Mo. Bot. Gard. 25:283. 1938. *(Geyer 505,* slopes of the high plains of the Kooskooskee [Clearwater River]) A form with puberulent fruits.
> LOMATIUM TRITERNATUM ssp. PLATYCARPUM (Torr.) Cronq. hoc loc. *Peucedanum triternatum* (as *"citernatum")* var. ? *platycarpum* Torr. in Stansb. Expl. Utah 389. 1852. *L. platycarpum* Coult. & Rose, Contr. U.S. Nat. Herb. 7:226. 1900. *Cogswellia platycarpa* M. E. Jones, Contr. West. Bot. 12:32. 1908. *(Stansbury,* Great Salt Lake, Utah)
> *Peucedanum simplex* Nutt. ex Wats. Bot. King Exp. 129. 1871. *Cogswellia simplex* M. E. Jones, Bull. U. Mont. Biol. 15:41. 1910. *L. simplex* Macbr. Contr. Gray Herb. n.s. 56:34. 1918. *(Nuttall,* Rocky Mts.) = ssp. *platycarpum.*
> *Peucedanum triternatum* var. *macrocarpum* Coult. & Rose, Rev. N. Am. Umbell. 70. 1888. *Peucedanum triternatum* var. *robustius* Coult. & Rose ex Holz. Contr. U.S. Nat. Herb. 3:228. 1895. *L. robustius* Coult. & Rose, Contr. U.S. Nat. Herb. 7:228. 1900. *Cogswellia triternata* var. *robustior* M. E. Jones, Contr. West. Bot. 12:32. 1908. *Cogswellia robustior* Coult. & Rose, Contr. U.S. Nat. Herb. 12:451. 1909. *L. triternatum* var. *macrocarpum* Mathias, Ann. Mo. Bot. Gard. 25:286. 1938. *(Suksdorf 502,* western Klickitat Co., Wash.) An extreme form of var. *triternatum* with long narrow fruits.
> *Peucedanum triternatum* var. *brevifolium* Coult. & Rose, Rev. N. Am. Umbell. 70. 1888. *L. brevifolium* Coult. & Rose, Contr. U.S. Nat. Herb. 7:232. 1900. *Cogswellia brevifolia* M. E. Jones, Contr. West. Bot. 12:32. 1908. *L. triternatum* var. *brevifolium* Mathias, Ann. Mo. Bot. Gard. 25:286. 1938. *(Howell 379,* Klickitat Co., Wash.) An extreme form of var. *triternatum* with relatively numerous, rather crowded, short leaf segments mostly about 1 cm. long.
> *Peucedanum triternatum* var. *alatum* Coult. & Rose, Rev. N. Am. Umbell. 70. 1888. *L. alatum* Coult. & Rose, Contr. U.S. Nat. Herb. 7:228. 1900. *Cogswellia triternata* var. *alata* M. E. Jones, Contr. West. Bot. 12:32. 1908. *Cogswellia alata* Coult. & Rose, Contr. U.S. Nat. Herb. 12:448. 1909. *L. triternatum* var. *alatum* Mathias, Ann. Mo. Bot. Gard. 25:287. 1938. *(Curran,* Folsom, Calif.) A form approaching ssp. *platycarpum* in its broadly winged fruit.
> *L. anomalum* M. E. Jones ex Coult. & Rose, Contr. U.S. Nat. Herb. 7:237. 1900. *Cogswellia anomala* M. E. Jones, Contr. West. Bot. 12:32. 1908. *L. triternatum* var. *anomalum* Mathias, Ann. Mo. Bot. Gard. 25:285. 1938. *(Jones,* Indian Valley, Washington [now Adams] Co., Ida.)
> *Cogswellia triternata* f. *lancifolia* St. John, Fl. S.E. Wash. 293. 1937. *L. triternatum* f. *lancifolium* St. John, Fl. S.E. Wash. 2nd ed. 548. 1956. *(Piper,* Spokane, Wash., in 1896) An extreme form of var. *triternatum.*

Taproot elongate and seldom much thickened, surmounted by a simple or occasionally few-branched crown or short caudex; stems or scapes solitary or few, more or less erect, mostly (1) 2-8 dm. tall at maturity; herbage finely hirtellous-puberulent throughout, or the leaves sometimes essentially glabrous; leaves chiefly or wholly basal or low-cauline, one or more fairly well developed cauline leaves often present on the middle or upper part of the stem but the stem not appearing very leafy; leaves ternately to ternate-pinnately (or at the base quinately) 2-3 times (seldom only once) cleft into long, usually narrow segments or leaflets 1-10 (20) cm. long (also relatively broad in one var.); rays of the umbel elongating unequally, the longer ones mostly 2-10 cm. long at maturity; fruits oblong to broadly elliptic, 7-15 (20) mm. long, narrowly to very broadly winged, glabrous to occasionally minutely puberulent or granular-scabrous; oil tubes solitary in the intervals, 2 on the commissure. N=11.

Open slopes and meadows, in dry to fairly moist soil, from the lowlands to moderate elevations in the mountains; s. Alta. and B.C. to Colo., Utah, and Calif. May-July.

This species consists of two major geographic races which have usually been treated as distinct species, under the names *L. triternatum* and *L. simplex*. The two are sufficiently well marked in their typical forms, but are so thoroughly intergradient in a broad contact zone that no clear line can be drawn between them. Occasional specimens from well within the range of one of these races approach or simulate the other in one or more respects. These two races are here treated as subspecies *triternatum* and subspecies *platycarpum*. The subspecies *triternatum* is highly variable, and several extreme forms have been described as varieties or segregate species. Only one of these variants, the var. *anomalum*, is sufficiently distinctive and geographically coherent to demand recognition. The differences among the taxa here accepted are summarized in the following key:

1 Fruit broadly elliptic, the wings nearly or fully as wide as (or wider than) the body; leaves fairly regularly dissected, tending to be more nearly ternate than pinnate, the ultimate segments always linear or nearly so, relatively few, less crowded, and less markedly unequal than in ssp. *triternatum;* plants more often scapose or nearly so than distinctly caulescent; c. and s.w. Colo., Utah, and n.e. Nev., n. to w. Mont. (chiefly e. of the continental divide, but extending also to Missoula Co. and even Flathead Co.), the drier parts of c. Ida. (but more common from the Snake R. plains southward), e. Oreg. (where mostly at lower elevations than ssp. *triternatum),* the drier parts of c. Wash., and the Okanogan Valley of s. B.C.   ssp. platycarpum (Torr.) Cronq.
1 Fruit usually relatively narrow, the wings seldom more than half as wide as the body; leaves tending to be less regularly and more nearly pinnately dissected than in ssp. *platycarpum,* although the first division is generally ternate or even quinate; ultimate segments of the leaves tending to be more numerous, more markedly unequal, and more crowded than in ssp. *platycarpum,* varying from linear to elliptic; plants distinctly caulescent, or less often scapose; s. Alta. and B.C. to n.w. Mont., c. Ida. (n. of the Snake R. plains), s. Oreg., and n. Calif.   ssp. triternatum
     2 Ultimate segments of the leaves, as in ssp. *platycarpum,* mostly linear or nearly
          so and more or less strongly acute; range of the subspecies   var. triternatum
     2 Ultimate segments of the leaves mostly broader than linear, obtuse or rounded to
          barely acute; Nez Perce Co. to Washington and Camas cos., Ida., and extending
          into adj. Oreg.   var. anomalum (M. E. Jones) Mathias

Lomatium tuberosum Hoover, Leafl. West. Bot. 4:39. 1944. (*Hoover 5726,* south of White Swan, Yakima Co., Wash.)

Root tuberous-thickened, vertical or horizontal, with a simple crown; bladeless basal sheaths very large and conspicuous, but the remnants from previous years not prominent as in *L. cuspidatum;* scape erect, 1-3 dm. tall; herbage glabrous and glaucous; leaves usually all basal, ternate-pinnately dissected into numerous small, crowded, blunt or barely apiculate ultimate segments that are mostly 2-8 mm. long; flowers purple; rays of the umbel mostly 5-13, up to 8 cm. long at maturity; involucel of a few inconspicuous narrow bractlets; fruit elliptic, 9-11 mm. long, the lateral wings thickened and less than 1 mm. wide.

Known to be represented in herbaria only by collections from rocky hillsides near Fort Simcoe and White Swan, Yakima Co., Wash. Mar.-Apr.

This imperfectly known species is suggestive of both *L. cuspidatum* and *L. minus,* but it does not seem to be properly referable to either of these taxa. It is probably most closely related to *L. minus,* likewise a little-known species, and it is possible that further collecting will show the two to be conspecific.

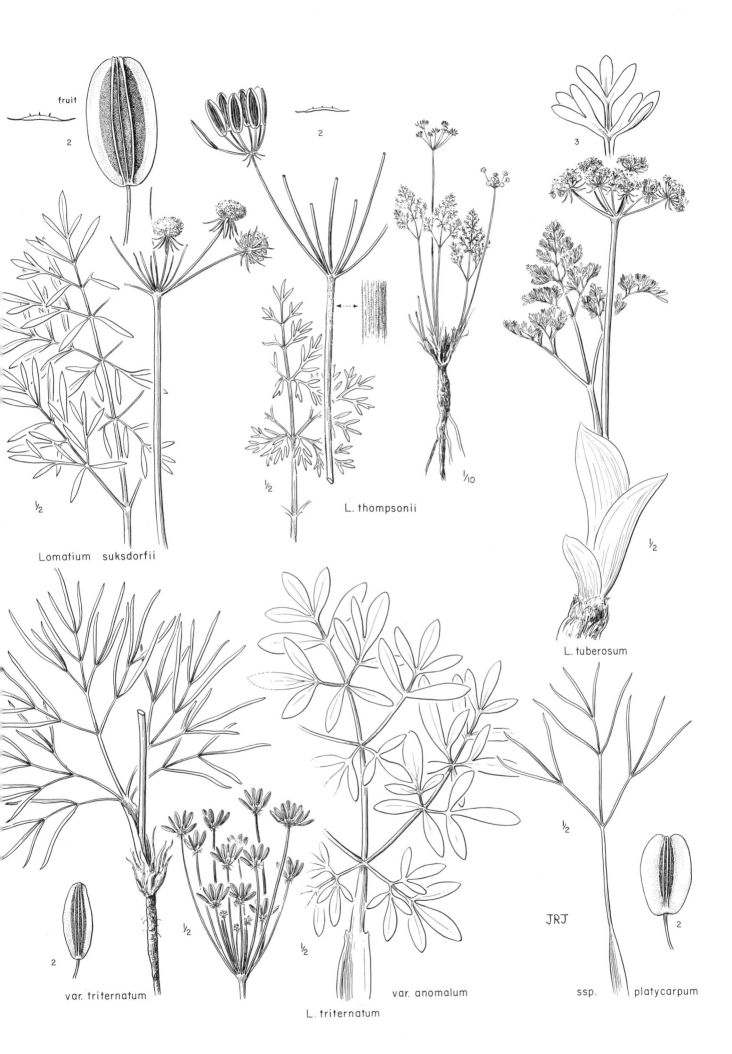

fruit

2

2

Lomatium suksdorfii

½

½

L. thompsonii

1/10

3

½

L. tuberosum

var. triternatum

½

2

½

L. triternatum

var. anomalum

½

ssp. platycarpum

½

JRJ

2

Lomatium utriculatum (Nutt.) Coult. & Rose, Contr. U.S. Nat. Herb. 7:215. 1900.

Peucedanum utriculatum Nutt. ex T. & G. Fl. N. Am. 1:628. 1840. *Cogswellia utriculata* M. E. Jones, Contr. West. Bot. 12:34. 1908. *(Nuttall,* near the confluence of the Wahlamet and Oregon rivers)

Taproot usually more or less elongate and seldom very much thickened, in any case not cormose; plants 1-6 dm. tall, glabrous or short-hairy, usually distinctly caulescent, the leaves often chiefly cauline, frequently some of them borne well above the middle; leaves soft, ternate-pinnately dissected, with slender, crowded, acute or acuminate segments generally well under 1 mm. wide and seldom over 5 (12) mm. long; rays of the umbel seldom as many as 15, elongating unequally, the longer ones mostly 2-7 cm. long at maturity; bractlets of the involucel well developed, subherbaceous to subscarious, 2-5 mm. long, obovate to elliptic or suborbicular, the tip often shallowly toothed or cleft; flowers yellow; pedicels 2-8 mm. long at maturity; fruit somewhat granular-roughened when young, usually glabrous at maturity, 5-11 mm. long, 3-6 mm. wide, the wings from a little narrower to a little broader than the body, the dorsal ribs prominent and slightly raised; oil tubes 1-4 in the intervals, 2-6 on the commissure, sometimes obscure. N=11.

Prairies and other open, often rocky places w. of the Cascades; s. B.C. to c. Calif. and apparently less often to s. Calif. Apr.-June.

Many of the Californian plants usually referred to *L. utriculatum* have broader or longer leaf segments and properly belong to other species such as *L. vaseyi* Coult. & Rose.

Lomatium vaginatum Coult. & Rose. Contr. U.S. Nat. Herb. 7:223. 1900.

Cogswellia vaginata M. E. Jones, Contr. West. Bot. 12:34. 1908. *(Cusick 1697,* Logan Valley, Union Co., Oreg.)

Root elongate and seldom much thickened; plants glabrous or granular-scaberulous, 1.5-4 dm. tall, distinctly caulescent, the leaves sometimes chiefly cauline, but mostly borne below the middle; petioles of the cauline leaves short and very conspicuously dilated; leaves flat or nearly so, ternate-pinnately dissected, the ultimate segments crowded, rather thick and firm, often blunt, seldom as much as 4 mm. long; umbels long-pedunculate, the rays elongating unequally, sometimes 6 cm. long at maturity; bractlets of the involucel oblanceolate to rather narrowly elliptic, acuminate, entire or with one or two large teeth, subscarious to more or less herbaceous, 3-6 mm. long; flowers yellow; pedicels slender, 5-15 mm. long at maturity; fruit granular-roughened when young, generally glabrous at maturity, elliptic, 8-12 mm. long, 5-8 mm. wide, the wings generally a little narrower than the body, the dorsal ribs sometimes narrowly raised; oil tubes 1-4 in the intervals, 4-5 on the commissure.

Open slopes and flats in the foothills and valleys, often on the heavy clays associated with volcanic tuff; Union, Grant, and Crook cos., Oreg., s. to n.e. Calif. Apr.-June.

Lomatium watsonii Coult. & Rose, Contr. U.S. Nat. Herb. 7:211. 1900.

Peucedanum watsoni Coult. & Rose, Bot. Gaz. 13:209. 1888. *Cogswellia watsoni* M. E. Jones, Contr. West. Bot. 12:33. 1908. *(Howell 830,* on denuded hilltops near Alkali, Oreg.)

Much like a small form of *L. cous;* taproot thickened and usually elongate; plants acaulescent or with a pseudoscape; leaf segments narrow, mostly 1-5 mm. long; flowers yellow; involucel more or less dimidiate, the broad bractlets usually connate below the middle or nearly to the tip; ovaries and fruit finely puberulent or occasionally glabrous, the mature fruit ovate, 6-7 mm. long and about half as wide, the wings less than half as wide as the body; oil tubes several in the intervals and on the commissure, but obscure.

Open hillsides, often with sagebrush; s.c. Wash. (s. Kittitas, Yakima, and Klickitat cos.) to n.c. Oreg. (Trout Creek, presumably the Trout Creek of Jefferson Co.). May.

Lomatium frenchii Math. & Const. Bull. Torrey Club 86:377. 1959, based on a series of collections made by David French in Warm Springs Indian Reservation, Jefferson and Wasco cos., Oreg., will key to *L. watsonii* because of its connate bractlets. In other respects, however, it more nearly resembles the *L. circumdatum* phase of *L. cous,* being caulescent, 1.5-2.5 dm. tall, with fruits 8-10 mm. long and with glabrous (rarely pubescent) foliage and ovaries. The thickened but elongate root is within the range of variation of both *L. watsonii* and *L. cous.* The geographical locality is peripheral to those of both *L. watsonii* and *L. cous,* but is more readily included with the former. The proper taxonomic status of these plants remains to be determined (Type: *French 640,* 1 mile northwest of Mill Creek, Warm Springs Indian Reservation, Jefferson Co. [actually in Wasco Co. acc. to verbal information from Dr. French], Oreg.).

Musineon Raf.

Inflorescence a compound umbel; involucre usually wanting; involucel of several distinct or basally connate bractlets; flowers yellow or white; calyx teeth well developed, ovate; styles slender, spreading, without a stylopodium; carpophore entire to deeply cleft; fruit ovoid to linear-oblong, laterally somewhat compressed, glabrous to scabrous, evidently ribbed; oil tubes 1-4 in the intervals and sometimes one in each rib, 2-6 on the commissure; low perennials from a thickened taproot, the leaves borne chiefly at or near the base (sometimes on a pseudoscape), pinnately or ternate-pinnately more or less dissected, with small ultimate segments.

Four species of w. U.S. (Name derived from the ancient Greek name for *Foeniculum* or some other umbellifer.)

1 Leaves mostly subopposite, deeply pinnatifid, the primary pinnae only shortly or scarcely
     stalked and generally appearing as more or less deeply cleft or toothed segments; taproot
     with a simple crown and plants often with a pseudoscape; species primarily of the high
     plains, extending w. to Custer Co., Ida., and to e. Nev.   M. DIVARICATUM
1 Leaves distinctly alternate, ternate-pinnately dissected into more or less linear ultimate
     segments, at least the lowest pair of primary segments distinctly slender-stalked; tap-
     root surmounted by a more or less branched caudex, and plants without a pseudoscape;
     Bridger Mts., Mont., to Bighorn Mts., Wyo.   M. VAGINATUM

Musineon divaricatum (Pursh) Nutt. ex T. & G. Fl. N. Am. 1:642. 1840.
    Seseli divaricatum Pursh, Fl. Am. Sept. 732. 1814. *Adorium divaricatum* Rydb. Bot. Surv. Nebr.
      3:37. 1894. *(Bradbury,* Upper Louisiana; bluffs of the Missouri R. at the mouth of the Niobrara
      R., fide Mathias)
    M. divaricatum var. *hookeri* T. & G. Fl. N. Am. 1:642. 1840. *M. hookeri* Nutt. ex Coult. & Rose,
      Bot. Gaz. 20:259. 1895. *Adorium hookeri* Rydb. Contr. U.S. Nat. Herb. 3:501. 1896. *(Nuttall,*
      plains of the Upper Platte, near the Rocky Mts.)
    M. trachyspermum Nutt. in T. & G. Fl. N. Am. 1:642. 1840. *(Nuttall,* plains of the Upper Platte,
      near the Rocky Mts.)
    M. angustifolium Nutt. in T. & G. Fl. N. Am. 1:642. 1840. *(Nuttall,* plains of the Upper Platte,
      within the Rocky Mts.)

Perennial from a stout, vertical, often deep-seated taproot and simple crown, often with a pseudoscape, 0.5-3 dm. tall, the stems clustered, frequently curved at the base; leaves mostly subopposite, more or less clustered toward the base, seldom as much as 12 cm. long (petiole included) and 3.5 cm. wide, somewhat bluish or glaucous, deeply pinnatifid, the primary segments sessile or nearly so and more or less deeply cleft or toothed; peduncles usually surpassed by the leaves at anthesis and slightly surpassing them at maturity; inflorescence 1-3 cm. wide at anthesis, slightly larger in fruit; flowers yellow; fruit more or less ovate, narrowed above, 3-6 mm. long.

Open flats and slopes in the plains, valleys, and foothills, tolerant of alkali; Sask. to Neb., w. to Alta., Mont., e. Custer Co., Ida., and e. Nev. Apr.-June.

The peduncles, rays of the umbel, fruits, and rachis of the leaves vary from glabrous to scabrous or granular-roughened. The more prominently scabrous or granular-roughened forms (including most or all of our plants) have been segregated as var. *hookeri* T. & G., but the distinction is only doubtfully significant.

Musineon vaginatum Rydb. Mem. N.Y. Bot. Gard. 1:288. 1900. *(Rydberg & Bessey 4626,* Bridger
    Mts., Mont.)

Perennial from a stout taproot that is generally surmounted by a short branched caudex; stems several, clustered, slender, commonly purple toward the base, as also the lower petioles; leaves basal and cauline, distinctly alternate, ternate-pinnately dissected into more or less linear ultimate segments, at least the lowest pair of primary segments distinctly slender-stalked; peduncles slender, equaling or surpassing the leaves at anthesis and becoming longer in fruit, the plant 1.5-3 dm. tall at maturity; inflorescence mostly 1-1.5 cm. wide at anthesis, slightly larger in fruit, the flowers reported to be white or yellowish; fruit ovate-oblong, 3-4 mm. long, glabrous to scaberulous or granular-roughened.

Rocky places in the mountains and foothills; Bridger Mts., Mont., to the Bighorn Mt. region of Wyo.; seldom collected. June-July.

This is a sharply distinct species.

Oenanthe L.

Inflorescence of compound umbels; involucre of narrow bracts, or wanting; involucel of numerous small, narrow bractlets; flowers white; calyx teeth evident, persistent; styles elongate and more or less erect, tending to be firmly persistent; stylopodium conic; carpophore wanting; fruit oblong or elliptic, truncate at the apex and much broader than the stylopodium, terete or slightly compressed laterally, glabrous, the prominent ribs corky-thickened and wider than the intervals; oil tubes usually solitary in the intervals, 1-2 on the commissure; glabrous perennials from fascicled fibrous or tuberous-thickened roots; leaves pinnately compound to dissected, with evident broad leaflets or small and narrow ultimate segments.

About 30 species, two in N. Am., the rest in the Old World. (The ancient Greek name for some other plant.)

Oenanthe sarmentosa Presl ex DC. Prodr. 4:138. 1830. *(Haenke,* Nootka Sound)
Fibrous-rooted, soft and weak, generally reclining or scrambling-ascending herbs, often rooting at the nodes, freely and loosely branched, up to 1 m. or more long; leaves mostly bipinnate or subtripinnate, with toothed to cleft leaflets commonly 1.5-6 cm. long and 7-50 mm. wide, the primary lateral veins of the leaflets directed to the marginal teeth; umbels pedunculate, leaf-opposed (i. e. terminal and the stem sympodial), the 10-20 rays mostly 1.5-3 cm. long at maturity; involucre of a few narrow leafy bracts, or more often wanting; involucel of evident narrow bractlets; fruit oblong, truncate, 2.5-3.5 mm. long, 1.5-2 mm. wide, the ribs broader than the narrow intervals. N=22.

Low wet places, in thickets and along streams and in marshes or around sloughs, west of the Cascades, extending up the Columbia R. to w. Klickitat Co., Wash., and in the Chilliwack Valley, B.C.; Alas. panhandle to c. Calif. June-Aug.

Orogenia Wats. Turkey Peas

Inflorescence a small compound umbel with few, spreading, unequal rays; involucre wanting; involucel minute or wanting; pedicels short or none; calyx teeth obsolete; petals white, obovate with a narrower inflexed apex; anthers and top of the ovary dark purplish; stylopodium and carpophore wanting; fruit oblong to oval, glabrous, subterete or slightly flattened laterally, the dorsal ribs evident or obscure, the lateral ones developed into inflexed corky wings, a corky riblike projection also running the length of the commissural face of each mericarp; oil tubes several in the intervals and on the commissure; delicate low vernal scapose perennials from a tuberous-thickened root; leaves petiolate, the petiole (especially of the reduced outermost leaves) more or less sheathing, the blade mostly once to thrice ternate, with elongate, narrow, mostly entire ultimate segments.

Two species of w. U.S., evidently related to some of the smaller white-flowered species of *Lomatium.* (Name from the Greek *oros,* mountain, and *genos,* race, referring to the habitat.)
1 Root globose; e. of the Cascade summits   O. LINEARIFOLIA
1 Root elongate, several times as long as thick; w. of the Cascade summits   O. FUSIFORMIS

Orogenia fusiformis Wats. Proc. Am. Acad. 22:474. 1887. *(Austin,* Plumas Co., Calif.)
Similar to *O. linearifolia,* differing chiefly in the more elongate, less-thickened root; leaves sometimes more numerous than in *O. linearifolia,* and the dorsal ribs of the fruit tending to be obscure.
Open places, from the valleys to moderate elev. in the mts.; Linn Co., Oreg., s. to n. Calif. May-July.

Orogenia linearifolia Wats. Bot. King Exp. 120. 1871. *(Watson 440,* north of Parley's Park, Wasatch Mts. of Utah)
Perennial from a globose root up to about 1.5 cm. thick that is generally seated 2-7 cm. below the ground surface; well-developed leaves mostly two or three, the blade only slightly elevated above the ground level, the ultimate segments mostly 1-4.5 cm. long and 0.5-4 mm. wide; scape ascending only a few cm. above the ground level; inflorescence at anthesis compact, 1-2 cm. long, longer than wide, more open in fruit; pedicels less than 2 mm. long; fruit 3-4 mm. long, oblong-elliptic, with evident dorsal ribs.

Open slopes and ridges from the lower foothills to moderate elevations in the mountains; s. Wash. (e. of the Cascades) and adj. Ida., e. to Ravalli Co., Mont., and s. to e. Oreg., s. Ida., Utah, and w. Colo., and to be expected in Nev. Mar.-May.

The species is often found near vernal snowbanks, blooming as soon as the ground is bared.

2
ucre bract

½

Lomatium vaginatum

Musineon divaricatum

fruit
4

involuce₁

4

½

Lomatium watsonii

½

involucre bracts
3

usineon vaginatum

½

Lomatium utriculatum

½

1.5

5

½

Oenanthe sarmentosa

JRJ

Osmorhiza Raf. Sweet Cicely

Inflorescence of loose compound umbels on terminal and often also axillary peduncles which generally surpass the leaves; involucre wanting, or of one or a few narrow, foliaceous bracts; involucel of several narrow foliaceous reflexed bractlets, or wanting; flowers white, yellow, purple, or pink; calyx teeth obsolete; stylopodium evidently conic to depressed; carpophore 2-cleft less than half its length; fruit narrow, linear or clavate, obtuse to constricted or short-beaked at the tip, rounded to caudate at the base, somewhat compressed laterally, bristly-hispid to glabrous, the ribs narrow, the oil tubes obscure or wanting; caulescent, thick-rooted perennials with petiolate leaves, the blade ternately to pinnately once to thrice compound, with lanceolate to orbicular, toothed to pinnatifid leaflets.

Eleven species, native to N. Am., the Andean region of S. Am., and e. Asia. (Name from the Greek *osme*, odor, and *rhiza*, root, from the pleasant odor of the original species.)

Reference:

Constance, L., and R. Shan. The genus Osmorhiza (Umbelliferae). U. Calif. Pub. Bot. 23:111-156. 1948.

1 Fruit glabrous, the base obtuse, exappendiculate; flowers yellow; stems clustered
O. OCCIDENTALIS

1 Fruit attenuate at the base into prominent, bristly appendages, the body often bristly as well; flowers greenish white, or sometimes pink or purple; stems mostly solitary

    2 Fruit concavely narrowed to the summit, the terminal (0. 5) 1-2 mm. more or less distinctly set off as a broadly beaklike apex (this condition more marked in *O. chilensis* than in *O. purpurea*); pedicels and rays of the umbel mostly ascending-spreading

        3 Fruit mostly 12-22 mm. long; stylopodium more or less conic, commonly at least as high as broad; flowers mostly greenish-white; common and widespread
O. CHILENSIS

        3 Fruit mostly 8-13 mm. long; stylopodium depressed, generally broader than high; flowers pink or purple to sometimes greenish-white; relatively uncommon in our range, and largely confined to the moister mountainous regions   O. PURPUREA

    2 Fruit convexly narrowed to the rounded or obtuse to merely acutish summit, the apex not at all beaklike; pedicels and rays of the umbel generally widely divaricate; widespread, but less common in our range than *O. chilensis*   O. DEPAUPERATA

Osmorhiza chilensis H. & A. Bot. Beechey Voy. 26. 1830. *(Lay & Collie,* Conception, Chile)
   O. divaricata Nutt. ex T. & G. Fl. N. Am. 1:639. 1840, nom. nud. ; Suksd. Allg. Bot. Zeits. 12:5. 1906. *Washingtonia divaricata* Britt. in Britt. & Brown, Ill. Fl. 2:531. 1897. *Scandix divaricata* K. -Pol. Bull. Soc. Nat. Mosc. II, 29:143. 1916. *O. nuda* var. *divaricata* Jeps. Madroño 1:119. 1923. *Uraspermum divaricatum* Farw. Am. Midl. Nat. 12:70. 1930. *(Nuttall,* Oregon woods)
   O. nuda Torr. Pac. R. R. Rep. 4:93. 1857. *Uraspermum nudum* Kuntze, Rev. Gen. 1:270. 1891. *Myrrhis nuda* Greene, Fl. Fran. 333. 1892. *Washingtonia nuda* Heller, Cat. N. Am. Pl. 5. 1898. *O. divaricata* var. *nuda* M. E. Jones, Bull. U. Mont. Biol. 15:42. 1910. *Scandix nuda* K. -Pol. Bull. Soc. Nat. Mosc. II, 29:143. 1916. *(Bigelow,* Napa Valley, Calif.)
   Washingtonia intermedia Rydb. Mem. N. Y. Bot. Gard. 1:289. 1900. *(Rydberg & Bessey 4595,* Bridger Mts. , Mont.)
   Washingtonia brevipes Coult. & Rose, Contr. U. S. Nat. Herb. 7:66. 1900. *O. brevipes* Suksd. Allg. Bot. Zeits. 12:5. 1906. *O. nuda* var. *brevipes* Jeps. Madroño 1:119. 1923. *Uraspermum brevipes* Farw. Am. Midl. Nat. 12:70. 1930. *(Palmer 2481,* Mount Shasta and vicinity, Siskiyou Co. , Calif.)

Perennial from a well-developed taproot which may be surmounted by a slightly branched caudex, not markedly odorous; stems solitary or sometimes 2 or 3, rather slender, 3-10 dm. tall; herbage more or less hirsute to occasionally essentially glabrous; leaves biternate, the leaflets thin, narrowly to very broadly ovate, coarsely toothed and sometimes incised or more or less tripartite, mostly 2-7 cm. long and 1-5. 5 cm. wide; basal leaves several, long-petiolate; cauline leaves 1-3, with shorter petioles or subsessile, but the blade fairly well developed; stem usually branched above and producing several umbels, these small, inconspicuous, and short-pedunculate at anthesis, becoming open and long-pedunculate at maturity, the peduncles 5-25 cm. long, the 3-8 rays ascending-spreading, 2-12 cm. long; involucre and involucel wanting or nearly so; flowers greenish-white, or reputedly rarely pinkish; stylopodium more or less conic and commonly as high as or higher than wide, together with the style 0.3-1.0 mm. long; fruit densely ascending-hispid toward the base, and generally sparsely so along the ribs above, linear-oblong, 12-22 mm. long including the caudate base, concavely narrowed toward the summit, the terminal 1-2 mm. more or less distinctly set off as a broadly beaklike apex. N=11.

Woodlands, from near sea level to moderate elevations in the mountains; s. Alas. (n. of the panhandle) to s. Calif., e. to s.w. Alta., w. S.D., Colo., and Ariz., reappearing in the Great Lakes region and from n. N.H. to Newf.; also in Chile and adj. Argentina. Common in all the forested parts of our range. Apr.-June.

The subapical constriction on the fruit which distinguishes this species from *O. depauperata* is evident well before maturity and can sometimes be observed with a lens even before the petals have fallen.

Osmorhiza depauperata Phil. Anal. Univ. Chile 85:726. 1894. *(Philippi,* Valle de las Nieblas, Chile)
   Washingtonia obtusa Coult. & Rose, Contr. U.S. Nat. Herb. 7:64. 1900. *O. obtusa* Fern. Rhodora 4: 154. 1902. *(Rose 476,* Ishawood Creek, northwestern Wyo.)
   Very similar in most respects to *O. chilensis;* pedicels and rays of the umbel more widely and stiffly divaricate; fruit mostly 10-15 mm. long, convexly narrowed to the rounded or obtuse to merely acutish summit, the apex not at all beaklike; stylopodium low-conic to depressed, as wide as or often wider than high.

Woodlands, from near sea level to moderate elevations in the mountains; s. Alas. (n. of the panhandle) to n.e. Calif. and s. Nev., e. to Sask., w. S.D., and N.M., reappearing in the Great Lakes region and from n. Vt. to Newf.; also in Chile and adj. Argentina. May-July.

Although the total ranges of *O. depauperata* and *O. chilensis* are very similar, their frequency within the range is quite different. *O. depauperata* is much less common in our region than *O. chilensis,* and apparently does not extend west of the Cascade summits; on the other hand it is abundant in the southern Rocky Mts. and Colorado Plateau, where *O. chilensis* is rare. The differences between the two species, though slight, are reasonably constant. Young fruits of *O. depauperata* are clavate and apically truncate, without the subapical constriction seen in *O. chilensis.*

The use of the name *O. depauperata* for this species, in place of the more familiar *O. obtusa,* follows Mathias & Constance, Contr. Fl. Nev. 44 *(Umbelliferae)*:11. 1957.

Osmorhiza occidentalis (Nutt.) Torr. Bot. Mex. Bound. 71. 1859.
   Glycosma occidentalis Nutt. ex T. & G. Fl. N. Am. 1:639. 1840. *Myrrhis occidentalis* Benth. & Hook. Gen. Pl. 1:897. 1867. *Washingtonia occidentalis* Coult. & Rose, Contr. U.S. Nat. Herb. 7: 67. 1910. *(Nuttall,* western side of the Blue Mts. of Oregon)
   Glycosma ambiguum Gray, Proc. Am. Acad. 8:386. 1872. *O. ambigua* Coult. & Rose, Rev. N. Am. Umbell. 119. 1888. *Myrrhis ambigua* Greene, Fl. Fran. 332. 1892. *Washingtonia ambigua* Coult. & Rose, Contr. U.S. Nat. Herb. 7:69. 1900. *(Hall 217,* foot of the Cascade Mts., Oreg.; Silver Creek, Marion Co., fide Constance & Shan)
   Plants, especially the roots, with a strong, heavy odor somewhat like that of licorice; stems stout, mostly 4-12 dm. tall, clustered on the summit of a caudex and stout, quickly deliquescent root; stem often villosulous just above the nodes, as also the basal part of the petiolar margins, the herbage otherwise hirtellous or glabrous; leaves once to thrice ternate or ternate-pinnate, the leaflets lanceolate or lance-elliptic to ovate, toothed and sometimes incised to trifid, mostly 2-10 cm. long and 0.5-5 cm. wide; basal leaves clustered and long-petiolate, the cauline ones usually several, shorter-petiolate or subsessile but otherwise well developed; stem commonly branched above and producing several pedunculate umbels, these well developed at anthesis, with 5-12 rays mostly 1-4 cm. long that elongate to as much as 7 or reputedly 13 cm. in fruit; involucre and involucel usually wanting; flowers yellow; stylopodium depressed or low-conic; fruit glabrous, 12-20 mm. long, linear-oblong, not appendiculate at the base, narrowed above to a short beaklike tip. N=11.

Thickets and open slopes, from the lowlands to moderate elevations in the mountains; s.w. Alta. to s. Colo., w. to s.e. B.C., n.w. Wash., and c. Calif. Apr.-July.

Osmorhiza occidentalis is so different from most other species of *Osmorhiza* that it has often been segregated as a distinct genus *Glycosma,* but the recently discovered *O. bipatriata* Constance & Shan, of Tex. and n. Mex., tends to link it with the remainder of the genus. Plants of *O. occidentalis* from west of the Cascade summits generally have thinner and more coarsely toothed leaflets than do those from east of the mountains, but the difference is rather tenuous and has been ignored by the latest monographers of the genus. The type of *Glycosma ambiguum* Gray presumably represents this thin-leaved western phase.

Osmorhiza purpurea (Coult. & Rose) Suksd. Allg. Bot. Zeits. 12:5. 1906.
Washingtonia purpurea Coult. & Rose, Contr. U.S. Nat. Herb. 7:67. 1900. *(Colville & Kearney 796,*
Sitka, Alas.)
Washingtonia leibergii Coult. & Rose, Contr. U.S. Nat. Herb. 7:66. 1900. *O. leibergii* Blank.
Mont. Agr. Coll. Stud. Bot. 1:93. 1905. *(Sandberg & Leiberg 666,* Nason Creek, Kittitas Co.,
Wash.)
Very similar in most respects to *O. chilensis;* flowers purplish or pinkish, or sometimes greenish-white; stylopodium generally depressed and broader than high; fruit mostly 8-13 mm. long, the beak-like apex often less sharply marked than in *O. chilensis* and sometimes only 0.5 mm. long.

Meadows, stream banks, and moist or wet open slopes in the mountains, often at rather high elevations, less commonly in woodlands; s. Alas. (n. of the panhandle) to the mountains of n.w. Mont., n. Ida., and n. Wash., and in the Cascade region to s. Oreg. and adj. Calif. June-July.

Osmorhiza purpurea is not always easily distinguished from either *O. chilensis* or *O. depauperata* in the herbarium, but the weak morphological differences are bolstered by the restricted distribution and usually different habitat. The frequently white flowers of *O. purpurea,* and the reported existence of purple flowers in *O. chilensis,* suggest the possibility that these two species may sometimes hybridize.

Pastinaca L. Parsnip

Inflorescence of compound umbels on terminal and lateral peduncles; involucre and involucel usually wanting; flowers yellow or red; calyx teeth obsolete; styles short, spreading, the stylopodium depressed-conic; carpophore bifid to the base; fruit glabrous, elliptic to obovate, strongly flattened dorsally, the dorsal ribs filiform, the lateral ones narrowly winged; oil tubes solitary in the intervals and visible to the naked eye through the pericarp, 2-4 on the commissure; stout, leafy-stemmed biennial and perennial herbs usually with a well-developed taproot; leaves pinnately compound, with broad, toothed to pinnatifid leaflets.

About 14 species, native to Eurasia. (The ancient Latin name of the parsnip.)

Pastinaca sativa L. Sp. Pl. 262. 1753.
Anathum pastinaca Wibel, Prim. Pl. Werth. 195. 1799. *Peucedanum pastinaca* Baillon, Hist. Pl. 7:
188. 1879. (Southern Europe)
Stout, aromatic biennial from a stout taproot, 3-10 dm. tall; basal leaves up to 5 dm. long and nearly half as wide, the leaflets up to 13 cm. long and 10 cm. wide, serrate and the lower sometimes also pinnately cleft, or some leaflets with a nearly separate large basal lobe; cauline leaves progressively reduced; rays of the umbel mostly 15-25, unequal, 2-10 cm. long; flowers yellow; fruit broadly elliptic, 5-6 mm. long and 4-5 mm. wide. N=11. Common parsnip, wild parsnip.

Roadsides, ditch banks, and other disturbed sites; native of Europe, now widely naturalized in N. Am. and occasionally found throughout our range. May-July.

The common cultivated parsnip and the similar wild plants with smaller roots are considered to belong to the same species, and some of the wild plants may actually be recent escapes from cultivation.

Perideridia Reichenb. Yampah

Inflorescence of compound umbels, the peduncles more or less elongate, terminal and lateral; involucre of few to numerous small and narrow, more or less scarious bracts; involucel mostly of scarious or colored bractlets, or obsolete; flowers white or pink; calyx teeth well developed; styles short, with a more or less conic stylopodium; carpophore bifid to the base; fruit linear-oblong to orbicular, laterally somewhat compressed, glabrous, with prominent to inconspicuous ribs; oil tubes 1-5 in the intervals, 2-8 on the commissure; slender glabrous perennials from more or less tuberous-thickened, edible, often fascicled roots; leaves cauline and basal, pinnately or ternate-pinnately compound or dissected, with mostly narrow ultimate segments.

Nine species of the U.S. and adj. Can., chiefly in the w. states, especially Calif. (Name unexplained.)

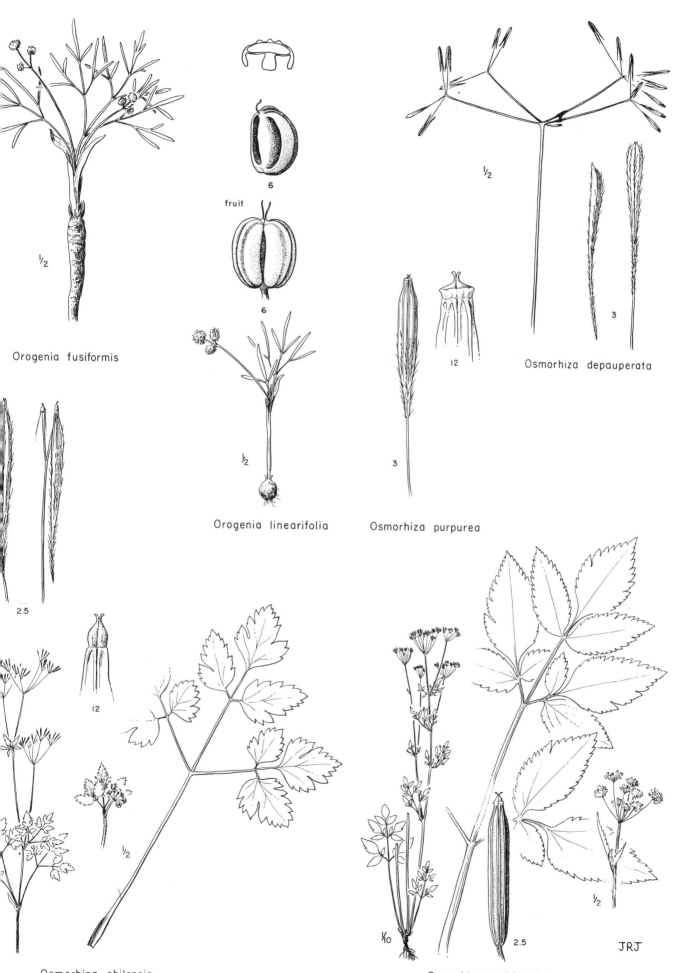

6

fruit

6

Orogenia fusiformis

12

Osmorhiza depauperata

1/2

3

Orogenia linearifolia

Osmorhiza purpurea

2.5

12

1/2

Osmorhiza chilensis

1/10

2.5

1/2

JRJ

Osmorhiza occidentalis

1 Principal leaves dissected, with numerous ultimate segments, these more or less dis-
    tinctly dimorphic, some much longer than others; petioles markedly dilated; oil tubes
    2-5 in the intervals   P. BOLANDERI
1 Principal leaves merely once or sometimes twice pinnate or ternate, the ultimate seg-
    ments few and not markedly dimorphic; petioles not much dilated; oil tubes solitary
    in the intervals
    2 Fruit orbicular or suborbicular, nearly or quite as wide as long; bractlets mostly
        setaceous, up to 0. 4 mm. wide, or obsolete; widespread   P. GAIRDNERI
    2 Fruit oblong-ovate, evidently longer than wide; bractlets broader and better de-
        deloped, mostly 0. 6-1 mm. wide; Cascades and westward   P. OREGANA

Perideridia bolanderi (Gray) Nels. & Macbr. Bot. Gaz. 61:33. 1916.
    Podosciadium bolanderi Gray, Proc. Am. Acad. 7:346. 1868. *Eulophus bolanderi* Coult. & Rose,
      Rev. N. Am. Umbell. 112. 1888. *Conopodium bolanderi* K.-Pol. Bull. Soc. Nat. Mosc. II, 29:205.
      1916. *(Bolander,* Mariposa trail, Yosemite, Calif.)
    Plants mostly 2-6 dm. tall, the solitary stem rather slender and delicately attached well below the
ground level to the fascicled root; lower leaves well developed and often rather crowded, the petiole
2-8 cm. long, conspicuously dilated, the blade 5-25 cm. long, ternate-pinnately dissected, with narrow and mostly rather numerous ultimate segments, these more or less distinctly dimorphic, some
much longer than others; middle and upper leaves mostly few and more or less strongly reduced, the
petiole remaining well developed, but the blade often reduced to a single elongate narrow segment;
umbels 1-several, rather long-pedunculate, commonly 1. 5-5 cm. wide at anthesis, slightly larger in
fruit, the rays up to 3. 5 cm. long at maturity; bractlets of the involucel well developed, scarious,
narrowly lanceolate to obovate, acuminate; fruit oblong, 3-5 mm. long; oil tubes 2-5 in the intervals,
6 on the commissure.
    Dry, open, often rocky hillsides, ridges, and dry washes in the foothills and high plains; w. Idaho
Co., Ida., and e. Wallowa Co., Oreg., s. through e. Oreg. and w. Ida. to n. e. Utah, Nev., and the
Sierra Nevada of Calif. ; an apparently outlying station in Jackson Hole, Wyo. May-July.

Perideridia gairdneri (H. & A.) Mathias, Brittonia 2:244. 1936.
    Atenia gairdneri H. & A. Bot. Beechey Voy. 349. 1838. *Edosmia gairdneri* T. & G. Fl. N. Am. 1:
      612. 1840. *Carum gairdneri* Gray, Proc. Am. Acad. 7:344. 1868. *(Douglas,* California)
    Carum montanum Blank. Mont. Agr. Coll. Stud. Bot. 1:91. 1905. *Atenia montana* Rydb. Bull.
      Torrey Club 40:67. 1913. *(Blankinship,* Bozeman, Mont., Aug. 11, 1898, is the first of three
      specimens cited)
    Plants mostly 4-12 dm. tall, the stem solitary, arising from a shallow or more deep-seated, soli-
tary or occasionally fascicled root; leaves several, well distributed along the stem, only gradually reduced upwards, the petiole not much expanded, the blade once or occasionally twice pinnate or ternate,
with more or less elongate, narrow ultimate segments; umbels 1-several, mostly 2. 5-7 cm. wide at
anthesis, slightly larger in fruit, the rays up to 6 cm. long at maturity; bractlets of the involucel
mostly setaceous, up to about 0. 4 mm. wide, or obsolete; fruit suborbicular, 2-3 mm. long and nearly
or quite as wide; oil tubes solitary in the intervals, 2 on the commissure.
    Woodlands and dry or wet meadows, from the lowlands to moderate elevations in the mountains; s.
B. C. to s. Calif., e. to Sask., S. D., and N. M. July-Aug.

Perideridia oregana (Wats.) Mathias, Brittonia 2:243. 1936.
    Edosmia oregana Nutt. ex T. & G. Fl. N. Am. 1:612. 1840, as a synonym. *Carum oreganum* Wats.
      Proc. Am. Acad. 20:368. 1885. *Ataenia oregana* Greene, Pitt. 1:274. 1889. *(Nuttall,* Wappatoo
      Island, at the mouth of the Willamette River, Oregon; this is now known as Sauvie's I.)
    Very similar to *P. gairdneri,* averaging smaller, mostly 3-6 dm. tall, and differing as indicated in
the key; roots more often fascicled; fruit 2. 5-4 mm. long.
    Moist or dry meadows and open slopes or flats, from the valleys to well up in the mountains; w.
Wash. to n. Calif., in and w. of the Cascade Mts. June-Aug.

Pimpinella L.

Inflorescence of compound umbels; involucre usually wanting; involucel of inconspicuous bractlets or
wanting; flowers mostly white or whitish, the outer petals often larger than the inner; calyx teeth mi-

fruit

4

5

Perideridia bolanderi

½

⅛

½

½

Pastinaca sativa

JRJ

nute or obsolete; stylopodium more or less conic; carpophore bifid to the middle or to the base; fruit small, oblong to orbicular, narrowed at the apex, rounded or cordate at the base, somewhat flattened laterally, glabrous or hairy, the ribs equal, filiform to very narrowly winged; oil tubes several in the intervals, or forming a circle around the seed; plants taprooted and mostly perennial, with branching stems; leaves variously simple or compound to dissected, often di- or trimorphic, the middle or upper more dissected than the lower.

Nearly 100 species of the Old World, centering in the Mediterranean region. (A late Latin name probably applied to a plant of this genus, said to be derived from *bipinnula,* referring to the leaves.)

Pimpinella saxifraga L. Sp. Pl. 263. 1753.

Tragoselinum saxifragum Moench. Meth. 99. 1794. *Carum saxifraga* Baillon, Hist. Pl. 7:178. 1879. *Selinum pimpinella* Krause in Sturm, Fl. Deuts. 2nd ed. 12:53. 1904. *Apium saxifragum* Calest. Webbia 1:178. 1905. (Europe)

P. nigra Mill. Gard. Dict. 1768. *P. saxifraga* ssp. *nigra* Gaudin, Fl. Helv. 2:440. 1828. *Selinum pimpinella nigra* Krause in Sturm, Fl. Deuts. 2nd ed. 12:55. 1904. (France?)

Plants 3-8 dm. tall, sparingly branched, shortly woolly-hirsute, especially on the stem, petioles, and lower surfaces of the leaves; leaves evidently di- or trimorphic, the lowermost ones pinnately compound, with mostly 7-13 leaflets 1-4 cm. long and up to 3 cm. wide, those next above generally more dissected, without well-defined leaflets, and the uppermost ones commonly reduced to petiolar sheaths with the blade vestigial or obsolete; involucre and involucel wanting or of a few small bracts or bractlets; rays of the umbel mostly 7-20, 2-4 cm. long at maturity; pedicels 3-8 mm. long at maturity; carpophore bifid to the base; fruit elliptic to orbicular, 2-2. 5 mm. long, glabrous, the ribs filiform and rather obscure; oil tubes mostly 3 in the intervals, 2-4 on the commissure.

Native of Eurasia, established as an occasional weed in the U.S. July-Sept.

In our area known only from the San Juan Is. of Wash. , where represented by the hairy form that is called ssp. *nigra* (Mill.) Gaudin. The nomenclaturally typical phase of the species is only inconspicuously or scarcely short-hairy.

Rhysopterus Coult. & Rose

Inflorescence of compact compound umbels on terminal peduncles; involucre wanting; involucel of well-developed, subcoriaceous or firm-scarious, partly connate bractlets, all on one side of the umbellet; flowers white; calyx teeth small but evident, broadly rounded, persistent and scarious; styles short, without a stylopodium; carpophore wanting; fruit ovoid to orbicular, somewhat flattened laterally, appearing to have strongly corrugated narrow wings when young, but at maturity the ribs merely raised and thickened, with or without an irregular vestige of a wing; oil tubes solitary in the intervals and in the apex of each rib, two on the commissure; low, taprooted perennials with a pseudoscape arising from the subterranean rootcrown, as in species of *Cymopterus;* leaves forming a rosette, not much dissected, the 3-5 primary segments broad and more or less confluent, merely toothed or lobed.

A single species, closely related to some more southern species of *Cymopterus,* but technically well marked by the absence of a carpophore and the virtual absence of wings from the mature fruit. (Name from the Greek *rhysos,* wrinkled, and *pteron,* wing, alluding to the strongly corrugated wings of the fruit of some of the original species which are now referred to *Cymopterus.)*

Rhysopterus plurijugus Coult. & Rose, Contr. U.S. Nat. Herb. 7:186. 1900.

Cymopterus plurijugus M. E. Jones, Contr. West. Bot. 12:25. 1908. *(Leiberg 2240,* Malheur Valley near Harper's Ranch, Oreg.)

Glabrous perennial with a slender pseudoscape arising to the ground level from the subterranean crown of a thickened taproot; leaves petiolate, spreading, firm, the blade mostly 1. 5-4 cm. long and nearly as wide; peduncles several, short, mostly 1-2 cm. long; fruiting umbel compact but scarcely headlike, the 6-17 rays mostly about 1 cm. long; fruit 3-4 mm. long.

Loose dry ground in Malheur, Harney, and e. Lake cos. , Oreg. , often associated with diatomite; barely or scarcely entering our range. May.

Sanicula L. Sanicle

Inflorescence of several or many compact, headlike ultimate umbels which may be arranged into a definite primary umbel or may be dichasially or rather irregularly arranged; flowers white or greenish white to yellow, purple, or blue, some staminate, others perfect, each ultimate umbel with both types, or some wholly staminate; perfect flowers sessile or nearly so, staminate flowers generally more or less pedicellate; involucre more or less foliaceous, often appearing as opposite, sessile leaves in forms with dichasially arranged umbellets; involucel of several or many prominent or inconspicuous bractlets; calyx teeth well developed, generally connate at least at the base; stylopodium flattened and disklike, or wanting; fruit oblong-ovoid to globose, somewhat compressed laterally, beset with prickles or tubercles, our spp. with uncinate prickles; ribs of the fruit usually obsolete; mericarps varying from terete to laterally or dorsally somewhat compressed; glabrous or subglabrous biennials or more often perennials with variously cleft or dissected leaves. X=8.

About 40 species, nearly cosmopolitan in distribution, characteristically found in temperate regions and mountainous parts of the tropics, absent from Australia and New Zealand. (Name from the Latin *sanare*, to heal.)

References:

Bell, C. R. The Sanicula crassicaulis complex (Umbelliferae). U. Calif. Pub. Bot. 27:133-230. 1954.

Shan, R., and L. Constance. The genus Sanicula (Umbelliferae) in the Old World and the New. U. Calif. Pub. Bot. 25:1-78. 1951.

1 Plants with a cluster of fibrous roots from a short simple caudex or crown, relatively robust, seldom less than 4 dm. tall, with palmately cleft to palmately compound leaves and toothed leaflets or segments; flowers greenish white; wholly e. of the Cascades
S. MARILANDICA

1. Plants more or less distinctly taprooted, often smaller than *S. marilandica*, and, except for *S. graveolens*, not extending much e. of the Cascades; flowers light yellowish to yellow or purple

  2 Plants prostrate or ascending, maritime; involucel conspicuous, generally surpassing the heads
S. ARCTOPOIDES

  2 Plants erect, not maritime, or only casually so; involucel inconspicuous

    3 Principal leaves once or twice pinnatifid, with a distinctly toothed rachis; flowers purple
S. BIPINNATIFIDA

    3 Principal leaves palmately or pinnipalmately lobed or divided to ternate-pinnate, without a toothed rachis; flowers yellow or yellowish except sometimes in *S. crassicaulis*

      4 Leaves palmately or pinnipalmately lobed or divided, without a narrow rachis, the primary divisions merely serrate or lobed; w. of the Cascades, extending to the e. end of the Columbia R. Gorge
S. CRASSICAULIS

      4 Leaves more or less ternate-pinnate, the primary divisions tending to be pinnatifid, the lowest pair of primary divisions separated from the terminal segment or segments by a narrow, entire rachis; widespread
S. GRAVEOLENS

Sanicula arctopoides H. & A. Bot. Beechey Voy. 141. 1832. *(Menzies,* Northwest coast of America) *S. howellii* Coult. & Rose, Bot. Gaz. 13:81. 1888. *S. crassicaulis* var. *howellii* Mathias, Brittonia 2:242. 1936. *(Howell 16,* Tillamook Bay and Ocean Beach, Oreg.)

Taprooted perennial; stems much branched at the base into prostrate or ascending branches 5-30 cm. long; leaves somewhat succulent, often yellowish, the basal ones rosette-forming, broadly petioled, the blade 2.5-6 cm. long and 2.5-9 cm. wide, 3-cleft, the primary segments irregularly laciniate-toothed or cleft and the teeth softly bristle-tipped, the terminal segment often larger and more dissected than the two lateral ones, which point forward at an angle; cauline leaves few and often more or less reduced, becoming sessile, commonly opposite and each subtending a branch; ultimate umbels terminating the branches, yellow, about 1 cm. wide at anthesis, each subtended by a prominent involucel of 8-17 oblanceolate, entire or trilobed, basally slightly connate bractlets mostly 5-15 mm. long; ultimate umbels mostly 20- to 25-flowered, with about equal numbers of perfect and staminate flowers, both types of flowers with soft, ovate, acute calyx lobes 1-2 mm. long and connate below the middle; fruits ovoid to subglobose, 2-5 mm. long and 2-3 mm. broad, covered upwards with stout, uncinate prickles, nearly smooth or merely tuberculate below; mericarps terete in cross-section. N=8. Footsteps of spring.

Coastal bluffs, from Santa Barbara Co., Calif., to Wash. and the s. tip of Vancouver I., not entering the Puget Sound area; seldom collected in our region. Mar.-May.

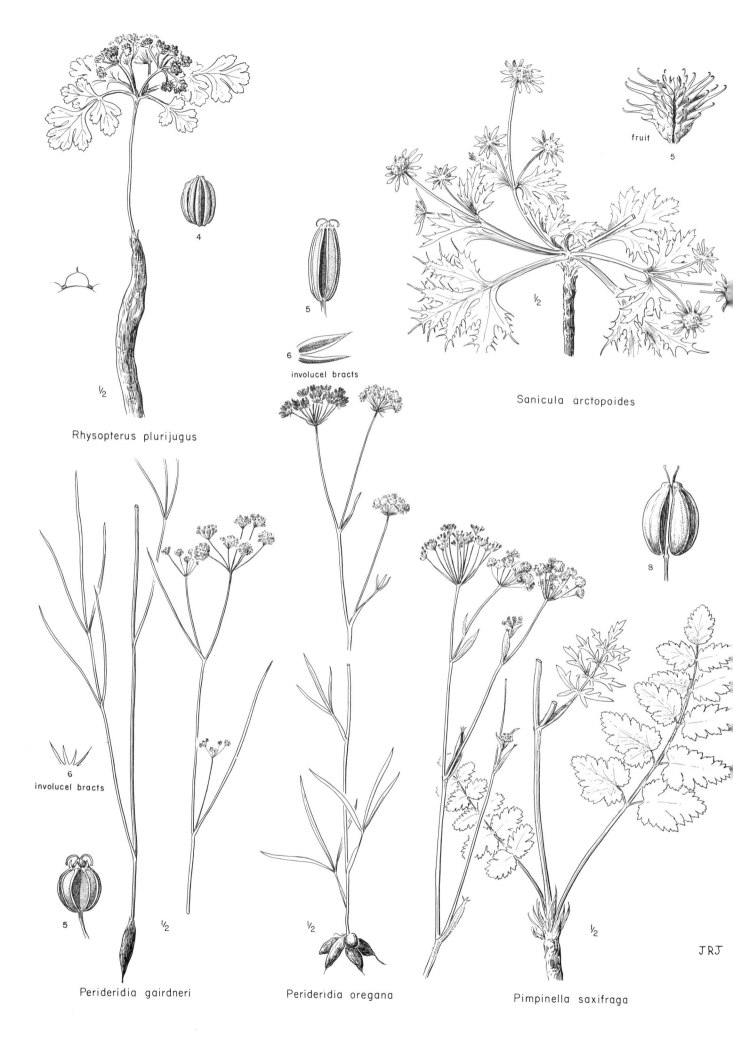

4

1/2

Rhysopterus plurijugus

fruit

5

1/2

Sanicula arctopoides

5

6

involucel bracts

8

6

involucel bracts

5

1/2

Perideridia gairdneri

1/2

Perideridia oregana

1/2

Pimpinella saxifraga

JRJ

Sanicula bipinnatifida Dougl. ex Hook. Fl. Bor. Am. 1:258. 1832. *(Douglas,* Fort Vancouver, on the Columbia)

Taprooted perennial 1-6 dm. tall, with solitary or few erect branching stems from the base; leaves alternate or partly opposite, the well-developed lower ones often clustered on a short pseudoscape at the base, the blade commonly 4-13 cm. long and 3-12 cm. wide, on a petiole of about equal length, variously pinnatifid or bipinnatifid, with a winged and evidently toothed rachis, the segments also toothed or cleft; cauline leaves few and more or less reduced, becoming sessile; primary umbels few-rayed and often irregular, subtended by leafy bracts; ultimate umbels 1 cm. wide or less at anthesis, dark purple (yellow only in some plants from south of our range), subtended by an inconspicuous involucel of 6-8 lanceolate bractlets mostly less than 4 mm. long; ultimate umbels about 20-flowered, with about equal numbers of staminate and perfect flowers, both types of flowers with soft, lance-ovate, acute calyx lobes about 1 mm. long or less and shortly connate at the base; fruits ovoid to subglobose, 3-6 mm. long and 2-4 mm. broad, covered with stout uncinate prickles; mericarps subterete in cross-section. N=8.

Open or sparsely wooded slopes and drier meadows; s. Vancouver I. and w. Wash. and Oreg. to Calif. and n. Baja Calif., wholly w. of the Cascade Mts. in our range. May-June.

Sanicula crassicaulis Poepp. ex DC. Prodr. 4:84. 1830.

Aulosolena crassicaulis K.-Pol. Bull. Soc. Nat. Mosc. II, 29:156. 1916. *(Poeppig 92,* Chile)

S. menziesii H. & A. Bot. Beechey Voy. 142. 1832. *S. crassicaulis* var. *menziesii* H. Wolff, Pflanzenr. IV, 228 (Heft 61):70. 1913. *Aulosolena menziesii* K.-Pol. Bull. Soc. Nat. Mosc. II, 29:156. 1916. *(Menzies,* northwest coast of America) = var. *crassicaulis.*

S. tripartita Suksd. Allg. Bot. Zeits. 12:5. 1906. *S. crassicaulis* var. *tripartita* H. Wolff, Pflanzenr. IV, 228 (Heft 61):70. 1913. *(Suksdorf 2650,* Bingen, Klickitat Co., Wash.)

S. diversiloba Suksd. Werdenda 1:29. 1927. *(Suksdorf 10015,* Bingen, Klickitat Co., Wash.) = var. *crassicaulis.*

More or less distinctly taprooted perennial, 2.5-12 dm. tall, with a solitary erect stem; leaves alternate, the basal and lower cauline ones long-petiolate, with palmately or pinnipalmately 3- to 5-lobed or -cleft blade mostly 3-13 cm. long and 2-18 cm. broad, the primary segments rather broad, more or less toothed and sometimes with a few shallow lobes; middle and upper cauline leaves few and reduced, becoming sessile; primary umbels few-rayed and often irregular, subtended by leafy bracts; ultimate umbels mostly less than 1 cm. wide at anthesis, yellow or sometimes (var. *tripartita)* more or less purplish, subtended by an inconspicuous involucel of about 5 narrow bractlets only 1-2 mm. long; ultimate umbels mostly 8- to 13-flowered, commonly with more perfect than staminate flowers, both types of flowers with soft, lanceolate or lance-ovate, basally connate calyx lobes less than 1 mm. long; fruits subglobose, 2-5 mm. long and 2-4 mm. broad, covered (except sometimes at the base) with stout, uncinate prickles; mericarps subterete in cross-section. N=16, 24, 32.

Moist or dry woods w. of the Cascades, and extending through the Columbia R. Gorge to Klickitat Co., Wash.; s. B.C. to n. Baja Calif., and in Chile. May-June.

Most of our material of *S. crassicaulis* is tetraploid or hexaploid on a base of X = 8, and has the blade of the basal leaves palmately lobed or cleft and about as wide as long, or a little wider than long. This represents the var. *crassicaulis.* To the south of our range octoploids of var. *crassicaulis* also occur. An octoploid occurring from the islands of Puget Sound to the Willamette Valley of Oreg., and e. through the Columbia R. Gorge to Klickitat Co., Wash., has the blade of the basal leaves more pinnipalmately cleft and often a little longer than wide, the central lobe being more or less elongate; the flowers often show a touch of purple. This is the var. *tripartita* (Suksd.) H. Wolff. It is believed that the var. *tripartita* has 2 genomes of *S. bipinnatifida* added to 6 genomes of *S. crassicaulis.* *S. crassicaulis* itself is believed to be an allopolyploid derived from two diploid Californian species. It will be noted that the range of *S. crassicaulis* var. *tripartita* lies entirely within the range of *S. crassicaulis* var. *crassicaulis,* and largely within the range of *S. bipinnatifida* as well.

Sanicula graveolens Poepp. ex DC. Prodr. 4:85. 1830. *(Poeppig 93,* Chile)

S. nevadensis Wats. Proc. Am. Acad. 11:139. 1876. *(Ames s.n.,* Indian Valley, Plumas Co., Calif., in 1872)

S. septentrionalis Greene, Erythea 1:6. 1893. *S. nevadensis* var. *septentrionalis* Mathias, Brittonia 2:241. 1936. *S. graveolens* var. *septentrionalis* St. John, Fl. S.E. Wash. 2nd ed. 549. 1956. *(Macoun n.s.,* Cedar Hill, Vancouver I., in 1887)

S. apiifolia Greene, Leafl. 2:46. 1910. *(Williams 982,* Columbia Falls, Flathead Co., Mont.)

Taprooted perennial 0.5-5 dm. tall, the single stem often branched near the base and tending to be irregularly branched upwards; leaves alternate, the lowermost cauline ones well developed, petiolate, often attached below the ground level and thus seeming to arise separately from the stem, the blade 1.5-4 cm. long and 2-3.5 cm. wide, more or less ternate-pinnate, the primary segments tending to be pinnatifid, and the lowest pair of primary segments separated from the upper by a narrow, entire rachis; middle and upper cauline leaves few and more or less reduced, often becoming sessile; ultimate umbels less than 1 cm. wide and often rather closely aggregated at anthesis, light yellowish, subtended by an inconspicuous involucel of 6-10 oblong-ovate to lance-linear, pointed, basally connate bractlets about 1 mm. long; ultimate umbels mostly 10- to 15-flowered, the staminate flowers more numerous than the pistillate, both types of flowers with ovate, acute calyx lobes 1 mm. long or less and connate below the middle; fruits ovoid-globose, 3-5 mm. long, 2-4 mm. broad, covered at least upwards with stout uncinate prickles. N=8.

Open or lightly wooded slopes and flats from the lowlands to moderate elevations in the mountains; s. B.C. to s. Calif., w. to w. Mont. and n.w. Wyo., though rarely collected in Ida.; also in Chile and Argentina. May-July.

Sanicula marilandica L. Sp. Pl. 235. 1753.

Caucalis marilandica Crantz, Class. Umbell. 110. 1767. *S. canadensis* var. *marylandica* Hitchc. Trans. Acad. St. Louis 5:497. 1891. *(Clayton 28* in part, Virginia)

Perennial with a cluster of fibrous roots from a short simple caudex or crown; stem solitary, erect, mostly 4-12 dm. tall, generally branched only above; basal and lowermost cauline leaves long-petiolate, the blade 6-15 cm. wide, palmately 5- to 7-parted or palmately compound, the segments or leaflets sharply toothed and sometimes shallowly lobed, or some of them (especially the two lateral ones in a 5-parted leaf) often more or less deeply bifid; cauline leaves usually several, gradually reduced upwards and becoming sessile; ultimate umbels about 1 cm. wide or less at anthesis, greenish-white, subtended by a few minute narrow bractlets, mostly 15- to 25-flowered, the staminate flowers more numerous than the perfect ones, or some umbellets all staminate; calyx lobes firm, lance-triangular, attenuate, slightly connate at the base; styles elongate, often persistent and longer than the prickles of the fruit; fruits ovoid, 4-6 mm. long, 3-5 mm. wide, covered with numerous uncinate, basally thickened prickles, the lower ones rudimentary; mericarps subterete in cross-section. N=8.

Moist low ground, less often on moist wooded slopes; Newf. to Fla., w. to e. B.C., n. Ida. and probably extreme n.e. Wash., s.w. Mont., Wyo., Colo., and n. N.M. June.

Scandix L.

Inflorescence of compound or simple umbels, the ultimate umbels aggregated into a usually few-rayed primary umbel or more or less scattered, often paired; involucre wanting, or a leafy bract; involucel of several lobed or dissected bractlets; flowers white; calyx teeth minute or obsolete; styles very short, the stylopodium depressed; carpophore entire, or bifid at the apex; fruit more or less linear, hispid or scabrous, the more or less quadrangular to subcylindric, evidently ribbed body tipped by a long, subcylindric or dorsally compressed, nearly ribless beak; oil tubes solitary in the intervals, very small or obsolete; taprooted annuals with branching, leafy stems, the leaves petiolate, pinnately dissected, with more or less linear, small and rather short ultimate segments.

About 10 species of the Old World. (The ancient Greek name of *Chaerophyllum bulbosum,* another umbelliferous plant, which is now cultivated under the common name chervil.)

Scandix pecten-veneris L. Sp. Pl. 256. 1753. (Germany and southern Europe)

Plants 1-3 dm. tall, generally branched from the base, more or less hirsute (especially the stems) to subglabrous; leaves several, well distributed along the stems, the petiole equaling or shorter than the blade, the blade mostly 1.5-9 (15) cm. long and 1-6 cm. wide, with ultimate segments 1-3 mm. long and 0.5-1 mm. wide; umbellets compact, less than 1 cm. wide at anthesis and not much expanded at maturity, often paired, less often borne singly or in threes, the peduncles or rays mostly 1-4 cm. long at maturity, or longer when solitary; bractlets of the involucel several, evident, generally bifid above; fruiting pedicels stout, erect or nearly so, 2-5 mm. long; fruit with a more or less quadrangular body 6-15 mm. long and 1-2 mm. wide, tipped by a stout, dorsally somewhat flattened beak 2-7 cm. long. N=8. Venus' comb, lady's comb, shepherd's needle.

fruit

5

Sanicula marilandica

½

5

var. tripartita

½

var. crassicaulis

Sanicula crassicaulis

Sanicula graveolens

½

½

Sanicula bipinnatifida

4

½

½

2

Scandix pecten-veneris

½

JRJ

A weed in fields and waste places, native of Eurasia, and sparingly introduced here and there in the U.S.; reported from Vancouver I. to s. Calif. Apr.-July.

The name is sometimes incorrectly written *Scandix pecten.*

Sium L.

Inflorescence of compound umbels; involucre of subfoliaceous, entire or incised, often reflexed bracts; involucel of well-developed narrow bractlets; flowers white; calyx teeth minute or obsolete, often unequal; styles short, reflexed, the stylopodium depressed or rarely conic; carpophore bifid to the base, the halves in our species inconspicuous and adnate to the mericarps; fruits glabrous, elliptic to orbicular, slightly compressed laterally and somewhat constricted at the commissure, the subequal ribs prominent and corky but scarcely winged; oil tubes 1-3 in the intervals, 2-6 on the commissural face; glabrous, leafy-stemmed perennials from fascicles of fibrous roots, the leaves mostly pinnately compound or decompound, with evident, toothed to pinnatifid leaflets.

About 8 species, native to the north temperate regions and to South Africa, three in the U.S. (Name from *Sion,* the ancient Greek name of some aquatic umbellifer, possibly a species of the present genus *Sium.)*

Sium suave Walt. Fl. Carol. 115. 1788. *(Walter,* Carolina)

  S. cicutaefolium Schrank, Baier Fl. 1:558. 1789. *Apium cicutaefolium* Benth. & Hook. ex Forbes & Hemsl. Journ. Linn. Soc. 23:328. 1887. (Ob River, in western Siberia)

Perennial from a very short erect crown, the fibrous roots sometimes originating at one or two lower nodes of the stem as well as from the crown; stem solitary, stout, ribbed-striate, mostly 5-12 dm. tall, generally branched above; leaves basal and cauline or all cauline, the lower long-petiolate, the upper with progressively shorter, winged petioles, all pinnately once compound with 7-13 sessile, merely serrulate to deeply pinnatisect and again cleft leaflets, these 2-9 cm. long, 1.5-10 (20) mm. wide, typically lance-linear; primary lateral veins of the leaflets mostly branched and inconspicuous, not bearing any obvious relation to the marginal teeth or sinuses; involucre of 6-10 lanceolate or linear bracts 3-15 mm. long, entire or incised, unequal, reflexed; involucel of 4-8 inconspicuous narrow bractlets 1-3 mm. long; rays of the umbels mostly 10-30, 1.5-3.5 cm. long at maturity; fruit broadly elliptic to orbicular, 2-3 mm. long.

Swampy places and in shallow water, in the valleys and foothills; s. B.C. to c. Calif., e. to Newf. and Va., found throughout our range. July-Aug.

A similar or identical form occurs in eastern Asia *(S. cicutaefolium).*

This species has several times been reported to be poisonous to livestock, but the reports may be due to confusion with *Cicuta.* Of the two species of *Cicuta* in our range, *C. bulbifera* is easily distinguished by its axillary bulbils, while *C. douglasii* differs in its thickened, septate stem base, thickened roots, and the venation pattern of the leaflets.

Sphenosciadium Gray   Swamp White-heads

Inflorescence of loose compound umbels, the umbellets compact and capitate with sessile flowers; involucre wanting; involucel of numerous slender tomentose bractlets; flowers white or occasionally purplish; calyx teeth obsolete; stylopodium small, broadly conic, the styles slender and spreading; carpophore bifid to the base; fruit tomentose, cuneate, strongly flattened dorsally; ribs of the fruit prominent, the dorsal ones narrowly winged, the lateral ones more prominently so; oil tubes small, solitary in the intervals, 2 on the commissure; perennial from a simple or compactly branched caudex which sometimes surmounts a stout taproot; leaves petiolate, pinnately or ternate-pinnately once or twice compound, with irregularly toothed or cleft leaflets.

A monotypic genus. (Name from the Greek *sphen,* wedge, and *skias,* umbrella, presumably referring to the fruits and inflorescence.)

Sphenosciadium capitellatum Gray, Proc. Am. Acad. 6:537. 1865.

  Selinum capitellatum Wats. Bot. King Exp. 126. 1871. *(Brewer,* Sierra Nevada near Ebbett's Pass, Alpine Co., Calif.)

Plants rather stout, 5-18 dm. tall, the foliage scabrous to subglabrous, the inflorescence tomentose; leaflets more or less lanceolate, mostly 2.5-8 cm. long and 0.5-2 cm. wide; umbels 1-several, the 4-18 rays mostly 1-5 cm. long; umbellets well separated, subglobose, 6-12 mm. wide; fruit 5-8 mm. long, 3-5 mm. wide at the top, the wings progressively wider toward the top of the fruit.

Wet meadows, swamps, stream banks, and other moist low places, from the foothills to moderate elevations in the mountains; c. Ida. to the Wallowa and Strawberry mts. of e. Oreg., s. irregularly to the Sierra Nevada of Calif. and the mts. of w. Nev. July-Aug.

A distinctive species, readily recognized by its well-separated, subglobose umbellets.

Tauschia Schlecht. Nom. Conserv.

Inflorescence a compound umbel, usually without an involucre but with an involucel; calyx teeth prominent to obsolete; petals yellow, white, or purplish, oblanceolate to obovate with a narrower in-flexed apex; stylopodium nearly or quite wanting; carpophore 2-cleft to the middle or below; fruit linear-oblong to subglobose, slightly flattened laterally, glabrous, the ribs more or less evident but not winged; low, acaulescent or shortly caulescent perennials from a taproot or thickened tuberous root, our spp. glabrous; leaves petiolate, pinnately to ternately cleft or dissected, rarely entire.

About two dozen species of w. N. Am. to C. Am. (Named for Ignaz Friedrich Tausch, 1793-1848, European botanist.)

Both of our species are somewhat anomalous in the genus, and *T. hooveri* is suggestive of the genus *Musineon* in some respects.

1 Leaf segments linear, mostly 1-2 mm. wide; fruit linear-oblong, 5-7 mm. long; flowers
   white; desert scablands in Yakima Co., Wash.   T. HOOVERI
1 Leaf segments lanceolate to ovate or elliptic, mostly 4-10 mm. wide; fruits subglobose or
   broadly ellipsoid, 2-3 mm. long; flowers yellow; Mt. Rainier, Wash.   T. STRICKLANDII

Tauschia hooveri Math. & Const. Madroño 7:65. 1943. *(Hoover 5689, near Cowiche, Yakima Co., Wash.)*

Perennial from a deep-seated globose tuberous root about 1.5 cm. thick, glabrous and glaucous throughout, the foliage leaves and the slender peduncles arising at the ground level from a pseudo-scape; leaves petiolate, the blade ternate or pinnate to partly bipinnate, with linear and entire ulti-mate segments mostly 1.5-3.5 cm. long and 1-2 mm. wide; peduncles 2-4 cm. long; inflorescence few-rayed, compact, about 1 cm. wide at anthesis; involucre and involucel wanting; calyx teeth obso-lete; flowers white with purple anthers; carpophore cleft about halfway or nearly to the base; fruit linear-oblong, 5-7 mm. long and about 2 mm. wide, glaucous, with narrow but evident ribs.

Sagebrush scablands in Yakima Co., Wash. Feb.-Apr.

This species is habitally suggestive of some of the cormose rooted, white-flowered Lomatiums of similar sites.

Tauschia stricklandii (Coult. & Rose) Math. & Const. Bull. Torrey Club 68:121. 1941.

Hesperogenia stricklandii Coult. & Rose, Contr. U.S. Nat. Herb. 5:203. 1899. *(Strickland s.n., Mt. Rainier, Wash., in 1896)*

Perennial from a thickened taproot and more or less branched caudex, glabrous throughout, essen-tially acaulescent, the scape 5-20 cm. tall, naked or with 1 or 2 reduced leaves; basal leaves clus-tered, rather long-petiolate, trifoliolate or with 5-7 pinnately or subpalmately disposed leaflets, these rather firm, lanceolate to ovate or elliptic, mostly 1-3 cm. long and 4-10 mm. wide, the 3 terminal ones often somewhat confluent; inflorescence only about 1 cm. wide at anthesis, more open in fruit, the several rays elongating unequally; involucre none; involucel of several slender bractlets 1-2 mm. long; calyx teeth minute; flowers yellow; carpophore cleft nearly to the base; fruit subglobose to broadly ellipsoid, 2-3 mm. long and nearly or quite as wide, with narrow, obscure nerves, the sur-face longitudinally ribbed and sulcate, the oil tubes visible beneath the sulci.

Meadows and moist slopes at 5000 to 6500 feet altitude on Mt. Rainier, Wash. Aug.

Zizia Koch

Inflorescence of compound umbels; involucre wanting; involucel of a few inconspicuous bractlets; pedicels short and stout, the central flower of each umbellet sessile or nearly so; flowers bright yel-low; calyx teeth well developed; styles slender, erect or spreading, without a stylopodium; carpophore bifid about half its length; fruit oblong or broadly elliptic, glabrous, somewhat compressed laterally, the ribs prominent but not much raised; oil tubes solitary in the intervals, two on the commissure; glabrous or subglabrous perennials from a short caudex and a cluster of fleshy-fibrous roots; leaves cauline and basal, petiolate, simple or ternately compound, the blade of the leaf or leaflets toothed. Four species of N. Am. (Named for Johann Baptist Ziz, 1779-1829, German botanist.)

Zizia aptera (Gray) Fern. Rhodora 41:441. 1939.
Thaspium trifoliatum var. *apterum* Gray, Man. 2nd ed. 156. 1856. (New York and New Jersey)
Z. cordata (Walt.) Koch, misapplied. See Fern. Rhodora 41:441-444. 1939.
Z. aptera var. *occidentalis* Fern. Rhodora 41:444. 1939. *(Cusick 2401*, Wallowa River, Oreg.)

Stems generally clustered on a compact caudex, 2-6 dm. tall; basal leaves rather long-petiolate, with broad, cordate, crenate or serrate blade mostly 2. 5-10 cm. long and 1. 5-8 cm. wide, or occasionally some (rarely all?) of them trifoliolate with more elliptic, scarcely cordate leaflets; cauline leaves few, shorter-petiolate, mostly trifoliolate, the upper (in our variety) tending to be more coarsely and irregularly and often more sharply toothed than the basal and lower cauline ones; inflorescence compact and densely flowered, mostly 1. 5-4 cm. wide at anthesis, a little larger in fruit, the spreading-ascending rays up to 3. 5 cm. long at maturity; fruit 2-4 mm. long. N=11.

Moist or wet meadows, stream banks, and moist low ground, tolerant of alkali; throughout most of e. and c. U.S. and adj. Can., and in the cordilleran region from s. Alta. to Colo., w. to e. Wash., n. w. Oreg., and n. e. Nev. May-July.

The cordilleran plants, constituting the var. *occidentalis* Fern., may be distinguished with some difficulty from the more eastern var. *aptera* by their more coarsely and irregularly toothed upper leaves. The var. *aptera* also tends to have thicker leaves than var. *occidentalis,* and is not so strictly confined to moist or wet habitats.

CORNACEAE Dogwood Family

Flowers small, regular, polypetalous, paniculate, capitate, or cymose, often conspicuously involucrate, perfect or imperfect and the plants monoecious to dioecious; sepals 4 (ours) or 5, usually very small; petals 4 (5 or 0); stamens usually as many as the petals and alternate with them, but sometimes twice as many, the anthers introrse; pistil 1, the carpels 2-4, style usually 1; ovary inferior, 1- to 4-locular, each locule with 1 pendulous ovule, the placentation mostly axile; fruit a drupe or berry; trees or shrubs (sometimes only suffrutescent) with opposite or (less commonly) alternate (rarely whorled), simple, petiolate, exstipulate leaves.

About 10 genera and 80 species, of widespread temperate and tropical distribution.

Cornus and *Aucuba* are genera of the family that include several popular ornamentals.

Cornus L. Dogwood

Flowers in open to congested cymes or heads, ebracteate or subtended by 4-8 conspicuous white or pinkish bracts; sepals 4, very small; petals 4, whitish to greenish, often purplish-tipped, valvate; stamens 4; pistil 2-carpellary, style 1, stigma more or less capitate; ovary 2-locular; fruit a fleshy, 2-seeded drupe; suffrutescent to decidedly woody, more or less malpighiaceous-strigillose shrubs or trees with opposite (or in *C. canadensis* terminally whorled), petiolate leaves with prominent pinnate venation.

Perhaps 30 species, chiefly of the North temperate region, including several species of much ornamental value. (Latin, meaning horn or antler, hence often supposed to refer to the hard wood, but *cornus* is also the name for the ornamental knobs or tips affixed to the ends of the cylinder on which ancient manuscripts were rolled, which might have suggested the capitate inflorescence of some species of the dogwood.)

1 Low trailing subshrub less than 3 dm. tall, the leaves whorled at the tip of the stem
C. CANADENSIS
1 Woody trees or shrubs with opposite leaves throughout
2 Plant arborescent; flowers capitate, conspicuously bracteate   C. NUTTALLII
2 Plant shrubby; flowers borne in open cymes, ebracteate   C. STOLONIFERA

Cornus canadensis L. Sp. Pl. 118. 1753.
Chamaepericlymenum canadense Aschers. & Graebn. Fl. Nordd. Flachl. 539. 1898. *Cornella canadensis* Rydb. Bull. Torrey Club 33:147. 1906. *Cynoxylon canadense* Schaffner, Cat. Ohio Vasc. Pl. 222. 1914. (Canada)
Cornus unalaschkensis Ledeb. Fl. Ross. 2:378. 1844-46. *Cornella unalaschkensis* Rydb. Bull.

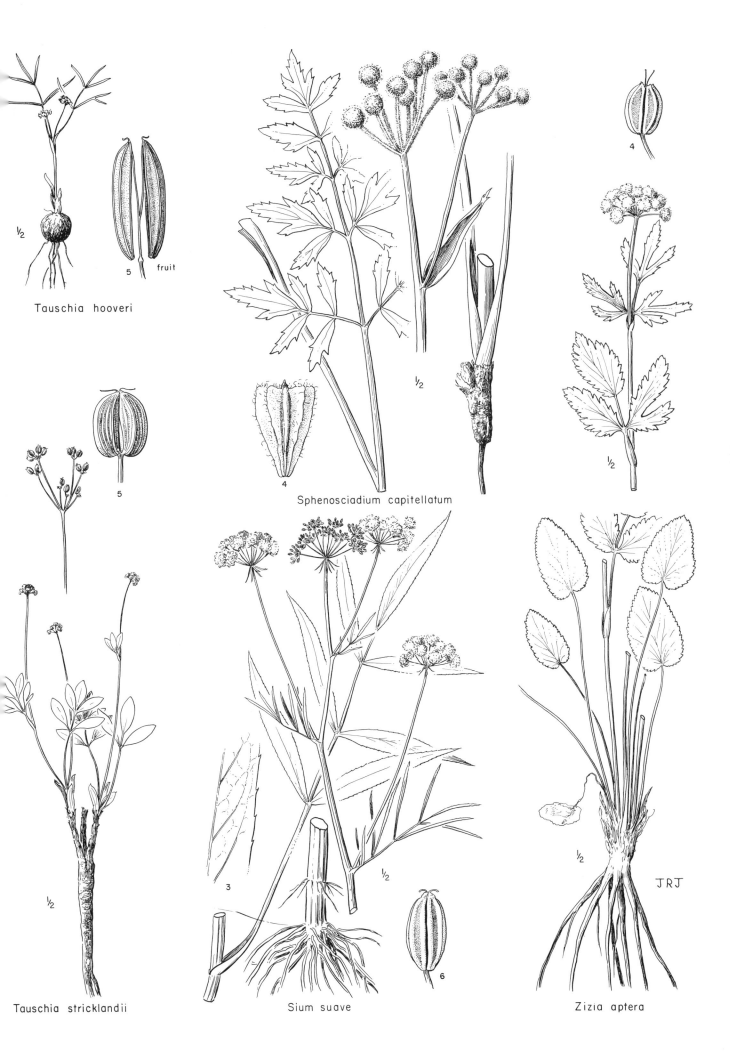

Tauschia hooveri

½

5   fruit

4

Tauschia stricklandii

5

Sphenosciadium capitellatum

½

4

½

½

3

½

6

Sium suave

Zizia aptera

½

JRJ

Torrey Club 33:147. 1906. *Chamaepericlymenum unalaschkense* Rydb. Fl. Rocky Mts. 635, 1065. 1917. *(Eschscholtz, Chamisso* [separate collections], Unalaschka)

Chamaepericlymenum canadense f. *purpurascens* Miyabe & Tatewaki, Trans. Sapporo Nat. Hist. Soc. 15:43. 1937. *Cornus canadensis* f. *purpurascens* Hara, Rhodora 44:20. 1942. *(Shirusaka,* Tonnai, Distr. Toyohara, Japan)

Cornus canadensis f. *rosea* Fern. Rhodora 43:156. 1941. *(Fernald et al. 25935,* Mt. Mattaouisse, Matane Co., Que., July 14, 1923)

Low, widely rhizomatous subshrubs 5-20 cm. tall, the erect, largely herbaceous, greenish to reddish, strigillose stems leafless or bracteate below, the bracts often becoming enlarged or foliar upward; true leaves 4-7 in a terminal whorl, elliptic, ovate- or rhombic-elliptic, nearly sessile, 2-8 cm. long, sparsely malpighiaceous-strigillose to subglabrous and green on the upper surface, paler and more or less glaucous beneath; flowers in a solitary, pedunculate, greatly condensed and semi-capitate cyme, subtended by 4 white, pinkish, or purplish-tinged, broadly to narrowly ovate bracts 1-2 (2.5) cm. long; petals 1-1.5 mm. long, greenish-white, often with a deep purplish tinge; drupes bright coral red, 6-8 mm. long; pericarp smooth. Bunchberry, dwarf cornel, puddingberry.

In moist woods from Alas. to Greenl., s. to Pa. and Minn., and in the mts. to N. M. and Calif.; e. Asia. June-Aug.

It would be difficult to overestimate the ornamental value of this charming little plant, which can be introduced readily by taking up a small amount of sod; it spreads readily, but not too aggressively. It is almost equally attractive in flower and in fruit. Occasional plants (sporadic in occurrence) with purplish or reddish bracts have been designated as forma *purpurascens* (Miyabe & Tatewaki) Hara or forma *rosea* Fern. These names, superfluous from our point of view, will probably be maintained by the plant grower.

Cornus suecica L., a closely related species, that differs among other ways in having several pairs of opposite foliage leaves below the terminal whorl, has been reported for Vancouver I., but is not believed to extend southward as far as our northern limit.

Cornus nuttallii Aud. ex T. & G. Fl. N. Am. 1:652. 1840.

Cynoxylon nuttallii Shafer in Britt. & Shafer, N.Am. Trees 746, fig. 684. 1908. ("Oregon, *Dr. Scouler, Mr. Tolmie, Nuttall")*

Shrublike to rather stately trees 2-20 m. tall; bark smooth, brownish, the younger branches gray-ish-purplish, usually abundantly strigillose, eventually glabrate; leaves with petioles 5-10 mm. long, the blades ovate-elliptic to rhombic-ovate or elliptic-obovate, usually shortly acuminate, 4-10 cm. long, malpighiaceous-strigillose and sometimes sparsely pilose on the upper surface, strigillose to pilose-lanate beneath; flowers numerous in hemispheric heads 1.5-2 cm. broad, subtended by 4-7 very conspicuous white or pinkish-tinged, elliptic to rhombic-obovate bracts 2-7 cm. long, the heads developing in the fall and beginning to flower early in the spring as the leaves expand; petals greenish-white, usually purplish-tipped, about 2.5 mm. long; drupes bright red, about 10 mm. long. Pacific, mountain, or western flowering dogwood.

British Columbia southward, in and w. of the Cascades in Wash. and Oreg., to s. Calif., also in the Selway-Lochsa area, Idaho Co., Ida.; along streams and in open to fairly dense forest. Apr.-June.

A beautiful and highly prized native ornamental tree, showing considerable variation in the flowering habit, some specimens flowering profusely in mid- or late summer in contrast to the majority that blossom in fairly early spring.

Cornus stolonifera Michx. Fl. Bor. Am. 1:92. 1803.

Suida stolonifera Rydb. Bull. Torrey Club 31:572. 1904. *C. alba* ssp. *stolonifera* Wang. in Engl. Pflanzenr. 41:53. 1910. *C. sericea* ssp. *stolonifera* and f. *stolonifera* Fosberg, Bull. Torrey Club 69:587. 1942. ("ad ripas amnium rivorumque Canadae, Novae Angliae")

C. sericea L. Mant. 2:199. 1771, in part, but the application uncertain and apparently the name misapplied by most American authors. (North America)

C. sanguinea sensu Pursh, Fl. Am. Sept. 109. 1814, but not of L. *C. purshii* G. Don, Gen. Syst. 3:399. 1834. (North America, near the lakes of Canada and New York)

C. pubescens Nutt. Sylva 3:54. 1849, not of Willd. in 1827. *Svida pubescens* Standl. Smith. Misc. Coll. 56:3. 1912. *C. californica* var. *pubescens* Macbr. Contr. Gray Herb. n. s. 56:54. 1918. (Borders of the Oregon and Wahlamet) = var. *occidentalis.*

CORNUS STOLONIFERA var. OCCIDENTALIS (T. & G.) C. L. Hitchc. hoc loc. *C. sericea* var.

occidentalis T. & G. Fl. N. Am. 1:652. 1840. *C. occidentalis* Cov. Contr. U.S. Nat. Herb. 4:
117. 1893. *C. sericea* ssp. *occidentalis* and f. *occidentalis* Fosberg, Bull. Torrey Club 69:589.
1942. *(Douglas,* "N.W. Coast," is the first collection cited)

CORNUS STOLONIFERA var. OCCIDENTALIS f. CALIFORNICA (Meyer) C. L. Hitchc. hoc loc. *C.*
californica Meyer, Bull. Phys. Math. Acad. Pétersb. 3:373. 1845. *C. pubescens* var. *californica*
Coult. & Evans, Bot. Gaz. 15:37. 1890. *Svida californica* Abrams, Bull. N. Y. Bot. Gard. 6:429.
1910. *C. stolonifera* var. *californica* McMinn, Ill. Man. Calif. Shrubs 377. 1939. *C. sericea* f.
californica Fosberg, Bull. Torrey Club 69:589. 1942. (Near San Francisco and Ft. Ross, Calif.)
= var. *occidentalis*.

C. alba var. *coloradense* Koehne, Mitt. Deuts. Dendrol. Ges. 1903:39. 1903. *(Purpus 460,* Grand
Mesa, Colo.) = var. *stolonifera*.

Suida interior Rydb. Bull. Torrey Club 31:572. 1904. *C. interior* Peters. Fl. Neb. 163. 1912.
Ossea interior Lunell, Am. Midl. Nat. 5:239. 1918. *C. stolonifera* var. *interior* St. John, Fl. S.
E. Wash. 303. 1937. *C. sericea* f. *interior* Fosberg, Bull. Torrey Club 69:588. 1942. *(Rydberg*
1414, Dismal River, Neb., in 1893) = var. *stolonifera*.

Suida stolonifera var. *riparia* Rydb. Bull. Torrey Club 31:573. 1904. *C. instoloneus* A. Nels. Bot.
Gaz. 53:224. 1912. *C. stolonifera riparia* Visher, Geogr. Geol. & Biol. S. Dak. 101. 1912. *Ossea*
instolonea Nieuwl. & Lunell, Am. Midl. Nat. 4:487. 1916. *Svida instolonea* Rydb. Fl. Rocky Mts.
635, 1065. 1917. Not *C. riparia* Raf. *(M. S. Baker 257,* Crystal Creek, Colo., in 1901) = var.
stolonifera.

C. californica var. *nevadensis* Jeps. Man. Fl. Pl. Calif. 733. 1925. *(Jepson 10629,* Rich Pt.,
Middle Fork Feather R., Calif.) = var. *occidentalis*.

Many-stemmed shrub 2-6 m. tall, usually freely spreading by the layering of decumbent or prostrate stems that simulate stolons, the younger branches usually bright reddish to reddish-purple,
later turning grayish-green, from subglabrous to more commonly malpighiaceous-strigose or spread-
ing- or crisp-pubescent; leaf blades ovate to elliptic-ovate or somewhat obovate, usually more or
less acuminate, 4-12 cm. long, sparsely strigillose and greenish on the upper surface, much paler
and from malpighiaceous-strigillose to spreading- or crisp-pubescent beneath; flowers in flat-topped
cymes, not showy-bracteate, the peduncle and branches from malpighiaceous-strigose to conspicuously spreading-pubescent; petals white, 2-4 mm. long; styles 1-3 mm. long; drupes white to somewhat bluish, glabrous to spreading-pubescent, 7-9 mm. long, the stone somewhat flattened, smooth
to rather conspicuously longitudinally several-ridged. Creek dogwood, red osier.

Usually in moist soil, especially along streams, over much of N. Am. May-July.

This plant is a rather popular ornamental shrub, chiefly because of the bright red stems and autumn coloration of the foliage and it is propagated under several horticultural varietal names.

The taxonomy of the native forms is very controversial. All workers have agreed that the red
osiers of e. and w. N. Am. are closely related, if not conspecific. It is also rather generally agreed
that there is almost complete transition between the two, so that it is not possible to delineate clean-
cut taxa at whatever level they are treated. In two rather intensive recent studies of the complex
(Fosberg, F. R., Bull. Torrey Club 69:583-589. 1942 and Rickett, H. W., Brittonia 5:149-159. 1944)
opposite interpretations are presented, the former author considering that there are two completely intergradient subspecies, with 5 formae, in the complex which he feels must be called *C.*
sericea. Rickett, on the other hand, believes that there are two distinct species involved, *C. stolo-*
nifera and *C. occidentalis* (he rejects the name *C. sericea* on the ground that it will be a permanent
source of confusion or error), which have so widely hybridized that specific limits are almost oblit-
erated. The disposition of taxa that follows below more nearly approaches that of Fosberg even though
the name *C. sericea* is discarded in favor of *C. stolonifera:*

1 Stone smooth; petals in general not over 3 (2-3) mm. long; styles 1-2 mm. long; pubescence almost entirely strigose; Alas. to Newf., s. to Pa., Mo., and in the Rocky Mts.
to Mex.; the common phase in Mont., and abundant in Ida., occasional in n. e. Wash.;
not uniform; a form centering in Neb., but occasional in Mont., with more spreading
or woolly hairs (at least on the peduncles) is often distinguished at one of several levels
under the epithet *interior*   var. stolonifera
1 Stone grooved lengthwise; petals mostly at least 3 mm. long; styles 2-3 mm. long;
pubescence various but often very conspicuously spreading or curled; Alas. to s.
Calif., e. to Ida. and Nev., occasional in w. Mont.   var. occidentalis (T. & G.) C. L. Hitchc.

Our West Coast plants vary from sparsely strigillose to copiously spreading or crisp-hairy, the
two forms being almost totally sympatric and completely intergradient, although the more hairy forma

occidentalis appears to be somewhat more abundant w. of the Cascades than is forma *californica* (Meyer) C. L. Hitchc., which is subglabrous to thinly appressed- or spreading-hairy and characteristically appressed- rather than spreading-hairy in the inflorescence.

GARRYACEAE Silk-Tassel Family

Flowers small, apetalous, imperfect, borne in pendent catkinlike racemes, in 3's (less commonly singly) in the axils of decussate, connate, more or less cupulate bracts, the central flower with the longest pedicel; staminate flowers of 4 elongate, usually apically connate bractlike sepals and 4 alternate stamens; pistillate flowers (usually reduced to one per axil at least toward the tip of the raceme) with the calyx adherent to the inferior ovary, but the lobes greatly reduced and only 2, or lacking entirely; pistil 1, carpels 2, the styles 2, spreading, persistent, the ovary 1-locular with 2 pendulous ovules; fruit apparently baccate but drying by maturity and pseudocapsular, 1- or 2-seeded; dioecious evergreen shrubs (sometimes more or less arborescent) with opposite, leathery, simple, subentire, exstipulate leaves, the petioles slightly connate in a basal line.

Only the one genus.

The phylogenetic position of the family is not agreed upon although it has classically been placed near such families as the Salicaceae and Juglandaceae in the Amentiferae, because of the catkinlike inflorescence and unisexual flowers. Most recent students consider it to have its closest affinities with the Cornaceae, chiefly because of the opposite leaves, 4-merous flowers, reduced calyx, and inferior ovary with pendulous ovules.

Garrya Dougl. Silk Tassel

About 15 species ranging from s. w. Wash. southward and in much of s. w. U.S. and n. Mex., with one species in Jamaica. (Named for Nicholas Garry, a personal friend of Douglas.)

The plants are beautiful evergreen ornamental shrubs, valued because of their mostly deep green foliage and unusual drooping racemes of long-persistent silky flowers that finally (in the pistillate plants) are replaced by the almost as attractive fruits. The fact that they blossom in the middle of winter enhances their ornamental value.

1 Leaves distinctly undulate-margined, thickly close-tomentose on the lower surface

G. ELLIPTICA

1 Leaves not undulate-margined, strigillose or glabrate on the lower surface   G. FREMONTII

Garrya elliptica Dougl. in Lindl. Bot. Reg. 20:pl. 1686. 1834. *(Douglas,* "Northern California")

Large shrubs 2-7 m. tall; young branches thickly hairy, becoming glabrous and brownish and later rough-barked; leaves with stout petioles 5-10 mm. long, the blades thick and leathery, usually markedly undulate, deep green and glabrous on the upper surface, rather thickly close-tomentose beneath, oblong-ovate to -obovate, mostly 5-8 cm. long; staminate racemes usually several per fascicle, 8-14 cm. long, the bracts densely sericeous; pistillate racemes with silky bracts, the densely aggregated fruits tomentose-villous, 7-10 mm. thick. N=11.

On coastal bluffs and hills from s. Lincoln Co., Oreg., to San Luis Obispo Co., Calif. Jan. - Apr., dependent upon elevation.

Garrya fremontii Torr. Pac. R. R. Rep. 4:136. 1857. *(Fremont,* Upper Sacramento, "above the Great Canyon," in 1846)

  G. rigida Eastw. Bot. Gaz. 36:461. 1903. (Mt. Tamalpais, Calif., probably collected by Eastwood)

  G. fremontii var. *laxa* Eastw. Bot. Gaz. 36:462. 1903. *(Eastwood,* Twin Lakes, Trinity Co., Calif., July 10, 1901)

Shrub 1-3 m. tall, the young branches strigillose, soon glabrate and more or less brownish-purple; young leaves strigillose on the lower surface, soon glabrous and rather yellow-green, the blades elliptic-ovate to -oblong, mostly 4-8 cm. long, the petioles 7-14 mm. long; staminate racemes 1-several per axil, lax and drooping, unbranched, 3-9 cm. long, the bracts sericeous or pilose (to subglabrous), 4-6 mm. long; pistillate racemes with densely silky bracts; berries purple, nearly globose, 5-6 mm. thick, hairy when young, usually becoming glabrous, very closely to loosely clustered.

In chaparral and woodland; along the Columbia R. in Wash. and Oreg., reappearing along the w. side of the Cascade Mts. in n. Lane Co., southward and becoming more abundant in s. w. Oreg. and

fruit

1.5

Cornus nuttallii

Cornus canadensis

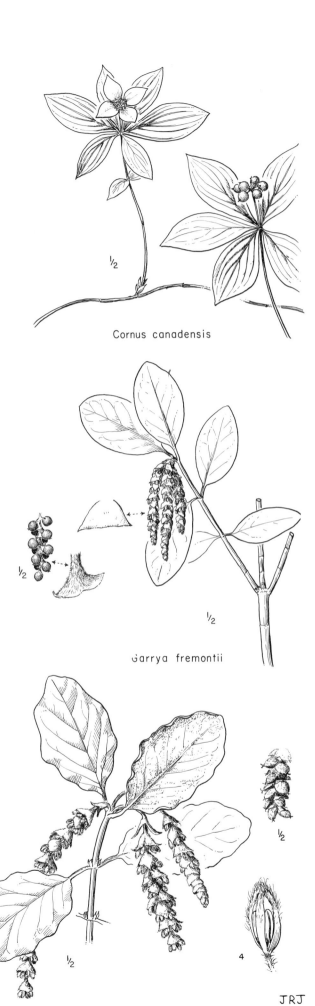

Garrya fremontii

Cornus stolonifera

Garrya elliptica

JRJ

n. Calif., commonly in the coastal mts. to Monterey Co., and in the Sierra Nevada; also in s. Calif. Jan.-May, dependent upon elevation.

The species is notably variable in several respects, but especially in the amount of pubescence on the leaves and fruits, and in the relative compactness of the fruiting catkins. Plants from the Columbia R. Gorge have perhaps the least hairy leaves, fruits, and inflorescences as well as the longest, least compact fruiting racemes. In contrast, plants of coastal Calif. have thick, crowded, heavily pubescent fruiting racemes. As the species occurs in the Sierras and in n. and n. c. Calif., it is more or less intermediate between the two extremes, but much more like the plants of the Columbia R. Gorge and the Oregon Cascades than like those of coastal and s. Calif. The name var. *laxa* Eastw. was proposed for plants from Trinity Co., Calif., characterized by loose fruiting racemes, and is available for our plants should badly needed field study prove that the recognition of geographic races is warranted.

Index and Partial List of Synonyms

Those names listed in the left-hand column which are followed by page reference in the right-hand column include well-known common names, accepted names of families and genera, and certain species incidentally mentioned but not formally described. Names in italics are not followed by page citation as they are considered to be synonyms referable to other taxa and are referred to the genus (and often the particular species) under which they may be found in the text. It has not in general been considered necessary to include in this second category the synonymy for genera that have fewer than six species in our area unless it happens to be particularly involved. Accepted species will be found in alphabetical sequence under the genus.

593